DERIVATIVES AND INTEGRALS

Basic Differentiation Rules

1. $\dfrac{d}{dx}[cu] = cu'$
2. $\dfrac{d}{dx}[u \pm v] = u' \pm v'$
3. $\dfrac{d}{dx}[uv] = uv' + vu'$
4. $\dfrac{d}{dx}\left[\dfrac{u}{v}\right] = \dfrac{vu' - uv'}{v^2}$
5. $\dfrac{d}{dx}[c] = 0$
6. $\dfrac{d}{dx}[u^n] = nu^{n-1}u'$
7. $\dfrac{d}{dx}[x] = 1$
8. $\dfrac{d}{dx}[|u|] = \dfrac{u}{|u|}(u'), \quad u \neq 0$
9. $\dfrac{d}{dx}[\ln u] = \dfrac{u'}{u}$
10. $\dfrac{d}{dx}[e^u] = e^u u'$
11. $\dfrac{d}{dx}[\log_a u] = \dfrac{u'}{(\ln a)u}$
12. $\dfrac{d}{dx}[a^u] = (\ln a)a^u u'$
13. $\dfrac{d}{dx}[\sin u] = (\cos u)u'$
14. $\dfrac{d}{dx}[\cos u] = -(\sin u)u'$
15. $\dfrac{d}{dx}[\tan u] = (\sec^2 u)u'$
16. $\dfrac{d}{dx}[\cot u] = -(\csc^2 u)u'$
17. $\dfrac{d}{dx}[\sec u] = (\sec u \tan u)u'$
18. $\dfrac{d}{dx}[\csc u] = -(\csc u \cot u)u'$
19. $\dfrac{d}{dx}[\arcsin u] = \dfrac{u'}{\sqrt{1 - u^2}}$
20. $\dfrac{d}{dx}[\arccos u] = \dfrac{-u'}{\sqrt{1 - u^2}}$
21. $\dfrac{d}{dx}[\arctan u] = \dfrac{u'}{1 + u^2}$
22. $\dfrac{d}{dx}[\operatorname{arccot} u] = \dfrac{-u'}{1 + u^2}$
23. $\dfrac{d}{dx}[\operatorname{arcsec} u] = \dfrac{u'}{|u|\sqrt{u^2 - 1}}$
24. $\dfrac{d}{dx}[\operatorname{arccsc} u] = \dfrac{-u'}{|u|\sqrt{u^2 - 1}}$
25. $\dfrac{d}{dx}[\sinh u] = (\cosh u)u'$
26. $\dfrac{d}{dx}[\cosh u] = (\sinh u)u'$
27. $\dfrac{d}{dx}[\tanh u] = (\operatorname{sech}^2 u)u'$
28. $\dfrac{d}{dx}[\coth u] = -(\operatorname{csch}^2 u)u'$
29. $\dfrac{d}{dx}[\operatorname{sech} u] = -(\operatorname{sech} u \tanh u)u'$
30. $\dfrac{d}{dx}[\operatorname{csch} u] = -(\operatorname{csch} u \coth u)u'$
31. $\dfrac{d}{dx}[\sinh^{-1} u] = \dfrac{u'}{\sqrt{u^2 + 1}}$
32. $\dfrac{d}{dx}[\cosh^{-1} u] = \dfrac{u'}{\sqrt{u^2 - 1}}$
33. $\dfrac{d}{dx}[\tanh^{-1} u] = \dfrac{u'}{1 - u^2}$
34. $\dfrac{d}{dx}[\coth^{-1} u] = \dfrac{u'}{1 - u^2}$
35. $\dfrac{d}{dx}[\operatorname{sech}^{-1} u] = \dfrac{-u'}{u\sqrt{1 - u^2}}$
36. $\dfrac{d}{dx}[\operatorname{csch}^{-1} u] = \dfrac{-u'}{|u|\sqrt{1 + u^2}}$

Basic Integration Formulas

1. $\int kf(u)\,du = k\int f(u)\,du$
2. $\int [f(u) \pm g(u)]\,du = \int f(u)\,du \pm \int g(u)\,du$
3. $\int du = u + C$
4. $\int a^u\,du = \left(\dfrac{1}{\ln a}\right)a^u + C$
5. $\int e^u\,du = e^u + C$
6. $\int \sin u\,du = -\cos u + C$
7. $\int \cos u\,du = \sin u + C$
8. $\int \tan u\,du = -\ln|\cos u| + C$
9. $\int \cot u\,du = \ln|\sin u| + C$
10. $\int \sec u\,du = \ln|\sec u + \tan u| + C$
11. $\int \csc u\,du = -\ln|\csc u + \cot u| + C$
12. $\int \sec^2 u\,du = \tan u + C$
13. $\int \csc^2 u\,du = -\cot u + C$
14. $\int \sec u \tan u\,du = \sec u + C$
15. $\int \csc u \cot u\,du = -\csc u + C$
16. $\int \dfrac{du}{\sqrt{a^2 - u^2}} = \arcsin\dfrac{u}{a} + C$
17. $\int \dfrac{du}{a^2 + u^2} = \dfrac{1}{a}\arctan\dfrac{u}{a} + C$
18. $\int \dfrac{du}{u\sqrt{u^2 - a^2}} = \dfrac{1}{a}\operatorname{arcsec}\dfrac{|u|}{a} + C$

TRIGONOMETRY

Definition of the Six Trigonometric Functions

Right triangle definitions, where $0 < \theta < \pi/2$.

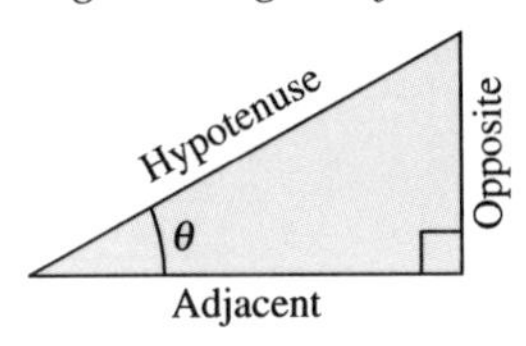

$$\sin\theta = \frac{\text{opp}}{\text{hyp}} \qquad \csc\theta = \frac{\text{hyp}}{\text{opp}}$$

$$\cos\theta = \frac{\text{adj}}{\text{hyp}} \qquad \sec\theta = \frac{\text{hyp}}{\text{adj}}$$

$$\tan\theta = \frac{\text{opp}}{\text{adj}} \qquad \cot\theta = \frac{\text{adj}}{\text{opp}}$$

Circular function definitions, where θ is any angle.

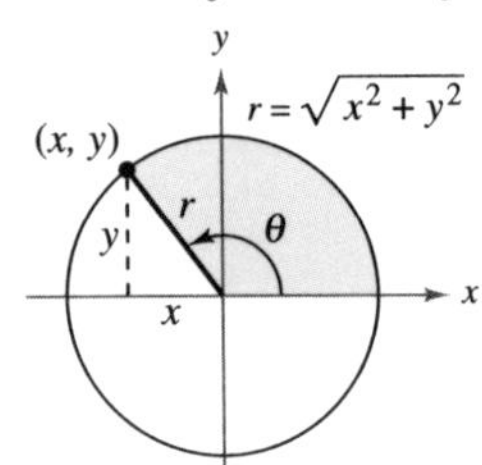

$$\sin\theta = \frac{y}{r} \qquad \csc\theta = \frac{r}{y}$$

$$\cos\theta = \frac{x}{r} \qquad \sec\theta = \frac{r}{x}$$

$$\tan\theta = \frac{y}{x} \qquad \cot\theta = \frac{x}{y}$$

Reciprocal Identities

$$\sin x = \frac{1}{\csc x} \qquad \sec x = \frac{1}{\cos x} \qquad \tan x = \frac{1}{\cot x}$$

$$\csc x = \frac{1}{\sin x} \qquad \cos x = \frac{1}{\sec x} \qquad \cot x = \frac{1}{\tan x}$$

Tangent and Cotangent Identities

$$\tan x = \frac{\sin x}{\cos x} \qquad \cot x = \frac{\cos x}{\sin x}$$

Pythagorean Identities

$$\sin^2 x + \cos^2 x = 1$$

$$1 + \tan^2 x = \sec^2 x \qquad 1 + \cot^2 x = \csc^2 x$$

Cofunction Identities

$$\sin\left(\frac{\pi}{2} - x\right) = \cos x \qquad \cos\left(\frac{\pi}{2} - x\right) = \sin x$$

$$\csc\left(\frac{\pi}{2} - x\right) = \sec x \qquad \tan\left(\frac{\pi}{2} - x\right) = \cot x$$

$$\sec\left(\frac{\pi}{2} - x\right) = \csc x \qquad \cot\left(\frac{\pi}{2} - x\right) = \tan x$$

Reduction Formulas

$$\sin(-x) = -\sin x \qquad \cos(-x) = \cos x$$

$$\csc(-x) = -\csc x \qquad \tan(-x) = -\tan x$$

$$\sec(-x) = \sec x \qquad \cot(-x) = -\cot x$$

Sum and Difference Formulas

$$\sin(u \pm v) = \sin u \cos v \pm \cos u \sin v$$

$$\cos(u \pm v) = \cos u \cos v \mp \sin u \sin v$$

$$\tan(u \pm v) = \frac{\tan u \pm \tan v}{1 \mp \tan u \tan v}$$

Double-Angle Formulas

$$\sin 2u = 2 \sin u \cos u$$

$$\cos 2u = \cos^2 u - \sin^2 u = 2\cos^2 u - 1 = 1 - 2\sin^2 u$$

$$\tan 2u = \frac{2 \tan u}{1 - \tan^2 u}$$

Power-Reducing Formulas

$$\sin^2 u = \frac{1 - \cos 2u}{2}$$

$$\cos^2 u = \frac{1 + \cos 2u}{2}$$

$$\tan^2 u = \frac{1 - \cos 2u}{1 + \cos 2u}$$

Sum-to-Product Formulas

$$\sin u + \sin v = 2 \sin\left(\frac{u + v}{2}\right)\cos\left(\frac{u - v}{2}\right)$$

$$\sin u - \sin v = 2 \cos\left(\frac{u + v}{2}\right)\sin\left(\frac{u - v}{2}\right)$$

$$\cos u + \cos v = 2 \cos\left(\frac{u + v}{2}\right)\cos\left(\frac{u - v}{2}\right)$$

$$\cos u - \cos v = -2 \sin\left(\frac{u + v}{2}\right)\sin\left(\frac{u - v}{2}\right)$$

Product-to-Sum Formulas

$$\sin u \sin v = \frac{1}{2}[\cos(u - v) - \cos(u + v)]$$

$$\cos u \cos v = \frac{1}{2}[\cos(u - v) + \cos(u + v)]$$

$$\sin u \cos v = \frac{1}{2}[\sin(u + v) + \sin(u - v)]$$

$$\cos u \sin v = \frac{1}{2}[\sin(u + v) - \sin(u - v)]$$

Multivariable Calculus

Ninth Edition

Ron Larson
The Pennsylvania State University
The Behrend College

Bruce H. Edwards
University of Florida

Australia • Brazil • Japan • Korea • Mexico • Singapore • Spain • United Kingdom • United States

Multivariable Calculus, Ninth Edition
Larson/Edwards

VP/Editor-in-Chief: Michelle Julet
Publisher: Richard Stratton
Senior Sponsoring Editor: Cathy Cantin
Development Editor: Peter Galuardi
Associate Editor: Jeannine Lawless
Editorial Assistant: Amy Haines
Media Editor: Peter Galuardi
Senior Content Manager: Maren Kunert
Senior Marketing Manager: Jennifer Jones
Marketing Communications Manager: Mary Anne Payumo
Senior Content Project Manager, Editorial Production: Tamela Ambush
Art and Design Manager: Jill Haber
Senior Manufacturing Coordinator: Diane Gibbons
Permissions Editor: Katie Huha
Text Designer: Nesbitt Graphics
Art Editor: Larson Texts, Inc.
Senior Photo Editor: Jennifer Meyer Dare
Illustrator: Larson Texts, Inc.
Cover Designer: Harold Burch
Cover Image: © Richard Edelman/Woodstock Graphics Studio
Compositor: Larson Texts, Inc.

Library of Congress Control Number: 2008939232
Student Edition:
ISBN-13: 978-0-547-20997-5
ISBN-10: 0-547-20997-5

Brooks/Cole
10 Davis Drive
Belmont, CA 94002-3098
USA

Cengage learning is a leading provider of customized learning solutions with office locations around the globe, including Singapore, the United Kingdom, Australia, Mexico, Brazil, and Japan. Locate your local office at: **international.cengage.com/region**

Cengage learning products are represented in Canada by Nelson Education, Ltd.

For your course and learning solutions, visit **www.cengage.com**

Purchase any of our products at your local college store or at our preferred online store **www.ichapters.com**

Printed in the United States of America
1 2 3 4 5 6 7 12 11 10 09 08

Contents

A Word from the Authors

Welcome to the Ninth Edition of *Multivariable Calculus*! We are proud to offer you a new and revised version of our textbook. Much has changed since we wrote the first edition over 35 years ago. With each edition we have listened to you, our users, and have incorporated many of your suggestions for improvement.

6th

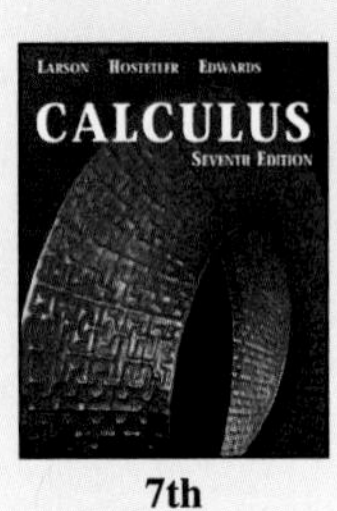

7th

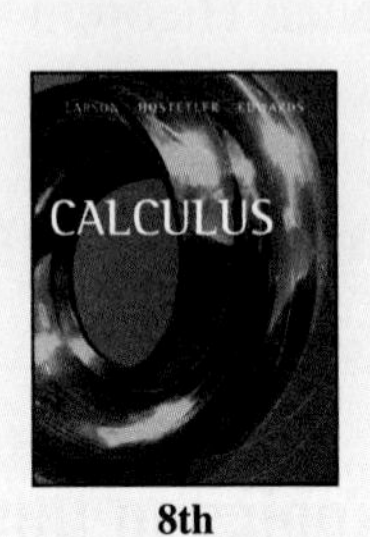

8th

9th

Throughout the years, our objective has always been to write in a precise, readable manner with the fundamental concepts and rules of calculus clearly defined and demonstrated. When writing for students, we strive to offer features and materials that enable mastery by all types of learners. For the instructors, we aim to provide a comprehensive teaching instrument that employs proven pedagogical techniques, freeing instructors to make the most efficient use of classroom time.

This revision brings us to a new level of change and improvement. For the past several years, we've maintained an independent website—**CalcChat.com**—that provides free solutions to all odd-numbered exercises in the text. Thousands of students using our textbooks have visited the site for practice and help with their homework. With the Ninth Edition, we were able to use information from CalcChat.com, including which solutions students accessed most often, to help guide the revision of the exercises. This edition of *Calculus* will be the first calculus textbook to use actual data from students.

We have also added a new feature called *Capstone* exercises to this edition. These conceptual problems synthesize key topics and provide students with a better understanding of each section's concepts. *Capstone* exercises are excellent for classroom discussion or test prep, and instructors may find value in integrating these problems into their review of the section. These and other new features join our time-tested pedagogy, with the goal of enabling students and instructors to make the best use of this text.

We hope you will enjoy the Ninth Edition of *Multivariable Calculus*. As always, we welcome comments and suggestions for continued improvements.

Ron Larson Bruce H. Edwards

5th

4th

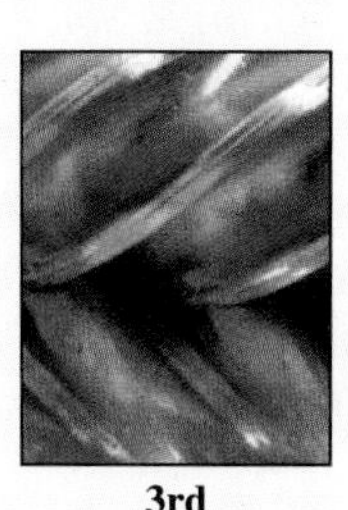
3rd

2nd

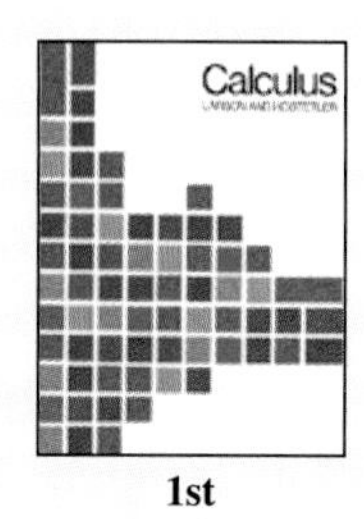

1st

Acknowledgments

We would like to thank the many people who have helped us at various stages of this project over the last 35 years. Their encouragement, criticisms, and suggestions have been invaluable to us.

Reviewers of the Ninth Edition

Ray Cannon, *Baylor University*
Sadeq Elbaneh, *Buffalo State College*
J. Fasteen, *Portland State University*
Audrey Gillant, *Binghamton University*
Sudhir Goel, *Valdosta State University*
Marcia Kemen, *Wentworth Institute of Technology*
Ibrahima Khalil Kaba, *Embry Riddle Aeronautical University*
Jean-Baptiste Meilhan, *University of California Riverside*
Catherine Moushon, *Elgin Community College*
Charles Odion, *Houston Community College*
Greg Oman, *The Ohio State University*
Dennis Pence, *Western Michigan University*
Jonathan Prewett, *University of Wyoming*
Lori Dunlop Pyle, *University of Central Florida*
Aaron Robertson, *Colgate University*
Matthew D. Sosa, *The Pennsylvania State University*
William T. Trotter, *Georgia Institute of Technology*
Dr. Draga Vidakovic, *Georgia State University*
Jay Wiestling, *Palomar College*
Jianping Zhu, *University of Texas at Arlington*

Ninth Edition Advisory Board Members

Jim Braselton, *Georgia Southern University;* Sien Deng, *Northern Illinois University;* Dimitar Grantcharov, *University of Texas, Arlington;* Dale Hughes, *Johnson County Community College;* Dr. Philippe B. Laval, *Kennesaw State University;* Kouok Law, *Georgia Perimeter College, Clarkson Campus;* Mara D. Neusel, *Texas Tech University;* Charlotte Newsom, *Tidewater Community College, Virginia Beach Campus;* Donald W. Orr, *Miami Dade College, Kendall Campus;* Jude Socrates, *Pasadena City College;* Betty Travis, *University of Texas at San Antonio;* Kuppalapalle Vajravelu, *University of Central Florida*

Reviewers of Previous Editions

Stan Adamski, *Owens Community College*; Alexander Arhangelskii, *Ohio University;* Seth G. Armstrong, *Southern Utah University;* Jim Ball, *Indiana State University;* Marcelle Bessman, *Jacksonville University;* Linda A. Bolte, *Eastern Washington University;* James Braselton, *Georgia Southern University;* Harvey Braverman, *Middlesex County College;* Tim Chappell, *Penn Valley Community College;* Oiyin Pauline Chow, *Harrisburg Area Community College;* Julie M. Clark, *Hollins University;* P.S. Crooke, *Vanderbilt University;* Jim Dotzler, *Nassau Community College;* Murray Eisenberg, *University of Massachusetts at Amherst;* Donna Flint, *South Dakota State University;* Michael Frantz, *University of La Verne;* Sudhir Goel, *Valdosta State University;* Arek Goetz, *San Francisco State University;* Donna J. Gorton, *Butler County Community College;* John Gosselin, *University of Georgia;* Shahryar Heydari, *Piedmont College;* Guy Hogan, *Norfolk State University;* Ashok Kumar, *Valdosta State University;* Kevin J. Leith, *Albuquerque Community College;* Douglas B. Meade, *University of South Carolina;* Teri Murphy, *University of Oklahoma;* Darren Narayan, *Rochester Institute of Technology;* Susan A. Natale, *The Ursuline School, NY;* Terence H. Perciante, *Wheaton College;* James Pommersheim, *Reed College;* Leland E. Rogers, *Pepperdine University;* Paul Seeburger, *Monroe Community College;* Edith A. Silver, *Mercer County Community College;* Howard Speier, *Chandler-Gilbert Community College;* Desmond Stephens, *Florida A&M University;* Jianzhong Su, *University of Texas at Arlington;* Patrick Ward, *Illinois Central College;* Diane Zych, *Erie Community College*

Many thanks to Robert Hostetler, The Behrend College, The Pennsylvania State University, and David Heyd, The Behrend College, The Pennsylvania State University, for their significant contributions to previous editions of this text.

A special note of thanks goes to the instructors who responded to our survey and to the over 2 million students who have used earlier editions of the text.

We would also like to thank the staff at Larson Texts, Inc., who assisted in preparing the manuscript, rendering the art package, typesetting, and proofreading the pages and supplements.

On a personal level, we are grateful to our wives, Deanna Gilbert Larson and Consuelo Edwards, for their love, patience, and support. Also, a special note of thanks goes out to R. Scott O'Neil.

If you have suggestions for improving this text, please feel free to write to us. Over the years we have received many useful comments from both instructors and students, and we value these very much.

Ron Larson

Bruce H. Edwards

Your Course. Your Way.

Calculus Textbook Options

The Ninth Edition of *Calculus* is available in a variety of textbook configurations to address the different ways instructors teach—and students take—their classes.

It is available in a comprehensive three-semester version or as single-variable and multivariable versions. The book can also be customized to meet your individual needs and is available through iChapters —www.ichapters.com.

TOPICS COVERED	APPROACH			
	Late Transcendental Functions	**Early Transcendental Functions**	**Accelerated coverage**	**Late Trigonometry**
3-semester	Calculus 9e 	Calculus: Early Transcendental Functions 4e 	Essential Calculus 	Calculus with Late Trigonometry 
Single Variable Only	Calculus 9e Single Variable 	Calculus: Early Transcendental Functions 4e Single Variable 		
Multivariable	Calculus 9e Multivariable 	Calculus 9e Multivariable 		
Custom All of these textbook choices can be customized to fit the individual needs of your course.	Calculus 9e 	Calculus: Early Transcendental Functions 4e 	Essential Calculus 	Calculus with Late Trigonometry 

Textbook Features

Tools to Build Mastery

CAPSTONE

70. Use the graph of f' shown in the figure to answer the following, given that $f(0) = -4$.

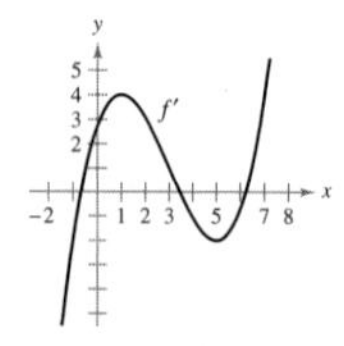

(a) Approximate the slope of f at $x = 4$. Explain.

(b) Is it possible that $f(2) = -1$? Explain.

(c) Is $f(5) - f(4) > 0$? Explain.

(d) Approximate the value of x where f is maximum. Explain.

(e) Approximate any intervals in which the graph of f is concave upward and any intervals in which it is concave downward. Approximate the x-coordinates of any points of inflection.

(f) Approximate the x-coordinate of the minimum of $f''(x)$.

(g) Sketch an approximate graph of f. To print an enlarged copy of the graph, go to the website *www.mathgraphs.com*.

CAPSTONES

NEW! Capstone exercises now appear in every section. These exercises synthesize the main concepts of each section and show students how the topics relate. They are often multipart problems that contain conceptual and noncomputational parts, and can be used for classroom discussion or test prep.

WRITING ABOUT CONCEPTS

59. The graph of f is shown in the figure.

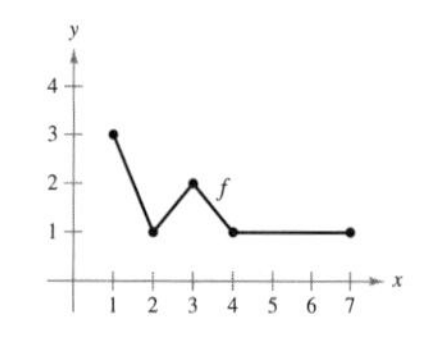

(a) Evaluate $\int_1^7 f(x)\,dx$.

(b) Determine the average value of f on the interval $[1, 7]$.

(c) Determine the answers to parts (a) and (b) if the graph is translated two units upward.

60. If $r'(t)$ represents the rate of growth of a dog in pounds per year, what does $r(t)$ represent? What does $\int_2^6 r'(t)\,dt$ represent about the dog?

WRITING ABOUT CONCEPTS

These writing exercises are questions designed to test students' understanding of basic concepts in each section. The exercises encourage students to verbalize and write answers, promoting technical communication skills that will be invaluable in their future careers.

STUDY TIPS

The devil is in the details. Study Tips help point out some of the troublesome common mistakes, indicate special cases that can cause confusion, or expand on important concepts. These tips provide students with valuable information, similar to what an instructor might comment on in class.

STUDY TIP Because integration is usually more difficult than differentiation, you should always check your answer to an integration problem by differentiating. For instance, in Example 4 you should differentiate $\frac{1}{3}(2x -$ ... that you obtain the ...

STUDY TIP Later in this chapter, you will learn convenient methods for calculating $\int_a^b f(x)\,dx$ for continuous ... or now, you must use the ... ion.

STUDY TIP Remember that you can check your answer by differentiating.

EXAMPLE 6 Evaluation of a Definite Integral

Evaluate $\int_1^3 (-x^2 + 4x - 3)\,dx$ using each of the following values.

$$\int_1^3 x^2\,dx = \frac{26}{3}, \quad \int_1^3 x\,dx = 4, \quad \int_1^3 dx = 2$$

Solution

$$\begin{aligned}\int_1^3 (-x^2 + 4x - 3)\,dx &= \int_1^3 (-x^2)\,dx + \int_1^3 4x\,dx + \int_1^3 (-3)\,dx\\ &= -\int_1^3 x^2\,dx + 4\int_1^3 x\,dx - 3\int_1^3 dx\\ &= -\left(\frac{26}{3}\right) + 4(4) - 3(2)\\ &= \frac{4}{3}\end{aligned}$$

EXAMPLES

Throughout the text, examples are worked out step-by-step. These worked examples demonstrate the procedures and techniques for solving problems, and give students an increased understanding of the concepts of calculus.

EXERCISES

Practice makes perfect. Exercises are often the first place students turn to in a textbook. The authors have spent a great deal of time analyzing and revising the exercises, and the result is a comprehensive and robust set of exercises at the end of every section. A variety of exercise types and levels of difficulty are included to accommodate students with all learning styles.

In addition to the exercises in the book, 3,000 algorithmic exercises appear in the WebAssign® course that accompanies Calculus.

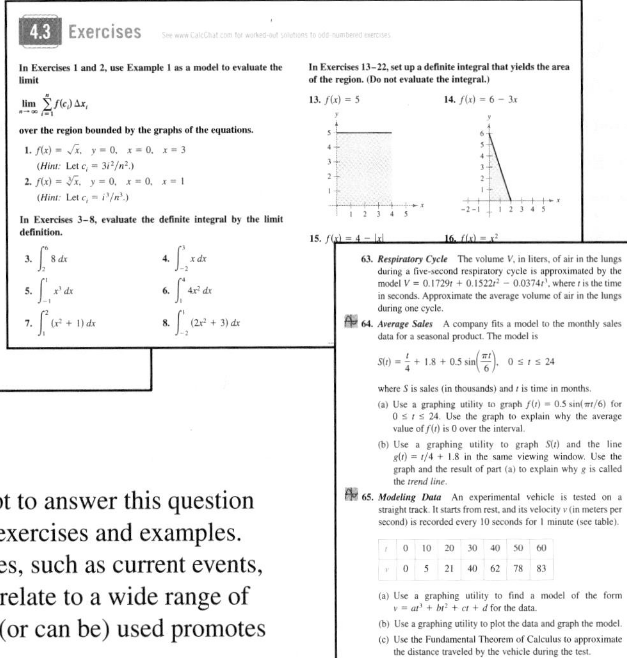

4.3 Exercises See www.CalcChat.com for worked-out solutions to odd-numbered exercises.

In Exercises 1 and 2, use Example 1 as a model to evaluate the limit

$$\lim_{n\to\infty}\sum_{i=1}^{n} f(c_i)\,\Delta x_i$$

over the region bounded by the graphs of the equations.

1. $f(x) = \sqrt{x},\ y = 0,\ x = 0,\ x = 3$
   (*Hint:* Let $c_i = 3i^2/n^2$.)
2. $f(x) = \sqrt[3]{x},\ y = 0,\ x = 0,\ x = 1$
   (*Hint:* Let $c_i = i^3/n^3$.)

In Exercises 3–8, evaluate the definite integral by the limit definition.

3. $\int_2^6 8\,dx$
4. $\int_{-2}^{3} x\,dx$
5. $\int_{-1}^{1} x^3\,dx$
6. $\int_1^4 4x^2\,dx$
7. $\int_1^2 (x^2 + 1)\,dx$
8. $\int_{-2}^{1} (2x^2 + 3)\,dx$

In Exercises 13–22, set up a definite integral that yields the area of the region. (Do not evaluate the integral.)

13. $f(x) = 5$
14. $f(x) = 6 - 3x$
15. $f(x) = 4 - |x|$
16. $f(x) = x^2$

63. ***Respiratory Cycle*** The volume V, in liters, of air in the lungs during a five-second respiratory cycle is approximated by the model $V = 0.1729t + 0.1522t^2 - 0.0374t^3$, where t is the time in seconds. Approximate the average volume of air in the lungs during one cycle.

64. ***Average Sales*** A company fits a model to the monthly sales data for a seasonal product. The model is

$$S(t) = \frac{t}{4} + 1.8 + 0.5\sin\left(\frac{\pi t}{6}\right),\quad 0 \le t \le 24$$

where S is sales (in thousands) and t is time in months.

(a) Use a graphing utility to graph $f(t) = 0.5\sin(\pi t/6)$ for $0 \le t \le 24$. Use the graph to explain why the average value of $f(t)$ is 0 over the interval.

(b) Use a graphing utility to graph $S(t)$ and the line $g(t) = t/4 + 1.8$ in the same viewing window. Use the graph and the result of part (a) to explain why g is called the *trend line*.

65. ***Modeling Data*** An experimental vehicle is tested on a straight track. It starts from rest, and its velocity v (in meters per second) is recorded every 10 seconds for 1 minute (see table).

t	0	10	20	30	40	50	60
v	0	5	21	40	62	78	83

(a) Use a graphing utility to find a model of the form $v = at^3 + bt^2 + ct + d$ for the data.

(b) Use a graphing utility to plot the data and graph the model.

(c) Use the Fundamental Theorem of Calculus to approximate the distance traveled by the vehicle during the test.

APPLICATIONS

"When will I use this?" The authors attempt to answer this question for students with carefully chosen applied exercises and examples. Applications are pulled from diverse sources, such as current events, world data, industry trends, and more, and relate to a wide range of interests. Understanding where calculus is (or can be) used promotes fuller understanding of the material.

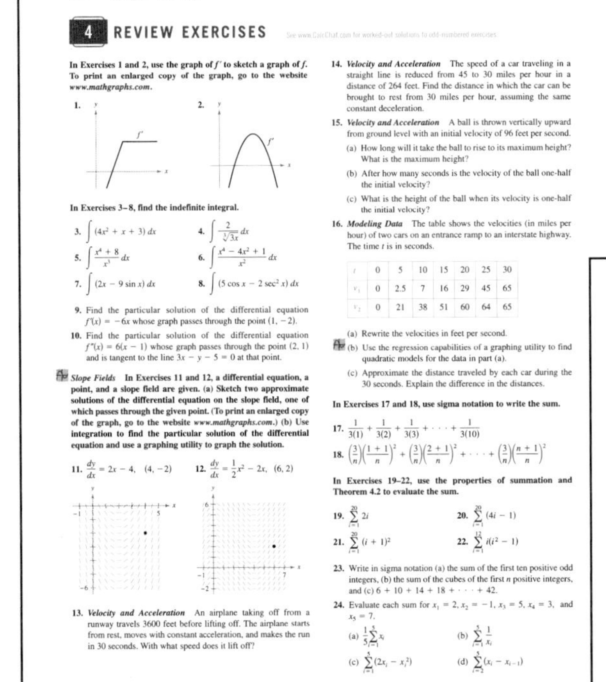

318 Chapter 4 Integration

4 REVIEW EXERCISES See www.CalcChat.com for worked-out solutions to odd-numbered exercises.

In Exercises 1 and 2, use the graph of f' to sketch a graph of f. To print an enlarged copy of the graph, go to the website *www.mathgraphs.com*.

1. 2.

In Exercises 3–8, find the indefinite integral.

3. $\int (4x^2 + x + 3)\,dx$
4. $\int \frac{2}{\sqrt[3]{3x}}\,dx$
5. $\int \frac{x^4 + 8}{x^3}\,dx$
6. $\int \frac{x^4 - 4x^2 + 1}{x^2}\,dx$
7. $\int (2x - 9\sin x)\,dx$
8. $\int (5\cos x - 2\sec^2 x)\,dx$

9. Find the particular solution of the differential equation $f'(x) = -6x$ whose graph passes through the point $(1, -2)$.
10. Find the particular solution of the differential equation $f''(x) = 6(x - 1)$ whose graph passes through the point $(2, 1)$ and is tangent to the line $3x - y - 5 = 0$ at that point.

Slope Fields **In Exercises 11 and 12, a differential equation, a point, and a slope field are given. (a) Sketch two approximate solutions of the differential equation on the slope field, one of which passes through the given point. (To print an enlarged copy of the graph, go to the website *www.mathgraphs.com*.) (b) Use integration to find the particular solution of the differential equation and use a graphing utility to graph the solution.**

11. $\frac{dy}{dx} = 2x - 4,\ (4, -2)$
12. $\frac{dy}{dx} = \frac{1}{2}x^2 - 2x,\ (6, 2)$

13. ***Velocity and Acceleration*** An airplane taking off from a runway travels 3600 feet before lifting off. The airplane starts from rest, moves with constant acceleration, and makes the run in 30 seconds. With what speed does it lift off?

14. ***Velocity and Acceleration*** The speed of a car traveling in a straight line is reduced from 45 to 30 miles per hour in a distance of 264 feet. Find the distance in which the car can be brought to rest from 30 miles per hour, assuming the same constant deceleration.

15. ***Velocity and Acceleration*** A ball is thrown vertically upward from ground level with an initial velocity of 96 feet per second.
    (a) How long will it take the ball to rise to its maximum height? What is the maximum height?
    (b) After how many seconds is the velocity of the ball one-half the initial velocity?
    (c) What is the height of the ball when its velocity is one-half the initial velocity?

16. ***Modeling Data*** The table shows the velocities (in miles per hour) of two cars on an entrance ramp to an interstate highway. The time t is in seconds.

t	0	5	10	15	20	25	30
v_1	0	2.5	7	16	29	45	65
v_2	0	21	38	51	60	64	65

(a) Rewrite the velocities in feet per second.
(b) Use the regression capabilities of a graphing utility to find quadratic models for the data in part (a).
(c) Approximate the distance traveled by each car during the 30 seconds. Explain the difference in the distances.

In Exercises 17 and 18, use sigma notation to write the sum.

17. $\frac{1}{3(1)} + \frac{1}{3(2)} + \frac{1}{3(3)} + \cdots + \frac{1}{3(10)}$
18. $\left(\frac{3}{n}\right)\left(\frac{1+1}{n}\right)^2 + \left(\frac{3}{n}\right)\left(\frac{2+1}{n}\right)^2 + \cdots + \left(\frac{3}{n}\right)\left(\frac{n+1}{n}\right)^2$

In Exercises 19–22, use the properties of summation and Theorem 4.2 to evaluate the sum.

19. $\sum_{i=1}^{20} 2i$
20. $\sum_{i=1}^{20} (4i - 1)$
21. $\sum_{i=1}^{20} (i + 1)^2$
22. $\sum_{i=1}^{12} i(i^2 - 1)$

23. Write in sigma notation (a) the sum of the first ten positive odd integers, (b) the sum of the cubes of the first n positive integers, and (c) $6 + 10 + 14 + 18 + \cdots + 42$.
24. Evaluate each sum for $x_1 = 2, x_2 = -1, x_3 = 5, x_4 = 3$, and $x_5 = 7$.
    (a) $\frac{1}{5}\sum_{i=1}^{5} x_i$ (b) $\sum_{i=1}^{5} \frac{1}{x_i}$
    (c) $\sum_{i=1}^{5} (2x_i - x_i^2)$ (d) $\sum_{i=2}^{5} (x_i - x_{i-1})$

REVIEW EXERCISES

Review Exercises at the end of each chapter provide more practice for students. These exercise sets provide a comprehensive review of the chapter's concepts and are an excellent way for students to prepare for an exam.

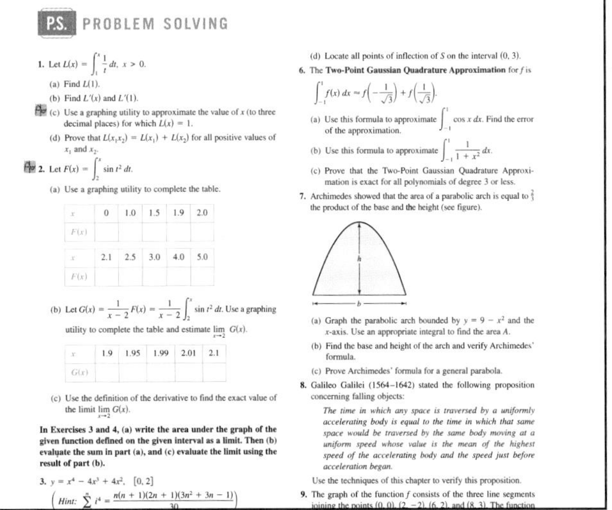

P.S. PROBLEM SOLVING

1. Let $L(x) = \int_1^x \frac{1}{t}\,dt,\ x > 0$.
   (a) Find $L(1)$.
   (b) Find $L'(x)$ and $L'(1)$.
   (c) Use a graphing utility to approximate the value of x (to three decimal places) for which $L(x) = 1$.
   (d) Prove that $L(x_1x_2) = L(x_1) + L(x_2)$ for all positive values of x_1 and x_2.
2. Let $F(x) = \int_2^x \sin t^2\,dt$.
   (a) Use a graphing utility to complete the table.

x	0	1.0	1.5	1.9	2.0
$F(x)$					

x	2.1	2.5	3.0	4.0	5.0
$F(x)$					

   (b) Let $G(x) = \frac{1}{x-2}F(x) = \frac{1}{x-2}\int_2^x \sin t^2\,dt$. Use a graphing utility to complete the table and estimate $\lim_{x\to 2} G(x)$.

x	1.9	1.95	1.99	2.01	2.1
$G(x)$					

   (c) Use the definition of the derivative to find the exact value of the limit $\lim_{x\to 2} G(x)$.

In Exercises 3 and 4, (a) write the area under the graph of the given function defined on the given interval as a limit. Then (b) evaluate the sum in part (a), and (c) evaluate the limit using the result of part (b).

3. $y = x^4 - 4x^3 + 4x^2,\ [0, 2]$
   $\left(\textit{Hint: } \sum_{i=1}^{n} i^4 = \frac{n(n+1)(2n+1)(3n^2+3n-1)}{30}\right)$

   (d) Locate all points of inflection of S on the interval $(0, 3)$.
6. The **Two-Point Gaussian Quadrature Approximation** for f is
   $\int_{-1}^{1} f(x)\,dx \approx f\left(-\frac{1}{\sqrt{3}}\right) + f\left(\frac{1}{\sqrt{3}}\right)$.
   (a) Use this formula to approximate $\int_{-1}^{1} \cos x\,dx$. Find the error of the approximation.
   (b) Use this formula to approximate $\int_{-1}^{1} \frac{1}{1+x^2}\,dx$.
   (c) Prove that the Two-Point Gaussian Quadrature Approximation is exact for all polynomials of degree 3 or less.
7. Archimedes showed that the area of a parabolic arch is equal to $\frac{2}{3}$ the product of the base and the height (see figure).
   (a) Graph the parabolic arch bounded by $y = 9 - x^2$ and the x-axis. Use an appropriate integral to find the area A.
   (b) Find the base and height of the arch and verify Archimedes' formula.
   (c) Prove Archimedes' formula for a general parabola.
8. Galileo Galilei (1564–1642) stated the following proposition concerning falling objects:
   The time in which any space is traversed by a uniformly accelerating body is equal to the time in which that same space would be traversed by the same body moving at a uniform speed whose value is the mean of the highest speed of the accelerating body and the speed just before acceleration began.
   Use the techniques of this chapter to verify this proposition.
9. The graph of the function f consists of the three line segments joining the points (0, 0), (2, −2), (6, 2), and (8, 3). The function

P.S. PROBLEM SOLVING

These sets of exercises at the end of each chapter test students' abilities with challenging, thought-provoking questions.

Classic Calculus with Contemporary Relevance

THEOREMS

Theorems provide the conceptual framework for calculus. Theorems are clearly stated and separated from the rest of the text by boxes for quick visual reference. Key proofs often follow the theorem, and other proofs are provided in an in-text appendix.

THEOREM 4.9 THE FUNDAMENTAL THEOREM OF CALCULUS

If a function f is continuous on the closed interval $[a, b]$ and F is an antiderivative of f on the interval $[a, b]$, then

$$\int_a^b f(x)\,dx = F(b) - F(a).$$

DEFINITIONS

As with the theorems, definitions are clearly stated using precise, formal wording and are separated from the text by boxes for quick visual reference.

DEFINITION OF DEFINITE INTEGRAL

If f is defined on the closed interval $[a, b]$ and the limit of Riemann sums over partitions Δ

$$\lim_{\|\Delta\| \to 0} \sum_{i=1}^{n} f(c_i)\,\Delta x_i$$

exists (as described above), then f is said to be **integrable** on $[a, b]$ and the limit is denoted by

$$\lim_{\|\Delta\| \to 0} \sum_{i=1}^{n} f(c_i)\,\Delta x_i = \int_a^b f(x)\,dx.$$

The limit is called the **definite integral** of f from a to b. **lower limit** of integration, and the number b is the **uppe**

PROCEDURES

Formal procedures are set apart from the text for easy reference. The procedures provide students with step-by-step instructions that will help them solve problems quickly and efficiently.

To complete the change of variables in Example 5, you solved for x in terms of u. Sometimes this is very difficult. Fortunately it is not always necessary, as shown in the next example.

EXAMPLE 6 Change of Variables

Find $\int \sin^2 3x \cos 3x\,dx$.

Solution Because $\sin^2 3x = (\sin 3x)^2$, you can let $u = \sin 3x$. Then

$$du = (\cos 3x)(3)\,dx.$$

Now, because $\cos 3x\,dx$ is part of the original integral, you can write

$$\frac{du}{3} = \cos 3x\,dx.$$

Substituting u and $du/3$ in the original integral yields

$$\begin{aligned}\int \sin^2 3x \cos 3x\,dx &= \int u^2 \frac{du}{3}\\ &= \frac{1}{3}\int u^2\,du\\ &= \frac{1}{3}\left(\frac{u^3}{3}\right) + C\\ &= \frac{1}{9}\sin^3 3x + C.\end{aligned}$$

You can check this by differentiating.

$$\begin{aligned}\frac{d}{dx}\left[\frac{1}{9}\sin^3 3x\right] &= \left(\frac{1}{9}\right)(3)(\sin 3x)^2(\cos 3x)(3)\\ &= \sin^2 3x \cos 3x\end{aligned}$$

that you have ■

NOTES

Notes provide additional details about theorems, definitions, and examples. They offer additional insight, or important generalizations that students might not immediately see. Like the study tips, notes can be invaluable to students.

NOTE There are two important points that should be made concerning the Trapezoidal Rule (or the Midpoint Rule). First, the approximation tends to become more accurate as n increases. For instance, in Example 1, if $n = 16$, the Trapezoidal Rule yields an approximation of 1.994. Second, although you could have used the Fundamental Theorem to evaluate the integral in Example 1, this theorem cannot be used to evaluate an integral as simple as $\int_0^{\pi} \sin x^2\,dx$ because $\sin x^2$ has no elementary antiderivative. Yet, the Trapezoidal Rule can be applied easily to estimate this integral. ■

Expanding the Experience of Calculus

CHAPTER OPENERS

Chapter Openers provide initial motivation for the upcoming chapter material. Along with a map of the chapter objectives, an important concept in the chapter is related to an application of the topic in the real world. Students are encouraged to see the real-life relevance of calculus.

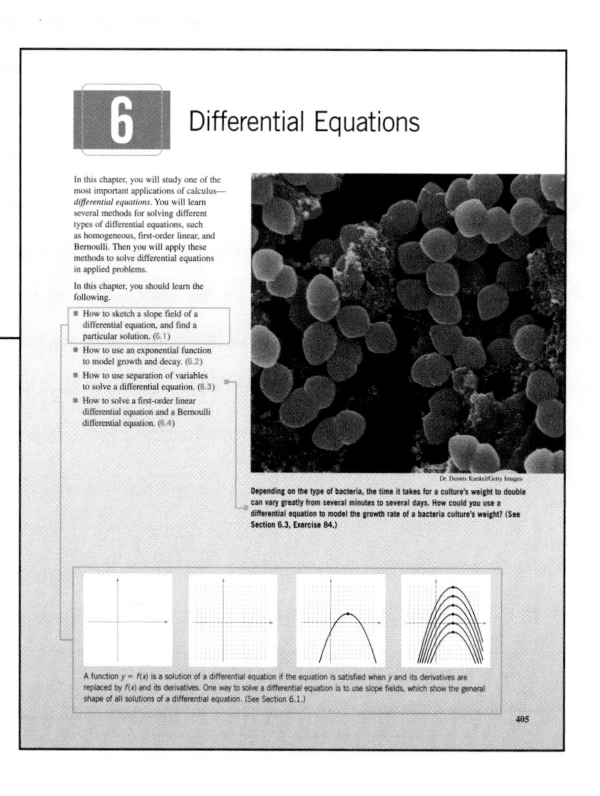
6 Differential Equations

In this chapter, you will study one of the most important applications of calculus—*differential equations*. You will learn several methods for solving different types of differential equations, such as homogeneous, first-order linear, and Bernoulli. Then you will apply these methods to solve differential equations in applied problems.

In this chapter, you should learn the following.

- How to sketch a slope field of a differential equation, and find a particular solution. (6.1)
- How to use an exponential function to model growth and decay. (6.2)
- How to use separation of variables to solve a differential equation. (6.3)
- How to solve a first-order linear differential equation and a Bernoulli differential equation. (6.4)

Dr. Dennis Kunkel/Getty Images

Depending on the type of bacteria, the time it takes for a culture's weight to double can vary greatly from several minutes to several days. How could you use a differential equation to model the growth rate of a bacteria culture's weight? (See Section 6.3, Exercise 84.)

A function $y = f(x)$ is a solution of a differential equation if the equation is satisfied when y and its derivatives are replaced by $f(x)$ and its derivatives. One way to solve a differential equation is to use slope fields, which show the general shape of all solutions of a differential equation. (See Section 6.1.)

405

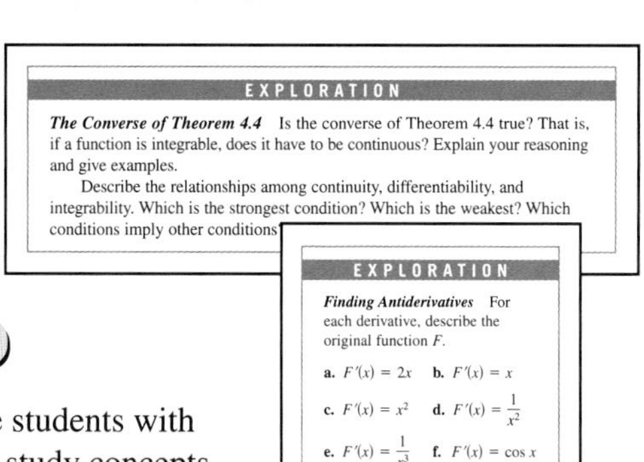
EXPLORATION

The Converse of Theorem 4.4 Is the converse of Theorem 4.4 true? That is, if a function is integrable, does it have to be continuous? Explain your reasoning and give examples.

Describe the relationships among continuity, differentiability, and integrability. Which is the strongest condition? Which is the weakest? Which conditions imply other conditions?

EXPLORATION

Finding Antiderivatives For each derivative, describe the original function F.

a. $F'(x) = 2x$ b. $F'(x) = x$

c. $F'(x) = x^2$ d. $F'(x) = \frac{1}{x^2}$

e. $F'(x) = \frac{1}{x^3}$ f. $F'(x) = \cos x$

What strategy did you use to find F?

EXPLORATIONS

Explorations provide students with unique challenges to study concepts that have not yet been formally covered. They allow students to learn by discovery and introduce topics related to ones they are presently studying. By exploring topics in this way, students are encouraged to think outside the box.

HISTORICAL NOTES AND BIOGRAPHIES

Historical Notes provide students with background information on the foundations of calculus, and Biographies help humanize calculus and teach students about the people who contributed to its formal creation.

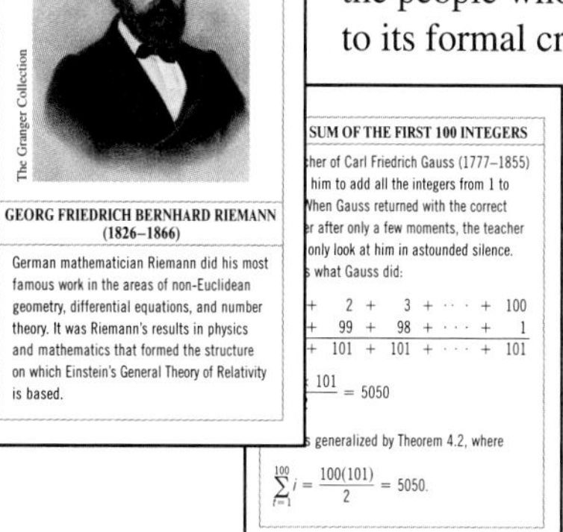
The Granger Collection

GEORG FRIEDRICH BERNHARD RIEMANN (1826–1866)

German mathematician Riemann did his most famous work in the areas of non-Euclidean geometry, differential equations, and number theory. It was Riemann's results in physics and mathematics that formed the structure on which Einstein's General Theory of Relativity is based.

SUM OF THE FIRST 100 INTEGERS

…her of Carl Friedrich Gauss (1777–1855) … him to add all the integers from 1 to … When Gauss returned with the correct … after only a few moments, the teacher … only look at him in astounded silence. … what Gauss did:

$$\begin{array}{l} \;\;\;2 + \;\;3 + \cdots + 100 \\ +\;\;99 + \;98 + \cdots + \;\;\;1 \\ \hline +\;101 + 101 + \cdots + 101 \end{array}$$

$\frac{101}{} = 5050$

… generalized by Theorem 4.2, where

$$\sum_{i=1}^{100} i = \frac{100(101)}{2} = 5050.$$

PUTNAM EXAM CHALLENGES

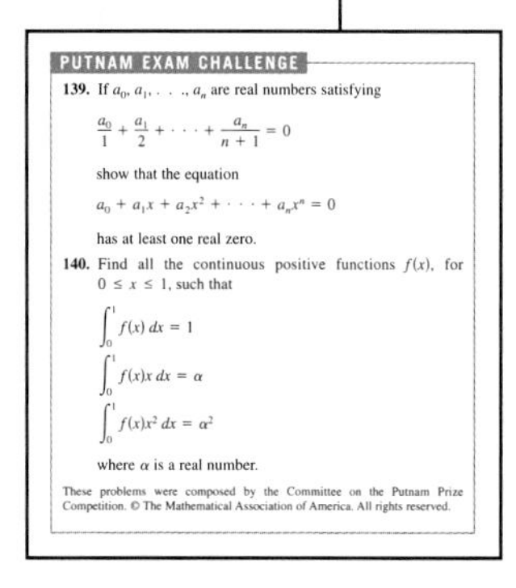
PUTNAM EXAM CHALLENGE

139. If $a_0, a_1, \ldots, a_n$ are real numbers satisfying

$$\frac{a_0}{1} + \frac{a_1}{2} + \cdots + \frac{a_n}{n+1} = 0$$

show that the equation

$$a_0 + a_1 x + a_2 x^2 + \cdots + a_n x^n = 0$$

has at least one real zero.

140. Find all the continuous positive functions $f(x)$, for $0 \le x \le 1$, such that

$$\int_0^1 f(x)\,dx = 1$$

$$\int_0^1 f(x)x\,dx = \alpha$$

$$\int_0^1 f(x)x^2\,dx = \alpha^2$$

where α is a real number.

These problems were composed by the Committee on the Putnam Prize Competition. © The Mathematical Association of America. All rights reserved.

Putnam Exam questions appear in selected sections and are drawn from actual Putnam Exams. These exercises will push the limits of students' understanding of calculus and provide extra challenges for motivated students.

SECTION PROJECTS

Projects appear in selected sections and more deeply explore applications related to the topics being studied. They provide an interesting and engaging way for students to work and investigate ideas collaboratively.

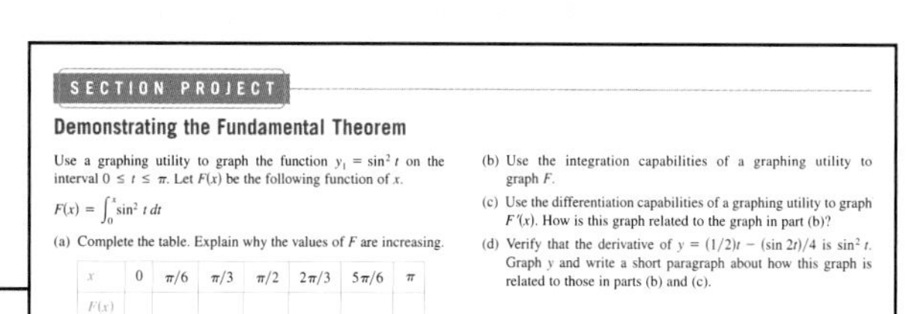
SECTION PROJECT

Demonstrating the Fundamental Theorem

Use a graphing utility to graph the function $y_1 = \sin^2 t$ on the interval $0 \le t \le \pi$. Let $F(x)$ be the following function of x.

$$F(x) = \int_0^x \sin^2 t\,dt$$

(a) Complete the table. Explain why the values of F are increasing.

x	0	$\pi/6$	$\pi/3$	$\pi/2$	$2\pi/3$	$5\pi/6$	π
$F(x)$							

(b) Use the integration capabilities of a graphing utility to graph F.

(c) Use the differentiation capabilities of a graphing utility to graph $F'(x)$. How is this graph related to the graph in part (b)?

(d) Verify that the derivative of $y = (1/2)t - (\sin 2t)/4$ is $\sin^2 t$. Graph y and write a short paragraph about how this graph is related to those in parts (b) and (c).

Integrated Technology for Today's World

 EXAMPLE 5 **Change of Variables**

Find $\int x\sqrt{2x-1}\,dx$.

Solution As in the previous example, let $u = 2x - 1$ and obtain $dx = du/2$. Because the integrand contains a factor of x, you must also solve for x in terms of u, as shown.

$$u = 2x - 1 \implies x = (u+1)/2 \qquad \text{Solve for } x \text{ in terms of } u.$$

Now, using substitution, you obtain

$$\begin{aligned}\int x\sqrt{2x-1}\,dx &= \int \left(\frac{u+1}{2}\right) u^{1/2} \left(\frac{du}{2}\right)\\ &= \frac{1}{4}\int (u^{3/2} + u^{1/2})\,du\\ &= \frac{1}{4}\left(\frac{u^{5/2}}{5/2} + \frac{u^{3/2}}{3/2}\right) + C\\ &= \frac{1}{10}(2x-1)^{5/2} + \frac{1}{6}(2x-1)^{3/2} + C.\end{aligned}$$

CAS INVESTIGATIONS

Examples throughout the book are accompanied by CAS Investigations. These investigations are linked explorations that use a computer algebra system (e.g., Maple®) to further explore a related example in the book. They allow students to explore calculus by manipulating functions, graphs, etc. and observing the results. (Formerly called Open Explorations)

Slope Fields **In Exercises 55 and 56, (a) use a graphing utility to graph a slope field for the differential equation, (b) use integration and the given point to find the particular solution of the differential equation, and (c) graph the solution and the slope field in the same viewing window.**

55. $\dfrac{dy}{dx} = 2x,\ (-2, -2)$ **56.** $\dfrac{dy}{dx} = 2\sqrt{x},\ (4, 12)$

GRAPHING TECH EXERCISES

Understanding is often enhanced by using a graph or visualization. Graphing Tech Exercises are exercises that ask students to make use of a graphing utility to help find a solution. These exercises are marked with a special icon.

TECHNOLOGY

Throughout the book, technology boxes give students a glimpse of how technology may be used to help solve problems and explore the concepts of calculus. They provide discussions of not only where technology succeeds, but also where it may fail.

CAS **49.** ***Investigation*** Consider the function

$$f(x, y) = x^2 - y^2$$

at the point $(4, -3, 7)$.

(a) Use a computer algebra system to graph the surface represented by the function.

(b) Determine the directional derivative $D_{\mathbf{u}}f(4, -3)$ as a function of θ, where $\mathbf{u} = \cos\theta\,\mathbf{i} + \sin\theta\,\mathbf{j}$. Use a computer algebra system to graph the function on the interval $[0, 2\pi)$.

(c) Approximate the zeros of the function in part (b) and interpret each in the context of the problem.

(d) Approximate the critical numbers of the function in part (b) and interpret each in the context of the problem.

(e) Find $\|\nabla f(4, -3)\|$ and explain its relationship to your answers in part (d).

(f) Use a computer algebra syst… of the function f at the level … the vector in the direction … relationship to the level curve.

CAS **In Exercises 79–82, use a computer algebra system to graph the plane.**

79. $2x + y - z = 6$ **80.** $x - 3z = 3$

81. $-5x + 4y - 6z = -8$ **82.** $2.1x - 4.7y - z = -3$

In Exercises 83–86, determine if any of the planes are parallel or identical.

CAS **In Exercises 21–24, use a computer algebra system to find u × v and a unit vector orthogonal to u and v.**

21. $\mathbf{u} = \langle 4, -3.5, 7\rangle$, $\mathbf{v} = \langle 2.5, 9, 3\rangle$ **22.** $\mathbf{u} = \langle -8, -6, 4\rangle$, $\mathbf{v} = \langle 10, -12, -2\rangle$

23. $\mathbf{u} = -3\mathbf{i} + 2\mathbf{j} - 5\mathbf{k}$, $\mathbf{v} = 0.4\mathbf{i} - 0.8\mathbf{j} + 0.2\mathbf{k}$ **24.** $\mathbf{u} = 0.7\mathbf{k}$, $\mathbf{v} = 1.5\mathbf{i} + 6.2\mathbf{k}$

TECHNOLOGY Most graphing utilities and computer algebra systems have built-in programs that can be used to approximate the value of a definite integral. Try using such a program to approximate the integral in Example 1. How close is your approximation?

When you use such a program, you need to be aware of its limitations. Often, you are given no indication of the degree of accuracy of the approximation. Other times, you may be given an approximation that is completely wrong. For instance, try using a built-in numerical integration program to evaluate

$$\int_{-1}^{2} \frac{1}{x}\,dx.$$

Your calculator should give an error message. Does yours?

CAS EXERCISES

NEW! Like the Graphing Tech Exercises, some exercises may best be solved using a computer algebra system. These CAS Exercises are new to this edition and are denoted by a special icon.

Additional Resources

Student Resources

- **Student Solutions Manual**—Need a leg up on your homework or help to prepare for an exam? The *Student Solutions Manual* contains worked-out solutions for all odd-numbered exercises in the text. It is a great resource to help you understand how to solve those tough problems.
- **Notetaking Guide**—This notebook organizer is designed to help you organize your notes, and provides section-by-section summaries of key topics and other helpful study tools. The Notetaking Guide is available for download on the book's website.
- **WebAssign®**—The most widely used homework system in higher education, WebAssign offers instant feedback and repeatable problems, everything you could ask for in an online homework system. WebAssign's homework system lets you practice and submit homework via the web. It is easy to use and loaded with extra resources. With this edition of Larson's *Calculus*, there are over 3,000 algorithmic homework exercises to use for practice and review.
- **DVD Lecture Series**—Comprehensive, instructional lecture presentations serve a number of uses. They are great if you need to catch up after missing a class, need to supplement online or hybrid instruction, or need material for self-study or review.
- **CalcLabs with Maple® and Mathematica®**—Working with *Maple* or *Mathematica* in class? Be sure to pick up one of these comprehensive manuals that will help you use each program efficiently.

Instructor Resources

- **WebAssign®**—Instant feedback, grading precision, and ease of use are just three reasons why WebAssign is the most widely used homework system in higher education. WebAssign's homework delivery system lets instructors deliver, collect, grade, and record assignments via the web. With this edition of Larson's *Calculus,* there are over 3,000 algorithmic homework exercises to choose from. These algorithmic exercises are based on the section exercises from the textbook to ensure alignment with course goals.
- **Instructor's Complete Solutions Manual**—This manual contains worked-out solutions for all exercises in the text. It also contains solutions for the special features in the text such as Explorations, Section Projects, etc. It is available on the *Instructor's Resource Center* at the book's website.
- **Instructor's Resource Manual**—This robust manual contains an abundance of resources keyed to the textbook by chapter and section, including chapter summaries and teaching strategies. New to this edition's manual are the authors' findings from CalcChat.com (*see* A Word from the Authors). They offer suggestions for exercises to cover in class, identify tricky exercises with tips on how best to use them, and explain what changes were made in the exercise set based on the research.
- **Power Lecture**—This comprehensive CD-ROM includes the *Instructor's Complete Solutions Manual*, PowerPoint® slides, and the computerized test bank featuring algorithmically created questions that can be used to create, deliver, and customize tests.
- **Computerized Test Bank**—Create, deliver, and customize tests and study guides in minutes with this easy to use assessment software on CD. The thousands of algorithmic questions in the test bank are derived from the textbook exercises, ensuring consistency between exams and the book.
- **JoinIn on TurningPoint**—Enhance your students' interactions with you, your lectures, and each other. Cengage Learning is now pleased to offer you book-specific content for Response Systems tailored to Larson's *Calculus*, allowing you to transform your classroom and assess your students' progress with instant in-class quizzes and polls.

11 Vectors and the Geometry of Space

This chapter introduces vectors and the three-dimensional coordinate system. Vectors are used to represent lines and planes, and are also used to represent quantities such as force and velocity. The three-dimensional coordinate system is used to represent surfaces such as ellipsoids and elliptical cones. Much of the material in the remaining chapters relies on an understanding of this system.

In this chapter, you should learn the following.

- How to write vectors, perform basic vector operations, and represent vectors graphically. (11.1)
- How to plot points in a three-dimensional coordinate system and analyze vectors in space. (11.2)
- How to find the dot product of two vectors (in the plane or in space). (11.3)
- How to find the cross product of two vectors (in space). (11.4)
- How to find equations of lines and planes in space, and how to sketch their graphs. (11.5)
- How to recognize and write equations of cylindrical and quadric surfaces and of surfaces of revolution. (11.6)
- How to use cylindrical and spherical coordinates to represent surfaces in space. (11.7)

Mark Hunt/Hunt Stock

Two tugboats are pushing an ocean liner, as shown above. Each boat is exerting a force of 400 pounds. What is the resultant force on the ocean liner? (See Section 11.1, Example 7.)

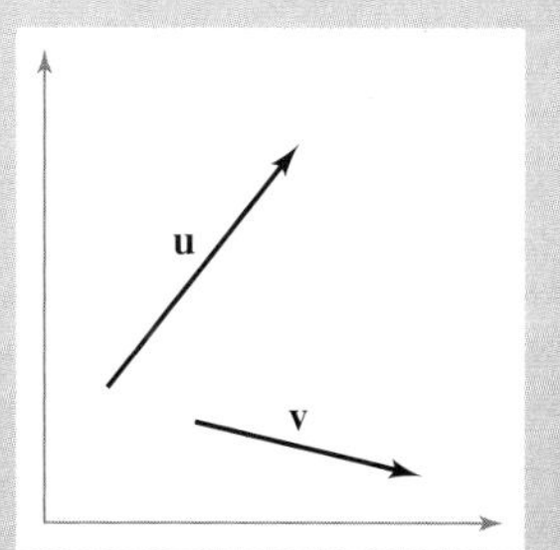

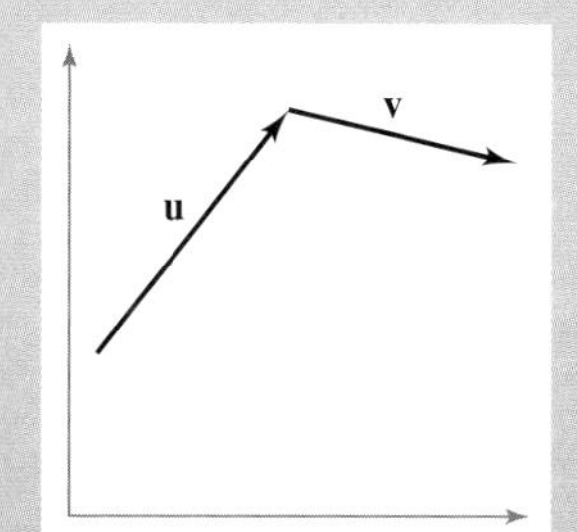

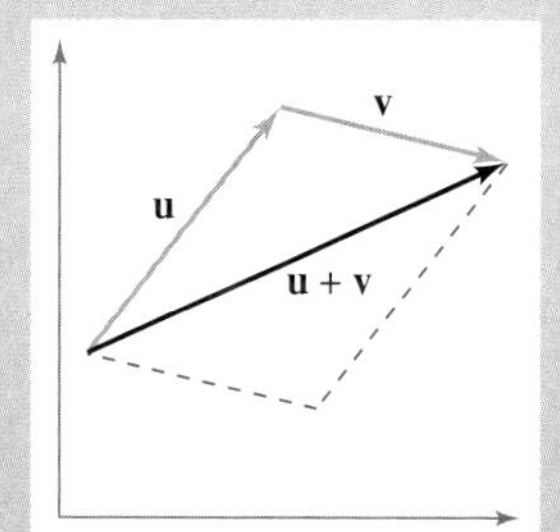

Vectors indicate quantities that involve both magnitude and direction. In Chapter 11, you will study operations of vectors in the plane and in space. You will also learn how to represent vector operations geometrically. For example, the graphs shown above represent vector addition in the plane.

11.1 Vectors in the Plane

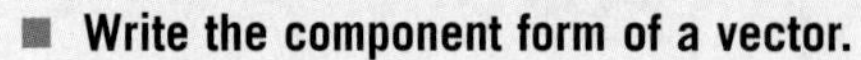

- **Write the component form of a vector.**
- **Perform vector operations and interpret the results geometrically.**
- **Write a vector as a linear combination of standard unit vectors.**
- **Use vectors to solve problems involving force or velocity.**

Component Form of a Vector

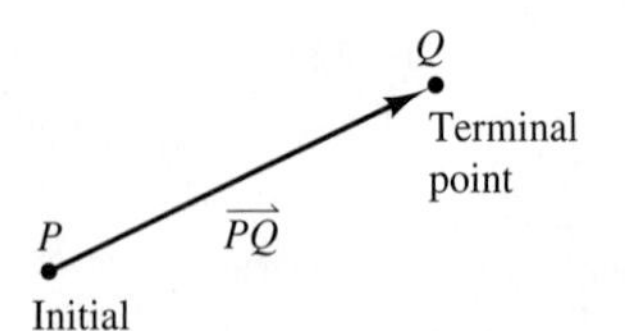

A directed line segment
Figure 11.1

Many quantities in geometry and physics, such as area, volume, temperature, mass, and time, can be characterized by a single real number scaled to appropriate units of measure. These are called **scalar quantities,** and the real number associated with each is called a **scalar.**

Other quantities, such as force, velocity, and acceleration, involve both magnitude and direction and cannot be characterized completely by a single real number. A **directed line segment** is used to represent such a quantity, as shown in Figure 11.1. The directed line segment $\overrightarrow{PQ}$ has **initial point** P and **terminal point** Q, and its **length** (or **magnitude**) is denoted by $\|\overrightarrow{PQ}\|$. Directed line segments that have the same length and direction are **equivalent,** as shown in Figure 11.2. The set of all directed line segments that are equivalent to a given directed line segment $\overrightarrow{PQ}$ is a **vector in the plane** and is denoted by $\mathbf{v} = \overrightarrow{PQ}$. In typeset material, vectors are usually denoted by lowercase, boldface letters such as **u**, **v**, and **w**. When written by hand, however, vectors are often denoted by letters with arrows above them, such as $\vec{u}$, $\vec{v}$, and $\vec{w}$.

Be sure you understand that a vector represents a *set* of directed line segments (each having the same length and direction). In practice, however, it is common not to distinguish between a vector and one of its representatives.

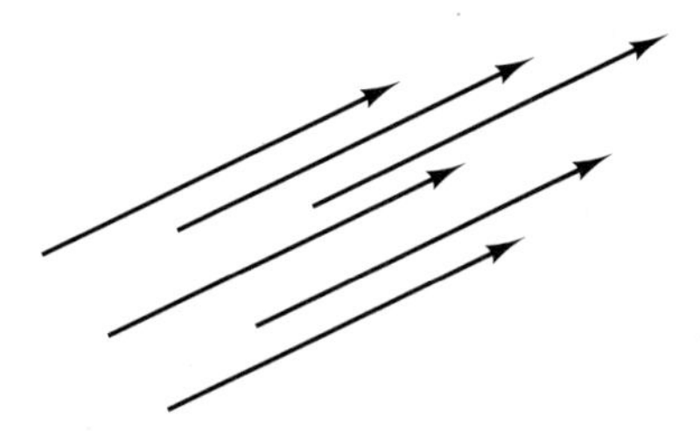

Equivalent directed line segments
Figure 11.2

EXAMPLE 1 Vector Representation by Directed Line Segments

Let **v** be represented by the directed line segment from $(0, 0)$ to $(3, 2)$, and let **u** be represented by the directed line segment from $(1, 2)$ to $(4, 4)$. Show that **v** and **u** are equivalent.

Solution Let $P(0, 0)$ and $Q(3, 2)$ be the initial and terminal points of **v**, and let $R(1, 2)$ and $S(4, 4)$ be the initial and terminal points of **u**, as shown in Figure 11.3. You can use the Distance Formula to show that $\overrightarrow{PQ}$ and $\overrightarrow{RS}$ have the *same length.*

$$\|\overrightarrow{PQ}\| = \sqrt{(3-0)^2 + (2-0)^2} = \sqrt{13} \qquad \text{Length of } \overrightarrow{PQ}$$
$$\|\overrightarrow{RS}\| = \sqrt{(4-1)^2 + (4-2)^2} = \sqrt{13} \qquad \text{Length of } \overrightarrow{RS}$$

Both line segments have the *same direction*, because they both are directed toward the upper right on lines having the same slope.

$$\text{Slope of } \overrightarrow{PQ} = \frac{2-0}{3-0} = \frac{2}{3}$$

and

$$\text{Slope of } \overrightarrow{RS} = \frac{4-2}{4-1} = \frac{2}{3}$$

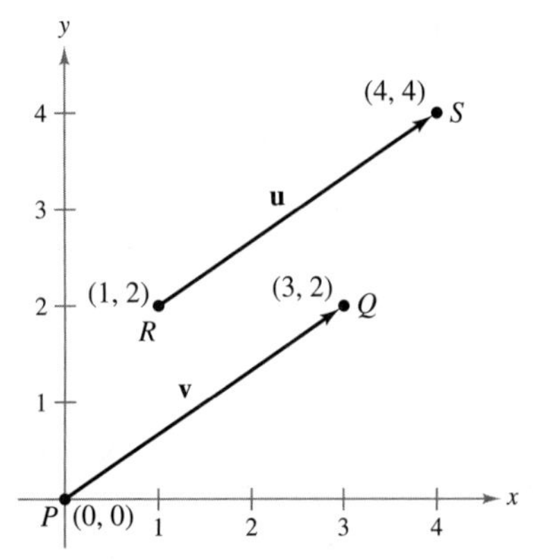

The vectors **u** and **v** are equivalent.
Figure 11.3

Because $\overrightarrow{PQ}$ and $\overrightarrow{RS}$ have the same length and direction, you can conclude that the two vectors are equivalent. That is, **v** and **u** are equivalent. ■

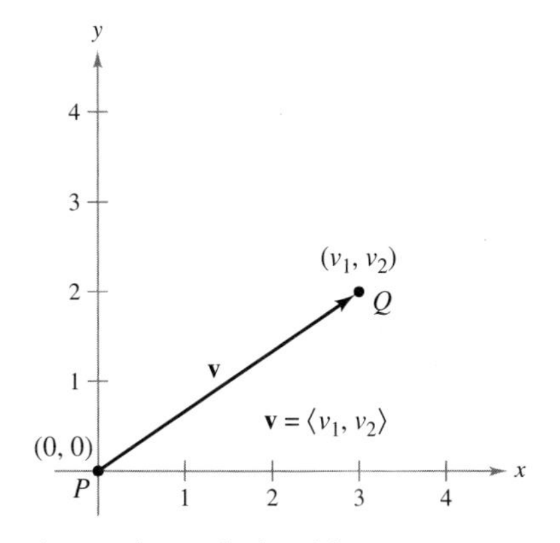

A vector in standard position
Figure 11.4

The directed line segment whose initial point is the origin is often the most convenient representative of a set of equivalent directed line segments such as those shown in Figure 11.3. This representation of **v** is said to be in **standard position.** A directed line segment whose initial point is the origin can be uniquely represented by the coordinates of its terminal point $Q(v_1, v_2)$, as shown in Figure 11.4.

> **DEFINITION OF COMPONENT FORM OF A VECTOR IN THE PLANE**
>
> If **v** is a vector in the plane whose initial point is the origin and whose terminal point is (v_1, v_2), then the **component form of v** is given by
>
> $$\mathbf{v} = \langle v_1, v_2 \rangle.$$
>
> The coordinates v_1 and v_2 are called the **components of v.** If both the initial point and the terminal point lie at the origin, then **v** is called the **zero vector** and is denoted by $\mathbf{0} = \langle 0, 0 \rangle$.

This definition implies that two vectors $\mathbf{u} = \langle u_1, u_2 \rangle$ and $\mathbf{v} = \langle v_1, v_2 \rangle$ are **equal** if and only if $u_1 = v_1$ and $u_2 = v_2$.

The following procedures can be used to convert directed line segments to component form or vice versa.

1. If $P(p_1, p_2)$ and $Q(q_1, q_2)$ are the initial and terminal points of a directed line segment, the component form of the vector **v** represented by $\overrightarrow{PQ}$ is $\langle v_1, v_2 \rangle = \langle q_1 - p_1, q_2 - p_2 \rangle$. Moreover, from the Distance Formula you can see that the **length** (or **magnitude**) **of v** is

$$\begin{aligned} \|\mathbf{v}\| &= \sqrt{(q_1 - p_1)^2 + (q_2 - p_2)^2} \\ &= \sqrt{v_1^2 + v_2^2}. \end{aligned}$$ Length of a vector

2. If $\mathbf{v} = \langle v_1, v_2 \rangle$, **v** can be represented by the directed line segment, in standard position, from $P(0, 0)$ to $Q(v_1, v_2)$.

The length of **v** is also called the **norm of v.** If $\|\mathbf{v}\| = 1$, **v** is a **unit vector.** Moreover, $\|\mathbf{v}\| = 0$ if and only if **v** is the zero vector **0**.

EXAMPLE 2 Finding the Component Form and Length of a Vector

Find the component form and length of the vector **v** that has initial point $(3, -7)$ and terminal point $(-2, 5)$.

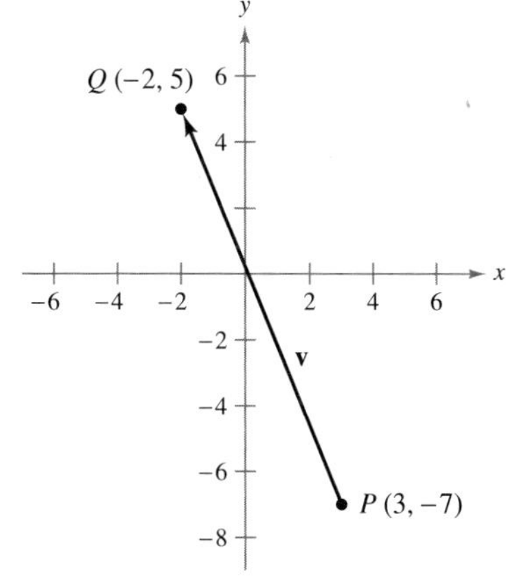

Component form of **v**: $\mathbf{v} = \langle -5, 12 \rangle$
Figure 11.5

Solution Let $P(3, -7) = (p_1, p_2)$ and $Q(-2, 5) = (q_1, q_2)$. Then the components of $\mathbf{v} = \langle v_1, v_2 \rangle$ are

$$v_1 = q_1 - p_1 = -2 - 3 = -5$$
$$v_2 = q_2 - p_2 = 5 - (-7) = 12.$$

So, as shown in Figure 11.5, $\mathbf{v} = \langle -5, 12 \rangle$, and the length of **v** is

$$\begin{aligned} \|\mathbf{v}\| &= \sqrt{(-5)^2 + 12^2} \\ &= \sqrt{169} \\ &= 13. \end{aligned}$$

■

Vector Operations

DEFINITIONS OF VECTOR ADDITION AND SCALAR MULTIPLICATION

Let $\mathbf{u} = \langle u_1, u_2 \rangle$ and $\mathbf{v} = \langle v_1, v_2 \rangle$ be vectors and let c be a scalar.

1. The **vector sum** of $\mathbf{u}$ and $\mathbf{v}$ is the vector $\mathbf{u} + \mathbf{v} = \langle u_1 + v_1, u_2 + v_2 \rangle$.
2. The **scalar multiple** of c and $\mathbf{u}$ is the vector $c\mathbf{u} = \langle cu_1, cu_2 \rangle$.
3. The **negative** of $\mathbf{v}$ is the vector

$$-\mathbf{v} = (-1)\mathbf{v} = \langle -v_1, -v_2 \rangle.$$

4. The **difference** of $\mathbf{u}$ and $\mathbf{v}$ is

$$\mathbf{u} - \mathbf{v} = \mathbf{u} + (-\mathbf{v}) = \langle u_1 - v_1, u_2 - v_2 \rangle.$$

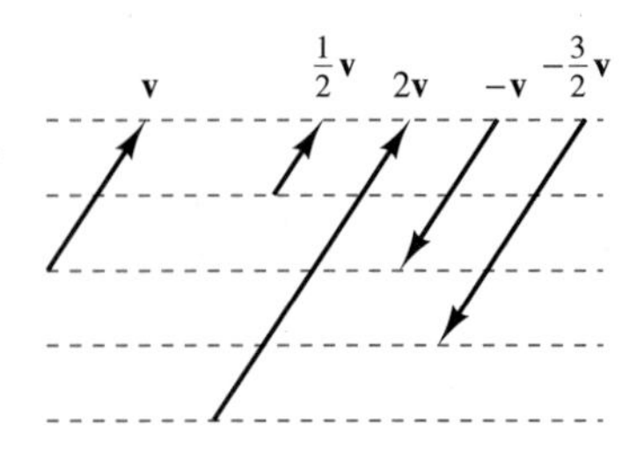

The scalar multiplication of v
Figure 11.6

Geometrically, the scalar multiple of a vector $\mathbf{v}$ and a scalar c is the vector that is $|c|$ times as long as $\mathbf{v}$, as shown in Figure 11.6. If c is positive, $c\mathbf{v}$ has the same direction as $\mathbf{v}$. If c is negative, $c\mathbf{v}$ has the opposite direction.

The sum of two vectors can be represented geometrically by positioning the vectors (without changing their magnitudes or directions) so that the initial point of one coincides with the terminal point of the other, as shown in Figure 11.7. The vector $\mathbf{u} + \mathbf{v}$, called the **resultant vector,** is the diagonal of a parallelogram having $\mathbf{u}$ and $\mathbf{v}$ as its adjacent sides.

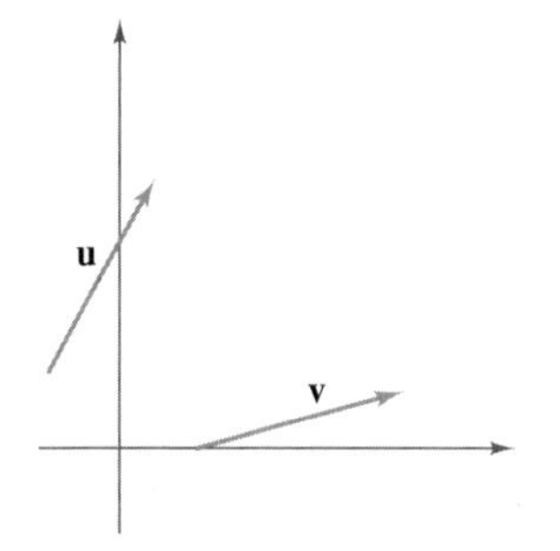

To find $\mathbf{u} + \mathbf{v}$,

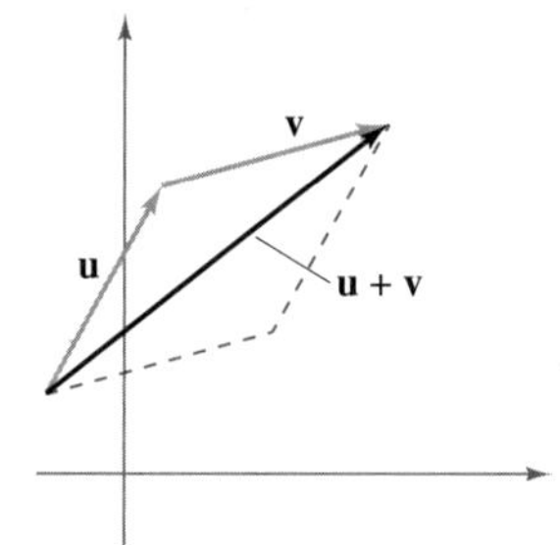

(1) move the initial point of v to the terminal point of u, or

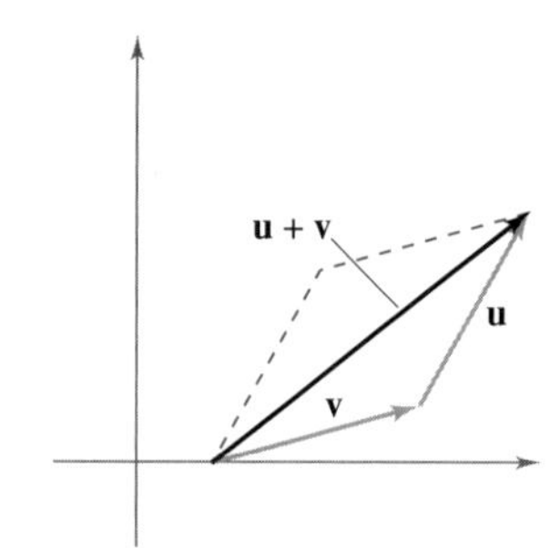

(2) move the initial point of u to the terminal point of v.

Figure 11.7

WILLIAM ROWAN HAMILTON (1805–1865)

Some of the earliest work with vectors was done by the Irish mathematician William Rowan Hamilton. Hamilton spent many years developing a system of vector-like quantities called *quaternions.* Although Hamilton was convinced of the benefits of quaternions, the operations he defined did not produce good models for physical phenomena. It wasn't until the latter half of the nineteenth century that the Scottish physicist James Maxwell (1831–1879) restructured Hamilton's quaternions in a form useful for representing physical quantities such as force, velocity, and acceleration.

Figure 11.8 shows the equivalence of the geometric and algebraic definitions of vector addition and scalar multiplication, and presents (at far right) a geometric interpretation of $\mathbf{u} - \mathbf{v}$.

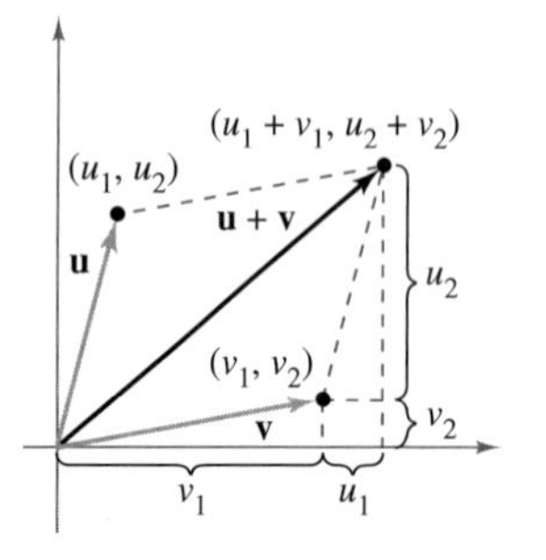

Vector addition

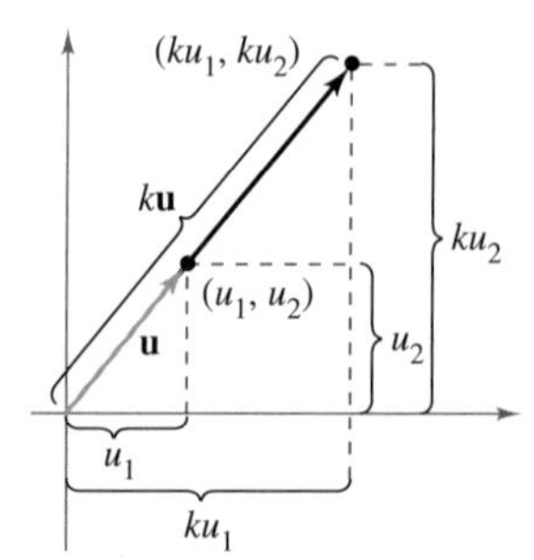

Scalar multiplication

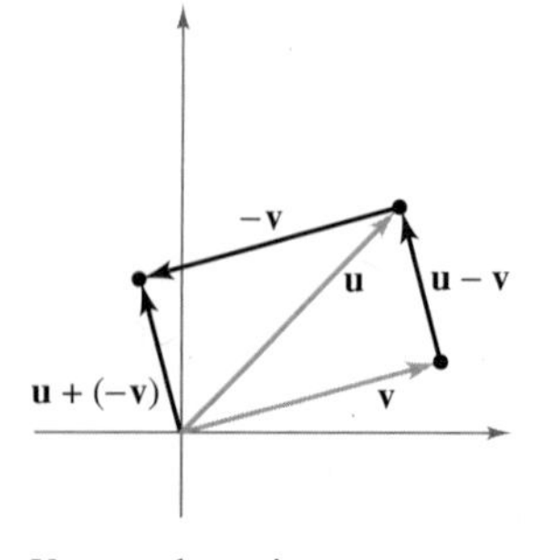

Vector subtraction

Figure 11.8

EXAMPLE 3 Vector Operations

Given $\mathbf{v} = \langle -2, 5 \rangle$ and $\mathbf{w} = \langle 3, 4 \rangle$, find each of the vectors.

a. $\frac{1}{2}\mathbf{v}$ **b.** $\mathbf{w} - \mathbf{v}$ **c.** $\mathbf{v} + 2\mathbf{w}$

Solution

a. $\frac{1}{2}\mathbf{v} = \left\langle \frac{1}{2}(-2), \frac{1}{2}(5) \right\rangle = \left\langle -1, \frac{5}{2} \right\rangle$

b. $\mathbf{w} - \mathbf{v} = \langle w_1 - v_1, w_2 - v_2 \rangle = \langle 3 - (-2), 4 - 5 \rangle = \langle 5, -1 \rangle$

c. Using $2\mathbf{w} = \langle 6, 8 \rangle$, you have

$$\begin{aligned}\mathbf{v} + 2\mathbf{w} &= \langle -2, 5 \rangle + \langle 6, 8 \rangle \\ &= \langle -2 + 6, 5 + 8 \rangle \\ &= \langle 4, 13 \rangle.\end{aligned}$$

■

Vector addition and scalar multiplication share many properties of ordinary arithmetic, as shown in the following theorem.

THEOREM 11.1 PROPERTIES OF VECTOR OPERATIONS

Let $\mathbf{u}$, $\mathbf{v}$, and $\mathbf{w}$ be vectors in the plane, and let c and d be scalars.

1. $\mathbf{u} + \mathbf{v} = \mathbf{v} + \mathbf{u}$	Commutative Property
2. $(\mathbf{u} + \mathbf{v}) + \mathbf{w} = \mathbf{u} + (\mathbf{v} + \mathbf{w})$	Associative Property
3. $\mathbf{u} + \mathbf{0} = \mathbf{u}$	Additive Identity Property
4. $\mathbf{u} + (-\mathbf{u}) = \mathbf{0}$	Additive Inverse Property
5. $c(d\mathbf{u}) = (cd)\mathbf{u}$	
6. $(c + d)\mathbf{u} = c\mathbf{u} + d\mathbf{u}$	Distributive Property
7. $c(\mathbf{u} + \mathbf{v}) = c\mathbf{u} + c\mathbf{v}$	Distributive Property
8. $1(\mathbf{u}) = \mathbf{u}, 0(\mathbf{u}) = \mathbf{0}$	

PROOF The proof of the *Associative Property* of vector addition uses the Associative Property of addition of real numbers.

$$\begin{aligned}(\mathbf{u} + \mathbf{v}) + \mathbf{w} &= [\langle u_1, u_2 \rangle + \langle v_1, v_2 \rangle] + \langle w_1, w_2 \rangle \\ &= \langle u_1 + v_1, u_2 + v_2 \rangle + \langle w_1, w_2 \rangle \\ &= \langle (u_1 + v_1) + w_1, (u_2 + v_2) + w_2 \rangle \\ &= \langle u_1 + (v_1 + w_1), u_2 + (v_2 + w_2) \rangle \\ &= \langle u_1, u_2 \rangle + \langle v_1 + w_1, v_2 + w_2 \rangle = \mathbf{u} + (\mathbf{v} + \mathbf{w})\end{aligned}$$

Similarly, the proof of the *Distributive Property* of vectors depends on the Distributive Property of real numbers.

$$\begin{aligned}(c + d)\mathbf{u} &= (c + d)\langle u_1, u_2 \rangle \\ &= \langle (c + d)u_1, (c + d)u_2 \rangle \\ &= \langle cu_1 + du_1, cu_2 + du_2 \rangle \\ &= \langle cu_1, cu_2 \rangle + \langle du_1, du_2 \rangle = c\mathbf{u} + d\mathbf{u}\end{aligned}$$

The other properties can be proved in a similar manner. ■

EMMY NOETHER (1882–1935)

One person who contributed to our knowledge of axiomatic systems was the German mathematician Emmy Noether. Noether is generally recognized as the leading woman mathematician in recent history.

Any set of vectors (with an accompanying set of scalars) that satisfies the eight properties given in Theorem 11.1 is a **vector space.*** The eight properties are the *vector space axioms*. So, this theorem states that the set of vectors in the plane (with the set of real numbers) forms a vector space.

THEOREM 11.2 LENGTH OF A SCALAR MULTIPLE

Let **v** be a vector and let c be a scalar. Then

$$\|c\mathbf{v}\| = |c|\,\|\mathbf{v}\|. \qquad |c| \text{ is the absolute value of } c.$$

PROOF Because $c\mathbf{v} = \langle cv_1, cv_2 \rangle$, it follows that

$$\begin{aligned} \|c\mathbf{v}\| = \|\langle cv_1, cv_2\rangle\| &= \sqrt{(cv_1)^2 + (cv_2)^2} \\ &= \sqrt{c^2v_1^2 + c^2v_2^2} \\ &= \sqrt{c^2(v_1^2 + v_2^2)} \\ &= |c|\sqrt{v_1^2 + v_2^2} \\ &= |c|\,\|\mathbf{v}\|. \end{aligned}$$

■

■ **FOR FURTHER INFORMATION** For more information on Emmy Noether, see the article "Emmy Noether, Greatest Woman Mathematician" by Clark Kimberling in *The Mathematics Teacher*. To view this article, go to the website *www.matharticles.com*.

In many applications of vectors, it is useful to find a unit vector that has the same direction as a given vector. The following theorem gives a procedure for doing this.

THEOREM 11.3 UNIT VECTOR IN THE DIRECTION OF v

If **v** is a nonzero vector in the plane, then the vector

$$\mathbf{u} = \frac{\mathbf{v}}{\|\mathbf{v}\|} = \frac{1}{\|\mathbf{v}\|}\mathbf{v}$$

has length 1 and the same direction as **v**.

PROOF Because $1/\|\mathbf{v}\|$ is positive and $\mathbf{u} = (1/\|\mathbf{v}\|)\mathbf{v}$, you can conclude that **u** has the same direction as **v**. To see that $\|\mathbf{u}\| = 1$, note that

$$\begin{aligned} \|\mathbf{u}\| &= \left\|\left(\frac{1}{\|\mathbf{v}\|}\right)\mathbf{v}\right\| \\ &= \left|\frac{1}{\|\mathbf{v}\|}\right|\,\|\mathbf{v}\| \\ &= \frac{1}{\|\mathbf{v}\|}\|\mathbf{v}\| \\ &= 1. \end{aligned}$$

So, **u** has length 1 and the same direction as **v**. ■

In Theorem 11.3, **u** is called a **unit vector in the direction of v.** The process of multiplying **v** by $1/\|\mathbf{v}\|$ to get a unit vector is called **normalization of v.**

---

* *For more information about vector spaces, see* Elementary Linear Algebra, *Sixth Edition, by Larson and Falvo (Boston: Houghton Mifflin Harcourt Publishing Company, 2009).*

EXAMPLE 4 Finding a Unit Vector

Find a unit vector in the direction of $\mathbf{v} = \langle -2, 5\rangle$ and verify that it has length 1.

Solution From Theorem 11.3, the unit vector in the direction of $\mathbf{v}$ is

$$\frac{\mathbf{v}}{\|\mathbf{v}\|} = \frac{\langle -2, 5\rangle}{\sqrt{(-2)^2 + (5)^2}} = \frac{1}{\sqrt{29}}\langle -2, 5\rangle = \left\langle \frac{-2}{\sqrt{29}}, \frac{5}{\sqrt{29}} \right\rangle.$$

This vector has length 1, because

$$\sqrt{\left(\frac{-2}{\sqrt{29}}\right)^2 + \left(\frac{5}{\sqrt{29}}\right)^2} = \sqrt{\frac{4}{29} + \frac{25}{29}} = \sqrt{\frac{29}{29}} = 1.$$

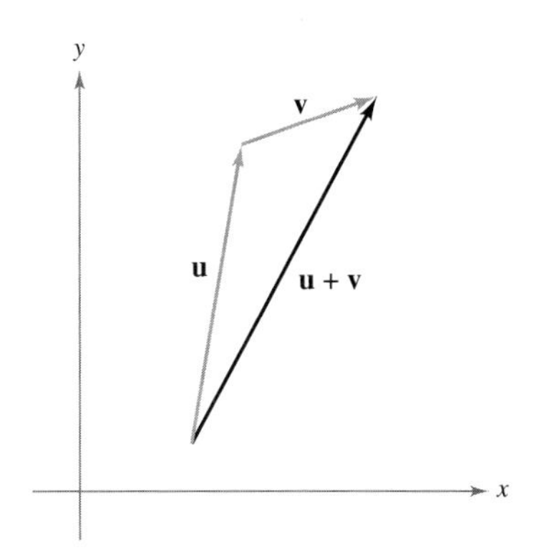

Triangle inequality
Figure 11.9

Generally, the length of the sum of two vectors is not equal to the sum of their lengths. To see this, consider the vectors $\mathbf{u}$ and $\mathbf{v}$ as shown in Figure 11.9. By considering $\mathbf{u}$ and $\mathbf{v}$ as two sides of a triangle, you can see that the length of the third side is $\|\mathbf{u} + \mathbf{v}\|$, and you have

$$\|\mathbf{u} + \mathbf{v}\| \le \|\mathbf{u}\| + \|\mathbf{v}\|.$$

Equality occurs only if the vectors $\mathbf{u}$ and $\mathbf{v}$ have the *same direction.* This result is called the **triangle inequality** for vectors. (You are asked to prove this in Exercise 91, Section 11.3.)

Standard Unit Vectors

The unit vectors $\langle 1, 0\rangle$ and $\langle 0, 1\rangle$ are called the **standard unit vectors** in the plane and are denoted by

$$\mathbf{i} = \langle 1, 0\rangle \quad \text{and} \quad \mathbf{j} = \langle 0, 1\rangle$$ Standard unit vectors

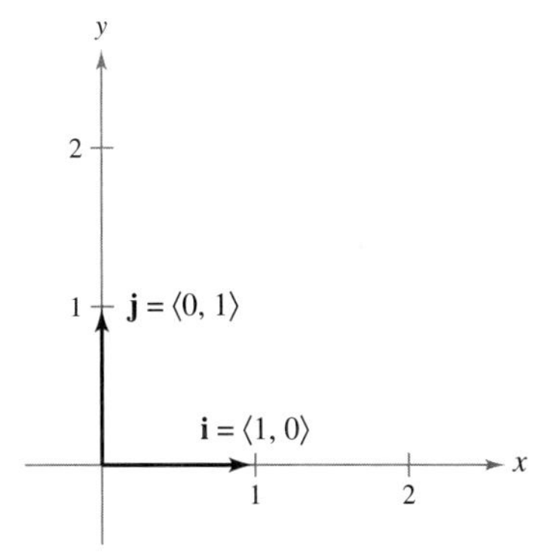

Standard unit vectors **i** and **j**
Figure 11.10

as shown in Figure 11.10. These vectors can be used to represent any vector uniquely, as follows.

$$\mathbf{v} = \langle v_1, v_2\rangle = \langle v_1, 0\rangle + \langle 0, v_2\rangle = v_1\langle 1, 0\rangle + v_2\langle 0, 1\rangle = v_1\mathbf{i} + v_2\mathbf{j}$$

The vector $\mathbf{v} = v_1\mathbf{i} + v_2\mathbf{j}$ is called a **linear combination** of $\mathbf{i}$ and $\mathbf{j}$. The scalars v_1 and v_2 are called the **horizontal** and **vertical components of v.**

EXAMPLE 5 Writing a Linear Combination of Unit Vectors

Let $\mathbf{u}$ be the vector with initial point $(2, -5)$ and terminal point $(-1, 3)$, and let $\mathbf{v} = 2\mathbf{i} - \mathbf{j}$. Write each vector as a linear combination of $\mathbf{i}$ and $\mathbf{j}$.

a. $\mathbf{u}$ **b.** $\mathbf{w} = 2\mathbf{u} - 3\mathbf{v}$

Solution

a. $\mathbf{u} = \langle q_1 - p_1, q_2 - p_2\rangle$
$= \langle -1 - 2, 3 - (-5)\rangle$
$= \langle -3, 8\rangle = -3\mathbf{i} + 8\mathbf{j}$

b. $\mathbf{w} = 2\mathbf{u} - 3\mathbf{v} = 2(-3\mathbf{i} + 8\mathbf{j}) - 3(2\mathbf{i} - \mathbf{j})$
$= -6\mathbf{i} + 16\mathbf{j} - 6\mathbf{i} + 3\mathbf{j}$
$= -12\mathbf{i} + 19\mathbf{j}$

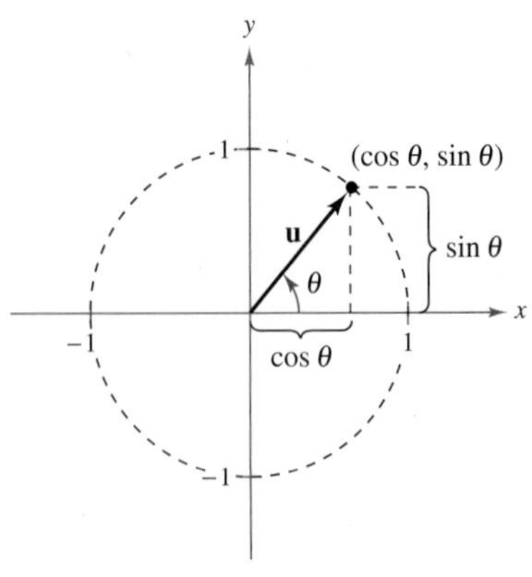

The angle θ from the positive x-axis to the vector $\mathbf{u}$
Figure 11.11

If $\mathbf{u}$ is a unit vector and θ is the angle (measured counterclockwise) from the positive x-axis to $\mathbf{u}$, then the terminal point of $\mathbf{u}$ lies on the unit circle, and you have

$$\mathbf{u} = \langle \cos\theta, \sin\theta\rangle = \cos\theta\mathbf{i} + \sin\theta\mathbf{j} \qquad \text{Unit vector}$$

as shown in Figure 11.11. Moreover, it follows that any other nonzero vector $\mathbf{v}$ making an angle θ with the positive x-axis has the same direction as $\mathbf{u}$, and you can write

$$\mathbf{v} = \|\mathbf{v}\|\langle \cos\theta, \sin\theta\rangle = \|\mathbf{v}\|\cos\theta\mathbf{i} + \|\mathbf{v}\|\sin\theta\mathbf{j}.$$

EXAMPLE 6 Writing a Vector of Given Magnitude and Direction

The vector $\mathbf{v}$ has a magnitude of 3 and makes an angle of $30° = \pi/6$ with the positive x-axis. Write $\mathbf{v}$ as a linear combination of the unit vectors $\mathbf{i}$ and $\mathbf{j}$.

Solution Because the angle between $\mathbf{v}$ and the positive x-axis is $\theta = \pi/6$, you can write the following.

$$\begin{aligned}\mathbf{v} &= \|\mathbf{v}\|\cos\theta\mathbf{i} + \|\mathbf{v}\|\sin\theta\mathbf{j}\\ &= 3\cos\frac{\pi}{6}\mathbf{i} + 3\sin\frac{\pi}{6}\mathbf{j}\\ &= \frac{3\sqrt{3}}{2}\mathbf{i} + \frac{3}{2}\mathbf{j}\end{aligned}$$

■

Applications of Vectors

Vectors have many applications in physics and engineering. One example is force. A vector can be used to represent force, because force has both magnitude and direction. If two or more forces are acting on an object, then the **resultant force** on the object is the vector sum of the vector forces.

EXAMPLE 7 Finding the Resultant Force

Two tugboats are pushing an ocean liner, as shown in Figure 11.12. Each boat is exerting a force of 400 pounds. What is the resultant force on the ocean liner?

Solution Using Figure 11.12, you can represent the forces exerted by the first and second tugboats as

$$\begin{aligned}\mathbf{F}_1 &= 400\langle \cos 20°, \sin 20°\rangle\\ &= 400\cos(20°)\mathbf{i} + 400\sin(20°)\mathbf{j}\\ \mathbf{F}_2 &= 400\langle \cos(-20°), \sin(-20°)\rangle\\ &= 400\cos(20°)\mathbf{i} - 400\sin(20°)\mathbf{j}.\end{aligned}$$

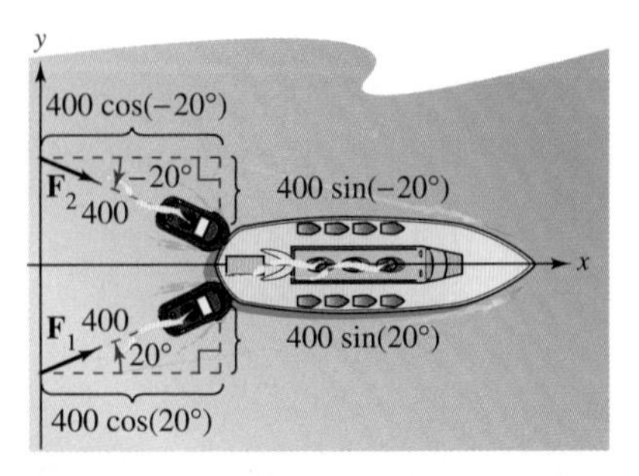

The resultant force on the ocean liner that is exerted by the two tugboats
Figure 11.12

The resultant force on the ocean liner is

$$\begin{aligned}\mathbf{F} &= \mathbf{F}_1 + \mathbf{F}_2\\ &= [400\cos(20°)\mathbf{i} + 400\sin(20°)\mathbf{j}] + [400\cos(20°)\mathbf{i} - 400\sin(20°)\mathbf{j}]\\ &= 800\cos(20°)\mathbf{i} \approx 752\mathbf{i}.\end{aligned}$$

So, the resultant force on the ocean liner is approximately 752 pounds in the direction of the positive x-axis. ■

In surveying and navigation, a **bearing** is a direction that measures the acute angle that a path or line of sight makes with a fixed north-south line. In air navigation, bearings are measured in degrees clockwise from north.

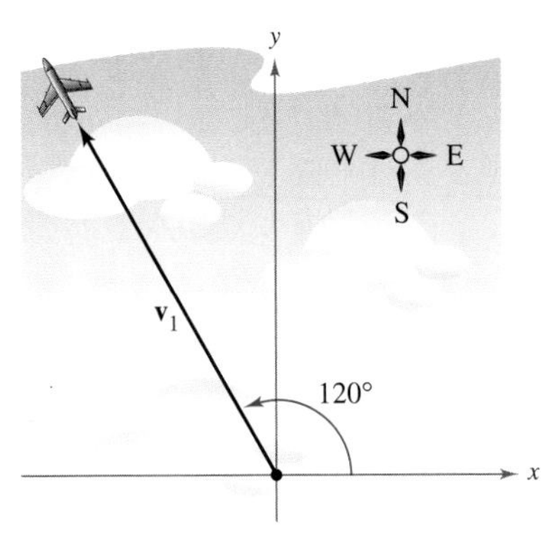

(a) Direction without wind

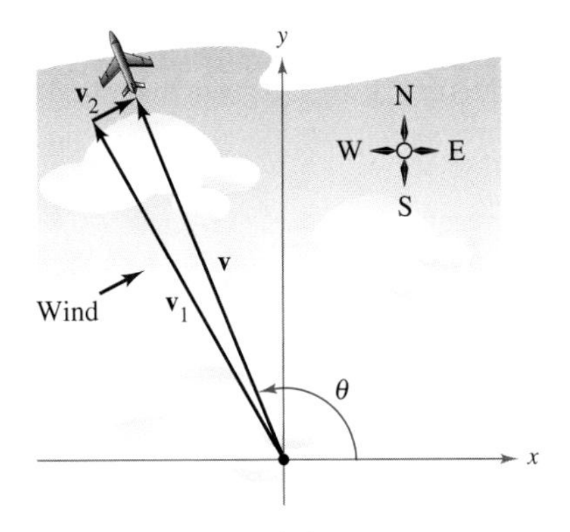

(b) Direction with wind

Figure 11.13

EXAMPLE 8 Finding a Velocity

An airplane is traveling at a fixed altitude with a negligible wind factor. The airplane is traveling at a speed of 500 miles per hour with a bearing of 330°, as shown in Figure 11.13(a). As the airplane reaches a certain point, it encounters wind with a velocity of 70 miles per hour in the direction N 45° E (45° east of north), as shown in Figure 11.13(b). What are the resultant speed and direction of the airplane?

Solution Using Figure 11.13(a), represent the velocity of the airplane (alone) as

$$\mathbf{v}_1 = 500\cos(120°)\mathbf{i} + 500\sin(120°)\mathbf{j}.$$

The velocity of the wind is represented by the vector

$$\mathbf{v}_2 = 70\cos(45°)\mathbf{i} + 70\sin(45°)\mathbf{j}.$$

The resultant velocity of the airplane (in the wind) is

$$\mathbf{v} = \mathbf{v}_1 + \mathbf{v}_2 = 500\cos(120°)\mathbf{i} + 500\sin(120°)\mathbf{j} + 70\cos(45°)\mathbf{i} + 70\sin(45°)\mathbf{j}$$
$$\approx -200.5\mathbf{i} + 482.5\mathbf{j}.$$

To find the resultant speed and direction, write $\mathbf{v} = \|\mathbf{v}\|(\cos\theta\,\mathbf{i} + \sin\theta\,\mathbf{j})$. Because $\|\mathbf{v}\| \approx \sqrt{(-200.5)^2 + (482.5)^2} \approx 522.5$, you can write

$$\mathbf{v} \approx 522.5\left(\frac{-200.5}{522.5}\mathbf{i} + \frac{482.5}{522.5}\mathbf{j}\right) \approx 522.5\left[\cos(112.6°)\mathbf{i} + \sin(112.6°)\mathbf{j}\right].$$

The new speed of the airplane, as altered by the wind, is approximately 522.5 miles per hour in a path that makes an angle of 112.6° with the positive x-axis. ■

11.1 Exercises

See www.CalcChat.com for worked-out solutions to odd-numbered exercises.

In Exercises 1–4, (a) find the component form of the vector v and (b) sketch the vector with its initial point at the origin.

1.
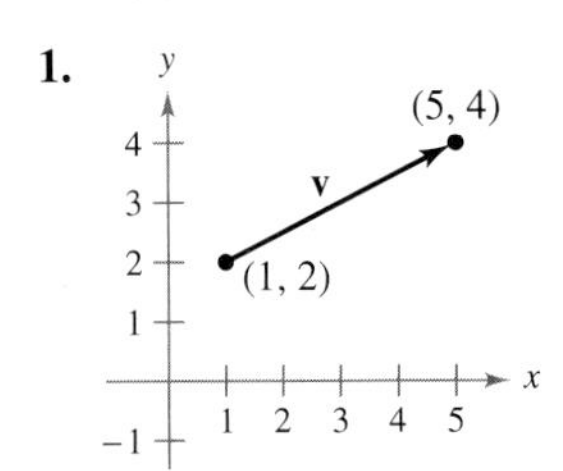

2.
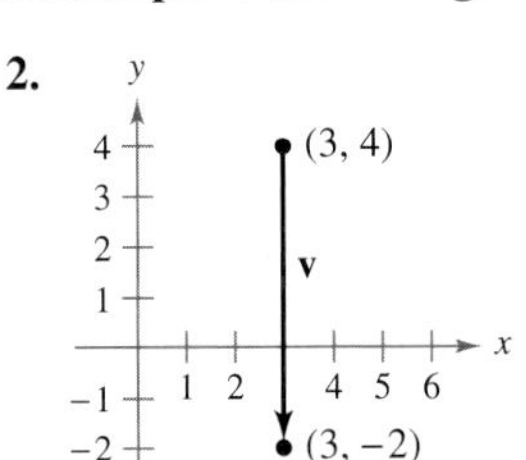

3.
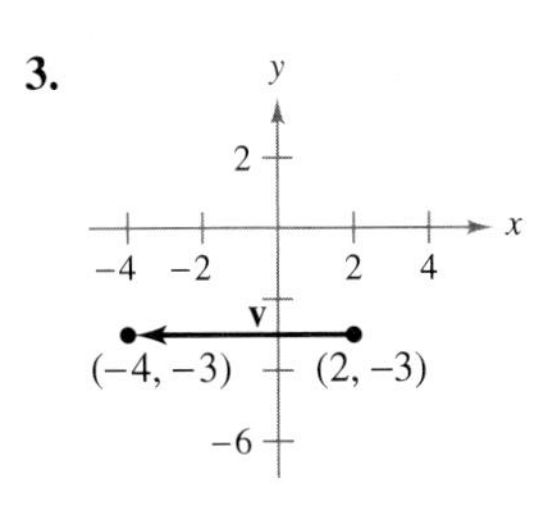

4.
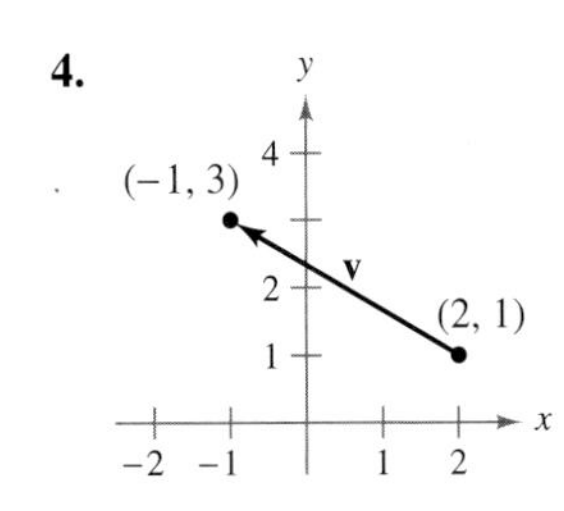

In Exercises 5–8, find the vectors u and v whose initial and terminal points are given. Show that u and v are equivalent.

5. **u**: (3, 2), (5, 6)
 v: (1, 4), (3, 8)

6. **u**: (−4, 0), (1, 8)
 v: (2, −1), (7, 7)

7. **u**: (0, 3), (6, −2)
 v: (3, 10), (9, 5)

8. **u**: (−4, −1), (11, −4)
 v: (10, 13), (25, 10)

In Exercises 9–16, the initial and terminal points of a vector v are given. (a) Sketch the given directed line segment, (b) write the vector in component form, (c) write the vector as the linear combination of the standard unit vectors i and j, and (d) sketch the vector with its initial point at the origin.

	Initial Point	*Terminal Point*		*Initial Point*	*Terminal Point*
9.	(2, 0)	(5, 5)	10.	(4, −6)	(3, 6)
11.	(8, 3)	(6, −1)	12.	(0, −4)	(−5, −1)

The icon indicates that you will find a CAS Investigation on the book's website. The CAS Investigation is a collaborative exploration of this example using the computer algebra systems Maple *and* Mathematica.

	Initial Point	*Terminal Point*		*Initial Point*	*Terminal Point*
13.	$(6, 2)$	$(6, 6)$	**14.**	$(7, -1)$	$(-3, -1)$
15.	$\left(\frac{3}{2}, \frac{4}{3}\right)$	$\left(\frac{1}{2}, 3\right)$	**16.**	$(0.12, 0.60)$	$(0.84, 1.25)$

In Exercises 17 and 18, sketch each scalar multiple of v.

17. $\mathbf{v} = \langle 3, 5 \rangle$

(a) $2\mathbf{v}$ (b) $-3\mathbf{v}$ (c) $\frac{7}{2}\mathbf{v}$ (d) $\frac{2}{3}\mathbf{v}$

18. $\mathbf{v} = \langle -2, 3 \rangle$

(a) $4\mathbf{v}$ (b) $-\frac{1}{2}\mathbf{v}$ (c) $0\mathbf{v}$ (d) $-6\mathbf{v}$

In Exercises 19–22, use the figure to sketch a graph of the vector. To print an enlarged copy of the graph, go to the website *www.mathgraphs.com*.

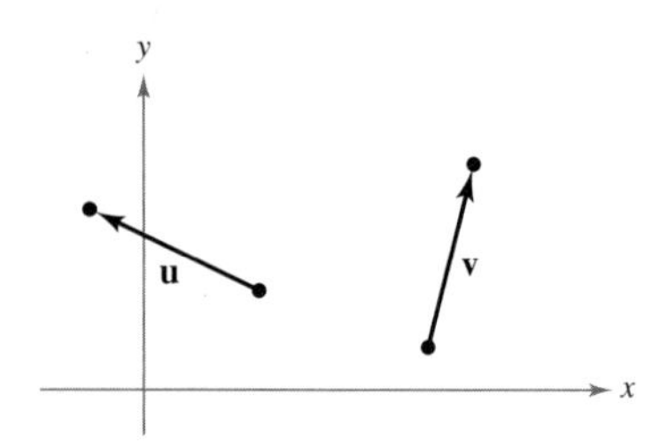

19. $-\mathbf{u}$ **20.** $2\mathbf{u}$

21. $\mathbf{u} - \mathbf{v}$ **22.** $\mathbf{u} + 2\mathbf{v}$

In Exercises 23 and 24, find (a) $\frac{2}{3}\mathbf{u}$, (b) $\mathbf{v} - \mathbf{u}$, and (c) $2\mathbf{u} + 5\mathbf{v}$.

23. $\mathbf{u} = \langle 4, 9 \rangle$, $\mathbf{v} = \langle 2, -5 \rangle$

24. $\mathbf{u} = \langle -3, -8 \rangle$, $\mathbf{v} = \langle 8, 25 \rangle$

In Exercises 25–28, find the vector v where $\mathbf{u} = \langle 2, -1 \rangle$ and $\mathbf{w} = \langle 1, 2 \rangle$. Illustrate the vector operations geometrically.

25. $\mathbf{v} = \frac{3}{2}\mathbf{u}$ **26.** $\mathbf{v} = \mathbf{u} + \mathbf{w}$

27. $\mathbf{v} = \mathbf{u} + 2\mathbf{w}$ **28.** $\mathbf{v} = 5\mathbf{u} - 3\mathbf{w}$

In Exercises 29 and 30, the vector v and its initial point are given. Find the terminal point.

29. $\mathbf{v} = \langle -1, 3 \rangle$; Initial point: $(4, 2)$

30. $\mathbf{v} = \langle 4, -9 \rangle$; Initial point: $(5, 3)$

In Exercises 31–36, find the magnitude of v.

31. $\mathbf{v} = 7\mathbf{i}$ **32.** $\mathbf{v} = -3\mathbf{i}$

33. $\mathbf{v} = \langle 4, 3 \rangle$ **34.** $\mathbf{v} = \langle 12, -5 \rangle$

35. $\mathbf{v} = 6\mathbf{i} - 5\mathbf{j}$ **36.** $\mathbf{v} = -10\mathbf{i} + 3\mathbf{j}$

In Exercises 37–40, find the unit vector in the direction of v and verify that it has length 1.

37. $\mathbf{v} = \langle 3, 12 \rangle$ **38.** $\mathbf{v} = \langle -5, 15 \rangle$

39. $\mathbf{v} = \left\langle \frac{3}{2}, \frac{5}{2} \right\rangle$ **40.** $\mathbf{v} = \langle -6.2, 3.4 \rangle$

In Exercises 41–44, find the following.

(a) $\|\mathbf{u}\|$ **(b)** $\|\mathbf{v}\|$ **(c)** $\|\mathbf{u} + \mathbf{v}\|$

(d) $\left\|\dfrac{\mathbf{u}}{\|\mathbf{u}\|}\right\|$ **(e)** $\left\|\dfrac{\mathbf{v}}{\|\mathbf{v}\|}\right\|$ **(f)** $\left\|\dfrac{\mathbf{u} + \mathbf{v}}{\|\mathbf{u} + \mathbf{v}\|}\right\|$

41. $\mathbf{u} = \langle 1, -1 \rangle$, $\mathbf{v} = \langle -1, 2 \rangle$

42. $\mathbf{u} = \langle 0, 1 \rangle$, $\mathbf{v} = \langle 3, -3 \rangle$

43. $\mathbf{u} = \left\langle 1, \frac{1}{2} \right\rangle$, $\mathbf{v} = \langle 2, 3 \rangle$

44. $\mathbf{u} = \langle 2, -4 \rangle$, $\mathbf{v} = \langle 5, 5 \rangle$

In Exercises 45 and 46, sketch a graph of u, v, and u + v. Then demonstrate the triangle inequality using the vectors u and v.

45. $\mathbf{u} = \langle 2, 1 \rangle$, $\mathbf{v} = \langle 5, 4 \rangle$ **46.** $\mathbf{u} = \langle -3, 2 \rangle$, $\mathbf{v} = \langle 1, -2 \rangle$

In Exercises 47–50, find the vector v with the given magnitude and the same direction as u.

	Magnitude	*Direction*
47.	$\|\mathbf{v}\| = 6$	$\mathbf{u} = \langle 0, 3 \rangle$
48.	$\|\mathbf{v}\| = 4$	$\mathbf{u} = \langle 1, 1 \rangle$
49.	$\|\mathbf{v}\| = 5$	$\mathbf{u} = \langle -1, 2 \rangle$
50.	$\|\mathbf{v}\| = 2$	$\mathbf{u} = \langle \sqrt{3}, 3 \rangle$

In Exercises 51–54, find the component form of v given its magnitude and the angle it makes with the positive *x*-axis.

51. $\|\mathbf{v}\| = 3, \quad \theta = 0°$ **52.** $\|\mathbf{v}\| = 5, \quad \theta = 120°$

53. $\|\mathbf{v}\| = 2, \quad \theta = 150°$ **54.** $\|\mathbf{v}\| = 4, \quad \theta = 3.5°$

In Exercises 55–58, find the component form of u + v given the lengths of u and v and the angles that u and v make with the positive *x*-axis.

55. $\|\mathbf{u}\| = 1, \quad \theta_{\mathbf{u}} = 0°$; $\|\mathbf{v}\| = 3, \quad \theta_{\mathbf{v}} = 45°$

56. $\|\mathbf{u}\| = 4, \quad \theta_{\mathbf{u}} = 0°$; $\|\mathbf{v}\| = 2, \quad \theta_{\mathbf{v}} = 60°$

57. $\|\mathbf{u}\| = 2, \quad \theta_{\mathbf{u}} = 4$; $\|\mathbf{v}\| = 1, \quad \theta_{\mathbf{v}} = 2$

58. $\|\mathbf{u}\| = 5, \quad \theta_{\mathbf{u}} = -0.5$; $\|\mathbf{v}\| = 5, \quad \theta_{\mathbf{v}} = 0.5$

WRITING ABOUT CONCEPTS

59. In your own words, state the difference between a scalar and a vector. Give examples of each.

60. Give geometric descriptions of the operations of addition of vectors and multiplication of a vector by a scalar.

61. Identify the quantity as a scalar or as a vector. Explain your reasoning.

(a) The muzzle velocity of a gun

(b) The price of a company's stock

62. Identify the quantity as a scalar or as a vector. Explain your reasoning.

(a) The air temperature in a room

(b) The weight of a car

In Exercises 63–68, find a and b such that $\mathbf{v} = a\mathbf{u} + b\mathbf{w}$, where $\mathbf{u} = \langle 1, 2\rangle$ and $\mathbf{w} = \langle 1, -1\rangle$.

63. $\mathbf{v} = \langle 2, 1\rangle$ **64.** $\mathbf{v} = \langle 0, 3\rangle$

65. $\mathbf{v} = \langle 3, 0\rangle$ **66.** $\mathbf{v} = \langle 3, 3\rangle$

67. $\mathbf{v} = \langle 1, 1\rangle$ **68.** $\mathbf{v} = \langle -1, 7\rangle$

In Exercises 69–74, find a unit vector (a) parallel to and (b) perpendicular to the graph of f at the given point. Then sketch the graph of f and sketch the vectors at the given point.

Function	*Point*
69. $f(x) = x^2$	$(3, 9)$
70. $f(x) = -x^2 + 5$	$(1, 4)$
71. $f(x) = x^3$	$(1, 1)$
72. $f(x) = x^3$	$(-2, -8)$
73. $f(x) = \sqrt{25 - x^2}$	$(3, 4)$
74. $f(x) = \tan x$	$\left(\frac{\pi}{4}, 1\right)$

In Exercises 75 and 76, find the component form of v given the magnitudes of u and u + v and the angles that u and u + v make with the positive x-axis.

75. $\|\mathbf{u}\| = 1, \theta = 45^\circ$
$\|\mathbf{u} + \mathbf{v}\| = \sqrt{2}, \theta = 90^\circ$

76. $\|\mathbf{u}\| = 4, \theta = 30^\circ$
$\|\mathbf{u} + \mathbf{v}\| = 6, \theta = 120^\circ$

77. ***Programming*** You are given the magnitudes of **u** and **v** and the angles that **u** and **v** make with the positive x-axis. Write a program for a graphing utility in which the output is the following.

(a) $\mathbf{u} + \mathbf{v}$ (b) $\|\mathbf{u} + \mathbf{v}\|$

(c) The angle that $\mathbf{u} + \mathbf{v}$ makes with the positive x-axis

(d) Use the program to find the magnitude and direction of the resultant of the vectors shown.

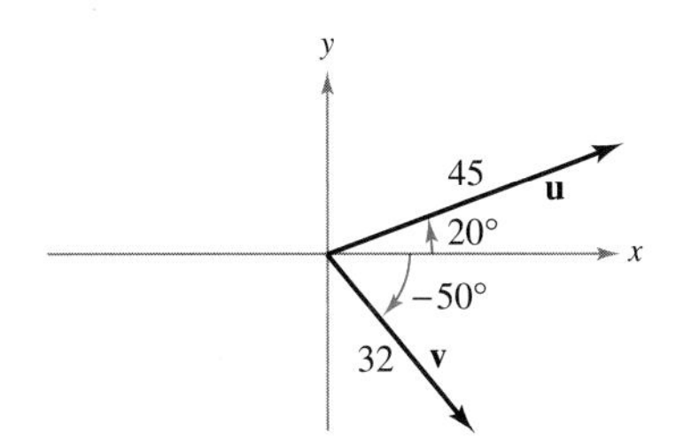

CAPSTONE

78. The initial and terminal points of vector **v** are $(3, -4)$ and $(9, 1)$, respectively.

(a) Write **v** in component form.

(b) Write **v** as the linear combination of the standard unit vectors **i** and **j**.

(c) Sketch **v** with its initial point at the origin.

(d) Find the magnitude of **v**.

In Exercises 79 and 80, use a graphing utility to find the magnitude and direction of the resultant of the vectors.

79.

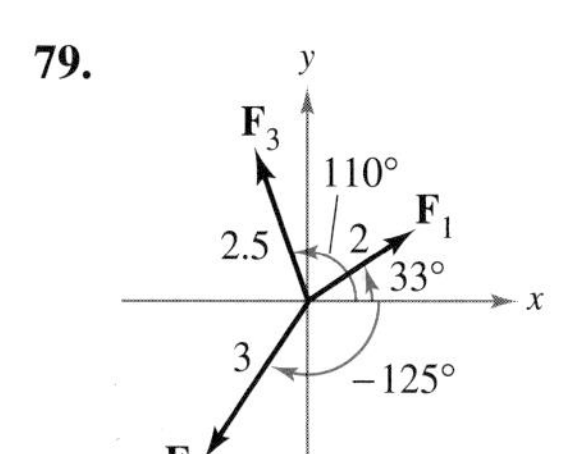

80.

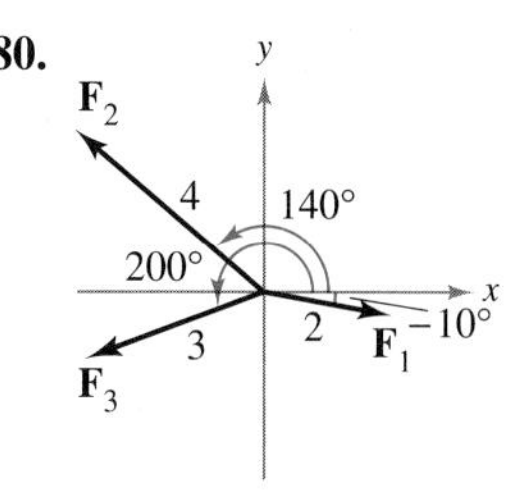

81. ***Resultant Force*** Forces with magnitudes of 500 pounds and 200 pounds act on a machine part at angles of 30° and −45°, respectively, with the x-axis (see figure). Find the direction and magnitude of the resultant force.

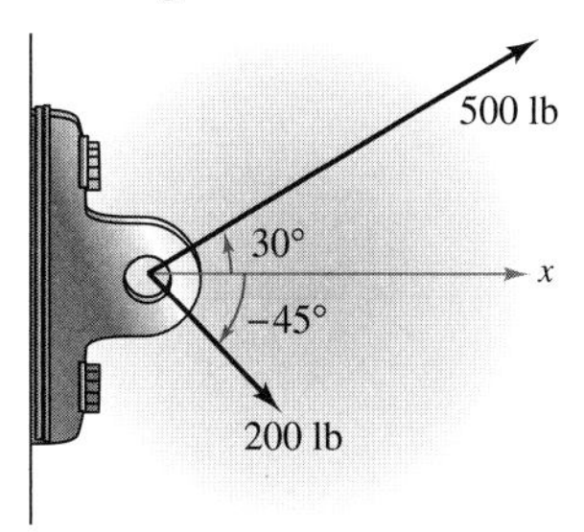

Figure for 81

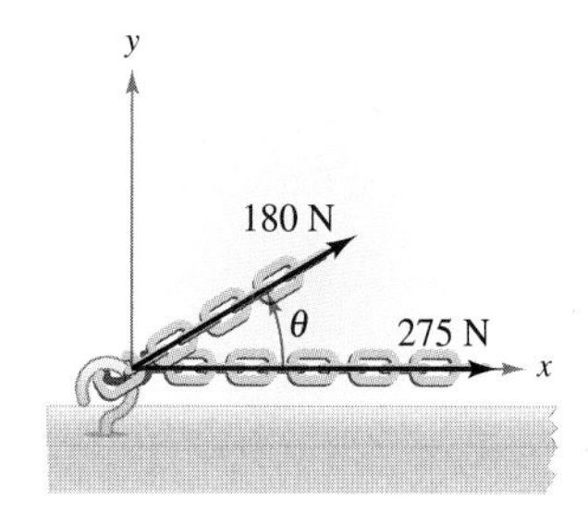

Figure for 82

82. ***Numerical and Graphical Analysis*** Forces with magnitudes of 180 newtons and 275 newtons act on a hook (see figure). The angle between the two forces is θ degrees.

(a) If $\theta = 30^\circ$, find the direction and magnitude of the resultant force.

(b) Write the magnitude M and direction α of the resultant force as functions of θ, where $0^\circ \leq \theta \leq 180^\circ$.

(c) Use a graphing utility to complete the table.

θ	0°	30°	60°	90°	120°	150°	180°
M							
α							

(d) Use a graphing utility to graph the two functions M and α.

(e) Explain why one of the functions decreases for increasing values of θ whereas the other does not.

83. ***Resultant Force*** Three forces with magnitudes of 75 pounds, 100 pounds, and 125 pounds act on an object at angles of 30°, 45°, and 120°, respectively, with the positive x-axis. Find the direction and magnitude of the resultant force.

84. ***Resultant Force*** Three forces with magnitudes of 400 newtons, 280 newtons, and 350 newtons act on an object at angles of −30°, 45°, and 135°, respectively, with the positive x-axis. Find the direction and magnitude of the resultant force.

85. ***Think About It*** Consider two forces of equal magnitude acting on a point.

(a) If the magnitude of the resultant is the sum of the magnitudes of the two forces, make a conjecture about the angle between the forces.

(b) If the resultant of the forces is **0**, make a conjecture about the angle between the forces.

(c) Can the magnitude of the resultant be greater than the sum of the magnitudes of the two forces? Explain.

86. ***Graphical Reasoning*** Consider two forces $\mathbf{F}_1 = \langle 20, 0\rangle$ and $\mathbf{F}_2 = 10\langle\cos\theta, \sin\theta\rangle$.

(a) Find $\|\mathbf{F}_1 + \mathbf{F}_2\|$.

(b) Determine the magnitude of the resultant as a function of θ. Use a graphing utility to graph the function for $0 \le \theta < 2\pi$.

(c) Use the graph in part (b) to determine the range of the function. What is its maximum and for what value of θ does it occur? What is its minimum and for what value of θ does it occur?

(d) Explain why the magnitude of the resultant is never 0.

87. Three vertices of a parallelogram are $(1, 2)$, $(3, 1)$, and $(8, 4)$. Find the three possible fourth vertices (see figure).

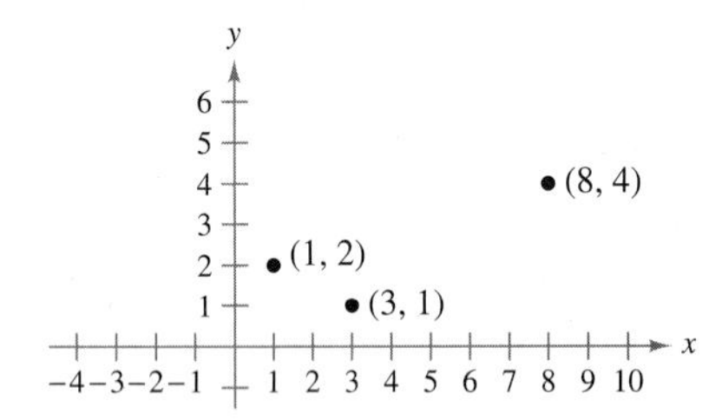

88. Use vectors to find the points of trisection of the line segment with endpoints $(1, 2)$ and $(7, 5)$.

Cable Tension **In Exercises 89 and 90, use the figure to determine the tension in each cable supporting the given load.**

89.

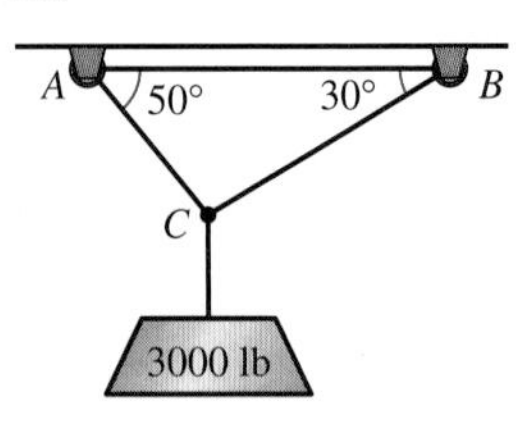

90.

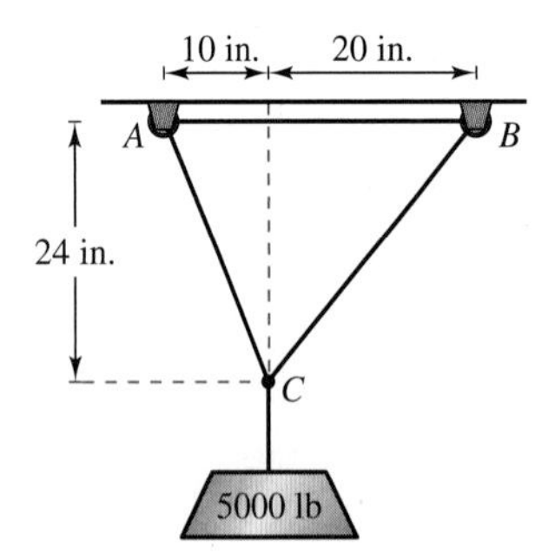

91. ***Projectile Motion*** A gun with a muzzle velocity of 1200 feet per second is fired at an angle of 6° above the horizontal. Find the vertical and horizontal components of the velocity.

92. ***Shared Load*** To carry a 100-pound cylindrical weight, two workers lift on the ends of short ropes tied to an eyelet on the top center of the cylinder. One rope makes a 20° angle away from the vertical and the other makes a 30° angle (see figure).

(a) Find each rope's tension if the resultant force is vertical.

(b) Find the vertical component of each worker's force.

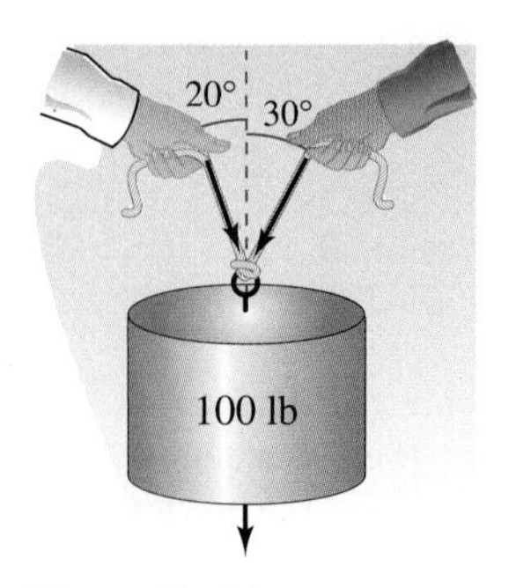

Figure for 92

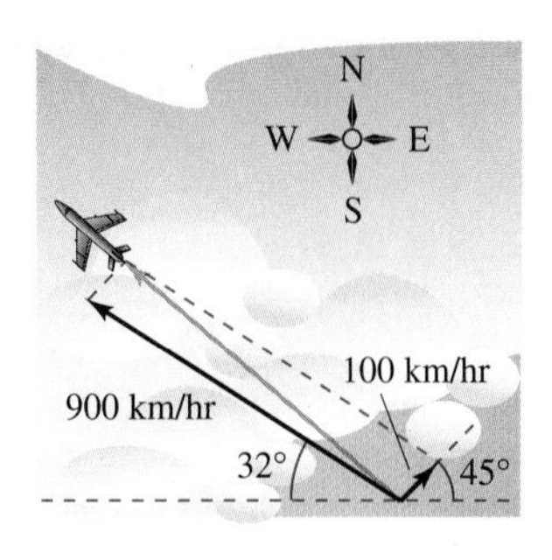

Figure for 93

93. ***Navigation*** A plane is flying with a bearing of 302°. Its speed with respect to the air is 900 kilometers per hour. The wind at the plane's altitude is from the southwest at 100 kilometers per hour (see figure). What is the true direction of the plane, and what is its speed with respect to the ground?

94. ***Navigation*** A plane flies at a constant groundspeed of 400 miles per hour due east and encounters a 50-mile-per-hour wind from the northwest. Find the airspeed and compass direction that will allow the plane to maintain its groundspeed and eastward direction.

True or False? **In Exercises 95–100, determine whether the statement is true or false. If it is false, explain why or give an example that shows it is false.**

95. If **u** and **v** have the same magnitude and direction, then **u** and **v** are equivalent.

96. If **u** is a unit vector in the direction of **v**, then $\mathbf{v} = \|\mathbf{v}\|\,\mathbf{u}$.

97. If $\mathbf{u} = a\mathbf{i} + b\mathbf{j}$ is a unit vector, then $a^2 + b^2 = 1$.

98. If $\mathbf{v} = a\mathbf{i} + b\mathbf{j} = \mathbf{0}$, then $a = -b$.

99. If $a = b$, then $\|a\mathbf{i} + b\mathbf{j}\| = \sqrt{2}a$.

100. If **u** and **v** have the same magnitude but opposite directions, then $\mathbf{u} + \mathbf{v} = \mathbf{0}$.

101. Prove that $\mathbf{u} = (\cos\theta)\mathbf{i} - (\sin\theta)\mathbf{j}$ and $\mathbf{v} = (\sin\theta)\mathbf{i} + (\cos\theta)\mathbf{j}$ are unit vectors for any angle θ.

102. ***Geometry*** Using vectors, prove that the line segment joining the midpoints of two sides of a triangle is parallel to, and one-half the length of, the third side.

103. ***Geometry*** Using vectors, prove that the diagonals of a parallelogram bisect each other.

104. Prove that the vector $\mathbf{w} = \|\mathbf{u}\|\mathbf{v} + \|\mathbf{v}\|\mathbf{u}$ bisects the angle between **u** and **v**.

105. Consider the vector $\mathbf{u} = \langle x, y\rangle$. Describe the set of all points (x, y) such that $\|\mathbf{u}\| = 5$.

PUTNAM EXAM CHALLENGE

106. A coast artillery gun can fire at any angle of elevation between 0° and 90° in a fixed vertical plane. If air resistance is neglected and the muzzle velocity is constant ($= v_0$), determine the set H of points in the plane and above the horizontal which can be hit.

This problem was composed by the Committee on the Putnam Prize Competition.

11.2 Space Coordinates and Vectors in Space

- **Understand the three-dimensional rectangular coordinate system.**
- **Analyze vectors in space.**
- **Use three-dimensional vectors to solve real-life problems.**

Coordinates in Space

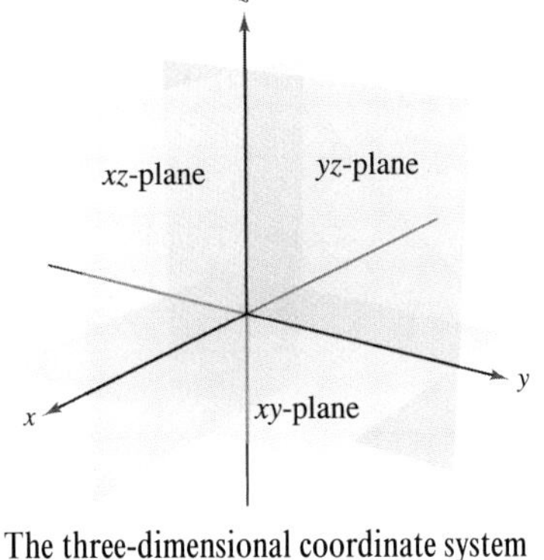

The three-dimensional coordinate system
Figure 11.14

Up to this point in the text, you have been primarily concerned with the two-dimensional coordinate system. Much of the remaining part of your study of calculus will involve the three-dimensional coordinate system.

Before extending the concept of a vector to three dimensions, you must be able to identify points in the **three-dimensional coordinate system.** You can construct this system by passing a z-axis perpendicular to both the x- and y-axes at the origin. Figure 11.14 shows the positive portion of each coordinate axis. Taken as pairs, the axes determine three **coordinate planes:** the ***xy*-plane,** the ***xz*-plane,** and the ***yz*-plane.** These three coordinate planes separate three-space into eight **octants.** The first octant is the one for which all three coordinates are positive. In this three-dimensional system, a point P in space is determined by an ordered triple (x, y, z) where x, y, and z are as follows.

$x =$ directed distance from yz-plane to P

$y =$ directed distance from xz-plane to P

$z =$ directed distance from xy-plane to P

Several points are shown in Figure 11.15.

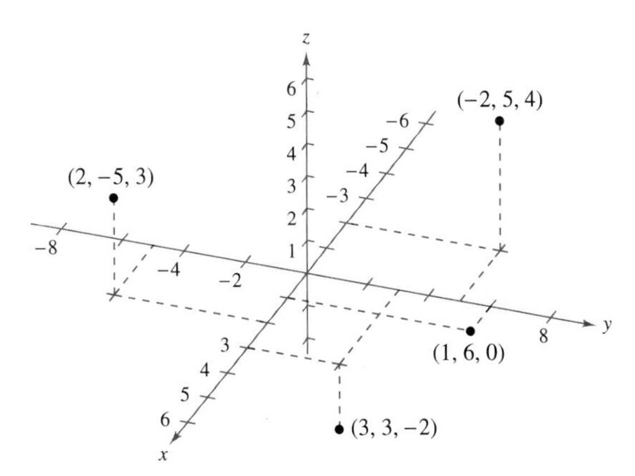

Points in the three-dimensional coordinate system are represented by ordered triples.
Figure 11.15

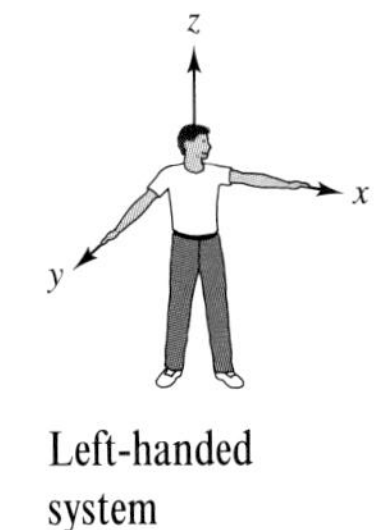

Right-handed system
Left-handed system
Figure 11.16

A three-dimensional coordinate system can have either a **left-handed** or a **right-handed** orientation. To determine the orientation of a system, imagine that you are standing at the origin, with your arms pointing in the direction of the positive x- and y-axes, and with the z-axis pointing up, as shown in Figure 11.16. The system is right-handed or left-handed depending on which hand points along the x-axis. In this text, you will work exclusively with the right-handed system.

NOTE The three-dimensional rotatable graphs that are available in the premium eBook for this text will help you visualize points or objects in a three-dimensional coordinate system. ■

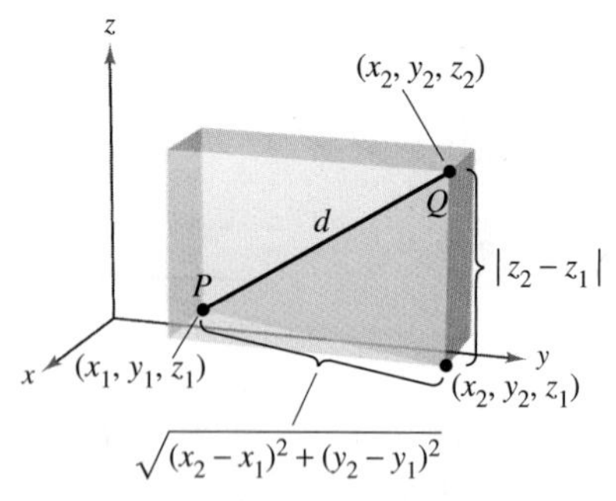

The distance between two points in space
Figure 11.17

Many of the formulas established for the two-dimensional coordinate system can be extended to three dimensions. For example, to find the distance between two points in space, you can use the Pythagorean Theorem twice, as shown in Figure 11.17. By doing this, you will obtain the formula for the distance between the points (x_1, y_1, z_1) and (x_2, y_2, z_2).

$$d = \sqrt{(x_2 - x_1)^2 + (y_2 - y_1)^2 + (z_2 - z_1)^2} \qquad \text{Distance Formula}$$

EXAMPLE 1 Finding the Distance Between Two Points in Space

The distance between the points $(2, -1, 3)$ and $(1, 0, -2)$ is

$$\begin{aligned} d &= \sqrt{(1-2)^2 + (0+1)^2 + (-2-3)^2} && \text{Distance Formula} \\ &= \sqrt{1 + 1 + 25} \\ &= \sqrt{27} \\ &= 3\sqrt{3}. \end{aligned}$$

■

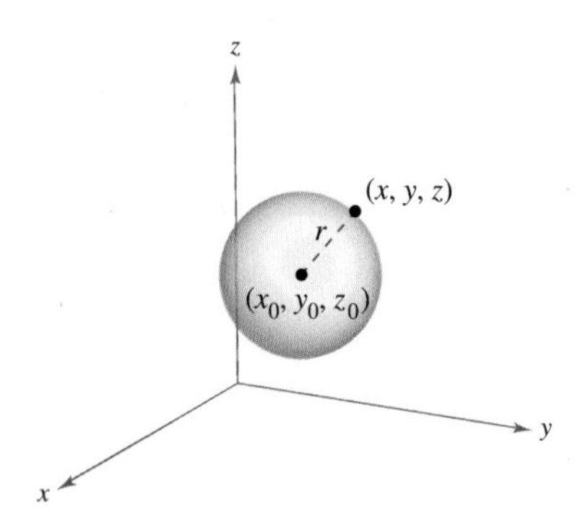

Figure 11.18

A **sphere** with center at (x_0, y_0, z_0) and radius r is defined to be the set of all points (x, y, z) such that the distance between (x, y, z) and (x_0, y_0, z_0) is r. You can use the Distance Formula to find the **standard equation of a sphere** of radius r, centered at (x_0, y_0, z_0). If (x, y, z) is an arbitrary point on the sphere, the equation of the sphere is

$$(x - x_0)^2 + (y - y_0)^2 + (z - z_0)^2 = r^2 \qquad \text{Equation of sphere}$$

as shown in Figure 11.18. Moreover, the midpoint of the line segment joining the points (x_1, y_1, z_1) and (x_2, y_2, z_2) has coordinates

$$\left(\frac{x_1 + x_2}{2}, \frac{y_1 + y_2}{2}, \frac{z_1 + z_2}{2}\right). \qquad \text{Midpoint Formula}$$

EXAMPLE 2 Finding the Equation of a Sphere

Find the standard equation of the sphere that has the points $(5, -2, 3)$ and $(0, 4, -3)$ as endpoints of a diameter.

Solution Using the Midpoint Formula, the center of the sphere is

$$\left(\frac{5+0}{2}, \frac{-2+4}{2}, \frac{3-3}{2}\right) = \left(\frac{5}{2}, 1, 0\right). \qquad \text{Midpoint Formula}$$

By the Distance Formula, the radius is

$$r = \sqrt{\left(0 - \frac{5}{2}\right)^2 + (4-1)^2 + (-3-0)^2} = \sqrt{\frac{97}{4}} = \frac{\sqrt{97}}{2}.$$

Therefore, the standard equation of the sphere is

$$\left(x - \frac{5}{2}\right)^2 + (y-1)^2 + z^2 = \frac{97}{4}. \qquad \text{Equation of sphere}$$

■

Vectors in Space

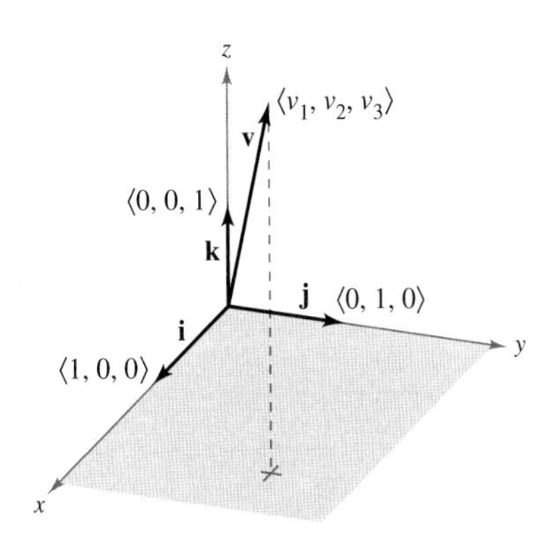

The standard unit vectors in space
Figure 11.19

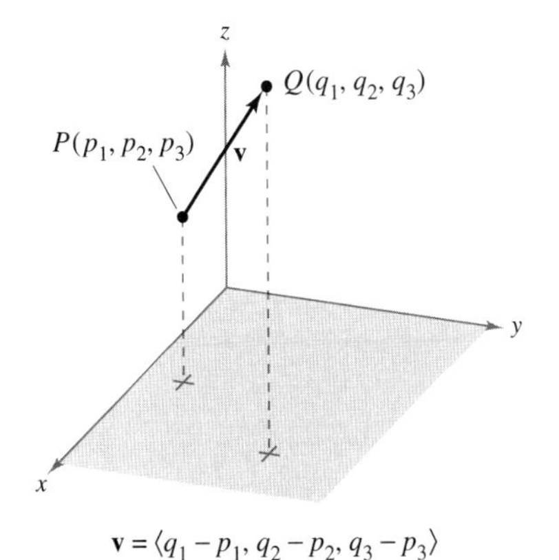

Figure 11.20

In space, vectors are denoted by ordered triples $\mathbf{v} = \langle v_1, v_2, v_3\rangle$. The **zero vector** is denoted by $\mathbf{0} = \langle 0, 0, 0\rangle$. Using the unit vectors $\mathbf{i} = \langle 1, 0, 0\rangle$, $\mathbf{j} = \langle 0, 1, 0\rangle$, and $\mathbf{k} = \langle 0, 0, 1\rangle$ in the direction of the positive z-axis, the **standard unit vector notation** for $\mathbf{v}$ is

$$\mathbf{v} = v_1\mathbf{i} + v_2\mathbf{j} + v_3\mathbf{k}$$

as shown in Figure 11.19. If $\mathbf{v}$ is represented by the directed line segment from $P(p_1, p_2, p_3)$ to $Q(q_1, q_2, q_3)$, as shown in Figure 11.20, the component form of $\mathbf{v}$ is given by subtracting the coordinates of the initial point from the coordinates of the terminal point, as follows.

$$\mathbf{v} = \langle v_1, v_2, v_3\rangle = \langle q_1 - p_1, q_2 - p_2, q_3 - p_3\rangle$$

VECTORS IN SPACE

Let $\mathbf{u} = \langle u_1, u_2, u_3\rangle$ and $\mathbf{v} = \langle v_1, v_2, v_3\rangle$ be vectors in space and let c be a scalar.

1. *Equality of Vectors:* $\mathbf{u} = \mathbf{v}$ if and only if $u_1 = v_1$, $u_2 = v_2$, and $u_3 = v_3$.
2. *Component Form:* If $\mathbf{v}$ is represented by the directed line segment from $P(p_1, p_2, p_3)$ to $Q(q_1, q_2, q_3)$, then
$$\mathbf{v} = \langle v_1, v_2, v_3\rangle = \langle q_1 - p_1, q_2 - p_2, q_3 - p_3\rangle.$$
3. *Length:* $\|\mathbf{v}\| = \sqrt{v_1^2 + v_2^2 + v_3^2}$
4. *Unit Vector in the Direction of* $\mathbf{v}$*:* $\dfrac{\mathbf{v}}{\|\mathbf{v}\|} = \left(\dfrac{1}{\|\mathbf{v}\|}\right)\langle v_1, v_2, v_3\rangle, \quad \mathbf{v} \neq \mathbf{0}$
5. *Vector Addition:* $\mathbf{v} + \mathbf{u} = \langle v_1 + u_1, v_2 + u_2, v_3 + u_3\rangle$
6. *Scalar Multiplication:* $c\mathbf{v} = \langle cv_1, cv_2, cv_3\rangle$

NOTE The properties of vector addition and scalar multiplication given in Theorem 11.1 are also valid for vectors in space. ■

EXAMPLE 3 Finding the Component Form of a Vector in Space

Find the component form and magnitude of the vector $\mathbf{v}$ having initial point $(-2, 3, 1)$ and terminal point $(0, -4, 4)$. Then find a unit vector in the direction of $\mathbf{v}$.

Solution The component form of $\mathbf{v}$ is

$$\begin{aligned}\mathbf{v} = \langle q_1 - p_1, q_2 - p_2, q_3 - p_3\rangle &= \langle 0 - (-2), -4 - 3, 4 - 1\rangle \\ &= \langle 2, -7, 3\rangle\end{aligned}$$

which implies that its magnitude is

$$\|\mathbf{v}\| = \sqrt{2^2 + (-7)^2 + 3^2} = \sqrt{62}.$$

The unit vector in the direction of $\mathbf{v}$ is

$$\mathbf{u} = \frac{\mathbf{v}}{\|\mathbf{v}\|} = \frac{1}{\sqrt{62}}\langle 2, -7, 3\rangle = \left\langle \frac{2}{\sqrt{62}}, \frac{-7}{\sqrt{62}}, \frac{3}{\sqrt{62}}\right\rangle.$$

■

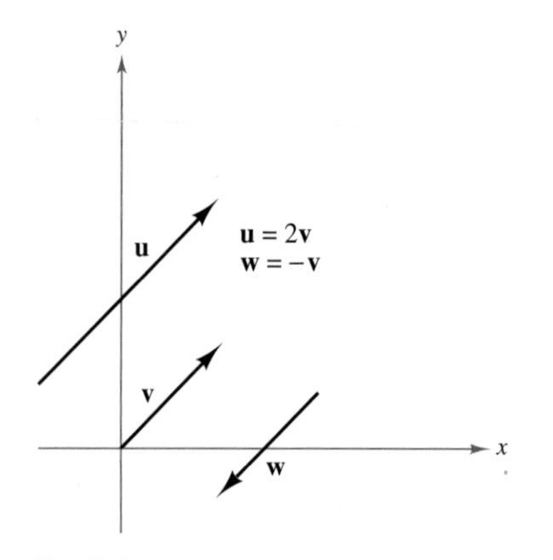

Parallel vectors
Figure 11.21

Recall from the definition of scalar multiplication that positive scalar multiples of a nonzero vector $\mathbf{v}$ have the same direction as $\mathbf{v}$, whereas negative multiples have the direction opposite of $\mathbf{v}$. In general, two nonzero vectors $\mathbf{u}$ and $\mathbf{v}$ are **parallel** if there is some scalar c such that $\mathbf{u} = c\mathbf{v}$.

DEFINITION OF PARALLEL VECTORS

Two nonzero vectors $\mathbf{u}$ and $\mathbf{v}$ are **parallel** if there is some scalar c such that $\mathbf{u} = c\mathbf{v}$.

For example, in Figure 11.21, the vectors $\mathbf{u}$, $\mathbf{v}$, and $\mathbf{w}$ are parallel because $\mathbf{u} = 2\mathbf{v}$ and $\mathbf{w} = -\mathbf{v}$.

EXAMPLE 4 Parallel Vectors

Vector $\mathbf{w}$ has initial point $(2, -1, 3)$ and terminal point $(-4, 7, 5)$. Which of the following vectors is parallel to $\mathbf{w}$?

a. $\mathbf{u} = \langle 3, -4, -1 \rangle$

b. $\mathbf{v} = \langle 12, -16, 4 \rangle$

Solution Begin by writing $\mathbf{w}$ in component form.

$$\mathbf{w} = \langle -4 - 2, 7 - (-1), 5 - 3 \rangle = \langle -6, 8, 2 \rangle$$

a. Because $\mathbf{u} = \langle 3, -4, -1 \rangle = -\frac{1}{2}\langle -6, 8, 2 \rangle = -\frac{1}{2}\mathbf{w}$, you can conclude that $\mathbf{u}$ is parallel to $\mathbf{w}$.

b. In this case, you want to find a scalar c such that

$$\langle 12, -16, 4 \rangle = c\langle -6, 8, 2 \rangle.$$

$$\begin{aligned} 12 &= -6c \rightarrow c = -2 \\ -16 &= 8c \rightarrow c = -2 \\ 4 &= 2c \rightarrow c = 2 \end{aligned}$$

Because there is no c for which the equation has a solution, the vectors are not parallel.

EXAMPLE 5 Using Vectors to Determine Collinear Points

Determine whether the points $P(1, -2, 3)$, $Q(2, 1, 0)$, and $R(4, 7, -6)$ are collinear.

Solution The component forms of $\overrightarrow{PQ}$ and $\overrightarrow{PR}$ are

$$\overrightarrow{PQ} = \langle 2 - 1, 1 - (-2), 0 - 3 \rangle = \langle 1, 3, -3 \rangle$$

and

$$\overrightarrow{PR} = \langle 4 - 1, 7 - (-2), -6 - 3 \rangle = \langle 3, 9, -9 \rangle.$$

These two vectors have a common initial point. So, P, Q, and R lie on the same line if and only if $\overrightarrow{PQ}$ and $\overrightarrow{PR}$ are parallel—which they are because $\overrightarrow{PR} = 3\,\overrightarrow{PQ}$, as shown in Figure 11.22. ■

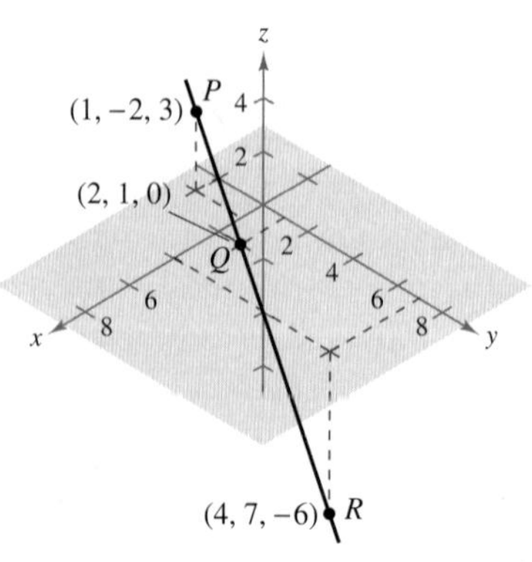

The points P, Q, and R lie on the same line.
Figure 11.22

EXAMPLE 6 Standard Unit Vector Notation

a. Write the vector $\mathbf{v} = 4\mathbf{i} - 5\mathbf{k}$ in component form.

b. Find the terminal point of the vector $\mathbf{v} = 7\mathbf{i} - \mathbf{j} + 3\mathbf{k}$, given that the initial point is $P(-2, 3, 5)$.

Solution

a. Because $\mathbf{j}$ is missing, its component is 0 and

$$\mathbf{v} = 4\mathbf{i} - 5\mathbf{k} = \langle 4, 0, -5 \rangle.$$

b. You need to find $Q(q_1, q_2, q_3)$ such that $\mathbf{v} = \overrightarrow{PQ} = 7\mathbf{i} - \mathbf{j} + 3\mathbf{k}$. This implies that $q_1 - (-2) = 7$, $q_2 - 3 = -1$, and $q_3 - 5 = 3$. The solution of these three equations is $q_1 = 5$, $q_2 = 2$, and $q_3 = 8$. Therefore, Q is $(5, 2, 8)$. ■

Application

EXAMPLE 7 Measuring Force

A television camera weighing 120 pounds is supported by a tripod, as shown in Figure 11.23. Represent the force exerted on each leg of the tripod as a vector.

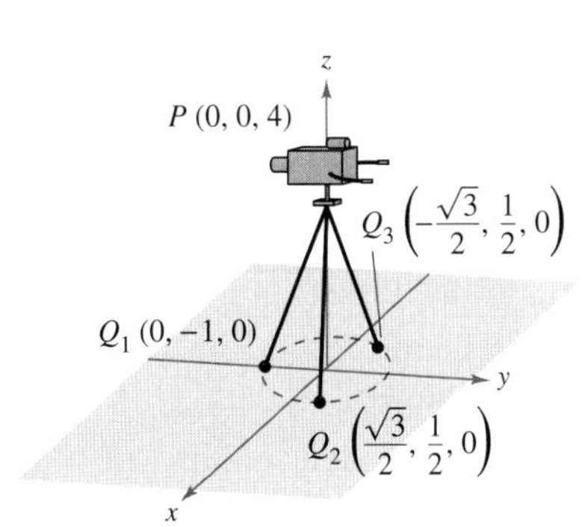

Figure 11.23

Solution Let the vectors $\mathbf{F}_1$, $\mathbf{F}_2$, and $\mathbf{F}_3$ represent the forces exerted on the three legs. From Figure 11.23, you can determine the directions of $\mathbf{F}_1$, $\mathbf{F}_2$, and $\mathbf{F}_3$ to be as follows.

$$\overrightarrow{PQ_1} = \langle 0 - 0, -1 - 0, 0 - 4 \rangle = \langle 0, -1, -4 \rangle$$

$$\overrightarrow{PQ_2} = \left\langle \frac{\sqrt{3}}{2} - 0, \frac{1}{2} - 0, 0 - 4 \right\rangle = \left\langle \frac{\sqrt{3}}{2}, \frac{1}{2}, -4 \right\rangle$$

$$\overrightarrow{PQ_3} = \left\langle -\frac{\sqrt{3}}{2} - 0, \frac{1}{2} - 0, 0 - 4 \right\rangle = \left\langle -\frac{\sqrt{3}}{2}, \frac{1}{2}, -4 \right\rangle$$

Because each leg has the same length, and the total force is distributed equally among the three legs, you know that $\|\mathbf{F}_1\| = \|\mathbf{F}_2\| = \|\mathbf{F}_3\|$. So, there exists a constant c such that

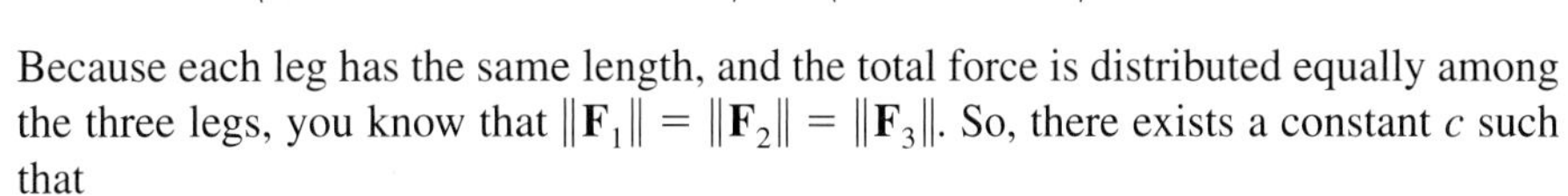

$$\mathbf{F}_1 = c\langle 0, -1, -4 \rangle, \quad \mathbf{F}_2 = c\left\langle \frac{\sqrt{3}}{2}, \frac{1}{2}, -4 \right\rangle, \quad \text{and} \quad \mathbf{F}_3 = c\left\langle -\frac{\sqrt{3}}{2}, \frac{1}{2}, -4 \right\rangle.$$

Let the total force exerted by the object be given by $\mathbf{F} = \langle 0, 0, -120 \rangle$. Then, using the fact that

$$\mathbf{F} = \mathbf{F}_1 + \mathbf{F}_2 + \mathbf{F}_3$$

you can conclude that $\mathbf{F}_1$, $\mathbf{F}_2$, and $\mathbf{F}_3$ all have a vertical component of -40. This implies that $c(-4) = -40$ and $c = 10$. Therefore, the forces exerted on the legs can be represented by

$$\mathbf{F}_1 = \langle 0, -10, -40 \rangle$$
$$\mathbf{F}_2 = \langle 5\sqrt{3}, 5, -40 \rangle$$
$$\mathbf{F}_3 = \langle -5\sqrt{3}, 5, -40 \rangle.$$

■

11.2 Exercises

See www.CalcChat.com for worked-out solutions to odd-numbered exercises.

In Exercises 1 and 2, approximate the coordinates of the points.

1.

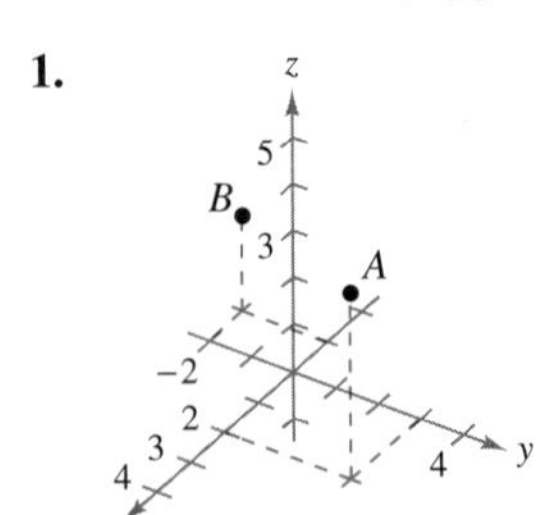

2.

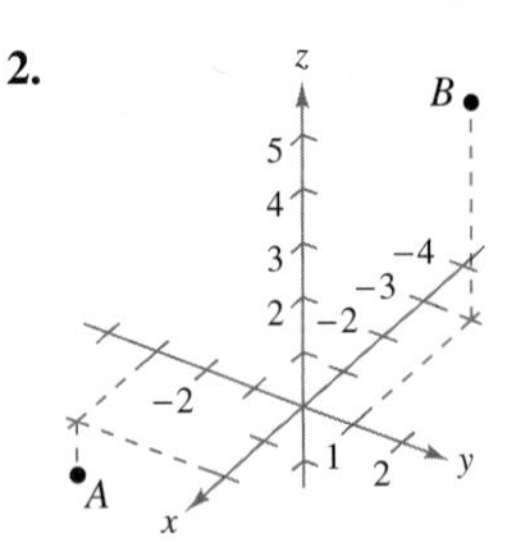

In Exercises 3–6, plot the points on the same three-dimensional coordinate system.

3. (a) $(2, 1, 3)$ (b) $(-1, 2, 1)$

4. (a) $(3, -2, 5)$ (b) $\left(\frac{3}{2}, 4, -2\right)$

5. (a) $(5, -2, 2)$ (b) $(5, -2, -2)$

6. (a) $(0, 4, -5)$ (b) $(4, 0, 5)$

In Exercises 7–10, find the coordinates of the point.

7. The point is located three units behind the yz-plane, four units to the right of the xz-plane, and five units above the xy-plane.

8. The point is located seven units in front of the yz-plane, two units to the left of the xz-plane, and one unit below the xy-plane.

9. The point is located on the x-axis, 12 units in front of the yz-plane.

10. The point is located in the yz-plane, three units to the right of the xz-plane, and two units above the xy-plane.

11. ***Think About It*** What is the z-coordinate of any point in the xy-plane?

12. ***Think About It*** What is the x-coordinate of any point in the yz-plane?

In Exercises 13–24, determine the location of a point (x, y, z) that satisfies the condition(s).

13. $z = 6$ **14.** $y = 2$

15. $x = -3$ **16.** $z = -\frac{5}{2}$

17. $y < 0$ **18.** $x > 0$

19. $|y| \le 3$ **20.** $|x| > 4$

21. $xy > 0, \quad z = -3$ **22.** $xy < 0, \quad z = 4$

23. $xyz < 0$ **24.** $xyz > 0$

In Exercises 25–28, find the distance between the points.

25. $(0, 0, 0), \quad (-4, 2, 7)$

26. $(-2, 3, 2), \quad (2, -5, -2)$

27. $(1, -2, 4), \quad (6, -2, -2)$

28. $(2, 2, 3), \quad (4, -5, 6)$

In Exercises 29–32, find the lengths of the sides of the triangle with the indicated vertices, and determine whether the triangle is a right triangle, an isosceles triangle, or neither.

29. $(0, 0, 4), (2, 6, 7), (6, 4, -8)$

30. $(3, 4, 1), (0, 6, 2), (3, 5, 6)$

31. $(-1, 0, -2), (-1, 5, 2), (-3, -1, 1)$

32. $(4, -1, -1), (2, 0, -4), (3, 5, -1)$

33. ***Think About It*** The triangle in Exercise 29 is translated five units upward along the z-axis. Determine the coordinates of the translated triangle.

34. ***Think About It*** The triangle in Exercise 30 is translated three units to the right along the y-axis. Determine the coordinates of the translated triangle.

In Exercises 35 and 36, find the coordinates of the midpoint of the line segment joining the points.

35. $(5, -9, 7), (-2, 3, 3)$ **36.** $(4, 0, -6), (8, 8, 20)$

In Exercises 37–40, find the standard equation of the sphere.

37. Center: $(0, 2, 5)$
Radius: 2

38. Center: $(4, -1, 1)$
Radius: 5

39. Endpoints of a diameter: $(2, 0, 0), (0, 6, 0)$

40. Center: $(-3, 2, 4)$, tangent to the yz-plane

In Exercises 41–44, complete the square to write the equation of the sphere in standard form. Find the center and radius.

41. $x^2 + y^2 + z^2 - 2x + 6y + 8z + 1 = 0$

42. $x^2 + y^2 + z^2 + 9x - 2y + 10z + 19 = 0$

43. $9x^2 + 9y^2 + 9z^2 - 6x + 18y + 1 = 0$

44. $4x^2 + 4y^2 + 4z^2 - 24x - 4y + 8z - 23 = 0$

In Exercises 45–48, describe the solid satisfying the condition.

45. $x^2 + y^2 + z^2 \le 36$ **46.** $x^2 + y^2 + z^2 > 4$

47. $x^2 + y^2 + z^2 < 4x - 6y + 8z - 13$

48. $x^2 + y^2 + z^2 > -4x + 6y - 8z - 13$

In Exercises 49–52, (a) find the component form of the vector v, (b) write the vector using standard unit vector notation, and (c) sketch the vector with its initial point at the origin.

49. **50.**

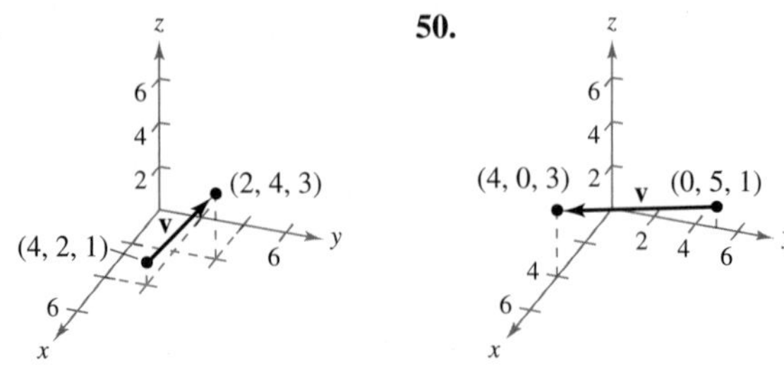

51.

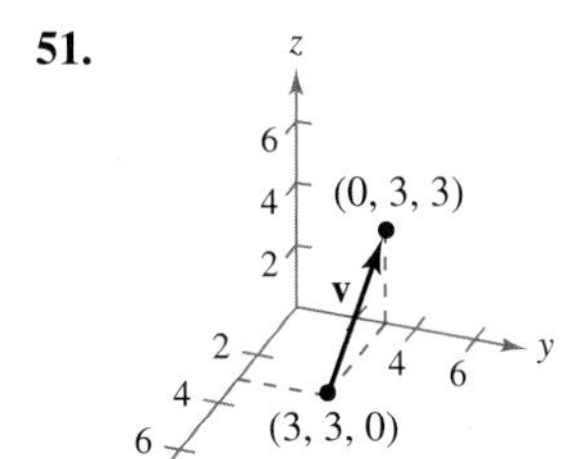

52.

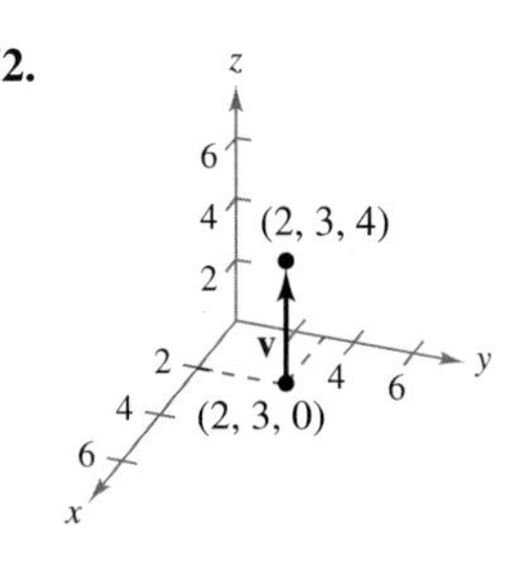

In Exercises 53–56, find the component form and magnitude of the vector v with the given initial and terminal points. Then find a unit vector in the direction of v.

	Initial Point	*Terminal Point*
53.	$(3, 2, 0)$	$(4, 1, 6)$
54.	$(4, -5, 2)$	$(-1, 7, -3)$
55.	$(-4, 3, 1)$	$(-5, 3, 0)$
56.	$(1, -2, 4)$	$(2, 4, -2)$

In Exercises 57 and 58, the initial and terminal points of a vector v are given. (a) Sketch the directed line segment, (b) find the component form of the vector, (c) write the vector using standard unit vector notation, and (d) sketch the vector with its initial point at the origin.

57. Initial point: $(-1, 2, 3)$
Terminal point: $(3, 3, 4)$

58. Initial point: $(2, -1, -2)$
Terminal point: $(-4, 3, 7)$

In Exercises 59 and 60, the vector v and its initial point are given. Find the terminal point.

59. $\mathbf{v} = \langle 3, -5, 6 \rangle$
Initial point: $(0, 6, 2)$

60. $\mathbf{v} = \left\langle 1, -\frac{2}{3}, \frac{1}{2} \right\rangle$
Initial point: $\left(0, 2, \frac{5}{2}\right)$

In Exercises 61 and 62, find each scalar multiple of v and sketch its graph.

61. $\mathbf{v} = \langle 1, 2, 2 \rangle$
(a) $2\mathbf{v}$ (b) $-\mathbf{v}$
(c) $\frac{3}{2}\mathbf{v}$ (d) $0\mathbf{v}$

62. $\mathbf{v} = \langle 2, -2, 1 \rangle$
(a) $-\mathbf{v}$ (b) $2\mathbf{v}$
(c) $\frac{1}{2}\mathbf{v}$ (d) $\frac{5}{2}\mathbf{v}$

In Exercises 63–68, find the vector z, given that $\mathbf{u} = \langle 1, 2, 3 \rangle$, $\mathbf{v} = \langle 2, 2, -1 \rangle$, and $\mathbf{w} = \langle 4, 0, -4 \rangle$.

63. $\mathbf{z} = \mathbf{u} - \mathbf{v}$

64. $\mathbf{z} = \mathbf{u} - \mathbf{v} + 2\mathbf{w}$

65. $\mathbf{z} = 2\mathbf{u} + 4\mathbf{v} - \mathbf{w}$

66. $\mathbf{z} = 5\mathbf{u} - 3\mathbf{v} - \frac{1}{2}\mathbf{w}$

67. $2\mathbf{z} - 3\mathbf{u} = \mathbf{w}$

68. $2\mathbf{u} + \mathbf{v} - \mathbf{w} + 3\mathbf{z} = \mathbf{0}$

In Exercises 69–72, determine which of the vectors is (are) parallel to z. Use a graphing utility to confirm your results.

69. $\mathbf{z} = \langle 3, 2, -5 \rangle$
(a) $\langle -6, -4, 10 \rangle$
(b) $\left\langle 2, \frac{4}{3}, -\frac{10}{3} \right\rangle$
(c) $\langle 6, 4, 10 \rangle$
(d) $\langle 1, -4, 2 \rangle$

70. $\mathbf{z} = \frac{1}{2}\mathbf{i} - \frac{2}{3}\mathbf{j} + \frac{3}{4}\mathbf{k}$
(a) $6\mathbf{i} - 4\mathbf{j} + 9\mathbf{k}$
(b) $-\mathbf{i} + \frac{4}{3}\mathbf{j} - \frac{3}{2}\mathbf{k}$
(c) $12\mathbf{i} + 9\mathbf{k}$
(d) $\frac{3}{4}\mathbf{i} - \mathbf{j} + \frac{9}{8}\mathbf{k}$

71. $\mathbf{z}$ has initial point $(1, -1, 3)$ and terminal point $(-2, 3, 5)$.
(a) $-6\mathbf{i} + 8\mathbf{j} + 4\mathbf{k}$ (b) $4\mathbf{j} + 2\mathbf{k}$

72. $\mathbf{z}$ has initial point $(5, 4, 1)$ and terminal point $(-2, -4, 4)$.
(a) $\langle 7, 6, 2 \rangle$ (b) $\langle 14, 16, -6 \rangle$

In Exercises 73–76, use vectors to determine whether the points are collinear.

73. $(0, -2, -5), (3, 4, 4), (2, 2, 1)$

74. $(4, -2, 7), (-2, 0, 3), (7, -3, 9)$

75. $(1, 2, 4), (2, 5, 0), (0, 1, 5)$

76. $(0, 0, 0), (1, 3, -2), (2, -6, 4)$

In Exercises 77 and 78, use vectors to show that the points form the vertices of a parallelogram.

77. $(2, 9, 1), (3, 11, 4), (0, 10, 2), (1, 12, 5)$

78. $(1, 1, 3), (9, -1, -2), (11, 2, -9), (3, 4, -4)$

In Exercises 79–84, find the magnitude of v.

79. $\mathbf{v} = \langle 0, 0, 0 \rangle$

80. $\mathbf{v} = \langle 1, 0, 3 \rangle$

81. $\mathbf{v} = 3\mathbf{j} - 5\mathbf{k}$

82. $\mathbf{v} = 2\mathbf{i} + 5\mathbf{j} - \mathbf{k}$

83. $\mathbf{v} = \mathbf{i} - 2\mathbf{j} - 3\mathbf{k}$

84. $\mathbf{v} = -4\mathbf{i} + 3\mathbf{j} + 7\mathbf{k}$

In Exercises 85–88, find a unit vector (a) in the direction of v and (b) in the direction opposite of v.

85. $\mathbf{v} = \langle 2, -1, 2 \rangle$

86. $\mathbf{v} = \langle 6, 0, 8 \rangle$

87. $\mathbf{v} = \langle 3, 2, -5 \rangle$

88. $\mathbf{v} = \langle 8, 0, 0 \rangle$

89. ***Programming*** You are given the component forms of the vectors **u** and **v**. Write a program for a graphing utility in which the output is (a) the component form of $\mathbf{u} + \mathbf{v}$, (b) $\|\mathbf{u} + \mathbf{v}\|$, (c) $\|\mathbf{u}\|$, and (d) $\|\mathbf{v}\|$. (e) Run the program for the vectors $\mathbf{u} = \langle -1, 3, 4 \rangle$ and $\mathbf{v} = \langle 5, 4.5, -6 \rangle$.

CAPSTONE

90. Consider the two nonzero vectors **u** and **v**, and let s and t be real numbers. Describe the geometric figure generated by the terminal points of the three vectors $t\mathbf{v}$, $\mathbf{u} + t\mathbf{v}$, and $s\mathbf{u} + t\mathbf{v}$.

In Exercises 91 and 92, determine the values of c that satisfy the equation. Let $\mathbf{u} = -\mathbf{i} + 2\mathbf{j} + 3\mathbf{k}$ and $\mathbf{v} = 2\mathbf{i} + 2\mathbf{j} - k$.

91. $\|c\mathbf{v}\| = 7$

92. $\|c\mathbf{u}\| = 4$

In Exercises 93–96, find the vector v with the given magnitude and direction u.

	Magnitude	*Direction*
93.	10	$\mathbf{u} = \langle 0, 3, 3 \rangle$
94.	3	$\mathbf{u} = \langle 1, 1, 1 \rangle$
95.	$\frac{3}{2}$	$\mathbf{u} = \langle 2, -2, 1 \rangle$
96.	7	$\mathbf{u} = \langle -4, 6, 2 \rangle$

In Exercises 97 and 98, sketch the vector v and write its component form.

97. $\mathbf{v}$ lies in the yz-plane, has magnitude 2, and makes an angle of 30° with the positive y-axis.

98. $\mathbf{v}$ lies in the xz-plane, has magnitude 5, and makes an angle of 45° with the positive z-axis.

In Exercises 99 and 100, use vectors to find the point that lies two-thirds of the way from P to Q.

99. $P(4, 3, 0), \quad Q(1, -3, 3)$ **100.** $P(1, 2, 5), \quad Q(6, 8, 2)$

101. Let $\mathbf{u} = \mathbf{i} + \mathbf{j}$, $\mathbf{v} = \mathbf{j} + \mathbf{k}$, and $\mathbf{w} = a\mathbf{u} + b\mathbf{v}$.

(a) Sketch $\mathbf{u}$ and $\mathbf{v}$.

(b) If $\mathbf{w} = \mathbf{0}$, show that a and b must both be zero.

(c) Find a and b such that $\mathbf{w} = \mathbf{i} + 2\mathbf{j} + \mathbf{k}$.

(d) Show that no choice of a and b yields $\mathbf{w} = \mathbf{i} + 2\mathbf{j} + 3\mathbf{k}$.

102. ***Writing*** The initial and terminal points of the vector $\mathbf{v}$ are (x_1, y_1, z_1) and (x, y, z). Describe the set of all points (x, y, z) such that $\|\mathbf{v}\| = 4$.

WRITING ABOUT CONCEPTS

103. A point in the three-dimensional coordinate system has coordinates (x_0, y_0, z_0). Describe what each coordinate measures.

104. Give the formula for the distance between the points (x_1, y_1, z_1) and (x_2, y_2, z_2).

105. Give the standard equation of a sphere of radius r, centered at (x_0, y_0, z_0).

106. State the definition of parallel vectors.

107. Let A, B, and C be vertices of a triangle. Find $\overrightarrow{AB} + \overrightarrow{BC} + \overrightarrow{CA}$.

108. Let $\mathbf{r} = \langle x, y, z \rangle$ and $\mathbf{r}_0 = \langle 1, 1, 1 \rangle$. Describe the set of all points (x, y, z) such that $\|\mathbf{r} - \mathbf{r}_0\| = 2$.

109. ***Numerical, Graphical, and Analytic Analysis*** The lights in an auditorium are 24-pound discs of radius 18 inches. Each disc is supported by three equally spaced cables that are L inches long (see figure).

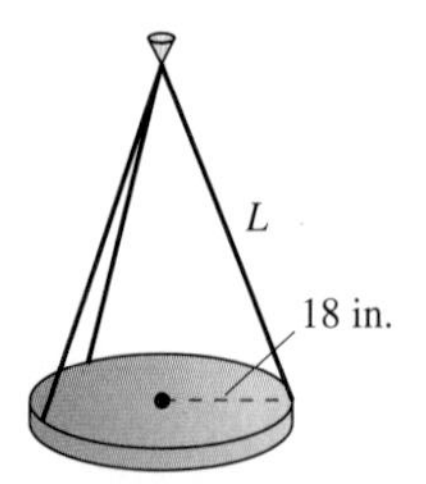

(a) Write the tension T in each cable as a function of L. Determine the domain of the function.

(b) Use a graphing utility and the function in part (a) to complete the table.

L	20	25	30	35	40	45	50
T							

(c) Use a graphing utility to graph the function in part (a). Determine the asymptotes of the graph.

(d) Confirm the asymptotes of the graph in part (c) analytically.

(e) Determine the minimum length of each cable if a cable is designed to carry a maximum load of 10 pounds.

110. ***Think About It*** Suppose the length of each cable in Exercise 109 has a fixed length $L = a$, and the radius of each disc is r_0 inches. Make a conjecture about the limit $\lim_{r_0 \to a^-} T$ and give a reason for your answer.

111. ***Diagonal of a Cube*** Find the component form of the unit vector $\mathbf{v}$ in the direction of the diagonal of the cube shown in the figure.

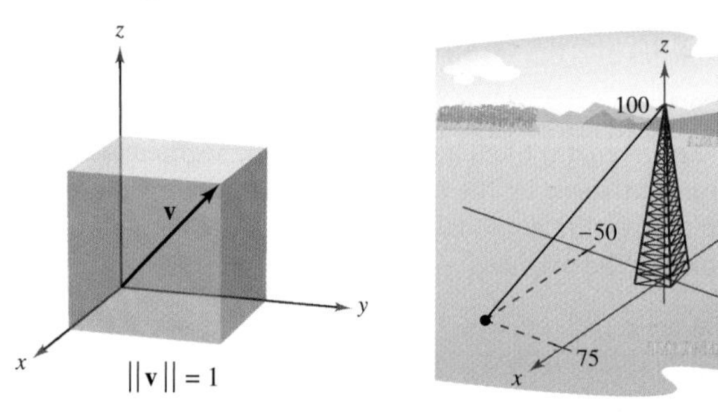

Figure for 111 **Figure for 112**

112. ***Tower Guy Wire*** The guy wire supporting a 100-foot tower has a tension of 550 pounds. Using the distances shown in the figure, write the component form of the vector $\mathbf{F}$ representing the tension in the wire.

113. ***Load Supports*** Find the tension in each of the supporting cables in the figure if the weight of the crate is 500 newtons.

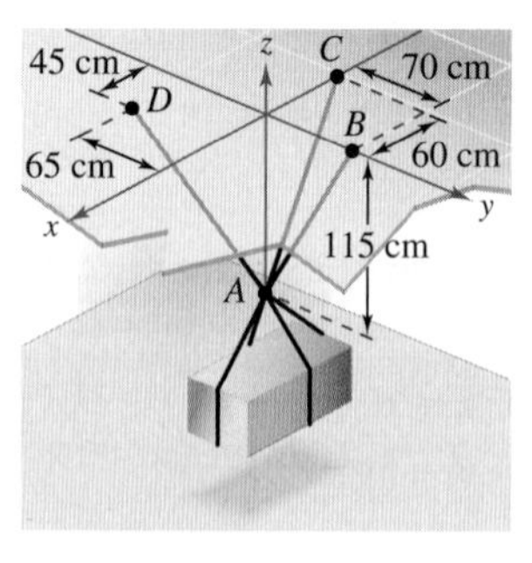

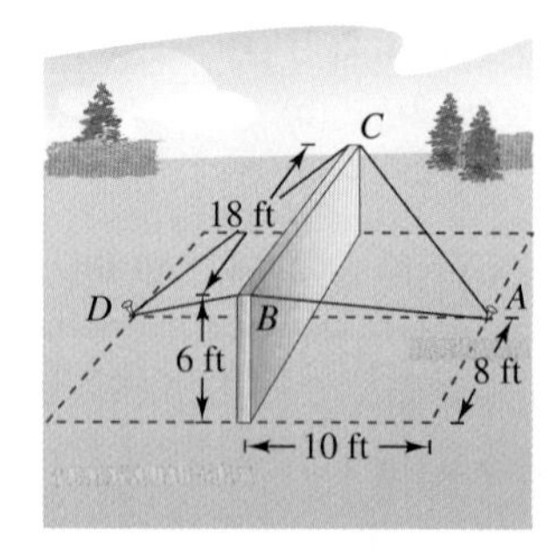

Figure for 113 **Figure for 114**

114. ***Construction*** A precast concrete wall is temporarily kept in its vertical position by ropes (see figure). Find the total force exerted on the pin at position A. The tensions in AB and AC are 420 pounds and 650 pounds.

115. Write an equation whose graph consists of the set of points $P(x, y, z)$ that are twice as far from $A(0, -1, 1)$ as from $B(1, 2, 0)$.

11.3 The Dot Product of Two Vectors

- Use properties of the dot product of two vectors.
- Find the angle between two vectors using the dot product.
- Find the direction cosines of a vector in space.
- Find the projection of a vector onto another vector.
- Use vectors to find the work done by a constant force.

EXPLORATION

Interpreting a Dot Product
Several vectors are shown below on the unit circle. Find the dot products of several pairs of vectors. Then find the angle between each pair that you used. Make a conjecture about the relationship between the dot product of two vectors and the angle between the vectors.

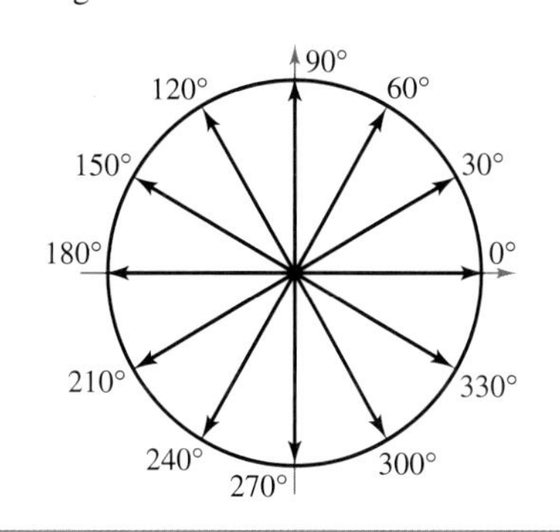

The Dot Product

So far you have studied two operations with vectors—vector addition and multiplication by a scalar—each of which yields another vector. In this section you will study a third vector operation, called the **dot product.** This product yields a scalar, rather than a vector.

DEFINITION OF DOT PRODUCT

The **dot product** of $\mathbf{u} = \langle u_1, u_2 \rangle$ and $\mathbf{v} = \langle v_1, v_2 \rangle$ is

$$\mathbf{u} \cdot \mathbf{v} = u_1v_1 + u_2v_2.$$

The **dot product** of $\mathbf{u} = \langle u_1, u_2, u_3 \rangle$ and $\mathbf{v} = \langle v_1, v_2, v_3 \rangle$ is

$$\mathbf{u} \cdot \mathbf{v} = u_1v_1 + u_2v_2 + u_3v_3.$$

NOTE Because the dot product of two vectors yields a scalar, it is also called the **scalar product** (or **inner product**) of the two vectors. ■

THEOREM 11.4 PROPERTIES OF THE DOT PRODUCT

Let $\mathbf{u}$, $\mathbf{v}$, and $\mathbf{w}$ be vectors in the plane or in space and let c be a scalar.

1. $\mathbf{u} \cdot \mathbf{v} = \mathbf{v} \cdot \mathbf{u}$ — Commutative Property
2. $\mathbf{u} \cdot (\mathbf{v} + \mathbf{w}) = \mathbf{u} \cdot \mathbf{v} + \mathbf{u} \cdot \mathbf{w}$ — Distributive Property
3. $c(\mathbf{u} \cdot \mathbf{v}) = c\mathbf{u} \cdot \mathbf{v} = \mathbf{u} \cdot c\mathbf{v}$
4. $\mathbf{0} \cdot \mathbf{v} = 0$
5. $\mathbf{v} \cdot \mathbf{v} = \|\mathbf{v}\|^2$

PROOF To prove the first property, let $\mathbf{u} = \langle u_1, u_2, u_3 \rangle$ and $\mathbf{v} = \langle v_1, v_2, v_3 \rangle$. Then

$$\begin{aligned}\mathbf{u} \cdot \mathbf{v} &= u_1v_1 + u_2v_2 + u_3v_3\\ &= v_1u_1 + v_2u_2 + v_3u_3\\ &= \mathbf{v} \cdot \mathbf{u}.\end{aligned}$$

For the fifth property, let $\mathbf{v} = \langle v_1, v_2, v_3 \rangle$. Then

$$\begin{aligned}\mathbf{v} \cdot \mathbf{v} &= v_1^2 + v_2^2 + v_3^2\\ &= \left(\sqrt{v_1^2 + v_2^2 + v_3^2}\right)^2\\ &= \|\mathbf{v}\|^2.\end{aligned}$$

Proofs of the other properties are left to you. ■

EXAMPLE 1 Finding Dot Products

Given $\mathbf{u} = \langle 2, -2\rangle$, $\mathbf{v} = \langle 5, 8\rangle$, and $\mathbf{w} = \langle -4, 3\rangle$, find each of the following.

a. $\mathbf{u} \cdot \mathbf{v}$ **b.** $(\mathbf{u} \cdot \mathbf{v})\mathbf{w}$

c. $\mathbf{u} \cdot (2\mathbf{v})$ **d.** $\|\mathbf{w}\|^2$

Solution

a. $\mathbf{u} \cdot \mathbf{v} = \langle 2, -2\rangle \cdot \langle 5, 8\rangle = 2(5) + (-2)(8) = -6$

b. $(\mathbf{u} \cdot \mathbf{v})\mathbf{w} = -6\langle -4, 3\rangle = \langle 24, -18\rangle$

c. $\mathbf{u} \cdot (2\mathbf{v}) = 2(\mathbf{u} \cdot \mathbf{v}) = 2(-6) = -12$ Theorem 11.4

d. $\|\mathbf{w}\|^2 = \mathbf{w} \cdot \mathbf{w}$ Theorem 11.4

$= \langle -4, 3\rangle \cdot \langle -4, 3\rangle$ Substitute $\langle -4, 3\rangle$ for $\mathbf{w}$.

$= (-4)(-4) + (3)(3)$ Definition of dot product

$= 25$ Simplify.

Notice that the result of part (b) is a *vector* quantity, whereas the results of the other three parts are *scalar* quantities. ■

Angle Between Two Vectors

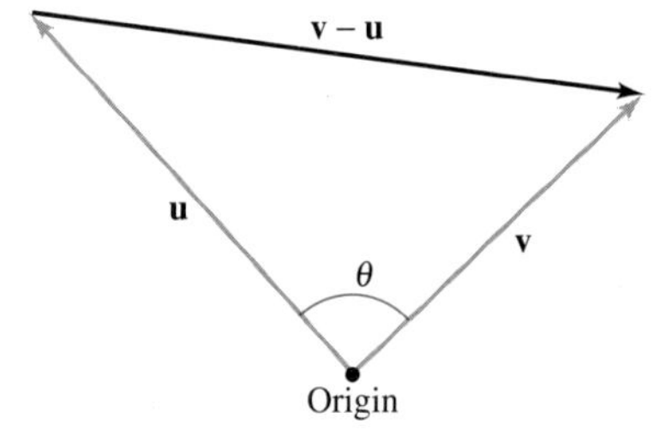

The angle between two vectors
Figure 11.24

The **angle between two nonzero vectors** is the angle θ, $0 \le \theta \le \pi$, between their respective standard position vectors, as shown in Figure 11.24. The next theorem shows how to find this angle using the dot product. (Note that the angle between the zero vector and another vector is not defined here.)

THEOREM 11.5 ANGLE BETWEEN TWO VECTORS

If θ is the angle between two nonzero vectors $\mathbf{u}$ and $\mathbf{v}$, then

$$\cos\theta = \frac{\mathbf{u} \cdot \mathbf{v}}{\|\mathbf{u}\|\,\|\mathbf{v}\|}.$$

PROOF Consider the triangle determined by vectors $\mathbf{u}$, $\mathbf{v}$, and $\mathbf{v} - \mathbf{u}$, as shown in Figure 11.24. By the Law of Cosines, you can write

$$\|\mathbf{v} - \mathbf{u}\|^2 = \|\mathbf{u}\|^2 + \|\mathbf{v}\|^2 - 2\|\mathbf{u}\|\,\|\mathbf{v}\|\cos\theta.$$

Using the properties of the dot product, the left side can be rewritten as

$$\begin{aligned}\|\mathbf{v} - \mathbf{u}\|^2 &= (\mathbf{v} - \mathbf{u}) \cdot (\mathbf{v} - \mathbf{u})\\ &= (\mathbf{v} - \mathbf{u}) \cdot \mathbf{v} - (\mathbf{v} - \mathbf{u}) \cdot \mathbf{u}\\ &= \mathbf{v} \cdot \mathbf{v} - \mathbf{u} \cdot \mathbf{v} - \mathbf{v} \cdot \mathbf{u} + \mathbf{u} \cdot \mathbf{u}\\ &= \|\mathbf{v}\|^2 - 2\mathbf{u} \cdot \mathbf{v} + \|\mathbf{u}\|^2\end{aligned}$$

and substitution back into the Law of Cosines yields

$$\begin{aligned}\|\mathbf{v}\|^2 - 2\mathbf{u} \cdot \mathbf{v} + \|\mathbf{u}\|^2 &= \|\mathbf{u}\|^2 + \|\mathbf{v}\|^2 - 2\|\mathbf{u}\|\,\|\mathbf{v}\|\cos\theta\\ -2\mathbf{u} \cdot \mathbf{v} &= -2\|\mathbf{u}\|\,\|\mathbf{v}\|\cos\theta\\ \cos\theta &= \frac{\mathbf{u} \cdot \mathbf{v}}{\|\mathbf{u}\|\,\|\mathbf{v}\|}.\end{aligned}$$

■

If the angle between two vectors is known, rewriting Theorem 11.5 in the form

$$\mathbf{u} \cdot \mathbf{v} = \|\mathbf{u}\| \|\mathbf{v}\| \cos \theta$$ Alternative form of dot product

produces an alternative way to calculate the dot product. From this form, you can see that because $\|\mathbf{u}\|$ and $\|\mathbf{v}\|$ are always positive, $\mathbf{u} \cdot \mathbf{v}$ and $\cos \theta$ will always have the same sign. Figure 11.25 shows the possible orientations of two vectors.

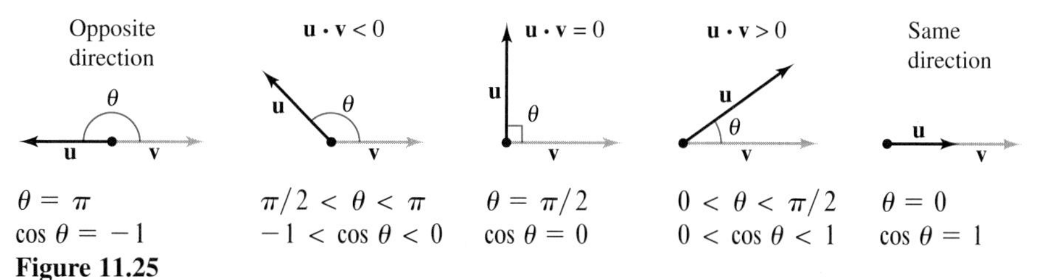

Figure 11.25

From Theorem 11.5, you can see that two nonzero vectors meet at a right angle if and only if their dot product is zero. Two such vectors are said to be **orthogonal.**

DEFINITION OF ORTHOGONAL VECTORS

The vectors **u** and **v** are orthogonal if $\mathbf{u} \cdot \mathbf{v} = 0$.

NOTE The terms "perpendicular," "orthogonal," and "normal" all mean essentially the same thing—meeting at right angles. However, it is common to say that two vectors are *orthogonal*, two lines or planes are *perpendicular*, and a vector is *normal* to a given line or plane. ■

From this definition, it follows that the zero vector is orthogonal to every vector **u**, because $\mathbf{0} \cdot \mathbf{u} = 0$. Moreover, for $0 \le \theta \le \pi$, you know that $\cos \theta = 0$ if and only if $\theta = \pi/2$. So, you can use Theorem 11.5 to conclude that two *nonzero* vectors are orthogonal if and only if the angle between them is $\pi/2$.

EXAMPLE 2 Finding the Angle Between Two Vectors

For $\mathbf{u} = \langle 3, -1, 2\rangle$, $\mathbf{v} = \langle -4, 0, 2\rangle$, $\mathbf{w} = \langle 1, -1, -2\rangle$, and $\mathbf{z} = \langle 2, 0, -1\rangle$, find the angle between each pair of vectors.

a. **u** and **v** **b.** **u** and **w** **c.** **v** and **z**

Solution

a. $\cos \theta = \dfrac{\mathbf{u} \cdot \mathbf{v}}{\|\mathbf{u}\| \|\mathbf{v}\|} = \dfrac{-12 + 0 + 4}{\sqrt{14}\sqrt{20}} = \dfrac{-8}{2\sqrt{14}\sqrt{5}} = \dfrac{-4}{\sqrt{70}}$

Because $\mathbf{u} \cdot \mathbf{v} < 0$, $\theta = \arccos \dfrac{-4}{\sqrt{70}} \approx 2.069$ radians.

b. $\cos \theta = \dfrac{\mathbf{u} \cdot \mathbf{w}}{\|\mathbf{u}\| \|\mathbf{w}\|} = \dfrac{3 + 1 - 4}{\sqrt{14}\sqrt{6}} = \dfrac{0}{\sqrt{84}} = 0$

Because $\mathbf{u} \cdot \mathbf{w} = 0$, **u** and **w** are *orthogonal*. So, $\theta = \pi/2$.

c. $\cos \theta = \dfrac{\mathbf{v} \cdot \mathbf{z}}{\|\mathbf{v}\| \|\mathbf{z}\|} = \dfrac{-8 + 0 - 2}{\sqrt{20}\sqrt{5}} = \dfrac{-10}{\sqrt{100}} = -1$

Consequently, $\theta = \pi$. Note that **v** and **z** are parallel, with $\mathbf{v} = -2\mathbf{z}$. ■

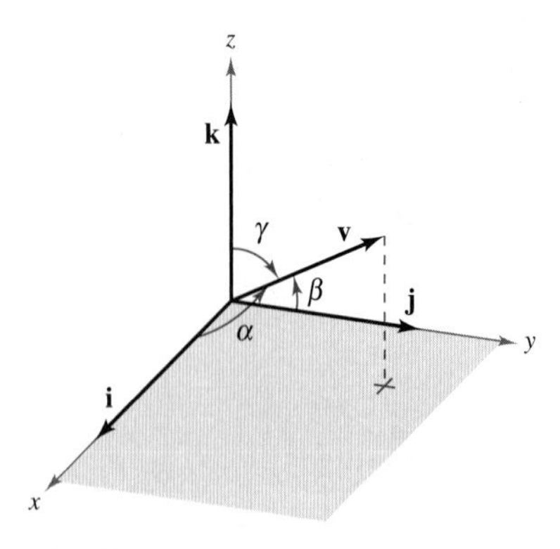

Direction angles
Figure 11.26

Direction Cosines

For a vector in the plane, you have seen that it is convenient to measure direction in terms of the angle, measured counterclockwise, *from* the positive x-axis *to* the vector. In space it is more convenient to measure direction in terms of the angles *between* the nonzero vector $\mathbf{v}$ and the three unit vectors $\mathbf{i}$, $\mathbf{j}$, and $\mathbf{k}$, as shown in Figure 11.26. The angles α, β, and γ are the **direction angles of v,** and $\cos\alpha$, $\cos\beta$, and $\cos\gamma$ are the **direction cosines of v.** Because

$$\mathbf{v}\cdot\mathbf{i} = \|\mathbf{v}\|\,\|\mathbf{i}\|\cos\alpha = \|\mathbf{v}\|\cos\alpha$$

and

$$\mathbf{v}\cdot\mathbf{i} = \langle v_1, v_2, v_3\rangle \cdot \langle 1, 0, 0\rangle = v_1$$

it follows that $\cos\alpha = v_1/\|\mathbf{v}\|$. By similar reasoning with the unit vectors $\mathbf{j}$ and $\mathbf{k}$, you have

$$\cos\alpha = \frac{v_1}{\|\mathbf{v}\|} \qquad \alpha \text{ is the angle between } \mathbf{v} \text{ and } \mathbf{i}.$$

$$\cos\beta = \frac{v_2}{\|\mathbf{v}\|} \qquad \beta \text{ is the angle between } \mathbf{v} \text{ and } \mathbf{j}.$$

$$\cos\gamma = \frac{v_3}{\|\mathbf{v}\|}. \qquad \gamma \text{ is the angle between } \mathbf{v} \text{ and } \mathbf{k}.$$

Consequently, any nonzero vector $\mathbf{v}$ in space has the normalized form

$$\frac{\mathbf{v}}{\|\mathbf{v}\|} = \frac{v_1}{\|\mathbf{v}\|}\mathbf{i} + \frac{v_2}{\|\mathbf{v}\|}\mathbf{j} + \frac{v_3}{\|\mathbf{v}\|}\mathbf{k} = \cos\alpha\,\mathbf{i} + \cos\beta\,\mathbf{j} + \cos\gamma\,\mathbf{k}$$

and because $\mathbf{v}/\|\mathbf{v}\|$ is a unit vector, it follows that

$$\cos^2\alpha + \cos^2\beta + \cos^2\gamma = 1.$$

EXAMPLE 3 Finding Direction Angles

Find the direction cosines and angles for the vector $\mathbf{v} = 2\mathbf{i} + 3\mathbf{j} + 4\mathbf{k}$, and show that $\cos^2\alpha + \cos^2\beta + \cos^2\gamma = 1$.

Solution Because $\|\mathbf{v}\| = \sqrt{2^2 + 3^2 + 4^2} = \sqrt{29}$, you can write the following.

$$\cos\alpha = \frac{v_1}{\|\mathbf{v}\|} = \frac{2}{\sqrt{29}} \implies \alpha \approx 68.2^\circ \qquad \text{Angle between } \mathbf{v} \text{ and } \mathbf{i}$$

$$\cos\beta = \frac{v_2}{\|\mathbf{v}\|} = \frac{3}{\sqrt{29}} \implies \beta \approx 56.1^\circ \qquad \text{Angle between } \mathbf{v} \text{ and } \mathbf{j}$$

$$\cos\gamma = \frac{v_3}{\|\mathbf{v}\|} = \frac{4}{\sqrt{29}} \implies \gamma \approx 42.0^\circ \qquad \text{Angle between } \mathbf{v} \text{ and } \mathbf{k}$$

Furthermore, the sum of the squares of the direction cosines is

$$\begin{aligned}\cos^2\alpha + \cos^2\beta + \cos^2\gamma &= \frac{4}{29} + \frac{9}{29} + \frac{16}{29} \\ &= \frac{29}{29} \\ &= 1.\end{aligned}$$

See Figure 11.27. ■

α = angle between $\mathbf{v}$ and $\mathbf{i}$
β = angle between $\mathbf{v}$ and $\mathbf{j}$
γ = angle between $\mathbf{v}$ and $\mathbf{k}$

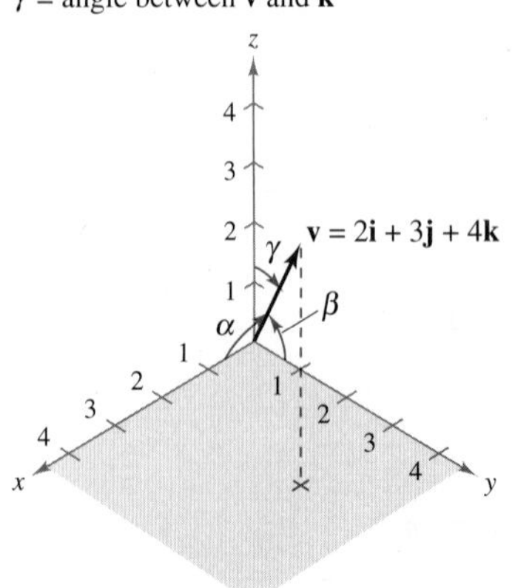

The direction angles of $\mathbf{v}$
Figure 11.27

Projections and Vector Components

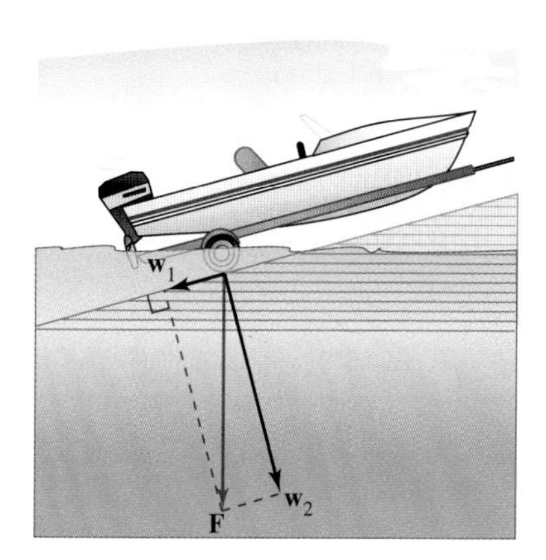

The force due to gravity pulls the boat against the ramp and down the ramp.
Figure 11.28

You have already seen applications in which two vectors are added to produce a resultant vector. Many applications in physics and engineering pose the reverse problem—decomposing a given vector into the sum of two **vector components.** The following physical example enables you to see the usefulness of this procedure.

Consider a boat on an inclined ramp, as shown in Figure 11.28. The force $\mathbf{F}$ due to gravity pulls the boat *down* the ramp and *against* the ramp. These two forces, $\mathbf{w}_1$ and $\mathbf{w}_2$, are orthogonal—they are called the vector components of $\mathbf{F}$.

$$\mathbf{F} = \mathbf{w}_1 + \mathbf{w}_2 \qquad \text{Vector components of } \mathbf{F}$$

The forces $\mathbf{w}_1$ and $\mathbf{w}_2$ help you analyze the effect of gravity on the boat. For example, $\mathbf{w}_1$ indicates the force necessary to keep the boat from rolling down the ramp, whereas $\mathbf{w}_2$ indicates the force that the tires must withstand.

DEFINITIONS OF PROJECTION AND VECTOR COMPONENTS

Let $\mathbf{u}$ and $\mathbf{v}$ be nonzero vectors. Moreover, let $\mathbf{u} = \mathbf{w}_1 + \mathbf{w}_2$, where $\mathbf{w}_1$ is parallel to $\mathbf{v}$, and $\mathbf{w}_2$ is orthogonal to $\mathbf{v}$, as shown in Figure 11.29.

1. $\mathbf{w}_1$ is called the **projection of u onto v** or the **vector component of u along v,** and is denoted by $\mathbf{w}_1 = \text{proj}_{\mathbf{v}}\mathbf{u}$.
2. $\mathbf{w}_2 = \mathbf{u} - \mathbf{w}_1$ is called the **vector component of u orthogonal to v.**

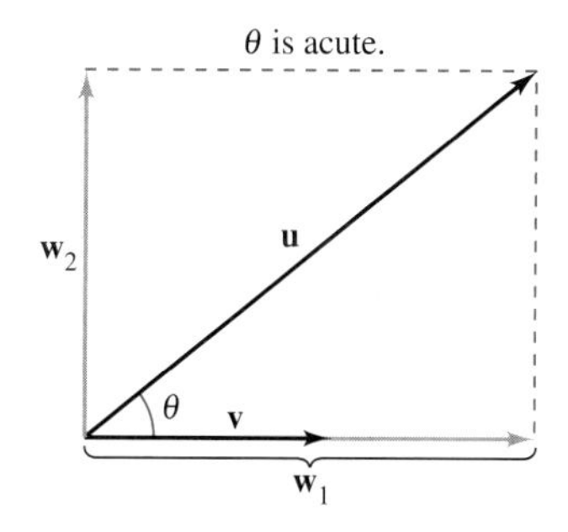

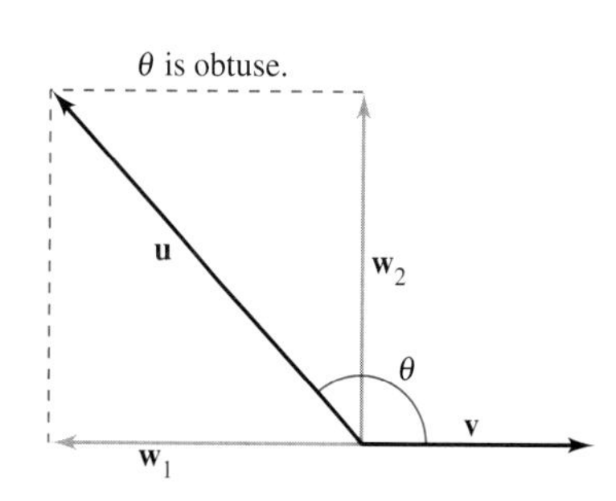

$\mathbf{w}_1 = \text{proj}_{\mathbf{v}}\mathbf{u} =$ projection of $\mathbf{u}$ onto $\mathbf{v}$ = vector component of $\mathbf{u}$ along $\mathbf{v}$
$\mathbf{w}_2 =$ vector component of $\mathbf{u}$ orthogonal to $\mathbf{v}$
Figure 11.29

EXAMPLE 4 Finding a Vector Component of u Orthogonal to v

Find the vector component of $\mathbf{u} = \langle 5, 10 \rangle$ that is orthogonal to $\mathbf{v} = \langle 4, 3 \rangle$, given that $\mathbf{w}_1 = \text{proj}_{\mathbf{v}}\mathbf{u} = \langle 8, 6 \rangle$ and

$$\mathbf{u} = \langle 5, 10 \rangle = \mathbf{w}_1 + \mathbf{w}_2.$$

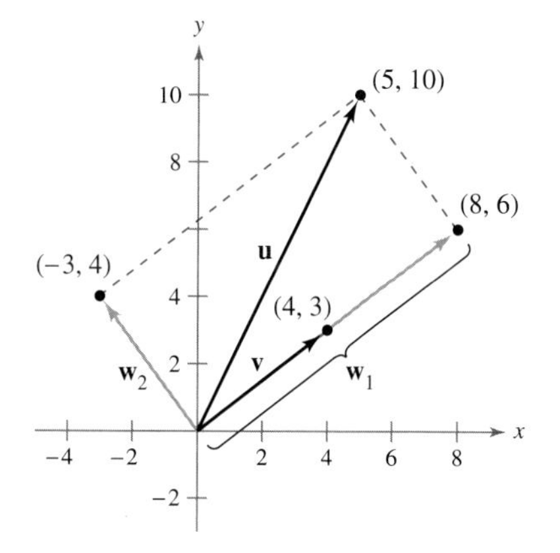

$\mathbf{u} = \mathbf{w}_1 + \mathbf{w}_2$
Figure 11.30

Solution Because $\mathbf{u} = \mathbf{w}_1 + \mathbf{w}_2$, where $\mathbf{w}_1$ is parallel to $\mathbf{v}$, it follows that $\mathbf{w}_2$ is the vector component of $\mathbf{u}$ orthogonal to $\mathbf{v}$. So, you have

$$\begin{aligned} \mathbf{w}_2 &= \mathbf{u} - \mathbf{w}_1 \\ &= \langle 5, 10 \rangle - \langle 8, 6 \rangle \\ &= \langle -3, 4 \rangle. \end{aligned}$$

Check to see that $\mathbf{w}_2$ is orthogonal to $\mathbf{v}$, as shown in Figure 11.30. ■

From Example 4, you can see that it is easy to find the vector component $\mathbf{w}_2$ once you have found the projection, $\mathbf{w}_1$, of $\mathbf{u}$ onto $\mathbf{v}$. To find this projection, use the dot product given in the theorem below, which you will prove in Exercise 92.

NOTE Note the distinction between the terms "component" and "vector component." For example, using the standard unit vectors with $\mathbf{u} = u_1\mathbf{i} + u_2\mathbf{j}$, u_1 is the *component* of $\mathbf{u}$ in the direction of $\mathbf{i}$ and $u_1\mathbf{i}$ is the *vector component* in the direction of $\mathbf{i}$.

THEOREM 11.6 PROJECTION USING THE DOT PRODUCT

If $\mathbf{u}$ and $\mathbf{v}$ are nonzero vectors, then the projection of $\mathbf{u}$ onto $\mathbf{v}$ is given by

$$\text{proj}_{\mathbf{v}}\mathbf{u} = \left(\frac{\mathbf{u} \cdot \mathbf{v}}{\|\mathbf{v}\|^2}\right)\mathbf{v}.$$

The projection of $\mathbf{u}$ onto $\mathbf{v}$ can be written as a scalar multiple of a unit vector in the direction of $\mathbf{v}$. That is,

$$\left(\frac{\mathbf{u} \cdot \mathbf{v}}{\|\mathbf{v}\|^2}\right)\mathbf{v} = \left(\frac{\mathbf{u} \cdot \mathbf{v}}{\|\mathbf{v}\|}\right)\frac{\mathbf{v}}{\|\mathbf{v}\|} = (k)\frac{\mathbf{v}}{\|\mathbf{v}\|} \quad \Longrightarrow \quad k = \frac{\mathbf{u} \cdot \mathbf{v}}{\|\mathbf{v}\|} = \|\mathbf{u}\| \cos \theta.$$

The scalar k is called the **component of u in the direction of v.**

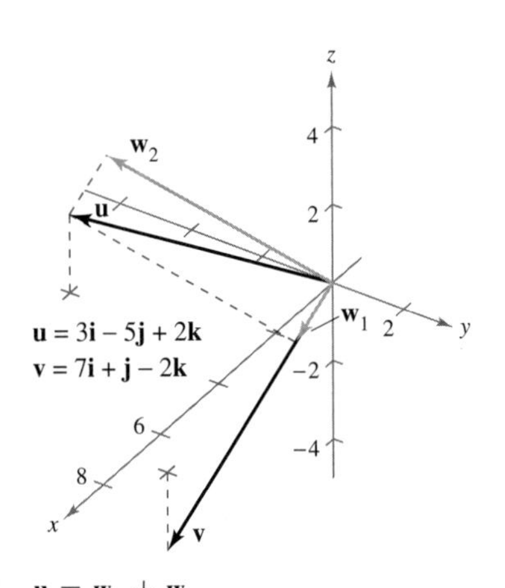

$\mathbf{u} = \mathbf{w}_1 + \mathbf{w}_2$
Figure 11.31

EXAMPLE 5 Decomposing a Vector into Vector Components

Find the projection of $\mathbf{u}$ onto $\mathbf{v}$ and the vector component of $\mathbf{u}$ orthogonal to $\mathbf{v}$ for the vectors $\mathbf{u} = 3\mathbf{i} - 5\mathbf{j} + 2\mathbf{k}$ and $\mathbf{v} = 7\mathbf{i} + \mathbf{j} - 2\mathbf{k}$ shown in Figure 11.31.

Solution The projection of $\mathbf{u}$ onto $\mathbf{v}$ is

$$\mathbf{w}_1 = \left(\frac{\mathbf{u} \cdot \mathbf{v}}{\|\mathbf{v}\|^2}\right)\mathbf{v} = \left(\frac{12}{54}\right)(7\mathbf{i} + \mathbf{j} - 2\mathbf{k}) = \frac{14}{9}\mathbf{i} + \frac{2}{9}\mathbf{j} - \frac{4}{9}\mathbf{k}.$$

The vector component of $\mathbf{u}$ orthogonal to $\mathbf{v}$ is the vector

$$\mathbf{w}_2 = \mathbf{u} - \mathbf{w}_1 = (3\mathbf{i} - 5\mathbf{j} + 2\mathbf{k}) - \left(\frac{14}{9}\mathbf{i} + \frac{2}{9}\mathbf{j} - \frac{4}{9}\mathbf{k}\right) = \frac{13}{9}\mathbf{i} - \frac{47}{9}\mathbf{j} + \frac{22}{9}\mathbf{k}.$$

EXAMPLE 6 Finding a Force

A 600-pound boat sits on a ramp inclined at 30°, as shown in Figure 11.32. What force is required to keep the boat from rolling down the ramp?

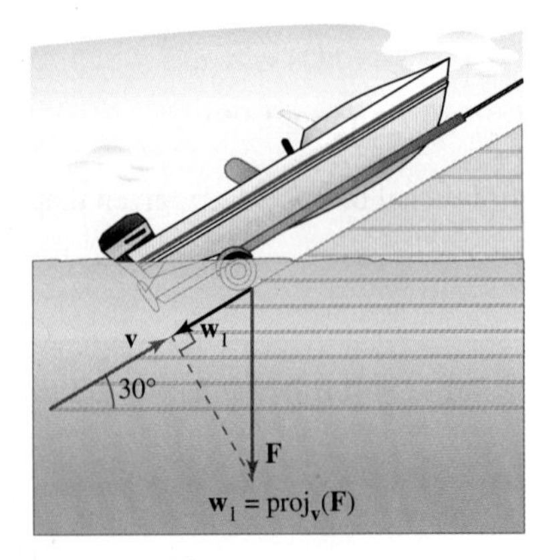

Figure 11.32

Solution Because the force due to gravity is vertical and downward, you can represent the gravitational force by the vector $\mathbf{F} = -600\mathbf{j}$. To find the force required to keep the boat from rolling down the ramp, project $\mathbf{F}$ onto a unit vector $\mathbf{v}$ in the direction of the ramp, as follows.

$$\mathbf{v} = \cos 30°\mathbf{i} + \sin 30°\mathbf{j} = \frac{\sqrt{3}}{2}\mathbf{i} + \frac{1}{2}\mathbf{j} \qquad \text{Unit vector along ramp}$$

Therefore, the projection of $\mathbf{F}$ onto $\mathbf{v}$ is given by

$$\mathbf{w}_1 = \text{proj}_{\mathbf{v}}\mathbf{F} = \left(\frac{\mathbf{F} \cdot \mathbf{v}}{\|\mathbf{v}\|^2}\right)\mathbf{v} = (\mathbf{F} \cdot \mathbf{v})\mathbf{v} = (-600)\left(\frac{1}{2}\right)\mathbf{v} = -300\left(\frac{\sqrt{3}}{2}\mathbf{i} + \frac{1}{2}\mathbf{j}\right).$$

The magnitude of this force is 300, and therefore a force of 300 pounds is required to keep the boat from rolling down the ramp. ■

Work $= \|\mathbf{F}\| \|\overrightarrow{PQ}\|$

(a) Force acts along the line of motion.

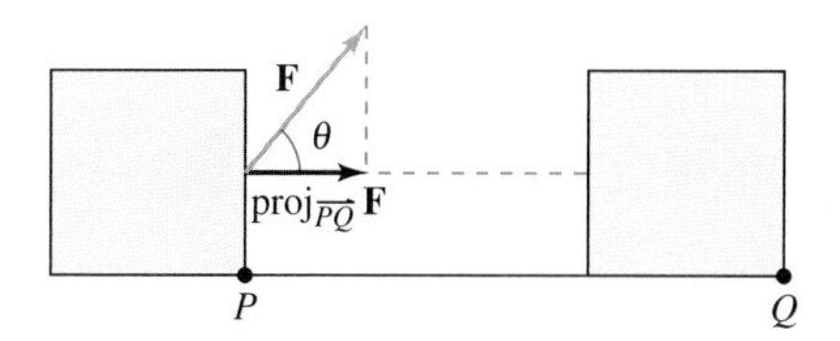

Work $= \|\text{proj}_{\overrightarrow{PQ}}\mathbf{F}\| \|\overrightarrow{PQ}\|$

(b) Force acts at angle θ with the line of motion.
Figure 11.33

Work

The work W done by the constant force $\mathbf{F}$ acting along the line of motion of an object is given by

$$W = (\text{magnitude of force})(\text{distance}) = \|\mathbf{F}\| \, \|\overrightarrow{PQ}\|$$

as shown in Figure 11.33(a). If the constant force $\mathbf{F}$ is not directed along the line of motion, you can see from Figure 11.33(b) that the work W done by the force is

$$W = \|\text{proj}_{\overrightarrow{PQ}}\mathbf{F}\| \, \|\overrightarrow{PQ}\| = (\cos\theta)\|\mathbf{F}\| \, \|\overrightarrow{PQ}\| = \mathbf{F} \cdot \overrightarrow{PQ}.$$

This notion of work is summarized in the following definition.

DEFINITION OF WORK

The work W done by a constant force $\mathbf{F}$ as its point of application moves along the vector $\overrightarrow{PQ}$ is given by either of the following.

1. $W = \|\text{proj}_{\overrightarrow{PQ}}\mathbf{F}\| \, \|\overrightarrow{PQ}\|$ — Projection form

2. $W = \mathbf{F} \cdot \overrightarrow{PQ}$ — Dot product form

Figure 11.34

EXAMPLE 7 Finding Work

To close a sliding door, a person pulls on a rope with a constant force of 50 pounds at a constant angle of 60°, as shown in Figure 11.34. Find the work done in moving the door 12 feet to its closed position.

Solution Using a projection, you can calculate the work as follows.

$$\begin{aligned} W &= \|\text{proj}_{\overrightarrow{PQ}}\mathbf{F}\| \, \|\overrightarrow{PQ}\| && \text{Projection form for work} \\ &= \cos(60°) \, \|\mathbf{F}\| \, \|\overrightarrow{PQ}\| \\ &= \frac{1}{2}(50)(12) \\ &= 300 \text{ foot-pounds} \end{aligned}$$

■

11.3 Exercises

See www.CalcChat.com for worked-out solutions to odd-numbered exercises.

In Exercises 1–8, find (a) $\mathbf{u} \cdot \mathbf{v}$, (b) $\mathbf{u} \cdot \mathbf{u}$, (c) $\|\mathbf{u}\|^2$, (d) $(\mathbf{u} \cdot \mathbf{v})\mathbf{v}$, and (e) $\mathbf{u} \cdot (2\mathbf{v})$.

1. $\mathbf{u} = \langle 3, 4\rangle, \quad \mathbf{v} = \langle -1, 5\rangle$

2. $\mathbf{u} = \langle 4, 10\rangle, \quad \mathbf{v} = \langle -2, 3\rangle$

3. $\mathbf{u} = \langle 6, -4\rangle, \quad \mathbf{v} = \langle -3, 2\rangle$

4. $\mathbf{u} = \langle -4, 8\rangle, \quad \mathbf{v} = \langle 7, 5\rangle$

5. $\mathbf{u} = \langle 2, -3, 4\rangle, \quad \mathbf{v} = \langle 0, 6, 5\rangle$

6. $\mathbf{u} = \mathbf{i}, \quad \mathbf{v} = \mathbf{i}$

7. $\mathbf{u} = 2\mathbf{i} - \mathbf{j} + \mathbf{k}$, $\mathbf{v} = \mathbf{i} - \mathbf{k}$

8. $\mathbf{u} = 2\mathbf{i} + \mathbf{j} - 2\mathbf{k}$, $\mathbf{v} = \mathbf{i} - 3\mathbf{j} + 2\mathbf{k}$

In Exercises 9 and 10, find $\mathbf{u} \cdot \mathbf{v}$.

9. $\|\mathbf{u}\| = 8$, $\|\mathbf{v}\| = 5$, and the angle between $\mathbf{u}$ and $\mathbf{v}$ is $\pi/3$.

10. $\|\mathbf{u}\| = 40$, $\|\mathbf{v}\| = 25$, and the angle between $\mathbf{u}$ and $\mathbf{v}$ is $5\pi/6$.

In Exercises 11–18, find the angle θ between the vectors.

11. $\mathbf{u} = \langle 1, 1\rangle, \mathbf{v} = \langle 2, -2\rangle$

12. $\mathbf{u} = \langle 3, 1\rangle, \mathbf{v} = \langle 2, -1\rangle$

13. $\mathbf{u} = 3\mathbf{i} + \mathbf{j}, \mathbf{v} = -2\mathbf{i} + 4\mathbf{j}$

14. $\mathbf{u} = \cos\left(\frac{\pi}{6}\right)\mathbf{i} + \sin\left(\frac{\pi}{6}\right)\mathbf{j}$

$\mathbf{v} = \cos\left(\frac{3\pi}{4}\right)\mathbf{i} + \sin\left(\frac{3\pi}{4}\right)\mathbf{j}$

15. $\mathbf{u} = \langle 1, 1, 1\rangle$, $\mathbf{v} = \langle 2, 1, -1\rangle$

16. $\mathbf{u} = 3\mathbf{i} + 2\mathbf{j} + \mathbf{k}$, $\mathbf{v} = 2\mathbf{i} - 3\mathbf{j}$

17. $\mathbf{u} = 3\mathbf{i} + 4\mathbf{j}$, $\mathbf{v} = -2\mathbf{j} + 3\mathbf{k}$

18. $\mathbf{u} = 2\mathbf{i} - 3\mathbf{j} + \mathbf{k}$, $\mathbf{v} = \mathbf{i} - 2\mathbf{j} + \mathbf{k}$

In Exercises 19–26, determine whether u and v are orthogonal, parallel, or neither.

19. $\mathbf{u} = \langle 4, 0\rangle, \quad \mathbf{v} = \langle 1, 1\rangle$

20. $\mathbf{u} = \langle 2, 18\rangle, \quad \mathbf{v} = \left\langle \frac{3}{2}, -\frac{1}{6}\right\rangle$

21. $\mathbf{u} = \langle 4, 3 \rangle$
$\mathbf{v} = \langle \frac{1}{2}, -\frac{2}{3} \rangle$

22. $\mathbf{u} = -\frac{1}{3}(\mathbf{i} - 2\mathbf{j})$
$\mathbf{v} = 2\mathbf{i} - 4\mathbf{j}$

23. $\mathbf{u} = \mathbf{j} + 6\mathbf{k}$
$\mathbf{v} = \mathbf{i} - 2\mathbf{j} - \mathbf{k}$

24. $\mathbf{u} = -2\mathbf{i} + 3\mathbf{j} - \mathbf{k}$
$\mathbf{v} = 2\mathbf{i} + \mathbf{j} - \mathbf{k}$

25. $\mathbf{u} = \langle 2, -3, 1 \rangle$
$\mathbf{v} = \langle -1, -1, -1 \rangle$

26. $\mathbf{u} = \langle \cos\theta, \sin\theta, -1 \rangle$
$\mathbf{v} = \langle \sin\theta, -\cos\theta, 0 \rangle$

In Exercises 27–30, the vertices of a triangle are given. Determine whether the triangle is an acute triangle, an obtuse triangle, or a right triangle. Explain your reasoning.

27. $(1, 2, 0), (0, 0, 0), (-2, 1, 0)$

28. $(-3, 0, 0), (0, 0, 0), (1, 2, 3)$

29. $(2, 0, 1), (0, 1, 2), (-0.5, 1.5, 0)$

30. $(2, -7, 3), (-1, 5, 8), (4, 6, -1)$

In Exercises 31–34, find the direction cosines of u and demonstrate that the sum of the squares of the direction cosines is 1.

31. $\mathbf{u} = \mathbf{i} + 2\mathbf{j} + 2\mathbf{k}$

32. $\mathbf{u} = 5\mathbf{i} + 3\mathbf{j} - \mathbf{k}$

33. $\mathbf{u} = \langle 0, 6, -4 \rangle$

34. $\mathbf{u} = \langle a, b, c \rangle$

In Exercises 35–38, find the direction angles of the vector.

35. $\mathbf{u} = 3\mathbf{i} + 2\mathbf{j} - 2\mathbf{k}$

36. $\mathbf{u} = -4\mathbf{i} + 3\mathbf{j} + 5\mathbf{k}$

37. $\mathbf{u} = \langle -1, 5, 2 \rangle$

38. $\mathbf{u} = \langle -2, 6, 1 \rangle$

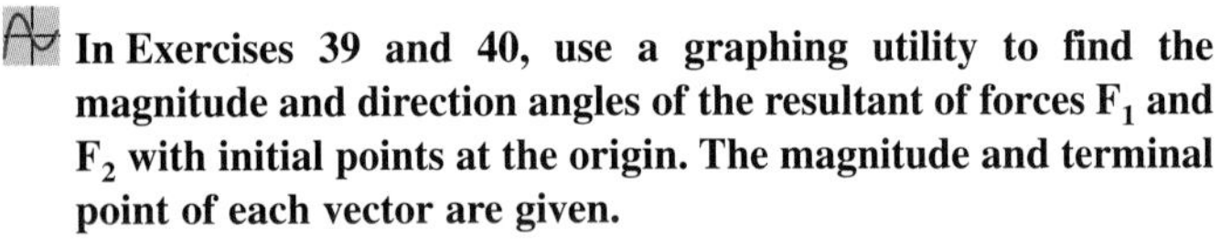
In Exercises 39 and 40, use a graphing utility to find the magnitude and direction angles of the resultant of forces $\mathbf{F}_1$ and $\mathbf{F}_2$ with initial points at the origin. The magnitude and terminal point of each vector are given.

	Vector	*Magnitude*	*Terminal Point*
39.	$\mathbf{F}_1$	50 lb	$(10, 5, 3)$
	$\mathbf{F}_2$	80 lb	$(12, 7, -5)$
40.	$\mathbf{F}_1$	300 N	$(-20, -10, 5)$
	$\mathbf{F}_2$	100 N	$(5, 15, 0)$

41. ***Load-Supporting Cables*** A load is supported by three cables, as shown in the figure. Find the direction angles of the load-supporting cable OA.

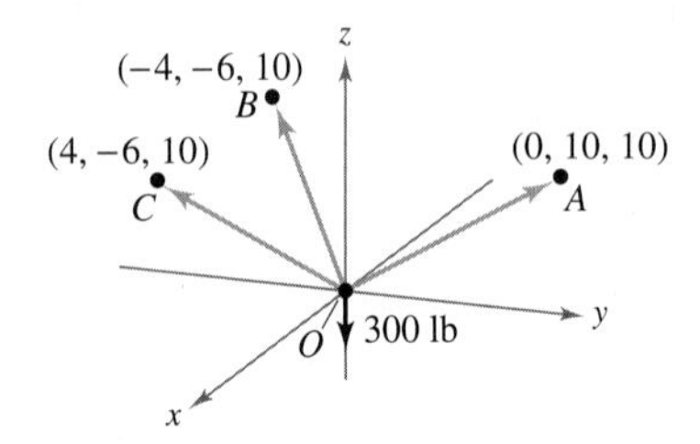

42. ***Load-Supporting Cables*** The tension in the cable OA in Exercise 41 is 200 newtons. Determine the weight of the load.

In Exercises 43–50, (a) find the projection of u onto v, and (b) find the vector component of u orthogonal to v.

43. $\mathbf{u} = \langle 6, 7 \rangle, \quad \mathbf{v} = \langle 1, 4 \rangle$

44. $\mathbf{u} = \langle 9, 7 \rangle, \quad \mathbf{v} = \langle 1, 3 \rangle$

45. $\mathbf{u} = 2\mathbf{i} + 3\mathbf{j}, \quad \mathbf{v} = 5\mathbf{i} + \mathbf{j}$

46. $\mathbf{u} = 2\mathbf{i} - 3\mathbf{j}, \quad \mathbf{v} = 3\mathbf{i} + 2\mathbf{j}$

47. $\mathbf{u} = \langle 0, 3, 3 \rangle, \quad \mathbf{v} = \langle -1, 1, 1 \rangle$

48. $\mathbf{u} = \langle 8, 2, 0 \rangle, \quad \mathbf{v} = \langle 2, 1, -1 \rangle$

49. $\mathbf{u} = 2\mathbf{i} + \mathbf{j} + 2\mathbf{k}, \quad \mathbf{v} = 3\mathbf{j} + 4\mathbf{k}$

50. $\mathbf{u} = \mathbf{i} + 4\mathbf{k}, \quad \mathbf{v} = 3\mathbf{i} + 2\mathbf{k}$

WRITING ABOUT CONCEPTS

51. Define the dot product of vectors **u** and **v**.

52. State the definition of orthogonal vectors. If vectors are neither parallel nor orthogonal, how do you find the angle between them? Explain.

53. Determine which of the following are defined for nonzero vectors **u**, **v**, and **w**. Explain your reasoning.

(a) $\mathbf{u} \cdot (\mathbf{v} + \mathbf{w})$ (b) $(\mathbf{u} \cdot \mathbf{v})\mathbf{w}$

(c) $\mathbf{u} \cdot \mathbf{v} + \mathbf{w}$ (d) $\|\mathbf{u}\| \cdot (\mathbf{v} + \mathbf{w})$

54. Describe direction cosines and direction angles of a vector **v**.

55. Give a geometric description of the projection of **u** onto **v**.

56. What can be said about the vectors **u** and **v** if (a) the projection of **u** onto **v** equals **u** and (b) the projection of **u** onto **v** equals **0**?

57. If the projection of **u** onto **v** has the same magnitude as the projection of **v** onto **u**, can you conclude that $\|\mathbf{u}\| = \|\mathbf{v}\|$? Explain.

CAPSTONE

58. What is known about θ, the angle between two nonzero vectors **u** and **v**, if

(a) $\mathbf{u} \cdot \mathbf{v} = 0$? (b) $\mathbf{u} \cdot \mathbf{v} > 0$? (c) $\mathbf{u} \cdot \mathbf{v} < 0$?

59. ***Revenue*** The vector $\mathbf{u} = \langle 3240, 1450, 2235 \rangle$ gives the numbers of hamburgers, chicken sandwiches, and cheeseburgers, respectively, sold at a fast-food restaurant in one week. The vector $\mathbf{v} = \langle 1.35, 2.65, 1.85 \rangle$ gives the prices (in dollars) per unit for the three food items. Find the dot product $\mathbf{u} \cdot \mathbf{v}$, and explain what information it gives.

60. ***Revenue*** Repeat Exercise 59 after increasing prices by 4%. Identify the vector operation used to increase prices by 4%.

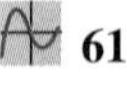
61. ***Programming*** Given vectors **u** and **v** in component form, write a program for a graphing utility in which the output is (a) $\|\mathbf{u}\|$, (b) $\|\mathbf{v}\|$, and (c) the angle between **u** and **v**.

62. ***Programming*** Use the program you wrote in Exercise 61 to find the angle between the vectors $\mathbf{u} = \langle 8, -4, 2 \rangle$ and $\mathbf{v} = \langle 2, 5, 2 \rangle$.

63. ***Programming*** Given vectors **u** and **v** in component form, write a program for a graphing utility in which the output is the component form of the projection of **u** onto **v**.

64. ***Programming*** Use the program you wrote in Exercise 63 to find the projection of **u** onto **v** for $\mathbf{u} = \langle 5, 6, 2 \rangle$ and $\mathbf{v} = \langle -1, 3, 4 \rangle$.

Think About It **In Exercises 65 and 66, use the figure to determine mentally the projection of u onto v. (The coordinates of the terminal points of the vectors in standard position are given.) Verify your results analytically.**

65.

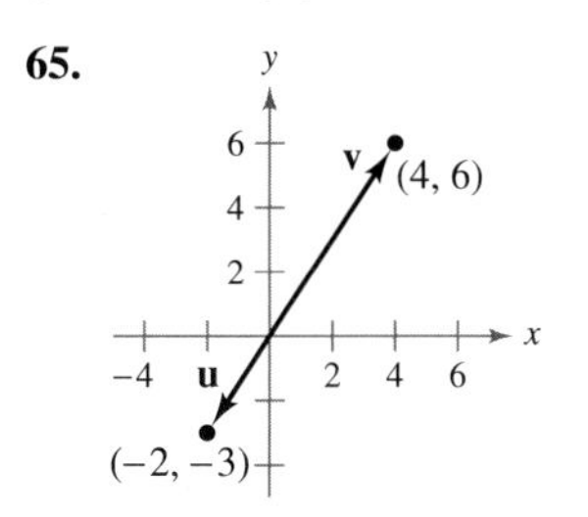

66.

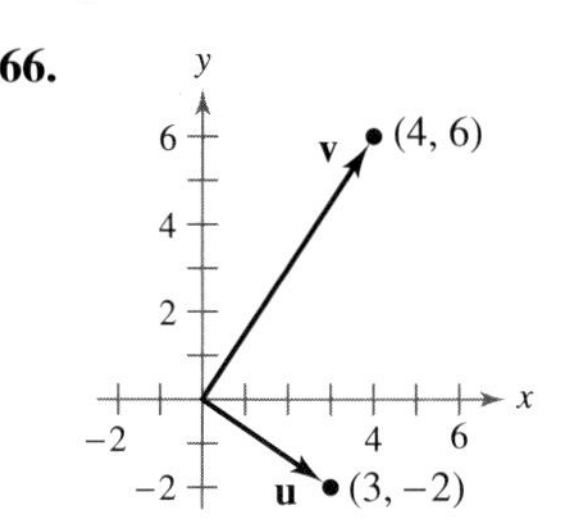

In Exercises 67–70, find two vectors in opposite directions that are orthogonal to the vector u. (The answers are not unique.)

67. $\mathbf{u} = -\frac{1}{4}\mathbf{i} + \frac{3}{2}\mathbf{j}$

68. $\mathbf{u} = 9\mathbf{i} - 4\mathbf{j}$

69. $\mathbf{u} = \langle 3, 1, -2 \rangle$

70. $\mathbf{u} = \langle 4, -3, 6 \rangle$

71. ***Braking Load*** A 48,000-pound truck is parked on a $10°$ slope (see figure). Assume the only force to overcome is that due to gravity. Find (a) the force required to keep the truck from rolling down the hill and (b) the force perpendicular to the hill.

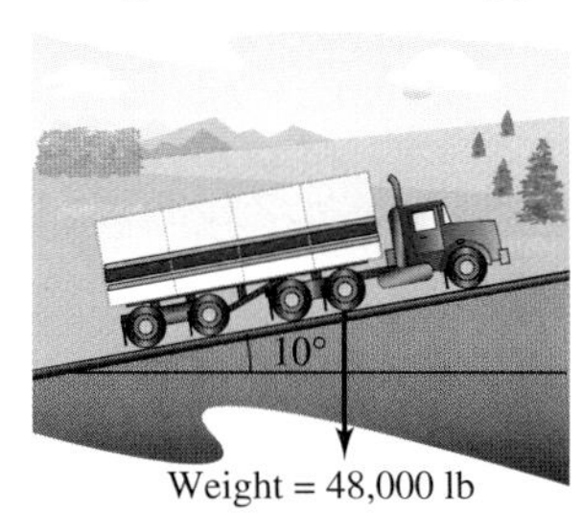

Figure for 71

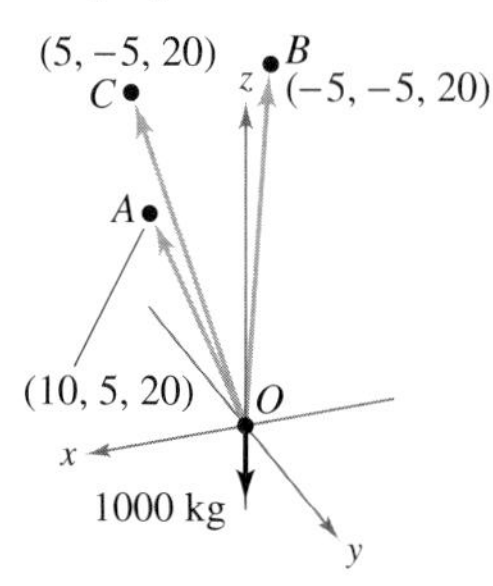

Figure for 72

72. ***Load-Supporting Cables*** Find the magnitude of the projection of the load-supporting cable OA onto the positive z-axis as shown in the figure.

73. ***Work*** An object is pulled 10 feet across a floor, using a force of 85 pounds. The direction of the force is $60°$ above the horizontal (see figure). Find the work done.

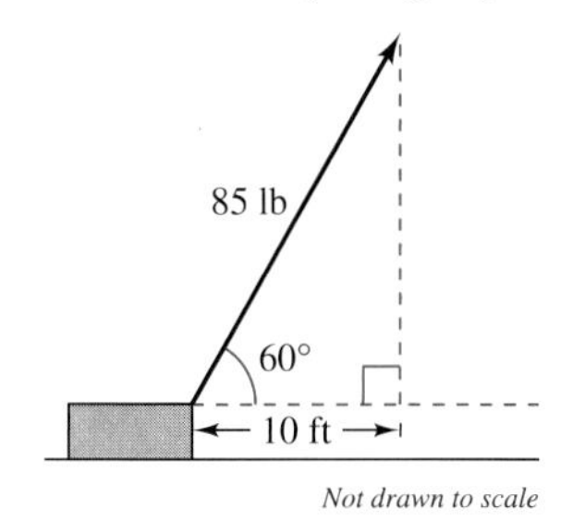

Figure for 73

Figure for 74

74. ***Work*** A toy wagon is pulled by exerting a force of 25 pounds on a handle that makes a $20°$ angle with the horizontal (see figure in left column). Find the work done in pulling the wagon 50 feet.

75. ***Work*** A car is towed using a force of 1600 newtons. The chain used to pull the car makes a $25°$ angle with the horizontal. Find the work done in towing the car 2 kilometers.

76. ***Work*** A sled is pulled by exerting a force of 100 newtons on a rope that makes a $25°$ angle with the horizontal. Find the work done in pulling the sled 40 meters.

True or False? **In Exercises 77 and 78, determine whether the statement is true or false. If it is false, explain why or give an example that shows it is false.**

77. If $\mathbf{u} \cdot \mathbf{v} = \mathbf{u} \cdot \mathbf{w}$ and $\mathbf{u} \neq \mathbf{0}$, then $\mathbf{v} = \mathbf{w}$.

78. If **u** and **v** are orthogonal to **w**, then $\mathbf{u} + \mathbf{v}$ is orthogonal to **w**.

79. Find the angle between a cube's diagonal and one of its edges.

80. Find the angle between the diagonal of a cube and the diagonal of one of its sides.

In Exercises 81–84, (a) find all points of intersection of the graphs of the two equations, (b) find the unit tangent vectors to each curve at their points of intersection, and (c) find the angles ($0° \leq \theta \leq 90°$) between the curves at their points of intersection.

81. $y = x^2, \quad y = x^{1/3}$

82. $y = x^3, \quad y = x^{1/3}$

83. $y = 1 - x^2, \quad y = x^2 - 1$

84. $(y + 1)^2 = x, \quad y = x^3 - 1$

85. Use vectors to prove that the diagonals of a rhombus are perpendicular.

86. Use vectors to prove that a parallelogram is a rectangle if and only if its diagonals are equal in length.

87. ***Bond Angle*** Consider a regular tetrahedron with vertices $(0, 0, 0)$, $(k, k, 0)$, $(k, 0, k)$, and $(0, k, k)$, where k is a positive real number.

    (a) Sketch the graph of the tetrahedron.

    (b) Find the length of each edge.

    (c) Find the angle between any two edges.

    (d) Find the angle between the line segments from the centroid $(k/2, k/2, k/2)$ to two vertices. This is the bond angle for a molecule such as CH_4 or $PbCl_4$, where the structure of the molecule is a tetrahedron.

88. Consider the vectors $\mathbf{u} = \langle \cos \alpha, \sin \alpha, 0 \rangle$ and $\mathbf{v} = \langle \cos \beta, \sin \beta, 0 \rangle$, where $\alpha > \beta$. Find the dot product of the vectors and use the result to prove the identity

    $$\cos(\alpha - \beta) = \cos \alpha \cos \beta + \sin \alpha \sin \beta.$$

89. Prove that $\|\mathbf{u} - \mathbf{v}\|^2 = \|\mathbf{u}\|^2 + \|\mathbf{v}\|^2 - 2\mathbf{u} \cdot \mathbf{v}$.

90. Prove the **Cauchy-Schwarz Inequality** $|\mathbf{u} \cdot \mathbf{v}| \leq \|\mathbf{u}\| \|\mathbf{v}\|$.

91. Prove the triangle inequality $\|\mathbf{u} + \mathbf{v}\| \leq \|\mathbf{u}\| + \|\mathbf{v}\|$.

92. Prove Theorem 11.6.

11.4 The Cross Product of Two Vectors in Space

- **Find the cross product of two vectors in space.**
- **Use the triple scalar product of three vectors in space.**

The Cross Product

Many applications in physics, engineering, and geometry involve finding a vector in space that is orthogonal to two given vectors. In this section you will study a product that will yield such a vector. It is called the **cross product,** and it is most conveniently defined and calculated using the standard unit vector form. Because the cross product yields a vector, it is also called the **vector product.**

> **EXPLORATION**
>
> ***Geometric Property of the Cross Product*** Three pairs of vectors are shown below. Use the definition to find the cross product of each pair. Sketch all three vectors in a three-dimensional system. Describe any relationships among the three vectors. Use your description to write a conjecture about **u**, **v**, and **u** × **v**.
>
> **a.** $\mathbf{u} = \langle 3, 0, 3 \rangle$, $\mathbf{v} = \langle 3, 0, -3 \rangle$
>
> 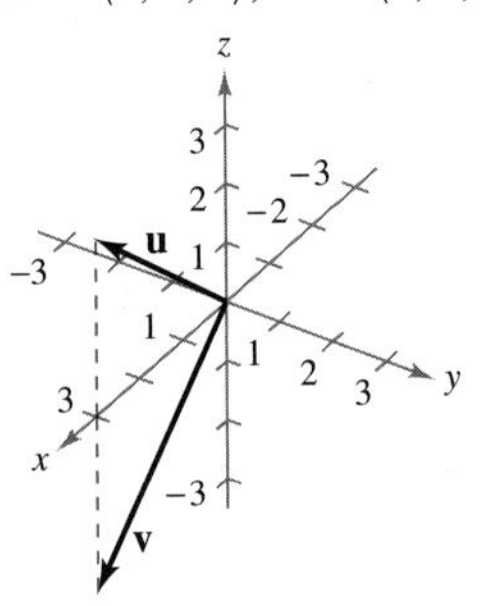
>
> **b.** $\mathbf{u} = \langle 0, 3, 3 \rangle$, $\mathbf{v} = \langle 0, -3, 3 \rangle$
>
> 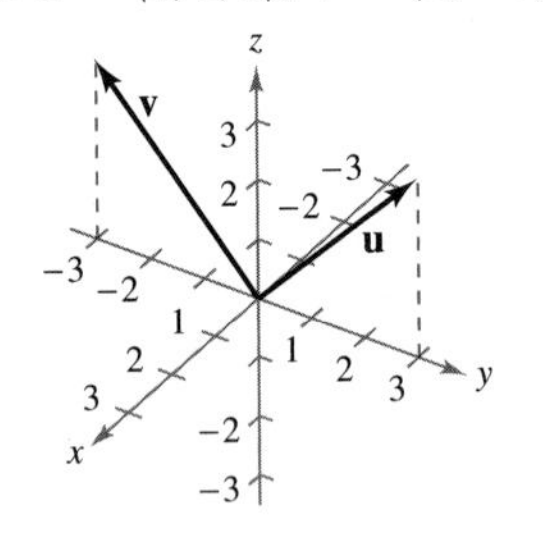
>
> **c.** $\mathbf{u} = \langle 3, 3, 0 \rangle$, $\mathbf{v} = \langle 3, -3, 0 \rangle$
>
> 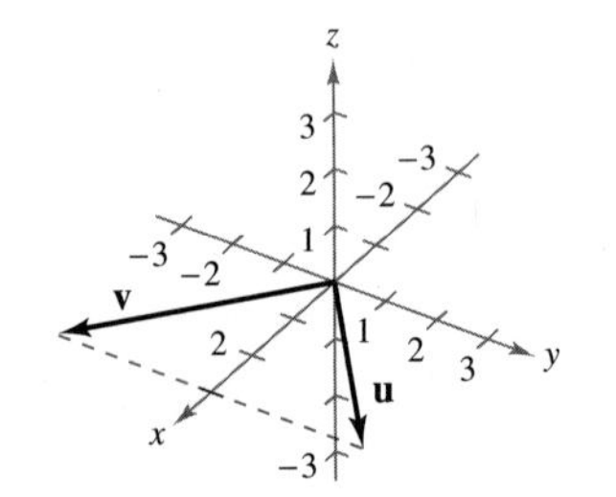

DEFINITION OF CROSS PRODUCT OF TWO VECTORS IN SPACE

Let

$$\mathbf{u} = u_1\mathbf{i} + u_2\mathbf{j} + u_3\mathbf{k} \quad \text{and} \quad \mathbf{v} = v_1\mathbf{i} + v_2\mathbf{j} + v_3\mathbf{k}$$

be vectors in space. The **cross product** of **u** and **v** is the vector

$$\mathbf{u} \times \mathbf{v} = (u_2v_3 - u_3v_2)\mathbf{i} - (u_1v_3 - u_3v_1)\mathbf{j} + (u_1v_2 - u_2v_1)\mathbf{k}.$$

NOTE Be sure you see that this definition applies only to three-dimensional vectors. The cross product is not defined for two-dimensional vectors. ■

A convenient way to calculate **u** × **v** is to use the following *determinant form* with cofactor expansion. (This 3 × 3 determinant form is used simply to help remember the formula for the cross product—it is technically not a determinant because the entries of the corresponding matrix are not all real numbers.)

$$\mathbf{u} \times \mathbf{v} = \begin{vmatrix} \mathbf{i} & \mathbf{j} & \mathbf{k} \\ u_1 & u_2 & u_3 \\ v_1 & v_2 & v_3 \end{vmatrix} \quad \begin{matrix} \\ \longleftarrow \text{Put "}\mathbf{u}\text{" in Row 2.} \\ \longleftarrow \text{Put "}\mathbf{v}\text{" in Row 3.} \end{matrix}$$

$$= \begin{vmatrix} \mathbf{i} & \mathbf{j} & \mathbf{k} \\ u_1 & u_2 & u_3 \\ v_1 & v_2 & v_3 \end{vmatrix}\mathbf{i} - \begin{vmatrix} \mathbf{i} & \mathbf{j} & \mathbf{k} \\ u_1 & u_2 & u_3 \\ v_1 & v_2 & v_3 \end{vmatrix}\mathbf{j} + \begin{vmatrix} \mathbf{i} & \mathbf{j} & \mathbf{k} \\ u_1 & u_2 & u_3 \\ v_1 & v_2 & v_3 \end{vmatrix}\mathbf{k}$$

$$= \begin{vmatrix} u_2 & u_3 \\ v_2 & v_3 \end{vmatrix}\mathbf{i} - \begin{vmatrix} u_1 & u_3 \\ v_1 & v_3 \end{vmatrix}\mathbf{j} + \begin{vmatrix} u_1 & u_2 \\ v_1 & v_2 \end{vmatrix}\mathbf{k}$$

$$= (u_2v_3 - u_3v_2)\mathbf{i} - (u_1v_3 - u_3v_1)\mathbf{j} + (u_1v_2 - u_2v_1)\mathbf{k}$$

Note the minus sign in front of the **j**-component. Each of the three 2 × 2 determinants can be evaluated by using the following diagonal pattern.

$$\begin{vmatrix} a & b \\ c & d \end{vmatrix} = ad - bc$$

Here are a couple of examples.

$$\begin{vmatrix} 2 & 4 \\ 3 & -1 \end{vmatrix} = (2)(-1) - (4)(3) = -2 - 12 = -14$$

$$\begin{vmatrix} 4 & 0 \\ -6 & 3 \end{vmatrix} = (4)(3) - (0)(-6) = 12$$

NOTATION FOR DOT AND CROSS PRODUCTS

The notation for the dot product and cross product of vectors was first introduced by the American physicist Josiah Willard Gibbs (1839–1903). In the early 1880s, Gibbs built a system to represent physical quantities called "vector analysis." The system was a departure from Hamilton's theory of quaternions.

EXAMPLE 1 Finding the Cross Product

Given $\mathbf{u} = \mathbf{i} - 2\mathbf{j} + \mathbf{k}$ and $\mathbf{v} = 3\mathbf{i} + \mathbf{j} - 2\mathbf{k}$, find each of the following.

a. $\mathbf{u} \times \mathbf{v}$ **b.** $\mathbf{v} \times \mathbf{u}$ **c.** $\mathbf{v} \times \mathbf{v}$

Solution

$$\textbf{a. } \mathbf{u} \times \mathbf{v} = \begin{vmatrix} \mathbf{i} & \mathbf{j} & \mathbf{k} \\ 1 & -2 & 1 \\ 3 & 1 & -2 \end{vmatrix} = \begin{vmatrix} -2 & 1 \\ 1 & -2 \end{vmatrix}\mathbf{i} - \begin{vmatrix} 1 & 1 \\ 3 & -2 \end{vmatrix}\mathbf{j} + \begin{vmatrix} 1 & -2 \\ 3 & 1 \end{vmatrix}\mathbf{k}$$

$$= (4 - 1)\mathbf{i} - (-2 - 3)\mathbf{j} + (1 + 6)\mathbf{k}$$

$$= 3\mathbf{i} + 5\mathbf{j} + 7\mathbf{k}$$

$$\textbf{b. } \mathbf{v} \times \mathbf{u} = \begin{vmatrix} \mathbf{i} & \mathbf{j} & \mathbf{k} \\ 3 & 1 & -2 \\ 1 & -2 & 1 \end{vmatrix} = \begin{vmatrix} 1 & -2 \\ -2 & 1 \end{vmatrix}\mathbf{i} - \begin{vmatrix} 3 & -2 \\ 1 & 1 \end{vmatrix}\mathbf{j} + \begin{vmatrix} 3 & 1 \\ 1 & -2 \end{vmatrix}\mathbf{k}$$

$$= (1 - 4)\mathbf{i} - (3 + 2)\mathbf{j} + (-6 - 1)\mathbf{k}$$

$$= -3\mathbf{i} - 5\mathbf{j} - 7\mathbf{k}$$

Note that this result is the negative of that in part (a).

$$\textbf{c. } \mathbf{v} \times \mathbf{v} = \begin{vmatrix} \mathbf{i} & \mathbf{j} & \mathbf{k} \\ 3 & 1 & -2 \\ 3 & 1 & -2 \end{vmatrix} = \mathbf{0}$$

■

The results obtained in Example 1 suggest some interesting *algebraic* properties of the cross product. For instance, $\mathbf{u} \times \mathbf{v} = -(\mathbf{v} \times \mathbf{u})$, and $\mathbf{v} \times \mathbf{v} = \mathbf{0}$. These properties, and several others, are summarized in the following theorem.

THEOREM 11.7 ALGEBRAIC PROPERTIES OF THE CROSS PRODUCT

Let $\mathbf{u}$, $\mathbf{v}$, and $\mathbf{w}$ be vectors in space, and let c be a scalar.

1. $\mathbf{u} \times \mathbf{v} = -(\mathbf{v} \times \mathbf{u})$
2. $\mathbf{u} \times (\mathbf{v} + \mathbf{w}) = (\mathbf{u} \times \mathbf{v}) + (\mathbf{u} \times \mathbf{w})$
3. $c(\mathbf{u} \times \mathbf{v}) = (c\mathbf{u}) \times \mathbf{v} = \mathbf{u} \times (c\mathbf{v})$
4. $\mathbf{u} \times \mathbf{0} = \mathbf{0} \times \mathbf{u} = \mathbf{0}$
5. $\mathbf{u} \times \mathbf{u} = \mathbf{0}$
6. $\mathbf{u} \cdot (\mathbf{v} \times \mathbf{w}) = (\mathbf{u} \times \mathbf{v}) \cdot \mathbf{w}$

PROOF To prove Property 1, let $\mathbf{u} = u_1\mathbf{i} + u_2\mathbf{j} + u_3\mathbf{k}$ and $\mathbf{v} = v_1\mathbf{i} + v_2\mathbf{j} + v_3\mathbf{k}$. Then,

$$\mathbf{u} \times \mathbf{v} = (u_2v_3 - u_3v_2)\mathbf{i} - (u_1v_3 - u_3v_1)\mathbf{j} + (u_1v_2 - u_2v_1)\mathbf{k}$$

and

$$\mathbf{v} \times \mathbf{u} = (v_2u_3 - v_3u_2)\mathbf{i} - (v_1u_3 - v_3u_1)\mathbf{j} + (v_1u_2 - v_2u_1)\mathbf{k}$$

which implies that $\mathbf{u} \times \mathbf{v} = -(\mathbf{v} \times \mathbf{u})$. Proofs of Properties 2, 3, 5, and 6 are left as exercises (see Exercises 59–62). ■

Note that Property 1 of Theorem 11.7 indicates that the cross product is *not commutative*. In particular, this property indicates that the vectors $\mathbf{u} \times \mathbf{v}$ and $\mathbf{v} \times \mathbf{u}$ have equal lengths but opposite directions. The following theorem lists some other *geometric* properties of the cross product of two vectors.

NOTE It follows from Properties 1 and 2 in Theorem 11.8 that if $\mathbf{n}$ is a unit vector orthogonal to both $\mathbf{u}$ and $\mathbf{v}$, then

$$\mathbf{u} \times \mathbf{v} = \pm(\|\mathbf{u}\|\ \|\mathbf{v}\| \sin\theta)\mathbf{n}.$$

THEOREM 11.8 GEOMETRIC PROPERTIES OF THE CROSS PRODUCT

Let $\mathbf{u}$ and $\mathbf{v}$ be nonzero vectors in space, and let θ be the angle between $\mathbf{u}$ and $\mathbf{v}$.

1. $\mathbf{u} \times \mathbf{v}$ is orthogonal to both $\mathbf{u}$ and $\mathbf{v}$.
2. $\|\mathbf{u} \times \mathbf{v}\| = \|\mathbf{u}\|\ \|\mathbf{v}\| \sin\theta$
3. $\mathbf{u} \times \mathbf{v} = \mathbf{0}$ if and only if $\mathbf{u}$ and $\mathbf{v}$ are scalar multiples of each other.
4. $\|\mathbf{u} \times \mathbf{v}\| =$ area of parallelogram having $\mathbf{u}$ and $\mathbf{v}$ as adjacent sides.

PROOF To prove Property 2, note because $\cos\theta = (\mathbf{u} \cdot \mathbf{v})/(\|\mathbf{u}\|\ \|\mathbf{v}\|)$, it follows that

$$\begin{aligned}
\|\mathbf{u}\|\ \|\mathbf{v}\| \sin\theta &= \|\mathbf{u}\|\ \|\mathbf{v}\|\sqrt{1-\cos^2\theta} \\
&= \|\mathbf{u}\|\ \|\mathbf{v}\|\sqrt{1-\frac{(\mathbf{u}\cdot\mathbf{v})^2}{\|\mathbf{u}\|^2\|\mathbf{v}\|^2}} \\
&= \sqrt{\|\mathbf{u}\|^2\|\mathbf{v}\|^2-(\mathbf{u}\cdot\mathbf{v})^2} \\
&= \sqrt{(u_1^2+u_2^2+u_3^2)(v_1^2+v_2^2+v_3^2)-(u_1v_1+u_2v_2+u_3v_3)^2} \\
&= \sqrt{(u_2v_3-u_3v_2)^2+(u_1v_3-u_3v_1)^2+(u_1v_2-u_2v_1)^2} \\
&= \|\mathbf{u}\times\mathbf{v}\|.
\end{aligned}$$

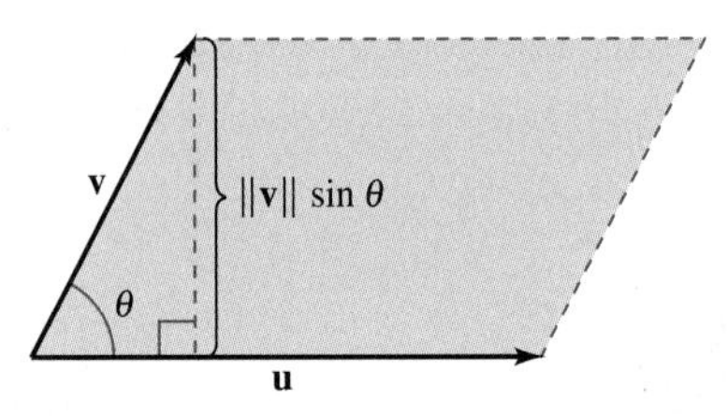

The vectors **u** and **v** form adjacent sides of a parallelogram.
Figure 11.35

To prove Property 4, refer to Figure 11.35, which is a parallelogram having $\mathbf{v}$ and $\mathbf{u}$ as adjacent sides. Because the height of the parallelogram is $\|\mathbf{v}\| \sin\theta$, the area is

$$\begin{aligned}
\text{Area} &= (\text{base})(\text{height}) \\
&= \|\mathbf{u}\|\ \|\mathbf{v}\| \sin\theta \\
&= \|\mathbf{u}\times\mathbf{v}\|.
\end{aligned}$$

Proofs of Properties 1 and 3 are left as exercises (see Exercises 63 and 64). ■

Both $\mathbf{u} \times \mathbf{v}$ and $\mathbf{v} \times \mathbf{u}$ are perpendicular to the plane determined by $\mathbf{u}$ and $\mathbf{v}$. One way to remember the orientations of the vectors $\mathbf{u}$, $\mathbf{v}$, and $\mathbf{u} \times \mathbf{v}$ is to compare them with the unit vectors $\mathbf{i}$, $\mathbf{j}$, and $\mathbf{k} = \mathbf{i} \times \mathbf{j}$, as shown in Figure 11.36. The three vectors $\mathbf{u}$, $\mathbf{v}$, and $\mathbf{u} \times \mathbf{v}$ form a *right-handed system*, whereas the three vectors $\mathbf{u}$, $\mathbf{v}$, and $\mathbf{v} \times \mathbf{u}$ form a *left-handed system*.

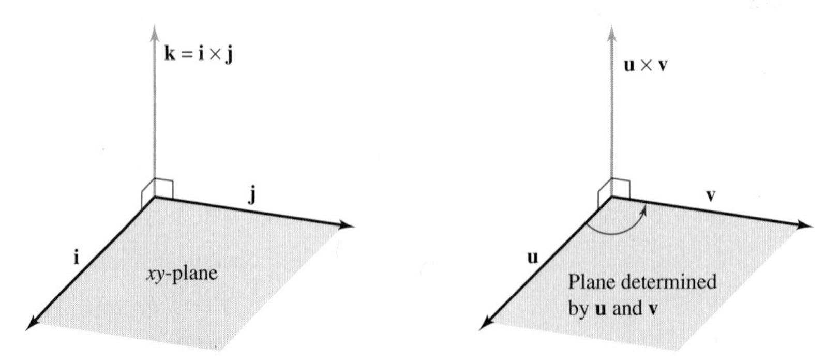

Right-handed systems
Figure 11.36

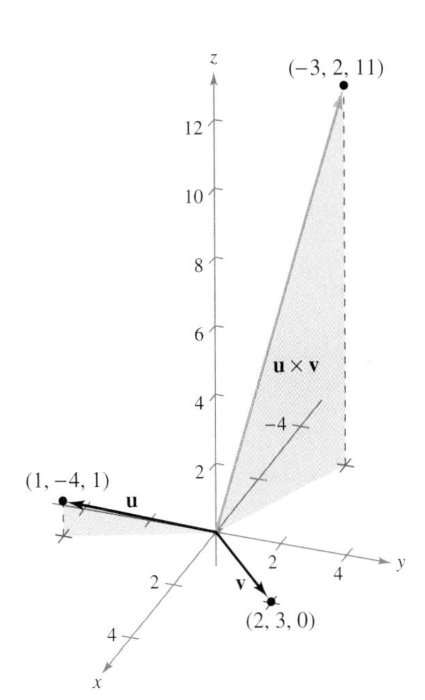

The vector $\mathbf{u} \times \mathbf{v}$ is orthogonal to both $\mathbf{u}$ and $\mathbf{v}$.
Figure 11.37

EXAMPLE 2 Using the Cross Product

Find a unit vector that is orthogonal to both

$$\mathbf{u} = \mathbf{i} - 4\mathbf{j} + \mathbf{k} \quad \text{and} \quad \mathbf{v} = 2\mathbf{i} + 3\mathbf{j}.$$

Solution The cross product $\mathbf{u} \times \mathbf{v}$, as shown in Figure 11.37, is orthogonal to both $\mathbf{u}$ and $\mathbf{v}$.

$$\mathbf{u} \times \mathbf{v} = \begin{vmatrix} \mathbf{i} & \mathbf{j} & \mathbf{k} \\ 1 & -4 & 1 \\ 2 & 3 & 0 \end{vmatrix} \qquad \text{Cross product}$$

$$= -3\mathbf{i} + 2\mathbf{j} + 11\mathbf{k}$$

Because

$$\|\mathbf{u} \times \mathbf{v}\| = \sqrt{(-3)^2 + 2^2 + 11^2} = \sqrt{134}$$

a unit vector orthogonal to both $\mathbf{u}$ and $\mathbf{v}$ is

$$\frac{\mathbf{u} \times \mathbf{v}}{\|\mathbf{u} \times \mathbf{v}\|} = -\frac{3}{\sqrt{134}}\mathbf{i} + \frac{2}{\sqrt{134}}\mathbf{j} + \frac{11}{\sqrt{134}}\mathbf{k}.$$

■

NOTE In Example 2, note that you could have used the cross product $\mathbf{v} \times \mathbf{u}$ to form a unit vector that is orthogonal to both $\mathbf{u}$ and $\mathbf{v}$. With that choice, you would have obtained the negative of the unit vector found in the example. ■

EXAMPLE 3 Geometric Application of the Cross Product

Show that the quadrilateral with vertices at the following points is a parallelogram, and find its area.

$$A = (5, 2, 0) \qquad B = (2, 6, 1)$$
$$C = (2, 4, 7) \qquad D = (5, 0, 6)$$

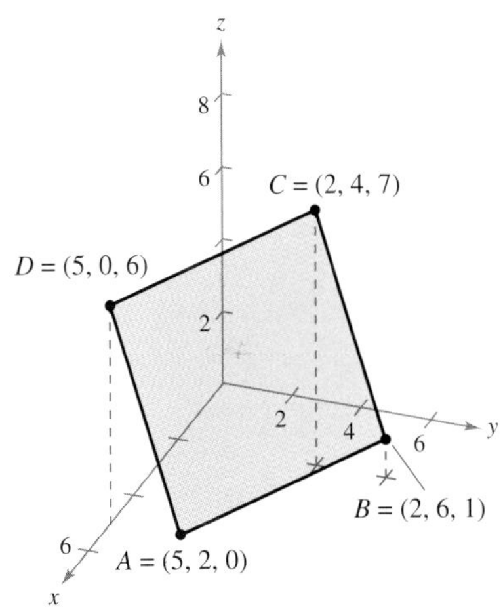

The area of the parallelogram is approximately 32.19.
Figure 11.38

Solution From Figure 11.38 you can see that the sides of the quadrilateral correspond to the following four vectors.

$$\overrightarrow{AB} = -3\mathbf{i} + 4\mathbf{j} + \mathbf{k} \qquad \overrightarrow{CD} = 3\mathbf{i} - 4\mathbf{j} - \mathbf{k} = -\overrightarrow{AB}$$
$$\overrightarrow{AD} = 0\mathbf{i} - 2\mathbf{j} + 6\mathbf{k} \qquad \overrightarrow{CB} = 0\mathbf{i} + 2\mathbf{j} - 6\mathbf{k} = -\overrightarrow{AD}$$

So, $\overrightarrow{AB}$ is parallel to $\overrightarrow{CD}$ and $\overrightarrow{AD}$ is parallel to $\overrightarrow{CB}$, and you can conclude that the quadrilateral is a parallelogram with $\overrightarrow{AB}$ and $\overrightarrow{AD}$ as adjacent sides. Moreover, because

$$\overrightarrow{AB} \times \overrightarrow{AD} = \begin{vmatrix} \mathbf{i} & \mathbf{j} & \mathbf{k} \\ -3 & 4 & 1 \\ 0 & -2 & 6 \end{vmatrix} \qquad \text{Cross product}$$

$$= 26\mathbf{i} + 18\mathbf{j} + 6\mathbf{k}$$

the area of the parallelogram is

$$\|\overrightarrow{AB} \times \overrightarrow{AD}\| = \sqrt{1036} \approx 32.19.$$

Is the parallelogram a rectangle? You can determine whether it is by finding the angle between the vectors $\overrightarrow{AB}$ and $\overrightarrow{AD}$. ■

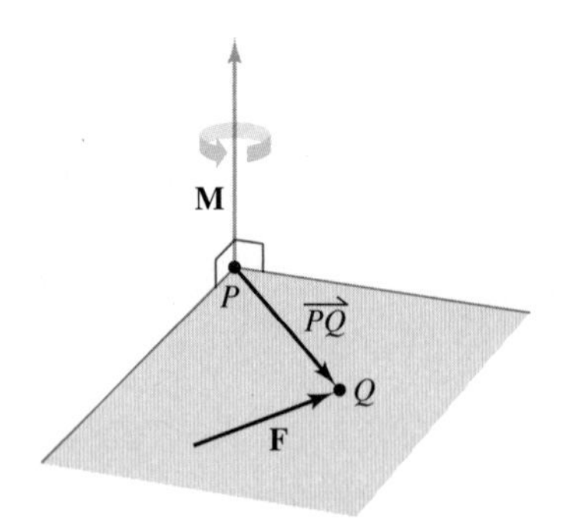

The moment of **F** about P
Figure 11.39

In physics, the cross product can be used to measure **torque**—the **moment M of a force F about a point *P*,** as shown in Figure 11.39. If the point of application of the force is Q, the moment of **F** about P is given by

$$\mathbf{M} = \overrightarrow{PQ} \times \mathbf{F}. \qquad \text{Moment of } \mathbf{F} \text{ about } P$$

The magnitude of the moment **M** measures the tendency of the vector $\overrightarrow{PQ}$ to rotate counterclockwise (using the right-hand rule) about an axis directed along the vector **M**.

EXAMPLE 4 An Application of the Cross Product

A vertical force of 50 pounds is applied to the end of a one-foot lever that is attached to an axle at point P, as shown in Figure 11.40. Find the moment of this force about the point P when $\theta = 60°$.

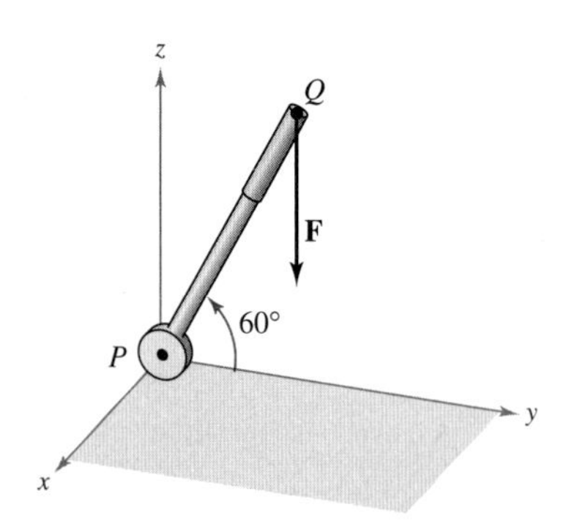

A vertical force of 50 pounds is applied at point Q.
Figure 11.40

Solution If you represent the 50-pound force as $\mathbf{F} = -50\mathbf{k}$ and the lever as

$$\overrightarrow{PQ} = \cos(60°)\mathbf{j} + \sin(60°)\mathbf{k} = \frac{1}{2}\mathbf{j} + \frac{\sqrt{3}}{2}\mathbf{k}$$

the moment of **F** about P is given by

$$\mathbf{M} = \overrightarrow{PQ} \times \mathbf{F} = \begin{vmatrix} \mathbf{i} & \mathbf{j} & \mathbf{k} \\ 0 & \frac{1}{2} & \frac{\sqrt{3}}{2} \\ 0 & 0 & -50 \end{vmatrix} = -25\mathbf{i}. \qquad \text{Moment of } \mathbf{F} \text{ about } P$$

The magnitude of this moment is 25 foot-pounds. ■

NOTE In Example 4, note that the moment (the tendency of the lever to rotate about its axle) is dependent on the angle θ. When $\theta = \pi/2$, the moment is 0. The moment is greatest when $\theta = 0$. ■

The Triple Scalar Product

For vectors **u**, **v**, and **w** in space, the dot product of **u** and $\mathbf{v} \times \mathbf{w}$

$$\mathbf{u} \cdot (\mathbf{v} \times \mathbf{w})$$

is called the **triple scalar product,** as defined in Theorem 11.9. The proof of this theorem is left as an exercise (see Exercise 67).

■ **FOR FURTHER INFORMATION** To see how the cross product is used to model the torque of the robot arm of a space shuttle, see the article "The Long Arm of Calculus" by Ethan Berkove and Rich Marchand in *The College Mathematics Journal*. To view this article, go to the website *www.matharticles.com*.

THEOREM 11.9 THE TRIPLE SCALAR PRODUCT

For $\mathbf{u} = u_1\mathbf{i} + u_2\mathbf{j} + u_3\mathbf{k}$, $\mathbf{v} = v_1\mathbf{i} + v_2\mathbf{j} + v_3\mathbf{k}$, and $\mathbf{w} = w_1\mathbf{i} + w_2\mathbf{j} + w_3\mathbf{k}$, the triple scalar product is given by

$$\mathbf{u} \cdot (\mathbf{v} \times \mathbf{w}) = \begin{vmatrix} u_1 & u_2 & u_3 \\ v_1 & v_2 & v_3 \\ w_1 & w_2 & w_3 \end{vmatrix}.$$

NOTE The value of a determinant is multiplied by -1 if two rows are interchanged. After two such interchanges, the value of the determinant will be unchanged. So, the following triple scalar products are equivalent.

$$\mathbf{u} \cdot (\mathbf{v} \times \mathbf{w}) = \mathbf{v} \cdot (\mathbf{w} \times \mathbf{u}) = \mathbf{w} \cdot (\mathbf{u} \times \mathbf{v})$$

■

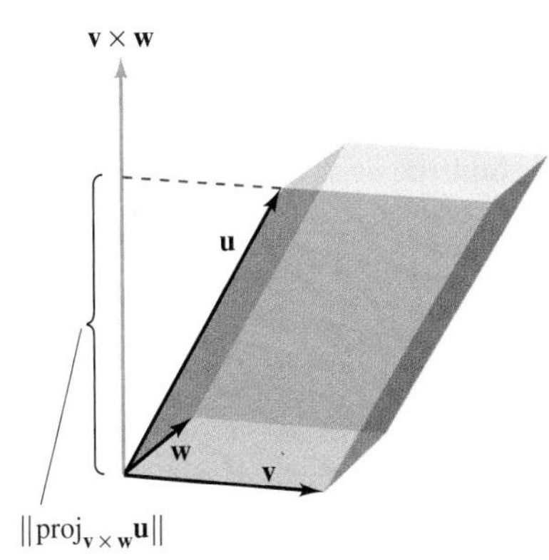

Area of base $= \|\mathbf{v} \times \mathbf{w}\|$
Volume of parallelepiped $= |\mathbf{u} \cdot (\mathbf{v} \times \mathbf{w})|$
Figure 11.41

If the vectors **u**, **v**, and **w** do not lie in the same plane, the triple scalar product $\mathbf{u} \cdot (\mathbf{v} \times \mathbf{w})$ can be used to determine the volume of the parallelepiped (a polyhedron, all of whose faces are parallelograms) with **u**, **v**, and **w** as adjacent edges, as shown in Figure 11.41. This is established in the following theorem.

THEOREM 11.10 GEOMETRIC PROPERTY OF THE TRIPLE SCALAR PRODUCT

The volume V of a parallelepiped with vectors **u**, **v**, and **w** as adjacent edges is given by

$$V = |\mathbf{u} \cdot (\mathbf{v} \times \mathbf{w})|.$$

PROOF In Figure 11.41, note that

$$\|\mathbf{v} \times \mathbf{w}\| = \text{area of base}$$

and

$$\|\text{proj}_{\mathbf{v} \times \mathbf{w}} \mathbf{u}\| = \text{height of parallelepiped.}$$

Therefore, the volume is

$$\begin{aligned} V = (\text{height})(\text{area of base}) &= \|\text{proj}_{\mathbf{v} \times \mathbf{w}} \mathbf{u}\| \|\mathbf{v} \times \mathbf{w}\| \\ &= \left| \frac{\mathbf{u} \cdot (\mathbf{v} \times \mathbf{w})}{\|\mathbf{v} \times \mathbf{w}\|} \right| \|\mathbf{v} \times \mathbf{w}\| \\ &= |\mathbf{u} \cdot (\mathbf{v} \times \mathbf{w})|. \end{aligned}$$

■

EXAMPLE 5 Volume by the Triple Scalar Product

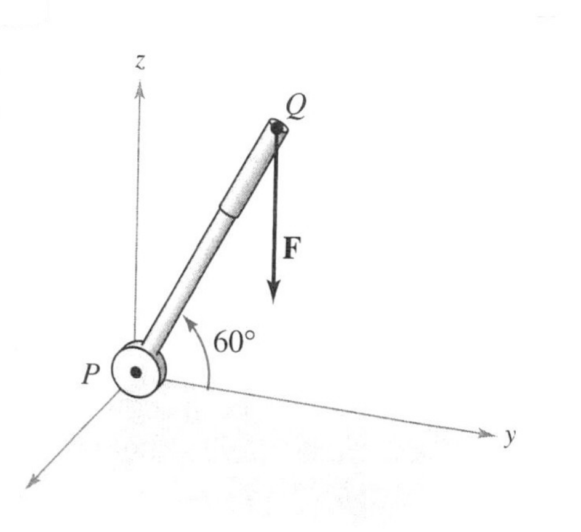

vertical force of 50 pounds is applied at oint Q.
igure 11.40

Find the volume of the parallelepiped shown in Figure 11.42 having $\mathbf{u} = 3\mathbf{i} - 5\mathbf{j} + \mathbf{k}$, $\mathbf{v} = 2\mathbf{j} - 2\mathbf{k}$, and $\mathbf{w} = 3\mathbf{i} + \mathbf{j} + \mathbf{k}$ as adjacent edges.

Solution By Theorem 11.10, you have

$$\begin{aligned} V &= |\mathbf{u} \cdot (\mathbf{v} \times \mathbf{w})| \qquad \text{Triple scalar product} \\ &= \begin{vmatrix} 3 & -5 & 1 \\ 0 & 2 & -2 \\ 3 & 1 & 1 \end{vmatrix} \\ &= 3\begin{vmatrix} 2 & -2 \\ 1 & 1 \end{vmatrix} - (-5)\begin{vmatrix} 0 & -2 \\ 3 & 1 \end{vmatrix} + (1)\begin{vmatrix} 0 & 2 \\ 3 & 1 \end{vmatrix} \\ &= 3(4) + 5(6) + 1(-6) \\ &= 36. \end{aligned}$$

■

A natural consequence of Theorem 11.10 is that the volume of the parallelepiped is 0 if and only if the three vectors are coplanar. That is, if the vectors $\mathbf{u} = \langle u_1, u_2, u_3 \rangle$, $\mathbf{v} = \langle v_1, v_2, v_3 \rangle$, and $\mathbf{w} = \langle w_1, w_2, w_3 \rangle$ have the same initial point, they lie in the same plane if and only if

$$\mathbf{u} \cdot (\mathbf{v} \times \mathbf{w}) = \begin{vmatrix} u_1 & u_2 & u_3 \\ v_1 & v_2 & v_3 \\ w_1 & w_2 & w_3 \end{vmatrix} = 0.$$

11.4 Exercises

See www.CalcChat.com for worked-out solutions to odd-numbered exercises.

In Exercises 1–6, find the cross product of the unit vectors and sketch your result.

1. $\mathbf{j} \times \mathbf{i}$
2. $\mathbf{i} \times \mathbf{j}$
3. $\mathbf{j} \times \mathbf{k}$
4. $\mathbf{k} \times \mathbf{j}$
5. $\mathbf{i} \times \mathbf{k}$
6. $\mathbf{k} \times \mathbf{i}$

In Exercises 7–10, find (a) u × v, (b) v × u, and (c) v × v.

7. $\mathbf{u} = -2\mathbf{i} + 4\mathbf{j}$, $\mathbf{v} = 3\mathbf{i} + 2\mathbf{j} + 5\mathbf{k}$
8. $\mathbf{u} = 3\mathbf{i} + 5\mathbf{k}$, $\mathbf{v} = 2\mathbf{i} + 3\mathbf{j} - 2\mathbf{k}$
9. $\mathbf{u} = \langle 7, 3, 2 \rangle$, $\mathbf{v} = \langle 1, -1, 5 \rangle$
10. $\mathbf{u} = \langle 3, -2, -2 \rangle$, $\mathbf{v} = \langle 1, 5, 1 \rangle$

In Exercises 11–16, find u × v and show that it is orthogonal to both u and v.

11. $\mathbf{u} = \langle 12, -3, 0 \rangle$, $\mathbf{v} = \langle -2, 5, 0 \rangle$
12. $\mathbf{u} = \langle -1, 1, 2 \rangle$, $\mathbf{v} = \langle 0, 1, 0 \rangle$
13. $\mathbf{u} = \langle 2, -3, 1 \rangle$, $\mathbf{v} = \langle 1, -2, 1 \rangle$
14. $\mathbf{u} = \langle -10, 0, 6 \rangle$, $\mathbf{v} = \langle 5, -3, 0 \rangle$
15. $\mathbf{u} = \mathbf{i} + \mathbf{j} + \mathbf{k}$, $\mathbf{v} = 2\mathbf{i} + \mathbf{j} - \mathbf{k}$
16. $\mathbf{u} = \mathbf{i} + 6\mathbf{j}$, $\mathbf{v} = -2\mathbf{i} + \mathbf{j} + \mathbf{k}$

Think About It **In Exercises 17–20, use the vectors u and v shown in the figure to sketch a vector in the direction of the indicated cross product in a right-handed system.**

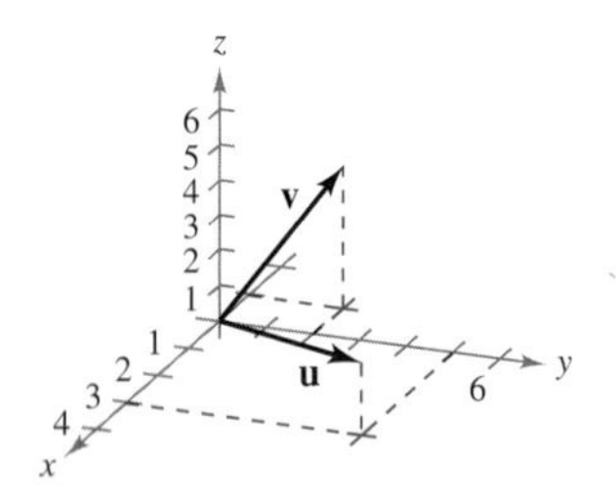

17. $\mathbf{u} \times \mathbf{v}$
18. $\mathbf{v} \times \mathbf{u}$
19. $(-\mathbf{v}) \times \mathbf{u}$
20. $\mathbf{u} \times (\mathbf{u} \times \mathbf{v})$

CAS **In Exercises 21–24, use a computer algebra system to find u × v and a unit vector orthogonal to u and v.**

21. $\mathbf{u} = \langle 4, -3.5, 7 \rangle$, $\mathbf{v} = \langle 2.5, 9, 3 \rangle$
22. $\mathbf{u} = \langle -8, -6, 4 \rangle$, $\mathbf{v} = \langle 10, -12, -2 \rangle$
23. $\mathbf{u} = -3\mathbf{i} + 2\mathbf{j} - 5\mathbf{k}$, $\mathbf{v} = 0.4\mathbf{i} - 0.8\mathbf{j} + 0.2\mathbf{k}$
24. $\mathbf{u} = 0.7\mathbf{k}$, $\mathbf{v} = 1.5\mathbf{i} + 6.2\mathbf{k}$

25. ***Programming*** Given the vectors **u** and **v** in component form, write a program for a graphing utility in which the output is $\mathbf{u} \times \mathbf{v}$ and $\|\mathbf{u} \times \mathbf{v}\|$.

26. ***Programming*** Use the program you wrote in Exercise 25 to find $\mathbf{u} \times \mathbf{v}$ and $\|\mathbf{u} \times \mathbf{v}\|$ for $\mathbf{u} = \langle -2, 6, 10 \rangle$ and $\mathbf{v} = \langle 3, 8, 5 \rangle$.

Area **In Exercises 27–30, find the area of the parallelogram that has the given vectors as adjacent sides. Use a computer algebra system or a graphing utility to verify your result.**

27. $\mathbf{u} = \mathbf{j}$, $\mathbf{v} = \mathbf{j} + \mathbf{k}$
28. $\mathbf{u} = \mathbf{i} + \mathbf{j} + \mathbf{k}$, $\mathbf{v} = \mathbf{j} + \mathbf{k}$
29. $\mathbf{u} = \langle 3, 2, -1 \rangle$, $\mathbf{v} = \langle 1, 2, 3 \rangle$
30. $\mathbf{u} = \langle 2, -1, 0 \rangle$, $\mathbf{v} = \langle -1, 2, 0 \rangle$

Area **In Exercises 31 and 32, verify that the points are the vertices of a parallelogram, and find its area.**

31. $A(0, 3, 2)$, $B(1, 5, 5)$, $C(6, 9, 5)$, $D(5, 7, 2)$
32. $A(2, -3, 1)$, $B(6, 5, -1)$, $C(7, 2, 2)$, $D(3, -6, 4)$

Area **In Exercises 33–36, find the area of the triangle with the given vertices. (*Hint:* $\frac{1}{2}\|\mathbf{u} \times \mathbf{v}\|$ is the area of the triangle having u and v as adjacent sides.)**

33. $A(0, 0, 0)$, $B(1, 0, 3)$, $C(-3, 2, 0)$
34. $A(2, -3, 4)$, $B(0, 1, 2)$, $C(-1, 2, 0)$
35. $A(2, -7, 3)$, $B(-1, 5, 8)$, $C(4, 6, -1)$
36. $A(1, 2, 0)$, $B(-2, 1, 0)$, $C(0, 0, 0)$

37. ***Torque*** A child applies the brakes on a bicycle by applying a downward force of 20 pounds on the pedal when the crank makes a 40° angle with the horizontal (see figure). The crank is 6 inches in length. Find the torque at P.

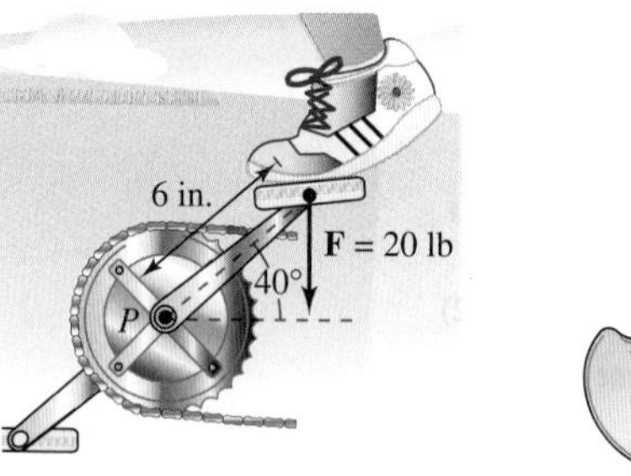

Figure for 37 **Figure for 38**

38. ***Torque*** Both the magnitude and the direction of the force on a crankshaft change as the crankshaft rotates. Find the torque on the crankshaft using the position and data shown in the figure.

39. ***Optimization*** A force of 56 pounds acts on the pipe wrench shown in the figure on the next page.
    (a) Find the magnitude of the moment about O by evaluating $\|\overrightarrow{OA} \times \mathbf{F}\|$. Use a graphing utility to graph the resulting function of θ.
    (b) Use the result of part (a) to determine the magnitude of the moment when $\theta = 45°$.
    (c) Use the result of part (a) to determine the angle θ when the magnitude of the moment is maximum. Is the answer what you expected? Why or why not?

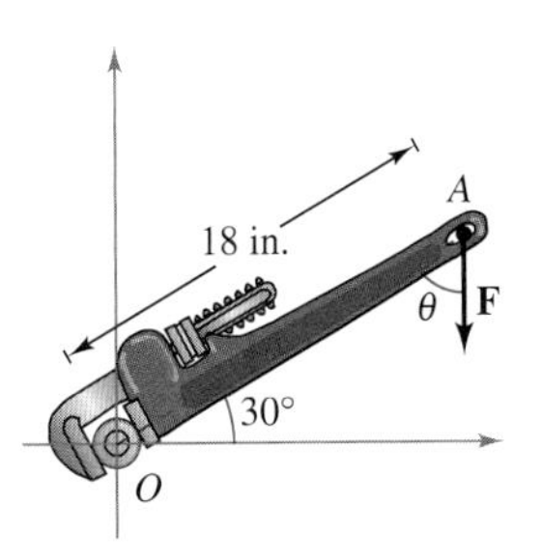

Figure for 39

Figure for 40

40. ***Optimization*** A force of 180 pounds acts on the bracket shown in the figure.

(a) Determine the vector $\overrightarrow{AB}$ and the vector $\mathbf{F}$ representing the force. ($\mathbf{F}$ will be in terms of θ.)

(b) Find the magnitude of the moment about A by evaluating $\|\overrightarrow{AB} \times \mathbf{F}\|$.

(c) Use the result of part (b) to determine the magnitude of the moment when $\theta = 30°$.

(d) Use the result of part (b) to determine the angle θ when the magnitude of the moment is maximum. At that angle, what is the relationship between the vectors $\mathbf{F}$ and $\overrightarrow{AB}$? Is it what you expected? Why or why not?

 (e) Use a graphing utility to graph the function for the magnitude of the moment about A for $0° \le \theta \le 180°$. Find the zero of the function in the given domain. Interpret the meaning of the zero in the context of the problem.

In Exercises 41–44, find $\mathbf{u} \cdot (\mathbf{v} \times \mathbf{w})$.

41. $\mathbf{u} = \mathbf{i}$, $\mathbf{v} = \mathbf{j}$, $\mathbf{w} = \mathbf{k}$

42. $\mathbf{u} = \langle 1, 1, 1\rangle$, $\mathbf{v} = \langle 2, 1, 0\rangle$, $\mathbf{w} = \langle 0, 0, 1\rangle$

43. $\mathbf{u} = \langle 2, 0, 1\rangle$, $\mathbf{v} = \langle 0, 3, 0\rangle$, $\mathbf{w} = \langle 0, 0, 1\rangle$

44. $\mathbf{u} = \langle 2, 0, 0\rangle$, $\mathbf{v} = \langle 1, 1, 1\rangle$, $\mathbf{w} = \langle 0, 2, 2\rangle$

Volume **In Exercises 45 and 46, use the triple scalar product to find the volume of the parallelepiped having adjacent edges u, v, and w.**

45. $\mathbf{u} = \mathbf{i} + \mathbf{j}$, $\mathbf{v} = \mathbf{j} + \mathbf{k}$, $\mathbf{w} = \mathbf{i} + \mathbf{k}$

46. $\mathbf{u} = \langle 1, 3, 1\rangle$, $\mathbf{v} = \langle 0, 6, 6\rangle$, $\mathbf{w} = \langle -4, 0, -4\rangle$

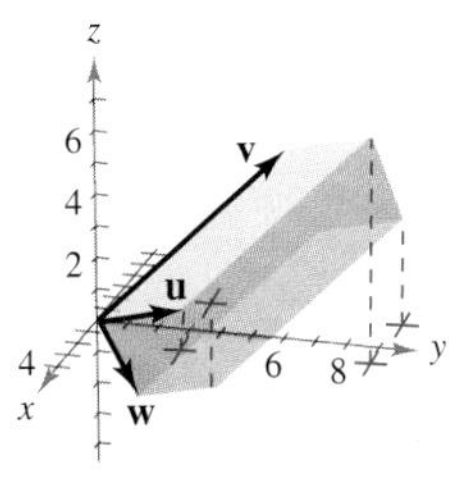

Volume **In Exercises 47 and 48, find the volume of the parallelepiped with the given vertices.**

47. $(0, 0, 0), (3, 0, 0), (0, 5, 1), (2, 0, 5)$
$(3, 5, 1), (5, 0, 5), (2, 5, 6), (5, 5, 6)$

48. $(0, 0, 0), (0, 4, 0), (-3, 0, 0), (-1, 1, 5)$
$(-3, 4, 0), (-1, 5, 5), (-4, 1, 5), (-4, 5, 5)$

49. If $\mathbf{u} \times \mathbf{v} = \mathbf{0}$ and $\mathbf{u} \cdot \mathbf{v} = 0$, what can you conclude about $\mathbf{u}$ and $\mathbf{v}$?

50. Identify the dot products that are equal. Explain your reasoning. (Assume $\mathbf{u}$, $\mathbf{v}$, and $\mathbf{w}$ are nonzero vectors.)

(a) $\mathbf{u} \cdot (\mathbf{v} \times \mathbf{w})$ **(b)** $(\mathbf{v} \times \mathbf{w}) \cdot \mathbf{u}$

(c) $(\mathbf{u} \times \mathbf{v}) \cdot \mathbf{w}$ **(d)** $(\mathbf{u} \times -\mathbf{w}) \cdot \mathbf{v}$

(e) $\mathbf{u} \cdot (\mathbf{w} \times \mathbf{v})$ **(f)** $\mathbf{w} \cdot (\mathbf{v} \times \mathbf{u})$

(g) $(-\mathbf{u} \times \mathbf{v}) \cdot \mathbf{w}$ **(h)** $(\mathbf{w} \times \mathbf{u}) \cdot \mathbf{v}$

WRITING ABOUT CONCEPTS

51. Define the cross product of vectors $\mathbf{u}$ and $\mathbf{v}$.

52. State the geometric properties of the cross product.

53. If the magnitudes of two vectors are doubled, how will the magnitude of the cross product of the vectors change? Explain.

CAPSTONE

54. The vertices of a triangle in space are (x_1, y_1, z_1), (x_2, y_2, z_2), and (x_3, y_3, z_3). Explain how to find a vector perpendicular to the triangle.

True or False? **In Exercises 55–58, determine whether the statement is true or false. If it is false, explain why or give an example that shows it is false.**

55. It is possible to find the cross product of two vectors in a two-dimensional coordinate system.

56. If $\mathbf{u}$ and $\mathbf{v}$ are vectors in space that are nonzero and nonparallel, then $\mathbf{u} \times \mathbf{v} = \mathbf{v} \times \mathbf{u}$.

57. If $\mathbf{u} \neq \mathbf{0}$ and $\mathbf{u} \times \mathbf{v} = \mathbf{u} \times \mathbf{w}$, then $\mathbf{v} = \mathbf{w}$.

58. If $\mathbf{u} \neq \mathbf{0}$, $\mathbf{u} \cdot \mathbf{v} = \mathbf{u} \cdot \mathbf{w}$, and $\mathbf{u} \times \mathbf{v} = \mathbf{u} \times \mathbf{w}$, then $\mathbf{v} = \mathbf{w}$.

In Exercises 59–66, prove the property of the cross product.

59. $\mathbf{u} \times (\mathbf{v} + \mathbf{w}) = (\mathbf{u} \times \mathbf{v}) + (\mathbf{u} \times \mathbf{w})$

60. $c(\mathbf{u} \times \mathbf{v}) = (c\mathbf{u}) \times \mathbf{v} = \mathbf{u} \times (c\mathbf{v})$

61. $\mathbf{u} \times \mathbf{u} = \mathbf{0}$

62. $\mathbf{u} \cdot (\mathbf{v} \times \mathbf{w}) = (\mathbf{u} \times \mathbf{v}) \cdot \mathbf{w}$

63. $\mathbf{u} \times \mathbf{v}$ is orthogonal to both $\mathbf{u}$ and $\mathbf{v}$.

64. $\mathbf{u} \times \mathbf{v} = \mathbf{0}$ if and only if $\mathbf{u}$ and $\mathbf{v}$ are scalar multiples of each other.

65. Prove that $\|\mathbf{u} \times \mathbf{v}\| = \|\mathbf{u}\|\ \|\mathbf{v}\|$ if $\mathbf{u}$ and $\mathbf{v}$ are orthogonal.

66. Prove that $\mathbf{u} \times (\mathbf{v} \times \mathbf{w}) = (\mathbf{u} \cdot \mathbf{w})\mathbf{v} - (\mathbf{u} \cdot \mathbf{v})\mathbf{w}$.

67. Prove Theorem 11.9.

11.5 Lines and Planes in Space

- **Write a set of parametric equations for a line in space.**
- **Write a linear equation to represent a plane in space.**
- **Sketch the plane given by a linear equation.**
- **Find the distances between points, planes, and lines in space.**

Lines in Space

In the plane, *slope* is used to determine an equation of a line. In space, it is more convenient to use *vectors* to determine the equation of a line.

In Figure 11.43, consider the line L through the point $P(x_1, y_1, z_1)$ and parallel to the vector $\mathbf{v} = \langle a, b, c\rangle$. The vector $\mathbf{v}$ is a **direction vector** for the line L, and a, b, and c are **direction numbers.** One way of describing the line L is to say that it consists of all points $Q(x, y, z)$ for which the vector $\overrightarrow{PQ}$ is parallel to $\mathbf{v}$. This means that $\overrightarrow{PQ}$ is a scalar multiple of $\mathbf{v}$, and you can write $\overrightarrow{PQ} = t\mathbf{v}$, where t is a scalar (a real number).

$$\overrightarrow{PQ} = \langle x - x_1, y - y_1, z - z_1\rangle = \langle at, bt, ct\rangle = t\mathbf{v}$$

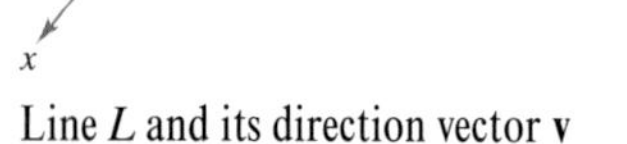

Line L and its direction vector $\mathbf{v}$
Figure 11.43

By equating corresponding components, you can obtain **parametric equations** of a line in space.

THEOREM 11.11 PARAMETRIC EQUATIONS OF A LINE IN SPACE

A line L parallel to the vector $\mathbf{v} = \langle a, b, c\rangle$ and passing through the point $P(x_1, y_1, z_1)$ is represented by the **parametric equations**

$$x = x_1 + at, \quad y = y_1 + bt, \quad \text{and} \quad z = z_1 + ct.$$

If the direction numbers a, b, and c are all nonzero, you can eliminate the parameter t to obtain **symmetric equations** of the line.

$$\frac{x - x_1}{a} = \frac{y - y_1}{b} = \frac{z - z_1}{c}$$ Symmetric equations

EXAMPLE 1 Finding Parametric and Symmetric Equations

Find parametric and symmetric equations of the line L that passes through the point $(1, -2, 4)$ and is parallel to $\mathbf{v} = \langle 2, 4, -4\rangle$.

Solution To find a set of parametric equations of the line, use the coordinates $x_1 = 1$, $y_1 = -2$, and $z_1 = 4$ and direction numbers $a = 2$, $b = 4$, and $c = -4$ (see Figure 11.44).

$$x = 1 + 2t, \quad y = -2 + 4t, \quad z = 4 - 4t$$ Parametric equations

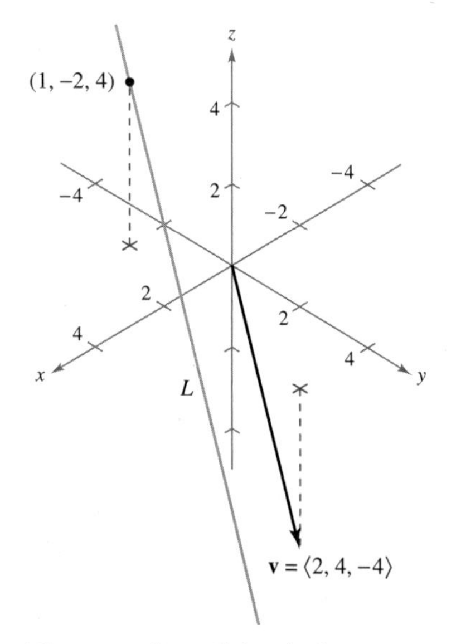

The vector $\mathbf{v}$ is parallel to the line L.
Figure 11.44

Because a, b, and c are all nonzero, a set of symmetric equations is

$$\frac{x - 1}{2} = \frac{y + 2}{4} = \frac{z - 4}{-4}.$$ Symmetric equations ■

Neither parametric equations nor symmetric equations of a given line are unique. For instance, in Example 1, by letting $t = 1$ in the parametric equations you would obtain the point $(3, 2, 0)$. Using this point with the direction numbers $a = 2$, $b = 4$, and $c = -4$ would produce a different set of parametric equations

$$x = 3 + 2t, \quad y = 2 + 4t, \quad \text{and} \quad z = -4t.$$

EXAMPLE 2 Parametric Equations of a Line Through Two Points

Find a set of parametric equations of the line that passes through the points $(-2, 1, 0)$ and $(1, 3, 5)$.

Solution Begin by using the points $P(-2, 1, 0)$ and $Q(1, 3, 5)$ to find a direction vector for the line passing through P and Q, given by

$$\mathbf{v} = \overrightarrow{PQ} = \langle 1 - (-2), 3 - 1, 5 - 0 \rangle = \langle 3, 2, 5 \rangle = \langle a, b, c \rangle.$$

Using the direction numbers $a = 3$, $b = 2$, and $c = 5$ with the point $P(-2, 1, 0)$, you can obtain the parametric equations

$$x = -2 + 3t, \quad y = 1 + 2t, \quad \text{and} \quad z = 5t.$$

■

NOTE As t varies over all real numbers, the parametric equations in Example 2 determine the points (x, y, z) on the line. In particular, note that $t = 0$ and $t = 1$ give the original points $(-2, 1, 0)$ and $(1, 3, 5)$. ■

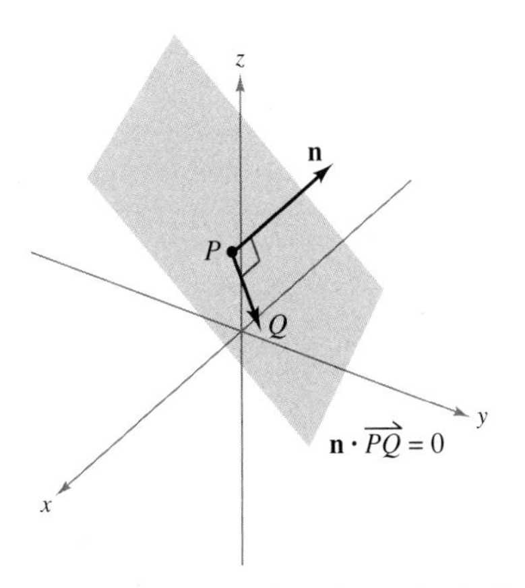

The normal vector $\mathbf{n}$ is orthogonal to each vector $\overrightarrow{PQ}$ in the plane.
Figure 11.45

Planes in Space

You have seen how an equation of a line in space can be obtained from a point on the line and a vector *parallel* to it. You will now see that an equation of a plane in space can be obtained from a point in the plane and a vector *normal* (perpendicular) to the plane.

Consider the plane containing the point $P(x_1, y_1, z_1)$ having a nonzero normal vector $\mathbf{n} = \langle a, b, c \rangle$, as shown in Figure 11.45. This plane consists of all points $Q(x, y, z)$ for which vector $\overrightarrow{PQ}$ is orthogonal to $\mathbf{n}$. Using the dot product, you can write the following.

$$\mathbf{n} \cdot \overrightarrow{PQ} = 0$$
$$\langle a, b, c \rangle \cdot \langle x - x_1, y - y_1, z - z_1 \rangle = 0$$
$$a(x - x_1) + b(y - y_1) + c(z - z_1) = 0$$

The third equation of the plane is said to be in **standard form.**

THEOREM 11.12 STANDARD EQUATION OF A PLANE IN SPACE

The plane containing the point (x_1, y_1, z_1) and having normal vector $\mathbf{n} = \langle a, b, c \rangle$ can be represented by the **standard form** of the equation of a plane

$$a(x - x_1) + b(y - y_1) + c(z - z_1) = 0.$$

By regrouping terms, you obtain the **general form** of the equation of a plane in space.

$$ax + by + cz + d = 0 \qquad \text{General form of equation of plane}$$

Given the general form of the equation of a plane, it is easy to find a normal vector to the plane. Simply use the coefficients of x, y, and z and write $\mathbf{n} = \langle a, b, c\rangle$.

EXAMPLE 3 Finding an Equation of a Plane in Three-Space

Find the general equation of the plane containing the points $(2, 1, 1)$, $(0, 4, 1)$, and $(-2, 1, 4)$.

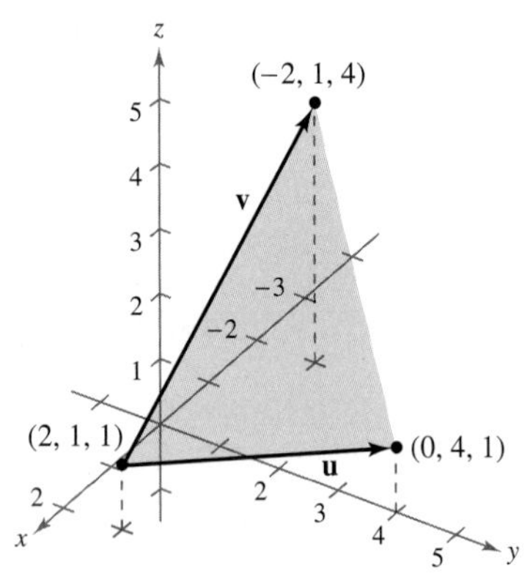

A plane determined by **u** and **v**
Figure 11.46

Solution To apply Theorem 11.12 you need a point in the plane and a vector that is normal to the plane. There are three choices for the point, but no normal vector is given. To obtain a normal vector, use the cross product of vectors **u** and **v** extending from the point $(2, 1, 1)$ to the points $(0, 4, 1)$ and $(-2, 1, 4)$, as shown in Figure 11.46. The component forms of **u** and **v** are

$$\mathbf{u} = \langle 0-2, 4-1, 1-1\rangle = \langle -2, 3, 0\rangle$$
$$\mathbf{v} = \langle -2-2, 1-1, 4-1\rangle = \langle -4, 0, 3\rangle$$

and it follows that

$$\begin{aligned}\mathbf{n} &= \mathbf{u}\times\mathbf{v}\\ &= \begin{vmatrix}\mathbf{i} & \mathbf{j} & \mathbf{k}\\ -2 & 3 & 0\\ -4 & 0 & 3\end{vmatrix}\\ &= 9\mathbf{i} + 6\mathbf{j} + 12\mathbf{k}\\ &= \langle a, b, c\rangle\end{aligned}$$

is normal to the given plane. Using the direction numbers for **n** and the point $(x_1, y_1, z_1) = (2, 1, 1)$, you can determine an equation of the plane to be

$$a(x-x_1) + b(y-y_1) + c(z-z_1) = 0$$
$$9(x-2) + 6(y-1) + 12(z-1) = 0 \qquad \text{Standard form}$$
$$9x + 6y + 12z - 36 = 0 \qquad \text{General form}$$
$$3x + 2y + 4z - 12 = 0. \qquad \text{Simplified general form}$$

■

NOTE In Example 3, check to see that each of the three original points satisfies the equation $3x + 2y + 4z - 12 = 0$. ■

Two distinct planes in three-space either are parallel or intersect in a line. If they intersect, you can determine the angle $(0 \le \theta \le \pi/2)$ between them from the angle between their normal vectors, as shown in Figure 11.47. Specifically, if vectors $\mathbf{n}_1$ and $\mathbf{n}_2$ are normal to two intersecting planes, the angle θ between the normal vectors is equal to the angle between the two planes and is given by

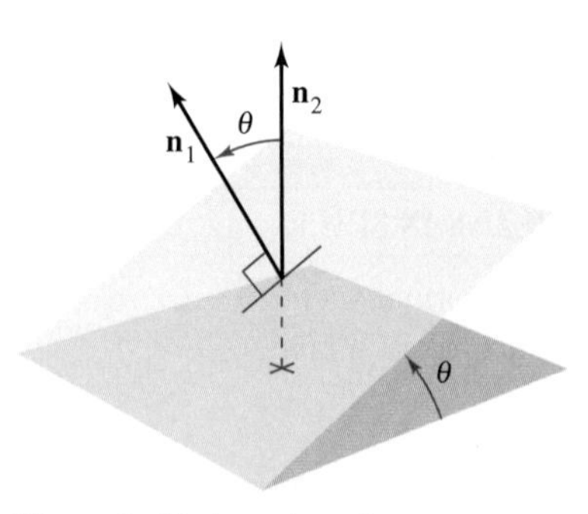

The angle θ between two planes
Figure 11.47

$$\cos\theta = \frac{|\mathbf{n}_1\cdot\mathbf{n}_2|}{\|\mathbf{n}_1\|\,\|\mathbf{n}_2\|}. \qquad \text{Angle between two planes}$$

Consequently, two planes with normal vectors $\mathbf{n}_1$ and $\mathbf{n}_2$ are

1. *perpendicular* if $\mathbf{n}_1\cdot\mathbf{n}_2 = 0$.
2. *parallel* if $\mathbf{n}_1$ is a scalar multiple of $\mathbf{n}_2$.

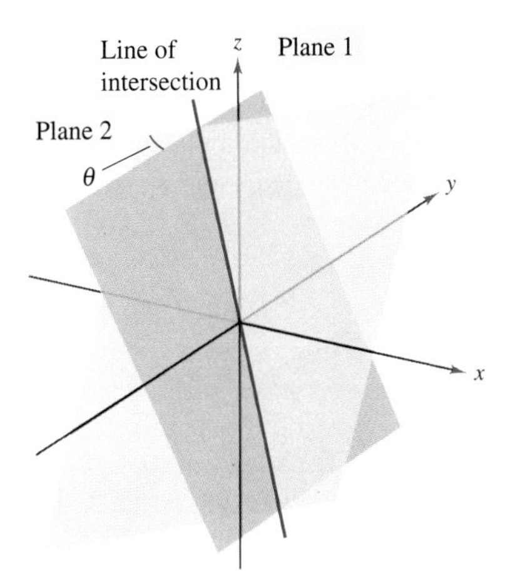

Figure 11.48

EXAMPLE 4 Finding the Line of Intersection of Two Planes

Find the angle between the two planes given by

$$x - 2y + z = 0 \qquad \text{Equation of plane 1}$$
$$2x + 3y - 2z = 0 \qquad \text{Equation of plane 2}$$

and find parametric equations of their line of intersection (see Figure 11.48).

Solution Normal vectors for the planes are $\mathbf{n}_1 = \langle 1, -2, 1 \rangle$ and $\mathbf{n}_2 = \langle 2, 3, -2 \rangle$. Consequently, the angle between the two planes is determined as follows.

$$\begin{aligned} \cos\theta &= \frac{|\mathbf{n}_1 \cdot \mathbf{n}_2|}{\|\mathbf{n}_1\|\,\|\mathbf{n}_2\|} && \text{Cosine of angle between } \mathbf{n}_1 \text{ and } \mathbf{n}_2 \\ &= \frac{|-6|}{\sqrt{6}\sqrt{17}} \\ &= \frac{6}{\sqrt{102}} \\ &\approx 0.59409 \end{aligned}$$

This implies that the angle between the two planes is $\theta \approx 53.55°$. You can find the line of intersection of the two planes by simultaneously solving the two linear equations representing the planes. One way to do this is to multiply the first equation by -2 and add the result to the second equation.

$$\begin{aligned} x - 2y + z &= 0 \quad \Rightarrow & -2x + 4y - 2z &= 0 \\ 2x + 3y - 2z &= 0 & 2x + 3y - 2z &= 0 \\ & & 7y - 4z &= 0 \quad \Rightarrow \quad y = \frac{4z}{7} \end{aligned}$$

Substituting $y = 4z/7$ back into one of the original equations, you can determine that $x = z/7$. Finally, by letting $t = z/7$, you obtain the parametric equations

$$x = t, \quad y = 4t, \quad \text{and} \quad z = 7t \qquad \text{Line of intersection}$$

which indicate that 1, 4, and 7 are direction numbers for the line of intersection. ■

Note that the direction numbers in Example 4 can be obtained from the cross product of the two normal vectors as follows.

$$\begin{aligned} \mathbf{n}_1 \times \mathbf{n}_2 &= \begin{vmatrix} \mathbf{i} & \mathbf{j} & \mathbf{k} \\ 1 & -2 & 1 \\ 2 & 3 & -2 \end{vmatrix} \\ &= \begin{vmatrix} -2 & 1 \\ 3 & -2 \end{vmatrix}\mathbf{i} - \begin{vmatrix} 1 & 1 \\ 2 & -2 \end{vmatrix}\mathbf{j} + \begin{vmatrix} 1 & -2 \\ 2 & 3 \end{vmatrix}\mathbf{k} \\ &= \mathbf{i} + 4\mathbf{j} + 7\mathbf{k} \end{aligned}$$

This means that the line of intersection of the two planes is parallel to the cross product of their normal vectors.

NOTE The three-dimensional rotatable graphs that are available in the premium eBook for this text can help you visualize surfaces such as those shown in Figure 11.48. If you have access to these graphs, you should use them to help your spatial intuition when studying this section and other sections in the text that deal with vectors, curves, or surfaces in space. ■

Sketching Planes in Space

If a plane in space intersects one of the coordinate planes, the line of intersection is called the **trace** of the given plane in the coordinate plane. To sketch a plane in space, it is helpful to find its points of intersection with the coordinate axes and its traces in the coordinate planes. For example, consider the plane given by

$$3x + 2y + 4z = 12. \qquad \text{Equation of plane}$$

You can find the xy-trace by letting $z = 0$ and sketching the line

$$3x + 2y = 12 \qquad xy\text{-trace}$$

in the xy-plane. This line intersects the x-axis at $(4, 0, 0)$ and the y-axis at $(0, 6, 0)$. In Figure 11.49, this process is continued by finding the yz-trace and the xz-trace, and then shading the triangular region lying in the first octant.

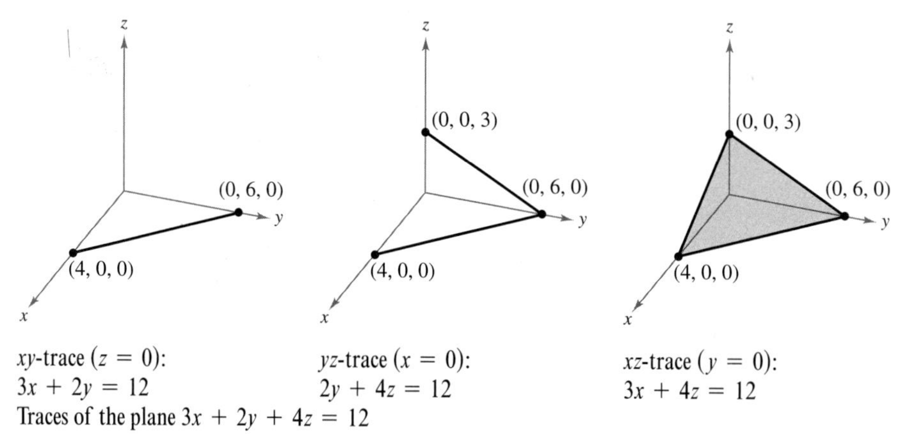

xy-trace $(z = 0)$: $3x + 2y = 12$ yz-trace $(x = 0)$: $2y + 4z = 12$ xz-trace $(y = 0)$: $3x + 4z = 12$
Traces of the plane $3x + 2y + 4z = 12$
Figure 11.49

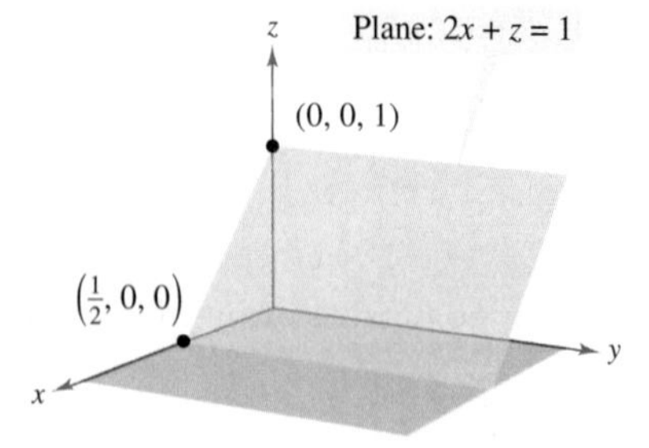

Plane $2x + z = 1$ is parallel to the y-axis.
Figure 11.50

If an equation of a plane has a missing variable, such as $2x + z = 1$, the plane must be *parallel to the axis* represented by the missing variable, as shown in Figure 11.50. If two variables are missing from an equation of a plane, it is *parallel to the coordinate plane* represented by the missing variables, as shown in Figure 11.51.

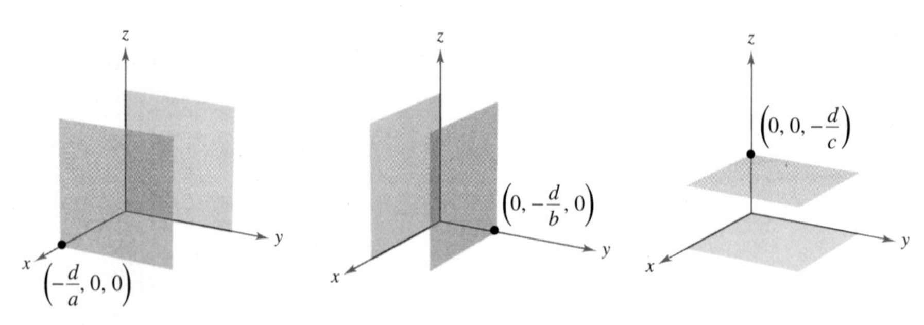

Plane $ax + d = 0$ is parallel to the yz-plane Plane $by + d = 0$ is parallel to the xz-plane Plane $cz + d = 0$ is parallel to the xy-plane
Figure 11.51

Distances Between Points, Planes, and Lines

This section is concluded with the following discussion of two basic types of problems involving distance in space.

1. Finding the distance between a point and a plane
2. Finding the distance between a point and a line

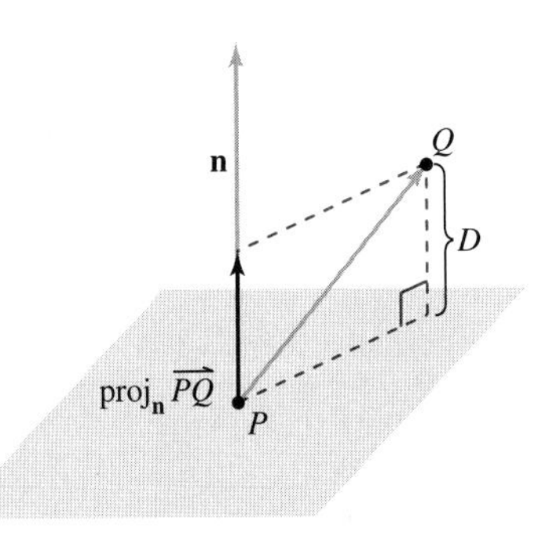

$D = \|\text{proj}_{\mathbf{n}} \overrightarrow{PQ}\|$

The distance between a point and a plane
Figure 11.52

The solutions of these problems illustrate the versatility and usefulness of vectors in coordinate geometry: the first problem uses the *dot product* of two vectors, and the second problem uses the *cross product.*

The distance D between a point Q and a plane is the length of the shortest line segment connecting Q to the plane, as shown in Figure 11.52. If P is *any* point in the plane, you can find this distance by projecting the vector $\overrightarrow{PQ}$ onto the normal vector $\mathbf{n}$. The length of this projection is the desired distance.

THEOREM 11.13 DISTANCE BETWEEN A POINT AND A PLANE

The distance between a plane and a point Q (not in the plane) is

$$D = \|\text{proj}_{\mathbf{n}}\overrightarrow{PQ}\| = \frac{|\overrightarrow{PQ} \cdot \mathbf{n}|}{\|\mathbf{n}\|}$$

where P is a point in the plane and $\mathbf{n}$ is normal to the plane.

To find a point in the plane given by $ax + by + cz + d = 0$ $(a \neq 0)$, let $y = 0$ and $z = 0$. Then, from the equation $ax + d = 0$, you can conclude that the point $(-d/a, 0, 0)$ lies in the plane.

EXAMPLE 5 Finding the Distance Between a Point and a Plane

Find the distance between the point $Q(1, 5, -4)$ and the plane given by

$$3x - y + 2z = 6.$$

Solution You know that $\mathbf{n} = \langle 3, -1, 2\rangle$ is normal to the given plane. To find a point in the plane, let $y = 0$ and $z = 0$, and obtain the point $P(2, 0, 0)$. The vector from P to Q is given by

$$\begin{aligned}\overrightarrow{PQ} &= \langle 1 - 2, 5 - 0, -4 - 0\rangle \\ &= \langle -1, 5, -4\rangle.\end{aligned}$$

Using the Distance Formula given in Theorem 11.13 produces

$$\begin{aligned}D = \frac{|\overrightarrow{PQ} \cdot \mathbf{n}|}{\|\mathbf{n}\|} &= \frac{|\langle -1, 5, -4\rangle \cdot \langle 3, -1, 2\rangle|}{\sqrt{9 + 1 + 4}} \\ &= \frac{|-3 - 5 - 8|}{\sqrt{14}} \\ &= \frac{16}{\sqrt{14}}.\end{aligned}$$

Distance between a point and a plane

■

NOTE The choice of the point P in Example 5 is arbitrary. Try choosing a different point in the plane to verify that you obtain the same distance. ■

From Theorem 11.13, you can determine that the distance between the point $Q(x_0, y_0, z_0)$ and the plane given by $ax + by + cz + d = 0$ is

$$D = \frac{|a(x_0 - x_1) + b(y_0 - y_1) + c(z_0 - z_1)|}{\sqrt{a^2 + b^2 + c^2}}$$

or

$$D = \frac{|ax_0 + by_0 + cz_0 + d|}{\sqrt{a^2 + b^2 + c^2}}$$ Distance between a point and a plane

where $P(x_1, y_1, z_1)$ is a point in the plane and $d = -(ax_1 + by_1 + cz_1)$.

EXAMPLE 6 Finding the Distance Between Two Parallel Planes

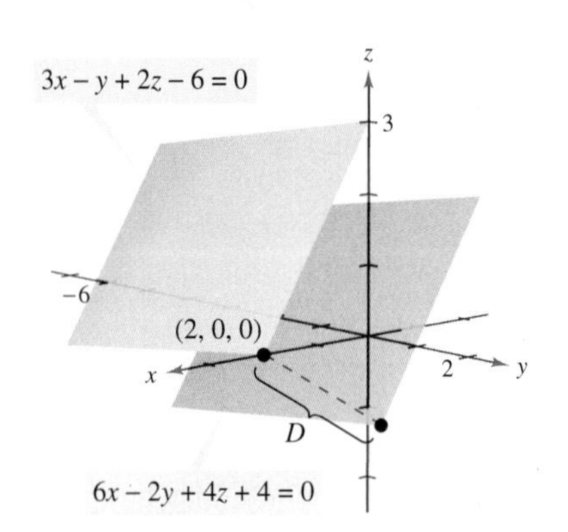

The distance between the parallel planes is approximately 2.14.
Figure 11.53

Find the distance between the two parallel planes given by

$$3x - y + 2z - 6 = 0 \quad \text{and} \quad 6x - 2y + 4z + 4 = 0.$$

Solution The two planes are shown in Figure 11.53. To find the distance between the planes, choose a point in the first plane, say $(x_0, y_0, z_0) = (2, 0, 0)$. Then, from the second plane, you can determine that $a = 6$, $b = -2$, $c = 4$, and $d = 4$, and conclude that the distance is

$$D = \frac{|ax_0 + by_0 + cz_0 + d|}{\sqrt{a^2 + b^2 + c^2}}$$ Distance between a point and a plane

$$= \frac{|6(2) + (-2)(0) + (4)(0) + 4|}{\sqrt{6^2 + (-2)^2 + 4^2}}$$

$$= \frac{16}{\sqrt{56}} = \frac{8}{\sqrt{14}} \approx 2.14.$$ ■

The formula for the distance between a point and a line in space resembles that for the distance between a point and a plane—except that you replace the dot product with the length of the cross product and the normal vector **n** with a direction vector for the line.

THEOREM 11.14 DISTANCE BETWEEN A POINT AND A LINE IN SPACE

The distance between a point Q and a line in space is given by

$$D = \frac{\|\overrightarrow{PQ} \times \mathbf{u}\|}{\|\mathbf{u}\|}$$

where **u** is a direction vector for the line and P is a point on the line.

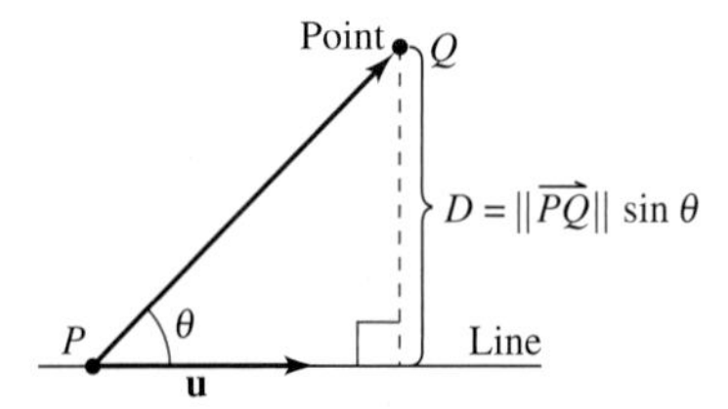

The distance between a point and a line
Figure 11.54

PROOF In Figure 11.54, let D be the distance between the point Q and the given line. Then $D = \|\overrightarrow{PQ}\| \sin\theta$, where θ is the angle between **u** and $\overrightarrow{PQ}$. By Property 2 of Theorem 11.8, you have

$$\|\mathbf{u}\| \|\overrightarrow{PQ}\| \sin\theta = \|\mathbf{u} \times \overrightarrow{PQ}\| = \|\overrightarrow{PQ} \times \mathbf{u}\|.$$

Consequently,

$$D = \|\overrightarrow{PQ}\| \sin\theta = \frac{\|\overrightarrow{PQ} \times \mathbf{u}\|}{\|\mathbf{u}\|}.$$ ■

EXAMPLE 7 Finding the Distance Between a Point and a Line

Find the distance between the point $Q(3, -1, 4)$ and the line given by

$$x = -2 + 3t, \quad y = -2t, \quad \text{and} \quad z = 1 + 4t.$$

Solution Using the direction numbers 3, -2, and 4, you know that a direction vector for the line is

$$\mathbf{u} = \langle 3, -2, 4 \rangle. \qquad \text{Direction vector for line}$$

To find a point on the line, let $t = 0$ and obtain

$$P = (-2, 0, 1). \qquad \text{Point on the line}$$

So,

$$\overrightarrow{PQ} = \langle 3 - (-2), -1 - 0, 4 - 1 \rangle = \langle 5, -1, 3 \rangle$$

and you can form the cross product

$$\overrightarrow{PQ} \times \mathbf{u} = \begin{vmatrix} \mathbf{i} & \mathbf{j} & \mathbf{k} \\ 5 & -1 & 3 \\ 3 & -2 & 4 \end{vmatrix} = 2\mathbf{i} - 11\mathbf{j} - 7\mathbf{k} = \langle 2, -11, -7 \rangle.$$

Finally, using Theorem 11.14, you can find the distance to be

$$\begin{aligned} D &= \frac{\|\overrightarrow{PQ} \times \mathbf{u}\|}{\|\mathbf{u}\|} \\ &= \frac{\sqrt{174}}{\sqrt{29}} \\ &= \sqrt{6} \approx 2.45. \end{aligned} \qquad \text{See Figure 11.55.}$$

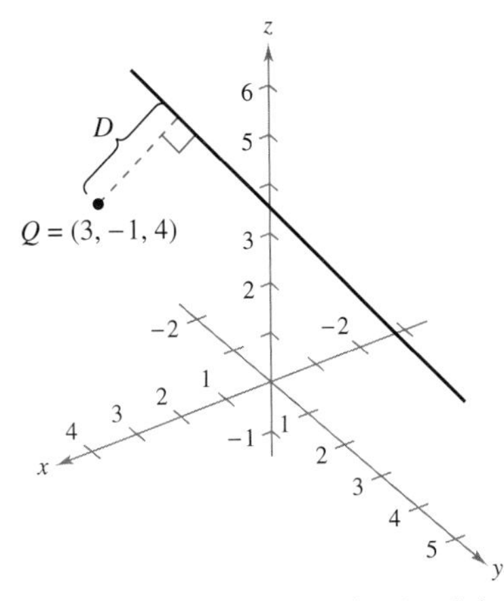

The distance between the point Q and the line is $\sqrt{6} \approx 2.45$.
Figure 11.55

11.5 Exercises

See www.CalcChat.com for worked-out solutions to odd-numbered exercises.

In Exercises 1 and 2, the figure shows the graph of a line given by the parametric equations. (a) Draw an arrow on the line to indicate its orientation. To print an enlarged copy of the graph, go to the website *www.mathgraphs.com*. (b) Find the coordinates of two points, P and Q, on the line. Determine the vector $\overrightarrow{PQ}$. What is the relationship between the components of the vector and the coefficients of t in the parametric equations? Why is this true? (c) Determine the coordinates of any points of intersection with the coordinate planes. If the line does not intersect a coordinate plane, explain why.

1. $x = 1 + 3t$
$y = 2 - t$
$z = 2 + 5t$

2. $x = 2 - 3t$
$y = 2$
$z = 1 - t$

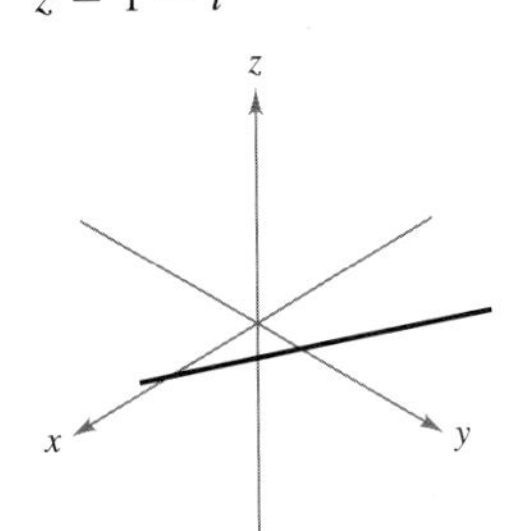

In Exercises 3 and 4, determine whether each point lies on the line.

3. $x = -2 + t, y = 3t, z = 4 + t$
(a) $(0, 6, 6)$ (b) $(2, 3, 5)$

4. $\dfrac{x-3}{2} = \dfrac{y-7}{8} = z + 2$
(a) $(7, 23, 0)$ (b) $(1, -1, -3)$

In Exercises 5–10, find sets of (a) parametric equations and (b) symmetric equations of the line through the point parallel to the given vector or line (if possible). (For each line, write the direction numbers as integers.)

	Point	*Parallel to*
5.	$(0, 0, 0)$	$\mathbf{v} = \langle 3, 1, 5 \rangle$
6.	$(0, 0, 0)$	$\mathbf{v} = \left\langle -2, \frac{5}{2}, 1 \right\rangle$
7.	$(-2, 0, 3)$	$\mathbf{v} = 2\mathbf{i} + 4\mathbf{j} - 2\mathbf{k}$
8.	$(-3, 0, 2)$	$\mathbf{v} = 6\mathbf{j} + 3\mathbf{k}$
9.	$(1, 0, 1)$	$x = 3 + 3t, y = 5 - 2t, z = -7 + t$
10.	$(-3, 5, 4)$	$\dfrac{x-1}{3} = \dfrac{y+1}{-2} = z - 3$

In Exercises 11–14, find sets of (a) parametric equations and (b) symmetric equations of the line through the two points (if possible). (For each line, write the direction numbers as integers.)

11. $(5, -3, -2), \left(-\frac{2}{3}, \frac{2}{3}, 1\right)$
12. $(0, 4, 3), (-1, 2, 5)$
13. $(7, -2, 6), (-3, 0, 6)$
14. $(0, 0, 25), (10, 10, 0)$

In Exercises 15–22, find a set of parametric equations of the line.

15. The line passes through the point $(2, 3, 4)$ and is parallel to the xz-plane and the yz-plane.
16. The line passes through the point $(-4, 5, 2)$ and is parallel to the xy-plane and the yz-plane.
17. The line passes through the point $(2, 3, 4)$ and is perpendicular to the plane given by $3x + 2y - z = 6$.
18. The line passes through the point $(-4, 5, 2)$ and is perpendicular to the plane given by $-x + 2y + z = 5$.
19. The line passes through the point $(5, -3, -4)$ and is parallel to $\mathbf{v} = \langle 2, -1, 3 \rangle$.
20. The line passes through the point $(-1, 4, -3)$ and is parallel to $\mathbf{v} = 5\mathbf{i} - \mathbf{j}$.
21. The line passes through the point $(2, 1, 2)$ and is parallel to the line $x = -t, y = 1 + t, z = -2 + t$.
22. The line passes through the point $(-6, 0, 8)$ and is parallel to the line $x = 5 - 2t, y = -4 + 2t, z = 0$.

In Exercises 23–26, find the coordinates of a point P on the line and a vector $\mathbf{v}$ parallel to the line.

23. $x = 3 - t, \quad y = -1 + 2t, \quad z = -2$
24. $x = 4t, \quad y = 5 - t, \quad z = 4 + 3t$
25. $\dfrac{x-7}{4} = \dfrac{y+6}{2} = z + 2$
26. $\dfrac{x+3}{5} = \dfrac{y}{8} = \dfrac{z-3}{6}$

In Exercises 27–30, determine if any of the lines are parallel or identical.

27. L_1: $x = 6 - 3t, \quad y = -2 + 2t, \quad z = 5 + 4t$
L_2: $x = 6t, \quad y = 2 - 4t, \quad z = 13 - 8t$
L_3: $x = 10 - 6t, \quad y = 3 + 4t, \quad z = 7 + 8t$
L_4: $x = -4 + 6t, \quad y = 3 + 4t, \quad z = 5 - 6t$
28. L_1: $x = 3 + 2t, \quad y = -6t, \quad z = 1 - 2t$
L_2: $x = 1 + 2t, \quad y = -1 - t, \quad z = 3t$
L_3: $x = -1 + 2t, \quad y = 3 - 10t, \quad z = 1 - 4t$
L_4: $x = 5 + 2t, \quad y = 1 - t, \quad z = 8 + 3t$
29. L_1: $\dfrac{x-8}{4} = \dfrac{y+5}{-2} = \dfrac{z+9}{3}$
L_2: $\dfrac{x+7}{2} = \dfrac{y-4}{1} = \dfrac{z+6}{5}$
L_3: $\dfrac{x+4}{-8} = \dfrac{y-1}{4} = \dfrac{z+18}{-6}$
L_4: $\dfrac{x-2}{-2} = \dfrac{y+3}{1} = \dfrac{z-4}{1.5}$
30. L_1: $\dfrac{x-3}{2} = \dfrac{y-2}{1} = \dfrac{z+2}{2}$
L_2: $\dfrac{x-1}{4} = \dfrac{y-1}{2} = \dfrac{z+3}{4}$
L_3: $\dfrac{x+2}{1} = \dfrac{y-1}{0.5} = \dfrac{z-3}{1}$
L_4: $\dfrac{x-3}{2} = \dfrac{y+1}{4} = \dfrac{z-2}{-1}$

In Exercises 31–34, determine whether the lines intersect, and if so, find the point of intersection and the cosine of the angle of intersection.

31. $x = 4t + 2, \quad y = 3, \quad z = -t + 1$
$x = 2s + 2, \quad y = 2s + 3, \quad z = s + 1$
32. $x = -3t + 1, y = 4t + 1, z = 2t + 4$
$x = 3s + 1, y = 2s + 4, z = -s + 1$
33. $\dfrac{x}{3} = \dfrac{y-2}{-1} = z + 1, \quad \dfrac{x-1}{4} = y + 2 = \dfrac{z+3}{-3}$
34. $\dfrac{x-2}{-3} = \dfrac{y-2}{6} = z - 3, \quad \dfrac{x-3}{2} = y + 5 = \dfrac{z+2}{4}$

CAS **In Exercises 35 and 36, use a computer algebra system to graph the pair of intersecting lines and find the point of intersection.**

35. $x = 2t + 3, y = 5t - 2, z = -t + 1$
$x = -2s + 7, y = s + 8, z = 2s - 1$
36. $x = 2t - 1, y = -4t + 10, z = t$
$x = -5s - 12, y = 3s + 11, z = -2s - 4$

Cross Product **In Exercises 37 and 38, (a) find the coordinates of three points P, Q, and R in the plane, and determine the vectors $\overrightarrow{PQ}$ and $\overrightarrow{PR}$. (b) Find $\overrightarrow{PQ} \times \overrightarrow{PR}$. What is the relationship between the components of the cross product and the coefficients of the equation of the plane? Why is this true?**

37. $4x - 3y - 6z = 6$

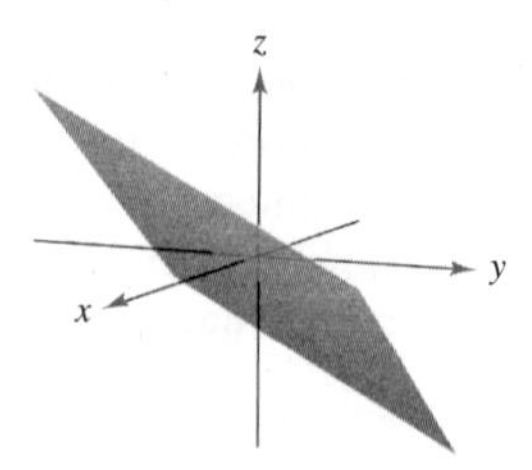

38. $2x + 3y + 4z = 4$

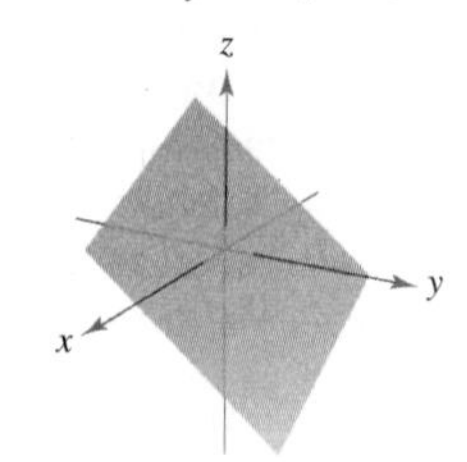

In Exercises 39 and 40, determine whether the plane passes through each point.

39. $x + 2y - 4z - 1 = 0$
(a) $(-7, 2, -1)$ (b) $(5, 2, 2)$
40. $2x + y + 3z - 6 = 0$
(a) $(3, 6, -2)$ (b) $(-1, 5, -1)$

In Exercises 41–46, find an equation of the plane passing through the point perpendicular to the given vector or line.

	Point	*Perpendicular to*
41.	$(1, 3, -7)$	$\mathbf{n} = \mathbf{j}$
42.	$(0, -1, 4)$	$\mathbf{n} = \mathbf{k}$
43.	$(3, 2, 2)$	$\mathbf{n} = 2\mathbf{i} + 3\mathbf{j} - \mathbf{k}$
44.	$(0, 0, 0)$	$\mathbf{n} = -3\mathbf{i} + 2\mathbf{k}$
45.	$(-1, 4, 0)$	$x = -1 + 2t, y = 5 - t, z = 3 - 2t$
46.	$(3, 2, 2)$	$\dfrac{x-1}{4} = y + 2 = \dfrac{z+3}{-3}$

In Exercises 47–58, find an equation of the plane.

47. The plane passes through $(0, 0, 0)$, $(2, 0, 3)$, and $(-3, -1, 5)$.

48. The plane passes through $(3, -1, 2)$, $(2, 1, 5)$, and $(1, -2, -2)$.

49. The plane passes through $(1, 2, 3)$, $(3, 2, 1)$, and $(-1, -2, 2)$.

50. The plane passes through the point $(1, 2, 3)$ and is parallel to the yz-plane.

51. The plane passes through the point $(1, 2, 3)$ and is parallel to the xy-plane.

52. The plane contains the y-axis and makes an angle of $\pi/6$ with the positive x-axis.

53. The plane contains the lines given by

$$\frac{x-1}{-2} = y - 4 = z \quad \text{and} \quad \frac{x-2}{-3} = \frac{y-1}{4} = \frac{z-2}{-1}.$$

$x = 2 - 3t$ $\quad y = 1 + 4t$ $\quad z = 2 - t$

$x = 1 - 2t$ $\quad y = 4 + t$ $\quad z = t$

54. The plane passes through the point $(2, 2, 1)$ and contains the line given by

$$\frac{x}{2} = \frac{y-4}{-1} = z.$$

55. The plane passes through the points $(2, 2, 1)$ and $(-1, 1, -1)$ and is perpendicular to the plane $2x - 3y + z = 3$.

56. The plane passes through the points $(3, 2, 1)$ and $(3, 1, -5)$ and is perpendicular to the plane $6x + 7y + 2z = 10$.

57. The plane passes through the points $(1, -2, -1)$ and $(2, 5, 6)$ and is parallel to the x-axis.

58. The plane passes through the points $(4, 2, 1)$ and $(-3, 5, 7)$ and is parallel to the z-axis.

In Exercises 59 and 60, sketch a graph of the line and find the points (if any) where the line intersects the *xy*-, *xz*-, and *yz*-planes.

59. $x = 1 - 2t, \quad y = -2 + 3t, \quad z = -4 + t$

60. $\dfrac{x-2}{3} = y + 1 = \dfrac{z-3}{2}$

In Exercises 61–64, find an equation of the plane that contains all the points that are equidistant from the given points.

61. $(2, 2, 0), \quad (0, 2, 2)$

62. $(1, 0, 2), \quad (2, 0, 1)$

63. $(-3, 1, 2), \quad (6, -2, 4)$

64. $(-5, 1, -3), \quad (2, -1, 6)$

In Exercises 65–70, determine whether the planes are parallel, orthogonal, or neither. If they are neither parallel nor orthogonal, find the angle of intersection.

65. $5x - 3y + z = 4$
$x + 4y + 7z = 1$

66. $3x + y - 4z = 3$
$-9x - 3y + 12z = 4$

67. $x - 3y + 6z = 4$
$5x + y - z = 4$

68. $3x + 2y - z = 7$
$x - 4y + 2z = 0$

69. $x - 5y - z = 1$
$5x - 25y - 5z = -3$

70. $2x - z = 1$
$4x + y + 8z = 10$

In Exercises 71–78, sketch a graph of the plane and label any intercepts.

71 $4x + 2y + 6z = 12$

72. $3x + 6y + 2z = 6$

73. $2x - y + 3z = 4$

74. $2x - y + z = 4$

75. $x + z = 6$

76. $2x + y = 8$

77. $x = 5$

78. $z = 8$

CAS **In Exercises 79–82, use a computer algebra system to graph the plane.**

79. $2x + y - z = 6$

80. $x - 3z = 3$

81. $-5x + 4y - 6z = -8$

82. $2.1x - 4.7y - z = -3$

In Exercises 83–86, determine if any of the planes are parallel or identical.

83. P_1: $15x - 6y + 24z = 17$
P_2: $-5x + 2y - 8z = 6$
P_3: $6x - 4y + 4z = 9$
P_4: $3x - 2y - 2z = 4$

84. P_1: $2x - y + 3z = 8$
P_2: $3x - 5y - 2z = 6$
P_3: $8x - 4y + 12z = 5$
P_4: $-4x - 2y + 6z = 11$

85. P_1: $3x - 2y + 5z = 10$
P_2: $-6x + 4y - 10z = 5$
P_3: $-3x + 2y + 5z = 8$
P_4: $75x - 50y + 125z = 250$

86. P_1: $-60x + 90y + 30z = 27$
P_2: $6x - 9y - 3z = 2$
P_3: $-20x + 30y + 10z = 9$
P_4: $12x - 18y + 6z = 5$

In Exercises 87–90, describe the family of planes represented by the equation, where *c* is any real number.

87. $x + y + z = c$

88. $x + y = c$

89. $cy + z = 0$

90. $x + cz = 0$

In Exercises 91 and 92, (a) find the angle between the two planes, and (b) find a set of parametric equations for the line of intersection of the planes.

91. $3x + 2y - z = 7$
$x - 4y + 2z = 0$

92. $6x - 3y + z = 5$
$-x + y + 5z = 5$

In Exercises 93–96, find the point(s) of intersection (if any) of the plane and the line. Also determine whether the line lies in the plane.

93. $2x - 2y + z = 12, \quad x - \frac{1}{2} = \frac{y + (3/2)}{-1} = \frac{z + 1}{2}$

94. $2x + 3y = -5, \quad \frac{x - 1}{4} = \frac{y}{2} = \frac{z - 3}{6}$

95. $2x + 3y = 10, \quad \frac{x - 1}{3} = \frac{y + 1}{-2} = z - 3$

96. $5x + 3y = 17, \quad \frac{x - 4}{2} = \frac{y + 1}{-3} = \frac{z + 2}{5}$

In Exercises 97–100, find the distance between the point and the plane.

97. $(0, 0, 0)$
$2x + 3y + z = 12$

98. $(0, 0, 0)$
$5x + y - z = 9$

99. $(2, 8, 4)$
$2x + y + z = 5$

100. $(1, 3, -1)$
$3x - 4y + 5z = 6$

In Exercises 101–104, verify that the two planes are parallel, and find the distance between the planes.

101. $x - 3y + 4z = 10$
$x - 3y + 4z = 6$

102. $4x - 4y + 9z = 7$
$4x - 4y + 9z = 18$

103. $-3x + 6y + 7z = 1$
$6x - 12y - 14z = 25$

104. $2x - 4z = 4$
$2x - 4z = 10$

In Exercises 105–108, find the distance between the point and the line given by the set of parametric equations.

105. $(1, 5, -2); \quad x = 4t - 2, \quad y = 3, \quad z = -t + 1$

106. $(1, -2, 4); \quad x = 2t, \quad y = t - 3, \quad z = 2t + 2$

107. $(-2, 1, 3); \quad x = 1 - t, \quad y = 2 + t, \quad z = -2t$

108. $(4, -1, 5); \quad x = 3, \quad y = 1 + 3t, \quad z = 1 + t$

In Exercises 109 and 110, verify that the lines are parallel, and find the distance between them.

109. L_1: $x = 2 - t, \quad y = 3 + 2t, \quad z = 4 + t$
L_2: $x = 3t, \quad y = 1 - 6t, \quad z = 4 - 3t$

110. L_1: $x = 3 + 6t, \quad y = -2 + 9t, \quad z = 1 - 12t$
L_2: $x = -1 + 4t, \quad y = 3 + 6t, \quad z = -8t$

WRITING ABOUT CONCEPTS

111. Give the parametric equations and the symmetric equations of a line in space. Describe what is required to find these equations.

112. Give the standard equation of a plane in space. Describe what is required to find this equation.

113. Describe a method of finding the line of intersection of two planes.

114. Describe each surface given by the equations $x = a$, $y = b$, and $z = c$.

WRITING ABOUT CONCEPTS (continued)

115. Describe a method for determining when two planes

$a_1x + b_1y + c_1z + d_1 = 0$ and

$a_2x + b_2y + c_2z + d_2 = 0$

are (a) parallel and (b) perpendicular. Explain your reasoning.

116. Let L_1 and L_2 be nonparallel lines that do not intersect. Is it possible to find a nonzero vector $\mathbf{v}$ such that $\mathbf{v}$ is perpendicular to both L_1 and L_2? Explain your reasoning.

117. Find an equation of the plane with x-intercept $(a, 0, 0)$, y-intercept $(0, b, 0)$, and z-intercept $(0, 0, c)$. (Assume a, b, and c are nonzero.)

CAPSTONE

118. Match the equation or set of equations with the description it represents.

(a) Set of parametric equations of a line

(b) Set of symmetric equations of a line

(c) Standard equation of a plane in space

(d) General form of an equation of a plane in space

(i) $(x - 6)/2 = (y + 1)/-3 = z/1$

(ii) $2x - 7y + 5z + 10 = 0$

(iii) $x = 4 + 7t, y = 3 + t, z = 3 - 3t$

(iv) $2(x - 1) + (y + 3) - 4(z - 5) = 0$

119. Describe and find an equation for the surface generated by all points (x, y, z) that are four units from the point $(3, -2, 5)$.

120. Describe and find an equation for the surface generated by all points (x, y, z) that are four units from the plane $4x - 3y + z = 10$.

121. ***Modeling Data*** Per capita consumptions (in gallons) of different types of milk in the United States from 1999 through 2005 are shown in the table. Consumptions of flavored milk, plain reduced-fat milk, and plain light and skim milks are represented by the variables x, y, and z, respectively. *(Source: U.S. Department of Agriculture)*

Year	1999	2000	2001	2002	2003	2004	2005
x	1.4	1.4	1.4	1.6	1.6	1.7	1.7
y	7.3	7.1	7.0	7.0	6.9	6.9	6.9
z	6.2	6.1	5.9	5.8	5.6	5.5	5.6

A model for the data is given by $0.92x - 1.03y + z = 0.02$.

(a) Complete a fourth row in the table using the model to approximate z for the given values of x and y. Compare the approximations with the actual values of z.

(b) According to this model, any increases in consumption of two types of milk will have what effect on the consumption of the third type?

122. ***Mechanical Design*** The figure shows a chute at the top of a grain elevator of a combine that funnels the grain into a bin. Find the angle between two adjacent sides.

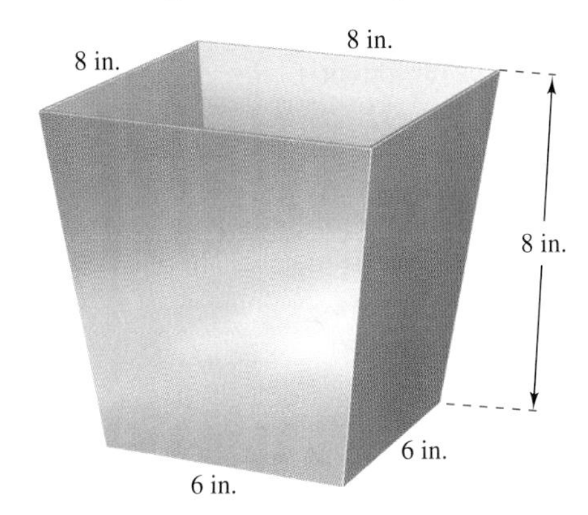

123. ***Distance*** Two insects are crawling along different lines in three-space. At time t (in minutes), the first insect is at the point (x, y, z) on the line $x = 6 + t$, $y = 8 - t$, $z = 3 + t$. Also, at time t, the second insect is at the point (x, y, z) on the line $x = 1 + t$, $y = 2 + t$, $z = 2t$.

Assume that distances are given in inches.

(a) Find the distance between the two insects at time $t = 0$.

 (b) Use a graphing utility to graph the distance between the insects from $t = 0$ to $t = 10$.

(c) Using the graph from part (b), what can you conclude about the distance between the insects?

(d) How close to each other do the insects get?

124. Find the standard equation of the sphere with center $(-3, 2, 4)$ that is tangent to the plane given by $2x + 4y - 3z = 8$.

125. Find the point of intersection of the plane $3x - y + 4z = 7$ and the line through $(5, 4, -3)$ that is perpendicular to this plane.

126. Show that the plane $2x - y - 3z = 4$ is parallel to the line $x = -2 + 2t$, $y = -1 + 4t$, $z = 4$, and find the distance between them.

127. Find the point of intersection of the line through $(1, -3, 1)$ and $(3, -4, 2)$, and the plane given by $x - y + z = 2$.

128. Find a set of parametric equations for the line passing through the point $(1, 0, 2)$ that is parallel to the plane given by $x + y + z = 5$, and perpendicular to the line $x = t$, $y = 1 + t$, $z = 1 + t$.

True or False? **In Exercises 129–134, determine whether the statement is true or false. If it is false, explain why or give an example that shows it is false.**

129. If $\mathbf{v} = a_1\mathbf{i} + b_1\mathbf{j} + c_1\mathbf{k}$ is any vector in the plane given by $a_2x + b_2y + c_2z + d_2 = 0$, then $a_1a_2 + b_1b_2 + c_1c_2 = 0$.

130. Every two lines in space are either intersecting or parallel.

131. Two planes in space are either intersecting or parallel.

132. If two lines L_1 and L_2 are parallel to a plane P, then L_1 and L_2 are parallel.

133. Two planes perpendicular to a third plane in space are parallel.

134. A plane and a line in space are either intersecting or parallel.

SECTION PROJECT

Distances in Space

You have learned two distance formulas in this section—the distance between a point and a plane, and the distance between a point and a line. In this project you will study a third distance problem—the distance between two skew lines. Two lines in space are *skew* if they are neither parallel nor intersecting (see figure).

(a) Consider the following two lines in space.

L_1: $x = 4 + 5t$, $y = 5 + 5t$, $z = 1 - 4t$

L_2: $x = 4 + s$, $y = -6 + 8s$, $z = 7 - 3s$

(i) Show that these lines are not parallel.

(ii) Show that these lines do not intersect, and therefore are skew lines.

(iii) Show that the two lines lie in parallel planes.

(iv) Find the distance between the parallel planes from part (iii). This is the distance between the original skew lines.

(b) Use the procedure in part (a) to find the distance between the lines.

L_1: $x = 2t$, $y = 4t$, $z = 6t$

L_2: $x = 1 - s$, $y = 4 + s$, $z = -1 + s$

(c) Use the procedure in part (a) to find the distance between the lines.

L_1: $x = 3t$, $y = 2 - t$, $z = -1 + t$

L_2: $x = 1 + 4s$, $y = -2 + s$, $z = -3 - 3s$

(d) Develop a formula for finding the distance between the skew lines.

L_1: $x = x_1 + a_1t$, $y = y_1 + b_1t$, $z = z_1 + c_1t$

L_2: $x = x_2 + a_2s$, $y = y_2 + b_2s$, $z = z_2 + c_2s$

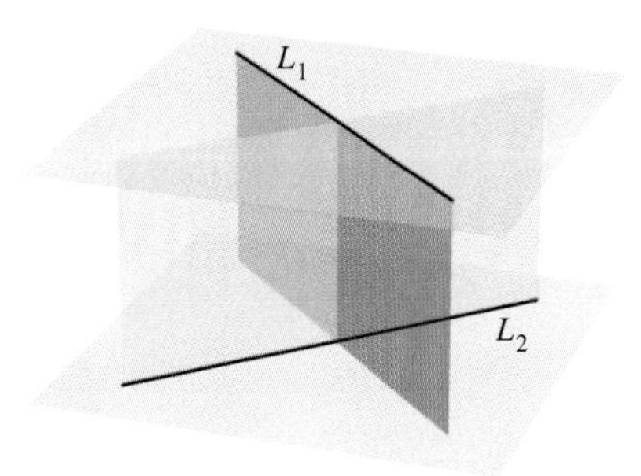

11.6 Surfaces in Space

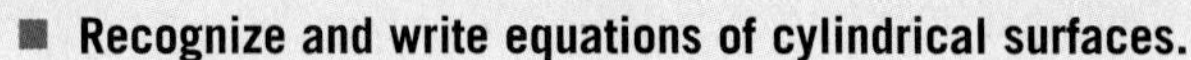

- **Recognize and write equations of cylindrical surfaces.**
- **Recognize and write equations of quadric surfaces.**
- **Recognize and write equations of surfaces of revolution.**

Cylindrical Surfaces

The first five sections of this chapter contained the vector portion of the preliminary work necessary to study vector calculus and the calculus of space. In this and the next section, you will study surfaces in space and alternative coordinate systems for space. You have already studied two special types of surfaces.

1. Spheres: $(x - x_0)^2 + (y - y_0)^2 + (z - z_0)^2 = r^2$ — Section 11.2
2. Planes: $ax + by + cz + d = 0$ — Section 11.5

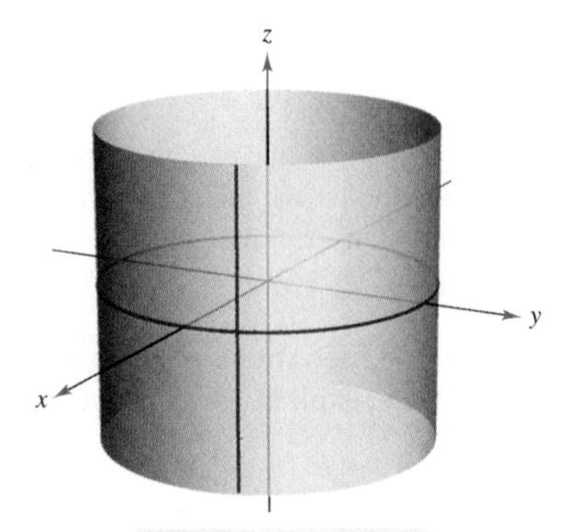

Right circular cylinder: $x^2 + y^2 = a^2$

Rulings are parallel to z-axis.
Figure 11.56

A third type of surface in space is called a **cylindrical surface,** or simply a **cylinder.** To define a cylinder, consider the familiar right circular cylinder shown in Figure 11.56. You can imagine that this cylinder is generated by a vertical line moving around the circle $x^2 + y^2 = a^2$ in the xy-plane. This circle is called a **generating curve** for the cylinder, as indicated in the following definition.

DEFINITION OF A CYLINDER

Let C be a curve in a plane and let L be a line not in a parallel plane. The set of all lines parallel to L and intersecting C is called a **cylinder.** C is called the **generating curve** (or **directrix**) of the cylinder, and the parallel lines are called **rulings.**

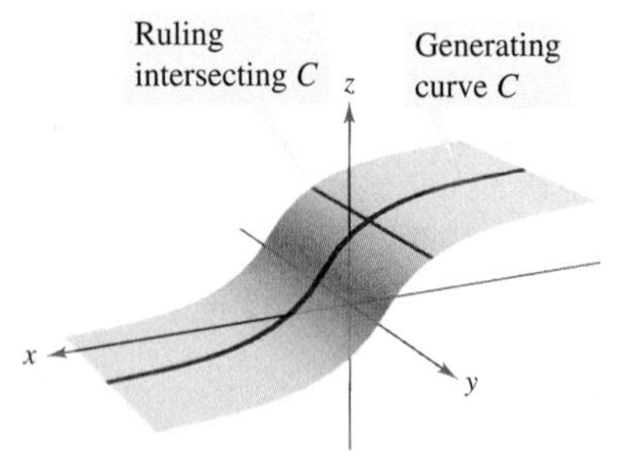

Cylinder: Rulings intersect C and are parallel to the given line.
Figure 11.57

NOTE Without loss of generality, you can assume that C lies in one of the three coordinate planes. Moreover, this text restricts the discussion to *right* cylinders—cylinders whose rulings are perpendicular to the coordinate plane containing C, as shown in Figure 11.57. ■

For the right circular cylinder shown in Figure 11.56, the equation of the generating curve is

$$x^2 + y^2 = a^2.$$ Equation of generating curve in xy-plane

To find an equation of the cylinder, note that you can generate any one of the rulings by fixing the values of x and y and then allowing z to take on all real values. In this sense, the value of z is arbitrary and is, therefore, not included in the equation. In other words, the equation of this cylinder is simply the equation of its generating curve.

$$x^2 + y^2 = a^2$$ Equation of cylinder in space

EQUATIONS OF CYLINDERS

The equation of a cylinder whose rulings are parallel to one of the coordinate axes contains only the variables corresponding to the other two axes.

EXAMPLE 1 Sketching a Cylinder

Sketch the surface represented by each equation.

a. $z = y^2$ **b.** $z = \sin x, \quad 0 \le x \le 2\pi$

Solution

a. The graph is a cylinder whose generating curve, $z = y^2$, is a parabola in the yz-plane. The rulings of the cylinder are parallel to the x-axis, as shown in Figure 11.58(a).

b. The graph is a cylinder generated by the sine curve in the xz-plane. The rulings are parallel to the y-axis, as shown in Figure 11.58(b).

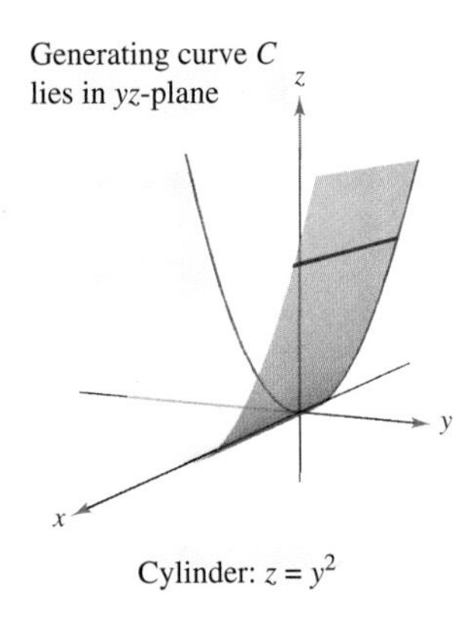

Cylinder: $z = y^2$

(a) Rulings are parallel to x-axis.

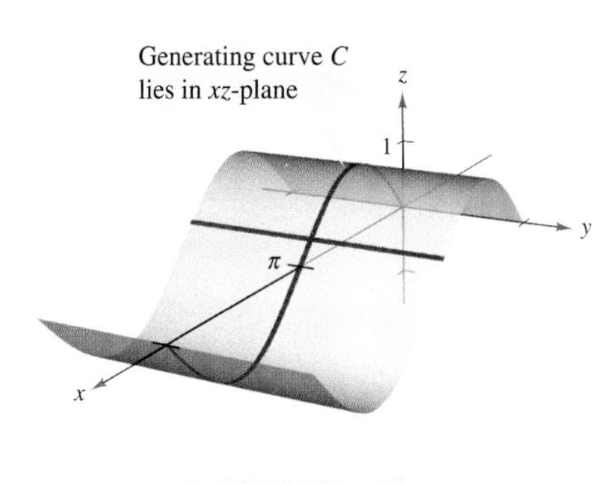

Cylinder: $z = \sin x$

(b) Rulings are parallel to y-axis.

Figure 11.58

■

Quadric Surfaces

STUDY TIP In the table on pages 814 and 815, only one of several orientations of each quadric surface is shown. If the surface is oriented along a different axis, its standard equation will change accordingly, as illustrated in Examples 2 and 3. The fact that the two types of paraboloids have one variable raised to the first power can be helpful in classifying quadric surfaces. The other four types of basic quadric surfaces have equations that are of *second degree* in all three variables.

The fourth basic type of surface in space is a **quadric surface.** Quadric surfaces are the three-dimensional analogs of conic sections.

QUADRIC SURFACE

The equation of a **quadric surface** in space is a second-degree equation in three variables. The **general form** of the equation is

$$Ax^2 + By^2 + Cz^2 + Dxy + Exz + Fyz + Gx + Hy + Iz + J = 0.$$

There are six basic types of quadric surfaces: **ellipsoid, hyperboloid of one sheet, hyperboloid of two sheets, elliptic cone, elliptic paraboloid, and hyperbolic paraboloid.**

The intersection of a surface with a plane is called the **trace of the surface** in the plane. To visualize a surface in space, it is helpful to determine its traces in some well-chosen planes. The traces of quadric surfaces are conics. These traces, together with the **standard form** of the equation of each quadric surface, are shown in the table on pages 814 and 815.

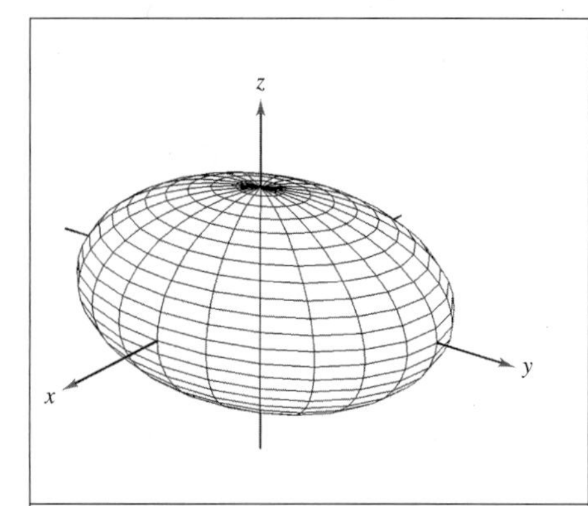

Ellipsoid

$$\frac{x^2}{a^2} + \frac{y^2}{b^2} + \frac{z^2}{c^2} = 1$$

Trace	*Plane*
Ellipse	Parallel to xy-plane
Ellipse	Parallel to xz-plane
Ellipse	Parallel to yz-plane

The surface is a sphere if $a = b = c \neq 0$.

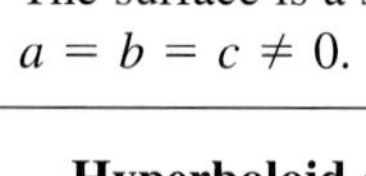

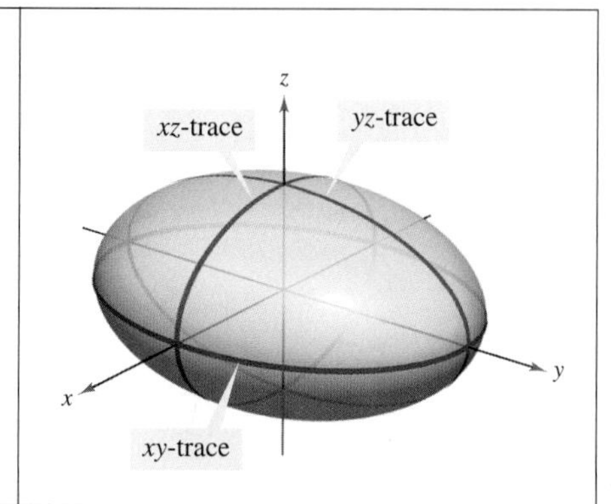

Hyperboloid of One Sheet

$$\frac{x^2}{a^2} + \frac{y^2}{b^2} - \frac{z^2}{c^2} = 1$$

Trace	*Plane*
Ellipse	Parallel to xy-plane
Hyperbola	Parallel to xz-plane
Hyperbola	Parallel to yz-plane

The axis of the hyperboloid corresponds to the variable whose coefficient is negative.

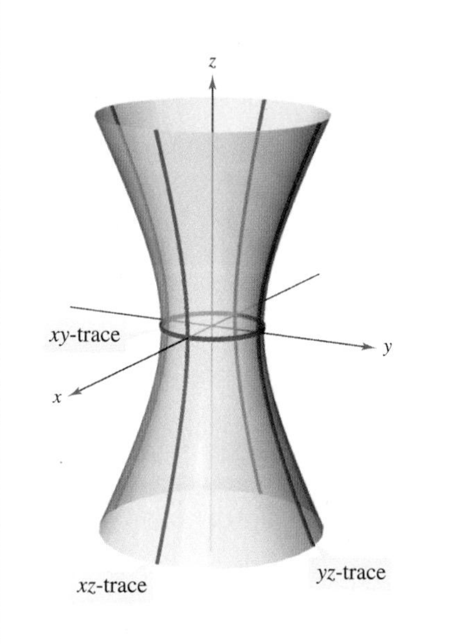

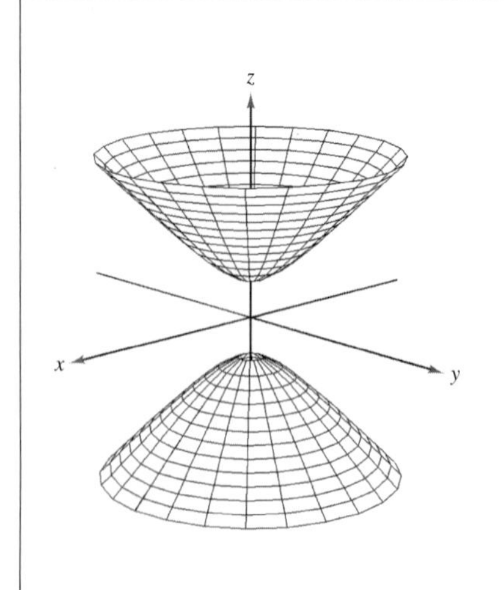

Hyperboloid of Two Sheets

$$\frac{z^2}{c^2} - \frac{x^2}{a^2} - \frac{y^2}{b^2} = 1$$

Trace	*Plane*
Ellipse	Parallel to xy-plane
Hyperbola	Parallel to xz-plane
Hyperbola	Parallel to yz-plane

The axis of the hyperboloid corresponds to the variable whose coefficient is positive. There is no trace in the coordinate plane perpendicular to this axis.

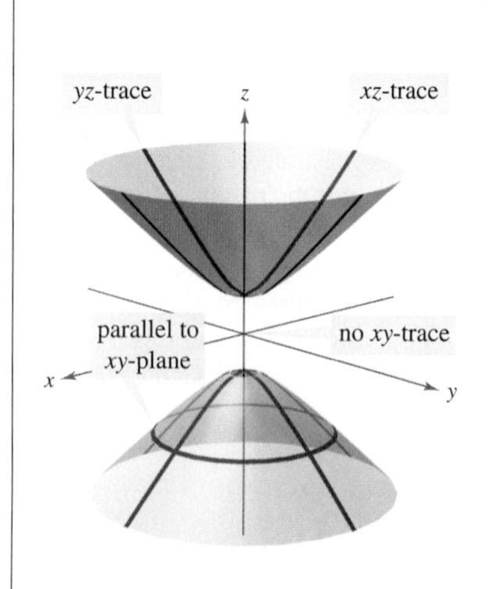

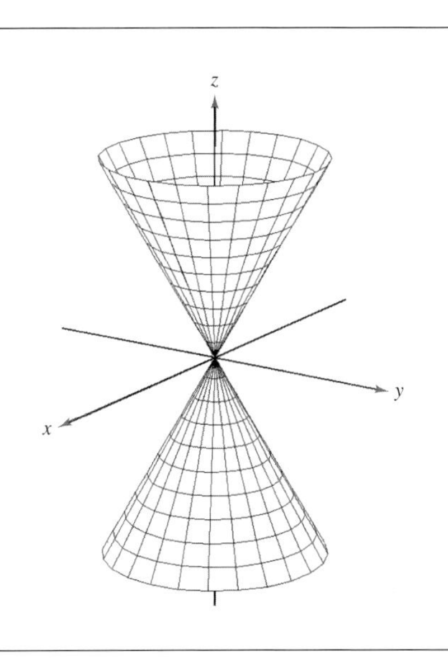

Elliptic Cone

$$\frac{x^2}{a^2} + \frac{y^2}{b^2} - \frac{z^2}{c^2} = 0$$

Trace	*Plane*
Ellipse	Parallel to xy-plane
Hyperbola	Parallel to xz-plane
Hyperbola	Parallel to yz-plane

The axis of the cone corresponds to the variable whose coefficient is negative. The traces in the coordinate planes parallel to this axis are intersecting lines.

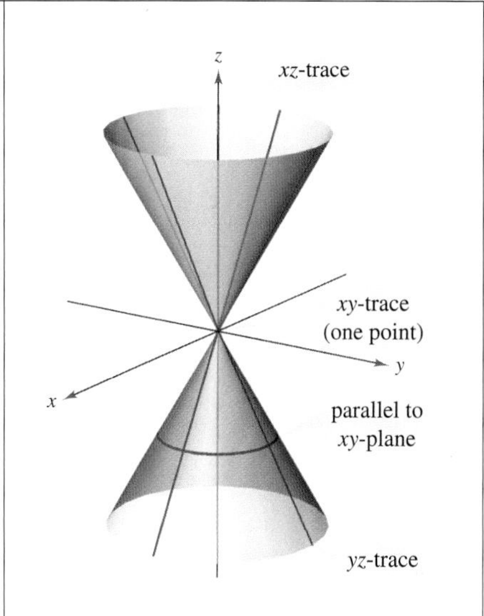

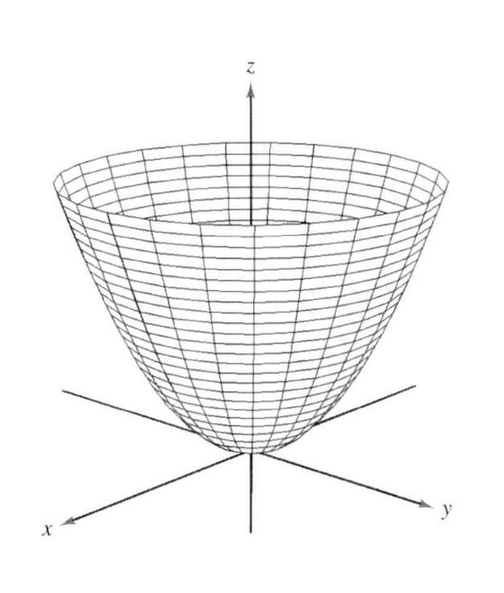

Elliptic Paraboloid

$$z = \frac{x^2}{a^2} + \frac{y^2}{b^2}$$

Trace	*Plane*
Ellipse	Parallel to xy-plane
Parabola	Parallel to xz-plane
Parabola	Parallel to yz-plane

The axis of the paraboloid corresponds to the variable raised to the first power.

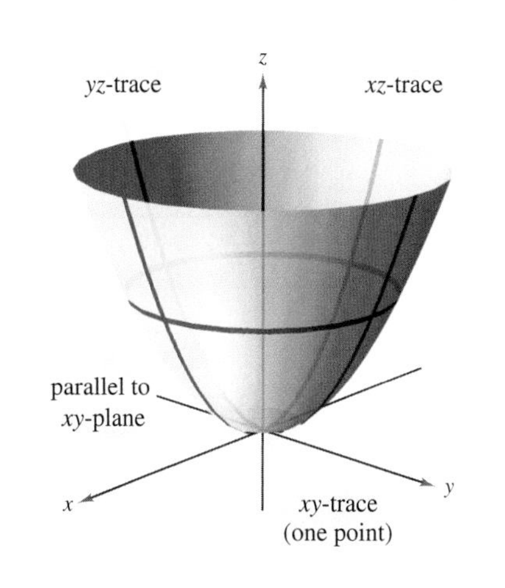

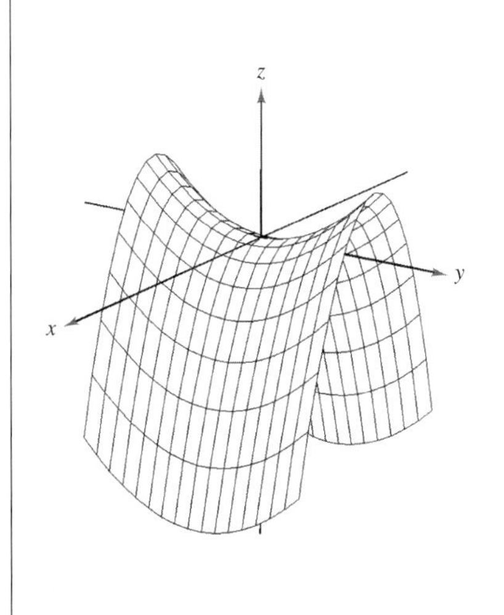

Hyperbolic Paraboloid

$$z = \frac{y^2}{b^2} - \frac{x^2}{a^2}$$

Trace	*Plane*
Hyperbola	Parallel to xy-plane
Parabola	Parallel to xz-plane
Parabola	Parallel to yz-plane

The axis of the paraboloid corresponds to the variable raised to the first power.

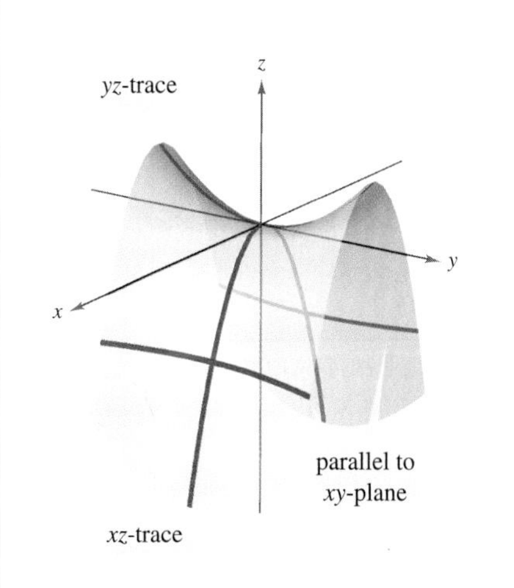

To classify a quadric surface, begin by writing the surface in standard form. Then, determine several traces taken in the coordinate planes *or* taken in planes that are parallel to the coordinate planes.

EXAMPLE 2 Sketching a Quadric Surface

Classify and sketch the surface given by $4x^2 - 3y^2 + 12z^2 + 12 = 0$.

Solution Begin by writing the equation in standard form.

$$4x^2 - 3y^2 + 12z^2 + 12 = 0 \qquad \text{Write original equation.}$$

$$\frac{x^2}{-3} + \frac{y^2}{4} - z^2 - 1 = 0 \qquad \text{Divide by } -12.$$

$$\frac{y^2}{4} - \frac{x^2}{3} - \frac{z^2}{1} = 1 \qquad \text{Standard form}$$

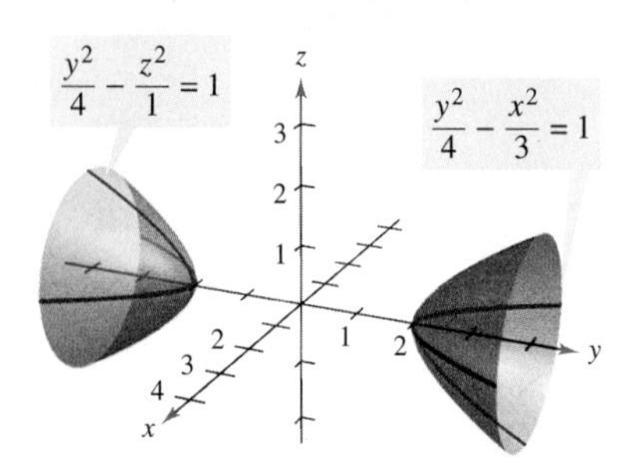

Figure 11.59

From the table on pages 814 and 815, you can conclude that the surface is a hyperboloid of two sheets with the y-axis as its axis. To sketch the graph of this surface, it helps to find the traces in the coordinate planes.

$$xy\text{-trace } (z = 0): \quad \frac{y^2}{4} - \frac{x^2}{3} = 1 \qquad \text{Hyperbola}$$

$$xz\text{-trace } (y = 0): \quad \frac{x^2}{3} + \frac{z^2}{1} = -1 \qquad \text{No trace}$$

$$yz\text{-trace } (x = 0): \quad \frac{y^2}{4} - \frac{z^2}{1} = 1 \qquad \text{Hyperbola}$$

The graph is shown in Figure 11.59.

EXAMPLE 3 Sketching a Quadric Surface

Classify and sketch the surface given by $x - y^2 - 4z^2 = 0$.

Solution Because x is raised only to the first power, the surface is a paraboloid. The axis of the paraboloid is the x-axis. In the standard form, the equation is

$$x = y^2 + 4z^2. \qquad \text{Standard form}$$

Some convenient traces are as follows.

$$xy\text{-trace } (z = 0): \quad x = y^2 \qquad \text{Parabola}$$

$$xz\text{-trace } (y = 0): \quad x = 4z^2 \qquad \text{Parabola}$$

$$\text{parallel to } yz\text{-plane } (x = 4): \quad \frac{y^2}{4} + \frac{z^2}{1} = 1 \qquad \text{Ellipse}$$

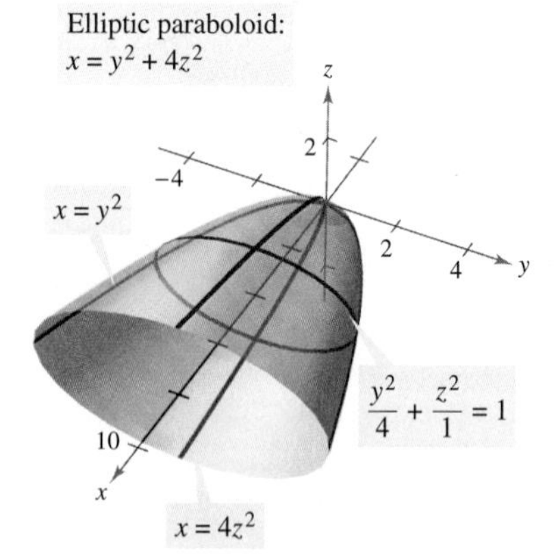

Figure 11.60

The surface is an *elliptic* paraboloid, as shown in Figure 11.60. ■

Some second-degree equations in x, y, and z do not represent any of the basic types of quadric surfaces. Here are two examples.

$$x^2 + y^2 + z^2 = 0 \qquad \text{Single point}$$

$$x^2 + y^2 = 1 \qquad \text{Right circular cylinder}$$

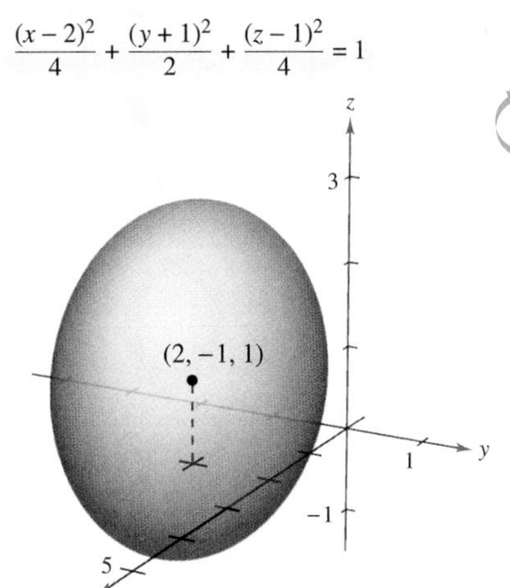

An ellipsoid centered at $(2, -1, 1)$
Figure 11.61

For a quadric surface not centered at the origin, you can form the standard equation by completing the square, as demonstrated in Example 4.

EXAMPLE 4 A Quadric Surface Not Centered at the Origin

Classify and sketch the surface given by

$$x^2 + 2y^2 + z^2 - 4x + 4y - 2z + 3 = 0.$$

Solution Completing the square for each variable produces the following.

$$(x^2 - 4x + \quad) + 2(y^2 + 2y + \quad) + (z^2 - 2z + \quad) = -3$$

$$(x^2 - 4x + 4) + 2(y^2 + 2y + 1) + (z^2 - 2z + 1) = -3 + 4 + 2 + 1$$

$$(x - 2)^2 + 2(y + 1)^2 + (z - 1)^2 = 4$$

$$\frac{(x - 2)^2}{4} + \frac{(y + 1)^2}{2} + \frac{(z - 1)^2}{4} = 1$$

From this equation, you can see that the quadric surface is an ellipsoid that is centered at $(2, -1, 1)$. Its graph is shown in Figure 11.61. ■

TECHNOLOGY A computer algebra system can help you visualize a surface in space.* Most of these computer algebra systems create three-dimensional illusions by sketching several traces of the surface and then applying a "hidden-line" routine that blocks out portions of the surface that lie behind other portions of the surface. Two examples of figures that were generated by *Mathematica* are shown below.

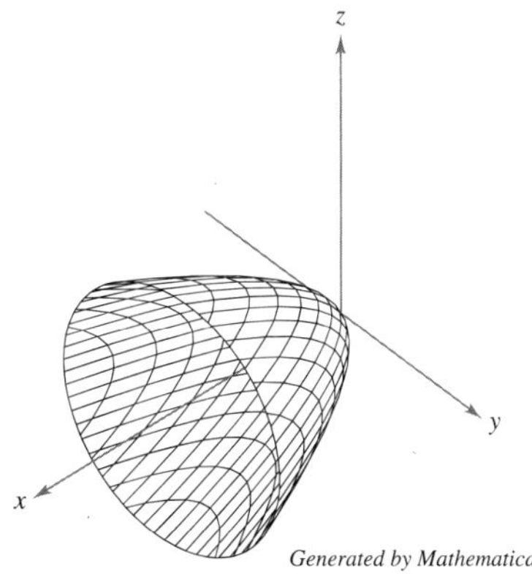

Elliptic paraboloid

$$x = \frac{y^2}{2} + \frac{z^2}{2}$$

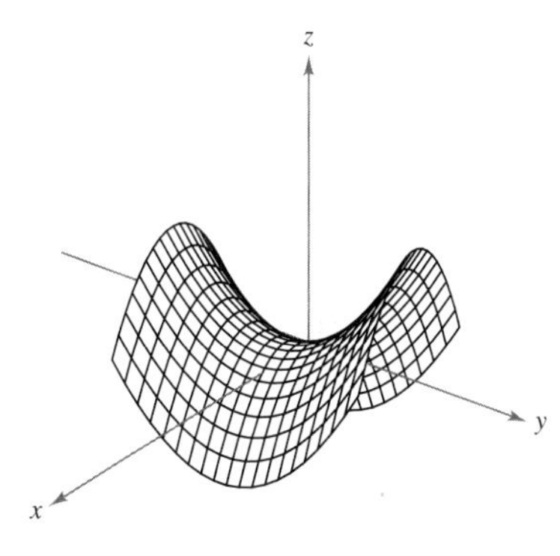

Hyperbolic paraboloid

$$z = \frac{y^2}{16} - \frac{x^2}{16}$$

Using a graphing utility to graph a surface in space requires practice. For one thing, you must know enough about the surface to be able to specify a *viewing window* that gives a representative view of the surface. Also, you can often improve the view of a surface by rotating the axes. For instance, note that the elliptic paraboloid in the figure is seen from a line of sight that is "higher" than the line of sight used to view the hyperbolic paraboloid.

**Some 3-D graphing utilities require surfaces to be entered with parametric equations. For a discussion of this technique, see Section 15.5.*

Surfaces of Revolution

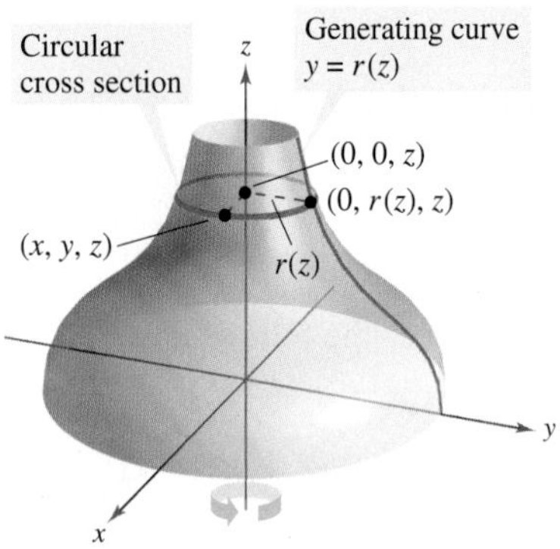

The fifth special type of surface you will study is called a **surface of revolution.** In Section 7.4, you studied a method for finding the *area* of such a surface. You will now look at a procedure for finding its *equation*. Consider the graph of the **radius function**

$$y = r(z) \qquad \text{Generating curve}$$

in the yz-plane. If this graph is revolved about the z-axis, it forms a surface of revolution, as shown in Figure 11.62. The trace of the surface in the plane $z = z_0$ is a circle whose radius is $r(z_0)$ and whose equation is

$$x^2 + y^2 = [r(z_0)]^2. \qquad \text{Circular trace in plane: } z = z_0$$

Replacing z_0 with z produces an equation that is valid for all values of z. In a similar manner, you can obtain equations for surfaces of revolution for the other two axes, and the results are summarized as follows.

Figure 11.62

SURFACE OF REVOLUTION

If the graph of a radius function r is revolved about one of the coordinate axes, the equation of the resulting surface of revolution has one of the following forms.

1. Revolved about the x-axis: $y^2 + z^2 = [r(x)]^2$
2. Revolved about the y-axis: $x^2 + z^2 = [r(y)]^2$
3. Revolved about the z-axis: $x^2 + y^2 = [r(z)]^2$

EXAMPLE 5 Finding an Equation for a Surface of Revolution

a. An equation for the surface of revolution formed by revolving the graph of

$$y = \frac{1}{z} \qquad \text{Radius function}$$

about the z-axis is

$$x^2 + y^2 = [r(z)]^2 \qquad \text{Revolved about the } z\text{-axis}$$

$$x^2 + y^2 = \left(\frac{1}{z}\right)^2. \qquad \text{Substitute } 1/z \text{ for } r(z).$$

b. To find an equation for the surface formed by revolving the graph of $9x^2 = y^3$ about the y-axis, solve for x in terms of y to obtain

$$x = \tfrac{1}{3}y^{3/2} = r(y). \qquad \text{Radius function}$$

So, the equation for this surface is

$$x^2 + z^2 = [r(y)]^2 \qquad \text{Revolved about the } y\text{-axis}$$

$$x^2 + z^2 = \left(\tfrac{1}{3}y^{3/2}\right)^2 \qquad \text{Substitute } \tfrac{1}{3}y^{3/2} \text{ for } r(y).$$

$$x^2 + z^2 = \tfrac{1}{9}y^3. \qquad \text{Equation of surface}$$

The graph is shown in Figure 11.63. ■

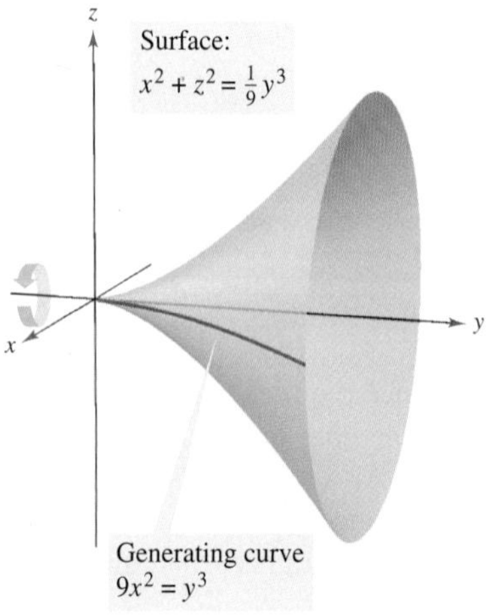

Figure 11.63

The generating curve for a surface of revolution is not unique. For instance, the surface

$$x^2 + z^2 = e^{-2y}$$

can be formed by revolving either the graph of $x = e^{-y}$ about the y-axis or the graph of $z = e^{-y}$ about the y-axis, as shown in Figure 11.64.

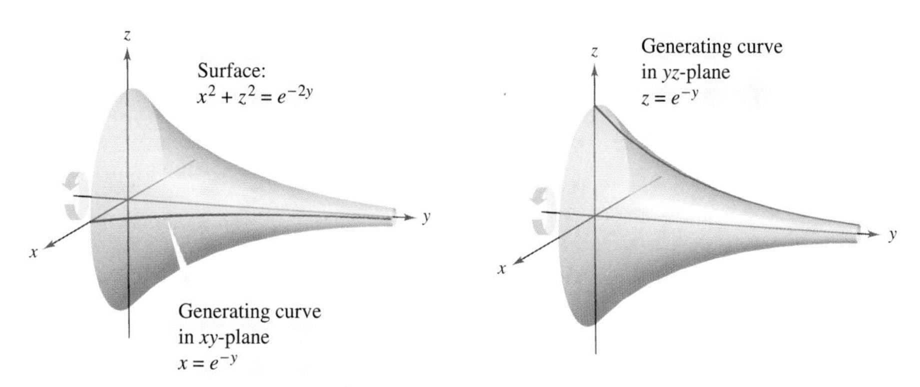

Figure 11.64

EXAMPLE 6 Finding a Generating Curve for a Surface of Revolution

Find a generating curve and the axis of revolution for the surface given by

$$x^2 + 3y^2 + z^2 = 9.$$

Solution You now know that the equation has one of the following forms.

$$x^2 + y^2 = [r(z)]^2 \quad \text{Revolved about } z\text{-axis}$$
$$y^2 + z^2 = [r(x)]^2 \quad \text{Revolved about } x\text{-axis}$$
$$x^2 + z^2 = [r(y)]^2 \quad \text{Revolved about } y\text{-axis}$$

Because the coefficients of x^2 and z^2 are equal, you should choose the third form and write

$$x^2 + z^2 = 9 - 3y^2.$$

The y-axis is the axis of revolution. You can choose a generating curve from either of the following traces.

$$x^2 = 9 - 3y^2 \quad \text{Trace in } xy\text{-plane}$$
$$z^2 = 9 - 3y^2 \quad \text{Trace in } yz\text{-plane}$$

For example, using the first trace, the generating curve is the semiellipse given by

$$x = \sqrt{9 - 3y^2}. \quad \text{Generating curve}$$

The graph of this surface is shown in Figure 11.65. ■

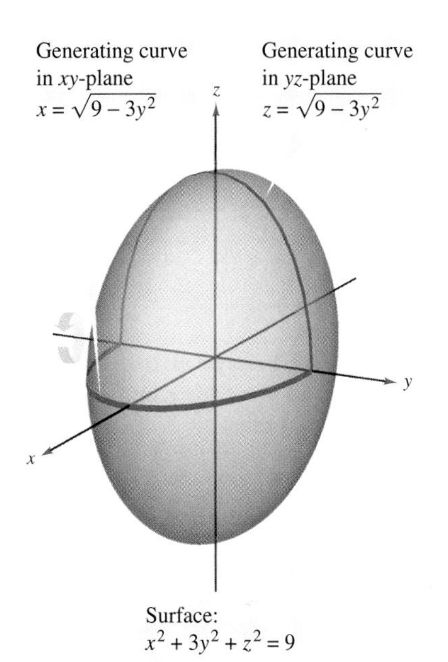

Figure 11.65

11.6 Exercises

See www.CalcChat.com for worked-out solutions to odd-numbered exercises.

In Exercises 1–6, match the equation with its graph. [The graphs are labeled (a), (b), (c), (d), (e), and (f).]

(a)

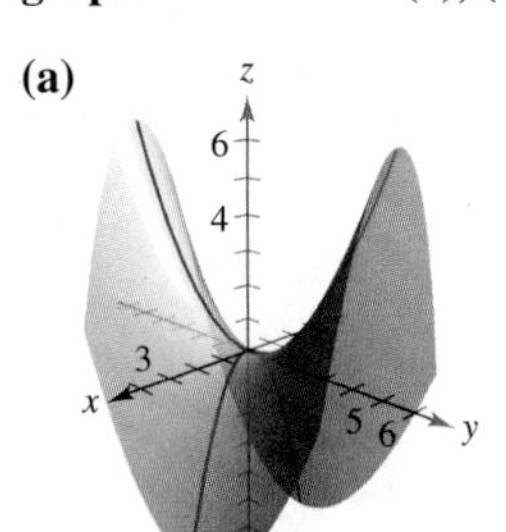

(b)

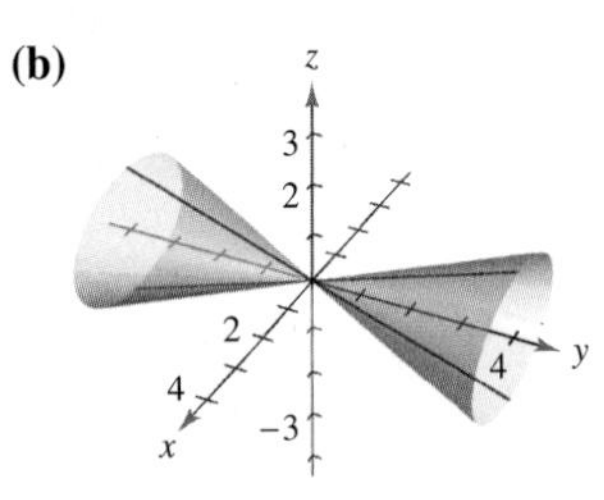

(c)

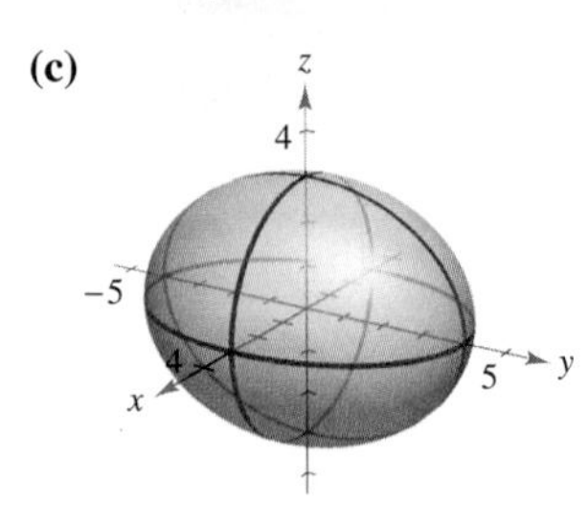

(d)

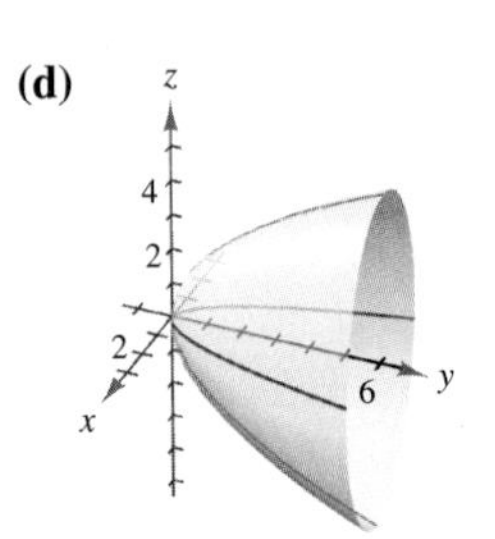

(e)

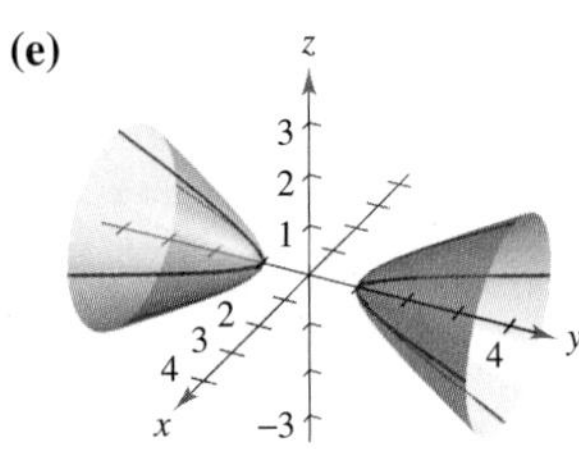

(f)

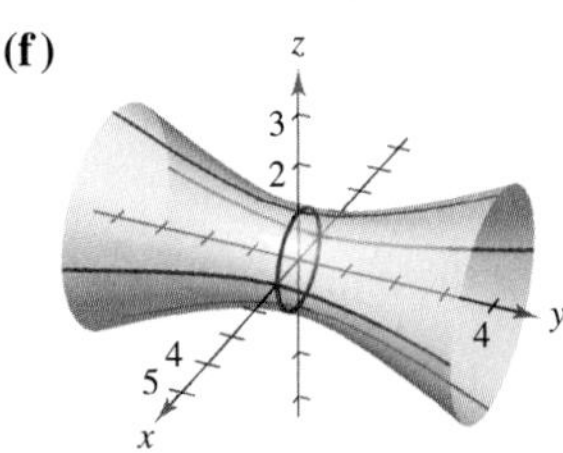

1. $\frac{x^2}{9} + \frac{y^2}{16} + \frac{z^2}{9} = 1$

2. $15x^2 - 4y^2 + 15z^2 = -4$

3. $4x^2 - y^2 + 4z^2 = 4$

4. $y^2 = 4x^2 + 9z^2$

5. $4x^2 - 4y + z^2 = 0$

6. $4x^2 - y^2 + 4z = 0$

In Exercises 7–16, describe and sketch the surface.

7. $y = 5$

8. $z = 2$

9. $y^2 + z^2 = 9$

10. $x^2 + z^2 = 25$

11. $x^2 - y = 0$

12. $y^2 + z = 6$

13. $4x^2 + y^2 = 4$

14. $y^2 - z^2 = 16$

15. $z - \sin y = 0$

16. $z - e^y = 0$

17. ***Think About It*** The four figures are graphs of the quadric surface $z = x^2 + y^2$. Match each of the four graphs with the point in space from which the paraboloid is viewed. The four points are $(0, 0, 20)$, $(0, 20, 0)$, $(20, 0, 0)$, and $(10, 10, 20)$.

(a)

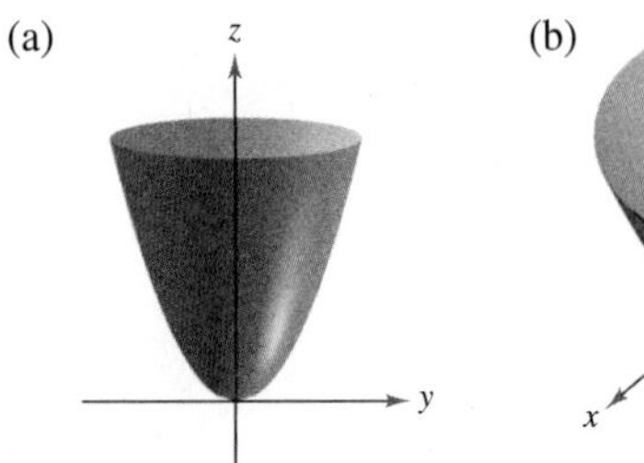

(b)

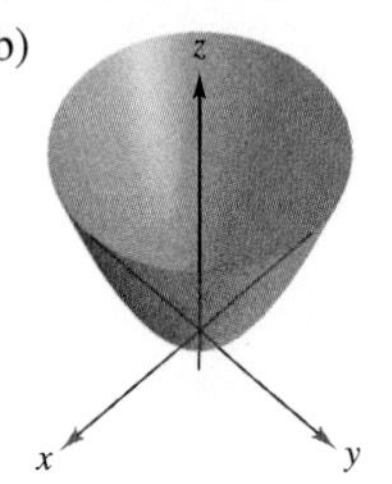

(c) (d)

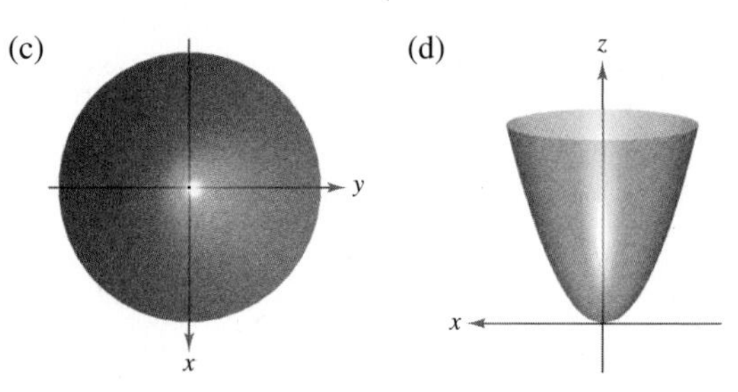

Figures for 17

18. Use a computer algebra system to graph a view of the cylinder $y^2 + z^2 = 4$ from each point.
    (a) $(10, 0, 0)$
    (b) $(0, 10, 0)$
    (c) $(10, 10, 10)$

In Exercises 19–32, identify and sketch the quadric surface. Use a computer algebra system to confirm your sketch.

19. $x^2 + \frac{y^2}{4} + z^2 = 1$

20. $\frac{x^2}{16} + \frac{y^2}{25} + \frac{z^2}{25} = 1$

21. $16x^2 - y^2 + 16z^2 = 4$

22. $-8x^2 + 18y^2 + 18z^2 = 2$

23. $4x^2 - y^2 - z^2 = 1$

24. $z^2 - x^2 - \frac{y^2}{4} = 1$

25. $x^2 - y + z^2 = 0$

26. $z = x^2 + 4y^2$

27. $x^2 - y^2 + z = 0$

28. $3z = -y^2 + x^2$

29. $z^2 = x^2 + \frac{y^2}{9}$

30. $x^2 = 2y^2 + 2z^2$

31. $16x^2 + 9y^2 + 16z^2 - 32x - 36y + 36 = 0$

32. $9x^2 + y^2 - 9z^2 - 54x - 4y - 54z + 4 = 0$

In Exercises 33–42, use a computer algebra system to graph the surface. (*Hint:* It may be necessary to solve for z and acquire two equations to graph the surface.)

33. $z = 2 \cos x$

34. $z = x^2 + 0.5y^2$

35. $z^2 = x^2 + 7.5y^2$

36. $3.25y = x^2 + z^2$

37. $x^2 + y^2 = \left(\frac{2}{z}\right)^2$

38. $x^2 + y^2 = e^{-z}$

39. $z = 10 - \sqrt{|xy|}$

40. $z = \frac{-x}{8 + x^2 + y^2}$

41. $6x^2 - 4y^2 + 6z^2 = -36$

42. $9x^2 + 4y^2 - 8z^2 = 72$

In Exercises 43–46, sketch the region bounded by the graphs of the equations.

43. $z = 2\sqrt{x^2 + y^2}, \; z = 2$

44. $z = \sqrt{4 - x^2}, \; y = \sqrt{4 - x^2}, \; x = 0, \; y = 0, \; z = 0$

45. $x^2 + y^2 = 1, \; x + z = 2, \; z = 0$

46. $z = \sqrt{4 - x^2 - y^2}, \; y = 2z, \; z = 0$

In Exercises 47–52, find an equation for the surface of revolution generated by revolving the curve in the indicated coordinate plane about the given axis.

	Equation of Curve	*Coordinate Plane*	*Axis of Revolution*
47.	$z^2 = 4y$	yz-plane	y-axis
48.	$z = 3y$	yz-plane	y-axis
49.	$z = 2y$	yz-plane	z-axis
50.	$2z = \sqrt{4 - x^2}$	xz-plane	x-axis
51.	$xy = 2$	xy-plane	x-axis
52.	$z = \ln y$	yz-plane	z-axis

In Exercises 53 and 54, find an equation of a generating curve given the equation of its surface of revolution.

53. $x^2 + y^2 - 2z = 0$ **54.** $x^2 + z^2 = \cos^2 y$

WRITING ABOUT CONCEPTS

55. State the definition of a cylinder.

56. What is meant by the trace of a surface? How do you find a trace?

57. Identify the six quadric surfaces and give the standard form of each.

CAPSTONE

58. What does the equation $z = x^2$ represent in the xz-plane? What does it represent in three-space?

In Exercises 59 and 60, use the shell method to find the volume of the solid below the surface of revolution and above the *xy*-plane.

59. The curve $z = 4x - x^2$ in the xz-plane is revolved about the z-axis.

60. The curve $z = \sin y$ $(0 \le y \le \pi)$ in the yz-plane is revolved about the z-axis.

In Exercises 61 and 62, analyze the trace when the surface

$$z = \tfrac{1}{2}x^2 + \tfrac{1}{4}y^2$$

is intersected by the indicated planes.

61. Find the lengths of the major and minor axes and the coordinates of the foci of the ellipse generated when the surface is intersected by the planes given by

(a) $z = 2$ and (b) $z = 8$.

62. Find the coordinates of the focus of the parabola formed when the surface is intersected by the planes given by

(a) $y = 4$ and (b) $x = 2$.

In Exercises 63 and 64, find an equation of the surface satisfying the conditions, and identify the surface.

63. The set of all points equidistant from the point $(0, 2, 0)$ and the plane $y = -2$

64. The set of all points equidistant from the point $(0, 0, 4)$ and the xy-plane

65. ***Geography*** Because of the forces caused by its rotation, Earth is an oblate ellipsoid rather than a sphere. The equatorial radius is 3963 miles and the polar radius is 3950 miles. Find an equation of the ellipsoid. (Assume that the center of Earth is at the origin and that the trace formed by the plane $z = 0$ corresponds to the equator.)

66. ***Machine Design*** The top of a rubber bushing designed to absorb vibrations in an automobile is the surface of revolution generated by revolving the curve $z = \frac{1}{2}y^2 + 1$ $(0 \le y \le 2)$ in the yz-plane about the z-axis.

(a) Find an equation for the surface of revolution.

(b) All measurements are in centimeters and the bushing is set on the xy-plane. Use the shell method to find its volume.

(c) The bushing has a hole of diameter 1 centimeter through its center and parallel to the axis of revolution. Find the volume of the rubber bushing.

67. Determine the intersection of the hyperbolic paraboloid $z = y^2/b^2 - x^2/a^2$ with the plane $bx + ay - z = 0$. (Assume $a, b > 0$.)

68. Explain why the curve of intersection of the surfaces $x^2 + 3y^2 - 2z^2 + 2y = 4$ and $2x^2 + 6y^2 - 4z^2 - 3x = 2$ lies in a plane.

True or False? **In Exercises 69–72, determine whether the statement is true or false. If it is false, explain why or give an example that shows it is false.**

69. A sphere is an ellipsoid.

70. The generating curve for a surface of revolution is unique.

71. All traces of an ellipsoid are ellipses.

72. All traces of a hyperboloid of one sheet are hyperboloids.

73. ***Think About It*** Three types of classic "topological" surfaces are shown below. The sphere and torus have both an "inside" and an "outside." Does the Klein bottle have both an inside and an outside? Explain.

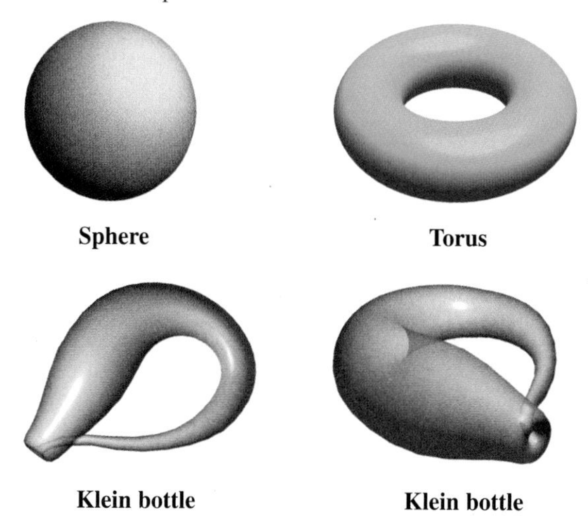

11.7 Cylindrical and Spherical Coordinates

- Use cylindrical coordinates to represent surfaces in space.
- Use spherical coordinates to represent surfaces in space.

Cylindrical Coordinates

You have already seen that some two-dimensional graphs are easier to represent in polar coordinates than in rectangular coordinates. A similar situation exists for surfaces in space. In this section, you will study two alternative space-coordinate systems. The first, the **cylindrical coordinate system,** is an extension of polar coordinates in the plane to three-dimensional space.

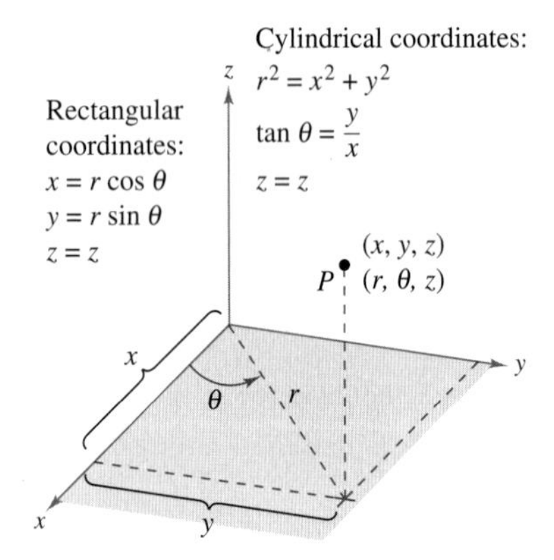

Figure 11.66

THE CYLINDRICAL COORDINATE SYSTEM

In a **cylindrical coordinate system,** a point P in space is represented by an ordered triple (r, θ, z).

1. (r, θ) is a polar representation of the projection of P in the xy-plane.
2. z is the directed distance from (r, θ) to P.

To convert from rectangular to cylindrical coordinates (or vice versa), use the following conversion guidelines for polar coordinates, as illustrated in Figure 11.66.

Cylindrical to rectangular:

$$x = r\cos\theta, \qquad y = r\sin\theta, \qquad z = z$$

Rectangular to cylindrical:

$$r^2 = x^2 + y^2, \qquad \tan\theta = \frac{y}{x}, \qquad z = z$$

The point $(0, 0, 0)$ is called the **pole.** Moreover, because the representation of a point in the polar coordinate system is not unique, it follows that the representation in the cylindrical coordinate system is also not unique.

EXAMPLE 1 Converting from Cylindrical to Rectangular Coordinates

Convert the point $(r, \theta, z) = \left(4, \frac{5\pi}{6}, 3\right)$ to rectangular coordinates.

Solution Using the cylindrical-to-rectangular conversion equations produces

$$x = 4\cos\frac{5\pi}{6} = 4\left(-\frac{\sqrt{3}}{2}\right) = -2\sqrt{3}$$

$$y = 4\sin\frac{5\pi}{6} = 4\left(\frac{1}{2}\right) = 2$$

$$z = 3.$$

So, in rectangular coordinates, the point is $(x, y, z) = \left(-2\sqrt{3}, 2, 3\right)$, as shown in Figure 11.67. ■

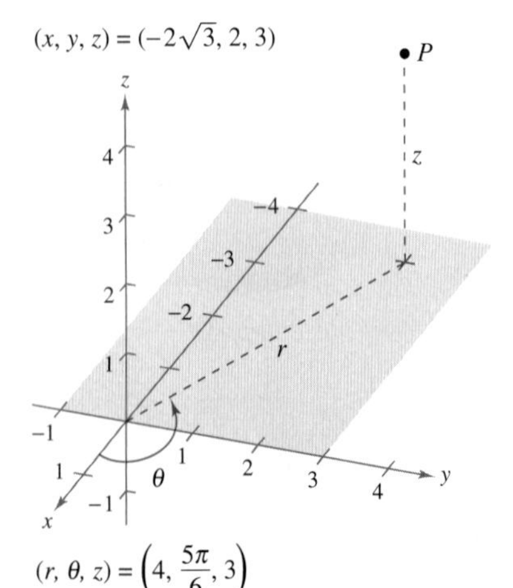

Figure 11.67

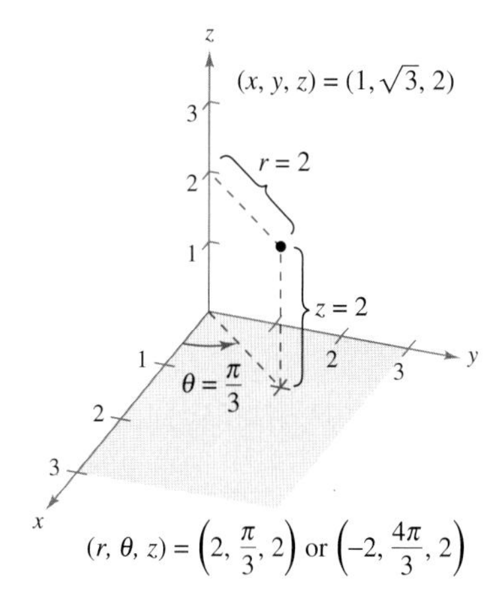

Figure 11.68

EXAMPLE 2 Converting from Rectangular to Cylindrical Coordinates

Convert the point $(x, y, z) = \left(1, \sqrt{3}, 2\right)$ to cylindrical coordinates.

Solution Use the rectangular-to-cylindrical conversion equations.

$$r = \pm\sqrt{1 + 3} = \pm 2$$

$$\tan \theta = \sqrt{3} \quad \Rightarrow \quad \theta = \arctan\left(\sqrt{3}\right) + n\pi = \frac{\pi}{3} + n\pi$$

$$z = 2$$

You have two choices for r and infinitely many choices for θ. As shown in Figure 11.68, two convenient representations of the point are

$$\left(2, \frac{\pi}{3}, 2\right) \qquad r > 0 \text{ and } \theta \text{ in Quadrant I}$$

$$\left(-2, \frac{4\pi}{3}, 2\right). \qquad r < 0 \text{ and } \theta \text{ in Quadrant III}$$

Cylindrical coordinates are especially convenient for representing cylindrical surfaces and surfaces of revolution with the z-axis as the axis of symmetry, as shown in Figure 11.69.

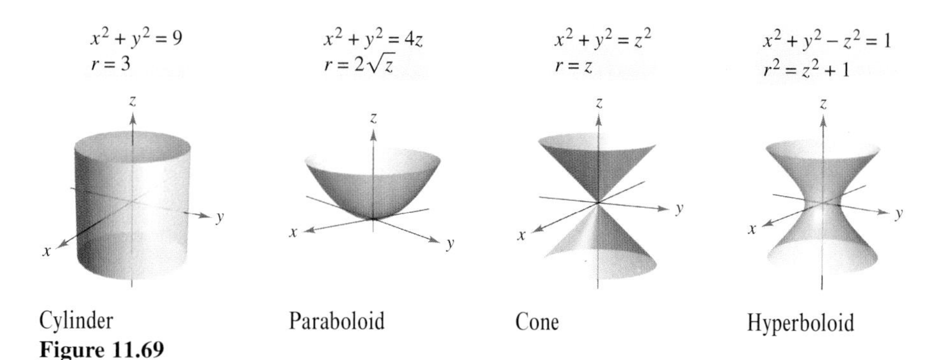

Cylinder Paraboloid Cone Hyperboloid

Figure 11.69

Vertical planes containing the z-axis and horizontal planes also have simple cylindrical coordinate equations, as shown in Figure 11.70.

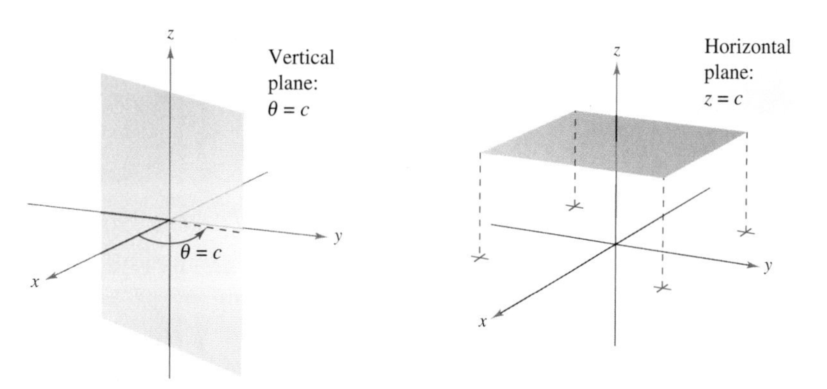

Figure 11.70

EXAMPLE 3 Rectangular-to-Cylindrical Conversion

Find an equation in cylindrical coordinates for the surface represented by each rectangular equation.

a. $x^2 + y^2 = 4z^2$

b. $y^2 = x$

Solution

a. From the preceding section, you know that the graph $x^2 + y^2 = 4z^2$ is an elliptic cone with its axis along the z-axis, as shown in Figure 11.71. If you replace $x^2 + y^2$ with r^2, the equation in cylindrical coordinates is

$$x^2 + y^2 = 4z^2 \quad \text{Rectangular equation}$$
$$r^2 = 4z^2. \quad \text{Cylindrical equation}$$

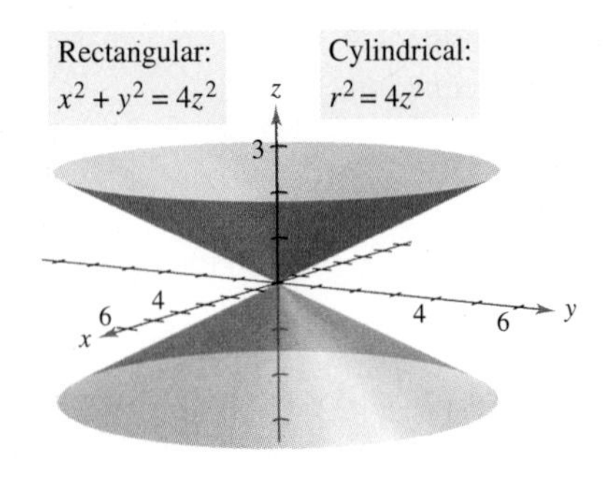

Figure 11.71

b. The graph of the surface $y^2 = x$ is a parabolic cylinder with rulings parallel to the z-axis, as shown in Figure 11.72. By replacing y^2 with $r^2 \sin^2 \theta$ and x with $r \cos \theta$, you obtain the following equation in cylindrical coordinates.

$$y^2 = x \quad \text{Rectangular equation}$$
$$r^2 \sin^2 \theta = r \cos \theta \quad \text{Substitute } r \sin \theta \text{ for } y \text{ and } r \cos \theta \text{ for } x.$$
$$r(r \sin^2 \theta - \cos \theta) = 0 \quad \text{Collect terms and factor.}$$
$$r \sin^2 \theta - \cos \theta = 0 \quad \text{Divide each side by } r.$$
$$r = \frac{\cos \theta}{\sin^2 \theta} \quad \text{Solve for } r.$$
$$r = \csc \theta \cot \theta \quad \text{Cylindrical equation}$$

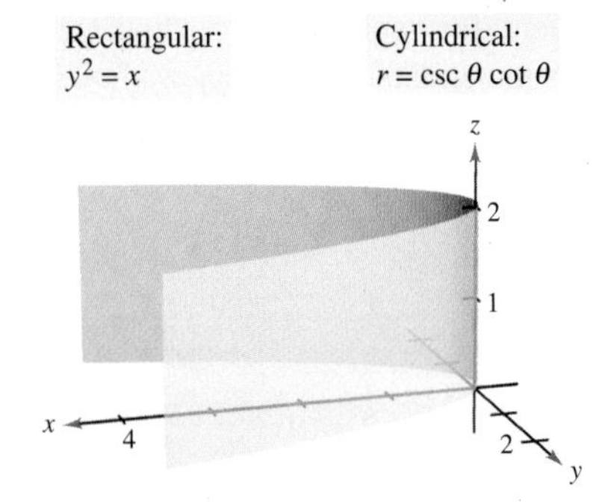

Figure 11.72

Note that this equation includes a point for which $r = 0$, so nothing was lost by dividing each side by the factor r. ■

Converting from cylindrical coordinates to rectangular coordinates is less straightforward than converting from rectangular coordinates to cylindrical coordinates, as demonstrated in Example 4.

EXAMPLE 4 Cylindrical-to-Rectangular Conversion

Find an equation in rectangular coordinates for the surface represented by the cylindrical equation

$$r^2 \cos 2\theta + z^2 + 1 = 0.$$

Solution

$$r^2 \cos 2\theta + z^2 + 1 = 0 \quad \text{Cylindrical equation}$$
$$r^2(\cos^2 \theta - \sin^2 \theta) + z^2 + 1 = 0 \quad \text{Trigonometric identity}$$
$$r^2 \cos^2 \theta - r^2 \sin^2 \theta + z^2 = -1$$
$$x^2 - y^2 + z^2 = -1 \quad \text{Replace } r \cos \theta \text{ with } x \text{ and } r \sin \theta \text{ with } y.$$
$$y^2 - x^2 - z^2 = 1 \quad \text{Rectangular equation}$$

This is a hyperboloid of two sheets whose axis lies along the y-axis, as shown in Figure 11.73. ■

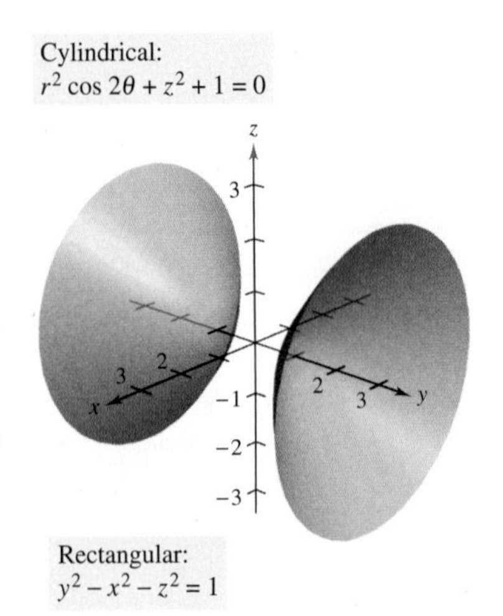

Figure 11.73

Spherical Coordinates

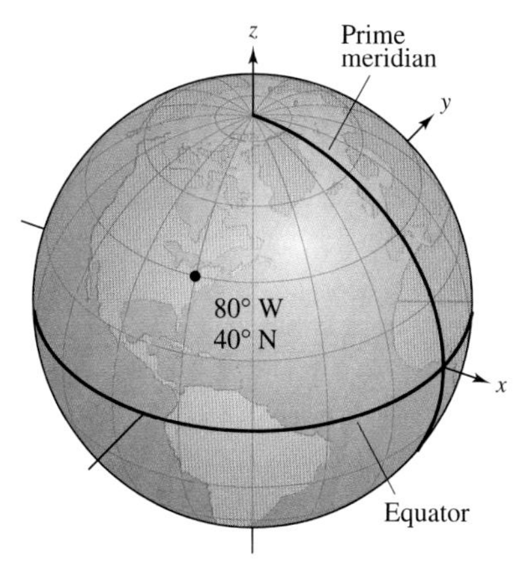

Figure 11.74

In the **spherical coordinate system,** each point is represented by an ordered triple: the first coordinate is a distance, and the second and third coordinates are angles. This system is similar to the latitude-longitude system used to identify points on the surface of Earth. For example, the point on the surface of Earth whose latitude is 40° North (of the equator) and whose longitude is 80° West (of the prime meridian) is shown in Figure 11.74. Assuming that the Earth is spherical and has a radius of 4000 miles, you would label this point as

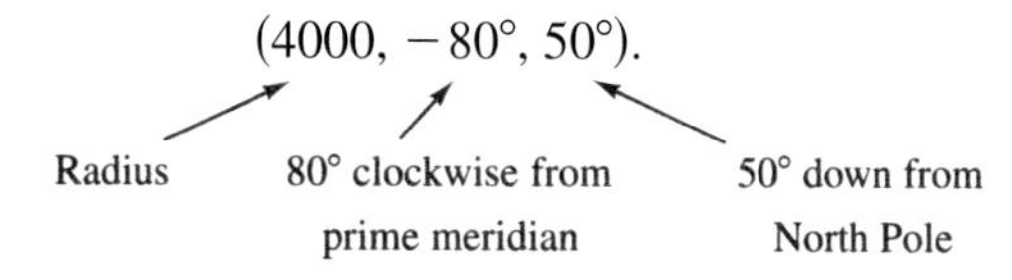

THE SPHERICAL COORDINATE SYSTEM

In a **spherical coordinate system,** a point P in space is represented by an ordered triple (ρ, θ, ϕ).

1. ρ is the distance between P and the origin, $\rho \geq 0$.
2. θ is the same angle used in cylindrical coordinates for $r \geq 0$.
3. ϕ is the angle *between* the positive z-axis and the line segment $\overrightarrow{OP}$, $0 \leq \phi \leq \pi$.

Note that the first and third coordinates, ρ and ϕ, are nonnegative. ρ is the lowercase Greek letter *rho*, and ϕ is the lowercase Greek letter *phi*.

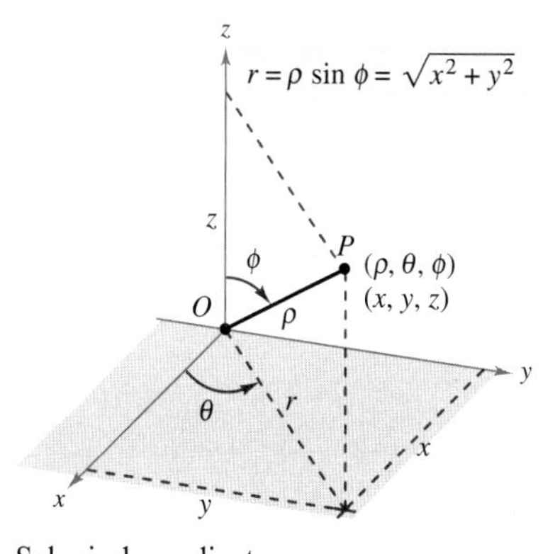

Spherical coordinates
Figure 11.75

The relationship between rectangular and spherical coordinates is illustrated in Figure 11.75. To convert from one system to the other, use the following.

Spherical to rectangular:

$$x = \rho \sin\phi \cos\theta, \qquad y = \rho \sin\phi \sin\theta, \qquad z = \rho \cos\phi$$

Rectangular to spherical:

$$\rho^2 = x^2 + y^2 + z^2, \qquad \tan\theta = \frac{y}{x}, \qquad \phi = \arccos\left(\frac{z}{\sqrt{x^2 + y^2 + z^2}}\right)$$

To change coordinates between the cylindrical and spherical systems, use the following.

Spherical to cylindrical ($r \geq 0$):

$$r^2 = \rho^2 \sin^2\phi, \qquad \theta = \theta, \qquad z = \rho \cos\phi$$

Cylindrical to spherical ($r \geq 0$):

$$\rho = \sqrt{r^2 + z^2}, \qquad \theta = \theta, \qquad \phi = \arccos\left(\frac{z}{\sqrt{r^2 + z^2}}\right)$$

The spherical coordinate system is useful primarily for surfaces in space that have a *point* or *center* of symmetry. For example, Figure 11.76 shows three surfaces with simple spherical equations.

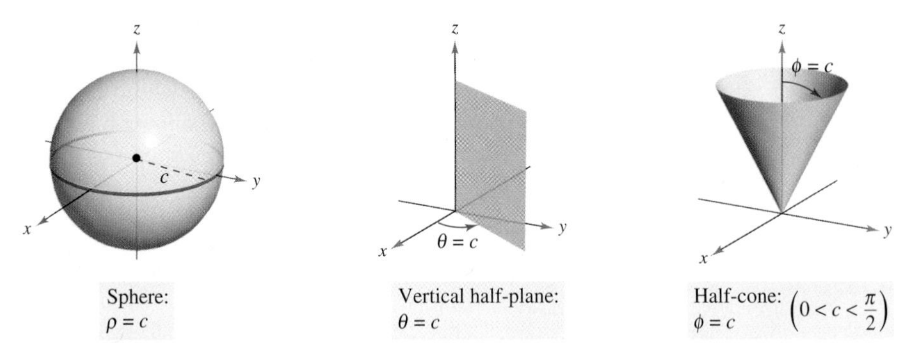

Figure 11.76

EXAMPLE 5 Rectangular-to-Spherical Conversion

Find an equation in spherical coordinates for the surface represented by each rectangular equation.

a. Cone: $x^2 + y^2 = z^2$

b. Sphere: $x^2 + y^2 + z^2 - 4z = 0$

Solution

a. Making the appropriate replacements for x, y, and z in the given equation yields the following.

$$x^2 + y^2 = z^2$$
$$\rho^2 \sin^2 \phi \cos^2 \theta + \rho^2 \sin^2 \phi \sin^2 \theta = \rho^2 \cos^2 \phi$$
$$\rho^2 \sin^2 \phi (\cos^2 \theta + \sin^2 \theta) = \rho^2 \cos^2 \phi$$
$$\rho^2 \sin^2 \phi = \rho^2 \cos^2 \phi$$
$$\frac{\sin^2 \phi}{\cos^2 \phi} = 1 \qquad \rho \geq 0$$
$$\tan^2 \phi = 1 \qquad \phi = \pi/4 \text{ or } \phi = 3\pi/4$$

The equation $\phi = \pi/4$ represents the *upper* half-cone, and the equation $\phi = 3\pi/4$ represents the *lower* half-cone.

b. Because $\rho^2 = x^2 + y^2 + z^2$ and $z = \rho \cos \phi$, the given equation has the following spherical form.

$$\rho^2 - 4\rho \cos \phi = 0 \quad \Longrightarrow \quad \rho(\rho - 4 \cos \phi) = 0$$

Temporarily discarding the possibility that $\rho = 0$, you have the spherical equation

$$\rho - 4 \cos \phi = 0 \qquad \text{or} \qquad \rho = 4 \cos \phi.$$

Note that the solution set for this equation includes a point for which $\rho = 0$, so nothing is lost by discarding the factor ρ. The sphere represented by the equation $\rho = 4 \cos \phi$ is shown in Figure 11.77. ■

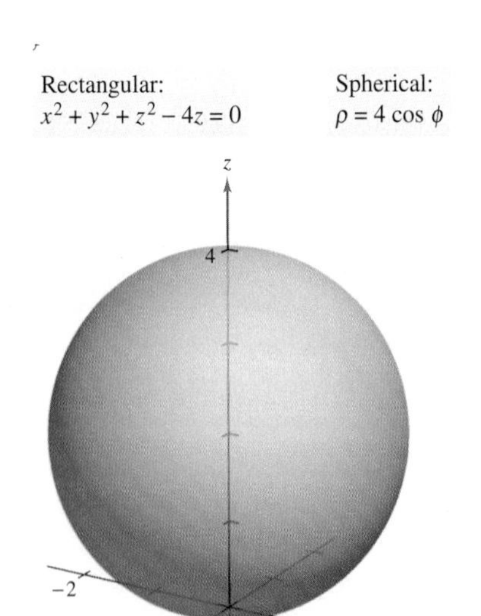

Figure 11.77

11.7 Exercises

See www.CalcChat.com for worked-out solutions to odd-numbered exercises.

In Exercises 1–6, convert the point from cylindrical coordinates to rectangular coordinates.

1. $(-7, 0, 5)$
2. $(2, -\pi, -4)$
3. $(3, \pi/4, 1)$
4. $(6, -\pi/4, 2)$
5. $(4, 7\pi/6, 3)$
6. $(-0.5, 4\pi/3, 8)$

In Exercises 7–12, convert the point from rectangular coordinates to cylindrical coordinates.

7. $(0, 5, 1)$
8. $(2\sqrt{2}, -2\sqrt{2}, 4)$
9. $(2, -2, -4)$
10. $(3, -3, 7)$
11. $(1, \sqrt{3}, 4)$
12. $(2\sqrt{3}, -2, 6)$

In Exercises 13–20, find an equation in cylindrical coordinates for the equation given in rectangular coordinates.

13. $z = 4$
14. $x = 9$
15. $x^2 + y^2 + z^2 = 17$
16. $z = x^2 + y^2 - 11$
17. $y = x^2$
18. $x^2 + y^2 = 8x$
19. $y^2 = 10 - z^2$
20. $x^2 + y^2 + z^2 - 3z = 0$

In Exercises 21–28, find an equation in rectangular coordinates for the equation given in cylindrical coordinates, and sketch its graph.

21. $r = 3$
22. $z = 2$
23. $\theta = \pi/6$
24. $r = \frac{1}{2}z$
25. $r^2 + z^2 = 5$
26. $z = r^2 \cos^2 \theta$
27. $r = 2 \sin \theta$
28. $r = 2 \cos \theta$

In Exercises 29–34, convert the point from rectangular coordinates to spherical coordinates.

29. $(4, 0, 0)$
30. $(-4, 0, 0)$
31. $(-2, 2\sqrt{3}, 4)$
32. $(2, 2, 4\sqrt{2})$
33. $(\sqrt{3}, 1, 2\sqrt{3})$
34. $(-1, 2, 1)$

In Exercises 35–40, convert the point from spherical coordinates to rectangular coordinates.

35. $(4, \pi/6, \pi/4)$
36. $(12, 3\pi/4, \pi/9)$
37. $(12, -\pi/4, 0)$
38. $(9, \pi/4, \pi)$
39. $(5, \pi/4, 3\pi/4)$
40. $(6, \pi, \pi/2)$

In Exercises 41–48, find an equation in spherical coordinates for the equation given in rectangular coordinates.

41. $y = 2$
42. $z = 6$
43. $x^2 + y^2 + z^2 = 49$
44. $x^2 + y^2 - 3z^2 = 0$
45. $x^2 + y^2 = 16$
46. $x = 13$
47. $x^2 + y^2 = 2z^2$
48. $x^2 + y^2 + z^2 - 9z = 0$

In Exercises 49–56, find an equation in rectangular coordinates for the equation given in spherical coordinates, and sketch its graph.

49. $\rho = 5$
50. $\theta = \dfrac{3\pi}{4}$
51. $\phi = \dfrac{\pi}{6}$
52. $\phi = \dfrac{\pi}{2}$
53. $\rho = 4 \cos \phi$
54. $\rho = 2 \sec \phi$
55. $\rho = \csc \phi$
56. $\rho = 4 \csc \phi \sec \theta$

In Exercises 57–64, convert the point from cylindrical coordinates to spherical coordinates.

57. $(4, \pi/4, 0)$
58. $(3, -\pi/4, 0)$
59. $(4, \pi/2, 4)$
60. $(2, 2\pi/3, -2)$
61. $(4, -\pi/6, 6)$
62. $(-4, \pi/3, 4)$
63. $(12, \pi, 5)$
64. $(4, \pi/2, 3)$

In Exercises 65–72, convert the point from spherical coordinates to cylindrical coordinates.

65. $(10, \pi/6, \pi/2)$
66. $(4, \pi/18, \pi/2)$
67. $(36, \pi, \pi/2)$
68. $(18, \pi/3, \pi/3)$
69. $(6, -\pi/6, \pi/3)$
70. $(5, -5\pi/6, \pi)$
71. $(8, 7\pi/6, \pi/6)$
72. $(7, \pi/4, 3\pi/4)$

CAS **In Exercises 73–88, use a computer algebra system or graphing utility to convert the point from one system to another among the rectangular, cylindrical, and spherical coordinate systems.**

	Rectangular	*Cylindrical*	*Spherical*
73.	$(4, 6, 3)$		
74.	$(6, -2, -3)$		
75.		$(5, \pi/9, 8)$	
76.		$(10, -0.75, 6)$	
77.			$(20, 2\pi/3, \pi/4)$
78.			$(7.5, 0.25, 1)$
79.	$(3, -2, 2)$		
80.	$(3\sqrt{2}, 3\sqrt{2}, -3)$		
81.	$(5/2, 4/3, -3/2)$		
82.	$(0, -5, 4)$		
83.		$(5, 3\pi/4, -5)$	
84.		$(-2, 11\pi/6, 3)$	
85.		$(-3.5, 2.5, 6)$	
86.		$(8.25, 1.3, -4)$	
87.			$(3, 3\pi/4, \pi/3)$
88.			$(8, -\pi/6, \pi)$

In Exercises 89–94, match the equation (written in terms of cylindrical or spherical coordinates) with its graph. [The graphs are labeled (a), (b), (c), (d), (e), and (f).]

(a)

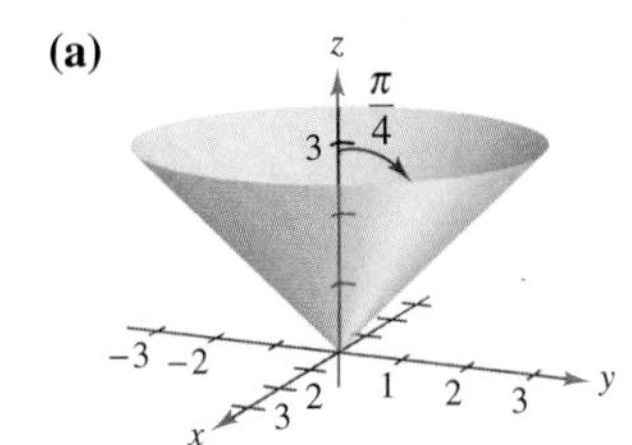

(b)

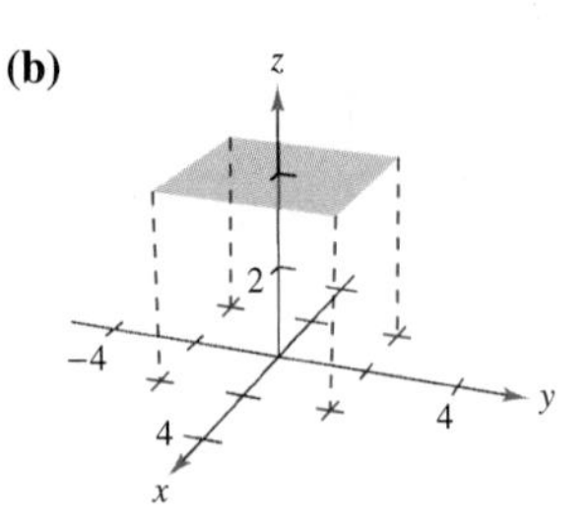

(c)

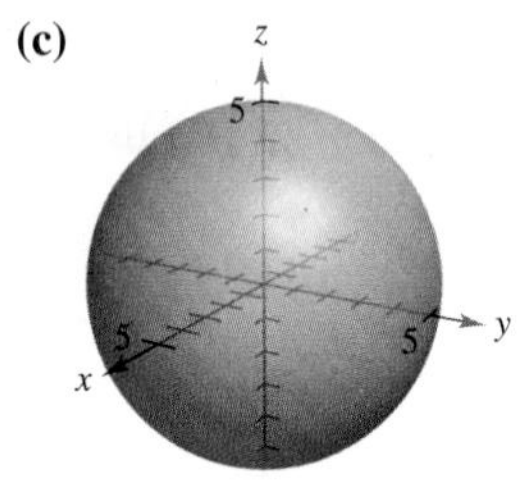

(d)

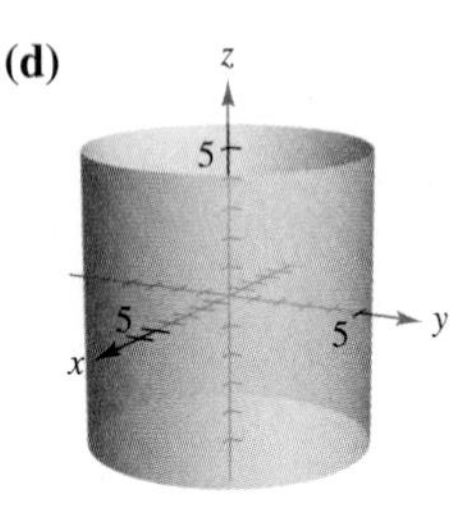

(e)

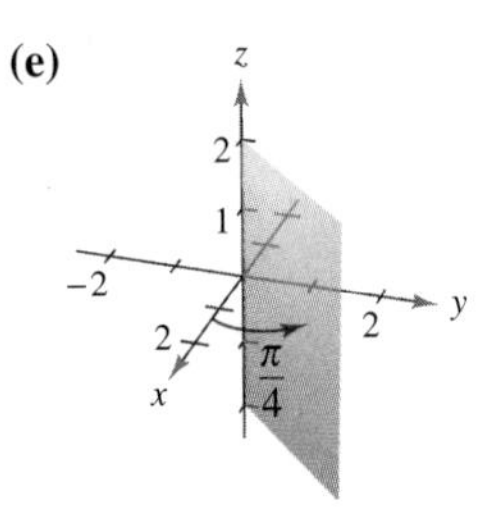

(f)

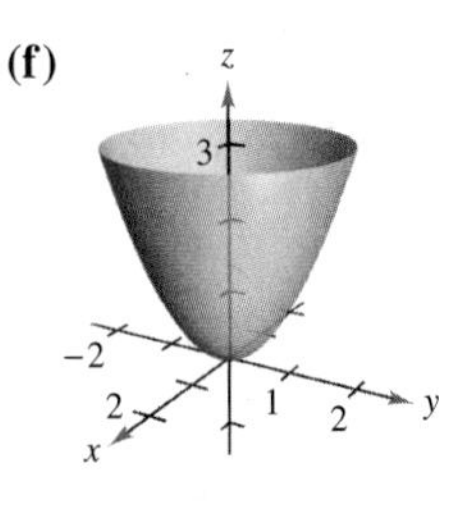

89. $r = 5$

90. $\theta = \dfrac{\pi}{4}$

91. $\rho = 5$

92. $\phi = \dfrac{\pi}{4}$

93. $r^2 = z$

94. $\rho = 4 \sec \phi$

WRITING ABOUT CONCEPTS

95. Give the equations for the coordinate conversion from rectangular to cylindrical coordinates and vice versa.

96. Explain why in spherical coordinates the graph of $\theta = c$ is a half-plane and not an entire plane.

97. Give the equations for the coordinate conversion from rectangular to spherical coordinates and vice versa.

CAPSTONE

98. (a) For constants a, b, and c, describe the graphs of the equations $r = a$, $\theta = b$, and $z = c$ in cylindrical coordinates.

(b) For constants a, b, and c, describe the graphs of the equations $\rho = a$, $\theta = b$, and $\phi = c$ in spherical coordinates.

In Exercises 99–106, convert the rectangular equation to an equation in (a) cylindrical coordinates and (b) spherical coordinates.

99. $x^2 + y^2 + z^2 = 25$

100. $4(x^2 + y^2) = z^2$

101. $x^2 + y^2 + z^2 - 2z = 0$

102. $x^2 + y^2 = z$

103. $x^2 + y^2 = 4y$

104. $x^2 + y^2 = 36$

105. $x^2 - y^2 = 9$

106. $y = 4$

In Exercises 107–110, sketch the solid that has the given description in cylindrical coordinates.

107. $0 \le \theta \le \pi/2,\ 0 \le r \le 2,\ 0 \le z \le 4$

108. $-\pi/2 \le \theta \le \pi/2,\ 0 \le r \le 3,\ 0 \le z \le r\cos\theta$

109. $0 \le \theta \le 2\pi,\ 0 \le r \le a,\ r \le z \le a$

110. $0 \le \theta \le 2\pi,\ 2 \le r \le 4,\ z^2 \le -r^2 + 6r - 8$

In Exercises 111–114, sketch the solid that has the given description in spherical coordinates.

111. $0 \le \theta \le 2\pi,\ 0 \le \phi \le \pi/6,\ 0 \le \rho \le a \sec\phi$

112. $0 \le \theta \le 2\pi,\ \pi/4 \le \phi \le \pi/2,\ 0 \le \rho \le 1$

113. $0 \le \theta \le \pi/2,\ 0 \le \phi \le \pi/2,\ 0 \le \rho \le 2$

114. $0 \le \theta \le \pi,\ 0 \le \phi \le \pi/2,\ 1 \le \rho \le 3$

Think About It **In Exercises 115–120, find inequalities that describe the solid, and state the coordinate system used. Position the solid on the coordinate system such that the inequalities are as simple as possible.**

115. A cube with each edge 10 centimeters long

116. A cylindrical shell 8 meters long with an inside diameter of 0.75 meter and an outside diameter of 1.25 meters

117. A spherical shell with inside and outside radii of 4 inches and 6 inches, respectively

118. The solid that remains after a hole 1 inch in diameter is drilled through the center of a sphere 6 inches in diameter

119. The solid inside both $x^2 + y^2 + z^2 = 9$ and $\left(x - \frac{3}{2}\right)^2 + y^2 = \frac{9}{4}$

120. The solid between the spheres $x^2 + y^2 + z^2 = 4$ and $x^2 + y^2 + z^2 = 9$, and inside the cone $z^2 = x^2 + y^2$

True or False? **In Exercises 121–124, determine whether the statement is true or false. If it is false, explain why or give an example that shows it is false.**

121. In cylindrical coordinates, the equation $r = z$ is a cylinder.

122. The equations $\rho = 2$ and $x^2 + y^2 + z^2 = 4$ represent the same surface.

123. The cylindrical coordinates of a point (x, y, z) are unique.

124. The spherical coordinates of a point (x, y, z) are unique.

125. Identify the curve of intersection of the surfaces (in cylindrical coordinates) $z = \sin\theta$ and $r = 1$.

126. Identify the curve of intersection of the surfaces (in spherical coordinates) $\rho = 2\sec\phi$ and $\rho = 4$.

11 REVIEW EXERCISES

See www.CalcChat.com for worked-out solutions to odd-numbered exercises.

In Exercises 1 and 2, let $\mathbf{u} = \overrightarrow{PQ}$ and $\mathbf{v} = \overrightarrow{PR}$, and (a) write u and v in component form, (b) write u as the linear combination of the standard unit vectors i and j, (c) find the magnitude of v, and (d) find $2\mathbf{u} + \mathbf{v}$.

1. $P = (1, 2), Q = (4, 1), R = (5, 4)$
2. $P = (-2, -1), Q = (5, -1), R = (2, 4)$

In Exercises 3 and 4, find the component form of v given its magnitude and the angle it makes with the positive x-axis.

3. $\|\mathbf{v}\| = 8,\ \theta = 60°$
4. $\|\mathbf{v}\| = \frac{1}{2},\ \theta = 225°$

5. Find the coordinates of the point in the xy-plane four units to the right of the xz-plane and five units behind the yz-plane.
6. Find the coordinates of the point located on the y-axis and seven units to the left of the xz-plane.

In Exercises 7 and 8, determine the location of a point (x, y, z) that satisfies the condition.

7. $yz > 0$
8. $xy < 0$

In Exercises 9 and 10, find the standard equation of the sphere.

9. Center: $(3, -2, 6)$; Diameter: 15
10. Endpoints of a diameter: $(0, 0, 4), (4, 6, 0)$

In Exercises 11 and 12, complete the square to write the equation of the sphere in standard form. Find the center and radius.

11. $x^2 + y^2 + z^2 - 4x - 6y + 4 = 0$
12. $x^2 + y^2 + z^2 - 10x + 6y - 4z + 34 = 0$

In Exercises 13 and 14, the initial and terminal points of a vector are given. (a) Sketch the directed line segment, (b) find the component form of the vector, (c) write the vector using standard unit vector notation, and (d) sketch the vector with its initial point at the origin.

13. Initial point: $(2, -1, 3)$
Terminal point: $(4, 4, -7)$
14. Initial point: $(6, 2, 0)$
Terminal point: $(3, -3, 8)$

In Exercises 15 and 16, use vectors to determine whether the points are collinear.

15. $(3, 4, -1),\ (-1, 6, 9),\ (5, 3, -6)$
16. $(5, -4, 7),\ (8, -5, 5),\ (11, 6, 3)$

17. Find a unit vector in the direction of $\mathbf{u} = \langle 2, 3, 5\rangle$.
18. Find the vector $\mathbf{v}$ of magnitude 8 in the direction $\langle 6, -3, 2\rangle$.

In Exercises 19 and 20, let $\mathbf{u} = \overrightarrow{PQ}$ and $\mathbf{v} = \overrightarrow{PR}$, and find (a) the component forms of u and v, (b) $\mathbf{u} \cdot \mathbf{v}$, and (c) $\mathbf{v} \cdot \mathbf{v}$.

19. $P = (5, 0, 0),\ Q = (4, 4, 0),\ R = (2, 0, 6)$
20. $P = (2, -1, 3),\ Q = (0, 5, 1),\ R = (5, 5, 0)$

In Exercises 21 and 22, determine whether u and v are orthogonal, parallel, or neither.

21. $\mathbf{u} = \langle 7, -2, 3\rangle$
$\mathbf{v} = \langle -1, 4, 5\rangle$
22. $\mathbf{u} = \langle -4, 3, -6\rangle$
$\mathbf{v} = \langle 16, -12, 24\rangle$

In Exercises 23–26, find the angle θ between the vectors.

23. $\mathbf{u} = 5[\cos(3\pi/4)\mathbf{i} + \sin(3\pi/4)\mathbf{j}]$
$\mathbf{v} = 2[\cos(2\pi/3)\mathbf{i} + \sin(2\pi/3)\mathbf{j}]$
24. $\mathbf{u} = 6\mathbf{i} + 2\mathbf{j} - 3\mathbf{k},\quad \mathbf{v} = -\mathbf{i} + 5\mathbf{j}$
25. $\mathbf{u} = \langle 10, -5, 15\rangle,\quad \mathbf{v} = \langle -2, 1, -3\rangle$
26. $\mathbf{u} = \langle 1, 0, -3\rangle,\quad \mathbf{v} = \langle 2, -2, 1\rangle$

27. Find two vectors in opposite directions that are orthogonal to the vector $\mathbf{u} = \langle 5, 6, -3\rangle$.
28. ***Work*** An object is pulled 8 feet across a floor using a force of 75 pounds. The direction of the force is 30° above the horizontal. Find the work done.

In Exercises 29–38, let $\mathbf{u} = \langle 3, -2, 1\rangle$, $\mathbf{v} = \langle 2, -4, -3\rangle$, and $\mathbf{w} = \langle -1, 2, 2\rangle$.

29. Show that $\mathbf{u} \cdot \mathbf{u} = \|\mathbf{u}\|^2$.
30. Find the angle between $\mathbf{u}$ and $\mathbf{v}$.
31. Determine the projection of $\mathbf{w}$ onto $\mathbf{u}$.
32. Find the work done in moving an object along the vector $\mathbf{u}$ if the applied force is $\mathbf{w}$.
33. Determine a unit vector perpendicular to the plane containing $\mathbf{v}$ and $\mathbf{w}$.
34. Show that $\mathbf{u} \times \mathbf{v} = -(\mathbf{v} \times \mathbf{u})$.
35. Find the volume of the solid whose edges are $\mathbf{u}$, $\mathbf{v}$, and $\mathbf{w}$.
36. Show that $\mathbf{u} \times (\mathbf{v} + \mathbf{w}) = (\mathbf{u} \times \mathbf{v}) + (\mathbf{u} \times \mathbf{w})$.
37. Find the area of the parallelogram with adjacent sides $\mathbf{u}$ and $\mathbf{v}$.
38. Find the area of the triangle with adjacent sides $\mathbf{v}$ and $\mathbf{w}$.

39. ***Torque*** The specifications for a tractor state that the torque on a bolt with head size $\frac{7}{8}$ inch cannot exceed 200 foot-pounds. Determine the maximum force $\|\mathbf{F}\|$ that can be applied to the wrench in the figure.

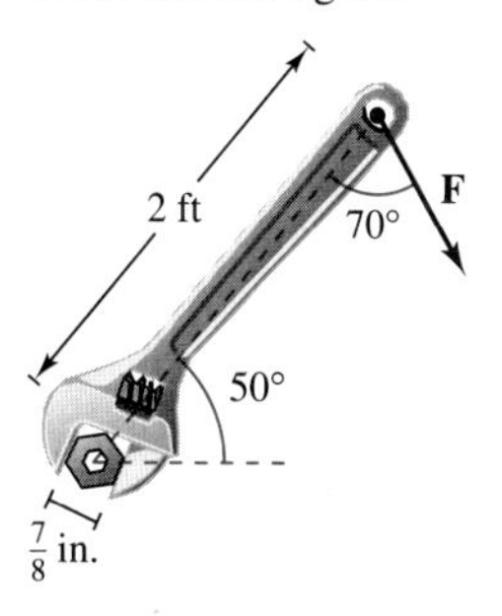

40. ***Volume*** Use the triple scalar product to find the volume of the parallelepiped having adjacent edges $\mathbf{u} = 2\mathbf{i} + \mathbf{j}$, $\mathbf{v} = 2\mathbf{j} + \mathbf{k}$, and $\mathbf{w} = -\mathbf{j} + 2\mathbf{k}$.

In Exercises 41 and 42, find sets of (a) parametric equations and (b) symmetric equations of the line through the two points. (For each line, write the direction numbers as integers.)

41. $(3, 0, 2)$, $(9, 11, 6)$

42. $(-1, 4, 3)$, $(8, 10, 5)$

In Exercises 43–46, (a) find a set of parametric equations for the line, (b) find a set of symmetric equations for the line, and (c) sketch a graph of the line.

43. The line passes through the point $(1, 2, 3)$ and is perpendicular to the xz-plane.

44. The line passes through the point $(1, 2, 3)$ and is parallel to the line given by $x = y = z$.

45. The intersection of the planes $3x - 3y - 7z = -4$ and $x - y + 2z = 3$.

46. The line passes through the point $(0, 1, 4)$ and is perpendicular to $\mathbf{u} = \langle 2, -5, 1\rangle$ and $\mathbf{v} = \langle -3, 1, 4\rangle$.

In Exercises 47–50, find an equation of the plane and sketch its graph.

47. The plane passes through

$(-3, -4, 2)$, $(-3, 4, 1)$, and $(1, 1, -2)$.

48. The plane passes through the point $(-2, 3, 1)$ and is perpendicular to $\mathbf{n} = 3\mathbf{i} - \mathbf{j} + \mathbf{k}$.

49. The plane contains the lines given by

$$\frac{x-1}{-2} = y = z + 1$$

and

$$\frac{x+1}{-2} = y - 1 = z - 2.$$

50. The plane passes through the points $(5, 1, 3)$ and $(2, -2, 1)$ and is perpendicular to the plane $2x + y - z = 4$.

51. Find the distance between the point $(1, 0, 2)$ and the plane $2x - 3y + 6z = 6$.

52. Find the distance between the point $(3, -2, 4)$ and the plane $2x - 5y + z = 10$.

53. Find the distance between the planes $5x - 3y + z = 2$ and $5x - 3y + z = -3$.

54. Find the distance between the point $(-5, 1, 3)$ and the line given by $x = 1 + t$, $y = 3 - 2t$, and $z = 5 - t$.

In Exercises 55–64, describe and sketch the surface.

55. $x + 2y + 3z = 6$

56. $y = z^2$

57. $y = \frac{1}{2}z$

58. $y = \cos z$

59. $\dfrac{x^2}{16} + \dfrac{y^2}{9} + z^2 = 1$

60. $16x^2 + 16y^2 - 9z^2 = 0$

61. $\dfrac{x^2}{16} - \dfrac{y^2}{9} + z^2 = -1$

62. $\dfrac{x^2}{25} + \dfrac{y^2}{4} - \dfrac{z^2}{100} = 1$

63. $x^2 + z^2 = 4$

64. $y^2 + z^2 = 16$

65. Find an equation of a generating curve of the surface of revolution $y^2 + z^2 - 4x = 0$.

66. Find an equation of a generating curve of the surface of revolution $x^2 + 2y^2 + z^2 = 3y$.

67. Find an equation for the surface of revolution generated by revolving the curve $z^2 = 2y$ in the yz-plane about the y-axis.

68. Find an equation for the surface of revolution generated by revolving the curve $2x + 3z = 1$ in the xz-plane about the x-axis.

In Exercises 69 and 70, convert the point from rectangular coordinates to (a) cylindrical coordinates and (b) spherical coordinates.

69. $\left(-2\sqrt{2}, 2\sqrt{2}, 2\right)$

70. $\left(\dfrac{\sqrt{3}}{4}, \dfrac{3}{4}, \dfrac{3\sqrt{3}}{2}\right)$

In Exercises 71 and 72, convert the point from cylindrical coordinates to spherical coordinates.

71. $\left(100, -\dfrac{\pi}{6}, 50\right)$

72. $\left(81, -\dfrac{5\pi}{6}, 27\sqrt{3}\right)$

In Exercises 73 and 74, convert the point from spherical coordinates to cylindrical coordinates.

73. $\left(25, -\dfrac{\pi}{4}, \dfrac{3\pi}{4}\right)$

74. $\left(12, -\dfrac{\pi}{2}, \dfrac{2\pi}{3}\right)$

In Exercises 75 and 76, convert the rectangular equation to an equation in (a) cylindrical coordinates and (b) spherical coordinates.

75. $x^2 - y^2 = 2z$

76. $x^2 + y^2 + z^2 = 16$

In Exercises 77 and 78, find an equation in rectangular coordinates for the equation given in cylindrical coordinates, and sketch its graph.

77. $r = 5 \cos \theta$

78. $z = 4$

In Exercises 79 and 80, find an equation in rectangular coordinates for the equation given in spherical coordinates, and sketch its graph.

79. $\theta = \dfrac{\pi}{4}$

80. $\rho = 3 \cos \phi$

P.S. PROBLEM SOLVING

1. Using vectors, prove the Law of Sines: If **a**, **b**, and **c** are the three sides of the triangle shown in the figure, then

$$\frac{\sin A}{\|\mathbf{a}\|} = \frac{\sin B}{\|\mathbf{b}\|} = \frac{\sin C}{\|\mathbf{c}\|}.$$

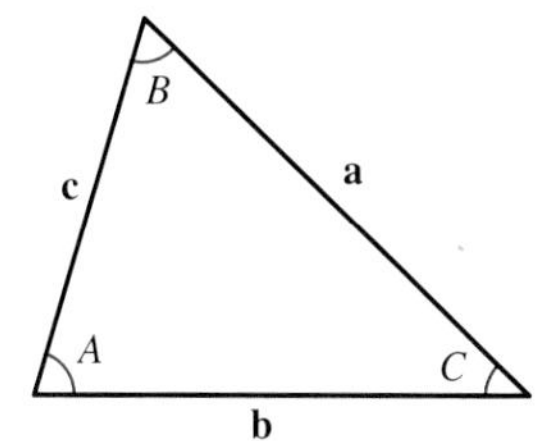

2. Consider the function $f(x) = \int_0^x \sqrt{t^4 + 1}\, dt$.

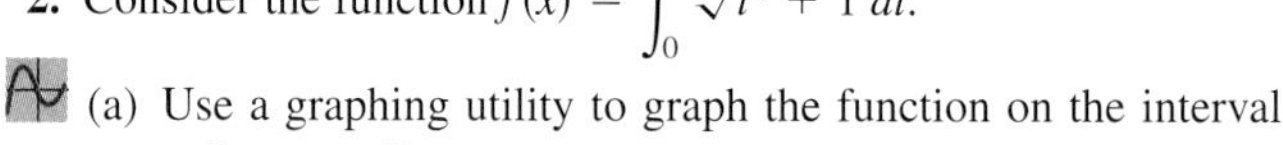

   (a) Use a graphing utility to graph the function on the interval $-2 \leq x \leq 2$.

   (b) Find a unit vector parallel to the graph of f at the point $(0, 0)$.

   (c) Find a unit vector perpendicular to the graph of f at the point $(0, 0)$.

   (d) Find the parametric equations of the tangent line to the graph of f at the point $(0, 0)$.

3. Using vectors, prove that the line segments joining the midpoints of the sides of a parallelogram form a parallelogram (see figure).

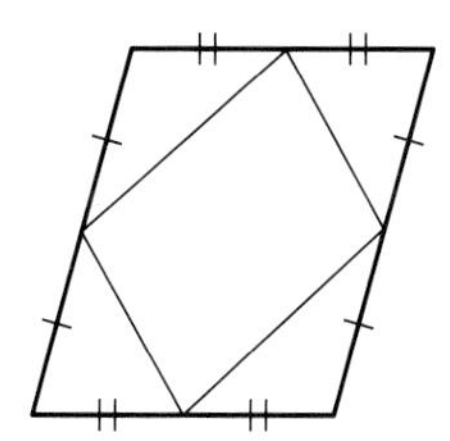

4. Using vectors, prove that the diagonals of a rhombus are perpendicular (see figure).

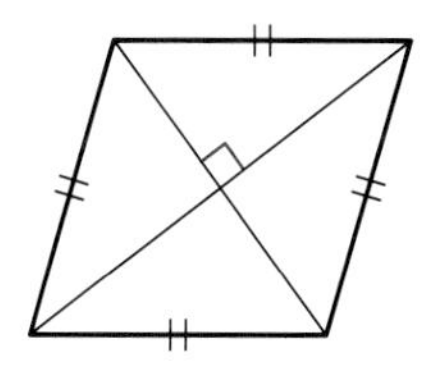

5. (a) Find the shortest distance between the point $Q(2, 0, 0)$ and the line determined by the points $P_1(0, 0, 1)$ and $P_2(0, 1, 2)$.

   (b) Find the shortest distance between the point $Q(2, 0, 0)$ and the line segment joining the points $P_1(0, 0, 1)$ and $P_2(0, 1, 2)$.

6. Let P_0 be a point in the plane with normal vector $\mathbf{n}$. Describe the set of points P in the plane for which $(\mathbf{n} + \overrightarrow{PP_0})$ is orthogonal to $(\mathbf{n} - \overrightarrow{PP_0})$.

7. (a) Find the volume of the solid bounded below by the paraboloid $z = x^2 + y^2$ and above by the plane $z = 1$.

   (b) Find the volume of the solid bounded below by the elliptic paraboloid $z = \frac{x^2}{a^2} + \frac{y^2}{b^2}$ and above by the plane $z = k$, where $k > 0$.

   (c) Show that the volume of the solid in part (b) is equal to one-half the product of the area of the base times the altitude, as shown in the figure.

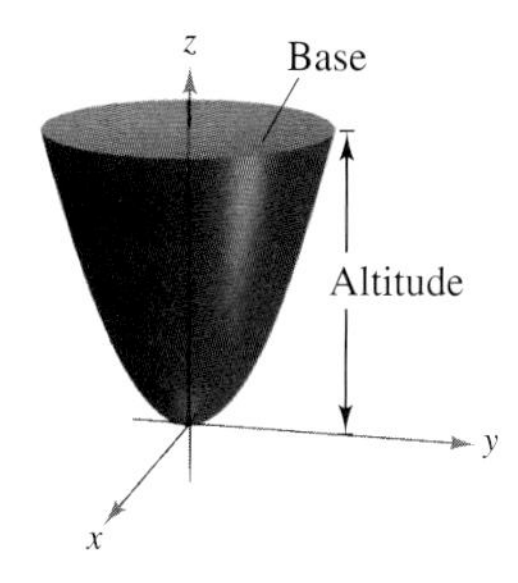

8. (a) Use the disk method to find the volume of the sphere $x^2 + y^2 + z^2 = r^2$.

   (b) Find the volume of the ellipsoid $\frac{x^2}{a^2} + \frac{y^2}{b^2} + \frac{z^2}{c^2} = 1$.

9. Sketch the graph of each equation given in spherical coordinates.

   (a) $\rho = 2 \sin \phi$

   (b) $\rho = 2 \cos \phi$

10. Sketch the graph of each equation given in cylindrical coordinates.

   (a) $r = 2 \cos \theta$

   (b) $z = r^2 \cos 2\theta$

11. Prove the following property of the cross product.

$$(\mathbf{u} \times \mathbf{v}) \times (\mathbf{w} \times \mathbf{z}) = (\mathbf{u} \times \mathbf{v} \cdot \mathbf{z})\mathbf{w} - (\mathbf{u} \times \mathbf{v} \cdot \mathbf{w})\mathbf{z}$$

12. Consider the line given by the parametric equations

$$x = -t + 3, \quad y = \tfrac{1}{2}t + 1, \quad z = 2t - 1$$

and the point $(4, 3, s)$ for any real number s.

   (a) Write the distance between the point and the line as a function of s.

   (b) Use a graphing utility to graph the function in part (a). Use the graph to find the value of s such that the distance between the point and the line is minimum.

   (c) Use the *zoom* feature of a graphing utility to zoom out several times on the graph in part (b). Does it appear that the graph has slant asymptotes? Explain. If it appears to have slant asymptotes, find them.

13. A tetherball weighing 1 pound is pulled outward from the pole by a horizontal force **u** until the rope makes an angle of θ degrees with the pole (see figure).

(a) Determine the resulting tension in the rope and the magnitude of **u** when $\theta = 30°$.

(b) Write the tension T in the rope and the magnitude of **u** as functions of θ. Determine the domains of the functions.

(c) Use a graphing utility to complete the table.

θ	0°	10°	20°	30°	40°	50°	60°
T							
$\|\mathbf{u}\|$							

(d) Use a graphing utility to graph the two functions for $0° \leq \theta \leq 60°$.

(e) Compare T and $\|\mathbf{u}\|$ as θ increases.

(f) Find (if possible) $\lim_{\theta \to \pi/2^-} T$ and $\lim_{\theta \to \pi/2^-} \|\mathbf{u}\|$. Are the results what you expected? Explain.

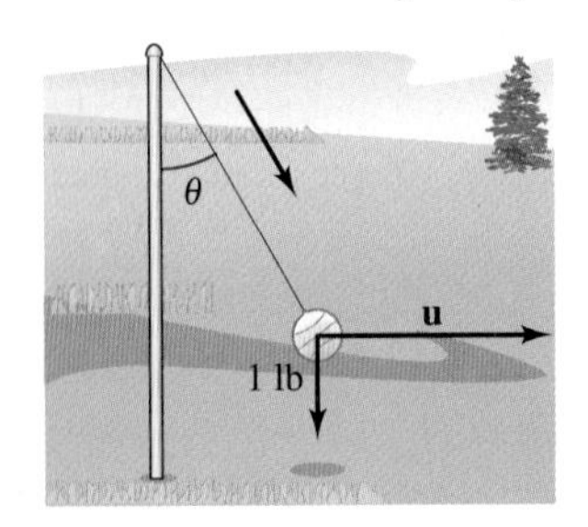

Figure for 13

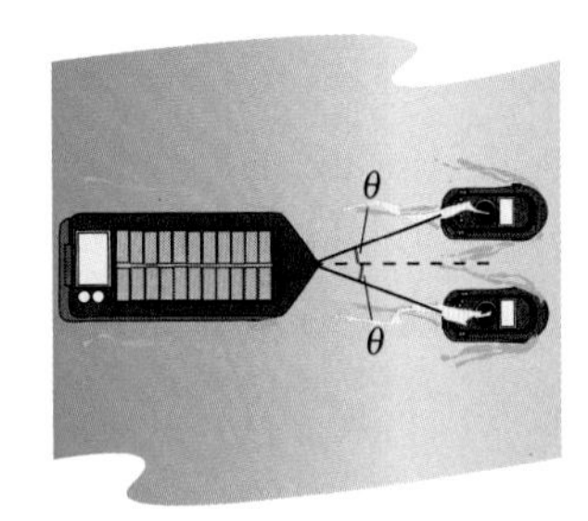

Figure for 14

14. A loaded barge is being towed by two tugboats, and the magnitude of the resultant is 6000 pounds directed along the axis of the barge (see figure). Each towline makes an angle of θ degrees with the axis of the barge.

(a) Find the tension in the towlines if $\theta = 20°$.

(b) Write the tension T of each line as a function of θ. Determine the domain of the function.

(c) Use a graphing utility to complete the table.

θ	10°	20°	30°	40°	50°	60°
T						

(d) Use a graphing utility to graph the tension function.

(e) Explain why the tension increases as θ increases.

15. Consider the vectors $\mathbf{u} = \langle \cos \alpha, \sin \alpha, 0 \rangle$ and $\mathbf{v} = \langle \cos \beta, \sin \beta, 0 \rangle$, where $\alpha > \beta$. Find the cross product of the vectors and use the result to prove the identity

$$\sin(\alpha - \beta) = \sin \alpha \cos \beta - \cos \alpha \sin \beta.$$

16. Los Angeles is located at 34.05° North latitude and 118.24° West longitude, and Rio de Janeiro, Brazil is located at 22.90° South latitude and 43.23° West longitude (see figure). Assume that Earth is spherical and has a radius of 4000 miles.

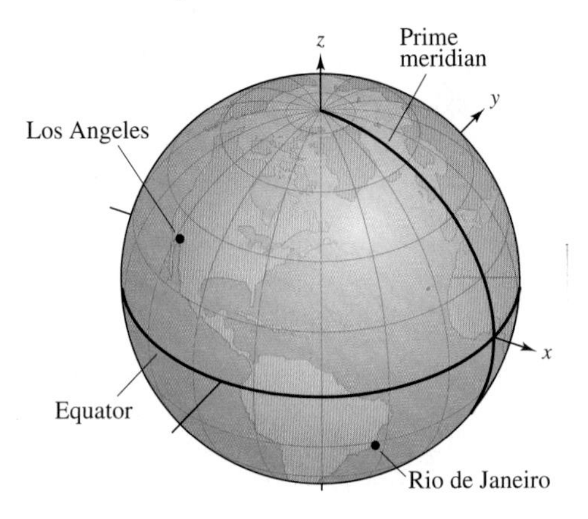

(a) Find the spherical coordinates for the location of each city.

(b) Find the rectangular coordinates for the location of each city.

(c) Find the angle (in radians) between the vectors from the center of Earth to the two cities.

(d) Find the great-circle distance s between the cities. (*Hint:* $s = r\theta$)

(e) Repeat parts (a)–(d) for the cities of Boston, located at 42.36° North latitude and 71.06° West longitude, and Honolulu, located at 21.31° North latitude and 157.86° West longitude.

17. Consider the plane that passes through the points P, R, and S. Show that the distance from a point Q to this plane is

$$\text{Distance} = \frac{|\mathbf{u} \cdot (\mathbf{v} \times \mathbf{w})|}{\|\mathbf{u} \times \mathbf{v}\|}$$

where $\mathbf{u} = \overrightarrow{PR}$, $\mathbf{v} = \overrightarrow{PS}$, and $\mathbf{w} = \overrightarrow{PQ}$.

18. Show that the distance between the parallel planes $ax + by + cz + d_1 = 0$ and $ax + by + cz + d_2 = 0$ is

$$\text{Distance} = \frac{|d_1 - d_2|}{\sqrt{a^2 + b^2 + c^2}}.$$

19. Show that the curve of intersection of the plane $z = 2y$ and the cylinder $x^2 + y^2 = 1$ is an ellipse.

20. Read the article "Tooth Tables: Solution of a Dental Problem by Vector Algebra" by Gary Hosler Meisters in *Mathematics Magazine*. (To view this article, go to the website *www.matharticles.com*.) Then write a paragraph explaining how vectors and vector algebra can be used in the construction of dental inlays.

12 Vector-Valued Functions

This chapter introduces the concept of vector-valued functions. Vector-valued functions can be used to study curves in the plane and in space. These functions can also be used to study the motion of an object along a curve.

In this chapter, you should learn the following.

- How to analyze and sketch a space curve represented by a vector-valued function. How to apply the concepts of limits and continuity to vector-valued functions. (12.1)
- How to differentiate and integrate vector-valued functions. (12.2)
- How to describe the velocity and acceleration associated with a vector-valued function and how to use a vector-valued function to analyze projectile motion. (12.3)
- How to find tangent vectors and normal vectors. (12.4)
- How to find the arc length and curvature of a curve. (12.5)

Jerry Driendl/Getty Images

A Ferris wheel is constructed using the basic principles of a bicycle wheel. You can use a vector-valued function to analyze the motion of a Ferris wheel, including its position and velocity. (See P.S. Problem Solving, Exercise 14.)

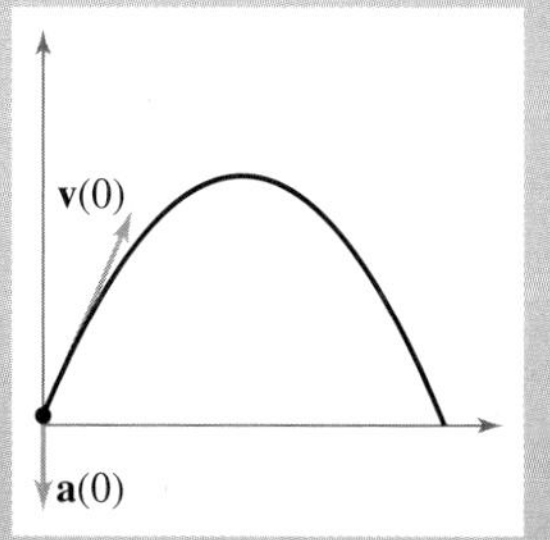

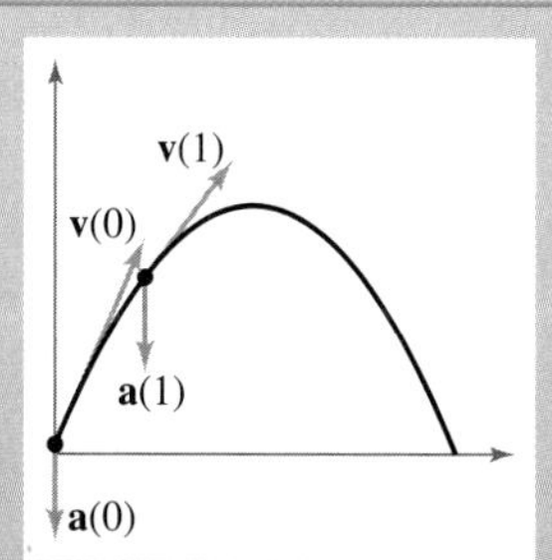

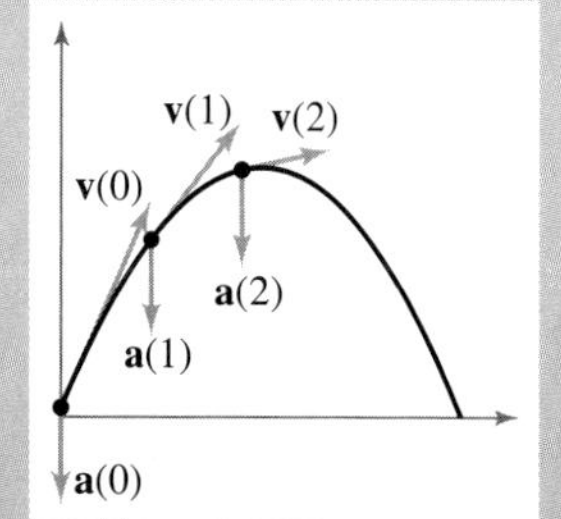

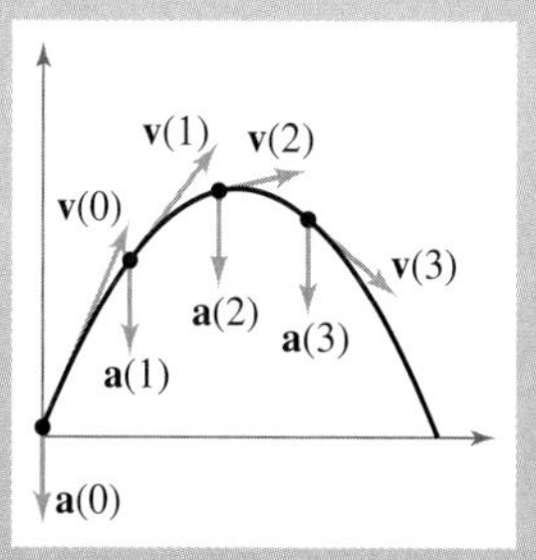

A *vector-valued function* maps real numbers to vectors. You can use a vector-valued function to represent the motion of a particle along a curve. In Section 12.3, you will use the first and second derivatives of a position vector to find a particle's velocity and acceleration.

12.1 Vector-Valued Functions

- **Analyze and sketch a space curve given by a vector-valued function.**
- **Extend the concepts of limits and continuity to vector-valued functions.**

Space Curves and Vector-Valued Functions

In Section 10.2, a *plane curve* was defined as the set of ordered pairs $(f(t), g(t))$ together with their defining parametric equations

$$x = f(t) \quad \text{and} \quad y = g(t)$$

where f and g are continuous functions of t on an interval I. This definition can be extended naturally to three-dimensional space as follows. A **space curve** C is the set of all ordered triples $(f(t), g(t), h(t))$ together with their defining parametric equations

$$x = f(t), \quad y = g(t), \quad \text{and} \quad z = h(t)$$

where f, g, and h are continuous functions of t on an interval I.

Before looking at examples of space curves, a new type of function, called a **vector-valued function,** is introduced. This type of function maps real numbers to vectors.

DEFINITION OF VECTOR-VALUED FUNCTION

A function of the form

$$\mathbf{r}(t) = f(t)\mathbf{i} + g(t)\mathbf{j} \qquad \text{Plane}$$

or

$$\mathbf{r}(t) = f(t)\mathbf{i} + g(t)\mathbf{j} + h(t)\mathbf{k} \qquad \text{Space}$$

is a **vector-valued function,** where the **component functions** f, g, and h are real-valued functions of the parameter t. Vector-valued functions are sometimes denoted as $\mathbf{r}(t) = \langle f(t), g(t)\rangle$ or $\mathbf{r}(t) = \langle f(t), g(t), h(t)\rangle$.

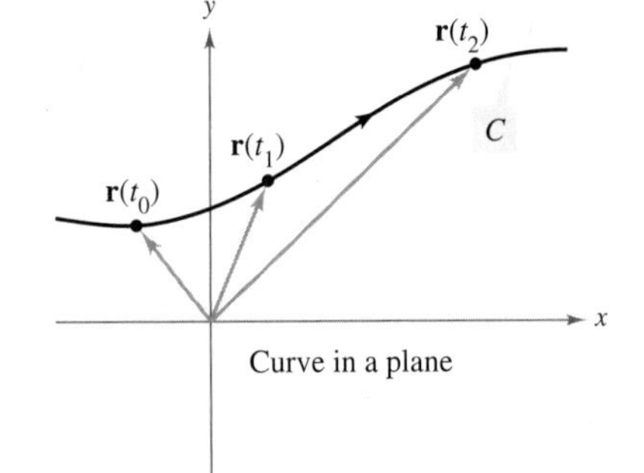

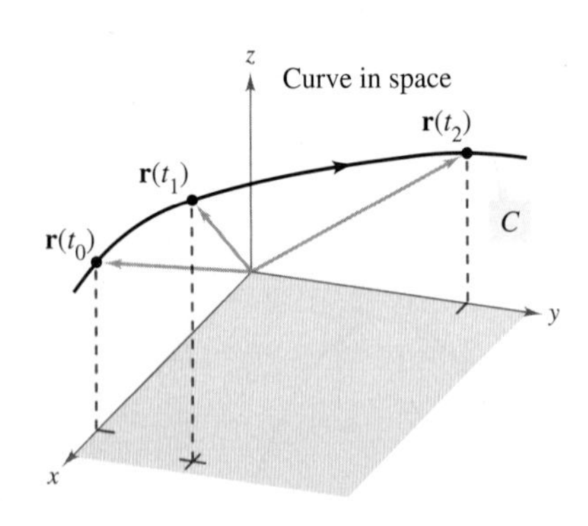

Curve C is traced out by the terminal point of position vector $\mathbf{r}(t)$.
Figure 12.1

Technically, a curve in the plane or in space consists of a collection of points and the defining parametric equations. Two different curves can have the same graph. For instance, each of the curves given by

$$\mathbf{r}(t) = \sin t\,\mathbf{i} + \cos t\,\mathbf{j} \quad \text{and} \quad \mathbf{r}(t) = \sin t^2\,\mathbf{i} + \cos t^2\,\mathbf{j}$$

has the unit circle as its graph, but these equations do not represent the same curve—because the circle is traced out in different ways on the graphs.

Be sure you see the distinction between the vector-valued function $\mathbf{r}$ and the real-valued functions f, g, and h. All are functions of the real variable t, but $\mathbf{r}(t)$ is a vector, whereas $f(t)$, $g(t)$, and $h(t)$ are real numbers (for each specific value of t).

Vector-valued functions serve dual roles in the representation of curves. By letting the parameter t represent time, you can use a vector-valued function to represent *motion* along a curve. Or, in the more general case, you can use a vector-valued function to *trace the graph* of a curve. In either case, the terminal point of the position vector $\mathbf{r}(t)$ coincides with the point (x, y) or (x, y, z) on the curve given by the parametric equations, as shown in Figure 12.1. The arrowhead on the curve indicates the curve's *orientation* by pointing in the direction of increasing values of t.

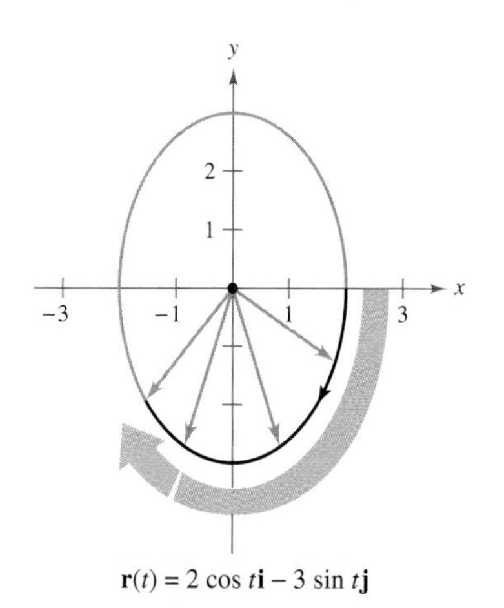

The ellipse is traced clockwise as t increases from 0 to 2π.
Figure 12.2

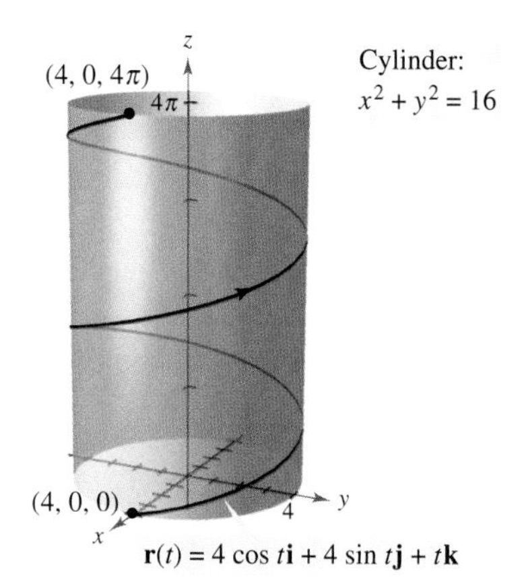

As t increases from 0 to 4π, two spirals on the helix are traced out.
Figure 12.3

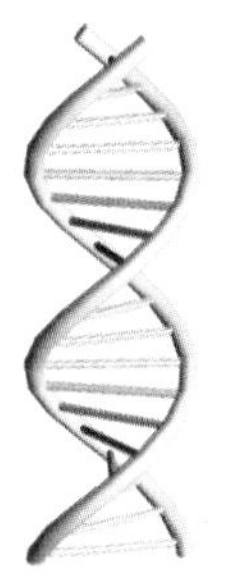

In 1953 Francis Crick and James D. Watson discovered the double helix structure of DNA.

Unless stated otherwise, the **domain** of a vector-valued function $\mathbf{r}$ is considered to be the intersection of the domains of the component functions f, g, and h. For instance, the domain of $\mathbf{r}(t) = \ln t\,\mathbf{i} + \sqrt{1-t}\,\mathbf{j} + t\mathbf{k}$ is the interval $(0, 1]$.

EXAMPLE 1 Sketching a Plane Curve

Sketch the plane curve represented by the vector-valued function

$$\mathbf{r}(t) = 2\cos t\mathbf{i} - 3\sin t\mathbf{j}, \quad 0 \le t \le 2\pi. \qquad \text{Vector-valued function}$$

Solution From the position vector $\mathbf{r}(t)$, you can write the parametric equations $x = 2\cos t$ and $y = -3\sin t$. Solving for $\cos t$ and $\sin t$ and using the identity $\cos^2 t + \sin^2 t = 1$ produces the rectangular equation

$$\frac{x^2}{2^2} + \frac{y^2}{3^2} = 1. \qquad \text{Rectangular equation}$$

The graph of this rectangular equation is the ellipse shown in Figure 12.2. The curve has a *clockwise* orientation. That is, as t increases from 0 to 2π, the position vector $\mathbf{r}(t)$ moves clockwise, and its terminal point traces the ellipse.

EXAMPLE 2 Sketching a Space Curve

Sketch the space curve represented by the vector-valued function

$$\mathbf{r}(t) = 4\cos t\mathbf{i} + 4\sin t\mathbf{j} + t\mathbf{k}, \quad 0 \le t \le 4\pi. \qquad \text{Vector-valued function}$$

Solution From the first two parametric equations $x = 4\cos t$ and $y = 4\sin t$, you can obtain

$$x^2 + y^2 = 16. \qquad \text{Rectangular equation}$$

This means that the curve lies on a right circular cylinder of radius 4, centered about the z-axis. To locate the curve on this cylinder, you can use the third parametric equation $z = t$. In Figure 12.3, note that as t increases from 0 to 4π, the point (x, y, z) spirals up the cylinder to produce a **helix.** A real-life example of a helix is shown in the drawing at the lower left. ■

In Examples 1 and 2, you were given a vector-valued function and were asked to sketch the corresponding curve. The next two examples address the reverse problem—finding a vector-valued function to represent a given graph. Of course, if the graph is described parametrically, representation by a vector-valued function is straightforward. For instance, to represent the line in space given by

$$x = 2 + t, \quad y = 3t, \quad \text{and} \quad z = 4 - t$$

you can simply use the vector-valued function given by

$$\mathbf{r}(t) = (2 + t)\mathbf{i} + 3t\mathbf{j} + (4 - t)\mathbf{k}.$$

If a set of parametric equations for the graph is not given, the problem of representing the graph by a vector-valued function boils down to finding a set of parametric equations.

The icon indicates that you will find a CAS Investigation on the book's website. The CAS Investigation is a collaborative exploration of this example using the computer algebra systems Maple *and* Mathematica.

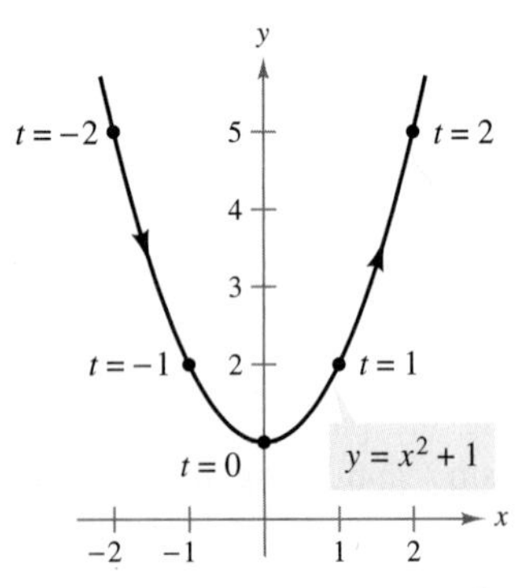

There are many ways to parametrize this graph. One way is to let $x = t$.
Figure 12.4

EXAMPLE 3 Representing a Graph by a Vector-Valued Function

Represent the parabola given by $y = x^2 + 1$ by a vector-valued function.

Solution Although there are many ways to choose the parameter t, a natural choice is to let $x = t$. Then $y = t^2 + 1$ and you have

$$\mathbf{r}(t) = t\mathbf{i} + (t^2 + 1)\mathbf{j}. \qquad \text{Vector-valued function}$$

Note in Figure 12.4 the orientation produced by this particular choice of parameter. Had you chosen $x = -t$ as the parameter, the curve would have been oriented in the opposite direction.

EXAMPLE 4 Representing a Graph by a Vector-Valued Function

Sketch the space curve C represented by the intersection of the semiellipsoid

$$\frac{x^2}{12} + \frac{y^2}{24} + \frac{z^2}{4} = 1, \quad z \geq 0$$

and the parabolic cylinder $y = x^2$. Then, find a vector-valued function to represent the graph.

Solution The intersection of the two surfaces is shown in Figure 12.5. As in Example 3, a natural choice of parameter is $x = t$. For this choice, you can use the given equation $y = x^2$ to obtain $y = t^2$. Then, it follows that

$$\frac{z^2}{4} = 1 - \frac{x^2}{12} - \frac{y^2}{24} = 1 - \frac{t^2}{12} - \frac{t^4}{24} = \frac{24 - 2t^2 - t^4}{24} = \frac{(6 + t^2)(4 - t^2)}{24}.$$

NOTE Curves in space can be specified in various ways. For instance, the curve in Example 4 is described as the intersection of two surfaces in space.

Because the curve lies above the xy-plane, you should choose the positive square root for z and obtain the following parametric equations.

$$x = t, \quad y = t^2, \quad \text{and} \quad z = \sqrt{\frac{(6 + t^2)(4 - t^2)}{6}}$$

The resulting vector-valued function is

$$\mathbf{r}(t) = t\mathbf{i} + t^2\mathbf{j} + \sqrt{\frac{(6 + t^2)(4 - t^2)}{6}}\,\mathbf{k}, \quad -2 \leq t \leq 2. \qquad \text{Vector-valued function}$$

(Note that the $\mathbf{k}$-component of $\mathbf{r}(t)$ implies $-2 \leq t \leq 2$.) From the points $(-2, 4, 0)$ and $(2, 4, 0)$ shown in Figure 12.5, you can see that the curve is traced as t increases from -2 to 2.

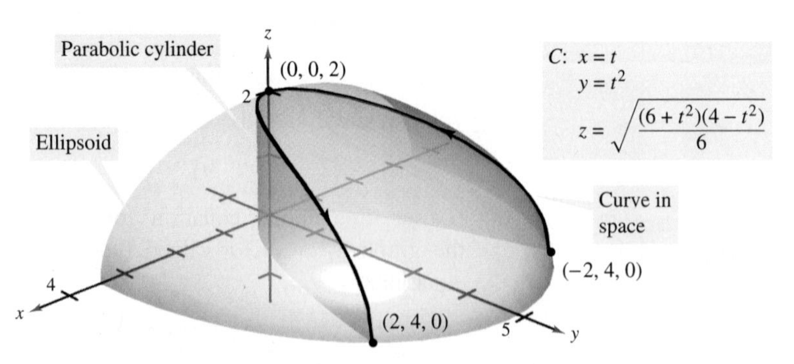

The curve C is the intersection of the semiellipsoid and the parabolic cylinder.
Figure 12.5

■

Limits and Continuity

Many techniques and definitions used in the calculus of real-valued functions can be applied to vector-valued functions. For instance, you can add and subtract vector-valued functions, multiply a vector-valued function by a scalar, take the limit of a vector-valued function, differentiate a vector-valued function, and so on. The basic approach is to capitalize on the linearity of vector operations by extending the definitions on a component-by-component basis. For example, to add or subtract two vector-valued functions (in the plane), you can write

$$\begin{aligned}\mathbf{r}_1(t) + \mathbf{r}_2(t) &= [f_1(t)\mathbf{i} + g_1(t)\mathbf{j}] + [f_2(t)\mathbf{i} + g_2(t)\mathbf{j}] \\ &= [f_1(t) + f_2(t)]\mathbf{i} + [g_1(t) + g_2(t)]\mathbf{j}\end{aligned}$$ Sum

$$\begin{aligned}\mathbf{r}_1(t) - \mathbf{r}_2(t) &= [f_1(t)\mathbf{i} + g_1(t)\mathbf{j}] - [f_2(t)\mathbf{i} + g_2(t)\mathbf{j}] \\ &= [f_1(t) - f_2(t)]\mathbf{i} + [g_1(t) - g_2(t)]\mathbf{j}.\end{aligned}$$ Difference

Similarly, to multiply and divide a vector-valued function by a scalar, you can write

$$\begin{aligned}c\mathbf{r}(t) &= c[f_1(t)\mathbf{i} + g_1(t)\mathbf{j}] \\ &= cf_1(t)\mathbf{i} + cg_1(t)\mathbf{j}\end{aligned}$$ Scalar multiplication

$$\begin{aligned}\frac{\mathbf{r}(t)}{c} &= \frac{[f_1(t)\mathbf{i} + g_1(t)\mathbf{j}]}{c}, \quad c \neq 0 \\ &= \frac{f_1(t)}{c}\mathbf{i} + \frac{g_1(t)}{c}\mathbf{j}.\end{aligned}$$ Scalar division

This component-by-component extension of operations with real-valued functions to vector-valued functions is further illustrated in the following definition of the limit of a vector-valued function.

DEFINITION OF THE LIMIT OF A VECTOR-VALUED FUNCTION

1. If $\mathbf{r}$ is a vector-valued function such that $\mathbf{r}(t) = f(t)\mathbf{i} + g(t)\mathbf{j}$, then

$$\lim_{t\to a} \mathbf{r}(t) = \left[\lim_{t\to a} f(t)\right]\mathbf{i} + \left[\lim_{t\to a} g(t)\right]\mathbf{j}$$ Plane

provided f and g have limits as $t \to a$.

2. If $\mathbf{r}$ is a vector-valued function such that $\mathbf{r}(t) = f(t)\mathbf{i} + g(t)\mathbf{j} + h(t)\mathbf{k}$, then

$$\lim_{t\to a} \mathbf{r}(t) = \left[\lim_{t\to a} f(t)\right]\mathbf{i} + \left[\lim_{t\to a} g(t)\right]\mathbf{j} + \left[\lim_{t\to a} h(t)\right]\mathbf{k}$$ Space

provided f, g, and h have limits as $t \to a$.

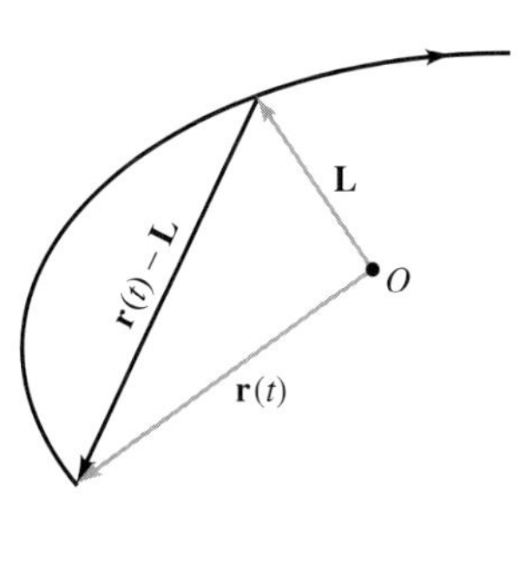

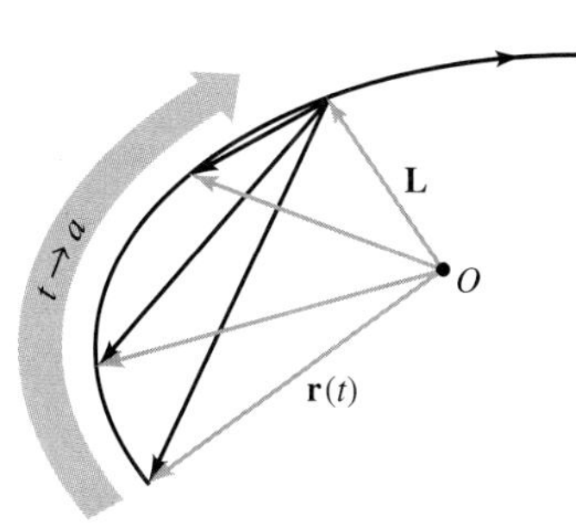

As t approaches a, $\mathbf{r}(t)$ approaches the limit $\mathbf{L}$. For the limit $\mathbf{L}$ to exist, it is not necessary that $\mathbf{r}(a)$ be defined or that $\mathbf{r}(a)$ be equal to $\mathbf{L}$.
Figure 12.6

If $\mathbf{r}(t)$ approaches the vector $\mathbf{L}$ as $t \to a$, the length of the vector $\mathbf{r}(t) - \mathbf{L}$ approaches 0. That is,

$$\|\mathbf{r}(t) - \mathbf{L}\| \to 0 \quad \text{as} \quad t \to a.$$

This is illustrated graphically in Figure 12.6. With this definition of the limit of a vector-valued function, you can develop vector versions of most of the limit theorems given in Chapter 1. For example, the limit of the sum of two vector-valued functions is the sum of their individual limits. Also, you can use the orientation of the curve $\mathbf{r}(t)$ to define one-sided limits of vector-valued functions. The next definition extends the notion of continuity to vector-valued functions.

DEFINITION OF CONTINUITY OF A VECTOR-VALUED FUNCTION

A vector-valued function $\mathbf{r}$ is **continuous at the point** given by $t = a$ if the limit of $\mathbf{r}(t)$ exists as $t \to a$ and

$$\lim_{t \to a} \mathbf{r}(t) = \mathbf{r}(a).$$

A vector-valued function $\mathbf{r}$ is **continuous on an interval** I if it is continuous at every point in the interval.

From this definition, it follows that a vector-valued function is continuous at $t = a$ if and only if each of its component functions is continuous at $t = a$.

EXAMPLE 5 Continuity of Vector-Valued Functions

Discuss the continuity of the vector-valued function given by

$$\mathbf{r}(t) = t\mathbf{i} + a\mathbf{j} + (a^2 - t^2)\mathbf{k} \qquad a \text{ is a constant.}$$

at $t = 0$.

Solution As t approaches 0, the limit is

$$\begin{aligned}\lim_{t \to 0} \mathbf{r}(t) &= \left[\lim_{t \to 0} t\right]\mathbf{i} + \left[\lim_{t \to 0} a\right]\mathbf{j} + \left[\lim_{t \to 0} (a^2 - t^2)\right]\mathbf{k} \\ &= 0\mathbf{i} + a\mathbf{j} + a^2\mathbf{k} \\ &= a\mathbf{j} + a^2\mathbf{k}.\end{aligned}$$

Because

$$\begin{aligned}\mathbf{r}(0) &= (0)\mathbf{i} + (a)\mathbf{j} + (a^2)\mathbf{k} \\ &= a\mathbf{j} + a^2\mathbf{k}\end{aligned}$$

you can conclude that $\mathbf{r}$ is continuous at $t = 0$. By similar reasoning, you can conclude that the vector-valued function $\mathbf{r}$ is continuous at all real-number values of t. ■

For each value of a, the curve represented by the vector-valued function in Example 5,

$$\mathbf{r}(t) = t\mathbf{i} + a\mathbf{j} + (a^2 - t^2)\mathbf{k} \qquad a \text{ is a constant.}$$

is a parabola. You can think of each parabola as the intersection of the vertical plane $y = a$ and the hyperbolic paraboloid

$$y^2 - x^2 = z$$

as shown in Figure 12.7.

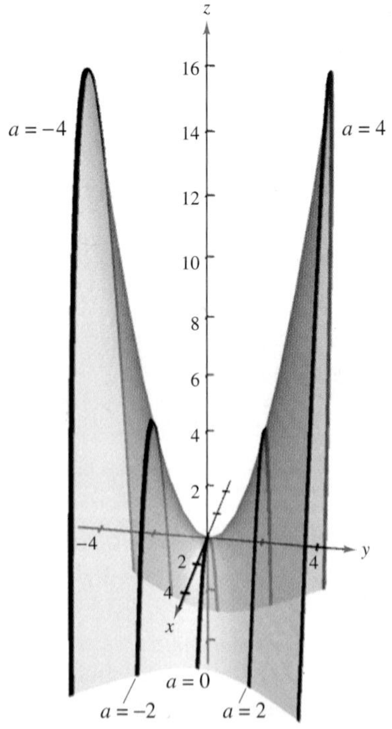

For each value of a, the curve represented by the vector-valued function $\mathbf{r}(t) = t\mathbf{i} + a\mathbf{j} + (a^2 - t^2)\mathbf{k}$ is a parabola.
Figure 12.7

TECHNOLOGY Almost any type of three-dimensional sketch is difficult to do by hand, but sketching curves in space is especially difficult. The problem is in trying to create the illusion of three dimensions. Graphing utilities use a variety of techniques to add "three-dimensionality" to graphs of space curves: one way is to show the curve on a surface, as in Figure 12.7.

12.1 Exercises

See www.CalcChat.com for worked-out solutions to odd-numbered exercises.

In Exercises 1–8, find the domain of the vector-valued function.

1. $\mathbf{r}(t) = \dfrac{1}{t+1}\mathbf{i} + \dfrac{t}{2}\mathbf{j} - 3t\mathbf{k}$

2. $\mathbf{r}(t) = \sqrt{4-t^2}\,\mathbf{i} + t^2\mathbf{j} - 6t\mathbf{k}$

3. $\mathbf{r}(t) = \ln t\,\mathbf{i} - e^t\,\mathbf{j} - t\mathbf{k}$

4. $\mathbf{r}(t) = \sin t\,\mathbf{i} + 4\cos t\,\mathbf{j} + t\mathbf{k}$

5. $\mathbf{r}(t) = \mathbf{F}(t) + \mathbf{G}(t)$ where

$\mathbf{F}(t) = \cos t\,\mathbf{i} - \sin t\,\mathbf{j} + \sqrt{t}\,\mathbf{k}, \quad \mathbf{G}(t) = \cos t\,\mathbf{i} + \sin t\,\mathbf{j}$

6. $\mathbf{r}(t) = \mathbf{F}(t) - \mathbf{G}(t)$ where

$\mathbf{F}(t) = \ln t\,\mathbf{i} + 5t\mathbf{j} - 3t^2\mathbf{k}, \quad \mathbf{G}(t) = \mathbf{i} + 4t\mathbf{j} - 3t^2\mathbf{k}$

7. $\mathbf{r}(t) = \mathbf{F}(t) \times \mathbf{G}(t)$ where

$\mathbf{F}(t) = \sin t\,\mathbf{i} + \cos t\,\mathbf{j}, \quad \mathbf{G}(t) = \sin t\,\mathbf{j} + \cos t\,\mathbf{k}$

8. $\mathbf{r}(t) = \mathbf{F}(t) \times \mathbf{G}(t)$ where

$\mathbf{F}(t) = t^3\mathbf{i} - t\mathbf{j} + t\mathbf{k}, \quad \mathbf{G}(t) = \sqrt[3]{t}\,\mathbf{i} + \dfrac{1}{t+1}\mathbf{j} + (t+2)\mathbf{k}$

In Exercises 9–12, evaluate (if possible) the vector-valued function at each given value of t.

9. $\mathbf{r}(t) = \frac{1}{2}t^2\mathbf{i} - (t-1)\mathbf{j}$

(a) $\mathbf{r}(1)$ (b) $\mathbf{r}(0)$ (c) $\mathbf{r}(s+1)$

(d) $\mathbf{r}(2+\Delta t) - \mathbf{r}(2)$

10. $\mathbf{r}(t) = \cos t\,\mathbf{i} + 2\sin t\,\mathbf{j}$

(a) $\mathbf{r}(0)$ (b) $\mathbf{r}(\pi/4)$ (c) $\mathbf{r}(\theta - \pi)$

(d) $\mathbf{r}(\pi/6 + \Delta t) - \mathbf{r}(\pi/6)$

11. $\mathbf{r}(t) = \ln t\,\mathbf{i} + \dfrac{1}{t}\mathbf{j} + 3t\mathbf{k}$

(a) $\mathbf{r}(2)$ (b) $\mathbf{r}(-3)$ (c) $\mathbf{r}(t-4)$

(d) $\mathbf{r}(1+\Delta t) - \mathbf{r}(1)$

12. $\mathbf{r}(t) = \sqrt{t}\,\mathbf{i} + t^{3/2}\mathbf{j} + e^{-t/4}\mathbf{k}$

(a) $\mathbf{r}(0)$ (b) $\mathbf{r}(4)$ (c) $\mathbf{r}(c+2)$

(d) $\mathbf{r}(9+\Delta t) - \mathbf{r}(9)$

In Exercises 13 and 14, find $\|\mathbf{r}(t)\|$.

13. $\mathbf{r}(t) = \sqrt{t}\,\mathbf{i} + 3t\mathbf{j} - 4t\mathbf{k}$

14. $\mathbf{r}(t) = \sin 3t\,\mathbf{i} + \cos 3t\,\mathbf{j} + t\mathbf{k}$

In Exercises 15–18, represent the line segment from P to Q by a vector-valued function and by a set of parametric equations.

15. $P(0, 0, 0), Q(3, 1, 2)$

16. $P(0, 2, -1), Q(4, 7, 2)$

17. $P(-2, 5, -3), Q(-1, 4, 9)$

18. $P(1, -6, 8), Q(-3, -2, 5)$

Think About It **In Exercises 19 and 20, find $\mathbf{r}(t) \cdot \mathbf{u}(t)$. Is the result a vector-valued function? Explain.**

19. $\mathbf{r}(t) = (3t-1)\mathbf{i} + \frac{1}{4}t^3\mathbf{j} + 4\mathbf{k}, \quad \mathbf{u}(t) = t^2\mathbf{i} - 8\mathbf{j} + t^3\mathbf{k}$

20. $\mathbf{r}(t) = \langle 3\cos t, 2\sin t, t-2\rangle, \quad \mathbf{u}(t) = \langle 4\sin t, -6\cos t, t^2\rangle$

In Exercises 21–24, match the equation with its graph. [The graphs are labeled (a), (b), (c), and (d).]

(a)

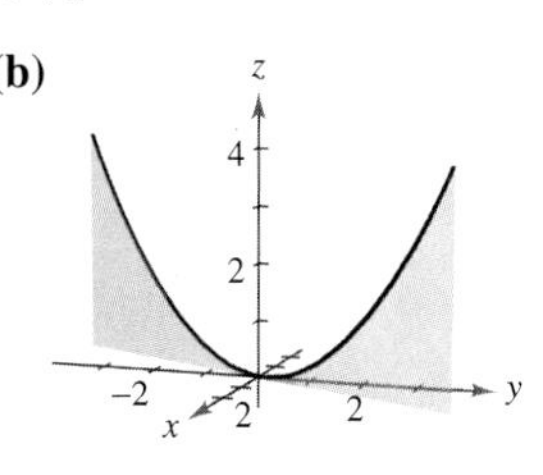

(c)

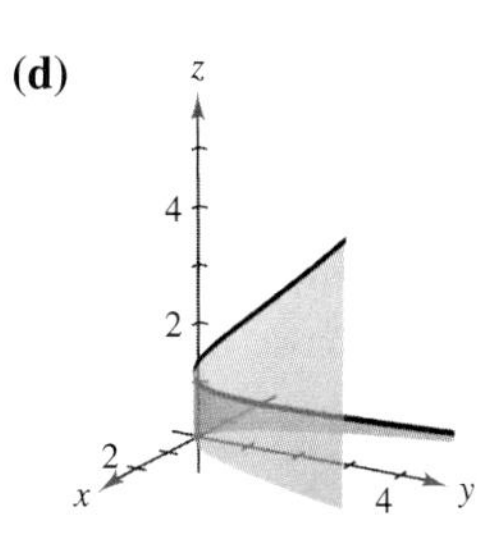

21. $\mathbf{r}(t) = t\mathbf{i} + 2t\mathbf{j} + t^2\mathbf{k}, \quad -2 \le t \le 2$

22. $\mathbf{r}(t) = \cos(\pi t)\mathbf{i} + \sin(\pi t)\mathbf{j} + t^2\mathbf{k}, \quad -1 \le t \le 1$

23. $\mathbf{r}(t) = t\mathbf{i} + t^2\mathbf{j} + e^{0.75t}\mathbf{k}, \quad -2 \le t \le 2$

24. $\mathbf{r}(t) = t\mathbf{i} + \ln t\,\mathbf{j} + \dfrac{2t}{3}\mathbf{k}, \quad 0.1 \le t \le 5$

25. ***Think About It*** The four figures below are graphs of the vector-valued function $\mathbf{r}(t) = 4\cos t\,\mathbf{i} + 4\sin t\,\mathbf{j} + (t/4)\mathbf{k}$. Match each of the four graphs with the point in space from which the helix is viewed. The four points are $(0, 0, 20)$, $(20, 0, 0)$, $(-20, 0, 0)$, and $(10, 20, 10)$.

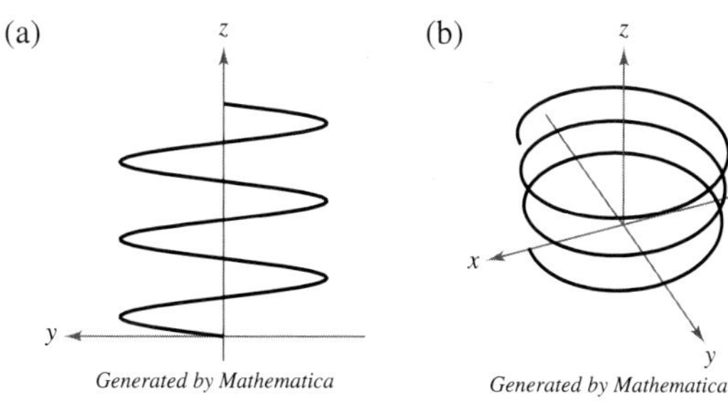

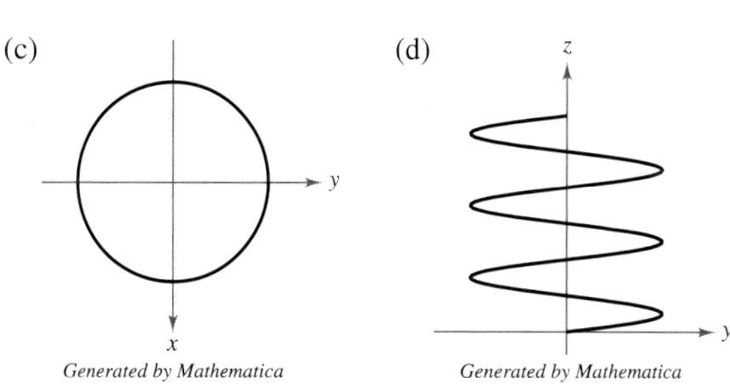

26. Sketch the three graphs of the vector-valued function $\mathbf{r}(t) = t\mathbf{i} + t\mathbf{j} + 2\mathbf{k}$ as viewed from each point.

(a) $(0, 0, 20)$ (b) $(10, 0, 0)$ (c) $(5, 5, 5)$

In Exercises 27–42, sketch the curve represented by the vector-valued function and give the orientation of the curve.

27. $\mathbf{r}(t) = \frac{t}{4}\mathbf{i} + (t - 1)\mathbf{j}$

28. $\mathbf{r}(t) = (5 - t)\mathbf{i} + \sqrt{t}\mathbf{j}$

29. $\mathbf{r}(t) = t^3\mathbf{i} + t^2\mathbf{j}$

30. $\mathbf{r}(t) = (t^2 + t)\mathbf{i} + (t^2 - t)\mathbf{j}$

31. $\mathbf{r}(\theta) = \cos\theta\mathbf{i} + 3\sin\theta\mathbf{j}$

32. $\mathbf{r}(t) = 2\cos t\mathbf{i} + 2\sin t\mathbf{j}$

33. $\mathbf{r}(\theta) = 3\sec\theta\mathbf{i} + 2\tan\theta\mathbf{j}$

34. $\mathbf{r}(t) = 2\cos^3 t\mathbf{i} + 2\sin^3 t\mathbf{j}$

35. $\mathbf{r}(t) = (-t + 1)\mathbf{i} + (4t + 2)\mathbf{j} + (2t + 3)\mathbf{k}$

36. $\mathbf{r}(t) = t\mathbf{i} + (2t - 5)\mathbf{j} + 3t\mathbf{k}$

37. $\mathbf{r}(t) = 2\cos t\mathbf{i} + 2\sin t\mathbf{j} + t\mathbf{k}$

38. $\mathbf{r}(t) = t\mathbf{i} + 3\cos t\mathbf{j} + 3\sin t\mathbf{k}$

39. $\mathbf{r}(t) = 2\sin t\mathbf{i} + 2\cos t\mathbf{j} + e^{-t}\mathbf{k}$

40. $\mathbf{r}(t) = t^2\mathbf{i} + 2t\mathbf{j} + \frac{3}{2}t\mathbf{k}$

41. $\mathbf{r}(t) = \left\langle t, t^2, \frac{2}{3}t^3\right\rangle$

42. $\mathbf{r}(t) = \langle\cos t + t\sin t, \sin t - t\cos t, t\rangle$

CAS In Exercises 43–46, use a computer algebra system to graph the vector-valued function and identify the common curve.

43. $\mathbf{r}(t) = -\frac{1}{2}t^2\mathbf{i} + t\mathbf{j} - \frac{\sqrt{3}}{2}t^2\mathbf{k}$

44. $\mathbf{r}(t) = t\mathbf{i} - \frac{\sqrt{3}}{2}t^2\mathbf{j} + \frac{1}{2}t^2\mathbf{k}$

45. $\mathbf{r}(t) = \sin t\mathbf{i} + \left(\frac{\sqrt{3}}{2}\cos t - \frac{1}{2}t\right)\mathbf{j} + \left(\frac{1}{2}\cos t + \frac{\sqrt{3}}{2}\right)\mathbf{k}$

46. $\mathbf{r}(t) = -\sqrt{2}\sin t\mathbf{i} + 2\cos t\mathbf{j} + \sqrt{2}\sin t\mathbf{k}$

CAS *Think About It* In Exercises 47 and 48, use a computer algebra system to graph the vector-valued function r(*t*). For each u(*t*), make a conjecture about the transformation (if any) of the graph of r(*t*). Use a computer algebra system to verify your conjecture.

47. $\mathbf{r}(t) = 2\cos t\mathbf{i} + 2\sin t\mathbf{j} + \frac{1}{2}t\mathbf{k}$
 (a) $\mathbf{u}(t) = 2(\cos t - 1)\mathbf{i} + 2\sin t\mathbf{j} + \frac{1}{2}t\mathbf{k}$
 (b) $\mathbf{u}(t) = 2\cos t\mathbf{i} + 2\sin t\mathbf{j} + 2t\mathbf{k}$
 (c) $\mathbf{u}(t) = 2\cos(-t)\mathbf{i} + 2\sin(-t)\mathbf{j} + \frac{1}{2}(-t)\mathbf{k}$
 (d) $\mathbf{u}(t) = \frac{1}{2}t\mathbf{i} + 2\sin t\mathbf{j} + 2\cos t\mathbf{k}$
 (e) $\mathbf{u}(t) = 6\cos t\mathbf{i} + 6\sin t\mathbf{j} + \frac{1}{2}t\mathbf{k}$

48. $\mathbf{r}(t) = t\mathbf{i} + t^2\mathbf{j} + \frac{1}{2}t^3\mathbf{k}$
 (a) $\mathbf{u}(t) = t\mathbf{i} + (t^2 - 2)\mathbf{j} + \frac{1}{2}t^3\mathbf{k}$
 (b) $\mathbf{u}(t) = t^2\mathbf{i} + t\mathbf{j} + \frac{1}{2}t^3\mathbf{k}$
 (c) $\mathbf{u}(t) = t\mathbf{i} + t^2\mathbf{j} + \left(\frac{1}{2}t^3 + 4\right)\mathbf{k}$
 (d) $\mathbf{u}(t) = t\mathbf{i} + t^2\mathbf{j} + \frac{1}{8}t^3\mathbf{k}$
 (e) $\mathbf{u}(t) = (-t)\mathbf{i} + (-t)^2\mathbf{j} + \frac{1}{2}(-t)^3\mathbf{k}$

In Exercises 49–56, represent the plane curve by a vector-valued function. (There are many correct answers.)

49. $y = x + 5$

50. $2x - 3y + 5 = 0$

51. $y = (x - 2)^2$

52. $y = 4 - x^2$

53. $x^2 + y^2 = 25$

54. $(x - 2)^2 + y^2 = 4$

55. $\frac{x^2}{16} - \frac{y^2}{4} = 1$

56. $\frac{x^2}{9} + \frac{y^2}{16} = 1$

In Exercises 57 and 58, find vector-valued functions forming the boundaries of the region in the figure. State the interval for the parameter of each function.

57.

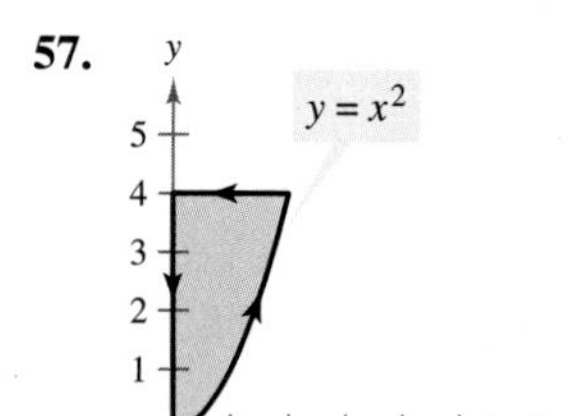

58.

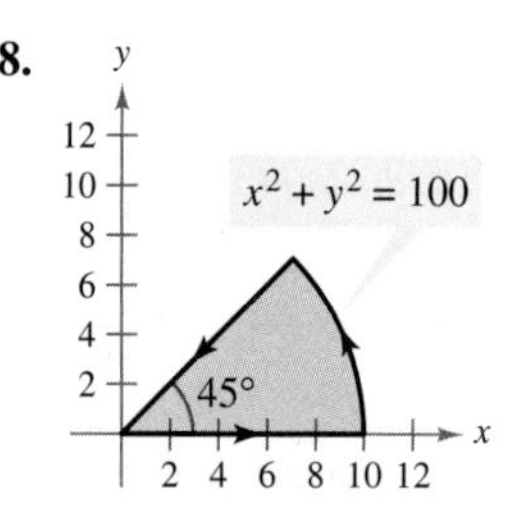

In Exercises 59–66, sketch the space curve represented by the intersection of the surfaces. Then represent the curve by a vector-valued function using the given parameter.

	Surfaces	*Parameter*
59.	$z = x^2 + y^2,\quad x + y = 0$	$x = t$
60.	$z = x^2 + y^2,\quad z = 4$	$x = 2\cos t$
61.	$x^2 + y^2 = 4,\quad z = x^2$	$x = 2\sin t$
62.	$4x^2 + 4y^2 + z^2 = 16,\quad x = z^2$	$z = t$
63.	$x^2 + y^2 + z^2 = 4,\quad x + z = 2$	$x = 1 + \sin t$
64.	$x^2 + y^2 + z^2 = 10,\quad x + y = 4$	$x = 2 + \sin t$
65.	$x^2 + z^2 = 4,\quad y^2 + z^2 = 4$	$x = t$ (first octant)
66.	$x^2 + y^2 + z^2 = 16,\quad xy = 4$	$x = t$ (first octant)

67. Show that the vector-valued function

$$\mathbf{r}(t) = t\mathbf{i} + 2t\cos t\mathbf{j} + 2t\sin t\mathbf{k}$$

lies on the cone $4x^2 = y^2 + z^2$. Sketch the curve.

68. Show that the vector-valued function

$$\mathbf{r}(t) = e^{-t}\cos t\mathbf{i} + e^{-t}\sin t\mathbf{j} + e^{-t}\mathbf{k}$$

lies on the cone $z^2 = x^2 + y^2$. Sketch the curve.

In Exercises 69–74, find the limit (if it exists).

69. $\lim_{t\to\pi}(t\mathbf{i} + \cos t\mathbf{j} + \sin t\mathbf{k})$

70. $\lim_{t\to 2}\left(3t\mathbf{i} + \frac{2}{t^2 - 1}\mathbf{j} + \frac{1}{t}\mathbf{k}\right)$

71. $\lim_{t\to 0}\left(t^2\mathbf{i} + 3t\mathbf{j} + \frac{1 - \cos t}{t}\mathbf{k}\right)$

72. $\lim_{t\to 1}\left(\sqrt{t}\mathbf{i} + \frac{\ln t}{t^2 - 1}\mathbf{j} + \frac{1}{t - 1}\mathbf{k}\right)$

73. $\lim_{t\to 0}\left(e^t\mathbf{i} + \frac{\sin t}{t}\mathbf{j} + e^{-t}\mathbf{k}\right)$

74. $\lim_{t\to\infty}\left(e^{-t}\mathbf{i} + \frac{1}{t}\mathbf{j} + \frac{t}{t^2 + 1}\mathbf{k}\right)$

In Exercises 75–80, determine the interval(s) on which the vector-valued function is continuous.

75. $\mathbf{r}(t) = t\mathbf{i} + \frac{1}{t}\mathbf{j}$

76. $\mathbf{r}(t) = \sqrt{t}\,\mathbf{i} + \sqrt{t-1}\,\mathbf{j}$

77. $\mathbf{r}(t) = t\mathbf{i} + \arcsin t\mathbf{j} + (t-1)\mathbf{k}$

78. $\mathbf{r}(t) = 2e^{-t}\mathbf{i} + e^{-t}\mathbf{j} + \ln(t-1)\mathbf{k}$

79. $\mathbf{r}(t) = \langle e^{-t}, t^2, \tan t \rangle$

80. $\mathbf{r}(t) = \langle 8, \sqrt{t}, \sqrt[3]{t} \rangle$

WRITING ABOUT CONCEPTS

81. Consider the vector-valued function

$$\mathbf{r}(t) = t^2\mathbf{i} + (t-3)\mathbf{j} + t\mathbf{k}.$$

Write a vector-valued function $\mathbf{s}(t)$ that is the specified transformation of $\mathbf{r}$.

(a) A vertical translation three units upward

(b) A horizontal translation two units in the direction of the negative x-axis

(c) A horizontal translation five units in the direction of the positive y-axis

82. State the definition of continuity of a vector-valued function. Give an example of a vector-valued function that is defined but not continuous at $t = 2$.

CAS **83.** The outer edge of a playground slide is in the shape of a helix of radius 1.5 meters. The slide has a height of 2 meters and makes one complete revolution from top to bottom. Find a vector-valued function for the helix. Use a computer algebra system to graph your function. (There are many correct answers.)

CAPSTONE

84. Which of the following vector-valued functions represent the same graph?

(a) $\mathbf{r}(t) = (-3\cos t + 1)\mathbf{i} + (5\sin t + 2)\mathbf{j} + 4\mathbf{k}$

(b) $\mathbf{r}(t) = 4\mathbf{i} + (-3\cos t + 1)\mathbf{j} + (5\sin t + 2)\mathbf{k}$

(c) $\mathbf{r}(t) = (3\cos t - 1)\mathbf{i} + (-5\sin t - 2)\mathbf{j} + 4\mathbf{k}$

(d) $\mathbf{r}(t) = (-3\cos 2t + 1)\mathbf{i} + (5\sin 2t + 2)\mathbf{j} + 4\mathbf{k}$

85. Let $\mathbf{r}(t)$ and $\mathbf{u}(t)$ be vector-valued functions whose limits exist as $t \to c$. Prove that

$$\lim_{t\to c} [\mathbf{r}(t) \times \mathbf{u}(t)] = \lim_{t\to c} \mathbf{r}(t) \times \lim_{t\to c} \mathbf{u}(t).$$

86. Let $\mathbf{r}(t)$ and $\mathbf{u}(t)$ be vector-valued functions whose limits exist as $t \to c$. Prove that

$$\lim_{t\to c} [\mathbf{r}(t) \cdot \mathbf{u}(t)] = \lim_{t\to c} \mathbf{r}(t) \cdot \lim_{t\to c} \mathbf{u}(t).$$

87. Prove that if $\mathbf{r}$ is a vector-valued function that is continuous at c, then $\|\mathbf{r}\|$ is continuous at c.

88. Verify that the converse of Exercise 87 is not true by finding a vector-valued function $\mathbf{r}$ such that $\|\mathbf{r}\|$ is continuous at c but $\mathbf{r}$ is not continuous at c.

In Exercises 89 and 90, two particles travel along the space curves r(*t*) and u(*t*). A collision will occur at the point of intersection *P* if both particles are at *P* at the same time. Do the particles collide? Do their paths intersect?

89. $\mathbf{r}(t) = t^2\mathbf{i} + (9t - 20)\mathbf{j} + t^2\mathbf{k}$

$\mathbf{u}(t) = (3t + 4)\mathbf{i} + t^2\mathbf{j} + (5t - 4)\mathbf{k}$

90. $\mathbf{r}(t) = t\mathbf{i} + t^2\mathbf{j} + t^3\mathbf{k}$

$\mathbf{u}(t) = (-2t + 3)\mathbf{i} + 8t\mathbf{j} + (12t + 2)\mathbf{k}$

Think About It **In Exercises 91 and 92, two particles travel along the space curves r(*t*) and u(*t*).**

91. If $\mathbf{r}(t)$ and $\mathbf{u}(t)$ intersect, will the particles collide?

92. If the particles collide, do their paths $\mathbf{r}(t)$ and $\mathbf{u}(t)$ intersect?

True or False? **In Exercises 93–96, determine whether the statement is true or false. If it is false, explain why or give an example that shows it is false.**

93. If f, g, and h are first-degree polynomial functions, then the curve given by $x = f(t)$, $y = g(t)$, and $z = h(t)$ is a line.

94. If the curve given by $x = f(t)$, $y = g(t)$, and $z = h(t)$ is a line, then f, g, and h are first-degree polynomial functions of t.

95. Two particles travel along the space curves $\mathbf{r}(t)$ and $\mathbf{u}(t)$. The intersection of their paths depends only on the curves traced out by $\mathbf{r}(t)$ and $\mathbf{u}(t)$, while collision depends on the parameterizations.

96. The vector-valued function $\mathbf{r}(t) = t^2\mathbf{i} + t\sin t\,\mathbf{j} + t\cos t\,\mathbf{k}$ lies on the paraboloid $x = y^2 + z^2$.

SECTION PROJECT

Witch of Agnesi

In Section 3.5, you studied a famous curve called the **Witch of Agnesi.** In this project you will take a closer look at this function.

Consider a circle of radius a centered on the y-axis at $(0, a)$. Let A be a point on the horizontal line $y = 2a$, let O be the origin, and let B be the point where the segment OA intersects the circle. A point P is on the Witch of Agnesi if P lies on the horizontal line through B and on the vertical line through A.

(a) Show that the point A is traced out by the vector-valued function

$$\mathbf{r}_A(\theta) = 2a\cot\theta\mathbf{i} + 2a\mathbf{j}, \quad 0 < \theta < \pi$$

where θ is the angle that OA makes with the positive x-axis.

(b) Show that the point B is traced out by the vector-valued function

$$\mathbf{r}_B(\theta) = a\sin 2\theta\mathbf{i} + a(1 - \cos 2\theta)\mathbf{j}, \quad 0 < \theta < \pi.$$

(c) Combine the results of parts (a) and (b) to find the vector-valued function $\mathbf{r}(\theta)$ for the Witch of Agnesi. Use a graphing utility to graph this curve for $a = 1$.

(d) Describe the limits $\lim_{\theta\to 0^+} \mathbf{r}(\theta)$ and $\lim_{\theta\to \pi^-} \mathbf{r}(\theta)$.

(e) Eliminate the parameter θ and determine the rectangular equation of the Witch of Agnesi. Use a graphing utility to graph this function for $a = 1$ and compare your graph with that obtained in part (c).

12.2 Differentiation and Integration of Vector-Valued Functions

- **Differentiate a vector-valued function.**
- **Integrate a vector-valued function.**

Differentiation of Vector-Valued Functions

In Sections 12.3–12.5, you will study several important applications involving the calculus of vector-valued functions. In preparation for that study, this section is devoted to the mechanics of differentiation and integration of vector-valued functions.

The definition of the derivative of a vector-valued function parallels the definition given for real-valued functions.

DEFINITION OF THE DERIVATIVE OF A VECTOR-VALUED FUNCTION

The **derivative of a vector-valued function r** is defined by

$$\mathbf{r}'(t) = \lim_{\Delta t \to 0} \frac{\mathbf{r}(t + \Delta t) - \mathbf{r}(t)}{\Delta t}$$

for all t for which the limit exists. If $\mathbf{r}'(t)$ exists, then $\mathbf{r}$ is **differentiable at *t*.** If $\mathbf{r}'(t)$ exists for all t in an open interval I, then $\mathbf{r}$ is **differentiable on the interval *I*.** Differentiability of vector-valued functions can be extended to closed intervals by considering one-sided limits.

NOTE In addition to $\mathbf{r}'(t)$, other notations for the derivative of a vector-valued function are

$$D_t[\mathbf{r}(t)], \quad \frac{d}{dt}[\mathbf{r}(t)], \quad \text{and} \quad \frac{d\mathbf{r}}{dt}.$$ ■

Differentiation of vector-valued functions can be done on a *component-by-component basis*. To see why this is true, consider the function given by

$$\mathbf{r}(t) = f(t)\mathbf{i} + g(t)\mathbf{j}.$$

Applying the definition of the derivative produces the following.

$$\begin{aligned}
\mathbf{r}'(t) &= \lim_{\Delta t \to 0} \frac{\mathbf{r}(t + \Delta t) - \mathbf{r}(t)}{\Delta t} \\
&= \lim_{\Delta t \to 0} \frac{f(t + \Delta t)\mathbf{i} + g(t + \Delta t)\mathbf{j} - f(t)\mathbf{i} - g(t)\mathbf{j}}{\Delta t} \\
&= \lim_{\Delta t \to 0} \left\{ \left[\frac{f(t + \Delta t) - f(t)}{\Delta t}\right]\mathbf{i} + \left[\frac{g(t + \Delta t) - g(t)}{\Delta t}\right]\mathbf{j} \right\} \\
&= \left\{ \lim_{\Delta t \to 0} \left[\frac{f(t + \Delta t) - f(t)}{\Delta t}\right] \right\}\mathbf{i} + \left\{ \lim_{\Delta t \to 0} \left[\frac{g(t + \Delta t) - g(t)}{\Delta t}\right] \right\}\mathbf{j} \\
&= f'(t)\mathbf{i} + g'(t)\mathbf{j}
\end{aligned}$$

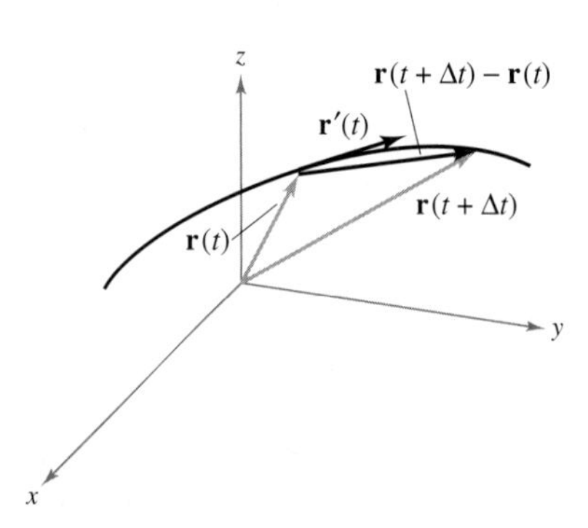

Figure 12.8

This important result is listed in the theorem on the next page. Note that the derivative of the vector-valued function $\mathbf{r}$ is itself a vector-valued function. You can see from Figure 12.8 that $\mathbf{r}'(t)$ is a vector tangent to the curve given by $\mathbf{r}(t)$ and pointing in the direction of increasing t-values.

THEOREM 12.1 DIFFERENTIATION OF VECTOR-VALUED FUNCTIONS

1. If $\mathbf{r}(t) = f(t)\mathbf{i} + g(t)\mathbf{j}$, where f and g are differentiable functions of t, then

$$\mathbf{r}'(t) = f'(t)\mathbf{i} + g'(t)\mathbf{j}. \qquad \text{Plane}$$

2. If $\mathbf{r}(t) = f(t)\mathbf{i} + g(t)\mathbf{j} + h(t)\mathbf{k}$, where f, g, and h are differentiable functions of t, then

$$\mathbf{r}'(t) = f'(t)\mathbf{i} + g'(t)\mathbf{j} + h'(t)\mathbf{k}. \qquad \text{Space}$$

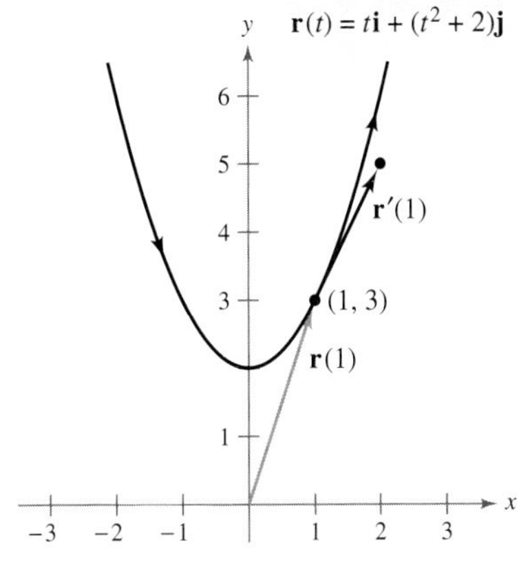

Figure 12.9

EXAMPLE 1 Differentiation of Vector-Valued Functions

For the vector-valued function given by $\mathbf{r}(t) = t\mathbf{i} + (t^2 + 2)\mathbf{j}$, find $\mathbf{r}'(t)$. Then sketch the plane curve represented by $\mathbf{r}(t)$, and the graphs of $\mathbf{r}(1)$ and $\mathbf{r}'(1)$.

Solution Differentiate on a component-by-component basis to obtain

$$\mathbf{r}'(t) = \mathbf{i} + 2t\mathbf{j}. \qquad \text{Derivative}$$

From the position vector $\mathbf{r}(t)$, you can write the parametric equations $x = t$ and $y = t^2 + 2$. The corresponding rectangular equation is $y = x^2 + 2$. When $t = 1$, $\mathbf{r}(1) = \mathbf{i} + 3\mathbf{j}$ and $\mathbf{r}'(1) = \mathbf{i} + 2\mathbf{j}$. In Figure 12.9, $\mathbf{r}(1)$ is drawn starting at the origin, and $\mathbf{r}'(1)$ is drawn starting at the terminal point of $\mathbf{r}(1)$. ■

Higher-order derivatives of vector-valued functions are obtained by successive differentiation of each component function.

EXAMPLE 2 Higher-Order Differentiation

For the vector-valued function given by $\mathbf{r}(t) = \cos t\mathbf{i} + \sin t\mathbf{j} + 2t\mathbf{k}$, find each of the following.

a. $\mathbf{r}'(t)$ **b.** $\mathbf{r}''(t)$

c. $\mathbf{r}'(t) \cdot \mathbf{r}''(t)$ **d.** $\mathbf{r}'(t) \times \mathbf{r}''(t)$

Solution

a. $\mathbf{r}'(t) = -\sin t\mathbf{i} + \cos t\mathbf{j} + 2\mathbf{k}$ First derivative

b. $\mathbf{r}''(t) = -\cos t\mathbf{i} - \sin t\mathbf{j} + 0\mathbf{k}$

$= -\cos t\mathbf{i} - \sin t\mathbf{j}$ Second derivative

c. $\mathbf{r}'(t) \cdot \mathbf{r}''(t) = \sin t \cos t - \sin t \cos t = 0$ Dot product

d. $\mathbf{r}'(t) \times \mathbf{r}''(t) = \begin{vmatrix} \mathbf{i} & \mathbf{j} & \mathbf{k} \\ -\sin t & \cos t & 2 \\ -\cos t & -\sin t & 0 \end{vmatrix}$ Cross product

$$= \begin{vmatrix} \cos t & 2 \\ -\sin t & 0 \end{vmatrix}\mathbf{i} - \begin{vmatrix} -\sin t & 2 \\ -\cos t & 0 \end{vmatrix}\mathbf{j} + \begin{vmatrix} -\sin t & \cos t \\ -\cos t & -\sin t \end{vmatrix}\mathbf{k}$$

$$= 2\sin t\mathbf{i} - 2\cos t\mathbf{j} + \mathbf{k}$$

Note that the dot product in part (c) is a *real-valued* function, not a vector-valued function. ■

The parametrization of the curve represented by the vector-valued function

$$\mathbf{r}(t) = f(t)\mathbf{i} + g(t)\mathbf{j} + h(t)\mathbf{k}$$

is **smooth on an open interval** I if f', g', and h' are continuous on I and $\mathbf{r}'(t) \neq \mathbf{0}$ for any value of t in the interval I.

EXAMPLE 3 Finding Intervals on Which a Curve Is Smooth

Find the intervals on which the epicycloid C given by

$$\mathbf{r}(t) = (5 \cos t - \cos 5t)\mathbf{i} + (5 \sin t - \sin 5t)\mathbf{j}, \quad 0 \le t \le 2\pi$$

is smooth.

Solution The derivative of $\mathbf{r}$ is

$$\mathbf{r}'(t) = (-5 \sin t + 5 \sin 5t)\mathbf{i} + (5 \cos t - 5 \cos 5t)\mathbf{j}.$$

In the interval $[0, 2\pi]$, the only values of t for which

$$\mathbf{r}'(t) = 0\mathbf{i} + 0\mathbf{j}$$

are $t = 0, \pi/2, \pi, 3\pi/2$, and 2π. Therefore, you can conclude that C is smooth in the intervals

$$\left(0, \frac{\pi}{2}\right), \left(\frac{\pi}{2}, \pi\right), \left(\pi, \frac{3\pi}{2}\right), \quad \text{and} \quad \left(\frac{3\pi}{2}, 2\pi\right)$$

as shown in Figure 12.10. ■

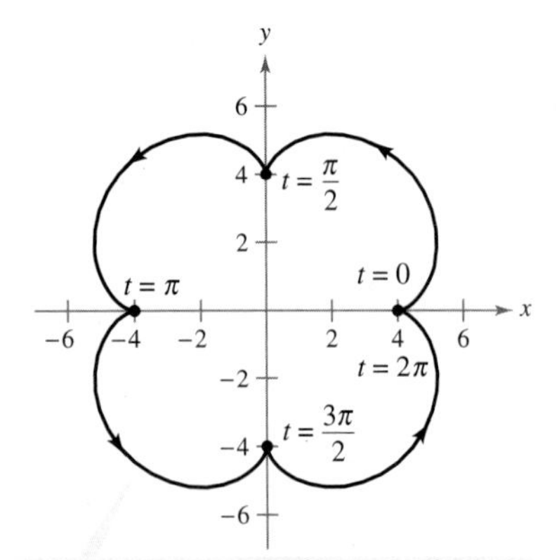

$\mathbf{r}(t) = (5 \cos t - \cos 5t)\mathbf{i} + (5 \sin t - \sin 5t)\mathbf{j}$

The epicycloid is not smooth at the points where it intersects the axes.
Figure 12.10

NOTE In Figure 12.10, note that the curve is not smooth at points at which the curve makes abrupt changes in direction. Such points are called **cusps** or **nodes.** ■

Most of the differentiation rules in Chapter 2 have counterparts for vector-valued functions, and several are listed in the following theorem. Note that the theorem contains three versions of "product rules." Property 3 gives the derivative of the product of a real-valued function w and a vector-valued function $\mathbf{r}$, Property 4 gives the derivative of the dot product of two vector-valued functions, and Property 5 gives the derivative of the cross product of two vector-valued functions (in space). Note that Property 5 applies only to three-dimensional vector-valued functions, because the cross product is not defined for two-dimensional vectors.

THEOREM 12.2 PROPERTIES OF THE DERIVATIVE

Let $\mathbf{r}$ and $\mathbf{u}$ be differentiable vector-valued functions of t, let w be a differentiable real-valued function of t, and let c be a scalar.

1. $D_t[c\mathbf{r}(t)] = c\mathbf{r}'(t)$
2. $D_t[\mathbf{r}(t) \pm \mathbf{u}(t)] = \mathbf{r}'(t) \pm \mathbf{u}'(t)$
3. $D_t[w(t)\mathbf{r}(t)] = w(t)\mathbf{r}'(t) + w'(t)\mathbf{r}(t)$
4. $D_t[\mathbf{r}(t) \cdot \mathbf{u}(t)] = \mathbf{r}(t) \cdot \mathbf{u}'(t) + \mathbf{r}'(t) \cdot \mathbf{u}(t)$
5. $D_t[\mathbf{r}(t) \times \mathbf{u}(t)] = \mathbf{r}(t) \times \mathbf{u}'(t) + \mathbf{r}'(t) \times \mathbf{u}(t)$
6. $D_t[\mathbf{r}(w(t))] = \mathbf{r}'(w(t))w'(t)$
7. If $\mathbf{r}(t) \cdot \mathbf{r}(t) = c$, then $\mathbf{r}(t) \cdot \mathbf{r}'(t) = 0$.

PROOF To prove Property 4, let

$$\mathbf{r}(t) = f_1(t)\mathbf{i} + g_1(t)\mathbf{j} \quad \text{and} \quad \mathbf{u}(t) = f_2(t)\mathbf{i} + g_2(t)\mathbf{j}$$

where f_1, f_2, g_1, and g_2 are differentiable functions of t. Then,

$$\mathbf{r}(t) \cdot \mathbf{u}(t) = f_1(t)f_2(t) + g_1(t)g_2(t)$$

and it follows that

$$\begin{aligned} D_t[\mathbf{r}(t) \cdot \mathbf{u}(t)] &= f_1(t)f_2'(t) + f_1'(t)f_2(t) + g_1(t)g_2'(t) + g_1'(t)g_2(t) \\ &= [f_1(t)f_2'(t) + g_1(t)g_2'(t)] + [f_1'(t)f_2(t) + g_1'(t)g_2(t)] \\ &= \mathbf{r}(t) \cdot \mathbf{u}'(t) + \mathbf{r}'(t) \cdot \mathbf{u}(t). \end{aligned}$$

Proofs of the other properties are left as exercises (see Exercises 77–81 and Exercise 84). ■

EXPLORATION

Let $\mathbf{r}(t) = \cos t\mathbf{i} + \sin t\mathbf{j}$. Sketch the graph of $\mathbf{r}(t)$. Explain why the graph is a circle of radius 1 centered at the origin. Calculate $\mathbf{r}(\pi/4)$ and $\mathbf{r}'(\pi/4)$. Position the vector $\mathbf{r}'(\pi/4)$ so that its initial point is at the terminal point of $\mathbf{r}(\pi/4)$. What do you observe? Show that $\mathbf{r}(t) \cdot \mathbf{r}(t)$ is constant and that $\mathbf{r}(t) \cdot \mathbf{r}'(t) = 0$ for all t. How does this example relate to Property 7 of Theorem 12.2?

EXAMPLE 4 Using Properties of the Derivative

For the vector-valued functions given by

$$\mathbf{r}(t) = \frac{1}{t}\mathbf{i} - \mathbf{j} + \ln t\mathbf{k} \quad \text{and} \quad \mathbf{u}(t) = t^2\mathbf{i} - 2t\mathbf{j} + \mathbf{k}$$

find

a. $D_t[\mathbf{r}(t) \cdot \mathbf{u}(t)]$ and **b.** $D_t[\mathbf{u}(t) \times \mathbf{u}'(t)]$.

Solution

a. Because $\mathbf{r}'(t) = -\frac{1}{t^2}\mathbf{i} + \frac{1}{t}\mathbf{k}$ and $\mathbf{u}'(t) = 2t\mathbf{i} - 2\mathbf{j}$, you have

$$\begin{aligned} D_t[\mathbf{r}(t) \cdot \mathbf{u}(t)] &= \mathbf{r}(t) \cdot \mathbf{u}'(t) + \mathbf{r}'(t) \cdot \mathbf{u}(t) \\ &= \left(\frac{1}{t}\mathbf{i} - \mathbf{j} + \ln t\mathbf{k}\right) \cdot (2t\mathbf{i} - 2\mathbf{j}) \\ &\quad + \left(-\frac{1}{t^2}\mathbf{i} + \frac{1}{t}\mathbf{k}\right) \cdot (t^2\mathbf{i} - 2t\mathbf{j} + \mathbf{k}) \\ &= 2 + 2 + (-1) + \frac{1}{t} \\ &= 3 + \frac{1}{t}. \end{aligned}$$

b. Because $\mathbf{u}'(t) = 2t\mathbf{i} - 2\mathbf{j}$ and $\mathbf{u}''(t) = 2\mathbf{i}$, you have

$$\begin{aligned} D_t[\mathbf{u}(t) \times \mathbf{u}'(t)] &= [\mathbf{u}(t) \times \mathbf{u}''(t)] + [\mathbf{u}'(t) \times \mathbf{u}'(t)] \\ &= \begin{vmatrix} \mathbf{i} & \mathbf{j} & \mathbf{k} \\ t^2 & -2t & 1 \\ 2 & 0 & 0 \end{vmatrix} + \mathbf{0} \\ &= \begin{vmatrix} -2t & 1 \\ 0 & 0 \end{vmatrix}\mathbf{i} - \begin{vmatrix} t^2 & 1 \\ 2 & 0 \end{vmatrix}\mathbf{j} + \begin{vmatrix} t^2 & -2t \\ 2 & 0 \end{vmatrix}\mathbf{k} \\ &= 0\mathbf{i} - (-2)\mathbf{j} + 4t\mathbf{k} \\ &= 2\mathbf{j} + 4t\mathbf{k}. \end{aligned}$$

■

NOTE Try reworking parts (a) and (b) in Example 4 by first forming the dot and cross products and then differentiating to see that you obtain the same results. ■

Integration of Vector-Valued Functions

The following definition is a rational consequence of the definition of the derivative of a vector-valued function.

DEFINITION OF INTEGRATION OF VECTOR-VALUED FUNCTIONS

1. If $\mathbf{r}(t) = f(t)\mathbf{i} + g(t)\mathbf{j}$, where f and g are continuous on $[a, b]$, then the **indefinite integral (antiderivative)** of $\mathbf{r}$ is

$$\int \mathbf{r}(t)\,dt = \left[\int f(t)\,dt\right]\mathbf{i} + \left[\int g(t)\,dt\right]\mathbf{j} \qquad \text{Plane}$$

and its **definite integral** over the interval $a \le t \le b$ is

$$\int_a^b \mathbf{r}(t)\,dt = \left[\int_a^b f(t)\,dt\right]\mathbf{i} + \left[\int_a^b g(t)\,dt\right]\mathbf{j}.$$

2. If $\mathbf{r}(t) = f(t)\mathbf{i} + g(t)\mathbf{j} + h(t)\mathbf{k}$, where f, g, and h are continuous on $[a, b]$, then the **indefinite integral (antiderivative)** of $\mathbf{r}$ is

$$\int \mathbf{r}(t)\,dt = \left[\int f(t)\,dt\right]\mathbf{i} + \left[\int g(t)\,dt\right]\mathbf{j} + \left[\int h(t)\,dt\right]\mathbf{k} \qquad \text{Space}$$

and its **definite integral** over the interval $a \le t \le b$ is

$$\int_a^b \mathbf{r}(t)\,dt = \left[\int_a^b f(t)\,dt\right]\mathbf{i} + \left[\int_a^b g(t)\,dt\right]\mathbf{j} + \left[\int_a^b h(t)\,dt\right]\mathbf{k}.$$

The antiderivative of a vector-valued function is a family of vector-valued functions all differing by a constant vector $\mathbf{C}$. For instance, if $\mathbf{r}(t)$ is a three-dimensional vector-valued function, then for the indefinite integral $\int \mathbf{r}(t)\,dt$, you obtain three constants of integration

$$\int f(t)\,dt = F(t) + C_1, \qquad \int g(t)\,dt = G(t) + C_2, \qquad \int h(t)\,dt = H(t) + C_3$$

where $F'(t) = f(t)$, $G'(t) = g(t)$, and $H'(t) = h(t)$. These three *scalar* constants produce one *vector* constant of integration,

$$\begin{aligned}\int \mathbf{r}(t)\,dt &= [F(t) + C_1]\mathbf{i} + [G(t) + C_2]\mathbf{j} + [H(t) + C_3]\mathbf{k} \\ &= [F(t)\mathbf{i} + G(t)\mathbf{j} + H(t)\mathbf{k}] + [C_1\mathbf{i} + C_2\mathbf{j} + C_3\mathbf{k}] \\ &= \mathbf{R}(t) + \mathbf{C}\end{aligned}$$

where $\mathbf{R}'(t) = \mathbf{r}(t)$.

EXAMPLE 5 Integrating a Vector-Valued Function

Find the indefinite integral

$$\int (t\,\mathbf{i} + 3\mathbf{j})\,dt.$$

Solution Integrating on a component-by-component basis produces

$$\int (t\,\mathbf{i} + 3\mathbf{j})\,dt = \frac{t^2}{2}\mathbf{i} + 3t\mathbf{j} + \mathbf{C}.$$

■

Example 6 shows how to evaluate the definite integral of a vector-valued function.

EXAMPLE 6 Definite Integral of a Vector-Valued Function

Evaluate the integral

$$\int_0^1 \mathbf{r}(t)\,dt = \int_0^1 \left(\sqrt[3]{t}\,\mathbf{i} + \frac{1}{t+1}\mathbf{j} + e^{-t}\mathbf{k}\right) dt.$$

Solution

$$\begin{aligned}\int_0^1 \mathbf{r}(t)\,dt &= \left(\int_0^1 t^{1/3}\,dt\right)\mathbf{i} + \left(\int_0^1 \frac{1}{t+1}\,dt\right)\mathbf{j} + \left(\int_0^1 e^{-t}\,dt\right)\mathbf{k}\\ &= \left[\left(\frac{3}{4}\right)t^{4/3}\right]_0^1 \mathbf{i} + \Big[\ln|t+1|\Big]_0^1 \mathbf{j} + \Big[-e^{-t}\Big]_0^1 \mathbf{k}\\ &= \frac{3}{4}\mathbf{i} + (\ln 2)\mathbf{j} + \left(1 - \frac{1}{e}\right)\mathbf{k}\end{aligned}$$

■

As with real-valued functions, you can narrow the family of antiderivatives of a vector-valued function $\mathbf{r}'$ down to a single antiderivative by imposing an initial condition on the vector-valued function $\mathbf{r}$. This is demonstrated in the next example.

EXAMPLE 7 The Antiderivative of a Vector-Valued Function

Find the antiderivative of

$$\mathbf{r}'(t) = \cos 2t\mathbf{i} - 2\sin t\mathbf{j} + \frac{1}{1+t^2}\mathbf{k}$$

that satisfies the initial condition $\mathbf{r}(0) = 3\mathbf{i} - 2\mathbf{j} + \mathbf{k}$.

Solution

$$\begin{aligned}\mathbf{r}(t) &= \int \mathbf{r}'(t)\,dt\\ &= \left(\int \cos 2t\,dt\right)\mathbf{i} + \left(\int -2\sin t\,dt\right)\mathbf{j} + \left(\int \frac{1}{1+t^2}\,dt\right)\mathbf{k}\\ &= \left(\frac{1}{2}\sin 2t + C_1\right)\mathbf{i} + (2\cos t + C_2)\mathbf{j} + (\arctan t + C_3)\mathbf{k}\end{aligned}$$

Letting $t = 0$ and using the fact that $\mathbf{r}(0) = 3\mathbf{i} - 2\mathbf{j} + \mathbf{k}$, you have

$$\begin{aligned}\mathbf{r}(0) &= (0 + C_1)\mathbf{i} + (2 + C_2)\mathbf{j} + (0 + C_3)\mathbf{k}\\ &= 3\mathbf{i} + (-2)\mathbf{j} + \mathbf{k}.\end{aligned}$$

Equating corresponding components produces

$$C_1 = 3, \qquad 2 + C_2 = -2, \qquad \text{and} \qquad C_3 = 1.$$

So, the antiderivative that satisfies the given initial condition is

$$\mathbf{r}(t) = \left(\frac{1}{2}\sin 2t + 3\right)\mathbf{i} + (2\cos t - 4)\mathbf{j} + (\arctan t + 1)\mathbf{k}.$$

■

12.2 Exercises

See www.CalcChat.com for worked-out solutions to odd-numbered exercises.

In Exercises 1–8, sketch the plane curve represented by the vector-valued function, and sketch the vectors $\mathbf{r}(t_0)$ and $\mathbf{r}'(t_0)$ for the given value of t_0. Position the vectors such that the initial point of $\mathbf{r}(t_0)$ is at the origin and the initial point of $\mathbf{r}'(t_0)$ is at the terminal point of $\mathbf{r}(t_0)$. What is the relationship between $\mathbf{r}'(t_0)$ and the curve?

1. $\mathbf{r}(t) = t^2\mathbf{i} + t\mathbf{j}, \quad t_0 = 2$
2. $\mathbf{r}(t) = t\mathbf{i} + (t^2 - 1)\mathbf{j}, \quad t_0 = 1$
3. $\mathbf{r}(t) = t^2\mathbf{i} + \dfrac{1}{t}\mathbf{j}, \quad t_0 = 2$
4. $\mathbf{r}(t) = (1 + t)\mathbf{i} + t^3\mathbf{j}, \quad t_0 = 1$
5. $\mathbf{r}(t) = \cos t\mathbf{i} + \sin t\mathbf{j}, \quad t_0 = \dfrac{\pi}{2}$
6. $\mathbf{r}(t) = 3 \sin t\mathbf{i} + 4 \cos t\mathbf{j}, \quad t_0 = \dfrac{\pi}{2}$
7. $\mathbf{r}(t) = \langle e^t, e^{2t} \rangle, \quad t_0 = 0$
8. $\mathbf{r}(t) = \langle e^{-t}, e^t \rangle, \quad t_0 = 0$

In Exercises 9 and 10, (a) sketch the space curve represented by the vector-valued function, and (b) sketch the vectors $\mathbf{r}(t_0)$ and $\mathbf{r}'(t_0)$ for the given value of t_0.

9. $\mathbf{r}(t) = 2 \cos t\mathbf{i} + 2 \sin t\mathbf{j} + t\mathbf{k}, \quad t_0 = \dfrac{3\pi}{2}$
10. $\mathbf{r}(t) = t\mathbf{i} + t^2\mathbf{j} + \frac{3}{2}\mathbf{k}, \quad t_0 = 2$

In Exercises 11–22, find $\mathbf{r}'(t)$.

11. $\mathbf{r}(t) = t^3\mathbf{i} - 3t\mathbf{j}$
12. $\mathbf{r}(t) = \sqrt{t}\,\mathbf{i} + (1 - t^3)\mathbf{j}$
13. $\mathbf{r}(t) = \langle 2 \cos t, 5 \sin t \rangle$
14. $\mathbf{r}(t) = \langle t \cos t, -2 \sin t \rangle$
15. $\mathbf{r}(t) = 6t\mathbf{i} - 7t^2\mathbf{j} + t^3\mathbf{k}$
16. $\mathbf{r}(t) = \dfrac{1}{t}\mathbf{i} + 16t\mathbf{j} + \dfrac{t^2}{2}\mathbf{k}$
17. $\mathbf{r}(t) = a \cos^3 t\mathbf{i} + a \sin^3 t\mathbf{j} + \mathbf{k}$
18. $\mathbf{r}(t) = 4\sqrt{t}\,\mathbf{i} + t^2\sqrt{t}\,\mathbf{j} + \ln t^2\mathbf{k}$
19. $\mathbf{r}(t) = e^{-t}\mathbf{i} + 4\mathbf{j} + 5te^t\mathbf{k}$
20. $\mathbf{r}(t) = \langle t^3, \cos 3t, \sin 3t \rangle$
21. $\mathbf{r}(t) = \langle t \sin t, t \cos t, t \rangle$
22. $\mathbf{r}(t) = \langle \arcsin t, \arccos t, 0 \rangle$

In Exercises 23–30, find (a) $\mathbf{r}'(t)$, (b) $\mathbf{r}''(t)$, and (c) $\mathbf{r}'(t) \cdot \mathbf{r}''(t)$.

23. $\mathbf{r}(t) = t^3\mathbf{i} + \frac{1}{2}t^2\mathbf{j}$
24. $\mathbf{r}(t) = (t^2 + t)\mathbf{i} + (t^2 - t)\mathbf{j}$
25. $\mathbf{r}(t) = 4 \cos t\mathbf{i} + 4 \sin t\mathbf{j}$
26. $\mathbf{r}(t) = 8 \cos t\mathbf{i} + 3 \sin t\mathbf{j}$
27. $\mathbf{r}(t) = \frac{1}{2}t^2\mathbf{i} - t\mathbf{j} + \frac{1}{6}t^3\mathbf{k}$
28. $\mathbf{r}(t) = t\mathbf{i} + (2t + 3)\mathbf{j} + (3t - 5)\mathbf{k}$
29. $\mathbf{r}(t) = \langle \cos t + t \sin t, \sin t - t \cos t, t \rangle$
30. $\mathbf{r}(t) = \langle e^{-t}, t^2, \tan t \rangle$

In Exercises 31 and 32, a vector-valued function and its graph are given. The graph also shows the unit vectors $\mathbf{r}'(t_0)/\|\mathbf{r}'(t_0)\|$ and $\mathbf{r}''(t_0)/\|\mathbf{r}''(t_0)\|$. Find these two unit vectors and identify them on the graph.

31. $\mathbf{r}(t) = \cos(\pi t)\mathbf{i} + \sin(\pi t)\mathbf{j} + t^2\mathbf{k}, \quad t_0 = -\frac{1}{4}$
32. $\mathbf{r}(t) = \frac{3}{2}t\mathbf{i} + t^2\mathbf{j} + e^{-t}\mathbf{k}, \quad t_0 = \frac{1}{4}$

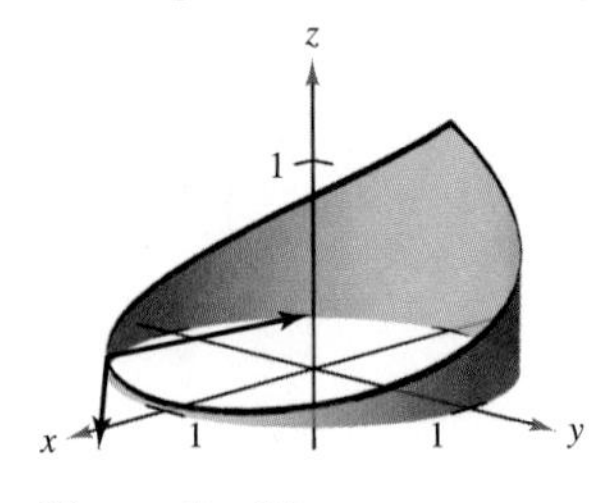

Figure for 31

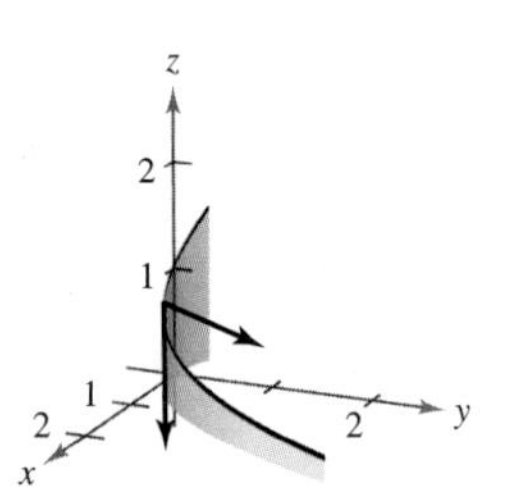

Figure for 32

In Exercises 33–42, find the open interval(s) on which the curve given by the vector-valued function is smooth.

33. $\mathbf{r}(t) = t^2\mathbf{i} + t^3\mathbf{j}$
34. $\mathbf{r}(t) = \dfrac{1}{t - 1}\mathbf{i} + 3t\mathbf{j}$
35. $\mathbf{r}(\theta) = 2 \cos^3 \theta\mathbf{i} + 3 \sin^3 \theta\mathbf{j}$
36. $\mathbf{r}(\theta) = (\theta + \sin \theta)\mathbf{i} + (1 - \cos \theta)\mathbf{j}$
37. $\mathbf{r}(\theta) = (\theta - 2 \sin \theta)\mathbf{i} + (1 - 2 \cos \theta)\mathbf{j}$
38. $\mathbf{r}(t) = \dfrac{2t}{8 + t^3}\mathbf{i} + \dfrac{2t^2}{8 + t^3}\mathbf{j}$
39. $\mathbf{r}(t) = (t - 1)\mathbf{i} + \dfrac{1}{t}\mathbf{j} - t^2\mathbf{k}$
40. $\mathbf{r}(t) = e^t\mathbf{i} - e^{-t}\mathbf{j} + 3t\mathbf{k}$
41. $\mathbf{r}(t) = t\mathbf{i} - 3t\mathbf{j} + \tan t\mathbf{k}$
42. $\mathbf{r}(t) = \sqrt{t}\,\mathbf{i} + (t^2 - 1)\mathbf{j} + \frac{1}{4}t\mathbf{k}$

In Exercises 43 and 44, use the properties of the derivative to find the following.

(a) $\mathbf{r}'(t)$ (b) $\mathbf{r}''(t)$ (c) $D_t[\mathbf{r}(t) \cdot \mathbf{u}(t)]$
(d) $D_t[3\mathbf{r}(t) - \mathbf{u}(t)]$ (e) $D_t[\mathbf{r}(t) \times \mathbf{u}(t)]$ (f) $D_t[\|\mathbf{r}(t)\|], \quad t > 0$

43. $\mathbf{r}(t) = t\mathbf{i} + 3t\mathbf{j} + t^2\mathbf{k}, \quad \mathbf{u}(t) = 4t\mathbf{i} + t^2\mathbf{j} + t^3\mathbf{k}$
44. $\mathbf{r}(t) = t\mathbf{i} + 2 \sin t\mathbf{j} + 2 \cos t\mathbf{k}$,
    $\mathbf{u}(t) = \dfrac{1}{t}\mathbf{i} + 2 \sin t\mathbf{j} + 2 \cos t\mathbf{k}$

In Exercises 45 and 46, find (a) $D_t[\mathbf{r}(t) \cdot \mathbf{u}(t)]$ and (b) $D_t[\mathbf{r}(t) \times \mathbf{u}(t)]$ in two different ways.

(i) Find the product first, then differentiate.

(ii) Apply the properties of Theorem 12.2.

45. $\mathbf{r}(t) = t\mathbf{i} + 2t^2\mathbf{j} + t^3\mathbf{k}, \quad \mathbf{u}(t) = t^4\mathbf{k}$
46. $\mathbf{r}(t) = \cos t\mathbf{i} + \sin t\mathbf{j} + t\mathbf{k}, \quad \mathbf{u}(t) = \mathbf{j} + t\mathbf{k}$

In Exercises 47 and 48, find the angle θ between $\mathbf{r}(t)$ and $\mathbf{r}'(t)$ as a function of t. Use a graphing utility to graph $\theta(t)$. Use the graph to find any extrema of the function. Find any values of t at which the vectors are orthogonal.

47. $\mathbf{r}(t) = 3 \sin t\mathbf{i} + 4 \cos t\mathbf{j}$
48. $\mathbf{r}(t) = t^2\mathbf{i} + t\mathbf{j}$

In Exercises 49–52, use the definition of the derivative to find $\mathbf{r}'(t)$.

49. $\mathbf{r}(t) = (3t + 2)\mathbf{i} + (1 - t^2)\mathbf{j}$

50. $\mathbf{r}(t) = \sqrt{t}\,\mathbf{i} + \frac{3}{t}\mathbf{j} - 2t\mathbf{k}$

51. $\mathbf{r}(t) = \langle t^2, 0, 2t\rangle$

52. $\mathbf{r}(t) = \langle 0, \sin t, 4t\rangle$

In Exercises 53–60, find the indefinite integral.

53. $\int (2t\mathbf{i} + \mathbf{j} + \mathbf{k})\, dt$

54. $\int \left(4t^3\,\mathbf{i} + 6t\mathbf{j} - 4\sqrt{t}\,\mathbf{k}\right) dt$

55. $\int \left(\frac{1}{t}\mathbf{i} + \mathbf{j} - t^{3/2}\,\mathbf{k}\right) dt$

56. $\int \left(\ln t\mathbf{i} + \frac{1}{t}\mathbf{j} + \mathbf{k}\right) dt$

57. $\int \left[(2t - 1)\mathbf{i} + 4t^3\mathbf{j} + 3\sqrt{t}\,\mathbf{k}\right] dt$

58. $\int (e^t\,\mathbf{i} + \sin t\mathbf{j} + \cos t\mathbf{k})\, dt$

59. $\int \left(\sec^2 t\mathbf{i} + \frac{1}{1 + t^2}\mathbf{j}\right) dt$

60. $\int (e^{-t}\sin t\mathbf{i} + e^{-t}\cos t\mathbf{j})\, dt$

In Exercises 61–66, evaluate the definite integral.

61. $\int_0^1 (8t\mathbf{i} + t\mathbf{j} - \mathbf{k})\, dt$

62. $\int_{-1}^1 \left(t\mathbf{i} + t^3\mathbf{j} + \sqrt[3]{t}\,\mathbf{k}\right) dt$

63. $\int_0^{\pi/2} \left[(a\cos t)\mathbf{i} + (a\sin t)\mathbf{j} + \mathbf{k}\right] dt$

64. $\int_0^{\pi/4} \left[(\sec t\tan t)\mathbf{i} + (\tan t)\mathbf{j} + (2\sin t\cos t)\mathbf{k}\right] dt$

65. $\int_0^2 (t\mathbf{i} + e^t\mathbf{j} - te^t\mathbf{k})\, dt$

66. $\int_0^3 \|t\mathbf{i} + t^2\mathbf{j}\|\, dt$

In Exercises 67–72, find $\mathbf{r}(t)$ for the given conditions.

67. $\mathbf{r}'(t) = 4e^{2t}\mathbf{i} + 3e^t\mathbf{j},\quad \mathbf{r}(0) = 2\mathbf{i}$

68. $\mathbf{r}'(t) = 3t^2\mathbf{j} + 6\sqrt{t}\,\mathbf{k},\quad \mathbf{r}(0) = \mathbf{i} + 2\mathbf{j}$

69. $\mathbf{r}''(t) = -32\mathbf{j},\quad \mathbf{r}'(0) = 600\sqrt{3}\mathbf{i} + 600\mathbf{j},\quad \mathbf{r}(0) = \mathbf{0}$

70. $\mathbf{r}''(t) = -4\cos t\mathbf{j} - 3\sin t\mathbf{k},\quad \mathbf{r}'(0) = 3\mathbf{k},\quad \mathbf{r}(0) = 4\mathbf{j}$

71. $\mathbf{r}'(t) = te^{-t^2}\mathbf{i} - e^{-t}\mathbf{j} + \mathbf{k},\quad \mathbf{r}(0) = \frac{1}{2}\mathbf{i} - \mathbf{j} + \mathbf{k}$

72. $\mathbf{r}'(t) = \frac{1}{1 + t^2}\mathbf{i} + \frac{1}{t^2}\mathbf{j} + \frac{1}{t}\mathbf{k},\quad \mathbf{r}(1) = 2\mathbf{i}$

WRITING ABOUT CONCEPTS

73. State the definition of the derivative of a vector-valued function. Describe how to find the derivative of a vector-valued function and give its geometric interpretation.

74. How do you find the integral of a vector-valued function?

75. The three components of the derivative of the vector-valued function $\mathbf{u}$ are positive at $t = t_0$. Describe the behavior of $\mathbf{u}$ at $t = t_0$.

76. The z-component of the derivative of the vector-valued function $\mathbf{u}$ is 0 for t in the domain of the function. What does this information imply about the graph of $\mathbf{u}$?

In Exercises 77–84, prove the property. In each case, assume $\mathbf{r}$, $\mathbf{u}$, and $\mathbf{v}$ are differentiable vector-valued functions of t in space, w is a differentiable real-valued function of t, and c is a scalar.

77. $D_t[c\mathbf{r}(t)] = c\mathbf{r}'(t)$

78. $D_t[\mathbf{r}(t) \pm \mathbf{u}(t)] = \mathbf{r}'(t) \pm \mathbf{u}'(t)$

79. $D_t[w(t)\mathbf{r}(t)] = w(t)\mathbf{r}'(t) + w'(t)\mathbf{r}(t)$

80. $D_t[\mathbf{r}(t) \times \mathbf{u}(t)] = \mathbf{r}(t) \times \mathbf{u}'(t) + \mathbf{r}'(t) \times \mathbf{u}(t)$

81. $D_t[\mathbf{r}(w(t))] = \mathbf{r}'(w(t))w'(t)$

82. $D_t[\mathbf{r}(t) \times \mathbf{r}'(t)] = \mathbf{r}(t) \times \mathbf{r}''(t)$

83. $D_t\{\mathbf{r}(t) \cdot [\mathbf{u}(t) \times \mathbf{v}(t)]\} = \mathbf{r}'(t) \cdot [\mathbf{u}(t) \times \mathbf{v}(t)] + \mathbf{r}(t) \cdot [\mathbf{u}'(t) \times \mathbf{v}(t)] + \mathbf{r}(t) \cdot [\mathbf{u}(t) \times \mathbf{v}'(t)]$

84. If $\mathbf{r}(t) \cdot \mathbf{r}(t)$ is a constant, then $\mathbf{r}(t) \cdot \mathbf{r}'(t) = 0$.

85. ***Particle Motion*** A particle moves in the xy-plane along the curve represented by the vector-valued function $\mathbf{r}(t) = (t - \sin t)\mathbf{i} + (1 - \cos t)\mathbf{j}$.
    (a) Use a graphing utility to graph $\mathbf{r}$. Describe the curve.
    (b) Find the minimum and maximum values of $\|\mathbf{r}'\|$ and $\|\mathbf{r}''\|$.

86. ***Particle Motion*** A particle moves in the yz-plane along the curve represented by the vector-valued function $\mathbf{r}(t) = (2\cos t)\mathbf{j} + (3\sin t)\mathbf{k}$.
    (a) Describe the curve.
    (b) Find the minimum and maximum values of $\|\mathbf{r}'\|$ and $\|\mathbf{r}''\|$.

87. Consider the vector-valued function
    $\mathbf{r}(t) = (e^t\sin t)\mathbf{i} + (e^t\cos t)\mathbf{j}$.
    Show that $\mathbf{r}(t)$ and $\mathbf{r}''(t)$ are always perpendicular to each other.

CAPSTONE

88. ***Investigation*** Consider the vector-valued function $\mathbf{r}(t) = t\mathbf{i} + (4 - t^2)\mathbf{j}$.
    (a) Sketch the graph of $\mathbf{r}(t)$. Use a graphing utility to verify your graph.
    (b) Sketch the vectors $\mathbf{r}(1)$, $\mathbf{r}(1.25)$, and $\mathbf{r}(1.25) - \mathbf{r}(1)$ on the graph in part (a).
    (c) Compare the vector $\mathbf{r}'(1)$ with the vector
    $$\frac{\mathbf{r}(1.25) - \mathbf{r}(1)}{1.25 - 1}.$$

True or False? **In Exercises 89–92, determine whether the statement is true or false. If it is false, explain why or give an example that shows it is false.**

89. If a particle moves along a sphere centered at the origin, then its derivative vector is always tangent to the sphere.

90. The definite integral of a vector-valued function is a real number.

91. $\frac{d}{dt}[\|\mathbf{r}(t)\|] = \|\mathbf{r}'(t)\|$

92. If $\mathbf{r}$ and $\mathbf{u}$ are differentiable vector-valued functions of t, then $D_t[\mathbf{r}(t) \cdot \mathbf{u}(t)] = \mathbf{r}'(t) \cdot \mathbf{u}'(t)$.

12.3 Velocity and Acceleration

- **Describe the velocity and acceleration associated with a vector-valued function.**
- **Use a vector-valued function to analyze projectile motion.**

> **EXPLORATION**
>
> ***Exploring Velocity*** Consider the circle given by
>
> $$\mathbf{r}(t) = (\cos \omega t)\mathbf{i} + (\sin \omega t)\mathbf{j}.$$
>
> Use a graphing utility in *parametric* mode to graph this circle for several values of ω. How does ω affect the velocity of the terminal point as it traces out the curve? For a given value of ω, does the speed appear constant? Does the acceleration appear constant? Explain your reasoning.
>
> 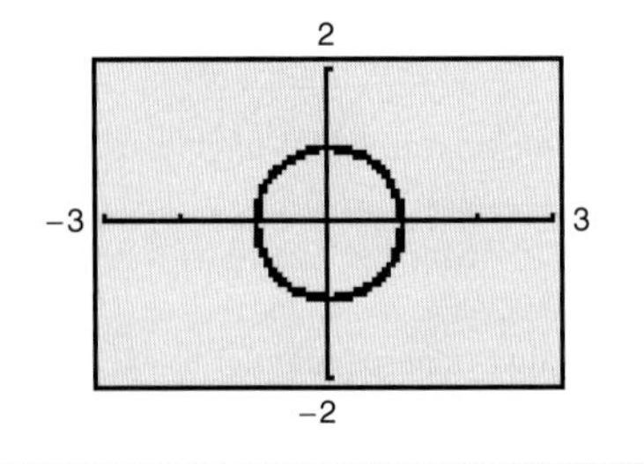
>

Velocity and Acceleration

You are now ready to combine your study of parametric equations, curves, vectors, and vector-valued functions to form a model for motion along a curve. You will begin by looking at the motion of an object in the plane. (The motion of an object in space can be developed similarly.)

As an object moves along a curve in the plane, the coordinates x and y of its center of mass are each functions of time t. Rather than using the letters f and g to represent these two functions, it is convenient to write $x = x(t)$ and $y = y(t)$. So, the position vector $\mathbf{r}(t)$ takes the form

$$\mathbf{r}(t) = x(t)\mathbf{i} + y(t)\mathbf{j}. \qquad \text{Position vector}$$

The beauty of this vector model for representing motion is that you can use the first and second derivatives of the vector-valued function $\mathbf{r}$ to find the object's velocity and acceleration. (Recall from the preceding chapter that velocity and acceleration are both vector quantities having magnitude and direction.) To find the velocity and acceleration vectors at a given time t, consider a point $Q(x(t + \Delta t), y(t + \Delta t))$ that is approaching the point $P(x(t), y(t))$ along the curve C given by $\mathbf{r}(t) = x(t)\mathbf{i} + y(t)\mathbf{j}$, as shown in Figure 12.11. As $\Delta t \to 0$, the direction of the vector $\overrightarrow{PQ}$ (denoted by $\Delta\mathbf{r}$) approaches the *direction of motion* at time t.

$$\Delta\mathbf{r} = \mathbf{r}(t + \Delta t) - \mathbf{r}(t)$$

$$\frac{\Delta\mathbf{r}}{\Delta t} = \frac{\mathbf{r}(t + \Delta t) - \mathbf{r}(t)}{\Delta t}$$

$$\lim_{\Delta t \to 0} \frac{\Delta\mathbf{r}}{\Delta t} = \lim_{\Delta t \to 0} \frac{\mathbf{r}(t + \Delta t) - \mathbf{r}(t)}{\Delta t}$$

If this limit exists, it is defined as the **velocity vector** or **tangent vector** to the curve at point P. Note that this is the same limit used to define $\mathbf{r}'(t)$. So, the direction of $\mathbf{r}'(t)$ gives the direction of motion at time t. Moreover, the magnitude of the vector $\mathbf{r}'(t)$

$$\|\mathbf{r}'(t)\| = \|x'(t)\mathbf{i} + y'(t)\mathbf{j}\| = \sqrt{[x'(t)]^2 + [y'(t)]^2}$$

gives the **speed** of the object at time t. Similarly, you can use $\mathbf{r}''(t)$ to find acceleration, as indicated in the definitions at the top of the next page.

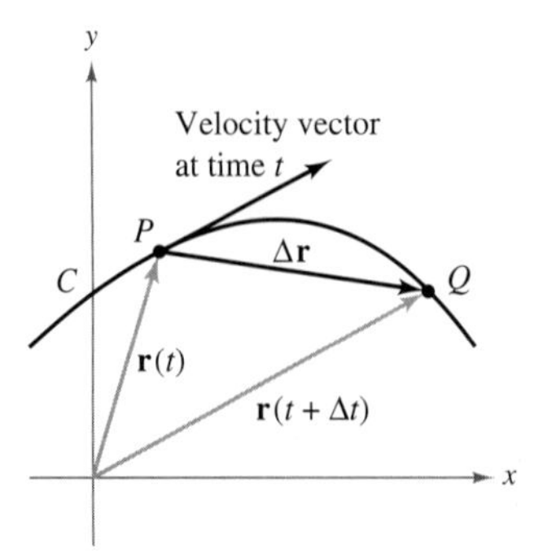

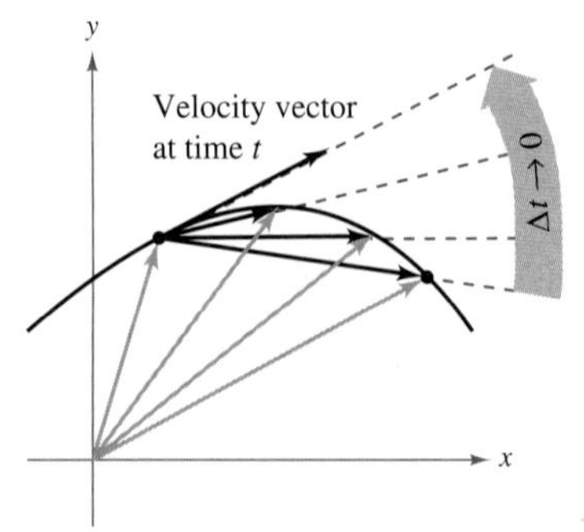

As $\Delta t \to 0$, $\dfrac{\Delta\mathbf{r}}{\Delta t}$ approaches the velocity vector.

Figure 12.11

DEFINITIONS OF VELOCITY AND ACCELERATION

If x and y are twice-differentiable functions of t, and $\mathbf{r}$ is a vector-valued function given by $\mathbf{r}(t) = x(t)\mathbf{i} + y(t)\mathbf{j}$, then the velocity vector, acceleration vector, and speed at time t are as follows.

$$\textbf{Velocity} = \mathbf{v}(t) = \mathbf{r}'(t) = x'(t)\mathbf{i} + y'(t)\mathbf{j}$$
$$\textbf{Acceleration} = \mathbf{a}(t) = \mathbf{r}''(t) = x''(t)\mathbf{i} + y''(t)\mathbf{j}$$
$$\textbf{Speed} = \|\mathbf{v}(t)\| = \|\mathbf{r}'(t)\| = \sqrt{[x'(t)]^2 + [y'(t)]^2}$$

For motion along a space curve, the definitions are similar. That is, if $\mathbf{r}(t) = x(t)\mathbf{i} + y(t)\mathbf{j} + z(t)\mathbf{k}$, you have

$$\textbf{Velocity} = \mathbf{v}(t) = \mathbf{r}'(t) = x'(t)\mathbf{i} + y'(t)\mathbf{j} + z'(t)\mathbf{k}$$
$$\textbf{Acceleration} = \mathbf{a}(t) = \mathbf{r}''(t) = x''(t)\mathbf{i} + y''(t)\mathbf{j} + z''(t)\mathbf{k}$$
$$\textbf{Speed} = \|\mathbf{v}(t)\| = \|\mathbf{r}'(t)\| = \sqrt{[x'(t)]^2 + [y'(t)]^2 + [z'(t)]^2}.$$

EXAMPLE 1 Finding Velocity and Acceleration Along a Plane Curve

NOTE In Example 1, note that the velocity and acceleration vectors are orthogonal at any point in time. This is characteristic of motion at a constant speed. (See Exercise 57.)

Find the velocity vector, speed, and acceleration vector of a particle that moves along the plane curve C described by

$$\mathbf{r}(t) = 2\sin\frac{t}{2}\mathbf{i} + 2\cos\frac{t}{2}\mathbf{j}. \qquad \text{Position vector}$$

Solution

The velocity vector is

$$\mathbf{v}(t) = \mathbf{r}'(t) = \cos\frac{t}{2}\mathbf{i} - \sin\frac{t}{2}\mathbf{j}. \qquad \text{Velocity vector}$$

The speed (at any time) is

$$\|\mathbf{r}'(t)\| = \sqrt{\cos^2\frac{t}{2} + \sin^2\frac{t}{2}} = 1. \qquad \text{Speed}$$

The acceleration vector is

$$\mathbf{a}(t) = \mathbf{r}''(t) = -\frac{1}{2}\sin\frac{t}{2}\mathbf{i} - \frac{1}{2}\cos\frac{t}{2}\mathbf{j}. \qquad \text{Acceleration vector}$$

■

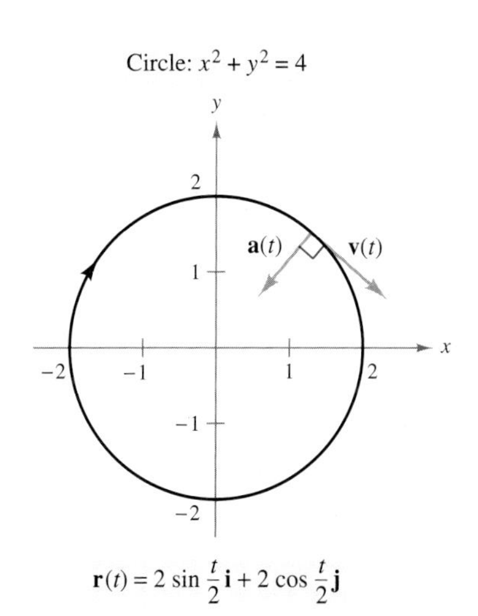

The particle moves around the circle at a constant speed.
Figure 12.12

The parametric equations for the curve in Example 1 are

$$x = 2\sin\frac{t}{2} \quad \text{and} \quad y = 2\cos\frac{t}{2}.$$

By eliminating the parameter t, you obtain the rectangular equation

$$x^2 + y^2 = 4. \qquad \text{Rectangular equation}$$

So, the curve is a circle of radius 2 centered at the origin, as shown in Figure 12.12. Because the velocity vector

$$\mathbf{v}(t) = \cos\frac{t}{2}\mathbf{i} - \sin\frac{t}{2}\mathbf{j}$$

has a constant magnitude but a changing direction as t increases, the particle moves around the circle at a constant speed.

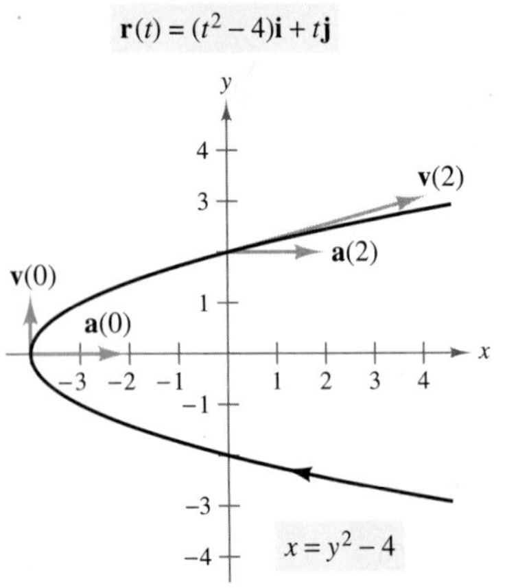

EXAMPLE 2 Sketching Velocity and Acceleration Vectors in the Plane

Sketch the path of an object moving along the plane curve given by

$$\mathbf{r}(t) = (t^2 - 4)\mathbf{i} + t\mathbf{j} \qquad \text{Position vector}$$

and find the velocity and acceleration vectors when $t = 0$ and $t = 2$.

Solution Using the parametric equations $x = t^2 - 4$ and $y = t$, you can determine that the curve is a parabola given by $x = y^2 - 4$, as shown in Figure 12.13. The velocity vector (at any time) is

$$\mathbf{v}(t) = \mathbf{r}'(t) = 2t\mathbf{i} + \mathbf{j} \qquad \text{Velocity vector}$$

and the acceleration vector (at any time) is

$$\mathbf{a}(t) = \mathbf{r}''(t) = 2\mathbf{i}. \qquad \text{Acceleration vector}$$

When $t = 0$, the velocity and acceleration vectors are given by

$$\mathbf{v}(0) = 2(0)\mathbf{i} + \mathbf{j} = \mathbf{j} \quad \text{and} \quad \mathbf{a}(0) = 2\mathbf{i}.$$

When $t = 2$, the velocity and acceleration vectors are given by

$$\mathbf{v}(2) = 2(2)\mathbf{i} + \mathbf{j} = 4\mathbf{i} + \mathbf{j} \quad \text{and} \quad \mathbf{a}(2) = 2\mathbf{i}.$$

■

$\mathbf{r}(t) = (t^2 - 4)\mathbf{i} + t\mathbf{j}$

At each point on the curve, the acceleration vector points to the right.
Figure 12.13

For the object moving along the path shown in Figure 12.13, note that the acceleration vector is constant (it has a magnitude of 2 and points to the right). This implies that the speed of the object is decreasing as the object moves toward the vertex of the parabola, and the speed is increasing as the object moves away from the vertex of the parabola.

This type of motion is *not* characteristic of comets that travel on parabolic paths through our solar system. For such comets, the acceleration vector always points to the origin (the sun), which implies that the comet's speed increases as it approaches the vertex of the path and decreases as it moves away from the vertex. (See Figure 12.14.)

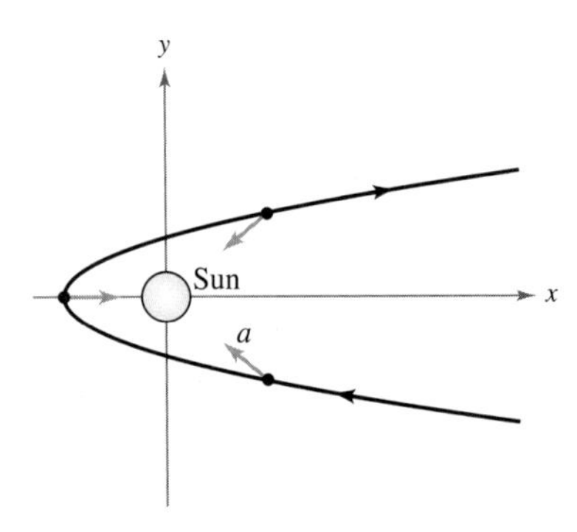

At each point in the comet's orbit, the acceleration vector points toward the sun.
Figure 12.14

EXAMPLE 3 Sketching Velocity and Acceleration Vectors in Space

Sketch the path of an object moving along the space curve C given by

$$\mathbf{r}(t) = t\mathbf{i} + t^3\mathbf{j} + 3t\mathbf{k}, \quad t \geq 0 \qquad \text{Position vector}$$

and find the velocity and acceleration vectors when $t = 1$.

Solution Using the parametric equations $x = t$ and $y = t^3$, you can determine that the path of the object lies on the cubic cylinder given by $y = x^3$. Moreover, because $z = 3t$, the object starts at $(0, 0, 0)$ and moves upward as t increases, as shown in Figure 12.15. Because $\mathbf{r}(t) = t\mathbf{i} + t^3\mathbf{j} + 3t\mathbf{k}$, you have

$$\mathbf{v}(t) = \mathbf{r}'(t) = \mathbf{i} + 3t^2\mathbf{j} + 3\mathbf{k} \qquad \text{Velocity vector}$$

and

$$\mathbf{a}(t) = \mathbf{r}''(t) = 6t\mathbf{j}. \qquad \text{Acceleration vector}$$

When $t = 1$, the velocity and acceleration vectors are given by

$$\mathbf{v}(1) = \mathbf{r}'(1) = \mathbf{i} + 3\mathbf{j} + 3\mathbf{k} \quad \text{and} \quad \mathbf{a}(1) = \mathbf{r}''(1) = 6\mathbf{j}.$$

■

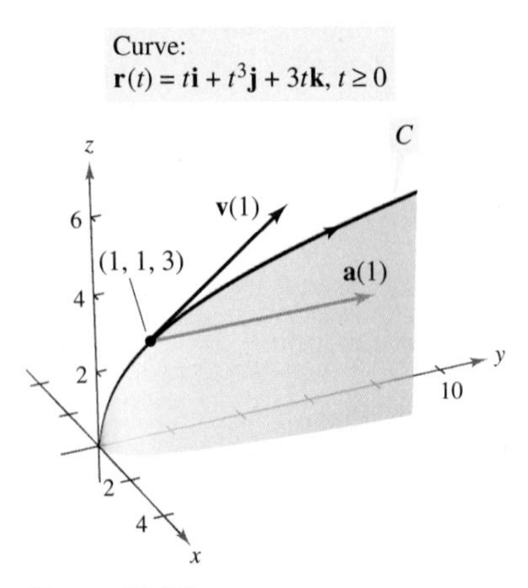

Figure 12.15

So far in this section, you have concentrated on finding the velocity and acceleration by differentiating the position function. Many practical applications involve the reverse problem—finding the position function for a given velocity or acceleration. This is demonstrated in the next example.

EXAMPLE 4 Finding a Position Function by Integration

An object starts from rest at the point $P(1, 2, 0)$ and moves with an acceleration of

$$\mathbf{a}(t) = \mathbf{j} + 2\mathbf{k} \qquad \text{Acceleration vector}$$

where $\|\mathbf{a}(t)\|$ is measured in feet per second per second. Find the location of the object after $t = 2$ seconds.

Solution From the description of the object's motion, you can deduce the following *initial conditions*. Because the object starts from rest, you have

$$\mathbf{v}(0) = \mathbf{0}.$$

Moreover, because the object starts at the point $(x, y, z) = (1, 2, 0)$, you have

$$\begin{aligned}\mathbf{r}(0) &= x(0)\mathbf{i} + y(0)\mathbf{j} + z(0)\mathbf{k}\\ &= 1\mathbf{i} + 2\mathbf{j} + 0\mathbf{k}\\ &= \mathbf{i} + 2\mathbf{j}.\end{aligned}$$

To find the position function, you should integrate twice, each time using one of the initial conditions to solve for the constant of integration. The velocity vector is

$$\begin{aligned}\mathbf{v}(t) = \int \mathbf{a}(t)\,dt &= \int (\mathbf{j} + 2\mathbf{k})\,dt\\ &= t\mathbf{j} + 2t\mathbf{k} + \mathbf{C}\end{aligned}$$

where $\mathbf{C} = C_1\mathbf{i} + C_2\mathbf{j} + C_3\mathbf{k}$. Letting $t = 0$ and applying the initial condition $\mathbf{v}(0) = \mathbf{0}$, you obtain

$$\mathbf{v}(0) = C_1\mathbf{i} + C_2\mathbf{j} + C_3\mathbf{k} = \mathbf{0} \quad \Longrightarrow \quad C_1 = C_2 = C_3 = 0.$$

So, the *velocity* at any time t is

$$\mathbf{v}(t) = t\mathbf{j} + 2t\mathbf{k}. \qquad \text{Velocity vector}$$

Integrating once more produces

$$\begin{aligned}\mathbf{r}(t) = \int \mathbf{v}(t)\,dt &= \int (t\mathbf{j} + 2t\mathbf{k})\,dt\\ &= \frac{t^2}{2}\mathbf{j} + t^2\mathbf{k} + \mathbf{C}\end{aligned}$$

where $\mathbf{C} = C_4\mathbf{i} + C_5\mathbf{j} + C_6\mathbf{k}$. Letting $t = 0$ and applying the initial condition $\mathbf{r}(0) = \mathbf{i} + 2\mathbf{j}$, you have

$$\mathbf{r}(0) = C_4\mathbf{i} + C_5\mathbf{j} + C_6\mathbf{k} = \mathbf{i} + 2\mathbf{j} \quad \Longrightarrow \quad C_4 = 1, C_5 = 2, C_6 = 0.$$

So, the *position* vector is

$$\mathbf{r}(t) = \mathbf{i} + \left(\frac{t^2}{2} + 2\right)\mathbf{j} + t^2\mathbf{k}. \qquad \text{Position vector}$$

The location of the object after $t = 2$ seconds is given by $\mathbf{r}(2) = \mathbf{i} + 4\mathbf{j} + 4\mathbf{k}$, as shown in Figure 12.16. ■

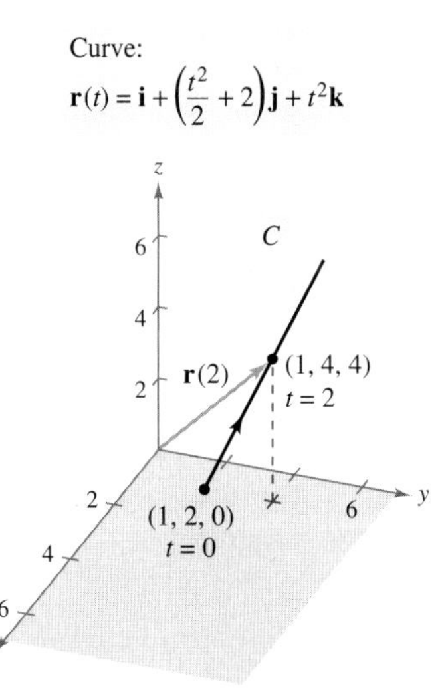

The object takes 2 seconds to move from point $(1, 2, 0)$ to point $(1, 4, 4)$ along C.
Figure 12.16

Projectile Motion

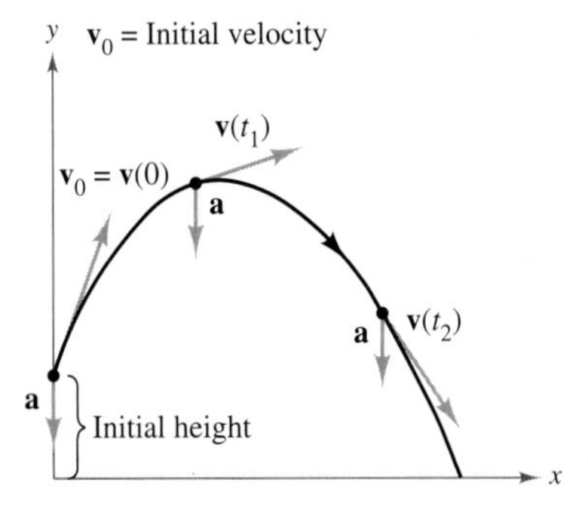

Figure 12.17

You now have the machinery to derive the parametric equations for the path of a projectile. Assume that gravity is the only force acting on the projectile after it is launched. So, the motion occurs in a vertical plane, which can be represented by the xy-coordinate system with the origin as a point on Earth's surface, as shown in Figure 12.17. For a projectile of mass m, the force due to gravity is

$$\mathbf{F} = -mg\mathbf{j} \qquad \text{Force due to gravity}$$

where the acceleration due to gravity is $g = 32$ feet per second per second, or 9.81 meters per second per second. By **Newton's Second Law of Motion,** this same force produces an acceleration $\mathbf{a} = \mathbf{a}(t)$, and satisfies the equation $\mathbf{F} = m\mathbf{a}$. Consequently, the acceleration of the projectile is given by $m\mathbf{a} = -mg\mathbf{j}$, which implies that

$$\mathbf{a} = -g\mathbf{j}. \qquad \text{Acceleration of projectile}$$

EXAMPLE 5 Derivation of the Position Function for a Projectile

A projectile of mass m is launched from an initial position $\mathbf{r}_0$ with an initial velocity $\mathbf{v}_0$. Find its position vector as a function of time.

Solution Begin with the acceleration $\mathbf{a}(t) = -g\mathbf{j}$ and integrate twice.

$$\mathbf{v}(t) = \int \mathbf{a}(t)\,dt = \int -g\mathbf{j}\,dt = -gt\mathbf{j} + \mathbf{C}_1$$

$$\mathbf{r}(t) = \int \mathbf{v}(t)\,dt = \int (-gt\mathbf{j} + \mathbf{C}_1)\,dt = -\frac{1}{2}gt^2\mathbf{j} + \mathbf{C}_1t + \mathbf{C}_2$$

You can use the facts that $\mathbf{v}(0) = \mathbf{v}_0$ and $\mathbf{r}(0) = \mathbf{r}_0$ to solve for the constant vectors $\mathbf{C}_1$ and $\mathbf{C}_2$. Doing this produces $\mathbf{C}_1 = \mathbf{v}_0$ and $\mathbf{C}_2 = \mathbf{r}_0$. Therefore, the position vector is

$$\mathbf{r}(t) = -\frac{1}{2}gt^2\mathbf{j} + t\mathbf{v}_0 + \mathbf{r}_0. \qquad \text{Position vector}$$

■

In many projectile problems, the constant vectors $\mathbf{r}_0$ and $\mathbf{v}_0$ are not given explicitly. Often you are given the initial height h, the initial speed v_0, and the angle θ at which the projectile is launched, as shown in Figure 12.18. From the given height, you can deduce that $\mathbf{r}_0 = h\mathbf{j}$. Because the speed gives the magnitude of the initial velocity, it follows that $v_0 = \|\mathbf{v}_0\|$ and you can write

$$\begin{aligned}\mathbf{v}_0 &= x\mathbf{i} + y\mathbf{j}\\ &= (\|\mathbf{v}_0\| \cos\theta)\mathbf{i} + (\|\mathbf{v}_0\| \sin\theta)\mathbf{j}\\ &= v_0\cos\theta\mathbf{i} + v_0\sin\theta\mathbf{j}.\end{aligned}$$

So, the position vector can be written in the form

$$\begin{aligned}\mathbf{r}(t) &= -\frac{1}{2}gt^2\mathbf{j} + t\mathbf{v}_0 + \mathbf{r}_0 \qquad \text{Position vector}\\ &= -\frac{1}{2}gt^2\mathbf{j} + tv_0\cos\theta\mathbf{i} + tv_0\sin\theta\mathbf{j} + h\mathbf{j}\\ &= (v_0\cos\theta)t\mathbf{i} + \left[h + (v_0\sin\theta)t - \frac{1}{2}gt^2\right]\mathbf{j}.\end{aligned}$$

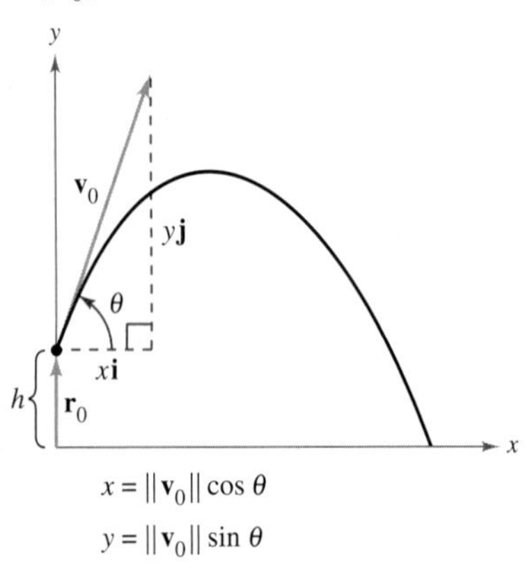

Figure 12.18

THEOREM 12.3 POSITION FUNCTION FOR A PROJECTILE

Neglecting air resistance, the path of a projectile launched from an initial height h with initial speed v_0 and angle of elevation θ is described by the vector function

$$\mathbf{r}(t) = (v_0 \cos \theta)t\mathbf{i} + \left[h + (v_0 \sin \theta)t - \frac{1}{2}gt^2\right]\mathbf{j}$$

where g is the acceleration due to gravity.

EXAMPLE 6 Describing the Path of a Baseball

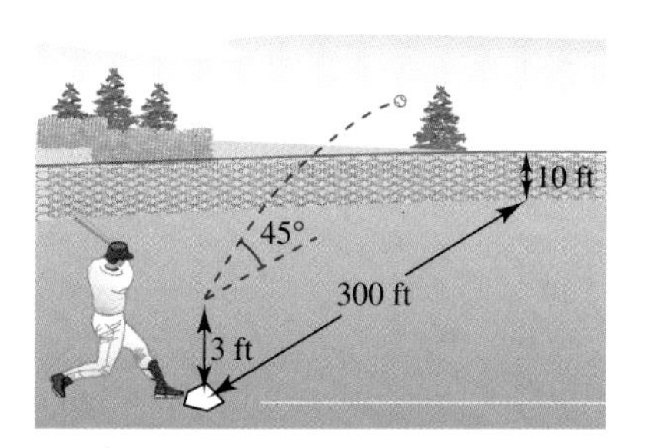

Figure 12.19

A baseball is hit 3 feet above ground level at 100 feet per second and at an angle of 45° with respect to the ground, as shown in Figure 12.19. Find the maximum height reached by the baseball. Will it clear a 10-foot-high fence located 300 feet from home plate?

Solution You are given $h = 3$, $v_0 = 100$, and $\theta = 45°$. So, using $g = 32$ feet per second per second produces

$$\begin{aligned}\mathbf{r}(t) &= \left(100 \cos \frac{\pi}{4}\right)t\mathbf{i} + \left[3 + \left(100 \sin \frac{\pi}{4}\right)t - 16t^2\right]\mathbf{j}\\ &= \left(50\sqrt{2}\,t\right)\mathbf{i} + \left(3 + 50\sqrt{2}\,t - 16t^2\right)\mathbf{j}\\ \mathbf{v}(t) &= \mathbf{r}'(t) = 50\sqrt{2}\,\mathbf{i} + \left(50\sqrt{2} - 32t\right)\mathbf{j}.\end{aligned}$$

The maximum height occurs when

$$y'(t) = 50\sqrt{2} - 32t = 0$$

which implies that

$$\begin{aligned}t &= \frac{25\sqrt{2}}{16}\\ &\approx 2.21 \text{ seconds.}\end{aligned}$$

So, the maximum height reached by the ball is

$$\begin{aligned}y &= 3 + 50\sqrt{2}\left(\frac{25\sqrt{2}}{16}\right) - 16\left(\frac{25\sqrt{2}}{16}\right)^2\\ &= \frac{649}{8}\\ &\approx 81 \text{ feet.}\end{aligned}$$

Maximum height when $t \approx 2.21$ seconds

The ball is 300 feet from where it was hit when

$$300 = x(t) = 50\sqrt{2}\,t.$$

Solving this equation for t produces $t = 3\sqrt{2} \approx 4.24$ seconds. At this time, the height of the ball is

$$\begin{aligned}y &= 3 + 50\sqrt{2}\left(3\sqrt{2}\right) - 16\left(3\sqrt{2}\right)^2\\ &= 303 - 288\\ &= 15 \text{ feet.}\end{aligned}$$

Height when $t \approx 4.24$ seconds

Therefore, the ball clears the 10-foot fence for a home run. ■

12.3 Exercises

See www.CalcChat.com for worked-out solutions to odd-numbered exercises.

In Exercises 1–10, the position vector r describes the path of an object moving in the xy-plane. Sketch a graph of the path and sketch the velocity and acceleration vectors at the given point.

Position Function	Point
1. $\mathbf{r}(t) = 3t\mathbf{i} + (t - 1)\mathbf{j}$	$(3, 0)$
2. $\mathbf{r}(t) = (6 - t)\mathbf{i} + t\mathbf{j}$	$(3, 3)$
3. $\mathbf{r}(t) = t^2\mathbf{i} + t\mathbf{j}$	$(4, 2)$
4. $\mathbf{r}(t) = t\mathbf{i} + (-t^2 + 4)\mathbf{j}$	$(1, 3)$
5. $\mathbf{r}(t) = t^2\mathbf{i} + t^3\mathbf{j}$	$(1, 1)$
6. $\mathbf{r}(t) = \left(\frac{1}{4}t^3 + 1\right)\mathbf{i} + t\mathbf{j}$	$(3, 2)$
7. $\mathbf{r}(t) = 2\cos t\mathbf{i} + 2\sin t\mathbf{j}$	$(\sqrt{2}, \sqrt{2})$
8. $\mathbf{r}(t) = 3\cos t\mathbf{i} + 2\sin t\mathbf{j}$	$(3, 0)$
9. $\mathbf{r}(t) = \langle t - \sin t, 1 - \cos t\rangle$	$(\pi, 2)$
10. $\mathbf{r}(t) = \langle e^{-t}, e^t\rangle$	$(1, 1)$

In Exercises 11–20, the position vector r describes the path of an object moving in space. Find the velocity, speed, and acceleration of the object.

11. $\mathbf{r}(t) = t\mathbf{i} + 5t\mathbf{j} + 3t\mathbf{k}$

12. $\mathbf{r}(t) = 4t\mathbf{i} + 4t\mathbf{j} + 2t\mathbf{k}$

13. $\mathbf{r}(t) = t\mathbf{i} + t^2\mathbf{j} + \dfrac{t^2}{2}\mathbf{k}$

14. $\mathbf{r}(t) = 3t\mathbf{i} + t\mathbf{j} + \dfrac{1}{4}t^2\mathbf{k}$

15. $\mathbf{r}(t) = t\mathbf{i} + t\mathbf{j} + \sqrt{9 - t^2}\,\mathbf{k}$

16. $\mathbf{r}(t) = t^2\mathbf{i} + t\mathbf{j} + 2t^{3/2}\mathbf{k}$

17. $\mathbf{r}(t) = \langle 4t, 3\cos t, 3\sin t\rangle$

18. $\mathbf{r}(t) = \langle 2\cos t, 2\sin t, t^2\rangle$

19. $\mathbf{r}(t) = \langle e^t\cos t, e^t\sin t, e^t\rangle$

20. $\mathbf{r}(t) = \left\langle \ln t, \dfrac{1}{t}, t^4\right\rangle$

Linear Approximation **In Exercises 21 and 22, the graph of the vector-valued function $\mathbf{r}(t)$ and a tangent vector to the graph at $t = t_0$ are given.**

(a) Find a set of parametric equations for the tangent line to the graph at $t = t_0$.

(b) Use the equations for the line to approximate $\mathbf{r}(t_0 + 0.1)$.

21. $\mathbf{r}(t) = \left\langle t, -t^2, \frac{1}{4}t^3\right\rangle, \quad t_0 = 1$

22. $\mathbf{r}(t) = \left\langle t, \sqrt{25 - t^2}, \sqrt{25 - t^2}\right\rangle, \quad t_0 = 3$

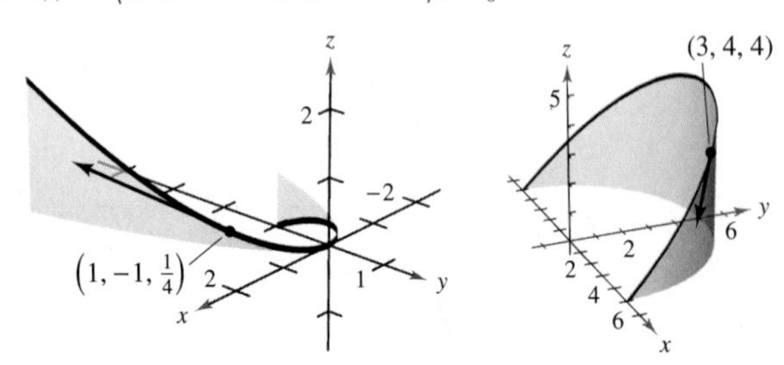

Figure for 21

Figure for 22

In Exercises 23–28, use the given acceleration function to find the velocity and position vectors. Then find the position at time $t = 2$.

23. $\mathbf{a}(t) = \mathbf{i} + \mathbf{j} + \mathbf{k}$
$\mathbf{v}(0) = \mathbf{0}, \quad \mathbf{r}(0) = \mathbf{0}$

24. $\mathbf{a}(t) = 2\mathbf{i} + 3\mathbf{k}$
$\mathbf{v}(0) = 4\mathbf{j}, \quad \mathbf{r}(0) = \mathbf{0}$

25. $\mathbf{a}(t) = t\mathbf{j} + t\mathbf{k}$
$\mathbf{v}(1) = 5\mathbf{j}, \quad \mathbf{r}(1) = \mathbf{0}$

26. $\mathbf{a}(t) = -32\mathbf{k}$
$\mathbf{v}(0) = 3\mathbf{i} - 2\mathbf{j} + \mathbf{k}, \quad \mathbf{r}(0) = 5\mathbf{j} + 2\mathbf{k}$

27. $\mathbf{a}(t) = -\cos t\mathbf{i} - \sin t\mathbf{j}$
$\mathbf{v}(0) = \mathbf{j} + \mathbf{k}, \quad \mathbf{r}(0) = \mathbf{i}$

28. $\mathbf{a}(t) = e^t\mathbf{i} - 8\mathbf{k}$
$\mathbf{v}(0) = 2\mathbf{i} + 3\mathbf{j} + \mathbf{k}, \quad \mathbf{r}(0) = \mathbf{0}$

Projectile Motion **In Exercises 29–44, use the model for projectile motion, assuming there is no air resistance.**

29. Find the vector-valued function for the path of a projectile launched at a height of 10 feet above the ground with an initial velocity of 88 feet per second and at an angle of 30° above the horizontal. Use a graphing utility to graph the path of the projectile.

30. Determine the maximum height and range of a projectile fired at a height of 3 feet above the ground with an initial velocity of 900 feet per second and at an angle of 45° above the horizontal.

31. A baseball, hit 3 feet above the ground, leaves the bat at an angle of 45° and is caught by an outfielder 3 feet above the ground and 300 feet from home plate. What is the initial speed of the ball, and how high does it rise?

32. A baseball player at second base throws a ball 90 feet to the player at first base. The ball is released at a point 5 feet above the ground with an initial velocity of 50 miles per hour and at an angle of 15° above the horizontal. At what height does the player at first base catch the ball?

33. Eliminate the parameter t from the position function for the motion of a projectile to show that the rectangular equation is

$$y = -\frac{16\sec^2\theta}{v_0^2}x^2 + (\tan\theta)x + h.$$

34. The path of a ball is given by the rectangular equation

$$y = x - 0.005x^2.$$

Use the result of Exercise 33 to find the position function. Then find the speed and direction of the ball at the point at which it has traveled 60 feet horizontally.

35. *Modeling Data* After the path of a ball thrown by a baseball player is videotaped, it is analyzed on a television set with a grid covering the screen. The tape is paused three times and the positions of the ball are measured. The coordinates are approximately (0, 6.0), (15, 10.6), and (30, 13.4). (The x-coordinate measures the horizontal distance from the player in feet and the y-coordinate measures the height in feet.)
   (a) Use a graphing utility to find a quadratic model for the data.
   (b) Use a graphing utility to plot the data and graph the model.
   (c) Determine the maximum height of the ball.
   (d) Find the initial velocity of the ball and the angle at which it was thrown.

36. A baseball is hit from a height of 2.5 feet above the ground with an initial velocity of 140 feet per second and at an angle of 22° above the horizontal. Use a graphing utility to graph the path of the ball and determine whether it will clear a 10-foot-high fence located 375 feet from home plate.

37. The Rogers Centre in Toronto, Ontario has a center field fence that is 10 feet high and 400 feet from home plate. A ball is hit 3 feet above the ground and leaves the bat at a speed of 100 miles per hour.
   (a) The ball leaves the bat at an angle of $\theta = \theta_0$ with the horizontal. Write the vector-valued function for the path of the ball.
   (b) Use a graphing utility to graph the vector-valued function for $\theta_0 = 10°$, $\theta_0 = 15°$, $\theta_0 = 20°$, and $\theta_0 = 25°$. Use the graphs to approximate the minimum angle required for the hit to be a home run.
   (c) Determine analytically the minimum angle required for the hit to be a home run.

38. The quarterback of a football team releases a pass at a height of 7 feet above the playing field, and the football is caught by a receiver 30 yards directly downfield at a height of 4 feet. The pass is released at an angle of 35° with the horizontal.
   (a) Find the speed of the football when it is released.
   (b) Find the maximum height of the football.
   (c) Find the time the receiver has to reach the proper position after the quarterback releases the football.

39. A bale ejector consists of two variable-speed belts at the end of a baler. Its purpose is to toss bales into a trailing wagon. In loading the back of a wagon, a bale must be thrown to a position 8 feet above and 16 feet behind the ejector.
   (a) Find the minimum initial speed of the bale and the corresponding angle at which it must be ejected from the baler.
   (b) The ejector has a fixed angle of 45°. Find the initial speed required.

40. A bomber is flying at an altitude of 30,000 feet at a speed of 540 miles per hour (see figure). When should the bomb be released for it to hit the target? (Give your answer in terms of the angle of depression from the plane to the target.) What is the speed of the bomb at the time of impact?

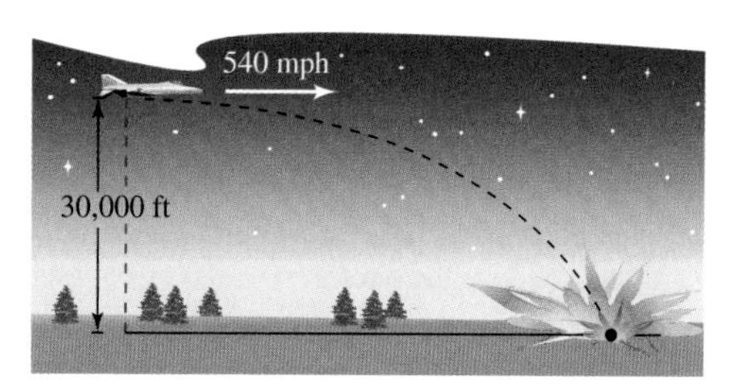

Figure for 40

41. A shot fired from a gun with a muzzle velocity of 1200 feet per second is to hit a target 3000 feet away. Determine the minimum angle of elevation of the gun.

42. A projectile is fired from ground level at an angle of 12° with the horizontal. The projectile is to have a range of 200 feet. Find the minimum initial velocity necessary.

43. Use a graphing utility to graph the paths of a projectile for the given values of θ and v_0. For each case, use the graph to approximate the maximum height and range of the projectile. (Assume that the projectile is launched from ground level.)
   (a) $\theta = 10°$, $v_0 = 66$ ft/sec (b) $\theta = 10°$, $v_0 = 146$ ft/sec
   (c) $\theta = 45°$, $v_0 = 66$ ft/sec (d) $\theta = 45°$, $v_0 = 146$ ft/sec
   (e) $\theta = 60°$, $v_0 = 66$ ft/sec (f) $\theta = 60°$, $v_0 = 146$ ft/sec

44. Find the angles at which an object must be thrown to obtain (a) the maximum range and (b) the maximum height.

Projectile Motion **In Exercises 45 and 46, use the model for projectile motion, assuming there is no air resistance. [$a(t) = -9.8$ meters per second per second]**

45. Determine the maximum height and range of a projectile fired at a height of 1.5 meters above the ground with an initial velocity of 100 meters per second and at an angle of 30° above the horizontal.

46. A projectile is fired from ground level at an angle of 8° with the horizontal. The projectile is to have a range of 50 meters. Find the minimum initial velocity necessary.

Cycloidal Motion **In Exercises 47 and 48, consider the motion of a point (or particle) on the circumference of a rolling circle. As the circle rolls, it generates the cycloid $\mathbf{r}(t) = b(\omega t - \sin \omega t)\mathbf{i} + b(1 - \cos \omega t)\mathbf{j}$, where ω is the constant angular velocity of the circle and b is the radius of the circle.**

47. Find the velocity and acceleration vectors of the particle. Use the results to determine the times at which the speed of the particle will be (a) zero and (b) maximized.

48. Find the maximum speed of a point on the circumference of an automobile tire of radius 1 foot when the automobile is traveling at 60 miles per hour. Compare this speed with the speed of the automobile.

Circular Motion **In Exercises 49–52, consider a particle moving on a circular path of radius b described by $\mathbf{r}(t) = b \cos \omega t\,\mathbf{i} + b \sin \omega t\,\mathbf{j}$, where $\omega = du/dt$ is the constant angular velocity.**

49. Find the velocity vector and show that it is orthogonal to $\mathbf{r}(t)$.

50. (a) Show that the speed of the particle is $b\omega$.

(b) Use a graphing utility in *parametric* mode to graph the circle for $b = 6$. Try different values of ω. Does the graphing utility draw the circle faster for greater values of ω?

51. Find the acceleration vector and show that its direction is always toward the center of the circle.

52. Show that the magnitude of the acceleration vector is $b\omega^2$.

Circular Motion **In Exercises 53 and 54, use the results of Exercises 49–52.**

53. A stone weighing 1 pound is attached to a two-foot string and is whirled horizontally (see figure). The string will break under a force of 10 pounds. Find the maximum speed the stone can attain without breaking the string. $\left(\text{Use } \mathbf{F} = m\mathbf{a}, \text{ where } m = \frac{1}{32}.\right)$

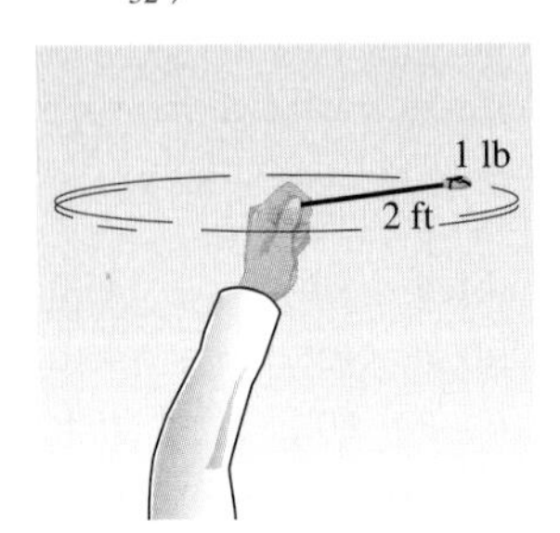

Figure for 53

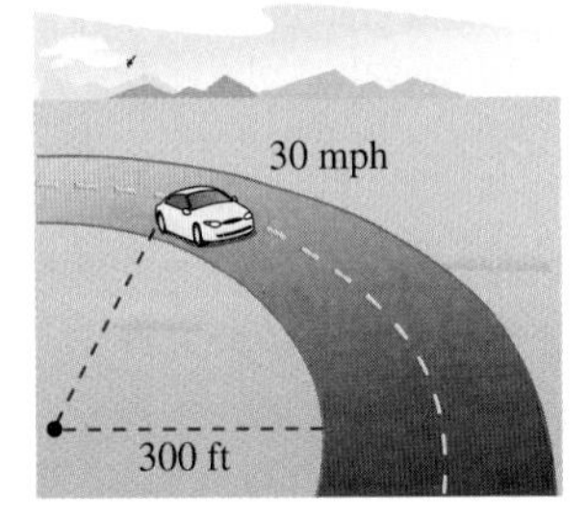

Figure for 54

54. A 3400-pound automobile is negotiating a circular interchange of radius 300 feet at 30 miles per hour (see figure). Assuming the roadway is level, find the force between the tires and the road such that the car stays on the circular path and does not skid. (Use $\mathbf{F} = m\mathbf{a}$, where $m = 3400/32$.) Find the angle at which the roadway should be banked so that no lateral frictional force is exerted on the tires of the automobile.

55. ***Shot-Put Throw*** The path of a shot thrown at an angle θ is

$$\mathbf{r}(t) = (v_0 \cos \theta)t\mathbf{i} + \left[h + (v_0 \sin \theta)t - \frac{1}{2}gt^2\right]\mathbf{j}$$

where v_0 is the initial speed, h is the initial height, t is the time in seconds, and g is the acceleration due to gravity. Verify that the shot will remain in the air for a total of

$$t = \frac{v_0 \sin \theta + \sqrt{v_0^2 \sin^2 \theta + 2gh}}{g} \text{ seconds}$$

and will travel a horizontal distance of

$$\frac{v_0^2 \cos \theta}{g}\left(\sin \theta + \sqrt{\sin^2 \theta + \frac{2gh}{v_0^2}}\right) \text{ feet.}$$

56. ***Shot-Put Throw*** A shot is thrown from a height of $h = 6$ feet with an initial speed of $v_0 = 45$ feet per second and at an angle of $\theta = 42.5°$ with the horizontal. Find the total time of travel and the total horizontal distance traveled.

57. Prove that if an object is traveling at a constant speed, its velocity and acceleration vectors are orthogonal.

58. Prove that an object moving in a straight line at a constant speed has an acceleration of 0.

59. ***Investigation*** A particle moves on an elliptical path given by the vector-valued function $\mathbf{r}(t) = 6 \cos t\mathbf{i} + 3 \sin t\mathbf{j}$.

(a) Find $\mathbf{v}(t)$, $\|\mathbf{v}(t)\|$, and $\mathbf{a}(t)$.

(b) Use a graphing utility to complete the table.

t	0	$\frac{\pi}{4}$	$\frac{\pi}{2}$	$\frac{2\pi}{3}$	π
Speed					

(c) Graph the elliptical path and the velocity and acceleration vectors at the values of t given in the table in part (b).

(d) Use the results of parts (b) and (c) to describe the geometric relationship between the velocity and acceleration vectors when the speed of the particle is increasing, and when it is decreasing.

CAPSTONE

60. Consider a particle moving on an elliptical path described by $\mathbf{r}(t) = a \cos \omega t\mathbf{i} + b \sin \omega t\mathbf{j}$, where $\omega = d\theta/dt$ is the constant angular velocity.

(a) Find the velocity vector. What is the speed of the particle?

(b) Find the acceleration vector and show that its direction is always toward the center of the ellipse.

WRITING ABOUT CONCEPTS

61. In your own words, explain the difference between the velocity of an object and its speed.

62. What is known about the speed of an object if the angle between the velocity and acceleration vectors is (a) acute and (b) obtuse?

63. Consider a particle that is moving on the path $\mathbf{r}_1(t) = x(t)\mathbf{i} + y(t)\mathbf{j} + z(t)\mathbf{k}$.

(a) Discuss any changes in the position, velocity, or acceleration of the particle if its position is given by the vector-valued function $\mathbf{r}_2(t) = \mathbf{r}_1(2t)$.

(b) Generalize the results for the position function $\mathbf{r}_3(t) = \mathbf{r}_1(\omega t)$.

64. When $t = 0$, an object is at the point $(0, 1)$ and has a velocity vector $\mathbf{v}(0) = -\mathbf{i}$. It moves with an acceleration of $\mathbf{a}(t) = \sin t\mathbf{i} - \cos t\mathbf{j}$. Show that the path of the object is a circle.

True or False? **In Exercises 65–68, determine whether the statement is true or false. If it is false, explain why or give an example that shows it is false.**

65. The acceleration of an object is the derivative of the speed.

66. The velocity of an object is the derivative of the position.

67. The velocity vector points in the direction of motion.

68. If a particle moves along a straight line, then the velocity and acceleration vectors are orthogonal.

12.4 Tangent Vectors and Normal Vectors

- **Find a unit tangent vector at a point on a space curve.**
- **Find the tangential and normal components of acceleration.**

Tangent Vectors and Normal Vectors

In the preceding section, you learned that the velocity vector points in the direction of motion. This observation leads to the following definition, which applies to any smooth curve—not just to those for which the parameter represents time.

DEFINITION OF UNIT TANGENT VECTOR

Let C be a smooth curve represented by $\mathbf{r}$ on an open interval I. The **unit tangent vector** $\mathbf{T}(t)$ at t is defined as

$$\mathbf{T}(t) = \frac{\mathbf{r}'(t)}{\|\mathbf{r}'(t)\|}, \quad \mathbf{r}'(t) \neq \mathbf{0}.$$

Recall that a curve is *smooth* on an interval if $\mathbf{r}'$ is continuous and nonzero on the interval. So, "smoothness" is sufficient to guarantee that a curve has a unit tangent vector.

EXAMPLE 1 Finding the Unit Tangent Vector

Find the unit tangent vector to the curve given by

$$\mathbf{r}(t) = t\mathbf{i} + t^2\mathbf{j}$$

when $t = 1$.

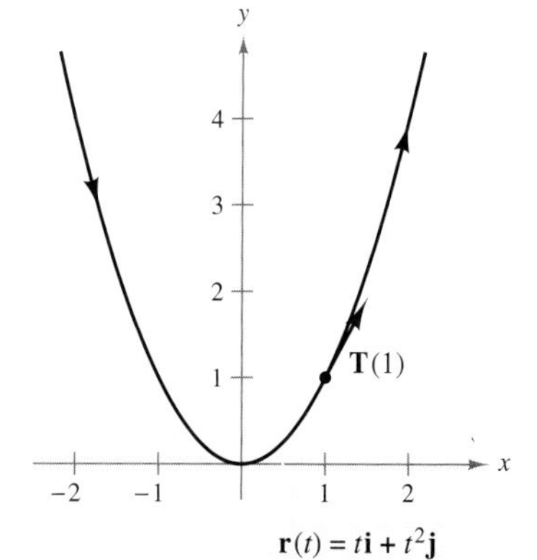

The direction of the unit tangent vector depends on the orientation of the curve.
Figure 12.20

Solution The derivative of $\mathbf{r}(t)$ is

$$\mathbf{r}'(t) = \mathbf{i} + 2t\mathbf{j}. \qquad \text{Derivative of } \mathbf{r}(t)$$

So, the unit tangent vector is

$$\mathbf{T}(t) = \frac{\mathbf{r}'(t)}{\|\mathbf{r}'(t)\|} \qquad \text{Definition of } \mathbf{T}(t)$$

$$= \frac{1}{\sqrt{1 + 4t^2}}(\mathbf{i} + 2t\mathbf{j}). \qquad \text{Substitute for } \mathbf{r}'(t).$$

When $t = 1$, the unit tangent vector is

$$\mathbf{T}(1) = \frac{1}{\sqrt{5}}(\mathbf{i} + 2\mathbf{j})$$

as shown in Figure 12.20. ■

NOTE In Example 1, note that the direction of the unit tangent vector depends on the orientation of the curve. For instance, if the parabola in Figure 12.20 were given by

$$\mathbf{r}(t) = -(t - 2)\mathbf{i} + (t - 2)^2\mathbf{j}$$

$\mathbf{T}(1)$ would still represent the unit tangent vector at the point $(1, 1)$, but it would point in the opposite direction. Try verifying this. ■

The **tangent line to a curve** at a point is the line that passes through the point and is parallel to the unit tangent vector. In Example 2, the unit tangent vector is used to find the tangent line at a point on a helix.

EXAMPLE 2 Finding the Tangent Line at a Point on a Curve

Find $\mathbf{T}(t)$ and then find a set of parametric equations for the tangent line to the helix given by

$$\mathbf{r}(t) = 2\cos t\mathbf{i} + 2\sin t\mathbf{j} + t\mathbf{k}$$

at the point $\left(\sqrt{2}, \sqrt{2}, \frac{\pi}{4}\right)$.

Solution The derivative of $\mathbf{r}(t)$ is $\mathbf{r}'(t) = -2\sin t\mathbf{i} + 2\cos t\mathbf{j} + \mathbf{k}$, which implies that $\|\mathbf{r}'(t)\| = \sqrt{4\sin^2 t + 4\cos^2 t + 1} = \sqrt{5}$. Therefore, the unit tangent vector is

$$\begin{aligned}\mathbf{T}(t) &= \frac{\mathbf{r}'(t)}{\|\mathbf{r}'(t)\|} \\ &= \frac{1}{\sqrt{5}}(-2\sin t\mathbf{i} + 2\cos t\mathbf{j} + \mathbf{k}).\end{aligned}$$ Unit tangent vector

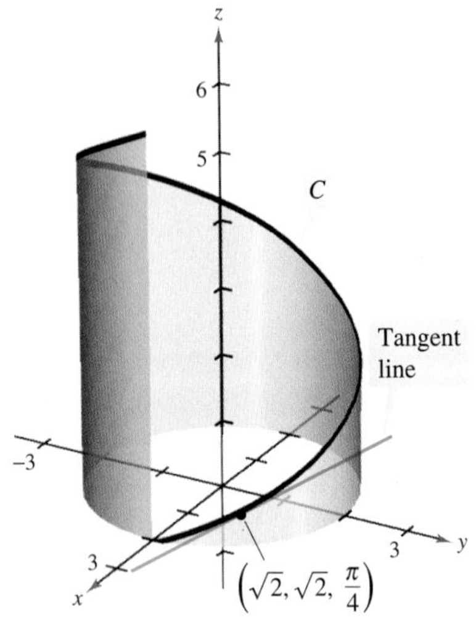

The tangent line to a curve at a point is determined by the unit tangent vector at the point.
Figure 12.21

At the point $\left(\sqrt{2}, \sqrt{2}, \pi/4\right)$, $t = \pi/4$ and the unit tangent vector is

$$\begin{aligned}\mathbf{T}\left(\frac{\pi}{4}\right) &= \frac{1}{\sqrt{5}}\left(-2\frac{\sqrt{2}}{2}\mathbf{i} + 2\frac{\sqrt{2}}{2}\mathbf{j} + \mathbf{k}\right) \\ &= \frac{1}{\sqrt{5}}\left(-\sqrt{2}\,\mathbf{i} + \sqrt{2}\,\mathbf{j} + \mathbf{k}\right).\end{aligned}$$

Using the direction numbers $a = -\sqrt{2}$, $b = \sqrt{2}$, and $c = 1$, and the point $(x_1, y_1, z_1) = \left(\sqrt{2}, \sqrt{2}, \pi/4\right)$, you can obtain the following parametric equations (given with parameter s).

$$\begin{aligned}x &= x_1 + as = \sqrt{2} - \sqrt{2}s \\ y &= y_1 + bs = \sqrt{2} + \sqrt{2}s \\ z &= z_1 + cs = \frac{\pi}{4} + s\end{aligned}$$

This tangent line is shown in Figure 12.21. ■

In Example 2, there are infinitely many vectors that are orthogonal to the tangent vector $\mathbf{T}(t)$. One of these is the vector $\mathbf{T}'(t)$. This follows from Property 7 of Theorem 12.2. That is,

$$\mathbf{T}(t)\cdot\mathbf{T}(t) = \|\mathbf{T}(t)\|^2 = 1 \quad \Longrightarrow \quad \mathbf{T}(t)\cdot\mathbf{T}'(t) = 0.$$

By normalizing the vector $\mathbf{T}'(t)$, you obtain a special vector called the **principal unit normal vector,** as indicated in the following definition.

DEFINITION OF PRINCIPAL UNIT NORMAL VECTOR

Let C be a smooth curve represented by $\mathbf{r}$ on an open interval I. If $\mathbf{T}'(t) \neq \mathbf{0}$, then the **principal unit normal vector** at t is defined as

$$\mathbf{N}(t) = \frac{\mathbf{T}'(t)}{\|\mathbf{T}'(t)\|}.$$

EXAMPLE 3 Finding the Principal Unit Normal Vector

Find $\mathbf{N}(t)$ and $\mathbf{N}(1)$ for the curve represented by

$$\mathbf{r}(t) = 3t\mathbf{i} + 2t^2\mathbf{j}.$$

Solution By differentiating, you obtain

$$\mathbf{r}'(t) = 3\mathbf{i} + 4t\mathbf{j} \quad \text{and} \quad \|\mathbf{r}'(t)\| = \sqrt{9 + 16t^2}$$

which implies that the unit tangent vector is

$$\mathbf{T}(t) = \frac{\mathbf{r}'(t)}{\|\mathbf{r}'(t)\|} = \frac{1}{\sqrt{9 + 16t^2}}(3\mathbf{i} + 4t\mathbf{j}). \qquad \text{Unit tangent vector}$$

Using Theorem 12.2, differentiate $\mathbf{T}(t)$ with respect to t to obtain

$$\mathbf{T}'(t) = \frac{1}{\sqrt{9 + 16t^2}}(4\mathbf{j}) - \frac{16t}{(9 + 16t^2)^{3/2}}(3\mathbf{i} + 4t\mathbf{j}) = \frac{12}{(9 + 16t^2)^{3/2}}(-4t\mathbf{i} + 3\mathbf{j})$$

$$\|\mathbf{T}'(t)\| = 12\sqrt{\frac{9 + 16t^2}{(9 + 16t^2)^3}} = \frac{12}{9 + 16t^2}.$$

Therefore, the principal unit normal vector is

$$\mathbf{N}(t) = \frac{\mathbf{T}'(t)}{\|\mathbf{T}'(t)\|} = \frac{1}{\sqrt{9 + 16t^2}}(-4t\mathbf{i} + 3\mathbf{j}). \qquad \text{Principal unit normal vector}$$

When $t = 1$, the principal unit normal vector is

$$\mathbf{N}(1) = \frac{1}{5}(-4\mathbf{i} + 3\mathbf{j})$$

as shown in Figure 12.22. ■

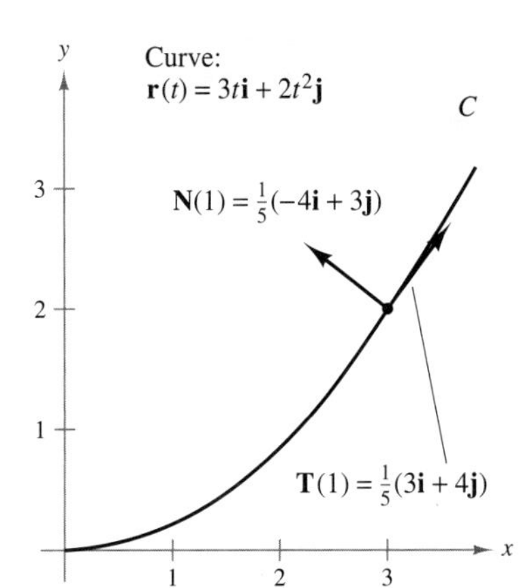

The principal unit normal vector points toward the concave side of the curve.
Figure 12.22

The principal unit normal vector can be difficult to evaluate algebraically. For plane curves, you can simplify the algebra by finding

$$\mathbf{T}(t) = x(t)\mathbf{i} + y(t)\mathbf{j} \qquad \text{Unit tangent vector}$$

and observing that $\mathbf{N}(t)$ must be either

$$\mathbf{N}_1(t) = y(t)\mathbf{i} - x(t)\mathbf{j} \qquad \text{or} \qquad \mathbf{N}_2(t) = -y(t)\mathbf{i} + x(t)\mathbf{j}.$$

Because $\sqrt{[x(t)]^2 + [y(t)]^2} = 1$, it follows that both $\mathbf{N}_1(t)$ and $\mathbf{N}_2(t)$ are unit normal vectors. The *principal* unit normal vector $\mathbf{N}$ is the one that points toward the concave side of the curve, as shown in Figure 12.22 (see Exercise 94). This also holds for curves in space. That is, for an object moving along a curve C in space, the vector $\mathbf{T}(t)$ points in the direction the object is moving, whereas the vector $\mathbf{N}(t)$ is orthogonal to $\mathbf{T}(t)$ and points in the direction in which the object is turning, as shown in Figure 12.23.

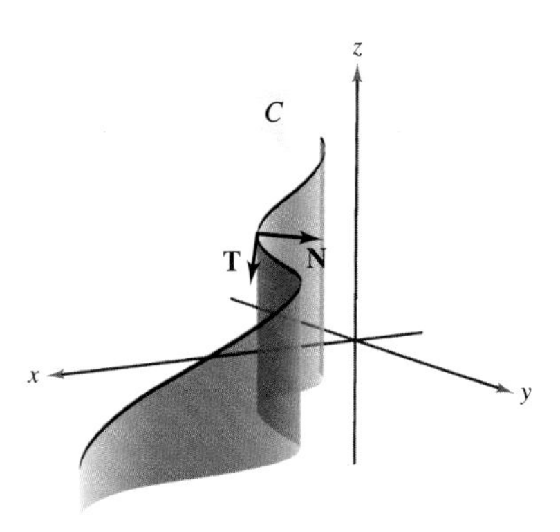

At any point on a curve, a unit normal vector is orthogonal to the unit tangent vector. The *principal* unit normal vector points in the direction in which the curve is turning.
Figure 12.23

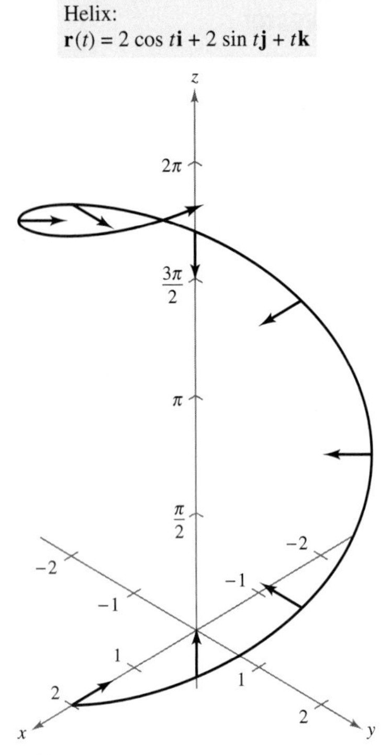

$\mathbf{N}(t)$ is horizontal and points toward the z-axis.
Figure 12.24

EXAMPLE 4 Finding the Principal Unit Normal Vector

Find the principal unit normal vector for the helix given by

$$\mathbf{r}(t) = 2\cos t\mathbf{i} + 2\sin t\mathbf{j} + t\mathbf{k}.$$

Solution From Example 2, you know that the unit tangent vector is

$$\mathbf{T}(t) = \frac{1}{\sqrt{5}}(-2\sin t\mathbf{i} + 2\cos t\mathbf{j} + \mathbf{k}). \qquad \text{Unit tangent vector}$$

So, $\mathbf{T}'(t)$ is given by

$$\mathbf{T}'(t) = \frac{1}{\sqrt{5}}(-2\cos t\mathbf{i} - 2\sin t\mathbf{j}).$$

Because $\|\mathbf{T}'(t)\| = 2/\sqrt{5}$, it follows that the principal unit normal vector is

$$\begin{aligned}\mathbf{N}(t) &= \frac{\mathbf{T}'(t)}{\|\mathbf{T}'(t)\|}\\ &= \frac{1}{2}(-2\cos t\mathbf{i} - 2\sin t\mathbf{j})\\ &= -\cos t\mathbf{i} - \sin t\mathbf{j}. \qquad \text{Principal unit normal vector}\end{aligned}$$

Note that this vector is horizontal and points toward the z-axis, as shown in Figure 12.24. ■

Tangential and Normal Components of Acceleration

Let's return to the problem of describing the motion of an object along a curve. In the preceding section, you saw that for an object traveling at a *constant speed*, the velocity and acceleration vectors are perpendicular. This seems reasonable, because the speed would not be constant if any acceleration were acting in the direction of motion. You can verify this observation by noting that

$$\mathbf{r}''(t) \cdot \mathbf{r}'(t) = 0$$

if $\|\mathbf{r}'(t)\|$ is a constant. (See Property 7 of Theorem 12.2.)

However, for an object traveling at a *variable speed*, the velocity and acceleration vectors are not necessarily perpendicular. For instance, you saw that the acceleration vector for a projectile always points down, regardless of the direction of motion.

In general, part of the acceleration (the tangential component) acts in the line of motion, and part (the normal component) acts perpendicular to the line of motion. In order to determine these two components, you can use the unit vectors $\mathbf{T}(t)$ and $\mathbf{N}(t)$, which serve in much the same way as do $\mathbf{i}$ and $\mathbf{j}$ in representing vectors in the plane. The following theorem states that the acceleration vector lies in the plane determined by $\mathbf{T}(t)$ and $\mathbf{N}(t)$.

THEOREM 12.4 ACCELERATION VECTOR

If $\mathbf{r}(t)$ is the position vector for a smooth curve C and $\mathbf{N}(t)$ exists, then the acceleration vector $\mathbf{a}(t)$ lies in the plane determined by $\mathbf{T}(t)$ and $\mathbf{N}(t)$.

PROOF To simplify the notation, write $\mathbf{T}$ for $\mathbf{T}(t)$, $\mathbf{T}'$ for $\mathbf{T}'(t)$, and so on. Because $\mathbf{T} = \mathbf{r}'/\|\mathbf{r}'\| = \mathbf{v}/\|\mathbf{v}\|$, it follows that

$$\mathbf{v} = \|\mathbf{v}\|\mathbf{T}.$$

By differentiating, you obtain

$$\begin{aligned}\mathbf{a} = \mathbf{v}' &= D_t[\|\mathbf{v}\|]\mathbf{T} + \|\mathbf{v}\|\mathbf{T}' && \text{Product Rule}\\ &= D_t[\|\mathbf{v}\|]\mathbf{T} + \|\mathbf{v}\|\mathbf{T}'\left(\frac{\|\mathbf{T}'\|}{\|\mathbf{T}'\|}\right)\\ &= D_t[\|\mathbf{v}\|]\mathbf{T} + \|\mathbf{v}\|\,\|\mathbf{T}'\|\,\mathbf{N}. && \mathbf{N} = \mathbf{T}'/\|\mathbf{T}'\|\end{aligned}$$

Because $\mathbf{a}$ is written as a linear combination of $\mathbf{T}$ and $\mathbf{N}$, it follows that $\mathbf{a}$ lies in the plane determined by $\mathbf{T}$ and $\mathbf{N}$. ■

The coefficients of $\mathbf{T}$ and $\mathbf{N}$ in the proof of Theorem 12.4 are called the **tangential and normal components of acceleration** and are denoted by $a_{\mathbf{T}} = D_t[\|\mathbf{v}\|]$ and $a_{\mathbf{N}} = \|\mathbf{v}\|\,\|\mathbf{T}'\|$. So, you can write

$$\mathbf{a}(t) = a_{\mathbf{T}}\mathbf{T}(t) + a_{\mathbf{N}}\mathbf{N}(t).$$

The following theorem gives some convenient formulas for $a_{\mathbf{N}}$ and $a_{\mathbf{T}}$.

THEOREM 12.5 TANGENTIAL AND NORMAL COMPONENTS OF ACCELERATION

If $\mathbf{r}(t)$ is the position vector for a smooth curve C [for which $\mathbf{N}(t)$ exists], then the tangential and normal components of acceleration are as follows.

$$a_{\mathbf{T}} = D_t[\|\mathbf{v}\|] = \mathbf{a}\cdot\mathbf{T} = \frac{\mathbf{v}\cdot\mathbf{a}}{\|\mathbf{v}\|}$$

$$a_{\mathbf{N}} = \|\mathbf{v}\|\,\|\mathbf{T}'\| = \mathbf{a}\cdot\mathbf{N} = \frac{\|\mathbf{v}\times\mathbf{a}\|}{\|\mathbf{v}\|} = \sqrt{\|\mathbf{a}\|^2 - a_{\mathbf{T}}^2}$$

Note that $a_{\mathbf{N}} \geq 0$. The normal component of acceleration is also called the **centripetal component of acceleration.**

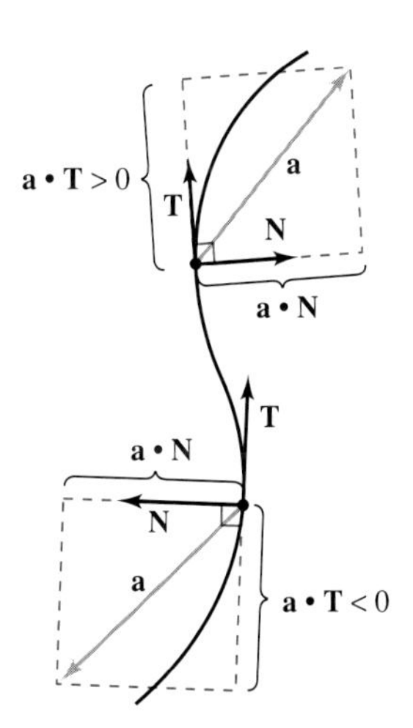

The tangential and normal components of acceleration are obtained by projecting **a** onto **T** and **N**.
Figure 12.25

PROOF Note that a lies in the plane of $\mathbf{T}$ and $\mathbf{N}$. So, you can use Figure 12.25 to conclude that, for any time t, the components of the projection of the acceleration vector onto $\mathbf{T}$ and onto $\mathbf{N}$ are given by $a_{\mathbf{T}} = \mathbf{a}\cdot\mathbf{T}$ and $a_{\mathbf{N}} = \mathbf{a}\cdot\mathbf{N}$, respectively. Moreover, because $\mathbf{a} = \mathbf{v}'$ and $\mathbf{T} = \mathbf{v}/\|\mathbf{v}\|$, you have

$$\begin{aligned}a_{\mathbf{T}} &= \mathbf{a}\cdot\mathbf{T}\\ &= \mathbf{T}\cdot\mathbf{a}\\ &= \frac{\mathbf{v}}{\|\mathbf{v}\|}\cdot\mathbf{a}\\ &= \frac{\mathbf{v}\cdot\mathbf{a}}{\|\mathbf{v}\|}.\end{aligned}$$

In Exercises 96 and 97, you are asked to prove the other parts of the theorem. ■

NOTE The formulas from Theorem 12.5, together with several other formulas from this chapter, are summarized on page 877. ■

EXAMPLE 5 Tangential and Normal Components of Acceleration

Find the tangential and normal components of acceleration for the position vector given by $\mathbf{r}(t) = 3t\mathbf{i} - t\mathbf{j} + t^2\mathbf{k}$.

Solution Begin by finding the velocity, speed, and acceleration.

$$\mathbf{v}(t) = \mathbf{r}'(t) = 3\mathbf{i} - \mathbf{j} + 2t\mathbf{k}$$
$$\|\mathbf{v}(t)\| = \sqrt{9 + 1 + 4t^2} = \sqrt{10 + 4t^2}$$
$$\mathbf{a}(t) = \mathbf{r}''(t) = 2\mathbf{k}$$

By Theorem 12.5, the tangential component of acceleration is

$$a_{\mathbf{T}} = \frac{\mathbf{v} \cdot \mathbf{a}}{\|\mathbf{v}\|} = \frac{4t}{\sqrt{10 + 4t^2}} \qquad \text{Tangential component of acceleration}$$

and because

$$\mathbf{v} \times \mathbf{a} = \begin{vmatrix} \mathbf{i} & \mathbf{j} & \mathbf{k} \\ 3 & -1 & 2t \\ 0 & 0 & 2 \end{vmatrix} = -2\mathbf{i} - 6\mathbf{j}$$

the normal component of acceleration is

$$a_{\mathbf{N}} = \frac{\|\mathbf{v} \times \mathbf{a}\|}{\|\mathbf{v}\|} = \frac{\sqrt{4 + 36}}{\sqrt{10 + 4t^2}} = \frac{2\sqrt{10}}{\sqrt{10 + 4t^2}}. \qquad \text{Normal component of acceleration}$$

■

NOTE In Example 5, you could have used the alternative formula for $a_{\mathbf{N}}$ as follows.

$$a_{\mathbf{N}} = \sqrt{\|\mathbf{a}\|^2 - a_{\mathbf{T}}^2} = \sqrt{(2)^2 - \frac{16t^2}{10 + 4t^2}} = \frac{2\sqrt{10}}{\sqrt{10 + 4t^2}}$$

■

EXAMPLE 6 Finding $a_{\mathbf{T}}$ and $a_{\mathbf{N}}$ for a Circular Helix

Find the tangential and normal components of acceleration for the helix given by $\mathbf{r}(t) = b \cos t\mathbf{i} + b \sin t\mathbf{j} + ct\mathbf{k}$, $b > 0$.

Solution

$$\mathbf{v}(t) = \mathbf{r}'(t) = -b \sin t\mathbf{i} + b \cos t\mathbf{j} + c\mathbf{k}$$
$$\|\mathbf{v}(t)\| = \sqrt{b^2 \sin^2 t + b^2 \cos^2 t + c^2} = \sqrt{b^2 + c^2}$$
$$\mathbf{a}(t) = \mathbf{r}''(t) = -b \cos t\mathbf{i} - b \sin t\mathbf{j}$$

By Theorem 12.5, the tangential component of acceleration is

$$a_{\mathbf{T}} = \frac{\mathbf{v} \cdot \mathbf{a}}{\|\mathbf{v}\|} = \frac{b^2 \sin t \cos t - b^2 \sin t \cos t + 0}{\sqrt{b^2 + c^2}} = 0. \qquad \text{Tangential component of acceleration}$$

Moreover, because $\|\mathbf{a}\| = \sqrt{b^2 \cos^2 t + b^2 \sin^2 t} = b$, you can use the alternative formula for the normal component of acceleration to obtain

$$a_{\mathbf{N}} = \sqrt{\|\mathbf{a}\|^2 - a_{\mathbf{T}}^2} = \sqrt{b^2 - 0^2} = b. \qquad \text{Normal component of acceleration}$$

Note that the normal component of acceleration is equal to the magnitude of the acceleration. In other words, because the speed is constant, the acceleration is perpendicular to the velocity. See Figure 12.26. ■

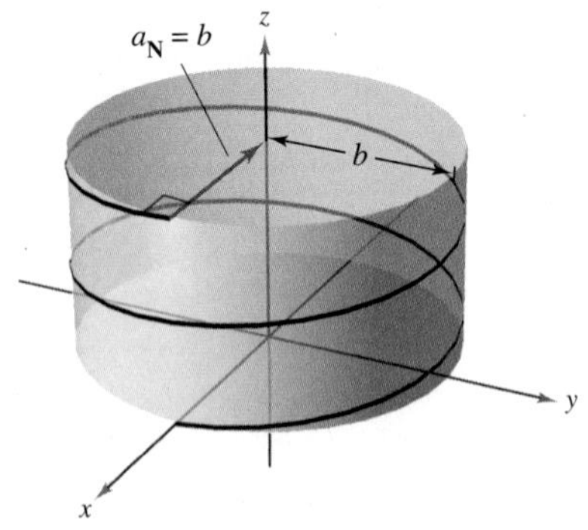

The normal component of acceleration is equal to the radius of the cylinder around which the helix is spiraling.
Figure 12.26

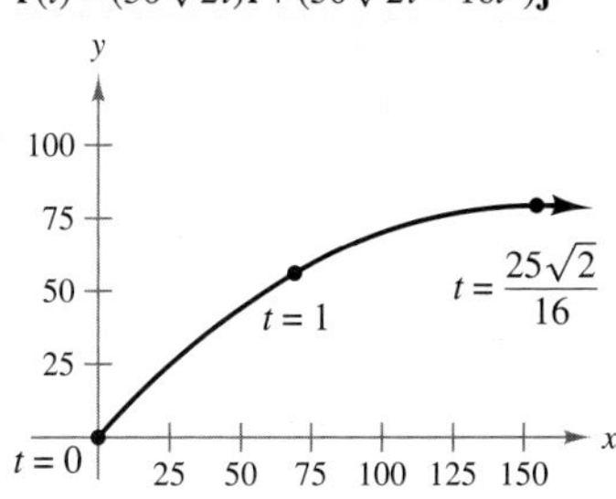

The path of a projectile
Figure 12.27

EXAMPLE 7 Projectile Motion

The position vector for the projectile shown in Figure 12.27 is given by

$$\mathbf{r}(t) = \left(50\sqrt{2}\,t\right)\mathbf{i} + \left(50\sqrt{2}\,t - 16t^2\right)\mathbf{j}. \qquad \text{Position vector}$$

Find the tangential components of acceleration when $t = 0$, 1, and $25\sqrt{2}/16$.

Solution

$$\mathbf{v}(t) = 50\sqrt{2}\,\mathbf{i} + \left(50\sqrt{2} - 32t\right)\mathbf{j} \qquad \text{Velocity vector}$$

$$\|\mathbf{v}(t)\| = 2\sqrt{50^2 - 16(50)\sqrt{2}t + 16^2t^2} \qquad \text{Speed}$$

$$\mathbf{a}(t) = -32\mathbf{j} \qquad \text{Acceleration vector}$$

The tangential component of acceleration is

$$a_{\mathbf{T}}(t) = \frac{\mathbf{v}(t) \cdot \mathbf{a}(t)}{\|\mathbf{v}(t)\|} = \frac{-32(50\sqrt{2} - 32t)}{2\sqrt{50^2 - 16(50)\sqrt{2}t + 16^2t^2}}. \qquad \text{Tangential component of acceleration}$$

At the specified times, you have

$$a_{\mathbf{T}}(0) = \frac{-32\left(50\sqrt{2}\right)}{100} = -16\sqrt{2} \approx -22.6$$

$$a_{\mathbf{T}}(1) = \frac{-32\left(50\sqrt{2} - 32\right)}{2\sqrt{50^2 - 16(50)\sqrt{2} + 16^2}} \approx -15.4$$

$$a_{\mathbf{T}}\left(\frac{25\sqrt{2}}{16}\right) = \frac{-32\left(50\sqrt{2} - 50\sqrt{2}\right)}{50\sqrt{2}} = 0.$$

You can see from Figure 12.27 that, at the maximum height, when $t = 25\sqrt{2}/16$, the tangential component is 0. This is reasonable because the direction of motion is horizontal at the point and the tangential component of the acceleration is equal to the horizontal component of the acceleration. ■

12.4 Exercises

See www.CalcChat.com for worked-out solutions to odd-numbered exercises.

In Exercises 1–4, sketch the unit tangent and normal vectors at the given points. To print an enlarged copy of the graph, go to the website *www.mathgraphs.com*.

1.

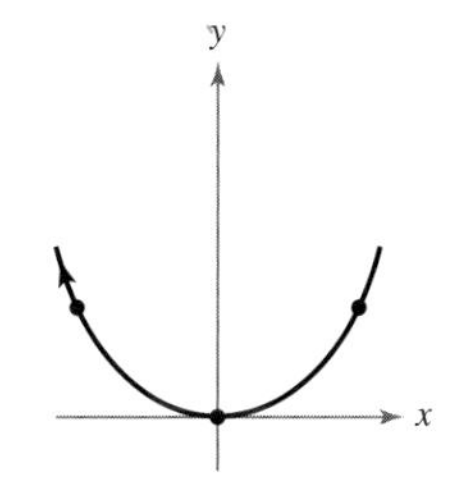

2.

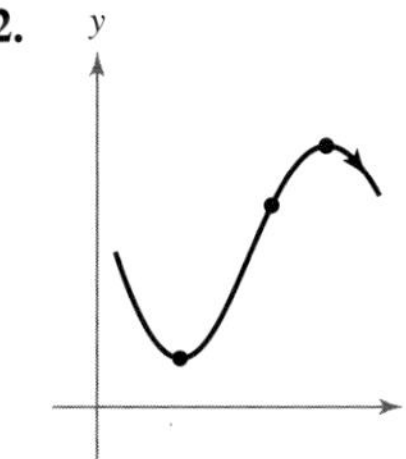

3.

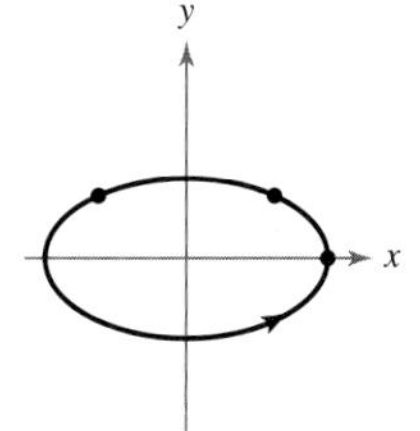

4.

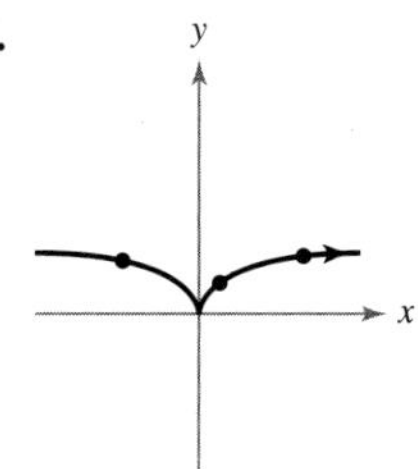

In Exercises 5–10, find the unit tangent vector to the curve at the specified value of the parameter.

5. $\mathbf{r}(t) = t^2\mathbf{i} + 2t\mathbf{j}, \quad t = 1$

6. $\mathbf{r}(t) = t^3\mathbf{i} + 2t^2\mathbf{j}, \quad t = 1$

7. $\mathbf{r}(t) = 4\cos t\mathbf{i} + 4\sin t\mathbf{j}, \quad t = \dfrac{\pi}{4}$

8. $\mathbf{r}(t) = 6\cos t\mathbf{i} + 2\sin t\mathbf{j}, \quad t = \dfrac{\pi}{3}$

9. $\mathbf{r}(t) = 3t\mathbf{i} - \ln t\mathbf{j}, \quad t = e$

10. $\mathbf{r}(t) = e^t\cos t\mathbf{i} + e^t\mathbf{j}, \quad t = 0$

In Exercises 11–16, find the unit tangent vector $\mathbf{T}(t)$ and find a set of parametric equations for the line tangent to the space curve at point P.

11. $\mathbf{r}(t) = t\mathbf{i} + t^2\mathbf{j} + t\mathbf{k}, \quad P(0, 0, 0)$

12. $\mathbf{r}(t) = t^2\mathbf{i} + t\mathbf{j} + \frac{4}{3}\mathbf{k}, \quad P\left(1, 1, \frac{4}{3}\right)$

13. $\mathbf{r}(t) = 3\cos t\mathbf{i} + 3\sin t\mathbf{j} + t\mathbf{k}, \quad P(3, 0, 0)$

14. $\mathbf{r}(t) = \left\langle t, t, \sqrt{4 - t^2}\right\rangle, \quad P\left(1, 1, \sqrt{3}\right)$

15. $\mathbf{r}(t) = \langle 2\cos t, 2\sin t, 4\rangle, \quad P\left(\sqrt{2}, \sqrt{2}, 4\right)$

16. $\mathbf{r}(t) = \langle 2\sin t, 2\cos t, 4\sin^2 t\rangle, \quad P\left(1, \sqrt{3}, 1\right)$

CAS **In Exercises 17 and 18, use a computer algebra system to graph the space curve. Then find $\mathbf{T}(t)$ and find a set of parametric equations for the line tangent to the space curve at point P. Graph the tangent line.**

17. $\mathbf{r}(t) = \langle t, t^2, 2t^3/3 \rangle, \quad P(3, 9, 18)$

18. $\mathbf{r}(t) = 3\cos t\mathbf{i} + 4\sin t\mathbf{j} + \frac{1}{2}t\mathbf{k}, \quad P(0, 4, \pi/4)$

Linear Approximation **In Exercises 19 and 20, find a set of parametric equations for the tangent line to the graph at $t = t_0$ and use the equations for the line to approximate $\mathbf{r}(t_0 + 0.1)$.**

19. $\mathbf{r}(t) = \langle t, \ln t, \sqrt{t} \rangle, \quad t_0 = 1$

20. $\mathbf{r}(t) = \langle e^{-t}, 2\cos t, 2\sin t \rangle, \quad t_0 = 0$

In Exercises 21 and 22, verify that the space curves intersect at the given values of the parameters. Find the angle between the tangent vectors to the curves at the point of intersection.

21. $\mathbf{r}(t) = \langle t - 2, t^2, \frac{1}{2}t \rangle, \quad t = 4$
$\mathbf{u}(s) = \langle \frac{1}{4}s, 2s, \sqrt[3]{s} \rangle, \quad s = 8$

22. $\mathbf{r}(t) = \langle t, \cos t, \sin t \rangle, \quad t = 0$
$\mathbf{u}(s) = \langle -\frac{1}{2}\sin^2 s - \sin s, 1 - \frac{1}{2}\sin^2 s - \sin s,$
$\frac{1}{2}\sin s \cos s + \frac{1}{2}s \rangle, \quad s = 0$

In Exercises 23–30, find the principal unit normal vector to the curve at the specified value of the parameter.

23. $\mathbf{r}(t) = t\mathbf{i} + \frac{1}{2}t^2\mathbf{j}, \quad t = 2$

24. $\mathbf{r}(t) = t\mathbf{i} + \dfrac{6}{t}\mathbf{j}, \quad t = 3$

25. $\mathbf{r}(t) = \ln t\mathbf{i} + (t + 1)\mathbf{j}, \quad t = 2$

26. $\mathbf{r}(t) = \pi\cos t\mathbf{i} + \pi\sin t\mathbf{j}, \quad t = \dfrac{\pi}{6}$

27. $\mathbf{r}(t) = t\mathbf{i} + t^2\mathbf{j} + \ln t\mathbf{k}, \quad t = 1$

28. $\mathbf{r}(t) = \sqrt{2}t\mathbf{i} + e^t\mathbf{j} + e^{-t}\mathbf{k}, \quad t = 0$

29. $\mathbf{r}(t) = 6\cos t\mathbf{i} + 6\sin t\mathbf{j} + \mathbf{k}, \quad t = \dfrac{3\pi}{4}$

30. $\mathbf{r}(t) = \cos 3t\mathbf{i} + 2\sin 3t\mathbf{j} + \mathbf{k}, \quad t = \pi$

In Exercises 31–34, find $\mathbf{v}(t)$, $\mathbf{a}(t)$, $\mathbf{T}(t)$, and $\mathbf{N}(t)$ (if it exists) for an object moving along the path given by the vector-valued function $\mathbf{r}(t)$. Use the results to determine the form of the path. Is the speed of the object constant or changing?

31. $\mathbf{r}(t) = 4t\mathbf{i}$

32. $\mathbf{r}(t) = 4t\mathbf{i} - 2t\mathbf{j}$

33. $\mathbf{r}(t) = 4t^2\mathbf{i}$

34. $\mathbf{r}(t) = t^2\mathbf{j} + \mathbf{k}$

In Exercises 35–44, find $\mathbf{T}(t)$, $\mathbf{N}(t)$, $a_{\mathbf{T}}$, and $a_{\mathbf{N}}$ at the given time t for the plane curve $\mathbf{r}(t)$.

35. $\mathbf{r}(t) = t\mathbf{i} + \dfrac{1}{t}\mathbf{j}, \quad t = 1$

36. $\mathbf{r}(t) = t^2\mathbf{i} + 2t\mathbf{j}, \quad t = 1$

37. $\mathbf{r}(t) = (t - t^3)\mathbf{i} + 2t^2\mathbf{j}, \quad t = 1$

38. $\mathbf{r}(t) = (t^3 - 4t)\mathbf{i} + (t^2 - 1)\mathbf{j}, \quad t = 0$

39. $\mathbf{r}(t) = e^t\mathbf{i} + e^{-2t}\mathbf{j}, \quad t = 0$

40. $\mathbf{r}(t) = e^t\mathbf{i} + e^{-t}\mathbf{j} + t\mathbf{k}, \quad t = 0$

41. $\mathbf{r}(t) = e^t\cos t\mathbf{i} + e^t\sin t\mathbf{j}, \quad t = \dfrac{\pi}{2}$

42. $\mathbf{r}(t) = a\cos\omega t\mathbf{i} + b\sin\omega t\mathbf{j}, \quad t = 0$

43. $\mathbf{r}(t) = \langle \cos\omega t + \omega t\sin\omega t, \sin\omega t - \omega t\cos\omega t \rangle, \quad t = t_0$

44. $\mathbf{r}(t) = \langle \omega t - \sin\omega t, 1 - \cos\omega t \rangle, \quad t = t_0$

Circular Motion **In Exercises 45–48, consider an object moving according to the position function**

$$\mathbf{r}(t) = a\cos\omega t\,\mathbf{i} + a\sin\omega t\,\mathbf{j}.$$

45. Find $\mathbf{T}(t)$, $\mathbf{N}(t)$, $a_{\mathbf{T}}$, and $a_{\mathbf{N}}$.

46. Determine the directions of $\mathbf{T}$ and $\mathbf{N}$ relative to the position function $\mathbf{r}$.

47. Determine the speed of the object at any time t and explain its value relative to the value of $a_{\mathbf{T}}$.

48. If the angular velocity ω is halved, by what factor is $a_{\mathbf{N}}$ changed?

In Exercises 49–54, sketch the graph of the plane curve given by the vector-valued function, and, at the point on the curve determined by $\mathbf{r}(t_0)$, sketch the vectors $\mathbf{T}$ and $\mathbf{N}$. Note that $\mathbf{N}$ points toward the concave side of the curve.

Function	*Time*
49. $\mathbf{r}(t) = t\mathbf{i} + \dfrac{1}{t}\mathbf{j}$	$t_0 = 2$
50. $\mathbf{r}(t) = t^3\mathbf{i} + t\mathbf{j}$	$t_0 = 1$
51. $\mathbf{r}(t) = 4t\mathbf{i} + 4t^2\mathbf{j}$	$t_0 = \dfrac{1}{4}$
52. $\mathbf{r}(t) = (2t + 1)\mathbf{i} - t^2\mathbf{j}$	$t_0 = 2$
53. $\mathbf{r}(t) = 2\cos t\mathbf{i} + 2\sin t\mathbf{j}$	$t_0 = \dfrac{\pi}{4}$
54. $\mathbf{r}(t) = 3\cos t\mathbf{i} + 2\sin t\mathbf{j}$	$t_0 = \pi$

In Exercises 55–62, find $\mathbf{T}(t)$, $\mathbf{N}(t)$, $a_{\mathbf{T}}$, and $a_{\mathbf{N}}$ at the given time t for the space curve $\mathbf{r}(t)$. [*Hint:* Find $\mathbf{a}(t)$, $\mathbf{T}(t)$, $a_{\mathbf{T}}$, and $a_{\mathbf{N}}$. Solve for $\mathbf{N}$ in the equation $\mathbf{a}(t) = a_{\mathbf{T}}\mathbf{T} + a_{\mathbf{N}}\mathbf{N}$.]

Function	*Time*
55. $\mathbf{r}(t) = t\mathbf{i} + 2t\mathbf{j} - 3t\mathbf{k}$	$t = 1$
56. $\mathbf{r}(t) = 4t\mathbf{i} - 4t\mathbf{j} + 2t\mathbf{k}$	$t = 2$
57. $\mathbf{r}(t) = \cos t\mathbf{i} + \sin t\mathbf{j} + 2t\mathbf{k}$	$t = \dfrac{\pi}{3}$
58. $\mathbf{r}(t) = 3t\mathbf{i} - t\mathbf{j} + t^2\mathbf{k}$	$t = -1$
59. $\mathbf{r}(t) = t\mathbf{i} + t^2\mathbf{j} + \dfrac{t^2}{2}\mathbf{k}$	$t = 1$
60. $\mathbf{r}(t) = (2t - 1)\mathbf{i} + t^2\mathbf{j} - 4t\mathbf{k}$	$t = 2$
61. $\mathbf{r}(t) = e^t\sin t\mathbf{i} + e^t\cos t\mathbf{j} + e^t\mathbf{k}$	$t = 0$
62. $\mathbf{r}(t) = e^t\mathbf{i} + 2t\mathbf{j} + e^{-t}\mathbf{k}$	$t = 0$

CAS **In Exercises 63–66, use a computer algebra system to graph the space curve. Then find $\mathbf{T}(t)$, $\mathbf{N}(t)$, $a_{\mathbf{T}}$, and $a_{\mathbf{N}}$ at the given time t. Sketch $\mathbf{T}(t)$ and $\mathbf{N}(t)$ on the space curve.**

Function	*Time*
63. $\mathbf{r}(t) = 4t\mathbf{i} + 3\cos t\mathbf{j} + 3\sin t\mathbf{k}$	$t = \frac{\pi}{2}$
64. $\mathbf{r}(t) = (2 + \cos t)\mathbf{i} + (1 - \sin t)\mathbf{j} + \frac{t}{3}\mathbf{k}$	$t = \pi$
65. $\mathbf{r}(t) = t\mathbf{i} + 3t^2\mathbf{j} + \frac{t^2}{2}\mathbf{k}$	$t = 2$
66. $\mathbf{r}(t) = t^2\mathbf{i} + \mathbf{j} + 2t\mathbf{k}$	$t = 1$

WRITING ABOUT CONCEPTS

67. Define the unit tangent vector, the principal unit normal vector, and the tangential and normal components of acceleration.

68. How is the unit tangent vector related to the orientation of a curve? Explain.

69. (a) Describe the motion of a particle if the normal component of acceleration is 0.

(b) Describe the motion of a particle if the tangential component of acceleration is 0.

CAPSTONE

70. An object moves along the path given by

$$\mathbf{r}(t) = 3t\mathbf{i} + 4t\mathbf{j}.$$

Find $v(t)$, $a(t)$, $\mathbf{T}(t)$, and $\mathbf{N}(t)$ (if it exists). What is the form of the path? Is the speed of the object constant or changing?

71. ***Cycloidal Motion*** The figure shows the path of a particle modeled by the vector-valued function

$$\mathbf{r}(t) = \langle \pi t - \sin \pi t, 1 - \cos \pi t \rangle.$$

The figure also shows the vectors $\mathbf{v}(t)/\|\mathbf{v}(t)\|$ and $\mathbf{a}(t)/\|\mathbf{a}(t)\|$ at the indicated values of t.

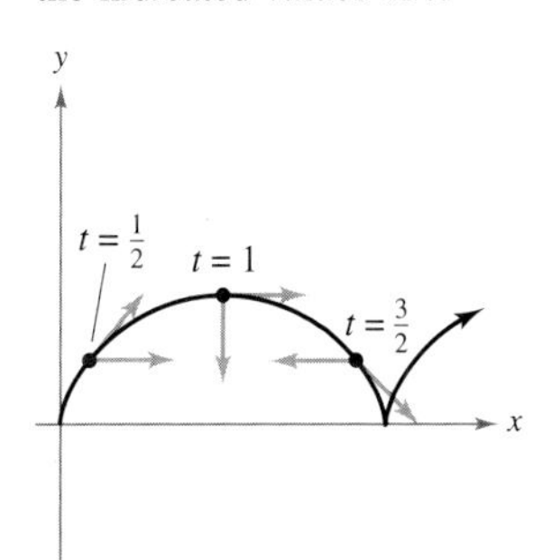

(a) Find $a_{\mathbf{T}}$ and $a_{\mathbf{N}}$ at $t = \frac{1}{2}$, $t = 1$, and $t = \frac{3}{2}$.

(b) Determine whether the speed of the particle is increasing or decreasing at each of the indicated values of t. Give reasons for your answers.

72. ***Motion Along an Involute of a Circle*** The figure shows a particle moving along a path modeled by $\mathbf{r}(t) = \langle \cos \pi t + \pi t \sin \pi t, \sin \pi t - \pi t \cos \pi t \rangle$. The figure also shows the vectors $\mathbf{v}(t)$ and $\mathbf{a}(t)$ for $t = 1$ and $t = 2$.

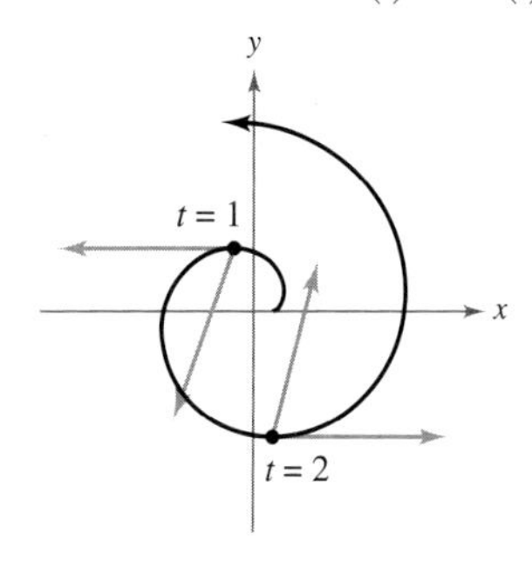

(a) Find $a_{\mathbf{T}}$ and $a_{\mathbf{N}}$ at $t = 1$ and $t = 2$.

(b) Determine whether the speed of the particle is increasing or decreasing at each of the indicated values of t. Give reasons for your answers.

In Exercises 73–78, find the vectors T and N, and the unit binormal vector $\mathbf{B} = \mathbf{T} \times \mathbf{N}$, for the vector-valued function $\mathbf{r}(t)$ at the given value of t.

73. $\mathbf{r}(t) = 2\cos t\mathbf{i} + 2\sin t\mathbf{j} + \frac{t}{2}\mathbf{k}$

$t_0 = \frac{\pi}{2}$

74. $\mathbf{r}(t) = t\mathbf{i} + t^2\mathbf{j} + \frac{t^3}{3}\mathbf{k}$

$t_0 = 1$

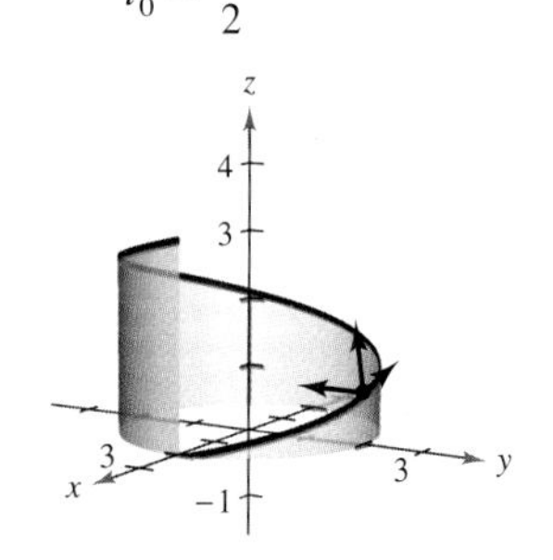

Figure for 73

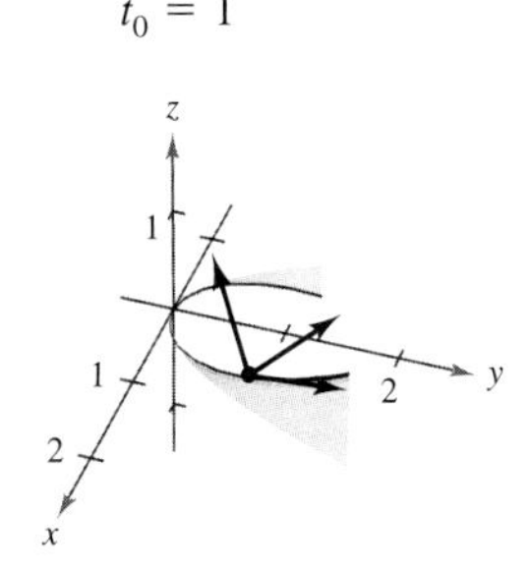

Figure for 74

75. $\mathbf{r}(t) = \mathbf{i} + \sin t\mathbf{j} + \cos t\mathbf{k}, \quad t_0 = \frac{\pi}{4}$

76. $\mathbf{r}(t) = 2e^t\mathbf{i} + e^t\cos t\mathbf{j} + e^t\sin t\mathbf{k}, \quad t_0 = 0$

77. $\mathbf{r}(t) = 4\sin t\mathbf{i} + 4\cos t\mathbf{j} + 2t\mathbf{k}, \quad t_0 = \frac{\pi}{3}$

78. $\mathbf{r}(t) = 3\cos 2t\mathbf{i} + 3\sin 2t\mathbf{j} + t\mathbf{k}, \quad t_0 = \frac{\pi}{4}$

79. ***Projectile Motion*** Find the tangential and normal components of acceleration for a projectile fired at an angle θ with the horizontal at an initial speed of v_0. What are the components when the projectile is at its maximum height?

80. ***Projectile Motion*** Use your results from Exercise 79 to find the tangential and normal components of acceleration for a projectile fired at an angle of 45° with the horizontal at an initial speed of 150 feet per second. What are the components when the projectile is at its maximum height?

81. ***Projectile Motion*** A projectile is launched with an initial velocity of 120 feet per second at a height of 5 feet and at an angle of 30° with the horizontal.

(a) Determine the vector-valued function for the path of the projectile.

(b) Use a graphing utility to graph the path and approximate the maximum height and range of the projectile.

(c) Find $\mathbf{v}(t)$, $\|\mathbf{v}(t)\|$, and $\mathbf{a}(t)$.

(d) Use a graphing utility to complete the table.

t	0.5	1.0	1.5	2.0	2.5	3.0
Speed						

(e) Use a graphing utility to graph the scalar functions $a_{\mathbf{T}}$ and $a_{\mathbf{N}}$. How is the speed of the projectile changing when $a_{\mathbf{T}}$ and $a_{\mathbf{N}}$ have opposite signs?

82. ***Projectile Motion*** A projectile is launched with an initial velocity of 220 feet per second at a height of 4 feet and at an angle of 45° with the horizontal.

(a) Determine the vector-valued function for the path of the projectile.

(b) Use a graphing utility to graph the path and approximate the maximum height and range of the projectile.

(c) Find $\mathbf{v}(t)$, $\|\mathbf{v}(t)\|$, and $\mathbf{a}(t)$.

(d) Use a graphing utility to complete the table.

t	0.5	1.0	1.5	2.0	2.5	3.0
Speed						

83. ***Air Traffic Control*** Because of a storm, ground controllers instruct the pilot of a plane flying at an altitude of 4 miles to make a 90° turn and climb to an altitude of 4.2 miles. The model for the path of the plane during this maneuver is

$$\mathbf{r}(t) = \langle 10\cos 10\pi t, 10\sin 10\pi t, 4 + 4t\rangle, \quad 0 \le t \le \tfrac{1}{20}$$

where t is the time in hours and $\mathbf{r}$ is the distance in miles.

(a) Determine the speed of the plane.

CAS (b) Use a computer algebra system to calculate $a_{\mathbf{T}}$ and $a_{\mathbf{N}}$. Why is one of these equal to 0?

84. ***Projectile Motion*** A plane flying at an altitude of 36,000 feet at a speed of 600 miles per hour releases a bomb. Find the tangential and normal components of acceleration acting on the bomb.

85. ***Centripetal Acceleration*** An object is spinning at a constant speed on the end of a string, according to the position function given in Exercises 45–48.

(a) If the angular velocity ω is doubled, how is the centripetal component of acceleration changed?

(b) If the angular velocity is unchanged but the length of the string is halved, how is the centripetal component of acceleration changed?

86. ***Centripetal Force*** An object of mass m moves at a constant speed v in a circular path of radius r. The force required to produce the centripetal component of acceleration is called the *centripetal force* and is given by $F = mv^2/r$. Newton's Law of Universal Gravitation is given by $F = GMm/d^2$, where d is the distance between the centers of the two bodies of masses M and m, and G is a gravitational constant. Use this law to show that the speed required for circular motion is $v = \sqrt{GM/r}$.

Orbital Speed **In Exercises 87–90, use the result of Exercise 86 to find the speed necessary for the given circular orbit around Earth. Let $GM = 9.56 \times 10^4$ cubic miles per second per second, and assume the radius of Earth is 4000 miles.**

87. The orbit of a space shuttle 115 miles above the surface of Earth

88. The orbit of a space shuttle 245 miles above the surface of Earth

89. The orbit of a heat capacity mapping satellite 385 miles above the surface of Earth

90. The orbit of a communications satellite r miles above the surface of Earth that is in geosynchronous orbit [The satellite completes one orbit per sidereal day (approximately 23 hours, 56 minutes), and therefore appears to remain stationary above a point on Earth.]

True or False? **In Exercises 91 and 92, determine whether the statement is true or false. If it is false, explain why or give an example that shows it is false.**

91. If a car's speedometer is constant, then the car cannot be accelerating.

92. If $a_{\mathbf{N}} = 0$ for a moving object, then the object is moving in a straight line.

93. A particle moves along a path modeled by $\mathbf{r}(t) = \cosh(bt)\mathbf{i} + \sinh(bt)\mathbf{j}$, where b is a positive constant.

(a) Show that the path of the particle is a hyperbola.

(b) Show that $\mathbf{a}(t) = b^2\,\mathbf{r}(t)$.

94. Prove that the principal unit normal vector $\mathbf{N}$ points toward the concave side of a plane curve.

95. Prove that the vector $\mathbf{T}'(t)$ is $\mathbf{0}$ for an object moving in a straight line.

96. Prove that $a_{\mathbf{N}} = \dfrac{\|\mathbf{v} \times \mathbf{a}\|}{\|\mathbf{v}\|}$.

97. Prove that $a_{\mathbf{N}} = \sqrt{\|\mathbf{a}\|^2 - a_{\mathbf{T}}^2}$.

PUTNAM EXAM CHALLENGE

98. A particle of unit mass moves on a straight line under the action of a force which is a function $f(v)$ of the velocity v of the particle, but the form of this function is not known. A motion is observed, and the distance x covered in time t is found to be connected with t by the formula $x = at + bt^2 + ct^3$, where a, b, and c have numerical values determined by observation of the motion. Find the function $f(v)$ for the range of v covered by the experiment.

12.5 Arc Length and Curvature

- Find the arc length of a space curve.
- Use the arc length parameter to describe a plane curve or space curve.
- Find the curvature of a curve at a point on the curve.
- Use a vector-valued function to find frictional force.

Arc Length

In Section 10.3, you saw that the arc length of a smooth *plane* curve C given by the parametric equations $x = x(t)$ and $y = y(t)$, $a \le t \le b$, is

$$s = \int_a^b \sqrt{[x'(t)]^2 + [y'(t)]^2}\, dt.$$

In vector form, where C is given by $\mathbf{r}(t) = x(t)\mathbf{i} + y(t)\mathbf{j}$, you can rewrite this equation for arc length as

$$s = \int_a^b \|\mathbf{r}'(t)\|\, dt.$$

The formula for the arc length of a plane curve has a natural extension to a smooth curve in *space*, as stated in the following theorem.

> **EXPLORATION**
>
> ***Arc Length Formula*** The formula for the arc length of a space curve is given in terms of the parametric equations used to represent the curve. Does this mean that the arc length of the curve depends on the parameter being used? Would you want this to be true? Explain your reasoning.
>
> Here is a different parametric representation of the curve in Example 1.
>
> $$\mathbf{r}(t) = t^2\mathbf{i} + \frac{4}{3}t^3\mathbf{j} + \frac{1}{2}t^4\mathbf{k}$$
>
> Find the arc length from $t = 0$ to $t = \sqrt{2}$ and compare the result with that found in Example 1.

THEOREM 12.6 ARC LENGTH OF A SPACE CURVE

If C is a smooth curve given by $\mathbf{r}(t) = x(t)\mathbf{i} + y(t)\mathbf{j} + z(t)\mathbf{k}$, on an interval $[a, b]$, then the arc length of C on the interval is

$$s = \int_a^b \sqrt{[x'(t)]^2 + [y'(t)]^2 + [z'(t)]^2}\, dt = \int_a^b \|\mathbf{r}'(t)\|\, dt.$$

EXAMPLE 1 Finding the Arc Length of a Curve in Space

Find the arc length of the curve given by

$$\mathbf{r}(t) = t\mathbf{i} + \frac{4}{3}t^{3/2}\mathbf{j} + \frac{1}{2}t^2\mathbf{k}$$

from $t = 0$ to $t = 2$, as shown in Figure 12.28.

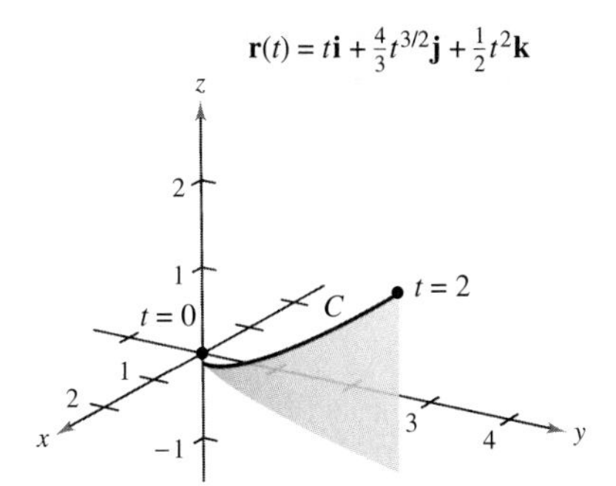

As t increases from 0 to 2, the vector $\mathbf{r}(t)$ traces out a curve.
Figure 12.28

Solution Using $x(t) = t$, $y(t) = \frac{4}{3}t^{3/2}$, and $z(t) = \frac{1}{2}t^2$, you obtain $x'(t) = 1$, $y'(t) = 2t^{1/2}$, and $z'(t) = t$. So, the arc length from $t = 0$ to $t = 2$ is given by

$$s = \int_0^2 \sqrt{[x'(t)]^2 + [y'(t)]^2 + [z'(t)]^2}\, dt \qquad \text{Formula for arc length}$$

$$= \int_0^2 \sqrt{1 + 4t + t^2}\, dt$$

$$= \int_0^2 \sqrt{(t+2)^2 - 3}\, dt \qquad \text{Integration tables (Appendix B), Formula 26}$$

$$= \left[\frac{t+2}{2}\sqrt{(t+2)^2 - 3} - \frac{3}{2}\ln\left|(t+2) + \sqrt{(t+2)^2 - 3}\right|\right]_0^2$$

$$= 2\sqrt{13} - \frac{3}{2}\ln\left(4 + \sqrt{13}\right) - 1 + \frac{3}{2}\ln 3 \approx 4.816.$$

■

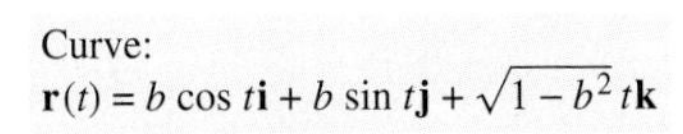

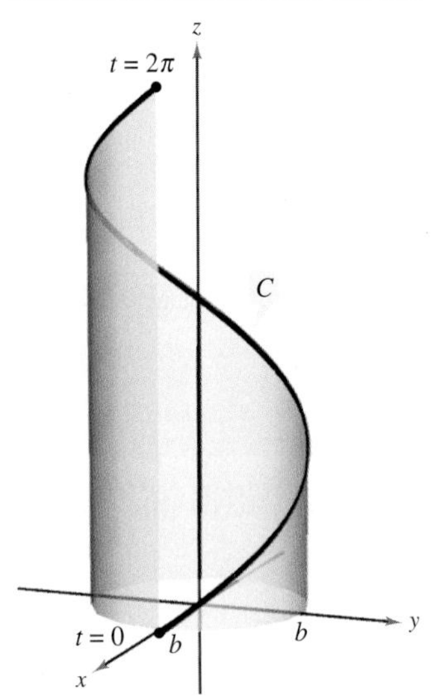

One turn of a helix
Figure 12.29

EXAMPLE 2 Finding the Arc Length of a Helix

Find the length of one turn of the helix given by

$$\mathbf{r}(t) = b\cos t\mathbf{i} + b\sin t\mathbf{j} + \sqrt{1-b^2}\,t\mathbf{k}$$

as shown in Figure 12.29.

Solution Begin by finding the derivative.

$$\mathbf{r}'(t) = -b\sin t\mathbf{i} + b\cos t\mathbf{j} + \sqrt{1-b^2}\mathbf{k} \qquad \text{Derivative}$$

Now, using the formula for arc length, you can find the length of one turn of the helix by integrating $\|\mathbf{r}'(t)\|$ from 0 to 2π.

$$\begin{aligned} s &= \int_0^{2\pi} \|\mathbf{r}'(t)\|\,dt && \text{Formula for arc length} \\ &= \int_0^{2\pi} \sqrt{b^2(\sin^2 t + \cos^2 t) + (1-b^2)}\,dt \\ &= \int_0^{2\pi} dt \\ &= t\Big]_0^{2\pi} = 2\pi \end{aligned}$$

So, the length is 2π units. ■

Arc Length Parameter

$$s(t) = \int_a^t \sqrt{[x'(u)]^2 + [y'(u)]^2 + [z'(u)]^2}\,du$$

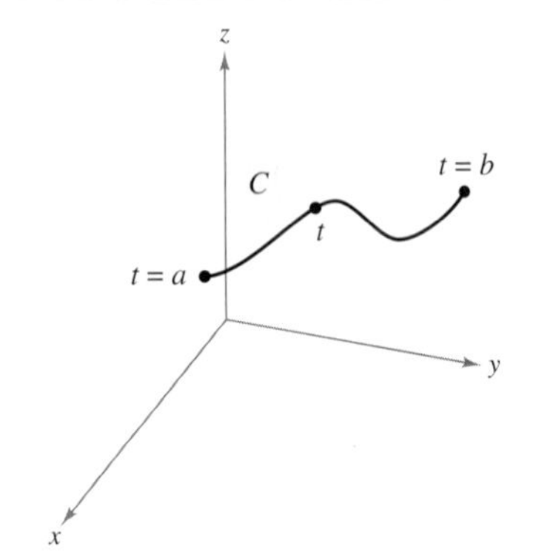

Figure 12.30

You have seen that curves can be represented by vector-valued functions in different ways, depending on the choice of parameter. For *motion* along a curve, the convenient parameter is time t. However, for studying the *geometric properties* of a curve, the convenient parameter is often arc length s.

DEFINITION OF ARC LENGTH FUNCTION

Let C be a smooth curve given by $\mathbf{r}(t)$ defined on the closed interval $[a, b]$. For $a \le t \le b$, the **arc length function** is given by

$$s(t) = \int_a^t \|\mathbf{r}'(u)\|\,du = \int_a^t \sqrt{[x'(u)]^2 + [y'(u)]^2 + [z'(u)]^2}\,du.$$

The arc length s is called the **arc length parameter.** (See Figure 12.30.)

NOTE The arc length function s is *nonnegative*. It measures the distance along C from the initial point $(x(a), y(a), z(a))$ to the point $(x(t), y(t), z(t))$. ■

Using the definition of the arc length function and the Second Fundamental Theorem of Calculus, you can conclude that

$$\frac{ds}{dt} = \|\mathbf{r}'(t)\|. \qquad \text{Derivative of arc length function}$$

In differential form, you can write

$$ds = \|\mathbf{r}'(t)\|\,dt.$$

EXAMPLE 3 Finding the Arc Length Function for a Line

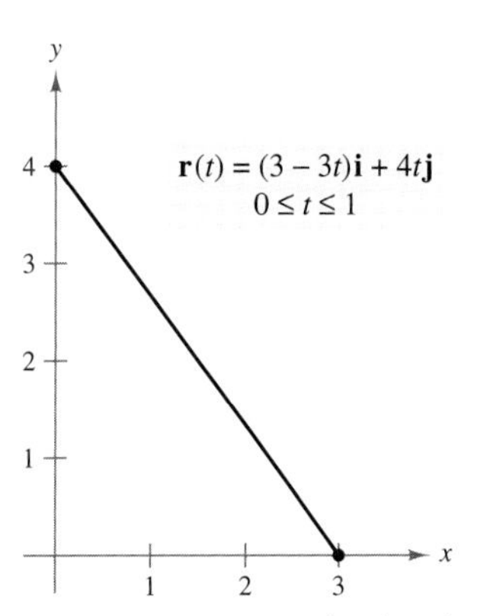

The line segment from (3, 0) to (0, 4) can be parametrized using the arc length parameter s.
Figure 12.31

Find the arc length function $s(t)$ for the line segment given by

$$\mathbf{r}(t) = (3 - 3t)\mathbf{i} + 4t\mathbf{j}, \quad 0 \le t \le 1$$

and write $\mathbf{r}$ as a function of the parameter s. (See Figure 12.31.)

Solution Because $\mathbf{r}'(t) = -3\mathbf{i} + 4\mathbf{j}$ and

$$\|\mathbf{r}'(t)\| = \sqrt{(-3)^2 + 4^2} = 5$$

you have

$$\begin{aligned} s(t) &= \int_0^t \|\mathbf{r}'(u)\|\, du \\ &= \int_0^t 5\, du \\ &= 5t. \end{aligned}$$

Using $s = 5t$ (or $t = s/5$), you can rewrite $\mathbf{r}$ using the arc length parameter as follows.

$$\mathbf{r}(s) = \left(3 - \tfrac{3}{5}s\right)\mathbf{i} + \tfrac{4}{5}s\mathbf{j}, \quad 0 \le s \le 5$$ ■

One of the advantages of writing a vector-valued function in terms of the arc length parameter is that $\|\mathbf{r}'(s)\| = 1$. For instance, in Example 3, you have

$$\|\mathbf{r}'(s)\| = \sqrt{\left(-\frac{3}{5}\right)^2 + \left(\frac{4}{5}\right)^2} = 1.$$

So, for a smooth curve C represented by $\mathbf{r}(s)$, where s is the arc length parameter, the arc length between a and b is

$$\begin{aligned} \text{Length of arc} &= \int_a^b \|\mathbf{r}'(s)\|\, ds \\ &= \int_a^b ds \\ &= b - a \\ &= \text{length of interval.} \end{aligned}$$

Furthermore, if t is *any* parameter such that $\|\mathbf{r}'(t)\| = 1$, then t must be the arc length parameter. These results are summarized in the following theorem, which is stated without proof.

THEOREM 12.7 ARC LENGTH PARAMETER

If C is a smooth curve given by

$$\mathbf{r}(s) = x(s)\mathbf{i} + y(s)\mathbf{j} \quad \text{or} \quad \mathbf{r}(s) = x(s)\mathbf{i} + y(s)\mathbf{j} + z(s)\mathbf{k}$$

where s is the arc length parameter, then

$$\|\mathbf{r}'(s)\| = 1.$$

Moreover, if t is *any* parameter for the vector-valued function $\mathbf{r}$ such that $\|\mathbf{r}'(t)\| = 1$, then t must be the arc length parameter.

Curvature

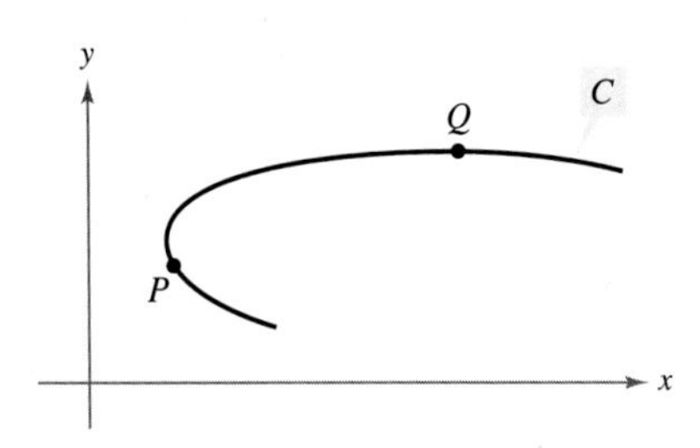

Curvature at P is greater than at Q.
Figure 12.32

An important use of the arc length parameter is to find **curvature**—the measure of how sharply a curve bends. For instance, in Figure 12.32 the curve bends more sharply at P than at Q, and you can say that the curvature is greater at P than at Q. You can calculate curvature by calculating the magnitude of the rate of change of the unit tangent vector $\mathbf{T}$ with respect to the arc length s, as shown in Figure 12.33.

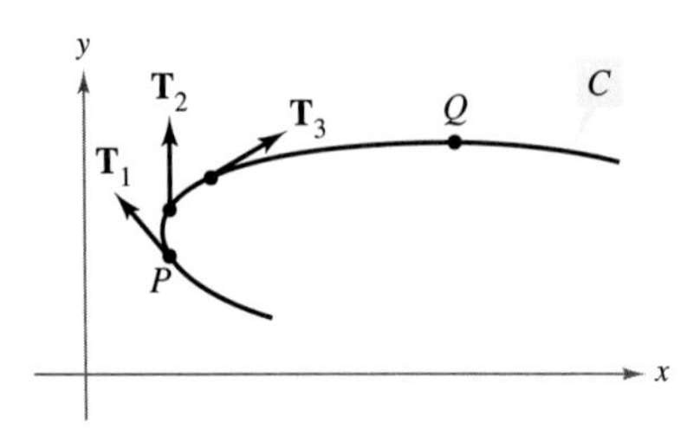

The magnitude of the rate of change of $\mathbf{T}$ with respect to the arc length is the curvature of a curve.
Figure 12.33

DEFINITION OF CURVATURE

Let C be a smooth curve (in the plane *or* in space) given by $\mathbf{r}(s)$, where s is the arc length parameter. The **curvature** K at s is given by

$$K = \left\| \frac{d\mathbf{T}}{ds} \right\| = \|\mathbf{T}'(s)\|.$$

A circle has the same curvature at any point. Moreover, the curvature and the radius of the circle are inversely related. That is, a circle with a large radius has a small curvature, and a circle with a small radius has a large curvature. This inverse relationship is made explicit in the following example.

EXAMPLE 4 Finding the Curvature of a Circle

Show that the curvature of a circle of radius r is $K = 1/r$.

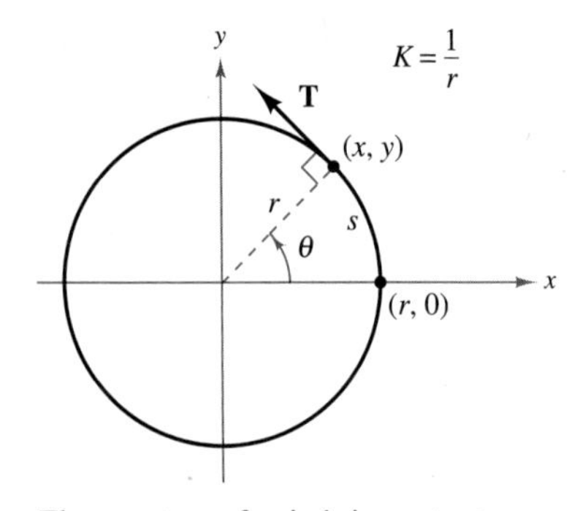

The curvature of a circle is constant.
Figure 12.34

Solution Without loss of generality you can consider the circle to be centered at the origin. Let (x, y) be any point on the circle and let s be the length of the arc from $(r, 0)$ to (x, y), as shown in Figure 12.34. By letting θ be the central angle of the circle, you can represent the circle by

$$\mathbf{r}(\theta) = r \cos \theta \mathbf{i} + r \sin \theta \mathbf{j}. \qquad \theta \text{ is the parameter.}$$

Using the formula for the length of a circular arc $s = r\theta$, you can rewrite $\mathbf{r}(\theta)$ in terms of the arc length parameter as follows.

$$\mathbf{r}(s) = r \cos \frac{s}{r} \mathbf{i} + r \sin \frac{s}{r} \mathbf{j} \qquad \text{Arc length } s \text{ is the parameter.}$$

So, $\mathbf{r}'(s) = -\sin \frac{s}{r} \mathbf{i} + \cos \frac{s}{r} \mathbf{j}$, and it follows that $\|\mathbf{r}'(s)\| = 1$, which implies that the unit tangent vector is

$$\mathbf{T}(s) = \frac{\mathbf{r}'(s)}{\|\mathbf{r}'(s)\|} = -\sin \frac{s}{r} \mathbf{i} + \cos \frac{s}{r} \mathbf{j}$$

and the curvature is given by

$$K = \|\mathbf{T}'(s)\| = \left\| -\frac{1}{r} \cos \frac{s}{r} \mathbf{i} - \frac{1}{r} \sin \frac{s}{r} \mathbf{j} \right\| = \frac{1}{r}$$

at every point on the circle. ■

NOTE Because a straight line doesn't curve, you would expect its curvature to be 0. Try checking this by finding the curvature of the line given by

$$\mathbf{r}(s) = \left(3 - \frac{3}{5}s\right)\mathbf{i} + \frac{4}{5}s\mathbf{j}.$$

■

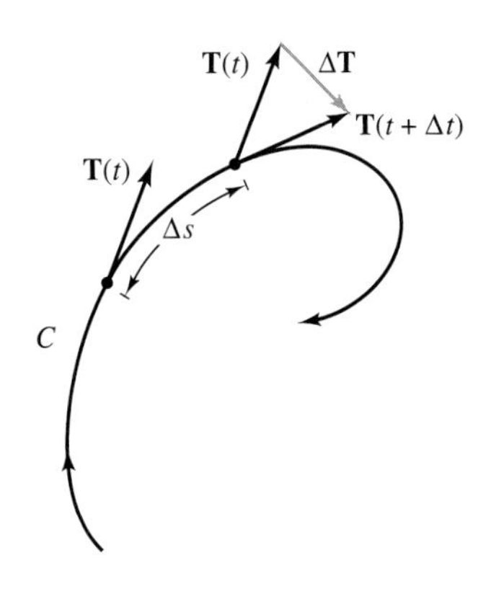

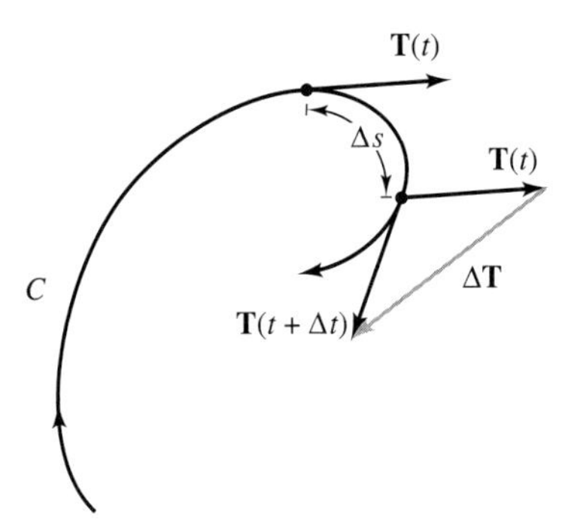

Figure 12.35

In Example 4, the curvature was found by applying the definition directly. This requires that the curve be written in terms of the arc length parameter s. The following theorem gives two other formulas for finding the curvature of a curve written in terms of an arbitrary parameter t. The proof of this theorem is left as an exercise [see Exercise 100, parts (a) and (b)].

THEOREM 12.8 FORMULAS FOR CURVATURE

If C is a smooth curve given by $\mathbf{r}(t)$, then the curvature K of C at t is given by

$$K = \frac{\|\mathbf{T}'(t)\|}{\|\mathbf{r}'(t)\|} = \frac{\|\mathbf{r}'(t) \times \mathbf{r}''(t)\|}{\|\mathbf{r}'(t)\|^3}.$$

Because $\|\mathbf{r}'(t)\| = ds/dt$, the first formula implies that curvature is the ratio of the rate of change in the tangent vector $\mathbf{T}$ to the rate of change in arc length. To see that this is reasonable, let Δt be a "small number." Then,

$$\frac{\mathbf{T}'(t)}{ds/dt} \approx \frac{[\mathbf{T}(t + \Delta t) - \mathbf{T}(t)]/\Delta t}{[s(t + \Delta t) - s(t)]/\Delta t} = \frac{\mathbf{T}(t + \Delta t) - \mathbf{T}(t)}{s(t + \Delta t) - s(t)} = \frac{\Delta \mathbf{T}}{\Delta s}.$$

In other words, for a given Δs, the greater the length of $\Delta \mathbf{T}$, the more the curve bends at t, as shown in Figure 12.35.

EXAMPLE 5 Finding the Curvature of a Space Curve

Find the curvature of the curve given by $\mathbf{r}(t) = 2t\mathbf{i} + t^2\mathbf{j} - \frac{1}{3}t^3\mathbf{k}$.

Solution It is not apparent whether this parameter represents arc length, so you should use the formula $K = \|\mathbf{T}'(t)\|/\|\mathbf{r}'(t)\|$.

$$\mathbf{r}'(t) = 2\mathbf{i} + 2t\mathbf{j} - t^2\mathbf{k}$$

$$\|\mathbf{r}'(t)\| = \sqrt{4 + 4t^2 + t^4} = t^2 + 2 \qquad \text{Length of } \mathbf{r}'(t)$$

$$\mathbf{T}(t) = \frac{\mathbf{r}'(t)}{\|\mathbf{r}'(t)\|} = \frac{2\mathbf{i} + 2t\mathbf{j} - t^2\mathbf{k}}{t^2 + 2}$$

$$\mathbf{T}'(t) = \frac{(t^2 + 2)(2\mathbf{j} - 2t\mathbf{k}) - (2t)(2\mathbf{i} + 2t\mathbf{j} - t^2\mathbf{k})}{(t^2 + 2)^2}$$

$$= \frac{-4t\mathbf{i} + (4 - 2t^2)\mathbf{j} - 4t\mathbf{k}}{(t^2 + 2)^2}$$

$$\|\mathbf{T}'(t)\| = \frac{\sqrt{16t^2 + 16 - 16t^2 + 4t^4 + 16t^2}}{(t^2 + 2)^2}$$

$$= \frac{2(t^2 + 2)}{(t^2 + 2)^2}$$

$$= \frac{2}{t^2 + 2} \qquad \text{Length of } \mathbf{T}'(t)$$

Therefore,

$$K = \frac{\|\mathbf{T}'(t)\|}{\|\mathbf{r}'(t)\|} = \frac{2}{(t^2 + 2)^2}. \qquad \text{Curvature}$$

■

The following theorem presents a formula for calculating the curvature of a plane curve given by $y = f(x)$.

THEOREM 12.9 CURVATURE IN RECTANGULAR COORDINATES

If C is the graph of a twice-differentiable function given by $y = f(x)$, then the curvature K at the point (x, y) is given by

$$K = \frac{|y''|}{[1 + (y')^2]^{3/2}}.$$

PROOF By representing the curve C by $\mathbf{r}(x) = x\mathbf{i} + f(x)\mathbf{j} + 0\mathbf{k}$ (where x is the parameter), you obtain $\mathbf{r}'(x) = \mathbf{i} + f'(x)\mathbf{j}$,

$$\|\mathbf{r}'(x)\| = \sqrt{1 + [f'(x)]^2}$$

and $\mathbf{r}''(x) = f''(x)\mathbf{j}$. Because $\mathbf{r}'(x) \times \mathbf{r}''(x) = f''(x)\mathbf{k}$, it follows that the curvature is

$$\begin{aligned} K &= \frac{\|\mathbf{r}'(x) \times \mathbf{r}''(x)\|}{\|\mathbf{r}'(x)\|^3} \\ &= \frac{|f''(x)|}{\{1 + [f'(x)]^2\}^{3/2}} \\ &= \frac{|y''|}{[1 + (y')^2]^{3/2}}. \end{aligned}$$

■

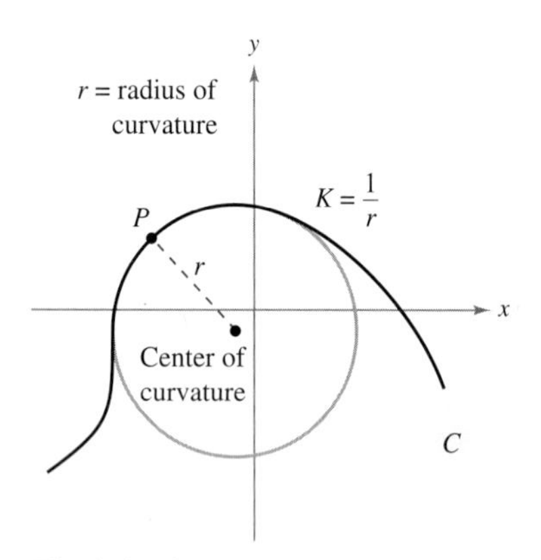

The circle of curvature
Figure 12.36

Let C be a curve with curvature K at point P. The circle passing through point P with radius $r = 1/K$ is called the **circle of curvature** if the circle lies on the concave side of the curve and shares a common tangent line with the curve at point P. The radius is called the **radius of curvature** at P, and the center of the circle is called the **center of curvature.**

The circle of curvature gives you a nice way to estimate graphically the curvature K at a point P on a curve. Using a compass, you can sketch a circle that lies against the concave side of the curve at point P, as shown in Figure 12.36. If the circle has a radius of r, you can estimate the curvature to be $K = 1/r$.

EXAMPLE 6 Finding Curvature in Rectangular Coordinates

Find the curvature of the parabola given by $y = x - \frac{1}{4}x^2$ at $x = 2$. Sketch the circle of curvature at $(2, 1)$.

Solution The curvature at $x = 2$ is as follows.

$$y' = 1 - \frac{x}{2} \qquad y' = 0$$

$$y'' = -\frac{1}{2} \qquad y'' = -\frac{1}{2}$$

$$K = \frac{|y''|}{[1 + (y')^2]^{3/2}} \qquad K = \frac{1}{2}$$

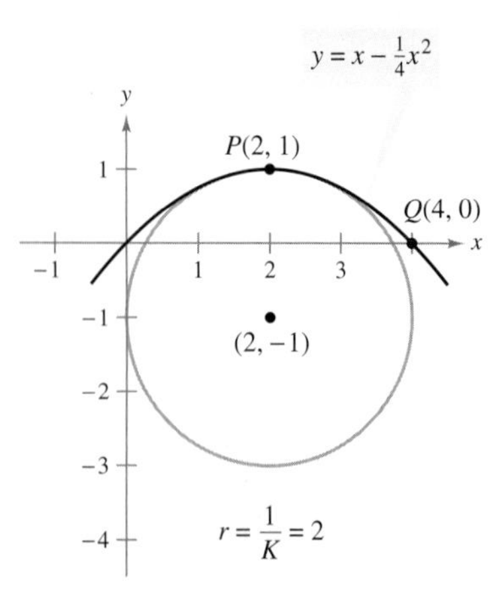

The circle of curvature
Figure 12.37

Because the curvature at $P(2, 1)$ is $\frac{1}{2}$, it follows that the radius of the circle of curvature at that point is 2. So, the center of curvature is $(2, -1)$, as shown in Figure 12.37. [In the figure, note that the curve has the greatest curvature at P. Try showing that the curvature at $Q(4, 0)$ is $1/2^{5/2} \approx 0.177$.] ■

The amount of thrust felt by passengers in a car that is turning depends on two things—the speed of the car and the sharpness of the turn.
Figure 12.38

Arc length and curvature are closely related to the tangential and normal components of acceleration. The tangential component of acceleration is the rate of change of the speed, which in turn is the rate of change of the arc length. This component is negative as a moving object slows down and positive as it speeds up—regardless of whether the object is turning or traveling in a straight line. So, the tangential component is solely a function of the arc length and is independent of the curvature.

On the other hand, the normal component of acceleration is a function of *both* speed and curvature. This component measures the acceleration acting perpendicular to the direction of motion. To see why the normal component is affected by both speed and curvature, imagine that you are driving a car around a turn, as shown in Figure 12.38. If your speed is high and the turn is sharp, you feel yourself thrown against the car door. By lowering your speed *or* taking a more gentle turn, you are able to lessen this sideways thrust.

The next theorem explicitly states the relationships among speed, curvature, and the components of acceleration.

NOTE Note that Theorem 12.10 gives additional formulas for $a_{\mathbf{T}}$ and $a_{\mathbf{N}}$.

THEOREM 12.10 ACCELERATION, SPEED, AND CURVATURE

If $\mathbf{r}(t)$ is the position vector for a smooth curve C, then the acceleration vector is given by

$$\mathbf{a}(t) = \frac{d^2s}{dt^2}\mathbf{T} + K\left(\frac{ds}{dt}\right)^2\mathbf{N}$$

where K is the curvature of C and ds/dt is the speed.

PROOF For the position vector $\mathbf{r}(t)$, you have

$$\begin{aligned}\mathbf{a}(t) &= a_{\mathbf{T}}\mathbf{T} + a_{\mathbf{N}}\mathbf{N}\\ &= D_t[\|\mathbf{v}\|]\mathbf{T} + \|\mathbf{v}\|\,\|\mathbf{T}'\|\mathbf{N}\\ &= \frac{d^2s}{dt^2}\mathbf{T} + \frac{ds}{dt}(\|\mathbf{v}\|K)\mathbf{N}\\ &= \frac{d^2s}{dt^2}\mathbf{T} + K\left(\frac{ds}{dt}\right)^2\mathbf{N}.\end{aligned}$$

■

EXAMPLE 7 Tangential and Normal Components of Acceleration

Find $a_{\mathbf{T}}$ and $a_{\mathbf{N}}$ for the curve given by

$$\mathbf{r}(t) = 2t\mathbf{i} + t^2\mathbf{j} - \tfrac{1}{3}t^3\mathbf{k}.$$

Solution From Example 5, you know that

$$\frac{ds}{dt} = \|\mathbf{r}'(t)\| = t^2 + 2 \quad \text{and} \quad K = \frac{2}{(t^2+2)^2}.$$

Therefore,

$$a_{\mathbf{T}} = \frac{d^2s}{dt^2} = 2t \qquad \text{Tangential component}$$

and

$$a_{\mathbf{N}} = K\left(\frac{ds}{dt}\right)^2 = \frac{2}{(t^2+2)^2}(t^2+2)^2 = 2. \qquad \text{Normal component}$$

■

Application

There are many applications in physics and engineering dynamics that involve the relationships among speed, arc length, curvature, and acceleration. One such application concerns frictional force.

A moving object with mass m is in contact with a stationary object. The total force required to produce an acceleration $\mathbf{a}$ along a given path is

$$\begin{aligned}\mathbf{F} = m\mathbf{a} &= m\left(\frac{d^2s}{dt^2}\right)\mathbf{T} + mK\left(\frac{ds}{dt}\right)^2\mathbf{N}\\ &= ma_{\mathbf{T}}\mathbf{T} + ma_{\mathbf{N}}\mathbf{N}.\end{aligned}$$

The portion of this total force that is supplied by the stationary object is called the **force of friction.** For example, if a car moving with constant speed is rounding a turn, the roadway exerts a frictional force that keeps the car from sliding off the road. If the car is not sliding, the frictional force is perpendicular to the direction of motion and has magnitude equal to the normal component of acceleration, as shown in Figure 12.39. The potential frictional force of a road around a turn can be increased by banking the roadway.

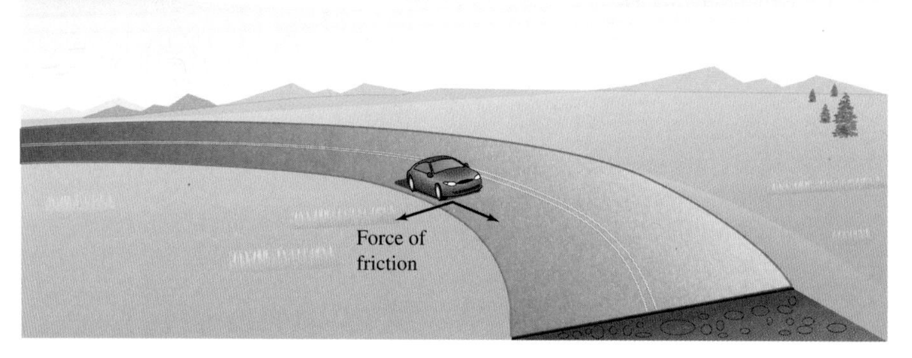

The force of friction is perpendicular to the direction of motion.
Figure 12.39

EXAMPLE 8 Frictional Force

A 360-kilogram go-cart is driven at a speed of 60 kilometers per hour around a circular racetrack of radius 12 meters, as shown in Figure 12.40. To keep the cart from skidding off course, what frictional force must the track surface exert on the tires?

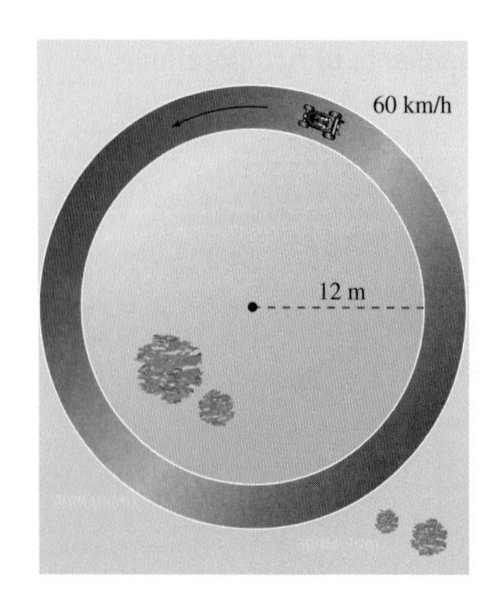

Figure 12.40

Solution The frictional force must equal the mass times the normal component of acceleration. For this circular path, you know that the curvature is

$$K = \frac{1}{12}. \qquad \text{Curvature of circular racetrack}$$

Therefore, the frictional force is

$$\begin{aligned}ma_{\mathbf{N}} &= mK\left(\frac{ds}{dt}\right)^2\\ &= (360 \text{ kg})\left(\frac{1}{12 \text{ m}}\right)\left(\frac{60{,}000 \text{ m}}{3600 \text{ sec}}\right)^2\\ &\approx 8333 \text{ (kg)(m)/sec}^2.\end{aligned}$$

SUMMARY OF VELOCITY, ACCELERATION, AND CURVATURE

Let C be a curve (in the plane or in space) given by the position function

$\mathbf{r}(t) = x(t)\mathbf{i} + y(t)\mathbf{j}$ — Curve in the plane

$\mathbf{r}(t) = x(t)\mathbf{i} + y(t)\mathbf{j} + z(t)\mathbf{k}.$ — Curve in space

Velocity vector, speed, and acceleration vector:

$\mathbf{v}(t) = \mathbf{r}'(t)$ — Velocity vector

$\|\mathbf{v}(t)\| = \dfrac{ds}{dt} = \|\mathbf{r}'(t)\|$ — Speed

$\mathbf{a}(t) = \mathbf{r}''(t) = a_{\mathbf{T}}\mathbf{T}(t) + a_{\mathbf{N}}\mathbf{N}(t)$ — Acceleration vector

Unit tangent vector and principal unit normal vector:

$$\mathbf{T}(t) = \frac{\mathbf{r}'(t)}{\|\mathbf{r}'(t)\|} \quad \text{and} \quad \mathbf{N}(t) = \frac{\mathbf{T}'(t)}{\|\mathbf{T}'(t)\|}$$

Components of acceleration:

$$a_{\mathbf{T}} = \mathbf{a} \cdot \mathbf{T} = \frac{\mathbf{v} \cdot \mathbf{a}}{\|\mathbf{v}\|} = \frac{d^2s}{dt^2}$$

$$a_{\mathbf{N}} = \mathbf{a} \cdot \mathbf{N} = \frac{\|\mathbf{v} \times \mathbf{a}\|}{\|\mathbf{v}\|} = \sqrt{\|\mathbf{a}\|^2 - a_{\mathbf{T}}^2} = K\left(\frac{ds}{dt}\right)^2$$

Formulas for curvature in the plane:

$K = \dfrac{|y''|}{[1 + (y')^2]^{3/2}}$ — C given by $y = f(x)$

$K = \dfrac{|x'y'' - y'x''|}{[(x')^2 + (y')^2]^{3/2}}$ — C given by $x = x(t)$, $y = y(t)$

Formulas for curvature in the plane or in space:

$K = \|\mathbf{T}'(s)\| = \|\mathbf{r}''(s)\|$ — s is arc length parameter.

$K = \dfrac{\|\mathbf{T}'(t)\|}{\|\mathbf{r}'(t)\|} = \dfrac{\|\mathbf{r}'(t) \times \mathbf{r}''(t)\|}{\|\mathbf{r}'(t)\|^3}$ — t is general parameter.

$$K = \frac{\mathbf{a}(t) \cdot \mathbf{N}(t)}{\|\mathbf{v}(t)\|^2}$$

Cross product formulas apply only to curves in space.

12.5 Exercises

See www.CalcChat.com for worked-out solutions to odd-numbered exercises.

In Exercises 1–6, sketch the plane curve and find its length over the given interval.

1. $\mathbf{r}(t) = 3t\mathbf{i} - t\mathbf{j}$, $[0, 3]$
2. $\mathbf{r}(t) = t\mathbf{i} + t^2\mathbf{j}$, $[0, 4]$
3. $\mathbf{r}(t) = t^3\mathbf{i} + t^2\mathbf{j}$, $[0, 1]$
4. $\mathbf{r}(t) = (t + 1)\mathbf{i} + t^2\mathbf{j}$, $[0, 6]$
5. $\mathbf{r}(t) = a\cos^3 t\,\mathbf{i} + a\sin^3 t\,\mathbf{j}$, $[0, 2\pi]$
6. $\mathbf{r}(t) = a\cos t\,\mathbf{i} + a\sin t\,\mathbf{j}$, $[0, 2\pi]$

7. ***Projectile Motion*** A baseball is hit 3 feet above the ground at 100 feet per second and at an angle of 45° with respect to the ground.
   (a) Find the vector-valued function for the path of the baseball.
   (b) Find the maximum height.
   (c) Find the range.
   (d) Find the arc length of the trajectory.

8. ***Projectile Motion*** An object is launched from ground level. Determine the angle of the launch to obtain (a) the maximum height, (b) the maximum range, and (c) the maximum length of the trajectory. For part (c), let $v_0 = 96$ feet per second.

In Exercises 9–14, sketch the space curve and find its length over the given interval.

	Function	*Interval*
9.	$\mathbf{r}(t) = -t\mathbf{i} + 4t\mathbf{j} + 3t\mathbf{k}$	$[0, 1]$
10.	$\mathbf{r}(t) = \mathbf{i} + t^2\mathbf{j} + t^3\mathbf{k}$	$[0, 2]$
11.	$\mathbf{r}(t) = \langle 4t, -\cos t, \sin t\rangle$	$\left[0, \frac{3\pi}{2}\right]$
12.	$\mathbf{r}(t) = \langle 2\sin t, 5t, 2\cos t\rangle$	$[0, \pi]$
13.	$\mathbf{r}(t) = a\cos t\,\mathbf{i} + a\sin t\,\mathbf{j} + bt\,\mathbf{k}$	$[0, 2\pi]$
14.	$\mathbf{r}(t) = \langle \cos t + t\sin t, \sin t - t\cos t, t^2\rangle$	$\left[0, \frac{\pi}{2}\right]$

In Exercises 15 and 16, use the integration capabilities of a graphing utility to approximate the length of the space curve over the given interval.

Function	*Interval*
15. $\mathbf{r}(t) = t^2\mathbf{i} + t\mathbf{j} + \ln t\mathbf{k}$	$1 \le t \le 3$
16. $\mathbf{r}(t) = \sin \pi t\mathbf{i} + \cos \pi t\mathbf{j} + t^3\mathbf{k}$	$0 \le t \le 2$

17. ***Investigation*** Consider the graph of the vector-valued function $\mathbf{r}(t) = t\mathbf{i} + (4 - t^2)\mathbf{j} + t^3\mathbf{k}$ on the interval $[0, 2]$.

(a) Approximate the length of the curve by finding the length of the line segment connecting its endpoints.

(b) Approximate the length of the curve by summing the lengths of the line segments connecting the terminal points of the vectors $\mathbf{r}(0)$, $\mathbf{r}(0.5)$, $\mathbf{r}(1)$, $\mathbf{r}(1.5)$, and $\mathbf{r}(2)$.

(c) Describe how you could obtain a more accurate approximation by continuing the processes in parts (a) and (b).

(d) Use the integration capabilities of a graphing utility to approximate the length of the curve. Compare this result with the answers in parts (a) and (b).

18. ***Investigation*** Repeat Exercise 17 for the vector-valued function $\mathbf{r}(t) = 6\cos(\pi t/4)\mathbf{i} + 2\sin(\pi t/4)\mathbf{j} + t\mathbf{k}$.

19. ***Investigation*** Consider the helix represented by the vector-valued function $\mathbf{r}(t) = \langle 2\cos t, 2\sin t, t\rangle$.

(a) Write the length of the arc s on the helix as a function of t by evaluating the integral

$$s = \int_0^t \sqrt{[x'(u)]^2 + [y'(u)]^2 + [z'(u)]^2}\, du.$$

(b) Solve for t in the relationship derived in part (a), and substitute the result into the original set of parametric equations. This yields a parametrization of the curve in terms of the arc length parameter s.

(c) Find the coordinates of the point on the helix for arc lengths $s = \sqrt{5}$ and $s = 4$.

(d) Verify that $\|\mathbf{r}'(s)\| = 1$.

20. ***Investigation*** Repeat Exercise 19 for the curve represented by the vector-valued function

$$\mathbf{r}(t) = \left\langle 4(\sin t - t\cos t), 4(\cos t + t\sin t), \tfrac{3}{2}t^2\right\rangle.$$

In Exercises 21–24, find the curvature K of the curve, where s is the arc length parameter.

21. $\mathbf{r}(s) = \left(1 + \dfrac{\sqrt{2}}{2}s\right)\mathbf{i} + \left(1 - \dfrac{\sqrt{2}}{2}s\right)\mathbf{j}$

22. $\mathbf{r}(s) = (3 + s)\mathbf{i} + \mathbf{j}$

23. Helix in Exercise 19: $\mathbf{r}(t) = \langle 2\cos t, 2\sin t, t\rangle$

24. Curve in Exercise 20:
$\mathbf{r}(t) = \left\langle 4(\sin t - t\cos t), 4(\cos t + t\sin t), \tfrac{3}{2}t^2\right\rangle$

In Exercises 25–30, find the curvature K of the plane curve at the given value of the parameter.

25. $\mathbf{r}(t) = 4t\mathbf{i} - 2t\mathbf{j}, \quad t = 1$

26. $\mathbf{r}(t) = t^2\mathbf{i} + \mathbf{j}, \quad t = 2$

27. $\mathbf{r}(t) = t\mathbf{i} + \dfrac{1}{t}\mathbf{j}, \quad t = 1$

28. $\mathbf{r}(t) = t\mathbf{i} + \dfrac{1}{9}t^3\mathbf{j}, \quad t = 2$

29. $\mathbf{r}(t) = \langle t, \sin t\rangle, \quad t = \dfrac{\pi}{2}$

30. $\mathbf{r}(t) = \langle 5\cos t, 4\sin t\rangle, \quad t = \dfrac{\pi}{3}$

In Exercises 31–40, find the curvature K of the curve.

31. $\mathbf{r}(t) = 4\cos 2\pi t\mathbf{i} + 4\sin 2\pi t\mathbf{j}$

32. $\mathbf{r}(t) = 2\cos \pi t\mathbf{i} + \sin \pi t\mathbf{j}$

33. $\mathbf{r}(t) = a\cos \omega t\mathbf{i} + a\sin \omega t\mathbf{j}$

34. $\mathbf{r}(t) = a\cos \omega t\mathbf{i} + b\sin \omega t\mathbf{j}$

35. $\mathbf{r}(t) = \langle a(\omega t - \sin \omega t), a(1 - \cos \omega t)\rangle$

36. $\mathbf{r}(t) = \langle \cos \omega t + \omega t\sin \omega t, \sin \omega t - \omega t\cos \omega t\rangle$

37. $\mathbf{r}(t) = t\mathbf{i} + t^2\mathbf{j} + \dfrac{t^2}{2}\mathbf{k}$

38. $\mathbf{r}(t) = 2t^2\mathbf{i} + t\mathbf{j} + \dfrac{1}{2}t^2\mathbf{k}$

39. $\mathbf{r}(t) = 4t\mathbf{i} + 3\cos t\mathbf{j} + 3\sin t\mathbf{k}$

40. $\mathbf{r}(t) = e^{2t}\mathbf{i} + e^{2t}\cos t\mathbf{j} + e^{2t}\sin t\mathbf{k}$

In Exercises 41–44, find the curvature K of the curve at the point P.

41. $\mathbf{r}(t) = 3t\mathbf{i} + 2t^2\mathbf{j}, \quad P(-3, 2)$

42. $\mathbf{r}(t) = e^t\mathbf{i} + 4t\mathbf{j}, \quad P(1, 0)$

43. $\mathbf{r}(t) = t\mathbf{i} + t^2\mathbf{j} + \dfrac{t^3}{4}\mathbf{k}, \quad P(2, 4, 2)$

44. $\mathbf{r}(t) = e^t\cos t\mathbf{i} + e^t\sin t\mathbf{j} + e^t\mathbf{k}, \quad P(1, 0, 1)$

In Exercises 45–54, find the curvature and radius of curvature of the plane curve at the given value of x.

45. $y = 3x - 2, \quad x = a$

46. $y = mx + b, \quad x = a$

47. $y = 2x^2 + 3, \quad x = -1$

48. $y = 2x + \dfrac{4}{x}, \quad x = 1$

49. $y = \cos 2x, \quad x = 2\pi$

50. $y = e^{3x}, \quad x = 0$

51. $y = \sqrt{a^2 - x^2}, \quad x = 0$

52. $y = \frac{3}{4}\sqrt{16 - x^2}, \quad x = 0$

53. $y = x^3, \quad x = 2$

54. $y = x^n, \quad x = 1, \quad n \ge 2$

Writing **In Exercises 55 and 56, two circles of curvature to the graph of the function are given. (a) Find the equation of the smaller circle, and (b) write a short paragraph explaining why the circles have different radii.**

55. $f(x) = \sin x$

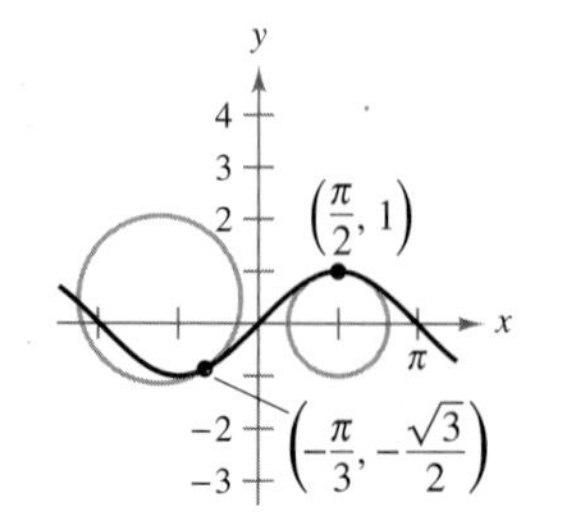

56. $f(x) = 4x^2/(x^2 + 3)$

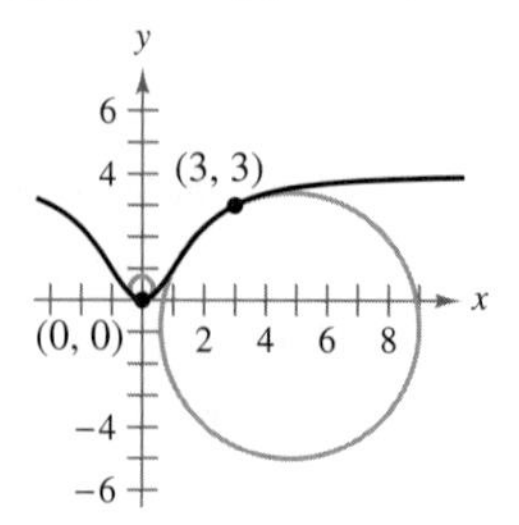

In Exercises 57–60, use a graphing utility to graph the function. In the same viewing window, graph the circle of curvature to the graph at the given value of x.

57. $y = x + \frac{1}{x}, \quad x = 1$

58. $y = \ln x, \quad x = 1$

59. $y = e^x, \quad x = 0$

60. $y = \frac{1}{3}x^3, \quad x = 1$

Evolute **An evolute is the curve formed by the set of centers of curvature of a curve. In Exercises 61 and 62, a curve and its evolute are given. Use a compass to sketch the circles of curvature with centers at points A and B. To print an enlarged copy of the graph, go to the website *www.mathgraphs.com*.**

61. Cycloid: $x = t - \sin t$
$y = 1 - \cos t$
Evolute: $x = \sin t + t$
$y = \cos t - 1$

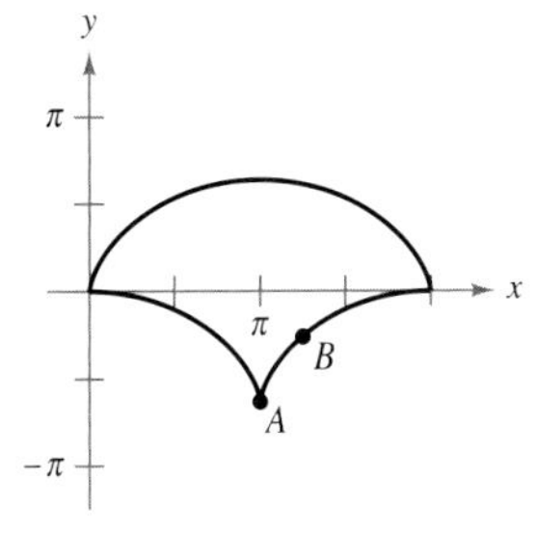

62. Ellipse: $x = 3\cos t$
$y = 2\sin t$
Evolute: $x = \frac{5}{3}\cos^3 t$
$y = \frac{5}{2}\sin^3 t$

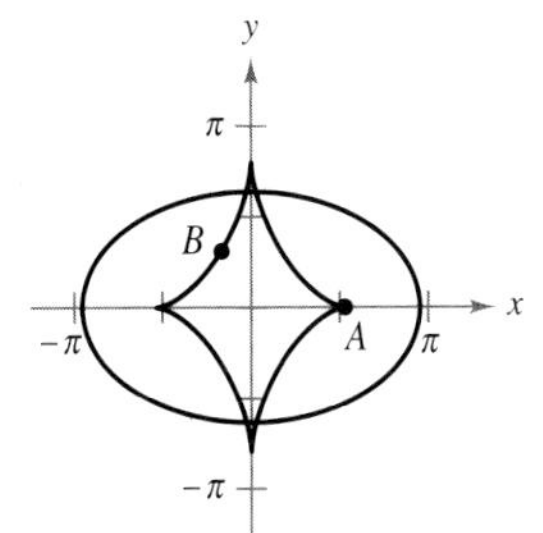

In Exercises 63–70, (a) find the point on the curve at which the curvature K is a maximum and (b) find the limit of K as $x \to \infty$.

63. $y = (x - 1)^2 + 3$

64. $y = x^3$

65. $y = x^{2/3}$

66. $y = \frac{1}{x}$

67. $y = \ln x$

68. $y = e^x$

69. $y = \sinh x$

70. $y = \cosh x$

In Exercises 71–74, find all points on the graph of the function such that the curvature is zero.

71. $y = 1 - x^3$

72. $y = (x - 1)^3 + 3$

73. $y = \cos x$

74. $y = \sin x$

WRITING ABOUT CONCEPTS

75. (a) Give the formula for the arc length of a smooth curve in space.
(b) Give the formulas for curvature in the plane and in space.

76. Describe the graph of a vector-valued function for which the curvature is 0 for all values of t in its domain.

WRITING ABOUT CONCEPTS (continued)

77. Given a twice-differentiable function $y = f(x)$, determine its curvature at a relative extremum. Can the curvature ever be greater than it is at a relative extremum? Why or why not?

CAPSTONE

78. A particle moves along the plane curve C described by $\mathbf{r}(t) = t\mathbf{i} + t^2\mathbf{j}$.
(a) Find the length of C on the interval $0 \le t \le 2$.
(b) Find the curvature K of the plane curve at $t = 0$, $t = 1$, and $t = 2$.
(c) Describe the curvature of C as t changes from $t = 0$ to $t = 2$.

79. Show that the curvature is greatest at the endpoints of the major axis, and is least at the endpoints of the minor axis, for the ellipse given by $x^2 + 4y^2 = 4$.

80. ***Investigation*** Find all a and b such that the two curves given by

$$y_1 = ax(b - x) \quad \text{and} \quad y_2 = \frac{x}{x + 2}$$

intersect at only one point and have a common tangent line and equal curvature at that point. Sketch a graph for each set of values of a and b.

CAS 81. ***Investigation*** Consider the function $f(x) = x^4 - x^2$.
(a) Use a computer algebra system to find the curvature K of the curve as a function of x.
(b) Use the result of part (a) to find the circles of curvature to the graph of f when $x = 0$ and $x = 1$. Use a computer algebra system to graph the function and the two circles of curvature.
(c) Graph the function $K(x)$ and compare it with the graph of $f(x)$. For example, do the extrema of f and K occur at the same critical numbers? Explain your reasoning.

82. ***Investigation*** The surface of a goblet is formed by revolving the graph of the function

$$y = \frac{1}{4}x^{8/5}, \quad 0 \le x \le 5$$

about the y-axis. The measurements are given in centimeters.
CAS (a) Use a computer algebra system to graph the surface.
(b) Find the volume of the goblet.
(c) Find the curvature K of the generating curve as a function of x. Use a graphing utility to graph K.
(d) If a spherical object is dropped into the goblet, is it possible for it to touch the bottom? Explain.

83. A sphere of radius 4 is dropped into the paraboloid given by $z = x^2 + y^2$.
(a) How close will the sphere come to the vertex of the paraboloid?
(b) What is the radius of the largest sphere that will touch the vertex?

84. *Speed* The smaller the curvature of a bend in a road, the faster a car can travel. Assume that the maximum speed around a turn is inversely proportional to the square root of the curvature. A car moving on the path $y = \frac{1}{3}x^3$ (x and y are measured in miles) can safely go 30 miles per hour at $\left(1, \frac{1}{3}\right)$. How fast can it go at $\left(\frac{3}{2}, \frac{9}{8}\right)$?

85. Let C be a curve given by $y = f(x)$. Let K be the curvature $(K \neq 0)$ at the point $P(x_0, y_0)$ and let

$$z = \frac{1 + f'(x_0)^2}{f''(x_0)}.$$

Show that the coordinates (α, β) of the center of curvature at P are $(\alpha, \beta) = (x_0 - f'(x_0)z, y_0 + z)$.

86. Use the result of Exercise 85 to find the center of curvature for the curve at the given point.

(a) $y = e^x, \quad (0, 1)$ (b) $y = \dfrac{x^2}{2}, \quad \left(1, \dfrac{1}{2}\right)$ (c) $y = x^2, \quad (0, 0)$

87. A curve C is given by the polar equation $r = f(\theta)$. Show that the curvature K at the point (r, θ) is

$$K = \frac{\left|2(r')^2 - rr'' + r^2\right|}{\left[(r')^2 + r^2\right]^{3/2}}$$

[*Hint:* Represent the curve by $\mathbf{r}(\theta) = r\cos\theta\mathbf{i} + r\sin\theta\mathbf{j}$.]

88. Use the result of Exercise 87 to find the curvature of each polar curve.

(a) $r = 1 + \sin\theta$ (b) $r = \theta$ (c) $r = a\sin\theta$ (d) $r = e^{\theta}$

89. Given the polar curve $r = e^{a\theta}$, $a > 0$, find the curvature K and determine the limit of K as (a) $\theta \to \infty$ and (b) $a \to \infty$.

90. Show that the formula for the curvature of a polar curve $r = f(\theta)$ given in Exercise 87 reduces to $K = 2/|r'|$ for the curvature *at the pole*.

In Exercises 91 and 92, use the result of Exercise 90 to find the curvature of the rose curve at the pole.

91. $r = 4\sin 2\theta$

92. $r = 6\cos 3\theta$

93. For a smooth curve given by the parametric equations $x = f(t)$ and $y = g(t)$, prove that the curvature is given by

$$K = \frac{|f'(t)g''(t) - g'(t)f''(t)|}{\{[f'(t)]^2 + [g'(t)]^2\}^{3/2}}.$$

94. Use the result of Exercise 93 to find the curvature K of the curve represented by the parametric equations $x(t) = t^3$ and $y(t) = \frac{1}{2}t^2$. Use a graphing utility to graph K and determine any horizontal asymptotes. Interpret the asymptotes in the context of the problem.

95. Use the result of Exercise 93 to find the curvature K of the cycloid represented by the parametric equations

$$x(\theta) = a(\theta - \sin\theta) \quad \text{and} \quad y(\theta) = a(1 - \cos\theta).$$

What are the minimum and maximum values of K?

96. Use Theorem 12.10 to find $a_{\mathbf{T}}$ and $a_{\mathbf{N}}$ for each curve given by the vector-valued function.

(a) $\mathbf{r}(t) = 3t^2\mathbf{i} + (3t - t^3)\mathbf{j}$ (b) $\mathbf{r}(t) = t\mathbf{i} + t^2\mathbf{j} + \frac{1}{2}t^2\mathbf{k}$

97. *Frictional Force* A 5500-pound vehicle is driven at a speed of 30 miles per hour on a circular interchange of radius 100 feet. To keep the vehicle from skidding off course, what frictional force must the road surface exert on the tires?

98. *Frictional Force* A 6400-pound vehicle is driven at a speed of 35 miles per hour on a circular interchange of radius 250 feet. To keep the vehicle from skidding off course, what frictional force must the road surface exert on the tires?

99. Verify that the curvature at any point (x, y) on the graph of $y = \cosh x$ is $1/y^2$.

100. Use the definition of curvature in space, $K = \|\mathbf{T}'(s)\| = \|\mathbf{r}''(s)\|$, to verify each formula.

(a) $K = \dfrac{\|\mathbf{T}'(t)\|}{\|\mathbf{r}'(t)\|}$

(b) $K = \dfrac{\|\mathbf{r}'(t) \times \mathbf{r}''(t)\|}{\|\mathbf{r}'(t)\|^3}$

(c) $K = \dfrac{\mathbf{a}(t) \cdot \mathbf{N}(t)}{\|\mathbf{v}(t)\|^2}$

True or False? **In Exercises 101–104, determine whether the statement is true or false. If it is false, explain why or give an example that shows it is false.**

101. The arc length of a space curve depends on the parametrization.

102. The curvature of a circle is the same as its radius.

103. The curvature of a line is 0.

104. The normal component of acceleration is a function of both speed and curvature.

Kepler's Laws **In Exercises 105–112, you are asked to verify Kepler's Laws of Planetary Motion. For these exercises, assume that each planet moves in an orbit given by the vector-valued function r. Let $r = \|\mathbf{r}\|$, let G represent the universal gravitational constant, let M represent the mass of the sun, and let m represent the mass of the planet.**

105. Prove that $\mathbf{r} \cdot \mathbf{r}' = r\dfrac{dr}{dt}$.

106. Using Newton's Second Law of Motion, $\mathbf{F} = m\mathbf{a}$, and Newton's Second Law of Gravitation, $\mathbf{F} = -(GmM/r^3)\mathbf{r}$, show that $\mathbf{a}$ and $\mathbf{r}$ are parallel, and that $\mathbf{r}(t) \times \mathbf{r}'(t) = \mathbf{L}$ is a constant vector. So, $\mathbf{r}(t)$ moves in a *fixed plane*, orthogonal to $\mathbf{L}$.

107. Prove that $\dfrac{d}{dt}\left[\dfrac{\mathbf{r}}{r}\right] = \dfrac{1}{r^3}\{[\mathbf{r} \times \mathbf{r}'] \times \mathbf{r}\}$.

108. Show that $\dfrac{\mathbf{r}'}{GM} \times \mathbf{L} - \dfrac{\mathbf{r}}{r} = \mathbf{e}$ is a constant vector.

109. Prove Kepler's First Law: Each planet moves in an elliptical orbit with the sun as a focus.

110. Assume that the elliptical orbit $r = ed/(1 + e\cos\theta)$ is in the xy-plane, with $\mathbf{L}$ along the z-axis. Prove that $\|\mathbf{L}\| = r^2\,d\theta/dt$.

111. Prove Kepler's Second Law: Each ray from the sun to a planet sweeps out equal areas of the ellipse in equal times.

112. Prove Kepler's Third Law: The square of the period of a planet's orbit is proportional to the cube of the mean distance between the planet and the sun.

12 REVIEW EXERCISES

See www.CalcChat.com for worked-out solutions to odd-numbered exercises.

In Exercises 1–4, (a) find the domain of r and (b) determine the values (if any) of t for which the function is continuous.

1. $\mathbf{r}(t) = \tan t\mathbf{i} + \mathbf{j} + t\mathbf{k}$
2. $\mathbf{r}(t) = \sqrt{t}\mathbf{i} + \dfrac{1}{t-4}\mathbf{j} + \mathbf{k}$
3. $\mathbf{r}(t) = \ln t\mathbf{i} + t\mathbf{j} + t\mathbf{k}$
4. $\mathbf{r}(t) = (2t+1)\mathbf{i} + t^2\mathbf{j} + t\mathbf{k}$

In Exercises 5 and 6, evaluate (if possible) the vector-valued function at each given value of t.

5. $\mathbf{r}(t) = (2t+1)\mathbf{i} + t^2\mathbf{j} - \sqrt{t+2}\,\mathbf{k}$
   (a) $\mathbf{r}(0)$ (b) $\mathbf{r}(-2)$ (c) $\mathbf{r}(c-1)$ (d) $\mathbf{r}(1+\Delta t) - \mathbf{r}(1)$
6. $\mathbf{r}(t) = 3\cos t\mathbf{i} + (1-\sin t)\mathbf{j} - t\mathbf{k}$
   (a) $\mathbf{r}(0)$ (b) $\mathbf{r}\left(\dfrac{\pi}{2}\right)$ (c) $\mathbf{r}(s-\pi)$ (d) $\mathbf{r}(\pi+\Delta t) - \mathbf{r}(\pi)$

In Exercises 7 and 8, sketch the plane curve represented by the vector-valued function and give the orientation of the curve.

7. $\mathbf{r}(t) = \langle \pi\cos t, \pi\sin t\rangle$
8. $\mathbf{r}(t) = \langle t+2, t^2-1\rangle$

CAS **In Exercises 9–14, use a computer algebra system to graph the space curve represented by the vector-valued function.**

9. $\mathbf{r}(t) = \mathbf{i} + t\mathbf{j} + t^2\mathbf{k}$
10. $\mathbf{r}(t) = t^2\mathbf{i} + 3t\mathbf{j} + t^3\mathbf{k}$
11. $\mathbf{r}(t) = \langle 1, \sin t, 1\rangle$
12. $\mathbf{r}(t) = \langle 2\cos t, t, 2\sin t\rangle$
13. $\mathbf{r}(t) = \left\langle t, \ln t, \frac{1}{2}t^2\right\rangle$
14. $\mathbf{r}(t) = \left\langle \frac{1}{2}t, \sqrt{t}, \frac{1}{4}t^3\right\rangle$

In Exercises 15 and 16, find vector-valued functions forming the boundaries of the region in the figure.

15.

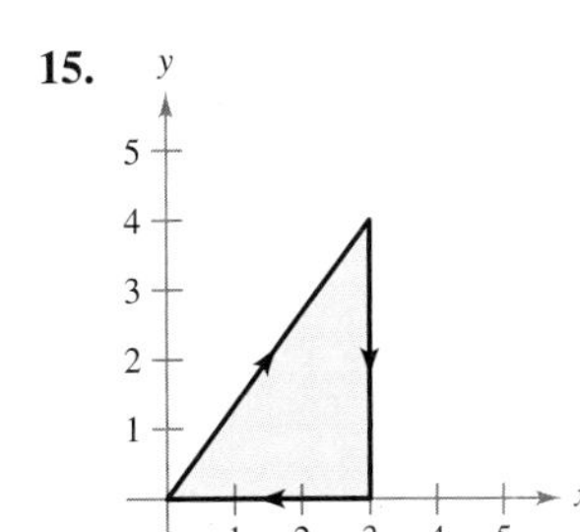

16.

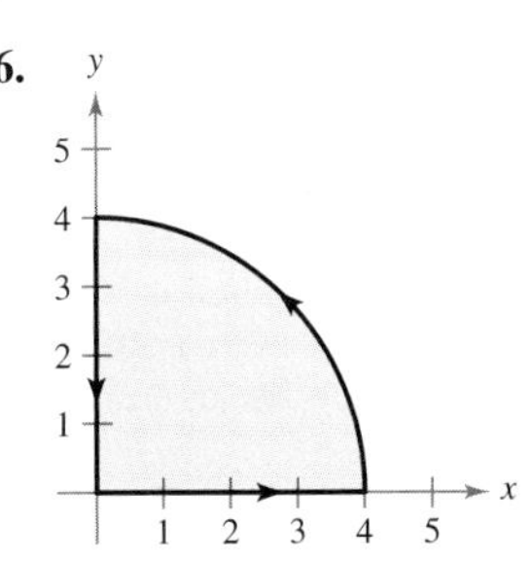

17. A particle moves on a straight-line path that passes through the points $(-2, -3, 8)$ and $(5, 1, -2)$. Find a vector-valued function for the path. (There are many correct answers.)
18. The outer edge of a spiral staircase is in the shape of a helix of radius 2 meters. The staircase has a height of 2 meters and is three-fourths of one complete revolution from bottom to top. Find a vector-valued function for the helix. (There are many correct answers.)

In Exercises 19 and 20, sketch the space curve represented by the intersection of the surfaces. Use the parameter $x = t$ to find a vector-valued function for the space curve.

19. $z = x^2 + y^2, \quad x + y = 0$
20. $x^2 + z^2 = 4, \quad x - y = 0$

In Exercises 21 and 22, evaluate the limit.

21. $\displaystyle\lim_{t\to 4^-} \left(t\mathbf{i} + \sqrt{4-t}\,\mathbf{j} + \mathbf{k}\right)$
22. $\displaystyle\lim_{t\to 0} \left(\frac{\sin 2t}{t}\mathbf{i} + e^{-t}\mathbf{j} + e^t\mathbf{k}\right)$

In Exercises 23 and 24, find the following.

(a) $\mathbf{r}'(t)$ **(b)** $\mathbf{r}''(t)$ **(c)** $D_t[\mathbf{r}(t)\cdot\mathbf{u}(t)]$
(d) $D_t[\mathbf{u}(t) - 2\mathbf{r}(t)]$ **(e)** $D_t[\|\mathbf{r}(t)\|],\ t > 0$ **(f)** $D_t[\mathbf{r}(t)\times\mathbf{u}(t)]$

23. $\mathbf{r}(t) = 3t\mathbf{i} + (t-1)\mathbf{j}, \quad \mathbf{u}(t) = t\mathbf{i} + t^2\mathbf{j} + \frac{2}{3}t^3\mathbf{k}$
24. $\mathbf{r}(t) = \sin t\mathbf{i} + \cos t\mathbf{j} + t\mathbf{k}, \quad \mathbf{u}(t) = \sin t\mathbf{i} + \cos t\mathbf{j} + \dfrac{1}{t}\mathbf{k}$

25. ***Writing*** The x- and y-components of the derivative of the vector-valued function $\mathbf{u}$ are positive at $t = t_0$, and the z-component is negative. Describe the behavior of $\mathbf{u}$ at $t = t_0$.
26. ***Writing*** The x-component of the derivative of the vector-valued function $\mathbf{u}$ is 0 for t in the domain of the function. What does this information imply about the graph of $\mathbf{u}$?

In Exercises 27–30, find the indefinite integral.

27. $\displaystyle\int (\cos t\mathbf{i} + t\cos t\mathbf{j})\,dt$
28. $\displaystyle\int (\ln t\mathbf{i} + t\ln t\mathbf{j} + \mathbf{k})\,dt$
29. $\displaystyle\int \|\cos t\mathbf{i} + \sin t\mathbf{j} + t\mathbf{k}\|\,dt$
30. $\displaystyle\int (t\mathbf{j} + t^2\mathbf{k}) \times (\mathbf{i} + t\mathbf{j} + t\mathbf{k})\,dt$

In Exercises 31–34, evaluate the definite integral.

31. $\displaystyle\int_{-2}^{2} (3t\mathbf{i} + 2t^2\mathbf{j} - t^3\mathbf{k})\,dt$
32. $\displaystyle\int_0^1 \left(\sqrt{t}\,\mathbf{j} + t\sin t\mathbf{k}\right)dt$
33. $\displaystyle\int_0^2 (e^{t/2}\,\mathbf{i} - 3t^2\mathbf{j} - \mathbf{k})\,dt$
34. $\displaystyle\int_{-1}^{1} (t^3\mathbf{i} + \arcsin t\mathbf{j} - t^2\mathbf{k})\,dt$

In Exercises 35 and 36, find r(t) for the given conditions.

35. $\mathbf{r}'(t) = 2t\mathbf{i} + e^t\mathbf{j} + e^{-t}\mathbf{k}, \quad \mathbf{r}(0) = \mathbf{i} + 3\mathbf{j} - 5\mathbf{k}$
36. $\mathbf{r}'(t) = \sec t\mathbf{i} + \tan t\mathbf{j} + t^2\mathbf{k}, \quad \mathbf{r}(0) = 3\mathbf{k}$

In Exercises 37–40, the position vector r describes the path of an object moving in space. Find the velocity, speed, and acceleration of the object.

37. $\mathbf{r}(t) = 4t\mathbf{i} + t^3\mathbf{j} - t\mathbf{k}$
38. $\mathbf{r}(t) = \sqrt{t}\,\mathbf{i} + 5t\mathbf{j} + 2t^2\mathbf{k}$
39. $\mathbf{r}(t) = \langle \cos^3 t, \sin^3 t, 3t\rangle$
40. $\mathbf{r}(t) = \langle t, -\tan t, e^t\rangle$

Linear Approximation **In Exercises 41 and 42, find a set of parametric equations for the tangent line to the graph of the vector-valued function at $t = t_0$. Use the equations for the line to approximate r(t_0 + 0.1).**

41. $\mathbf{r}(t) = \ln(t-3)\mathbf{i} + t^2\mathbf{j} + \frac{1}{2}t\mathbf{k}, \quad t_0 = 4$
42. $\mathbf{r}(t) = 3\cosh t\mathbf{i} + \sinh t\mathbf{j} - 2t\mathbf{k}, \quad t_0 = 0$

Projectile Motion **In Exercises 43–46, use the model for projectile motion, assuming there is no air resistance. [$a(t) = -32$ feet per second per second or $a(t) = -9.8$ meters per second per second.]**

43. A projectile is fired from ground level with an initial velocity of 84 feet per second at an angle of 30° with the horizontal. Find the range of the projectile.

44. The center of a truck bed is 6 feet below and 4 feet horizontally from the end of a horizontal conveyor that is discharging gravel (see figure). Determine the speed ds/dt at which the conveyor belt should be moving so that the gravel falls onto the center of the truck bed.

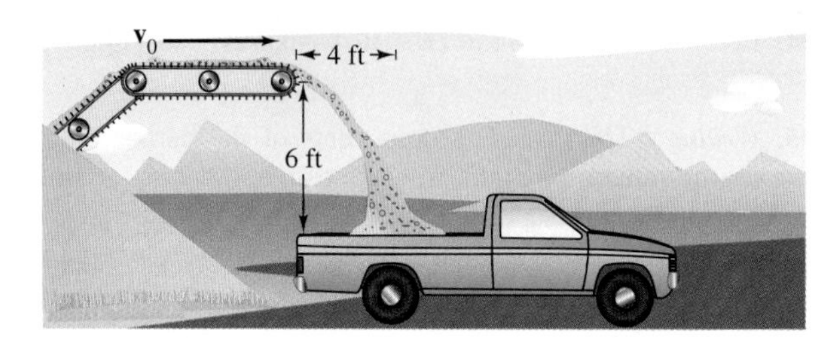

45. A projectile is fired from ground level at an angle of 20° with the horizontal. The projectile has a range of 95 meters. Find the minimum initial velocity.

46. Use a graphing utility to graph the paths of a projectile for $v_0 = 20$ meters per second, $h = 0$ and (a) $\theta = 30°$, (b) $\theta = 45°$, and (c) $\theta = 60°$. Use the graphs to approximate the maximum height and range of the projectile for each case.

In Exercises 47–54, find the velocity, speed, and acceleration at time t. Then find $\mathbf{a} \cdot \mathbf{T}$ and $\mathbf{a} \cdot \mathbf{N}$ at time t.

47. $\mathbf{r}(t) = (2 - t)\mathbf{i} + 3t\mathbf{j}$

48. $\mathbf{r}(t) = (1 + 4t)\mathbf{i} + (2 - 3t)\mathbf{j}$

49. $\mathbf{r}(t) = t\mathbf{i} + \sqrt{t}\,\mathbf{j}$

50. $\mathbf{r}(t) = 2(t + 1)\mathbf{i} + \dfrac{2}{t + 1}\mathbf{j}$

51. $\mathbf{r}(t) = e^t\mathbf{i} + e^{-t}\mathbf{j}$

52. $\mathbf{r}(t) = t\cos t\,\mathbf{i} + t\sin t\,\mathbf{j}$

53. $\mathbf{r}(t) = t\mathbf{i} + t^2\mathbf{j} + \dfrac{1}{2}t^2\mathbf{k}$

54. $\mathbf{r}(t) = (t - 1)\mathbf{i} + t\mathbf{j} + \dfrac{1}{t}\mathbf{k}$

In Exercises 55 and 56, find a set of parametric equations for the line tangent to the space curve at the given value of the parameter.

55. $\mathbf{r}(t) = 2\cos t\,\mathbf{i} + 2\sin t\,\mathbf{j} + t\mathbf{k}, \quad t = \dfrac{\pi}{3}$

56. $\mathbf{r}(t) = t\mathbf{i} + t^2\mathbf{j} + \frac{2}{3}t^3\mathbf{k}, \quad t = 2$

57. ***Satellite Orbit*** Find the speed necessary for a satellite to maintain a circular orbit 550 miles above the surface of Earth.

58. ***Centripetal Force*** An automobile in a circular traffic exchange is traveling at twice the posted speed. By what factor is the centripetal force increased over that which would occur at the posted speed?

In Exercises 59–62, sketch the plane curve and find its length over the given interval.

	Function	*Interval*
59.	$\mathbf{r}(t) = 2t\mathbf{i} - 3t\mathbf{j}$	$[0, 5]$
60.	$\mathbf{r}(t) = t^2\mathbf{i} + 2t\mathbf{k}$	$[0, 3]$
61.	$\mathbf{r}(t) = 10\cos^3 t\,\mathbf{i} + 10\sin^3 t\,\mathbf{j}$	$[0, 2\pi]$
62.	$\mathbf{r}(t) = 10\cos t\,\mathbf{i} + 10\sin t\,\mathbf{j}$	$[0, 2\pi]$

In Exercises 63–66, sketch the space curve and find its length over the given interval.

	Function	*Interval*
63.	$\mathbf{r}(t) = -3t\mathbf{i} + 2t\mathbf{j} + 4t\mathbf{k}$	$[0, 3]$
64.	$\mathbf{r}(t) = t\mathbf{i} + t^2\mathbf{j} + 2t\mathbf{k}$	$[0, 2]$
65.	$\mathbf{r}(t) = \langle 8\cos t, 8\sin t, t\rangle$	$[0, \pi/2]$
66.	$\mathbf{r}(t) = \langle 2(\sin t - t\cos t), 2(\cos t + t\sin t), t\rangle$	$[0, \pi/2]$

In Exercises 67–70, find the curvature K of the curve.

67. $\mathbf{r}(t) = 3t\mathbf{i} + 2t\mathbf{j}$

68. $\mathbf{r}(t) = 2\sqrt{t}\,\mathbf{i} + 3t\mathbf{j}$

69. $\mathbf{r}(t) = 2t\mathbf{i} + \frac{1}{2}t^2\mathbf{j} + t^2\mathbf{k}$

70. $\mathbf{r}(t) = 2t\mathbf{i} + 5\cos t\mathbf{j} + 5\sin t\mathbf{k}$

In Exercises 71 and 72, find the curvature K of the curve at point P.

71. $\mathbf{r}(t) = \dfrac{1}{2}t^2\mathbf{i} + t\mathbf{j} + \dfrac{1}{3}t^3\mathbf{k}, \quad P\left(\dfrac{1}{2}, 1, \dfrac{1}{3}\right)$

72. $\mathbf{r}(t) = 4\cos t\mathbf{i} + 3\sin t\mathbf{j} + t\mathbf{k}, \quad P(-4, 0, \pi)$

In Exercises 73–76, find the curvature and radius of curvature of the plane curve at the given value of x.

73. $y = \frac{1}{2}x^2 + 2, \quad x = 4$

74. $y = e^{-x/2}, \quad x = 0$

75. $y = \ln x, \quad x = 1$

76. $y = \tan x, \quad x = \dfrac{\pi}{4}$

77. ***Writing*** A civil engineer designs a highway as shown in the figure. BC is an arc of the circle. AB and CD are straight lines tangent to the circular arc. Criticize the design.

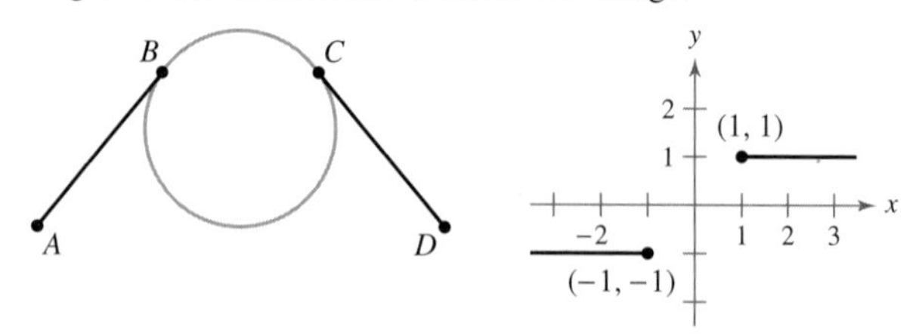

Figure for 77 **Figure for 78**

78. A line segment extends horizontally to the left from the point $(-1, -1)$, and another line segment extends horizontally to the right from the point $(1, 1)$, as shown in the figure. Find a curve of the form

$$y = ax^5 + bx^3 + cx$$

that connects the points $(-1, -1)$ and $(1, 1)$ so that the slope and curvature of the curve are zero at the endpoints.

P.S. PROBLEM SOLVING

1. The **cornu spiral** is given by

$$x(t) = \int_0^t \cos\left(\frac{\pi u^2}{2}\right) du \quad \text{and} \quad y(t) = \int_0^t \sin\left(\frac{\pi u^2}{2}\right) du.$$

The spiral shown in the figure was plotted over the interval $-\pi \leq t \leq \pi$.

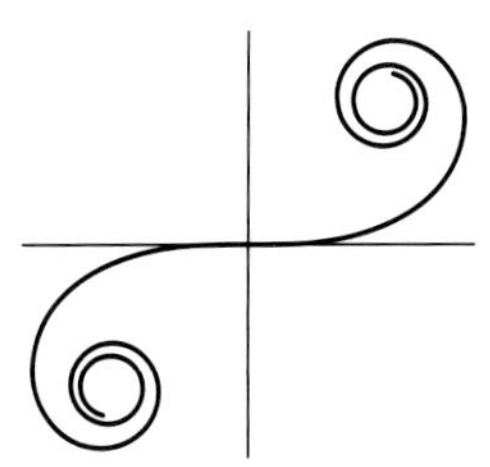

Generated by Mathematica

(a) Find the arc length of this curve from $t = 0$ to $t = a$.

(b) Find the curvature of the graph when $t = a$.

(c) The cornu spiral was discovered by James Bernoulli. He found that the spiral has an amazing relationship between curvature and arc length. What is this relationship?

2. Let T be the tangent line at the point $P(x, y)$ to the graph of the curve $x^{2/3} + y^{2/3} = a^{2/3}$, $a > 0$, as shown in the figure. Show that the radius of curvature at P is three times the distance from the origin to the tangent line T.

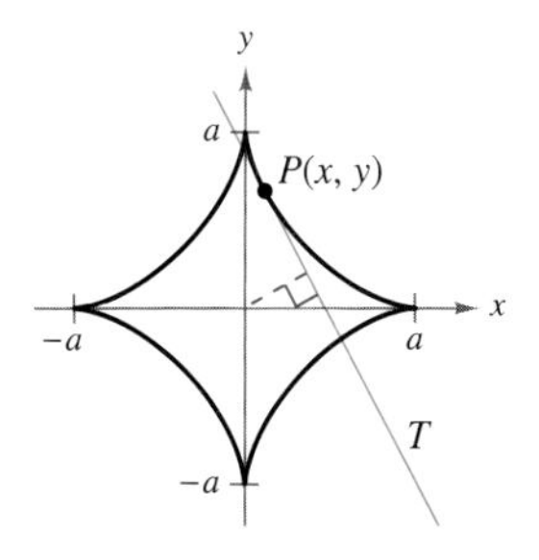

3. A bomber is flying horizontally at an altitude of 3200 feet with a velocity of 400 feet per second when it releases a bomb. A projectile is launched 5 seconds later from a cannon at a site facing the bomber and 5000 feet from the point that was directly beneath the bomber when the bomb was released, as shown in the figure. The projectile is to intercept the bomb at an altitude of 1600 feet. Determine the required initial speed and angle of inclination of the projectile. (Ignore air resistance.)

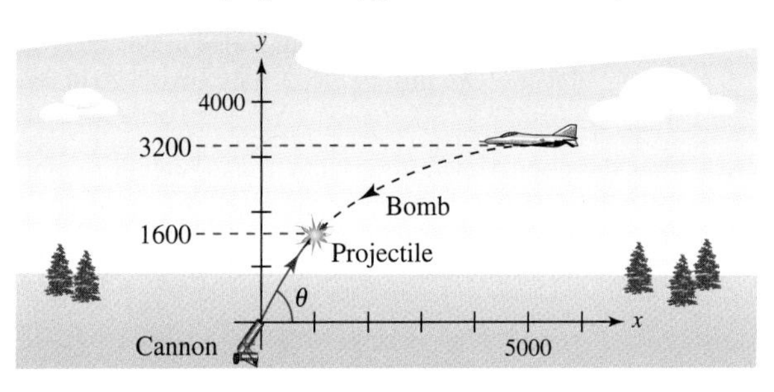

4. Repeat Exercise 3 for the case in which the bomber is facing *away* from the launch site, as shown in the figure.

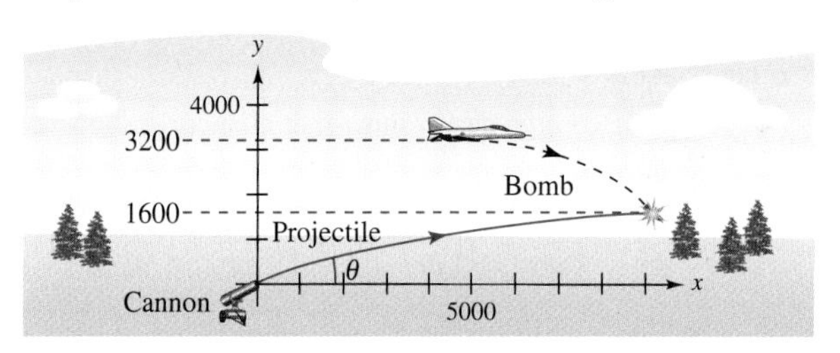

5. Consider one arch of the cycloid

$$\mathbf{r}(\theta) = (\theta - \sin\theta)\mathbf{i} + (1 - \cos\theta)\mathbf{j}, \quad 0 \leq \theta \leq 2\pi$$

as shown in the figure. Let $s(\theta)$ be the arc length from the highest point on the arch to the point $(x(\theta), y(\theta))$, and let $\rho(\theta) = \dfrac{1}{K}$ be the radius of curvature at the point $(x(\theta), y(\theta))$. Show that s and ρ are related by the equation $s^2 + \rho^2 = 16$. (This equation is called a *natural equation* for the curve.)

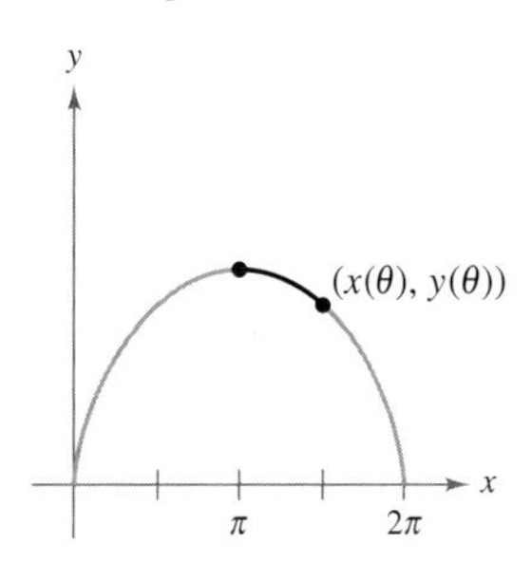

6. Consider the cardioid $r = 1 - \cos\theta$, $0 \leq \theta \leq 2\pi$, as shown in the figure. Let $s(\theta)$ be the arc length from the point $(2, \pi)$ on the cardioid to the point (r, θ), and let $\rho(\theta) = \dfrac{1}{K}$ be the radius of curvature at the point (r, θ). Show that s and ρ are related by the equation $s^2 + 9\rho^2 = 16$. (This equation is called a *natural equation* for the curve.)

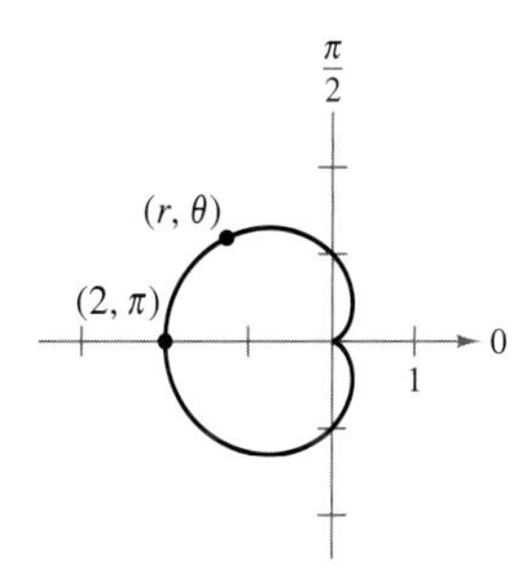

7. If $\mathbf{r}(t)$ is a nonzero differentiable function of t, prove that

$$\frac{d}{dt}(\|\mathbf{r}(t)\|) = \frac{1}{\|\mathbf{r}(t)\|}\mathbf{r}(t) \cdot \mathbf{r}'(t).$$

8. A communications satellite moves in a circular orbit around Earth at a distance of 42,000 kilometers from the center of Earth. The angular velocity

$$\frac{d\theta}{dt} = \omega = \frac{\pi}{12} \text{ radian per hour}$$

is constant.

(a) Use polar coordinates to show that the acceleration vector is given by

$$\mathbf{a} = \frac{d^2\mathbf{r}}{dt^2} = \left[\frac{d^2r}{dt^2} - r\left(\frac{d\theta}{dt}\right)^2\right]\mathbf{u}_\mathbf{r} + \left[r\frac{d^2\theta}{dt^2} + 2\frac{dr}{dt}\frac{d\theta}{dt}\right]\mathbf{u}_\theta$$

where $\mathbf{u}_\mathbf{r} = \cos\theta\mathbf{i} + \sin\theta\mathbf{j}$ is the unit vector in the radial direction and $\mathbf{u}_\theta = -\sin\theta\mathbf{i} + \cos\theta\mathbf{j}$.

(b) Find the radial and angular components of acceleration for the satellite.

In Exercises 9–11, use the binormal vector defined by the equation $\mathbf{B} = \mathbf{T} \times \mathbf{N}$.

9. Find the unit tangent, unit normal, and binormal vectors for the helix $\mathbf{r}(t) = 4\cos t\mathbf{i} + 4\sin t\mathbf{j} + 3t\mathbf{k}$ at $t = \frac{\pi}{2}$. Sketch the helix together with these three mutually orthogonal unit vectors.

10. Find the unit tangent, unit normal, and binormal vectors for the curve $\mathbf{r}(t) = \cos t\mathbf{i} + \sin t\mathbf{j} - \mathbf{k}$ at $t = \frac{\pi}{4}$. Sketch the curve together with these three mutually orthogonal unit vectors.

11. (a) Prove that there exists a scalar τ, called the **torsion,** such that $d\mathbf{B}/ds = -\tau\mathbf{N}$.

(b) Prove that $\dfrac{d\mathbf{N}}{ds} = -K\mathbf{T} + \tau\mathbf{B}$.

(The three equations $d\mathbf{T}/ds = K\mathbf{N}$, $d\mathbf{N}/ds = -K\mathbf{T} + \tau\mathbf{B}$, and $d\mathbf{B}/ds = -\tau\mathbf{N}$ are called the *Frenet-Serret formulas.*)

12. A highway has an exit ramp that begins at the origin of a coordinate system and follows the curve $y = \frac{1}{32}x^{5/2}$ to the point $(4, 1)$ (see figure). Then it follows a circular path whose curvature is that given by the curve at $(4, 1)$. What is the radius of the circular arc? Explain why the curve and the circular arc should have the same curvature at $(4, 1)$.

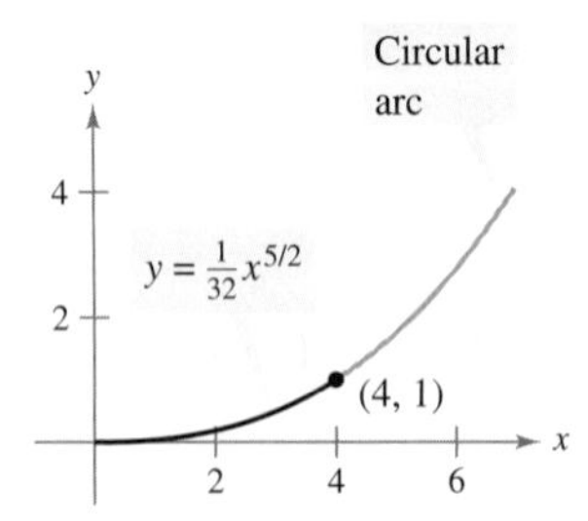

13. Consider the vector-valued function $\mathbf{r}(t) = \langle t\cos\pi t, t\sin\pi t\rangle$, $0 \le t \le 2$.

(a) Use a graphing utility to graph the function.

(b) Find the length of the arc in part (a).

(c) Find the curvature K as a function of t. Find the curvatures for t-values of 0, 1, and 2.

(d) Use a graphing utility to graph the function K.

(e) Find (if possible) $\lim_{t\to\infty} K$.

(f) Using the result of part (e), make a conjecture about the graph of $\mathbf{r}$ as $t \to \infty$.

14. You want to toss an object to a friend who is riding a Ferris wheel (see figure). The following parametric equations give the path of the friend $\mathbf{r}_1(t)$ and the path of the object $\mathbf{r}_2(t)$. Distance is measured in meters and time is measured in seconds.

$$\mathbf{r}_1(t) = 15\left(\sin\frac{\pi t}{10}\right)\mathbf{i} + \left(16 - 15\cos\frac{\pi t}{10}\right)\mathbf{j}$$

$$\mathbf{r}_2(t) = [22 - 8.03(t - t_0)]\mathbf{i} + [1 + 11.47(t - t_0) - 4.9(t - t_0)^2]\mathbf{j}$$

(a) Locate your friend's position on the Ferris wheel at time $t = 0$.

(b) Determine the number of revolutions per minute of the Ferris wheel.

(c) What are the speed and angle of inclination (in degrees) at which the object is thrown at time $t = t_0$?

(d) Use a graphing utility to graph the vector-valued functions using a value of t_0 that allows your friend to be within reach of the object. (Do this by trial and error.) Explain the significance of t_0.

(e) Find the approximate time your friend should be able to catch the object. Approximate the speeds of your friend and the object at that time.

13 Functions of Several Variables

In this chapter, you will study functions of more than one independent variable. Many of the concepts presented are extensions of familiar ideas from earlier chapters.

In this chapter, you should learn the following.

- How to sketch a graph, level curves, and level surfaces. (13.1)
- How to find a limit and determine continuity. (13.2)
- How to find and use a partial derivative. (13.3)
- How to find and use a total differential and determine differentiability. (13.4)
- How to use the Chain Rules and find a partial derivative implicitly. (13.5)
- How to find and use a directional derivative and a gradient. (13.6)
- How to find an equation of a tangent plane and an equation of a normal line to a surface, and how to find the angle of inclination of a plane. (13.7)
- How to find absolute and relative extrema. (13.8)
- How to solve an optimization problem, including constrained optimization using a Lagrange multiplier, and how to use the method of least squares. (13.9, 13.10)

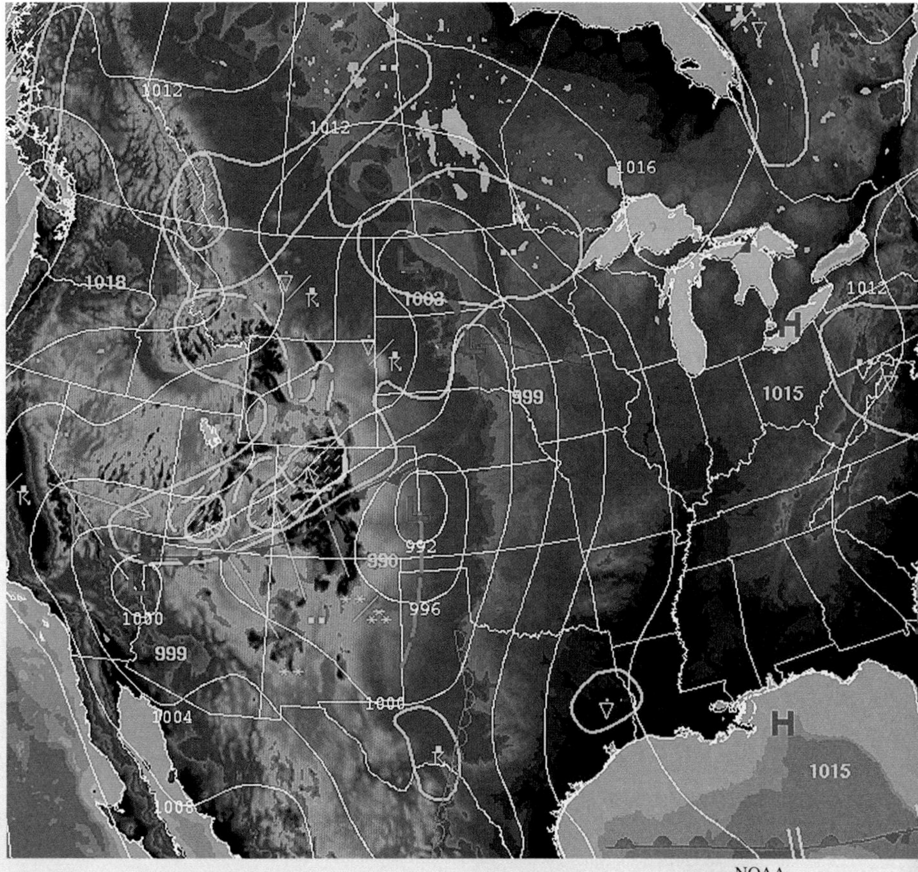

NOAA

Meteorologists use maps that show curves of equal atmospheric pressure, called *isobars*, to predict weather patterns. How can you use pressure gradients to determine the area of the country that has the greatest wind speed? (See Section 13.6, Exercise 68.)

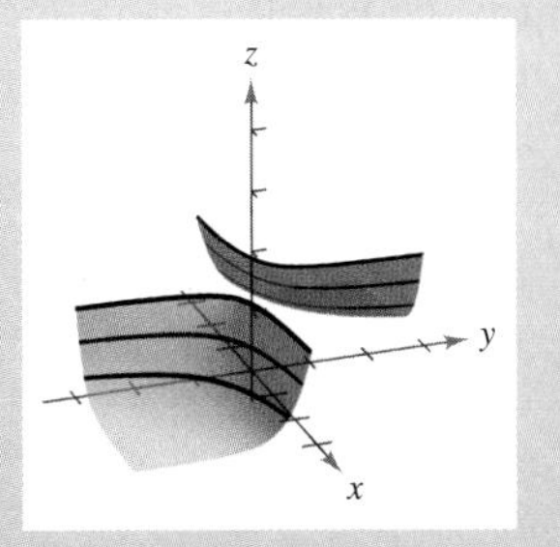

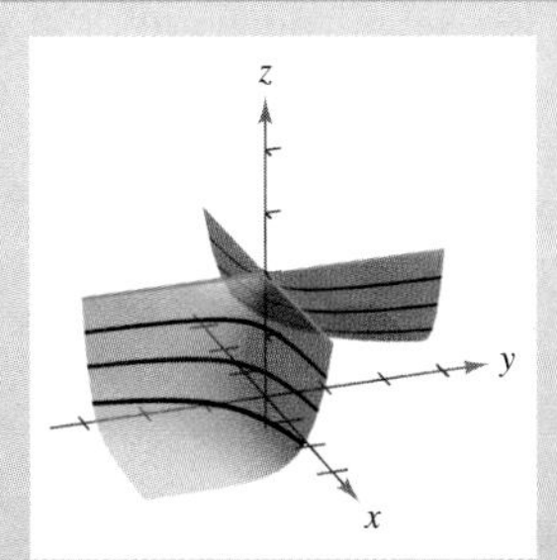

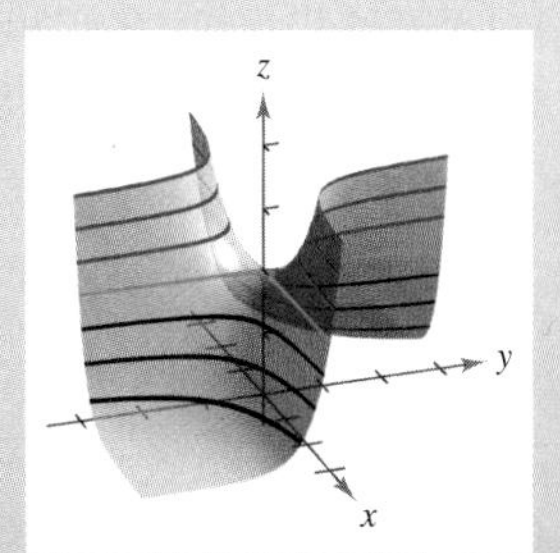

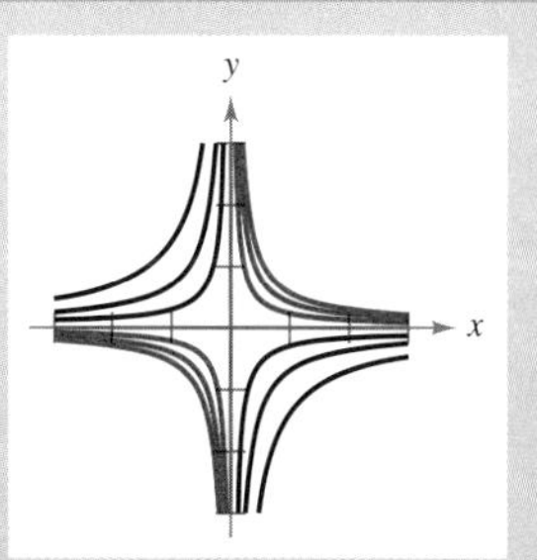

Many real-life quantities are functions of two or more variables. In Section 13.1, you will learn how to graph a function of two variables, like the one shown above. The first three graphs show cut-away views of the surface at various traces. Another way to visualize this surface is to project the traces onto the *xy*-plane, as shown in the fourth graph.

13.1 Introduction to Functions of Several Variables

- **Understand the notation for a function of several variables.**
- **Sketch the graph of a function of two variables.**
- **Sketch level curves for a function of two variables.**
- **Sketch level surfaces for a function of three variables.**
- **Use computer graphics to graph a function of two variables.**

Functions of Several Variables

EXPLORATION

Comparing Dimensions
Without using a graphing utility, describe the graph of each function of two variables.

a. $z = x^2 + y^2$
b. $z = x + y$
c. $z = x^2 + y$
d. $z = \sqrt{x^2 + y^2}$
e. $z = \sqrt{1 - x^2 + y^2}$

So far in this text, you have dealt only with functions of a single (independent) variable. Many familiar quantities, however, are functions of two or more variables. For instance, the work done by a force $(W = FD)$ and the volume of a right circular cylinder $(V = \pi r^2 h)$ are both functions of two variables. The volume of a rectangular solid $(V = lwh)$ is a function of three variables. The notation for a function of two or more variables is similar to that for a function of a single variable. Here are two examples.

$$z = f(\underbrace{x, y}_{\text{2 variables}}) = x^2 + xy \qquad \text{Function of two variables}$$

and

$$w = f(\underbrace{x, y, z}_{\text{3 variables}}) = x + 2y - 3z \qquad \text{Function of three variables}$$

DEFINITION OF A FUNCTION OF TWO VARIABLES

Let D be a set of ordered pairs of real numbers. If to each ordered pair (x, y) in D there corresponds a unique real number $f(x, y)$, then f is called a **function of x and y.** The set D is the **domain** of f, and the corresponding set of values for $f(x, y)$ is the **range** of f.

For the function given by $z = f(x, y)$, x and y are called the **independent variables** and z is called the **dependent variable.**

Similar definitions can be given for functions of three, four, or n variables, where the domains consist of ordered triples (x_1, x_2, x_3), quadruples (x_1, x_2, x_3, x_4), and n-tuples $(x_1, x_2, \ldots, x_n)$. In all cases, the range is a set of real numbers. In this chapter, you will study only functions of two or three variables.

As with functions of one variable, the most common way to describe a function of several variables is with an *equation*, and unless otherwise restricted, you can assume that the domain is the set of all points for which the equation is defined. For instance, the domain of the function given by

$$f(x, y) = x^2 + y^2$$

is assumed to be the entire xy-plane. Similarly, the domain of

$$f(x, y) = \ln xy$$

is the set of all points (x, y) in the plane for which $xy > 0$. This consists of all points in the first and third quadrants.

MARY FAIRFAX SOMERVILLE (1780–1872)

Somerville was interested in the problem of creating geometric models for functions of several variables. Her most well-known book, *The Mechanics of the Heavens*, was published in 1831.

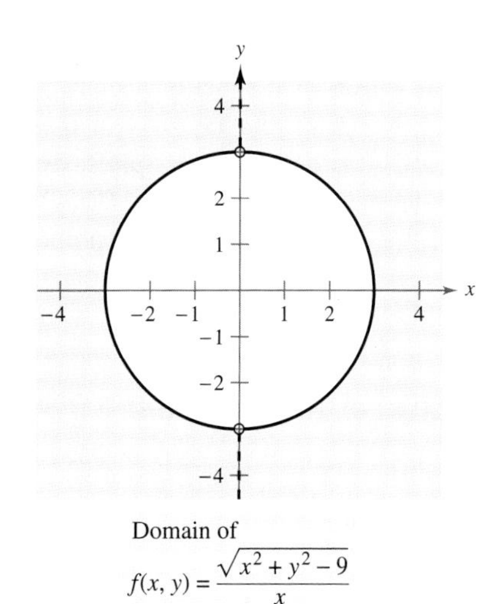

Domain of

$f(x, y) = \dfrac{\sqrt{x^2 + y^2 - 9}}{x}$

Figure 13.1

EXAMPLE 1 Domains of Functions of Several Variables

Find the domain of each function.

a. $f(x, y) = \dfrac{\sqrt{x^2 + y^2 - 9}}{x}$ **b.** $g(x, y, z) = \dfrac{x}{\sqrt{9 - x^2 - y^2 - z^2}}$

Solution

a. The function f is defined for all points (x, y) such that $x \neq 0$ and

$$x^2 + y^2 \geq 9.$$

So, the domain is the set of all points lying on or outside the circle $x^2 + y^2 = 9$, *except* those points on the y-axis, as shown in Figure 13.1.

b. The function g is defined for all points (x, y, z) such that

$$x^2 + y^2 + z^2 < 9.$$

Consequently, the domain is the set of all points (x, y, z) lying inside a sphere of radius 3 that is centered at the origin. ■

Functions of several variables can be combined in the same ways as functions of single variables. For instance, you can form the sum, difference, product, and quotient of two functions of two variables as follows.

$(f \pm g)(x, y) = f(x, y) \pm g(x, y)$	Sum or difference
$(fg)(x, y) = f(x, y)g(x, y)$	Product
$\dfrac{f}{g}(x, y) = \dfrac{f(x, y)}{g(x, y)}, \quad g(x, y) \neq 0$	Quotient

You cannot form the composite of two functions of several variables. However, if h is a function of several variables and g is a function of a single variable, you can form the **composite** function $(g \circ h)(x, y)$ as follows.

$$(g \circ h)(x, y) = g(h(x, y)) \qquad \text{Composition}$$

The domain of this composite function consists of all (x, y) in the domain of h such that $h(x, y)$ is in the domain of g. For example, the function given by

$$f(x, y) = \sqrt{16 - 4x^2 - y^2}$$

can be viewed as the composite of the function of two variables given by $h(x, y) = 16 - 4x^2 - y^2$ and the function of a single variable given by $g(u) = \sqrt{u}$. The domain of this function is the set of all points lying on or inside the ellipse given by $4x^2 + y^2 = 16$.

A function that can be written as a sum of functions of the form $cx^m y^n$ (where c is a real number and m and n are nonnegative integers) is called a **polynomial function** of two variables. For instance, the functions given by

$$f(x, y) = x^2 + y^2 - 2xy + x + 2 \quad \text{and} \quad g(x, y) = 3xy^2 + x - 2$$

are polynomial functions of two variables. A **rational function** is the quotient of two polynomial functions. Similar terminology is used for functions of more than two variables.

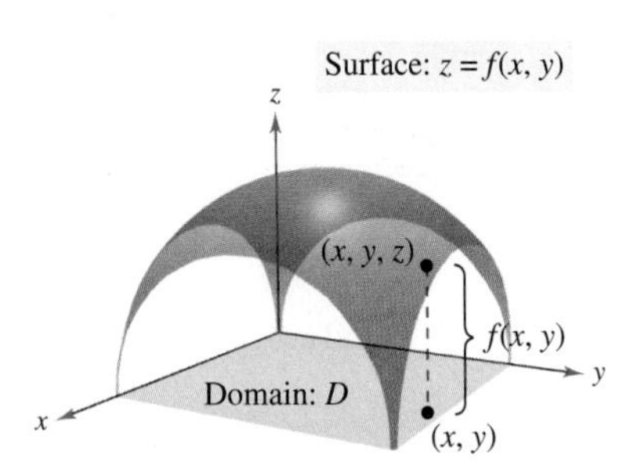

Figure 13.2

The Graph of a Function of Two Variables

As with functions of a single variable, you can learn a lot about the behavior of a function of two variables by sketching its graph. The **graph** of a function f of two variables is the set of all points (x, y, z) for which $z = f(x, y)$ and (x, y) is in the domain of f. This graph can be interpreted geometrically as a *surface in space*, as discussed in Sections 11.5 and 11.6. In Figure 13.2, note that the graph of $z = f(x, y)$ is a surface whose projection onto the xy-plane is D, the domain of f. To each point (x, y) in D there corresponds a point (x, y, z) on the surface, and, conversely, to each point (x, y, z) on the surface there corresponds a point (x, y) in D.

EXAMPLE 2 Describing the Graph of a Function of Two Variables

What is the range of $f(x, y) = \sqrt{16 - 4x^2 - y^2}$? Describe the graph of f.

Solution The domain D implied by the equation of f is the set of all points (x, y) such that $16 - 4x^2 - y^2 \geq 0$. So, D is the set of all points lying on or inside the ellipse given by

$$\frac{x^2}{4} + \frac{y^2}{16} = 1. \qquad \text{Ellipse in the } xy\text{-plane}$$

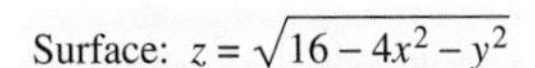

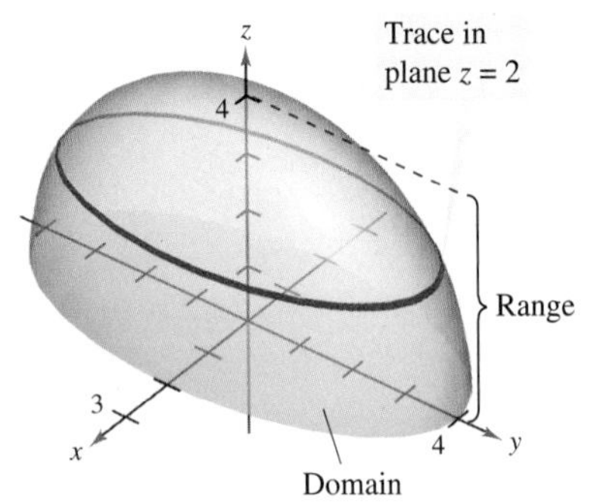

The graph of $f(x, y) = \sqrt{16 - 4x^2 - y^2}$ is the upper half of an ellipsoid.
Figure 13.3

The range of f is all values $z = f(x, y)$ such that $0 \leq z \leq \sqrt{16}$ or

$$0 \leq z \leq 4. \qquad \text{Range of } f$$

A point (x, y, z) is on the graph of f if and only if

$$z = \sqrt{16 - 4x^2 - y^2}$$
$$z^2 = 16 - 4x^2 - y^2$$
$$4x^2 + y^2 + z^2 = 16$$
$$\frac{x^2}{4} + \frac{y^2}{16} + \frac{z^2}{16} = 1, \qquad 0 \leq z \leq 4.$$

From Section 11.6, you know that the graph of f is the upper half of an ellipsoid, as shown in Figure 13.3. ■

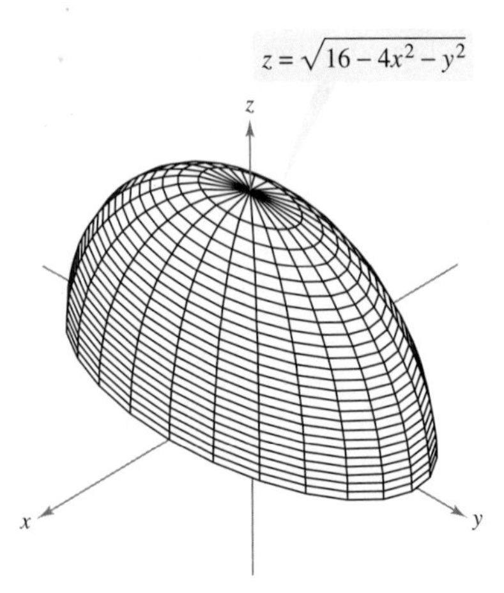

Figure 13.4

To sketch a surface in space *by hand,* it helps to use traces in planes parallel to the coordinate planes, as shown in Figure 13.3. For example, to find the trace of the surface in the plane $z = 2$, substitute $z = 2$ in the equation $z = \sqrt{16 - 4x^2 - y^2}$ and obtain

$$2 = \sqrt{16 - 4x^2 - y^2} \quad \Longrightarrow \quad \frac{x^2}{3} + \frac{y^2}{12} = 1.$$

So, the trace is an ellipse centered at the point $(0, 0, 2)$ with major and minor axes of lengths $4\sqrt{3}$ and $2\sqrt{3}$.

Traces are also used with most three-dimensional graphing utilities. For instance, Figure 13.4 shows a computer-generated version of the surface given in Example 2. For this graph, the computer took 25 traces parallel to the xy-plane and 12 traces in vertical planes.

If you have access to a three-dimensional graphing utility, use it to graph several surfaces.

Level Curves

A second way to visualize a function of two variables is to use a **scalar field** in which the scalar $z = f(x, y)$ is assigned to the point (x, y). A scalar field can be characterized by **level curves** (or **contour lines**) along which the value of $f(x, y)$ is constant. For instance, the weather map in Figure 13.5 shows level curves of equal pressure called **isobars.** In weather maps for which the level curves represent points of equal temperature, the level curves are called **isotherms,** as shown in Figure 13.6. Another common use of level curves is in representing electric potential fields. In this type of map, the level curves are called **equipotential lines.**

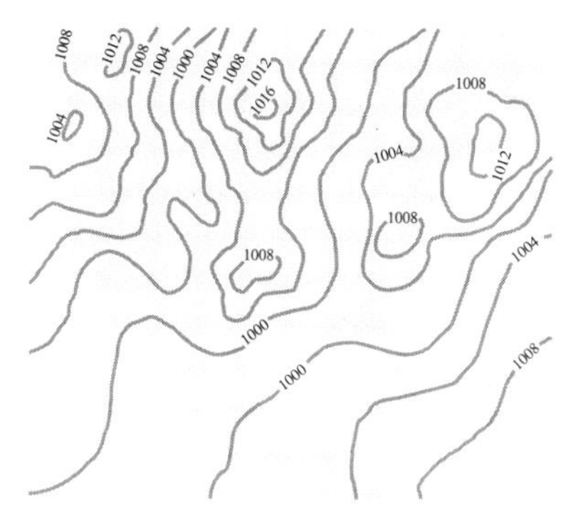

Level curves show the lines of equal pressure (isobars) measured in millibars.

Figure 13.5

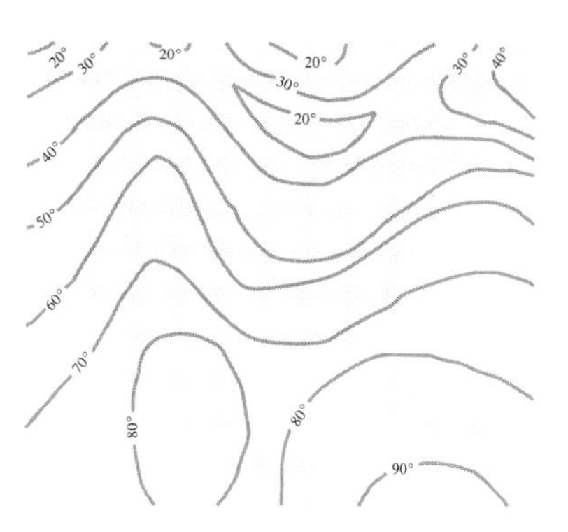

Level curves show the lines of equal temperature (isotherms) measured in degrees Fahrenheit.

Figure 13.6

Contour maps are commonly used to show regions on Earth's surface, with the level curves representing the height above sea level. This type of map is called a **topographic map.** For example, the mountain shown in Figure 13.7 is represented by the topographic map in Figure 13.8.

A contour map depicts the variation of z with respect to x and y by the spacing between level curves. Much space between level curves indicates that z is changing slowly, whereas little space indicates a rapid change in z. Furthermore, to produce a good three-dimensional illusion in a contour map, it is important to choose c-values that are *evenly spaced.*

Alfred B. Thomas/Earth Scenes

Figure 13.7

USGS

Figure 13.8

EXAMPLE 3 Sketching a Contour Map

The hemisphere given by $f(x, y) = \sqrt{64 - x^2 - y^2}$ is shown in Figure 13.9. Sketch a contour map of this surface using level curves corresponding to $c = 0, 1, 2, \ldots, 8$.

Solution For each value of c, the equation given by $f(x, y) = c$ is a circle (or point) in the xy-plane. For example, when $c_1 = 0$, the level curve is

$$x^2 + y^2 = 64 \qquad \text{Circle of radius 8}$$

which is a circle of radius 8. Figure 13.10 shows the nine level curves for the hemisphere.

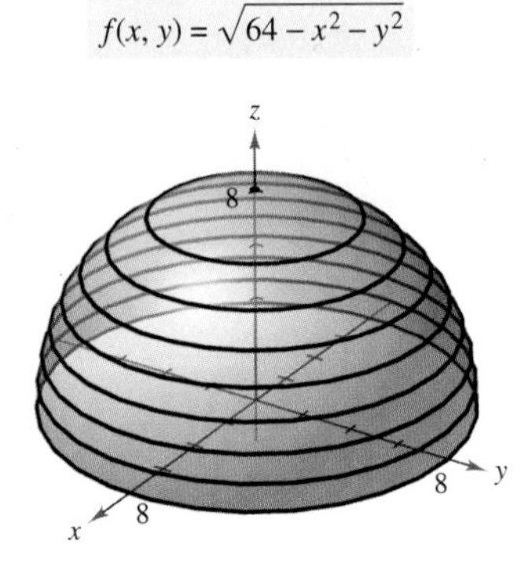

Hemisphere
Figure 13.9

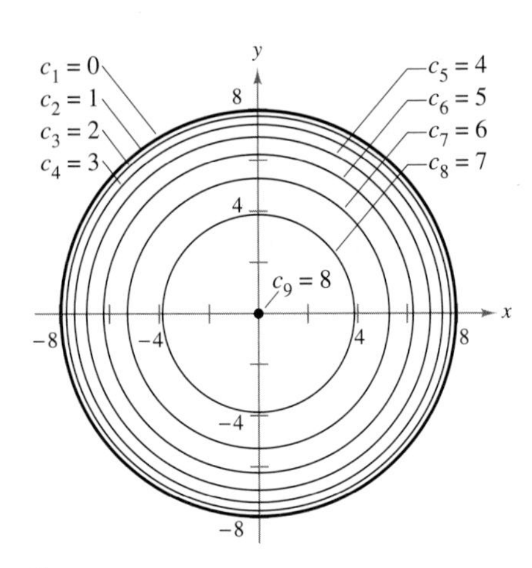

Contour map
Figure 13.10

EXAMPLE 4 Sketching a Contour Map

The hyperbolic paraboloid given by

$$z = y^2 - x^2$$

is shown in Figure 13.11. Sketch a contour map of this surface.

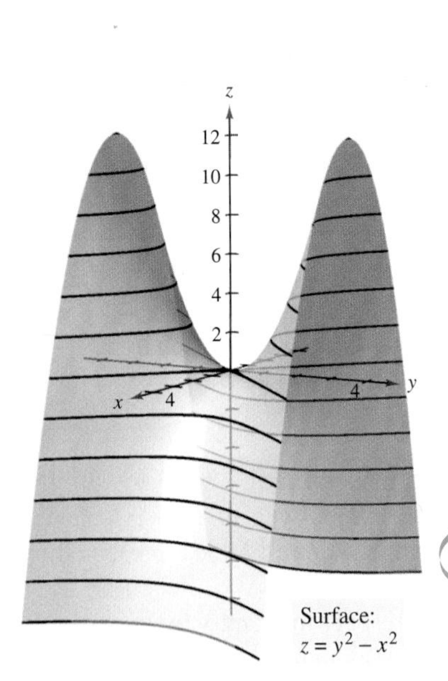

Hyperbolic paraboloid
Figure 13.11

Solution For each value of c, let $f(x, y) = c$ and sketch the resulting level curve in the xy-plane. For this function, each of the level curves $(c \neq 0)$ is a hyperbola whose asymptotes are the lines $y = \pm x$. If $c < 0$, the transverse axis is horizontal. For instance, the level curve for $c = -4$ is given by

$$\frac{x^2}{2^2} - \frac{y^2}{2^2} = 1. \qquad \text{Hyperbola with horizontal transverse axis}$$

If $c > 0$, the transverse axis is vertical. For instance, the level curve for $c = 4$ is given by

$$\frac{y^2}{2^2} - \frac{x^2}{2^2} = 1. \qquad \text{Hyperbola with vertical transverse axis}$$

If $c = 0$, the level curve is the degenerate conic representing the intersecting asymptotes, as shown in Figure 13.12. ■

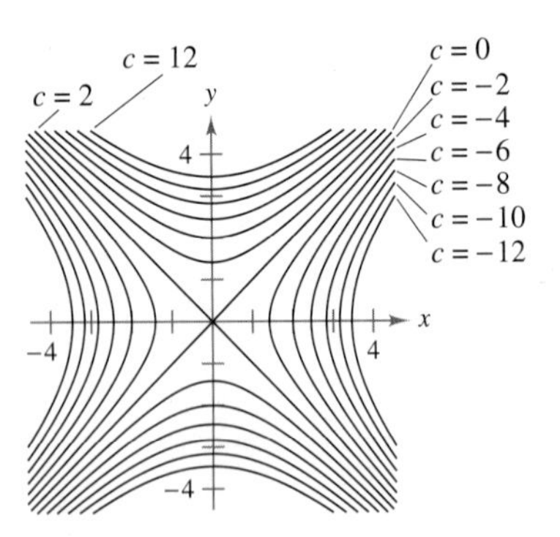

Hyperbolic level curves (at increments of 2)
Figure 13.12

The icon indicates that you will find a CAS Investigation on the book's website. The CAS Investigation is a collaborative exploration of this example using the computer algebra systems Maple *and* Mathematica.

One example of a function of two variables used in economics is the **Cobb-Douglas production function.** This function is used as a model to represent the numbers of units produced by varying amounts of labor and capital. If x measures the units of labor and y measures the units of capital, the number of units produced is given by

$$f(x, y) = Cx^a y^{1-a}$$

where C and a are constants with $0 < a < 1$.

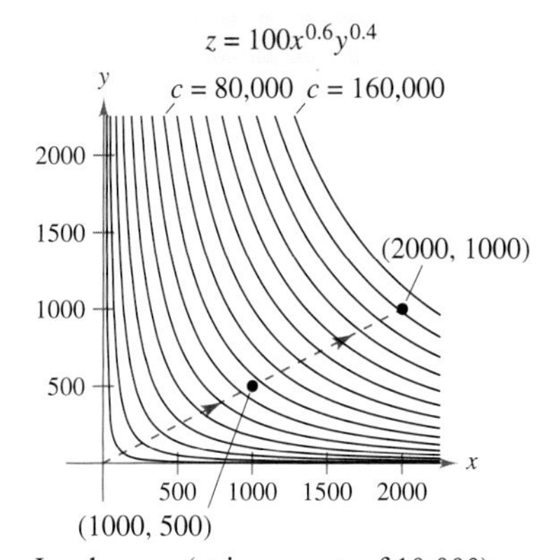

Level curves (at increments of 10,000)
Figure 13.13

EXAMPLE 5 The Cobb-Douglas Production Function

A toy manufacturer estimates a production function to be $f(x, y) = 100x^{0.6}y^{0.4}$, where x is the number of units of labor and y is the number of units of capital. Compare the production level when $x = 1000$ and $y = 500$ with the production level when $x = 2000$ and $y = 1000$.

Solution When $x = 1000$ and $y = 500$, the production level is

$$f(1000, 500) = 100(1000^{0.6})(500^{0.4}) \approx 75{,}786.$$

When $x = 2000$ and $y = 1000$, the production level is

$$f(2000, 1000) = 100(2000^{0.6})(1000^{0.4}) = 151{,}572.$$

The level curves of $z = f(x, y)$ are shown in Figure 13.13. Note that by doubling both x and y, you double the production level (see Exercise 79). ■

Level Surfaces

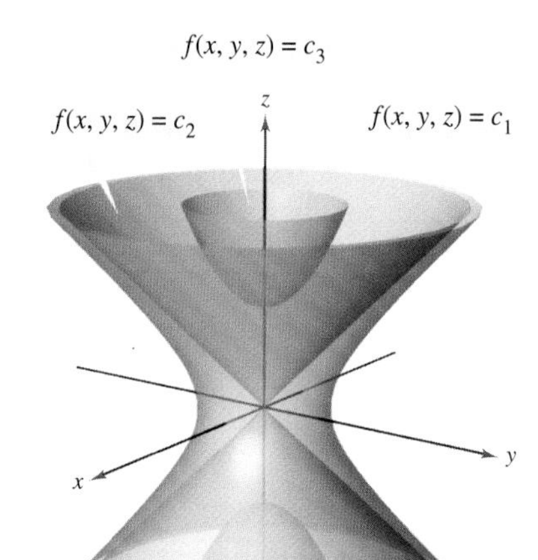

Level surfaces of f
Figure 13.14

The concept of a level curve can be extended by one dimension to define a **level surface.** If f is a function of three variables and c is a constant, the graph of the equation $f(x, y, z) = c$ is a **level surface** of the function f, as shown in Figure 13.14.

With computers, engineers and scientists have developed other ways to view functions of three variables. For instance, Figure 13.15 shows a computer simulation that uses color to represent the temperature distribution of fluid inside a pipe fitting.

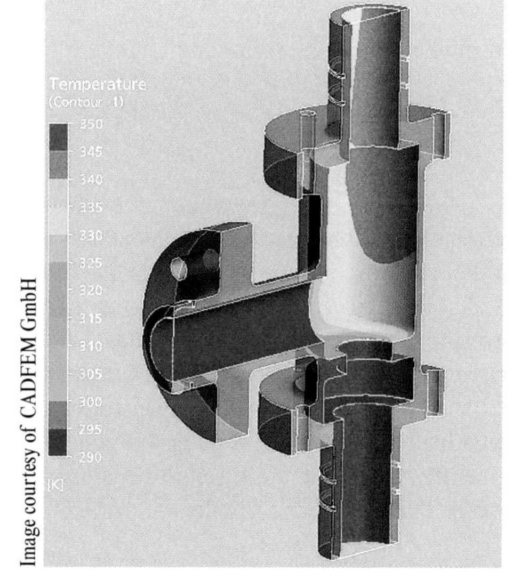

One-way coupling of ANSYS CFX™ and ANSYS Mechanical™ for thermal stress analysis
Figure 13.15

EXAMPLE 6 Level Surfaces

Describe the level surfaces of the function

$$f(x, y, z) = 4x^2 + y^2 + z^2.$$

Solution Each level surface has an equation of the form

$$4x^2 + y^2 + z^2 = c. \qquad \text{Equation of level surface}$$

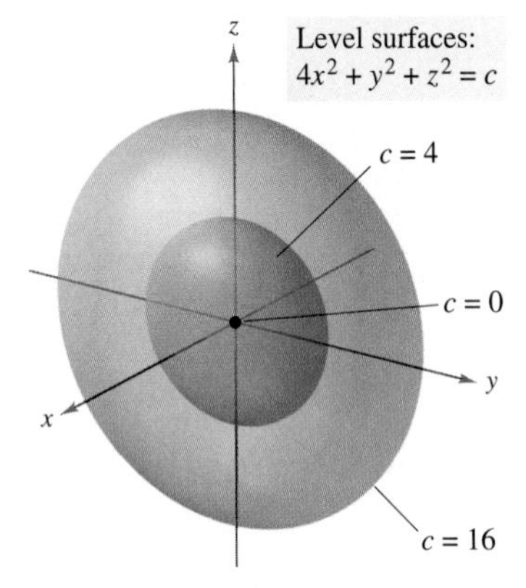

Figure 13.16

So, the level surfaces are ellipsoids (whose cross sections parallel to the yz-plane are circles). As c increases, the radii of the circular cross sections increase according to the square root of c. For example, the level surfaces corresponding to the values $c = 0$, $c = 4$, and $c = 16$ are as follows.

$$4x^2 + y^2 + z^2 = 0 \qquad \text{Level surface for } c = 0 \text{ (single point)}$$

$$\frac{x^2}{1} + \frac{y^2}{4} + \frac{z^2}{4} = 1 \qquad \text{Level surface for } c = 4 \text{ (ellipsoid)}$$

$$\frac{x^2}{4} + \frac{y^2}{16} + \frac{z^2}{16} = 1 \qquad \text{Level surface for } c = 16 \text{ (ellipsoid)}$$

These level surfaces are shown in Figure 13.16. ■

NOTE If the function in Example 6 represented the *temperature* at the point (x, y, z), the level surfaces shown in Figure 13.16 would be called **isothermal surfaces.** ■

Computer Graphics

The problem of sketching the graph of a surface in space can be simplified by using a computer. Although there are several types of three-dimensional graphing utilities, most use some form of trace analysis to give the illusion of three dimensions. To use such a graphing utility, you usually need to enter the equation of the surface, the region in the xy-plane over which the surface is to be plotted, and the number of traces to be taken. For instance, to graph the surface given by

$$f(x, y) = (x^2 + y^2)e^{1-x^2-y^2}$$

you might choose the following bounds for x, y, and z.

$$-3 \le x \le 3 \qquad \text{Bounds for } x$$

$$-3 \le y \le 3 \qquad \text{Bounds for } y$$

$$0 \le z \le 3 \qquad \text{Bounds for } z$$

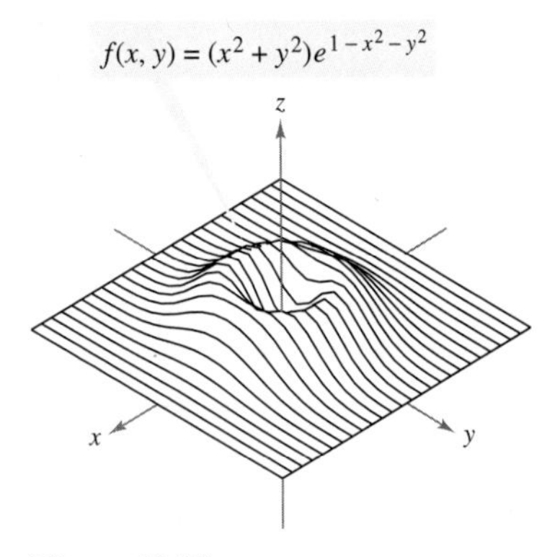

Figure 13.17

Figure 13.17 shows a computer-generated graph of this surface using 26 traces taken parallel to the yz-plane. To heighten the three-dimensional effect, the program uses a "hidden line" routine. That is, it begins by plotting the traces in the foreground (those corresponding to the largest x-values), and then, as each new trace is plotted, the program determines whether all or only part of the next trace should be shown.

The graphs on page 893 show a variety of surfaces that were plotted by computer. If you have access to a computer drawing program, use it to reproduce these surfaces. Remember also that the three-dimensional graphics in this text can be viewed and rotated. These rotatable graphs are available in the premium eBook for this text.

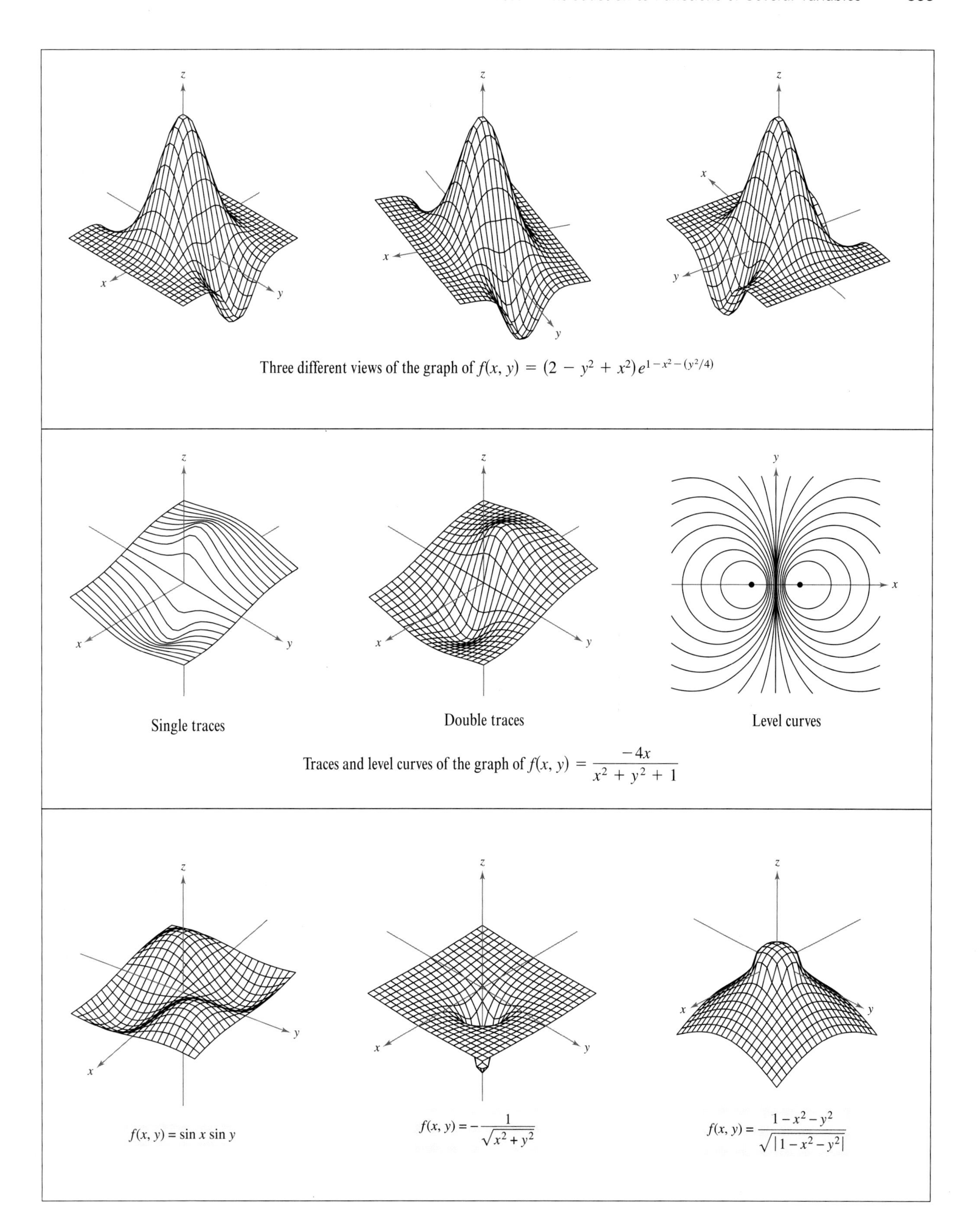

Three different views of the graph of $f(x, y) = (2 - y^2 + x^2)e^{1 - x^2 - (y^2/4)}$

Traces and level curves of the graph of $f(x, y) = \dfrac{-4x}{x^2 + y^2 + 1}$

$f(x, y) = \sin x \sin y$

$f(x, y) = -\dfrac{1}{\sqrt{x^2 + y^2}}$

$f(x, y) = \dfrac{1 - x^2 - y^2}{\sqrt{|1 - x^2 - y^2|}}$

13.1 Exercises

See www.CalcChat.com for worked-out solutions to odd-numbered exercises.

In Exercises 1 and 2, use the graph to determine whether z is a function of x and y. Explain.

1.

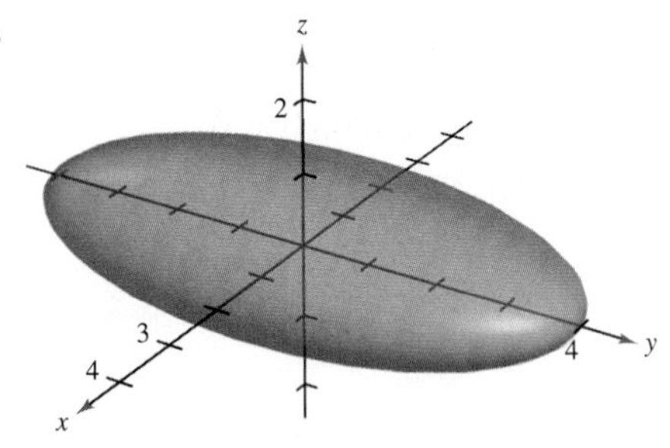

2.

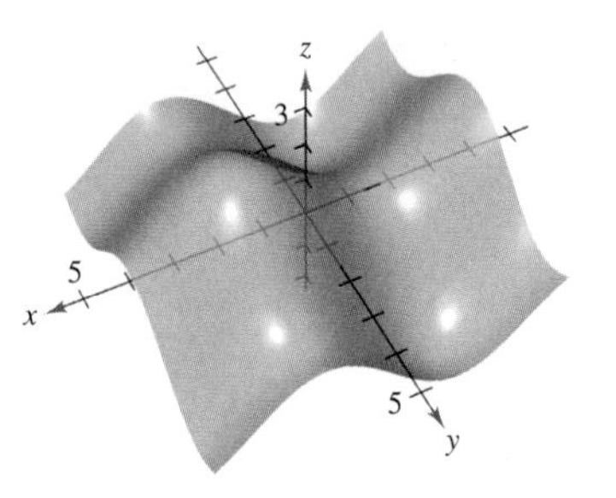

In Exercises 3–6, determine whether z is a function of x and y.

3. $x^2z + 3y^2 - xy = 10$

4. $xz^2 + 2xy - y^2 = 4$

5. $\dfrac{x^2}{4} + \dfrac{y^2}{9} + z^2 = 1$

6. $z + x \ln y - 8yz = 0$

In Exercises 7–18, find and simplify the function values.

7. $f(x, y) = xy$
   (a) $(3, 2)$ (b) $(-1, 4)$ (c) $(30, 5)$
   (d) $(5, y)$ (e) $(x, 2)$ (f) $(5, t)$

8. $f(x, y) = 4 - x^2 - 4y^2$
   (a) $(0, 0)$ (b) $(0, 1)$ (c) $(2, 3)$
   (d) $(1, y)$ (e) $(x, 0)$ (f) $(t, 1)$

9. $f(x, y) = xe^y$
   (a) $(5, 0)$ (b) $(3, 2)$ (c) $(2, -1)$
   (d) $(5, y)$ (e) $(x, 2)$ (f) (t, t)

10. $g(x, y) = \ln|x + y|$
    (a) $(1, 0)$ (b) $(0, -1)$ (c) $(0, e)$
    (d) $(1, 1)$ (e) $(e, e/2)$ (f) $(2, 5)$

11. $h(x, y, z) = \dfrac{xy}{z}$
    (a) $(2, 3, 9)$ (b) $(1, 0, 1)$ (c) $(-2, 3, 4)$ (d) $(5, 4, -6)$

12. $f(x, y, z) = \sqrt{x + y + z}$
    (a) $(0, 5, 4)$ (b) $(6, 8, -3)$
    (c) $(4, 6, 2)$ (d) $(10, -4, -3)$

13. $f(x, y) = x \sin y$
    (a) $(2, \pi/4)$ (b) $(3, 1)$ (c) $(-3, \pi/3)$ (d) $(4, \pi/2)$

14. $V(r, h) = \pi r^2 h$
    (a) $(3, 10)$ (b) $(5, 2)$ (c) $(4, 8)$ (d) $(6, 4)$

15. $g(x, y) = \displaystyle\int_x^y (2t - 3)\, dt$
    (a) $(4, 0)$ (b) $(4, 1)$ (c) $\left(4, \frac{3}{2}\right)$ (d) $\left(\frac{3}{2}, 0\right)$

16. $g(x, y) = \displaystyle\int_x^y \frac{1}{t}\, dt$
    (a) $(4, 1)$ (b) $(6, 3)$ (c) $(2, 5)$ (d) $\left(\frac{1}{2}, 7\right)$

17. $f(x, y) = 2x + y^2$
    (a) $\dfrac{f(x + \Delta x, y) - f(x, y)}{\Delta x}$
    (b) $\dfrac{f(x, y + \Delta y) - f(x, y)}{\Delta y}$

18. $f(x, y) = 3x^2 - 2y$
    (a) $\dfrac{f(x + \Delta x, y) - f(x, y)}{\Delta x}$
    (b) $\dfrac{f(x, y + \Delta y) - f(x, y)}{\Delta y}$

In Exercises 19–30, describe the domain and range of the function.

19. $f(x, y) = x^2 + y^2$

20. $f(x, y) = e^{xy}$

21. $g(x, y) = x\sqrt{y}$

22. $g(x, y) = \dfrac{y}{\sqrt{x}}$

23. $z = \dfrac{x + y}{xy}$

24. $z = \dfrac{xy}{x - y}$

25. $f(x, y) = \sqrt{4 - x^2 - y^2}$

26. $f(x, y) = \sqrt{4 - x^2 - 4y^2}$

27. $f(x, y) = \arccos(x + y)$

28. $f(x, y) = \arcsin(y/x)$

29. $f(x, y) = \ln(4 - x - y)$

30. $f(x, y) = \ln(xy - 6)$

31. ***Think About It*** The graphs labeled (a), (b), (c), and (d) are graphs of the function $f(x, y) = -4x/(x^2 + y^2 + 1)$. Match the four graphs with the points in space from which the surface is viewed. The four points are $(20, 15, 25)$, $(-15, 10, 20)$, $(20, 20, 0)$, and $(20, 0, 0)$.

(a)

z
y

Generated by Maple

(b)

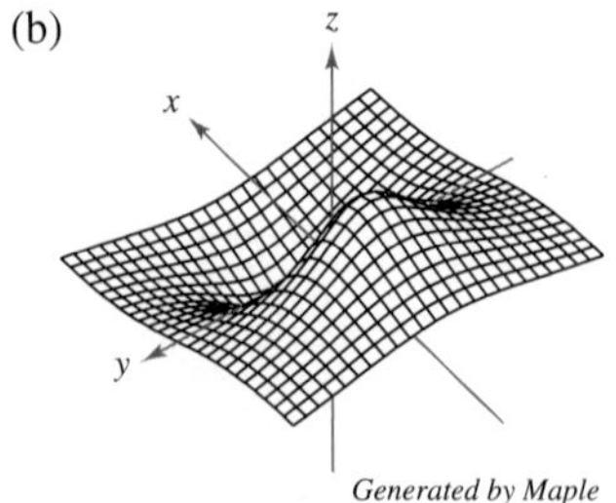

Generated by Maple

(c)

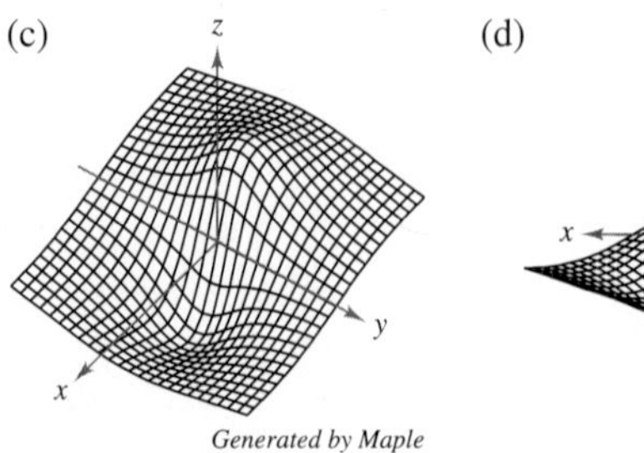

Generated by Maple

(d)

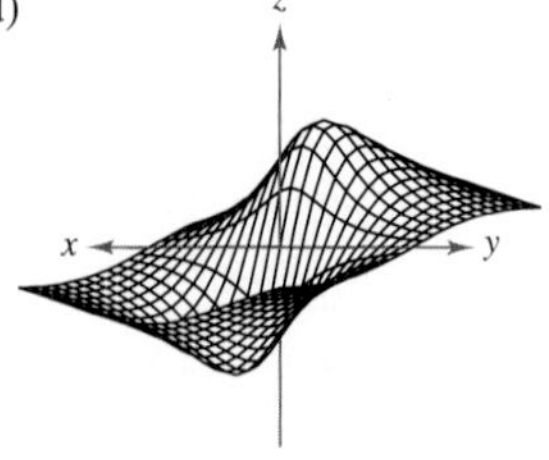

Generated by Maple

32. ***Think About It*** Use the function given in Exercise 31.
   (a) Find the domain and range of the function.
   (b) Identify the points in the xy-plane at which the function value is 0.
   (c) Does the surface pass through all the octants of the rectangular coordinate system? Give reasons for your answer.

In Exercises 33–40, sketch the surface given by the function.

33. $f(x, y) = 4$
34. $f(x, y) = 6 - 2x - 3y$
35. $f(x, y) = y^2$
36. $g(x, y) = \frac{1}{2}y$
37. $z = -x^2 - y^2$
38. $z = \frac{1}{2}\sqrt{x^2 + y^2}$
39. $f(x, y) = e^{-x}$
40. $f(x, y) = \begin{cases} xy, & x \ge 0, y \ge 0 \\ 0, & x < 0 \text{ or } y < 0 \end{cases}$

CAS **In Exercises 41–44, use a computer algebra system to graph the function.**

41. $z = y^2 - x^2 + 1$
42. $z = \frac{1}{12}\sqrt{144 - 16x^2 - 9y^2}$
43. $f(x, y) = x^2e^{(-xy/2)}$
44. $f(x, y) = x \sin y$

In Exercises 45–48, match the graph of the surface with one of the contour maps. [The contour maps are labeled (a), (b), (c), and (d).]

(a)

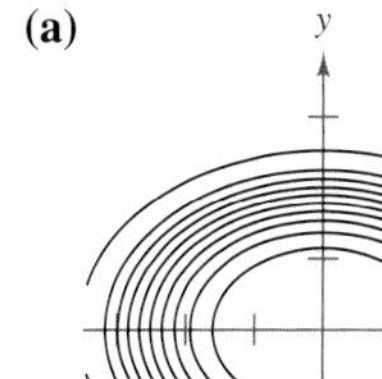

(b)

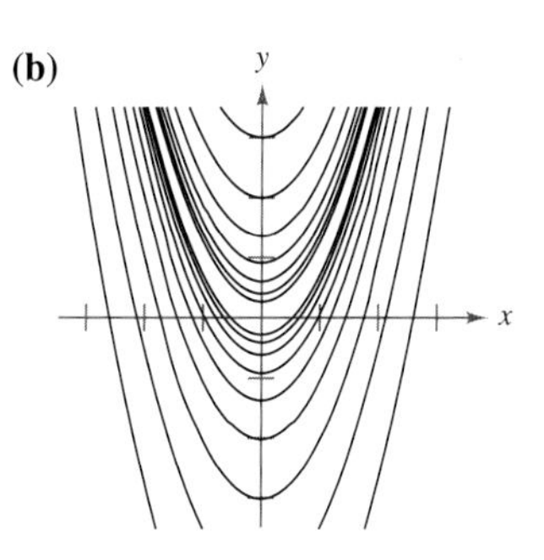

(c)

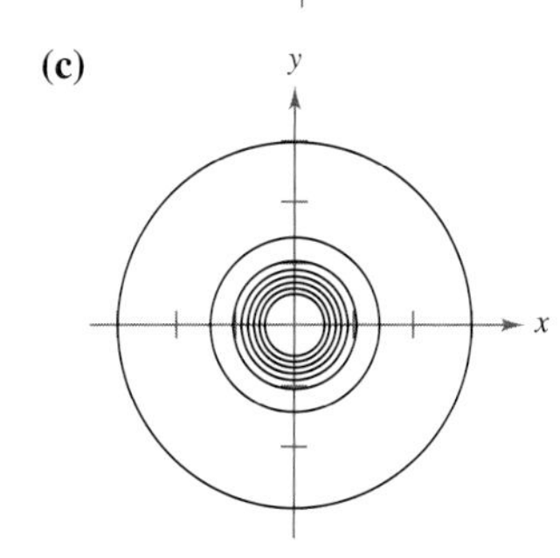

(d)

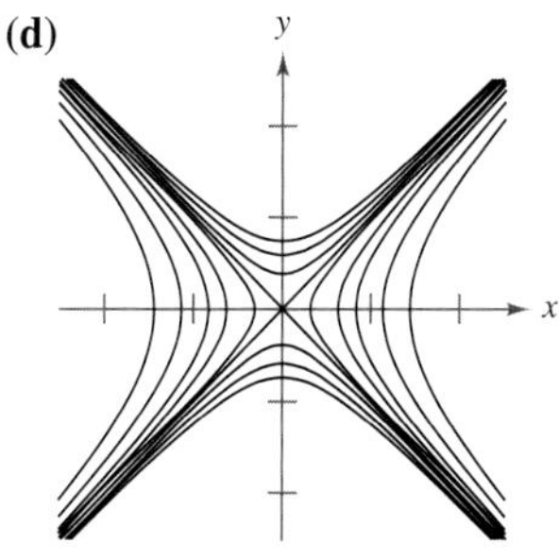

45. 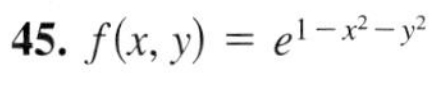 $f(x, y) = e^{1 - x^2 - y^2}$

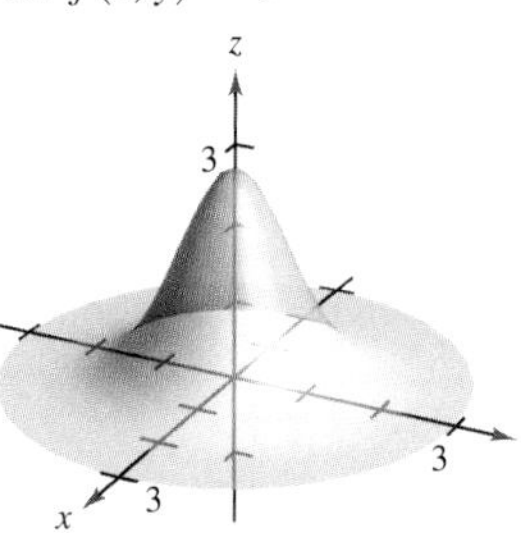

46. $f(x, y) = e^{1 - x^2 + y^2}$

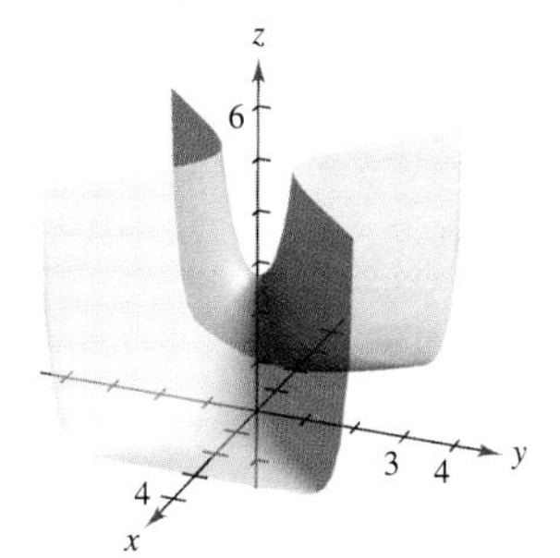

47. $f(x, y) = \ln|y - x^2|$

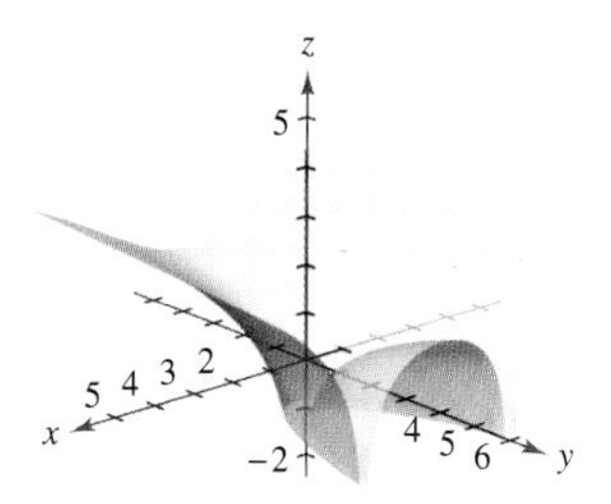

48. 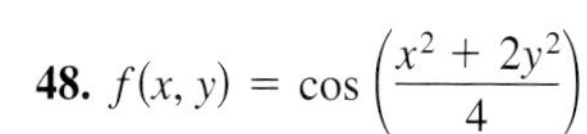 $f(x, y) = \cos\left(\frac{x^2 + 2y^2}{4}\right)$

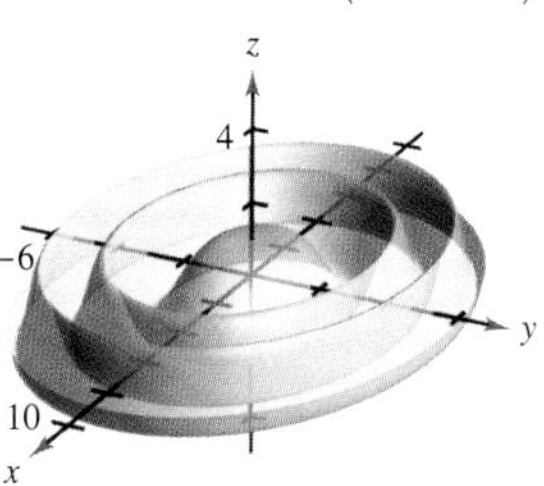

In Exercises 49–56, describe the level curves of the function. Sketch the level curves for the given c-values.

49. $z = x + y, \quad c = -1, 0, 2, 4$
50. $z = 6 - 2x - 3y, \quad c = 0, 2, 4, 6, 8, 10$
51. $z = x^2 + 4y^2, \quad c = 0, 1, 2, 3, 4$
52. $f(x, y) = \sqrt{9 - x^2 - y^2}, \quad c = 0, 1, 2, 3$
53. $f(x, y) = xy, \quad c = \pm 1, \pm 2, \ldots, \pm 6$
54. $f(x, y) = e^{xy/2}, \quad c = 2, 3, 4, \frac{1}{2}, \frac{1}{3}, \frac{1}{4}$
55. $f(x, y) = x/(x^2 + y^2), \quad c = \pm\frac{1}{2}, \pm 1, \pm\frac{3}{2}, \pm 2$
56. $f(x, y) = \ln(x - y), \quad c = 0, \pm\frac{1}{2}, \pm 1, \pm\frac{3}{2}, \pm 2$

In Exercises 57–60, use a graphing utility to graph six level curves of the function.

57. $f(x, y) = x^2 - y^2 + 2$
58. $f(x, y) = |xy|$
59. $g(x, y) = \dfrac{8}{1 + x^2 + y^2}$
60. $h(x, y) = 3 \sin(|x| + |y|)$

WRITING ABOUT CONCEPTS

61. What is a graph of a function of two variables? How is it interpreted geometrically? Describe level curves.
62. All of the level curves of the surface given by $z = f(x, y)$ are concentric circles. Does this imply that the graph of f is a hemisphere? Illustrate your answer with an example.
63. Construct a function whose level curves are lines passing through the origin.

CAPSTONE

64. Consider the function $f(x, y) = xy$, for $x \ge 0$ and $y \ge 0$.
   (a) Sketch the graph of the surface given by f.
   (b) Make a conjecture about the relationship between the graphs of f and $g(x, y) = f(x, y) - 3$. Explain your reasoning.
   (c) Make a conjecture about the relationship between the graphs of f and $g(x, y) = -f(x, y)$. Explain your reasoning.
   (d) Make a conjecture about the relationship between the graphs of f and $g(x, y) = \frac{1}{2}f(x, y)$. Explain your reasoning.
   (e) On the surface in part (a), sketch the graph of $z = f(x, x)$.

Writing **In Exercises 65 and 66, use the graphs of the level curves (c-values evenly spaced) of the function f to write a description of a possible graph of f. Is the graph of f unique? Explain.**

65. **66.**

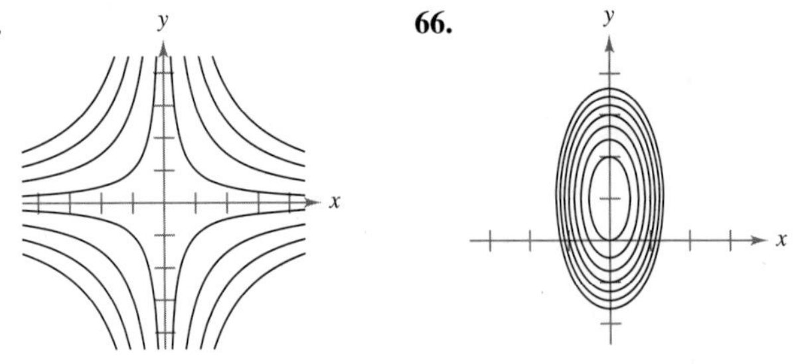

67. ***Investment*** In 2009, an investment of \$1000 was made in a bond earning 6% compounded annually. Assume that the buyer pays tax at rate R and the annual rate of inflation is I. In the year 2019, the value V of the investment in constant 2009 dollars is

$$V(I, R) = 1000\left[\frac{1 + 0.06(1 - R)}{1 + I}\right]^{10}.$$

Use this function of two variables to complete the table.

	Inflation Rate		
Tax Rate	0	0.03	0.05
0			
0.28			
0.35			

68. ***Investment*** A principal of \$5000 is deposited in a savings account that earns interest at a rate of r (written as a decimal), compounded continuously. The amount $A(r, t)$ after t years is $A(r, t) = 5000e^{rt}$. Use this function of two variables to complete the table.

	Number of Years			
Rate	5	10	15	20
0.02				
0.03				
0.04				
0.05				

In Exercises 69–74, sketch the graph of the level surface $f(x, y, z) = c$ at the given value of c.

69. $f(x, y, z) = x - y + z, \quad c = 1$

70. $f(x, y, z) = 4x + y + 2z, \quad c = 4$

71. $f(x, y, z) = x^2 + y^2 + z^2, \quad c = 9$

72. $f(x, y, z) = x^2 + \frac{1}{4}y^2 - z, \quad c = 1$

73. $f(x, y, z) = 4x^2 + 4y^2 - z^2, \quad c = 0$

74. $f(x, y, z) = \sin x - z, \quad c = 0$

75. ***Forestry*** The **Doyle Log Rule** is one of several methods used to determine the lumber yield of a log (in board-feet) in terms of its diameter d (in inches) and its length L (in feet). The number of board-feet is

$$N(d, L) = \left(\frac{d - 4}{4}\right)^2 L.$$

(a) Find the number of board-feet of lumber in a log 22 inches in diameter and 12 feet in length.

(b) Find $N(30, 12)$.

76. ***Queuing Model*** The average length of time that a customer waits in line for service is

$$W(x, y) = \frac{1}{x - y}, \quad x > y$$

where y is the average arrival rate, written as the number of customers per unit of time, and x is the average service rate, written in the same units. Evaluate each of the following.

(a) $W(15, 9)$ (b) $W(15, 13)$ (c) $W(12, 7)$ (d) $W(5, 2)$

77. ***Temperature Distribution*** The temperature T (in degrees Celsius) at any point (x, y) in a circular steel plate of radius 10 meters is $T = 600 - 0.75x^2 - 0.75y^2$, where x and y are measured in meters. Sketch some of the isothermal curves.

78. ***Electric Potential*** The electric potential V at any point (x, y) is

$$V(x, y) = \frac{5}{\sqrt{25 + x^2 + y^2}}.$$

Sketch the equipotential curves for $V = \frac{1}{2}$, $V = \frac{1}{3}$, and $V = \frac{1}{4}$.

79. ***Cobb-Douglas Production Function*** Use the Cobb-Douglas production function (see Example 5) to show that if the number of units of labor and the number of units of capital are doubled, the production level is also doubled.

80. ***Cobb-Douglas Production Function*** Show that the Cobb-Douglas production function $z = Cx^a y^{1-a}$ can be rewritten as

$$\ln\frac{z}{y} = \ln C + a\ln\frac{x}{y}.$$

81. ***Construction Cost*** A rectangular box with an open top has a length of x feet, a width of y feet, and a height of z feet. It costs \$1.20 per square foot to build the base and \$0.75 per square foot to build the sides. Write the cost C of constructing the box as a function of x, y, and z.

82. ***Volume*** A propane tank is constructed by welding hemispheres to the ends of a right circular cylinder. Write the volume V of the tank as a function of r and l, where r is the radius of the cylinder and hemispheres, and l is the length of the cylinder.

83. ***Ideal Gas Law*** According to the Ideal Gas Law, $PV = kT$, where P is pressure, V is volume, T is temperature (in Kelvins), and k is a constant of proportionality. A tank contains 2000 cubic inches of nitrogen at a pressure of 26 pounds per square inch and a temperature of 300 K.

(a) Determine k.

(b) Write P as a function of V and T and describe the level curves.

84. ***Modeling Data*** The table shows the net sales x (in billions of dollars), the total assets y (in billions of dollars), and the shareholder's equity z (in billions of dollars) for Wal-Mart for the years 2002 through 2007. (*Source: 2007 Annual Report for Wal-Mart*)

Year	2002	2003	2004	2005	2006	2007
x	201.2	226.5	252.8	281.5	208.9	345.0
y	79.3	90.2	102.5	117.1	135.6	151.2
z	35.2	39.5	43.6	49.4	53.2	61.6

A model for these data is

$$z = f(x, y) = 0.026x + 0.316y + 5.04.$$

(a) Use a graphing utility and the model to approximate z for the given values of x and y.

(b) Which of the two variables in this model has the greater influence on shareholder's equity?

(c) Simplify the expression for $f(x, 95)$ and interpret its meaning in the context of the problem.

85. ***Meteorology*** Meteorologists measure the atmospheric pressure in millibars. From these observations they create weather maps on which the curves of equal atmospheric pressure (isobars) are drawn (see figure). On the map, the closer the isobars the higher the wind speed. Match points A, B, and C with (a) highest pressure, (b) lowest pressure, and (c) highest wind velocity.

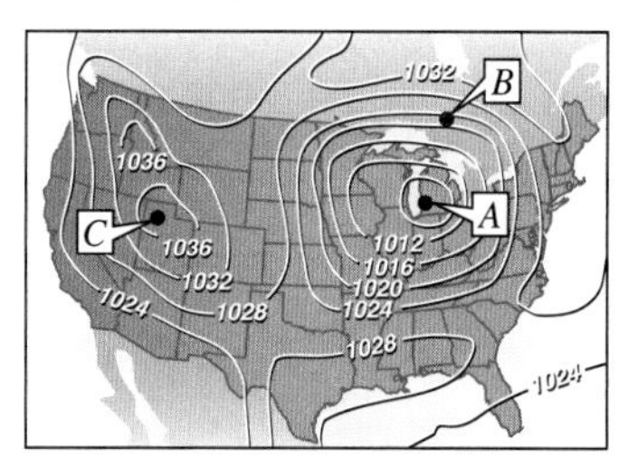

Figure for 85

Figure for 86

86. ***Acid Rain*** The acidity of rainwater is measured in units called pH. A pH of 7 is neutral, smaller values are increasingly acidic, and larger values are increasingly alkaline. The map shows curves of equal pH and gives evidence that downwind of heavily industrialized areas the acidity has been increasing. Using the level curves on the map, determine the direction of the prevailing winds in the northeastern United States.

87. ***Atmosphere*** The contour map shown in the figure was computer generated using data collected by satellite instrumentation. Color is used to show the "ozone hole" in Earth's atmosphere. The purple and blue areas represent the lowest levels of ozone and the green areas represent the highest levels. (*Source: National Aeronautics and Space Administration*)

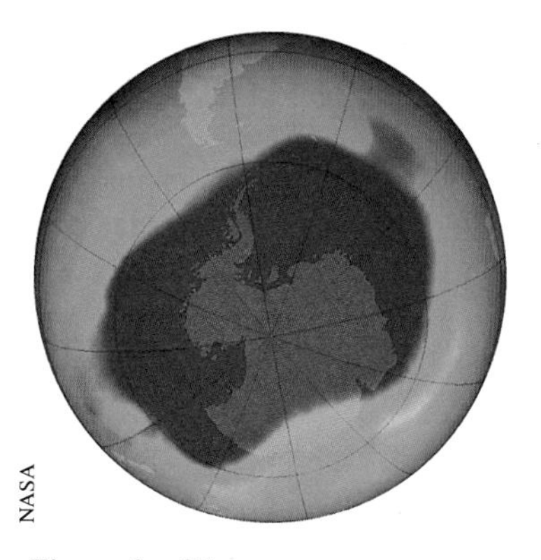

NASA

Figure for 87

(a) Do the level curves correspond to equally spaced ozone levels? Explain.

(b) Describe how to obtain a more detailed contour map.

88. ***Geology*** The contour map in the figure represents color-coded seismic amplitudes of a fault horizon and a projected contour map, which is used in earthquake studies. (*Source: Adapted from Shipman/Wilson/Todd,* An Introduction to Physical Science, *Tenth Edition*)

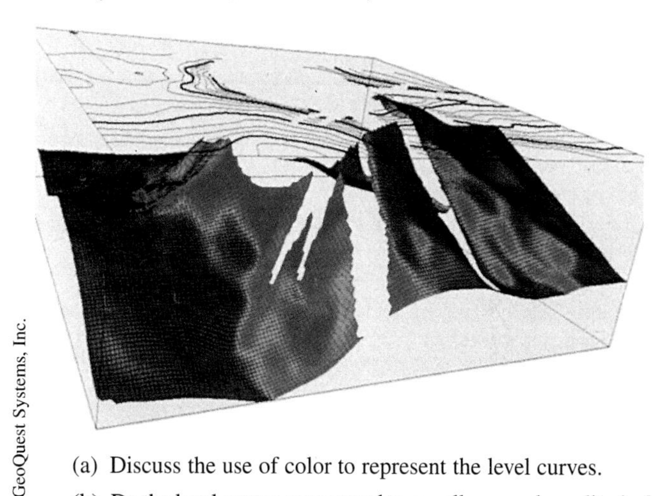

GeoQuest Systems, Inc.

(a) Discuss the use of color to represent the level curves.

(b) Do the level curves correspond to equally spaced amplitudes? Explain.

True or False? **In Exercises 89–92, determine whether the statement is true or false. If it is false, explain why or give an example that shows it is false.**

89. If $f(x_0, y_0) = f(x_1, y_1)$, then $x_0 = x_1$ and $y_0 = y_1$.

90. If f is a function, then $f(ax, ay) = a^2 f(x, y)$.

91. A vertical line can intersect the graph of $z = f(x, y)$ at most once.

92. Two different level curves of the graph of $z = f(x, y)$ can intersect.

13.2 Limits and Continuity

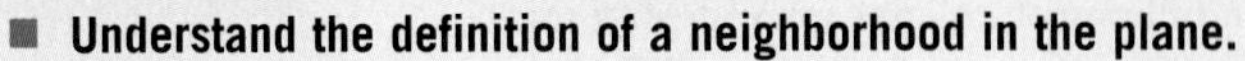

- Understand the definition of a neighborhood in the plane.
- Understand and use the definition of the limit of a function of two variables.
- Extend the concept of continuity to a function of two variables.
- Extend the concept of continuity to a function of three variables.

Neighborhoods in the Plane

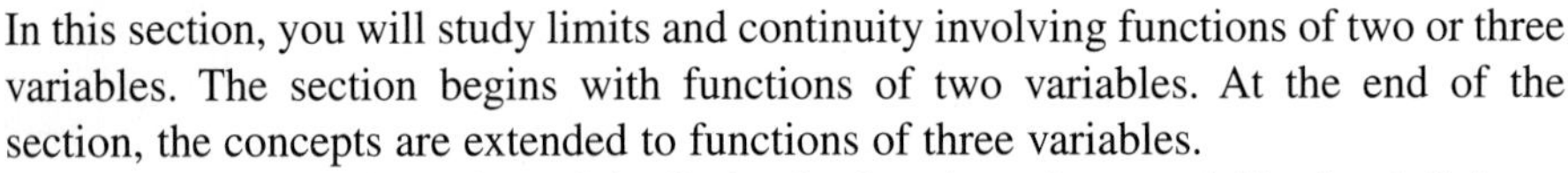

In this section, you will study limits and continuity involving functions of two or three variables. The section begins with functions of two variables. At the end of the section, the concepts are extended to functions of three variables.

We begin our discussion of the limit of a function of two variables by defining a two-dimensional analog to an interval on the real number line. Using the formula for the distance between two points (x, y) and (x_0, y_0) in the plane, you can define the **δ-neighborhood** about (x_0, y_0) to be the **disk** centered at (x_0, y_0) with radius $\delta > 0$

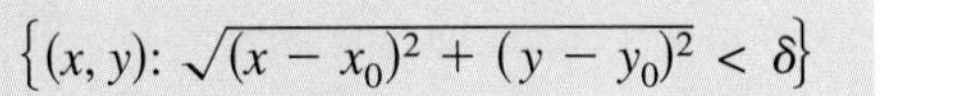

$$\left\{(x, y): \sqrt{(x - x_0)^2 + (y - y_0)^2} < \delta\right\} \qquad \text{Open disk}$$

as shown in Figure 13.18. When this formula contains the *less than* inequality sign, <, the disk is called **open,** and when it contains the *less than or equal to* inequality sign, ≤, the disk is called **closed.** This corresponds to the use of < and ≤ to define open and closed intervals.

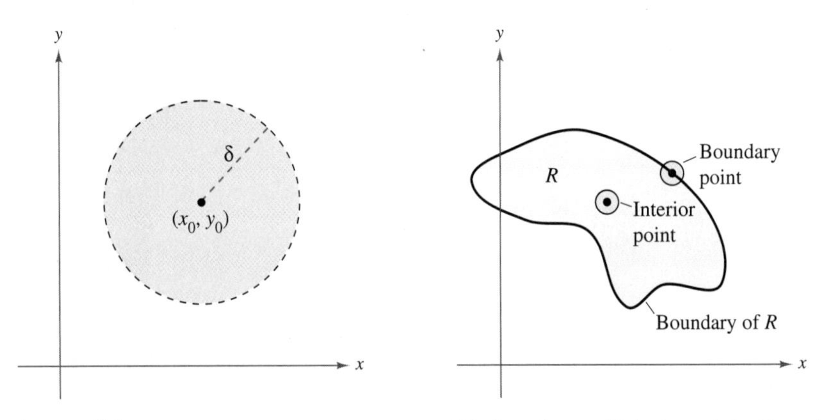

An open disk
Figure 13.18

The boundary and interior points of a region R
Figure 13.19

The Granger Collection

SONYA KOVALEVSKY (1850–1891)

Much of the terminology used to define limits and continuity of a function of two or three variables was introduced by the German mathematician Karl Weierstrass (1815–1897). Weierstrass's rigorous approach to limits and other topics in calculus gained him the reputation as the "father of modern analysis." Weierstrass was a gifted teacher. One of his best-known students was the Russian mathematician Sonya Kovalevsky, who applied many of Weierstrass's techniques to problems in mathematical physics and became one of the first women to gain acceptance as a research mathematician.

A point (x_0, y_0) in a plane region R is an **interior point** of R if there exists a δ-neighborhood about (x_0, y_0) that lies entirely in R, as shown in Figure 13.19. If every point in R is an interior point, then R is an **open region.** A point (x_0, y_0) is a **boundary point** of R if every open disk centered at (x_0, y_0) contains points inside R *and* points outside R. By definition, a region must contain its interior points, but it need not contain its boundary points. If a region contains all its boundary points, the region is **closed.** A region that contains some but not all of its boundary points is neither open nor closed.

■ **FOR FURTHER INFORMATION** For more information on Sonya Kovalevsky, see the article "S. Kovalevsky: A Mathematical Lesson" by Karen D. Rappaport in *The American Mathematical Monthly*. To view this article, go to the website *www.matharticles.com*. ■

Limit of a Function of Two Variables

DEFINITION OF THE LIMIT OF A FUNCTION OF TWO VARIABLES

Let f be a function of two variables defined, except possibly at (x_0, y_0), on an open disk centered at (x_0, y_0), and let L be a real number. Then

$$\lim_{(x, y)\to(x_0, y_0)} f(x, y) = L$$

if for each $\varepsilon > 0$ there corresponds a $\delta > 0$ such that

$$|f(x, y) - L| < \varepsilon \quad \text{whenever} \quad 0 < \sqrt{(x - x_0)^2 + (y - y_0)^2} < \delta.$$

NOTE Graphically, this definition of a limit implies that for any point $(x, y) \neq (x_0, y_0)$ in the disk of radius δ, the value $f(x, y)$ lies between $L + \varepsilon$ and $L - \varepsilon$, as shown in Figure 13.20. ■

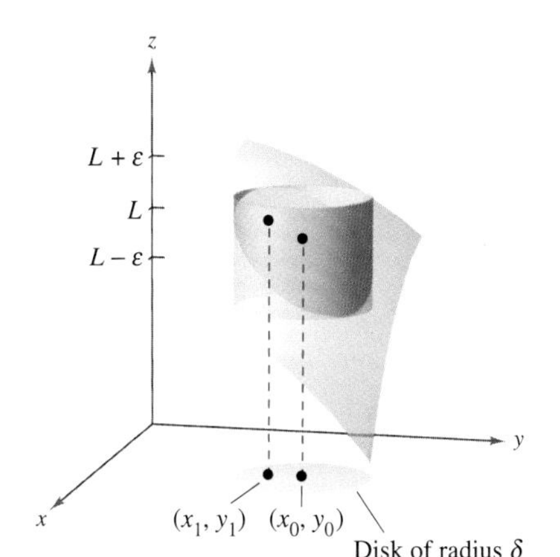

For any (x, y) in the disk of radius δ, the value $f(x, y)$ lies between $L + \varepsilon$ and $L - \varepsilon$.
Figure 13.20

The definition of the limit of a function of two variables is similar to the definition of the limit of a function of a single variable, yet there is a critical difference. To determine whether a function of a single variable has a limit, you need only test the approach from two directions—from the right and from the left. If the function approaches the same limit from the right and from the left, you can conclude that the limit exists. However, for a function of two variables, the statement

$$(x, y) \to (x_0, y_0)$$

means that the point (x, y) is allowed to approach (x_0, y_0) from any direction. If the value of

$$\lim_{(x, y)\to(x_0, y_0)} f(x, y)$$

is not the same for all possible approaches, or **paths,** to (x_0, y_0), the limit does not exist.

EXAMPLE 1 Verifying a Limit by the Definition

Show that

$$\lim_{(x, y)\to(a, b)} x = a.$$

Solution Let $f(x, y) = x$ and $L = a$. You need to show that for each $\varepsilon > 0$, there exists a δ-neighborhood about (a, b) such that

$$|f(x, y) - L| = |x - a| < \varepsilon$$

whenever $(x, y) \neq (a, b)$ lies in the neighborhood. You can first observe that from

$$0 < \sqrt{(x - a)^2 + (y - b)^2} < \delta$$

it follows that

$$\begin{aligned} |f(x, y) - a| &= |x - a| \\ &= \sqrt{(x - a)^2} \\ &\le \sqrt{(x - a)^2 + (y - b)^2} \\ &< \delta. \end{aligned}$$

So, you can choose $\delta = \varepsilon$, and the limit is verified. ■

Limits of functions of several variables have the same properties regarding sums, differences, products, and quotients as do limits of functions of single variables. (See Theorem 1.2 in Section 1.3.) Some of these properties are used in the next example.

EXAMPLE 2 Verifying a Limit

Evaluate $\displaystyle\lim_{(x, y)\to(1, 2)} \frac{5x^2y}{x^2 + y^2}$.

Solution By using the properties of limits of products and sums, you obtain

$$\begin{aligned}\lim_{(x, y)\to(1, 2)} 5x^2y &= 5(1^2)(2)\\ &= 10\end{aligned}$$

and

$$\begin{aligned}\lim_{(x, y)\to(1, 2)} (x^2 + y^2) &= (1^2 + 2^2)\\ &= 5.\end{aligned}$$

Because the limit of a quotient is equal to the quotient of the limits (and the denominator is not 0), you have

$$\begin{aligned}\lim_{(x, y)\to(1, 2)} \frac{5x^2y}{x^2 + y^2} &= \frac{10}{5}\\ &= 2.\end{aligned}$$

EXAMPLE 3 Verifying a Limit

Evaluate $\displaystyle\lim_{(x, y)\to(0, 0)} \frac{5x^2y}{x^2 + y^2}$.

Solution In this case, the limits of the numerator and of the denominator are both 0, and so you cannot determine the existence (or nonexistence) of a limit by taking the limits of the numerator and denominator separately and then dividing. However, from the graph of f in Figure 13.21, it seems reasonable that the limit might be 0. So, you can try applying the definition to $L = 0$. First, note that

$$|y| \le \sqrt{x^2 + y^2} \qquad \text{and} \qquad \frac{x^2}{x^2 + y^2} \le 1.$$

Then, in a δ-neighborhood about $(0, 0)$, you have $0 < \sqrt{x^2 + y^2} < \delta$, and it follows that, for $(x, y) \neq (0, 0)$,

$$\begin{aligned}|f(x, y) - 0| &= \left|\frac{5x^2y}{x^2 + y^2}\right|\\ &= 5|y|\left(\frac{x^2}{x^2 + y^2}\right)\\ &\le 5|y|\\ &\le 5\sqrt{x^2 + y^2}\\ &< 5\delta.\end{aligned}$$

So, you can choose $\delta = \varepsilon/5$ and conclude that

$$\lim_{(x, y)\to(0, 0)} \frac{5x^2y}{x^2 + y^2} = 0.$$

■

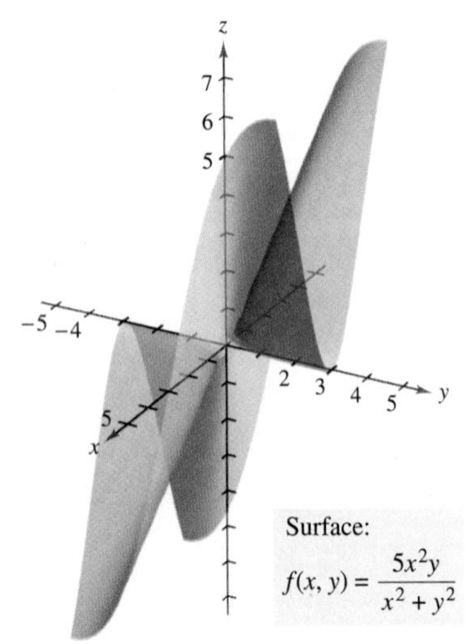

Figure 13.21

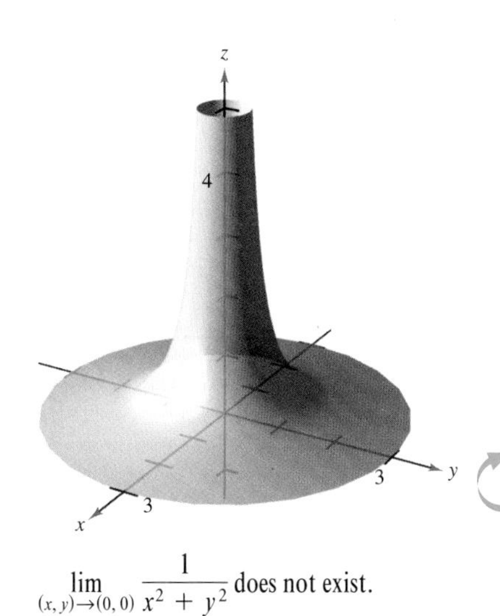

$\lim_{(x, y)\to(0, 0)} \frac{1}{x^2 + y^2}$ does not exist.

Figure 13.22

For some functions, it is easy to recognize that a limit does not exist. For instance, it is clear that the limit

$$\lim_{(x, y)\to(0, 0)} \frac{1}{x^2 + y^2}$$

does not exist because the values of $f(x, y)$ increase without bound as (x, y) approaches $(0, 0)$ along *any* path (see Figure 13.22).

For other functions, it is not so easy to recognize that a limit does not exist. For instance, the next example describes a limit that does not exist because the function approaches different values along different paths.

EXAMPLE 4 A Limit That Does Not Exist

Show that the following limit does not exist.

$$\lim_{(x, y)\to(0, 0)} \left(\frac{x^2 - y^2}{x^2 + y^2}\right)^2$$

Solution The domain of the function given by

$$f(x, y) = \left(\frac{x^2 - y^2}{x^2 + y^2}\right)^2$$

consists of all points in the xy-plane except for the point $(0, 0)$. To show that the limit as (x, y) approaches $(0, 0)$ does not exist, consider approaching $(0, 0)$ along two different "paths," as shown in Figure 13.23. Along the x-axis, every point is of the form $(x, 0)$, and the limit along this approach is

$$\lim_{(x, 0)\to(0, 0)} \left(\frac{x^2 - 0^2}{x^2 + 0^2}\right)^2 = \lim_{(x, 0)\to(0, 0)} 1^2 = 1. \qquad \text{Limit along } x\text{-axis}$$

However, if (x, y) approaches $(0, 0)$ along the line $y = x$, you obtain

$$\lim_{(x, x)\to(0, 0)} \left(\frac{x^2 - x^2}{x^2 + x^2}\right)^2 = \lim_{(x, x)\to(0, 0)} \left(\frac{0}{2x^2}\right)^2 = 0. \qquad \text{Limit along line } y = x$$

This means that in any open disk centered at $(0, 0)$, there are points (x, y) at which f takes on the value 1, and other points at which f takes on the value 0. For instance, $f(x, y) = 1$ at the points $(1, 0)$, $(0.1, 0)$, $(0.01, 0)$, and $(0.001, 0)$, and $f(x, y) = 0$ at the points $(1, 1)$, $(0.1, 0.1)$, $(0.01, 0.01)$, and $(0.001, 0.001)$. So, f does not have a limit as $(x, y) \to (0, 0)$.

NOTE In Example 4, you could conclude that the limit does not exist because you found two approaches that produced different limits. If two approaches had produced the same limit, you still could not have concluded that the limit exists. To form such a conclusion, you must show that the limit is the same along *all* possible approaches.

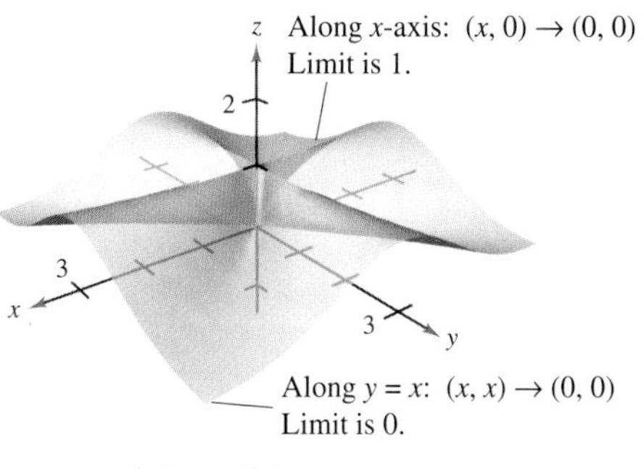

$\lim_{(x, y)\to(0, 0)} \left(\frac{x^2 - y^2}{x^2 + y^2}\right)^2$ does not exist.

Figure 13.23

■

Continuity of a Function of Two Variables

Notice in Example 2 that the limit of $f(x, y) = 5x^2y/(x^2 + y^2)$ as $(x, y) \to (1, 2)$ can be evaluated by direct substitution. That is, the limit is $f(1, 2) = 2$. In such cases the function f is said to be **continuous** at the point $(1, 2)$.

NOTE This definition of continuity can be extended to *boundary points* of the open region R by considering a special type of limit in which (x, y) is allowed to approach (x_0, y_0) along paths lying in the region R. This notion is similar to that of one-sided limits, as discussed in Chapter 1.

DEFINITION OF CONTINUITY OF A FUNCTION OF TWO VARIABLES

A function f of two variables is **continuous at a point (x_0, y_0)** in an open region R if $f(x_0, y_0)$ is equal to the limit of $f(x, y)$ as (x, y) approaches (x_0, y_0). That is,

$$\lim_{(x, y)\to(x_0, y_0)} f(x, y) = f(x_0, y_0).$$

The function f is **continuous in the open region R** if it is continuous at every point in R.

In Example 3, it was shown that the function

$$f(x, y) = \frac{5x^2y}{x^2 + y^2}$$

is not continuous at $(0, 0)$. However, because the limit at this point exists, you can remove the discontinuity by defining f at $(0, 0)$ as being equal to its limit there. Such a discontinuity is called **removable.** In Example 4, the function

$$f(x, y) = \left(\frac{x^2 - y^2}{x^2 + y^2}\right)^2$$

was also shown not to be continuous at $(0, 0)$, but this discontinuity is **nonremovable.**

THEOREM 13.1 CONTINUOUS FUNCTIONS OF TWO VARIABLES

If k is a real number and f and g are continuous at (x_0, y_0), then the following functions are continuous at (x_0, y_0).

1. Scalar multiple: kf
2. Sum and difference: $f \pm g$
3. Product: fg
4. Quotient: f/g, if $g(x_0, y_0) \neq 0$

Theorem 13.1 establishes the continuity of *polynomial* and *rational* functions at every point in their domains. Furthermore, the continuity of other types of functions can be extended naturally from one to two variables. For instance, the functions whose graphs are shown in Figures 13.24 and 13.25 are continuous at every point in the plane.

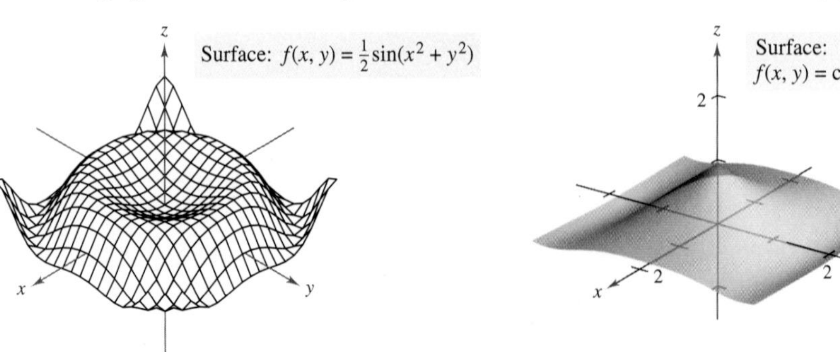

The function f is continuous at every point in the plane.
Figure 13.24

The function f is continuous at every point in the plane.
Figure 13.25

EXPLORATION

Hold a spoon a foot or so from your eyes. Look at your image in the spoon. It should be upside down. Now, move the spoon closer and closer to one eye. At some point, your image will be right side up. Could it be that your image is being continuously deformed? Talk about this question and the general meaning of continuity with other members of your class. (This exploration was suggested by Irvin Roy Hentzel, Iowa State University.)

The next theorem states conditions under which a composite function is continuous.

THEOREM 13.2 CONTINUITY OF A COMPOSITE FUNCTION

If h is continuous at (x_0, y_0) and g is continuous at $h(x_0, y_0)$, then the composite function given by $(g \circ h)(x, y) = g(h(x, y))$ is continuous at (x_0, y_0). That is,

$$\lim_{(x, y)\to(x_0, y_0)} g(h(x, y)) = g(h(x_0, y_0)).$$

NOTE Note in Theorem 13.2 that h is a function of two variables and g is a function of one variable.

EXAMPLE 5 Testing for Continuity

Discuss the continuity of each function.

a. $f(x, y) = \dfrac{x - 2y}{x^2 + y^2}$ **b.** $g(x, y) = \dfrac{2}{y - x^2}$

Solution

a. Because a rational function is continuous at every point in its domain, you can conclude that f is continuous at each point in the xy-plane except at $(0, 0)$, as shown in Figure 13.26.

b. The function given by $g(x, y) = 2/(y - x^2)$ is continuous except at the points at which the denominator is 0, $y - x^2 = 0$. So, you can conclude that the function is continuous at all points except those lying on the parabola $y = x^2$. Inside this parabola, you have $y > x^2$, and the surface represented by the function lies above the xy-plane, as shown in Figure 13.27. Outside the parabola, $y < x^2$, and the surface lies below the xy-plane.

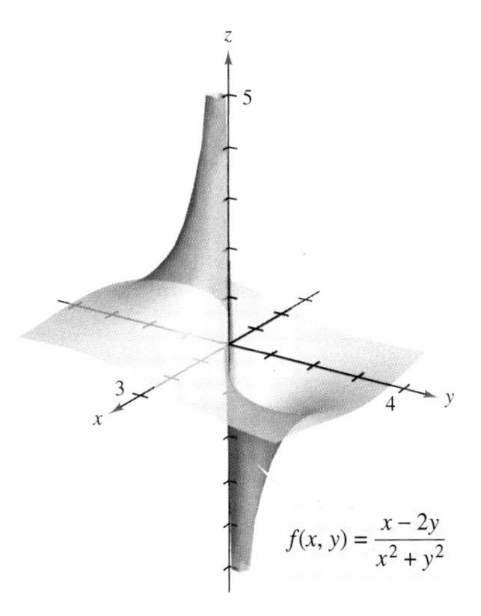

The function f is not continuous at $(0, 0)$.

Figure 13.26

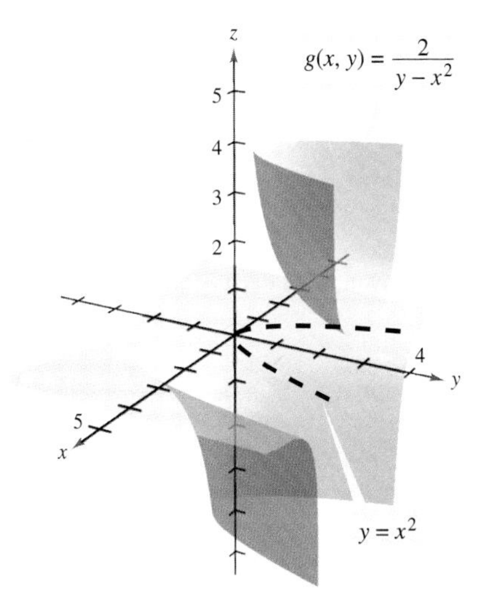

The function g is not continuous on the parabola $y = x^2$.

Figure 13.27

Continuity of a Function of Three Variables

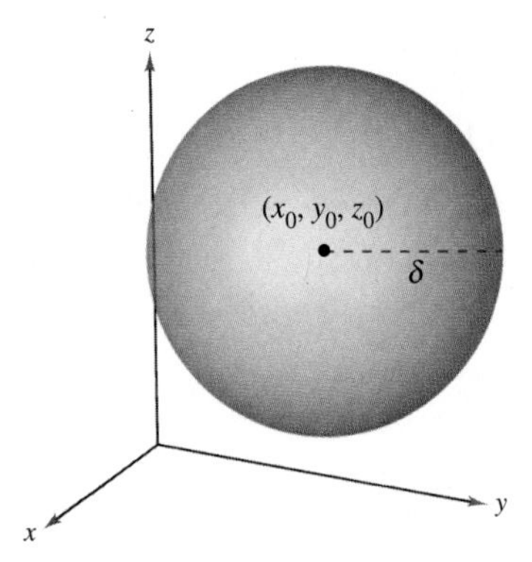

Open sphere in space
Figure 13.28

The preceding definitions of limits and continuity can be extended to functions of three variables by considering points (x, y, z) within the *open sphere*

$$(x - x_0)^2 + (y - y_0)^2 + (z - z_0)^2 < \delta^2. \quad \text{Open sphere}$$

The radius of this sphere is δ, and the sphere is centered at (x_0, y_0, z_0), as shown in Figure 13.28. A point (x_0, y_0, z_0) in a region R in space is an **interior point** of R if there exists a δ-sphere about (x_0, y_0, z_0) that lies entirely in R. If every point in R is an interior point, then R is called **open.**

DEFINITION OF CONTINUITY OF A FUNCTION OF THREE VARIABLES

A function f of three variables is **continuous at a point (x_0, y_0, z_0)** in an open region R if $f(x_0, y_0, z_0)$ is defined and is equal to the limit of $f(x, y, z)$ as (x, y, z) approaches (x_0, y_0, z_0). That is,

$$\lim_{(x, y, z)\to(x_0, y_0, z_0)} f(x, y, z) = f(x_0, y_0, z_0).$$

The function f is **continuous in the open region R** if it is continuous at every point in R.

EXAMPLE 6 Testing Continuity of a Function of Three Variables

The function

$$f(x, y, z) = \frac{1}{x^2 + y^2 - z}$$

is continuous at each point in space except at the points on the paraboloid given by $z = x^2 + y^2$. ■

13.2 Exercises

See www.CalcChat.com for worked-out solutions to odd-numbered exercises.

In Exercises 1–4, use the definition of the limit of a function of two variables to verify the limit.

1. $\lim_{(x, y)\to(1, 0)} x = 1$
2. $\lim_{(x, y)\to(4, -1)} x = 4$
3. $\lim_{(x, y)\to(1, -3)} y = -3$
4. $\lim_{(x, y)\to(a, b)} y = b$

In Exercises 5–8, find the indicated limit by using the limits

$$\lim_{(x,y)\to(a,b)} f(x, y) = 4 \quad \text{and} \quad \lim_{(x,y)\to(a,b)} g(x, y) = 3.$$

5. $\lim_{(x, y)\to(a, b)} [f(x, y) - g(x, y)]$
6. $\lim_{(x, y)\to(a, b)} \left[\frac{5f(x, y)}{g(x, y)}\right]$
7. $\lim_{(x, y)\to(a, b)} [f(x, y)g(x, y)]$
8. $\lim_{(x, y)\to(a, b)} \left[\frac{f(x, y) + g(x, y)}{f(x, y)}\right]$

In Exercises 9–22, find the limit and discuss the continuity of the function.

9. $\lim_{(x, y)\to(2, 1)} (2x^2 + y)$
10. $\lim_{(x, y)\to(0, 0)} (x + 4y + 1)$
11. $\lim_{(x, y)\to(1, 2)} e^{xy}$
12. $\lim_{(x, y)\to(2, 4)} \frac{x + y}{x^2 + 1}$
13. $\lim_{(x, y)\to(0, 2)} \frac{x}{y}$
14. $\lim_{(x, y)\to(-1, 2)} \frac{x + y}{x - y}$
15. $\lim_{(x, y)\to(1, 1)} \frac{xy}{x^2 + y^2}$
16. $\lim_{(x, y)\to(1, 1)} \frac{x}{\sqrt{x + y}}$
17. $\lim_{(x, y)\to(\pi/4, 2)} y \cos xy$
18. $\lim_{(x, y)\to(2\pi, 4)} \sin \frac{x}{y}$
19. $\lim_{(x, y)\to(0, 1)} \frac{\arcsin xy}{1 - xy}$
20. $\lim_{(x, y)\to(0, 1)} \frac{\arccos (x/y)}{1 + xy}$
21. $\lim_{(x, y, z)\to(1, 3, 4)} \sqrt{x + y + z}$
22. $\lim_{(x, y, z)\to(-2, 1, 0)} xe^{yz}$

In Exercises 23–36, find the limit (if it exists). If the limit does not exist, explain why.

23. $\displaystyle\lim_{(x,y)\to(1,1)} \frac{xy-1}{1+xy}$

24. $\displaystyle\lim_{(x,y)\to(1,-1)} \frac{x^2y}{1+xy^2}$

25. $\displaystyle\lim_{(x,y)\to(0,0)} \frac{1}{x+y}$

26. $\displaystyle\lim_{(x,y)\to(0,0)} \frac{1}{x^2y^2}$

27. $\displaystyle\lim_{(x,y)\to(2,2)} \frac{x^2-y^2}{x-y}$

28. $\displaystyle\lim_{(x,y)\to(0,0)} \frac{x^4-4y^4}{x^2+2y^2}$

29. $\displaystyle\lim_{(x,y)\to(0,0)} \frac{x-y}{\sqrt{x}-\sqrt{y}}$

30. $\displaystyle\lim_{(x,y)\to(2,1)} \frac{x-y-1}{\sqrt{x-y}-1}$

31. $\displaystyle\lim_{(x,y)\to(0,0)} \frac{x+y}{x^2+y}$

32. $\displaystyle\lim_{(x,y)\to(0,0)} \frac{x}{x^2-y^2}$

33. $\displaystyle\lim_{(x,y)\to(0,0)} \frac{x^2}{(x^2+1)(y^2+1)}$

34. $\displaystyle\lim_{(x,y)\to(0,0)} \ln(x^2+y^2)$

35. $\displaystyle\lim_{(x,y,z)\to(0,0,0)} \frac{xy+yz+xz}{x^2+y^2+z^2}$

36. $\displaystyle\lim_{(x,y,z)\to(0,0,0)} \frac{xy+yz^2+xz^2}{x^2+y^2+z^2}$

In Exercises 37 and 38, discuss the continuity of the function and evaluate the limit of $f(x, y)$ (if it exists) as $(x, y) \to (0, 0)$.

37. $f(x, y) = e^{xy}$

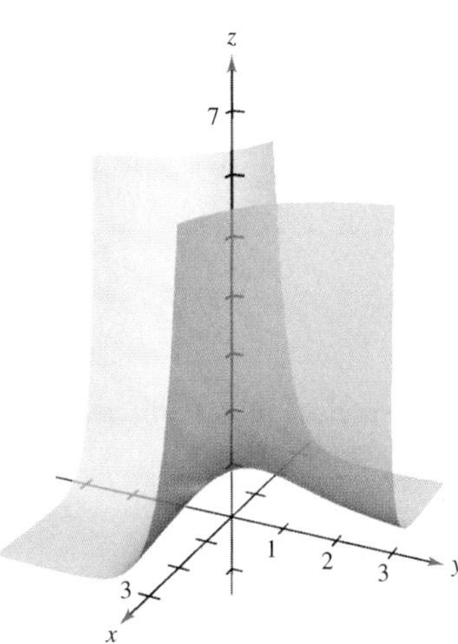

38. $f(x, y) = 1 - \dfrac{\cos(x^2+y^2)}{x^2+y^2}$

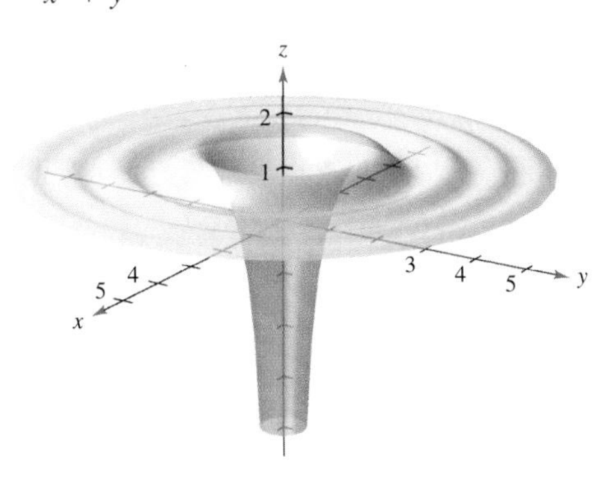

In Exercises 39–42, use a graphing utility to make a table showing the values of $f(x, y)$ at the given points for each path. Use the result to make a conjecture about the limit of $f(x, y)$ as $(x, y) \to (0, 0)$. Determine whether the limit exists analytically and discuss the continuity of the function.

39. $f(x, y) = \dfrac{xy}{x^2+y^2}$

Path: $y = 0$

Points: (1, 0), (0.5, 0), (0.1, 0), (0.01, 0), (0.001, 0)

Path: $y = x$

Points: (1, 1), (0.5, 0.5), (0.1, 0.1), (0.01, 0.01), (0.001, 0.001)

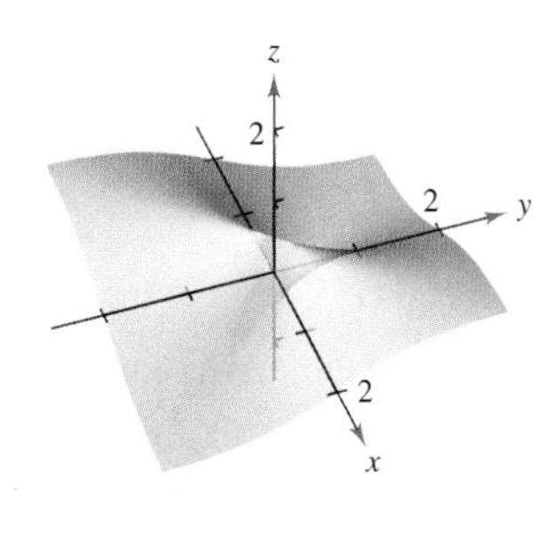

40. $f(x, y) = \dfrac{y}{x^2+y^2}$

Path: $y = 0$

Points: (1, 0), (0.5, 0), (0.1, 0), (0.01, 0), (0.001, 0)

Path: $y = x$

Points: (1, 1), (0.5, 0.5), (0.1, 0.1), (0.01, 0.01), (0.001, 0.001)

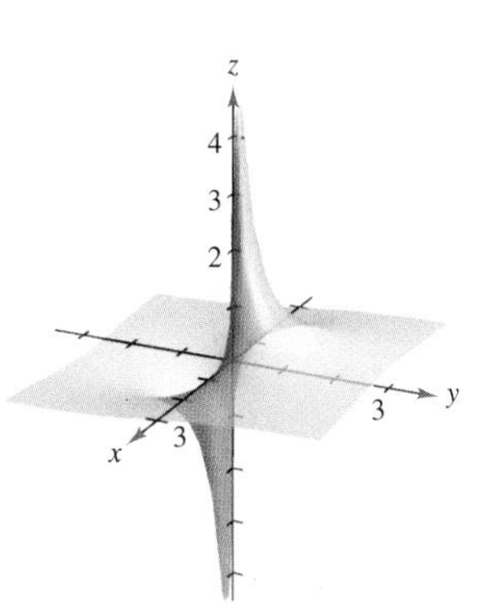

41. $f(x, y) = -\dfrac{xy^2}{x^2+y^4}$

Path: $x = y^2$

Points: (1, 1), (0.25, 0.5), (0.01, 0.1), (0.0001, 0.01), (0.000001, 0.001)

Path: $x = -y^2$

Points: (−1, 1), (−0.25, 0.5), (−0.01, 0.1), (−0.0001, 0.01), (−0.000001, 0.001)

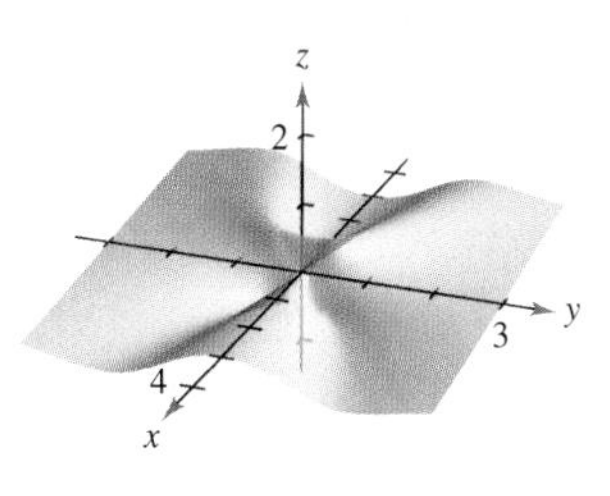

42. $f(x, y) = \dfrac{2x - y^2}{2x^2 + y}$

Path: $y = 0$

Points: $(1, 0)$, $(0.25, 0)$, $(0.01, 0)$, $(0.001, 0)$, $(0.000001, 0)$

Path: $y = x$

Points: $(1, 1)$, $(0.25, 0.25)$, $(0.01, 0.01)$, $(0.001, 0.001)$, $(0.0001, 0.0001)$

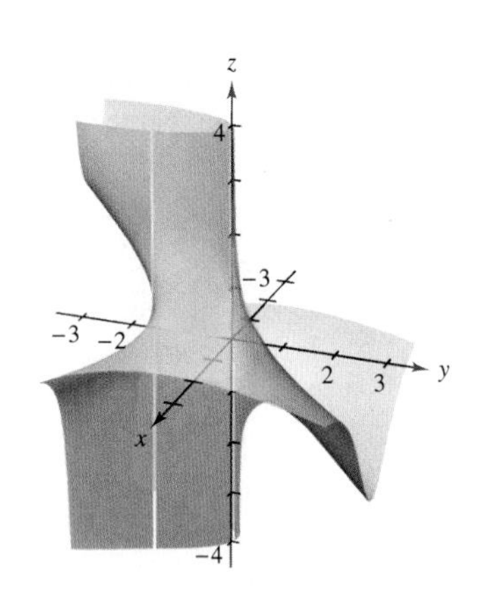

In Exercises 43–46, discuss the continuity of the functions f and g. Explain any differences.

43. $f(x, y) = \begin{cases} \dfrac{x^4 - y^4}{x^2 + y^2}, & (x, y) \neq (0, 0) \\ 0, & (x, y) = (0, 0) \end{cases}$

$g(x, y) = \begin{cases} \dfrac{x^4 - y^4}{x^2 + y^2}, & (x, y) \neq (0, 0) \\ 1, & (x, y) = (0, 0) \end{cases}$

44. $f(x, y) = \begin{cases} \dfrac{4x^4 - y^4}{2x^2 + y^2}, & (x, y) \neq (0, 0) \\ -1, & (x, y) = (0, 0) \end{cases}$

$g(x, y) = \begin{cases} \dfrac{4x^4 - y^4}{2x^2 + y^2}, & (x, y) \neq (0, 0) \\ 0, & (x, y) = (0, 0) \end{cases}$

45. $f(x, y) = \begin{cases} \dfrac{4x^2y^2}{x^2 + y^2}, & (x, y) \neq (0, 0) \\ 0, & (x, y) = (0, 0) \end{cases}$

$g(x, y) = \begin{cases} \dfrac{4x^2y^2}{x^2 + y^2}, & (x, y) \neq (0, 0) \\ 2, & (x, y) = (0, 0) \end{cases}$

46. $f(x, y) = \begin{cases} \dfrac{x^2 + 2xy^2 + y^2}{x^2 + y^2}, & (x, y) \neq (0, 0) \\ 0, & (x, y) = (0, 0) \end{cases}$

$g(x, y) = \begin{cases} \dfrac{x^2 + 2xy^2 + y^2}{x^2 + y^2}, & (x, y) \neq (0, 0) \\ 1, & (x, y) = (0, 0) \end{cases}$

CAS **In Exercises 47–52, use a computer algebra system to graph the function and find $\lim_{(x, y)\to(0, 0)} f(x, y)$ (if it exists).**

47. $f(x, y) = \sin x + \sin y$

48. $f(x, y) = \sin \dfrac{1}{x} + \cos \dfrac{1}{x}$

49. $f(x, y) = \dfrac{x^2y}{x^4 + 2y^2}$

50. $f(x, y) = \dfrac{x^2 + y^2}{x^2y}$

51. $f(x, y) = \dfrac{5xy}{x^2 + 2y^2}$

52. $f(x, y) = \dfrac{6xy}{x^2 + y^2 + 1}$

In Exercises 53–58, use polar coordinates to find the limit. [*Hint:* Let $x = r \cos \theta$ and $y = r \sin \theta$, and note that $(x, y) \to (0, 0)$ implies $r \to 0$.]

53. $\lim_{(x, y)\to(0, 0)} \dfrac{xy^2}{x^2 + y^2}$

54. $\lim_{(x, y)\to(0, 0)} \dfrac{x^3 + y^3}{x^2 + y^2}$

55. $\lim_{(x, y)\to(0, 0)} \dfrac{x^2y^2}{x^2 + y^2}$

56. $\lim_{(x, y)\to(0, 0)} \dfrac{x^2 - y^2}{\sqrt{x^2 + y^2}}$

57. $\lim_{(x, y)\to(0, 0)} \cos(x^2 + y^2)$

58. $\lim_{(x, y)\to(0, 0)} \sin\sqrt{x^2 + y^2}$

In Exercises 59–62, use polar coordinates and L'Hôpital's Rule to find the limit.

59. $\lim_{(x, y)\to(0, 0)} \dfrac{\sin\sqrt{x^2 + y^2}}{\sqrt{x^2 + y^2}}$

60. $\lim_{(x, y)\to(0, 0)} \dfrac{\sin(x^2 + y^2)}{x^2 + y^2}$

61. $\lim_{(x, y)\to(0, 0)} \dfrac{1 - \cos(x^2 + y^2)}{x^2 + y^2}$

62. $\lim_{(x, y)\to(0, 0)} (x^2 + y^2)\ln(x^2 + y^2)$

In Exercises 63–68, discuss the continuity of the function.

63. $f(x, y, z) = \dfrac{1}{\sqrt{x^2 + y^2 + z^2}}$

64. $f(x, y, z) = \dfrac{z}{x^2 + y^2 - 4}$

65. $f(x, y, z) = \dfrac{\sin z}{e^x + e^y}$

66. $f(x, y, z) = xy \sin z$

67. $f(x, y) = \begin{cases} \dfrac{\sin xy}{xy}, & xy \neq 0 \\ 1, & xy = 0 \end{cases}$

68. $f(x, y) = \begin{cases} \dfrac{\sin(x^2 - y^2)}{x^2 - y^2}, & x^2 \neq y^2 \\ 1, & x^2 = y^2 \end{cases}$

In Exercises 69–72, discuss the continuity of the composite function $f \circ g$.

69. $f(t) = t^2$

$g(x, y) = 2x - 3y$

70. $f(t) = \dfrac{1}{t}$

$g(x, y) = x^2 + y^2$

71. $f(t) = \dfrac{1}{t}$

$g(x, y) = 2x - 3y$

72. $f(t) = \dfrac{1}{1 - t}$

$g(x, y) = x^2 + y^2$

In Exercises 73–78, find each limit.

(a) $\lim_{\Delta x\to 0} \dfrac{f(x + \Delta x, y) - f(x, y)}{\Delta x}$

(b) $\lim_{\Delta y\to 0} \dfrac{f(x, y + \Delta y) - f(x, y)}{\Delta y}$

73. $f(x, y) = x^2 - 4y$

74. $f(x, y) = x^2 + y^2$

75. $f(x, y) = \dfrac{x}{y}$

76. $f(x, y) = \dfrac{1}{x + y}$

77. $f(x, y) = 3x + xy - 2y$

78. $f(x, y) = \sqrt{y}\,(y + 1)$

True or False? **In Exercises 79–82, determine whether the statement is true or false. If it is false, explain why or give an example that shows it is false.**

79. If $\lim_{(x, y)\to(0, 0)} f(x, y) = 0$, then $\lim_{x\to 0} f(x, 0) = 0$.

80. If $\lim_{(x, y)\to(0, 0)} f(0, y) = 0$, then $\lim_{(x, y)\to(0, 0)} f(x, y) = 0$.

81. If f is continuous for all nonzero x and y, and $f(0, 0) = 0$, then $\lim_{(x, y)\to(0, 0)} f(x, y) = 0$.

82. If g and h are continuous functions of x and y, and $f(x, y) = g(x) + h(y)$, then f is continuous.

83. Consider $\lim_{(x, y)\to(0, 0)} \dfrac{x^2 + y^2}{xy}$ (see figure).

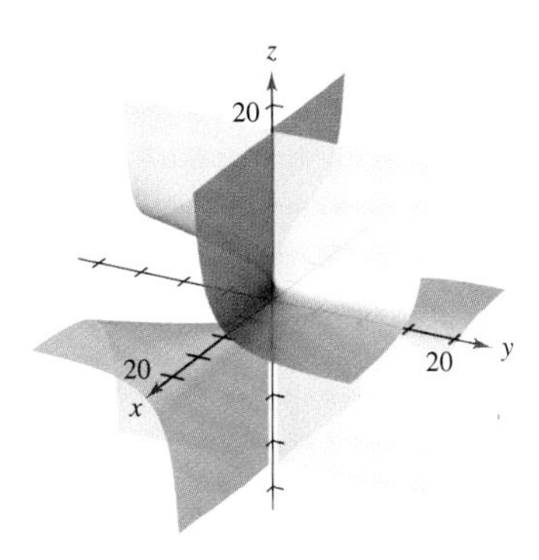

(a) Determine (if possible) the limit along any line of the form $y = ax$.

(b) Determine (if possible) the limit along the parabola $y = x^2$.

(c) Does the limit exist? Explain.

84. Consider $\lim_{(x, y)\to(0, 0)} \dfrac{x^2 y}{x^4 + y^2}$ (see figure).

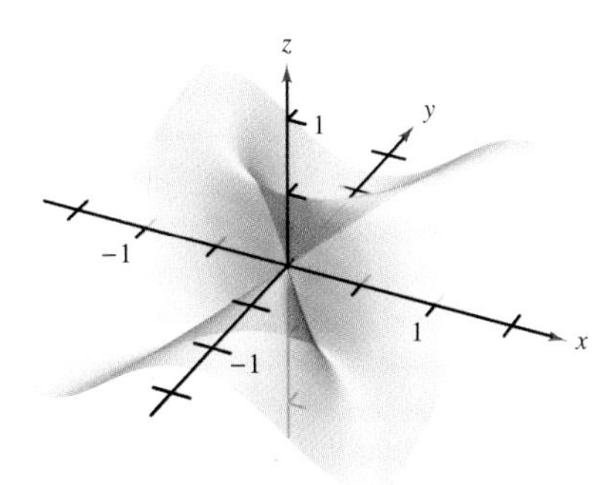

(a) Determine (if possible) the limit along any line of the form $y = ax$.

(b) Determine (if possible) the limit along the parabola $y = x^2$.

(c) Does the limit exist? Explain.

In Exercises 85 and 86, use spherical coordinates to find the limit. [*Hint:* Let $x = \rho \sin\phi \cos\theta$, $y = \rho \sin\phi \sin\theta$, and $z = \rho \cos\phi$, and note that $(x, y, z) \to (0, 0, 0)$ implies $\rho \to 0^+$.]

85. $\lim_{(x, y, z)\to(0, 0, 0)} \dfrac{xyz}{x^2 + y^2 + z^2}$

86. $\lim_{(x, y, z)\to(0, 0, 0)} \tan^{-1}\left[\dfrac{1}{x^2 + y^2 + z^2}\right]$

87. Find the following limit.

$$\lim_{(x, y)\to(0, 1)} \tan^{-1}\left[\frac{x^2 + 1}{x^2 + (y - 1)^2}\right]$$

88. For the function

$$f(x, y) = xy\left(\frac{x^2 - y^2}{x^2 + y^2}\right)$$

define $f(0, 0)$ such that f is continuous at the origin.

89. Prove that

$$\lim_{(x, y)\to(a, b)} [f(x, y) + g(x, y)] = L_1 + L_2$$

where $f(x, y)$ approaches L_1 and $g(x, y)$ approaches L_2 as $(x, y) \to (a, b)$.

90. Prove that if f is continuous and $f(a, b) < 0$, there exists a δ-neighborhood about (a, b) such that $f(x, y) < 0$ for every point (x, y) in the neighborhood.

WRITING ABOUT CONCEPTS

91. Define the limit of a function of two variables. Describe a method for showing that

$$\lim_{(x, y)\to(x_0, y_0)} f(x, y)$$

does not exist.

92. State the definition of continuity of a function of two variables.

93. Determine whether each of the following statements is true or false. Explain your reasoning.

(a) If $\lim_{(x, y)\to(2, 3)} f(x, y) = 4$, then $\lim_{x\to 2} f(x, 3) = 4$.

(b) If $\lim_{x\to 2} f(x, 3) = 4$, then $\lim_{(x, y)\to(2, 3)} f(x, y) = 4$.

(c) If $\lim_{x\to 2} f(x, 3) = \lim_{y\to 3} f(2, y) = 4$, then $\lim_{(x, y)\to(2, 3)} f(x, y) = 4$.

(d) If $\lim_{(x, y)\to(0, 0)} f(x, y) = 0$, then for any real number k, $\lim_{(x, y)\to(0, 0)} f(kx, y) = 0$.

CAPSTONE

94. (a) If $f(2, 3) = 4$, can you conclude anything about $\lim_{(x, y)\to(2, 3)} f(x, y)$? Give reasons for your answer.

(b) If $\lim_{(x, y)\to(2, 3)} f(x, y) = 4$, can you conclude anything about $f(2, 3)$? Give reasons for your answer.

13.3 Partial Derivatives

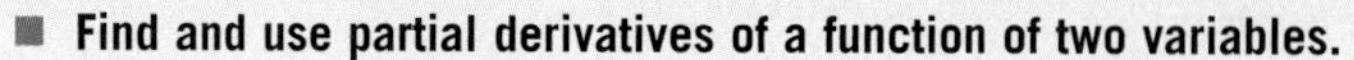

- Find and use partial derivatives of a function of two variables.
- Find and use partial derivatives of a function of three or more variables.
- Find higher-order partial derivatives of a function of two or three variables.

Partial Derivatives of a Function of Two Variables

Mary Evans Picture Library

JEAN LE ROND D'ALEMBERT (1717–1783)

The introduction of partial derivatives followed Newton's and Leibniz's work in calculus by several years. Between 1730 and 1760, Leonhard Euler and Jean Le Rond d'Alembert separately published several papers on dynamics, in which they established much of the theory of partial derivatives. These papers used functions of two or more variables to study problems involving equilibrium, fluid motion, and vibrating strings.

In applications of functions of several variables, the question often arises, "How will the value of a function be affected by a change in one of its independent variables?" You can answer this by considering the independent variables one at a time. For example, to determine the effect of a catalyst in an experiment, a chemist could conduct the experiment several times using varying amounts of the catalyst, while keeping constant other variables such as temperature and pressure. You can use a similar procedure to determine the rate of change of a function f with respect to one of its several independent variables. This process is called **partial differentiation,** and the result is referred to as the **partial derivative** of f with respect to the chosen independent variable.

DEFINITION OF PARTIAL DERIVATIVES OF A FUNCTION OF TWO VARIABLES

If $z = f(x, y)$, then the **first partial derivatives** of f with respect to x and y are the functions f_x and f_y defined by

$$f_x(x, y) = \lim_{\Delta x \to 0} \frac{f(x + \Delta x, y) - f(x, y)}{\Delta x}$$

$$f_y(x, y) = \lim_{\Delta y \to 0} \frac{f(x, y + \Delta y) - f(x, y)}{\Delta y}$$

provided the limits exist.

This definition indicates that if $z = f(x, y)$, then to find f_x you *consider y constant* and differentiate with respect to x. Similarly, to find f_y, you *consider x constant* and differentiate with respect to y.

EXAMPLE 1 Finding Partial Derivatives

Find the partial derivatives f_x and f_y for the function

$$f(x, y) = 3x - x^2y^2 + 2x^3y.$$

Solution Considering y to be constant and differentiating with respect to x produces

$$f(x, y) = 3x - x^2y^2 + 2x^3y \qquad \text{Write original function.}$$

$$f_x(x, y) = 3 - 2xy^2 + 6x^2y. \qquad \text{Partial derivative with respect to } x$$

Considering x to be constant and differentiating with respect to y produces

$$f(x, y) = 3x - x^2y^2 + 2x^3y \qquad \text{Write original function.}$$

$$f_y(x, y) = -2x^2y + 2x^3. \qquad \text{Partial derivative with respect to } y$$

■

NOTATION FOR FIRST PARTIAL DERIVATIVES

For $z = f(x, y)$, the partial derivatives f_x and f_y are denoted by

$$\frac{\partial}{\partial x}f(x, y) = f_x(x, y) = z_x = \frac{\partial z}{\partial x}$$

and

$$\frac{\partial}{\partial y}f(x, y) = f_y(x, y) = z_y = \frac{\partial z}{\partial y}.$$

The first partials evaluated at the point (a, b) are denoted by

$$\left.\frac{\partial z}{\partial x}\right|_{(a, b)} = f_x(a, b) \quad \text{and} \quad \left.\frac{\partial z}{\partial y}\right|_{(a, b)} = f_y(a, b).$$

EXAMPLE 2 Finding and Evaluating Partial Derivatives

For $f(x, y) = xe^{x^2y}$, find f_x and f_y, and evaluate each at the point $(1, \ln 2)$.

Solution Because

$$f_x(x, y) = xe^{x^2y}(2xy) + e^{x^2y} \qquad \text{Partial derivative with respect to } x$$

the partial derivative of f with respect to x at $(1, \ln 2)$ is

$$\begin{aligned} f_x(1, \ln 2) &= e^{\ln 2}(2 \ln 2) + e^{\ln 2} \\ &= 4 \ln 2 + 2. \end{aligned}$$

Because

$$\begin{aligned} f_y(x, y) &= xe^{x^2y}(x^2) \\ &= x^3e^{x^2y} \end{aligned} \qquad \text{Partial derivative with respect to } y$$

the partial derivative of f with respect to y at $(1, \ln 2)$ is

$$\begin{aligned} f_y(1, \ln 2) &= e^{\ln 2} \\ &= 2. \end{aligned}$$

■

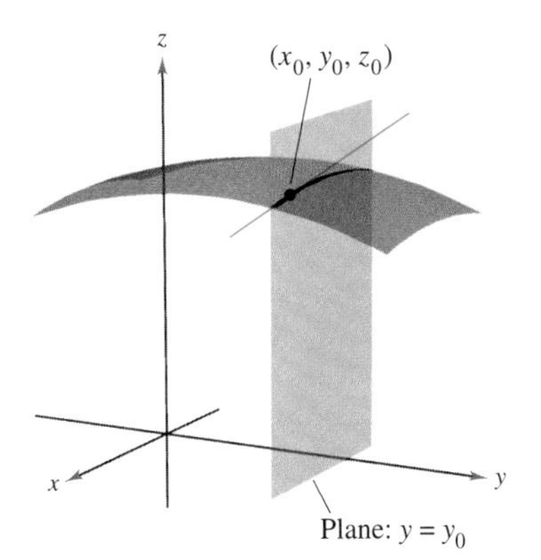

$\frac{\partial f}{\partial x}$ = slope in x-direction

Figure 13.29

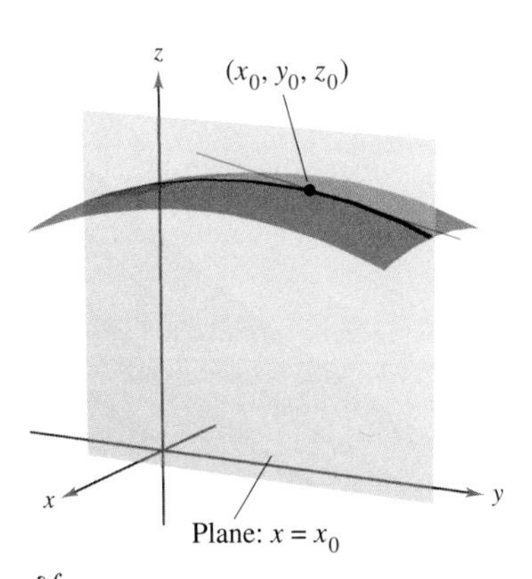

$\frac{\partial f}{\partial y}$ = slope in y-direction

Figure 13.30

The partial derivatives of a function of two variables, $z = f(x, y)$, have a useful geometric interpretation. If $y = y_0$, then $z = f(x, y_0)$ represents the curve formed by intersecting the surface $z = f(x, y)$ with the plane $y = y_0$, as shown in Figure 13.29. Therefore,

$$f_x(x_0, y_0) = \lim_{\Delta x \to 0} \frac{f(x_0 + \Delta x, y_0) - f(x_0, y_0)}{\Delta x}$$

represents the slope of this curve at the point $(x_0, y_0, f(x_0, y_0))$. Note that both the curve and the tangent line lie in the plane $y = y_0$. Similarly,

$$f_y(x_0, y_0) = \lim_{\Delta y \to 0} \frac{f(x_0, y_0 + \Delta y) - f(x_0, y_0)}{\Delta y}$$

represents the slope of the curve given by the intersection of $z = f(x, y)$ and the plane $x = x_0$ at $(x_0, y_0, f(x_0, y_0))$, as shown in Figure 13.30.

Informally, the values of $\partial f/\partial x$ and $\partial f/\partial y$ at the point (x_0, y_0, z_0) denote the **slopes of the surface in the x- and y-directions,** respectively.

EXAMPLE 3 Finding the Slopes of a Surface in the x- and y-Directions

Find the slopes in the x-direction and in the y-direction of the surface given by

$$f(x, y) = -\frac{x^2}{2} - y^2 + \frac{25}{8}$$

at the point $\left(\frac{1}{2}, 1, 2\right)$.

Solution The partial derivatives of f with respect to x and y are

$$f_x(x, y) = -x \quad \text{and} \quad f_y(x, y) = -2y. \qquad \text{Partial derivatives}$$

So, in the x-direction, the slope is

$$f_x\left(\frac{1}{2}, 1\right) = -\frac{1}{2} \qquad \text{Figure 13.31(a)}$$

and in the y-direction, the slope is

$$f_y\left(\frac{1}{2}, 1\right) = -2. \qquad \text{Figure 13.31(b)}$$

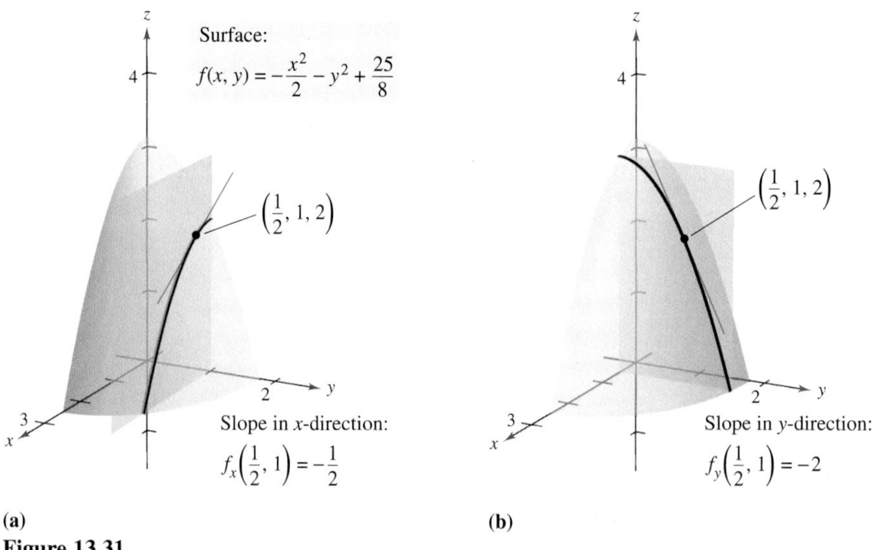

Figure 13.31

EXAMPLE 4 Finding the Slopes of a Surface in the x- and y-Directions

Find the slopes of the surface given by

$$f(x, y) = 1 - (x - 1)^2 - (y - 2)^2$$

at the point $(1, 2, 1)$ in the x-direction and in the y-direction.

Solution The partial derivatives of f with respect to x and y are

$$f_x(x, y) = -2(x - 1) \quad \text{and} \quad f_y(x, y) = -2(y - 2). \qquad \text{Partial derivatives}$$

So, at the point $(1, 2, 1)$, the slopes in the x- and y-directions are

$$f_x(1, 2) = -2(1 - 1) = 0 \quad \text{and} \quad f_y(1, 2) = -2(2 - 2) = 0$$

as shown in Figure 13.32. ■

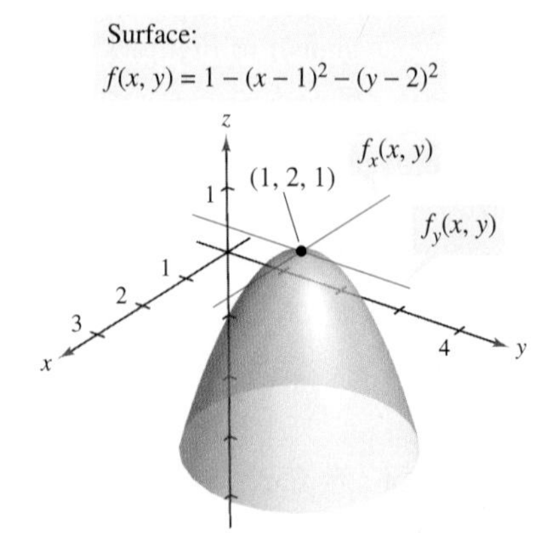

Figure 13.32

No matter how many variables are involved, partial derivatives can be interpreted as *rates of change*.

EXAMPLE 5 Using Partial Derivatives to Find Rates of Change

The area of a parallelogram with adjacent sides a and b and included angle θ is given by $A = ab \sin \theta$, as shown in Figure 13.33.

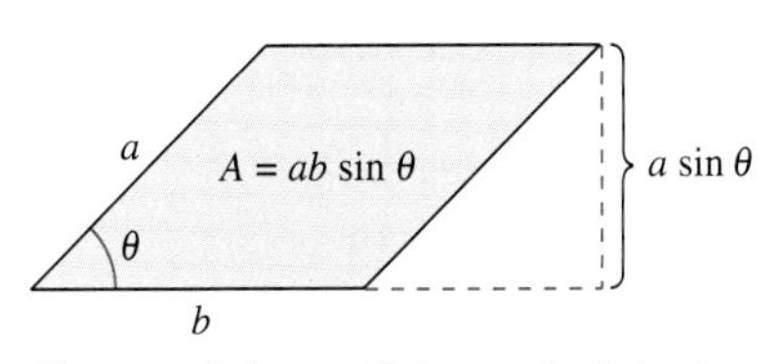

The area of the parallelogram is $ab \sin \theta$.
Figure 13.33

a. Find the rate of change of A with respect to a for $a = 10$, $b = 20$, and $\theta = \dfrac{\pi}{6}$.

b. Find the rate of change of A with respect to θ for $a = 10$, $b = 20$, and $\theta = \dfrac{\pi}{6}$.

Solution

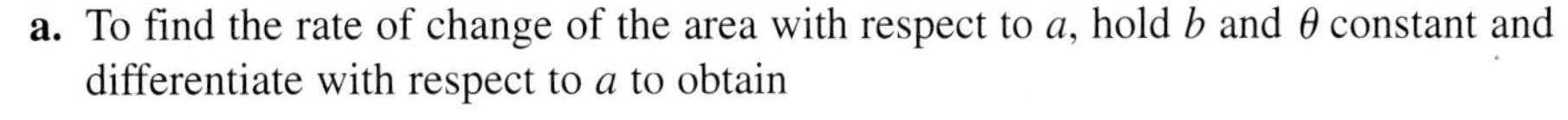

a. To find the rate of change of the area with respect to a, hold b and θ constant and differentiate with respect to a to obtain

$$\frac{\partial A}{\partial a} = b \sin \theta \qquad \text{Find partial with respect to } a.$$

$$\frac{\partial A}{\partial a} = 20 \sin \frac{\pi}{6} = 10. \qquad \text{Substitute for } b \text{ and } \theta.$$

b. To find the rate of change of the area with respect to θ, hold a and b constant and differentiate with respect to θ to obtain

$$\frac{\partial A}{\partial \theta} = ab \cos \theta \qquad \text{Find partial with respect to } \theta.$$

$$\frac{\partial A}{\partial \theta} = 200 \cos \frac{\pi}{6} = 100\sqrt{3}. \qquad \text{Substitute for } a, b, \text{ and } \theta.$$

■

Partial Derivatives of a Function of Three or More Variables

The concept of a partial derivative can be extended naturally to functions of three or more variables. For instance, if $w = f(x, y, z)$, there are three partial derivatives, each of which is formed by holding two of the variables constant. That is, to define the partial derivative of w with respect to x, consider y and z to be constant and differentiate with respect to x. A similar process is used to find the derivatives of w with respect to y and with respect to z.

$$\frac{\partial w}{\partial x} = f_x(x, y, z) = \lim_{\Delta x \to 0} \frac{f(x + \Delta x, y, z) - f(x, y, z)}{\Delta x}$$

$$\frac{\partial w}{\partial y} = f_y(x, y, z) = \lim_{\Delta y \to 0} \frac{f(x, y + \Delta y, z) - f(x, y, z)}{\Delta y}$$

$$\frac{\partial w}{\partial z} = f_z(x, y, z) = \lim_{\Delta z \to 0} \frac{f(x, y, z + \Delta z) - f(x, y, z)}{\Delta z}$$

In general, if $w = f(x_1, x_2, \ldots, x_n)$, there are n partial derivatives denoted by

$$\frac{\partial w}{\partial x_k} = f_{x_k}(x_1, x_2, \ldots, x_n), \quad k = 1, 2, \ldots, n.$$

To find the partial derivative with respect to one of the variables, hold the other variables constant and differentiate with respect to the given variable.

EXAMPLE 6 Finding Partial Derivatives

a. To find the partial derivative of $f(x, y, z) = xy + yz^2 + xz$ with respect to z, consider x and y to be constant and obtain

$$\frac{\partial}{\partial z}[xy + yz^2 + xz] = 2yz + x.$$

b. To find the partial derivative of $f(x, y, z) = z \sin(xy^2 + 2z)$ with respect to z, consider x and y to be constant. Then, using the Product Rule, you obtain

$$\begin{aligned}\frac{\partial}{\partial z}[z \sin(xy^2 + 2z)] &= (z)\frac{\partial}{\partial z}[\sin(xy^2 + 2z)] + \sin(xy^2 + 2z)\frac{\partial}{\partial z}[z]\\ &= (z)[\cos(xy^2 + 2z)](2) + \sin(xy^2 + 2z)\\ &= 2z\cos(xy^2 + 2z) + \sin(xy^2 + 2z).\end{aligned}$$

c. To find the partial derivative of $f(x, y, z, w) = (x + y + z)/w$ with respect to w, consider x, y, and z to be constant and obtain

$$\frac{\partial}{\partial w}\left[\frac{x + y + z}{w}\right] = -\frac{x + y + z}{w^2}.$$

■

Higher-Order Partial Derivatives

As is true for ordinary derivatives, it is possible to take second, third, and higher-order partial derivatives of a function of several variables, provided such derivatives exist. Higher-order derivatives are denoted by the order in which the differentiation occurs. For instance, the function $z = f(x, y)$ has the following second partial derivatives.

1. Differentiate twice with respect to x:

$$\frac{\partial}{\partial x}\left(\frac{\partial f}{\partial x}\right) = \frac{\partial^2 f}{\partial x^2} = f_{xx}.$$

2. Differentiate twice with respect to y:

$$\frac{\partial}{\partial y}\left(\frac{\partial f}{\partial y}\right) = \frac{\partial^2 f}{\partial y^2} = f_{yy}.$$

3. Differentiate first with respect to x and then with respect to y:

$$\frac{\partial}{\partial y}\left(\frac{\partial f}{\partial x}\right) = \frac{\partial^2 f}{\partial y \partial x} = f_{xy}.$$

4. Differentiate first with respect to y and then with respect to x:

$$\frac{\partial}{\partial x}\left(\frac{\partial f}{\partial y}\right) = \frac{\partial^2 f}{\partial x \partial y} = f_{yx}.$$

The third and fourth cases are called **mixed partial derivatives.**

NOTE Note that the two types of notation for mixed partials have different conventions for indicating the order of differentiation.

$$\frac{\partial}{\partial y}\left(\frac{\partial f}{\partial x}\right) = \frac{\partial^2 f}{\partial y \partial x} \qquad \text{Right-to-left order}$$

$$(f_x)_y = f_{xy} \qquad \text{Left-to-right order}$$

You can remember the order by observing that in both notations you differentiate first with respect to the variable "nearest" f.

EXAMPLE 7 Finding Second Partial Derivatives

Find the second partial derivatives of $f(x, y) = 3xy^2 - 2y + 5x^2y^2$, and determine the value of $f_{xy}(-1, 2)$.

Solution Begin by finding the first partial derivatives with respect to x and y.

$$f_x(x, y) = 3y^2 + 10xy^2 \quad \text{and} \quad f_y(x, y) = 6xy - 2 + 10x^2y$$

Then, differentiate each of these with respect to x and y.

$$f_{xx}(x, y) = 10y^2 \quad \text{and} \quad f_{yy}(x, y) = 6x + 10x^2$$

$$f_{xy}(x, y) = 6y + 20xy \quad \text{and} \quad f_{yx}(x, y) = 6y + 20xy$$

At $(-1, 2)$, the value of f_{xy} is $f_{xy}(-1, 2) = 12 - 40 = -28$. ■

NOTE Notice in Example 7 that the two mixed partials are equal. Sufficient conditions for this occurrence are given in Theorem 13.3. ■

THEOREM 13.3 EQUALITY OF MIXED PARTIAL DERIVATIVES

If f is a function of x and y such that f_{xy} and f_{yx} are continuous on an open disk R, then, for every (x, y) in R,

$$f_{xy}(x, y) = f_{yx}(x, y).$$

Theorem 13.3 also applies to a function f of *three or more variables* so long as all second partial derivatives are continuous. For example, if $w = f(x, y, z)$ and all the second partial derivatives are continuous in an open region R, then at each point in R the order of differentiation in the mixed second partial derivatives is irrelevant. If the third partial derivatives of f are also continuous, the order of differentiation of the mixed third partial derivatives is irrelevant.

EXAMPLE 8 Finding Higher-Order Partial Derivatives

Show that $f_{xz} = f_{zx}$ and $f_{xzz} = f_{zxz} = f_{zzx}$ for the function given by

$$f(x, y, z) = ye^x + x \ln z.$$

Solution

First partials:

$$f_x(x, y, z) = ye^x + \ln z, \quad f_z(x, y, z) = \frac{x}{z}$$

Second partials (note that the first two are equal):

$$f_{xz}(x, y, z) = \frac{1}{z}, \quad f_{zx}(x, y, z) = \frac{1}{z}, \quad f_{zz}(x, y, z) = -\frac{x}{z^2}$$

Third partials (note that all three are equal):

$$f_{xzz}(x, y, z) = -\frac{1}{z^2}, \quad f_{zxz}(x, y, z) = -\frac{1}{z^2}, \quad f_{zzx}(x, y, z) = -\frac{1}{z^2}$$ ■

13.3 Exercises

See www.CalcChat.com for worked-out solutions to odd-numbered exercises.

Think About It **In Exercises 1–4, use the graph of the surface to determine the sign of the indicated partial derivative.**

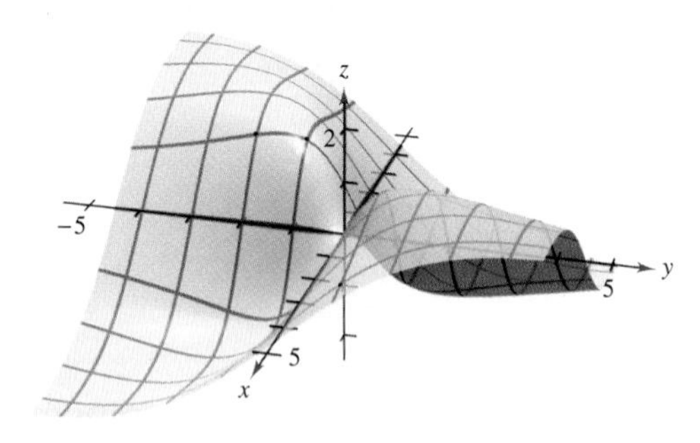

1. $f_x(4, 1)$
2. $f_y(-1, -2)$
3. $f_y(4, 1)$
4. $f_x(-1, -1)$

In Exercises 5–8, explain whether the Quotient Rule should be used to find the partial derivative. Do not differentiate.

5. $\dfrac{\partial}{\partial y}\left(\dfrac{x - y}{x^2 + 1}\right)$
6. $\dfrac{\partial}{\partial x}\left(\dfrac{x - y}{x^2 + 1}\right)$
7. $\dfrac{\partial}{\partial x}\left(\dfrac{xy}{x^2 + 1}\right)$
8. $\dfrac{\partial}{\partial y}\left(\dfrac{xy}{x^2 + 1}\right)$

In Exercises 9–40, find both first partial derivatives.

9. $f(x, y) = 2x - 5y + 3$
10. $f(x, y) = x^2 - 2y^2 + 4$
11. $f(x, y) = x^2y^3$
12. $f(x, y) = 4x^3y^{-2}$
13. $z = x\sqrt{y}$
14. $z = 2y^2\sqrt{x}$
15. $z = x^2 - 4xy + 3y^2$
16. $z = y^3 - 2xy^2 - 1$
17. $z = e^{xy}$
18. $z = e^{x/y}$
19. $z = x^2e^{2y}$
20. $z = ye^{y/x}$
21. $z = \ln \dfrac{x}{y}$
22. $z = \ln \sqrt{xy}$
23. $z = \ln(x^2 + y^2)$
24. $z = \ln \dfrac{x + y}{x - y}$
25. $z = \dfrac{x^2}{2y} + \dfrac{3y^2}{x}$
26. $z = \dfrac{xy}{x^2 + y^2}$
27. $h(x, y) = e^{-(x^2+y^2)}$
28. $g(x, y) = \ln \sqrt{x^2 + y^2}$
29. $f(x, y) = \sqrt{x^2 + y^2}$
30. $f(x, y) = \sqrt{2x + y^3}$
31. $z = \cos xy$
32. $z = \sin(x + 2y)$
33. $z = \tan(2x - y)$
34. $z = \sin 5x \cos 5y$
35. $z = e^y \sin xy$
36. $z = \cos(x^2 + y^2)$
37. $z = \sinh(2x + 3y)$
38. $z = \cosh xy^2$
39. $f(x, y) = \displaystyle\int_x^y (t^2 - 1)\, dt$
40. $f(x, y) = \displaystyle\int_x^y (2t + 1)\, dt + \int_y^x (2t - 1)\, dt$

In Exercises 41–44, use the limit definition of partial derivatives to find $f_x(x, y)$ and $f_y(x, y)$.

41. $f(x, y) = 3x + 2y$
42. $f(x, y) = x^2 - 2xy + y^2$
43. $f(x, y) = \sqrt{x + y}$
44. $f(x, y) = \dfrac{1}{x + y}$

In Exercises 45–52, evaluate f_x and f_y at the given point.

45. $f(x, y) = e^y \sin x,\quad (\pi, 0)$
46. $f(x, y) = e^{-x} \cos y,\quad (0, 0)$
47. $f(x, y) = \cos(2x - y),\quad \left(\dfrac{\pi}{4}, \dfrac{\pi}{3}\right)$
48. $f(x, y) = \sin xy,\quad \left(2, \dfrac{\pi}{4}\right)$
49. $f(x, y) = \arctan \dfrac{y}{x},\quad (2, -2)$
50. $f(x, y) = \arccos xy,\quad (1, 1)$
51. $f(x, y) = \dfrac{xy}{x - y},\quad (2, -2)$
52. $f(x, y) = \dfrac{2xy}{\sqrt{4x^2 + 5y^2}},\quad (1, 1)$

In Exercises 53 and 54, find the slopes of the surface in the x- and y-directions at the given point.

53. $g(x, y) = 4 - x^2 - y^2$
$(1, 1, 2)$
54. $h(x, y) = x^2 - y^2$
$(-2, 1, 3)$

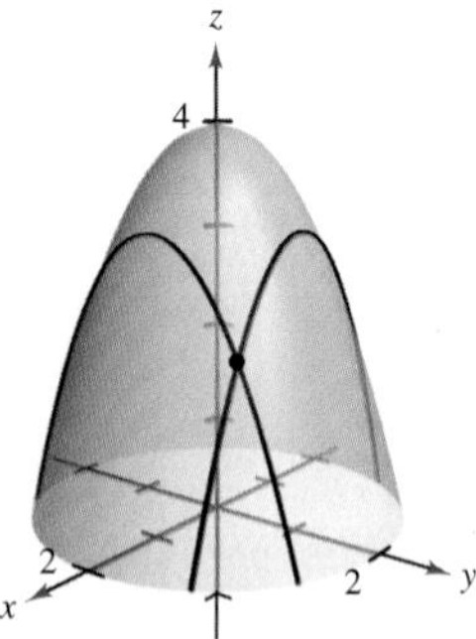

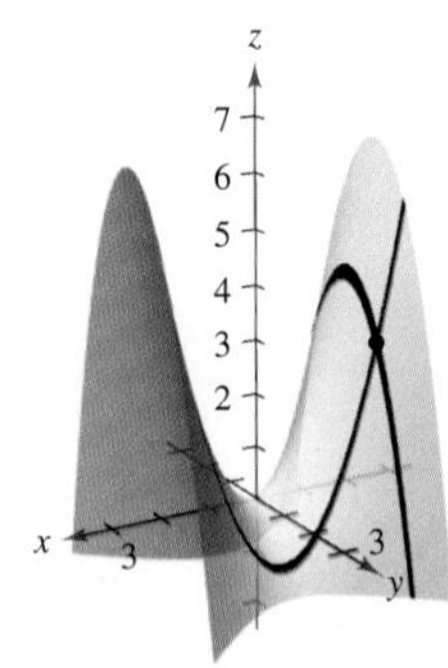

CAS In Exercises 55–58, use a computer algebra system to graph the curve formed by the intersection of the surface and the plane. Find the slope of the curve at the given point.

	Surface	*Plane*	*Point*
55.	$z = \sqrt{49 - x^2 - y^2}$	$x = 2$	$(2, 3, 6)$
56.	$z = x^2 + 4y^2$	$y = 1$	$(2, 1, 8)$
57.	$z = 9x^2 - y^2$	$y = 3$	$(1, 3, 0)$
58.	$z = 9x^2 - y^2$	$x = 1$	$(1, 3, 0)$

In Exercises 59–64, find the first partial derivatives with respect to x, y, and z.

59. $H(x, y, z) = \sin(x + 2y + 3z)$

60. $f(x, y, z) = 3x^2y - 5xyz + 10yz^2$

61. $w = \sqrt{x^2 + y^2 + z^2}$

62. $w = \dfrac{7xz}{x + y}$

63. $F(x, y, z) = \ln\sqrt{x^2 + y^2 + z^2}$

64. $G(x, y, z) = \dfrac{1}{\sqrt{1 - x^2 - y^2 - z^2}}$

In Exercises 65–70, evaluate f_x, f_y, and f_z at the given point.

65. $f(x, y, z) = x^3yz^2, \quad (1, 1, 1)$

66. $f(x, y, z) = x^2y^3 + 2xyz - 3yz, \quad (-2, 1, 2)$

67. $f(x, y, z) = \dfrac{x}{yz}, \quad (1, -1, -1)$

68. $f(x, y, z) = \dfrac{xy}{x + y + z}, \quad (3, 1, -1)$

69. $f(x, y, z) = z\sin(x + y), \quad \left(0, \dfrac{\pi}{2}, -4\right)$

70. $f(x, y, z) = \sqrt{3x^2 + y^2 - 2z^2}, \quad (1, -2, 1)$

In Exercises 71–80, find the four second partial derivatives. Observe that the second mixed partials are equal.

71. $z = 3xy^2$

72. $z = x^2 + 3y^2$

73. $z = x^2 - 2xy + 3y^2$

74. $z = x^4 - 3x^2y^2 + y^4$

75. $z = \sqrt{x^2 + y^2}$

76. $z = \ln(x - y)$

77. $z = e^x \tan y$

78. $z = 2xe^y - 3ye^{-x}$

79. $z = \cos xy$

80. $z = \arctan \dfrac{y}{x}$

In Exercises 81–88, for $f(x, y)$, find all values of x and y such that $f_x(x, y) = 0$ and $f_y(x, y) = 0$ simultaneously.

81. $f(x, y) = x^2 + xy + y^2 - 2x + 2y$

82. $f(x, y) = x^2 - xy + y^2 - 5x + y$

83. $f(x, y) = x^2 + 4xy + y^2 - 4x + 16y + 3$

84. $f(x, y) = x^2 - xy + y^2$

85. $f(x, y) = \dfrac{1}{x} + \dfrac{1}{y} + xy$

86. $f(x, y) = 3x^3 - 12xy + y^3$

87. $f(x, y) = e^{x^2 + xy + y^2}$

88. $f(x, y) = \ln(x^2 + y^2 + 1)$

CAS **In Exercises 89–92, use a computer algebra system to find the first and second partial derivatives of the function. Determine whether there exist values of x and y such that $f_x(x, y) = 0$ and $f_y(x, y) = 0$ simultaneously.**

89. $f(x, y) = x\sec y$

90. $f(x, y) = \sqrt{25 - x^2 - y^2}$

91. $f(x, y) = \ln\dfrac{x}{x^2 + y^2}$

92. $f(x, y) = \dfrac{xy}{x - y}$

In Exercises 93–96, show that the mixed partial derivatives f_{xyy}, f_{yxy}, and f_{yyx} are equal.

93. $f(x, y, z) = xyz$

94. $f(x, y, z) = x^2 - 3xy + 4yz + z^3$

95. $f(x, y, z) = e^{-x}\sin yz$

96. $f(x, y, z) = \dfrac{2z}{x + y}$

Laplace's Equation **In Exercises 97–100, show that the function satisfies Laplace's equation $\partial^2 z/\partial x^2 + \partial^2 z/\partial y^2 = 0$.**

97. $z = 5xy$

98. $z = \frac{1}{2}(e^y - e^{-y})\sin x$

99. $z = e^x \sin y$

100. $z = \arctan \dfrac{y}{x}$

Wave Equation **In Exercises 101–104, show that the function satisfies the wave equation $\partial^2 z/\partial t^2 = c^2(\partial^2 z/\partial x^2)$.**

101. $z = \sin(x - ct)$

102. $z = \cos(4x + 4ct)$

103. $z = \ln(x + ct)$

104. $z = \sin \omega ct \sin \omega x$

Heat Equation **In Exercises 105 and 106, show that the function satisfies the heat equation $\partial z/\partial t = c^2(\partial^2 z/\partial x^2)$.**

105. $z = e^{-t}\cos\dfrac{x}{c}$

106. $z = e^{-t}\sin\dfrac{x}{c}$

In Exercises 107 and 108, determine whether there exists a function $f(x, y)$ with the given partial derivatives. Explain your reasoning. If such a function exists, give an example.

107. $f_x(x, y) = -3\sin(3x - 2y), f_y(x, y) = 2\sin(3x - 2y)$

108. $f_x(x, y) = 2x + y, f_y(x, y) = x - 4y$

In Exercises 109 and 110, find the first partial derivative with respect to x.

109. $f(x, y, z) = (\tan y^2z)e^{z^2 + y^{-2}\sqrt{z}}$

110. $f(x, y, z) = x\left(\sinh\dfrac{y}{z}\right)(y^2 - 2\sqrt{y - 1})z$

WRITING ABOUT CONCEPTS

111. Let f be a function of two variables x and y. Describe the procedure for finding the first partial derivatives.

112. Sketch a surface representing a function f of two variables x and y. Use the sketch to give geometric interpretations of $\partial f/\partial x$ and $\partial f/\partial y$.

113. Sketch the graph of a function $z = f(x, y)$ whose derivative f_x is always negative and whose derivative f_y is always positive.

114. Sketch the graph of a function $z = f(x, y)$ whose derivatives f_x and f_y are always positive.

115. If f is a function of x and y such that f_{xy} and f_{yx} are continuous, what is the relationship between the mixed partial derivatives? Explain.

CAPSTONE

116. Find the four second partial derivatives of the function given by $f(x, y) = \sin(x - 2y)$. Show that the second mixed partial derivatives f_{xy} and f_{yx} are equal.

117. ***Marginal Revenue*** A pharmaceutical corporation has two plants that produce the same over-the-counter medicine. If x_1 and x_2 are the numbers of units produced at plant 1 and plant 2, respectively, then the total revenue for the product is given by $R = 200x_1 + 200x_2 - 4x_1^2 - 8x_1x_2 - 4x_2^2$. When $x_1 = 4$ and $x_2 = 12$, find (a) the marginal revenue for plant 1, $\partial R/\partial x_1$, and (b) the marginal revenue for plant 2, $\partial R/\partial x_2$.

118. ***Marginal Costs*** A company manufactures two types of wood-burning stoves: a freestanding model and a fireplace-insert model. The cost function for producing x freestanding and y fireplace-insert stoves is

$$C = 32\sqrt{xy} + 175x + 205y + 1050.$$

(a) Find the marginal costs ($\partial C/\partial x$ and $\partial C/\partial y$) when $x = 80$ and $y = 20$.

(b) When additional production is required, which model of stove results in the cost increasing at a higher rate? How can this be determined from the cost model?

119. ***Psychology*** Early in the twentieth century, an intelligence test called the *Stanford-Binet Test* (more commonly known as the *IQ test*) was developed. In this test, an individual's mental age M is divided by the individual's chronological age C and the quotient is multiplied by 100. The result is the individual's IQ.

$$IQ(M, C) = \frac{M}{C} \times 100$$

Find the partial derivatives of IQ with respect to M and with respect to C. Evaluate the partial derivatives at the point $(12, 10)$ and interpret the result. *(Source: Adapted from Bernstein/Clark-Stewart/Roy/Wickens, Psychology, Fourth Edition)*

120. ***Marginal Productivity*** Consider the Cobb-Douglas production function $f(x, y) = 200x^{0.7}y^{0.3}$. When $x = 1000$ and $y = 500$, find (a) the marginal productivity of labor, $\partial f/\partial x$, and (b) the marginal productivity of capital, $\partial f/\partial y$.

121. ***Think About It*** Let N be the number of applicants to a university, p the charge for food and housing at the university, and t the tuition. N is a function of p and t such that $\partial N/\partial p < 0$ and $\partial N/\partial t < 0$. What information is gained by noticing that both partials are negative?

122. ***Investment*** The value of an investment of \$1000 earning 6% compounded annually is

$$V(I, R) = 1000\left[\frac{1 + 0.06(1 - R)}{1 + I}\right]^{10}$$

where I is the annual rate of inflation and R is the tax rate for the person making the investment. Calculate $V_I(0.03, 0.28)$ and $V_R(0.03, 0.28)$. Determine whether the tax rate or the rate of inflation is the greater "negative" factor in the growth of the investment.

123. ***Temperature Distribution*** The temperature at any point (x, y) in a steel plate is $T = 500 - 0.6x^2 - 1.5y^2$, where x and y are measured in meters. At the point $(2, 3)$, find the rates of change of the temperature with respect to the distances moved along the plate in the directions of the x- and y-axes.

124. ***Apparent Temperature*** A measure of how hot weather feels to an average person is the Apparent Temperature Index. A model for this index is

$$A = 0.885t - 22.4h + 1.20th - 0.544$$

where A is the apparent temperature in degrees Celsius, t is the air temperature, and h is the relative humidity in decimal form. *(Source: The UMAP Journal)*

(a) Find $\partial A/\partial t$ and $\partial A/\partial h$ when $t = 30°$ and $h = 0.80$.

(b) Which has a greater effect on A, air temperature or humidity? Explain.

125. ***Ideal Gas Law*** The Ideal Gas Law states that $PV = nRT$, where P is pressure, V is volume, n is the number of moles of gas, R is a fixed constant (the gas constant), and T is absolute temperature. Show that

$$\frac{\partial T}{\partial P}\frac{\partial P}{\partial V}\frac{\partial V}{\partial T} = -1.$$

126. ***Marginal Utility*** The utility function $U = f(x, y)$ is a measure of the utility (or satisfaction) derived by a person from the consumption of two products x and y. Suppose the utility function is $U = -5x^2 + xy - 3y^2$.

(a) Determine the marginal utility of product x.

(b) Determine the marginal utility of product y.

(c) When $x = 2$ and $y = 3$, should a person consume one more unit of product x or one more unit of product y? Explain your reasoning.

CAS (d) Use a computer algebra system to graph the function. Interpret the marginal utilities of products x and y graphically.

127. ***Modeling Data*** Per capita consumptions (in gallons) of different types of milk in the United States from 1999 through 2005 are shown in the table. Consumption of flavored milk, plain reduced-fat milk, and plain light and skim milks are represented by the variables x, y, and z, respectively. *(Source: U.S. Department of Agriculture)*

Year	1999	2000	2001	2002	2003	2004	2005
x	1.4	1.4	1.4	1.6	1.6	1.7	1.7
y	7.3	7.1	7.0	7.0	6.9	6.9	6.9
z	6.2	6.1	5.9	5.8	5.6	5.5	5.6

A model for the data is given by $z = -0.92x + 1.03y + 0.02$.

(a) Find $\dfrac{\partial z}{\partial x}$ and $\dfrac{\partial z}{\partial y}$.

(b) Interpret the partial derivatives in the context of the problem.

128. *Modeling Data* The table shows the public medical expenditures (in billions of dollars) for worker's compensation x, public assistance y, and Medicare z from 2000 through 2005. *(Source: Centers for Medicare and Medicaid Services)*

Year	2000	2001	2002	2003	2004	2005
x	24.9	28.1	30.1	31.4	32.1	33.5
y	207.5	233.2	258.4	281.9	303.2	324.9
z	224.3	247.7	265.7	283.5	312.8	342.0

A model for the data is given by

$$z = -1.2225x^2 + 0.0096y^2 + 71.381x - 4.121y - 354.65.$$

(a) Find $\dfrac{\partial^2 z}{\partial x^2}$ and $\dfrac{\partial^2 z}{\partial y^2}$.

(b) Determine the concavity of traces parallel to the xz-plane. Interpret the result in the context of the problem.

(c) Determine the concavity of traces parallel to the yz-plane. Interpret the result in the context of the problem.

True or False? **In Exercises 129–132, determine whether the statement is true or false. If it is false, explain why or give an example that shows it is false.**

129. If $z = f(x, y)$ and $\partial z/\partial x = \partial z/\partial y$, then $z = c(x + y)$.

130. If $z = f(x)g(y)$, then $(\partial z/\partial x) + (\partial z/\partial y) = f'(x)g(y) + f(x)g'(y)$.

131. If $z = e^{xy}$, then $\dfrac{\partial^2 z}{\partial y \partial x} = (xy + 1)e^{xy}$.

132. If a cylindrical surface $z = f(x, y)$ has rulings parallel to the y-axis, then $\partial z/\partial y = 0$.

133. Consider the function defined by

$$f(x, y) = \begin{cases} \dfrac{xy(x^2 - y^2)}{x^2 + y^2}, & (x, y) \neq (0, 0) \\ 0, & (x, y) = (0, 0) \end{cases}.$$

(a) Find $f_x(x, y)$ and $f_y(x, y)$ for $(x, y) \neq (0, 0)$.

(b) Use the definition of partial derivatives to find $f_x(0, 0)$ and $f_y(0, 0)$.

$$\left[\text{Hint: } f_x(0, 0) = \lim_{\Delta x \to 0} \frac{f(\Delta x, 0) - f(0, 0)}{\Delta x}.\right]$$

(c) Use the definition of partial derivatives to find $f_{xy}(0, 0)$ and $f_{yx}(0, 0)$.

(d) Using Theorem 13.3 and the result of part (c), what can be said about f_{xy} or f_{yx}?

134. Let $f(x, y) = \displaystyle\int_x^y \sqrt{1 + t^3}\, dt$. Find $f_x(x, y)$ and $f_y(x, y)$.

135. Consider the function $f(x, y) = (x^3 + y^3)^{1/3}$.

(a) Find $f_x(0, 0)$ and $f_y(0, 0)$.

(b) Determine the points (if any) at which $f_x(x, y)$ or $f_y(x, y)$ fails to exist.

136. Consider the function $f(x, y) = (x^2 + y^2)^{2/3}$. Show that

$$f_x(x, y) = \begin{cases} \dfrac{4x}{3(x^2 + y^2)^{1/3}}, & (x, y) \neq (0, 0) \\ 0, & (x, y) = (0, 0) \end{cases}.$$

FOR FURTHER INFORMATION For more information about this problem, see the article "A Classroom Note on a Naturally Occurring Piecewise Defined Function" by Don Cohen in *Mathematics and Computer Education.*

SECTION PROJECT

Moiré Fringes

Read the article "Moiré Fringes and the Conic Sections" by Mike Cullen in *The College Mathematics Journal*. The article describes how two families of level curves given by

$$f(x, y) = a \quad \text{and} \quad g(x, y) = b$$

can form Moiré patterns. After reading the article, write a paper explaining how the expression

$$\frac{\partial f}{\partial x} \cdot \frac{\partial g}{\partial x} + \frac{\partial f}{\partial y} \cdot \frac{\partial g}{\partial y}$$

is related to the Moiré patterns formed by intersecting the two families of level curves. Use one of the following patterns as an example in your paper.

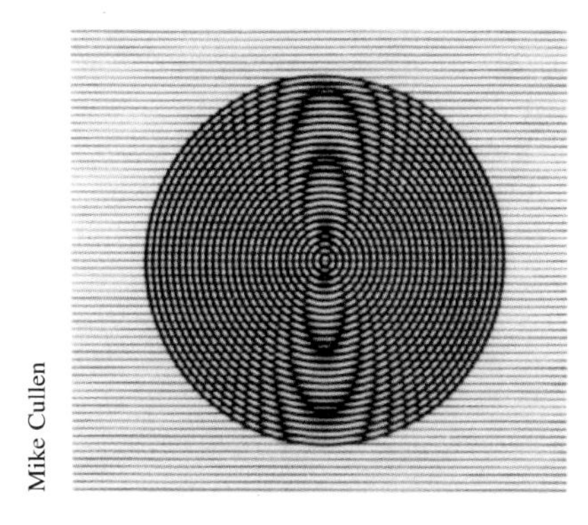

Mike Cullen

Mike Cullen

13.4 Differentials

- **Understand the concepts of increments and differentials.**
- **Extend the concept of differentiability to a function of two variables.**
- **Use a differential as an approximation.**

Increments and Differentials

In this section, the concepts of increments and differentials are generalized to functions of two or more variables. Recall from Section 3.9 that for $y = f(x)$, the differential of y was defined as

$$dy = f'(x)\,dx.$$

Similar terminology is used for a function of two variables, $z = f(x, y)$. That is, Δx and Δy are the **increments of x and y,** and the **increment of z** is given by

$$\Delta z = f(x + \Delta x, y + \Delta y) - f(x, y). \qquad \text{Increment of } z$$

DEFINITION OF TOTAL DIFFERENTIAL

If $z = f(x, y)$ and Δx and Δy are increments of x and y, then the **differentials** of the independent variables x and y are

$$dx = \Delta x \qquad \text{and} \qquad dy = \Delta y$$

and the **total differential** of the dependent variable z is

$$dz = \frac{\partial z}{\partial x}\,dx + \frac{\partial z}{\partial y}\,dy = f_x(x, y)\,dx + f_y(x, y)\,dy.$$

This definition can be extended to a function of three or more variables. For instance, if $w = f(x, y, z, u)$, then $dx = \Delta x$, $dy = \Delta y$, $dz = \Delta z$, $du = \Delta u$, and the total differential of w is

$$dw = \frac{\partial w}{\partial x}\,dx + \frac{\partial w}{\partial y}\,dy + \frac{\partial w}{\partial z}\,dz + \frac{\partial w}{\partial u}\,du.$$

EXAMPLE 1 Finding the Total Differential

Find the total differential for each function.

a. $z = 2x \sin y - 3x^2y^2$ **b.** $w = x^2 + y^2 + z^2$

Solution

a. The total differential dz for $z = 2x \sin y - 3x^2y^2$ is

$$dz = \frac{\partial z}{\partial x}\,dx + \frac{\partial z}{\partial y}\,dy \qquad \text{Total differential } dz$$
$$= (2 \sin y - 6xy^2)\,dx + (2x \cos y - 6x^2y)\,dy.$$

b. The total differential dw for $w = x^2 + y^2 + z^2$ is

$$dw = \frac{\partial w}{\partial x}\,dx + \frac{\partial w}{\partial y}\,dy + \frac{\partial w}{\partial z}\,dz \qquad \text{Total differential } dw$$
$$= 2x\,dx + 2y\,dy + 2z\,dz.$$

■

Differentiability

In Section 3.9, you learned that for a *differentiable* function given by $y = f(x)$, you can use the differential $dy = f'(x)\,dx$ as an approximation (for small Δx) to the value $\Delta y = f(x + \Delta x) - f(x)$. When a similar approximation is possible for a function of two variables, the function is said to be **differentiable.** This is stated explicitly in the following definition.

DEFINITION OF DIFFERENTIABILITY

A function f given by $z = f(x, y)$ is **differentiable** at (x_0, y_0) if Δz can be written in the form

$$\Delta z = f_x(x_0, y_0)\,\Delta x + f_y(x_0, y_0)\,\Delta y + \varepsilon_1 \Delta x + \varepsilon_2 \Delta y$$

where both ε_1 and $\varepsilon_2 \to 0$ as $(\Delta x, \Delta y) \to (0, 0)$. The function f is **differentiable in a region R** if it is differentiable at each point in R.

EXAMPLE 2 Showing That a Function Is Differentiable

Show that the function given by

$$f(x, y) = x^2 + 3y$$

is differentiable at every point in the plane.

Solution Letting $z = f(x, y)$, the increment of z at an arbitrary point (x, y) in the plane is

$$\begin{aligned}
\Delta z &= f(x + \Delta x, y + \Delta y) - f(x, y) && \text{Increment of } z\\
&= (x^2 + 2x\Delta x + \Delta x^2) + 3(y + \Delta y) - (x^2 + 3y)\\
&= 2x\Delta x + \Delta x^2 + 3\Delta y\\
&= 2x(\Delta x) + 3(\Delta y) + \Delta x(\Delta x) + 0(\Delta y)\\
&= f_x(x, y)\,\Delta x + f_y(x, y)\,\Delta y + \varepsilon_1 \Delta x + \varepsilon_2 \Delta y
\end{aligned}$$

where $\varepsilon_1 = \Delta x$ and $\varepsilon_2 = 0$. Because $\varepsilon_1 \to 0$ and $\varepsilon_2 \to 0$ as $(\Delta x, \Delta y) \to (0, 0)$, it follows that f is differentiable at every point in the plane. The graph of f is shown in Figure 13.34. ■

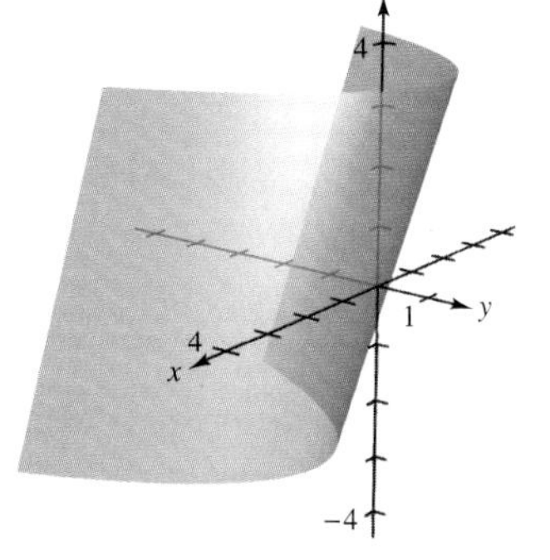

Figure 13.34

Be sure you see that the term "differentiable" is used differently for functions of two variables than for functions of one variable. A function of one variable is differentiable at a point if its derivative exists at the point. However, for a function of two variables, the existence of the partial derivatives f_x and f_y does not guarantee that the function is differentiable (see Example 5). The following theorem gives a *sufficient* condition for differentiability of a function of two variables. A proof of Theorem 13.4 is given in Appendix A.

THEOREM 13.4 SUFFICIENT CONDITION FOR DIFFERENTIABILITY

If f is a function of x and y, where f_x and f_y are continuous in an open region R, then f is differentiable on R.

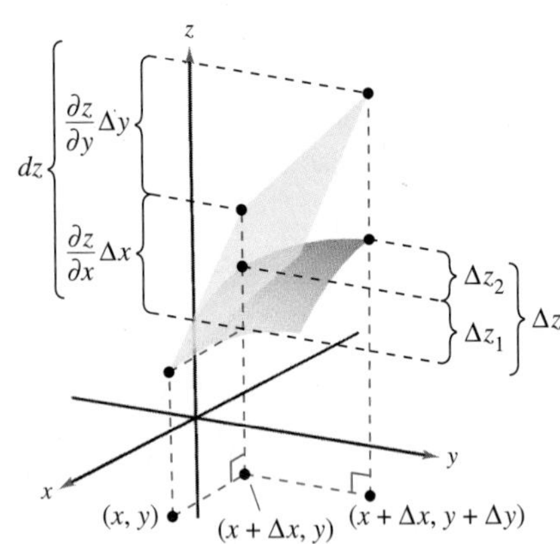

The exact change in z is Δz. This change can be approximated by the differential dz.
Figure 13.35

Approximation by Differentials

Theorem 13.4 tells you that you can choose $(x + \Delta x, y + \Delta y)$ close enough to (x, y) to make $\varepsilon_1 \Delta x$ and $\varepsilon_2 \Delta y$ insignificant. In other words, for small Δx and Δy, you can use the approximation

$$\Delta z \approx dz.$$

This approximation is illustrated graphically in Figure 13.35. Recall that the partial derivatives $\partial z/\partial x$ and $\partial z/\partial y$ can be interpreted as the slopes of the surface in the x- and y-directions. This means that

$$dz = \frac{\partial z}{\partial x}\Delta x + \frac{\partial z}{\partial y}\Delta y$$

represents the change in height of a plane that is tangent to the surface at the point $(x, y, f(x, y))$. Because a plane in space is represented by a linear equation in the variables x, y, and z, the approximation of Δz by dz is called a **linear approximation.** You will learn more about this geometric interpretation in Section 13.7.

EXAMPLE 3 Using a Differential as an Approximation

Use the differential dz to approximate the change in $z = \sqrt{4 - x^2 - y^2}$ as (x, y) moves from the point $(1, 1)$ to the point $(1.01, 0.97)$. Compare this approximation with the exact change in z.

Solution Letting $(x, y) = (1, 1)$ and $(x + \Delta x, y + \Delta y) = (1.01, 0.97)$ produces $dx = \Delta x = 0.01$ and $dy = \Delta y = -0.03$. So, the change in z can be approximated by

$$\Delta z \approx dz = \frac{\partial z}{\partial x}dx + \frac{\partial z}{\partial y}dy = \frac{-x}{\sqrt{4 - x^2 - y^2}}\Delta x + \frac{-y}{\sqrt{4 - x^2 - y^2}}\Delta y.$$

When $x = 1$ and $y = 1$, you have

$$\Delta z \approx -\frac{1}{\sqrt{2}}(0.01) - \frac{1}{\sqrt{2}}(-0.03) = \frac{0.02}{\sqrt{2}} = \sqrt{2}\,(0.01) \approx 0.0141.$$

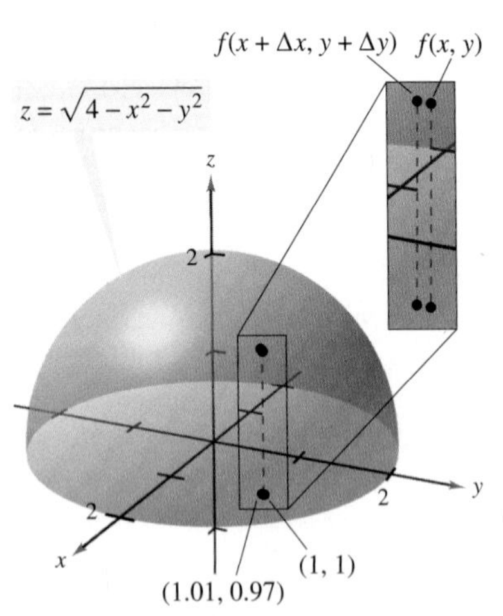

As (x, y) moves from $(1, 1)$ to the point $(1.01, 0.97)$, the value of $f(x, y)$ changes by about 0.0137.
Figure 13.36

In Figure 13.36, you can see that the exact change corresponds to the difference in the heights of two points on the surface of a hemisphere. This difference is given by

$$\begin{aligned}\Delta z &= f(1.01, 0.97) - f(1, 1)\\ &= \sqrt{4 - (1.01)^2 - (0.97)^2} - \sqrt{4 - 1^2 - 1^2} \approx 0.0137.\end{aligned}$$

■

A function of three variables $w = f(x, y, z)$ is called **differentiable** at (x, y, z) provided that

$$\Delta w = f(x + \Delta x, y + \Delta y, z + \Delta z) - f(x, y, z)$$

can be written in the form

$$\Delta w = f_x \Delta x + f_y \Delta y + f_z \Delta z + \varepsilon_1 \Delta x + \varepsilon_2 \Delta y + \varepsilon_3 \Delta z$$

where ε_1, ε_2, and $\varepsilon_3 \to 0$ as $(\Delta x, \Delta y, \Delta z) \to (0, 0, 0)$. With this definition of differentiability, Theorem 13.4 has the following extension for functions of three variables: If f is a function of x, y, and z, where f, f_x, f_y, and f_z are continuous in an open region R, then f is differentiable on R.

In Section 3.9, you used differentials to approximate the propagated error introduced by an error in measurement. This application of differentials is further illustrated in Example 4.

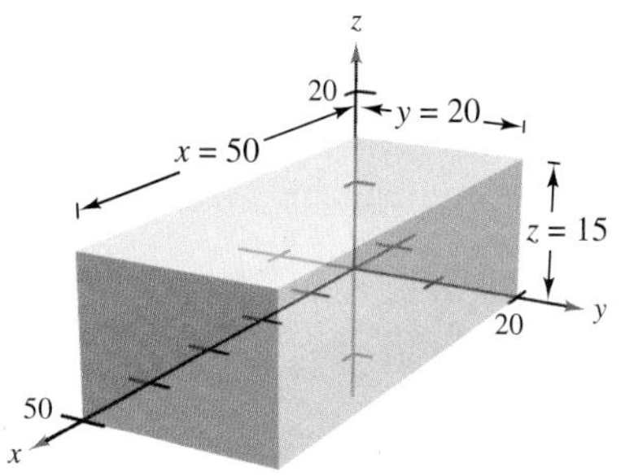

Volume $= xyz$
Figure 13.37

EXAMPLE 4 Error Analysis

The possible error involved in measuring each dimension of a rectangular box is ± 0.1 millimeter. The dimensions of the box are $x = 50$ centimeters, $y = 20$ centimeters, and $z = 15$ centimeters, as shown in Figure 13.37. Use dV to estimate the propagated error and the relative error in the calculated volume of the box.

Solution The volume of the box is given by $V = xyz$, and so

$$\begin{aligned} dV &= \frac{\partial V}{\partial x}dx + \frac{\partial V}{\partial y}dy + \frac{\partial V}{\partial z}dz \\ &= yz\,dx + xz\,dy + xy\,dz. \end{aligned}$$

Using 0.1 millimeter $= 0.01$ centimeter, you have $dx = dy = dz = \pm 0.01$, and the propagated error is approximately

$$\begin{aligned} dV &= (20)(15)(\pm 0.01) + (50)(15)(\pm 0.01) + (50)(20)(\pm 0.01) \\ &= 300(\pm 0.01) + 750(\pm 0.01) + 1000(\pm 0.01) \\ &= 2050(\pm 0.01) = \pm 20.5 \text{ cubic centimeters.} \end{aligned}$$

Because the measured volume is

$$V = (50)(20)(15) = 15{,}000 \text{ cubic centimeters,}$$

the relative error, $\Delta V/V$, is approximately

$$\frac{\Delta V}{V} \approx \frac{dV}{V} = \frac{20.5}{15{,}000} \approx 0.14\%.$$

■

As is true for a function of a single variable, if a function in two or more variables is differentiable at a point, it is also continuous there.

THEOREM 13.5 DIFFERENTIABILITY IMPLIES CONTINUITY

If a function of x and y is differentiable at (x_0, y_0), then it is continuous at (x_0, y_0).

PROOF Let f be differentiable at (x_0, y_0), where $z = f(x, y)$. Then

$$\Delta z = [f_x(x_0, y_0) + \varepsilon_1]\,\Delta x + [f_y(x_0, y_0) + \varepsilon_2]\,\Delta y$$

where both ε_1 and $\varepsilon_2 \to 0$ as $(\Delta x, \Delta y) \to (0, 0)$. However, by definition, you know that Δz is given by

$$\Delta z = f(x_0 + \Delta x, y_0 + \Delta y) - f(x_0, y_0).$$

Letting $x = x_0 + \Delta x$ and $y = y_0 + \Delta y$ produces

$$\begin{aligned} f(x, y) - f(x_0, y_0) &= [f_x(x_0, y_0) + \varepsilon_1]\,\Delta x + [f_y(x_0, y_0) + \varepsilon_2]\,\Delta y \\ &= [f_x(x_0, y_0) + \varepsilon_1](x - x_0) + [f_y(x_0, y_0) + \varepsilon_2](y - y_0). \end{aligned}$$

Taking the limit as $(x, y) \to (x_0, y_0)$, you have

$$\lim_{(x, y)\to(x_0, y_0)} f(x, y) = f(x_0, y_0)$$

which means that f is continuous at (x_0, y_0).

■

Remember that the existence of f_x and f_y is not sufficient to guarantee differentiability, as illustrated in the next example.

EXAMPLE 5 A Function That Is Not Differentiable

Show that $f_x(0, 0)$ and $f_y(0, 0)$ both exist, but that f is not differentiable at $(0, 0)$ where f is defined as

$$f(x, y) = \begin{cases} \dfrac{-3xy}{x^2 + y^2}, & \text{if } (x, y) \neq (0, 0) \\ 0, & \text{if } (x, y) = (0, 0) \end{cases}.$$

TECHNOLOGY Use a graphing utility to graph the function given in Example 5. For instance, the graph shown below was generated by *Mathematica*.

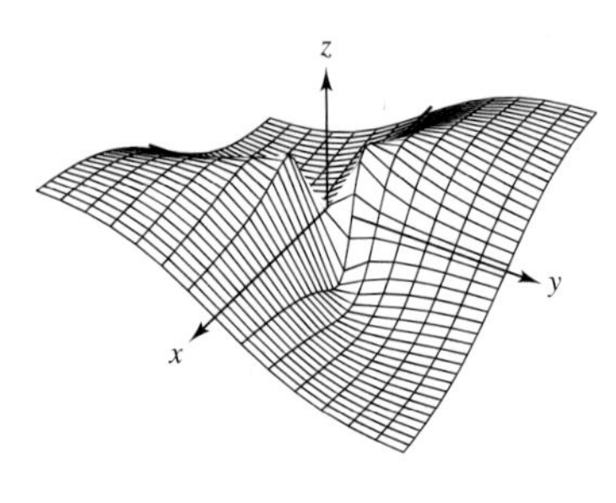
z
y
x

Generated by Mathematica

Solution You can show that f is not differentiable at $(0, 0)$ by showing that it is not continuous at this point. To see that f is not continuous at $(0, 0)$, look at the values of $f(x, y)$ along two different approaches to $(0, 0)$, as shown in Figure 13.38. Along the line $y = x$, the limit is

$$\lim_{(x, x)\to(0, 0)} f(x, y) = \lim_{(x, x)\to(0, 0)} \frac{-3x^2}{2x^2} = -\frac{3}{2}$$

whereas along $y = -x$ you have

$$\lim_{(x, -x)\to(0, 0)} f(x, y) = \lim_{(x, -x)\to(0, 0)} \frac{3x^2}{2x^2} = \frac{3}{2}.$$

So, the limit of $f(x, y)$ as $(x, y) \to (0, 0)$ does not exist, and you can conclude that f is not continuous at $(0, 0)$. Therefore, by Theorem 13.5, you know that f is not differentiable at $(0, 0)$. On the other hand, by the definition of the partial derivatives f_x and f_y, you have

$$f_x(0, 0) = \lim_{\Delta x\to 0} \frac{f(\Delta x, 0) - f(0, 0)}{\Delta x} = \lim_{\Delta x\to 0} \frac{0 - 0}{\Delta x} = 0$$

and

$$f_y(0, 0) = \lim_{\Delta y\to 0} \frac{f(0, \Delta y) - f(0, 0)}{\Delta y} = \lim_{\Delta y\to 0} \frac{0 - 0}{\Delta y} = 0.$$

So, the partial derivatives at $(0, 0)$ exist.

$$f(x, y) = \begin{cases} \dfrac{-3xy}{x^2 + y^2}, & (x, y) \neq (0, 0) \\ 0, & (x, y) = (0, 0) \end{cases}$$

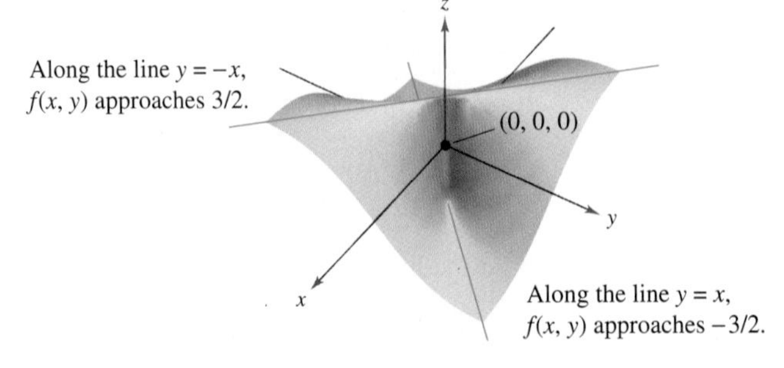
Along the line y = −x, f(x, y) approaches 3/2.
z
(0, 0, 0)
y
x
Along the line y = x, f(x, y) approaches −3/2.

Figure 13.38

■

13.4 Exercises

See www.CalcChat.com for worked-out solutions to odd-numbered exercises.

In Exercises 1–10, find the total differential.

1. $z = 2x^2y^3$
2. $z = \dfrac{x^2}{y}$
3. $z = \dfrac{-1}{x^2 + y^2}$
4. $w = \dfrac{x + y}{z - 3y}$
5. $z = x \cos y - y \cos x$
6. $z = \frac{1}{2}(e^{x^2+y^2} - e^{-x^2-y^2})$
7. $z = e^x \sin y$
8. $w = e^y \cos x + z^2$
9. $w = 2z^3y \sin x$
10. $w = x^2yz^2 + \sin yz$

In Exercises 11–16, (a) evaluate $f(2, 1)$ and $f(2.1, 1.05)$ and calculate Δz, and (b) use the total differential dz to approximate Δz.

11. $f(x, y) = 2x - 3y$
12. $f(x, y) = x^2 + y^2$
13. $f(x, y) = 16 - x^2 - y^2$
14. $f(x, y) = \dfrac{y}{x}$
15. $f(x, y) = ye^x$
16. $f(x, y) = x \cos y$

In Exercises 17–20, find $z = f(x, y)$ and use the total differential to approximate the quantity.

17. $(2.01)^2(9.02) - 2^2 \cdot 9$
18. $\sqrt{(5.05)^2 + (3.1)^2} - \sqrt{5^2 + 3^2}$
19. $\dfrac{1 - (3.05)^2}{(5.95)^2} - \dfrac{1 - 3^2}{6^2}$
20. $\sin[(1.05)^2 + (0.95)^2] - \sin(1^2 + 1^2)$

WRITING ABOUT CONCEPTS

21. Define the total differential of a function of two variables.
22. Describe the change in accuracy of dz as an approximation of Δz as Δx and Δy increase.
23. What is meant by a linear approximation of $z = f(x, y)$ at the point $P(x_0, y_0)$?
24. When using differentials, what is meant by the terms *propagated error* and *relative error*?

25. ***Area*** The area of the shaded rectangle in the figure is $A = lh$. The possible errors in the length and height are Δl and Δh, respectively. Find dA and identify the regions in the figure whose areas are given by the terms of dA. What region represents the difference between ΔA and dA?

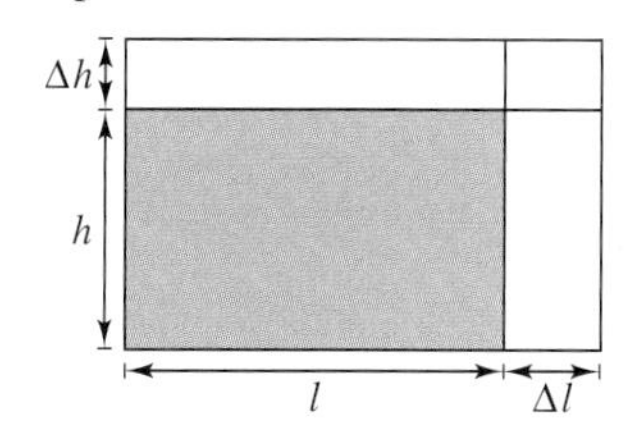

Figure for 25

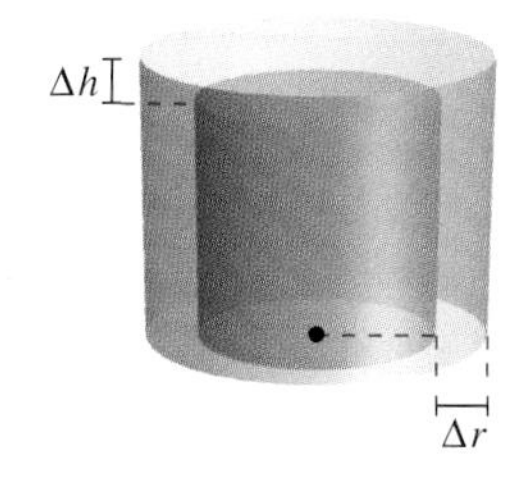

Figure for 26

26. ***Volume*** The volume of the red right circular cylinder in the figure is $V = \pi r^2 h$. The possible errors in the radius and the height are Δr and Δh, respectively. Find dV and identify the solids in the figure whose volumes are given by the terms of dV. What solid represents the difference between ΔV and dV?

27. ***Numerical Analysis*** A right circular cone of height $h = 8$ and radius $r = 4$ is constructed, and in the process errors Δr and Δh are made in the radius and height, respectively. Complete the table to show the relationship between ΔV and dV for the indicated errors.

Δr	Δh	dV or dS	ΔV or ΔS	$\Delta V - dV$ or $\Delta S - dS$
0.1	0.1			
0.1	−0.1			
0.001	0.002			
−0.0001	0.0002			

28. ***Numerical Analysis*** The height and radius of a right circular cone are measured as $h = 16$ meters and $r = 6$ meters. In the process of measuring, errors Δr and Δh are made. S is the lateral surface area of the cone. Complete the table above to show the relationship between ΔS and dS for the indicated errors.

29. ***Modeling Data*** Per capita consumptions (in gallons) of different types of plain milk in the United States from 1999 through 2005 are shown in the table. Consumption of flavored milk, plain reduced-fat milk, and plain light and skim milks are represented by the variables x, y, and z, respectively. *(Source: U.S. Department of Agriculture)*

Year	1999	2000	2001	2002	2003	2004	2005
x	1.4	1.4	1.4	1.6	1.6	1.7	1.7
y	7.3	7.1	7.0	7.0	6.9	6.9	6.9
z	6.2	6.1	5.9	5.8	5.6	5.5	5.6

A model for the data is given by $z = -0.92x + 1.03y + 0.02$.

(a) Find the total differential of the model.

(b) A dairy industry forecast for a future year is that per capita consumption of flavored milk will be 1.9 ± 0.25 gallons and that per capita consumption of plain reduced-fat milk will be 7.5 ± 0.25 gallons. Use dz to estimate the maximum possible propagated error and relative error in the prediction for the consumption of plain light and skim milks.

30. ***Rectangular to Polar Coordinates*** A rectangular coordinate system is placed over a map, and the coordinates of a point of interest are $(7.2, 2.5)$. There is a possible error of 0.05 in each coordinate. Approximate the maximum possible error in measuring the polar coordinates of the point.

31. ***Volume*** The radius r and height h of a right circular cylinder are measured with possible errors of 4% and 2%, respectively. Approximate the maximum possible percent error in measuring the volume.

32. ***Area*** A triangle is measured and two adjacent sides are found to be 3 inches and 4 inches long, with an included angle of $\pi/4$. The possible errors in measurement are $\frac{1}{16}$ inch for the sides and 0.02 radian for the angle. Approximate the maximum possible error in the computation of the area.

33. ***Wind Chill*** The formula for wind chill C (in degrees Fahrenheit) is given by

$$C = 35.74 + 0.6215T - 35.75v^{0.16} + 0.4275Tv^{0.16}$$

where v is the wind speed in miles per hour and T is the temperature in degrees Fahrenheit. The wind speed is 23 ± 3 miles per hour and the temperature is $8° \pm 1°$. Use dC to estimate the maximum possible propagated error and relative error in calculating the wind chill. *(Source: National Oceanic and Atmospheric Administration)*

34. ***Acceleration*** The centripetal acceleration of a particle moving in a circle is $a = v^2/r$, where v is the velocity and r is the radius of the circle. Approximate the maximum percent error in measuring the acceleration due to errors of 3% in v and 2% in r.

35. ***Volume*** A trough is 16 feet long (see figure). Its cross sections are isosceles triangles with each of the two equal sides measuring 18 inches. The angle between the two equal sides is θ.

(a) Write the volume of the trough as a function of θ and determine the value of θ such that the volume is a maximum.

(b) The maximum error in the linear measurements is one-half inch and the maximum error in the angle measure is 2°. Approximate the change in the maximum volume.

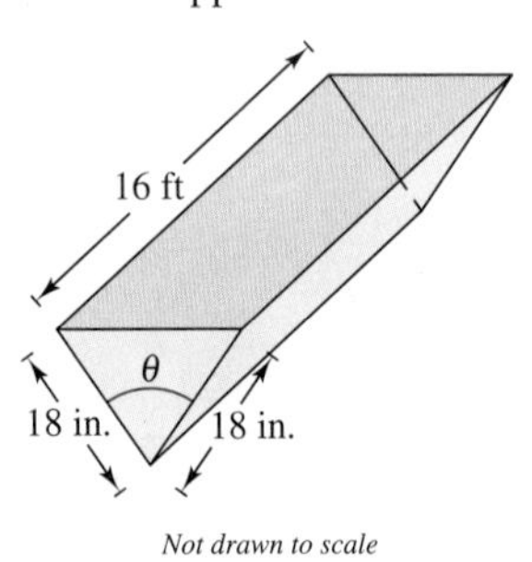

Figure for 35

Figure for 36

36. ***Sports*** A baseball player in center field is playing approximately 330 feet from a television camera that is behind home plate. A batter hits a fly ball that goes to the wall 420 feet from the camera (see figure).

(a) The camera turns 9° to follow the play. Approximate the number of feet that the center fielder has to run to make the catch.

(b) The position of the center fielder could be in error by as much as 6 feet and the maximum error in measuring the rotation of the camera is 1°. Approximate the maximum possible error in the result of part (a).

37. ***Power*** Electrical power P is given by $P = E^2/R$, where E is voltage and R is resistance. Approximate the maximum percent error in calculating power if 120 volts is applied to a 2000-ohm resistor and the possible percent errors in measuring E and R are 3% and 4%, respectively.

38. ***Resistance*** The total resistance R of two resistors connected in parallel is given by

$$\frac{1}{R} = \frac{1}{R_1} + \frac{1}{R_2}.$$

Approximate the change in R as R_1 is increased from 10 ohms to 10.5 ohms and R_2 is decreased from 15 ohms to 13 ohms.

39. ***Inductance*** The inductance L (in microhenrys) of a straight nonmagnetic wire in free space is

$$L = 0.00021\left(\ln\frac{2h}{r} - 0.75\right)$$

where h is the length of the wire in millimeters and r is the radius of a circular cross section. Approximate L when $r = 2 \pm \frac{1}{16}$ millimeters and $h = 100 \pm \frac{1}{100}$ millimeters.

40. ***Pendulum*** The period T of a pendulum of length L is $T = 2\pi\sqrt{L/g}$, where g is the acceleration due to gravity. A pendulum is moved from the Canal Zone, where $g = 32.09$ feet per second per second, to Greenland, where $g = 32.23$ feet per second per second. Because of the change in temperature, the length of the pendulum changes from 2.5 feet to 2.48 feet. Approximate the change in the period of the pendulum.

In Exercises 41–44, show that the function is differentiable by finding values of ε_1 and ε_2 as designated in the definition of differentiability, and verify that both ε_1 and $\varepsilon_2 \to 0$ as $(\Delta x, \Delta y) \to (0, 0)$.

41. $f(x, y) = x^2 - 2x + y$

42. $f(x, y) = x^2 + y^2$

43. $f(x, y) = x^2y$

44. $f(x, y) = 5x - 10y + y^3$

In Exercises 45 and 46, use the function to show that $f_x(0, 0)$ and $f_y(0, 0)$ both exist, but that f is not differentiable at $(0, 0)$.

45. $f(x, y) = \begin{cases} \dfrac{3x^2y}{x^4 + y^2}, & (x, y) \neq (0, 0) \\ 0, & (x, y) = (0, 0) \end{cases}$

46. $f(x, y) = \begin{cases} \dfrac{5x^2y}{x^3 + y^3}, & (x, y) \neq (0, 0) \\ 0, & (x, y) = (0, 0) \end{cases}$

47. Show that if $f(x, y)$ is differentiable at (x_0, y_0), then $f(x, y_0)$ is differentiable at $x = x_0$. Use this result to prove that $f(x, y) = \sqrt{x^2 + y^2}$ is not differentiable at $(0, 0)$.

CAPSTONE

48. Consider the function $f(x, y) = \sqrt{x^2 + y^2}$.

(a) Evaluate $f(3, 1)$ and $f(3.05, 1.1)$.

(b) Use the results of part (a) to calculate Δz.

(c) Use the total differential dz to approximate Δz. Compare your result with that of part (b).

13.5 Chain Rules for Functions of Several Variables

- **Use the Chain Rules for functions of several variables.**
- **Find partial derivatives implicitly.**

Chain Rules for Functions of Several Variables

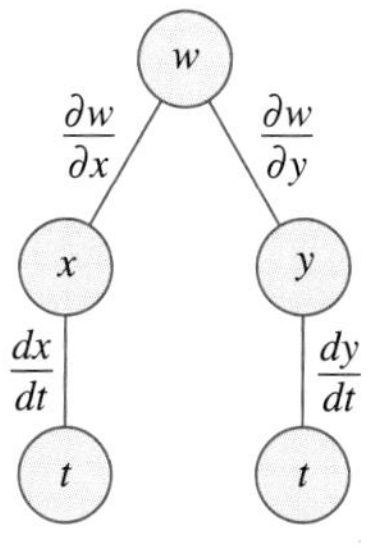

Chain Rule: one independent variable w is a function of x and y, which are each functions of t. This diagram represents the derivative of w with respect to t.
Figure 13.39

Your work with differentials in the preceding section provides the basis for the extension of the Chain Rule to functions of two variables. There are two cases—the first case involves w as a function of x and y, where x and y are functions of a single independent variable t, as shown in Theorem 13.6. (A proof of this theorem is given in Appendix A.)

THEOREM 13.6 CHAIN RULE: ONE INDEPENDENT VARIABLE

Let $w = f(x, y)$, where f is a differentiable function of x and y. If $x = g(t)$ and $y = h(t)$, where g and h are differentiable functions of t, then w is a differentiable function of t, and

$$\frac{dw}{dt} = \frac{\partial w}{\partial x}\frac{dx}{dt} + \frac{\partial w}{\partial y}\frac{dy}{dt}.$$

See Figure 13.39.

EXAMPLE 1 Using the Chain Rule with One Independent Variable

Let $w = x^2y - y^2$, where $x = \sin t$ and $y = e^t$. Find dw/dt when $t = 0$.

Solution By the Chain Rule for one independent variable, you have

$$\begin{aligned}\frac{dw}{dt} &= \frac{\partial w}{\partial x}\frac{dx}{dt} + \frac{\partial w}{\partial y}\frac{dy}{dt}\\ &= 2xy(\cos t) + (x^2 - 2y)e^t\\ &= 2(\sin t)(e^t)(\cos t) + (\sin^2 t - 2e^t)e^t\\ &= 2e^t \sin t \cos t + e^t \sin^2 t - 2e^{2t}.\end{aligned}$$

When $t = 0$, it follows that

$$\frac{dw}{dt} = -2.$$

■

The Chain Rules presented in this section provide alternative techniques for solving many problems in single-variable calculus. For instance, in Example 1, you could have used single-variable techniques to find dw/dt by first writing w as a function of t,

$$\begin{aligned}w &= x^2y - y^2\\ &= (\sin t)^2(e^t) - (e^t)^2\\ &= e^t \sin^2 t - e^{2t}\end{aligned}$$

and then differentiating as usual.

$$\frac{dw}{dt} = 2e^t \sin t \cos t + e^t \sin^2 t - 2e^{2t}$$

The Chain Rule in Theorem 13.6 can be extended to any number of variables. For example, if each x_i is a differentiable function of a single variable t, then for

$$w = f(x_1, x_2, \ldots, x_n)$$

you have

$$\frac{dw}{dt} = \frac{\partial w}{\partial x_1}\frac{dx_1}{dt} + \frac{\partial w}{\partial x_2}\frac{dx_2}{dt} + \cdots + \frac{\partial w}{\partial x_n}\frac{dx_n}{dt}.$$

EXAMPLE 2 An Application of a Chain Rule to Related Rates

Two objects are traveling in elliptical paths given by the following parametric equations.

$$x_1 = 4\cos t \quad \text{and} \quad y_1 = 2\sin t \qquad \text{First object}$$

$$x_2 = 2\sin 2t \quad \text{and} \quad y_2 = 3\cos 2t \qquad \text{Second object}$$

At what rate is the distance between the two objects changing when $t = \pi$?

Solution From Figure 13.40, you can see that the distance s between the two objects is given by

$$s = \sqrt{(x_2 - x_1)^2 + (y_2 - y_1)^2}$$

and that when $t = \pi$, you have $x_1 = -4$, $y_1 = 0$, $x_2 = 0$, $y_2 = 3$, and

$$s = \sqrt{(0 + 4)^2 + (3 - 0)^2} = 5.$$

When $t = \pi$, the partial derivatives of s are as follows.

$$\frac{\partial s}{\partial x_1} = \frac{-(x_2 - x_1)}{\sqrt{(x_2 - x_1)^2 + (y_2 - y_1)^2}} = -\frac{1}{5}(0 + 4) = -\frac{4}{5}$$

$$\frac{\partial s}{\partial y_1} = \frac{-(y_2 - y_1)}{\sqrt{(x_2 - x_1)^2 + (y_2 - y_1)^2}} = -\frac{1}{5}(3 - 0) = -\frac{3}{5}$$

$$\frac{\partial s}{\partial x_2} = \frac{(x_2 - x_1)}{\sqrt{(x_2 - x_1)^2 + (y_2 - y_1)^2}} = \frac{1}{5}(0 + 4) = \frac{4}{5}$$

$$\frac{\partial s}{\partial y_2} = \frac{(y_2 - y_1)}{\sqrt{(x_2 - x_1)^2 + (y_2 - y_1)^2}} = \frac{1}{5}(3 - 0) = \frac{3}{5}$$

When $t = \pi$, the derivatives of x_1, y_1, x_2, and y_2 are

$$\frac{dx_1}{dt} = -4\sin t = 0 \qquad \frac{dy_1}{dt} = 2\cos t = -2$$

$$\frac{dx_2}{dt} = 4\cos 2t = 4 \qquad \frac{dy_2}{dt} = -6\sin 2t = 0.$$

So, using the appropriate Chain Rule, you know that the distance is changing at a rate of

$$\frac{ds}{dt} = \frac{\partial s}{\partial x_1}\frac{dx_1}{dt} + \frac{\partial s}{\partial y_1}\frac{dy_1}{dt} + \frac{\partial s}{\partial x_2}\frac{dx_2}{dt} + \frac{\partial s}{\partial y_2}\frac{dy_2}{dt}$$

$$= \left(-\frac{4}{5}\right)(0) + \left(-\frac{3}{5}\right)(-2) + \left(\frac{4}{5}\right)(4) + \left(\frac{3}{5}\right)(0)$$

$$= \frac{22}{5}.$$

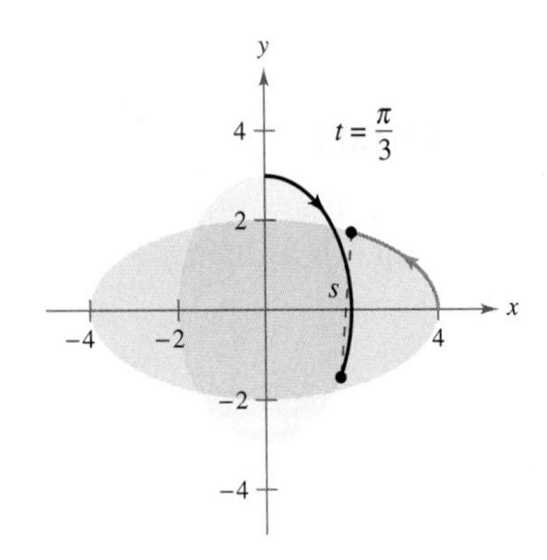

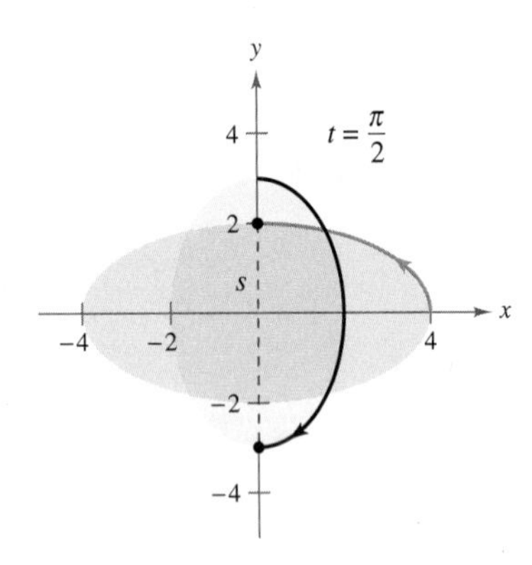

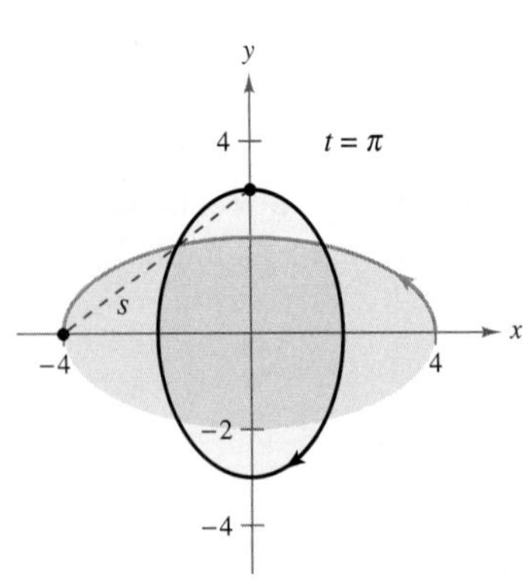

Paths of two objects traveling in elliptical orbits
Figure 13.40

In Example 2, note that s is the function of four *intermediate* variables, x_1, y_1, x_2, and y_2, each of which is a function of a single variable t. Another type of composite function is one in which the intermediate variables are themselves functions of more than one variable. For instance, if $w = f(x, y)$, where $x = g(s, t)$ and $y = h(s, t)$, it follows that w is a function of s and t, and you can consider the partial derivatives of w with respect to s and t. One way to find these partial derivatives is to write w as a function of s and t explicitly by substituting the equations $x = g(s, t)$ and $y = h(s, t)$ into the equation $w = f(x, y)$. Then you can find the partial derivatives in the usual way, as demonstrated in the next example.

EXAMPLE 3 Finding Partial Derivatives by Substitution

Find $\partial w/\partial s$ and $\partial w/\partial t$ for $w = 2xy$, where $x = s^2 + t^2$ and $y = s/t$.

Solution Begin by substituting $x = s^2 + t^2$ and $y = s/t$ into the equation $w = 2xy$ to obtain

$$w = 2xy = 2(s^2 + t^2)\left(\frac{s}{t}\right) = 2\left(\frac{s^3}{t} + st\right).$$

Then, to find $\partial w/\partial s$, hold t constant and differentiate with respect to s.

$$\frac{\partial w}{\partial s} = 2\left(\frac{3s^2}{t} + t\right)$$

$$= \frac{6s^2 + 2t^2}{t}$$

Similarly, to find $\partial w/\partial t$, hold s constant and differentiate with respect to t to obtain

$$\frac{\partial w}{\partial t} = 2\left(-\frac{s^3}{t^2} + s\right)$$

$$= 2\left(\frac{-s^3 + st^2}{t^2}\right)$$

$$= \frac{2st^2 - 2s^3}{t^2}.$$

■

Theorem 13.7 gives an alternative method for finding the partial derivatives in Example 3, without explicitly writing w as a function of s and t.

THEOREM 13.7 CHAIN RULE: TWO INDEPENDENT VARIABLES

Let $w = f(x, y)$, where f is a differentiable function of x and y. If $x = g(s, t)$ and $y = h(s, t)$ such that the first partials $\partial x/\partial s$, $\partial x/\partial t$, $\partial y/\partial s$, and $\partial y/\partial t$ all exist, then $\partial w/\partial s$ and $\partial w/\partial t$ exist and are given by

$$\frac{\partial w}{\partial s} = \frac{\partial w}{\partial x}\frac{\partial x}{\partial s} + \frac{\partial w}{\partial y}\frac{\partial y}{\partial s} \quad \text{and} \quad \frac{\partial w}{\partial t} = \frac{\partial w}{\partial x}\frac{\partial x}{\partial t} + \frac{\partial w}{\partial y}\frac{\partial y}{\partial t}.$$

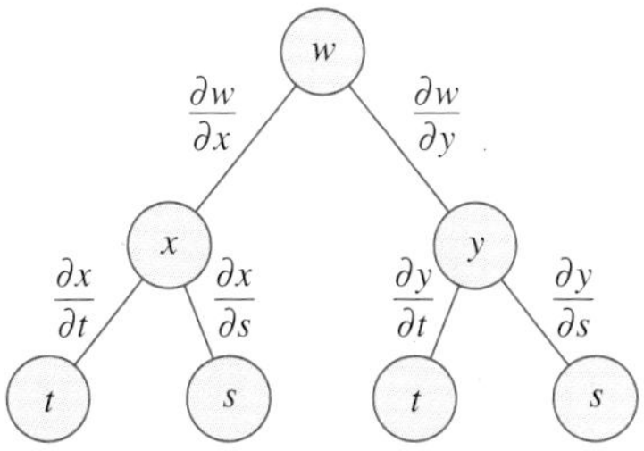

PROOF To obtain $\partial w/\partial s$, hold t constant and apply Theorem 13.6 to obtain the desired result. Similarly, for $\partial w/\partial t$, hold s constant and apply Theorem 13.6. ■

Chain Rule: two independent variables
Figure 13.41

NOTE The Chain Rule in Theorem 13.7 is shown schematically in Figure 13.41. ■

EXAMPLE 4 The Chain Rule with Two Independent Variables

Use the Chain Rule to find $\partial w/\partial s$ and $\partial w/\partial t$ for

$$w = 2xy$$

where $x = s^2 + t^2$ and $y = s/t$.

Solution Note that these same partials were found in Example 3. This time, using Theorem 13.7, you can hold t constant and differentiate with respect to s to obtain

$$\begin{aligned}
\frac{\partial w}{\partial s} &= \frac{\partial w}{\partial x}\frac{\partial x}{\partial s} + \frac{\partial w}{\partial y}\frac{\partial y}{\partial s}\\
&= 2y(2s) + 2x\left(\frac{1}{t}\right)\\
&= 2\left(\frac{s}{t}\right)(2s) + 2(s^2 + t^2)\left(\frac{1}{t}\right) && \text{Substitute } (s/t) \text{ for } y \text{ and } s^2 + t^2 \text{ for } x.\\
&= \frac{4s^2}{t} + \frac{2s^2 + 2t^2}{t}\\
&= \frac{6s^2 + 2t^2}{t}.
\end{aligned}$$

Similarly, holding s constant gives

$$\begin{aligned}
\frac{\partial w}{\partial t} &= \frac{\partial w}{\partial x}\frac{\partial x}{\partial t} + \frac{\partial w}{\partial y}\frac{\partial y}{\partial t}\\
&= 2y(2t) + 2x\left(\frac{-s}{t^2}\right)\\
&= 2\left(\frac{s}{t}\right)(2t) + 2(s^2 + t^2)\left(\frac{-s}{t^2}\right) && \text{Substitute } (s/t) \text{ for } y \text{ and } s^2 + t^2 \text{ for } x.\\
&= 4s - \frac{2s^3 + 2st^2}{t^2}\\
&= \frac{4st^2 - 2s^3 - 2st^2}{t^2}\\
&= \frac{2st^2 - 2s^3}{t^2}.
\end{aligned}$$

■

The Chain Rule in Theorem 13.7 can also be extended to any number of variables. For example, if w is a differentiable function of the n variables $x_1, x_2, \ldots, x_n$, where each x_i is a differentiable function of the m variables $t_1, t_2, \ldots, t_m$, then for

$$w = f(x_1, x_2, \ldots, x_n)$$

you obtain the following.

$$\begin{aligned}
\frac{\partial w}{\partial t_1} &= \frac{\partial w}{\partial x_1}\frac{\partial x_1}{\partial t_1} + \frac{\partial w}{\partial x_2}\frac{\partial x_2}{\partial t_1} + \cdots + \frac{\partial w}{\partial x_n}\frac{\partial x_n}{\partial t_1}\\
\frac{\partial w}{\partial t_2} &= \frac{\partial w}{\partial x_1}\frac{\partial x_1}{\partial t_2} + \frac{\partial w}{\partial x_2}\frac{\partial x_2}{\partial t_2} + \cdots + \frac{\partial w}{\partial x_n}\frac{\partial x_n}{\partial t_2}\\
&\vdots\\
\frac{\partial w}{\partial t_m} &= \frac{\partial w}{\partial x_1}\frac{\partial x_1}{\partial t_m} + \frac{\partial w}{\partial x_2}\frac{\partial x_2}{\partial t_m} + \cdots + \frac{\partial w}{\partial x_n}\frac{\partial x_n}{\partial t_m}
\end{aligned}$$

EXAMPLE 5 The Chain Rule for a Function of Three Variables

Find $\partial w/\partial s$ and $\partial w/\partial t$ when $s = 1$ and $t = 2\pi$ for the function given by

$$w = xy + yz + xz$$

where $x = s \cos t$, $y = s \sin t$, and $z = t$.

Solution By extending the result of Theorem 13.7, you have

$$\begin{aligned}\frac{\partial w}{\partial s} &= \frac{\partial w}{\partial x}\frac{\partial x}{\partial s} + \frac{\partial w}{\partial y}\frac{\partial y}{\partial s} + \frac{\partial w}{\partial z}\frac{\partial z}{\partial s}\\ &= (y + z)(\cos t) + (x + z)(\sin t) + (y + x)(0)\\ &= (y + z)(\cos t) + (x + z)(\sin t).\end{aligned}$$

When $s = 1$ and $t = 2\pi$, you have $x = 1$, $y = 0$, and $z = 2\pi$. So, $\partial w/\partial s = (0 + 2\pi)(1) + (1 + 2\pi)(0) = 2\pi$. Furthermore,

$$\begin{aligned}\frac{\partial w}{\partial t} &= \frac{\partial w}{\partial x}\frac{\partial x}{\partial t} + \frac{\partial w}{\partial y}\frac{\partial y}{\partial t} + \frac{\partial w}{\partial z}\frac{\partial z}{\partial t}\\ &= (y + z)(-s \sin t) + (x + z)(s \cos t) + (y + x)(1)\end{aligned}$$

and for $s = 1$ and $t = 2\pi$, it follows that

$$\begin{aligned}\frac{\partial w}{\partial t} &= (0 + 2\pi)(0) + (1 + 2\pi)(1) + (0 + 1)(1)\\ &= 2 + 2\pi.\end{aligned}$$

■

Implicit Partial Differentiation

This section concludes with an application of the Chain Rule to determine the derivative of a function defined *implicitly*. Suppose that x and y are related by the equation $F(x, y) = 0$, where it is assumed that $y = f(x)$ is a differentiable function of x. To find dy/dx, you could use the techniques discussed in Section 2.5. However, you will see that the Chain Rule provides a convenient alternative. If you consider the function given by

$$w = F(x, y) = F(x, f(x))$$

you can apply Theorem 13.6 to obtain

$$\frac{dw}{dx} = F_x(x, y)\frac{dx}{dx} + F_y(x, y)\frac{dy}{dx}.$$

Because $w = F(x, y) = 0$ for all x in the domain of f, you know that $dw/dx = 0$ and you have

$$F_x(x, y)\frac{dx}{dx} + F_y(x, y)\frac{dy}{dx} = 0.$$

Now, if $F_y(x, y) \neq 0$, you can use the fact that $dx/dx = 1$ to conclude that

$$\frac{dy}{dx} = -\frac{F_x(x, y)}{F_y(x, y)}.$$

A similar procedure can be used to find the partial derivatives of functions of several variables that are defined implicitly.

THEOREM 13.8 CHAIN RULE: IMPLICIT DIFFERENTIATION

If the equation $F(x, y) = 0$ defines y implicitly as a differentiable function of x, then

$$\frac{dy}{dx} = -\frac{F_x(x, y)}{F_y(x, y)}, \qquad F_y(x, y) \neq 0.$$

If the equation $F(x, y, z) = 0$ defines z implicitly as a differentiable function of x and y, then

$$\frac{\partial z}{\partial x} = -\frac{F_x(x, y, z)}{F_z(x, y, z)} \quad \text{and} \quad \frac{\partial z}{\partial y} = -\frac{F_y(x, y, z)}{F_z(x, y, z)}, \quad F_z(x, y, z) \neq 0.$$

This theorem can be extended to differentiable functions defined implicitly with any number of variables.

EXAMPLE 6 Finding a Derivative Implicitly

Find dy/dx, given $y^3 + y^2 - 5y - x^2 + 4 = 0$.

Solution Begin by defining a function F as

$$F(x, y) = y^3 + y^2 - 5y - x^2 + 4.$$

Then, using Theorem 13.8, you have

$$F_x(x, y) = -2x \quad \text{and} \quad F_y(x, y) = 3y^2 + 2y - 5$$

and it follows that

$$\frac{dy}{dx} = -\frac{F_x(x, y)}{F_y(x, y)} = \frac{-(-2x)}{3y^2 + 2y - 5} = \frac{2x}{3y^2 + 2y - 5}.$$

■

NOTE Compare the solution of Example 6 with the solution of Example 2 in Section 2.5. ■

EXAMPLE 7 Finding Partial Derivatives Implicitly

Find $\partial z/\partial x$ and $\partial z/\partial y$, given $3x^2z - x^2y^2 + 2z^3 + 3yz - 5 = 0$.

Solution To apply Theorem 13.8, let

$$F(x, y, z) = 3x^2z - x^2y^2 + 2z^3 + 3yz - 5.$$

Then

$$F_x(x, y, z) = 6xz - 2xy^2$$
$$F_y(x, y, z) = -2x^2y + 3z$$
$$F_z(x, y, z) = 3x^2 + 6z^2 + 3y$$

and you obtain

$$\frac{\partial z}{\partial x} = -\frac{F_x(x, y, z)}{F_z(x, y, z)} = \frac{2xy^2 - 6xz}{3x^2 + 6z^2 + 3y}$$
$$\frac{\partial z}{\partial y} = -\frac{F_y(x, y, z)}{F_z(x, y, z)} = \frac{2x^2y - 3z}{3x^2 + 6z^2 + 3y}.$$

■

13.5 Exercises

See www.CalcChat.com for worked-out solutions to odd-numbered exercises.

In Exercises 1–4, find dw/dt using the appropriate Chain Rule.

1. $w = x^2 + y^2$
   $x = 2t, \quad y = 3t$
2. $w = \sqrt{x^2 + y^2}$
   $x = \cos t, \quad y = e^t$
3. $w = x \sin y$
   $x = e^t, \quad y = \pi - t$
4. $w = \ln \dfrac{y}{x}$
   $x = \cos t, \quad y = \sin t$

In Exercises 5–10, find dw/dt (a) by using the appropriate Chain Rule and (b) by converting w to a function of t before differentiating.

5. $w = xy, \quad x = e^t, \quad y = e^{-2t}$
6. $w = \cos(x - y), \quad x = t^2, \quad y = 1$
7. $w = x^2 + y^2 + z^2, \quad x = \cos t, \quad y = \sin t, \quad z = e^t$
8. $w = xy \cos z, \quad x = t, \quad y = t^2, \quad z = \arccos t$
9. $w = xy + xz + yz, \quad x = t - 1, \quad y = t^2 - 1, \quad z = t$
10. $w = xy^2 + x^2z + yz^2, \quad x = t^2, \quad y = 2t, \quad z = 2$

Projectile Motion **In Exercises 11 and 12, the parametric equations for the paths of two projectiles are given. At what rate is the distance between the two objects changing at the given value of t?**

11. $x_1 = 10 \cos 2t, \ y_1 = 6 \sin 2t$ — First object
    $x_2 = 7 \cos t, \ y_2 = 4 \sin t$ — Second object
    $t = \pi/2$
12. $x_1 = 48\sqrt{2}\,t, \ y_1 = 48\sqrt{2}\,t - 16t^2$ — First object
    $x_2 = 48\sqrt{3}\,t, \ y_2 = 48t - 16t^2$ — Second object
    $t = 1$

In Exercises 13 and 14, find d^2w/dt^2 using the appropriate Chain Rule. Evaluate d^2w/dt^2 at the given value of t.

13. $w = \ln(x + y), \quad x = e^t, \quad y = e^{-t}, \quad t = 0$
14. $w = \dfrac{x^2}{y}, \quad x = t^2, \quad y = t + 1, \quad t = 1$

In Exercises 15–18, find $\partial w/\partial s$ and $\partial w/\partial t$ using the appropriate Chain Rule, and evaluate each partial derivative at the given values of s and t.

	Function	*Point*
15.	$w = x^2 + y^2$ $x = s + t, \quad y = s - t$	$s = 1, \quad t = 0$
16.	$w = y^3 - 3x^2y$ $x = e^s, \quad y = e^t$	$s = -1, \quad t = 2$
17.	$w = \sin(2x + 3y)$ $x = s + t, \quad y = s - t$	$s = 0, \quad t = \dfrac{\pi}{2}$
18.	$w = x^2 - y^2$ $x = s \cos t, \quad y = s \sin t$	$s = 3, \quad t = \dfrac{\pi}{4}$

In Exercises 19–22, find $\partial w/\partial r$ and $\partial w/\partial \theta$ (a) by using the appropriate Chain Rule and (b) by converting w to a function of r and θ before differentiating.

19. $w = \dfrac{yz}{x}, \quad x = \theta^2, \quad y = r + \theta, \quad z = r - \theta$
20. $w = x^2 - 2xy + y^2, \quad x = r + \theta, \quad y = r - \theta$
21. $w = \arctan \dfrac{y}{x}, \quad x = r \cos \theta, \quad y = r \sin \theta$
22. $w = \sqrt{25 - 5x^2 - 5y^2}, \quad x = r \cos \theta, \quad y = r \sin \theta$

In Exercises 23–26, find $\partial w/\partial s$ and $\partial w/\partial t$ by using the appropriate Chain Rule.

23. $w = xyz, \quad x = s + t, \quad y = s - t, \quad z = st^2$
24. $w = x^2 + y^2 + z^2, \quad x = t \sin s, \quad y = t \cos s, \quad z = st^2$
25. $w = ze^{xy}, \quad x = s - t, \quad y = s + t, \quad z = st$
26. $w = x \cos yz, \quad x = s^2, \quad y = t^2, \quad z = s - 2t$

In Exercises 27–30, differentiate implicitly to find dy/dx.

27. $x^2 - xy + y^2 - x + y = 0$
28. $\sec xy + \tan xy + 5 = 0$
29. $\ln \sqrt{x^2 + y^2} + x + y = 4$
30. $\dfrac{x}{x^2 + y^2} - y^2 = 6$

In Exercises 31–38, differentiate implicitly to find the first partial derivatives of z.

31. $x^2 + y^2 + z^2 = 1$
32. $xz + yz + xy = 0$
33. $x^2 + 2yz + z^2 = 1$
34. $x + \sin(y + z) = 0$
35. $\tan(x + y) + \tan(y + z) = 1$
36. $z = e^x \sin(y + z)$
37. $e^{xz} + xy = 0$
38. $x \ln y + y^2z + z^2 = 8$

In Exercises 39–42, differentiate implicitly to find the first partial derivatives of w.

39. $xy + yz - wz + wx = 5$
40. $x^2 + y^2 + z^2 - 5yw + 10w^2 = 2$
41. $\cos xy + \sin yz + wz = 20$
42. $w - \sqrt{x - y} - \sqrt{y - z} = 0$

Homogeneous Functions **A function f is *homogeneous of degree n* if $f(tx, ty) = t^n f(x, y)$. In Exercises 43–46, (a) show that the function is homogeneous and determine n, and (b) show that $xf_x(x, y) + yf_y(x, y) = nf(x, y)$.**

43. $f(x, y) = \dfrac{xy}{\sqrt{x^2 + y^2}}$
44. $f(x, y) = x^3 - 3xy^2 + y^3$
45. $f(x, y) = e^{x/y}$
46. $f(x, y) = \dfrac{x^2}{\sqrt{x^2 + y^2}}$

47. Let $w = f(x, y)$, $x = g(t)$, and $y = h(t)$, where f, g, and h are differentiable. Use the appropriate Chain Rule to find dw/dt when $t = 2$ given the following table of values.

$g(2)$	$h(2)$	$g'(2)$	$h'(2)$	$f_x(4, 3)$	$f_y(4, 3)$
4	3	-1	6	-5	7

48. Let $w = f(x, y)$, $x = g(s, t)$, and $y = h(s, t)$, where f, g, and h are differentiable. Use the appropriate Chain Rule to find $w_s(1, 2)$ and $w_t(1, 2)$ given the following table of values.

$g(1, 2)$	$h(1, 2)$	$g_s(1, 2)$	$h_s(1, 2)$
4	3	-3	5

$g_t(1, 2)$	$h_t(1, 2)$	$f_x(4, 3)$	$f_y(4, 3)$
-2	8	-5	7

WRITING ABOUT CONCEPTS

49. Let $w = f(x, y)$ be a function in which x and y are functions of a single variable t. Give the Chain Rule for finding dw/dt.

50. Let $w = f(x, y)$ be a function in which x and y are functions of two variables s and t. Give the Chain Rule for finding $\partial w/\partial s$ and $\partial w/\partial t$.

51. If $f(x, y) = 0$, give the rule for finding dy/dx implicitly. If $f(x, y, z) = 0$, give the rule for finding $\partial z/\partial x$ and $\partial z/\partial y$ implicitly.

CAPSTONE

52. Consider the function $f(x, y, z) = xyz$, where $x = t^2$, $y = 2t$, and $z = e^{-t}$.

(a) Use the appropriate Chain Rule to find df/dt.

(b) Write f as a function of t and then find df/dt. Explain why this result is the same as that of part (a).

53. ***Volume and Surface Area*** The radius of a right circular cylinder is increasing at a rate of 6 inches per minute, and the height is decreasing at a rate of 4 inches per minute. What are the rates of change of the volume and surface area when the radius is 12 inches and the height is 36 inches?

54. ***Volume and Surface Area*** Repeat Exercise 53 for a right circular cone.

55. ***Ideal Gas Law*** The Ideal Gas Law is $pV = mRT$, where R is a constant, m is a constant mass, and p and V are functions of time. Find dT/dt, the rate at which the temperature changes with respect to time.

56. ***Area*** Let θ be the angle between equal sides of an isosceles triangle and let x be the length of these sides. x is increasing at $\frac{1}{2}$ meter per hour and θ is increasing at $\pi/90$ radian per hour. Find the rate of increase of the area when $x = 6$ and $\theta = \pi/4$.

57. ***Moment of Inertia*** An annular cylinder has an inside radius of r_1 and an outside radius of r_2 (see figure). Its moment of inertia is $I = \frac{1}{2}m(r_1^2 + r_2^2)$, where m is the mass. The two radii are increasing at a rate of 2 centimeters per second. Find the rate at which I is changing at the instant the radii are 6 centimeters and 8 centimeters. (Assume mass is a constant.)

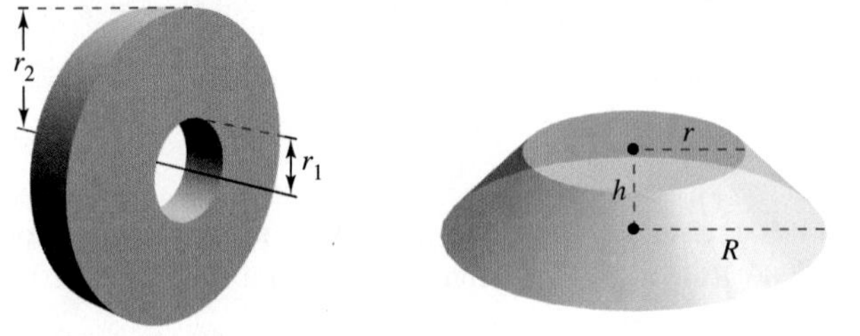

Figure for 57 **Figure for 58**

58. ***Volume and Surface Area*** The two radii of the frustum of a right circular cone are increasing at a rate of 4 centimeters per minute, and the height is increasing at a rate of 12 centimeters per minute (see figure). Find the rates at which the volume and surface area are changing when the two radii are 15 centimeters and 25 centimeters, and the height is 10 centimeters.

59. Show that $(\partial w/\partial u) + (\partial w/\partial v) = 0$ for $w = f(x, y)$, $x = u - v$, and $y = v - u$.

60. Demonstrate the result of Exercise 59 for

$w = (x - y)\sin(y - x)$.

61. Consider the function $w = f(x, y)$, where $x = r\cos\theta$ and $y = r\sin\theta$. Verify each of the following.

(a) $$\frac{\partial w}{\partial x} = \frac{\partial w}{\partial r}\cos\theta - \frac{\partial w}{\partial \theta}\frac{\sin\theta}{r}$$

$$\frac{\partial w}{\partial y} = \frac{\partial w}{\partial r}\sin\theta + \frac{\partial w}{\partial \theta}\frac{\cos\theta}{r}$$

(b) $$\left(\frac{\partial w}{\partial x}\right)^2 + \left(\frac{\partial w}{\partial y}\right)^2 = \left(\frac{\partial w}{\partial r}\right)^2 + \left(\frac{1}{r^2}\right)\left(\frac{\partial w}{\partial \theta}\right)^2$$

62. Demonstrate the result of Exercise 61(b) for $w = \arctan(y/x)$.

63. ***Cauchy-Riemann Equations*** Given the functions $u(x, y)$ and $v(x, y)$, verify that the **Cauchy-Riemann differential equations**

$$\frac{\partial u}{\partial x} = \frac{\partial v}{\partial y} \quad \text{and} \quad \frac{\partial u}{\partial y} = -\frac{\partial v}{\partial x}$$

can be written in polar coordinate form as

$$\frac{\partial u}{\partial r} = \frac{1}{r}\frac{\partial v}{\partial \theta} \quad \text{and} \quad \frac{\partial v}{\partial r} = -\frac{1}{r}\frac{\partial u}{\partial \theta}.$$

64. Demonstrate the result of Exercise 63 for the functions

$$u = \ln\sqrt{x^2 + y^2} \quad \text{and} \quad v = \arctan\frac{y}{x}.$$

65. Show that if $f(x,y)$ is homogeneous of degree n, then

$xf_x(x, y) + yf_y(x, y) = nf(x, y)$.

[*Hint:* Let $g(t) = f(tx, ty) = t^n f(x, y)$. Find $g'(t)$ and then let $t = 1$.]

13.6 Directional Derivatives and Gradients

- **Find and use directional derivatives of a function of two variables.**
- **Find the gradient of a function of two variables.**
- **Use the gradient of a function of two variables in applications.**
- **Find directional derivatives and gradients of functions of three variables.**

Directional Derivative

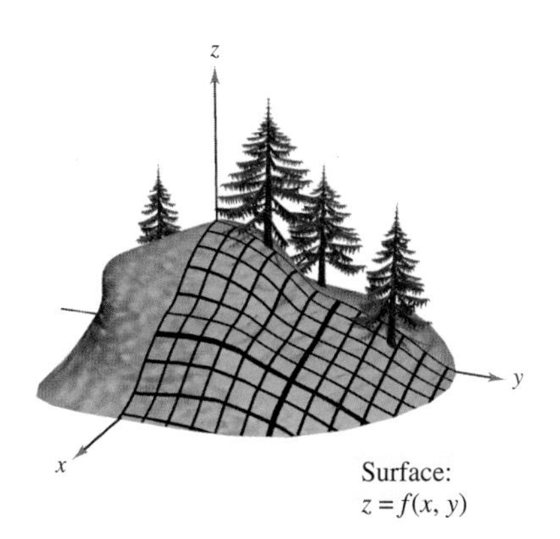

Figure 13.42

You are standing on the hillside pictured in Figure 13.42 and want to determine the hill's incline toward the z-axis. If the hill were represented by $z = f(x, y)$, you would already know how to determine the slopes in two different directions—the slope in the y-direction would be given by the partial derivative $f_y(x, y)$, and the slope in the x-direction would be given by the partial derivative $f_x(x, y)$. In this section, you will see that these two partial derivatives can be used to find the slope in *any* direction.

To determine the slope at a point on a surface, you will define a new type of derivative called a **directional derivative.** Begin by letting $z = f(x, y)$ be a *surface* and $P(x_0, y_0)$ be a *point* in the domain of f, as shown in Figure 13.43. The "direction" of the directional derivative is given by a unit vector

$$\mathbf{u} = \cos\theta\mathbf{i} + \sin\theta\mathbf{j}$$

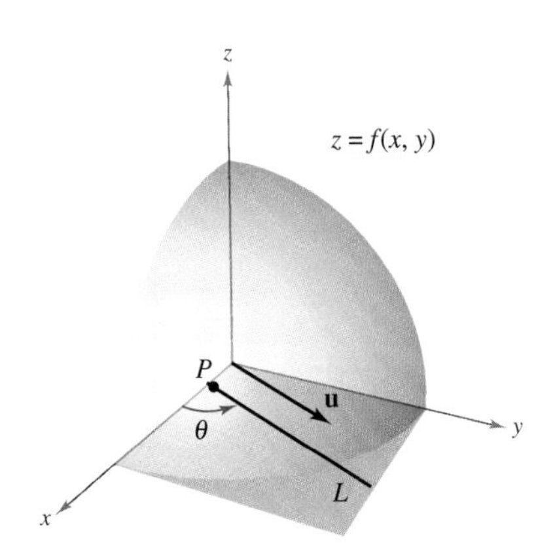

Figure 13.43

where θ is the angle the vector makes with the positive x-axis. To find the desired slope, reduce the problem to two dimensions by intersecting the surface with a vertical plane passing through the point P and parallel to $\mathbf{u}$, as shown in Figure 13.44. This vertical plane intersects the surface to form a curve C. The slope of the surface at $(x_0, y_0, f(x_0, y_0))$ in the direction of $\mathbf{u}$ is defined as the slope of the curve C at that point.

Informally, you can write the slope of the curve C as a limit that looks much like those used in single-variable calculus. The vertical plane used to form C intersects the xy-plane in a line L, represented by the parametric equations

$$x = x_0 + t\cos\theta$$

and

$$y = y_0 + t\sin\theta$$

so that for any value of t, the point $Q(x, y)$ lies on the line L. For each of the points P and Q, there is a corresponding point on the surface.

$$(x_0, y_0, f(x_0, y_0)) \qquad \text{Point above } P$$
$$(x, y, f(x, y)) \qquad \text{Point above } Q$$

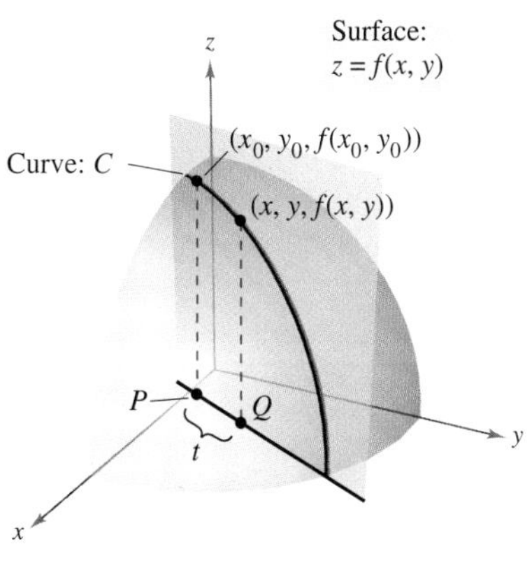

Figure 13.44

Moreover, because the distance between P and Q is

$$\sqrt{(x - x_0)^2 + (y - y_0)^2} = \sqrt{(t\cos\theta)^2 + (t\sin\theta)^2} = |t|$$

you can write the slope of the secant line through $(x_0, y_0, f(x_0, y_0))$ and $(x, y, f(x, y))$ as

$$\frac{f(x, y) - f(x_0, y_0)}{t} = \frac{f(x_0 + t\cos\theta, y_0 + t\sin\theta) - f(x_0, y_0)}{t}.$$

Finally, by letting t approach 0, you arrive at the following definition.

DEFINITION OF DIRECTIONAL DERIVATIVE

Let f be a function of two variables x and y and let $\mathbf{u} = \cos\theta\mathbf{i} + \sin\theta\mathbf{j}$ be a unit vector. Then the **directional derivative of f in the direction of u,** denoted by $D_{\mathbf{u}}f$, is

$$D_{\mathbf{u}}f(x, y) = \lim_{t\to 0} \frac{f(x + t\cos\theta, y + t\sin\theta) - f(x, y)}{t}$$

provided this limit exists.

Calculating directional derivatives by this definition is similar to finding the derivative of a function of one variable by the limit process (given in Section 2.1). A simpler "working" formula for finding directional derivatives involves the partial derivatives f_x and f_y.

THEOREM 13.9 DIRECTIONAL DERIVATIVE

If f is a differentiable function of x and y, then the directional derivative of f in the direction of the unit vector $\mathbf{u} = \cos\theta\mathbf{i} + \sin\theta\mathbf{j}$ is

$$D_{\mathbf{u}}f(x, y) = f_x(x, y)\cos\theta + f_y(x, y)\sin\theta.$$

PROOF For a fixed point (x_0, y_0), let $x = x_0 + t\cos\theta$ and let $y = y_0 + t\sin\theta$. Then, let $g(t) = f(x, y)$. Because f is differentiable, you can apply the Chain Rule given in Theorem 13.6 to obtain

$$g'(t) = f_x(x, y)x'(t) + f_y(x, y)y'(t) = f_x(x, y)\cos\theta + f_y(x, y)\sin\theta.$$

If $t = 0$, then $x = x_0$ and $y = y_0$, so

$$g'(0) = f_x(x_0, y_0)\cos\theta + f_y(x_0, y_0)\sin\theta.$$

By the definition of $g'(t)$, it is also true that

$$\begin{aligned} g'(0) &= \lim_{t\to 0} \frac{g(t) - g(0)}{t} \\ &= \lim_{t\to 0} \frac{f(x_0 + t\cos\theta, y_0 + t\sin\theta) - f(x_0, y_0)}{t}. \end{aligned}$$

Consequently, $D_{\mathbf{u}}f(x_0, y_0) = f_x(x_0, y_0)\cos\theta + f_y(x_0, y_0)\sin\theta$. ■

There are infinitely many directional derivatives of a surface at a given point—one for each direction specified by **u**, as shown in Figure 13.45. Two of these are the partial derivatives f_x and f_y.

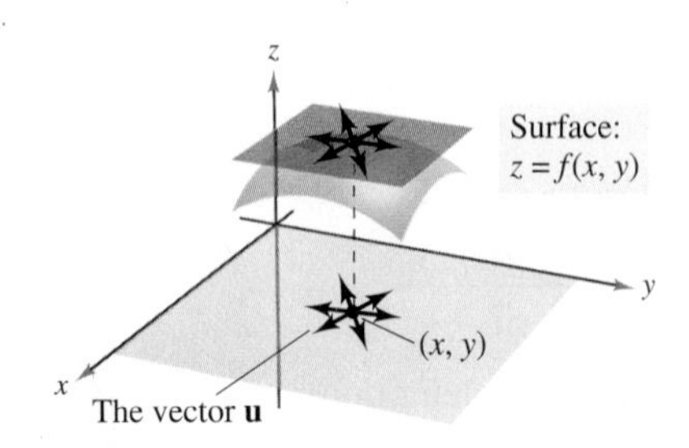

Figure 13.45

1. Direction of positive x-axis ($\theta = 0$): $\mathbf{u} = \cos 0\,\mathbf{i} + \sin 0\,\mathbf{j} = \mathbf{i}$

$$D_{\mathbf{i}}f(x, y) = f_x(x, y)\cos 0 + f_y(x, y)\sin 0 = f_x(x, y)$$

2. Direction of positive y-axis ($\theta = \pi/2$): $\mathbf{u} = \cos\frac{\pi}{2}\mathbf{i} + \sin\frac{\pi}{2}\mathbf{j} = \mathbf{j}$

$$D_{\mathbf{j}}f(x, y) = f_x(x, y)\cos\frac{\pi}{2} + f_y(x, y)\sin\frac{\pi}{2} = f_y(x, y)$$

EXAMPLE 1 Finding a Directional Derivative

Find the directional derivative of

$$f(x, y) = 4 - x^2 - \tfrac{1}{4}y^2 \qquad \text{Surface}$$

at $(1, 2)$ in the direction of

$$\mathbf{u} = \left(\cos\frac{\pi}{3}\right)\mathbf{i} + \left(\sin\frac{\pi}{3}\right)\mathbf{j}. \qquad \text{Direction}$$

Solution Because f_x and f_y are continuous, f is differentiable, and you can apply Theorem 13.9.

$$\begin{aligned} D_{\mathbf{u}}f(x, y) &= f_x(x, y)\cos\theta + f_y(x, y)\sin\theta \\ &= (-2x)\cos\theta + \left(-\frac{y}{2}\right)\sin\theta \end{aligned}$$

Evaluating at $\theta = \pi/3$, $x = 1$, and $y = 2$ produces

$$\begin{aligned} D_{\mathbf{u}}f(1, 2) &= (-2)\left(\frac{1}{2}\right) + (-1)\left(\frac{\sqrt{3}}{2}\right) \\ &= -1 - \frac{\sqrt{3}}{2} \\ &\approx -1.866. \end{aligned} \qquad \text{See Figure 13.46.}$$

Surface:
$f(x, y) = 4 - x^2 - \frac{1}{4}y^2$

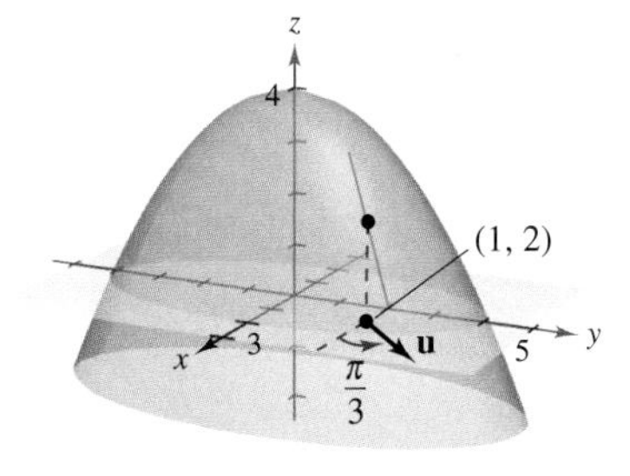

Figure 13.46

NOTE Note in Figure 13.46 that you can interpret the directional derivative as giving the slope of the surface at the point $(1, 2, 2)$ in the direction of the unit vector $\mathbf{u}$.

You have been specifying direction by a unit vector $\mathbf{u}$. If the direction is given by a vector whose length is not 1, you must normalize the vector before applying the formula in Theorem 13.9.

EXAMPLE 2 Finding a Directional Derivative

Find the directional derivative of

$$f(x, y) = x^2 \sin 2y \qquad \text{Surface}$$

at $(1, \pi/2)$ in the direction of

$$\mathbf{v} = 3\mathbf{i} - 4\mathbf{j}. \qquad \text{Direction}$$

Solution Because f_x and f_y are continuous, f is differentiable, and you can apply Theorem 13.9. Begin by finding a unit vector in the direction of $\mathbf{v}$.

$$\mathbf{u} = \frac{\mathbf{v}}{\|\mathbf{v}\|} = \frac{3}{5}\mathbf{i} - \frac{4}{5}\mathbf{j} = \cos\theta\,\mathbf{i} + \sin\theta\,\mathbf{j}$$

Using this unit vector, you have

$$\begin{aligned} D_{\mathbf{u}}f(x, y) &= (2x\sin 2y)(\cos\theta) + (2x^2\cos 2y)(\sin\theta) \\ D_{\mathbf{u}}f\left(1, \frac{\pi}{2}\right) &= (2\sin\pi)\left(\frac{3}{5}\right) + (2\cos\pi)\left(-\frac{4}{5}\right) \\ &= (0)\left(\frac{3}{5}\right) + (-2)\left(-\frac{4}{5}\right) \\ &= \frac{8}{5}. \end{aligned} \qquad \text{See Figure 13.47.}$$

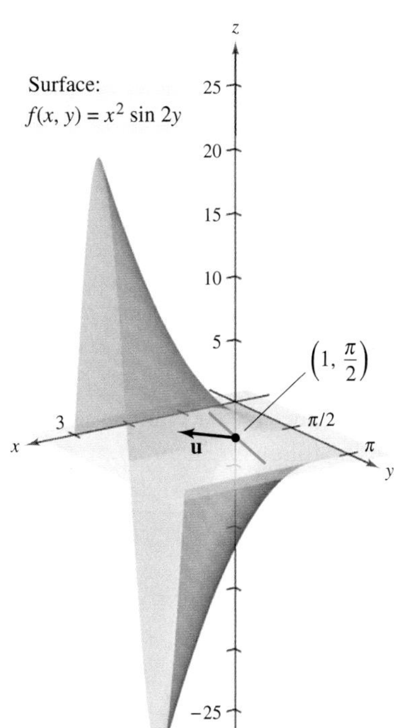

Figure 13.47

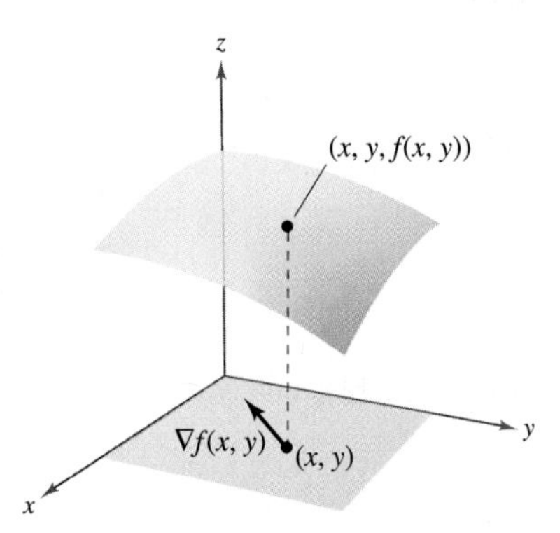

The gradient of f is a vector in the xy-plane.
Figure 13.48

The Gradient of a Function of Two Variables

The **gradient** of a function of two variables is a vector-valued function of two variables. This function has many important uses, some of which are described later in this section.

DEFINITION OF GRADIENT OF A FUNCTION OF TWO VARIABLES

Let $z = f(x, y)$ be a function of x and y such that f_x and f_y exist. Then the **gradient of *f***, denoted by $\nabla f(x, y)$, is the vector

$$\nabla f(x, y) = f_x(x, y)\mathbf{i} + f_y(x, y)\mathbf{j}.$$

∇f is read as "del f." Another notation for the gradient is **grad** $f(x, y)$. In Figure 13.48, note that for each (x, y), the gradient $\nabla f(x, y)$ is a vector in the plane (not a vector in space).

NOTE No value is assigned to the symbol ∇ by itself. It is an operator in the same sense that d/dx is an operator. When ∇ operates on $f(x, y)$, it produces the vector $\nabla f(x, y)$. ■

EXAMPLE 3 Finding the Gradient of a Function

Find the gradient of $f(x, y) = y \ln x + xy^2$ at the point $(1, 2)$.

Solution Using

$$f_x(x, y) = \frac{y}{x} + y^2 \quad \text{and} \quad f_y(x, y) = \ln x + 2xy$$

you have

$$\nabla f(x, y) = \left(\frac{y}{x} + y^2\right)\mathbf{i} + (\ln x + 2xy)\mathbf{j}.$$

At the point $(1, 2)$, the gradient is

$$\begin{aligned}\nabla f(1, 2) &= \left(\frac{2}{1} + 2^2\right)\mathbf{i} + [\ln 1 + 2(1)(2)]\mathbf{j}\\ &= 6\mathbf{i} + 4\mathbf{j}.\end{aligned}$$

■

Because the gradient of f is a vector, you can write the directional derivative of f in the direction of $\mathbf{u}$ as

$$D_{\mathbf{u}}f(x, y) = [f_x(x, y)\mathbf{i} + f_y(x, y)\mathbf{j}] \cdot [\cos\theta\mathbf{i} + \sin\theta\mathbf{j}].$$

In other words, the directional derivative is the dot product of the gradient and the direction vector. This useful result is summarized in the following theorem.

THEOREM 13.10 ALTERNATIVE FORM OF THE DIRECTIONAL DERIVATIVE

If f is a differentiable function of x and y, then the directional derivative of f in the direction of the unit vector $\mathbf{u}$ is

$$D_{\mathbf{u}}f(x, y) = \nabla f(x, y) \cdot \mathbf{u}.$$

EXAMPLE 4 Using $\nabla f(x, y)$ to Find a Directional Derivative

Find the directional derivative of

$$f(x, y) = 3x^2 - 2y^2$$

at $\left(-\frac{3}{4}, 0\right)$ in the direction from $P\left(-\frac{3}{4}, 0\right)$ to $Q(0, 1)$.

Solution Because the partials of f are continuous, f is differentiable and you can apply Theorem 13.10. A vector in the specified direction is

$$\overrightarrow{PQ} = \mathbf{v} = \left(0 + \frac{3}{4}\right)\mathbf{i} + (1 - 0)\mathbf{j}$$

$$= \frac{3}{4}\mathbf{i} + \mathbf{j}$$

and a unit vector in this direction is

$$\mathbf{u} = \frac{\mathbf{v}}{\|\mathbf{v}\|} = \frac{3}{5}\mathbf{i} + \frac{4}{5}\mathbf{j}. \qquad \text{Unit vector in direction of } \overrightarrow{PQ}$$

Because $\nabla f(x, y) = f_x(x, y)\mathbf{i} + f_y(x, y)\mathbf{j} = 6x\mathbf{i} - 4y\mathbf{j}$, the gradient at $\left(-\frac{3}{4}, 0\right)$ is

$$\nabla f\left(-\frac{3}{4}, 0\right) = -\frac{9}{2}\mathbf{i} + 0\mathbf{j}. \qquad \text{Gradient at } \left(-\tfrac{3}{4}, 0\right)$$

Consequently, at $\left(-\frac{3}{4}, 0\right)$ the directional derivative is

$$D_{\mathbf{u}} f\left(-\frac{3}{4}, 0\right) = \nabla f\left(-\frac{3}{4}, 0\right) \cdot \mathbf{u}$$

$$= \left(-\frac{9}{2}\mathbf{i} + 0\mathbf{j}\right) \cdot \left(\frac{3}{5}\mathbf{i} + \frac{4}{5}\mathbf{j}\right)$$

$$= -\frac{27}{10}. \qquad \text{Directional derivative at } \left(-\tfrac{3}{4}, 0\right)$$

See Figure 13.49. ■

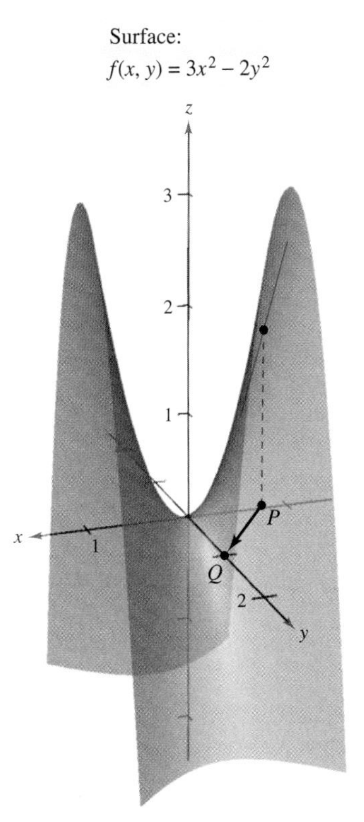

Figure 13.49

Applications of the Gradient

You have already seen that there are many directional derivatives at the point (x, y) on a surface. In many applications, you may want to know in which direction to move so that $f(x, y)$ increases most rapidly. This direction is called the direction of steepest ascent, and it is given by the gradient, as stated in the following theorem.

THEOREM 13.11 PROPERTIES OF THE GRADIENT

Let f be differentiable at the point (x, y).

1. If $\nabla f(x, y) = \mathbf{0}$, then $D_{\mathbf{u}} f(x, y) = 0$ for all $\mathbf{u}$.
2. The direction of *maximum* increase of f is given by $\nabla f(x, y)$. The maximum value of $D_{\mathbf{u}} f(x, y)$ is $\|\nabla f(x, y)\|$.
3. The direction of *minimum* increase of f is given by $-\nabla f(x, y)$. The minimum value of $D_{\mathbf{u}} f(x, y)$ is $-\|\nabla f(x, y)\|$.

NOTE Part 2 of Theorem 13.11 says that at the point (x, y), f increases most rapidly in the direction of the gradient, $\nabla f(x, y)$.

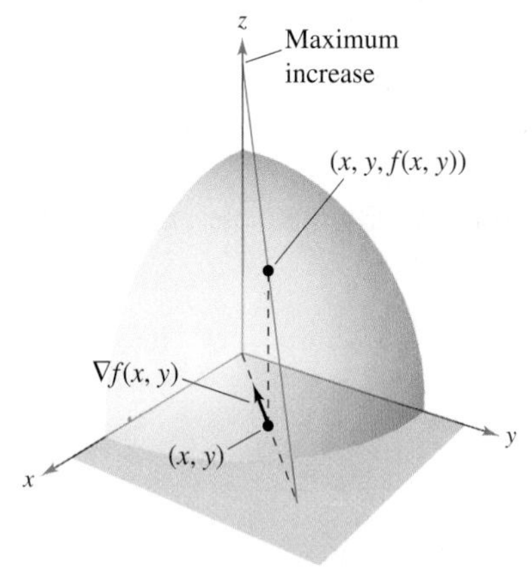

The gradient of f is a vector in the xy-plane that points in the direction of maximum increase on the surface given by $z = f(x, y)$.
Figure 13.50

PROOF If $\nabla f(x, y) = \mathbf{0}$, then for any direction (any $\mathbf{u}$), you have

$$\begin{aligned} D_{\mathbf{u}}f(x, y) &= \nabla f(x, y) \cdot \mathbf{u} \\ &= (0\mathbf{i} + 0\mathbf{j}) \cdot (\cos\theta\mathbf{i} + \sin\theta\mathbf{j}) \\ &= 0. \end{aligned}$$

If $\nabla f(x, y) \neq \mathbf{0}$, then let ϕ be the angle between $\nabla f(x, y)$ and a unit vector $\mathbf{u}$. Using the dot product, you can apply Theorem 11.5 to conclude that

$$\begin{aligned} D_{\mathbf{u}}f(x, y) &= \nabla f(x, y) \cdot \mathbf{u} \\ &= \|\nabla f(x, y)\| \, \|\mathbf{u}\| \cos\phi \\ &= \|\nabla f(x, y)\| \cos\phi \end{aligned}$$

and it follows that the maximum value of $D_{\mathbf{u}}f(x, y)$ will occur when $\cos\phi = 1$. So, $\phi = 0$, and the maximum value of the directional derivative occurs when $\mathbf{u}$ has the same direction as $\nabla f(x, y)$. Moreover, this largest value of $D_{\mathbf{u}}f(x, y)$ is precisely

$$\|\nabla f(x, y)\| \cos\phi = \|\nabla f(x, y)\|.$$

Similarly, the minimum value of $D_{\mathbf{u}}f(x, y)$ can be obtained by letting $\phi = \pi$ so that $\mathbf{u}$ points in the direction opposite that of $\nabla f(x, y)$, as shown in Figure 13.50. ■

To visualize one of the properties of the gradient, imagine a skier coming down a mountainside. If $f(x, y)$ denotes the altitude of the skier, then $-\nabla f(x, y)$ indicates the *compass direction* the skier should take to ski the path of steepest descent. (Remember that the gradient indicates direction in the xy-plane and does not itself point up or down the mountainside.)

As another illustration of the gradient, consider the temperature $T(x, y)$ at any point (x, y) on a flat metal plate. In this case, $\nabla T(x, y)$ gives the direction of greatest temperature increase at the point (x, y), as illustrated in the next example.

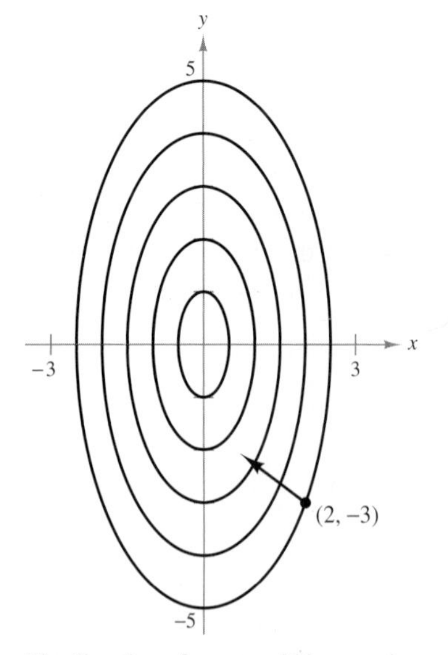

The direction of most rapid increase in temperature at $(2, -3)$ is given by $-16\mathbf{i} + 6\mathbf{j}$.
Figure 13.51

EXAMPLE 5 Finding the Direction of Maximum Increase

The temperature in degrees Celsius on the surface of a metal plate is

$$T(x, y) = 20 - 4x^2 - y^2$$

where x and y are measured in centimeters. In what direction from $(2, -3)$ does the temperature increase most rapidly? What is this rate of increase?

Solution The gradient is

$$\begin{aligned} \nabla T(x, y) &= T_x(x, y)\mathbf{i} + T_y(x, y)\mathbf{j} \\ &= -8x\mathbf{i} - 2y\mathbf{j}. \end{aligned}$$

It follows that the direction of maximum increase is given by

$$\nabla T(2, -3) = -16\mathbf{i} + 6\mathbf{j}$$

as shown in Figure 13.51, and the rate of increase is

$$\begin{aligned} \|\nabla T(2, -3)\| &= \sqrt{256 + 36} \\ &= \sqrt{292} \\ &\approx 17.09^\circ \text{ per centimeter.} \end{aligned}$$

■

The solution presented in Example 5 can be misleading. Although the gradient points in the direction of maximum temperature increase, it does not necessarily point toward the hottest spot on the plate. In other words, the gradient provides a local solution to finding an increase relative to the temperature at the point $(2, -3)$. *Once you leave that position, the direction of maximum increase may change.*

EXAMPLE 6 Finding the Path of a Heat-Seeking Particle

A heat-seeking particle is located at the point $(2, -3)$ on a metal plate whose temperature at (x, y) is

$$T(x, y) = 20 - 4x^2 - y^2.$$

Find the path of the particle as it continuously moves in the direction of maximum temperature increase.

Solution Let the path be represented by the position function

$$\mathbf{r}(t) = x(t)\mathbf{i} + y(t)\mathbf{j}.$$

A tangent vector at each point $(x(t), y(t))$ is given by

$$\mathbf{r}'(t) = \frac{dx}{dt}\mathbf{i} + \frac{dy}{dt}\mathbf{j}.$$

Because the particle seeks maximum temperature increase, the directions of $\mathbf{r}'(t)$ and $\nabla T(x, y) = -8x\mathbf{i} - 2y\mathbf{j}$ are the same at each point on the path. So,

$$-8x = k\frac{dx}{dt} \quad \text{and} \quad -2y = k\frac{dy}{dt}$$

where k depends on t. By solving each equation for dt/k and equating the results, you obtain

$$\frac{dx}{-8x} = \frac{dy}{-2y}.$$

The solution of this differential equation is $x = Cy^4$. Because the particle starts at the point $(2, -3)$, you can determine that $C = 2/81$. So, the path of the heat-seeking particle is

$$x = \frac{2}{81}y^4.$$

The path is shown in Figure 13.52. ■

Level curves: $T(x, y) = 20 - 4x^2 - y^2$

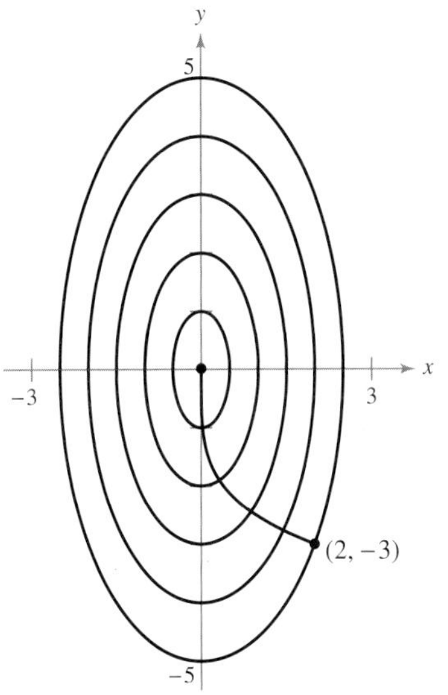

Path followed by a heat-seeking particle
Figure 13.52

In Figure 13.52, the path of the particle (determined by the gradient at each point) appears to be orthogonal to each of the level curves. This becomes clear when you consider that the temperature $T(x, y)$ is constant along a given level curve. So, at any point (x, y) on the curve, the rate of change of T in the direction of a unit tangent vector $\mathbf{u}$ is 0, and you can write

$$\nabla f(x, y) \cdot \mathbf{u} = D_{\mathbf{u}}T(x, y) = 0. \qquad \mathbf{u} \text{ is a unit tangent vector.}$$

Because the dot product of $\nabla f(x, y)$ and $\mathbf{u}$ is 0, you can conclude that they must be orthogonal. This result is stated in the following theorem.

THEOREM 13.12 GRADIENT IS NORMAL TO LEVEL CURVES

If f is differentiable at (x_0, y_0) and $\nabla f(x_0, y_0) \neq \mathbf{0}$, then $\nabla f(x_0, y_0)$ is normal to the level curve through (x_0, y_0).

EXAMPLE 7 Finding a Normal Vector to a Level Curve

Sketch the level curve corresponding to $c = 0$ for the function given by

$$f(x, y) = y - \sin x$$

and find a normal vector at several points on the curve.

Solution The level curve for $c = 0$ is given by

$$0 = y - \sin x$$
$$y = \sin x$$

as shown in Figure 13.53(a). Because the gradient vector of f at (x, y) is

$$\begin{aligned}\nabla f(x, y) &= f_x(x, y)\mathbf{i} + f_y(x, y)\mathbf{j} \\ &= -\cos x\mathbf{i} + \mathbf{j}\end{aligned}$$

you can use Theorem 13.12 to conclude that $\nabla f(x, y)$ is normal to the level curve at the point (x, y). Some gradient vectors are

$$\nabla f(-\pi, 0) = \mathbf{i} + \mathbf{j}$$
$$\nabla f\left(-\frac{2\pi}{3}, -\frac{\sqrt{3}}{2}\right) = \frac{1}{2}\mathbf{i} + \mathbf{j}$$
$$\nabla f\left(-\frac{\pi}{2}, -1\right) = \mathbf{j}$$
$$\nabla f\left(-\frac{\pi}{3}, -\frac{\sqrt{3}}{2}\right) = -\frac{1}{2}\mathbf{i} + \mathbf{j}$$
$$\nabla f(0, 0) = -\mathbf{i} + \mathbf{j}$$
$$\nabla f\left(\frac{\pi}{3}, \frac{\sqrt{3}}{2}\right) = -\frac{1}{2}\mathbf{i} + \mathbf{j}$$
$$\nabla f\left(\frac{\pi}{2}, 1\right) = \mathbf{j}.$$

These are shown in Figure 13.53(b).

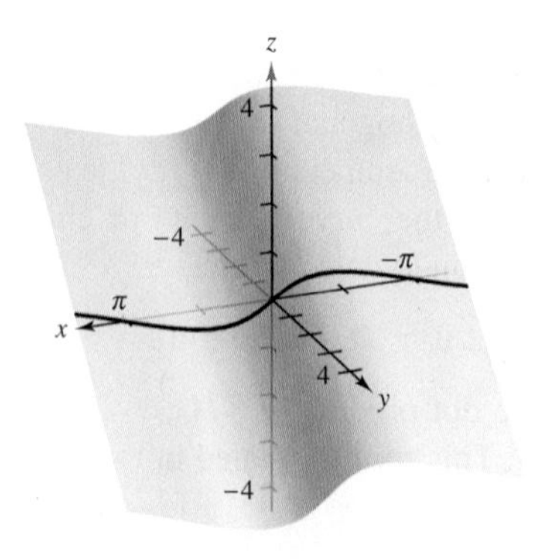

(a) The surface is given by $f(x, y) = y - \sin x$.

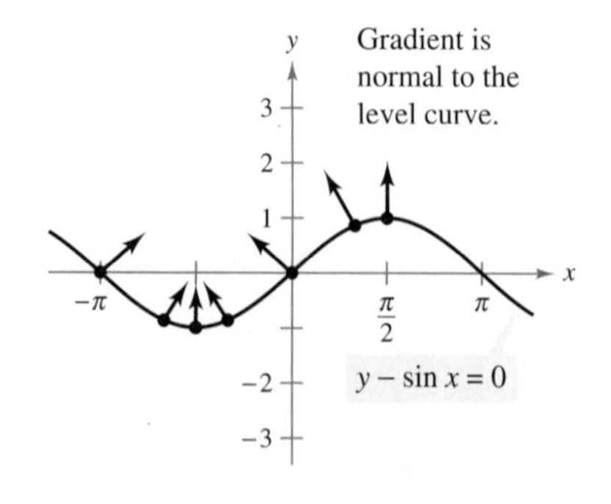

(b) The level curve is given by $f(x, y) = 0$.

Figure 13.53

Functions of Three Variables

The definitions of the directional derivative and the gradient can be extended naturally to functions of three or more variables. As often happens, some of the geometric interpretation is lost in the generalization from functions of two variables to those of three variables. For example, you cannot interpret the directional derivative of a function of three variables to represent slope.

The definitions and properties of the directional derivative and the gradient of a function of three variables are given in the following summary.

DIRECTIONAL DERIVATIVE AND GRADIENT FOR THREE VARIABLES

Let f be a function of x, y, and z, with continuous first partial derivatives. The **directional derivative of f** in the direction of a unit vector $\mathbf{u} = a\mathbf{i} + b\mathbf{j} + c\mathbf{k}$ is given by

$$D_{\mathbf{u}} f(x, y, z) = af_x(x, y, z) + bf_y(x, y, z) + cf_z(x, y, z).$$

The **gradient of f** is defined as

$$\nabla f(x, y, z) = f_x(x, y, z)\mathbf{i} + f_y(x, y, z)\mathbf{j} + f_z(x, y, z)\mathbf{k}.$$

Properties of the gradient are as follows.

1. $D_{\mathbf{u}} f(x, y, z) = \nabla f(x, y, z) \cdot \mathbf{u}$
2. If $\nabla f(x, y, z) = \mathbf{0}$, then $D_{\mathbf{u}} f(x, y, z) = 0$ for all $\mathbf{u}$.
3. The direction of *maximum* increase of f is given by $\nabla f(x, y, z)$. The maximum value of $D_{\mathbf{u}} f(x, y, z)$ is

   $\|\nabla f(x, y, z)\|$. Maximum value of $D_{\mathbf{u}} f(x, y, z)$
4. The direction of *minimum* increase of f is given by $-\nabla f(x, y, z)$. The minimum value of $D_{\mathbf{u}} f(x, y, z)$ is

   $-\|\nabla f(x, y, z)\|$. Minimum value of $D_{\mathbf{u}} f(x, y, z)$

NOTE You can generalize Theorem 13.12 to functions of three variables. Under suitable hypotheses,

$$\nabla f(x_0, y_0, z_0)$$

is normal to the level surface through (x_0, y_0, z_0). ■

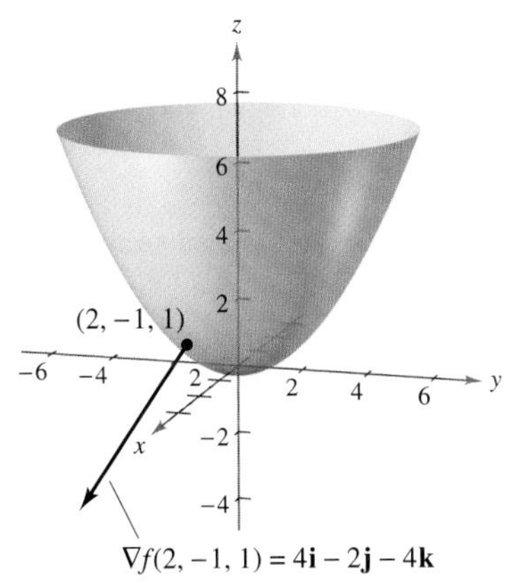

Level surface and gradient vector at $(2, -1, 1)$ for $f(x, y, z) = x^2 + y^2 - 4z$
Figure 13.54

EXAMPLE 8 Finding the Gradient for a Function of Three Variables

Find $\nabla f(x, y, z)$ for the function given by

$$f(x, y, z) = x^2 + y^2 - 4z$$

and find the direction of maximum increase of f at the point $(2, -1, 1)$.

Solution The gradient vector is given by

$$\begin{aligned}\nabla f(x, y, z) &= f_x(x, y, z)\mathbf{i} + f_y(x, y, z)\mathbf{j} + f_z(x, y, z)\mathbf{k}\\ &= 2x\mathbf{i} + 2y\mathbf{j} - 4\mathbf{k}.\end{aligned}$$

So, it follows that the direction of maximum increase at $(2, -1, 1)$ is

$$\nabla f(2, -1, 1) = 4\mathbf{i} - 2\mathbf{j} - 4\mathbf{k}.$$ See Figure 13.54. ■

13.6 Exercises

See www.CalcChat.com for worked-out solutions to odd-numbered exercises.

In Exercises 1–12, find the directional derivative of the function at P in the direction of v.

1. $f(x, y) = 3x - 4xy + 9y,\quad P(1, 2),\ \mathbf{v} = \frac{3}{5}\mathbf{i} + \frac{4}{5}\mathbf{j}$
2. $f(x, y) = x^3 - y^3,\quad P(4, 3),\ \mathbf{v} = \frac{\sqrt{2}}{2}(\mathbf{i} + \mathbf{j})$
3. $f(x, y) = xy,\quad P(0, -2),\ \mathbf{v} = \frac{1}{2}(\mathbf{i} + \sqrt{3}\mathbf{j})$
4. $f(x, y) = \frac{x}{y},\quad P(1, 1),\ \mathbf{v} = -\mathbf{j}$
5. $h(x, y) = e^x \sin y,\quad P\left(1, \frac{\pi}{2}\right),\ \mathbf{v} = -\mathbf{i}$
6. $g(x, y) = \arccos xy,\quad P(1, 0),\ \mathbf{v} = \mathbf{j}$
7. $g(x, y) = \sqrt{x^2 + y^2},\quad P(3, 4),\ \mathbf{v} = 3\mathbf{i} - 4\mathbf{j}$
8. $h(x, y) = e^{-(x^2+y^2)},\quad P(0, 0),\ \mathbf{v} = \mathbf{i} + \mathbf{j}$
9. $f(x, y, z) = x^2 + y^2 + z^2,\quad P(1, 1, 1),\ \mathbf{v} = \frac{\sqrt{3}}{3}(\mathbf{i} - \mathbf{j} + \mathbf{k})$
10. $f(x, y, z) = xy + yz + xz,\quad P(1, 2, -1),\ \mathbf{v} = 2\mathbf{i} + \mathbf{j} - \mathbf{k}$
11. $h(x, y, z) = xyz,\quad P(2, 1, 1),\ \mathbf{v} = \langle 2, 1, 2\rangle$
12. $h(x, y, z) = x \arctan yz,\quad P(4, 1, 1),\ \mathbf{v} = \langle 1, 2, -1\rangle$

In Exercises 13–16, find the directional derivative of the function in the direction of the unit vector $\mathbf{u} = \cos\theta\mathbf{i} + \sin\theta\mathbf{j}$.

13. $f(x, y) = x^2 + y^2,\quad \theta = \frac{\pi}{4}$
14. $f(x, y) = \frac{y}{x + y},\quad \theta = -\frac{\pi}{6}$
15. $f(x, y) = \sin(2x + y),\quad \theta = \frac{\pi}{3}$
16. $g(x, y) = xe^y,\quad \theta = \frac{2\pi}{3}$

In Exercises 17–20, find the directional derivative of the function at P in the direction of Q.

17. $f(x, y) = x^2 + 3y^2,\quad P(1, 1), Q(4, 5)$
18. $f(x, y) = \cos(x + y),\quad P(0, \pi), Q\left(\frac{\pi}{2}, 0\right)$
19. $g(x, y, z) = xye^z,\quad P(2, 4, 0), Q(0, 0, 0)$
20. $h(x, y, z) = \ln(x + y + z),\quad P(1, 0, 0), Q(4, 3, 1)$

In Exercises 21–26, find the gradient of the function at the given point.

21. $f(x, y) = 3x + 5y^2 + 1,\quad (2, 1)$
22. $g(x, y) = 2xe^{y/x},\quad (2, 0)$
23. $z = \ln(x^2 - y),\quad (2, 3)$
24. $z = \cos(x^2 + y^2),\quad (3, -4)$
25. $w = 3x^2 - 5y^2 + 2z^2,\quad (1, 1, -2)$
26. $w = x\tan(y + z),\quad (4, 3, -1)$

In Exercises 27–30, use the gradient to find the directional derivative of the function at P in the direction of Q.

27. $g(x, y) = x^2 + y^2 + 1,\quad P(1, 2),\ Q(2, 3)$
28. $f(x, y) = 3x^2 - y^2 + 4,\quad P(-1, 4),\ Q(3, 6)$
29. $f(x, y) = e^y \sin x,\quad P(0, 0),\ Q(2, 1)$
30. $f(x, y) = \sin 2x \cos y,\quad P(\pi, 0),\ Q\left(\frac{\pi}{2}, \pi\right)$

In Exercises 31–40, find the gradient of the function and the maximum value of the directional derivative at the given point.

	Function	*Point*
31.	$f(x, y) = x^2 + 2xy$	$(1, 0)$
32.	$f(x, y) = \frac{x + y}{y + 1}$	$(0, 1)$
33.	$h(x, y) = x\tan y$	$\left(2, \frac{\pi}{4}\right)$
34.	$h(x, y) = y\cos(x - y)$	$\left(0, \frac{\pi}{3}\right)$
35.	$g(x, y) = ye^{-x}$	$(0, 5)$
36.	$g(x, y) = \ln \sqrt[3]{x^2 + y^2}$	$(1, 2)$
37.	$f(x, y, z) = \sqrt{x^2 + y^2 + z^2}$	$(1, 4, 2)$
38.	$w = \frac{1}{\sqrt{1 - x^2 - y^2 - z^2}}$	$(0, 0, 0)$
39.	$w = xy^2z^2$	$(2, 1, 1)$
40.	$f(x, y, z) = xe^{yz}$	$(2, 0, -4)$

In Exercises 41–46, consider the function $f(x, y) = 3 - \frac{x}{3} - \frac{y}{2}$.

41. Sketch the graph of f in the first octant and plot the point $(3, 2, 1)$ on the surface.
42. Find $D_{\mathbf{u}}f(3, 2)$, where $\mathbf{u} = \cos\theta\mathbf{i} + \sin\theta\mathbf{j}$, using each given value of θ.
    (a) $\theta = \frac{\pi}{4}$ (b) $\theta = \frac{2\pi}{3}$
    (c) $\theta = \frac{4\pi}{3}$ (d) $\theta = -\frac{\pi}{6}$
43. Find $D_{\mathbf{u}}f(3, 2)$, where $\mathbf{u} = \frac{\mathbf{v}}{\|\mathbf{v}\|}$, using each given vector $\mathbf{v}$.
    (a) $\mathbf{v} = \mathbf{i} + \mathbf{j}$
    (b) $\mathbf{v} = -3\mathbf{i} - 4\mathbf{j}$
    (c) $\mathbf{v}$ is the vector from $(1, 2)$ to $(-2, 6)$.
    (d) $\mathbf{v}$ is the vector from $(3, 2)$ to $(4, 5)$.
44. Find $\nabla f(x, y)$.
45. Find the maximum value of the directional derivative at $(3, 2)$.
46. Find a unit vector $\mathbf{u}$ orthogonal to $\nabla f(3, 2)$ and calculate $D_{\mathbf{u}}f(3, 2)$. Discuss the geometric meaning of the result.

Investigation In Exercises 47 and 48, (a) use the graph to estimate the components of the vector in the direction of the maximum rate of increase in the function at the given point. (b) Find the gradient at the point and compare it with your estimate in part (a). (c) In what direction would the function be decreasing at the greatest rate? Explain.

47. $f(x, y) = \frac{1}{10}(x^2 - 3xy + y^2)$, $(1, 2)$ **48.** $f(x, y) = \frac{1}{2}y\sqrt{x}$, $(1, 2)$

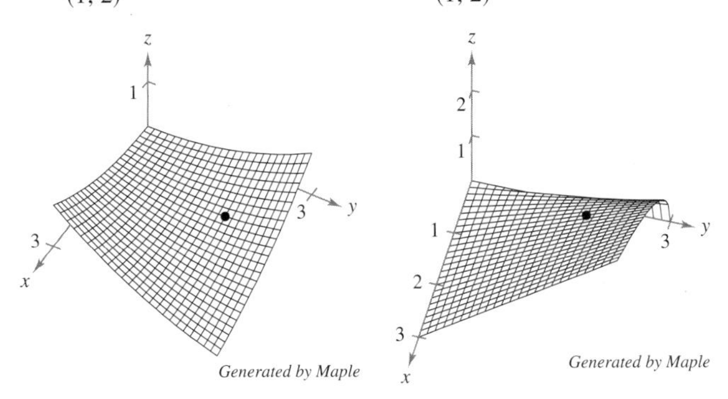

CAS **49.** ***Investigation*** Consider the function

$$f(x, y) = x^2 - y^2$$

at the point $(4, -3, 7)$.

(a) Use a computer algebra system to graph the surface represented by the function.

(b) Determine the directional derivative $D_{\mathbf{u}}f(4, -3)$ as a function of θ, where $\mathbf{u} = \cos\theta\,\mathbf{i} + \sin\theta\,\mathbf{j}$. Use a computer algebra system to graph the function on the interval $[0, 2\pi)$.

(c) Approximate the zeros of the function in part (b) and interpret each in the context of the problem.

(d) Approximate the critical numbers of the function in part (b) and interpret each in the context of the problem.

(e) Find $\|\nabla f(4, -3)\|$ and explain its relationship to your answers in part (d).

(f) Use a computer algebra system to graph the level curve of the function f at the level $c = 7$. On this curve, graph the vector in the direction of $\nabla f(4, -3)$, and state its relationship to the level curve.

50. ***Investigation*** Consider the function

$$f(x, y) = \frac{8y}{1 + x^2 + y^2}.$$

(a) Analytically verify that the level curve of $f(x, y)$ at the level $c = 2$ is a circle.

(b) At the point $(\sqrt{3}, 2)$ on the level curve for which $c = 2$, sketch the vector showing the direction of the greatest rate of increase of the function. (To print an enlarged copy of the graph, go to the website *www.mathgraphs.com.*)

(c) At the point $(\sqrt{3}, 2)$ on the level curve, sketch a vector such that the directional derivative is 0.

CAS (d) Use a computer algebra system to graph the surface to verify your answers in parts (a)–(c).

In Exercises 51–54, find a normal vector to the level curve $f(x, y) = c$ at P.

51. $f(x, y) = 6 - 2x - 3y$
$c = 6, \quad P(0, 0)$

52. $f(x, y) = x^2 + y^2$
$c = 25, \quad P(3, 4)$

53. $f(x, y) = xy$
$c = -3, \quad P(-1, 3)$

54. $f(x, y) = \dfrac{x}{x^2 + y^2}$
$c = \frac{1}{2}, \quad P(1, 1)$

In Exercises 55–58, (a) find the gradient of the function at P, (b) find a unit normal vector to the level curve $f(x, y) = c$ at P, (c) find the tangent line to the level curve $f(x, y) = c$ at P, and (d) sketch the level curve, the unit normal vector, and the tangent line in the xy-plane.

55. $f(x, y) = 4x^2 - y$
$c = 6, P(2, 10)$

56. $f(x, y) = x - y^2$
$c = 3, P(4, -1)$

57. $f(x, y) = 3x^2 - 2y^2$
$c = 1, P(1, 1)$

58. $f(x, y) = 9x^2 + 4y^2$
$c = 40, P(2, -1)$

WRITING ABOUT CONCEPTS

59. Define the derivative of the function $z = f(x, y)$ in the direction $\mathbf{u} = \cos\theta\,\mathbf{i} + \sin\theta\,\mathbf{j}$.

60. Write a paragraph describing the directional derivative of the function f in the direction $\mathbf{u} = \cos\theta\,\mathbf{i} + \sin\theta\,\mathbf{j}$ when (a) $\theta = 0°$ and (b) $\theta = 90°$.

61. Define the gradient of a function of two variables. State the properties of the gradient.

62. Sketch the graph of a surface and select a point P on the surface. Sketch a vector in the xy-plane giving the direction of steepest ascent on the surface at P.

63. Describe the relationship of the gradient to the level curves of a surface given by $z = f(x, y)$.

CAPSTONE

64. Consider the function $f(x, y) = 9 - x^2 - y^2$.

(a) Sketch the graph of f in the first octant and plot the point $(1, 2, 4)$ on the surface.

(b) Find $D_{\mathbf{u}}f(1, 2)$, where $\mathbf{u} = \cos\theta\,\mathbf{i} + \sin\theta\,\mathbf{j}$, for $\theta = -\pi/4$.

(c) Repeat part (b) for $\theta = \pi/3$.

(d) Find $\nabla f(1, 2)$ and $\|\nabla f(1, 2)\|$.

(e) Find a unit vector $\mathbf{u}$ orthogonal to $\nabla f(1, 2)$ and calculate $D_{\mathbf{u}}f(1, 2)$. Discuss the geometric meaning of the result.

65. ***Temperature Distribution*** The temperature at the point (x, y) on a metal plate is

$$T = \frac{x}{x^2 + y^2}.$$

Find the direction of greatest increase in heat from the point $(3, 4)$.

66. ***Topography*** The surface of a mountain is modeled by the equation $h(x, y) = 5000 - 0.001x^2 - 0.004y^2$. A mountain climber is at the point $(500, 300, 4390)$. In what direction should the climber move in order to ascend at the greatest rate?

67. ***Topography*** The figure shows a topographic map carried by a group of hikers. Sketch the paths of steepest descent if the hikers start at point A and if they start at point B. (To print an enlarged copy of the graph, go to the website *www.mathgraphs.com*.)

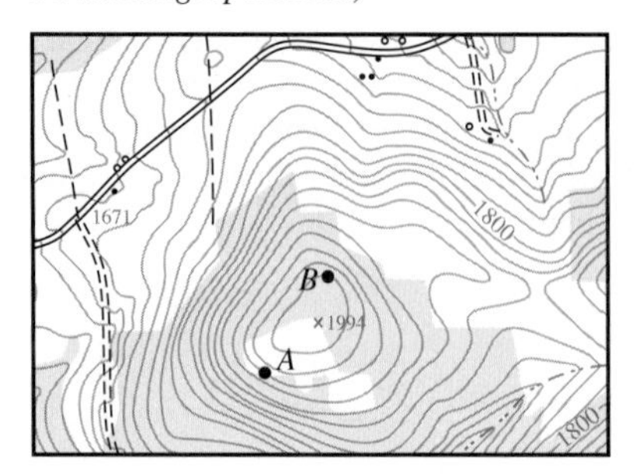

68. ***Meteorology*** Meteorologists measure the atmospheric pressure in units called millibars. From these observations they create weather maps on which the curves of equal atmospheric pressure (isobars) are drawn (see figure). These are level curves to the function $P(x, y)$ yielding the pressure at any point. Sketch the gradients to the isobars at the points A, B, and C. Although the magnitudes of the gradients are unknown, their lengths relative to each other can be estimated. At which of the three points is the wind speed greatest if the speed increases as the pressure gradient increases? (To print an enlarged copy of the graph, go to the website *www.mathgraphs.com*.)

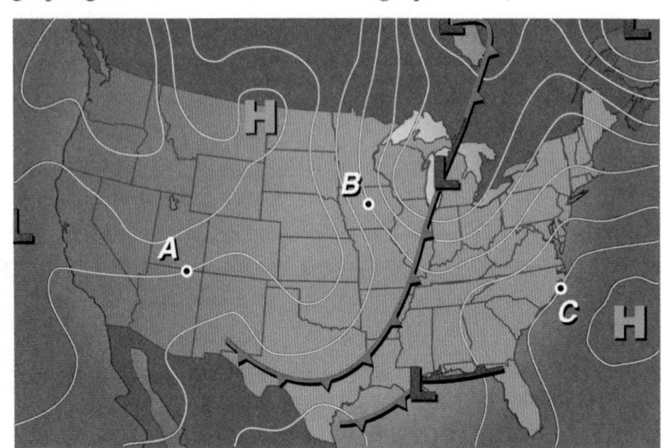

Heat-Seeking Path **In Exercises 69 and 70, find the path of a heat-seeking particle placed at point P on a metal plate with a temperature field $T(x, y)$.**

	Temperature Field	*Point*
69.	$T(x, y) = 400 - 2x^2 - y^2$	$P(10, 10)$
70.	$T(x, y) = 100 - x^2 - 2y^2$	$P(4, 3)$

71. ***Temperature*** The temperature at the point (x, y) on a metal plate is modeled by $T(x, y) = 400e^{-(x^2+y)/2}$, $x \geq 0$, $y \geq 0$.

CAS (a) Use a computer algebra system to graph the temperature distribution function.

(b) Find the directions of no change in heat on the plate from the point $(3, 5)$.

(c) Find the direction of greatest increase in heat from the point $(3, 5)$.

CAS **72.** ***Investigation*** A team of oceanographers is mapping the ocean floor to assist in the recovery of a sunken ship. Using sonar, they develop the model

$$D = 250 + 30x^2 + 50\sin\frac{\pi y}{2}, \quad 0 \leq x \leq 2, 0 \leq y \leq 2$$

where D is the depth in meters, and x and y are the distances in kilometers.

(a) Use a computer algebra system to graph the surface.

(b) Because the graph in part (a) is showing depth, it is not a map of the ocean floor. How could the model be changed so that the graph of the ocean floor could be obtained?

(c) What is the depth of the ship if it is located at the coordinates $x = 1$ and $y = 0.5$?

(d) Determine the steepness of the ocean floor in the positive x-direction from the position of the ship.

(e) Determine the steepness of the ocean floor in the positive y-direction from the position of the ship.

(f) Determine the direction of the greatest rate of change of depth from the position of the ship.

True or False? **In Exercises 73–76, determine whether the statement is true or false. If it is false, explain why or give an example that shows it is false.**

73. If $f(x, y) = \sqrt{1 - x^2 - y^2}$, then $D_{\mathbf{u}} f(0, 0) = 0$ for any unit vector $\mathbf{u}$.

74. If $f(x, y) = x + y$, then $-1 \leq D_{\mathbf{u}} f(x, y) \leq 1$.

75. If $D_{\mathbf{u}} f(x, y)$ exists, then $D_{\mathbf{u}} f(x, y) = -D_{-\mathbf{u}} f(x, y)$.

76. If $D_{\mathbf{u}} f(x_0, y_0) = c$ for any unit vector $\mathbf{u}$, then $c = 0$.

77. Find a function f such that $\nabla f = e^x \cos y\,\mathbf{i} - e^x \sin y\,\mathbf{j} + z\,\mathbf{k}$.

78. Consider the function

$$f(x, y) = \begin{cases} \dfrac{4xy}{x^2 + y^2}, & (x, y) \neq (0, 0) \\ 0, & (x, y) = (0, 0) \end{cases}$$

and the unit vector $\mathbf{u} = \dfrac{1}{\sqrt{2}}(\mathbf{i} + \mathbf{j})$.

Does the directional derivative of f at $P(0, 0)$ in the direction of $\mathbf{u}$ exist? If $f(0, 0)$ were defined as 2 instead of 0, would the directional derivative exist?

79. Consider the function $f(x, y) = \sqrt[3]{xy}$.

(a) Show that f is continuous at the origin.

(b) Show that f_x and f_y exist at the origin, but that the directional derivatives at the origin in all other directions do not exist.

CAS (c) Use a computer algebra system to graph f near the origin to verify your answers in parts (a) and (b). Explain.

13.7 Tangent Planes and Normal Lines

- **Find equations of tangent planes and normal lines to surfaces.**
- **Find the angle of inclination of a plane in space.**
- **Compare the gradients $\nabla f(x, y)$ and $\nabla F(x, y, z)$.**

EXPLORATION

Billiard Balls and Normal Lines
In each of the three figures below, the cue ball is about to strike a stationary ball at point P. Explain how you can use the normal line to the stationary ball at point P to describe the resulting motion of each of the two balls. Assuming that each cue ball has the same speed, which stationary ball will acquire the greatest speed? Which will acquire the least? Explain your reasoning.

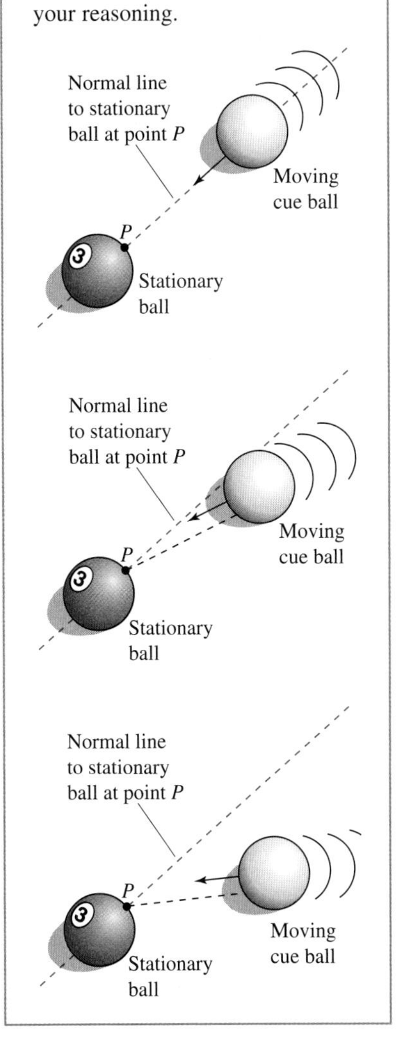

Tangent Plane and Normal Line to a Surface

So far you have represented surfaces in space primarily by equations of the form

$$z = f(x, y). \qquad \text{Equation of a surface } S$$

In the development to follow, however, it is convenient to use the more general representation $F(x, y, z) = 0$. For a surface S given by $z = f(x, y)$, you can convert to the general form by defining F as

$$F(x, y, z) = f(x, y) - z.$$

Because $f(x, y) - z = 0$, you can consider S to be the level surface of F given by

$$F(x, y, z) = 0. \qquad \text{Alternative equation of surface } S$$

EXAMPLE 1 Writing an Equation of a Surface

For the function given by

$$F(x, y, z) = x^2 + y^2 + z^2 - 4$$

describe the level surface given by $F(x, y, z) = 0$.

Solution The level surface given by $F(x, y, z) = 0$ can be written as

$$x^2 + y^2 + z^2 = 4$$

which is a sphere of radius 2 whose center is at the origin. ■

You have seen many examples of the usefulness of normal lines in applications involving curves. Normal lines are equally important in analyzing surfaces and solids. For example, consider the collision of two billiard balls. When a stationary ball is struck at a point P on its surface, it moves along the **line of impact** determined by P and the center of the ball. The impact can occur in *two* ways. If the cue ball is moving along the line of impact, it stops dead and imparts all of its momentum to the stationary ball, as shown in Figure 13.55. If the cue ball is not moving along the line of impact, it is deflected to one side or the other and retains part of its momentum. That part of the momentum that is transferred to the stationary ball occurs along the line of impact, *regardless* of the direction of the cue ball, as shown in Figure 13.56. This line of impact is called the **normal line** to the surface of the ball at the point P.

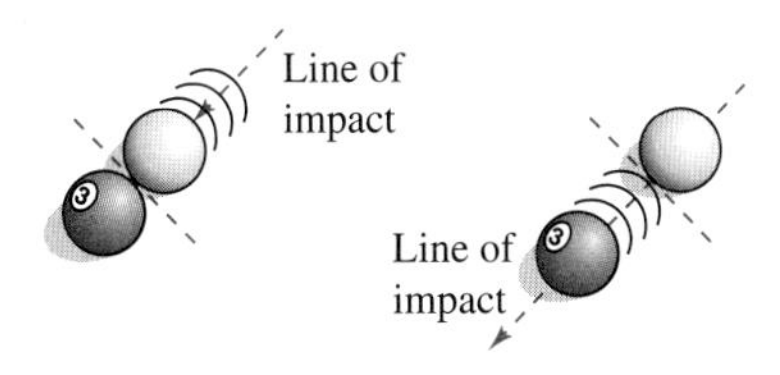

Figure 13.55

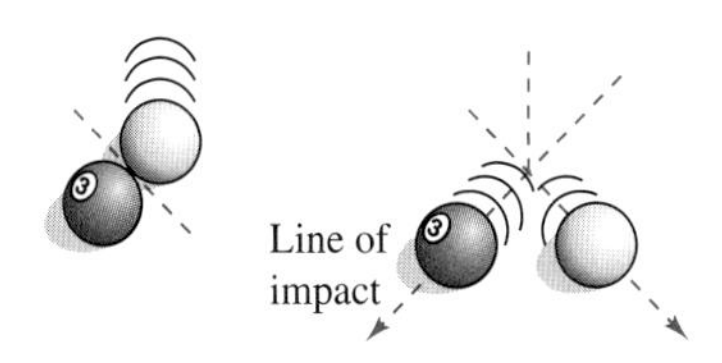

Figure 13.56

In the process of finding a normal line to a surface, you are also able to solve the problem of finding a **tangent plane** to the surface. Let S be a surface given by

$$F(x, y, z) = 0$$

and let $P(x_0, y_0, z_0)$ be a point on S. Let C be a curve on S through P that is defined by the vector-valued function

$$\mathbf{r}(t) = x(t)\mathbf{i} + y(t)\mathbf{j} + z(t)\mathbf{k}.$$

Then, for all t,

$$F(x(t),\ y(t),\ z(t)) = 0.$$

If F is differentiable and $x'(t)$, $y'(t)$, and $z'(t)$ all exist, it follows from the Chain Rule that

$$\begin{aligned} 0 &= F'(t) \\ &= F_x(x, y, z)x'(t) + F_y(x, y, z)y'(t) + F_z(x, y, z)z'(t). \end{aligned}$$

At (x_0, y_0, z_0), the equivalent vector form is

$$0 = \underbrace{\nabla F(x_0, y_0, z_0)}_{\text{Gradient}} \cdot \underbrace{\mathbf{r}'(t_0)}_{\text{Tangent vector}}.$$

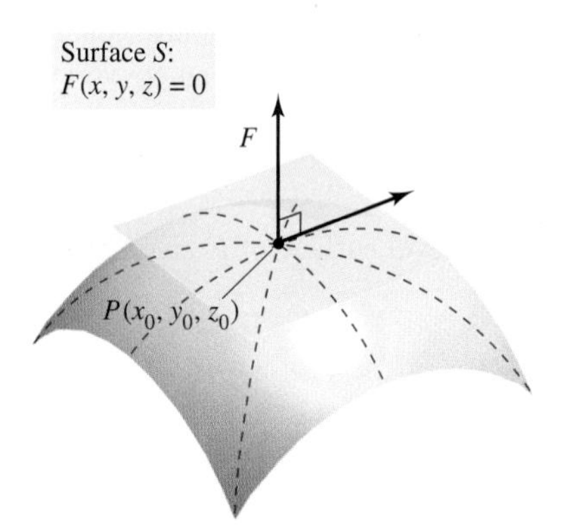

Tangent plane to surface S at P
Figure 13.57

This result means that the gradient at P is orthogonal to the tangent vector of every curve on S through P. So, all tangent lines on S lie in a plane that is normal to $\nabla F(x_0, y_0, z_0)$ and contains P, as shown in Figure 13.57.

DEFINITIONS OF TANGENT PLANE AND NORMAL LINE

Let F be differentiable at the point $P(x_0, y_0, z_0)$ on the surface S given by $F(x, y, z) = 0$ such that $\nabla F(x_0, y_0, z_0) \neq \mathbf{0}$.

1. The plane through P that is normal to $\nabla F(x_0, y_0, z_0)$ is called the **tangent plane to S at P.**
2. The line through P having the direction of $\nabla F(x_0, y_0, z_0)$ is called the **normal line to S at P.**

NOTE In the remainder of this section, assume $\nabla F(x_0, y_0, z_0)$ to be nonzero unless stated otherwise. ■

To find an equation for the tangent plane to S at (x_0, y_0, z_0), let (x, y, z) be an arbitrary point in the tangent plane. Then the vector

$$\mathbf{v} = (x - x_0)\mathbf{i} + (y - y_0)\mathbf{j} + (z - z_0)\mathbf{k}$$

lies in the tangent plane. Because $\nabla F(x_0, y_0, z_0)$ is normal to the tangent plane at (x_0, y_0, z_0), it must be orthogonal to every vector in the tangent plane, and you have $\nabla F(x_0, y_0, z_0) \cdot \mathbf{v} = 0$, which leads to the following theorem.

THEOREM 13.13 EQUATION OF TANGENT PLANE

If F is differentiable at (x_0, y_0, z_0), then an equation of the tangent plane to the surface given by $F(x, y, z) = 0$ at (x_0, y_0, z_0) is

$$F_x(x_0, y_0, z_0)(x - x_0) + F_y(x_0, y_0, z_0)(y - y_0) + F_z(x_0, y_0, z_0)(z - z_0) = 0.$$

EXAMPLE 2 Finding an Equation of a Tangent Plane

Find an equation of the tangent plane to the hyperboloid given by

$$z^2 - 2x^2 - 2y^2 = 12$$

at the point $(1, -1, 4)$.

Solution Begin by writing the equation of the surface as

$$z^2 - 2x^2 - 2y^2 - 12 = 0.$$

Then, considering

$$F(x, y, z) = z^2 - 2x^2 - 2y^2 - 12$$

you have

$$F_x(x, y, z) = -4x, \qquad F_y(x, y, z) = -4y, \quad \text{and} \quad F_z(x, y, z) = 2z.$$

At the point $(1, -1, 4)$ the partial derivatives are

$$F_x(1, -1, 4) = -4, \qquad F_y(1, -1, 4) = 4, \quad \text{and} \quad F_z(1, -1, 4) = 8.$$

So, an equation of the tangent plane at $(1, -1, 4)$ is

$$\begin{aligned} -4(x-1) + 4(y+1) + 8(z-4) &= 0 \\ -4x + 4 + 4y + 4 + 8z - 32 &= 0 \\ -4x + 4y + 8z - 24 &= 0 \\ x - y - 2z + 6 &= 0. \end{aligned}$$

Surface:
$z^2 - 2x^2 - 2y^2 - 12 = 0$

$F(1, -1, 4)$

Tangent plane to surface
Figure 13.58

Figure 13.58 shows a portion of the hyperboloid and tangent plane. ■

TECHNOLOGY Some three-dimensional graphing utilities are capable of graphing tangent planes to surfaces. Two examples are shown below.

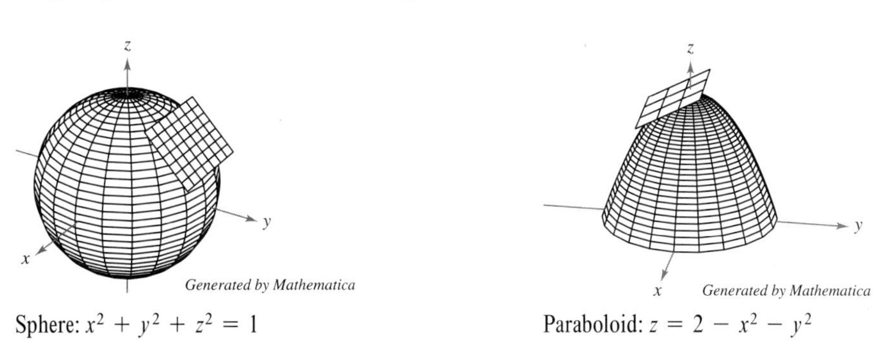

Sphere: $x^2 + y^2 + z^2 = 1$ Paraboloid: $z = 2 - x^2 - y^2$

To find the equation of the tangent plane at a point on a surface given by $z = f(x, y)$, you can define the function F by

$$F(x, y, z) = f(x, y) - z.$$

Then S is given by the level surface $F(x, y, z) = 0$, and by Theorem 13.13 an equation of the tangent plane to S at the point (x_0, y_0, z_0) is

$$f_x(x_0, y_0)(x - x_0) + f_y(x_0, y_0)(y - y_0) - (z - z_0) = 0.$$

EXAMPLE 3 Finding an Equation of the Tangent Plane

Find the equation of the tangent plane to the paraboloid

$$z = 1 - \frac{1}{10}(x^2 + 4y^2)$$

at the point $\left(1, 1, \frac{1}{2}\right)$.

Solution From $z = f(x, y) = 1 - \frac{1}{10}(x^2 + 4y^2)$, you obtain

$$f_x(x, y) = -\frac{x}{5} \quad \Longrightarrow \quad f_x(1, 1) = -\frac{1}{5}$$

and

$$f_y(x, y) = -\frac{4y}{5} \quad \Longrightarrow \quad f_y(1, 1) = -\frac{4}{5}.$$

So, an equation of the tangent plane at $\left(1, 1, \frac{1}{2}\right)$ is

$$f_x(1, 1)(x - 1) + f_y(1, 1)(y - 1) - \left(z - \frac{1}{2}\right) = 0$$

$$-\frac{1}{5}(x - 1) - \frac{4}{5}(y - 1) - \left(z - \frac{1}{2}\right) = 0$$

$$-\frac{1}{5}x - \frac{4}{5}y - z + \frac{3}{2} = 0.$$

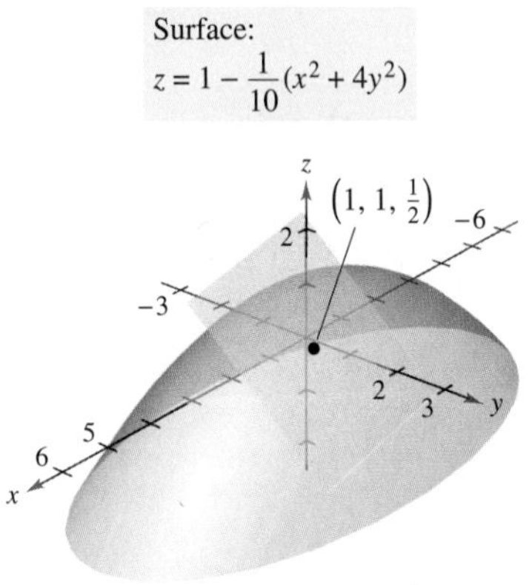

This tangent plane is shown in Figure 13.59. ■

Surface: $z = 1 - \frac{1}{10}(x^2 + 4y^2)$

Figure 13.59

The gradient $\nabla F(x, y, z)$ provides a convenient way to find equations of normal lines, as shown in Example 4.

EXAMPLE 4 Finding an Equation of a Normal Line to a Surface

Find a set of symmetric equations for the normal line to the surface given by $xyz = 12$ at the point $(2, -2, -3)$.

Solution Begin by letting

$$F(x, y, z) = xyz - 12.$$

Then, the gradient is given by

$$\begin{aligned}\nabla F(x, y, z) &= F_x(x, y, z)\mathbf{i} + F_y(x, y, z)\mathbf{j} + F_z(x, y, z)\mathbf{k} \\ &= yz\mathbf{i} + xz\mathbf{j} + xy\mathbf{k}\end{aligned}$$

and at the point $(2, -2, -3)$ you have

$$\begin{aligned}\nabla F(2, -2, -3) &= (-2)(-3)\mathbf{i} + (2)(-3)\mathbf{j} + (2)(-2)\mathbf{k} \\ &= 6\mathbf{i} - 6\mathbf{j} - 4\mathbf{k}.\end{aligned}$$

The normal line at $(2, -2, -3)$ has direction numbers 6, -6, and -4, and the corresponding set of symmetric equations is

$$\frac{x - 2}{6} = \frac{y + 2}{-6} = \frac{z + 3}{-4}.$$

See Figure 13.60. ■

Surface: $xyz = 12$

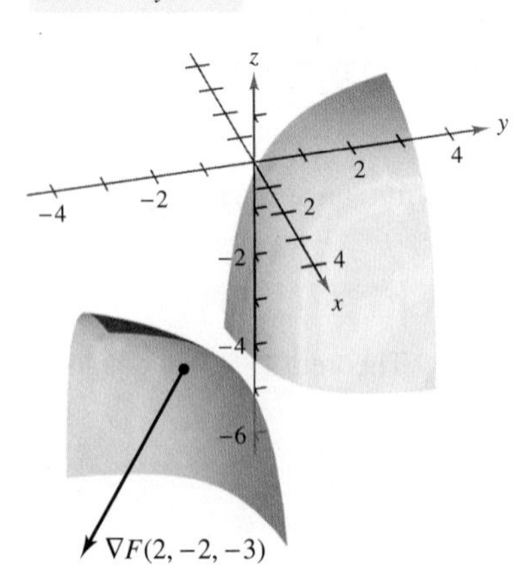

Figure 13.60

Knowing that the gradient $\nabla F(x, y, z)$ is normal to the surface given by $F(x, y, z) = 0$ allows you to solve a variety of problems dealing with surfaces and curves in space.

EXAMPLE 5 Finding the Equation of a Tangent Line to a Curve

Ellipsoid: $x^2 + 2y^2 + 2z^2 = 20$

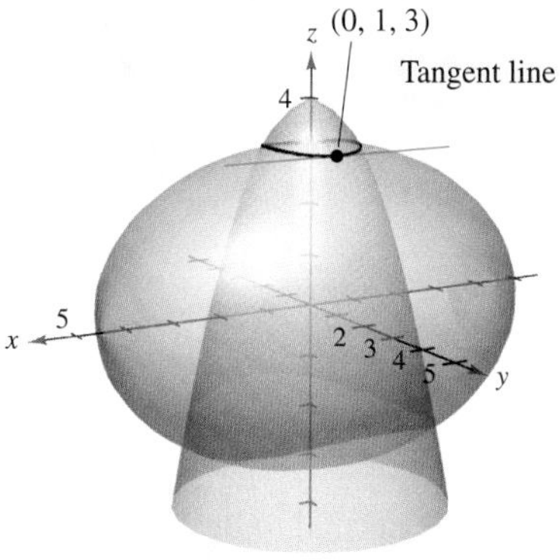

Paraboloid: $x^2 + y^2 + z = 4$

Figure 13.61

Describe the tangent line to the curve of intersection of the surfaces

$$x^2 + 2y^2 + 2z^2 = 20 \qquad \text{Ellipsoid}$$

$$x^2 + y^2 + z = 4 \qquad \text{Paraboloid}$$

at the point (0, 1, 3), as shown in Figure 13.61.

Solution Begin by finding the gradients to both surfaces at the point (0, 1, 3).

Ellipsoid

$$F(x, y, z) = x^2 + 2y^2 + 2z^2 - 20$$
$$\nabla F(x, y, z) = 2x\mathbf{i} + 4y\mathbf{j} + 4z\mathbf{k}$$
$$\nabla F(0, 1, 3) = 4\mathbf{j} + 12\mathbf{k}$$

Paraboloid

$$G(x, y, z) = x^2 + y^2 + z - 4$$
$$\nabla G(x, y, z) = 2x\mathbf{i} + 2y\mathbf{j} + \mathbf{k}$$
$$\nabla G(0, 1, 3) = 2\mathbf{j} + \mathbf{k}$$

The cross product of these two gradients is a vector that is tangent to both surfaces at the point (0, 1, 3).

$$\nabla F(0, 1, 3) \times \nabla G(0, 1, 3) = \begin{vmatrix} \mathbf{i} & \mathbf{j} & \mathbf{k} \\ 0 & 4 & 12 \\ 0 & 2 & 1 \end{vmatrix} = -20\mathbf{i}$$

So, the tangent line to the curve of intersection of the two surfaces at the point (0, 1, 3) is a line that is parallel to the x-axis and passes through the point (0, 1, 3). ■

The Angle of Inclination of a Plane

Another use of the gradient $\nabla F(x, y, z)$ is to determine the angle of inclination of the tangent plane to a surface. The **angle of inclination** of a plane is defined as the angle θ $(0 \leq \theta \leq \pi/2)$ between the given plane and the xy-plane, as shown in Figure 13.62. (The angle of inclination of a horizontal plane is defined as zero.) Because the vector $\mathbf{k}$ is normal to the xy-plane, you can use the formula for the cosine of the angle between two planes (given in Section 11.5) to conclude that the angle of inclination of a plane with normal vector $\mathbf{n}$ is given by

$$\cos \theta = \frac{|\mathbf{n} \cdot \mathbf{k}|}{\|\mathbf{n}\| \|\mathbf{k}\|} = \frac{|\mathbf{n} \cdot \mathbf{k}|}{\|\mathbf{n}\|}. \qquad \text{Angle of inclination of a plane}$$

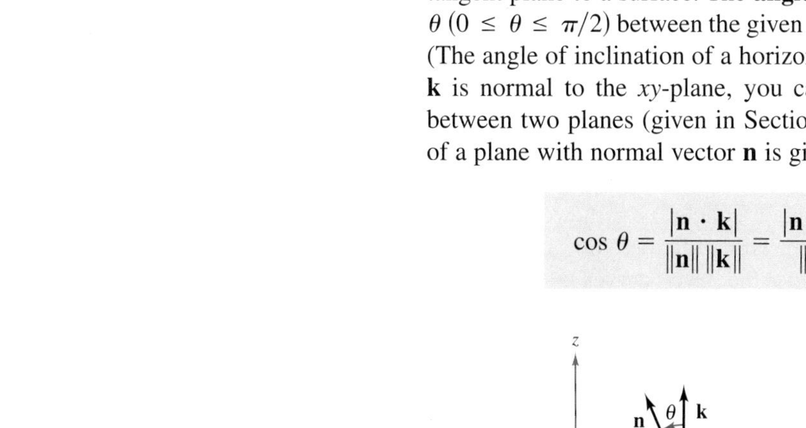

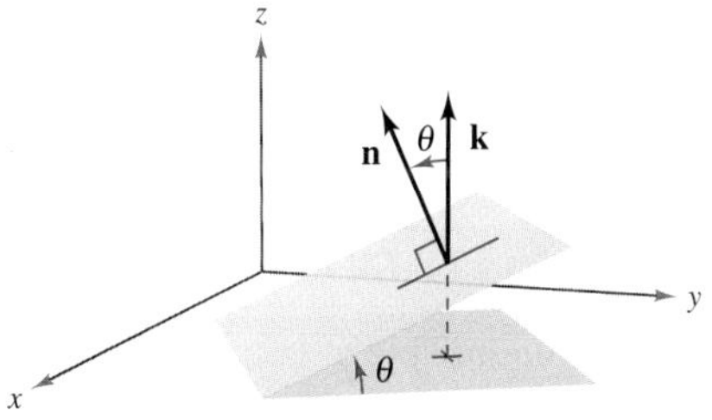

The angle of inclination
Figure 13.62

EXAMPLE 6 Finding the Angle of Inclination of a Tangent Plane

Find the angle of inclination of the tangent plane to the ellipsoid given by

$$\frac{x^2}{12} + \frac{y^2}{12} + \frac{z^2}{3} = 1$$

at the point $(2, 2, 1)$.

Solution If you let

$$F(x, y, z) = \frac{x^2}{12} + \frac{y^2}{12} + \frac{z^2}{3} - 1$$

the gradient of F at the point $(2, 2, 1)$ is given by

$$\nabla F(x, y, z) = \frac{x}{6}\mathbf{i} + \frac{y}{6}\mathbf{j} + \frac{2z}{3}\mathbf{k}$$

$$\nabla F(2, 2, 1) = \frac{1}{3}\mathbf{i} + \frac{1}{3}\mathbf{j} + \frac{2}{3}\mathbf{k}.$$

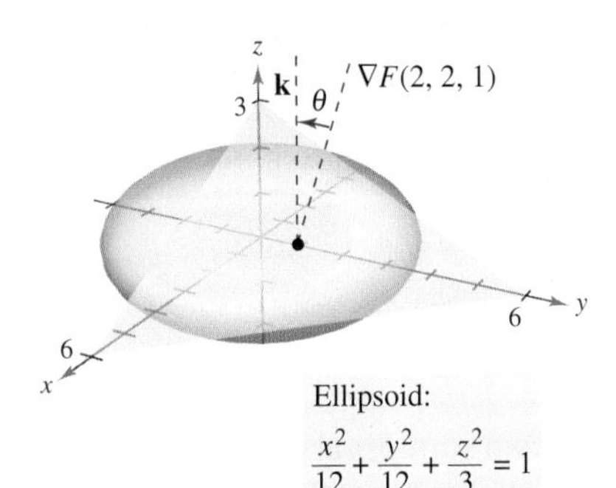

Figure 13.63

Because $\nabla F(2, 2, 1)$ is normal to the tangent plane and $\mathbf{k}$ is normal to the xy-plane, it follows that the angle of inclination of the tangent plane is given by

$$\cos\theta = \frac{|\nabla F(2, 2, 1) \cdot \mathbf{k}|}{\|\nabla F(2, 2, 1)\|} = \frac{2/3}{\sqrt{(1/3)^2 + (1/3)^2 + (2/3)^2}} = \sqrt{\frac{2}{3}}$$

which implies that

$$\theta = \arccos\sqrt{\frac{2}{3}} \approx 35.3^\circ,$$

as shown in Figure 13.63. ■

NOTE A special case of the procedure shown in Example 6 is worth noting. The angle of inclination θ of the tangent plane to the surface $z = f(x, y)$ at (x_0, y_0, z_0) is given by

$$\cos\theta = \frac{1}{\sqrt{[f_x(x_0, y_0)]^2 + [f_y(x_0, y_0)]^2 + 1}}.$$

Alternative formula for angle of inclination (See Exercise 77.) ■

A Comparison of the Gradients $\nabla f(x, y)$ and $\nabla F(x, y, z)$

This section concludes with a comparison of the gradients $\nabla f(x, y)$ and $\nabla F(x, y, z)$. In the preceding section, you saw that the gradient of a function f of two variables is normal to the level curves of f. Specifically, Theorem 13.12 states that if f is differentiable at (x_0, y_0) and $\nabla f(x_0, y_0) \neq \mathbf{0}$, then $\nabla f(x_0, y_0)$ is normal to the level curve through (x_0, y_0). Having developed normal lines to surfaces, you can now extend this result to a function of three variables. The proof of Theorem 13.14 is left as an exercise (see Exercise 78).

THEOREM 13.14 GRADIENT IS NORMAL TO LEVEL SURFACES

If F is differentiable at (x_0, y_0, z_0) and $\nabla F(x_0, y_0, z_0) \neq \mathbf{0}$, then $\nabla F(x_0, y_0, z_0)$ is normal to the level surface through (x_0, y_0, z_0).

When working with the gradients $\nabla f(x, y)$ and $\nabla F(x, y, z)$, be sure you remember that $\nabla f(x, y)$ is a vector in the xy-plane and $\nabla F(x, y, z)$ is a vector in space.

13.7 Exercises

See www.CalcChat.com for worked-out solutions to odd-numbered exercises.

In Exercises 1–4, describe the level surface $F(x, y, z) = 0$.

1. $F(x, y, z) = 3x - 5y + 3z - 15$

2. $F(x, y, z) = x^2 + y^2 + z^2 - 25$

3. $F(x, y, z) = 4x^2 + 9y^2 - 4z^2$

4. $F(x, y, z) = 16x^2 - 9y^2 + 36z$

In Exercises 5–16, find a unit normal vector to the surface at the given point. [*Hint:* Normalize the gradient vector $\nabla F(x, y, z)$.]

	Surface	*Point*
5.	$3x + 4y + 12z = 0$	$(0, 0, 0)$
6.	$x + y + z = 4$	$(2, 0, 2)$
7.	$x^2 + y^2 + z^2 = 6$	$(1, 1, 2)$
8.	$z = \sqrt{x^2 + y^2}$	$(3, 4, 5)$
9.	$z = x^3$	$(2, -1, 8)$
10.	$x^2y^4 - z = 0$	$(1, 2, 16)$
11.	$x^2 + 3y + z^3 = 9$	$(2, -1, 2)$
12.	$x^2y^3 - y^2z + 2xz^3 = 4$	$(-1, 1, -1)$
13.	$\ln\left(\frac{x}{y - z}\right) = 0$	$(1, 4, 3)$
14.	$ze^{x^2 - y^2} - 3 = 0$	$(2, 2, 3)$
15.	$z - x \sin y = 4$	$\left(6, \frac{\pi}{6}, 7\right)$
16.	$\sin(x - y) - z = 2$	$\left(\frac{\pi}{3}, \frac{\pi}{6}, -\frac{3}{2}\right)$

In Exercises 17–30, find an equation of the tangent plane to the surface at the given point.

17. $z = x^2 + y^2 + 3$

$(2, 1, 8)$

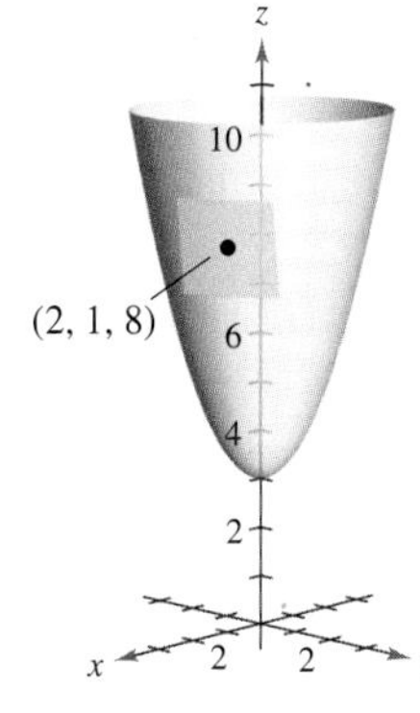

18. $f(x, y) = \frac{y}{x}$ 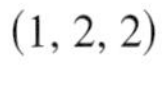

$(1, 2, 2)$

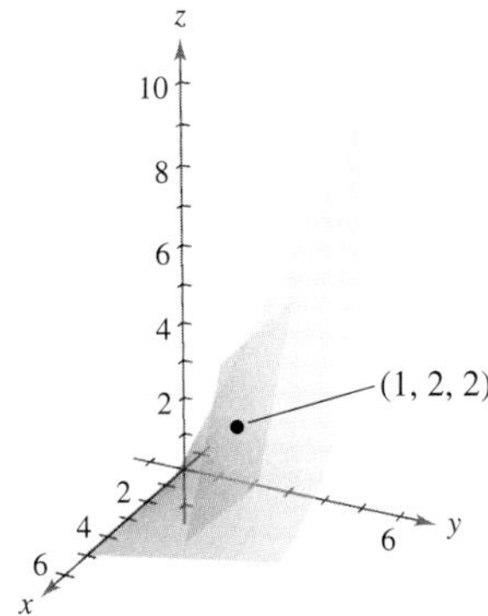

19. $z = \sqrt{x^2 + y^2}$, $(3, 4, 5)$

20. $g(x, y) = \arctan \frac{y}{x}$, $(1, 0, 0)$

21. $g(x, y) = x^2 + y^2$, $(1, -1, 2)$

22. $f(x, y) = x^2 - 2xy + y^2$, $(1, 2, 1)$

23. $z = 2 - \frac{2}{3}x - y$, $(3, -1, 1)$

24. $z = e^x(\sin y + 1)$, $\left(0, \frac{\pi}{2}, 2\right)$

25. $h(x, y) = \ln \sqrt{x^2 + y^2}$, $(3, 4, \ln 5)$

26. $h(x, y) = \cos y$, $\left(5, \frac{\pi}{4}, \frac{\sqrt{2}}{2}\right)$

27. $x^2 + 4y^2 + z^2 = 36$, $(2, -2, 4)$

28. $x^2 + 2z^2 = y^2$, $(1, 3, -2)$

29. $xy^2 + 3x - z^2 = 8$, $(1, -3, 2)$

30. $x = y(2z - 3)$, $(4, 4, 2)$

In Exercises 31–40, find an equation of the tangent plane and find symmetric equations of the normal line to the surface at the given point.

31. $x + y + z = 9$, $(3, 3, 3)$

32. $x^2 + y^2 + z^2 = 9$, $(1, 2, 2)$

33. $x^2 + y^2 + z = 9$, $(1, 2, 4)$

34. $z = 16 - x^2 - y^2$, $(2, 2, 8)$

35. $z = x^2 - y^2$, $(3, 2, 5)$

36. $xy - z = 0$, $(-2, -3, 6)$

37. $xyz = 10$, $(1, 2, 5)$

38. $z = ye^{2xy}$, $(0, 2, 2)$

39. $z = \arctan \frac{y}{x}$, $\left(1, 1, \frac{\pi}{4}\right)$

40. $y \ln xz^2 = 2$, $(e, 2, 1)$

In Exercises 41–46, (a) find symmetric equations of the tangent line to the curve of intersection of the surfaces at the given point, and (b) find the cosine of the angle between the gradient vectors at this point. State whether the surfaces are orthogonal at the point of intersection.

41. $x^2 + y^2 = 2$, $z = x$, $(1, 1, 1)$

42. $z = x^2 + y^2$, $z = 4 - y$, $(2, -1, 5)$

43. $x^2 + z^2 = 25$, $y^2 + z^2 = 25$, $(3, 3, 4)$

44. $z = \sqrt{x^2 + y^2}$, $5x - 2y + 3z = 22$, $(3, 4, 5)$

45. $x^2 + y^2 + z^2 = 14$, $x - y - z = 0$, $(3, 1, 2)$

46. $z = x^2 + y^2$, $x + y + 6z = 33$, $(1, 2, 5)$

In Exercises 47–50, find the angle of inclination θ of the tangent plane to the surface at the given point.

47. $3x^2 + 2y^2 - z = 15$, $(2, 2, 5)$

48. $2xy - z^3 = 0$, $(2, 2, 2)$

49. $x^2 - y^2 + z = 0$, $(1, 2, 3)$

50. $x^2 + y^2 = 5$, $(2, 1, 3)$

In Exercises 51–56, find the point(s) on the surface at which the tangent plane is horizontal.

51. $z = 3 - x^2 - y^2 + 6y$

52. $z = 3x^2 + 2y^2 - 3x + 4y - 5$

53. $z = x^2 - xy + y^2 - 2x - 2y$

54. $z = 4x^2 + 4xy - 2y^2 + 8x - 5y - 4$

55. $z = 5xy$

56. $z = xy + \frac{1}{x} + \frac{1}{y}$

In Exercises 57 and 58, show that the surfaces are tangent to each other at the given point by showing that the surfaces have the same tangent plane at this point.

57. $x^2 + 2y^2 + 3z^2 = 3,\ x^2 + y^2 + z^2 + 6x - 10y + 14 = 0,$ $(-1, 1, 0)$

58. $x^2 + y^2 + z^2 - 8x - 12y + 4z + 42 = 0,\ x^2 + y^2 + 2z = 7,$ $(2, 3, -3)$

In Exercises 59 and 60, (a) show that the surfaces intersect at the given point, and (b) show that the surfaces have perpendicular tangent planes at this point.

59. $z = 2xy^2,\ 8x^2 - 5y^2 - 8z = -13,\quad (1, 1, 2)$

60. $x^2 + y^2 + z^2 + 2x - 4y - 4z - 12 = 0,$
$4x^2 + y^2 + 16z^2 = 24,\quad (1, -2, 1)$

61. Find a point on the ellipsoid $x^2 + 4y^2 + z^2 = 9$ where the tangent plane is perpendicular to the line with parametric equations

$x = 2 - 4t,\ y = 1 + 8t,$ and $z = 3 - 2t.$

62. Find a point on the hyperboloid $x^2 + 4y^2 - z^2 = 1$ where the tangent plane is parallel to the plane $x + 4y - z = 0$.

WRITING ABOUT CONCEPTS

63. Give the standard form of the equation of the tangent plane to a surface given by $F(x, y, z) = 0$ at (x_0, y_0, z_0).

64. For some surfaces, the normal lines at any point pass through the same geometric object. What is the common geometric object for a sphere? What is the common geometric object for a right circular cylinder? Explain.

65. Discuss the relationship between the tangent plane to a surface and approximation by differentials.

CAPSTONE

66. Consider the elliptic cone given by

$x^2 - y^2 + z^2 = 0.$

(a) Find an equation of the tangent plane at the point $(5, 13, -12)$.

(b) Find symmetric equations of the normal line at the point $(5, 13, -12)$.

67. ***Investigation*** Consider the function

$$f(x, y) = \frac{4xy}{(x^2 + 1)(y^2 + 1)}$$

on the intervals $-2 \le x \le 2$ and $0 \le y \le 3$.

(a) Find a set of parametric equations of the normal line and an equation of the tangent plane to the surface at the point $(1, 1, 1)$.

(b) Repeat part (a) for the point $\left(-1, 2, -\frac{4}{5}\right)$.

CAS (c) Use a computer algebra system to graph the surface, the normal lines, and the tangent planes found in parts (a) and (b).

68. ***Investigation*** Consider the function

$$f(x, y) = \frac{\sin y}{x}$$

on the intervals $-3 \le x \le 3$ and $0 \le y \le 2\pi$.

(a) Find a set of parametric equations of the normal line and an equation of the tangent plane to the surface at the point

$$\left(2, \frac{\pi}{2}, \frac{1}{2}\right).$$

(b) Repeat part (a) for the point $\left(-\frac{2}{3}, \frac{3\pi}{2}, \frac{3}{2}\right)$.

CAS (c) Use a computer algebra system to graph the surface, the normal lines, and the tangent planes found in parts (a) and (b).

69. Consider the functions

$$f(x, y) = 6 - x^2 - y^2/4 \quad\text{and}\quad g(x, y) = 2x + y.$$

(a) Find a set of parametric equations of the tangent line to the curve of intersection of the surfaces at the point $(1, 2, 4)$, and find the angle between the gradient vectors.

CAS (b) Use a computer algebra system to graph the surfaces. Graph the tangent line found in part (a).

70. Consider the functions

$$f(x, y) = \sqrt{16 - x^2 - y^2 + 2x - 4y}$$

and

$$g(x, y) = \frac{\sqrt{2}}{2}\sqrt{1 - 3x^2 + y^2 + 6x + 4y}.$$

CAS (a) Use a computer algebra system to graph the first-octant portion of the surfaces represented by f and g.

(b) Find one first-octant point on the curve of intersection and show that the surfaces are orthogonal at this point.

(c) These surfaces are orthogonal along the curve of intersection. Does part (b) prove this fact? Explain.

In Exercises 71 and 72, show that the tangent plane to the quadric surface at the point (x_0, y_0, z_0) can be written in the given form.

71. Ellipsoid: $\frac{x^2}{a^2} + \frac{y^2}{b^2} + \frac{z^2}{c^2} = 1$

Plane: $\frac{x_0 x}{a^2} + \frac{y_0 y}{b^2} + \frac{z_0 z}{c^2} = 1$

72. Hyperboloid: $\frac{x^2}{a^2} + \frac{y^2}{b^2} - \frac{z^2}{c^2} = 1$

Plane: $\frac{x_0x}{a^2} + \frac{y_0y}{b^2} - \frac{z_0z}{c^2} = 1$

73. Show that any tangent plane to the cone

$z^2 = a^2x^2 + b^2y^2$

passes through the origin.

74. Let f be a differentiable function and consider the surface $z = xf(y/x)$. Show that the tangent plane at any point $P(x_0, y_0, z_0)$ on the surface passes through the origin.

75. ***Approximation*** Consider the following approximations for a function $f(x, y)$ centered at $(0, 0)$.

Linear approximation:

$P_1(x, y) = f(0, 0) + f_x(0, 0)x + f_y(0, 0)y$

Quadratic approximation:

$P_2(x, y) = f(0, 0) + f_x(0, 0)x + f_y(0, 0)y +$
$\frac{1}{2}f_{xx}(0, 0)x^2 + f_{xy}(0, 0)xy + \frac{1}{2}f_{yy}(0, 0)y^2$

[Note that the linear approximation is the tangent plane to the surface at $(0, 0, f(0, 0))$.]

(a) Find the linear approximation of $f(x, y) = e^{(x-y)}$ centered at $(0, 0)$.

(b) Find the quadratic approximation of $f(x, y) = e^{(x-y)}$ centered at $(0, 0)$.

(c) If $x = 0$ in the quadratic approximation, you obtain the second-degree Taylor polynomial for what function? Answer the same question for $y = 0$.

(d) Complete the table.

x	y	$f(x, y)$	$P_1(x, y)$	$P_2(x, y)$
0	0			
0	0.1			
0.2	0.1			
0.2	0.5			
1	0.5			

CAS (e) Use a computer algebra system to graph the surfaces $z = f(x, y)$, $z = P_1(x, y)$, and $z = P_2(x, y)$.

76. ***Approximation*** Repeat Exercise 75 for the function $f(x, y) = \cos(x + y)$.

77. Prove that the angle of inclination θ of the tangent plane to the surface $z = f(x, y)$ at the point (x_0, y_0, z_0) is given by

$$\cos \theta = \frac{1}{\sqrt{[f_x(x_0, y_0)]^2 + [f_y(x_0, y_0)]^2 + 1}}.$$

78. Prove Theorem 13.14.

SECTION PROJECT

Wildflowers

The diversity of wildflowers in a meadow can be measured by counting the numbers of daisies, buttercups, shooting stars, and so on. If there are n types of wildflowers, each with a proportion p_i of the total population, it follows that $p_1 + p_2 + \cdot \cdot \cdot + p_n = 1$. The measure of diversity of the population is defined as

$$H = -\sum_{i=1}^{n} p_i \log_2 p_i.$$

In this definition, it is understood that $p_i \log_2 p_i = 0$ when $p_i = 0$. The tables show proportions of wildflowers in a meadow in May, June, August, and September.

May

Flower type	1	2	3	4
Proportion	$\frac{5}{16}$	$\frac{5}{16}$	$\frac{5}{16}$	$\frac{1}{16}$

June

Flower type	1	2	3	4
Proportion	$\frac{1}{4}$	$\frac{1}{4}$	$\frac{1}{4}$	$\frac{1}{4}$

August

Flower type	1	2	3	4
Proportion	$\frac{1}{4}$	0	$\frac{1}{4}$	$\frac{1}{2}$

September

Flower type	1	2	3	4
Proportion	0	0	0	1

(a) Determine the wildflower diversity for each month. How would you interpret September's diversity? Which month had the greatest diversity?

(b) If the meadow contains 10 types of wildflowers in roughly equal proportions, is the diversity of the population greater than or less than the diversity of a similar distribution of 4 types of flowers? What type of distribution (of 10 types of wildflowers) would produce maximum diversity?

(c) Let H_n represent the maximum diversity of n types of wildflowers. Does H_n approach a limit as $n \to \infty$?

FOR FURTHER INFORMATION Biologists use the concept of diversity to measure the proportions of different types of organisms within an environment. For more information on this technique, see the article "Information Theory and Biological Diversity" by Steven Kolmes and Kevin Mitchell in the *UMAP Modules*.

13.8 Extrema of Functions of Two Variables

- Find absolute and relative extrema of a function of two variables.
- Use the Second Partials Test to find relative extrema of a function of two variables.

Absolute Extrema and Relative Extrema

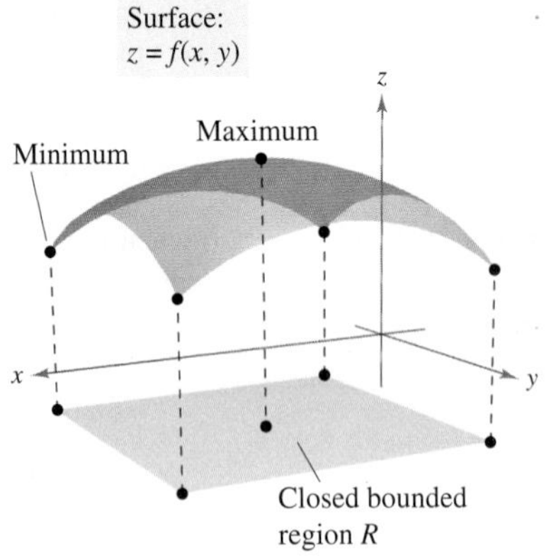

R contains point(s) at which $f(x, y)$ is a minimum and point(s) at which $f(x, y)$ is a maximum.
Figure 13.64

In Chapter 3, you studied techniques for finding the extreme values of a function of a single variable. In this section, you will extend these techniques to functions of two variables. For example, in Theorem 13.15 below, the Extreme Value Theorem for a function of a single variable is extended to a function of two variables.

Consider the continuous function f of two variables, defined on a closed bounded region R. The values $f(a, b)$ and $f(c, d)$ such that

$$f(a, b) \leq f(x, y) \leq f(c, d) \qquad (a, b) \text{ and } (c, d) \text{ are in } R.$$

for all (x, y) in R are called the **minimum** and **maximum** of f in the region R, as shown in Figure 13.64. Recall from Section 13.2 that a region in the plane is *closed* if it contains all of its boundary points. The Extreme Value Theorem deals with a region in the plane that is both closed and *bounded*. A region in the plane is called **bounded** if it is a subregion of a closed disk in the plane.

THEOREM 13.15 EXTREME VALUE THEOREM

Let f be a continuous function of two variables x and y defined on a closed bounded region R in the xy-plane.

1. There is at least one point in R at which f takes on a minimum value.
2. There is at least one point in R at which f takes on a maximum value.

A minimum is also called an **absolute minimum** and a maximum is also called an **absolute maximum.** As in single-variable calculus, there is a distinction made between absolute extrema and **relative extrema.**

DEFINITION OF RELATIVE EXTREMA

Let f be a function defined on a region R containing (x_0, y_0).

1. The function f has a **relative minimum** at (x_0, y_0) if

$$f(x, y) \geq f(x_0, y_0)$$

for all (x, y) in an *open* disk containing (x_0, y_0).

2. The function f has a **relative maximum** at (x_0, y_0) if

$$f(x, y) \leq f(x_0, y_0)$$

for all (x, y) in an *open* disk containing (x_0, y_0).

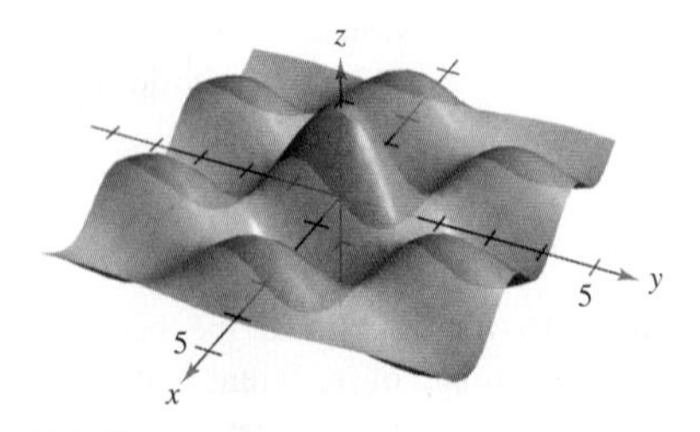

Relative extrema
Figure 13.65

To say that f has a relative maximum at (x_0, y_0) means that the point (x_0, y_0, z_0) is at least as high as all nearby points on the graph of $z = f(x, y)$. Similarly, f has a relative minimum at (x_0, y_0) if (x_0, y_0, z_0) is at least as low as all nearby points on the graph. (See Figure 13.65.)

KARL WEIERSTRASS (1815–1897)

Although the Extreme Value Theorem had been used by earlier mathematicians, the first to provide a rigorous proof was the German mathematician Karl Weierstrass. Weierstrass also provided rigorous justifications for many other mathematical results already in common use. We are indebted to him for much of the logical foundation on which modern calculus is built.

To locate relative extrema of f, you can investigate the points at which the gradient of f is **0** or the points at which one of the partial derivatives does not exist. Such points are called **critical points** of f.

DEFINITION OF CRITICAL POINT

Let f be defined on an open region R containing (x_0, y_0). The point (x_0, y_0) is a **critical point** of f if one of the following is true.

1. $f_x(x_0, y_0) = 0$ and $f_y(x_0, y_0) = 0$
2. $f_x(x_0, y_0)$ or $f_y(x_0, y_0)$ does not exist.

Recall from Theorem 13.11 that if f is differentiable and

$$\begin{aligned}\nabla f(x_0, y_0) &= f_x(x_0, y_0)\mathbf{i} + f_y(x_0, y_0)\mathbf{j}\\ &= 0\mathbf{i} + 0\mathbf{j}\end{aligned}$$

then every directional derivative at (x_0, y_0) must be 0. This implies that the function has a horizontal tangent plane at the point (x_0, y_0), as shown in Figure 13.66. It appears that such a point is a likely location of a relative extremum. This is confirmed by Theorem 13.16.

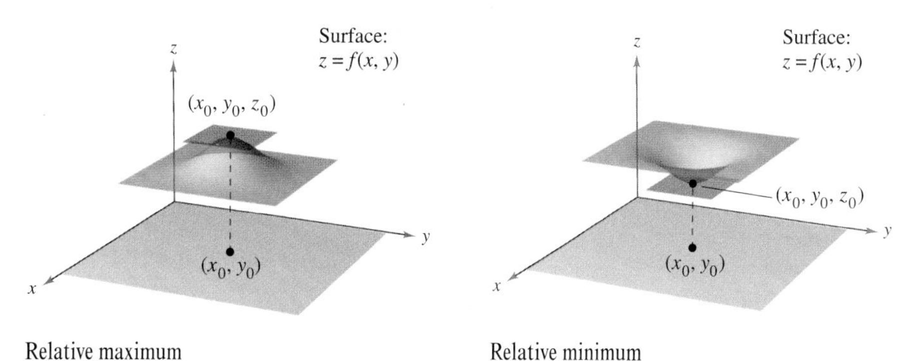

Relative maximum Relative minimum
Figure 13.66

THEOREM 13.16 RELATIVE EXTREMA OCCUR ONLY AT CRITICAL POINTS

If f has a relative extremum at (x_0, y_0) on an open region R, then (x_0, y_0) is a critical point of f.

EXPLORATION

Use a graphing utility to graph

$$z = x^3 - 3xy + y^3$$

using the bounds $0 \le x \le 3$, $0 \le y \le 3$, and $-3 \le z \le 3$. This view makes it appear as though the surface has an absolute minimum. But does it?

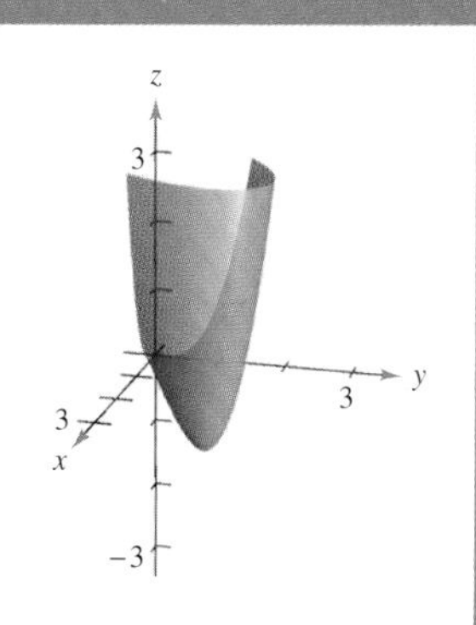

EXAMPLE 1 Finding a Relative Extremum

Determine the relative extrema of

$$f(x, y) = 2x^2 + y^2 + 8x - 6y + 20.$$

Solution Begin by finding the critical points of f. Because

$$f_x(x, y) = 4x + 8 \qquad \text{Partial with respect to } x$$

and

$$f_y(x, y) = 2y - 6 \qquad \text{Partial with respect to } y$$

are defined for all x and y, the only critical points are those for which both first partial derivatives are 0. To locate these points, set $f_x(x, y)$ and $f_y(x, y)$ equal to 0, and solve the equations

$$4x + 8 = 0 \quad \text{and} \quad 2y - 6 = 0$$

to obtain the critical point $(-2, 3)$. By completing the square, you can conclude that for all $(x, y) \neq (-2, 3)$

$$f(x, y) = 2(x + 2)^2 + (y - 3)^2 + 3 > 3.$$

So, a relative *minimum* of f occurs at $(-2, 3)$. The value of the relative minimum is $f(-2, 3) = 3$, as shown in Figure 13.67. ■

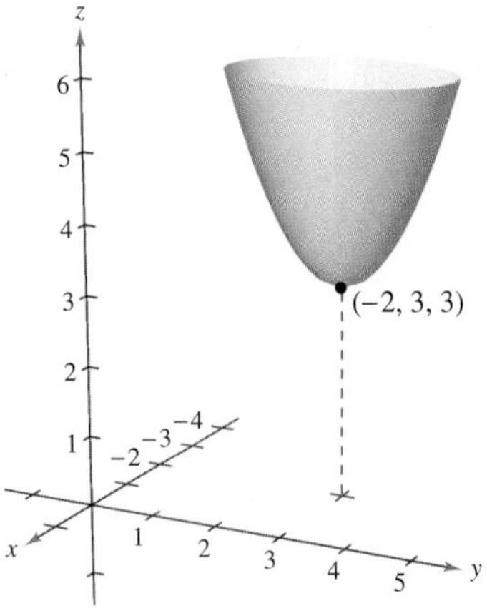

Surface:
$f(x, y) = 2x^2 + y^2 + 8x - 6y + 20$

The function $z = f(x, y)$ has a relative minimum at $(-2, 3)$.
Figure 13.67

Example 1 shows a relative minimum occurring at one type of critical point—the type for which both $f_x(x, y)$ and $f_y(x, y)$ are 0. The next example concerns a relative maximum that occurs at the other type of critical point—the type for which either $f_x(x, y)$ or $f_y(x, y)$ does not exist.

EXAMPLE 2 Finding a Relative Extremum

Determine the relative extrema of $f(x, y) = 1 - (x^2 + y^2)^{1/3}$.

Solution Because

$$f_x(x, y) = -\frac{2x}{3(x^2 + y^2)^{2/3}} \qquad \text{Partial with respect to } x$$

and

$$f_y(x, y) = -\frac{2y}{3(x^2 + y^2)^{2/3}} \qquad \text{Partial with respect to } y$$

it follows that both partial derivatives exist for all points in the xy-plane except for $(0, 0)$. Moreover, because the partial derivatives cannot both be 0 unless both x and y are 0, you can conclude that $(0, 0)$ is the only critical point. In Figure 13.68, note that $f(0, 0)$ is 1. For all other (x, y) it is clear that

$$f(x, y) = 1 - (x^2 + y^2)^{1/3} < 1.$$

So, f has a relative *maximum* at $(0, 0)$. ■

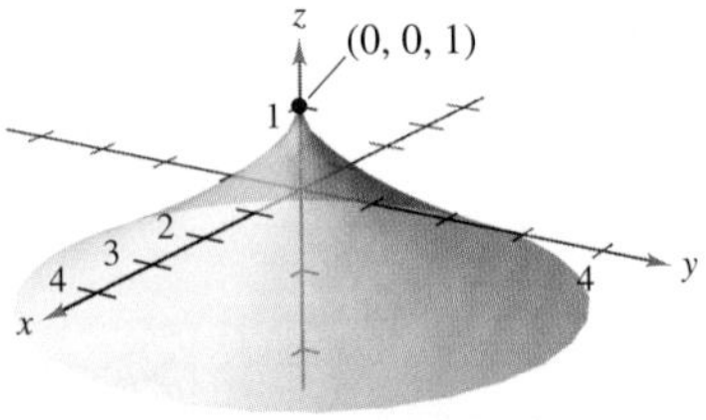

Surface:
$f(x, y) = 1 - (x^2 + y^2)^{1/3}$

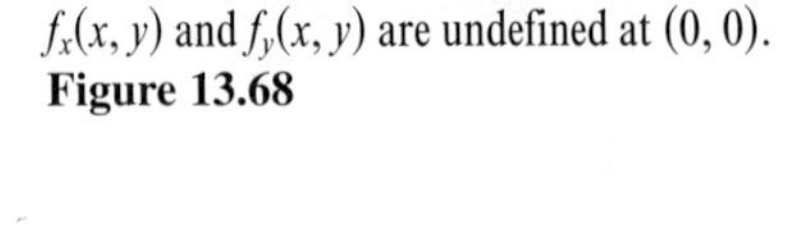

$f_x(x, y)$ and $f_y(x, y)$ are undefined at $(0, 0)$.
Figure 13.68

NOTE In Example 2, $f_x(x, y) = 0$ for every point on the y-axis other than $(0, 0)$. However, because $f_y(x, y)$ is nonzero, these are not critical points. Remember that *one* of the partials must not exist or *both* must be 0 in order to yield a critical point. ■

The Second Partials Test

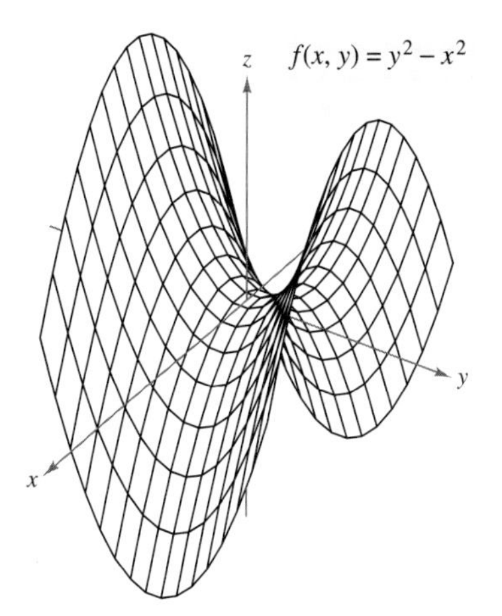

Saddle point at $(0, 0, 0)$:
$f_x(0, 0) = f_y(0, 0) = 0$

Figure 13.69

Theorem 13.16 tells you that to find relative extrema you need only examine values of $f(x, y)$ at critical points. However, as is true for a function of one variable, the critical points of a function of two variables do not always yield relative maxima or minima. Some critical points yield **saddle points,** which are neither relative maxima nor relative minima.

As an example of a critical point that does not yield a relative extremum, consider the surface given by

$$f(x, y) = y^2 - x^2 \qquad \text{Hyperbolic paraboloid}$$

as shown in Figure 13.69. At the point $(0, 0)$, both partial derivatives are 0. The function f does not, however, have a relative extremum at this point because in any open disk centered at $(0, 0)$ the function takes on both negative values (along the x-axis) *and* positive values (along the y-axis). So, the point $(0, 0, 0)$ is a saddle point of the surface. (The term "saddle point" comes from the fact that surfaces such as the one shown in Figure 13.69 resemble saddles.)

For the functions in Examples 1 and 2, it was relatively easy to determine the relative extrema, because each function was either given, or able to be written, in completed square form. For more complicated functions, algebraic arguments are less convenient and it is better to rely on the analytic means presented in the following Second Partials Test. This is the two-variable counterpart of the Second Derivative Test for functions of one variable. The proof of this theorem is best left to a course in advanced calculus.

THEOREM 13.17 SECOND PARTIALS TEST

Let f have continuous second partial derivatives on an open region containing a point (a, b) for which

$$f_x(a, b) = 0 \quad \text{and} \quad f_y(a, b) = 0.$$

To test for relative extrema of f, consider the quantity

$$d = f_{xx}(a, b)f_{yy}(a, b) - [f_{xy}(a, b)]^2.$$

1. If $d > 0$ and $f_{xx}(a, b) > 0$, then f has a **relative minimum** at (a, b).
2. If $d > 0$ and $f_{xx}(a, b) < 0$, then f has a **relative maximum** at (a, b).
3. If $d < 0$, then $(a, b, f(a, b))$ is a **saddle point.**
4. The test is inconclusive if $d = 0$.

NOTE If $d > 0$, then $f_{xx}(a, b)$ and $f_{yy}(a, b)$ must have the same sign. This means that $f_{xx}(a, b)$ can be replaced by $f_{yy}(a, b)$ in the first two parts of the test. ■

A convenient device for remembering the formula for d in the Second Partials Test is given by the 2×2 determinant

$$d = \begin{vmatrix} f_{xx}(a, b) & f_{xy}(a, b) \\ f_{yx}(a, b) & f_{yy}(a, b) \end{vmatrix}$$

where $f_{xy}(a, b) = f_{yx}(a, b)$ by Theorem 13.3.

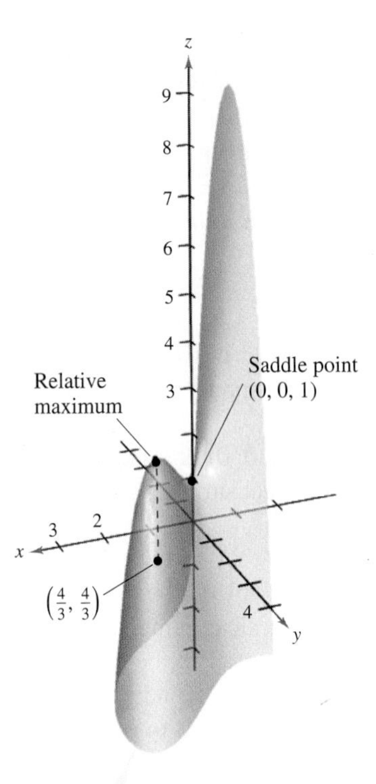

Figure 13.70

EXAMPLE 3 Using the Second Partials Test

Find the relative extrema of $f(x, y) = -x^3 + 4xy - 2y^2 + 1$.

Solution Begin by finding the critical points of f. Because

$$f_x(x, y) = -3x^2 + 4y \quad \text{and} \quad f_y(x, y) = 4x - 4y$$

exist for all x and y, the only critical points are those for which both first partial derivatives are 0. To locate these points, set $f_x(x, y)$ and $f_y(x, y)$ equal to 0 to obtain $-3x^2 + 4y = 0$ and $4x - 4y = 0$. From the second equation you know that $x = y$, and, by substitution into the first equation, you obtain two solutions: $y = x = 0$ and $y = x = \frac{4}{3}$. Because

$$f_{xx}(x, y) = -6x, \quad f_{yy}(x, y) = -4, \quad \text{and} \quad f_{xy}(x, y) = 4$$

it follows that, for the critical point $(0, 0)$,

$$d = f_{xx}(0, 0)f_{yy}(0, 0) - [f_{xy}(0, 0)]^2 = 0 - 16 < 0$$

and, by the Second Partials Test, you can conclude that $(0, 0, 1)$ is a saddle point of f. Furthermore, for the critical point $\left(\frac{4}{3}, \frac{4}{3}\right)$,

$$\begin{aligned} d &= f_{xx}\left(\tfrac{4}{3}, \tfrac{4}{3}\right) f_{yy}\left(\tfrac{4}{3}, \tfrac{4}{3}\right) - \left[f_{xy}\left(\tfrac{4}{3}, \tfrac{4}{3}\right)\right]^2 \\ &= -8(-4) - 16 \\ &= 16 \\ &> 0 \end{aligned}$$

and because $f_{xx}\left(\frac{4}{3}, \frac{4}{3}\right) = -8 < 0$ you can conclude that f has a relative maximum at $\left(\frac{4}{3}, \frac{4}{3}\right)$, as shown in Figure 13.70. ■

The Second Partials Test can fail to find relative extrema in two ways. If either of the first partial derivatives does not exist, you cannot use the test. Also, if

$$d = f_{xx}(a, b)f_{yy}(a, b) - [f_{xy}(a, b)]^2 = 0$$

the test fails. In such cases, you can try a sketch or some other approach, as demonstrated in the next example.

EXAMPLE 4 Failure of the Second Partials Test

Find the relative extrema of $f(x, y) = x^2y^2$.

Solution Because $f_x(x, y) = 2xy^2$ and $f_y(x, y) = 2x^2y$, you know that both partial derivatives are 0 if $x = 0$ or $y = 0$. That is, every point along the x- or y-axis is a critical point. Moreover, because

$$f_{xx}(x, y) = 2y^2, \quad f_{yy}(x, y) = 2x^2, \quad \text{and} \quad f_{xy}(x, y) = 4xy$$

you know that if either $x = 0$ or $y = 0$, then

$$\begin{aligned} d &= f_{xx}(x, y)f_{yy}(x, y) - [f_{xy}(x, y)]^2 \\ &= 4x^2y^2 - 16x^2y^2 = -12x^2y^2 = 0. \end{aligned}$$

So, the Second Partials Test fails. However, because $f(x, y) = 0$ for every point along the x- or y-axis and $f(x, y) = x^2y^2 > 0$ for all other points, you can conclude that each of these critical points yields an absolute minimum, as shown in Figure 13.71. ■

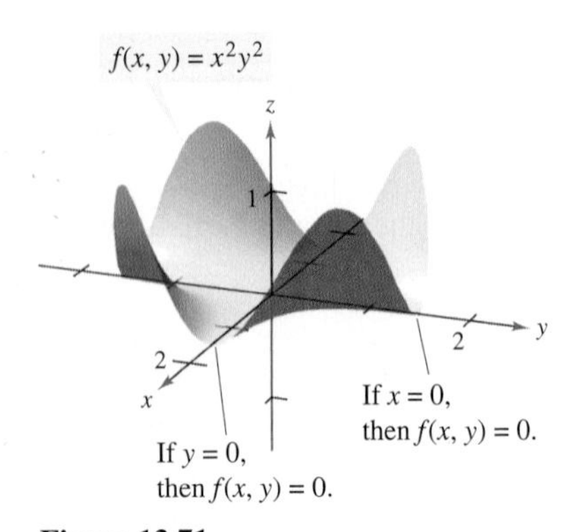

Figure 13.71

Absolute extrema of a function can occur in two ways. First, some relative extrema also happen to be absolute extrema. For instance, in Example 1, $f(-2, 3)$ is an absolute minimum of the function. (On the other hand, the relative maximum found in Example 3 is not an absolute maximum of the function.) Second, absolute extrema can occur at a boundary point of the domain. This is illustrated in Example 5.

EXAMPLE 5 Finding Absolute Extrema

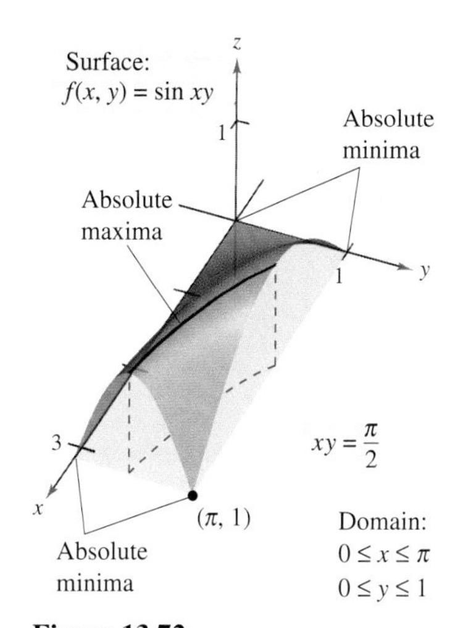

Figure 13.72

Find the absolute extrema of the function

$$f(x, y) = \sin xy$$

on the closed region given by $0 \le x \le \pi$ and $0 \le y \le 1$.

Solution From the partial derivatives

$$f_x(x, y) = y \cos xy \quad \text{and} \quad f_y(x, y) = x \cos xy$$

you can see that each point lying on the hyperbola given by $xy = \pi/2$ is a critical point. These points each yield the value

$$f(x, y) = \sin \frac{\pi}{2} = 1$$

which you know is the absolute maximum, as shown in Figure 13.72. The only other critical point of f *lying in the given region* is $(0, 0)$. It yields an absolute minimum of 0, because

$$0 \le xy \le \pi$$

implies that

$$0 \le \sin xy \le 1.$$

To locate other absolute extrema, you should consider the four boundaries of the region formed by taking traces with the vertical planes $x = 0$, $x = \pi$, $y = 0$, and $y = 1$. In doing this, you will find that $\sin xy = 0$ at all points on the x-axis, at all points on the y-axis, and at the point $(\pi, 1)$. Each of these points yields an absolute minimum for the surface, as shown in Figure 13.72. ■

The concepts of relative extrema and critical points can be extended to functions of three or more variables. If all first partial derivatives of

$$w = f(x_1, x_2, x_3, \ldots, x_n)$$

exist, it can be shown that a relative maximum or minimum can occur at $(x_1, x_2, x_3, \ldots, x_n)$ only if every first partial derivative is 0 at that point. This means that the critical points are obtained by solving the following system of equations.

$$\begin{aligned} f_{x_1}(x_1, x_2, x_3, \ldots, x_n) &= 0 \\ f_{x_2}(x_1, x_2, x_3, \ldots, x_n) &= 0 \\ &\vdots \\ f_{x_n}(x_1, x_2, x_3, \ldots, x_n) &= 0 \end{aligned}$$

The extension of Theorem 13.17 to three or more variables is also possible, although you will not consider such an extension in this text.

13.8 Exercises

See www.CalcChat.com for worked-out solutions to odd-numbered exercises.

In Exercises 1–6, identify any extrema of the function by recognizing its given form or its form after completing the square. Verify your results by using the partial derivatives to locate any critical points and test for relative extrema.

1. $g(x, y) = (x - 1)^2 + (y - 3)^2$
2. $g(x, y) = 5 - (x - 3)^2 - (y + 2)^2$
3. $f(x, y) = \sqrt{x^2 + y^2 + 1}$
4. $f(x, y) = \sqrt{25 - (x - 2)^2 - y^2}$
5. $f(x, y) = x^2 + y^2 + 2x - 6y + 6$
6. $f(x, y) = -x^2 - y^2 + 10x + 12y - 64$

In Exercises 7–16, examine the function for relative extrema.

7. $f(x, y) = 3x^2 + 2y^2 - 6x - 4y + 16$
8. $f(x, y) = -3x^2 - 2y^2 + 3x - 4y + 5$
9. $f(x, y) = -x^2 - 5y^2 + 10x - 10y - 28$
10. $f(x, y) = 2x^2 + 2xy + y^2 + 2x - 3$
11. $z = x^2 + xy + \frac{1}{2}y^2 - 2x + y$
12. $z = -5x^2 + 4xy - y^2 + 16x + 10$
13. $f(x, y) = \sqrt{x^2 + y^2}$
14. $h(x, y) = (x^2 + y^2)^{1/3} + 2$
15. $g(x, y) = 4 - |x| - |y|$
16. $f(x, y) = |x + y| - 2$

CAS **In Exercises 17–20, use a computer algebra system to graph the surface and locate any relative extrema and saddle points.**

17. $z = \dfrac{-4x}{x^2 + y^2 + 1}$
18. $f(x, y) = y^3 - 3yx^2 - 3y^2 - 3x^2 + 1$
19. $z = (x^2 + 4y^2)e^{1 - x^2 - y^2}$
20. $z = e^{xy}$

In Exercises 21–28, examine the function for relative extrema and saddle points.

21. $h(x, y) = 80x + 80y - x^2 - y^2$
22. $g(x, y) = x^2 - y^2 - x - y$
23. $g(x, y) = xy$
24. $h(x, y) = x^2 - 3xy - y^2$
25. $f(x, y) = x^2 - xy - y^2 - 3x - y$

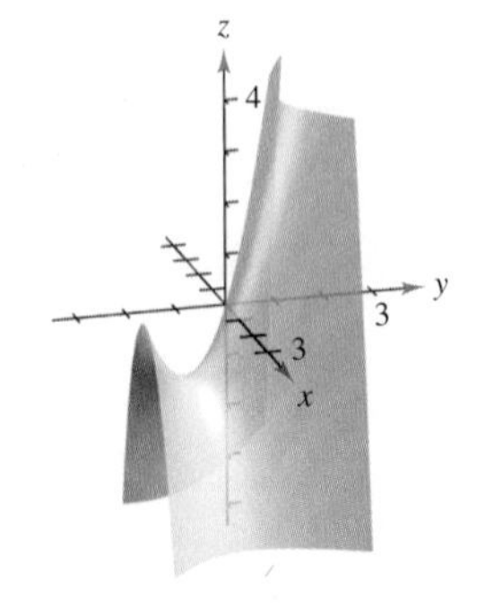

26. $f(x, y) = 2xy - \frac{1}{2}(x^4 + y^4) + 1$

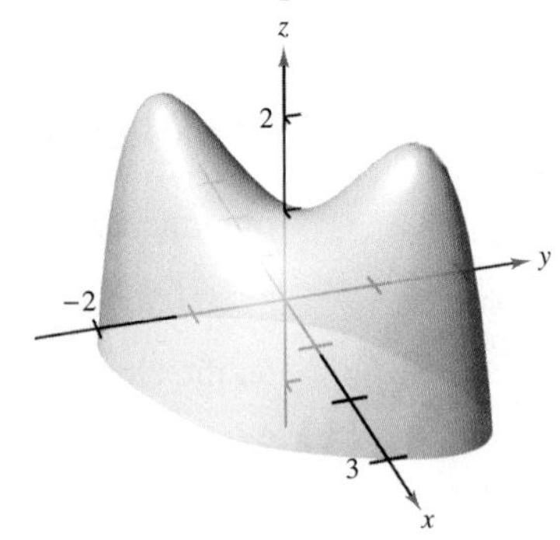

27. $z = e^{-x} \sin y$

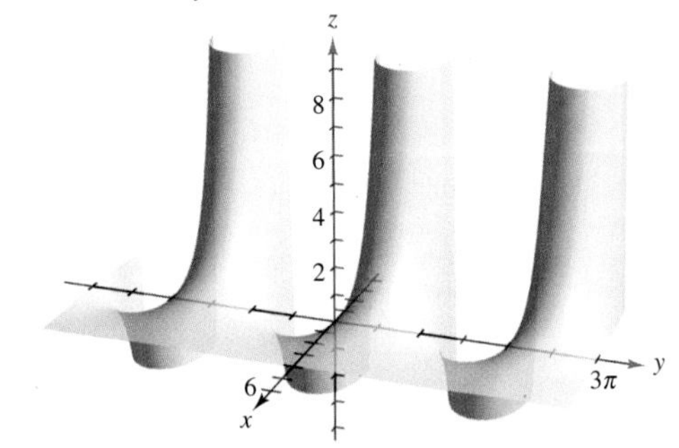

28. $z = \left(\frac{1}{2} - x^2 + y^2\right)e^{1 - x^2 - y^2}$

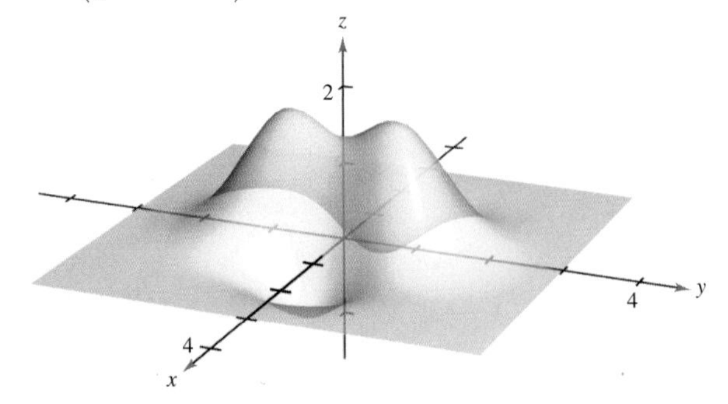

CAS **In Exercises 29 and 30, examine the function for extrema without using the derivative tests, and use a computer algebra system to graph the surface. (*Hint:* By observation, determine if it is possible for z to be negative. When is z equal to 0?)**

29. $z = \dfrac{(x - y)^4}{x^2 + y^2}$
30. $z = \dfrac{(x^2 - y^2)^2}{x^2 + y^2}$

Think About It **In Exercises 31–34, determine whether there is a relative maximum, a relative minimum, a saddle point, or insufficient information to determine the nature of the function $f(x, y)$ at the critical point (x_0, y_0).**

31. $f_{xx}(x_0, y_0) = 9,\quad f_{yy}(x_0, y_0) = 4,\quad f_{xy}(x_0, y_0) = 6$
32. $f_{xx}(x_0, y_0) = -3,\quad f_{yy}(x_0, y_0) = -8,\quad f_{xy}(x_0, y_0) = 2$
33. $f_{xx}(x_0, y_0) = -9,\quad f_{yy}(x_0, y_0) = 6,\quad f_{xy}(x_0, y_0) = 10$
34. $f_{xx}(x_0, y_0) = 25,\quad f_{yy}(x_0, y_0) = 8,\quad f_{xy}(x_0, y_0) = 10$

35. A function f has continuous second partial derivatives on an open region containing the critical point $(3, 7)$. The function has a minimum at $(3, 7)$, and $d > 0$ for the Second Partials Test. Determine the interval for $f_{xy}(3, 7)$ if $f_{xx}(3, 7) = 2$ and $f_{yy}(3, 7) = 8$.

36. A function f has continuous second partial derivatives on an open region containing the critical point (a, b). If $f_{xx}(a, b)$ and $f_{yy}(a, b)$ have opposite signs, what is implied? Explain.

CAS **In Exercises 37–42, (a) find the critical points, (b) test for relative extrema, (c) list the critical points for which the Second Partials Test fails, and (d) use a computer algebra system to graph the function, labeling any extrema and saddle points.**

37. $f(x, y) = x^3 + y^3$

38. $f(x, y) = x^3 + y^3 - 6x^2 + 9y^2 + 12x + 27y + 19$

39. $f(x, y) = (x - 1)^2(y + 4)^2$

40. $f(x, y) = \sqrt{(x - 1)^2 + (y + 2)^2}$

41. $f(x, y) = x^{2/3} + y^{2/3}$

42. $f(x, y) = (x^2 + y^2)^{2/3}$

In Exercises 43 and 44, find the critical points of the function and, from the form of the function, determine whether a relative maximum or a relative minimum occurs at each point.

43. $f(x, y, z) = x^2 + (y - 3)^2 + (z + 1)^2$

44. $f(x, y, z) = 9 - [x(y - 1)(z + 2)]^2$

In Exercises 45–54, find the absolute extrema of the function over the region R. (In each case, R contains the boundaries.) Use a computer algebra system to confirm your results.

45. $f(x, y) = x^2 - 4xy + 5$

$R = \{(x, y) : 1 \le x \le 4, 0 \le y \le 2\}$

46. $f(x, y) = x^2 + xy, \quad R = \{(x, y) : |x| \le 2, |y| \le 1\}$

47. $f(x, y) = 12 - 3x - 2y$

R: The triangular region in the xy-plane with vertices $(2, 0)$, $(0, 1)$, and $(1, 2)$

48. $f(x, y) = (2x - y)^2$

R: The triangular region in the xy-plane with vertices $(2, 0)$, $(0, 1)$, and $(1, 2)$

49. $f(x, y) = 3x^2 + 2y^2 - 4y$

R: The region in the xy-plane bounded by the graphs of $y = x^2$ and $y = 4$

50. $f(x, y) = 2x - 2xy + y^2$

R: The region in the xy-plane bounded by the graphs of $y = x^2$ and $y = 1$

51. $f(x, y) = x^2 + 2xy + y^2, \quad R = \{(x, y) : |x| \le 2, |y| \le 1\}$

52. $f(x, y) = x^2 + 2xy + y^2, \quad R = \{(x, y) : x^2 + y^2 \le 8\}$

53. $f(x, y) = \dfrac{4xy}{(x^2 + 1)(y^2 + 1)}$

$R = \{(x, y) : 0 \le x \le 1, 0 \le y \le 1\}$

54. $f(x, y) = \dfrac{4xy}{(x^2 + 1)(y^2 + 1)}$

$R = \{(x, y) : x \ge 0, y \ge 0, x^2 + y^2 \le 1\}$

WRITING ABOUT CONCEPTS

55. The figure shows the level curves for an unknown function $f(x, y)$. What, if any, information can be given about f at the point A? Explain your reasoning.

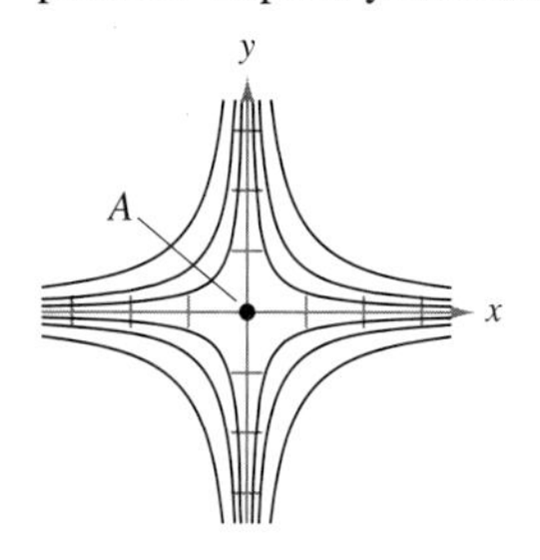

Figure for 55

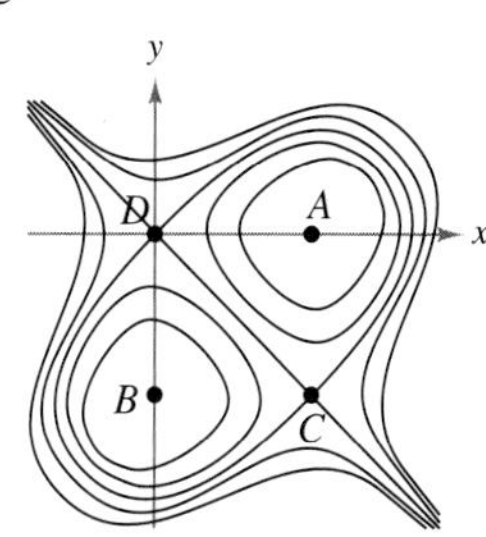

Figure for 56

56. The figure shows the level curves for an unknown function $f(x, y)$. What, if any, information can be given about f at the points A, B, C, and D? Explain your reasoning.

In Exercises 57–59, sketch the graph of an arbitrary function f satisfying the given conditions. State whether the function has any extrema or saddle points. (There are many correct answers.)

57. $f_x(x, y) > 0$ and $f_y(x, y) < 0$ for all (x, y).

58. All of the first and second partial derivatives of f are 0.

59. $f_x(0, 0) = 0, \ f_y(0, 0) = 0$

$f_x(x, y)\begin{cases} < 0, & x < 0 \\ > 0, & x > 0 \end{cases}, \quad f_y(x, y)\begin{cases} > 0, & y < 0 \\ < 0, & y > 0 \end{cases}$

$f_{xx}(x, y) > 0, f_{yy}(x, y) < 0$, and $f_{xy}(x, y) = 0$ for all (x, y).

CAPSTONE

60. Consider the functions

$f(x, y) = x^2 - y^2$ and $g(x, y) = x^2 + y^2$.

(a) Show that both functions have a critical point at $(0, 0)$.

(b) Explain how f and g behave differently at this critical point.

True or False? **In Exercises 61–64, determine whether the statement is true or false. If it is false, explain why or give an example that shows it is false.**

61. If f has a relative maximum at (x_0, y_0, z_0), then $f_x(x_0, y_0) = f_y(x_0, y_0) = 0$.

62. If $f_x(x_0, y_0) = f_y(x_0, y_0) = 0$, then f has a relative maximum at (x_0, y_0, z_0).

63. Between any two relative minima of f, there must be at least one relative maximum of f.

64. If f is continuous for all x and y and has two relative minima, then f must have at least one relative maximum.

13.9 Applications of Extrema of Functions of Two Variables

- **Solve optimization problems involving functions of several variables.**
- **Use the method of least squares.**

Applied Optimization Problems

In this section, you will survey a few of the many applications of extrema of functions of two (or more) variables.

EXAMPLE 1 Finding Maximum Volume

A rectangular box is resting on the xy-plane with one vertex at the origin. The opposite vertex lies in the plane

$$6x + 4y + 3z = 24$$

as shown in Figure 13.73. Find the maximum volume of such a box.

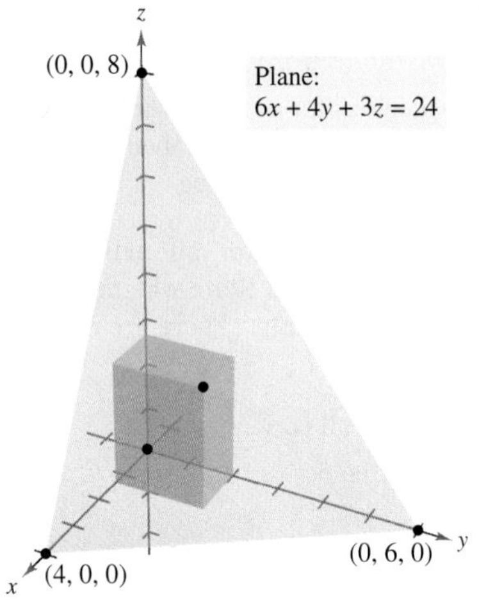

Figure 13.73

Solution Let x, y, and z represent the length, width, and height of the box. Because one vertex of the box lies in the plane $6x + 4y + 3z = 24$, you know that $z = \frac{1}{3}(24 - 6x - 4y)$, and you can write the volume xyz of the box as a function of two variables.

$$\begin{aligned} V(x, y) &= (x)(y)\left[\tfrac{1}{3}(24 - 6x - 4y)\right] \\ &= \tfrac{1}{3}(24xy - 6x^2y - 4xy^2) \end{aligned}$$

By setting the first partial derivatives equal to 0

$$V_x(x, y) = \frac{1}{3}(24y - 12xy - 4y^2) = \frac{y}{3}(24 - 12x - 4y) = 0$$

$$V_y(x, y) = \frac{1}{3}(24x - 6x^2 - 8xy) = \frac{x}{3}(24 - 6x - 8y) = 0$$

you obtain the critical points $(0, 0)$ and $\left(\frac{4}{3}, 2\right)$. At $(0, 0)$ the volume is 0, so that point does not yield a maximum volume. At the point $\left(\frac{4}{3}, 2\right)$, you can apply the Second Partials Test.

$$V_{xx}(x, y) = -4y$$

$$V_{yy}(x, y) = \frac{-8x}{3}$$

$$V_{xy}(x, y) = \frac{1}{3}(24 - 12x - 8y)$$

NOTE In many applied problems, the domain of the function to be optimized is a closed bounded region. To find minimum or maximum points, you must not only test critical points, but also consider the values of the function at points on the boundary.

Because

$$V_{xx}\left(\tfrac{4}{3}, 2\right)V_{yy}\left(\tfrac{4}{3}, 2\right) - \left[V_{xy}\left(\tfrac{4}{3}, 2\right)\right]^2 = (-8)\left(-\tfrac{32}{9}\right) - \left(-\tfrac{8}{3}\right)^2 = \tfrac{64}{3} > 0$$

and

$$V_{xx}\left(\tfrac{4}{3}, 2\right) = -8 < 0$$

you can conclude from the Second Partials Test that the maximum volume is

$$\begin{aligned} V\left(\tfrac{4}{3}, 2\right) &= \tfrac{1}{3}\left[24\left(\tfrac{4}{3}\right)(2) - 6\left(\tfrac{4}{3}\right)^2(2) - 4\left(\tfrac{4}{3}\right)(2^2)\right] \\ &= \tfrac{64}{9} \text{ cubic units.} \end{aligned}$$

Note that the volume is 0 at the boundary points of the triangular domain of V. ■

Applications of extrema in economics and business often involve more than one independent variable. For instance, a company may produce several models of one type of product. The price per unit and profit per unit are usually different for each model. Moreover, the demand for each model is often a function of the prices of the other models (as well as its own price). The next example illustrates an application involving two products.

EXAMPLE 2 Finding the Maximum Profit

An electronics manufacturer determines that the profit P (in dollars) obtained by producing and selling x units of a DVD player and y units of a DVD recorder is approximated by the model

$$P(x, y) = 8x + 10y - (0.001)(x^2 + xy + y^2) - 10{,}000.$$

Find the production level that produces a maximum profit. What is the maximum profit?

Solution The partial derivatives of the profit function are

$$P_x(x, y) = 8 - (0.001)(2x + y) \qquad \text{and} \qquad P_y(x, y) = 10 - (0.001)(x + 2y).$$

By setting these partial derivatives equal to 0, you obtain the following system of equations.

$$\begin{aligned} 8 - (0.001)(2x + y) &= 0 \\ 10 - (0.001)(x + 2y) &= 0 \end{aligned}$$

After simplifying, this system of linear equations can be written as

$$\begin{aligned} 2x + y &= 8000 \\ x + 2y &= 10{,}000. \end{aligned}$$

Solving this system produces $x = 2000$ and $y = 4000$. The second partial derivatives of P are

$$\begin{aligned} P_{xx}(2000, 4000) &= -0.002 \\ P_{yy}(2000, 4000) &= -0.002 \\ P_{xy}(2000, 4000) &= -0.001. \end{aligned}$$

Because $P_{xx} < 0$ and

$$P_{xx}(2000, 4000)P_{yy}(2000, 4000) - [P_{xy}(2000, 4000)]^2 = (-0.002)^2 - (-0.001)^2 > 0$$

you can conclude that the production level of $x = 2000$ units and $y = 4000$ units yields a *maximum* profit. The maximum profit is

$$\begin{aligned} P(2000, 4000) &= 8(2000) + 10(4000) - \\ &\qquad (0.001)[2000^2 + 2000(4000) + 4000^2)] - 10{,}000 \\ &= \$18{,}000. \end{aligned}$$

■ **FOR FURTHER INFORMATION** For more information on the use of mathematics in economics, see the article "Mathematical Methods of Economics" by Joel Franklin in *The American Mathematical Monthly*. To view this article, go to the website *www.matharticles.com*.

NOTE In Example 2, it was assumed that the manufacturing plant is able to produce the required number of units to yield a maximum profit. In actual practice, the production would be bounded by physical constraints. You will study such constrained optimization problems in the next section.

The Method of Least Squares

Many of the examples in this text have involved **mathematical models.** For instance, Example 2 involves a quadratic model for profit. There are several ways to develop such models; one is called the **method of least squares.**

In constructing a model to represent a particular phenomenon, the goals are simplicity and accuracy. Of course, these goals often conflict. For instance, a simple linear model for the points in Figure 13.74 is

$$y = 1.8566x - 5.0246.$$

However, Figure 13.75 shows that by choosing the slightly more complicated quadratic model*

$$y = 0.1996x^2 - 0.7281x + 1.3749$$

you can achieve greater accuracy.

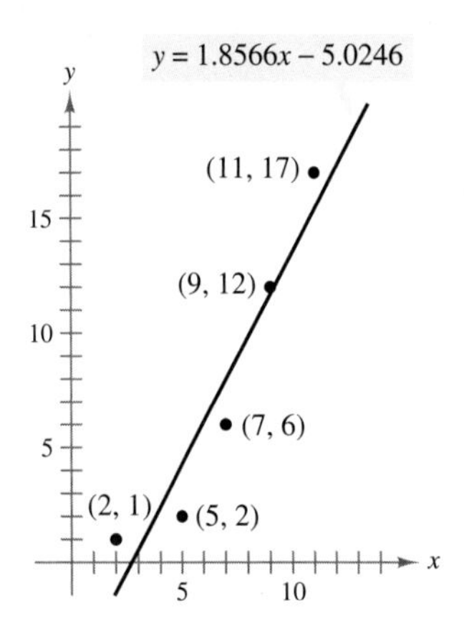

Figure 13.74

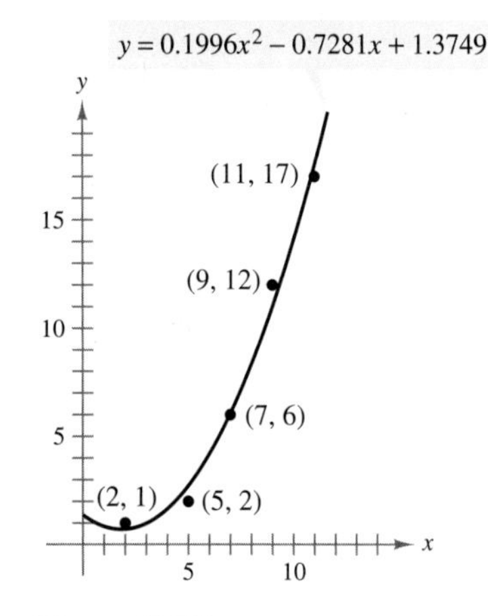

Figure 13.75

As a measure of how well the model $y = f(x)$ fits the collection of points

$$\{(x_1, y_1), (x_2, y_2), (x_3, y_3), \ldots, (x_n, y_n)\}$$

you can add the squares of the differences between the actual y-values and the values given by the model to obtain the **sum of the squared errors**

$$S = \sum_{i=1}^{n} [f(x_i) - y_i]^2. \qquad \text{Sum of the squared errors}$$

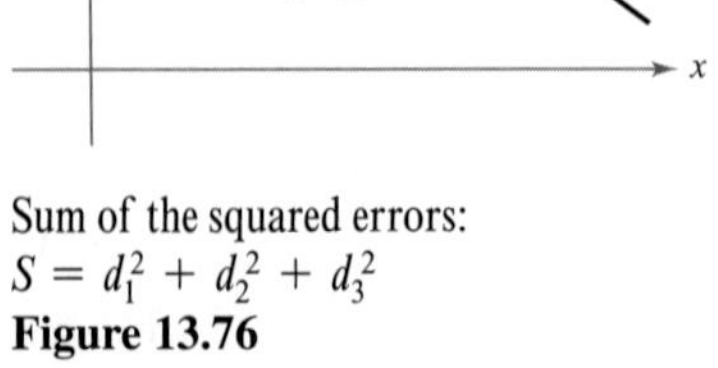

Sum of the squared errors: $S = d_1^2 + d_2^2 + d_3^2$

Figure 13.76

Graphically, S can be interpreted as the sum of the squares of the vertical distances between the graph of f and the given points in the plane, as shown in Figure 13.76. If the model is perfect, then $S = 0$. However, when perfection is not feasible, you can settle for a model that minimizes S. For instance, the sum of the squared errors for the linear model in Figure 13.74 is $S \approx 17$. Statisticians call the *linear model* that minimizes S the **least squares regression line.** The proof that this line actually minimizes S involves the minimizing of a function of two variables.

---

* *A method for finding the least squares quadratic model for a collection of data is described in Exercise 37.*

ADRIEN-MARIE LEGENDRE (1752–1833)

The method of least squares was introduced by the French mathematician Adrien-Marie Legendre. Legendre is best known for his work in geometry. In fact, his text Elements of Geometry was so popular in the United States that it continued to be used for 33 editions, spanning a period of more than 100 years.

THEOREM 13.18 LEAST SQUARES REGRESSION LINE

The **least squares regression line** for $\{(x_1, y_1), (x_2, y_2), \ldots, (x_n, y_n)\}$ is given by $f(x) = ax + b$, where

$$a = \frac{n\sum_{i=1}^{n} x_i y_i - \sum_{i=1}^{n} x_i \sum_{i=1}^{n} y_i}{n\sum_{i=1}^{n} x_i^2 - \left(\sum_{i=1}^{n} x_i\right)^2} \quad \text{and} \quad b = \frac{1}{n}\left(\sum_{i=1}^{n} y_i - a\sum_{i=1}^{n} x_i\right).$$

PROOF Let $S(a, b)$ represent the sum of the squared errors for the model $f(x) = ax + b$ and the given set of points. That is,

$$\begin{aligned} S(a, b) &= \sum_{i=1}^{n} [f(x_i) - y_i]^2 \\ &= \sum_{i=1}^{n} (ax_i + b - y_i)^2 \end{aligned}$$

where the points (x_i, y_i) represent constants. Because S is a function of a and b, you can use the methods discussed in the preceding section to find the minimum value of S. Specifically, the first partial derivatives of S are

$$\begin{aligned} S_a(a, b) &= \sum_{i=1}^{n} 2x_i(ax_i + b - y_i) \\ &= 2a\sum_{i=1}^{n} x_i^2 + 2b\sum_{i=1}^{n} x_i - 2\sum_{i=1}^{n} x_i y_i \\ S_b(a, b) &= \sum_{i=1}^{n} 2(ax_i + b - y_i) \\ &= 2a\sum_{i=1}^{n} x_i + 2nb - 2\sum_{i=1}^{n} y_i. \end{aligned}$$

By setting these two partial derivatives equal to 0, you obtain the values of a and b that are listed in the theorem. It is left to you to apply the Second Partials Test (see Exercise 47) to verify that these values of a and b yield a minimum. ■

If the x-values are symmetrically spaced about the y-axis, then $\sum x_i = 0$ and the formulas for a and b simplify to

$$a = \frac{\sum_{i=1}^{n} x_i y_i}{\sum_{i=1}^{n} x_i^2}$$

and

$$b = \frac{1}{n}\sum_{i=1}^{n} y_i.$$

This simplification is often possible with a translation of the x-values. For instance, if the x-values in a data collection consist of the years 2005, 2006, 2007, 2008, and 2009, you could let 2007 be represented by 0.

EXAMPLE 3 Finding the Least Squares Regression Line

Find the least squares regression line for the points $(-3, 0)$, $(-1, 1)$, $(0, 2)$, and $(2, 3)$.

Solution The table shows the calculations involved in finding the least squares regression line using $n = 4$.

x	y	xy	x^2
-3	0	0	9
-1	1	-1	1
0	2	0	0
2	3	6	4
$\sum_{i=1}^{n} x_i = -2$	$\sum_{i=1}^{n} y_i = 6$	$\sum_{i=1}^{n} x_i y_i = 5$	$\sum_{i=1}^{n} x_i^2 = 14$

TECHNOLOGY Many calculators have "built-in" least squares regression programs. If your calculator has such a program, use it to duplicate the results of Example 3.

Applying Theorem 13.18 produces

$$a = \frac{n\sum_{i=1}^{n} x_i y_i - \sum_{i=1}^{n} x_i \sum_{i=1}^{n} y_i}{n\sum_{i=1}^{n} x_i^2 - \left(\sum_{i=1}^{n} x_i\right)^2} = \frac{4(5) - (-2)(6)}{4(14) - (-2)^2} = \frac{8}{13}$$

and

$$b = \frac{1}{n}\left(\sum_{i=1}^{n} y_i - a\sum_{i=1}^{n} x_i\right) = \frac{1}{4}\left[6 - \frac{8}{13}(-2)\right] = \frac{47}{26}.$$

The least squares regression line is $f(x) = \frac{8}{13}x + \frac{47}{26}$, as shown in Figure 13.77. ■

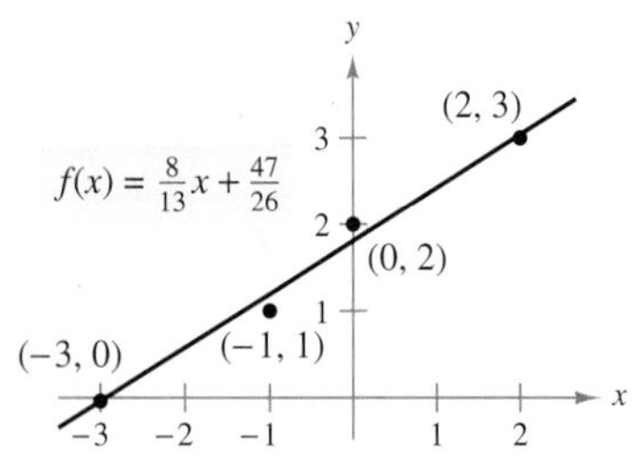

Least squares regression line
Figure 13.77

13.9 Exercises

See www.CalcChat.com for worked-out solutions to odd-numbered exercises.

In Exercises 1 and 2, find the minimum distance from the point to the plane $x - y + z = 3$. (*Hint:* To simplify the computations, minimize the square of the distance.)

1. $(0, 0, 0)$

2. $(1, 2, 3)$

In Exercises 3 and 4, find the minimum distance from the point to the surface $z = \sqrt{1 - 2x - 2y}$. (*Hint:* In Exercise 4, use the *root* feature of a graphing utility.)

3. $(-2, -2, 0)$

4. $(0, 0, 2)$

In Exercises 5–8, find three positive integers x, y, and z that satisfy the given conditions.

5. The product is 27 and the sum is a minimum.

6. The sum is 32 and $P = xy^2z$ is a maximum.

7. The sum is 30 and the sum of the squares is a minimum.

8. The product is 1 and the sum of the squares is a minimum.

9. ***Cost*** A home improvement contractor is painting the walls and ceiling of a rectangular room. The volume of the room is 668.25 cubic feet. The cost of wall paint is \$0.06 per square foot and the cost of ceiling paint is \$0.11 per square foot. Find the room dimensions that result in a minimum cost for the paint. What is the minimum cost for the paint?

10. ***Maximum Volume*** The material for constructing the base of an open box costs 1.5 times as much per unit area as the material for constructing the sides. For a fixed amount of money C, find the dimensions of the box of largest volume that can be made.

11. ***Maximum Volume*** The volume of an ellipsoid

$$\frac{x^2}{a^2} + \frac{y^2}{b^2} + \frac{z^2}{c^2} = 1$$

is $4\pi abc/3$. For a fixed sum $a + b + c$, show that the ellipsoid of maximum volume is a sphere.

12. ***Maximum Volume*** Show that the rectangular box of maximum volume inscribed in a sphere of radius r is a cube.

13. ***Volume and Surface Area*** Show that a rectangular box of given volume and minimum surface area is a cube.

14. ***Area*** A trough with trapezoidal cross sections is formed by turning up the edges of a 30-inch-wide sheet of aluminum (see figure). Find the cross section of maximum area.

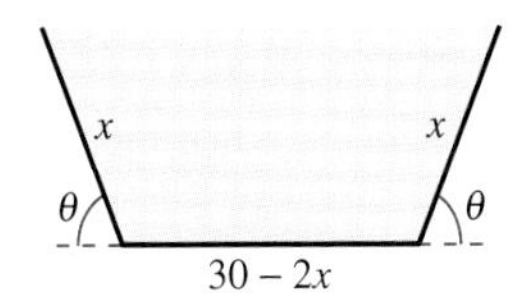

15. ***Maximum Revenue*** A company manufactures two types of sneakers, running shoes and basketball shoes. The total revenue from x_1 units of running shoes and x_2 units of basketball shoes is $R = -5x_1^2 - 8x_2^2 - 2x_1x_2 + 42x_1 + 102x_2$, where x_1 and x_2 are in thousands of units. Find x_1 and x_2 so as to maximize the revenue.

16. ***Maximum Profit*** A corporation manufactures candles at two locations. The cost of producing x_1 units at location 1 is

$$C_1 = 0.02x_1^2 + 4x_1 + 500$$

and the cost of producing x_2 units at location 2 is

$$C_2 = 0.05x_2^2 + 4x_2 + 275.$$

The candles sell for \$15 per unit. Find the quantity that should be produced at each location to maximize the profit $P = 15(x_1 + x_2) - C_1 - C_2$.

17. ***Hardy-Weinberg Law*** Common blood types are determined genetically by three alleles A, B, and O. (An allele is any of a group of possible mutational forms of a gene.) A person whose blood type is AA, BB, or OO is homozygous. A person whose blood type is AB, AO, or BO is heterozygous. The Hardy-Weinberg Law states that the proportion P of heterozygous individuals in any given population is

$$P(p, q, r) = 2pq + 2pr + 2qr$$

where p represents the percent of allele A in the population, q represents the percent of allele B in the population, and r represents the percent of allele O in the population. Use the fact that $p + q + r = 1$ to show that the maximum proportion of heterozygous individuals in any population is $\frac{2}{3}$.

18. ***Shannon Diversity Index*** One way to measure species diversity is to use the Shannon diversity index H. If a habitat consists of three species, A, B, and C, its Shannon diversity index is

$$H = -x \ln x - y \ln y - z \ln z$$

where x is the percent of species A in the habitat, y is the percent of species B in the habitat, and z is the percent of species C in the habitat.

(a) Use the fact that $x + y + z = 1$ to show that the maximum value of H occurs when $x = y = z = \frac{1}{3}$.

(b) Use the results of part (a) to show that the maximum value of H in this habitat is ln 3.

19. ***Minimum Cost*** A water line is to be built from point P to point S and must pass through regions where construction costs differ (see figure). The cost per kilometer in dollars is $3k$ from P to Q, $2k$ from Q to R, and k from R to S. Find x and y such that the total cost C will be minimized.

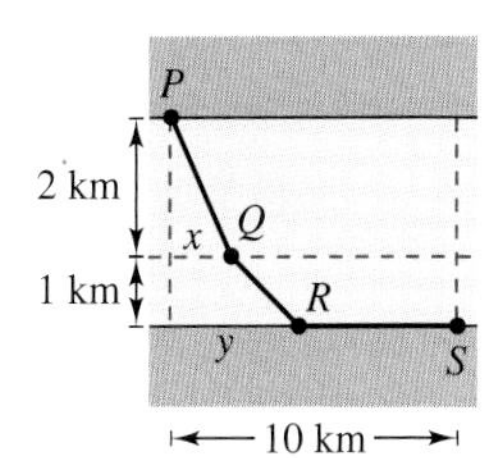

20. ***Distance*** A company has retail outlets located at the points (0, 0), (2, 2), and (−2, 2) (see figure). Management plans to build a distribution center located such that the sum of the distances S from the center to the outlets is minimum. From the symmetry of the problem it is clear that the distribution center will be located on the y-axis, and therefore S is a function of the single variable y. Using techniques presented in Chapter 3, find the required value of y.

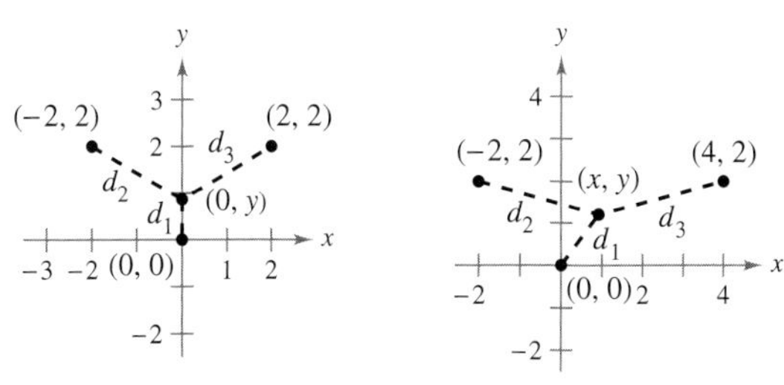

Figure for 20 **Figure for 21**

CAS **21.** ***Investigation*** The retail outlets described in Exercise 20 are located at (0, 0), (4, 2), and (−2, 2) (see figure). The location of the distribution center is (x, y), and therefore the sum of the distances S is a function of x and y.

(a) Write the expression giving the sum of the distances S. Use a computer algebra system to graph S. Does the surface have a minimum?

(b) Use a computer algebra system to obtain S_x and S_y. Observe that solving the system $S_x = 0$ and $S_y = 0$ is very difficult. So, approximate the location of the distribution center.

(c) An initial estimate of the critical point is $(x_1, y_1) = (1, 1)$. Calculate $-\nabla S(1, 1)$ with components $-S_x(1, 1)$ and $-S_y(1, 1)$. What direction is given by the vector $-\nabla S(1, 1)$?

(d) The second estimate of the critical point is

$$(x_2, y_2) = (x_1 - S_x(x_1, y_1)t, y_1 - S_y(x_1, y_1)t).$$

If these coordinates are substituted into $S(x, y)$, then S becomes a function of the single variable t. Find the value of t that minimizes S. Use this value of t to estimate (x_2, y_2).

(e) Complete two more iterations of the process in part (d) to obtain (x_4, y_4). For this location of the distribution center, what is the sum of the distances to the retail outlets?

(f) Explain why $-\nabla S(x, y)$ was used to approximate the minimum value of S. In what types of problems would you use $\nabla S(x, y)$?

22. ***Investigation*** Repeat Exercise 21 for retail outlets located at the points $(-4, 0)$, $(1, 6)$, and $(12, 2)$.

WRITING ABOUT CONCEPTS

23. In your own words, state the problem-solving strategy for applied minimum and maximum problems.

24. In your own words, describe the method of least squares for finding mathematical models.

In Exercises 25–28, (a) find the least squares regression line and (b) calculate *S*, the sum of the squared errors. Use the regression capabilities of a graphing utility to verify your results.

25.

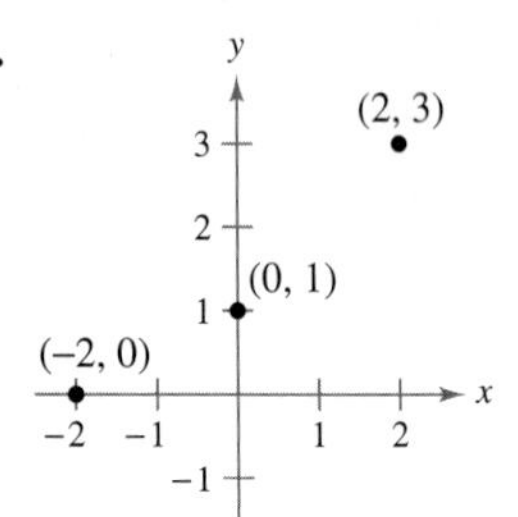

26.

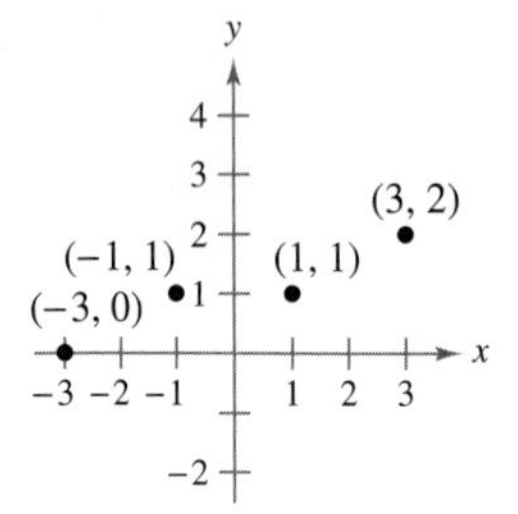

27.

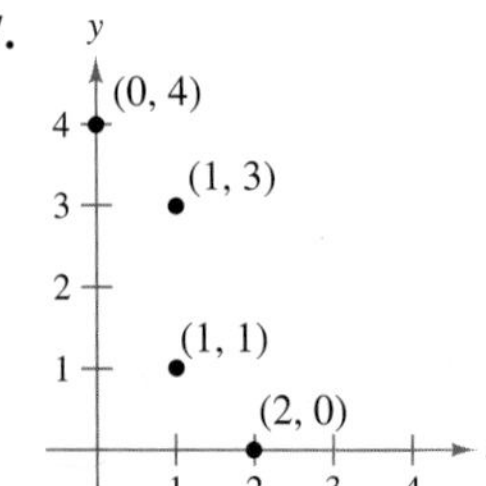

28.

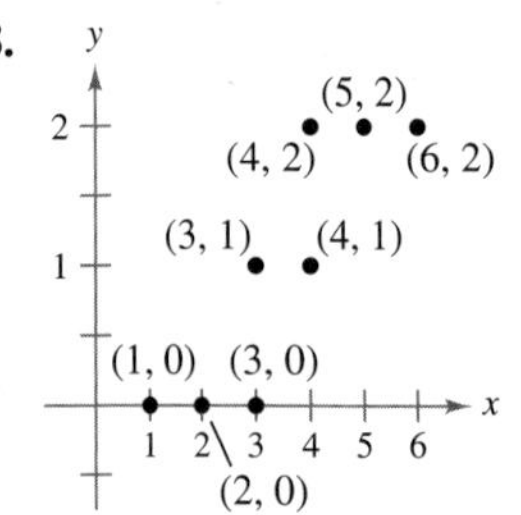

In Exercises 29–32, find the least squares regression line for the points. Use the regression capabilities of a graphing utility to verify your results. Use the graphing utility to plot the points and graph the regression line.

29. $(0, 0)$, $(1, 1)$, $(3, 4)$, $(4, 2)$, $(5, 5)$

30. $(1, 0)$, $(3, 3)$, $(5, 6)$

31. $(0, 6)$, $(4, 3)$, $(5, 0)$, $(8, -4)$, $(10, -5)$

32. $(6, 4)$, $(1, 2)$, $(3, 3)$, $(8, 6)$, $(11, 8)$, $(13, 8)$

33. ***Modeling Data*** The ages x (in years) and systolic blood pressures y of seven men are shown in the table.

Age, x	16	25	39	45	49	64	70
Systolic Blood Pressure, y	109	122	150	165	159	183	199

(a) Use the regression capabilities of a graphing utility to find the least squares regression line for the data.

(b) Use a graphing utility to plot the data and graph the model.

(c) Use the model to approximate the change in systolic blood pressure for each one-year increase in age.

34. ***Modeling Data*** A store manager wants to know the demand y for an energy bar as a function of price x. The daily sales for three different prices of the energy bar are shown in the table.

Price, x	\$1.29	\$1.49	\$1.69
Demand, y	450	375	330

(a) Use the regression capabilities of a graphing utility to find the least squares regression line for the data.

(b) Use the model to estimate the demand when the price is \$1.59.

35. ***Modeling Data*** An agronomist used four test plots to determine the relationship between the wheat yield y (in bushels per acre) and the amount of fertilizer x (in hundreds of pounds per acre). The results are shown in the table.

Fertilizer, x	1.0	1.5	2.0	2.5
Yield, y	32	41	48	53

Use the regression capabilities of a graphing utility to find the least squares regression line for the data, and estimate the yield for a fertilizer application of 160 pounds per acre.

36. ***Modeling Data*** The table shows the percents x and numbers y (in millions) of women in the work force for selected years. *(Source: U.S. Bureau of Labor Statistics)*

Year	1970	1975	1980	1985
Percent, x	43.3	46.3	51.5	54.5
Number, y	31.5	37.5	45.5	51.1

Year	1990	1995	2000	2005
Percent, x	57.5	58.9	59.9	59.3
Number, y	56.8	60.9	66.3	69.3

(a) Use the regression capabilities of a graphing utility to find the least squares regression line for the data.

(b) According to this model, approximately how many women enter the labor force for each one-point increase in the percent of women in the labor force?

37. Find a system of equations whose solution yields the coefficients a, b, and c for the least squares regression quadratic $y = ax^2 + bx + c$ for the points $(x_1, y_1), (x_2, y_2), \ldots, (x_n, y_n)$ by minimizing the sum

$$S(a, b, c) = \sum_{i=1}^{n} (y_i - ax_i^2 - bx_i - c)^2.$$

CAPSTONE

38. The sum of the length and the girth (perimeter of a cross section) of a package carried by a delivery service cannot exceed 108 inches. Find the dimensions of the rectangular package of largest volume that may be sent.

 **In Exercises 39–42, use the result of Exercise 37 to find the least squares regression quadratic for the given points. Use the regression capabilities of a graphing utility to confirm your results. Use the graphing utility to plot the points and graph the least squares regression quadratic.**

39. $(-2, 0), (-1, 0), (0, 1), (1, 2), (2, 5)$

40. $(-4, 5), (-2, 6), (2, 6), (4, 2)$

41. $(0, 0), (2, 2), (3, 6), (4, 12)$ **42.** $(0, 10), (1, 9), (2, 6), (3, 0)$

 **43.** ***Modeling Data*** After a new turbocharger for an automobile engine was developed, the following experimental data were obtained for speed y in miles per hour at two-second time intervals x.

Time, x	0	2	4	6	8	10
Speed, y	0	15	30	50	65	70

(a) Find a least squares regression quadratic for the data. Use a graphing utility to confirm your results.

(b) Use a graphing utility to plot the points and graph the model.

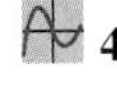 **44.** ***Modeling Data*** The table shows the world populations y (in billions) for five different years. Let $x = 8$ represent the year 1998. *(Source: U.S. Census Bureau, International Data Base)*

Year, x	1998	2000	2002	2004	2006
Population, y	5.9	6.1	6.2	6.4	6.5

(a) Use the regression capabilities of a graphing utility to find the least squares regression line for the data.

(b) Use the regression capabilities of a graphing utility to find the least squares regression quadratic for the data.

(c) Use a graphing utility to plot the data and graph the models.

(d) Use both models to forecast the world population for the year 2014. How do the two models differ as you extrapolate into the future?

 **45.** ***Modeling Data*** A meteorologist measures the atmospheric pressure P (in kilograms per square meter) at altitude h (in kilometers). The data are shown below.

Altitude, h	0	5	10	15	20
Pressure, P	10,332	5583	2376	1240	517

(a) Use the regression capabilities of a graphing utility to find a least squares regression line for the points $(h, \ln P)$.

(b) The result in part (a) is an equation of the form $\ln P = ah + b$. Write this logarithmic form in exponential form.

(c) Use a graphing utility to plot the original data and graph the exponential model in part (b).

(d) If your graphing utility can fit logarithmic models to data, use it to verify the result in part (b).

 **46.** ***Modeling Data*** The endpoints of the interval over which distinct vision is possible are called the near point and far point of the eye. With increasing age, these points normally change. The table shows the approximate near points y (in inches) for various ages x (in years). *(Source: Ophthalmology & Physiological Optics)*

Age, x	16	32	44	50	60
Near Point, y	3.0	4.7	9.8	19.7	39.4

(a) Find a rational model for the data by taking the reciprocals of the near points to generate the points $(x, 1/y)$. Use the regression capabilities of a graphing utility to find a least squares regression line for the revised data. The resulting line has the form $1/y = ax + b$. Solve for y.

(b) Use a graphing utility to plot the data and graph the model.

(c) Do you think the model can be used to predict the near point for a person who is 70 years old? Explain.

47. Use the Second Partials Test to verify that the formulas for a and b given in Theorem 13.18 yield a minimum.

$\left[\textit{Hint:} \text{ Use the fact that } n\sum_{i=1}^{n} x_i^2 \geq \left(\sum_{i=1}^{n} x_i\right)^2.\right]$

SECTION PROJECT

Building a Pipeline

An oil company wishes to construct a pipeline from its offshore facility A to its refinery B. The offshore facility is 2 miles from shore, and the refinery is 1 mile inland. Furthermore, A and B are 5 miles apart, as shown in the figure.

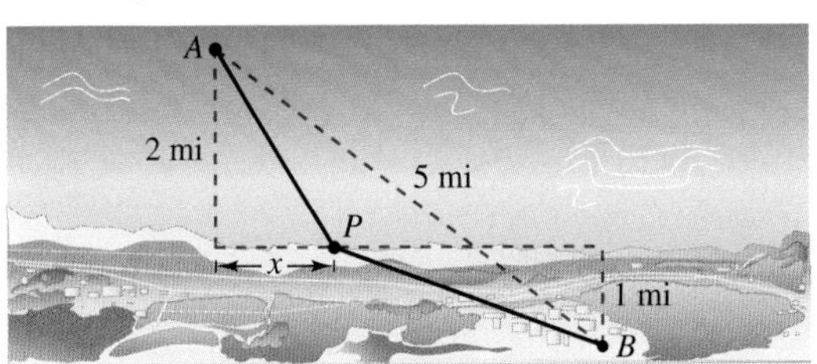

The cost of building the pipeline is \$3 million per mile in the water and \$4 million per mile on land. So, the cost of the pipeline depends on the location of point P, where it meets the shore. What would be the most economical route of the pipeline?

Imagine that you are to write a report to the oil company about this problem. Let x be the distance shown in the figure. Determine the cost of building the pipeline from A to P, and the cost from P to B. Analyze some sample pipeline routes and their corresponding costs. For instance, what is the cost of the most direct route? Then use calculus to determine the route of the pipeline that minimizes the cost. Explain all steps of your development and include any relevant graphs.

13.10 Lagrange Multipliers

- Understand the Method of Lagrange Multipliers.
- Use Lagrange multipliers to solve constrained optimization problems.
- Use the Method of Lagrange Multipliers with two constraints.

Lagrange Multipliers

JOSEPH-LOUIS LAGRANGE (1736–1813)

The Method of Lagrange Multipliers is named after the French mathematician Joseph-Louis Lagrange. Lagrange first introduced the method in his famous paper on mechanics, written when he was just 19 years old.

Many optimization problems have restrictions, or **constraints,** on the values that can be used to produce the optimal solution. Such constraints tend to complicate optimization problems because the optimal solution can occur at a boundary point of the domain. In this section, you will study an ingenious technique for solving such problems. It is called the **Method of Lagrange Multipliers.**

To see how this technique works, suppose you want to find the rectangle of maximum area that can be inscribed in the ellipse given by

$$\frac{x^2}{3^2} + \frac{y^2}{4^2} = 1.$$

Let (x, y) be the vertex of the rectangle in the first quadrant, as shown in Figure 13.78. Because the rectangle has sides of lengths $2x$ and $2y$, its area is given by

$$f(x, y) = 4xy. \qquad \text{Objective function}$$

You want to find x and y such that $f(x, y)$ is a maximum. Your choice of (x, y) is restricted to first-quadrant points that lie on the ellipse

$$\frac{x^2}{3^2} + \frac{y^2}{4^2} = 1. \qquad \text{Constraint}$$

Now, consider the constraint equation to be a fixed level curve of

$$g(x, y) = \frac{x^2}{3^2} + \frac{y^2}{4^2}.$$

The level curves of f represent a family of hyperbolas $f(x, y) = 4xy = k$. In this family, the level curves that meet the given constraint correspond to the hyperbolas that intersect the ellipse. Moreover, to maximize $f(x, y)$, you want to find the hyperbola that just barely satisfies the constraint. The level curve that does this is the one that is *tangent* to the ellipse, as shown in Figure 13.79.

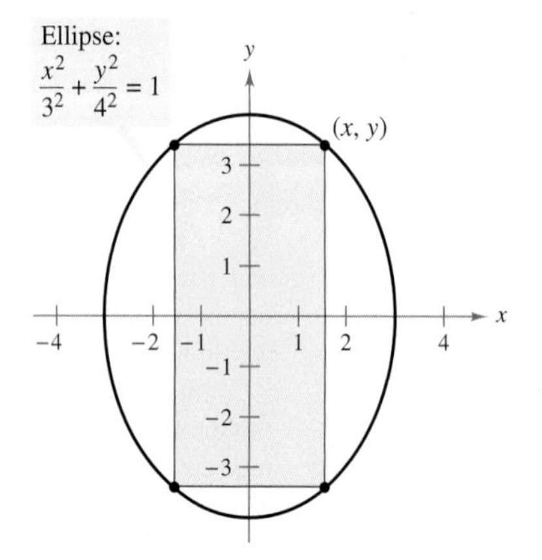

Objective function: $f(x, y) = 4xy$

Figure 13.78

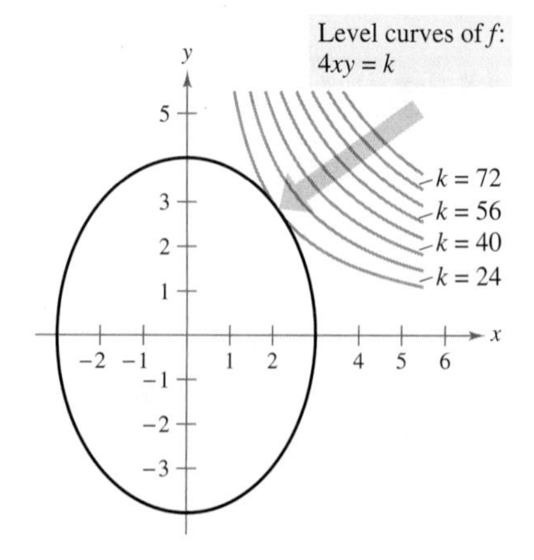

Constraint: $g(x, y) = \frac{x^2}{3^2} + \frac{y^2}{4^2} = 1$

Figure 13.79

To find the appropriate hyperbola, use the fact that two curves are tangent at a point if and only if their gradient vectors are parallel. This means that $\nabla f(x, y)$ must be a scalar multiple of $\nabla g(x, y)$ at the point of tangency. In the context of constrained optimization problems, this scalar is denoted by λ (the lowercase Greek letter lambda).

$$\nabla f(x, y) = \lambda \nabla g(x, y)$$

The scalar λ is called a **Lagrange multiplier.** Theorem 13.19 gives the necessary conditions for the existence of such multipliers.

THEOREM 13.19 LAGRANGE'S THEOREM

Let f and g have continuous first partial derivatives such that f has an extremum at a point (x_0, y_0) on the smooth constraint curve $g(x, y) = c$. If $\nabla g(x_0, y_0) \neq \mathbf{0}$, then there is a real number λ such that

$$\nabla f(x_0, y_0) = \lambda \nabla g(x_0, y_0).$$

PROOF To begin, represent the smooth curve given by $g(x, y) = c$ by the vector-valued function

$$\mathbf{r}(t) = x(t)\mathbf{i} + y(t)\mathbf{j}, \quad \mathbf{r}'(t) \neq \mathbf{0}$$

where x' and y' are continuous on an open interval I. Define the function h as $h(t) = f(x(t), y(t))$. Then, because $f(x_0, y_0)$ is an extreme value of f, you know that

$$h(t_0) = f(x(t_0), y(t_0)) = f(x_0, y_0)$$

is an extreme value of h. This implies that $h'(t_0) = 0$, and, by the Chain Rule,

$$h'(t_0) = f_x(x_0, y_0)x'(t_0) + f_y(x_0, y_0)y'(t_0) = \nabla f(x_0, y_0) \cdot \mathbf{r}'(t_0) = 0.$$

So, $\nabla f(x_0, y_0)$ is orthogonal to $\mathbf{r}'(t_0)$. Moreover, by Theorem 13.12, $\nabla g(x_0, y_0)$ is also orthogonal to $\mathbf{r}'(t_0)$. Consequently, the gradients $\nabla f(x_0, y_0)$ and $\nabla g(x_0, y_0)$ are parallel, and there must exist a scalar λ such that

$$\nabla f(x_0, y_0) = \lambda \nabla g(x_0, y_0).$$ ■

NOTE Lagrange's Theorem can be shown to be true for functions of three variables, using a similar argument with level surfaces and Theorem 13.14.

The Method of Lagrange Multipliers uses Theorem 13.19 to find the extreme values of a function f subject to a constraint.

METHOD OF LAGRANGE MULTIPLIERS

Let f and g satisfy the hypothesis of Lagrange's Theorem, and let f have a minimum or maximum subject to the constraint $g(x, y) = c$. To find the minimum or maximum of f, use the following steps.

1. Simultaneously solve the equations $\nabla f(x, y) = \lambda \nabla g(x, y)$ and $g(x, y) = c$ by solving the following system of equations.

$$f_x(x, y) = \lambda g_x(x, y)$$
$$f_y(x, y) = \lambda g_y(x, y)$$
$$g(x, y) = c$$

2. Evaluate f at each solution point obtained in the first step. The largest value yields the maximum of f subject to the constraint $g(x, y) = c$, and the smallest value yields the minimum of f subject to the constraint $g(x, y) = c$.

NOTE As you will see in Examples 1 and 2, the Method of Lagrange Multipliers requires solving systems of nonlinear equations. This often can require some tricky algebraic manipulation.

Constrained Optimization Problems

In the problem at the beginning of this section, you wanted to maximize the area of a rectangle that is inscribed in an ellipse. Example 1 shows how to use Lagrange multipliers to solve this problem.

EXAMPLE 1 Using a Lagrange Multiplier with One Constraint

Find the maximum value of $f(x, y) = 4xy$ where $x > 0$ and $y > 0$, subject to the constraint $(x^2/3^2) + (y^2/4^2) = 1$.

Solution To begin, let

$$g(x, y) = \frac{x^2}{3^2} + \frac{y^2}{4^2} = 1.$$

By equating $\nabla f(x, y) = 4y\mathbf{i} + 4x\mathbf{j}$ and $\lambda\nabla g(x, y) = (2\lambda x/9)\mathbf{i} + (\lambda y/8)\mathbf{j}$, you can obtain the following system of equations.

$$4y = \frac{2}{9}\lambda x \qquad f_x(x, y) = \lambda g_x(x, y)$$

$$4x = \frac{1}{8}\lambda y \qquad f_y(x, y) = \lambda g_y(x, y)$$

$$\frac{x^2}{3^2} + \frac{y^2}{4^2} = 1 \qquad \text{Constraint}$$

From the first equation, you obtain $\lambda = 18y/x$, and substitution into the second equation produces

$$4x = \frac{1}{8}\left(\frac{18y}{x}\right)y \quad \Longrightarrow \quad x^2 = \frac{9}{16}y^2.$$

Substituting this value for x^2 into the third equation produces

$$\frac{1}{9}\left(\frac{9}{16}y^2\right) + \frac{1}{16}y^2 = 1 \quad \Longrightarrow \quad y^2 = 8.$$

So, $y = \pm 2\sqrt{2}$. Because it is required that $y > 0$, choose the positive value and find that

$$x^2 = \frac{9}{16}y^2$$

$$= \frac{9}{16}(8) = \frac{9}{2}$$

$$x = \frac{3}{\sqrt{2}}.$$

So, the maximum value of f is

$$f\left(\frac{3}{\sqrt{2}}, 2\sqrt{2}\right) = 4xy = 4\left(\frac{3}{\sqrt{2}}\right)(2\sqrt{2}) = 24.$$ ■

Note that writing the constraint as

$$g(x, y) = \frac{x^2}{3^2} + \frac{y^2}{4^2} = 1 \quad \text{or} \quad g(x, y) = \frac{x^2}{3^2} + \frac{y^2}{4^2} - 1 = 0$$

does not affect the solution—the constant is eliminated when you form ∇g.

NOTE Example 1 can also be solved using the techniques you learned in Chapter 3. To see how, try to find the maximum value of $A = 4xy$ given that

$$\frac{x^2}{3^2} + \frac{y^2}{4^2} = 1.$$

To begin, solve the second equation for y to obtain

$$y = \tfrac{4}{3}\sqrt{9 - x^2}.$$

Then substitute into the first equation to obtain

$$A = 4x\left(\tfrac{4}{3}\sqrt{9 - x^2}\right).$$

Finally, use the techniques of Chapter 3 to maximize A.

EXAMPLE 2 A Business Application

The Cobb-Douglas production function (see Example 5, Section 13.1) for a software manufacturer is given by

$$f(x, y) = 100x^{3/4}y^{1/4} \qquad \text{Objective function}$$

where x represents the units of labor (at \$150 per unit) and y represents the units of capital (at \$250 per unit). The total cost of labor and capital is limited to \$50,000. Find the maximum production level for this manufacturer.

FOR FURTHER INFORMATION For more information on the use of Lagrange multipliers in economics, see the article "Lagrange Multiplier Problems in Economics" by John V. Baxley and John C. Moorhouse in *The American Mathematical Monthly.* To view this article, go to the website *www.matharticles.com.*

Solution From the given function, you have

$$\nabla f(x, y) = 75x^{-1/4}y^{1/4}\mathbf{i} + 25x^{3/4}y^{-3/4}\mathbf{j}.$$

The limit on the cost of labor and capital produces the constraint

$$g(x, y) = 150x + 250y = 50{,}000. \qquad \text{Constraint}$$

So, $\lambda\nabla g(x, y) = 150\lambda\mathbf{i} + 250\lambda\mathbf{j}$. This gives rise to the following system of equations.

$$75x^{-1/4}y^{1/4} = 150\lambda \qquad f_x(x, y) = \lambda g_x(x, y)$$

$$25x^{3/4}y^{-3/4} = 250\lambda \qquad f_y(x, y) = \lambda g_y(x, y)$$

$$150x + 250y = 50{,}000 \qquad \text{Constraint}$$

By solving for λ in the first equation

$$\lambda = \frac{75x^{-1/4}y^{1/4}}{150} = \frac{x^{-1/4}y^{1/4}}{2}$$

and substituting into the second equation, you obtain

$$25x^{3/4}y^{-3/4} = 250\left(\frac{x^{-1/4}y^{1/4}}{2}\right)$$

$$25x = 125y. \qquad \text{Multiply by } x^{1/4}y^{3/4}.$$

So, $x = 5y$. By substituting into the third equation, you have

$$150(5y) + 250y = 50{,}000$$

$$1000y = 50{,}000$$

$$y = 50 \text{ units of capital}$$

$$x = 250 \text{ units of labor.}$$

So, the maximum production level is

$$f(250, 50) = 100(250)^{3/4}(50)^{1/4}$$

$$\approx 16{,}719 \text{ product units.}$$

■

Economists call the Lagrange multiplier obtained in a production function the **marginal productivity of money.** For instance, in Example 2 the marginal productivity of money at $x = 250$ and $y = 50$ is

$$\lambda = \frac{x^{-1/4}y^{1/4}}{2} = \frac{(250)^{-1/4}(50)^{1/4}}{2} \approx 0.334$$

which means that for each additional dollar spent on production, an additional 0.334 unit of the product can be produced.

EXAMPLE 3 Lagrange Multipliers and Three Variables

Find the minimum value of

$$f(x, y, z) = 2x^2 + y^2 + 3z^2 \qquad \text{Objective function}$$

subject to the constraint $2x - 3y - 4z = 49$.

Solution Let $g(x, y, z) = 2x - 3y - 4z = 49$. Then, because

$$\nabla f(x, y, z) = 4x\mathbf{i} + 2y\mathbf{j} + 6z\mathbf{k} \quad \text{and} \quad \lambda\nabla g(x, y, z) = 2\lambda\mathbf{i} - 3\lambda\mathbf{j} - 4\lambda\mathbf{k}$$

you obtain the following system of equations.

$$\begin{aligned} 4x &= 2\lambda & \quad & f_x(x, y, z) = \lambda g_x(x, y, z) \\ 2y &= -3\lambda & & f_y(x, y, z) = \lambda g_y(x, y, z) \\ 6z &= -4\lambda & & f_z(x, y, z) = \lambda g_z(x, y, z) \\ 2x - 3y - 4z &= 49 & & \text{Constraint} \end{aligned}$$

The solution of this system is $x = 3$, $y = -9$, and $z = -4$. So, the optimum value of f is

$$\begin{aligned} f(3, -9, -4) &= 2(3)^2 + (-9)^2 + 3(-4)^2 \\ &= 147. \end{aligned}$$

From the original function and constraint, it is clear that $f(x, y, z)$ has no maximum. So, the optimum value of f determined above is a minimum. ■

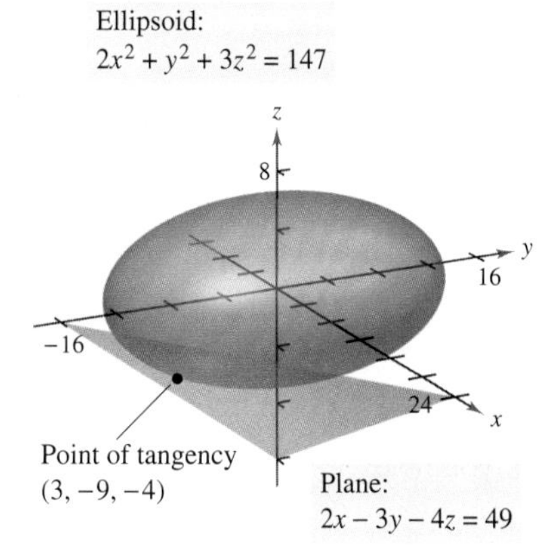

Figure 13.80

A graphical interpretation of constrained optimization problems in two variables was given at the beginning of this section. In three variables, the interpretation is similar, except that level surfaces are used instead of level curves. For instance, in Example 3, the level surfaces of f are ellipsoids centered at the origin, and the constraint

$$2x - 3y - 4z = 49$$

is a plane. The minimum value of f is represented by the ellipsoid that is tangent to the constraint plane, as shown in Figure 13.80.

EXAMPLE 4 Optimization Inside a Region

Find the extreme values of

$$f(x, y) = x^2 + 2y^2 - 2x + 3 \qquad \text{Objective function}$$

subject to the constraint $x^2 + y^2 \le 10$.

Solution To solve this problem, you can break the constraint into two cases.

a. For points *on the circle* $x^2 + y^2 = 10$, you can use Lagrange multipliers to find that the maximum value of $f(x, y)$ is 24—this value occurs at $(-1, 3)$ and at $(-1, -3)$. In a similar way, you can determine that the minimum value of $f(x, y)$ is approximately 6.675—this value occurs at $(\sqrt{10}, 0)$.

b. For points *inside the circle*, you can use the techniques discussed in Section 13.8 to conclude that the function has a relative minimum of 2 at the point $(1, 0)$.

By combining these two results, you can conclude that f has a maximum of 24 at $(-1, \pm 3)$ and a minimum of 2 at $(1, 0)$, as shown in Figure 13.81. ■

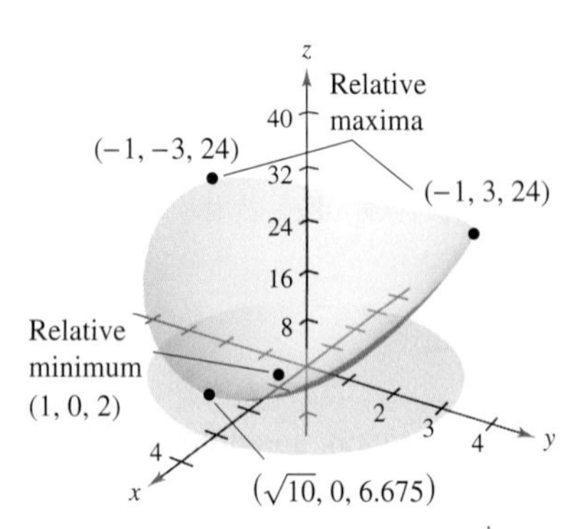

Figure 13.81

The Method of Lagrange Multipliers with Two Constraints

For optimization problems involving *two* constraint functions g and h, you can introduce a second Lagrange multiplier, μ (the lowercase Greek letter mu), and then solve the equation

$$\nabla f = \lambda \nabla g + \mu \nabla h$$

where the gradient vectors are not parallel, as illustrated in Example 5.

EXAMPLE 5 Optimization with Two Constraints

Let $T(x, y, z) = 20 + 2x + 2y + z^2$ represent the temperature at each point on the sphere $x^2 + y^2 + z^2 = 11$. Find the extreme temperatures on the curve formed by the intersection of the plane $x + y + z = 3$ and the sphere.

Solution The two constraints are

$$g(x, y, z) = x^2 + y^2 + z^2 = 11 \quad \text{and} \quad h(x, y, z) = x + y + z = 3.$$

Using

$$\nabla T(x, y, z) = 2\mathbf{i} + 2\mathbf{j} + 2z\mathbf{k}$$
$$\lambda \nabla g(x, y, z) = 2\lambda x\mathbf{i} + 2\lambda y\mathbf{j} + 2\lambda z\mathbf{k}$$

and

$$\mu \nabla h(x, y, z) = \mu\mathbf{i} + \mu\mathbf{j} + \mu\mathbf{k}$$

you can write the following system of equations.

$$2 = 2\lambda x + \mu \qquad T_x(x, y, z) = \lambda g_x(x, y, z) + \mu h_x(x, y, z)$$
$$2 = 2\lambda y + \mu \qquad T_y(x, y, z) = \lambda g_y(x, y, z) + \mu h_y(x, y, z)$$
$$2z = 2\lambda z + \mu \qquad T_z(x, y, z) = \lambda g_z(x, y, z) + \mu h_z(x, y, z)$$
$$x^2 + y^2 + z^2 = 11 \qquad \text{Constraint 1}$$
$$x + y + z = 3 \qquad \text{Constraint 2}$$

By subtracting the second equation from the first, you can obtain the following system.

$$\lambda(x - y) = 0$$
$$2z(1 - \lambda) - \mu = 0$$
$$x^2 + y^2 + z^2 = 11$$
$$x + y + z = 3$$

STUDY TIP The systems of equations that arise when the Method of Lagrange Multipliers is used are not, in general, linear systems, and finding the solutions often requires ingenuity.

From the first equation, you can conclude that $\lambda = 0$ or $x = y$. If $\lambda = 0$, you can show that the critical points are $(3, -1, 1)$ and $(-1, 3, 1)$. (Try doing this—it takes a little work.) If $\lambda \neq 0$, then $x = y$ and you can show that the critical points occur when $x = y = \left(3 \pm 2\sqrt{3}\right)/3$ and $z = \left(3 \mp 4\sqrt{3}\right)/3$. Finally, to find the optimal solutions, compare the temperatures at the four critical points.

$$T(3, -1, 1) = T(-1, 3, 1) = 25$$
$$T\left(\frac{3 - 2\sqrt{3}}{3}, \frac{3 - 2\sqrt{3}}{3}, \frac{3 + 4\sqrt{3}}{3}\right) = \frac{91}{3} \approx 30.33$$
$$T\left(\frac{3 + 2\sqrt{3}}{3}, \frac{3 + 2\sqrt{3}}{3}, \frac{3 - 4\sqrt{3}}{3}\right) = \frac{91}{3} \approx 30.33$$

So, $T = 25$ is the minimum temperature and $T = \frac{91}{3}$ is the maximum temperature on the curve. ■

13.10 Exercises

See www.CalcChat.com for worked-out solutions to odd-numbered exercises.

In Exercises 1–4, identify the constraint and level curves of the objective function shown in the figure. Use the figure to approximate the indicated extrema, assuming that x and y are positive. Use Lagrange multipliers to verify your result.

1. Maximize $z = xy$

   Constraint: $x + y = 10$

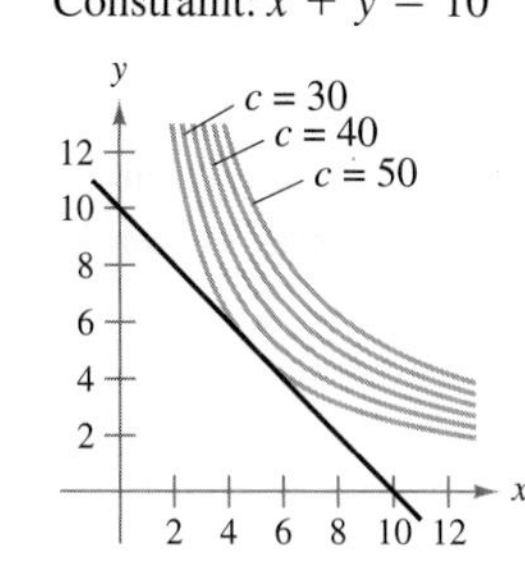

2. Maximize $z = xy$

   Constraint: $2x + y = 4$

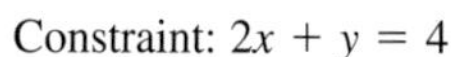

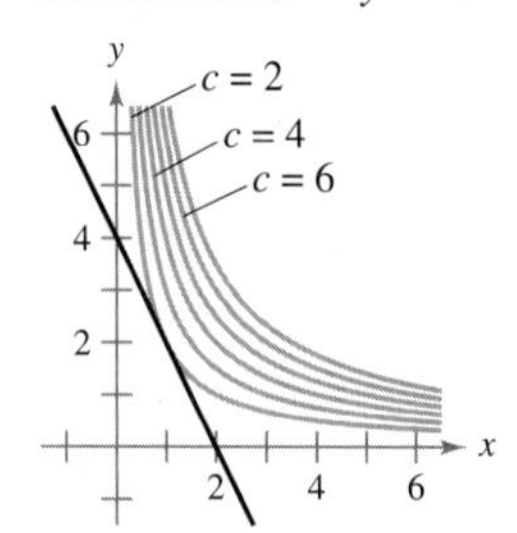

3. Minimize $z = x^2 + y^2$

   Constraint: $x + y - 4 = 0$

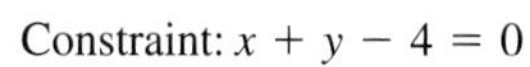

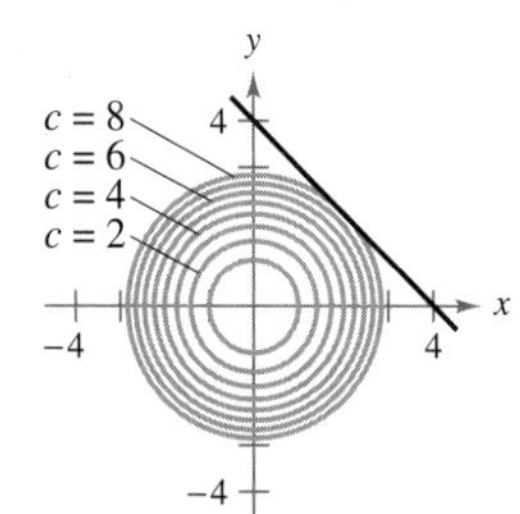

4. Minimize $z = x^2 + y^2$

   Constraint: $2x + 4y = 5$

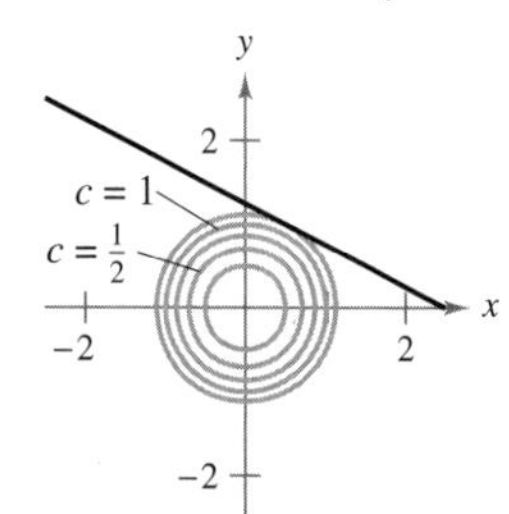

In Exercises 5–10, use Lagrange multipliers to find the indicated extrema, assuming that x and y are positive.

5. Minimize $f(x, y) = x^2 + y^2$

   Constraint: $x + 2y - 5 = 0$

6. Maximize $f(x, y) = x^2 - y^2$

   Constraint: $2y - x^2 = 0$

7. Maximize $f(x, y) = 2x + 2xy + y$

   Constraint: $2x + y = 100$

8. Minimize $f(x, y) = 3x + y + 10$

   Constraint: $x^2y = 6$

9. Maximize $f(x, y) = \sqrt{6 - x^2 - y^2}$

   Constraint: $x + y - 2 = 0$

10. Minimize $f(x, y) = \sqrt{x^2 + y^2}$

    Constraint: $2x + 4y - 15 = 0$

In Exercises 11–14, use Lagrange multipliers to find the indicated extrema, assuming that x, y, and z are positive.

11. Minimize $f(x, y, z) = x^2 + y^2 + z^2$

    Constraint: $x + y + z - 9 = 0$

12. Maximize $f(x, y, z) = xyz$

    Constraint: $x + y + z - 3 = 0$

13. Minimize $f(x, y, z) = x^2 + y^2 + z^2$

    Constraint: $x + y + z = 1$

14. Minimize $f(x, y) = x^2 - 10x + y^2 - 14y + 28$

    Constraint: $x + y = 10$

In Exercises 15 and 16, use Lagrange multipliers to find any extrema of the function subject to the constraint $x^2 + y^2 \le 1$.

15. $f(x, y) = x^2 + 3xy + y^2$

16. $f(x, y) = e^{-xy/4}$

In Exercises 17 and 18, use Lagrange multipliers to find the indicated extrema of f subject to two constraints. In each case, assume that x, y, and z are nonnegative.

17. Maximize $f(x, y, z) = xyz$

    Constraints: $x + y + z = 32, \quad x - y + z = 0$

18. Minimize $f(x, y, z) = x^2 + y^2 + z^2$

    Constraints: $x + 2z = 6, \quad x + y = 12$

In Exercises 19–28, use Lagrange multipliers to find the minimum distance from the curve or surface to the indicated point. [*Hints:* In Exercise 19, minimize $f(x, y) = x^2 + y^2$ subject to the constraint $x + y = 1$. In Exercise 25, use the *root* feature of a graphing utility.]

	Curve	*Point*
19.	Line: $x + y = 1$	$(0, 0)$
20.	Line: $2x + 3y = -1$	$(0, 0)$
21.	Line: $x - y = 4$	$(0, 2)$
22.	Line: $x + 4y = 3$	$(1, 0)$
23.	Parabola: $y = x^2$	$(0, 3)$
24.	Parabola: $y = x^2$	$(-3, 0)$
25.	Parabola: $y = x^2 + 1$	$(\frac{1}{2}, 1)$
26.	Circle: $(x - 4)^2 + y^2 = 4$	$(0, 10)$
	Surface	*Point*
27.	Plane: $x + y + z = 1$	$(2, 1, 1)$
28.	Cone: $z = \sqrt{x^2 + y^2}$	$(4, 0, 0)$

In Exercises 29 and 30, find the highest point on the curve of intersection of the surfaces.

29. Cone: $x^2 + y^2 - z^2 = 0$, Plane: $x + 2z = 4$

30. Sphere: $x^2 + y^2 + z^2 = 36$, Plane: $2x + y - z = 2$

WRITING ABOUT CONCEPTS

31. Explain what is meant by constrained optimization problems.

32. Explain the Method of Lagrange Multipliers for solving constrained optimization problems.

In Exercises 33–42, use Lagrange multipliers to solve the indicated exercise in Section 13.9.

33. Exercise 1

34. Exercise 2

35. Exercise 5

36. Exercise 6

37. Exercise 9

38. Exercise 10

39. Exercise 11

40. Exercise 12

41. Exercise 17

42. Exercise 18

43. ***Maximum Volume*** Use Lagrange multipliers to find the dimensions of a rectangular box of maximum volume that can be inscribed (with edges parallel to the coordinate axes) in the ellipsoid $(x^2/a^2) + (y^2/b^2) + (z^2/c^2) = 1$.

CAPSTONE

44. The sum of the length and the girth (perimeter of a cross section) of a package carried by a delivery service cannot exceed 108 inches.

(a) Determine whether Lagrange multipliers can be used to find the dimensions of the rectangular package of largest volume that may be sent. Explain your reasoning.

(b) If Lagrange multipliers can be used, find the dimensions. Compare your answer with that obtained in Exercise 38, Section 13.9.

45. ***Minimum Cost*** A cargo container (in the shape of a rectangular solid) must have a volume of 480 cubic feet. The bottom will cost \$5 per square foot to construct and the sides and the top will cost \$3 per square foot to construct. Use Lagrange multipliers to find the dimensions of the container of this size that has minimum cost.

46. ***Geometric and Arithmetic Means***

(a) Use Lagrange multipliers to prove that the product of three positive numbers x, y, and z, whose sum has the constant value S, is a maximum when the three numbers are equal. Use this result to prove that $\sqrt[3]{xyz} \le (x + y + z)/3$.

(b) Generalize the result of part (a) to prove that the product $x_1 x_2 x_3 \cdot \cdot \cdot x_n$ is a maximum when $x_1 = x_2 = x_3 = \cdot \cdot \cdot = x_n$, $\sum_{i=1}^{n} x_i = S$, and all $x_i \ge 0$. Then prove that

$$\sqrt[n]{x_1 x_2 x_3 \cdot \cdot \cdot x_n} \le \frac{x_1 + x_2 + x_3 + \cdot \cdot \cdot + x_n}{n}.$$

This shows that the geometric mean is never greater than the arithmetic mean.

47. ***Minimum Surface Area*** Use Lagrange multipliers to find the dimensions of a right circular cylinder with volume V_0 cubic units and minimum surface area.

48. ***Temperature Distribution*** Let $T(x, y, z) = 100 + x^2 + y^2$ represent the temperature at each point on the sphere $x^2 + y^2 + z^2 = 50$. Find the maximum temperature on the curve formed by the intersection of the sphere and the plane $x - z = 0$.

49. ***Refraction of Light*** When light waves traveling in a transparent medium strike the surface of a second transparent medium, they tend to "bend" in order to follow the path of minimum time. This tendency is called *refraction* and is described by **Snell's Law of Refraction,**

$$\frac{\sin \theta_1}{v_1} = \frac{\sin \theta_2}{v_2}$$

where θ_1 and θ_2 are the magnitudes of the angles shown in the figure, and v_1 and v_2 are the velocities of light in the two media. Use Lagrange multipliers to derive this law using $x + y = a$.

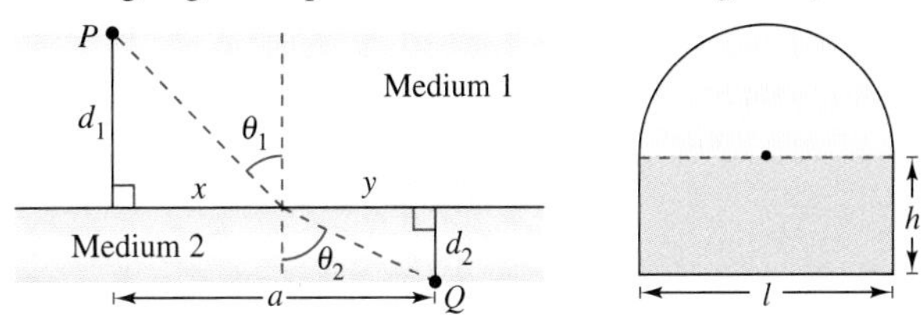

Figure for 49 **Figure for 50**

50. ***Area and Perimeter*** A semicircle is on top of a rectangle (see figure). If the area is fixed and the perimeter is a minimum, or if the perimeter is fixed and the area is a maximum, use Lagrange multipliers to verify that the length of the rectangle is twice its height.

Production Level **In Exercises 51 and 52, find the maximum production level P if the total cost of labor (at \$72 per unit) and capital (at \$60 per unit) is limited to \$250,000, where x is the number of units of labor and y is the number of units of capital.**

51. $P(x, y) = 100x^{0.25}y^{0.75}$

52. $P(x, y) = 100x^{0.4}y^{0.6}$

Cost **In Exercises 53 and 54, find the minimum cost of producing 50,000 units of a product, where x is the number of units of labor (at \$72 per unit) and y is the number of units of capital (at \$60 per unit).**

53. $P(x, y) = 100x^{0.25}y^{0.75}$

54. $P(x, y) = 100x^{0.6}y^{0.4}$

55. ***Investigation*** Consider the objective function $g(\alpha, \beta, \gamma) = \cos \alpha \cos \beta \cos \gamma$ subject to the constraint that α, β, and γ are the angles of a triangle.

(a) Use Lagrange multipliers to maximize g.

CAS (b) Use the constraint to reduce the function g to a function of two independent variables. Use a computer algebra system to graph the surface represented by g. Identify the maximum values on the graph.

PUTNAM EXAM CHALLENGE

56. A can buoy is to be made of three pieces, namely, a cylinder and two equal cones, the altitude of each cone being equal to the altitude of the cylinder. For a given area of surface, what shape will have the greatest volume?

This problem was composed by the Committee on the Putnam Prize Competition.

13 REVIEW EXERCISES

See www.CalcChat.com for worked-out solutions to odd-numbered exercises.

In Exercises 1 and 2, sketch the graph of the level surface $f(x, y, z) = c$ at the given value of c.

1. $f(x, y, z) = x^2 - y + z^2, \quad c = 2$
2. $f(x, y, z) = 4x^2 - y^2 + 4z^2, \quad c = 0$

3. *Conjecture* Consider the function $f(x, y) = x^2 + y^2$.
   (a) Sketch the graph of the surface given by f.
   (b) Make a conjecture about the relationship between the graphs of f and $g(x, y) = f(x, y) + 2$. Explain your reasoning.
   (c) Make a conjecture about the relationship between the graphs of f and $g(x, y) = f(x, y - 2)$. Explain your reasoning.
   (d) On the surface in part (a), sketch the graphs of $z = f(1, y)$ and $z = f(x, 1)$.

4. *Conjecture* Consider the function
   $f(x, y) = \sqrt{1 - x^2 - y^2}$.
   (a) Sketch the graph of the surface given by f.
   (b) Make a conjecture about the relationship between the graphs of f and $g(x, y) = f(x + 2, y)$. Explain your reasoning.
   (c) Make a conjecture about the relationship between the graphs of f and $g(x, y) = 4 - f(x, y)$. Explain your reasoning.
   (d) On the surface in part (a), sketch the graphs of $z = f(0, y)$ and $z = f(x, 0)$.

CAS **In Exercises 5–8, use a computer algebra system to graph several level curves of the function.**

5. $f(x, y) = e^{x^2+y^2}$
6. $f(x, y) = \ln xy$
7. $f(x, y) = x^2 - y^2$
8. $f(x, y) = \dfrac{x}{x + y}$

CAS **In Exercises 9 and 10, use a computer algebra system to graph the function.**

9. $f(x, y) = e^{-(x^2+y^2)}$
10. $g(x, y) = |y|^{1+|x|}$

In Exercises 11–14, find the limit and discuss the continuity of the function (if it exists).

11. $\displaystyle\lim_{(x, y)\to(1, 1)} \frac{xy}{x^2 + y^2}$
12. $\displaystyle\lim_{(x, y)\to(1, 1)} \frac{xy}{x^2 - y^2}$
13. $\displaystyle\lim_{(x, y)\to(0, 0)} \frac{y + xe^{-y^2}}{1 + x^2}$
14. $\displaystyle\lim_{(x, y)\to(0, 0)} \frac{x^2y}{x^4 + y^2}$

In Exercises 15–24, find all first partial derivatives.

15. $f(x, y) = e^x \cos y$
16. $f(x, y) = \dfrac{xy}{x + y}$
17. $z = e^{-y} + e^{-x}$
18. $z = \ln(x^2 + y^2 + 1)$
19. $g(x, y) = \dfrac{xy}{x^2 + y^2}$
20. $w = \sqrt{x^2 - y^2 - z^2}$
21. $f(x, y, z) = z \arctan \dfrac{y}{x}$
22. $f(x, y, z) = \dfrac{1}{\sqrt{1 + x^2 + y^2 + z^2}}$
23. $u(x, t) = ce^{-n^2t} \sin nx$
24. $u(x, t) = c \sin(akx) \cos kt$

25. *Think About It* Sketch a graph of a function $z = f(x, y)$ whose derivative f_x is always negative and whose derivative f_y is always negative.

26. Find the slopes of the surface $z = x^2 \ln(y + 1)$ in the x- and y-directions at the point $(2, 0, 0)$.

In Exercises 27–30, find all second partial derivatives and verify that the second mixed partials are equal.

27. $f(x, y) = 3x^2 - xy + 2y^3$
28. $h(x, y) = \dfrac{x}{x + y}$
29. $h(x, y) = x \sin y + y \cos x$
30. $g(x, y) = \cos(x - 2y)$

Laplace's Equation **In Exercises 31–34, show that the function satisfies Laplace's equation**

$$\frac{\partial^2 z}{\partial x^2} + \frac{\partial^2 z}{\partial y^2} = 0.$$

31. $z = x^2 - y^2$
32. $z = x^3 - 3xy^2$
33. $z = \dfrac{y}{x^2 + y^2}$
34. $z = e^y \sin x$

In Exercises 35 and 36, find the total differential.

35. $z = x \sin xy$
36. $z = \dfrac{xy}{\sqrt{x^2 + y^2}}$

37. *Error Analysis* The legs of a right triangle are measured to be 5 centimeters and 12 centimeters, with a possible error of $\frac{1}{2}$ centimeter. Approximate the maximum possible error in computing the length of the hypotenuse. Approximate the maximum percent error.

38. *Error Analysis* To determine the height of a tower, the angle of elevation to the top of the tower is measured from a point 100 feet $\pm \frac{1}{2}$ foot from the base. The angle is measured at 33°, with a possible error of 1°. Assuming that the ground is horizontal, approximate the maximum error in determining the height of the tower.

39. *Volume* A right circular cone is measured, and the radius and height are found to be 2 inches and 5 inches, respectively. The possible error in measurement is $\frac{1}{8}$ inch. Approximate the maximum possible error in the computation of the volume.

40. *Lateral Surface Area* Approximate the error in the computation of the lateral surface area of the cone in Exercise 39. (The lateral surface area is given by $A = \pi r\sqrt{r^2 + h^2}$.)

In Exercises 41–44, find the indicated derivatives (a) using the appropriate Chain Rule and (b) using substitution before differentiating.

41. $w = \ln(x^2 + y), \quad \dfrac{dw}{dt}$

$x = 2t, \quad y = 4 - t$

42. $u = y^2 - x, \quad \dfrac{du}{dt}$

$x = \cos t, \quad y = \sin t$

43. $w = \dfrac{xy}{z}, \quad \dfrac{\partial w}{\partial r}, \dfrac{\partial w}{\partial t}$

$x = 2r + t, \quad y = rt, \quad z = 2r - t$

44. $u = x^2 + y^2 + z^2, \quad \dfrac{\partial u}{\partial r}, \dfrac{\partial u}{\partial t}$

$x = r\cos t, \quad y = r\sin t, \quad z = t$

In Exercises 45 and 46, differentiate implicitly to find the first partial derivatives of z.

45. $x^2 + xy + y^2 + yz + z^2 = 0$

46. $xz^2 - y\sin z = 0$

In Exercises 47–50, find the directional derivative of the function at P in the direction of v.

47. $f(x, y) = x^2y, \quad (-5, 5), \quad \mathbf{v} = 3\mathbf{i} - 4\mathbf{j}$

48. $f(x, y) = \frac{1}{4}y^2 - x^2, \quad (1, 4), \quad \mathbf{v} = 2\mathbf{i} + \mathbf{j}$

49. $w = y^2 + xz, \quad (1, 2, 2), \quad \mathbf{v} = 2\mathbf{i} - \mathbf{j} + 2\mathbf{k}$

50. $w = 5x^2 + 2xy - 3y^2z, \quad (1, 0, 1), \quad \mathbf{v} = \mathbf{i} + \mathbf{j} - \mathbf{k}$

In Exercises 51–54, find the gradient of the function and the maximum value of the directional derivative at the given point.

51. $z = x^2y, \quad (2, 1)$

52. $z = e^{-x}\cos y, \quad \left(0, \dfrac{\pi}{4}\right)$

53. $z = \dfrac{y}{x^2 + y^2}, \quad (1, 1)$

54. $z = \dfrac{x^2}{x - y}, \quad (2, 1)$

In Exercises 55 and 56, (a) find the gradient of the function at P, (b) find a unit normal vector to the level curve $f(x, y) = c$ at P, (c) find the tangent line to the level curve $f(x, y) = c$ at P, and (d) sketch the level curve, the unit normal vector, and the tangent line in the xy-plane.

55. $f(x, y) = 9x^2 - 4y^2$

$c = 65, \ P(3, 2)$

56. $f(x, y) = 4y\sin x - y$

$c = 3, \ P\left(\dfrac{\pi}{2}, 1\right)$

In Exercises 57–60, find an equation of the tangent plane and parametric equations of the normal line to the surface at the given point.

Surface	*Point*
57. $f(x, y) = x^2y$	$(2, 1, 4)$
58. $f(x, y) = \sqrt{25 - y^2}$	$(2, 3, 4)$
59. $z = -9 + 4x - 6y - x^2 - y^2$	$(2, -3, 4)$
60. $z = \sqrt{9 - x^2 - y^2}$	$(1, 2, 2)$

In Exercises 61 and 62, find symmetric equations of the tangent line to the curve of intersection of the surfaces at the given point.

Surfaces	*Point*
61. $z = 9 - y^2, \ y = x$	$(2, 2, 5)$
62. $z = x^2 - y^2, \ z = 3$	$(2, 1, 3)$

63. Find the angle of inclination θ of the tangent plane to the surface $x^2 + y^2 + z^2 = 14$ at the point $(2, 1, 3)$.

64. ***Approximation*** Consider the following approximations for a function $f(x, y)$ centered at $(0, 0)$.

Linear approximation:

$P_1(x, y) = f(0, 0) + f_x(0, 0)x + f_y(0, 0)y$

Quadratic approximation:

$P_2(x, y) = f(0, 0) + f_x(0, 0)x + f_y(0, 0)y + \frac{1}{2}f_{xx}(0, 0)x^2 + f_{xy}(0, 0)xy + \frac{1}{2}f_{yy}(0, 0)y^2$

[Note that the linear approximation is the tangent plane to the surface at $(0, 0, f(0, 0))$.]

(a) Find the linear approximation of $f(x, y) = \cos x + \sin y$ centered at $(0, 0)$.

(b) Find the quadratic approximation of $f(x, y) = \cos x + \sin y$ centered at $(0, 0)$.

(c) If $y = 0$ in the quadratic approximation, you obtain the second-degree Taylor polynomial for what function?

(d) Complete the table.

x	y	$f(x, y)$	$P_1(x, y)$	$P_2(x, y)$
0	0			
0	0.1			
0.2	0.1			
0.5	0.3			
1	0.5			

CAS (e) Use a computer algebra system to graph the surfaces $z = f(x, y)$, $z = P_1(x, y)$, and $z = P_2(x, y)$. How does the accuracy of the approximations change as the distance from $(0, 0)$ increases?

CAS **In Exercises 65–68, examine the function for relative extrema and saddle points. Use a computer algebra system to graph the function and confirm your results.**

65. $f(x, y) = 2x^2 + 6xy + 9y^2 + 8x + 14$

66. $f(x, y) = x^2 + 3xy + y^2 - 5x$

67. $f(x, y) = xy + \dfrac{1}{x} + \dfrac{1}{y}$

68. $z = 50(x + y) - (0.1x^3 + 20x + 150) - (0.05y^3 + 20.6y + 125)$

Writing **In Exercises 69 and 70, write a short paragraph about the surface whose level curves (*c*-values evenly spaced) are shown. Comment on possible extrema, saddle points, the magnitude of the gradient, etc.**

69.

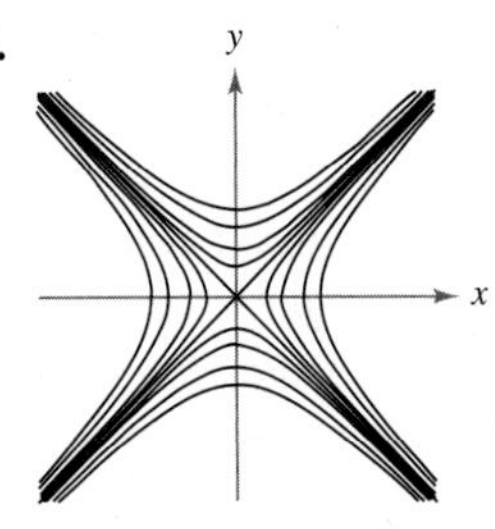

70.

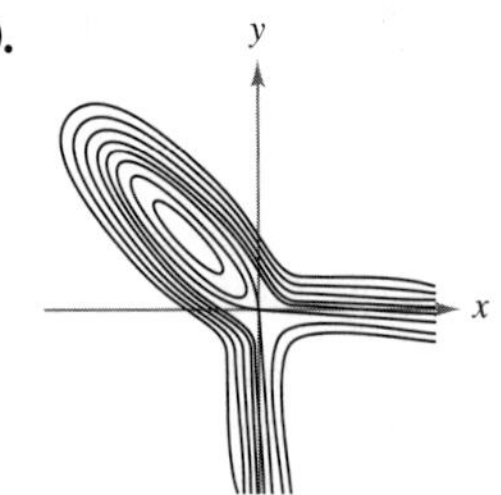

71. ***Maximum Profit*** A corporation manufactures digital cameras at two locations. The cost functions for producing x_1 units at location 1 and x_2 units at location 2 are

$$C_1 = 0.05x_1^2 + 15x_1 + 5400$$
$$C_2 = 0.03x_2^2 + 15x_2 + 6100$$

and the total revenue function is

$$R = [225 - 0.4(x_1 + x_2)](x_1 + x_2).$$

Find the production levels at the two locations that will maximize the profit $P(x_1, x_2) = R - C_1 - C_2$.

72. ***Minimum Cost*** A manufacturer has an order for 1000 units of wooden benches that can be produced at two locations. Let x_1 and x_2 be the numbers of units produced at the two locations. The cost function is

$$C = 0.25x_1^2 + 10x_1 + 0.15x_2^2 + 12x_2.$$

Find the number that should be produced at each location to meet the order and minimize cost.

73. ***Production Level*** The production function for a candy manufacturer is

$$f(x, y) = 4x + xy + 2y$$

where x is the number of units of labor and y is the number of units of capital. Assume that the total amount available for labor and capital is \$2000, and that units of labor and capital cost \$20 and \$4, respectively. Find the maximum production level for this manufacturer.

74. Find the minimum distance from the point $(2, 2, 0)$ to the surface $z = x^2 + y^2$.

75. ***Modeling Data*** The table shows the drag force y in kilograms for a motor vehicle at indicated speeds x in kilometers per hour.

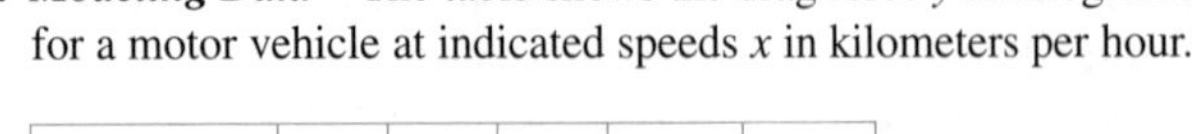

Speed, x	25	50	75	100	125
Drag, y	24	34	50	71	98

(a) Use the regression capabilities of a graphing utility to find the least squares regression quadratic for the data.

(b) Use the model to estimate the total drag when the vehicle is moving at 80 kilometers per hour.

76. ***Modeling Data*** The data in the table show the yield y (in milligrams) of a chemical reaction after t minutes.

Minutes, t	1	2	3	4
Yield, y	1.2	7.1	9.9	13.1

Minutes, t	5	6	7	8
Yield, y	15.5	16.0	17.9	18.0

(a) Use the regression capabilities of a graphing utility to find the least squares regression line for the data. Then use the graphing utility to plot the data and graph the model.

(b) Use a graphing utility to plot the points $(\ln t, y)$. Do these points appear to follow a linear pattern more closely than the plot of the given data in part (a)?

(c) Use the regression capabilities of a graphing utility to find the least squares regression line for the points $(\ln t, y)$ and obtain the logarithmic model $y = a + b \ln t$.

(d) Use a graphing utility to plot the data and graph the linear and logarithmic models. Which is a better model? Explain.

In Exercises 77 and 78, use Lagrange multipliers to locate and classify any extrema of the function.

77. $w = xy + yz + xz$

Constraint: $x + y + z = 1$

78. $z = x^2y$

Constraint: $x + 2y = 2$

79. ***Minimum Cost*** A water line is to be built from point P to point S and must pass through regions where construction costs differ (see figure). The cost per kilometer in dollars is $3k$ from P to Q, $2k$ from Q to R, and k from R to S. For simplicity, let $k = 1$. Use Lagrange multipliers to find x, y, and z such that the total cost C will be minimized.

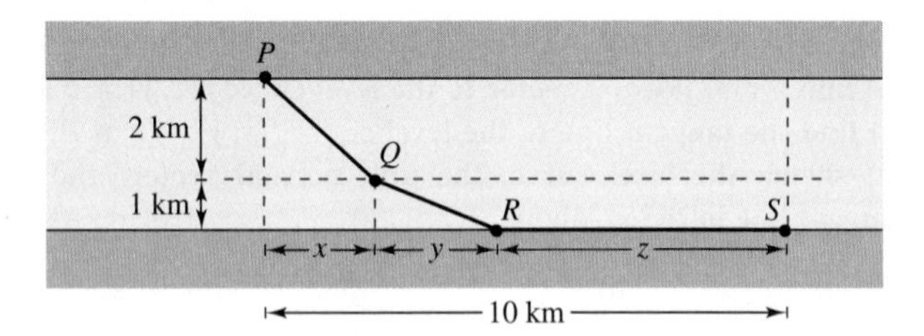

80. ***Investigation*** Consider the objective function $f(x, y) = ax + by$ subject to the constraint $x^2/64 + y^2/36 = 1$. Assume that x and y are positive.

(a) Use a computer algebra system to graph the constraint. For $a = 4$ and $b = 3$, use the computer algebra system to graph the level curves of the objective function. By trial and error, find the level curve that appears to be tangent to the ellipse. Use the result to approximate the maximum of f subject to the constraint.

(b) Repeat part (a) for $a = 4$ and $b = 9$.

P.S. PROBLEM SOLVING

1. **Heron's Formula** states that the area of a triangle with sides of lengths a, b, and c is given by

$$A = \sqrt{s(s - a)(s - b)(s - c)}$$

where $s = \dfrac{a + b + c}{2}$, as shown in the figure.

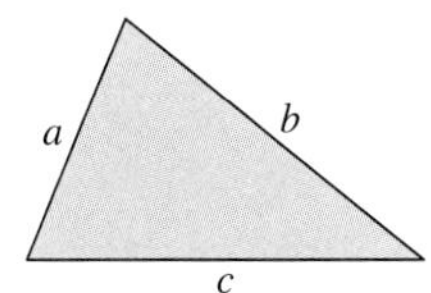

(a) Use Heron's Formula to find the area of the triangle with vertices $(0, 0)$, $(3, 4)$, and $(6, 0)$.

(b) Show that among all triangles having a fixed perimeter, the triangle with the largest area is an equilateral triangle.

(c) Show that among all triangles having a fixed area, the triangle with the smallest perimeter is an equilateral triangle.

2. An industrial container is in the shape of a cylinder with hemispherical ends, as shown in the figure. The container must hold 1000 liters of fluid. Determine the radius r and length h that minimize the amount of material used in the construction of the tank.

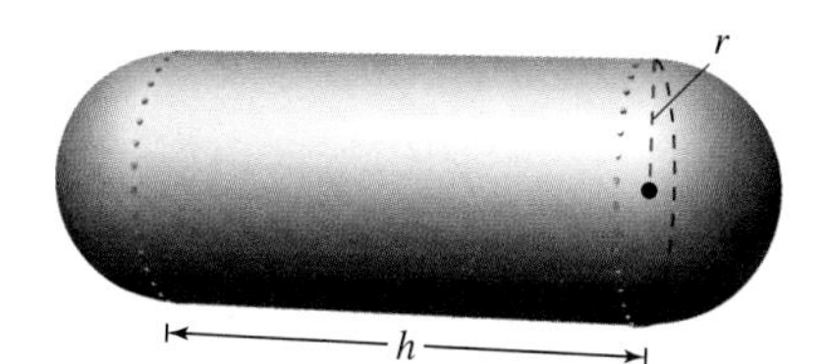

3. Let $P(x_0, y_0, z_0)$ be a point in the first octant on the surface $xyz = 1$.

(a) Find the equation of the tangent plane to the surface at the point P.

(b) Show that the volume of the tetrahedron formed by the three coordinate planes and the tangent plane is constant, independent of the point of tangency (see figure).

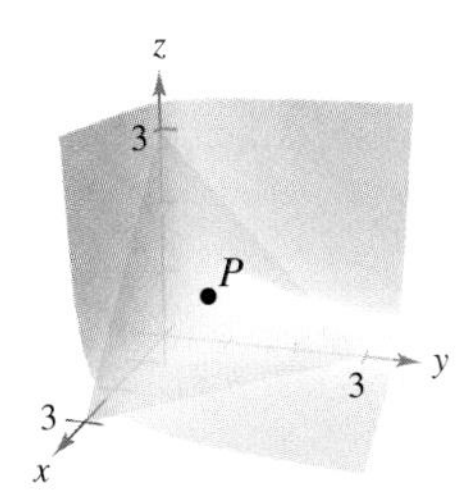

4. Use a graphing utility to graph the functions $f(x) = \sqrt[3]{x^3 - 1}$ and $g(x) = x$ in the same viewing window.

(a) Show that

$$\lim_{x \to \infty} [f(x) - g(x)] = 0 \text{ and } \lim_{x \to -\infty} [f(x) - g(x)] = 0.$$

(b) Find the point on the graph of f that is farthest from the graph of g.

5. (a) Let $f(x, y) = x - y$ and $g(x, y) = x^2 + y^2 = 4$. Graph various level curves of f and the constraint g in the xy-plane. Use the graph to determine the maximum value of f subject to the constraint $g = 4$. Then verify your answer using Lagrange multipliers.

(b) Let $f(x, y) = x - y$ and $g(x, y) = x^2 + y^2 = 0$. Find the maximum and minimum values of f subject to the constraint $g = 0$. Does the method of Lagrange Multipliers work in this case? Explain.

6. A heated storage room has the shape of a rectangular box and has a volume of 1000 cubic feet, as shown in the figure. Because warm air rises, the heat loss per unit of area through the ceiling is five times as great as the heat loss through the floor. The heat loss through the four walls is three times as great as the heat loss through the floor. Determine the room dimensions that will minimize heat loss and therefore minimize heating costs.

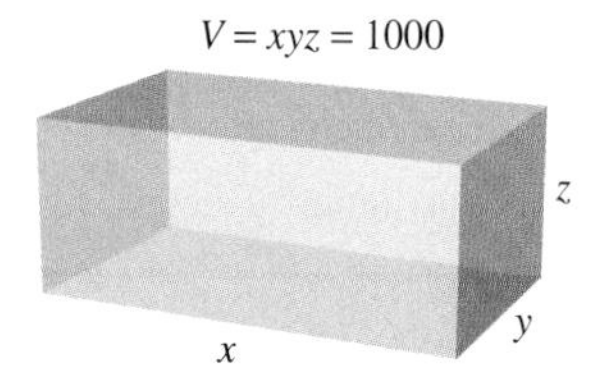

7. Repeat Exercise 6 assuming that the heat loss through the walls and ceiling remain the same, but the floor is insulated so that there is no heat loss through the floor.

8. Consider a circular plate of radius 1 given by $x^2 + y^2 \le 1$, as shown in the figure. The temperature at any point $P(x, y)$ on the plate is $T(x, y) = 2x^2 + y^2 - y + 10$.

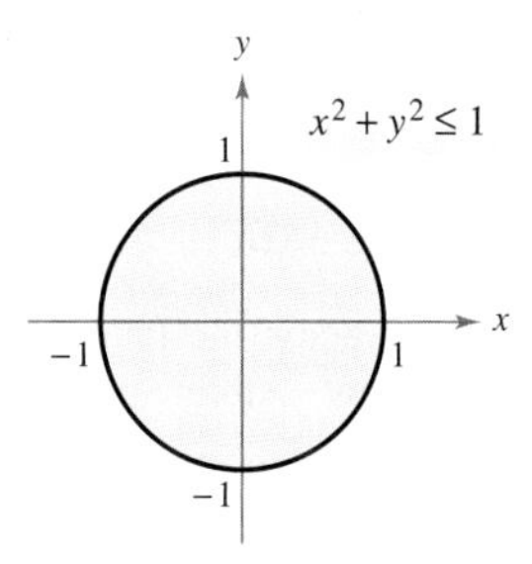

(a) Sketch the isotherm $T(x, y) = 10$. To print an enlarged copy of the graph, go to the website *www.mathgraphs.com*.

(b) Find the hottest and coldest points on the plate.

9. Consider the Cobb-Douglas production function

$$f(x, y) = Cx^a y^{1-a}, \quad 0 < a < 1.$$

(a) Show that f satisfies the equation $x\dfrac{\partial f}{\partial x} + y\dfrac{\partial f}{\partial y} = f$.

(b) Show that $f(tx, ty) = tf(x, y)$.

10. Rewrite Laplace's equation $\dfrac{\partial^2 u}{\partial x^2} + \dfrac{\partial^2 u}{\partial y^2} + \dfrac{\partial^2 u}{\partial z^2} = 0$ in cylindrical coordinates.

11. A projectile is launched at an angle of 45° with the horizontal and with an initial velocity of 64 feet per second. A television camera is located in the plane of the path of the projectile 50 feet behind the launch site (see figure).

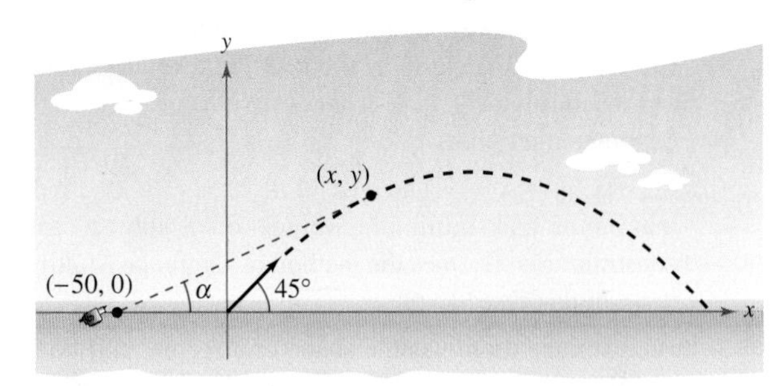

(a) Find parametric equations for the path of the projectile in terms of the parameter t representing time.

(b) Write the angle α that the camera makes with the horizontal in terms of x and y and in terms of t.

(c) Use the results of part (b) to find $d\alpha/dt$.

(d) Use a graphing utility to graph α in terms of t. Is the graph symmetric to the axis of the parabolic arch of the projectile? At what time is the rate of change of α greatest?

(e) At what time is the angle α maximum? Does this occur when the projectile is at its greatest height?

12. Consider the distance d between the launch site and the projectile in Exercise 11.

(a) Write the distance d in terms of x and y and in terms of the parameter t.

(b) Use the results of part (a) to find the rate of change of d.

(c) Find the rate of change of the distance when $t = 2$.

(d) When is the rate of change of d minimum during the flight of the projectile? Does this occur at the time when the projectile reaches its maximum height?

CAS **13.** Consider the function

$$f(x, y) = (\alpha x^2 + \beta y^2)e^{-(x^2+y^2)}, \quad 0 < |\alpha| < \beta.$$

(a) Use a computer algebra system to graph the function for $\alpha = 1$ and $\beta = 2$, and identify any extrema or saddle points.

(b) Use a computer algebra system to graph the function for $\alpha = -1$ and $\beta = 2$, and identify any extrema or saddle points.

(c) Generalize the results in parts (a) and (b) for the function f.

14. Prove that if f is a differentiable function such that

$$\nabla f(x_0, y_0) = \mathbf{0}$$

then the tangent plane at (x_0, y_0) is horizontal.

15. The figure shows a rectangle that is approximately $l = 6$ centimeters long and $h = 1$ centimeter high.

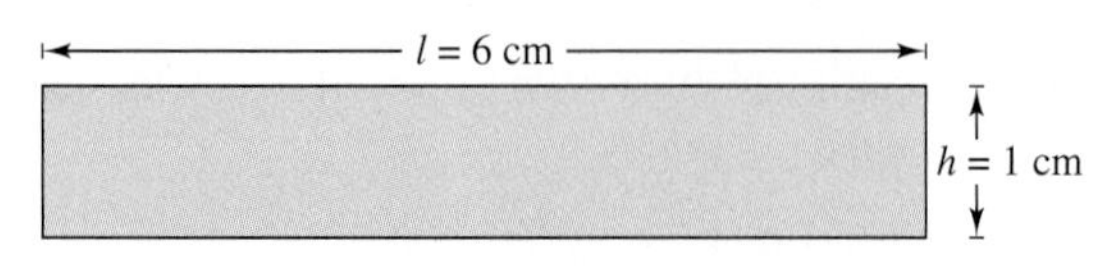

(a) Draw a rectangular strip along the rectangular region showing a small increase in length.

(b) Draw a rectangular strip along the rectangular region showing a small increase in height.

(c) Use the results in parts (a) and (b) to identify the measurement that has more effect on the area A of the rectangle.

(d) Verify your answer in part (c) analytically by comparing the value of dA when $dl = 0.01$ and when $dh = 0.01$.

16. Consider converting a point $(5 \pm 0.05, \pi/18 \pm 0.05)$ in polar coordinates to rectangular coordinates (x, y).

(a) Use a geometric argument to determine whether the accuracy in x is more dependent on the accuracy in r or on the accuracy in θ. Explain. Verify your answer analytically.

(b) Use a geometric argument to determine whether the accuracy in y is more dependent on the accuracy in r or on the accuracy in θ. Explain. Verify your answer analytically.

17. Let f be a differentiable function of one variable. Show that all tangent planes to the surface $z = yf(x/y)$ intersect in a common point.

18. Consider the ellipse

$$\frac{x^2}{a^2} + \frac{y^2}{b^2} = 1$$

that encloses the circle $x^2 + y^2 = 2x$. Find values of a and b that minimize the area of the ellipse.

19. Show that

$$u(x, t) = \frac{1}{2}[\sin(x - t) + \sin(x + t)]$$

is a solution to the one-dimensional wave equation

$$\frac{\partial^2 u}{\partial t^2} = \frac{\partial^2 u}{\partial x^2}.$$

20. Show that

$$u(x, t) = \frac{1}{2}[f(x - ct) + f(x + ct)]$$

is a solution to the one-dimensional wave equation

$$\frac{\partial^2 u}{\partial t^2} = c^2\frac{\partial^2 u}{\partial x^2}.$$

(This equation describes the small transverse vibration of an elastic string such as those on certain musical instruments.)

14 Multiple Integration

This chapter introduces the concepts of double integrals over regions in the plane and triple integrals over regions in space.

In this chapter, you should learn the following.

- How to evaluate an iterated integral and find the area of a plane region. (14.1)
- How to use a double integral to find the volume of a solid region. (14.2)
- How to write and evaluate double integrals in polar coordinates. (14.3)
- How to find the mass of a planar lamina, the center of mass of a planar lamina, and moments of inertia using double integrals. (14.4)
- How to use a double integral to find the area of a surface. (14.5)
- How to use a triple integral to find the volume, center of mass, and moments of inertia of a solid region. (14.6)
- How to write and evaluate triple integrals in cylindrical and spherical coordinates. (14.7)
- How to use a Jacobian to change variables in a double integral. (14.8)

The center of pressure on a sail is that point at which the total aerodynamic force may be assumed to act. Letting the sail be represented by a plane region, how can you use double integrals to find the center of pressure on a sail? (See Section 14.4, Section Project.)

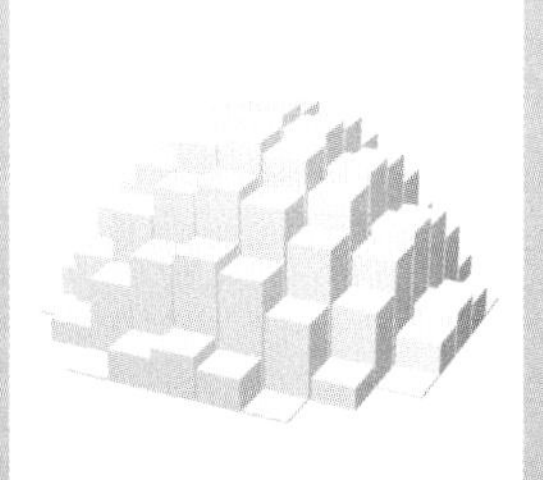

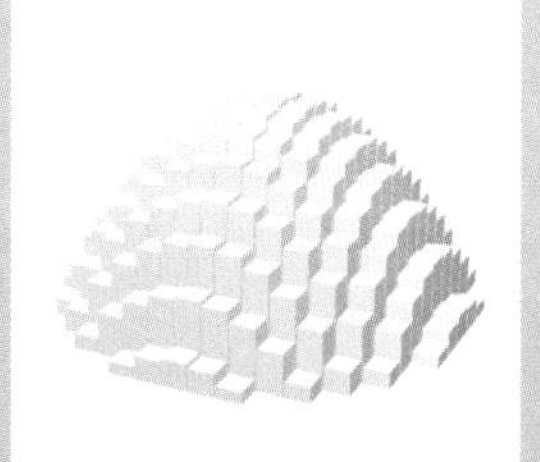

You can approximate the volume of a solid region by finding the sum of the volumes of representative rectangular prisms. As you increase the number of rectangular prisms, the approximation tends to become more and more accurate. In Chapter 14, you will learn how to use multiple integrals to find the volume of a solid region.

14.1 Iterated Integrals and Area in the Plane

- **Evaluate an iterated integral.**
- **Use an iterated integral to find the area of a plane region.**

Iterated Integrals

NOTE In Chapters 14 and 15, you will study several applications of integration involving functions of several variables. Chapter 14 is much like Chapter 7 in that it surveys the use of integration to find plane areas, volumes, surface areas, moments, and centers of mass.

In Chapter 13, you saw that it is meaningful to differentiate functions of several variables with respect to one variable while holding the other variables constant. You can *integrate* functions of several variables by a similar procedure. For example, if you are given the partial derivative

$$f_x(x, y) = 2xy$$

then, by considering y constant, you can integrate with respect to x to obtain

$$\begin{aligned} f(x, y) &= \int f_x(x, y)\, dx && \text{Integrate with respect to } x. \\ &= \int 2xy\, dx && \text{Hold } y \text{ constant.} \\ &= y \int 2x\, dx && \text{Factor out constant } y. \\ &= y(x^2) + C(y) && \text{Antiderivative of } 2x \text{ is } x^2. \\ &= x^2y + C(y). && C(y) \text{ is a function of } y. \end{aligned}$$

The "constant" of integration, $C(y)$, is a function of y. In other words, by integrating with respect to x, you are able to recover $f(x, y)$ only partially. The total recovery of a function of x and y from its partial derivatives is a topic you will study in Chapter 15. For now, we are more concerned with extending definite integrals to functions of several variables. For instance, by considering y constant, you can apply the Fundamental Theorem of Calculus to evaluate

$$\int_1^{2y} 2xy\, dx = x^2y\Big]_1^{2y} = (2y)^2y - (1)^2y = 4y^3 - y.$$

x is the variable of integration and y is fixed.

Replace x by the limits of integration.

The result is a function of y.

Similarly, you can integrate with respect to y by holding x fixed. Both procedures are summarized as follows.

$$\int_{h_1(y)}^{h_2(y)} f_x(x, y)\, dx = f(x, y)\Big]_{h_1(y)}^{h_2(y)} = f(h_2(y), y) - f(h_1(y), y) \qquad \text{With respect to } x$$

$$\int_{g_1(x)}^{g_2(x)} f_y(x, y)\, dy = f(x, y)\Big]_{g_1(x)}^{g_2(x)} = f(x, g_2(x)) - f(x, g_1(x)) \qquad \text{With respect to } y$$

Note that the variable of integration cannot appear in either limit of integration. For instance, it makes no sense to write

$$\int_0^x y\, dx.$$

EXAMPLE 1 Integrating with Respect to y

Evaluate $\displaystyle\int_1^x (2x^2y^{-2} + 2y)\, dy$.

Solution Considering x to be constant and integrating with respect to y produces

$$\begin{aligned}\int_1^x (2x^2y^{-2} + 2y)\, dy &= \left[\frac{-2x^2}{y} + y^2\right]_1^x && \text{Integrate with respect to } y.\\ &= \left(\frac{-2x^2}{x} + x^2\right) - \left(\frac{-2x^2}{1} + 1\right)\\ &= 3x^2 - 2x - 1.\end{aligned}$$

■

Notice in Example 1 that the integral defines a function of x and can *itself* be integrated, as shown in the next example.

EXAMPLE 2 The Integral of an Integral

Evaluate $\displaystyle\int_1^2 \left[\int_1^x (2x^2y^{-2} + 2y)\, dy\right] dx$.

Solution Using the result of Example 1, you have

$$\begin{aligned}\int_1^2 \left[\int_1^x (2x^2y^{-2} + 2y)\, dy\right] dx &= \int_1^2 (3x^2 - 2x - 1)\, dx\\ &= \Big[x^3 - x^2 - x\Big]_1^2 && \text{Integrate with respect to } x.\\ &= 2 - (-1)\\ &= 3.\end{aligned}$$

■

The integral in Example 2 is an **iterated integral.** The brackets used in Example 2 are normally not written. Instead, iterated integrals are usually written simply as

$$\int_a^b \int_{g_1(x)}^{g_2(x)} f(x, y)\, dy\, dx \quad \text{and} \quad \int_c^d \int_{h_1(y)}^{h_2(y)} f(x, y)\, dx\, dy.$$

The **inside limits of integration** can be variable with respect to the outer variable of integration. However, the **outside limits of integration** *must be* constant with respect to both variables of integration. After performing the inside integration, you obtain a "standard" definite integral, and the second integration produces a real number. The limits of integration for an iterated integral identify two sets of boundary intervals for the variables. For instance, in Example 2, the outside limits indicate that x lies in the interval $1 \le x \le 2$ and the inside limits indicate that y lies in the interval $1 \le y \le x$. Together, these two intervals determine the **region of integration *R*** of the iterated integral, as shown in Figure 14.1.

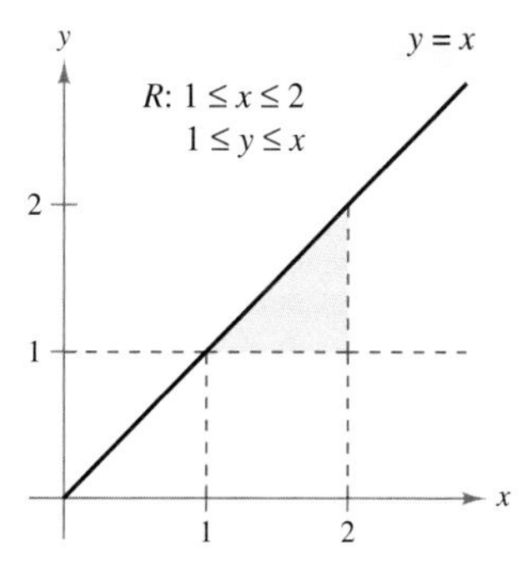

The region of integration for $\displaystyle\int_1^2 \int_1^x f(x, y)\, dy\, dx$

Figure 14.1

Because an iterated integral is just a special type of definite integral—one in which the integrand is also an integral—you can use the properties of definite integrals to evaluate iterated integrals.

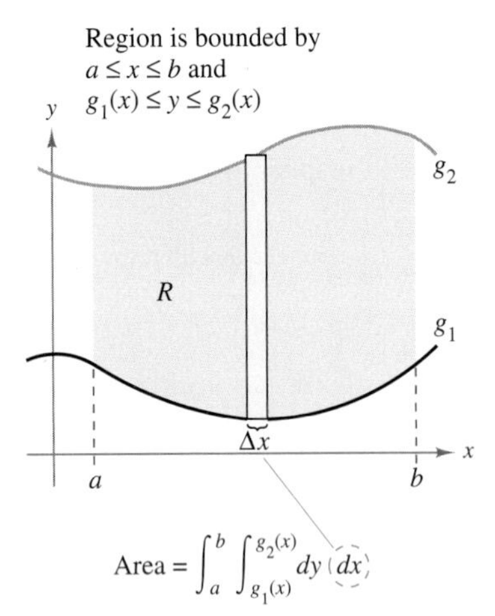

Vertically simple region
Figure 14.2

Area of a Plane Region

In the remainder of this section, you will take a new look at an old problem—that of finding the area of a plane region. Consider the plane region R bounded by $a \le x \le b$ and $g_1(x) \le y \le g_2(x)$, as shown in Figure 14.2. The area of R is given by the definite integral

$$\int_a^b [g_2(x) - g_1(x)]\, dx. \qquad \text{Area of } R$$

Using the Fundamental Theorem of Calculus, you can rewrite the integrand $g_2(x) - g_1(x)$ as a definite integral. Specifically, if you consider x to be fixed and let y vary from $g_1(x)$ to $g_2(x)$, you can write

$$\int_{g_1(x)}^{g_2(x)} dy = y\Big]_{g_1(x)}^{g_2(x)} = g_2(x) - g_1(x).$$

Combining these two integrals, you can write the area of the region R as an iterated integral

$$\int_a^b \int_{g_1(x)}^{g_2(x)} dy\, dx = \int_a^b y\Big]_{g_1(x)}^{g_2(x)} dx \qquad \text{Area of } R$$

$$= \int_a^b [g_2(x) - g_1(x)]\, dx.$$

Placing a representative rectangle in the region R helps determine both the order and the limits of integration. A vertical rectangle implies the order $dy\,dx$, with the inside limits corresponding to the upper and lower bounds of the rectangle, as shown in Figure 14.2. This type of region is called **vertically simple,** because the outside limits of integration represent the vertical lines $x = a$ and $x = b$.

Similarly, a horizontal rectangle implies the order $dx\,dy$, with the inside limits determined by the left and right bounds of the rectangle, as shown in Figure 14.3. This type of region is called **horizontally simple,** because the outside limits represent the horizontal lines $y = c$ and $y = d$. The iterated integrals used for these two types of simple regions are summarized as follows.

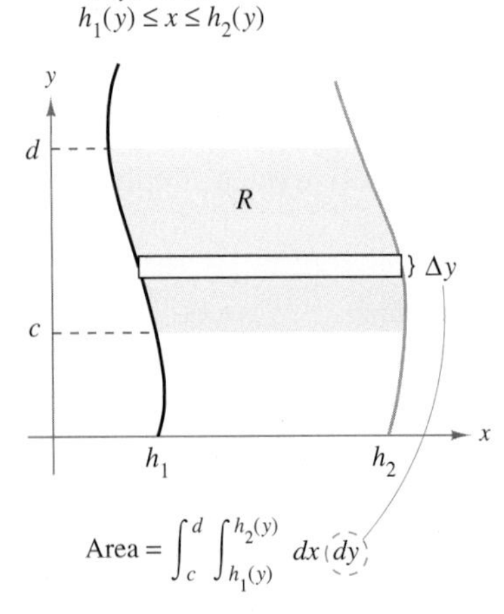

Horizontally simple region
Figure 14.3

AREA OF A REGION IN THE PLANE

1. If R is defined by $a \le x \le b$ and $g_1(x) \le y \le g_2(x)$, where g_1 and g_2 are continuous on $[a, b]$, then the area of R is given by

$$A = \int_a^b \int_{g_1(x)}^{g_2(x)} dy\, dx. \qquad \text{Figure 14.2 (vertically simple)}$$

2. If R is defined by $c \le y \le d$ and $h_1(y) \le x \le h_2(y)$, where h_1 and h_2 are continuous on $[c, d]$, then the area of R is given by

$$A = \int_c^d \int_{h_1(y)}^{h_2(y)} dx\, dy. \qquad \text{Figure 14.3 (horizontally simple)}$$

NOTE Be sure you see that the orders of integration of these two integrals are different—the order $dy\,dx$ corresponds to a vertically simple region, and the order $dx\,dy$ corresponds to a horizontally simple region. ■

If all four limits of integration happen to be constants, the region of integration is rectangular, as shown in Example 3.

EXAMPLE 3 The Area of a Rectangular Region

Use an iterated integral to represent the area of the rectangle shown in Figure 14.4.

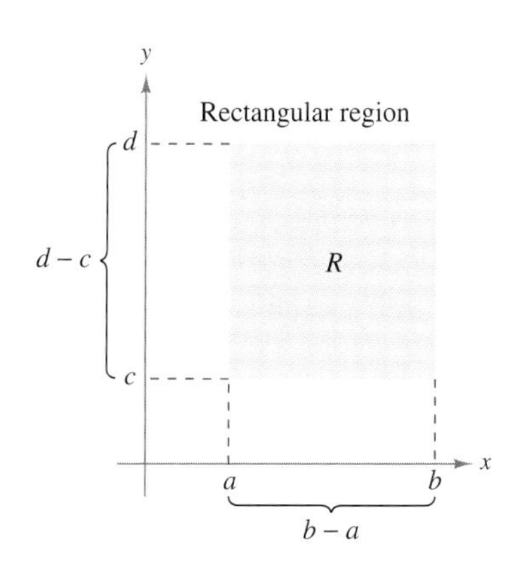

Figure 14.4

Solution The region shown in Figure 14.4 is both vertically simple and horizontally simple, so you can use either order of integration. By choosing the order $dy\,dx$, you obtain the following.

$$\begin{aligned}\int_a^b \int_c^d dy\,dx &= \int_a^b \left[y \right]_c^d dx && \text{Integrate with respect to } y.\\ &= \int_a^b (d-c)\,dx \\ &= \Big[(d-c)x \Big]_a^b && \text{Integrate with respect to } x.\\ &= (d-c)(b-a)\end{aligned}$$

Notice that this answer is consistent with what you know from geometry.

EXAMPLE 4 Finding Area by an Iterated Integral

Use an iterated integral to find the area of the region bounded by the graphs of

$$f(x) = \sin x \qquad \text{Sine curve forms upper boundary.}$$
$$g(x) = \cos x \qquad \text{Cosine curve forms lower boundary.}$$

between $x = \pi/4$ and $x = 5\pi/4$.

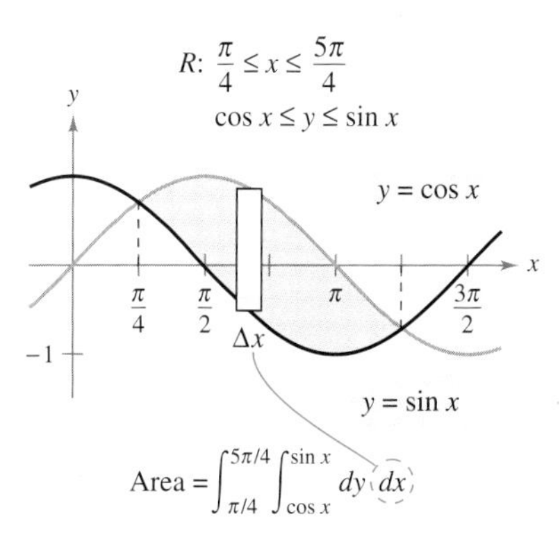

Figure 14.5

Solution Because f and g are given as functions of x, a vertical representative rectangle is convenient, and you can choose $dy\,dx$ as the order of integration, as shown in Figure 14.5. The outside limits of integration are $\pi/4 \le x \le 5\pi/4$. Moreover, because the rectangle is bounded above by $f(x) = \sin x$ and below by $g(x) = \cos x$, you have

$$\begin{aligned}\text{Area of } R &= \int_{\pi/4}^{5\pi/4} \int_{\cos x}^{\sin x} dy\,dx \\ &= \int_{\pi/4}^{5\pi/4} \left[y \right]_{\cos x}^{\sin x} dx && \text{Integrate with respect to } y.\\ &= \int_{\pi/4}^{5\pi/4} (\sin x - \cos x)\,dx \\ &= \Big[-\cos x - \sin x \Big]_{\pi/4}^{5\pi/4} && \text{Integrate with respect to } x.\\ &= 2\sqrt{2}.\end{aligned}$$

■

NOTE The region of integration of an iterated integral need not have any straight lines as boundaries. For instance, the region of integration shown in Figure 14.5 is *vertically simple* even though it has no vertical lines as left and right boundaries. The quality that makes the region vertically simple is that it is bounded above and below by the graphs of *functions of x*.

■

One order of integration will often produce a simpler integration problem than the other order. For instance, try reworking Example 4 with the order $dx\,dy$—you may be surprised to see that the task is formidable. However, if you succeed, you will see that the answer is the same. In other words, the order of integration affects the ease of integration, but not the value of the integral.

EXAMPLE 5 Comparing Different Orders of Integration

Sketch the region whose area is represented by the integral

$$\int_0^2 \int_{y^2}^4 dx\,dy.$$

Then find another iterated integral using the order $dy\,dx$ to represent the same area and show that both integrals yield the same value.

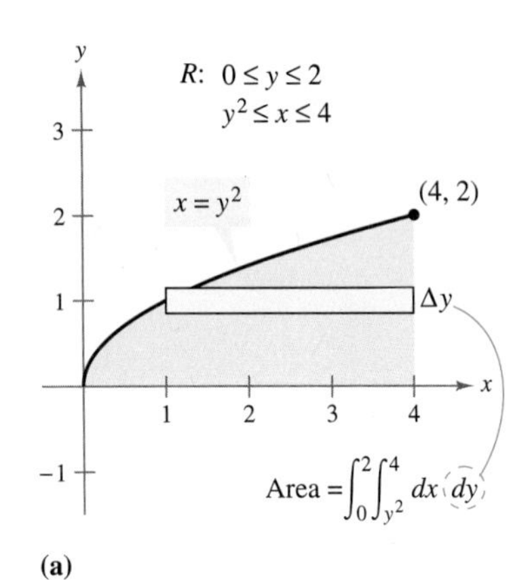

(a)

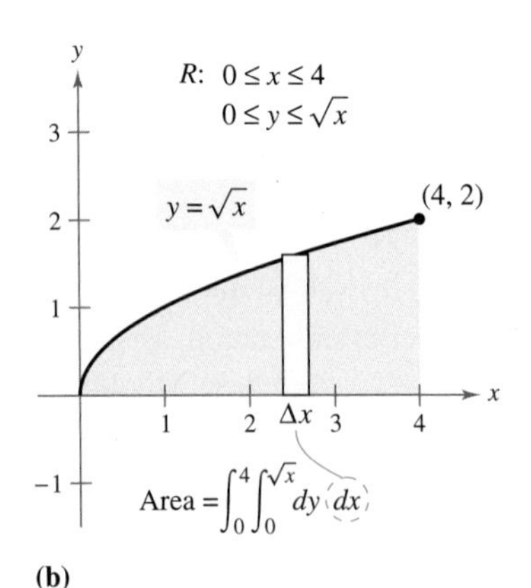

(b)

Figure 14.6

Solution From the given limits of integration, you know that

$$y^2 \le x \le 4 \qquad \text{Inner limits of integration}$$

which means that the region R is bounded on the left by the parabola $x = y^2$ and on the right by the line $x = 4$. Furthermore, because

$$0 \le y \le 2 \qquad \text{Outer limits of integration}$$

you know that R is bounded below by the x-axis, as shown in Figure 14.6(a). The value of this integral is

$$\int_0^2 \int_{y^2}^4 dx\,dy = \int_0^2 x\Big]_{y^2}^4 dy \qquad \text{Integrate with respect to } x.$$

$$= \int_0^2 (4 - y^2)\,dy$$

$$= \left[4y - \frac{y^3}{3}\right]_0^2 = \frac{16}{3}. \qquad \text{Integrate with respect to } y.$$

To change the order of integration to $dy\,dx$, place a vertical rectangle in the region, as shown in Figure 14.6(b). From this you can see that the constant bounds $0 \le x \le 4$ serve as the outer limits of integration. By solving for y in the equation $x = y^2$, you can conclude that the inner bounds are $0 \le y \le \sqrt{x}$. So, the area of the region can also be represented by

$$\int_0^4 \int_0^{\sqrt{x}} dy\,dx.$$

By evaluating this integral, you can see that it has the same value as the original integral.

$$\int_0^4 \int_0^{\sqrt{x}} dy\,dx = \int_0^4 y\Big]_0^{\sqrt{x}} dx \qquad \text{Integrate with respect to } y.$$

$$= \int_0^4 \sqrt{x}\,dx$$

$$= \frac{2}{3}x^{3/2}\Big]_0^4 = \frac{16}{3} \qquad \text{Integrate with respect to } x.$$

■

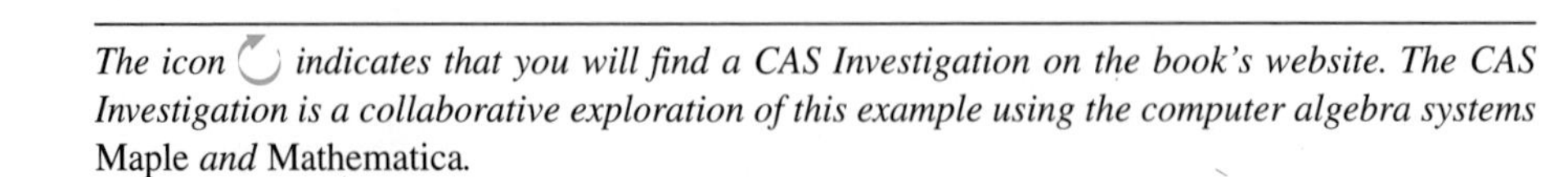

The icon indicates that you will find a CAS Investigation on the book's website. The CAS Investigation is a collaborative exploration of this example using the computer algebra systems Maple *and* Mathematica.

Sometimes it is not possible to calculate the area of a region with a single iterated integral. In these cases you can divide the region into subregions such that the area of each subregion can be calculated by an iterated integral. The total area is then the sum of the iterated integrals.

TECHNOLOGY Some computer software can perform symbolic integration for integrals such as those in Example 6. If you have access to such software, use it to evaluate the integrals in the exercises and examples given in this section.

EXAMPLE 6 An Area Represented by Two Iterated Integrals

Find the area of the region R that lies below the parabola

$$y = 4x - x^2 \qquad \text{Parabola forms upper boundary.}$$

above the x-axis, and above the line

$$y = -3x + 6. \qquad \text{Line and } x\text{-axis form lower boundary.}$$

Solution Begin by dividing R into the two subregions R_1 and R_2 shown in Figure 14.7.

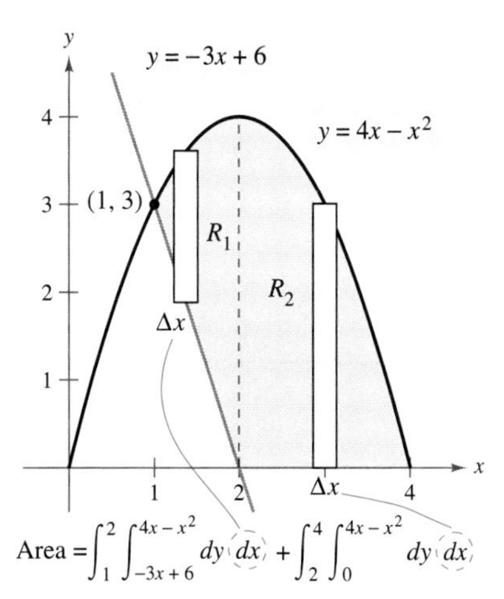

Figure 14.7

In both regions, it is convenient to use vertical rectangles, and you have

$$\begin{aligned}
\text{Area} &= \int_1^2 \int_{-3x+6}^{4x-x^2} dy\,dx + \int_2^4 \int_0^{4x-x^2} dy\,dx \\
&= \int_1^2 (4x - x^2 + 3x - 6)\,dx + \int_2^4 (4x - x^2)\,dx \\
&= \left[\frac{7x^2}{2} - \frac{x^3}{3} - 6x\right]_1^2 + \left[2x^2 - \frac{x^3}{3}\right]_2^4 \\
&= \left(14 - \frac{8}{3} - 12 - \frac{7}{2} + \frac{1}{3} + 6\right) + \left(32 - \frac{64}{3} - 8 + \frac{8}{3}\right) = \frac{15}{2}.
\end{aligned}$$

The area of the region is $15/2$ square units. Try checking this using the procedure for finding the area between two curves, as presented in Section 7.1. ■

At this point you may be wondering why you would need iterated integrals. After all, you already know how to use conventional integration to find the area of a region in the plane. (For instance, compare the solution of Example 4 in this section with that given in Example 3 in Section 7.1.) The need for iterated integrals will become clear in the next section. In this section, primary attention is given to procedures for finding the limits of integration of the region of an iterated integral, and the following exercise set is designed to develop skill in this important procedure.

NOTE In Examples 3 to 6, be sure you see the benefit of sketching the region of integration. You should develop the habit of making sketches to help you determine the limits of integration for all iterated integrals in this chapter.

14.1 Exercises

See www.CalcChat.com for worked-out solutions to odd-numbered exercises.

In Exercises 1–10, evaluate the integral.

1. $\int_0^x (x + 2y)\, dy$
2. $\int_x^{x^2} \frac{y}{x}\, dy$
3. $\int_1^{2y} \frac{y}{x}\, dx, \quad y > 0$
4. $\int_0^{\cos y} y\, dx$
5. $\int_0^{\sqrt{4-x^2}} x^2y\, dy$
6. $\int_{x^3}^{\sqrt{x}} (x^2 + 3y^2)\, dy$
7. $\int_{e^y}^{y} \frac{y \ln x}{x}\, dx, \quad y > 0$
8. $\int_{-\sqrt{1-y^2}}^{\sqrt{1-y^2}} (x^2 + y^2)\, dx$
9. $\int_0^{x^3} ye^{-y/x}\, dy$
10. $\int_y^{\pi/2} \sin^3 x \cos y\, dx$

In Exercises 11–30, evaluate the iterated integral.

11. $\int_0^1 \int_0^2 (x + y)\, dy\, dx$
12. $\int_{-1}^1 \int_{-2}^2 (x^2 - y^2)\, dy\, dx$
13. $\int_1^2 \int_0^4 (x^2 - 2y^2)\, dx\, dy$
14. $\int_{-1}^2 \int_1^3 (x + y^2)\, dx\, dy$
15. $\int_0^{\pi/2} \int_0^1 y \cos x\, dy\, dx$
16. $\int_0^{\ln 4} \int_0^{\ln 3} e^{x+y}\, dy\, dx$
17. $\int_0^{\pi} \int_0^{\sin x} (1 + \cos x)\, dy\, dx$
18. $\int_1^4 \int_1^{\sqrt{x}} 2ye^{-x}\, dy\, dx$
19. $\int_0^1 \int_0^x \sqrt{1 - x^2}\, dy\, dx$
20. $\int_{-4}^4 \int_0^{x^2} \sqrt{64 - x^3}\, dy\, dx$
21. $\int_{-1}^5 \int_0^{3y} \left(3 + x^2 + \frac{1}{4}y^2\right) dx\, dy$
22. $\int_0^2 \int_y^{2y} (10 + 2x^2 + 2y^2)\, dx\, dy$
23. $\int_0^1 \int_0^{\sqrt{1-y^2}} (x + y)\, dx\, dy$
24. $\int_0^2 \int_{3y^2-6y}^{2y-y^2} 3y\, dx\, dy$
25. $\int_0^2 \int_0^{\sqrt{4-y^2}} \frac{2}{\sqrt{4 - y^2}}\, dx\, dy$
26. $\int_1^3 \int_0^y \frac{4}{x^2 + y^2}\, dx\, dy$
27. $\int_0^{\pi/2} \int_0^{2\cos\theta} r\, dr\, d\theta$
28. $\int_0^{\pi/4} \int_{\sqrt{3}}^{\sqrt{3}\cos\theta} r\, dr\, d\theta$
29. $\int_0^{\pi/2} \int_0^{\sin\theta} \theta r\, dr\, d\theta$
30. $\int_0^{\pi/4} \int_0^{\cos\theta} 3r^2 \sin\theta\, dr\, d\theta$

In Exercises 31–34, evaluate the improper iterated integral.

31. $\int_1^{\infty} \int_0^{1/x} y\, dy\, dx$
32. $\int_0^3 \int_0^{\infty} \frac{x^2}{1 + y^2}\, dy\, dx$
33. $\int_1^{\infty} \int_1^{\infty} \frac{1}{xy}\, dx\, dy$
34. $\int_0^{\infty} \int_0^{\infty} xye^{-(x^2+y^2)}\, dx\, dy$

In Exercises 35–38, use an iterated integral to find the area of the region.

35.
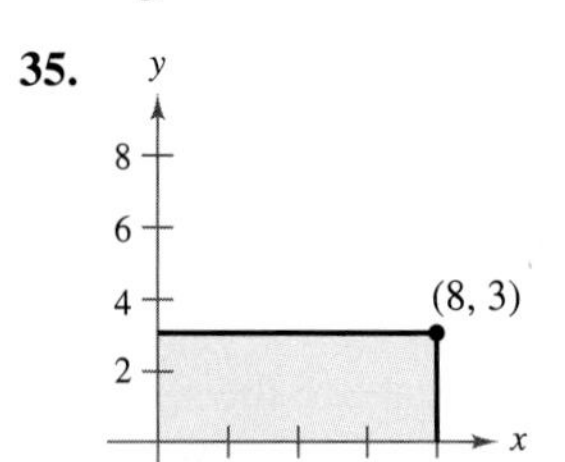

36.
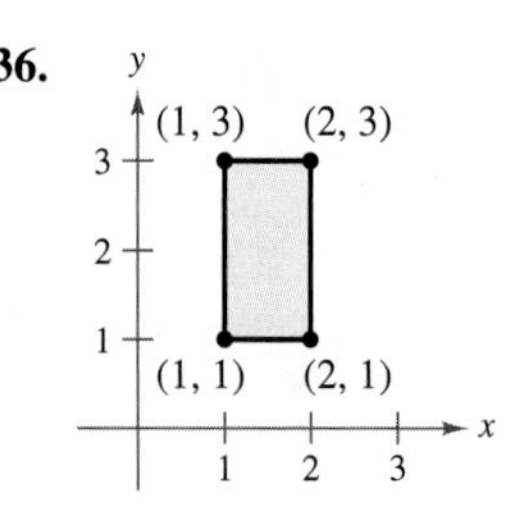

37.
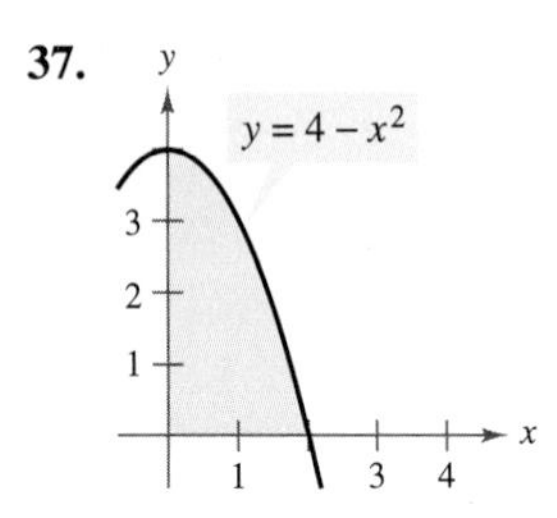

38.
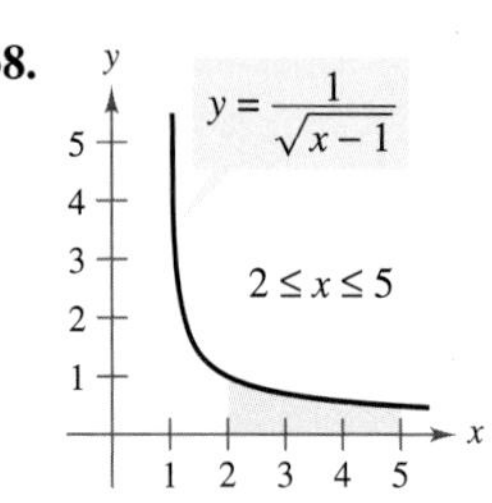

In Exercises 39–46, use an iterated integral to find the area of the region bounded by the graphs of the equations.

39. $\sqrt{x} + \sqrt{y} = 2, \quad x = 0, \quad y = 0$
40. $y = x^{3/2}, \quad y = 2x$
41. $2x - 3y = 0, \quad x + y = 5, \quad y = 0$
42. $xy = 9, \quad y = x, \quad y = 0, \quad x = 9$
43. $\frac{x^2}{a^2} + \frac{y^2}{b^2} = 1$
44. $y = x, \quad y = 2x, \quad x = 2$
45. $y = 4 - x^2, \quad y = x + 2$
46. $x^2 + y^2 = 4, \quad x = 0, \quad y = 0$

In Exercises 47–54, sketch the region R of integration and switch the order of integration.

47. $\int_0^4 \int_0^y f(x, y)\, dx\, dy$
48. $\int_0^4 \int_{\sqrt{y}}^2 f(x, y)\, dx\, dy$
49. $\int_{-2}^2 \int_0^{\sqrt{4-x^2}} f(x, y)\, dy\, dx$
50. $\int_0^2 \int_0^{4-x^2} f(x, y)\, dy\, dx$
51. $\int_1^{10} \int_0^{\ln y} f(x, y)\, dx\, dy$
52. $\int_{-1}^2 \int_0^{e^{-x}} f(x, y)\, dy\, dx$
53. $\int_{-1}^1 \int_{x^2}^1 f(x, y)\, dy\, dx$
54. $\int_{-\pi/2}^{\pi/2} \int_0^{\cos x} f(x, y)\, dy\, dx$

In Exercises 55–64, sketch the region R whose area is given by the iterated integral. Then switch the order of integration and show that both orders yield the same area.

55. $\int_0^1 \int_0^2 dy\, dx$
56. $\int_1^2 \int_2^4 dx\, dy$
57. $\int_0^1 \int_{-\sqrt{1-y^2}}^{\sqrt{1-y^2}} dx\, dy$
58. $\int_{-2}^2 \int_{-\sqrt{4-x^2}}^{\sqrt{4-x^2}} dy\, dx$

59. $\int_0^2 \int_0^x dy\,dx + \int_2^4 \int_0^{4-x} dy\,dx$

60. $\int_0^4 \int_0^{x/2} dy\,dx + \int_4^6 \int_0^{6-x} dy\,dx$

61. $\int_0^2 \int_{x/2}^1 dy\,dx$

62. $\int_0^9 \int_{\sqrt{x}}^3 dy\,dx$

63. $\int_0^1 \int_{y^2}^{\sqrt[3]{y}} dx\,dy$

64. $\int_{-2}^2 \int_0^{4-y^2} dx\,dy$

65. ***Think About It*** Give a geometric argument for the equality. Verify the equality analytically.

$$\int_0^5 \int_x^{\sqrt{50-x^2}} x^2y^2\,dy\,dx =$$

$$\int_0^5 \int_0^y x^2y^2\,dx\,dy + \int_5^{5\sqrt{2}} \int_0^{\sqrt{50-y^2}} x^2y^2\,dx\,dy$$

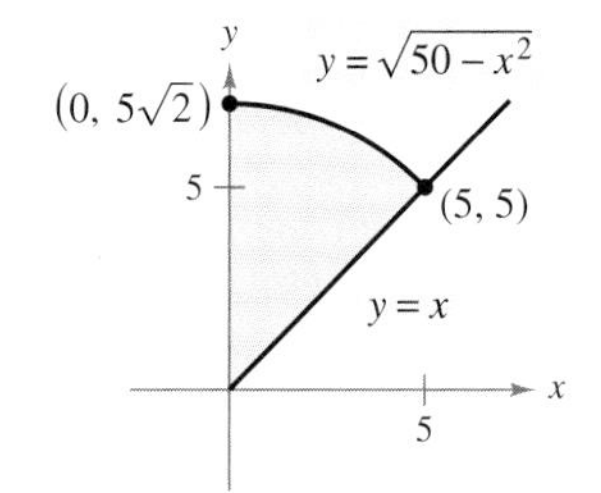

CAPSTONE

66. ***Think About It*** Complete the iterated integrals so that each one represents the area of the region R (see figure). Then show that both integrals yield the same area.

(a) Area $= \int\int dx\,dy$ (b) Area $= \int\int dy\,dx$

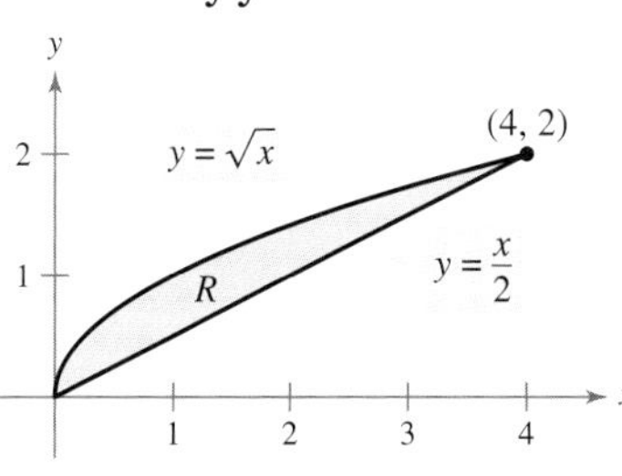

In Exercises 67–72, sketch the region of integration. Then evaluate the iterated integral. (Note that it is necessary to switch the order of integration.)

67. $\int_0^2 \int_x^2 x\sqrt{1 + y^3}\,dy\,dx$

68. $\int_0^4 \int_{\sqrt{x}}^2 \frac{3}{2 + y^3}\,dy\,dx$

69. $\int_0^1 \int_{2x}^2 4e^{y^2}\,dy\,dx$

70. $\int_0^2 \int_x^2 e^{-y^2}\,dy\,dx$

71. $\int_0^1 \int_y^1 \sin x^2\,dx\,dy$

72. $\int_0^2 \int_{y^2}^4 \sqrt{x}\sin x\,dx\,dy$

CAS **In Exercises 73–76, use a computer algebra system to evaluate the iterated integral.**

73. $\int_0^2 \int_{x^2}^{2x} (x^3 + 3y^2)\,dy\,dx$

74. $\int_0^1 \int_y^{2y} \sin(x + y)\,dx\,dy$

75. $\int_0^4 \int_0^y \frac{2}{(x + 1)(y + 1)}\,dx\,dy$

76. $\int_0^a \int_0^{a-x} (x^2 + y^2)\,dy\,dx$

CAS **In Exercises 77 and 78, (a) sketch the region of integration, (b) switch the order of integration, and (c) use a computer algebra system to show that both orders yield the same value.**

77. $\int_0^2 \int_{y^3}^{4\sqrt{2y}} (x^2y - xy^2)\,dx\,dy$

78. $\int_0^2 \int_{\sqrt{4-x^2}}^{4-x^2/4} \frac{xy}{x^2 + y^2 + 1}\,dy\,dx$

CAS **In Exercises 79–82, use a computer algebra system to approximate the iterated integral.**

79. $\int_0^2 \int_0^{4-x^2} e^{xy}\,dy\,dx$

80. $\int_0^2 \int_x^2 \sqrt{16 - x^3 - y^3}\,dy\,dx$

81. $\int_0^{2\pi} \int_0^{1+\cos\theta} 6r^2\cos\theta\,dr\,d\theta$

82. $\int_0^{\pi/2} \int_0^{1+\sin\theta} 15\theta r\,dr\,d\theta$

WRITING ABOUT CONCEPTS

83. Explain what is meant by an iterated integral. How is it evaluated?

84. Describe regions that are vertically simple and regions that are horizontally simple.

85. Give a geometric description of the region of integration if the inside and outside limits of integration are constants.

86. Explain why it is sometimes an advantage to change the order of integration.

True or False? **In Exercises 87 and 88, determine whether the statement is true or false. If it is false, explain why or give an example that shows it is false.**

87. $\int_a^b \int_c^d f(x, y)\,dy\,dx = \int_c^d \int_a^b f(x, y)\,dx\,dy$

88. $\int_0^1 \int_0^x f(x, y)\,dy\,dx = \int_0^1 \int_0^y f(x, y)\,dx\,dy$

14.2 Double Integrals and Volume

- **Use a double integral to represent the volume of a solid region.**
- **Use properties of double integrals.**
- **Evaluate a double integral as an iterated integral.**
- **Find the average value of a function over a region.**

Double Integrals and Volume of a Solid Region

You already know that a definite integral over an *interval* uses a limit process to assign measures to quantities such as area, volume, arc length, and mass. In this section, you will use a similar process to define the **double integral** of a function of two variables over a *region in the plane*.

Consider a continuous function f such that $f(x, y) \geq 0$ for all (x, y) in a region R in the xy-plane. The goal is to find the volume of the solid region lying between the surface given by

$$z = f(x, y) \qquad \text{Surface lying above the } xy\text{-plane}$$

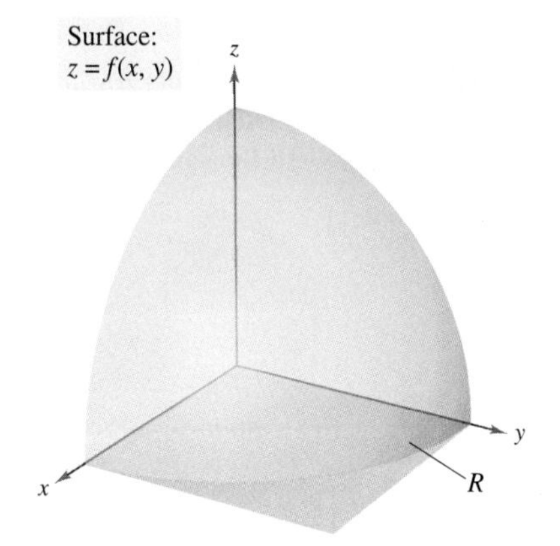

Figure 14.8

and the xy-plane, as shown in Figure 14.8. You can begin by superimposing a rectangular grid over the region, as shown in Figure 14.9. The rectangles lying entirely within R form an **inner partition** Δ, whose **norm** $\|\Delta\|$ is defined as the length of the longest diagonal of the n rectangles. Next, choose a point (x_i, y_i) in each rectangle and form the rectangular prism whose height is $f(x_i, y_i)$, as shown in Figure 14.10. Because the area of the ith rectangle is

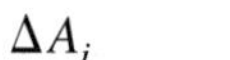

$$\Delta A_i \qquad \text{Area of } i\text{th rectangle}$$

it follows that the volume of the ith prism is

$$f(x_i, y_i)\,\Delta A_i \qquad \text{Volume of } i\text{th prism}$$

and you can approximate the volume of the solid region by the Riemann sum of the volumes of all n prisms,

$$\sum_{i=1}^{n} f(x_i, y_i)\,\Delta A_i \qquad \text{Riemann sum}$$

as shown in Figure 14.11. This approximation can be improved by tightening the mesh of the grid to form smaller and smaller rectangles, as shown in Example 1.

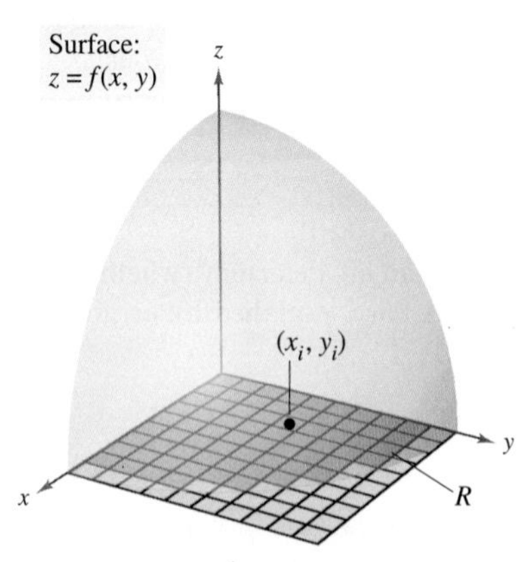

The rectangles lying within R form an inner partition of R.
Figure 14.9

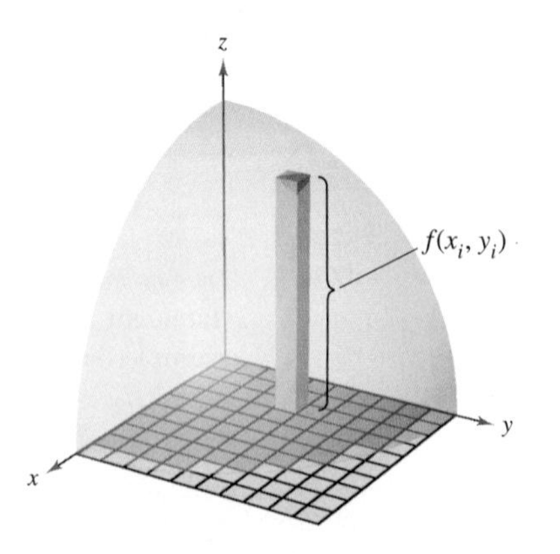

Rectangular prism whose base has an area of ΔA_i and whose height is $f(x_i, y_i)$
Figure 14.10

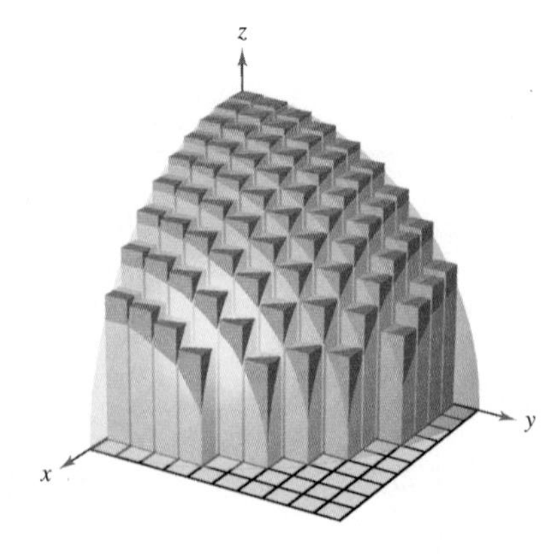

Volume approximated by rectangular prisms
Figure 14.11

EXAMPLE 1 Approximating the Volume of a Solid

Approximate the volume of the solid lying between the paraboloid

$$f(x, y) = 1 - \frac{1}{2}x^2 - \frac{1}{2}y^2$$

and the square region R given by $0 \le x \le 1, 0 \le y \le 1$. Use a partition made up of squares whose sides have a length of $\frac{1}{4}$.

Solution Begin by forming the specified partition of R. For this partition, it is convenient to choose the centers of the subregions as the points at which to evaluate $f(x, y)$.

$$\begin{array}{cccc} \left(\frac{1}{8}, \frac{1}{8}\right) & \left(\frac{1}{8}, \frac{3}{8}\right) & \left(\frac{1}{8}, \frac{5}{8}\right) & \left(\frac{1}{8}, \frac{7}{8}\right) \\ \left(\frac{3}{8}, \frac{1}{8}\right) & \left(\frac{3}{8}, \frac{3}{8}\right) & \left(\frac{3}{8}, \frac{5}{8}\right) & \left(\frac{3}{8}, \frac{7}{8}\right) \\ \left(\frac{5}{8}, \frac{1}{8}\right) & \left(\frac{5}{8}, \frac{3}{8}\right) & \left(\frac{5}{8}, \frac{5}{8}\right) & \left(\frac{5}{8}, \frac{7}{8}\right) \\ \left(\frac{7}{8}, \frac{1}{8}\right) & \left(\frac{7}{8}, \frac{3}{8}\right) & \left(\frac{7}{8}, \frac{5}{8}\right) & \left(\frac{7}{8}, \frac{7}{8}\right) \end{array}$$

Because the area of each square is $\Delta A_i = \frac{1}{16}$, you can approximate the volume by the sum

$$\sum_{i=1}^{16} f(x_i, y_i)\,\Delta A_i = \sum_{i=1}^{16}\left(1 - \frac{1}{2}x_i^2 - \frac{1}{2}y_i^2\right)\left(\frac{1}{16}\right) \approx 0.672.$$

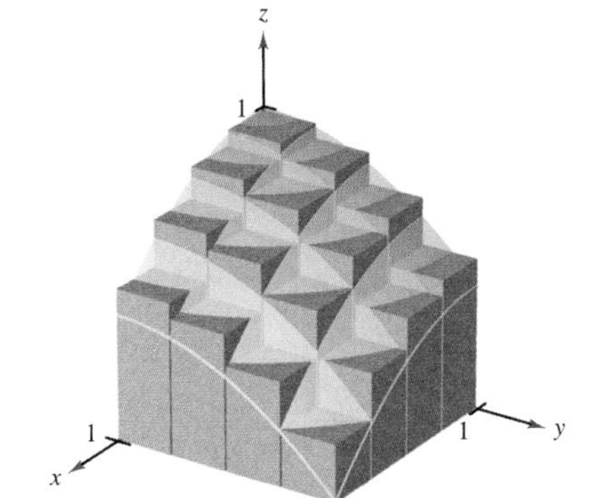

Surface:
$f(x, y) = 1 - \frac{1}{2}x^2 - \frac{1}{2}y^2$

Figure 14.12

This approximation is shown graphically in Figure 14.12. The exact volume of the solid is $\frac{2}{3}$ (see Example 2). You can obtain a better approximation by using a finer partition. For example, with a partition of squares with sides of length $\frac{1}{10}$, the approximation is 0.668. ■

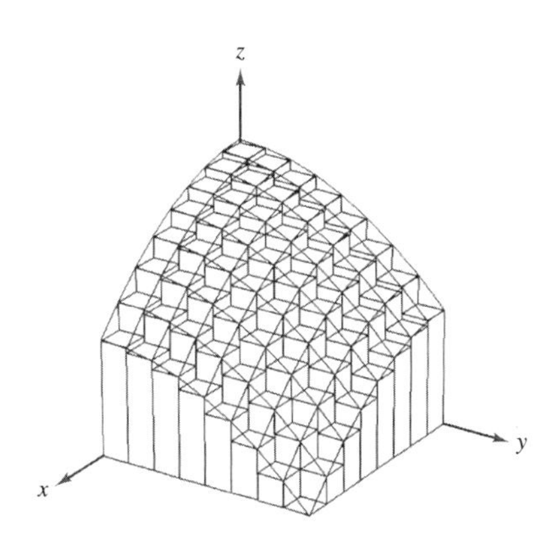

Figure 14.13

TECHNOLOGY Some three-dimensional graphing utilities are capable of graphing figures such as that shown in Figure 14.12. For instance, the graph shown in Figure 14.13 was drawn with a computer program. In this graph, note that each of the rectangular prisms lies within the solid region.

In Example 1, note that by using finer partitions, you obtain better approximations of the volume. This observation suggests that you could obtain the exact volume by taking a limit. That is,

$$\text{Volume} = \lim_{\|\Delta\| \to 0} \sum_{i=1}^{n} f(x_i, y_i)\,\Delta A_i.$$

The precise meaning of this limit is that the limit is equal to L if for every $\varepsilon > 0$ there exists a $\delta > 0$ such that

$$\left| L - \sum_{i=1}^{n} f(x_i, y_i)\,\Delta A_i \right| < \varepsilon$$

for all partitions Δ of the plane region R (that satisfy $\|\Delta\| < \delta$) and for all possible choices of x_i and y_i in the ith region.

Using the limit of a Riemann sum to define volume is a special case of using the limit to define a **double integral.** The general case, however, does not require that the function be positive or continuous.

EXPLORATION

The entries in the table represent the depths (in 10-yard units) of earth at the centers of the squares in the figure below.

x \ y	1	2	3
1	10	9	7
2	7	7	4
3	5	5	4
4	4	5	3

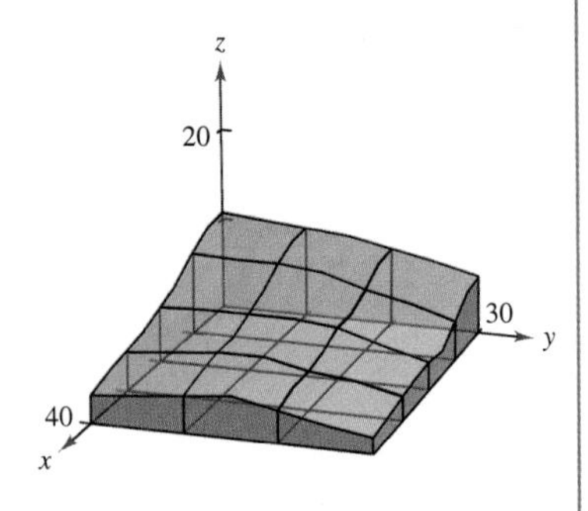

Approximate the number of cubic yards of earth in the first octant. (This exploration was submitted by Robert Vojack, Ridgewood High School, Ridgewood, NJ.)

DEFINITION OF DOUBLE INTEGRAL

If f is defined on a closed, bounded region R in the xy-plane, then the **double integral of f over R** is given by

$$\int_R\int f(x, y)\, dA = \lim_{\|\Delta\|\to 0} \sum_{i=1}^{n} f(x_i, y_i)\, \Delta A_i$$

provided the limit exists. If the limit exists, then f is **integrable** over R.

NOTE Having defined a double integral, you will see that a definite integral is occasionally referred to as a **single integral.**

Sufficient conditions for the double integral of f on the region R to exist are that R can be written as a union of a finite number of nonoverlapping subregions (see Figure 14.14) that are vertically or horizontally simple *and* that f is continuous on the region R.

A double integral can be used to find the volume of a solid region that lies between the xy-plane and the surface given by $z = f(x, y)$.

VOLUME OF A SOLID REGION

If f is integrable over a plane region R and $f(x, y) \geq 0$ for all (x, y) in R, then the volume of the solid region that lies above R and below the graph of f is defined as

$$V = \int_R\int f(x, y)\, dA.$$

Properties of Double Integrals

Double integrals share many properties of single integrals.

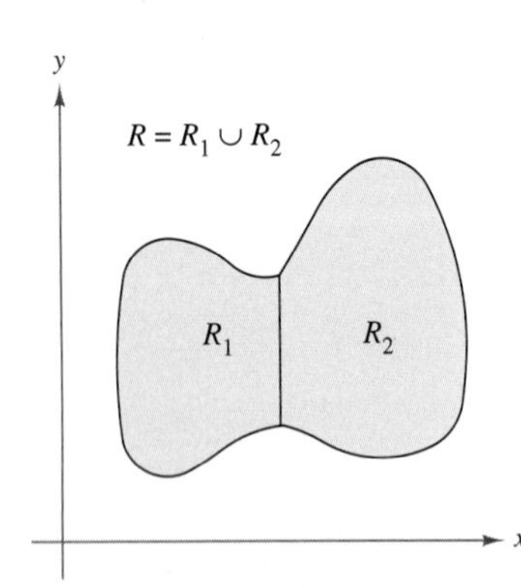

Two regions are nonoverlapping if their intersection is a set that has an area of 0. In this figure, the area of the line segment that is common to R_1 and R_2 is 0.
Figure 14.14

THEOREM 14.1 PROPERTIES OF DOUBLE INTEGRALS

Let f and g be continuous over a closed, bounded plane region R, and let c be a constant.

1. $\displaystyle\int_R\int cf(x, y)\, dA = c\int_R\int f(x, y)\, dA$
2. $\displaystyle\int_R\int [f(x, y) \pm g(x, y)]\, dA = \int_R\int f(x, y)\, dA \pm \int_R\int g(x, y)\, dA$
3. $\displaystyle\int_R\int f(x, y)\, dA \geq 0,\qquad \text{if } f(x, y) \geq 0$
4. $\displaystyle\int_R\int f(x, y)\, dA \geq \int_R\int g(x, y)\, dA,\qquad \text{if } f(x, y) \geq g(x, y)$
5. $\displaystyle\int_R\int f(x, y)\, dA = \int_{R_1}\int f(x, y)\, dA + \int_{R_2}\int f(x, y)\, dA$, where R is the union of two nonoverlapping subregions R_1 and R_2.

Evaluation of Double Integrals

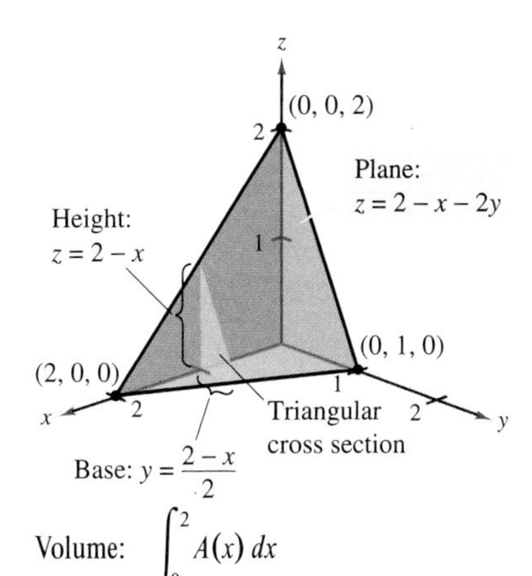

Figure 14.15

Normally, the first step in evaluating a double integral is to rewrite it as an iterated integral. To show how this is done, a geometric model of a double integral is used as the volume of a solid.

Consider the solid region bounded by the plane $z = f(x, y) = 2 - x - 2y$ and the three coordinate planes, as shown in Figure 14.15. Each vertical cross section taken parallel to the yz-plane is a triangular region whose base has a length of $y = (2 - x)/2$ and whose height is $z = 2 - x$. This implies that for a fixed value of x, the area of the triangular cross section is

$$A(x) = \frac{1}{2}(\text{base})(\text{height}) = \frac{1}{2}\left(\frac{2-x}{2}\right)(2 - x) = \frac{(2-x)^2}{4}.$$

By the formula for the volume of a solid with known cross sections (Section 7.2), the volume of the solid is

$$\begin{aligned}\text{Volume} &= \int_a^b A(x)\,dx\\ &= \int_0^2 \frac{(2-x)^2}{4}\,dx\\ &= -\frac{(2-x)^3}{12}\bigg]_0^2 = \frac{2}{3}.\end{aligned}$$

$z = 2 - x - 2y$

$y = 0$ $\quad$ $y = \frac{2-x}{2}$

Triangular cross section
Figure 14.16

This procedure works no matter how $A(x)$ is obtained. In particular, you can find $A(x)$ by integration, as shown in Figure 14.16. That is, you consider x to be constant, and integrate $z = 2 - x - 2y$ from 0 to $(2 - x)/2$ to obtain

$$\begin{aligned}A(x) &= \int_0^{(2-x)/2} (2 - x - 2y)\,dy\\ &= \Big[(2 - x)y - y^2\Big]_0^{(2-x)/2}\\ &= \frac{(2-x)^2}{4}.\end{aligned}$$

Combining these results, you have the *iterated integral*

$$\text{Volume} = \iint_R f(x, y)\,dA = \int_0^2 \int_0^{(2-x)/2} (2 - x - 2y)\,dy\,dx.$$

To understand this procedure better, it helps to imagine the integration as two sweeping motions. For the inner integration, a vertical line sweeps out the area of a cross section. For the outer integration, the triangular cross section sweeps out the volume, as shown in Figure 14.17.

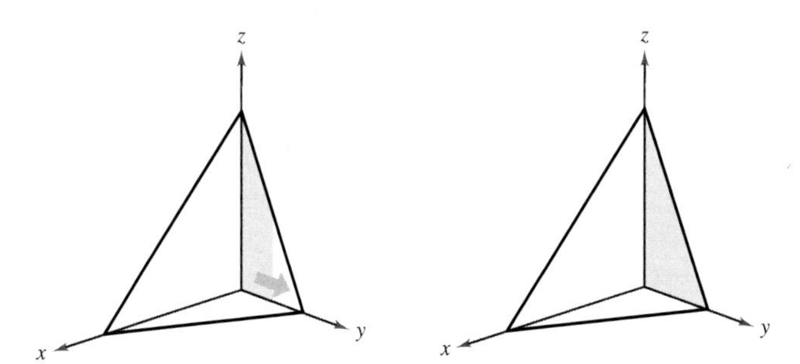

Integrate with respect to y to obtain the area of the cross section.

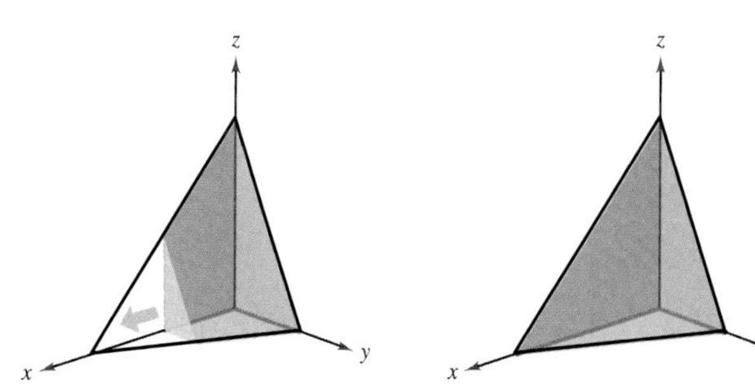

Integrate with respect to x to obtain the volume of the solid.

Figure 14.17

The following theorem was proved by the Italian mathematician Guido Fubini (1879–1943). The theorem states that if R is a vertically or horizontally simple region and f is continuous on R, the double integral of f on R is equal to an iterated integral.

THEOREM 14.2 FUBINI'S THEOREM

Let f be continuous on a plane region R.

1. If R is defined by $a \leq x \leq b$ and $g_1(x) \leq y \leq g_2(x)$, where g_1 and g_2 are continuous on $[a, b]$, then

$$\int_R\int f(x, y)\, dA = \int_a^b \int_{g_1(x)}^{g_2(x)} f(x, y)\, dy\, dx.$$

2. If R is defined by $c \leq y \leq d$ and $h_1(y) \leq x \leq h_2(y)$, where h_1 and h_2 are continuous on $[c, d]$, then

$$\int_R\int f(x, y)\, dA = \int_c^d \int_{h_1(y)}^{h_2(y)} f(x, y)\, dx\, dy.$$

EXAMPLE 2 Evaluating a Double Integral as an Iterated Integral

Evaluate

$$\int_R\int \left(1 - \frac{1}{2}x^2 - \frac{1}{2}y^2\right) dA$$

where R is the region given by $0 \leq x \leq 1$, $0 \leq y \leq 1$.

Solution Because the region R is a square, it is both vertically and horizontally simple, and you can use either order of integration. Choose $dy\,dx$ by placing a vertical representative rectangle in the region, as shown in Figure 14.18. This produces the following.

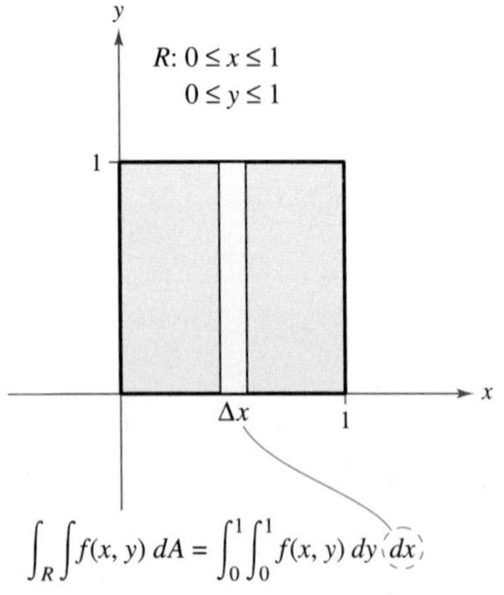

The volume of the solid region is $\frac{2}{3}$.
Figure 14.18

$$\begin{aligned}
\int_R\int \left(1 - \frac{1}{2}x^2 - \frac{1}{2}y^2\right) dA &= \int_0^1\int_0^1 \left(1 - \frac{1}{2}x^2 - \frac{1}{2}y^2\right) dy\, dx \\
&= \int_0^1 \left[\left(1 - \frac{1}{2}x^2\right)y - \frac{y^3}{6}\right]_0^1 dx \\
&= \int_0^1 \left(\frac{5}{6} - \frac{1}{2}x^2\right) dx \\
&= \left[\frac{5}{6}x - \frac{x^3}{6}\right]_0^1 \\
&= \frac{2}{3}
\end{aligned}$$

The double integral evaluated in Example 2 represents the volume of the solid region approximated in Example 1. Note that the approximation obtained in Example 1 is quite good $\left(0.672 \text{ vs. } \frac{2}{3}\right)$, even though you used a partition consisting of only 16 squares. The error resulted because the centers of the square subregions were used as the points in the approximation. This is comparable to the Midpoint Rule approximation of a single integral.

EXPLORATION

Volume of a Paraboloid Sector
The solid in Example 3 has an elliptical (not a circular) base. Consider the region bounded by the circular paraboloid

$$z = a^2 - x^2 - y^2, \quad a > 0$$

and the xy-plane. How many ways of finding the volume of this solid do you now know? For instance, you could use the disk method to find the volume as a solid of revolution. Does each method involve integration?

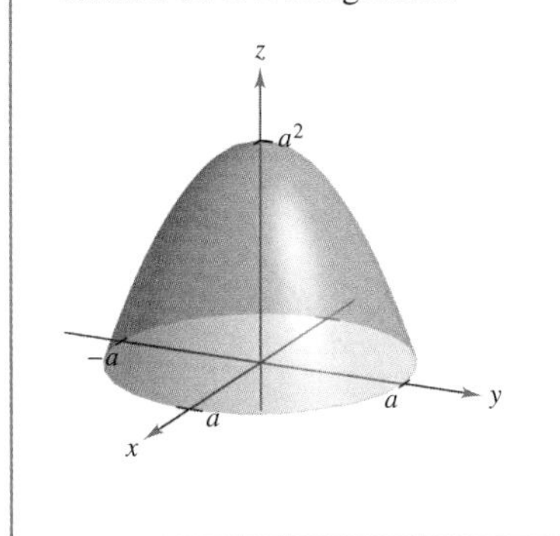

The difficulty of evaluating a single integral $\int_a^b f(x)\,dx$ usually depends on the function f, and not on the interval $[a, b]$. This is a major difference between single and double integrals. In the next example, you will integrate a function similar to the one in Examples 1 and 2. Notice that a change in the region R produces a much more difficult integration problem.

EXAMPLE 3 Finding Volume by a Double Integral

Find the volume of the solid region bounded by the paraboloid $z = 4 - x^2 - 2y^2$ and the xy-plane.

Solution By letting $z = 0$, you can see that the base of the region in the xy-plane is the ellipse $x^2 + 2y^2 = 4$, as shown in Figure 14.19(a). This plane region is both vertically and horizontally simple, so the order $dy\,dx$ is appropriate.

Variable bounds for y: $-\sqrt{\dfrac{(4 - x^2)}{2}} \le y \le \sqrt{\dfrac{(4 - x^2)}{2}}$

Constant bounds for x: $-2 \le x \le 2$

The volume is given by

$$\begin{aligned}
V &= \int_{-2}^{2}\int_{-\sqrt{(4-x^2)/2}}^{\sqrt{(4-x^2)/2}} (4 - x^2 - 2y^2)\,dy\,dx && \text{See Figure 14.19(b).}\\
&= \int_{-2}^{2}\left[(4 - x^2)y - \frac{2y^3}{3}\right]_{-\sqrt{(4-x^2)/2}}^{\sqrt{(4-x^2)/2}} dx\\
&= \frac{4}{3\sqrt{2}}\int_{-2}^{2}(4 - x^2)^{3/2}\,dx\\
&= \frac{4}{3\sqrt{2}}\int_{-\pi/2}^{\pi/2} 16\cos^4\theta\,d\theta && x = 2\sin\theta\\
&= \frac{64}{3\sqrt{2}}(2)\int_{0}^{\pi/2}\cos^4\theta\,d\theta\\
&= \frac{128}{3\sqrt{2}}\left(\frac{3\pi}{16}\right) && \text{Wallis's Formula}\\
&= 4\sqrt{2}\pi.
\end{aligned}$$

NOTE In Example 3, note the usefulness of Wallis's Formula to evaluate $\int_0^{\pi/2}\cos^n\theta\,d\theta$. You may want to review this formula in Section 8.3.

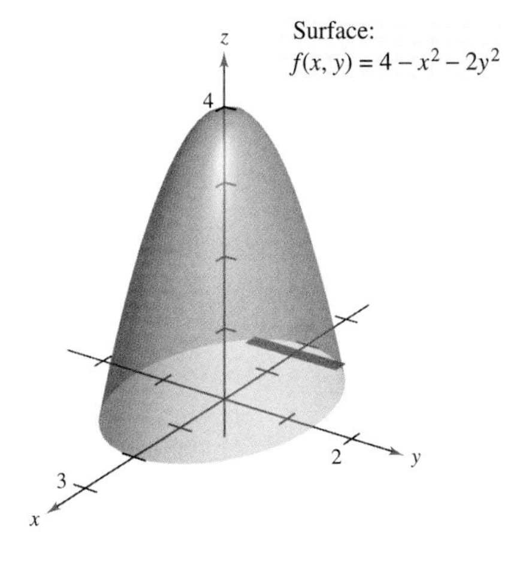

(a)

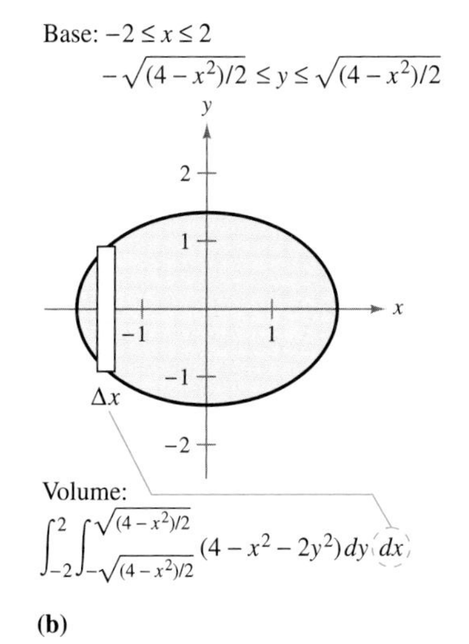

(b)

Figure 14.19

In Examples 2 and 3, the problems could be solved with either order of integration because the regions were both vertically and horizontally simple. Moreover, had you used the order $dx\,dy$, you would have obtained integrals of comparable difficulty. There are, however, some occasions in which one order of integration is much more convenient than the other. Example 4 shows such a case.

EXAMPLE 4 Comparing Different Orders of Integration

Find the volume of the solid region R bounded by the surface

$$f(x, y) = e^{-x^2} \qquad \text{Surface}$$

and the planes $z = 0$, $y = 0$, $y = x$, and $x = 1$, as shown in Figure 14.20.

Surface: $f(x, y) = e^{-x^2}$

Base is bounded by $y = 0$, $y = x$, and $x = 1$.

Figure 14.20

Solution The base of R in the xy-plane is bounded by the lines $y = 0$, $x = 1$, and $y = x$. The two possible orders of integration are shown in Figure 14.21.

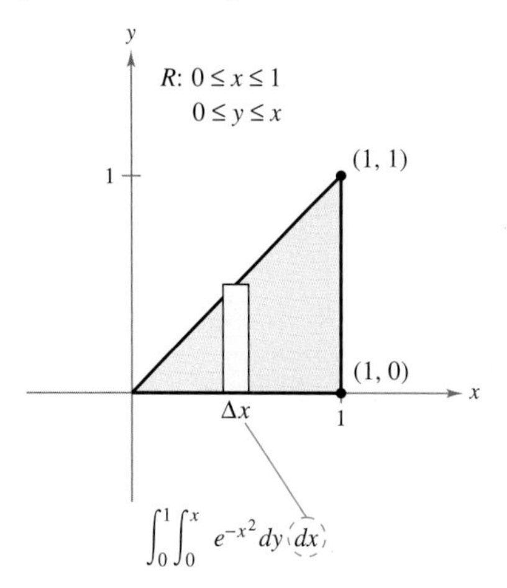

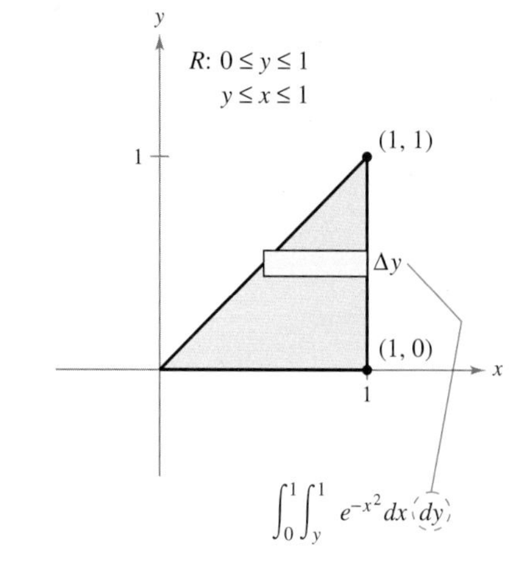

Figure 14.21

By setting up the corresponding iterated integrals, you can see that the order $dx\,dy$ requires the antiderivative $\int e^{-x^2}\,dx$, which is not an elementary function. On the other hand, the order $dy\,dx$ produces the integral

$$\begin{aligned}
\int_0^1 \int_0^x e^{-x^2}\,dy\,dx &= \int_0^1 e^{-x^2} y \Big]_0^x \,dx \\
&= \int_0^1 x e^{-x^2}\,dx \\
&= -\frac{1}{2} e^{-x^2} \Big]_0^1 \\
&= -\frac{1}{2}\left(\frac{1}{e} - 1\right) \\
&= \frac{e - 1}{2e} \\
&\approx 0.316.
\end{aligned}$$

■

NOTE Try using a symbolic integration utility to evaluate the integral in Example 4. ■

EXAMPLE 5 Volume of a Region Bounded by Two Surfaces

Find the volume of the solid region R bounded above by the paraboloid $z = 1 - x^2 - y^2$ and below by the plane $z = 1 - y$, as shown in Figure 14.22.

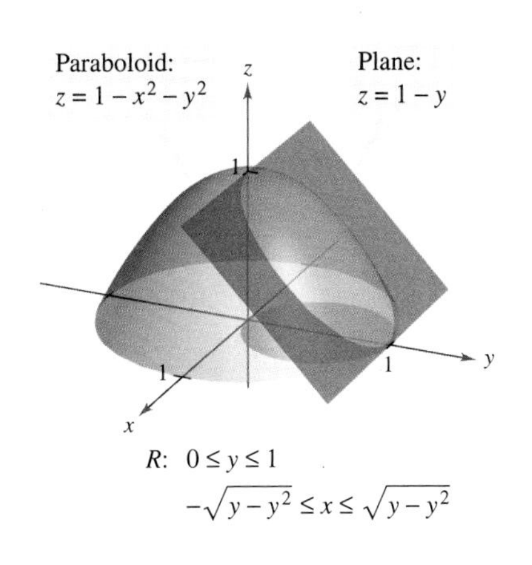

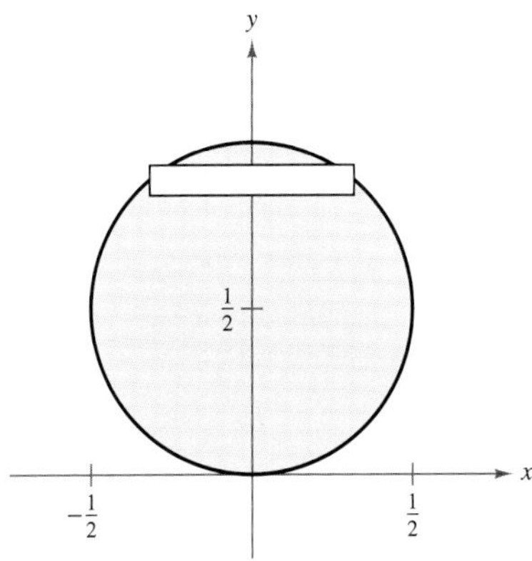

Figure 14.22

Solution Equating z-values, you can determine that the intersection of the two surfaces occurs on the right circular cylinder given by

$$1 - y = 1 - x^2 - y^2 \quad \Longrightarrow \quad x^2 = y - y^2.$$

Because the volume of R is the difference between the volume under the paraboloid and the volume under the plane, you have

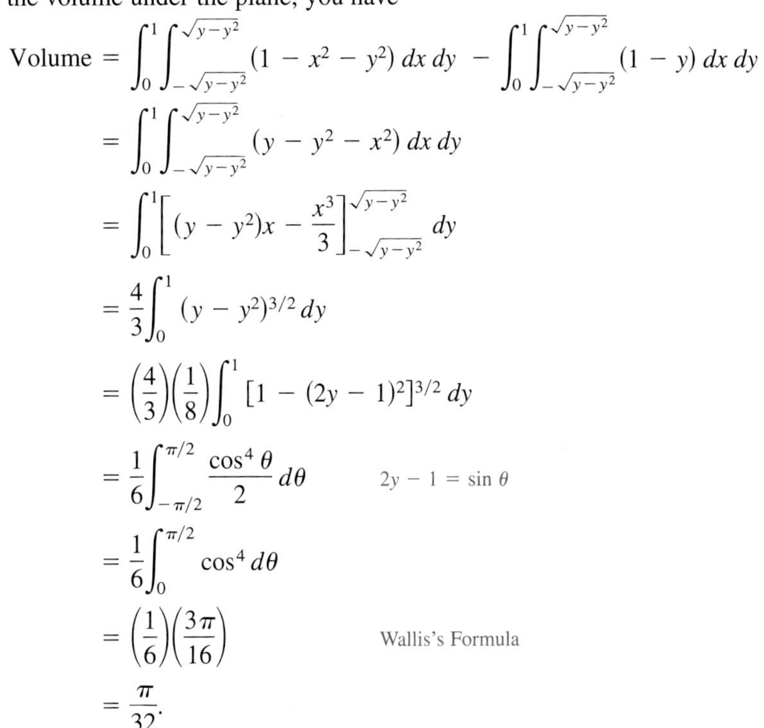

$$\begin{aligned}
\text{Volume} &= \int_0^1 \int_{-\sqrt{y-y^2}}^{\sqrt{y-y^2}} (1 - x^2 - y^2)\, dx\, dy - \int_0^1 \int_{-\sqrt{y-y^2}}^{\sqrt{y-y^2}} (1 - y)\, dx\, dy \\
&= \int_0^1 \int_{-\sqrt{y-y^2}}^{\sqrt{y-y^2}} (y - y^2 - x^2)\, dx\, dy \\
&= \int_0^1 \left[(y - y^2)x - \frac{x^3}{3} \right]_{-\sqrt{y-y^2}}^{\sqrt{y-y^2}} dy \\
&= \frac{4}{3} \int_0^1 (y - y^2)^{3/2}\, dy \\
&= \left(\frac{4}{3}\right)\left(\frac{1}{8}\right) \int_0^1 [1 - (2y - 1)^2]^{3/2}\, dy \\
&= \frac{1}{6} \int_{-\pi/2}^{\pi/2} \frac{\cos^4 \theta}{2}\, d\theta \qquad 2y - 1 = \sin\theta \\
&= \frac{1}{6} \int_0^{\pi/2} \cos^4 d\theta \\
&= \left(\frac{1}{6}\right)\left(\frac{3\pi}{16}\right) \qquad \text{Wallis's Formula} \\
&= \frac{\pi}{32}.
\end{aligned}$$

■

Average Value of a Function

Recall from Section 4.4 that for a function f in one variable, the average value of f on $[a, b]$ is

$$\frac{1}{b - a} \int_a^b f(x)\, dx.$$

Given a function f in two variables, you can find the average value of f over the region R as shown in the following definition.

DEFINITION OF THE AVERAGE VALUE OF A FUNCTION OVER A REGION

If f is integrable over the plane region R, then the **average value** of f over R is

$$\frac{1}{A} \int_R \int f(x, y)\, dA$$

where A is the area of R.

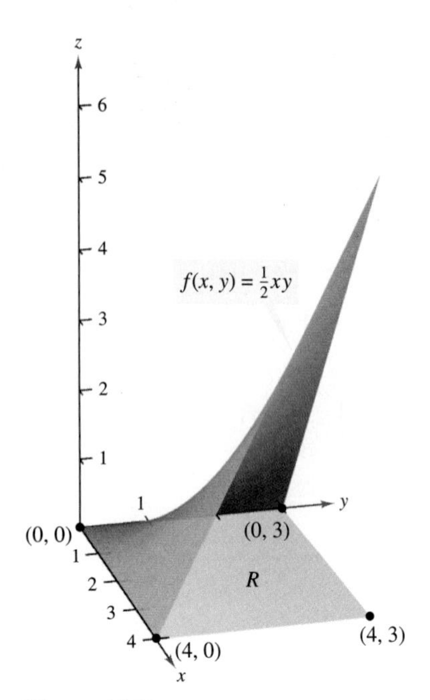

Figure 14.23

EXAMPLE 6 Finding the Average Value of a Function

Find the average value of $f(x, y) = \frac{1}{2}xy$ over the region R, where R is a rectangle with vertices $(0, 0)$, $(4, 0)$, $(4, 3)$, and $(0, 3)$.

Solution The area of the rectangular region R is $A = 12$ (see Figure 14.23). The average value is given by

$$\begin{aligned}
\frac{1}{A}\int_R\int f(x, y)\, dA &= \frac{1}{12}\int_0^4\int_0^3 \frac{1}{2}xy\, dy\, dx \\
&= \frac{1}{12}\int_0^4 \left[\frac{1}{4}xy^2\right]_0^3 dx \\
&= \left(\frac{1}{12}\right)\left(\frac{9}{4}\right)\int_0^4 x\, dx \\
&= \frac{3}{16}\left[\frac{1}{2}x^2\right]_0^4 \\
&= \left(\frac{3}{16}\right)(8) \\
&= \frac{3}{2}.
\end{aligned}$$

14.2 Exercises

See www.CalcChat.com for worked-out solutions to odd-numbered exercises.

Approximation **In Exercises 1–4, approximate the integral $\int_R\int f(x, y)\, dA$ by dividing the rectangle R with vertices $(0, 0)$, $(4, 0)$, $(4, 2)$, and $(0, 2)$ into eight equal squares and finding the sum $\sum_{i=1}^{8} f(x_i, y_i)\,\Delta A_i$ where (x_i, y_i) is the center of the ith square. Evaluate the iterated integral and compare it with the approximation.**

1. $\int_0^4\int_0^2 (x + y)\, dy\, dx$
2. $\frac{1}{2}\int_0^4\int_0^2 x^2y\, dy\, dx$
3. $\int_0^4\int_0^2 (x^2 + y^2)\, dy\, dx$
4. $\int_0^4\int_0^2 \frac{1}{(x + 1)(y + 1)}\, dy\, dx$

5. ***Approximation*** The table shows values of a function f over a square region R. Divide the region into 16 equal squares and select (x_i, y_i) to be the point in the ith square closest to the origin. Compare this approximation with that obtained by using the point in the ith square farthest from the origin.

$\int_0^4\int_0^4 f(x, y)\, dy\, dx$

x \ y	0	1	2	3	4
0	32	31	28	23	16
1	31	30	27	22	15
2	28	27	24	19	12
3	23	22	19	14	7
4	16	15	12	7	0

6. ***Approximation*** The figure shows the level curves for a function f over a square region R. Approximate the integral using four squares, selecting the midpoint of each square as (x_i, y_i).

$\int_0^2\int_0^2 f(x, y)\, dy\, dx$

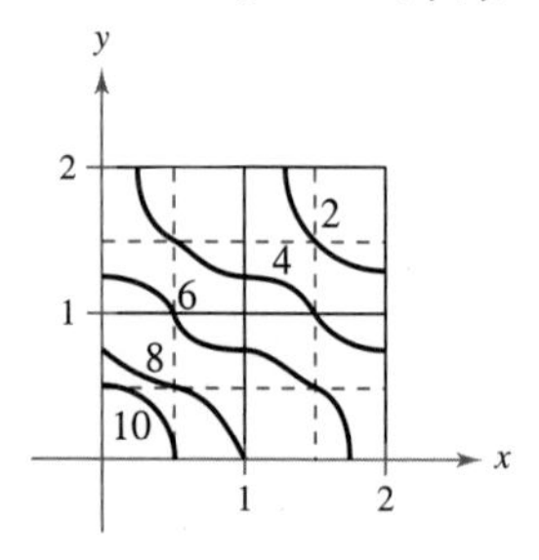

In Exercises 7–12, sketch the region R and evaluate the iterated integral $\int_R\int f(x, y)\, dA$.

7. $\int_0^2\int_0^1 (1 + 2x + 2y)\, dy\, dx$
8. $\int_0^{\pi}\int_0^{\pi/2} \sin^2 x \cos^2 y\, dy\, dx$
9. $\int_0^6\int_{y/2}^3 (x + y)\, dx\, dy$
10. $\int_0^4\int_{\frac{1}{2}y}^{\sqrt{y}} x^2y^2\, dx\, dy$
11. $\int_{-a}^{a}\int_{-\sqrt{a^2-x^2}}^{\sqrt{a^2-x^2}} (x + y)\, dy\, dx$
12. $\int_0^1\int_{y-1}^{0} e^{x+y}\, dx\, dy + \int_0^1\int_0^{1-y} e^{x+y}\, dx\, dy$

In Exercises 13–20, set up integrals for both orders of integration, and use the more convenient order to evaluate the integral over the region R.

13. $\displaystyle\int\!\!\int_R xy\,dA$

R: rectangle with vertices $(0, 0)$, $(0, 5)$, $(3, 5)$, $(3, 0)$

14. $\displaystyle\int\!\!\int_R \sin x \sin y\,dA$

R: rectangle with vertices $(-\pi, 0)$, $(\pi, 0)$, $(\pi, \pi/2)$, $(-\pi, \pi/2)$

15. $\displaystyle\int\!\!\int_R \frac{y}{x^2 + y^2}\,dA$

R: trapezoid bounded by $y = x$, $y = 2x$, $x = 1$, $x = 2$

16. $\displaystyle\int\!\!\int_R xe^y\,dA$

R: triangle bounded by $y = 4 - x$, $y = 0$, $x = 0$

17. $\displaystyle\int\!\!\int_R -2y\,dA$

R: region bounded by $y = 4 - x^2$, $y = 4 - x$

18. $\displaystyle\int\!\!\int_R \frac{y}{1 + x^2}\,dA$

R: region bounded by $y = 0$, $y = \sqrt{x}$, $x = 4$

19. $\displaystyle\int\!\!\int_R x\,dA$

R: sector of a circle in the first quadrant bounded by $y = \sqrt{25 - x^2}$, $3x - 4y = 0$, $y = 0$

20. $\displaystyle\int\!\!\int_R (x^2 + y^2)\,dA$

R: semicircle bounded by $y = \sqrt{4 - x^2}$, $y = 0$

In Exercises 21–30, use a double integral to find the volume of the indicated solid.

21.

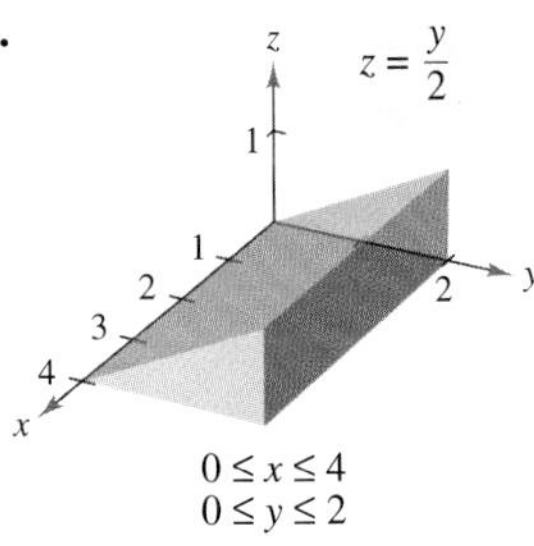

22.

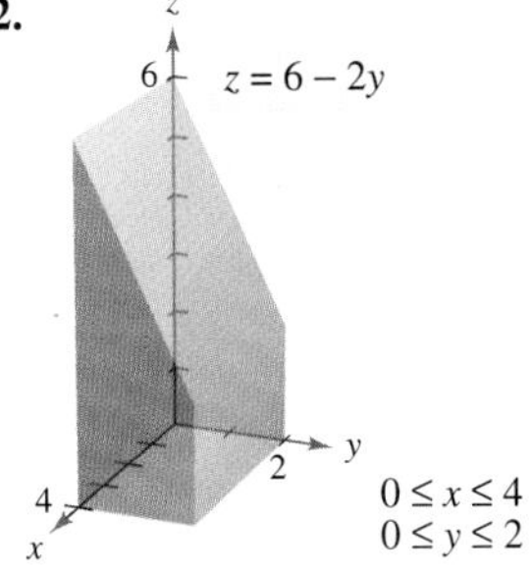

23.

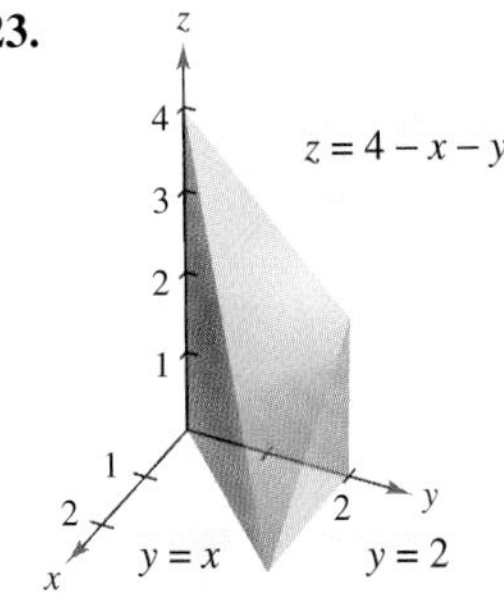

24.

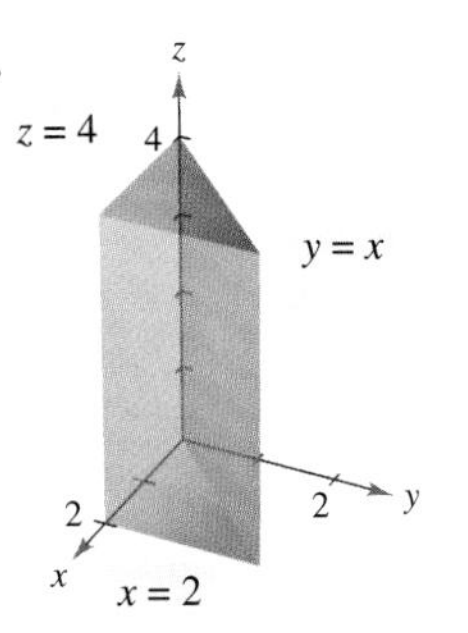

25.

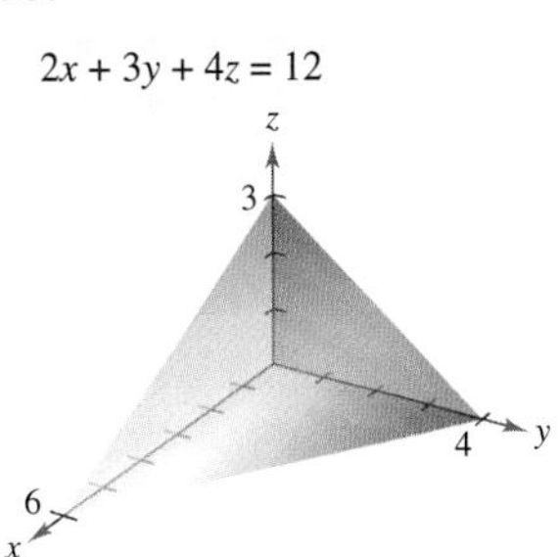

26.

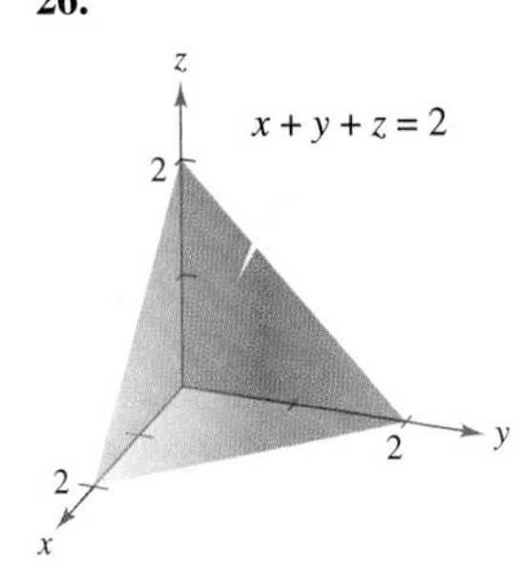

27.

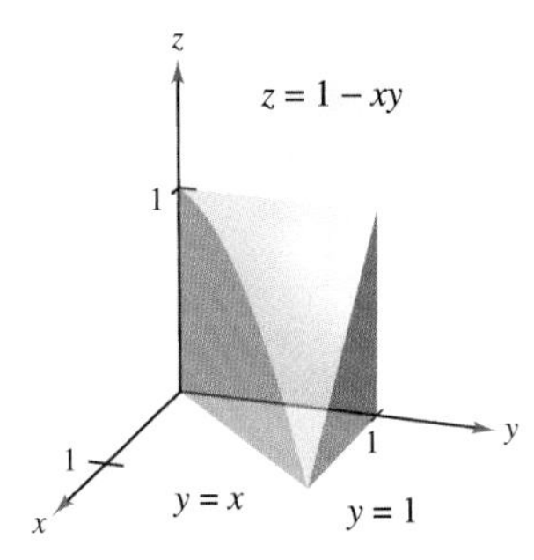

28.

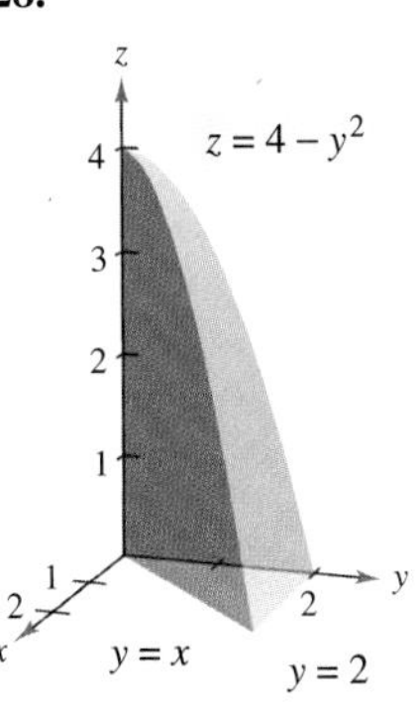

29. Improper integral

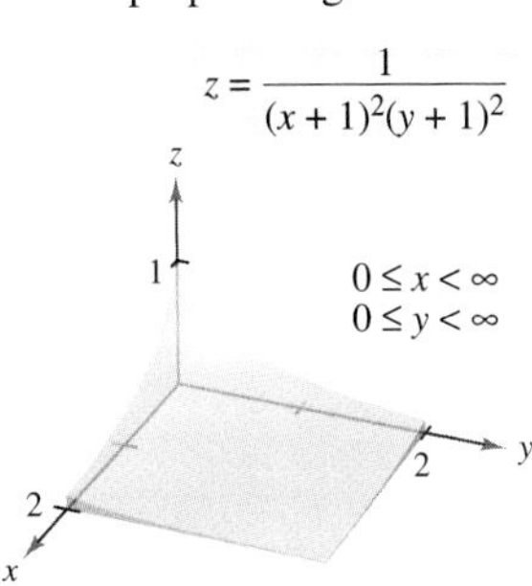

30. Improper integral

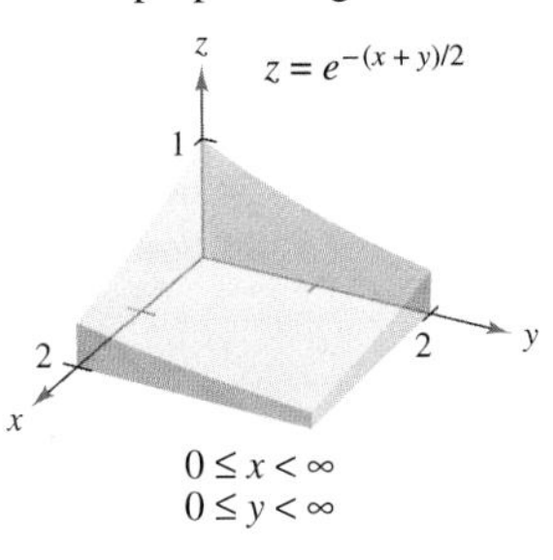

CAS **In Exercises 31 and 32, use a computer algebra system to find the volume of the solid.**

31.

$z = \sqrt{8 - x^2 - y^2}$

32.

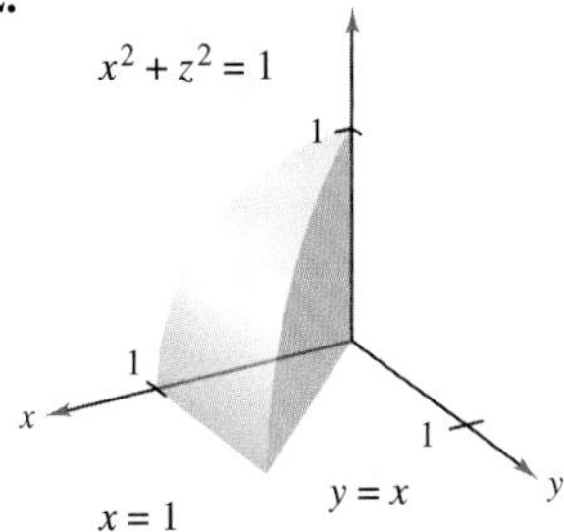

In Exercises 33–40, set up and evaluate a double integral to find the volume of the solid bounded by the graphs of the equations.

33. $z = xy$, $z = 0$, $y = x$, $x = 1$, first octant

34. $y = 0$, $z = 0$, $y = x$, $z = x$, $x = 0$, $x = 5$

35. $z = 0$, $z = x^2$, $x = 0$, $x = 2$, $y = 0$, $y = 4$

36. $x^2 + y^2 + z^2 = r^2$

37. $x^2 + z^2 = 1,\ y^2 + z^2 = 1,$ first octant

38. $y = 4 - x^2,\ z = 4 - x^2,$ first octant

39. $z = x + y,\ x^2 + y^2 = 4,$ first octant

40. $z = \dfrac{1}{1 + y^2},\ x = 0,\ x = 2,\ y \geq 0$

In Exercises 41–46, set up a double integral to find the volume of the solid region bounded by the graphs of the equations. Do not evaluate the integral.

41.

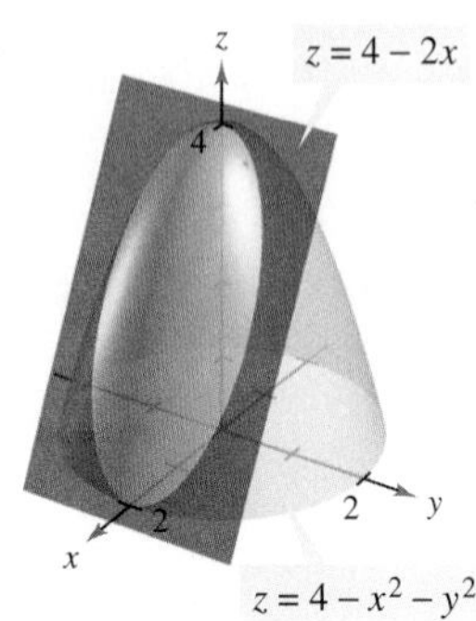

42.

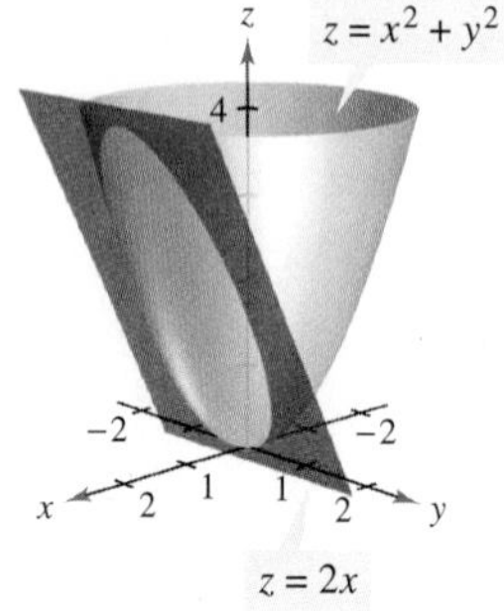

43. $z = x^2 + y^2,\ x^2 + y^2 = 4,\ z = 0$

44. $z = \sin^2 x,\ z = 0,\ 0 \leq x \leq \pi,\ 0 \leq y \leq 5$

45. $z = x^2 + 2y^2,\ z = 4y$

46. $z = x^2 + y^2,\ z = 18 - x^2 - y^2$

CAS **In Exercises 47–50, use a computer algebra system to find the volume of the solid bounded by the graphs of the equations.**

47. $z = 9 - x^2 - y^2,\ z = 0$

48. $x^2 = 9 - y,\ z^2 = 9 - y,$ first octant

49. $z = \dfrac{2}{1 + x^2 + y^2},\ z = 0,\ y = 0,\ x = 0,\ y = -0.5x + 1$

50. $z = \ln(1 + x + y),\ z = 0,\ y = 0,\ x = 0,\ x = 4 - \sqrt{y}$

51. If f is a continuous function such that $0 \leq f(x, y) \leq 1$ over a region R of area 1, prove that $0 \leq \int_R\int f(x, y)\, dA \leq 1$.

52. Find the volume of the solid in the first octant bounded by the coordinate planes and the plane $(x/a) + (y/b) + (z/c) = 1$, where $a > 0$, $b > 0$, and $c > 0$.

In Exercises 53–58, sketch the region of integration. Then evaluate the iterated integral, switching the order of integration if necessary.

53. $\displaystyle\int_0^1 \int_{y/2}^{1/2} e^{-x^2}\, dx\, dy$

54. $\displaystyle\int_0^{\ln 10} \int_{e^x}^{10} \frac{1}{\ln y}\, dy\, dx$

55. $\displaystyle\int_{-2}^{2} \int_{-\sqrt{4-x^2}}^{\sqrt{4-x^2}} \sqrt{4 - y^2}\, dy\, dx$

56. $\displaystyle\int_0^3 \int_{y/3}^{1} \frac{1}{1 + x^4}\, dx\, dy$

57. $\displaystyle\int_0^1 \int_0^{\arccos y} \sin x\sqrt{1 + \sin^2 x}\, dx\, dy$

58. $\displaystyle\int_0^2 \int_{(1/2)x^2}^{2} \sqrt{y} \cos y\, dy\, dx$

Average Value **In Exercises 59–64, find the average value of $f(x, y)$ over the region R.**

59. $f(x, y) = x$

R: rectangle with vertices $(0, 0), (4, 0), (4, 2), (0, 2)$

60. $f(x, y) = 2xy$

R: rectangle with vertices $(0, 0), (5, 0), (5, 3), (0, 3)$

61. $f(x, y) = x^2 + y^2$

R: square with vertices $(0, 0), (2, 0), (2, 2), (0, 2)$

62. $f(x, y) = \dfrac{1}{x + y}$

R: triangle with vertices $(0, 0), (1, 0), (1, 1)$

63. $f(x, y) = e^{x+y}$

R: triangle with vertices $(0, 0), (0, 1), (1, 1)$

64. $f(x, y) = \sin(x + y)$

R: rectangle with vertices $(0, 0), (\pi, 0), (\pi, \pi), (0, \pi)$

65. ***Average Production*** The Cobb-Douglas production function for an automobile manufacturer is $f(x, y) = 100x^{0.6}y^{0.4}$, where x is the number of units of labor and y is the number of units of capital. Estimate the average production level if the number of units of labor x varies between 200 and 250 and the number of units of capital y varies between 300 and 325.

66. ***Average Temperature*** The temperature in degrees Celsius on the surface of a metal plate is $T(x, y) = 20 - 4x^2 - y^2$, where x and y are measured in centimeters. Estimate the average temperature if x varies between 0 and 2 centimeters and y varies between 0 and 4 centimeters.

WRITING ABOUT CONCEPTS

67. State the definition of a double integral. If the integrand is a nonnegative function over the region of integration, give the geometric interpretation of a double integral.

68. Let R be a region in the xy-plane whose area is B. If $f(x, y) = k$ for every point (x, y) in R, what is the value of $\int_R\int f(x, y)\, dA$? Explain.

69. Let R represent a county in the northern part of the United States, and let $f(x, y)$ represent the total annual snowfall at the point (x, y) in R. Interpret each of the following.

(a) $\displaystyle\int_R\int f(x, y)\, dA$ (b) $\dfrac{\displaystyle\int_R\int f(x, y)\, dA}{\displaystyle\int_R\int dA}$

70. Identify the expression that is invalid. Explain your reasoning.

(a) $\displaystyle\int_0^2 \int_0^3 f(x, y)\, dy\, dx$ (b) $\displaystyle\int_0^2 \int_0^y f(x, y)\, dy\, dx$

(c) $\displaystyle\int_0^2 \int_x^3 f(x, y)\, dy\, dx$ (d) $\displaystyle\int_0^2 \int_0^x f(x, y)\, dy\, dx$

71. Let the plane region R be a unit circle and let the maximum value of f on R be 6. Is the greatest possible value of $\int_R\int f(x, y)\, dy\, dx$ equal to 6? Why or why not? If not, what is the greatest possible value?

CAPSTONE

72. The following iterated integrals represent the solution to the same problem. Which iterated integral is easier to evaluate? Explain your reasoning.

$$\int_0^4 \int_{x/2}^2 \sin y^2 \, dy \, dx = \int_0^2 \int_0^{2y} \sin y^2 \, dx \, dy$$

Probability **A joint density function of the continuous random variables x and y is a function $f(x, y)$ satisfying the following properties.**

(a) $f(x, y) \geq 0$ for all (x, y) **(b) $\int_{-\infty}^{\infty} \int_{-\infty}^{\infty} f(x, y) \, dA = 1$**

(c) $P[(x, y) \in R] = \int_R \int f(x, y) \, dA$

In Exercises 73–76, show that the function is a joint density function and find the required probability.

73. $f(x, y) = \begin{cases} \frac{1}{10}, & 0 \leq x \leq 5, \ 0 \leq y \leq 2 \\ 0, & \text{elsewhere} \end{cases}$

$P(0 \leq x \leq 2, \ 1 \leq y \leq 2)$

74. $f(x, y) = \begin{cases} \frac{1}{4}xy, & 0 \leq x \leq 2, \ 0 \leq y \leq 2 \\ 0, & \text{elsewhere} \end{cases}$

$P(0 \leq x \leq 1, \ 1 \leq y \leq 2)$

75. $f(x, y) = \begin{cases} \frac{1}{27}(9 - x - y), & 0 \leq x \leq 3, \ 3 \leq y \leq 6 \\ 0, & \text{elsewhere} \end{cases}$

$P(0 \leq x \leq 1, \ 4 \leq y \leq 6)$

76. $f(x, y) = \begin{cases} e^{-x-y}, & x \geq 0, \ y \geq 0 \\ 0, & \text{elsewhere} \end{cases}$

$P(0 \leq x \leq 1, \ x \leq y \leq 1)$

77. ***Approximation*** The base of a pile of sand at a cement plant is rectangular with approximate dimensions of 20 meters by 30 meters. If the base is placed on the xy-plane with one vertex at the origin, the coordinates on the surface of the pile are $(5, 5, 3)$, $(15, 5, 6)$, $(25, 5, 4)$, $(5, 15, 2)$, $(15, 15, 7)$, and $(25, 15, 3)$. Approximate the volume of sand in the pile.

78. ***Programming*** Consider a continuous function $f(x, y)$ over the rectangular region R with vertices (a, c), (b, c), (a, d), and (b, d), where $a < b$ and $c < d$. Partition the intervals $[a, b]$ and $[c, d]$ into m and n subintervals, so that the subintervals in a given direction are of equal length. Write a program for a graphing utility to compute the sum

$$\sum_{i=1}^{n} \sum_{j=1}^{m} f(x_i, y_j) \, \Delta A_i \approx \int_a^b \int_c^d f(x, y) \, dA$$

where (x_i, y_j) is the center of a representative rectangle in R.

CAS ***Approximation*** **In Exercises 79–82, (a) use a computer algebra system to approximate the iterated integral, and (b) use the program in Exercise 78 to approximate the iterated integral for the given values of m and n.**

79. $\int_0^1 \int_0^2 \sin \sqrt{x + y} \, dy \, dx$

$m = 4, \ n = 8$

80. $\int_0^2 \int_0^4 20e^{-x^3/8} dy \, dx$

$m = 10, \ n = 20$

81. $\int_4^6 \int_0^2 y \cos \sqrt{x} \, dx \, dy$

$m = 4, \ n = 8$

82. $\int_1^4 \int_1^2 \sqrt{x^3 + y^3} \, dx \, dy$

$m = 6, \ n = 4$

Approximation **In Exercises 83 and 84, determine which value best approximates the volume of the solid between the xy-plane and the function over the region. (Make your selection on the basis of a sketch of the solid and *not* by performing any calculations.)**

83. $f(x, y) = 4x$

R: square with vertices $(0, 0)$, $(4, 0)$, $(4, 4)$, $(0, 4)$

(a) -200 (b) 600 (c) 50 (d) 125 (e) 1000

84. $f(x, y) = \sqrt{x^2 + y^2}$

R: circle bounded by $x^2 + y^2 = 9$

(a) 50 (b) 500 (c) -500 (d) 5 (e) 5000

True or False? **In Exercises 85 and 86, determine whether the statement is true or false. If it is false, explain why or give an example that shows it is false.**

85. The volume of the sphere $x^2 + y^2 + z^2 = 1$ is given by the integral

$$V = 8 \int_0^1 \int_0^1 \sqrt{1 - x^2 - y^2} \, dx \, dy.$$

86. If $f(x, y) \leq g(x, y)$ for all (x, y) in R, and both f and g are continuous over R, then $\int_R \int f(x, y) \, dA \leq \int_R \int g(x, y) \, dA$.

87. Let $f(x) = \int_1^x e^{t^2} dt$. Find the average value of f on the interval $[0, 1]$.

88. Find $\int_0^\infty \frac{e^{-x} - e^{-2x}}{x} dx$. $\left(\textit{Hint:} \text{ Evaluate } \int_1^2 e^{-xy} \, dy.\right)$

89. Determine the region R in the xy-plane that maximizes the value of $\int_R \int (9 - x^2 - y^2) \, dA$.

90. Determine the region R in the xy-plane that minimizes the value of $\int_R \int (x^2 + y^2 - 4) \, dA$.

91. Find $\int_0^2 [\arctan(\pi x) - \arctan x] \, dx$. (*Hint:* Convert the integral to a double integral.)

92. Use a geometric argument to show that

$$\int_0^3 \int_0^{\sqrt{9-y^2}} \sqrt{9 - x^2 - y^2} \, dx \, dy = \frac{9\pi}{2}.$$

PUTNAM EXAM CHALLENGE

93. Evaluate $\int_0^a \int_0^b e^{\max\{b^2x^2, a^2y^2\}} \, dy \, dx$, where a and b are positive.

94. Show that if $\lambda > \frac{1}{2}$ there does not exist a real-valued function u such that for all x in the closed interval $0 \leq x \leq 1$, $u(x) = 1 + \lambda \int_x^1 u(y) u(y - x) \, dy$.

14.3 Change of Variables: Polar Coordinates

- **Write and evaluate double integrals in polar coordinates.**

Double Integrals in Polar Coordinates

Some double integrals are *much* easier to evaluate in polar form than in rectangular form. This is especially true for regions such as circles, cardioids, and rose curves, and for integrands that involve $x^2 + y^2$.

In Section 10.4, you learned that the polar coordinates (r, θ) of a point are related to the rectangular coordinates (x, y) of the point as follows.

$$x = r\cos\theta \quad \text{and} \quad y = r\sin\theta$$

$$r^2 = x^2 + y^2 \quad \text{and} \quad \tan\theta = \frac{y}{x}$$

EXAMPLE 1 Using Polar Coordinates to Describe a Region

Use polar coordinates to describe each region shown in Figure 14.24.

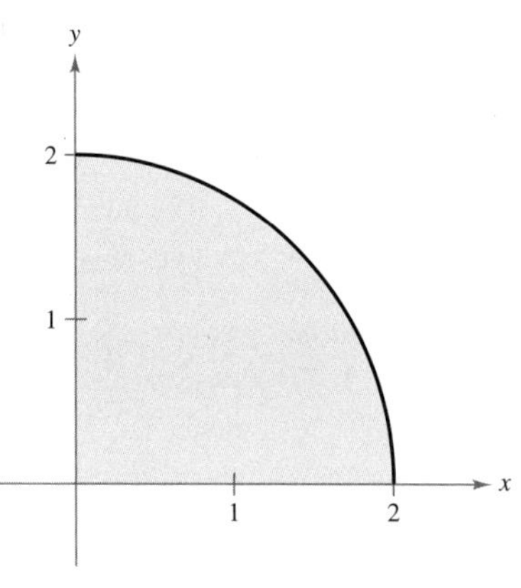

(a)

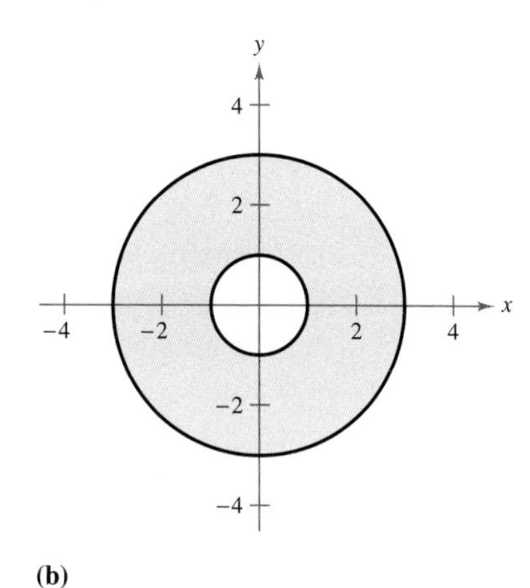

(b)

Figure 14.24

Solution

a. The region R is a quarter circle of radius 2. It can be described in polar coordinates as

$$R = \{(r, \theta): 0 \le r \le 2, \quad 0 \le \theta \le \pi/2\}.$$

b. The region R consists of all points between concentric circles of radii 1 and 3. It can be described in polar coordinates as

$$R = \{(r, \theta): 1 \le r \le 3, \quad 0 \le \theta \le 2\pi\}.$$ ■

The regions in Example 1 are special cases of **polar sectors**

$$R = \{(r, \theta): r_1 \le r \le r_2, \quad \theta_1 \le \theta \le \theta_2\} \qquad \text{Polar sector}$$

as shown in Figure 14.25.

Polar sector
Figure 14.25

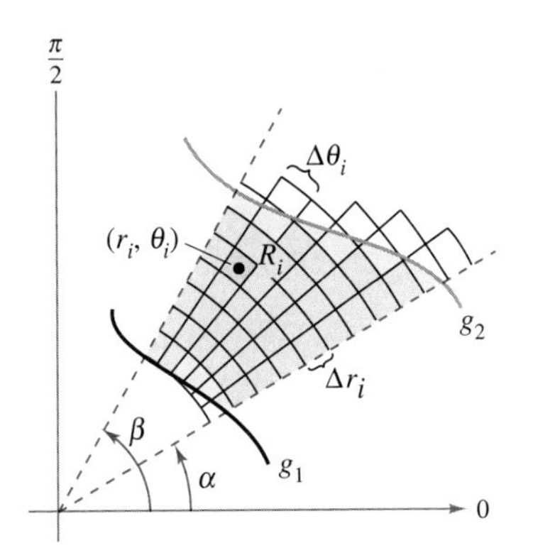

Polar grid superimposed over region R
Figure 14.26

To define a double integral of a continuous function $z = f(x, y)$ in polar coordinates, consider a region R bounded by the graphs of $r = g_1(\theta)$ and $r = g_2(\theta)$ and the lines $\theta = \alpha$ and $\theta = \beta$. Instead of partitioning R into small rectangles, use a partition of small polar sectors. On R, superimpose a polar grid made of rays and circular arcs, as shown in Figure 14.26. The polar sectors R_i lying entirely within R form an **inner polar partition** Δ, whose **norm** $\|\Delta\|$ is the length of the longest diagonal of the n polar sectors.

Consider a specific polar sector R_i, as shown in Figure 14.27. It can be shown (see Exercise 75) that the area of R_i is

$$\Delta A_i = r_i \Delta r_i \Delta\theta_i \qquad \text{Area of } R_i$$

where $\Delta r_i = r_2 - r_1$ and $\Delta\theta_i = \theta_2 - \theta_1$. This implies that the volume of the solid of height $f(r_i \cos\theta_i, r_i \sin\theta_i)$ above R_i is approximately

$$f(r_i \cos\theta_i, r_i \sin\theta_i) r_i \Delta r_i \Delta\theta_i$$

and you have

$$\int_R\int f(x, y)\, dA \approx \sum_{i=1}^{n} f(r_i \cos\theta_i, r_i \sin\theta_i) r_i \Delta r_i \Delta\theta_i.$$

The sum on the right can be interpreted as a Riemann sum for $f(r\cos\theta, r\sin\theta)r$. The region R corresponds to a *horizontally simple* region S in the $r\theta$-plane, as shown in Figure 14.28. The polar sectors R_i correspond to rectangles S_i, and the area ΔA_i of S_i is $\Delta r_i \Delta\theta_i$. So, the right-hand side of the equation corresponds to the double integral

$$\int_S\int f(r\cos\theta, r\sin\theta) r\, dA.$$

From this, you can apply Theorem 14.2 to write

$$\begin{aligned}\int_R\int f(x, y)\, dA &= \int_S\int f(r\cos\theta, r\sin\theta) r\, dA\\ &= \int_{\alpha}^{\beta}\int_{g_1(\theta)}^{g_2(\theta)} f(r\cos\theta, r\sin\theta) r\, dr\, d\theta.\end{aligned}$$

This suggests the following theorem, the proof of which is discussed in Section 14.8.

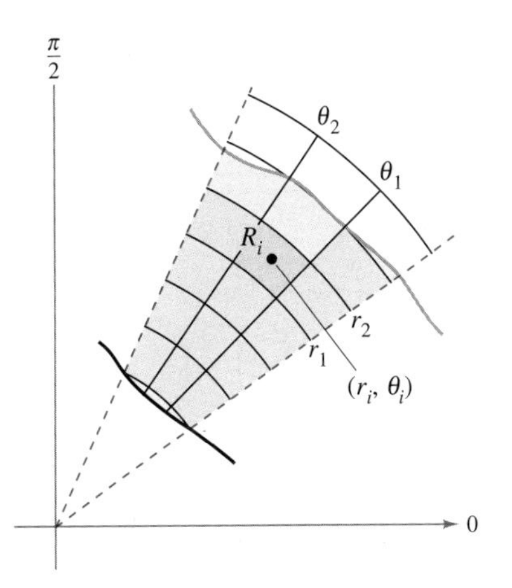

The polar sector R_i is the set of all points (r, θ) such that $r_1 \le r \le r_2$ and $\theta_1 \le \theta \le \theta_2$.
Figure 14.27

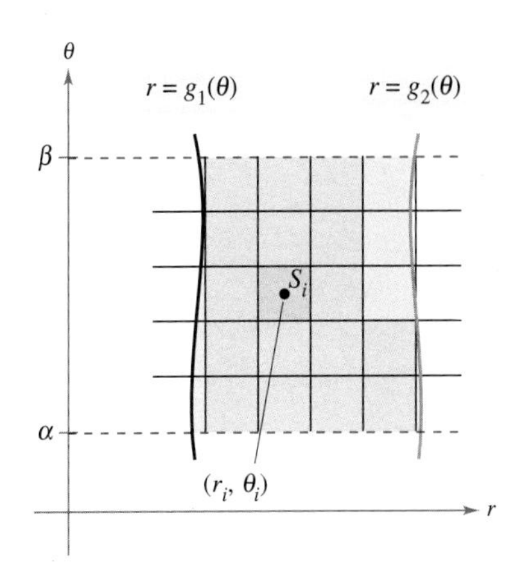

Horizontally simple region S
Figure 14.28

THEOREM 14.3 CHANGE OF VARIABLES TO POLAR FORM

Let R be a plane region consisting of all points $(x, y) = (r\cos\theta, r\sin\theta)$ satisfying the conditions $0 \le g_1(\theta) \le r \le g_2(\theta)$, $\alpha \le \theta \le \beta$, where $0 \le (\beta - \alpha) \le 2\pi$. If g_1 and g_2 are continuous on $[\alpha, \beta]$ and f is continuous on R, then

$$\int_R\int f(x, y)\,dA = \int_\alpha^\beta \int_{g_1(\theta)}^{g_2(\theta)} f(r\cos\theta, r\sin\theta)r\,dr\,d\theta.$$

EXPLORATION

Volume of a Paraboloid Sector
In the Exploration feature on page 997, you were asked to summarize the different ways you know of finding the volume of the solid bounded by the paraboloid

$$z = a^2 - x^2 - y^2, \quad a > 0$$

and the xy-plane. You now know another way. Use it to find the volume of the solid.

NOTE If $z = f(x, y)$ is nonnegative on R, then the integral in Theorem 14.3 can be interpreted as the volume of the solid region between the graph of f and the region R. When using the integral in Theorem 14.3, be certain not to omit the extra factor of r in the integrand. ■

The region R is restricted to two basic types, ***r*-simple** regions and ***θ*-simple** regions, as shown in Figure 14.29.

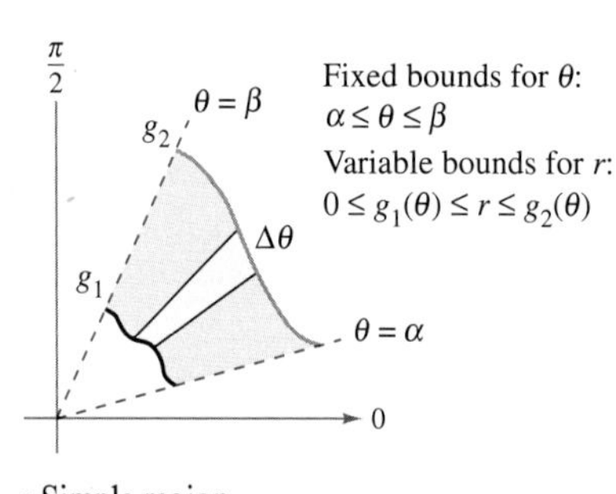

r-Simple region

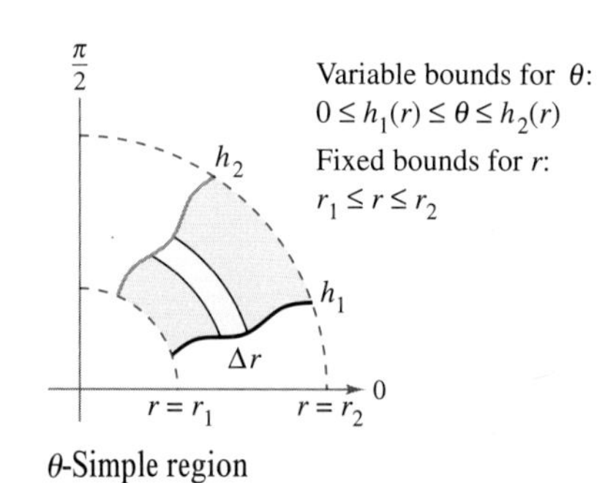

θ-Simple region

Figure 14.29

EXAMPLE 2 Evaluating a Double Polar Integral

Let R be the annular region lying between the two circles $x^2 + y^2 = 1$ and $x^2 + y^2 = 5$. Evaluate the integral $\int_R\int (x^2 + y)\,dA$.

Solution The polar boundaries are $1 \le r \le \sqrt{5}$ and $0 \le \theta \le 2\pi$, as shown in Figure 14.30. Furthermore, $x^2 = (r\cos\theta)^2$ and $y = r\sin\theta$. So, you have

$$\begin{aligned}
\int_R\int (x^2 + y)\,dA &= \int_0^{2\pi}\int_1^{\sqrt{5}} (r^2\cos^2\theta + r\sin\theta)r\,dr\,d\theta \\
&= \int_0^{2\pi}\int_1^{\sqrt{5}} (r^3\cos^2\theta + r^2\sin\theta)\,dr\,d\theta \\
&= \int_0^{2\pi} \left(\frac{r^4}{4}\cos^2\theta + \frac{r^3}{3}\sin\theta\right)\Big]_1^{\sqrt{5}}\,d\theta \\
&= \int_0^{2\pi} \left(6\cos^2\theta + \frac{5\sqrt{5} - 1}{3}\sin\theta\right)d\theta \\
&= \int_0^{2\pi} \left(3 + 3\cos 2\theta + \frac{5\sqrt{5} - 1}{3}\sin\theta\right)d\theta \\
&= \left(3\theta + \frac{3\sin 2\theta}{2} - \frac{5\sqrt{5} - 1}{3}\cos\theta\right)\Big]_0^{2\pi} \\
&= 6\pi.
\end{aligned}$$

■

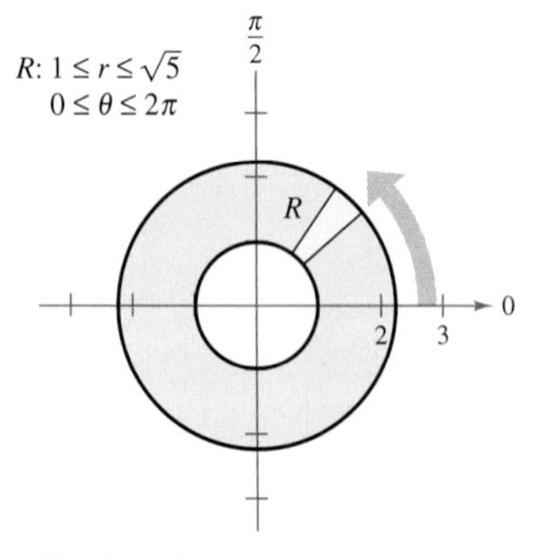

r-Simple region
Figure 14.30

In Example 2, be sure to notice the extra factor of r in the integrand. This comes from the formula for the area of a polar sector. In differential notation, you can write

$$dA = r\,dr\,d\theta$$

which indicates that the area of a polar sector increases as you move away from the origin.

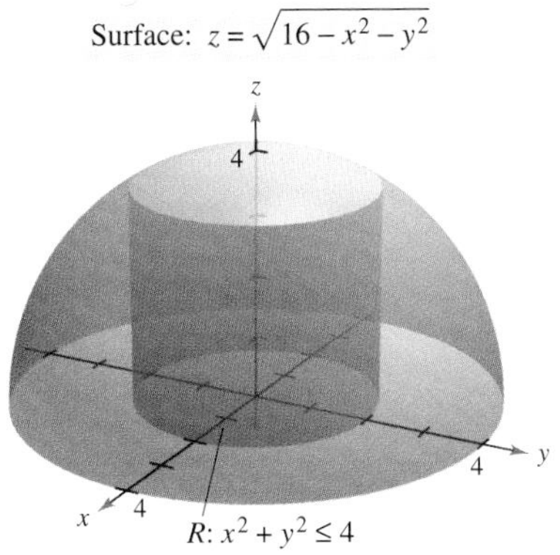

Surface: $z = \sqrt{16 - x^2 - y^2}$

R: $x^2 + y^2 \le 4$

Figure 14.31

EXAMPLE 3 Change of Variables to Polar Coordinates

Use polar coordinates to find the volume of the solid region bounded above by the hemisphere

$$z = \sqrt{16 - x^2 - y^2} \qquad \text{Hemisphere forms upper surface.}$$

and below by the circular region R given by

$$x^2 + y^2 \le 4 \qquad \text{Circular region forms lower surface.}$$

as shown in Figure 14.31.

Solution In Figure 14.31, you can see that R has the bounds

$$-\sqrt{4 - y^2} \le x \le \sqrt{4 - y^2}, \quad -2 \le y \le 2$$

and that $0 \le z \le \sqrt{16 - x^2 - y^2}$. In polar coordinates, the bounds are

$$0 \le r \le 2 \quad \text{and} \quad 0 \le \theta \le 2\pi$$

with height $z = \sqrt{16 - x^2 - y^2} = \sqrt{16 - r^2}$. Consequently, the volume V is given by

$$\begin{aligned} V = \int_R\!\!\int f(x, y)\,dA &= \int_0^{2\pi}\!\!\int_0^2 \sqrt{16 - r^2}\,r\,dr\,d\theta \\ &= -\frac{1}{3}\int_0^{2\pi} (16 - r^2)^{3/2}\Big]_0^2\,d\theta \\ &= -\frac{1}{3}\int_0^{2\pi} \left(24\sqrt{3} - 64\right) d\theta \\ &= -\frac{8}{3}\left(3\sqrt{3} - 8\right)\theta\Big]_0^{2\pi} \\ &= \frac{16\pi}{3}\left(8 - 3\sqrt{3}\right) \approx 46.979. \end{aligned}$$

■

NOTE To see the benefit of polar coordinates in Example 3, you should try to evaluate the corresponding rectangular iterated integral

$$\int_{-2}^{2}\int_{-\sqrt{4-y^2}}^{\sqrt{4-y^2}} \sqrt{16 - x^2 - y^2}\,dx\,dy.$$

TECHNOLOGY Any computer algebra system that can handle double integrals in rectangular coordinates can also handle double integrals in polar coordinates. The reason this is true is that once you have formed the iterated integral, its value is not changed by using different variables. In other words, if you use a computer algebra system to evaluate

$$\int_0^{2\pi}\!\!\int_0^2 \sqrt{16 - x^2}\,x\,dx\,dy$$

you should obtain the same value as that obtained in Example 3.

Just as with rectangular coordinates, the double integral

$$\int_R\!\!\int dA$$

can be used to find the area of a region in the plane.

EXAMPLE 4 Finding Areas of Polar Regions

Use a double integral to find the area enclosed by the graph of $r = 3\cos 3\theta$.

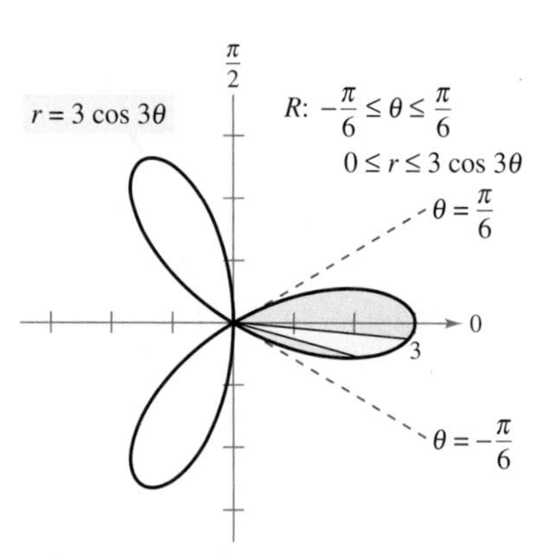

Figure 14.32

Solution Let R be one petal of the curve shown in Figure 14.32. This region is r-simple, and the boundaries are as follows.

$$-\frac{\pi}{6} \le \theta \le \frac{\pi}{6} \qquad \text{Fixed bounds on } \theta$$

$$0 \le r \le 3\cos 3\theta \qquad \text{Variable bounds on } r$$

So, the area of one petal is

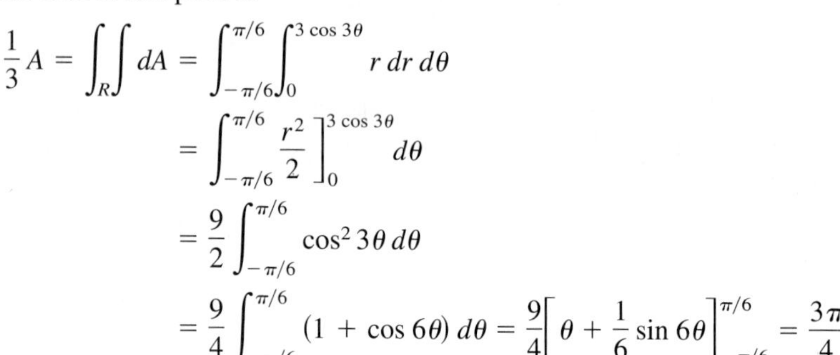

$$\frac{1}{3}A = \iint_R dA = \int_{-\pi/6}^{\pi/6}\int_0^{3\cos 3\theta} r\,dr\,d\theta$$

$$= \int_{-\pi/6}^{\pi/6} \left[\frac{r^2}{2}\right]_0^{3\cos 3\theta} d\theta$$

$$= \frac{9}{2}\int_{-\pi/6}^{\pi/6} \cos^2 3\theta\,d\theta$$

$$= \frac{9}{4}\int_{-\pi/6}^{\pi/6} (1 + \cos 6\theta)\,d\theta = \frac{9}{4}\left[\theta + \frac{1}{6}\sin 6\theta\right]_{-\pi/6}^{\pi/6} = \frac{3\pi}{4}.$$

So, the total area is $A = 9\pi/4$. ■

As illustrated in Example 4, the area of a region in the plane can be represented by

$$A = \int_\alpha^\beta \int_{g_1(\theta)}^{g_2(\theta)} r\,dr\,d\theta.$$

If $g_1(\theta) = 0$, you obtain

$$A = \int_\alpha^\beta \int_0^{g_2(\theta)} r\,dr\,d\theta = \int_\alpha^\beta \left[\frac{r^2}{2}\right]_0^{g_2(\theta)} d\theta = \int_\alpha^\beta \frac{1}{2}(g_2(\theta))^2\,d\theta$$

which agrees with Theorem 10.13.

So far in this section, all of the examples of iterated integrals in polar form have been of the form

$$\int_\alpha^\beta \int_{g_1(\theta)}^{g_2(\theta)} f(r\cos\theta, r\sin\theta)r\,dr\,d\theta$$

in which the order of integration is with respect to r first. Sometimes you can obtain a simpler integration problem by switching the order of integration, as illustrated in the next example.

EXAMPLE 5 Changing the Order of Integration

Find the area of the region bounded above by the spiral $r = \pi/(3\theta)$ and below by the polar axis, between $r = 1$ and $r = 2$.

Solution The region is shown in Figure 14.33. The polar boundaries for the region are

$$1 \le r \le 2 \quad \text{and} \quad 0 \le \theta \le \frac{\pi}{3r}.$$

So, the area of the region can be evaluated as follows.

$$A = \int_1^2 \int_0^{\pi/(3r)} r\,d\theta\,dr = \int_1^2 \left[r\theta\right]_0^{\pi/(3r)} dr = \int_1^2 \frac{\pi}{3}\,dr = \left[\frac{\pi r}{3}\right]_1^2 = \frac{\pi}{3}$$

■

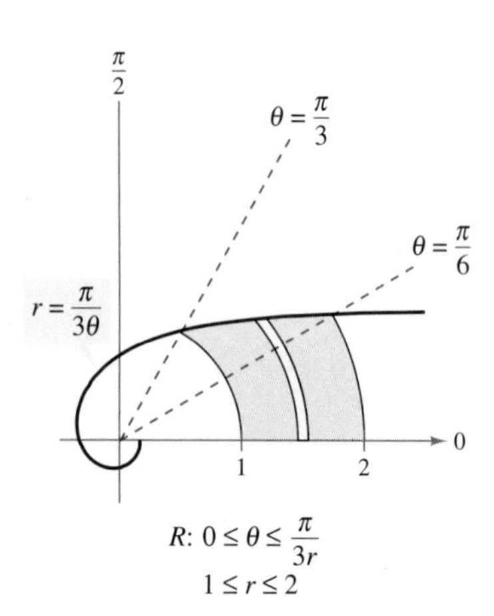

θ-Simple region
Figure 14.33

14.3 Exercises

See www.CalcChat.com for worked-out solutions to odd-numbered exercises.

In Exercises 1–4, the region R for the integral $\int_R\int f(x, y)\, dA$ is shown. State whether you would use rectangular or polar coordinates to evaluate the integral.

1.

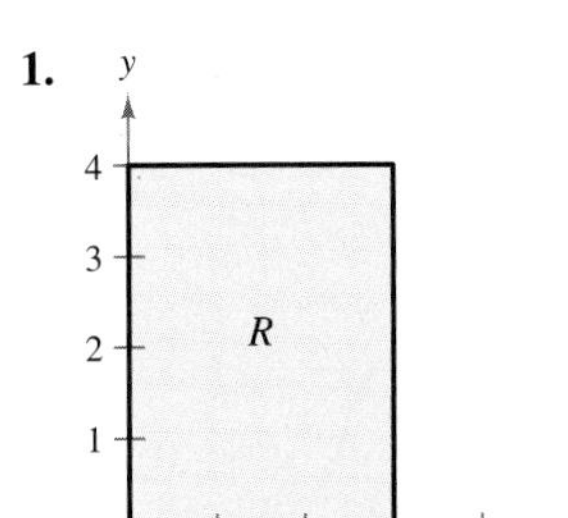

2.

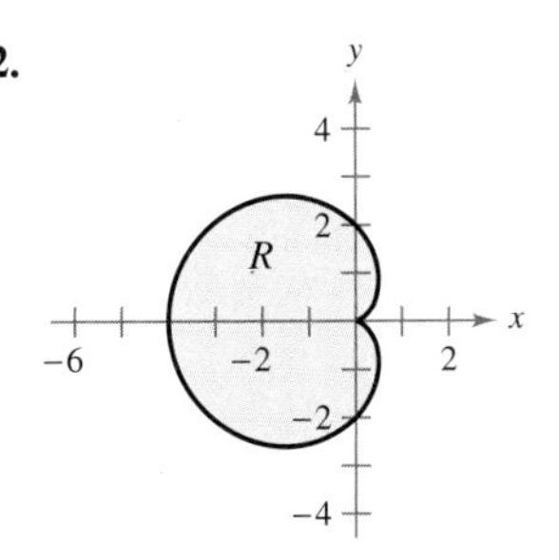

3.

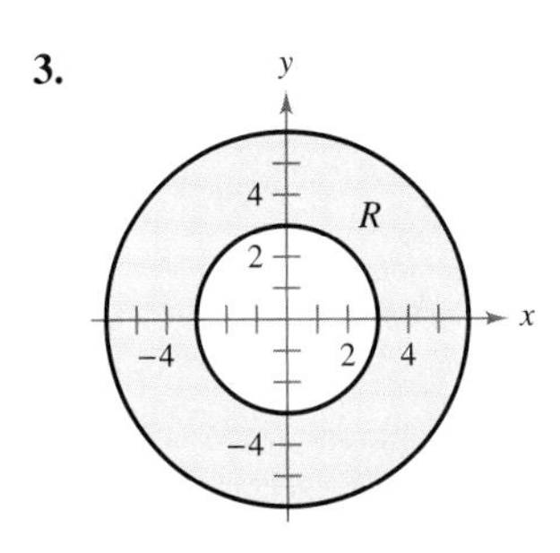

4.

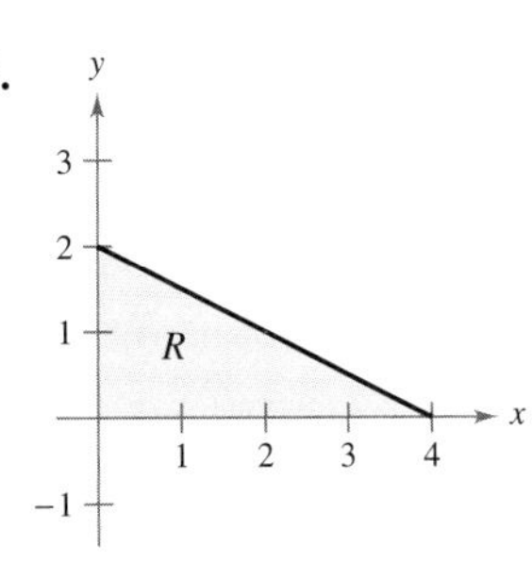

In Exercises 5–8, use polar coordinates to describe the region shown.

5.

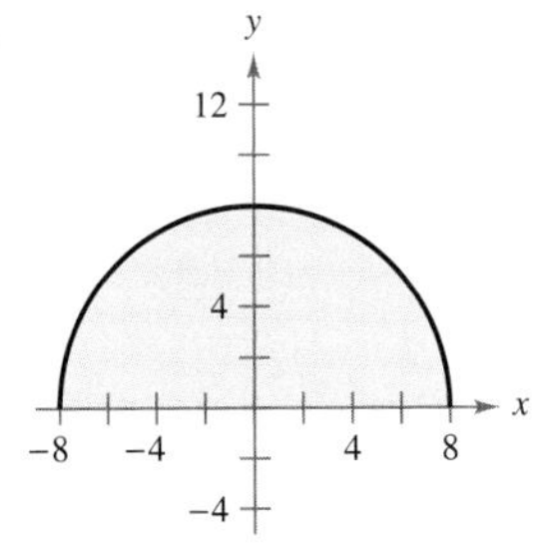

6.

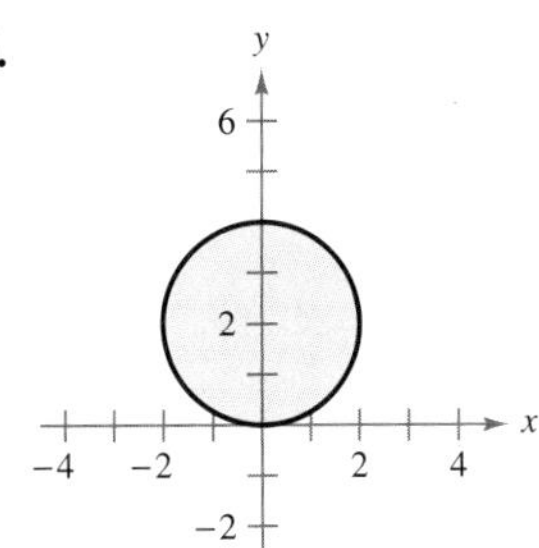

7.

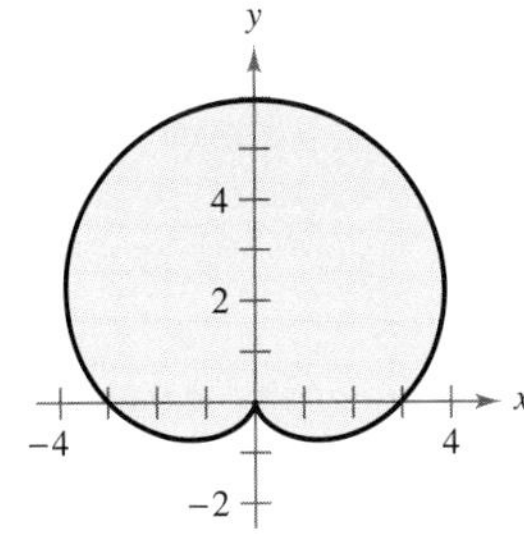

8.

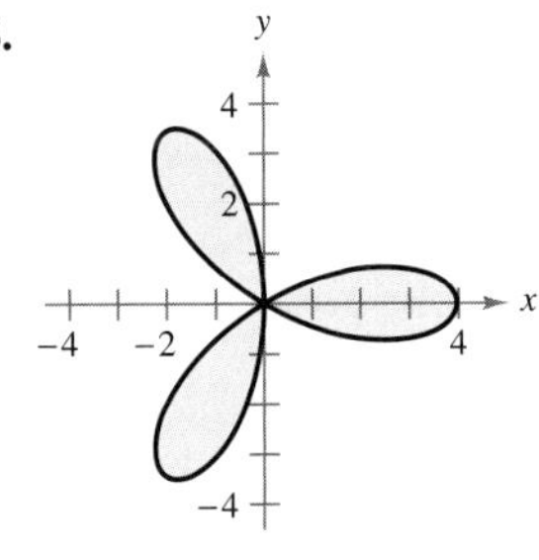

In Exercises 9–16, evaluate the double integral $\int_R\int f(r, \theta)\, dA$, and sketch the region R.

9. $\displaystyle\int_0^{\pi}\int_0^{\cos\theta} r\, dr\, d\theta$

10. $\displaystyle\int_0^{\pi}\int_0^{\sin\theta} r^2\, dr\, d\theta$

11. $\displaystyle\int_0^{2\pi}\int_0^{6} 3r^2 \sin\theta\, dr\, d\theta$

12. $\displaystyle\int_0^{\pi/4}\int_0^{4} r^2 \sin\theta\cos\theta\, dr\, d\theta$

13. $\displaystyle\int_0^{\pi/2}\int_2^{3} \sqrt{9 - r^2}\, r\, dr\, d\theta$

14. $\displaystyle\int_0^{\pi/2}\int_0^{3} re^{-r^2}\, dr\, d\theta$

15. $\displaystyle\int_0^{\pi/2}\int_0^{1+\sin\theta} \theta r\, dr\, d\theta$

16. $\displaystyle\int_0^{\pi/2}\int_0^{1-\cos\theta} (\sin\theta) r\, dr\, d\theta$

In Exercises 17–26, evaluate the iterated integral by converting to polar coordinates.

17. $\displaystyle\int_0^{a}\int_0^{\sqrt{a^2-y^2}} y\, dx\, dy$

18. $\displaystyle\int_0^{a}\int_0^{\sqrt{a^2-x^2}} x\, dy\, dx$

19. $\displaystyle\int_{-2}^{2}\int_0^{\sqrt{4-x^2}} (x^2 + y^2)\, dy\, dx$

20. $\displaystyle\int_0^{1}\int_{-\sqrt{x-x^2}}^{\sqrt{x-x^2}} (x^2 + y^2)\, dy\, dx$

21. $\displaystyle\int_0^{3}\int_0^{\sqrt{9-x^2}} (x^2 + y^2)^{3/2}\, dy\, dx$

22. $\displaystyle\int_0^{2}\int_y^{\sqrt{8-y^2}} \sqrt{x^2 + y^2}\, dx\, dy$

23. $\displaystyle\int_0^{2}\int_0^{\sqrt{2x-x^2}} xy\, dy\, dx$

24. $\displaystyle\int_0^{4}\int_0^{\sqrt{4y-y^2}} x^2\, dx\, dy$

25. $\displaystyle\int_{-1}^{1}\int_0^{\sqrt{1-x^2}} \cos(x^2 + y^2)\, dy\, dx$

26. $\displaystyle\int_0^{2}\int_0^{\sqrt{4-x^2}} \sin\sqrt{x^2 + y^2}\, dy\, dx$

In Exercises 27 and 28, combine the sum of the two iterated integrals into a single iterated integral by converting to polar coordinates. Evaluate the resulting iterated integral.

27. $\displaystyle\int_0^{2}\int_0^{x} \sqrt{x^2 + y^2}\, dy\, dx + \int_2^{2\sqrt{2}}\int_0^{\sqrt{8-x^2}} \sqrt{x^2 + y^2}\, dy\, dx$

28. $\displaystyle\int_0^{5\sqrt{2}/2}\int_0^{x} xy\, dy\, dx + \int_{5\sqrt{2}/2}^{5}\int_0^{\sqrt{25-x^2}} xy\, dy\, dx$

In Exercises 29–32, use polar coordinates to set up and evaluate the double integral $\int_R\int f(x, y)\, dA$.

29. $f(x, y) = x + y$, R: $x^2 + y^2 \le 4$, $x \ge 0$, $y \ge 0$

30. $f(x, y) = e^{-(x^2+y^2)/2}$, R: $x^2 + y^2 \le 25$, $x \ge 0$

31. $f(x, y) = \arctan\dfrac{y}{x}$, R: $x^2 + y^2 \ge 1$, $x^2 + y^2 \le 4$, $0 \le y \le x$

32. $f(x, y) = 9 - x^2 - y^2$, R: $x^2 + y^2 \le 9$, $x \ge 0$, $y \ge 0$

Volume **In Exercises 33–38, use a double integral in polar coordinates to find the volume of the solid bounded by the graphs of the equations.**

33. $z = xy$, $x^2 + y^2 = 1$, first octant

34. $z = x^2 + y^2 + 3$, $z = 0$, $x^2 + y^2 = 1$

35. $z = \sqrt{x^2 + y^2}$, $z = 0$, $x^2 + y^2 = 25$

36. $z = \ln(x^2 + y^2)$, $z = 0$, $x^2 + y^2 \ge 1$, $x^2 + y^2 \le 4$

37. Inside the hemisphere $z = \sqrt{16 - x^2 - y^2}$ and inside the cylinder $x^2 + y^2 - 4x = 0$

38. Inside the hemisphere $z = \sqrt{16 - x^2 - y^2}$ and outside the cylinder $x^2 + y^2 = 1$

39. ***Volume*** Find a such that the volume inside the hemisphere $z = \sqrt{16 - x^2 - y^2}$ and outside the cylinder $x^2 + y^2 = a^2$ is one-half the volume of the hemisphere.

40. ***Volume*** Use a double integral in polar coordinates to find the volume of a sphere of radius a.

41. ***Volume*** Determine the diameter of a hole that is drilled vertically through the center of the solid bounded by the graphs of the equations $z = 25e^{-(x^2+y^2)/4}$, $z = 0$, and $x^2 + y^2 = 16$ if one-tenth of the volume of the solid is removed.

CAS **42.** ***Machine Design*** The surfaces of a double-lobed cam are modeled by the inequalities $\frac{1}{4} \leq r \leq \frac{1}{2}(1 + \cos^2 \theta)$ and

$$\frac{-9}{4(x^2 + y^2 + 9)} \leq z \leq \frac{9}{4(x^2 + y^2 + 9)}$$

where all measurements are in inches.

(a) Use a computer algebra system to graph the cam.

(b) Use a computer algebra system to approximate the perimeter of the polar curve

$$r = \tfrac{1}{2}(1 + \cos^2 \theta).$$

This is the distance a roller must travel as it runs against the cam through one revolution of the cam.

(c) Use a computer algebra system to find the volume of steel in the cam.

Area **In Exercises 43–48, use a double integral to find the area of the shaded region.**

43.

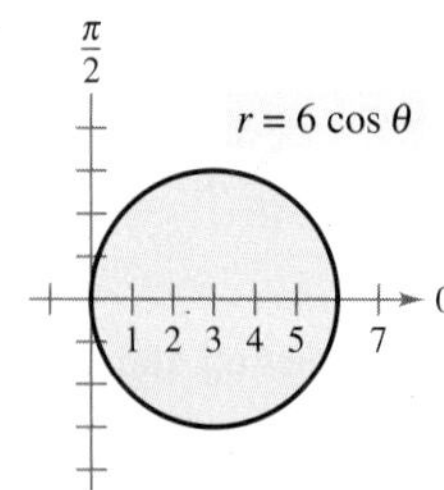

44.

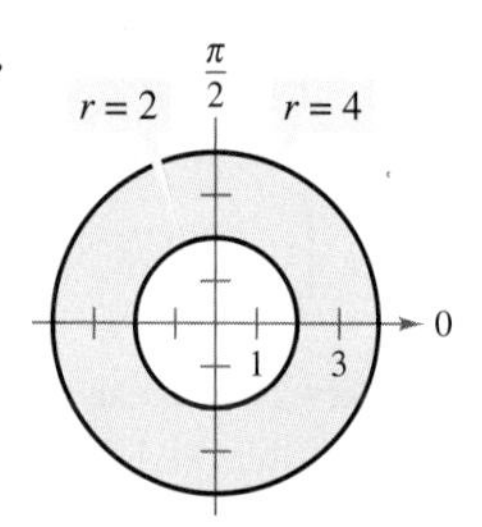

45.

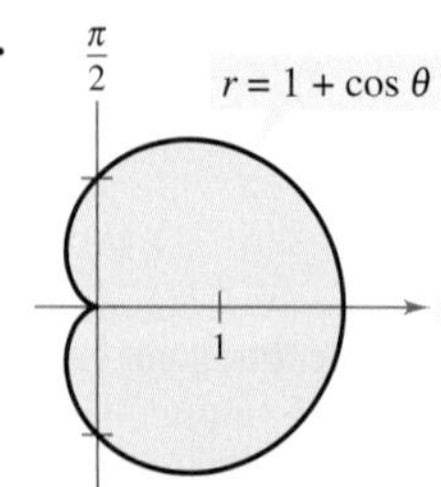

46.

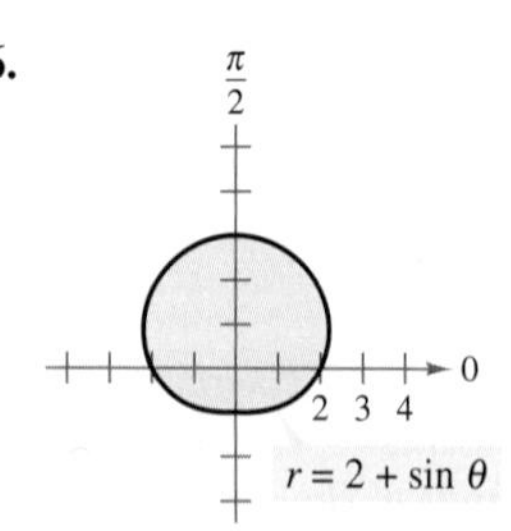

47.

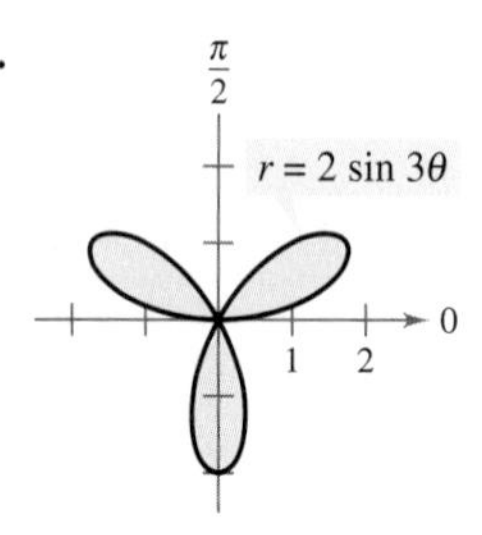

48.

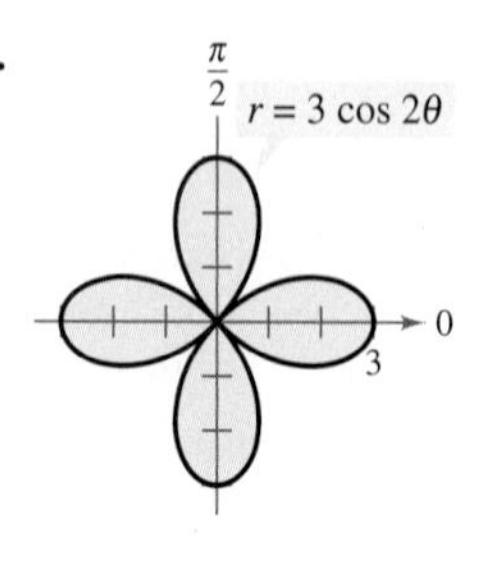

Area **In Exercises 49–54, sketch a graph of the region bounded by the graphs of the equations. Then use a double integral to find the area of the region.**

49. Inside the circle $r = 2 \cos \theta$ and outside the circle $r = 1$

50. Inside the cardioid $r = 2 + 2 \cos \theta$ and outside the circle $r = 1$

51. Inside the circle $r = 3 \cos \theta$ and outside the cardioid $r = 1 + \cos \theta$

52. Inside the cardioid $r = 1 + \cos \theta$ and outside the circle $r = 3 \cos \theta$

53. Inside the rose curve $r = 4 \sin 3\theta$ and outside the circle $r = 2$

54. Inside the circle $r = 2$ and outside the cardioid $r = 2 - 2 \cos \theta$

WRITING ABOUT CONCEPTS

55. Describe the partition of the region R of integration in the xy-plane when polar coordinates are used to evaluate a double integral.

56. Explain how to change from rectangular coordinates to polar coordinates in a double integral.

57. In your own words, describe r-simple regions and θ-simple regions.

58. Each figure shows a region of integration for the double integral $\int_R\int f(x, y)\, dA$. For each region, state whether horizontal representative elements, vertical representative elements, or polar sectors would yield the easiest method for obtaining the limits of integration. Explain your reasoning.

(a) (b)

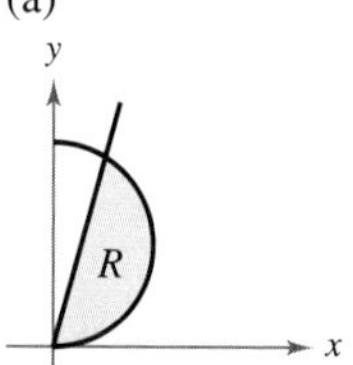

(c)

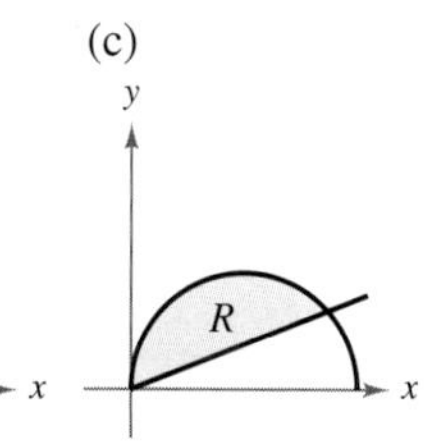

59. Let R be the region bounded by the circle $x^2 + y^2 = 9$.

(a) Set up the integral $\int_R\int f(x, y)\, dA$.

(b) Convert the integral in part (a) to polar coordinates.

(c) Which integral would you choose to evaluate? Why?

CAPSTONE

60. ***Think About It*** Without performing any calculations, identify the double integral that represents the integral of $f(x) = x^2 + y^2$ over a circle of radius 4. Explain your reasoning.

(a) $\displaystyle\int_0^{2\pi}\int_0^4 r^2\, dr\, d\theta$

(b) $\displaystyle\int_0^4\int_0^{2\pi} r^3\, dr\, d\theta$

(c) $\displaystyle\int_0^{2\pi}\int_0^4 r^3\, dr\, d\theta$

(d) $\displaystyle\int_0^{2\pi}\int_{-4}^4 r^3\, dr\, d\theta$

61. ***Think About It*** Consider the program you wrote to approximate double integrals in rectangular coordinates in Exercise 78, in Section 14.2. If the program is used to approximate the double integral

$$\int_R\int f(r, \theta)\, dA$$

in polar coordinates, how will you modify f when it is entered into the program? Because the limits of integration are constants, describe the plane region of integration.

62. ***Approximation*** Horizontal cross sections of a piece of ice that broke from a glacier are in the shape of a quarter of a circle with a radius of approximately 50 feet. The base is divided into 20 subregions, as shown in the figure. At the center of each subregion, the height of the ice is measured, yielding the following points in cylindrical coordinates.

$\left(5, \frac{\pi}{16}, 7\right), \left(15, \frac{\pi}{16}, 8\right), \left(25, \frac{\pi}{16}, 10\right), \left(35, \frac{\pi}{16}, 12\right), \left(45, \frac{\pi}{16}, 9\right),$
$\left(5, \frac{3\pi}{16}, 9\right), \left(15, \frac{3\pi}{16}, 10\right), \left(25, \frac{3\pi}{16}, 14\right), \left(35, \frac{3\pi}{16}, 15\right), \left(45, \frac{3\pi}{16}, 10\right),$
$\left(5, \frac{5\pi}{16}, 9\right), \left(15, \frac{5\pi}{16}, 11\right), \left(25, \frac{5\pi}{16}, 15\right), \left(35, \frac{5\pi}{16}, 18\right), \left(45, \frac{5\pi}{16}, 14\right),$
$\left(5, \frac{7\pi}{16}, 5\right), \left(15, \frac{7\pi}{16}, 8\right), \left(25, \frac{7\pi}{16}, 11\right), \left(35, \frac{7\pi}{16}, 16\right), \left(45, \frac{7\pi}{16}, 12\right)$

(a) Approximate the volume of the solid.

(b) Ice weighs approximately 57 pounds per cubic foot. Approximate the weight of the solid.

(c) There are 7.48 gallons of water per cubic foot. Approximate the number of gallons of water in the solid.

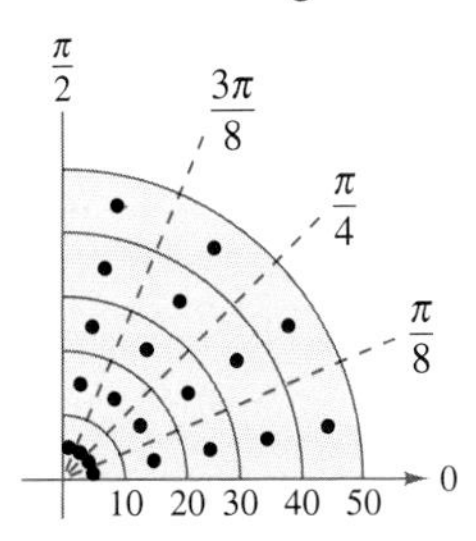

CAS ***Approximation*** **In Exercises 63 and 64, use a computer algebra system to approximate the iterated integral.**

63. $\displaystyle\int_{\pi/4}^{\pi/2}\int_0^5 r\sqrt{1 + r^3}\sin\sqrt{\theta}\, dr\, d\theta$

64. $\displaystyle\int_0^{\pi/4}\int_0^4 5re^{\sqrt{r\theta}}\, dr\, d\theta$

Approximation **In Exercises 65 and 66, determine which value best approximates the volume of the solid between the *xy*-plane and the function over the region. (Make your selection on the basis of a sketch of the solid and *not* by performing any calculations.)**

65. $f(x, y) = 15 - 2y$; R: semicircle: $x^2 + y^2 = 16, y \geq 0$

(a) 100 (b) 200 (c) 300 (d) -200 (e) 800

66. $f(x, y) = xy + 2$; R: quarter circle: $x^2 + y^2 = 9, x \geq 0, y \geq 0$

(a) 25 (b) 8 (c) 100 (d) 50 (e) -30

True or False? **In Exercises 67 and 68, determine whether the statement is true or false. If it is false, explain why or give an example that shows it is false.**

67. If $\int_R\int f(r, \theta)\, dA > 0$, then $f(r, \theta) > 0$ for all (r, θ) in R.

68. If $f(r, \theta)$ is a constant function and the area of the region S is twice that of the region R, then $2\int_R\int f(r, \theta)\, dA = \int_S\int f(r, \theta)\, dA$.

69. ***Probability*** The value of the integral $I = \int_{-\infty}^{\infty} e^{-x^2/2}\, dx$ is required in the development of the normal probability density function.

(a) Use polar coordinates to evaluate the improper integral.

$$I^2 = \left(\int_{-\infty}^{\infty} e^{-x^2/2}\, dx\right)\left(\int_{-\infty}^{\infty} e^{-y^2/2}\, dy\right)$$
$$= \int_{-\infty}^{\infty}\int_{-\infty}^{\infty} e^{-(x^2+y^2)/2}\, dA$$

(b) Use the result of part (a) to determine I.

FOR FURTHER INFORMATION For more information on this problem, see the article "Integrating e^{-x^2} Without Polar Coordinates" by William Dunham in *Mathematics Teacher*. To view this article, go to the website *www.matharticles.com*.

70. Use the result of Exercise 69 and a change of variables to evaluate each integral. No integration is required.

(a) $\displaystyle\int_{-\infty}^{\infty} e^{-x^2}\, dx$ (b) $\displaystyle\int_{-\infty}^{\infty} e^{-4x^2}\, dx$

71. ***Population*** The population density of a city is approximated by the model $f(x, y) = 4000e^{-0.01(x^2+y^2)}$, $x^2 + y^2 \leq 49$, where x and y are measured in miles. Integrate the density function over the indicated circular region to approximate the population of the city.

72. ***Probability*** Find k such that the function

$$f(x, y) = \begin{cases} ke^{-(x^2+y^2)}, & x \geq 0, y \geq 0 \\ 0, & \text{elsewhere} \end{cases}$$

is a probability density function.

73. ***Think About It*** Consider the region bounded by the graphs of $y = 2$, $y = 4$, $y = x$, and $y = \sqrt{3}x$ and the double integral $\int_R\int f\, dA$. Determine the limits of integration if the region R is divided into (a) horizontal representative elements, (b) vertical representative elements, and (c) polar sectors.

74. Repeat Exercise 73 for a region R bounded by the graph of the equation $(x - 2)^2 + y^2 = 4$.

75. Show that the area A of the polar sector R (see figure) is $A = r\Delta r\Delta\theta$, where $r = (r_1 + r_2)/2$ is the average radius of R.

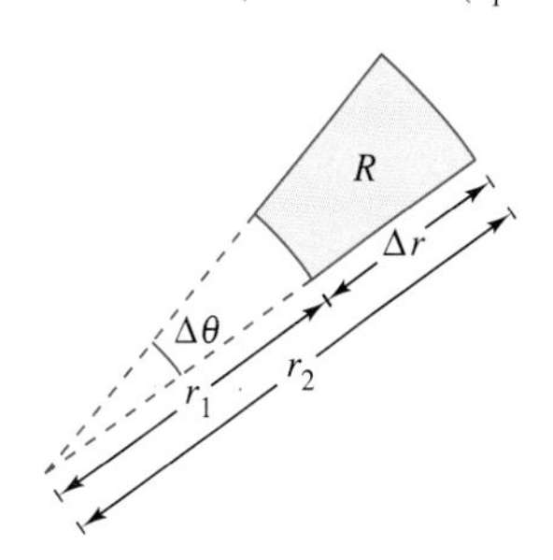

14.4 Center of Mass and Moments of Inertia

- **Find the mass of a planar lamina using a double integral.**
- **Find the center of mass of a planar lamina using double integrals.**
- **Find moments of inertia using double integrals.**

Mass

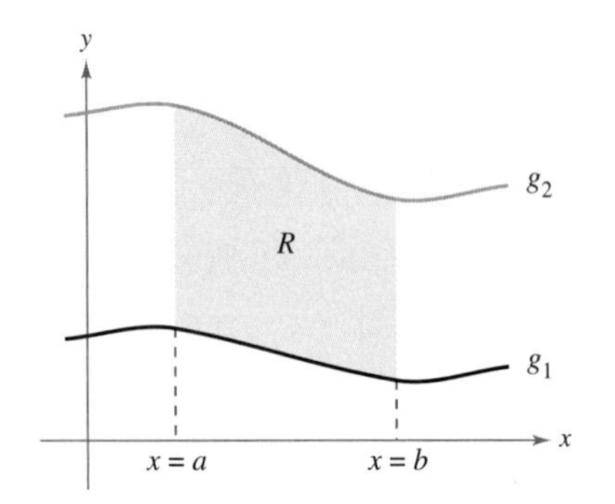

Lamina of constant density ρ
Figure 14.34

Section 7.6 discussed several applications of integration involving a lamina of *constant* density ρ. For example, if the lamina corresponding to the region R, as shown in Figure 14.34, has a constant density ρ, then the mass of the lamina is given by

$$\text{Mass} = \rho A = \rho \int_R \int dA = \int_R \int \rho \, dA. \quad \text{Constant density}$$

If not otherwise stated, a lamina is assumed to have a constant density. In this section, however, you will extend the definition of the term *lamina* to include thin plates of *variable* density. Double integrals can be used to find the mass of a lamina of variable density, where the density at (x, y) is given by the **density function ρ.**

DEFINITION OF MASS OF A PLANAR LAMINA OF VARIABLE DENSITY

If ρ is a continuous density function on the lamina corresponding to a plane region R, then the mass m of the lamina is given by

$$m = \int_R \int \rho(x, y) \, dA. \quad \text{Variable density}$$

NOTE Density is normally expressed as mass per unit volume. For a planar lamina, however, density is mass per unit surface area. ■

EXAMPLE 1 Finding the Mass of a Planar Lamina

Find the mass of the triangular lamina with vertices $(0, 0)$, $(0, 3)$, and $(2, 3)$, given that the density at (x, y) is $\rho(x, y) = 2x + y$.

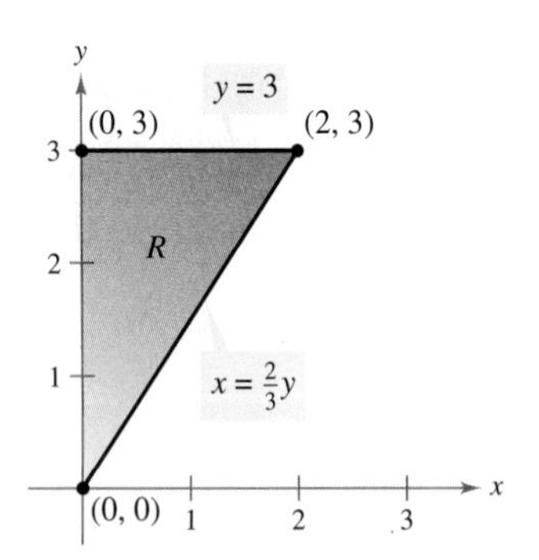

Lamina of variable density $\rho(x, y) = 2x + y$
Figure 14.35

Solution As shown in Figure 14.35, region R has the boundaries $x = 0$, $y = 3$, and $y = 3x/2$ (or $x = 2y/3$). Therefore, the mass of the lamina is

$$\begin{aligned}
m = \int_R \int (2x + y) \, dA &= \int_0^3 \int_0^{2y/3} (2x + y) \, dx \, dy \\
&= \int_0^3 \Big[x^2 + xy \Big]_0^{2y/3} dy \\
&= \frac{10}{9} \int_0^3 y^2 \, dy \\
&= \frac{10}{9} \left[\frac{y^3}{3} \right]_0^3 \\
&= 10.
\end{aligned}$$

■

NOTE In Figure 14.35, note that the planar lamina is shaded so that the darkest shading corresponds to the densest part. ■

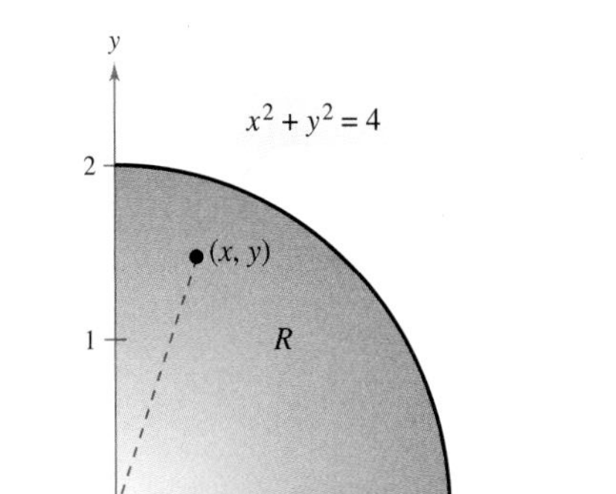

Density at (x, y): $\rho(x, y) = k\sqrt{x^2 + y^2}$
Figure 14.36

EXAMPLE 2 Finding Mass by Polar Coordinates

Find the mass of the lamina corresponding to the first-quadrant portion of the circle

$$x^2 + y^2 = 4$$

where the density at the point (x, y) is proportional to the distance between the point and the origin, as shown in Figure 14.36.

Solution At any point (x, y), the density of the lamina is

$$\begin{aligned}\rho(x, y) &= k\sqrt{(x-0)^2 + (y-0)^2}\\ &= k\sqrt{x^2 + y^2}.\end{aligned}$$

Because $0 \leq x \leq 2$ and $0 \leq y \leq \sqrt{4 - x^2}$, the mass is given by

$$\begin{aligned}m &= \int_R\!\!\int k\sqrt{x^2 + y^2}\,dA\\ &= \int_0^2\int_0^{\sqrt{4-x^2}} k\sqrt{x^2 + y^2}\,dy\,dx.\end{aligned}$$

To simplify the integration, you can convert to polar coordinates, using the bounds $0 \leq \theta \leq \pi/2$ and $0 \leq r \leq 2$. So, the mass is

$$\begin{aligned}m = \int_R\!\!\int k\sqrt{x^2 + y^2}\,dA &= \int_0^{\pi/2}\int_0^2 k\sqrt{r^2}\,r\,dr\,d\theta\\ &= \int_0^{\pi/2}\int_0^2 kr^2\,dr\,d\theta\\ &= \int_0^{\pi/2} \frac{kr^3}{3}\Big]_0^2\,d\theta\\ &= \frac{8k}{3}\int_0^{\pi/2} d\theta\\ &= \frac{8k}{3}\Big[\theta\Big]_0^{\pi/2}\\ &= \frac{4\pi k}{3}.\end{aligned}$$

■

TECHNOLOGY On many occasions, this text has mentioned the benefits of computer programs that perform symbolic integration. Even if you use such a program regularly, you should remember that its greatest benefit comes only in the hands of a knowledgeable user. For instance, notice how much simpler the integral in Example 2 becomes when it is converted to polar form.

Rectangular Form	*Polar Form*
$\int_0^2\int_0^{\sqrt{4-x^2}} k\sqrt{x^2 + y^2}\,dy\,dx$	$\int_0^{\pi/2}\int_0^2 kr^2\,dr\,d\theta$

If you have access to software that performs symbolic integration, use it to evaluate both integrals. Some software programs cannot handle the first integral, but any program that can handle double integrals can evaluate the second integral.

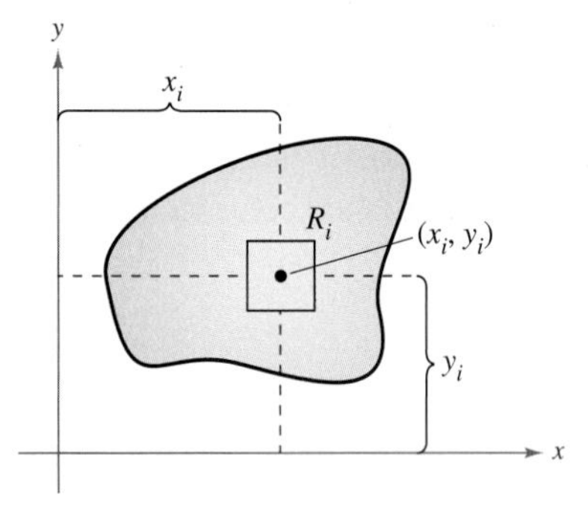

$M_x = (\text{mass})(y_i)$
$M_y = (\text{mass})(x_i)$
Figure 14.37

Moments and Center of Mass

For a lamina of variable density, moments of mass are defined in a manner similar to that used for the uniform density case. For a partition Δ of a lamina corresponding to a plane region R, consider the ith rectangle R_i of one area ΔA_i, as shown in Figure 14.37. Assume that the mass of R_i is concentrated at one of its interior points (x_i, y_i). The moment of mass of R_i with respect to the x-axis can be approximated by

$$(\text{Mass})(y_i) \approx [\rho(x_i, y_i)\,\Delta A_i](y_i).$$

Similarly, the moment of mass with respect to the y-axis can be approximated by

$$(\text{Mass})(x_i) \approx [\rho(x_i, y_i)\,\Delta A_i](x_i).$$

By forming the Riemann sum of all such products and taking the limits as the norm of Δ approaches 0, you obtain the following definitions of moments of mass with respect to the x- and y-axes.

MOMENTS AND CENTER OF MASS OF A VARIABLE DENSITY PLANAR LAMINA

Let ρ be a continuous density function on the planar lamina R. The **moments of mass** with respect to the x- and y-axes are

$$M_x = \int_R\!\!\int y\rho(x, y)\,dA \quad \text{and} \quad M_y = \int_R\!\!\int x\rho(x, y)\,dA.$$

If m is the mass of the lamina, then the **center of mass** is

$$(\bar{x}, \bar{y}) = \left(\frac{M_y}{m}, \frac{M_x}{m}\right).$$

If R represents a simple plane region rather than a lamina, the point $(\bar{x}, \bar{y})$ is called the **centroid** of the region.

For some planar laminas with a constant density ρ, you can determine the center of mass (or one of its coordinates) using symmetry rather than using integration. For instance, consider the laminas of constant density shown in Figure 14.38. Using symmetry, you can see that $\bar{y} = 0$ for the first lamina and $\bar{x} = 0$ for the second lamina.

R: $0 \le x \le 1$
$-\sqrt{1-x^2} \le y \le \sqrt{1-x^2}$

Lamina of constant density that is symmetric with respect to the x-axis

R: $-\sqrt{1-y^2} \le x \le \sqrt{1-y^2}$
$0 \le y \le 1$

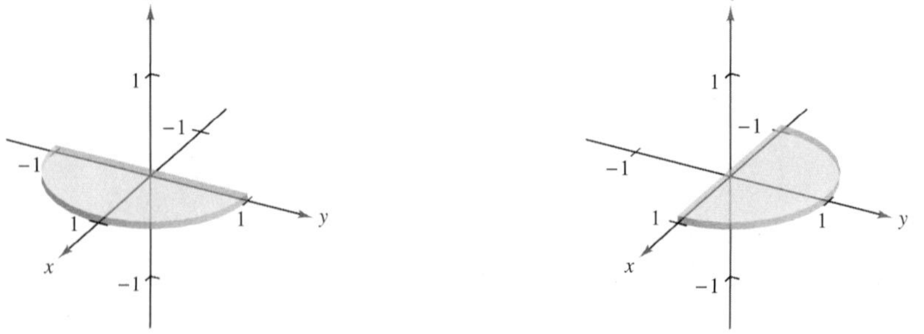

Lamina of constant density that is symmetric with respect to the y-axis

Figure 14.38

EXAMPLE 3 Finding the Center of Mass

Find the center of mass of the lamina corresponding to the parabolic region

$$0 \le y \le 4 - x^2 \qquad \text{Parabolic region}$$

where the density at the point (x, y) is proportional to the distance between (x, y) and the x-axis, as shown in Figure 14.39.

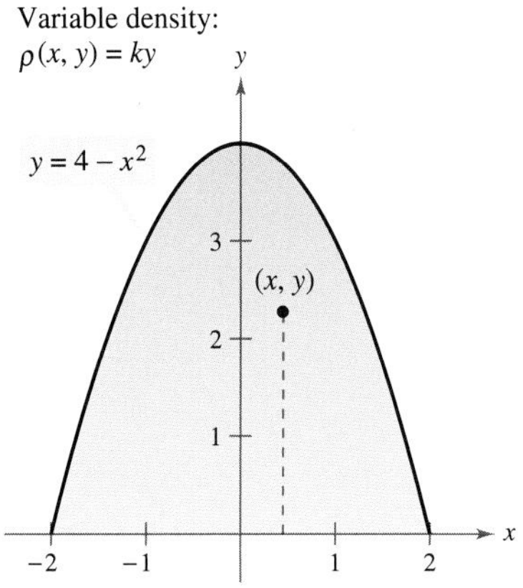

Parabolic region of variable density
Figure 14.39

Solution Because the lamina is symmetric with respect to the y-axis and

$$\rho(x, y) = ky$$

the center of mass lies on the y-axis. So, $\bar{x} = 0$. To find $\bar{y}$, first find the mass of the lamina.

$$\begin{aligned}
\text{Mass} = \int_{-2}^{2}\int_{0}^{4-x^2} ky\,dy\,dx &= \frac{k}{2}\int_{-2}^{2} y^2\Big]_0^{4-x^2}\,dx \\
&= \frac{k}{2}\int_{-2}^{2} (16 - 8x^2 + x^4)\,dx \\
&= \frac{k}{2}\left[16x - \frac{8x^3}{3} + \frac{x^5}{5}\right]_{-2}^{2} \\
&= k\left(32 - \frac{64}{3} + \frac{32}{5}\right) \\
&= \frac{256k}{15}
\end{aligned}$$

Next, find the moment about the x-axis.

$$\begin{aligned}
M_x = \int_{-2}^{2}\int_{0}^{4-x^2} (y)(ky)\,dy\,dx &= \frac{k}{3}\int_{-2}^{2} y^3\Big]_0^{4-x^2}\,dx \\
&= \frac{k}{3}\int_{-2}^{2} (64 - 48x^2 + 12x^4 - x^6)\,dx \\
&= \frac{k}{3}\left[64x - 16x^3 + \frac{12x^5}{5} - \frac{x^7}{7}\right]_{-2}^{2} \\
&= \frac{4096k}{105}
\end{aligned}$$

So,

$$\bar{y} = \frac{M_x}{m} = \frac{4096k/105}{256k/15} = \frac{16}{7}$$

and the center of mass is $\left(0, \frac{16}{7}\right)$. ■

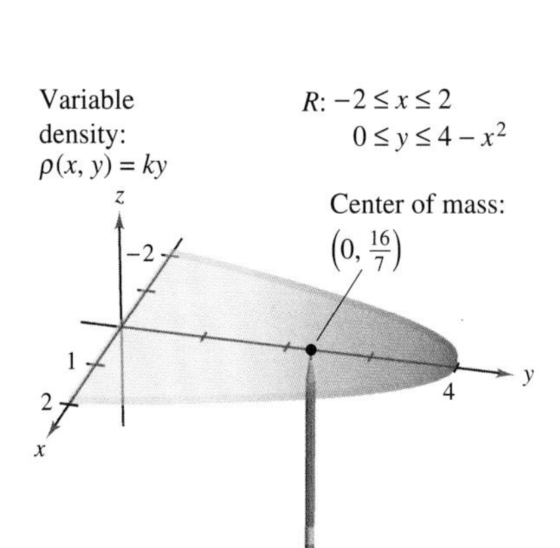

Figure 14.40

Although you can think of the moments M_x and M_y as measuring the tendency to rotate about the x- or y-axis, the calculation of moments is usually an intermediate step toward a more tangible goal. The use of the moments M_x and M_y is typical—to find the center of mass. Determination of the center of mass is useful in a variety of applications that allow you to treat a lamina as if its mass were concentrated at just one point. Intuitively, you can think of the center of mass as the balancing point of the lamina. For instance, the lamina in Example 3 should balance on the point of a pencil placed at $\left(0, \frac{16}{7}\right)$, as shown in Figure 14.40.

Moments of Inertia

The moments of M_x and M_y used in determining the center of mass of a lamina are sometimes called the **first moments** about the x- and y-axes. In each case, the moment is the product of a mass times a distance.

$$M_x = \int_R\int \underbrace{(y)}_{\text{Distance to } x\text{-axis}}\underbrace{\rho(x, y)\, dA}_{\text{Mass}} \qquad M_y = \int_R\int \underbrace{(x)}_{\text{Distance to } y\text{-axis}}\underbrace{\rho(x, y)\, dA}_{\text{Mass}}$$

You will now look at another type of moment—the **second moment,** or the **moment of inertia** of a lamina about a line. In the same way that mass is a measure of the tendency of matter to resist a change in straight-line motion, the moment of inertia about a line is a *measure of the tendency of matter to resist a change in rotational motion.* For example, if a particle of mass m is a distance d from a fixed line, its moment of inertia about the line is defined as

$$I = md^2 = (\text{mass})(\text{distance})^2.$$

As with moments of mass, you can generalize this concept to obtain the moments of inertia about the x- and y-axes of a lamina of variable density. These second moments are denoted by I_x and I_y, and in each case the moment is the product of a mass times the square of a distance.

$$I_x = \int_R\int \underbrace{(y^2)}_{\text{Square of distance to } x\text{-axis}}\underbrace{\rho(x, y)\, dA}_{\text{Mass}} \qquad I_y = \int_R\int \underbrace{(x^2)}_{\text{Square of distance to } y\text{-axis}}\underbrace{\rho(x, y)\, dA}_{\text{Mass}}$$

The sum of the moments I_x and I_y is called the **polar moment of inertia** and is denoted by I_0.

NOTE For a lamina in the xy-plane, I_0 represents the moment of inertia of the lamina about the z-axis. The term "polar moment of inertia" stems from the fact that the square of the polar distance r is used in the calculation.

$$I_0 = \int_R\int (x^2 + y^2)\rho(x, y)\, dA = \int_R\int r^2\rho(x, y)\, dA$$

EXAMPLE 4 Finding the Moment of Inertia

Find the moment of inertia about the x-axis of the lamina in Example 3.

Solution From the definition of moment of inertia, you have

$$\begin{aligned} I_x &= \int_{-2}^{2}\int_{0}^{4-x^2} y^2(ky)\, dy\, dx \\ &= \frac{k}{4}\int_{-2}^{2} y^4\Big]_0^{4-x^2} dx \\ &= \frac{k}{4}\int_{-2}^{2} (256 - 256x^2 + 96x^4 - 16x^6 + x^8)\, dx \\ &= \frac{k}{4}\left[256x - \frac{256x^3}{3} + \frac{96x^5}{5} - \frac{16x^7}{7} + \frac{x^9}{9}\right]_{-2}^{2} \\ &= \frac{32{,}768k}{315}. \end{aligned}$$

■

The moment of inertia I of a revolving lamina can be used to measure its kinetic energy. For example, suppose a planar lamina is revolving about a line with an **angular speed** of ω radians per second, as shown in Figure 14.41. The kinetic energy E of the revolving lamina is

$$E = \frac{1}{2}I\omega^2. \qquad \text{Kinetic energy for rotational motion}$$

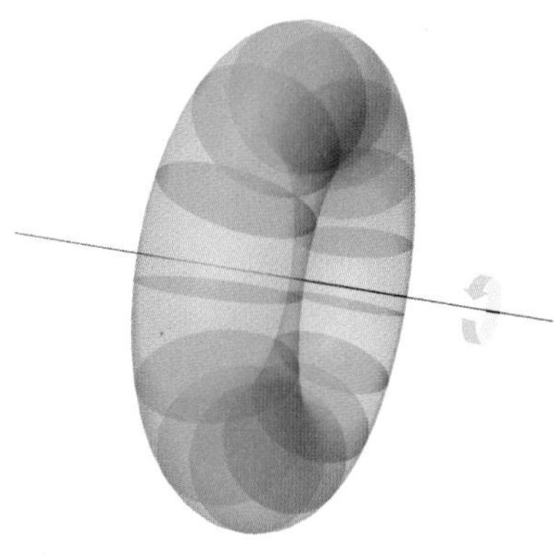

Planar lamina revolving at ω radians per second
Figure 14.41

On the other hand, the kinetic energy E of a mass m moving in a straight line at a velocity v is

$$E = \frac{1}{2}mv^2. \qquad \text{Kinetic energy for linear motion}$$

So, the kinetic energy of a mass moving in a straight line is proportional to its mass, but the kinetic energy of a mass revolving about an axis is proportional to its moment of inertia.

The **radius of gyration** $\bar{\bar{r}}$ of a revolving mass m with moment of inertia I is defined as

$$\bar{\bar{r}} = \sqrt{\frac{I}{m}}. \qquad \text{Radius of gyration}$$

If the entire mass were located at a distance $\bar{\bar{r}}$ from its axis of revolution, it would have the same moment of inertia and, consequently, the same kinetic energy. For instance, the radius of gyration of the lamina in Example 4 about the x-axis is given by

$$\bar{\bar{y}} = \sqrt{\frac{I_x}{m}} = \sqrt{\frac{32{,}768k/315}{256k/15}} = \sqrt{\frac{128}{21}} \approx 2.469.$$

EXAMPLE 5 Finding the Radius of Gyration

Find the radius of gyration about the y-axis for the lamina corresponding to the region R: $0 \le y \le \sin x$, $0 \le x \le \pi$, where the density at (x, y) is given by $\rho(x, y) = x$.

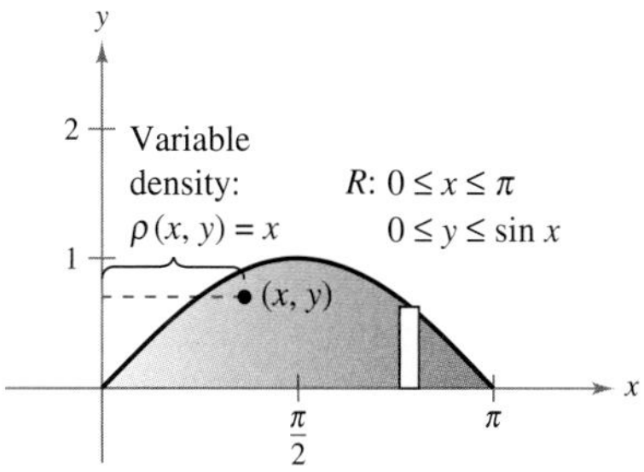

Solution The region R is shown in Figure 14.42. By integrating $\rho(x, y) = x$ over the region R, you can determine that the mass of the region is π. The moment of inertia about the y-axis is

$$\begin{aligned}
I_y &= \int_0^{\pi}\int_0^{\sin x} x^3\,dy\,dx \\
&= \int_0^{\pi} x^3 y\Big]_0^{\sin x}\,dx \\
&= \int_0^{\pi} x^3 \sin x\,dx \\
&= \Big[(3x^2 - 6)(\sin x) - (x^3 - 6x)(\cos x)\Big]_0^{\pi} \\
&= \pi^3 - 6\pi.
\end{aligned}$$

Variable density: $\rho(x, y) = x$
R: $0 \le x \le \pi$, $0 \le y \le \sin x$
(x, y)

Figure 14.42

So, the radius of gyration about the y-axis is

$$\begin{aligned}
\bar{\bar{x}} &= \sqrt{\frac{I_y}{m}} \\
&= \sqrt{\frac{\pi^3 - 6\pi}{\pi}} \\
&= \sqrt{\pi^2 - 6} \approx 1.967.
\end{aligned}$$

■

14.4 Exercises

See www.CalcChat.com for worked-out solutions to odd-numbered exercises.

In Exercises 1–4, find the mass of the lamina described by the inequalities, given that its density is $\rho(x, y) = xy$. (*Hint:* Some of the integrals are simpler in polar coordinates.)

1. $0 \le x \le 2,\ 0 \le y \le 2$
2. $0 \le x \le 3,\ 0 \le y \le 9 - x^2$
3. $0 \le x \le 1,\ 0 \le y \le \sqrt{1 - x^2}$
4. $x \ge 0,\ 3 \le y \le 3 + \sqrt{9 - x^2}$

In Exercises 5–8, find the mass and center of mass of the lamina for each density.

5. R: square with vertices $(0, 0), (a, 0), (0, a), (a, a)$
   (a) $\rho = k$ (b) $\rho = ky$ (c) $\rho = kx$
6. R: rectangle with vertices $(0, 0), (a, 0), (0, b), (a, b)$
   (a) $\rho = kxy$ (b) $\rho = k(x^2 + y^2)$
7. R: triangle with vertices $(0, 0), (0, a), (a, a)$
   (a) $\rho = k$ (b) $\rho = ky$ (c) $\rho = kx$
8. R: triangle with vertices $(0, 0), (a/2, a), (a, 0)$
   (a) $\rho = k$ (b) $\rho = kxy$

9. ***Translations in the Plane*** Translate the lamina in Exercise 5 to the right five units and determine the resulting center of mass.
10. ***Conjecture*** Use the result of Exercise 9 to make a conjecture about the change in the center of mass when a lamina of constant density is translated c units horizontally or d units vertically. Is the conjecture true if the density is not constant? Explain.

In Exercises 11–22, find the mass and center of mass of the lamina bounded by the graphs of the equations for the given density or densities. (*Hint:* Some of the integrals are simpler in polar coordinates.)

11. $y = \sqrt{x},\ y = 0,\ x = 1,\ \rho = ky$
12. $y = x^2,\ y = 0,\ x = 2,\ \rho = kxy$
13. $y = 4/x,\ y = 0,\ x = 1,\ x = 4,\ \rho = kx^2$
14. $y = \dfrac{1}{1 + x^2},\ y = 0,\ x = -1,\ x = 1,\ \rho = k$
15. $y = e^x,\ y = 0,\ x = 0,\ x = 1$
    (a) $\rho = k$ (b) $\rho = ky$
16. $y = e^{-x},\ y = 0,\ x = 0,\ x = 1$
    (a) $\rho = ky$ (b) $\rho = ky^2$
17. $y = 4 - x^2,\ y = 0,\ \rho = ky$
18. $x = 9 - y^2,\ x = 0,\ \rho = kx$
19. $y = \sin \dfrac{\pi x}{L},\ y = 0,\ x = 0,\ x = L,\ \rho = k$
20. $y = \cos \dfrac{\pi x}{L},\ y = 0,\ x = 0,\ x = \dfrac{L}{2},\ \rho = ky$
21. $y = \sqrt{a^2 - x^2},\ 0 \le y \le x,\ \rho = k$
22. $x^2 + y^2 = a^2,\ 0 \le x,\ 0 \le y,\ \rho = k(x^2 + y^2)$

CAS **In Exercises 23–26, use a computer algebra system to find the mass and center of mass of the lamina bounded by the graphs of the equations for the given density.**

23. $y = e^{-x},\ y = 0,\ x = 0,\ x = 2,\ \rho = kxy$
24. $y = \ln x,\ y = 0,\ x = 1,\ x = e,\ \rho = k/x$
25. $r = 2 \cos 3\theta,\ -\pi/6 \le \theta \le \pi/6,\ \rho = k$
26. $r = 1 + \cos \theta,\ \rho = k$

In Exercises 27–32, verify the given moment(s) of inertia and find $\bar{x}$ and $\bar{y}$. Assume that each lamina has a density of $\rho = 1$ gram per square centimeter. (These regions are common shapes used in engineering.)

27. Rectangle

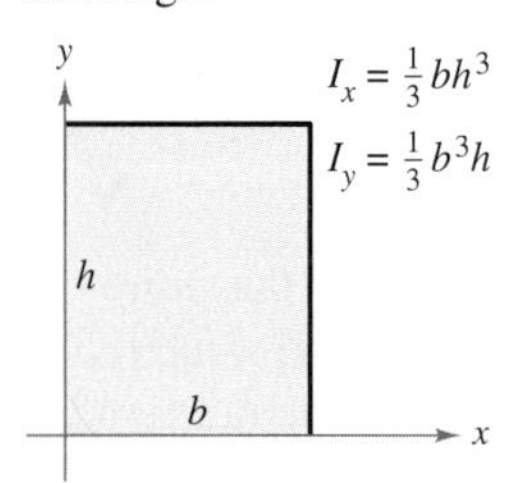

28. Right triangle

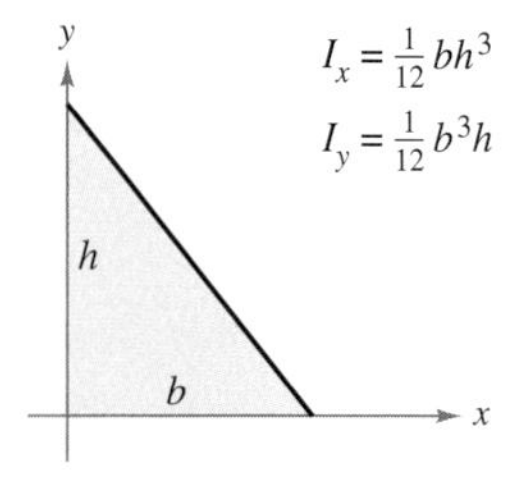

29. Circle

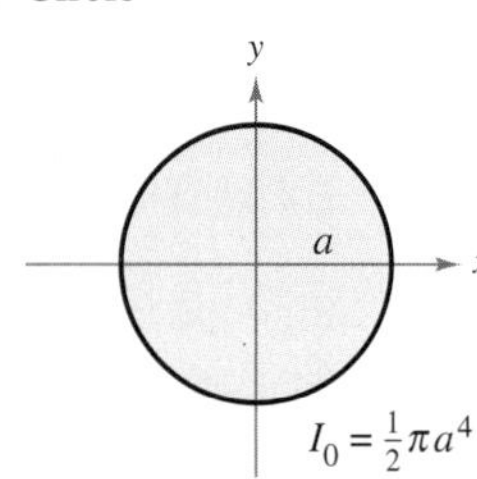

30. Semicircle

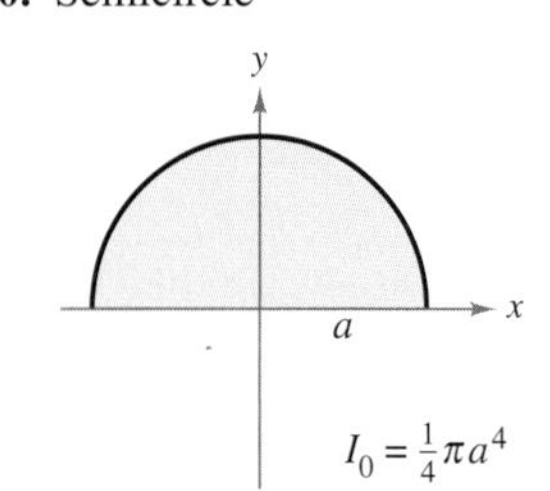

31. Quarter circle

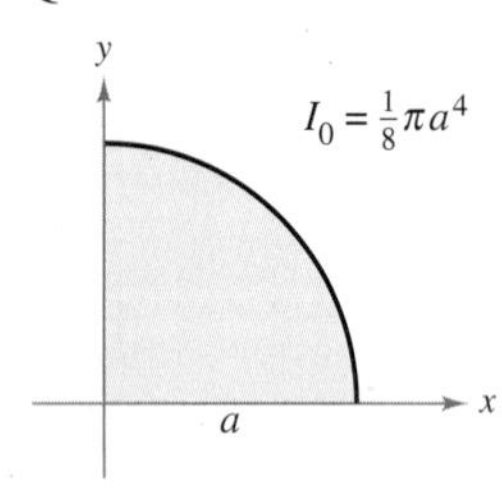

32. Ellipse

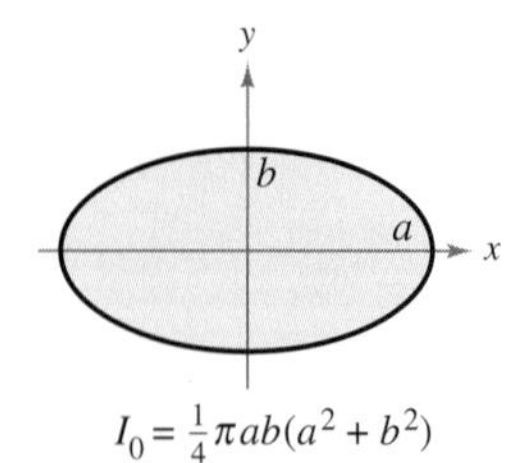

CAS **In Exercises 33–40, find $I_x, I_y, I_0, \bar{x}$, and $\bar{y}$ for the lamina bounded by the graphs of the equations. Use a computer algebra system to evaluate the double integrals.**

33. $y = 0,\ y = b,\ x = 0,\ x = a,\ \rho = ky$
34. $y = \sqrt{a^2 - x^2},\ y = 0,\ \rho = ky$
35. $y = 4 - x^2,\ y = 0,\ x > 0,\ \rho = kx$
36. $y = x,\ y = x^2,\ \rho = kxy$
37. $y = \sqrt{x},\ y = 0,\ x = 4,\ \rho = kxy$
38. $y = x^2,\ y^2 = x,\ \rho = x^2 + y^2$
39. $y = x^2,\ y^2 = x,\ \rho = kx$
40. $y = x^3,\ y = 4x,\ \rho = k|y|$

CAS In Exercises 41–46, set up the double integral required to find the moment of inertia I, about the given line, of the lamina bounded by the graphs of the equations. Use a computer algebra system to evaluate the double integral.

41. $x^2 + y^2 = b^2,\ \rho = k,$ line: $x = a\ (a > b)$

42. $y = 0,\ y = 2,\ x = 0,\ x = 4,\ \rho = k,$ line: $x = 6$

43. $y = \sqrt{x},\ y = 0,\ x = 4,\ \rho = kx,$ line: $x = 6$

44. $y = \sqrt{a^2 - x^2},\ y = 0,\ \rho = ky,$ line: $y = a$

45. $y = \sqrt{a^2 - x^2},\ y = 0,\ x \geq 0,\ \rho = k(a - y),$ line: $y = a$

46. $y = 4 - x^2,\ y = 0,\ \rho = k,$ line: $y = 2$

WRITING ABOUT CONCEPTS

47. Give the formulas for finding the moments and center of mass of a variable density planar lamina.

48. Give the formulas for finding the moments of inertia about the x- and y-axes for a variable density planar lamina.

49. In your own words, describe what the radius of gyration measures.

CAPSTONE

50. The center of mass of the lamina of constant density shown in the figure is $\left(2, \frac{8}{5}\right)$. Make a conjecture about how the center of mass $(\bar{x}, \bar{y})$ changes for each given nonconstant density $\rho(x, y)$. Explain. (Make your conjecture *without* performing any calculations.)

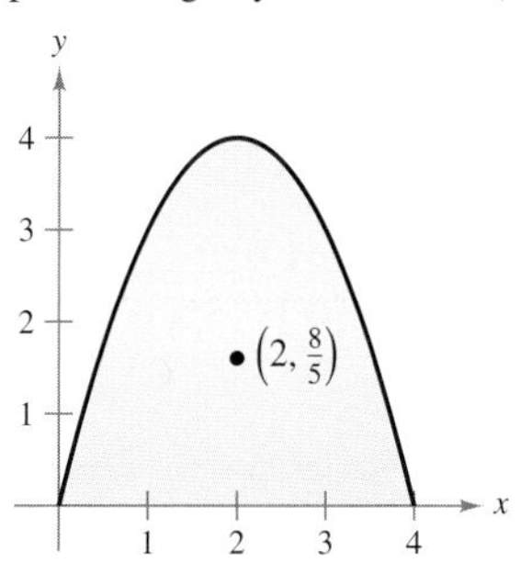

(a) $\rho(x, y) = ky$ (b) $\rho(x, y) = k|2 - x|$

(c) $\rho(x, y) = kxy$ (d) $\rho(x, y) = k(4 - x)(4 - y)$

Hydraulics **In Exercises 51–54, determine the location of the horizontal axis y_a at which a vertical gate in a dam is to be hinged so that there is no moment causing rotation under the indicated loading (see figure). The model for y_a is**

$$y_a = \bar{y} - \frac{I_{\bar{y}}}{hA}$$

where $\bar{y}$ is the y-coordinate of the centroid of the gate, $I_{\bar{y}}$ is the moment of inertia of the gate about the line $y = \bar{y}$, h is the depth of the centroid below the surface, and A is the area of the gate.

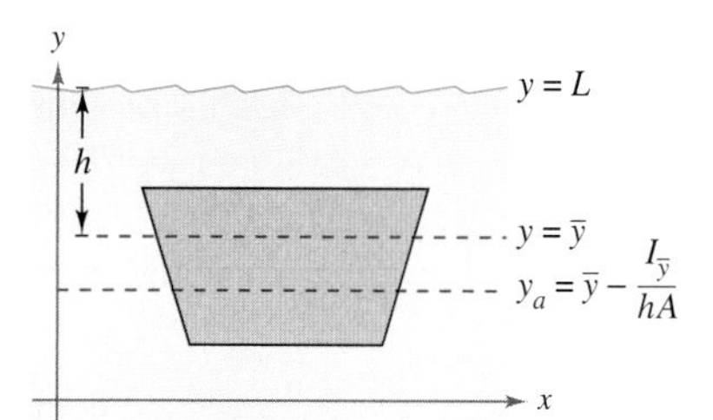

51.

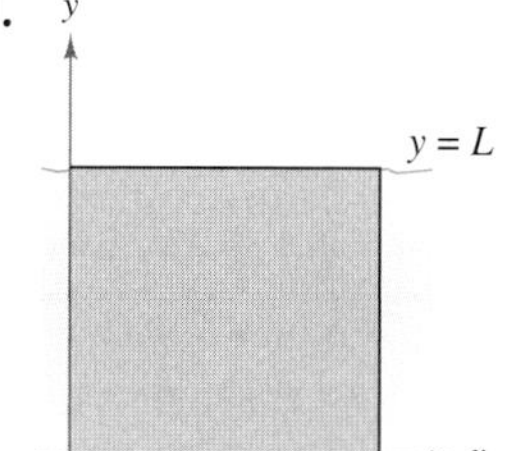

52.

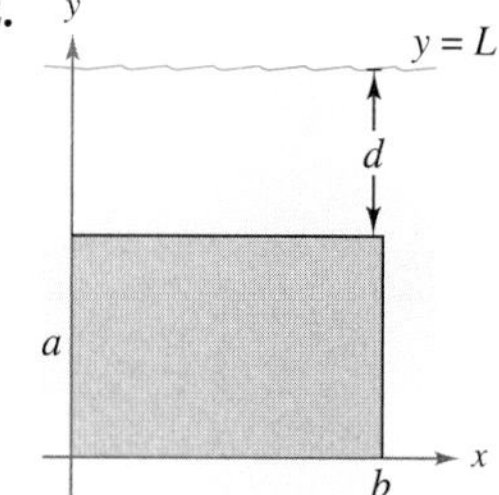

53.

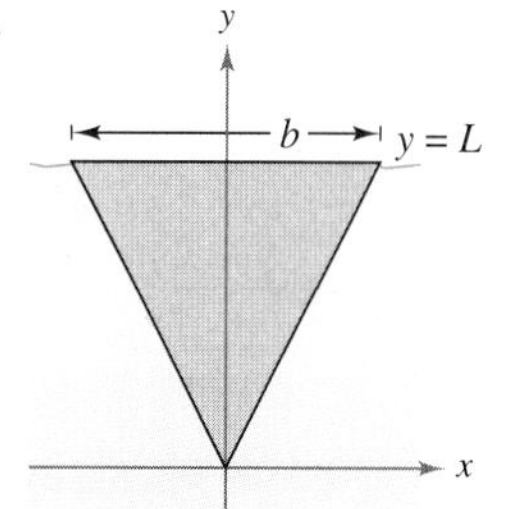

54.

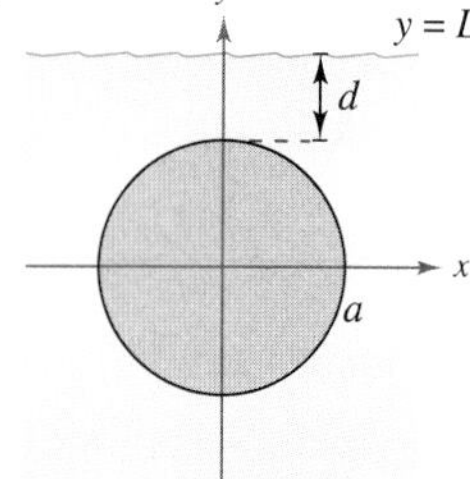

55. Prove the following Theorem of Pappus: Let R be a region in a plane and let L be a line in the same plane such that L does not intersect the interior of R. If r is the distance between the centroid of R and the line, then the volume V of the solid of revolution formed by revolving R about the line is given by $V = 2\pi rA$, where A is the area of R.

SECTION PROJECT

Center of Pressure on a Sail

The center of pressure on a sail is that point (x_p, y_p) at which the total aerodynamic force may be assumed to act. If the sail is represented by a plane region R, the center of pressure is

$$x_p = \frac{\int_R\int xy\,dA}{\int_R\int y\,dA} \quad \text{and} \quad y_p = \frac{\int_R\int y^2\,dA}{\int_R\int y\,dA}.$$

Consider a triangular sail with vertices at $(0, 0)$, $(2, 1)$, and $(0, 5)$. Verify the value of each integral.

(a) $\displaystyle\int_R\int y\,dA = 10$ (b) $\displaystyle\int_R\int xy\,dA = \frac{35}{6}$ (c) $\displaystyle\int_R\int y^2\,dA = \frac{155}{6}$

Calculate the coordinates (x_p, y_p) of the center of pressure. Sketch a graph of the sail and indicate the location of the center of pressure.

14.5 Surface Area

■ **Use a double integral to find the area of a surface.**

Surface Area

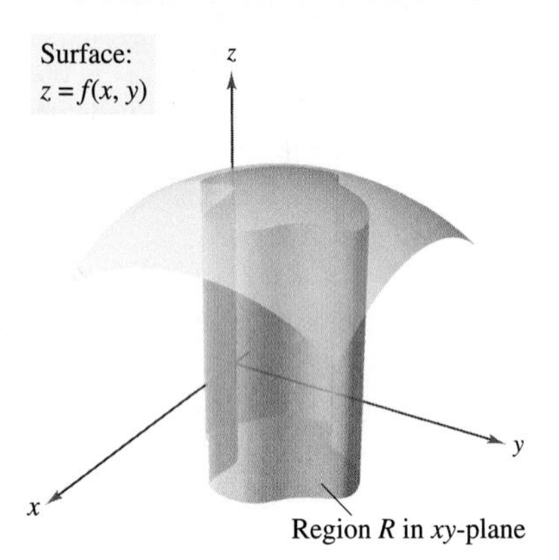

Figure 14.43

At this point you know a great deal about the solid region lying between a surface and a closed and bounded region R in the xy-plane, as shown in Figure 14.43. For example, you know how to find the extrema of f on R (Section 13.8), the area of the base R of the solid (Section 14.1), the volume of the solid (Section 14.2), and the centroid of the base R (Section 14.4).

In this section, you will learn how to find the upper **surface area** of the solid. Later, you will learn how to find the centroid of the solid (Section 14.6) and the lateral surface area (Section 15.2).

To begin, consider a surface S given by

$$z = f(x, y) \qquad \text{Surface defined over a region } R$$

defined over a region R. Assume that R is closed and bounded and that f has continuous first partial derivatives. To find the surface area, construct an inner partition of R consisting of n rectangles, where the area of the ith rectangle R_i is $\Delta A_i = \Delta x_i \Delta y_i$, as shown in Figure 14.44. In each R_i let (x_i, y_i) be the point that is closest to the origin. At the point $(x_i, y_i, z_i) = (x_i, y_i, f(x_i, y_i))$ on the surface S, construct a tangent plane T_i. The area of the portion of the tangent plane that lies directly above R_i is approximately equal to the area of the surface lying directly above R_i. That is, $\Delta T_i \approx \Delta S_i$. So, the surface area of S is given by

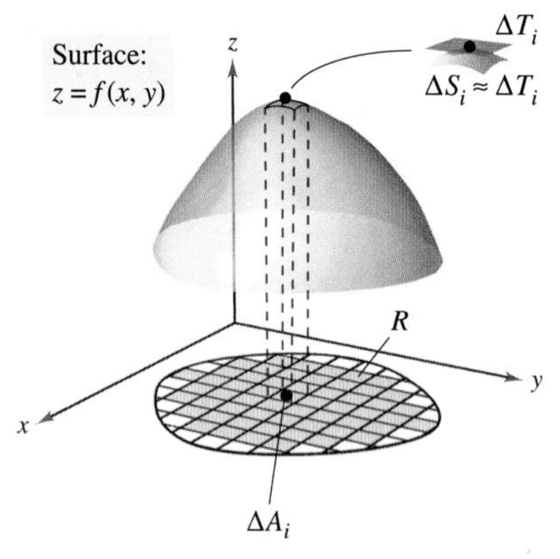

Figure 14.44

$$\sum_{i=1}^{n} \Delta S_i \approx \sum_{i=1}^{n} \Delta T_i.$$

To find the area of the parallelogram ΔT_i, note that its sides are given by the vectors

$$\mathbf{u} = \Delta x_i \mathbf{i} + f_x(x_i, y_i)\,\Delta x_i \mathbf{k}$$

and

$$\mathbf{v} = \Delta y_i \mathbf{j} + f_y(x_i, y_i)\,\Delta y_i \mathbf{k}.$$

From Theorem 11.8, the area of ΔT_i is given by $\|\mathbf{u} \times \mathbf{v}\|$, where

$$\begin{aligned} \mathbf{u} \times \mathbf{v} &= \begin{vmatrix} \mathbf{i} & \mathbf{j} & \mathbf{k} \\ \Delta x_i & 0 & f_x(x_i, y_i)\,\Delta x_i \\ 0 & \Delta y_i & f_y(x_i, y_i)\,\Delta y_i \end{vmatrix} \\ &= -f_x(x_i, y_i)\,\Delta x_i \Delta y_i \mathbf{i} - f_y(x_i, y_i)\,\Delta x_i \Delta y_i \mathbf{j} + \Delta x_i \Delta y_i \mathbf{k} \\ &= (-f_x(x_i, y_i)\mathbf{i} - f_y(x_i, y_i)\mathbf{j} + \mathbf{k})\,\Delta A_i. \end{aligned}$$

So, the area of ΔT_i is $\|\mathbf{u} \times \mathbf{v}\| = \sqrt{[f_x(x_i, y_i)]^2 + [f_y(x_i, y_i)]^2 + 1}\,\Delta A_i$, and

$$\begin{aligned} \text{Surface area of } S &\approx \sum_{i=1}^{n} \Delta S_i \\ &\approx \sum_{i=1}^{n} \sqrt{1 + [f_x(x_i, y_i)]^2 + [f_y(x_i, y_i)]^2}\,\Delta A_i. \end{aligned}$$

This suggests the following definition of surface area.

DEFINITION OF SURFACE AREA

If f and its first partial derivatives are continuous on the closed region R in the xy-plane, then the **area of the surface S** given by $z = f(x, y)$ over R is defined as

$$\text{Surface area} = \int\int_R dS$$
$$= \int\int_R \sqrt{1 + [f_x(x, y)]^2 + [f_y(x, y)]^2}\, dA.$$

As an aid to remembering the double integral for surface area, it is helpful to note its similarity to the integral for arc length.

Length on x-axis: $\int_a^b dx$

Arc length in xy-plane: $\int_a^b ds = \int_a^b \sqrt{1 + [f'(x)]^2}\, dx$

Area in xy-plane: $\int\int_R dA$

Surface area in space: $\int\int_R dS = \int\int_R \sqrt{1 + [f_x(x, y)]^2 + [f_y(x, y)]^2}\, dA$

Like integrals for arc length, integrals for surface area are often very difficult to evaluate. However, one type that is easily evaluated is demonstrated in the next example.

EXAMPLE 1 The Surface Area of a Plane Region

Find the surface area of the portion of the plane

$$z = 2 - x - y$$

that lies above the circle $x^2 + y^2 \le 1$ in the first quadrant, as shown in Figure 14.45.

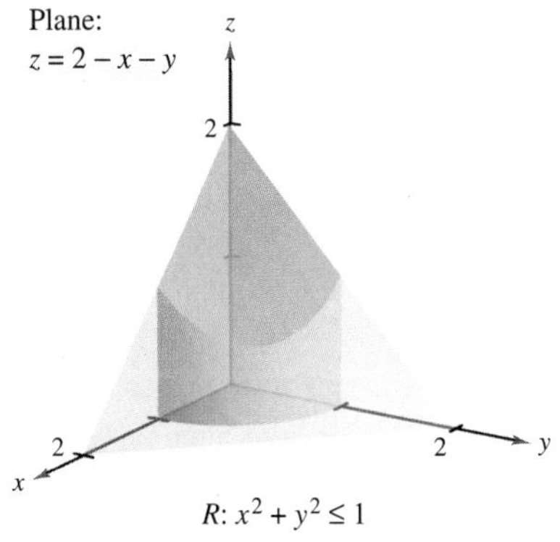

Figure 14.45

Solution Because $f_x(x, y) = -1$ and $f_y(x, y) = -1$, the surface area is given by

$$S = \int\int_R \sqrt{1 + [f_x(x, y)]^2 + [f_y(x, y)]^2}\, dA \qquad \text{Formula for surface area}$$
$$= \int\int_R \sqrt{1 + (-1)^2 + (-1)^2}\, dA \qquad \text{Substitute.}$$
$$= \int\int_R \sqrt{3}\, dA$$
$$= \sqrt{3}\int\int_R dA.$$

Note that the last integral is simply $\sqrt{3}$ times the area of the region R. R is a quarter circle of radius 1, with an area of $\frac{1}{4}\pi(1^2)$ or $\pi/4$. So, the area of S is

$$S = \sqrt{3}\,(\text{area of } R)$$
$$= \sqrt{3}\left(\frac{\pi}{4}\right)$$
$$= \frac{\sqrt{3}\,\pi}{4}.$$

■

EXAMPLE 2 Finding Surface Area

Find the area of the portion of the surface

$$f(x, y) = 1 - x^2 + y$$

that lies above the triangular region with vertices $(1, 0, 0)$, $(0, -1, 0)$, and $(0, 1, 0)$, as shown in Figure 14.46(a).

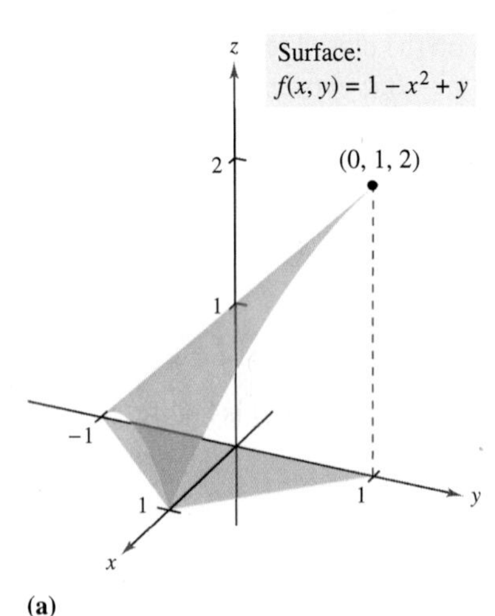

(a)

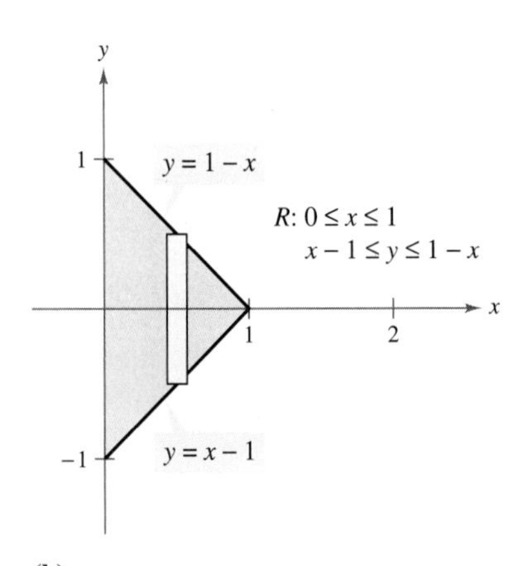

(b)

Figure 14.46

Solution Because $f_x(x, y) = -2x$ and $f_y(x, y) = 1$, you have

$$S = \iint_R \sqrt{1 + [f_x(x, y)]^2 + [f_y(x, y)]^2}\, dA = \iint_R \sqrt{1 + 4x^2 + 1}\, dA.$$

In Figure 14.46(b), you can see that the bounds for R are $0 \le x \le 1$ and $x - 1 \le y \le 1 - x$. So, the integral becomes

$$\begin{aligned}
S &= \int_0^1 \int_{x-1}^{1-x} \sqrt{2 + 4x^2}\, dy\, dx \\
&= \int_0^1 y\sqrt{2 + 4x^2}\Big]_{x-1}^{1-x}\, dx \\
&= \int_0^1 \Big[(1 - x)\sqrt{2 + 4x^2} - (x - 1)\sqrt{2 + 4x^2}\Big]\, dx \\
&= \int_0^1 \left(2\sqrt{2 + 4x^2} - 2x\sqrt{2 + 4x^2}\right) dx \qquad \text{Integration tables (Appendix B), Formula 26 and Power Rule} \\
&= \left[x\sqrt{2 + 4x^2} + \ln\left(2x + \sqrt{2 + 4x^2}\right) - \frac{(2 + 4x^2)^{3/2}}{6}\right]_0^1 \\
&= \sqrt{6} + \ln\left(2 + \sqrt{6}\right) - \sqrt{6} - \ln\sqrt{2} + \frac{1}{3}\sqrt{2} \approx 1.618.
\end{aligned}$$

EXAMPLE 3 Change of Variables to Polar Coordinates

Find the surface area of the paraboloid $z = 1 + x^2 + y^2$ that lies above the unit circle, as shown in Figure 14.47.

Solution Because $f_x(x, y) = 2x$ and $f_y(x, y) = 2y$, you have

$$S = \iint_R \sqrt{1 + [f_x(x, y)]^2 + [f_y(x, y)]^2}\, dA = \iint_R \sqrt{1 + 4x^2 + 4y^2}\, dA.$$

You can convert to polar coordinates by letting $x = r\cos\theta$ and $y = r\sin\theta$. Then, because the region R is bounded by $0 \le r \le 1$ and $0 \le \theta \le 2\pi$, you have

$$\begin{aligned}
S &= \int_0^{2\pi} \int_0^1 \sqrt{1 + 4r^2}\, r\, dr\, d\theta \\
&= \int_0^{2\pi} \frac{1}{12}(1 + 4r^2)^{3/2}\Big]_0^1\, d\theta \\
&= \int_0^{2\pi} \frac{5\sqrt{5} - 1}{12}\, d\theta \\
&= \frac{5\sqrt{5} - 1}{12}\,\theta\Big]_0^{2\pi} \\
&= \frac{\pi\left(5\sqrt{5} - 1\right)}{6} \\
&\approx 5.33.
\end{aligned}$$

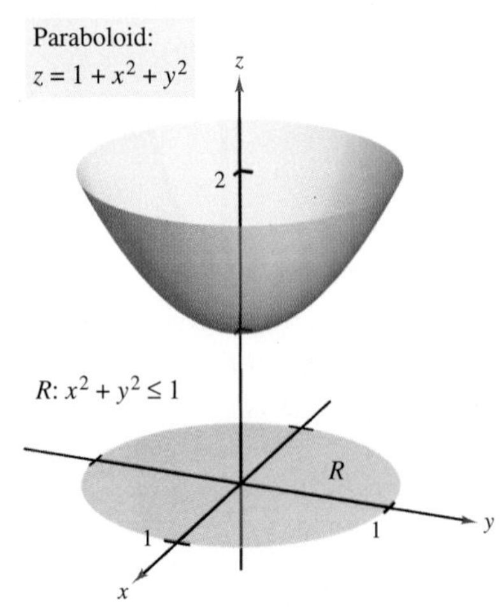

Figure 14.47

EXAMPLE 4 Finding Surface Area

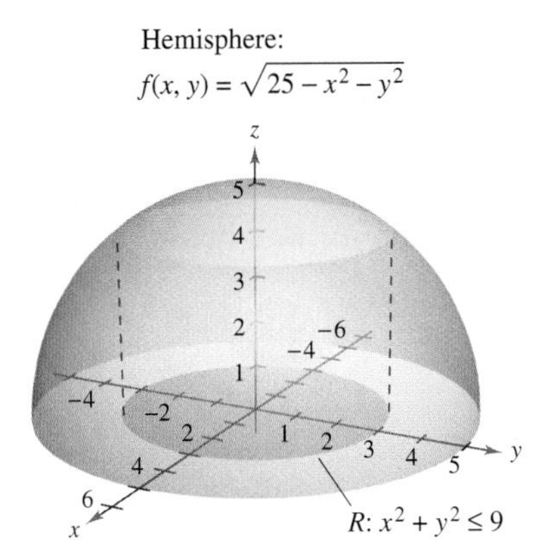

Figure 14.48

Find the surface area S of the portion of the hemisphere

$$f(x, y) = \sqrt{25 - x^2 - y^2} \qquad \text{Hemisphere}$$

that lies above the region R bounded by the circle $x^2 + y^2 \le 9$, as shown in Figure 14.48.

Solution The first partial derivatives of f are

$$f_x(x, y) = \frac{-x}{\sqrt{25 - x^2 - y^2}} \quad \text{and} \quad f_y(x, y) = \frac{-y}{\sqrt{25 - x^2 - y^2}}$$

and, from the formula for surface area, you have

$$\begin{aligned} dS &= \sqrt{1 + [f_x(x, y)]^2 + [f_y(x, y)]^2}\, dA \\ &= \sqrt{1 + \left(\frac{-x}{\sqrt{25 - x^2 - y^2}}\right)^2 + \left(\frac{-y}{\sqrt{25 - x^2 - y^2}}\right)^2}\, dA \\ &= \frac{5}{\sqrt{25 - x^2 - y^2}}\, dA. \end{aligned}$$

So, the surface area is

$$S = \iint_R \frac{5}{\sqrt{25 - x^2 - y^2}}\, dA.$$

You can convert to polar coordinates by letting $x = r \cos \theta$ and $y = r \sin \theta$. Then, because the region R is bounded by $0 \le r \le 3$ and $0 \le \theta \le 2\pi$, you obtain

$$\begin{aligned} S &= \int_0^{2\pi} \int_0^3 \frac{5}{\sqrt{25 - r^2}}\, r\, dr\, d\theta \\ &= 5 \int_0^{2\pi} -\sqrt{25 - r^2}\, \Big]_0^3\, d\theta \\ &= 5 \int_0^{2\pi} d\theta \\ &= 10\pi. \end{aligned}$$

The procedure used in Example 4 can be extended to find the surface area of a sphere by using the region R bounded by the circle $x^2 + y^2 \le a^2$, where $0 < a < 5$, as shown in Figure 14.49. The surface area of the portion of the hemisphere

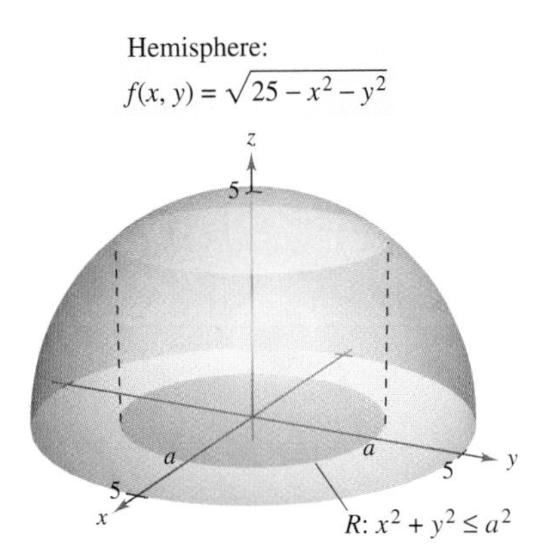

Figure 14.49

$$f(x, y) = \sqrt{25 - x^2 - y^2}$$

lying above the circular region can be shown to be

$$\begin{aligned} S &= \iint_R \frac{5}{\sqrt{25 - x^2 - y^2}}\, dA \\ &= \int_0^{2\pi} \int_0^a \frac{5}{\sqrt{25 - r^2}}\, r\, dr\, d\theta \\ &= 10\pi\left(5 - \sqrt{25 - a^2}\right). \end{aligned}$$

By taking the limit as a approaches 5 and doubling the result, you obtain a total area of 100π. (The surface area of a sphere of radius r is $S = 4\pi r^2$.)

You can use Simpson's Rule or the Trapezoidal Rule to approximate the value of a double integral, *provided* you can get through the first integration. This is demonstrated in the next example.

EXAMPLE 5 Approximating Surface Area by Simpson's Rule

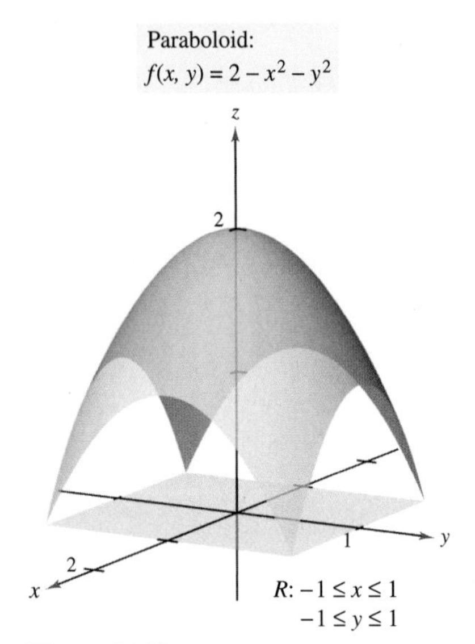

Figure 14.50

Find the area of the surface of the paraboloid

$$f(x, y) = 2 - x^2 - y^2 \qquad \text{Paraboloid}$$

that lies above the square region bounded by $-1 \le x \le 1$ and $-1 \le y \le 1$, as shown in Figure 14.50.

Solution Using the partial derivatives

$$f_x(x, y) = -2x \quad \text{and} \quad f_y(x, y) = -2y$$

you have a surface area of

$$\begin{aligned} S &= \int\!\!\int_R \sqrt{1 + [f_x(x, y)]^2 + [f_y(x, y)]^2}\, dA \\ &= \int\!\!\int_R \sqrt{1 + (-2x)^2 + (-2y)^2}\, dA \\ &= \int\!\!\int_R \sqrt{1 + 4x^2 + 4y^2}\, dA. \end{aligned}$$

In polar coordinates, the line $x = 1$ is given by $r \cos\theta = 1$ or $r = \sec\theta$, and you can determine from Figure 14.51 that one-fourth of the region R is bounded by

$$0 \le r \le \sec\theta \quad \text{and} \quad -\frac{\pi}{4} \le \theta \le \frac{\pi}{4}.$$

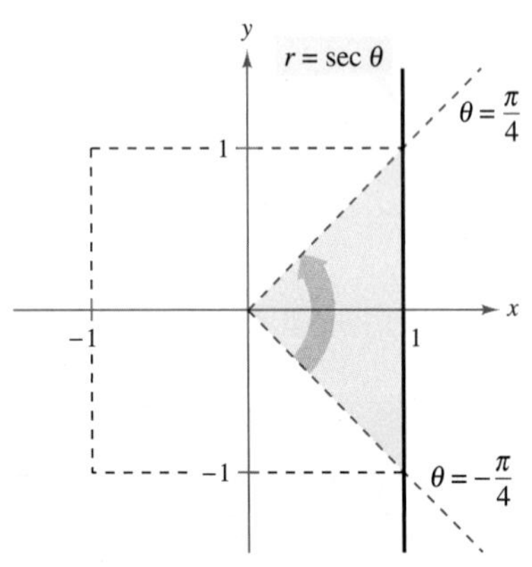

One-fourth of the region R is bounded by $0 \le r \le \sec\theta$ and $-\frac{\pi}{4} \le \theta \le \frac{\pi}{4}$.

Figure 14.51

Letting $x = r\cos\theta$ and $y = r\sin\theta$ produces

$$\begin{aligned} \frac{1}{4}S &= \frac{1}{4}\int\!\!\int_R \sqrt{1 + 4x^2 + 4y^2}\, dA \\ &= \int_{-\pi/4}^{\pi/4}\int_0^{\sec\theta} \sqrt{1 + 4r^2}\, r\, dr\, d\theta \\ &= \int_{-\pi/4}^{\pi/4} \frac{1}{12}(1 + 4r^2)^{3/2}\Big]_0^{\sec\theta}\, d\theta \\ &= \frac{1}{12}\int_{-\pi/4}^{\pi/4} \left[(1 + 4\sec^2\theta)^{3/2} - 1\right] d\theta. \end{aligned}$$

Finally, using Simpson's Rule with $n = 10$, you can approximate this single integral to be

$$\begin{aligned} S &= \frac{1}{3}\int_{-\pi/4}^{\pi/4} \left[(1 + 4\sec^2\theta)^{3/2} - 1\right] d\theta \\ &\approx 7.450. \end{aligned}$$

■

TECHNOLOGY Most computer programs that are capable of performing symbolic integration for multiple integrals are also capable of performing numerical approximation techniques. If you have access to such software, use it to approximate the value of the integral in Example 5.

14.5 Exercises

See www.CalcChat.com for worked-out solutions to odd-numbered exercises.

In Exercises 1–14, find the area of the surface given by $z = f(x, y)$ over the region R. (*Hint:* Some of the integrals are simpler in polar coordinates.)

1. $f(x, y) = 2x + 2y$
   R: triangle with vertices $(0, 0), (4, 0), (0, 4)$
2. $f(x, y) = 15 + 2x - 3y$
   R: square with vertices $(0, 0), (3, 0), (0, 3), (3, 3)$
3. $f(x, y) = 7 + 2x + 2y$
   $R = \{(x, y): x^2 + y^2 \leq 4\}$
4. $f(x, y) = 12 + 2x - 3y$
   $R = \{(x, y): x^2 + y^2 \leq 9\}$
5. $f(x, y) = 9 - x^2$
   R: square with vertices $(0, 0), (2, 0), (0, 2), (2, 2)$
6. $f(x, y) = y^2$
   R: square with vertices $(0, 0), (3, 0), (0, 3), (3, 3)$
7. $f(x, y) = 3 + x^{3/2}$
   R: rectangle with vertices $(0, 0), (0, 4), (3, 4), (3, 0)$
8. $f(x, y) = 2 + \frac{2}{3}y^{3/2}$
   $R = \{(x, y): 0 \leq x \leq 2, 0 \leq y \leq 2 - x\}$
9. $f(x, y) = \ln|\sec x|$
   $R = \left\{(x, y): 0 \leq x \leq \frac{\pi}{4}, 0 \leq y \leq \tan x\right\}$
10. $f(x, y) = 13 + x^2 - y^2$
    $R = \{(x, y): x^2 + y^2 \leq 4\}$
11. $f(x, y) = \sqrt{x^2 + y^2}$, $R = \{(x, y): 0 \leq f(x, y) \leq 1\}$
12. $f(x, y) = xy$, $R = \{(x, y): x^2 + y^2 \leq 16\}$
13. $f(x, y) = \sqrt{a^2 - x^2 - y^2}$
    $R = \{(x, y): x^2 + y^2 \leq b^2, 0 < b < a\}$
14. $f(x, y) = \sqrt{a^2 - x^2 - y^2}$
    $R = \{(x, y): x^2 + y^2 \leq a^2\}$

In Exercises 15–18, find the area of the surface.

15. The portion of the plane $z = 24 - 3x - 2y$ in the first octant
16. The portion of the paraboloid $z = 16 - x^2 - y^2$ in the first octant
17. The portion of the sphere $x^2 + y^2 + z^2 = 25$ inside the cylinder $x^2 + y^2 = 9$
18. The portion of the cone $z = 2\sqrt{x^2 + y^2}$ inside the cylinder $x^2 + y^2 = 4$

CAS **In Exercises 19–24, write a double integral that represents the surface area of $z = f(x, y)$ over the region R. Use a computer algebra system to evaluate the double integral.**

19. $f(x, y) = 2y + x^2$
    R: triangle with vertices $(0, 0), (1, 0), (1, 1)$
20. $f(x, y) = 2x + y^2$
    R: triangle with vertices $(0, 0), (2, 0), (2, 2)$
21. $f(x, y) = 9 - x^2 - y^2$
    $R = \{(x, y): 0 \leq f(x, y)\}$
22. $f(x, y) = x^2 + y^2$
    $R = \{(x, y): 0 \leq f(x, y) \leq 16\}$
23. $f(x, y) = 4 - x^2 - y^2$
    $R = \{(x, y): 0 \leq x \leq 1, 0 \leq y \leq 1\}$
24. $f(x, y) = \frac{2}{3}x^{3/2} + \cos x$
    $R = \{(x, y): 0 \leq x \leq 1, 0 \leq y \leq 1\}$

Approximation **In Exercises 25 and 26, determine which value best approximates the surface area of $z = f(x, y)$ over the region R. (Make your selection on the basis of a sketch of the surface and *not* by performing any calculations.)**

25. $f(x, y) = 10 - \frac{1}{2}y^2$
    R: square with vertices $(0, 0), (4, 0), (4, 4), (0, 4)$
    (a) 16 (b) 200 (c) -100 (d) 72 (e) 36
26. $f(x, y) = \frac{1}{4}\sqrt{x^2 + y^2}$
    R: circle bounded by $x^2 + y^2 = 9$
    (a) -100 (b) 150 (c) 9π (d) 55 (e) 500

CAS **In Exercises 27 and 28, use a computer algebra system to approximate the double integral that gives the surface area of the graph of f over the region $R = \{(x, y): 0 \leq x \leq 1, 0 \leq y \leq 1\}$.**

27. $f(x, y) = e^x$
28. $f(x, y) = \frac{2}{5}y^{5/2}$

In Exercises 29–34, set up a double integral that gives the area of the surface on the graph of f over the region R.

29. $f(x, y) = x^3 - 3xy + y^3$
    R: square with vertices $(1, 1), (-1, 1), (-1, -1), (1, -1)$
30. $f(x, y) = x^2 - 3xy - y^2$
    $R = \{(x, y): 0 \leq x \leq 4, 0 \leq y \leq x\}$
31. $f(x, y) = e^{-x} \sin y$
    $R = \{(x, y): x^2 + y^2 \leq 4\}$
32. $f(x, y) = \cos(x^2 + y^2)$
    $R = \left\{(x, y): x^2 + y^2 \leq \frac{\pi}{2}\right\}$
33. $f(x, y) = e^{xy}$
    $R = \{(x, y): 0 \leq x \leq 4, 0 \leq y \leq 10\}$
34. $f(x, y) = e^{-x} \sin y$
    $R = \{(x, y): 0 \leq x \leq 4, 0 \leq y \leq x\}$

WRITING ABOUT CONCEPTS

35. State the double integral definition of the area of a surface S given by $z = f(x, y)$ over the region R in the xy-plane.
36. Consider the surface $f(x, y) = x^2 + y^2$ and the surface area of f over each region R. Without integrating, order the surface areas from least to greatest. Explain your reasoning.
    (a) R: rectangle with vertices $(0, 0), (2, 0), (2, 2), (0, 2)$
    (b) R: triangle with vertices $(0, 0), (2, 0), (0, 2)$
    (c) $R = \{(x, y): x^2 + y^2 \leq 4$, first quadrant only$\}$
37. Will the surface area of the graph of a function $z = f(x, y)$ over a region R increase if the graph is shifted k units vertically? Why or why not?

CAPSTONE

38. Answer the following questions about the surface area S on a surface given by a positive function $z = f(x, y)$ over a region R in the xy-plane. Explain each answer.

(a) Is it possible for S to equal the area of R?

(b) Can S be greater than the area of R?

(c) Can S be less than the area of R?

39. Find the surface area of the solid of intersection of the cylinders $x^2 + z^2 = 1$ and $y^2 + z^2 = 1$ (see figure).

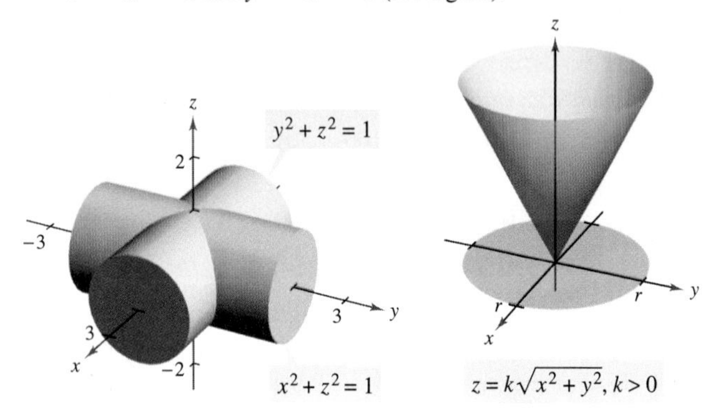

Figure for 39 **Figure for 40**

40. Show that the surface area of the cone $z = k\sqrt{x^2 + y^2}$, $k > 0$, over the circular region $x^2 + y^2 \le r^2$ in the xy-plane is $\pi r^2\sqrt{k^2 + 1}$ (see figure).

41. ***Product Design*** A company produces a spherical object of radius 25 centimeters. A hole of radius 4 centimeters is drilled through the center of the object. Find (a) the volume of the object and (b) the outer surface area of the object.

42. ***Modeling Data*** A rancher builds a barn with dimensions 30 feet by 50 feet. The symmetrical shape and selected heights of the roof are shown in the figure.

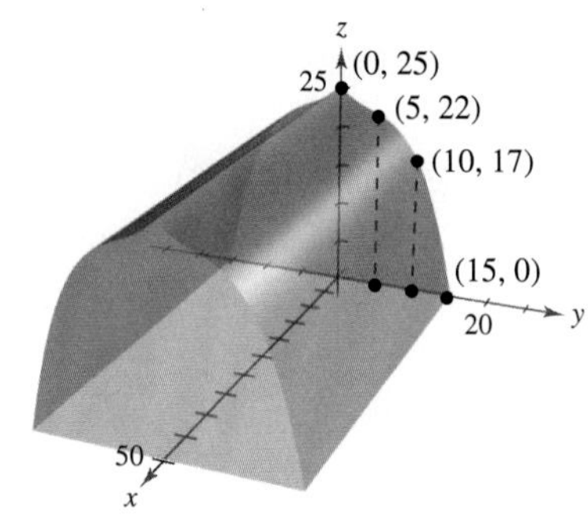

(a) Use the regression capabilities of a graphing utility to find a model of the form $z = ay^3 + by^2 + cy + d$ for the roof line.

(b) Use the numerical integration capabilities of a graphing utility and the model in part (a) to approximate the volume of storage space in the barn.

(c) Use the numerical integration capabilities of a graphing utility and the model in part (a) to approximate the surface area of the roof.

(d) Approximate the arc length of the roof line and find the surface area of the roof by multiplying the arc length by the length of the barn. Compare the results and the integrations with those found in part (c).

SECTION PROJECT

Capillary Action

A well-known property of liquids is that they will rise in narrow vertical channels—this property is called "capillary action." The figure shows two plates, which form a narrow wedge, in a container of liquid. The upper surface of the liquid follows a hyperbolic shape given by

$$z = \frac{k}{\sqrt{x^2 + y^2}}$$

where x, y, and z are measured in inches. The constant k depends on the angle of the wedge, the type of liquid, and the material that comprises the flat plates.

(a) Find the volume of the liquid that has risen in the wedge. (Assume $k = 1$.)

(b) Find the horizontal surface area of the liquid that has risen in the wedge.

Adaptation of Capillary Action problem from "Capillary Phenomena" by Thomas B. Greenslade, Jr., *Physics Teacher*, May 1992. By permission of the author.

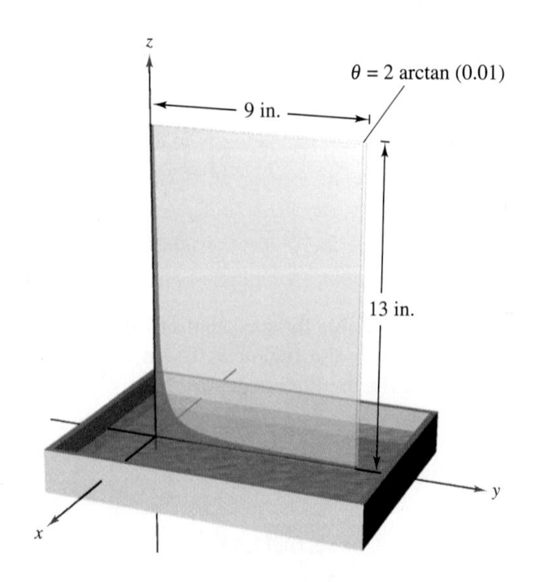

14.6 Triple Integrals and Applications

- **Use a triple integral to find the volume of a solid region.**
- **Find the center of mass and moments of inertia of a solid region.**

Triple Integrals

The procedure used to define a **triple integral** follows that used for double integrals. Consider a function f of three variables that is continuous over a bounded solid region Q. Then, encompass Q with a network of boxes and form the **inner partition** consisting of all boxes lying entirely within Q, as shown in Figure 14.52. The volume of the ith box is

$$\Delta V_i = \Delta x_i \Delta y_i \Delta z_i. \qquad \text{Volume of } i\text{th box}$$

The **norm** $\|\Delta\|$ of the partition is the length of the longest diagonal of the n boxes in the partition. Choose a point (x_i, y_i, z_i) in each box and form the Riemann sum

$$\sum_{i=1}^{n} f(x_i, y_i, z_i)\, \Delta V_i.$$

Taking the limit as $\|\Delta\| \to 0$ leads to the following definition.

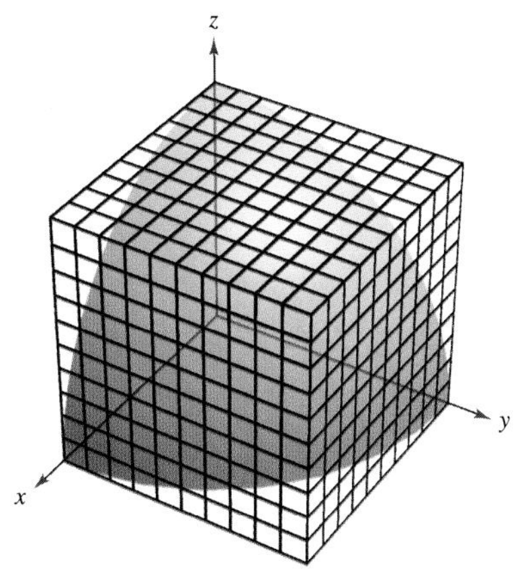

Solid region Q

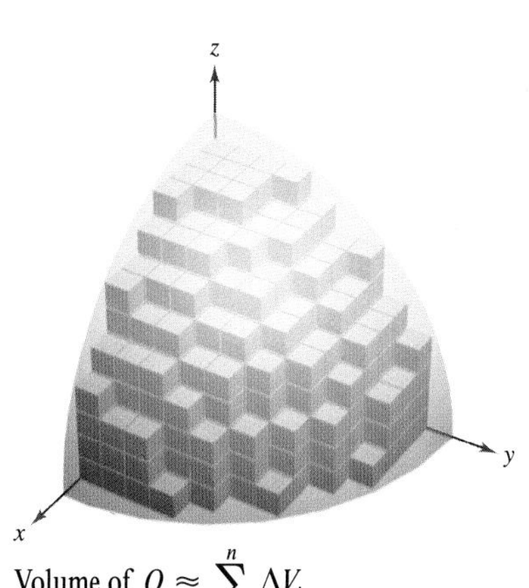

Volume of $Q \approx \sum_{i=1}^{n} \Delta V_i$

Figure 14.52

DEFINITION OF TRIPLE INTEGRAL

If f is continuous over a bounded solid region Q, then the **triple integral of f over Q** is defined as

$$\iiint_Q f(x, y, z)\, dV = \lim_{\|\Delta\| \to 0} \sum_{i=1}^{n} f(x_i, y_i, z_i)\, \Delta V_i$$

provided the limit exists. The **volume** of the solid region Q is given by

$$\text{Volume of } Q = \iiint_Q dV.$$

Some of the properties of double integrals in Theorem 14.1 can be restated in terms of triple integrals.

1. $\displaystyle\iiint_Q cf(x, y, z)\, dV = c\iiint_Q f(x, y, z)\, dV$

2. $\displaystyle\iiint_Q [f(x, y, z) \pm g(x, y, z)]\, dV = \iiint_Q f(x, y, z)\, dV \pm \iiint_Q g(x, y, z)\, dV$

3. $\displaystyle\iiint_Q f(x, y, z)\, dV = \iiint_{Q_1} f(x, y, z)\, dV + \iiint_{Q_2} f(x, y, z)\, dV$

In the properties above, Q is the union of two nonoverlapping solid subregions Q_1 and Q_2. If the solid region Q is simple, the triple integral $\iiint f(x, y, z)\, dV$ can be evaluated with an iterated integral using one of the six possible orders of integration:

$$dx\,dy\,dz \quad dy\,dx\,dz \quad dz\,dx\,dy \quad dx\,dz\,dy \quad dy\,dz\,dx \quad dz\,dy\,dx.$$

> **EXPLORATION**
>
> ***Volume of a Paraboloid Sector***
> On pages 997 and 1006, you were asked to summarize the different ways you know of finding the volume of the solid bounded by the paraboloid
>
> $$z = a^2 - x^2 - y^2, \quad a > 0$$
>
> and the xy-plane. You now know one more way. Use it to find the volume of the solid.

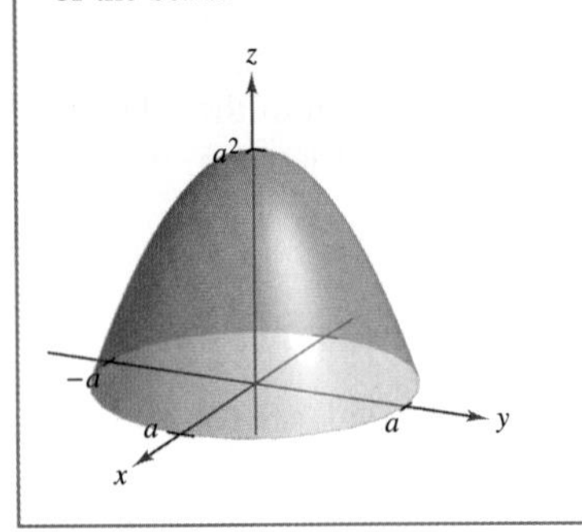

The following version of Fubini's Theorem describes a region that is considered simple with respect to the order $dz\,dy\,dx$. Similar descriptions can be given for the other five orders.

> **THEOREM 14.4 EVALUATION BY ITERATED INTEGRALS**
>
> Let f be continuous on a solid region Q defined by
>
> $$a \le x \le b, \quad h_1(x) \le y \le h_2(x), \quad g_1(x, y) \le z \le g_2(x, y)$$
>
> where h_1, h_2, g_1, and g_2 are continuous functions. Then,
>
> $$\iiint_Q f(x, y, z)\,dV = \int_a^b \int_{h_1(x)}^{h_2(x)} \int_{g_1(x, y)}^{g_2(x, y)} f(x, y, z)\,dz\,dy\,dx.$$

To evaluate a triple iterated integral in the order $dz\,dy\,dx$, hold *both* x and y constant for the innermost integration. Then, hold x constant for the second integration.

EXAMPLE 1 Evaluating a Triple Iterated Integral

Evaluate the triple iterated integral

$$\int_0^2 \int_0^x \int_0^{x+y} e^x(y + 2z)\,dz\,dy\,dx.$$

Solution For the first integration, hold x and y constant and integrate with respect to z.

$$\begin{aligned}\int_0^2 \int_0^x \int_0^{x+y} e^x(y + 2z)\,dz\,dy\,dx &= \int_0^2 \int_0^x e^x(yz + z^2)\Big]_0^{x+y}\,dy\,dx \\ &= \int_0^2 \int_0^x e^x(x^2 + 3xy + 2y^2)\,dy\,dx\end{aligned}$$

For the second integration, hold x constant and integrate with respect to y.

$$\begin{aligned}\int_0^2 \int_0^x e^x(x^2 + 3xy + 2y^2)\,dy\,dx &= \int_0^2 \left[e^x\left(x^2y + \frac{3xy^2}{2} + \frac{2y^3}{3}\right)\right]_0^x dx \\ &= \frac{19}{6}\int_0^2 x^3e^x\,dx\end{aligned}$$

Finally, integrate with respect to x.

$$\begin{aligned}\frac{19}{6}\int_0^2 x^3e^x\,dx &= \frac{19}{6}\Big[e^x(x^3 - 3x^2 + 6x - 6)\Big]_0^2 \\ &= 19\left(\frac{e^2}{3} + 1\right) \\ &\approx 65.797\end{aligned}$$

■

Example 1 demonstrates the integration order $dz\,dy\,dx$. For other orders, you can follow a similar procedure. For instance, to evaluate a triple iterated integral in the order $dx\,dy\,dz$, hold both y and z constant for the innermost integration and integrate with respect to x. Then, for the second integration, hold z constant and integrate with respect to y. Finally, for the third integration, integrate with respect to z.

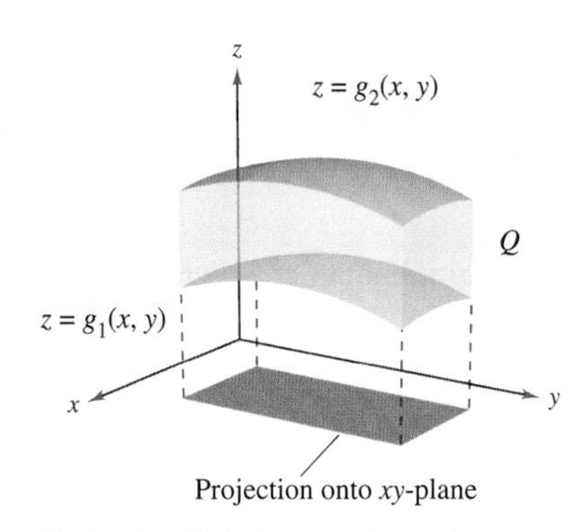

Solid region Q lies between two surfaces.
Figure 14.53

To find the limits for a particular order of integration, it is generally advisable first to determine the innermost limits, which may be functions of the outer two variables. Then, by projecting the solid Q onto the coordinate plane of the outer two variables, you can determine their limits of integration by the methods used for double integrals. For instance, to evaluate

$$\iiint_Q f(x, y, z)\, dz\, dy\, dx$$

first determine the limits for z, and then the integral has the form

$$\iint \left[\int_{g_1(x, y)}^{g_2(x, y)} f(x, y, z)\, dz \right] dy\, dx.$$

By projecting the solid Q onto the xy-plane, you can determine the limits for x and y as you did for double integrals, as shown in Figure 14.53.

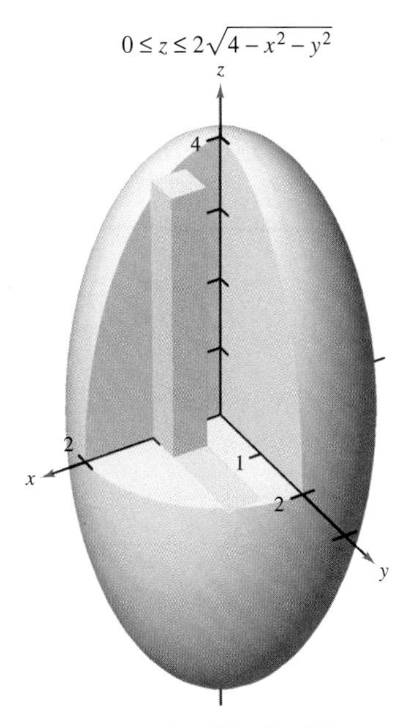

Ellipsoid: $4x^2 + 4y^2 + z^2 = 16$

Figure 14.54

EXAMPLE 2 Using a Triple Integral to Find Volume

Find the volume of the ellipsoid given by $4x^2 + 4y^2 + z^2 = 16$.

Solution Because x, y, and z play similar roles in the equation, the order of integration is probably immaterial, and you can arbitrarily choose $dz\, dy\, dx$. Moreover, you can simplify the calculation by considering only the portion of the ellipsoid lying in the first octant, as shown in Figure 14.54. From the order $dz\, dy\, dx$, you first determine the bounds for z.

$$0 \le z \le 2\sqrt{4 - x^2 - y^2}$$

In Figure 14.55, you can see that the boundaries for x and y are $0 \le x \le 2$ and $0 \le y \le \sqrt{4 - x^2}$, so the volume of the ellipsoid is

$$\begin{aligned}
V &= \iiint_Q dV \\
&= 8\int_0^2 \int_0^{\sqrt{4-x^2}} \int_0^{2\sqrt{4-x^2-y^2}} dz\, dy\, dx \\
&= 8\int_0^2 \int_0^{\sqrt{4-x^2}} \Big[z \Big]_0^{2\sqrt{4-x^2-y^2}} dy\, dx \\
&= 16\int_0^2 \int_0^{\sqrt{4-x^2}} \sqrt{(4 - x^2) - y^2}\, dy\, dx \qquad \text{Integration tables (Appendix B), Formula 37} \\
&= 8\int_0^2 \left[y\sqrt{4 - x^2 - y^2} + (4 - x^2)\arcsin\left(\frac{y}{\sqrt{4 - x^2}}\right) \right]_0^{\sqrt{4-x^2}} dx \\
&= 8\int_0^2 \left[0 + (4 - x^2)\arcsin(1) - 0 - 0 \right] dx \\
&= 8\int_0^2 (4 - x^2)\left(\frac{\pi}{2}\right) dx \\
&= 4\pi\left[4x - \frac{x^3}{3} \right]_0^2 \\
&= \frac{64\pi}{3}.
\end{aligned}$$

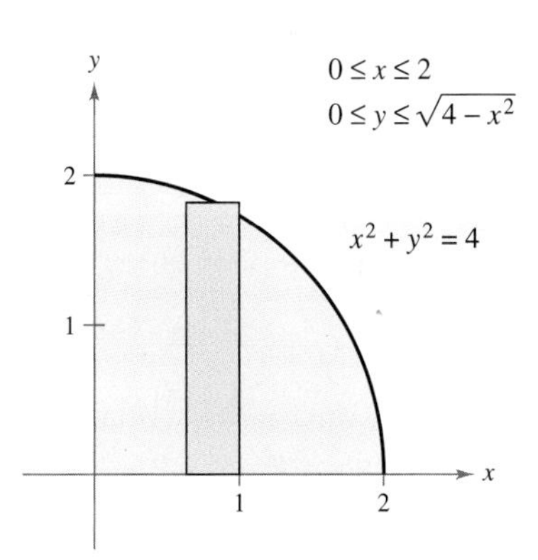

Figure 14.55

Example 2 is unusual in that all six possible orders of integration produce integrals of comparable difficulty. Try setting up some other possible orders of integration to find the volume of the ellipsoid. For instance, the order $dx\,dy\,dz$ yields the integral

$$V = 8\int_0^4\int_0^{\sqrt{16-z^2}/2}\int_0^{\sqrt{16-4y^2-z^2}/2} dx\,dy\,dz.$$

If you solve this integral, you will obtain the same volume obtained in Example 2. This is always the case—the order of integration does not affect the value of the integral. However, the order of integration often does affect the complexity of the integral. In Example 3, the given order of integration is not convenient, so you can change the order to simplify the problem.

EXAMPLE 3 Changing the Order of Integration

Evaluate $\displaystyle\int_0^{\sqrt{\pi/2}}\int_x^{\sqrt{\pi/2}}\int_1^3 \sin(y^2)\,dz\,dy\,dx.$

Solution Note that after one integration in the given order, you would encounter the integral $2\int \sin(y^2)\,dy$, which is not an elementary function. To avoid this problem, change the order of integration to $dz\,dx\,dy$, so that y is the outer variable. From Figure 14.56, you can see that the solid region Q is given by

$$0 \le x \le \sqrt{\frac{\pi}{2}},\quad x \le y \le \sqrt{\frac{\pi}{2}},\quad 1 \le z \le 3$$

and the projection of Q in the xy-plane yields the bounds

$$0 \le y \le \sqrt{\frac{\pi}{2}}\quad \text{and}\quad 0 \le x \le y.$$

So, evaluating the triple integral using the order $dz\,dx\,dy$ produces

$$\begin{aligned}
\int_0^{\sqrt{\pi/2}}\int_0^y\int_1^3 \sin(y^2)\,dz\,dx\,dy &= \int_0^{\sqrt{\pi/2}}\int_0^y z\sin(y^2)\Big]_1^3\,dx\,dy\\
&= 2\int_0^{\sqrt{\pi/2}}\int_0^y \sin(y^2)\,dx\,dy\\
&= 2\int_0^{\sqrt{\pi/2}} x\sin(y^2)\Big]_0^y\,dy\\
&= 2\int_0^{\sqrt{\pi/2}} y\sin(y^2)\,dy\\
&= -\cos(y^2)\Big]_0^{\sqrt{\pi/2}}\\
&= 1.
\end{aligned}$$

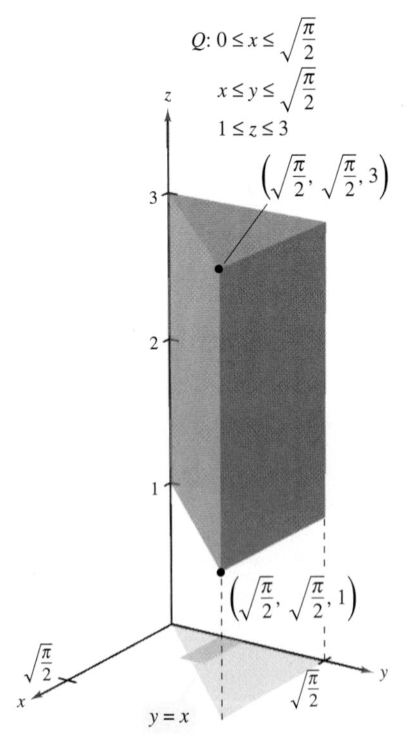

Figure 14.56

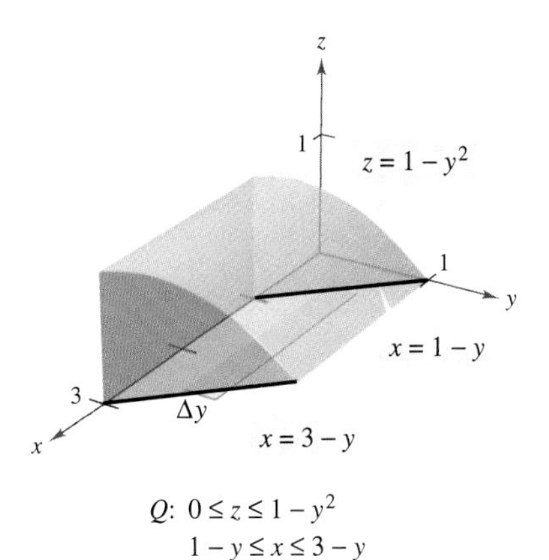

Q: $0 \le z \le 1 - y^2$
$1 - y \le x \le 3 - y$
$0 \le y \le 1$

Figure 14.57

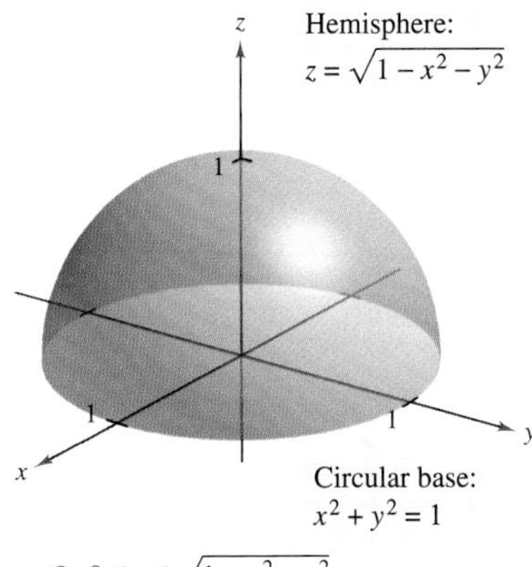

Q: $0 \le z \le \sqrt{1 - x^2 - y^2}$
$-\sqrt{1 - y^2} \le x \le \sqrt{1 - y^2}$
$-1 \le y \le 1$

Figure 14.58

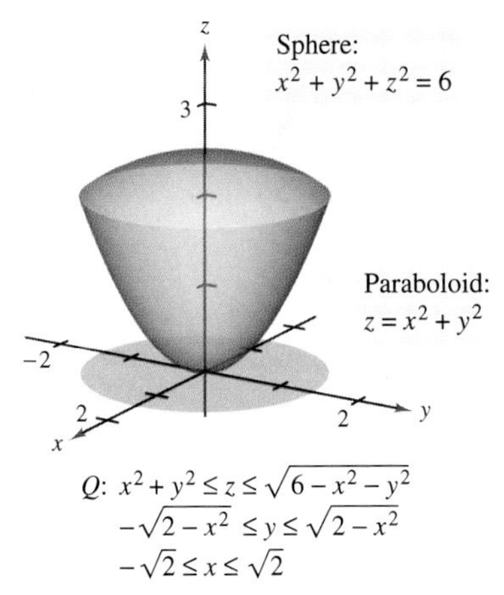

Q: $x^2 + y^2 \le z \le \sqrt{6 - x^2 - y^2}$
$-\sqrt{2 - x^2} \le y \le \sqrt{2 - x^2}$
$-\sqrt{2} \le x \le \sqrt{2}$

Figure 14.59

EXAMPLE 4 Determining the Limits of Integration

Set up a triple integral for the volume of each solid region.

a. The region in the first octant bounded above by the cylinder $z = 1 - y^2$ and lying between the vertical planes $x + y = 1$ and $x + y = 3$

b. The upper hemisphere given by $z = \sqrt{1 - x^2 - y^2}$

c. The region bounded below by the paraboloid $z = x^2 + y^2$ and above by the sphere $x^2 + y^2 + z^2 = 6$

Solution

a. In Figure 14.57, note that the solid is bounded below by the xy-plane ($z = 0$) and above by the cylinder $z = 1 - y^2$. So,

$$0 \le z \le 1 - y^2. \qquad \text{Bounds for } z$$

Projecting the region onto the xy-plane produces a parallelogram. Because two sides of the parallelogram are parallel to the x-axis, you have the following bounds:

$$1 - y \le x \le 3 - y \quad \text{and} \quad 0 \le y \le 1.$$

So, the volume of the region is given by

$$V = \iiint_Q dV = \int_0^1 \int_{1-y}^{3-y} \int_0^{1-y^2} dz\, dx\, dy.$$

b. For the upper hemisphere given by $z = \sqrt{1 - x^2 - y^2}$, you have

$$0 \le z \le \sqrt{1 - x^2 - y^2}. \qquad \text{Bounds for } z$$

In Figure 14.58, note that the projection of the hemisphere onto the xy-plane is the circle given by $x^2 + y^2 = 1$, and you can use either order $dx\,dy$ or $dy\,dx$. Choosing the first produces

$$-\sqrt{1 - y^2} \le x \le \sqrt{1 - y^2} \quad \text{and} \quad -1 \le y \le 1$$

which implies that the volume of the region is given by

$$V = \iiint_Q dV = \int_{-1}^1 \int_{-\sqrt{1-y^2}}^{\sqrt{1-y^2}} \int_0^{\sqrt{1-x^2-y^2}} dz\, dx\, dy.$$

c. For the region bounded below by the paraboloid $z = x^2 + y^2$ and above by the sphere $x^2 + y^2 + z^2 = 6$, you have

$$x^2 + y^2 \le z \le \sqrt{6 - x^2 - y^2}. \qquad \text{Bounds for } z$$

The sphere and the paraboloid intersect at $z = 2$. Moreover, you can see in Figure 14.59 that the projection of the solid region onto the xy-plane is the circle given by $x^2 + y^2 = 2$. Using the order $dy\,dx$ produces

$$-\sqrt{2 - x^2} \le y \le \sqrt{2 - x^2} \quad \text{and} \quad -\sqrt{2} \le x \le \sqrt{2}$$

which implies that the volume of the region is given by

$$V = \iiint_Q dV = \int_{-\sqrt{2}}^{\sqrt{2}} \int_{-\sqrt{2-x^2}}^{\sqrt{2-x^2}} \int_{x^2+y^2}^{\sqrt{6-x^2-y^2}} dz\, dy\, dx.$$

■

Center of Mass and Moments of Inertia

EXPLORATION

Sketch the solid (of uniform density) bounded by $z = 0$ and

$$z = \frac{1}{1 + x^2 + y^2}$$

where $x^2 + y^2 \le 1$. From your sketch, estimate the coordinates of the center of mass of the solid. Now use a computer algebra system to verify your estimate. What do you observe?

In the remainder of this section, two important engineering applications of triple integrals are discussed. Consider a solid region Q whose density is given by the **density function ρ.** The **center of mass** of a solid region Q of mass m is given by $(\bar{x}, \bar{y}, \bar{z})$, where

$$m = \iiint_Q \rho(x, y, z)\, dV \qquad \text{Mass of the solid}$$

$$M_{yz} = \iiint_Q x\rho(x, y, z)\, dV \qquad \text{First moment about } yz\text{-plane}$$

$$M_{xz} = \iiint_Q y\rho(x, y, z)\, dV \qquad \text{First moment about } xz\text{-plane}$$

$$M_{xy} = \iiint_Q z\rho(x, y, z)\, dV \qquad \text{First moment about } xy\text{-plane}$$

and

$$\bar{x} = \frac{M_{yz}}{m}, \quad \bar{y} = \frac{M_{xz}}{m}, \quad \bar{z} = \frac{M_{xy}}{m}.$$

The quantities M_{yz}, M_{xz}, and M_{xy} are called the **first moments** of the region Q about the yz-, xz-, and xy-planes, respectively.

The first moments for solid regions are taken about a plane, whereas the second moments for solids are taken about a line. The **second moments** (or **moments of inertia**) about the x-, y-, and z-axes are as follows.

NOTE In engineering and physics, the moment of inertia of a mass is used to find the time required for the mass to reach a given speed of rotation about an axis, as shown in Figure 14.60. The greater the moment of inertia, the longer a force must be applied for the mass to reach the given speed.

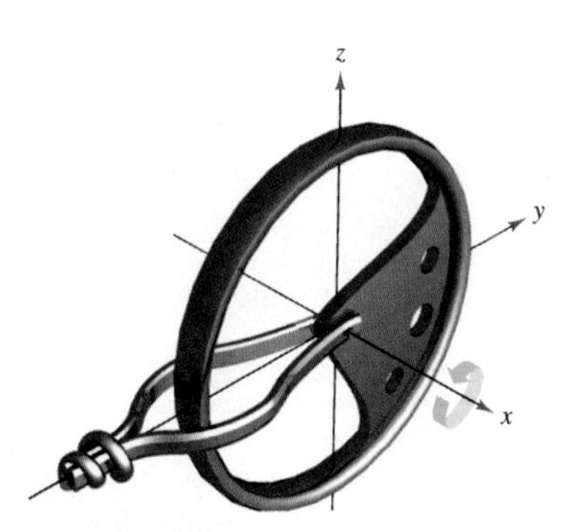

Figure 14.60

$$I_x = \iiint_Q (y^2 + z^2)\rho(x, y, z)\, dV \qquad \text{Moment of inertia about } x\text{-axis}$$

$$I_y = \iiint_Q (x^2 + z^2)\rho(x, y, z)\, dV \qquad \text{Moment of inertia about } y\text{-axis}$$

$$I_z = \iiint_Q (x^2 + y^2)\rho(x, y, z)\, dV \qquad \text{Moment of inertia about } z\text{-axis}$$

For problems requiring the calculation of all three moments, considerable effort can be saved by applying the additive property of triple integrals and writing

$$I_x = I_{xz} + I_{xy}, \quad I_y = I_{yz} + I_{xy}, \quad \text{and} \quad I_z = I_{yz} + I_{xz}$$

where I_{xy}, I_{xz}, and I_{yz} are as follows.

$$I_{xy} = \iiint_Q z^2\rho(x, y, z)\, dV$$

$$I_{xz} = \iiint_Q y^2\rho(x, y, z)\, dV$$

$$I_{yz} = \iiint_Q x^2\rho(x, y, z)\, dV$$

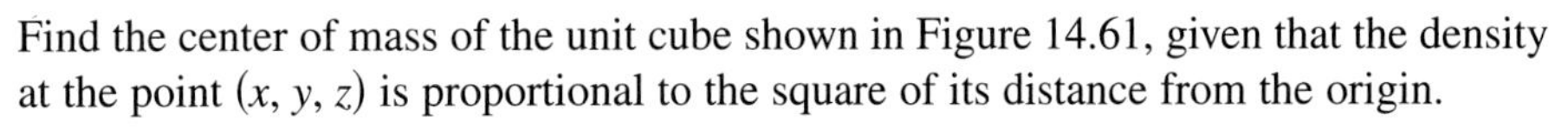

EXAMPLE 5 Finding the Center of Mass of a Solid Region

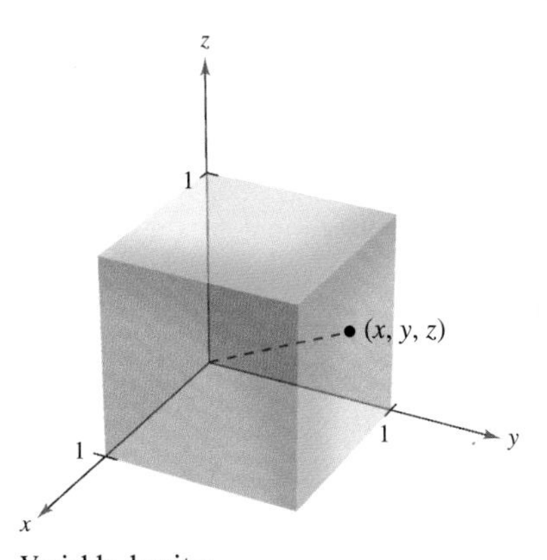

Variable density:
$\rho(x, y, z) = k(x^2 + y^2 + z^2)$
Figure 14.61

Find the center of mass of the unit cube shown in Figure 14.61, given that the density at the point (x, y, z) is proportional to the square of its distance from the origin.

Solution Because the density at (x, y, z) is proportional to the square of the distance between $(0, 0, 0)$ and (x, y, z), you have

$$\rho(x, y, z) = k(x^2 + y^2 + z^2).$$

You can use this density function to find the mass of the cube. Because of the symmetry of the region, any order of integration will produce an integral of comparable difficulty.

$$\begin{aligned}
m &= \int_0^1 \int_0^1 \int_0^1 k(x^2 + y^2 + z^2)\, dz\, dy\, dx \\
&= k \int_0^1 \int_0^1 \left[(x^2 + y^2)z + \frac{z^3}{3} \right]_0^1 dy\, dx \\
&= k \int_0^1 \int_0^1 \left(x^2 + y^2 + \frac{1}{3} \right) dy\, dx \\
&= k \int_0^1 \left[\left(x^2 + \frac{1}{3} \right) y + \frac{y^3}{3} \right]_0^1 dx \\
&= k \int_0^1 \left(x^2 + \frac{2}{3} \right) dx \\
&= k \left[\frac{x^3}{3} + \frac{2x}{3} \right]_0^1 = k
\end{aligned}$$

The first moment about the yz-plane is

$$\begin{aligned}
M_{yz} &= k \int_0^1 \int_0^1 \int_0^1 x(x^2 + y^2 + z^2)\, dz\, dy\, dx \\
&= k \int_0^1 x \left[\int_0^1 \int_0^1 (x^2 + y^2 + z^2)\, dz\, dy \right] dx.
\end{aligned}$$

Note that x can be factored out of the two inner integrals, because it is constant with respect to y and z. After factoring, the two inner integrals are the same as for the mass m. Therefore, you have

$$\begin{aligned}
M_{yz} &= k \int_0^1 x \left(x^2 + \frac{2}{3} \right) dx \\
&= k \left[\frac{x^4}{4} + \frac{x^2}{3} \right]_0^1 \\
&= \frac{7k}{12}.
\end{aligned}$$

So,

$$\bar{x} = \frac{M_{yz}}{m} = \frac{7k/12}{k} = \frac{7}{12}.$$

Finally, from the nature of ρ and the symmetry of x, y, and z in this solid region, you have $\bar{x} = \bar{y} = \bar{z}$, and the center of mass is $\left(\frac{7}{12}, \frac{7}{12}, \frac{7}{12}\right)$. ■

EXAMPLE 6 Moments of Inertia for a Solid Region

Find the moments of inertia about the x- and y-axes for the solid region lying between the hemisphere

$$z = \sqrt{4 - x^2 - y^2}$$

and the xy-plane, given that the density at (x, y, z) is proportional to the distance between (x, y, z) and the xy-plane.

$0 \le z \le \sqrt{4 - x^2 - y^2}$
$-\sqrt{4 - x^2} \le y \le \sqrt{4 - x^2}$
$-2 \le x \le 2$

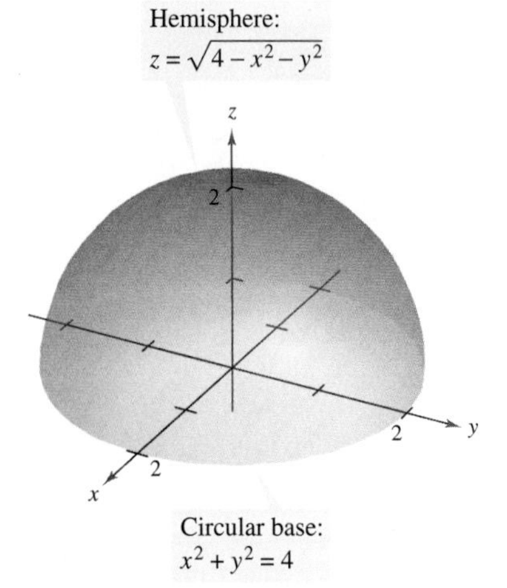

Variable density: $\rho(x, y, z) = kz$
Figure 14.62

Solution The density of the region is given by $\rho(x, y, z) = kz$. Considering the symmetry of this problem, you know that $I_x = I_y$, and you need to compute only one moment, say I_x. From Figure 14.62, choose the order $dz\,dy\,dx$ and write

$$\begin{aligned}
I_x &= \iiint_Q (y^2 + z^2)\rho(x, y, z)\,dV \\
&= \int_{-2}^{2}\int_{-\sqrt{4-x^2}}^{\sqrt{4-x^2}}\int_{0}^{\sqrt{4-x^2-y^2}} (y^2 + z^2)(kz)\,dz\,dy\,dx \\
&= k\int_{-2}^{2}\int_{-\sqrt{4-x^2}}^{\sqrt{4-x^2}} \left[\frac{y^2z^2}{2} + \frac{z^4}{4}\right]_0^{\sqrt{4-x^2-y^2}} dy\,dx \\
&= k\int_{-2}^{2}\int_{-\sqrt{4-x^2}}^{\sqrt{4-x^2}} \left[\frac{y^2(4 - x^2 - y^2)}{2} + \frac{(4 - x^2 - y^2)^2}{4}\right] dy\,dx \\
&= \frac{k}{4}\int_{-2}^{2}\int_{-\sqrt{4-x^2}}^{\sqrt{4-x^2}} [(4 - x^2)^2 - y^4]\,dy\,dx \\
&= \frac{k}{4}\int_{-2}^{2} \left[(4 - x^2)^2 y - \frac{y^5}{5}\right]_{-\sqrt{4-x^2}}^{\sqrt{4-x^2}} dx \\
&= \frac{k}{4}\int_{-2}^{2} \frac{8}{5}(4 - x^2)^{5/2}\,dx \\
&= \frac{4k}{5}\int_{0}^{2} (4 - x^2)^{5/2}\,dx && x = 2\sin\theta \\
&= \frac{4k}{5}\int_{0}^{\pi/2} 64\cos^6\theta\,d\theta \\
&= \left(\frac{256k}{5}\right)\left(\frac{5\pi}{32}\right) && \text{Wallis's Formula} \\
&= 8k\pi.
\end{aligned}$$

So, $I_x = 8k\pi = I_y$. ■

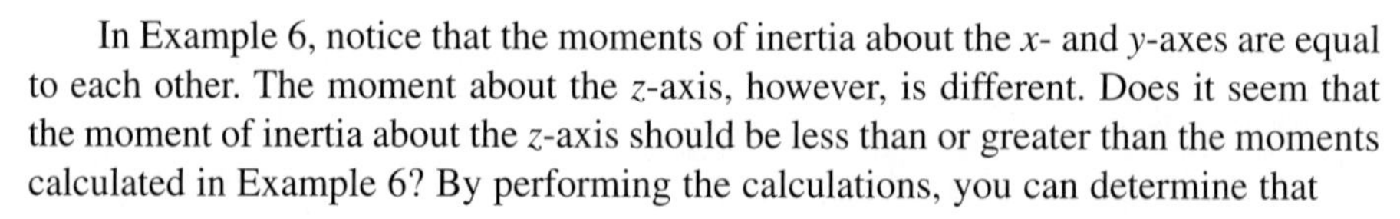

In Example 6, notice that the moments of inertia about the x- and y-axes are equal to each other. The moment about the z-axis, however, is different. Does it seem that the moment of inertia about the z-axis should be less than or greater than the moments calculated in Example 6? By performing the calculations, you can determine that

$$I_z = \frac{16}{3}k\pi.$$

This tells you that the solid shown in Figure 14.62 has a greater resistance to rotation about the x- or y-axis than about the z-axis.

14.6 Exercises

See www.CalcChat.com for worked-out solutions to odd-numbered exercises.

In Exercises 1–8, evaluate the iterated integral.

1. $\int_0^3 \int_0^2 \int_0^1 (x + y + z)\, dx\, dz\, dy$

2. $\int_{-1}^1 \int_{-1}^1 \int_{-1}^1 x^2y^2z^2\, dx\, dy\, dz$

3. $\int_0^1 \int_0^x \int_0^{xy} x\, dz\, dy\, dx$

4. $\int_0^9 \int_0^{y/3} \int_0^{\sqrt{y^2-9x^2}} z\, dz\, dx\, dy$

5. $\int_1^4 \int_0^1 \int_0^x 2ze^{-x^2}\, dy\, dx\, dz$

6. $\int_1^4 \int_1^{e^2} \int_0^{1/xz} \ln z\, dy\, dz\, dx$

7. $\int_0^4 \int_0^{\pi/2} \int_0^{1-x} x \cos y\, dz\, dy\, dx$

8. $\int_0^{\pi/2} \int_0^{y/2} \int_0^{1/y} \sin y\, dz\, dx\, dy$

CAS **In Exercises 9 and 10, use a computer algebra system to evaluate the iterated integral.**

9. $\int_0^3 \int_{-\sqrt{9-y^2}}^{\sqrt{9-y^2}} \int_0^{y^2} y\, dz\, dx\, dy$

10. $\int_0^{\sqrt{2}} \int_0^{\sqrt{2-x^2}} \int_{2x^2+y^2}^{4-y^2} y\, dz\, dy\, dx$

CAS **In Exercises 11 and 12, use a computer algebra system to approximate the iterated integral.**

11. $\int_0^2 \int_0^{\sqrt{4-x^2}} \int_1^4 \frac{x^2 \sin y}{z}\, dz\, dy\, dx$

12. $\int_0^3 \int_0^{2-(2y/3)} \int_0^{6-2y-3z} ze^{-x^2y^2}\, dx\, dz\, dy$

In Exercises 13–18, set up a triple integral for the volume of the solid.

13. The solid in the first octant bounded by the coordinate planes and the plane $z = 5 - x - y$

14. The solid bounded by $z = 9 - x^2$, $z = 0$, $y = 0$, and $y = 2x$

15. The solid bounded by $z = 6 - x^2 - y^2$ and $z = 0$

16. The solid bounded by $z = \sqrt{16 - x^2 - y^2}$ and $z = 0$

17. The solid that is the common interior below the sphere $x^2 + y^2 + z^2 = 80$ and above the paraboloid $z = \frac{1}{2}(x^2 + y^2)$

18. The solid bounded above by the cylinder $z = 4 - x^2$ and below by the paraboloid $z = x^2 + 3y^2$

Volume **In Exercises 19–22, use a triple integral to find the volume of the solid shown in the figure.**

19.

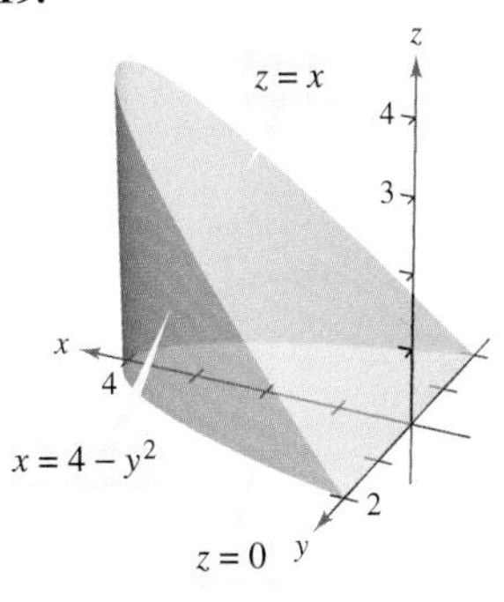

20.

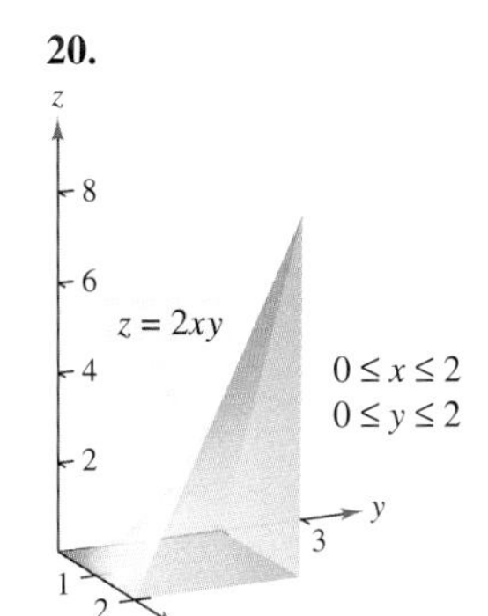

21.

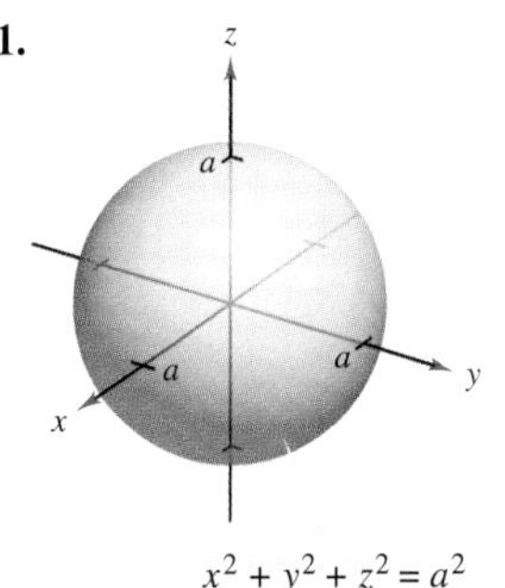

22.

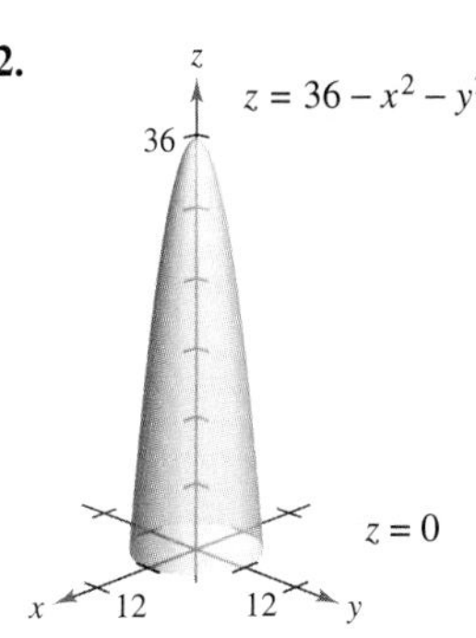

Volume **In Exercises 23–26, use a triple integral to find the volume of the solid bounded by the graphs of the equations.**

23. $z = 4 - x^2$, $y = 4 - x^2$, first octant

24. $z = 9 - x^3$, $y = -x^2 + 2$, $y = 0$, $z = 0$, $x \geq 0$

25. $z = 2 - y$, $z = 4 - y^2$, $x = 0$, $x = 3$, $y = 0$

26. $z = x$, $y = x + 2$, $y = x^2$, first octant

In Exercises 27–32, sketch the solid whose volume is given by the iterated integral and rewrite the integral using the indicated order of integration.

27. $\int_0^1 \int_{-1}^0 \int_0^{y^2} dz\, dy\, dx$

Rewrite using the order $dy\, dz\, dx$.

28. $\int_{-1}^1 \int_{y^2}^1 \int_0^{1-x} dz\, dx\, dy$

Rewrite using the order $dx\, dz\, dy$.

29. $\int_0^4 \int_0^{(4-x)/2} \int_0^{(12-3x-6y)/4} dz\, dy\, dx$

Rewrite using the order $dy\, dx\, dz$.

30. $\int_0^3 \int_0^{\sqrt{9-x^2}} \int_0^{6-x-y} dz\, dy\, dx$

Rewrite using the order $dz\, dx\, dy$.

31. $\int_0^1 \int_y^1 \int_0^{\sqrt{1-y^2}} dz\, dx\, dy$

Rewrite using the order $dz\, dy\, dx$.

32. $\int_0^2 \int_{2x}^4 \int_0^{\sqrt{y^2-4x^2}} dz\, dy\, dx$

Rewrite using the order $dx\, dy\, dz$.

In Exercises 33–36, list the six possible orders of integration for the triple integral over the solid region Q, $\iiint_Q xyz\, dV$.

33. $Q = \{(x, y, z): 0 \leq x \leq 1, 0 \leq y \leq x, 0 \leq z \leq 3\}$

34. $Q = \{(x, y, z): 0 \leq x \leq 2, x^2 \leq y \leq 4, 0 \leq z \leq 2 - x\}$

35. $Q = \{(x, y, z): x^2 + y^2 \leq 9, 0 \leq z \leq 4\}$

36. $Q = \{(x, y, z): 0 \leq x \leq 1, y \leq 1 - x^2, 0 \leq z \leq 6\}$

In Exercises 37 and 38, the figure shows the region of integration for the given integral. Rewrite the integral as an equivalent iterated integral in the five other orders.

37. $\displaystyle\int_0^1 \int_0^{1-y^2} \int_0^{1-y} dz\, dx\, dy$ **38.** $\displaystyle\int_0^3 \int_0^x \int_0^{9-x^2} dz\, dy\, dx$

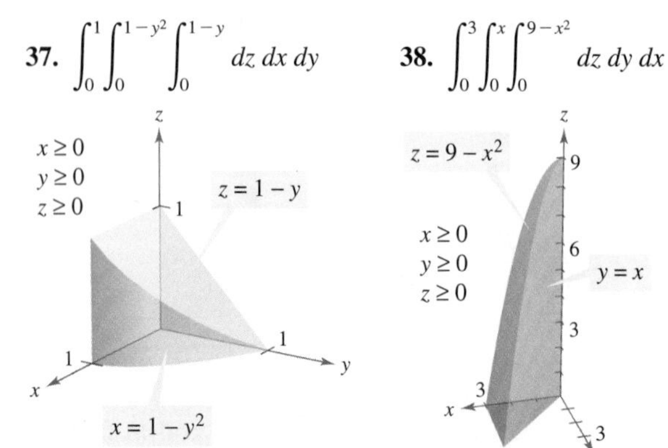

Mass and Center of Mass **In Exercises 39–42, find the mass and the indicated coordinates of the center of mass of the solid of given density bounded by the graphs of the equations.**

39. Find $\bar{x}$ using $\rho(x, y, z) = k$.

Q: $2x + 3y + 6z = 12, x = 0, y = 0, z = 0$

40. Find $\bar{y}$ using $\rho(x, y, z) = ky$.

Q: $3x + 3y + 5z = 15, x = 0, y = 0, z = 0$

41. Find $\bar{z}$ using $\rho(x, y, z) = kx$.

Q: $z = 4 - x, z = 0, y = 0, y = 4, x = 0$

42. Find $\bar{y}$ using $\rho(x, y, z) = k$.

Q: $\dfrac{x}{a} + \dfrac{y}{b} + \dfrac{z}{c} = 1\ (a, b, c > 0), x = 0, y = 0, z = 0$

Mass and Center of Mass **In Exercises 43 and 44, set up the triple integrals for finding the mass and the center of mass of the solid bounded by the graphs of the equations.**

43. $x = 0, x = b, y = 0, y = b, z = 0, z = b$

$\rho(x, y, z) = kxy$

44. $x = 0, x = a, y = 0, y = b, z = 0, z = c$

$\rho(x, y, z) = kz$

Think About It **The center of mass of a solid of constant density is shown in the figure. In Exercises 45–48, make a conjecture about how the center of mass $(\bar{x}, \bar{y}, \bar{z})$ will change for the nonconstant density $\rho(x, y, z)$. Explain.**

45. $\rho(x, y, z) = kx$

46. $\rho(x, y, z) = kz$

47. $\rho(x, y, z) = k(y + 2)$

48. $\rho(x, y, z) = kxz^2(y + 2)^2$

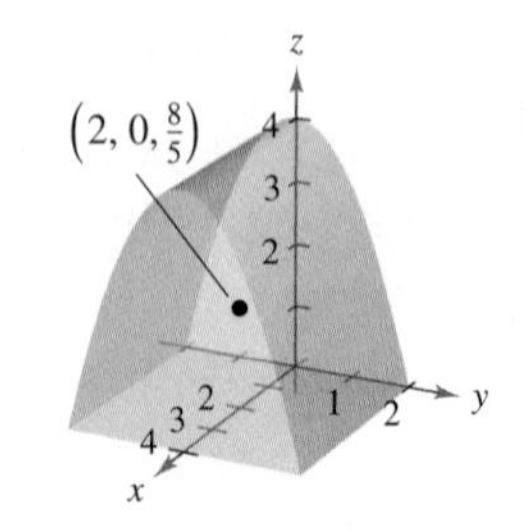

CAS *Centroid* **In Exercises 49–54, find the centroid of the solid region bounded by the graphs of the equations or described by the figure. Use a computer algebra system to evaluate the triple integrals. (Assume uniform density and find the center of mass.)**

49. $z = \dfrac{h}{r}\sqrt{x^2 + y^2}, z = h$ **50.** $y = \sqrt{9 - x^2}, z = y, z = 0$

51. $z = \sqrt{16 - x^2 - y^2}, z = 0$

52. $z = \dfrac{1}{y^2 + 1}, z = 0, x = -2, x = 2, y = 0, y = 1$

53.

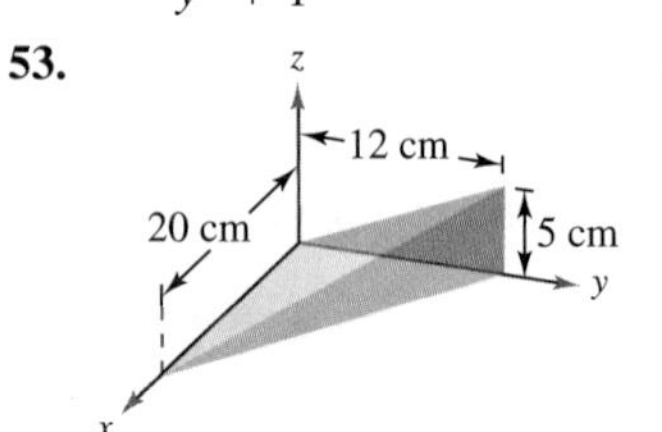

54.

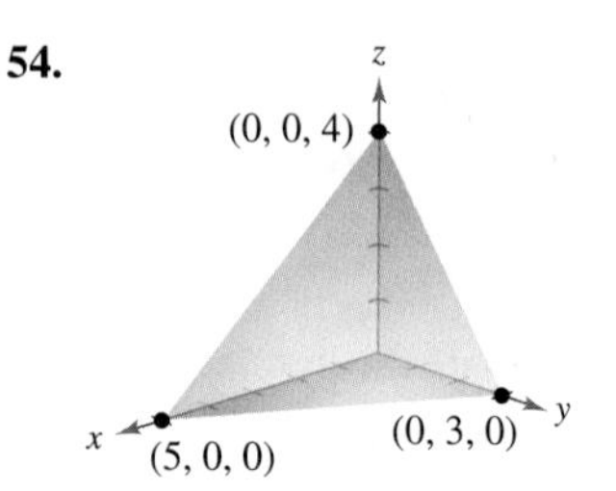

CAS *Moments of Inertia* **In Exercises 55–58, find I_x, I_y, and I_z for the solid of given density. Use a computer algebra system to evaluate the triple integrals.**

55. (a) $\rho = k$
(b) $\rho = kxyz$

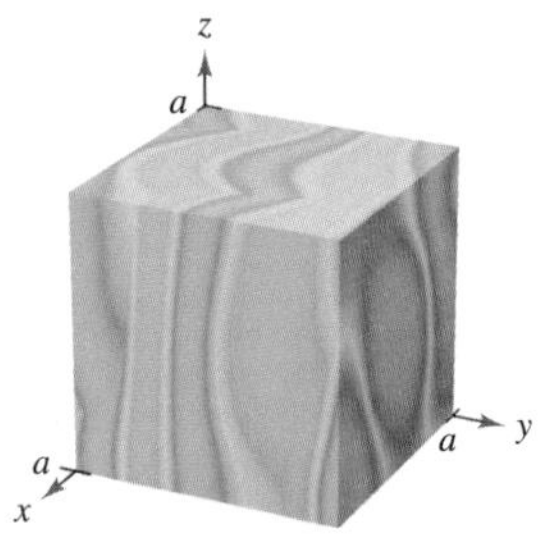

56. (a) $\rho(x, y, z) = k$

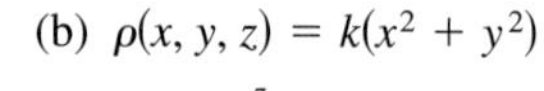

(b) $\rho(x, y, z) = k(x^2 + y^2)$

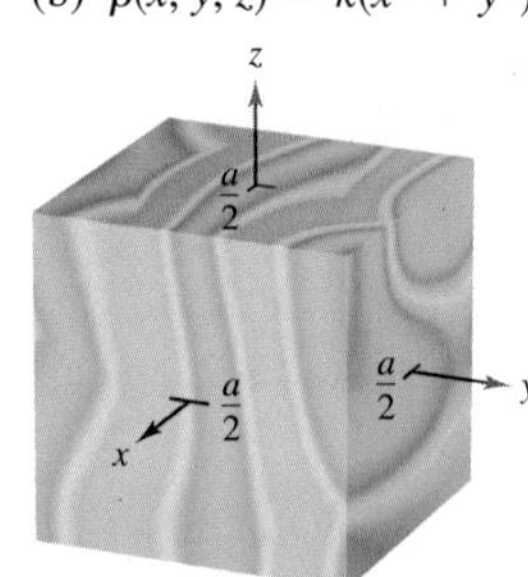

57. (a) $\rho(x, y, z) = k$
(b) $\rho = ky$

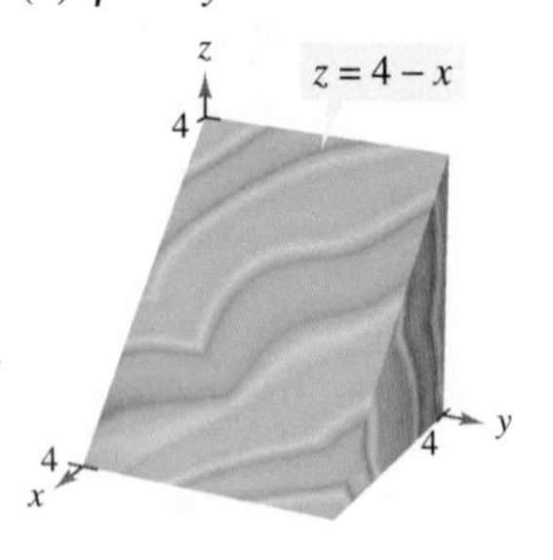

58. (a) $\rho = kz$
(b) $\rho = k(4 - z)$

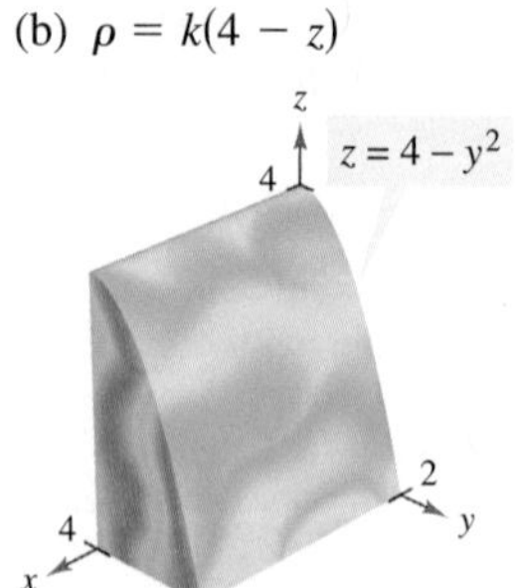

CAS *Moments of Inertia* **In Exercises 59 and 60, verify the moments of inertia for the solid of uniform density. Use a computer algebra system to evaluate the triple integrals.**

59. $I_x = \frac{1}{12}m(3a^2 + L^2)$

$I_y = \frac{1}{2}ma^2$

$I_z = \frac{1}{12}m(3a^2 + L^2)$

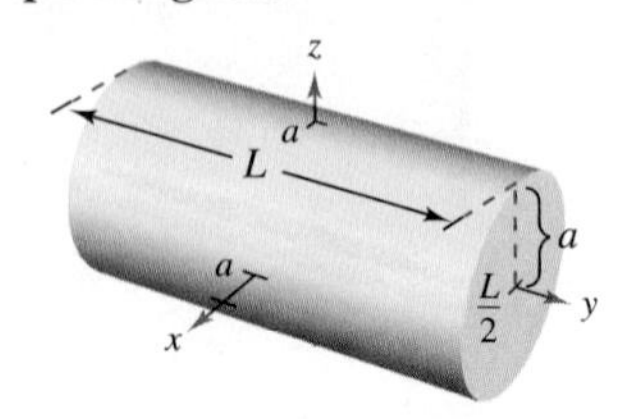

60. $I_x = \frac{1}{12}m(a^2 + b^2)$
$I_y = \frac{1}{12}m(b^2 + c^2)$
$I_z = \frac{1}{12}m(a^2 + c^2)$

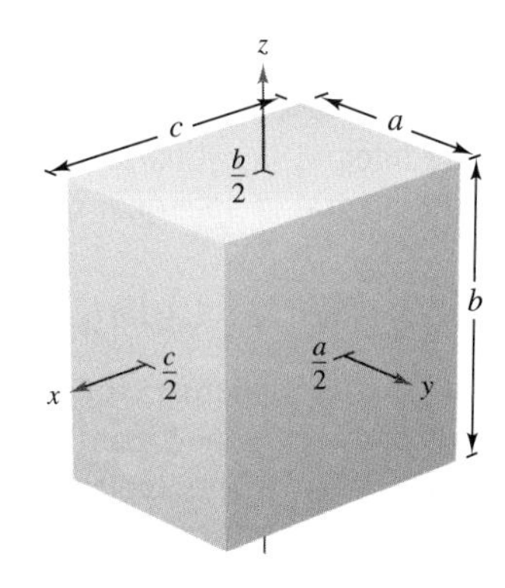

Moments of Inertia **In Exercises 61 and 62, set up a triple integral that gives the moment of inertia about the z-axis of the solid region Q of density ρ.**

61. $Q = \{(x, y, z): -1 \le x \le 1, -1 \le y \le 1, 0 \le z \le 1 - x\}$
$\rho = \sqrt{x^2 + y^2 + z^2}$

62. $Q = \{(x, y, z): x^2 + y^2 \le 1, 0 \le z \le 4 - x^2 - y^2\}$
$\rho = kx^2$

In Exercises 63 and 64, using the description of the solid region, set up the integral for (a) the mass, (b) the center of mass, and (c) the moment of inertia about the z-axis.

63. The solid bounded by $z = 4 - x^2 - y^2$ and $z = 0$ with density function $\rho = kz$

64. The solid in the first octant bounded by the coordinate planes and $x^2 + y^2 + z^2 = 25$ with density function $\rho = kxy$

WRITING ABOUT CONCEPTS

65. Define a triple integral and describe a method of evaluating a triple integral.

66. Determine whether the moment of inertia about the y-axis of the cylinder in Exercise 59 will increase or decrease for the nonconstant density $\rho(x, y, z) = \sqrt{x^2 + z^2}$ and $a = 4$.

67. Consider two solids, solid A and solid B, of equal weight as shown below.

(a) Because the solids have the same weight, which has the greater density?

(b) Which solid has the greater moment of inertia? Explain.

(c) The solids are rolled down an inclined plane. They are started at the same time and at the same height. Which will reach the bottom first? Explain.

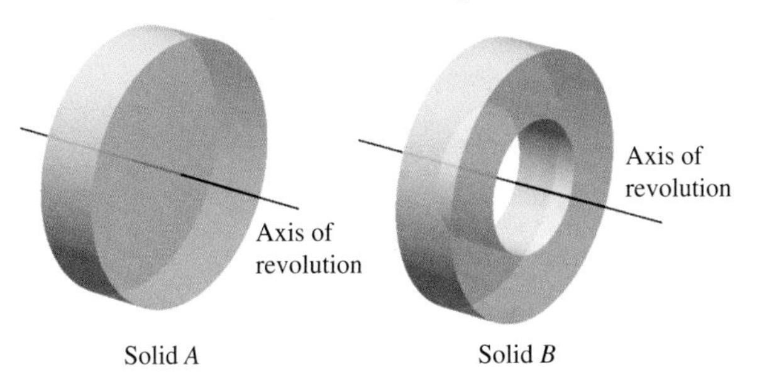

CAPSTONE

68. ***Think About It*** Of the integrals (a)–(c), which one is equal to $\int_1^3 \int_0^2 \int_{-1}^1 f(x, y, z)\, dz\, dy\, dx$? Explain.

(a) $\int_1^3 \int_0^2 \int_{-1}^1 f(x, y, z)\, dz\, dx\, dy$

(b) $\int_{-1}^1 \int_0^2 \int_1^3 f(x, y, z)\, dx\, dy\, dz$

(c) $\int_0^2 \int_1^3 \int_{-1}^1 f(x, y, z)\, dy\, dx\, dz$

Average Value **In Exercises 69–72, find the average value of the function over the given solid. The average value of a continuous function $f(x, y, z)$ over a solid region Q is**

$$\frac{1}{V}\iiint_Q f(x, y, z)\, dV$$

where V is the volume of the solid region Q.

69. $f(x, y, z) = z^2 + 4$ over the cube in the first octant bounded by the coordinate planes and the planes $x = 1$, $y = 1$, and $z = 1$

70. $f(x, y, z) = xyz$ over the cube in the first octant bounded by the coordinate planes and the planes $x = 4$, $y = 4$, and $z = 4$

71. $f(x, y, z) = x + y + z$ over the tetrahedron in the first octant with vertices $(0, 0, 0)$, $(2, 0, 0)$, $(0, 2, 0)$ and $(0, 0, 2)$

72. $f(x, y, z) = x + y$ over the solid bounded by the sphere $x^2 + y^2 + z^2 = 3$

CAS **73.** Find the solid region Q where the triple integral

$$\iiint_Q (1 - 2x^2 - y^2 - 3z^2)\, dV$$

is a maximum. Use a computer algebra system to approximate the maximum value. What is the exact maximum value?

CAS **74.** Find the solid region Q where the triple integral

$$\iiint_Q (1 - x^2 - y^2 - z^2)\, dV$$

is a maximum. Use a computer algebra system to approximate the maximum value. What is the exact maximum value?

75. Solve for a in the triple integral.

$$\int_0^1 \int_0^{3-a-y^2} \int_a^{4-x-y^2} dz\, dx\, dy = \frac{14}{15}$$

76. Determine the value of b such that the volume of the ellipsoid $x^2 + (y^2/b^2) + (z^2/9) = 1$ is 16π.

PUTNAM EXAM CHALLENGE

77. Evaluate

$$\lim_{n\to\infty} \int_0^1 \int_0^1 \cdots \int_0^1 \cos^2\left\{\frac{\pi}{2n}(x_1 + x_2 + \cdots + x_n)\right\} dx_1\, dx_2 \cdots dx_n.$$

This problem was composed by the Committee on the Putnam Prize Competition.

14.7 Triple Integrals in Cylindrical and Spherical Coordinates

- Write and evaluate a triple integral in cylindrical coordinates.
- Write and evaluate a triple integral in spherical coordinates.

PIERRE SIMON DE LAPLACE (1749–1827)

One of the first to use a cylindrical coordinate system was the French mathematician Pierre Simon de Laplace. Laplace has been called the "Newton of France," and he published many important works in mechanics, differential equations, and probability.

Triple Integrals in Cylindrical Coordinates

Many common solid regions such as spheres, ellipsoids, cones, and paraboloids can yield difficult triple integrals in rectangular coordinates. In fact, it is precisely this difficulty that led to the introduction of nonrectangular coordinate systems. In this section, you will learn how to use *cylindrical* and *spherical* coordinates to evaluate triple integrals.

Recall from Section 11.7 that the rectangular conversion equations for cylindrical coordinates are

$$x = r\cos\theta$$
$$y = r\sin\theta$$
$$z = z.$$

STUDY TIP An easy way to remember these conversions is to note that the equations for x and y are the same as in polar coordinates and z is unchanged.

In this coordinate system, the simplest solid region is a cylindrical block determined by

$$r_1 \le r \le r_2, \quad \theta_1 \le \theta \le \theta_2, \quad z_1 \le z \le z_2$$

as shown in Figure 14.63. To obtain the cylindrical coordinate form of a triple integral, suppose that Q is a solid region whose projection R onto the xy-plane can be described in polar coordinates. That is,

$$Q = \{(x, y, z): (x, y) \text{ is in } R, \quad h_1(x, y) \le z \le h_2(x, y)\}$$

and

$$R = \{(r, \theta): \theta_1 \le \theta \le \theta_2, \quad g_1(\theta) \le r \le g_2(\theta)\}.$$

If f is a continuous function on the solid Q, you can write the triple integral of f over Q as

$$\iiint_Q f(x, y, z)\, dV = \iint_R \left[\int_{h_1(x, y)}^{h_2(x, y)} f(x, y, z)\, dz\right] dA$$

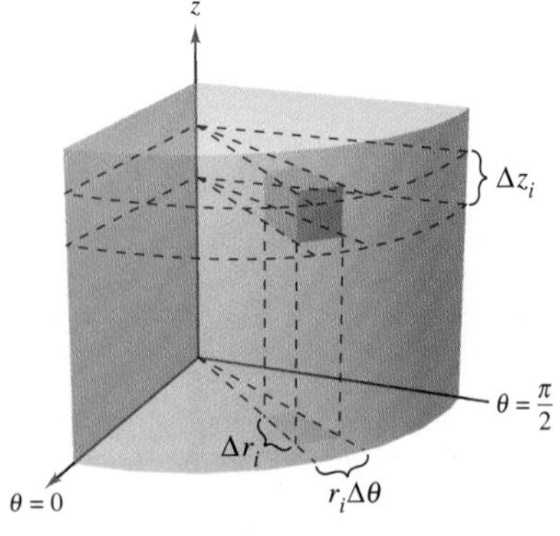

Volume of cylindrical block:
$\Delta V_i = r_i \Delta r_i \Delta\theta_i \Delta z_i$
Figure 14.63

where the double integral over R is evaluated in polar coordinates. That is, R is a plane region that is either r-simple or θ-simple. If R is r-simple, the iterated form of the triple integral in cylindrical form is

$$\iiint_Q f(x, y, z)\, dV = \int_{\theta_1}^{\theta_2}\int_{g_1(\theta)}^{g_2(\theta)}\int_{h_1(r\cos\theta,\, r\sin\theta)}^{h_2(r\cos\theta,\, r\sin\theta)} f(r\cos\theta, r\sin\theta, z) r\, dz\, dr\, d\theta.$$

NOTE This is only one of six possible orders of integration. The other five are $dz\,d\theta\,dr$, $dr\,dz\,d\theta$, $dr\,d\theta\,dz$, $d\theta\,dz\,dr$, and $d\theta\,dr\,dz$.

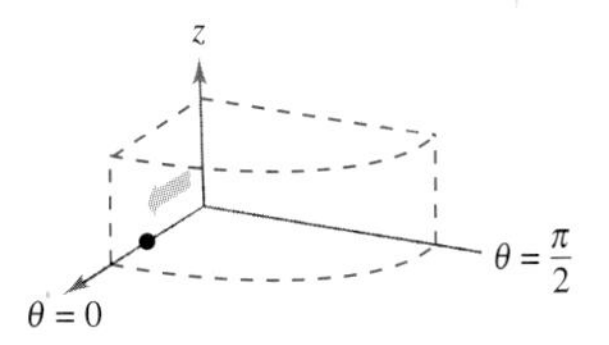

Integrate with respect to r.

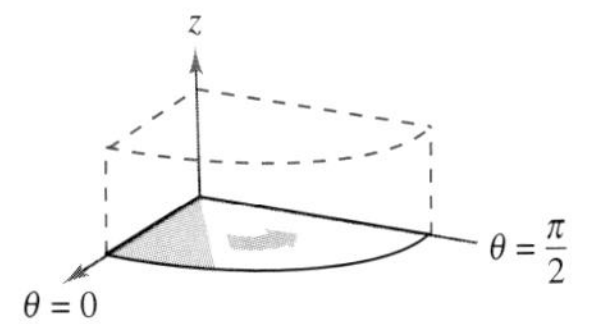

Integrate with respect to θ.

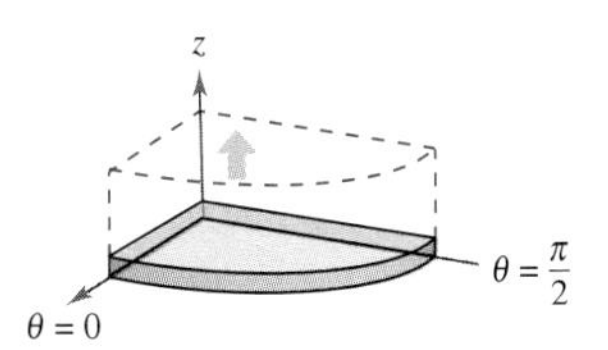

Integrate with respect to z.

Figure 14.64

To visualize a particular order of integration, it helps to view the iterated integral in terms of three sweeping motions—each adding another dimension to the solid. For instance, in the order $dr\,d\theta\,dz$, the first integration occurs in the r-direction as a point sweeps out a ray. Then, as θ increases, the line sweeps out a sector. Finally, as z increases, the sector sweeps out a solid wedge, as shown in Figure 14.64.

EXPLORATION

Volume of a Paraboloid Sector On pages 997, 1006, and 1028, you were asked to summarize the different ways you know of finding the volume of the solid bounded by the paraboloid

$$z = a^2 - x^2 - y^2, \quad a > 0$$

and the xy-plane. You now know one more way. Use it to find the volume of the solid. Compare the different methods. What are the advantages and disadvantages of each?

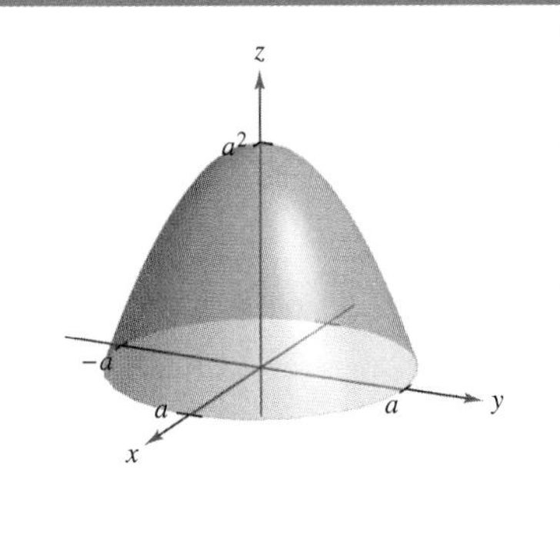

EXAMPLE 1 Finding Volume in Cylindrical Coordinates

Find the volume of the solid region Q cut from the sphere $x^2 + y^2 + z^2 = 4$ by the cylinder $r = 2 \sin \theta$, as shown in Figure 14.65.

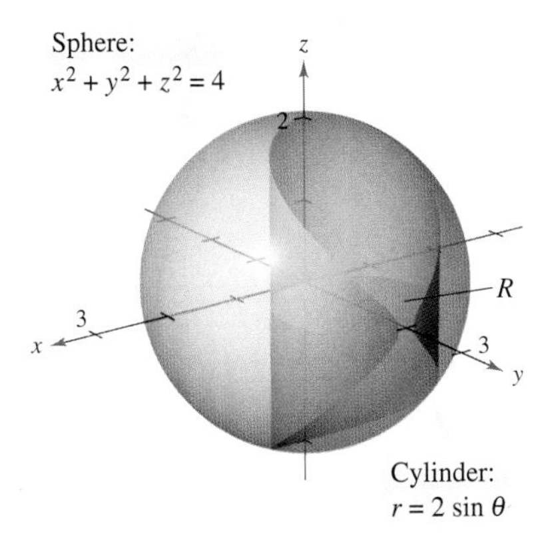

Figure 14.65

Solution Because $x^2 + y^2 + z^2 = r^2 + z^2 = 4$, the bounds on z are

$$-\sqrt{4 - r^2} \le z \le \sqrt{4 - r^2}.$$

Let R be the circular projection of the solid onto the $r\theta$-plane. Then the bounds on R are $0 \le r \le 2 \sin \theta$ and $0 \le \theta \le \pi$. So, the volume of Q is

$$\begin{aligned}
V &= \int_0^{\pi} \int_0^{2 \sin \theta} \int_{-\sqrt{4-r^2}}^{\sqrt{4-r^2}} r\,dz\,dr\,d\theta \\
&= 2 \int_0^{\pi/2} \int_0^{2 \sin \theta} \int_{-\sqrt{4-r^2}}^{\sqrt{4-r^2}} r\,dz\,dr\,d\theta \\
&= 2 \int_0^{\pi/2} \int_0^{2 \sin \theta} 2r\sqrt{4 - r^2}\,dr\,d\theta \\
&= 2 \int_0^{\pi/2} -\frac{2}{3}(4 - r^2)^{3/2} \Big]_0^{2 \sin \theta} d\theta \\
&= \frac{4}{3} \int_0^{\pi/2} (8 - 8 \cos^3 \theta)\,d\theta \\
&= \frac{32}{3} \int_0^{\pi/2} [1 - (\cos \theta)(1 - \sin^2 \theta)]\,d\theta \\
&= \frac{32}{3}\left[\theta - \sin \theta + \frac{\sin^3 \theta}{3}\right]_0^{\pi/2} \\
&= \frac{16}{9}(3\pi - 4) \\
&\approx 9.644.
\end{aligned}$$

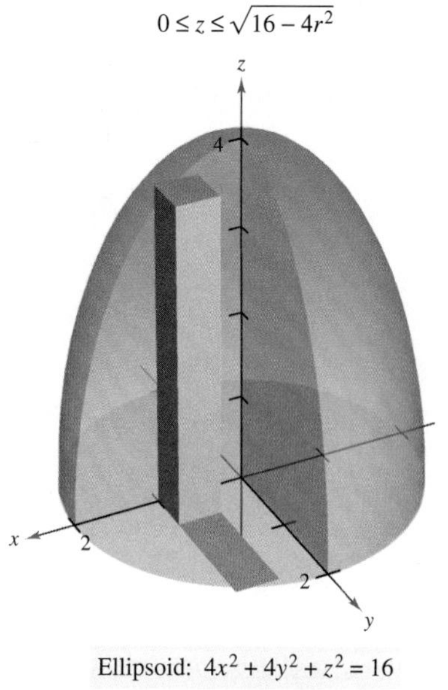

EXAMPLE 2 Finding Mass in Cylindrical Coordinates

Find the mass of the ellipsoid Q given by $4x^2 + 4y^2 + z^2 = 16$, lying above the xy-plane. The density at a point in the solid is proportional to the distance between the point and the xy-plane.

Solution The density function is $\rho(r, \theta, z) = kz$. The bounds on z are

$$0 \le z \le \sqrt{16 - 4x^2 - 4y^2} = \sqrt{16 - 4r^2}$$

where $0 \le r \le 2$ and $0 \le \theta \le 2\pi$, as shown in Figure 14.66. The mass of the solid is

$$\begin{aligned} m &= \int_0^{2\pi}\int_0^2\int_0^{\sqrt{16-4r^2}} kzr\,dz\,dr\,d\theta \\ &= \frac{k}{2}\int_0^{2\pi}\int_0^2 z^2r\Big]_0^{\sqrt{16-4r^2}} dr\,d\theta \\ &= \frac{k}{2}\int_0^{2\pi}\int_0^2 (16r - 4r^3)\,dr\,d\theta \\ &= \frac{k}{2}\int_0^{2\pi}\Big[8r^2 - r^4\Big]_0^2 d\theta \\ &= 8k\int_0^{2\pi} d\theta = 16\pi k. \end{aligned}$$

$0 \le z \le \sqrt{16 - 4r^2}$

Ellipsoid: $4x^2 + 4y^2 + z^2 = 16$

Figure 14.66

Integration in cylindrical coordinates is useful when factors involving $x^2 + y^2$ appear in the integrand, as illustrated in Example 3.

EXAMPLE 3 Finding a Moment of Inertia

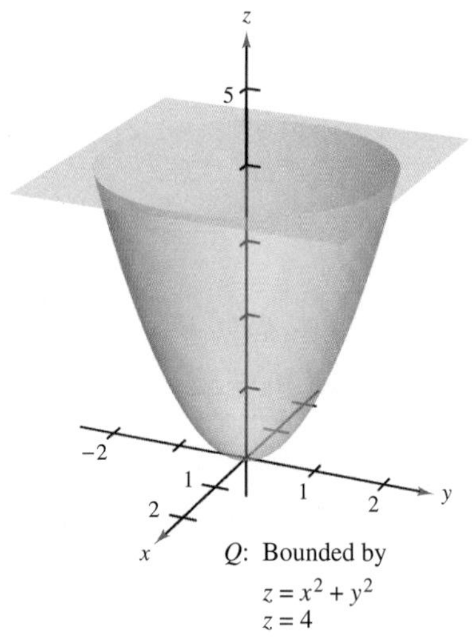

Find the moment of inertia about the axis of symmetry of the solid Q bounded by the paraboloid $z = x^2 + y^2$ and the plane $z = 4$, as shown in Figure 14.67. The density at each point is proportional to the distance between the point and the z-axis.

Solution Because the z-axis is the axis of symmetry, and $\rho(x, y, z) = k\sqrt{x^2 + y^2}$, it follows that

$$I_z = \iiint_Q k(x^2 + y^2)\sqrt{x^2 + y^2}\,dV.$$

In cylindrical coordinates, $0 \le r \le \sqrt{x^2 + y^2} = \sqrt{z}$. So, you have

$$\begin{aligned} I_z &= k\int_0^4\int_0^{2\pi}\int_0^{\sqrt{z}} r^2(r)r\,dr\,d\theta\,dz \\ &= k\int_0^4\int_0^{2\pi} \frac{r^5}{5}\Big]_0^{\sqrt{z}} d\theta\,dz \\ &= k\int_0^4\int_0^{2\pi} \frac{z^{5/2}}{5}\,d\theta\,dz \\ &= \frac{k}{5}\int_0^4 z^{5/2}(2\pi)\,dz \\ &= \frac{2\pi k}{5}\Big[\frac{2}{7}z^{7/2}\Big]_0^4 = \frac{512k\pi}{35}. \end{aligned}$$

Q: Bounded by $z = x^2 + y^2$, $z = 4$

Figure 14.67

Triple Integrals in Spherical Coordinates

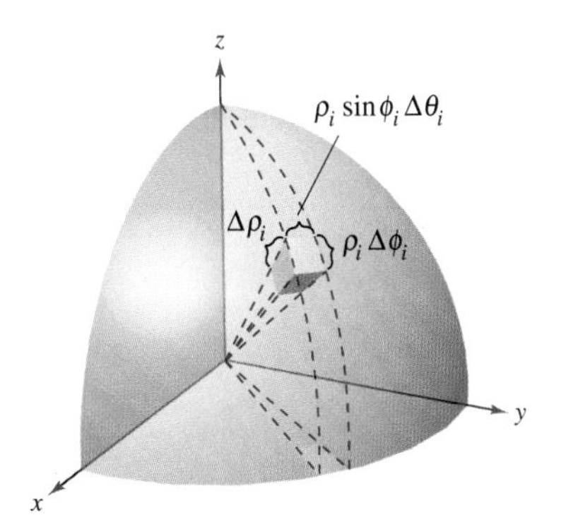

Spherical block:
$\Delta V_i \approx \rho_i^2 \sin\phi_i \Delta\rho_i \Delta\phi_i \Delta\theta_i$
Figure 14.68

Triple integrals involving spheres or cones are often easier to evaluate by converting to spherical coordinates. Recall from Section 11.7 that the rectangular conversion equations for spherical coordinates are

$$x = \rho \sin\phi \cos\theta$$
$$y = \rho \sin\phi \sin\theta$$
$$z = \rho \cos\phi.$$

In this coordinate system, the simplest region is a spherical block determined by

$$\{(\rho, \theta, \phi): \rho_1 \le \rho \le \rho_2, \quad \theta_1 \le \theta \le \theta_2, \quad \phi_1 \le \phi \le \phi_2\}$$

where $\rho_1 \ge 0$, $\theta_2 - \theta_1 \le 2\pi$, and $0 \le \phi_1 \le \phi_2 \le \pi$, as shown in Figure 14.68. If (ρ, θ, ϕ) is a point in the interior of such a block, then the volume of the block can be approximated by $\Delta V \approx \rho^2 \sin\phi \,\Delta\rho\, \Delta\phi\, \Delta\theta$ (see Exercise 18 in the Problem Solving exercises at the end of this chapter).

Using the usual process involving an inner partition, summation, and a limit, you can develop the following version of a triple integral in spherical coordinates for a continuous function f defined on the solid region Q.

$$\iiint_Q f(x, y, z)\, dV = \int_{\theta_1}^{\theta_2}\int_{\phi_1}^{\phi_2}\int_{\rho_1}^{\rho_2} f(\rho\sin\phi\cos\theta, \rho\sin\phi\sin\theta, \rho\cos\phi)\rho^2 \sin\phi \, d\rho\, d\phi\, d\theta$$

This formula can be modified for different orders of integration and generalized to include regions with variable boundaries.

Like triple integrals in cylindrical coordinates, triple integrals in spherical coordinates are evaluated with iterated integrals. As with cylindrical coordinates, you can visualize a particular order of integration by viewing the iterated integral in terms of three sweeping motions—each adding another dimension to the solid. For instance, the iterated integral

$$\int_0^{2\pi}\int_0^{\pi/4}\int_0^3 \rho^2 \sin\phi \, d\rho\, d\phi\, d\theta$$

(which is used in Example 4) is illustrated in Figure 14.69.

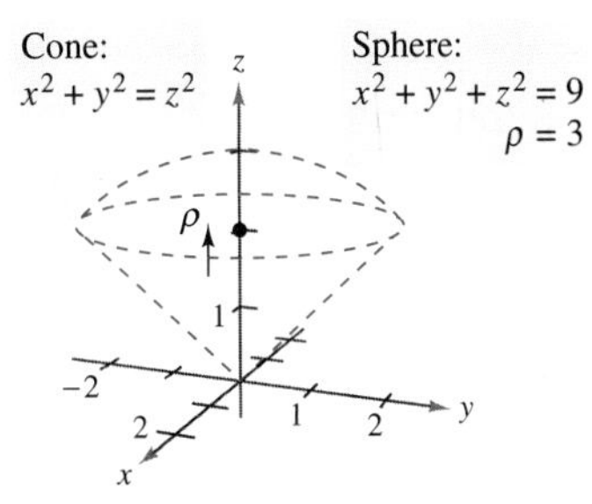

ρ varies from 0 to 3 with ϕ and θ held constant.

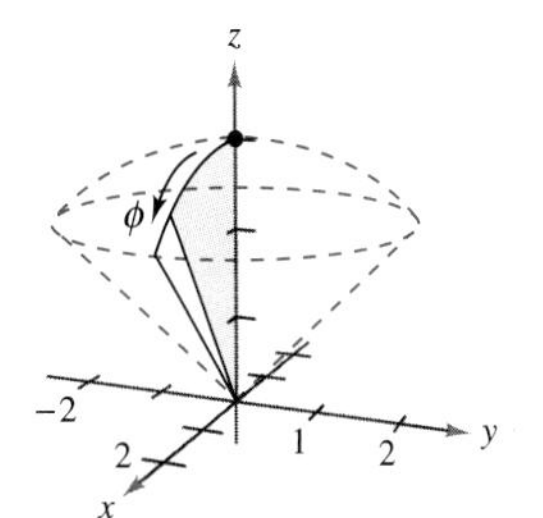

ϕ varies from 0 to $\pi/4$ with θ held constant.

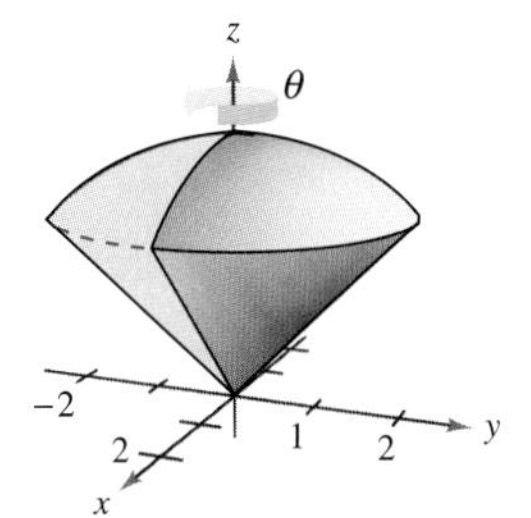

θ varies from 0 to 2π.

Figure 14.69

NOTE The Greek letter ρ used in spherical coordinates is not related to density. Rather, it is the three-dimensional analog of the r used in polar coordinates. For problems involving spherical coordinates and a density function, this text uses a different symbol to denote density. ■

EXAMPLE 4 Finding Volume in Spherical Coordinates

Find the volume of the solid region Q bounded below by the upper nappe of the cone $z^2 = x^2 + y^2$ and above by the sphere $x^2 + y^2 + z^2 = 9$, as shown in Figure 14.70.

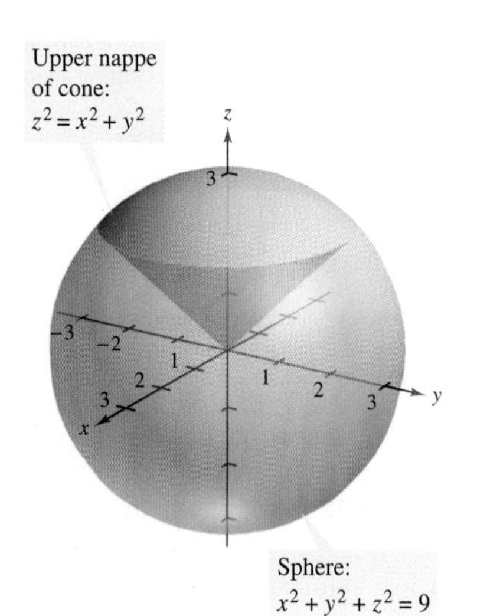

Figure 14.70

Solution In spherical coordinates, the equation of the sphere is

$$\rho^2 = x^2 + y^2 + z^2 = 9 \quad \Longrightarrow \quad \rho = 3.$$

Furthermore, the sphere and cone intersect when

$$(x^2 + y^2) + z^2 = (z^2) + z^2 = 9 \quad \Longrightarrow \quad z = \frac{3}{\sqrt{2}}$$

and, because $z = \rho \cos \phi$, it follows that

$$\left(\frac{3}{\sqrt{2}}\right)\left(\frac{1}{3}\right) = \cos \phi \quad \Longrightarrow \quad \phi = \frac{\pi}{4}.$$

Consequently, you can use the integration order $d\rho \, d\phi \, d\theta$, where $0 \le \rho \le 3$, $0 \le \phi \le \pi/4$, and $0 \le \theta \le 2\pi$. The volume is

$$\begin{aligned}
V = \iiint_Q dV &= \int_0^{2\pi} \int_0^{\pi/4} \int_0^3 \rho^2 \sin \phi \, d\rho \, d\phi \, d\theta \\
&= \int_0^{2\pi} \int_0^{\pi/4} 9 \sin \phi \, d\phi \, d\theta \\
&= 9 \int_0^{2\pi} -\cos \phi \Big]_0^{\pi/4} d\theta \\
&= 9 \int_0^{2\pi} \left(1 - \frac{\sqrt{2}}{2}\right) d\theta = 9\pi\left(2 - \sqrt{2}\right) \approx 16.563.
\end{aligned}$$

EXAMPLE 5 Finding the Center of Mass of a Solid Region

Find the center of mass of the solid region Q of uniform density, bounded below by the upper nappe of the cone $z^2 = x^2 + y^2$ and above by the sphere $x^2 + y^2 + z^2 = 9$.

Solution Because the density is uniform, you can consider the density at the point (x, y, z) to be k. By symmetry, the center of mass lies on the z-axis, and you need only calculate $\bar{z} = M_{xy}/m$, where $m = kV = 9k\pi\left(2 - \sqrt{2}\right)$ from Example 4. Because $z = \rho \cos \phi$, it follows that

$$\begin{aligned}
M_{xy} = \iiint_Q kz \, dV &= k \int_0^3 \int_0^{2\pi} \int_0^{\pi/4} (\rho \cos \phi)\rho^2 \sin \phi \, d\phi \, d\theta \, d\rho \\
&= k \int_0^3 \int_0^{2\pi} \rho^3 \frac{\sin^2 \phi}{2} \Big]_0^{\pi/4} d\theta \, d\rho \\
&= \frac{k}{4} \int_0^3 \int_0^{2\pi} \rho^3 \, d\theta \, d\rho = \frac{k\pi}{2} \int_0^3 \rho^3 \, d\rho = \frac{81k\pi}{8}.
\end{aligned}$$

So,

$$\bar{z} = \frac{M_{xy}}{m} = \frac{81k\pi/8}{9k\pi\left(2 - \sqrt{2}\right)} = \frac{9\left(2 + \sqrt{2}\right)}{16} \approx 1.920$$

and the center of mass is approximately $(0, 0, 1.92)$. ■

14.7 Exercises

See www.CalcChat.com for worked-out solutions to odd-numbered exercises.

In Exercises 1–6, evaluate the iterated integral.

1. $\displaystyle\int_{-1}^{5}\int_{0}^{\pi/2}\int_{0}^{3} r\cos\theta\, dr\, d\theta\, dz$

2. $\displaystyle\int_{0}^{\pi/4}\int_{0}^{6}\int_{0}^{6-r} rz\, dz\, dr\, d\theta$

3. $\displaystyle\int_{0}^{\pi/2}\int_{0}^{2\cos^2\theta}\int_{0}^{4-r^2} r\sin\theta\, dz\, dr\, d\theta$

4. $\displaystyle\int_{0}^{\pi/2}\int_{0}^{\pi}\int_{0}^{2} e^{-\rho^3}\rho^2\, d\rho\, d\theta\, d\phi$

5. $\displaystyle\int_{0}^{2\pi}\int_{0}^{\pi/4}\int_{0}^{\cos\phi} \rho^2\sin\phi\, d\rho\, d\phi\, d\theta$

6. $\displaystyle\int_{0}^{\pi/4}\int_{0}^{\pi/4}\int_{0}^{\cos\theta} \rho^2\sin\phi\cos\phi\, d\rho\, d\theta\, d\phi$

CAS In Exercises 7 and 8, use a computer algebra system to evaluate the iterated integral.

7. $\displaystyle\int_{0}^{4}\int_{0}^{z}\int_{0}^{\pi/2} re^{r}\, d\theta\, dr\, dz$

8. $\displaystyle\int_{0}^{\pi/2}\int_{0}^{\pi}\int_{0}^{\sin\theta} (2\cos\phi)\rho^2\, d\rho\, d\theta\, d\phi$

In Exercises 9–12, sketch the solid region whose volume is given by the iterated integral, and evaluate the iterated integral.

9. $\displaystyle\int_{0}^{\pi/2}\int_{0}^{3}\int_{0}^{e^{-r^2}} r\, dz\, dr\, d\theta$

10. $\displaystyle\int_{0}^{2\pi}\int_{0}^{\sqrt{5}}\int_{0}^{5-r^2} r\, dz\, dr\, d\theta$

11. $\displaystyle\int_{0}^{2\pi}\int_{\pi/6}^{\pi/2}\int_{0}^{4} \rho^2\sin\phi\, d\rho\, d\phi\, d\theta$

12. $\displaystyle\int_{0}^{2\pi}\int_{0}^{\pi}\int_{2}^{5} \rho^2\sin\phi\, d\rho\, d\phi\, d\theta$

In Exercises 13–16, convert the integral from rectangular coordinates to both cylindrical and spherical coordinates, and evaluate the simplest iterated integral.

13. $\displaystyle\int_{-2}^{2}\int_{-\sqrt{4-x^2}}^{\sqrt{4-x^2}}\int_{x^2+y^2}^{4} x\, dz\, dy\, dx$

14. $\displaystyle\int_{0}^{2}\int_{0}^{\sqrt{4-x^2}}\int_{0}^{\sqrt{16-x^2-y^2}} \sqrt{x^2+y^2}\, dz\, dy\, dx$

15. $\displaystyle\int_{-a}^{a}\int_{-\sqrt{a^2-x^2}}^{\sqrt{a^2-x^2}}\int_{a}^{a+\sqrt{a^2-x^2-y^2}} x\, dz\, dy\, dx$

16. $\displaystyle\int_{0}^{3}\int_{0}^{\sqrt{9-x^2}}\int_{0}^{\sqrt{9-x^2-y^2}} \sqrt{x^2+y^2+z^2}\, dz\, dy\, dx$

Volume **In Exercises 17–22, use cylindrical coordinates to find the volume of the solid.**

17. Solid inside both $x^2 + y^2 + z^2 = a^2$ and $(x - a/2)^2 + y^2 = (a/2)^2$

18. Solid inside $x^2 + y^2 + z^2 = 16$ and outside $z = \sqrt{x^2 + y^2}$

19. Solid bounded above by $z = 2x$ and below by $z = 2x^2 + 2y^2$

20. Solid bounded above by $z = 2 - x^2 - y^2$ and below by $z = x^2 + y^2$

21. Solid bounded by the graphs of the sphere $r^2 + z^2 = a^2$ and the cylinder $r = a\cos\theta$

22. Solid inside the sphere $x^2 + y^2 + z^2 = 4$ and above the upper nappe of the cone $z^2 = x^2 + y^2$

Mass **In Exercises 23 and 24, use cylindrical coordinates to find the mass of the solid Q.**

23. $Q = \{(x, y, z): 0 \le z \le 9 - x - 2y,\ x^2 + y^2 \le 4\}$
$\rho(x, y, z) = k\sqrt{x^2 + y^2}$

24. $Q = \{(x, y, z): 0 \le z \le 12e^{-(x^2+y^2)},\ x^2 + y^2 \le 4,\ x \ge 0,\ y \ge 0\}$
$\rho(x, y, z) = k$

In Exercises 25–30, use cylindrical coordinates to find the indicated characteristic of the cone shown in the figure.

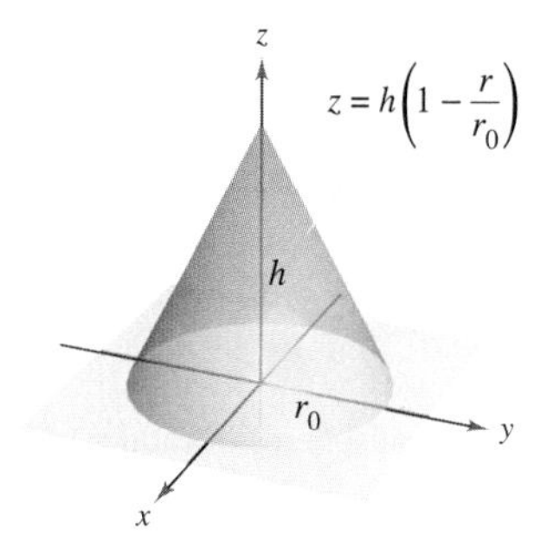

25. ***Volume*** Find the volume of the cone.

26. ***Centroid*** Find the centroid of the cone.

27. **CAS** ***Center of Mass*** Find the center of mass of the cone assuming that its density at any point is proportional to the distance between the point and the axis of the cone. Use a computer algebra system to evaluate the triple integral.

28. **CAS** ***Center of Mass*** Find the center of mass of the cone assuming that its density at any point is proportional to the distance between the point and the base. Use a computer algebra system to evaluate the triple integral.

29. ***Moment of Inertia*** Assume that the cone has uniform density and show that the moment of inertia about the z-axis is

$I_z = \frac{3}{10}mr_0^{\,2}.$

30. ***Moment of Inertia*** Assume that the density of the cone is $\rho(x, y, z) = k\sqrt{x^2 + y^2}$ and find the moment of inertia about the z-axis.

Moment of Inertia **In Exercises 31 and 32, use cylindrical coordinates to verify the given formula for the moment of inertia of the solid of uniform density.**

31. Cylindrical shell: $I_z = \frac{1}{2}m(a^2 + b^2)$
$0 < a \le r \le b,\quad 0 \le z \le h$

CAS **32.** Right circular cylinder: $I_z = \frac{3}{2}ma^2$

$r = 2a \sin \theta, \quad 0 \le z \le h$

Use a computer algebra system to evaluate the triple integral.

Volume **In Exercises 33–36, use spherical coordinates to find the volume of the solid.**

33. Solid inside $x^2 + y^2 + z^2 = 9$, outside $z = \sqrt{x^2 + y^2}$, and above the xy-plane

34. Solid bounded above by $x^2 + y^2 + z^2 = z$ and below by $z = \sqrt{x^2 + y^2}$

CAS **35.** The torus given by $\rho = 4 \sin \phi$ (Use a computer algebra system to evaluate the triple integral.)

36. The solid between the spheres $x^2 + y^2 + z^2 = a^2$ and $x^2 + y^2 + z^2 = b^2$, $b > a$, and inside the cone $z^2 = x^2 + y^2$

Mass **In Exercises 37 and 38, use spherical coordinates to find the mass of the sphere $x^2 + y^2 + z^2 = a^2$ with the given density.**

37. The density at any point is proportional to the distance between the point and the origin.

38. The density at any point is proportional to the distance of the point from the z-axis.

Center of Mass **In Exercises 39 and 40, use spherical coordinates to find the center of mass of the solid of uniform density.**

39. Hemispherical solid of radius r

40. Solid lying between two concentric hemispheres of radii r and R, where $r < R$

Moment of Inertia **In Exercises 41 and 42, use spherical coordinates to find the moment of inertia about the z-axis of the solid of uniform density.**

41. Solid bounded by the hemisphere $\rho = \cos \phi$, $\pi/4 \le \phi \le \pi/2$, and the cone $\phi = \pi/4$

42. Solid lying between two concentric hemispheres of radii r and R, where $r < R$

WRITING ABOUT CONCEPTS

43. Give the equations for conversion from rectangular to cylindrical coordinates and vice versa.

44. Give the equations for conversion from rectangular to spherical coordinates and vice versa.

45. Give the iterated form of the triple integral $\iiint\limits_Q f(x, y, z)\, dV$ in cylindrical form.

46. Give the iterated form of the triple integral $\iiint\limits_Q f(x, y, z)\, dV$ in spherical form.

47. Describe the surface whose equation is a coordinate equal to a constant for each of the coordinates in (a) the cylindrical coordinate system and (b) the spherical coordinate system.

CAPSTONE

48. Convert the integral from rectangular coordinates to both (a) cylindrical and (b) spherical coordinates. Without calculating, which integral appears to be the simplest to evaluate? Why?

$$\int_0^a \int_0^{\sqrt{a^2-x^2}} \int_0^{\sqrt{a^2-x^2-y^2}} \sqrt{x^2 + y^2 + z^2}\, dz\, dy\, dx$$

49. Find the "volume" of the "four-dimensional sphere"

$x^2 + y^2 + z^2 + w^2 = a^2$

by evaluating

$$16 \int_0^a \int_0^{\sqrt{a^2-x^2}} \int_0^{\sqrt{a^2-x^2-y^2}} \int_0^{\sqrt{a^2-x^2-y^2-z^2}} dw\, dz\, dy\, dx.$$

50. Use spherical coordinates to show that

$$\int_{-\infty}^{\infty} \int_{-\infty}^{\infty} \int_{-\infty}^{\infty} \sqrt{x^2 + y^2 + z^2}\, e^{-(x^2+y^2+z^2)}\, dx\, dy\, dz = 2\pi.$$

PUTNAM EXAM CHALLENGE

51. Find the volume of the region of points (x, y, z) such that $(x^2 + y^2 + z^2 + 8)^2 \le 36(x^2 + y^2)$.

This problem was composed by the Committee on the Putnam Prize Competition.

SECTION PROJECT

Wrinkled and Bumpy Spheres

In parts (a) and (b), find the volume of the wrinkled sphere or bumpy sphere. These solids are used as models for tumors.

(a) Wrinkled sphere

$\rho = 1 + 0.2 \sin 8\theta \sin \phi$

$0 \le \theta \le 2\pi, 0 \le \phi \le \pi$

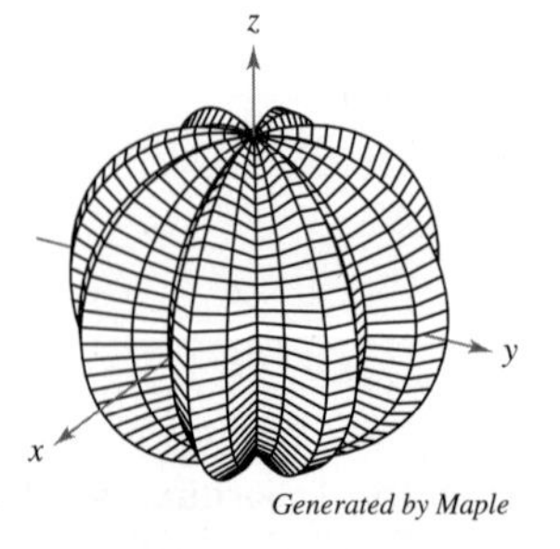

Generated by Maple

(b) Bumpy sphere

$\rho = 1 + 0.2 \sin 8\theta \sin 4\phi$

$0 \le \theta \le 2\pi, 0 \le \phi \le \pi$

z

y

x

Generated by Maple

■ **FOR FURTHER INFORMATION** For more information on these types of spheres, see the article "Heat Therapy for Tumors" by Leah Edelstein-Keshet in *The UMAP Journal*.

14.8 Change of Variables: Jacobians

- **Understand the concept of a Jacobian.**
- **Use a Jacobian to change variables in a double integral.**

Jacobians

> **CARL GUSTAV JACOBI (1804–1851)**
>
> The Jacobian is named after the German mathematician Carl Gustav Jacobi. Jacobi is known for his work in many areas of mathematics, but his interest in integration stemmed from the problem of finding the circumference of an ellipse.

For the single integral

$$\int_a^b f(x)\,dx$$

you can change variables by letting $x = g(u)$, so that $dx = g'(u)\,du$, and obtain

$$\int_a^b f(x)\,dx = \int_c^d f(g(u))g'(u)\,du$$

where $a = g(c)$ and $b = g(d)$. Note that the change of variables process introduces an additional factor $g'(u)$ into the integrand. This also occurs in the case of double integrals

$$\int_R\!\int f(x, y)\,dA = \int_S\!\int f(g(u, v), h(u, v))\underbrace{\left|\frac{\partial x}{\partial u}\frac{\partial y}{\partial v} - \frac{\partial y}{\partial u}\frac{\partial x}{\partial v}\right|}_{\text{Jacobian}}\,du\,dv$$

where the change of variables $x = g(u, v)$ and $y = h(u, v)$ introduces a factor called the **Jacobian** of x and y with respect to u and v. In defining the Jacobian, it is convenient to use the following determinant notation.

> **DEFINITION OF THE JACOBIAN**
>
> If $x = g(u, v)$ and $y = h(u, v)$, then the **Jacobian** of x and y with respect to u and v, denoted by $\partial(x, y)/\partial(u, v)$, is
>
> $$\frac{\partial(x, y)}{\partial(u, v)} = \begin{vmatrix} \dfrac{\partial x}{\partial u} & \dfrac{\partial x}{\partial v} \\[2mm] \dfrac{\partial y}{\partial u} & \dfrac{\partial y}{\partial v} \end{vmatrix} = \frac{\partial x}{\partial u}\frac{\partial y}{\partial v} - \frac{\partial y}{\partial u}\frac{\partial x}{\partial v}.$$

EXAMPLE 1 The Jacobian for Rectangular-to-Polar Conversion

Find the Jacobian for the change of variables defined by

$$x = r\cos\theta \qquad \text{and} \qquad y = r\sin\theta.$$

Solution From the definition of the Jacobian, you obtain

$$\begin{aligned}
\frac{\partial(x, y)}{\partial(r, \theta)} &= \begin{vmatrix} \dfrac{\partial x}{\partial r} & \dfrac{\partial x}{\partial \theta} \\[2mm] \dfrac{\partial y}{\partial r} & \dfrac{\partial y}{\partial \theta} \end{vmatrix} \\
&= \begin{vmatrix} \cos\theta & -r\sin\theta \\ \sin\theta & r\cos\theta \end{vmatrix} \\
&= r\cos^2\theta + r\sin^2\theta \\
&= r.
\end{aligned}$$

■

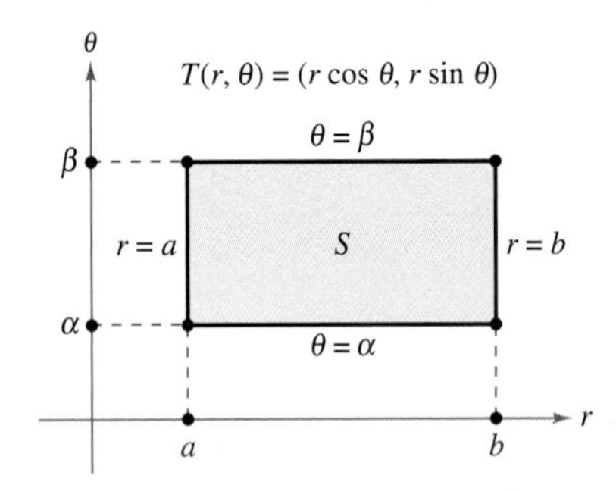

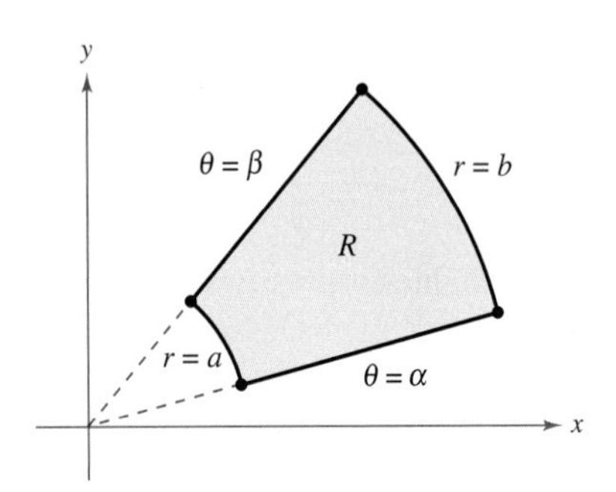

S is the region in the $r\theta$-plane that corresponds to R in the xy-plane.
Figure 14.71

Example 1 points out that the change of variables from rectangular to polar coordinates for a double integral can be written as

$$\int_R\int f(x, y)\, dA = \int_S\int f(r\cos\theta, r\sin\theta) r\, dr\, d\theta, \quad r > 0$$

$$= \int_S\int f(r\cos\theta, r\sin\theta)\left|\frac{\partial(x, y)}{\partial(r, \theta)}\right| dr\, d\theta$$

where S is the region in the $r\theta$-plane that corresponds to the region R in the xy-plane, as shown in Figure 14.71. This formula is similar to that found in Theorem 14.3 on page 1006.

In general, a change of variables is given by a one-to-one **transformation** T from a region S in the uv-plane to a region R in the xy-plane, to be given by

$$T(u, v) = (x, y) = (g(u, v), h(u, v))$$

where g and h have continuous first partial derivatives in the region S. Note that the point (u, v) lies in S and the point (x, y) lies in R. In most cases, you are hunting for a transformation in which the region S is simpler than the region R.

EXAMPLE 2 Finding a Change of Variables to Simplify a Region

Let R be the region bounded by the lines

$$x - 2y = 0, \quad x - 2y = -4, \quad x + y = 4, \quad \text{and} \quad x + y = 1$$

as shown in Figure 14.72. Find a transformation T from a region S to R such that S is a rectangular region (with sides parallel to the u- or v-axis).

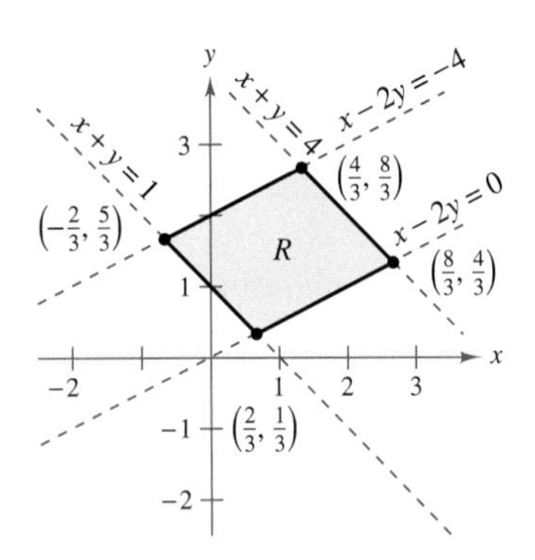

Region R in the xy-plane
Figure 14.72

Solution To begin, let $u = x + y$ and $v = x - 2y$. Solving this system of equations for x and y produces $T(u, v) = (x, y)$, where

$$x = \frac{1}{3}(2u + v) \quad \text{and} \quad y = \frac{1}{3}(u - v).$$

The four boundaries for R in the xy-plane give rise to the following bounds for S in the uv-plane.

Bounds in the xy-Plane		*Bounds in the uv-Plane*
$x + y = 1$	⇨	$u = 1$
$x + y = 4$	⇨	$u = 4$
$x - 2y = 0$	⇨	$v = 0$
$x - 2y = -4$	⇨	$v = -4$

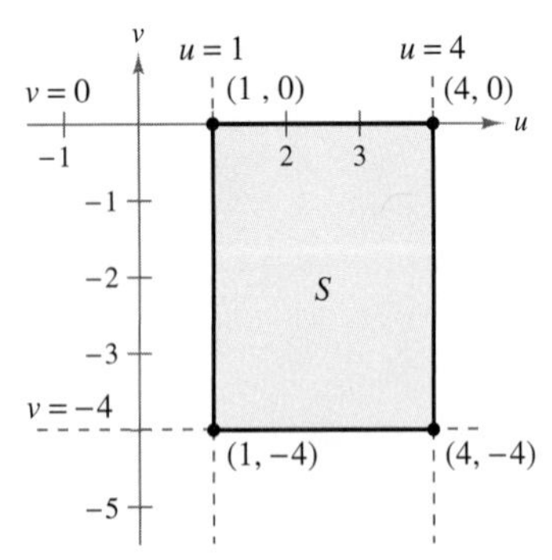

Region S in the uv-plane
Figure 14.73

The region S is shown in Figure 14.73. Note that the transformation

$$T(u, v) = (x, y) = \left(\frac{1}{3}[2u + v], \frac{1}{3}[u - v]\right)$$

maps the vertices of the region S onto the vertices of the region R. For instance,

$$T(1, 0) = \left(\tfrac{1}{3}[2(1) + 0], \tfrac{1}{3}[1 - 0]\right) = \left(\tfrac{2}{3}, \tfrac{1}{3}\right)$$

$$T(4, 0) = \left(\tfrac{1}{3}[2(4) + 0], \tfrac{1}{3}[4 - 0]\right) = \left(\tfrac{8}{3}, \tfrac{4}{3}\right)$$

$$T(4, -4) = \left(\tfrac{1}{3}[2(4) - 4], \tfrac{1}{3}[4 - (-4)]\right) = \left(\tfrac{4}{3}, \tfrac{8}{3}\right)$$

$$T(1, -4) = \left(\tfrac{1}{3}[2(1) - 4], \tfrac{1}{3}[1 - (-4)]\right) = \left(-\tfrac{2}{3}, \tfrac{5}{3}\right).$$

■

Change of Variables for Double Integrals

THEOREM 14.5 CHANGE OF VARIABLES FOR DOUBLE INTEGRALS

Let R be a vertically or horizontally simple region in the xy-plane, and let S be a vertically or horizontally simple region in the uv-plane. Let T from S to R be given by $T(u, v) = (x, y) = (g(u, v), h(u, v))$, where g and h have continuous first partial derivatives. Assume that T is one-to-one except possibly on the boundary of S. If f is continuous on R, and $\partial(x, y)/\partial(u, v)$ is nonzero on S, then

$$\int_R\int f(x, y)\,dx\,dy = \int_S\int f(g(u, v), h(u, v))\left|\frac{\partial(x, y)}{\partial(u, v)}\right| du\,dv.$$

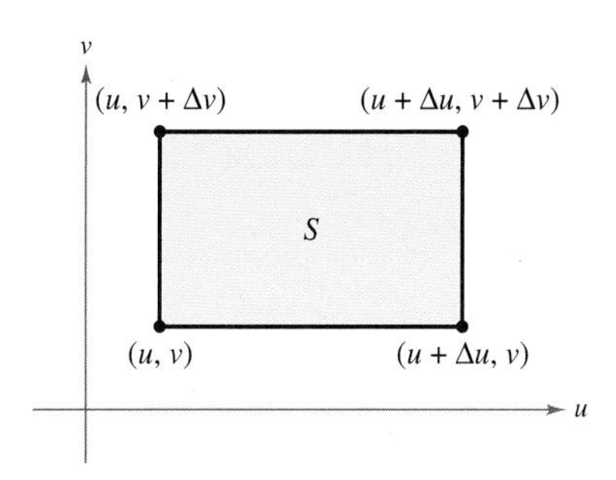

Area of $S = \Delta u\,\Delta v$
$\Delta u > 0,\ \Delta v > 0$
Figure 14.74

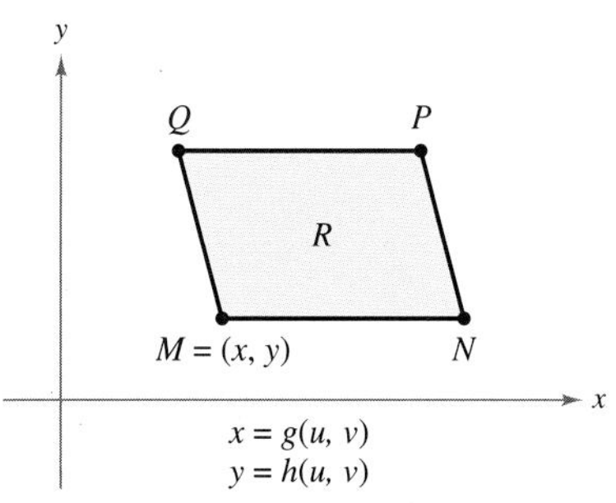

The vertices in the xy-plane are $M(g(u, v), h(u, v))$, $N(g(u + \Delta u, v), h(u + \Delta u, v))$, $P(g(u + \Delta u, v + \Delta v), h(u + \Delta u, v + \Delta v))$, and $Q(g(u, v + \Delta v), h(u, v + \Delta v))$.
Figure 14.75

PROOF Consider the case in which S is a rectangular region in the uv-plane with vertices (u, v), $(u + \Delta u, v)$, $(u + \Delta u, v + \Delta v)$, and $(u, v + \Delta v)$, as shown in Figure 14.74. The images of these vertices in the xy-plane are shown in Figure 14.75. If Δu and Δv are small, the continuity of g and h implies that R is approximately a parallelogram determined by the vectors $\overrightarrow{MN}$ and $\overrightarrow{MQ}$. So, the area of R is

$$\Delta A \approx \|\overrightarrow{MN} \times \overrightarrow{MQ}\|.$$

Moreover, for small Δu and Δv, the partial derivatives of g and h with respect to u can be approximated by

$$g_u(u, v) \approx \frac{g(u + \Delta u, v) - g(u, v)}{\Delta u} \quad \text{and} \quad h_u(u, v) \approx \frac{h(u + \Delta u, v) - h(u, v)}{\Delta u}.$$

Consequently,

$$\begin{aligned}\overrightarrow{MN} &= [g(u + \Delta u, v) - g(u, v)]\mathbf{i} + [h(u + \Delta u, v) - h(u, v)]\mathbf{j}\\ &\approx [g_u(u, v)\,\Delta u]\mathbf{i} + [h_u(u, v)\,\Delta u]\mathbf{j}\\ &= \frac{\partial x}{\partial u}\Delta u\mathbf{i} + \frac{\partial y}{\partial u}\Delta u\mathbf{j}.\end{aligned}$$

Similarly, you can approximate $\overrightarrow{MQ}$ by $\dfrac{\partial x}{\partial v}\Delta v\mathbf{i} + \dfrac{\partial y}{\partial v}\Delta v\mathbf{j}$, which implies that

$$\overrightarrow{MN} \times \overrightarrow{MQ} \approx \begin{vmatrix}\mathbf{i} & \mathbf{j} & \mathbf{k}\\ \dfrac{\partial x}{\partial u}\Delta u & \dfrac{\partial y}{\partial u}\Delta u & 0\\ \dfrac{\partial x}{\partial v}\Delta v & \dfrac{\partial y}{\partial v}\Delta v & 0\end{vmatrix} = \begin{vmatrix}\dfrac{\partial x}{\partial u} & \dfrac{\partial y}{\partial u}\\ \dfrac{\partial x}{\partial v} & \dfrac{\partial y}{\partial v}\end{vmatrix}\Delta u\,\Delta v\mathbf{k}.$$

It follows that, in Jacobian notation,

$$\Delta A \approx \|\overrightarrow{MN} \times \overrightarrow{MQ}\| \approx \left|\frac{\partial(x, y)}{\partial(u, v)}\right| \Delta u\,\Delta v.$$

Because this approximation improves as Δu and Δv approach 0, the limiting case can be written as

$$dA \approx \|\overrightarrow{MN} \times \overrightarrow{MQ}\| \approx \left|\frac{\partial(x, y)}{\partial(u, v)}\right| du\,dv.$$

So,

$$\int_R\int f(x, y)\,dx\,dy = \int_S\int f(g(u, v), h(u, v))\left|\frac{\partial(x, y)}{\partial(u, v)}\right| du\,dv.$$ ■

The next two examples show how a change of variables can simplify the integration process. The simplification can occur in various ways. You can make a change of variables to simplify either the *region R* or the *integrand f(x, y)*, or both.

EXAMPLE 3 Using a Change of Variables to Simplify a Region

Let R be the region bounded by the lines

$$x - 2y = 0, \quad x - 2y = -4, \quad x + y = 4, \quad \text{and} \quad x + y = 1$$

as shown in Figure 14.76. Evaluate the double integral

$$\int_R\int 3xy\, dA.$$

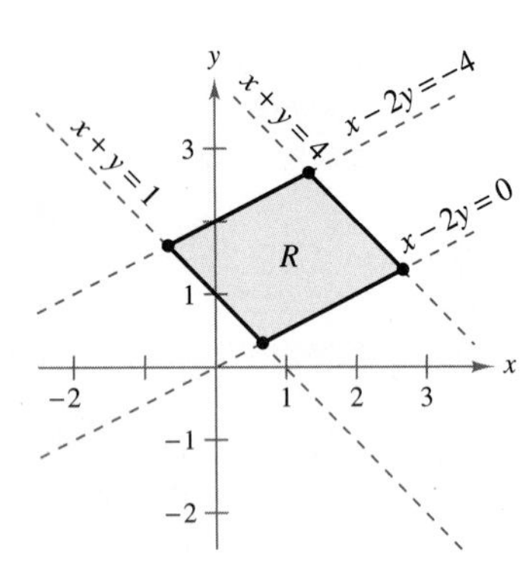

Figure 14.76

Solution From Example 2, you can use the following change of variables.

$$x = \frac{1}{3}(2u + v) \quad \text{and} \quad y = \frac{1}{3}(u - v)$$

The partial derivatives of x and y are

$$\frac{\partial x}{\partial u} = \frac{2}{3}, \quad \frac{\partial x}{\partial v} = \frac{1}{3}, \quad \frac{\partial y}{\partial u} = \frac{1}{3}, \quad \text{and} \quad \frac{\partial y}{\partial v} = -\frac{1}{3}$$

which implies that the Jacobian is

$$\begin{aligned}
\frac{\partial(x, y)}{\partial(u, v)} &= \begin{vmatrix} \frac{\partial x}{\partial u} & \frac{\partial x}{\partial v} \\ \frac{\partial y}{\partial u} & \frac{\partial y}{\partial v} \end{vmatrix} \\
&= \begin{vmatrix} \frac{2}{3} & \frac{1}{3} \\ \frac{1}{3} & -\frac{1}{3} \end{vmatrix} \\
&= -\frac{2}{9} - \frac{1}{9} \\
&= -\frac{1}{3}.
\end{aligned}$$

So, by Theorem 14.5, you obtain

$$\begin{aligned}
\int_R\int 3xy\, dA &= \int_S\int 3\left[\frac{1}{3}(2u + v)\frac{1}{3}(u - v)\right]\left|\frac{\partial(x, y)}{\partial(u, v)}\right| dv\, du \\
&= \int_1^4\int_{-4}^0 \frac{1}{9}(2u^2 - uv - v^2)\, dv\, du \\
&= \frac{1}{9}\int_1^4 \left[2u^2v - \frac{uv^2}{2} - \frac{v^3}{3}\right]_{-4}^0 du \\
&= \frac{1}{9}\int_1^4 \left(8u^2 + 8u - \frac{64}{3}\right) du \\
&= \frac{1}{9}\left[\frac{8u^3}{3} + 4u^2 - \frac{64}{3}u\right]_1^4 \\
&= \frac{164}{9}.
\end{aligned}$$

■

EXAMPLE 4 Using a Change of Variables to Simplify an Integrand

Let R be the region bounded by the square with vertices $(0, 1)$, $(1, 2)$, $(2, 1)$, and $(1, 0)$. Evaluate the integral

$$\int_R\int (x + y)^2 \sin^2(x - y)\, dA.$$

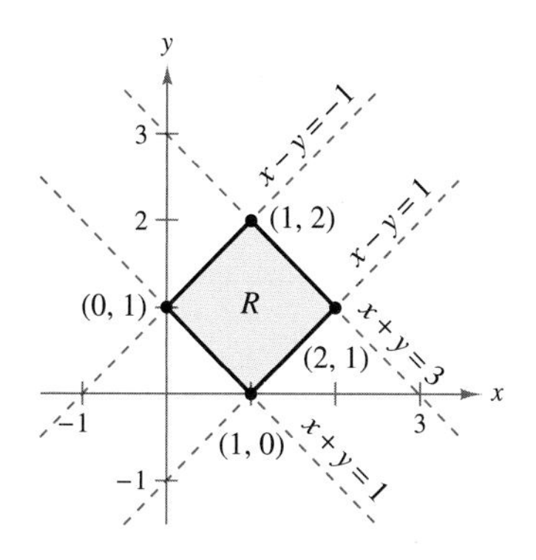

Region R in the xy-plane
Figure 14.77

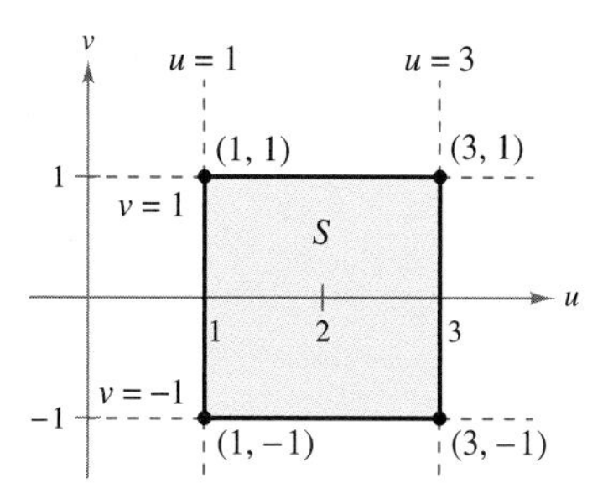

Region S in the uv-plane
Figure 14.78

Solution Note that the sides of R lie on the lines $x + y = 1$, $x - y = 1$, $x + y = 3$, and $x - y = -1$, as shown in Figure 14.77. Letting $u = x + y$ and $v = x - y$, you can determine the bounds for region S in the uv-plane to be

$$1 \le u \le 3 \quad \text{and} \quad -1 \le v \le 1$$

as shown in Figure 14.78. Solving for x and y in terms of u and v produces

$$x = \frac{1}{2}(u + v) \quad \text{and} \quad y = \frac{1}{2}(u - v).$$

The partial derivatives of x and y are

$$\frac{\partial x}{\partial u} = \frac{1}{2}, \quad \frac{\partial x}{\partial v} = \frac{1}{2}, \quad \frac{\partial y}{\partial u} = \frac{1}{2}, \quad \text{and} \quad \frac{\partial y}{\partial v} = -\frac{1}{2}$$

which implies that the Jacobian is

$$\frac{\partial(x, y)}{\partial(u, v)} = \begin{vmatrix} \dfrac{\partial x}{\partial u} & \dfrac{\partial x}{\partial v} \\ \dfrac{\partial y}{\partial u} & \dfrac{\partial y}{\partial v} \end{vmatrix} = \begin{vmatrix} \dfrac{1}{2} & \dfrac{1}{2} \\ \dfrac{1}{2} & -\dfrac{1}{2} \end{vmatrix} = -\frac{1}{4} - \frac{1}{4} = -\frac{1}{2}.$$

By Theorem 14.5, it follows that

$$\begin{aligned}
\int_R\int (x + y)^2 \sin^2(x - y)\, dA &= \int_{-1}^{1}\int_{1}^{3} u^2 \sin^2 v\left(\frac{1}{2}\right) du\, dv \\
&= \frac{1}{2}\int_{-1}^{1} (\sin^2 v) \frac{u^3}{3}\Big]_1^3\, dv \\
&= \frac{13}{3}\int_{-1}^{1} \sin^2 v\, dv \\
&= \frac{13}{6}\int_{-1}^{1} (1 - \cos 2v)\, dv \\
&= \frac{13}{6}\left[v - \frac{1}{2}\sin 2v\right]_{-1}^{1} \\
&= \frac{13}{6}\left[2 - \frac{1}{2}\sin 2 + \frac{1}{2}\sin(-2)\right] \\
&= \frac{13}{6}(2 - \sin 2) \\
&\approx 2.363.
\end{aligned}$$

■

In each of the change of variables examples in this section, the region S has been a rectangle with sides parallel to the u- or v-axis. Occasionally, a change of variables can be used for other types of regions. For instance, letting $T(u, v) = \left(x, \frac{1}{2}y\right)$ changes the circular region $u^2 + v^2 = 1$ to the elliptical region $x^2 + (y^2/4) = 1$.

14.8 Exercises

See www.CalcChat.com for worked-out solutions to odd-numbered exercises.

In Exercises 1–8, find the Jacobian $\partial(x, y)/\partial(u, v)$ for the indicated change of variables.

1. $x = -\frac{1}{2}(u - v),\ y = \frac{1}{2}(u + v)$
2. $x = au + bv,\ y = cu + dv$
3. $x = u - v^2,\ y = u + v$
4. $x = uv - 2u,\ y = uv$
5. $x = u\cos\theta - v\sin\theta,\ y = u\sin\theta + v\cos\theta$
6. $x = u + a,\ y = v + a$
7. $x = e^u \sin v,\ y = e^u \cos v$
8. $x = \dfrac{u}{v},\ y = u + v$

In Exercises 9–12, sketch the image S in the uv-plane of the region R in the xy-plane using the given transformations.

9. $x = 3u + 2v$
   $y = 3v$

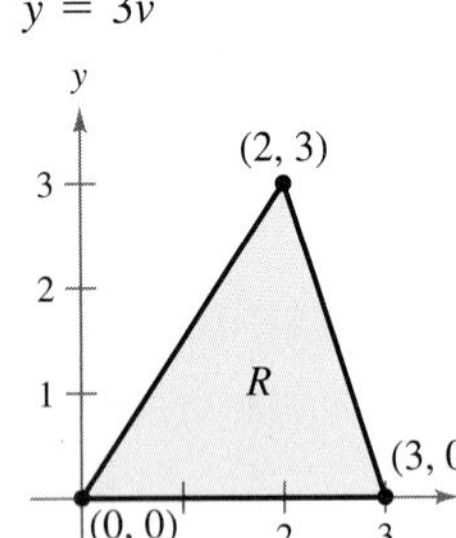

10. $x = \frac{1}{3}(4u - v)$
    $y = \frac{1}{3}(u - v)$

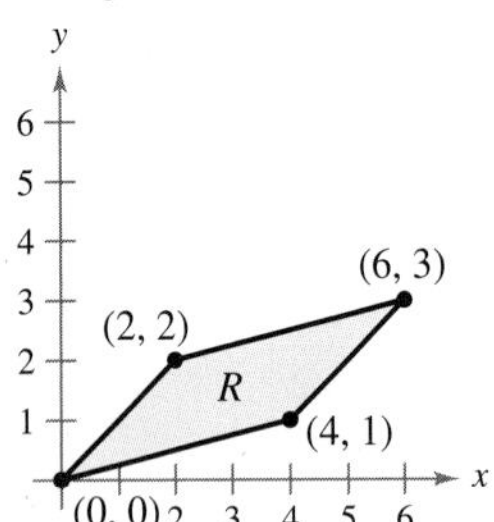

11. $x = \frac{1}{2}(u + v)$
    $y = \frac{1}{2}(u - v)$

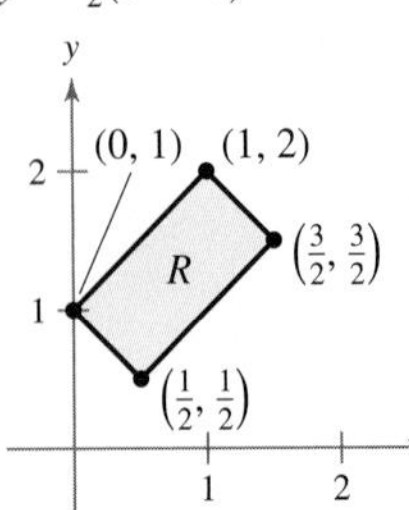

12. $x = \frac{1}{3}(v - u)$
    $y = \frac{1}{3}(2v + u)$

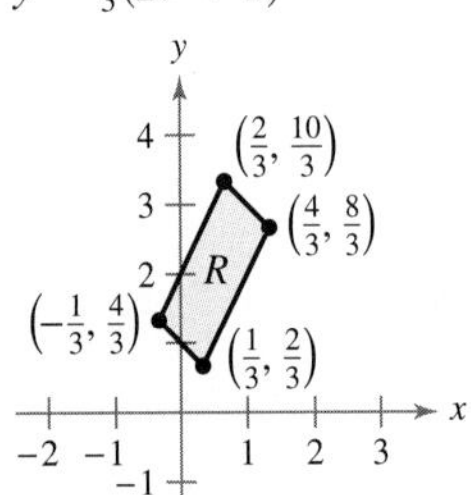

CAS In Exercises 13 and 14, verify the result of the indicated example by setting up the integral using $dy\,dx$ or $dx\,dy$ for dA. Then use a computer algebra system to evaluate the integral.

13. Example 3
14. Example 4

In Exercises 15–20, use the indicated change of variables to evaluate the double integral.

15. $\displaystyle\int_R\int 4(x^2 + y^2)\,dA$
    $x = \frac{1}{2}(u + v)$
    $y = \frac{1}{2}(u - v)$

16. $\displaystyle\int_R\int 60xy\,dA$
    $x = \frac{1}{2}(u + v)$
    $y = -\frac{1}{2}(u - v)$

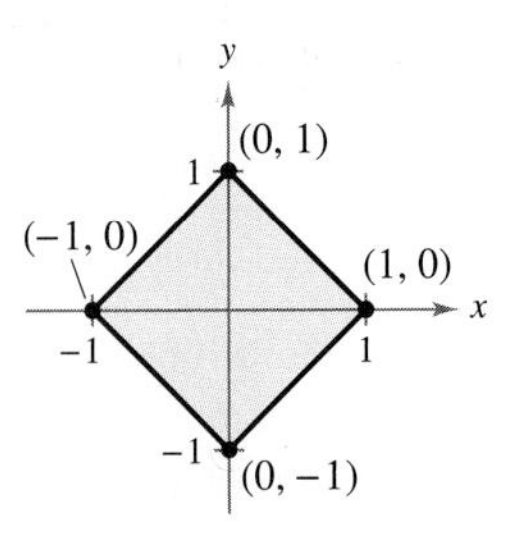

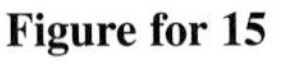

Figure for 15

Figure for 16

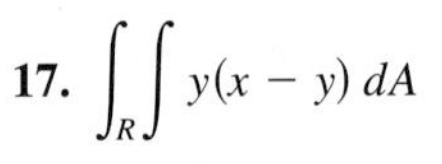

17. $\displaystyle\int_R\int y(x - y)\,dA$
    $x = u + v$
    $y = u$

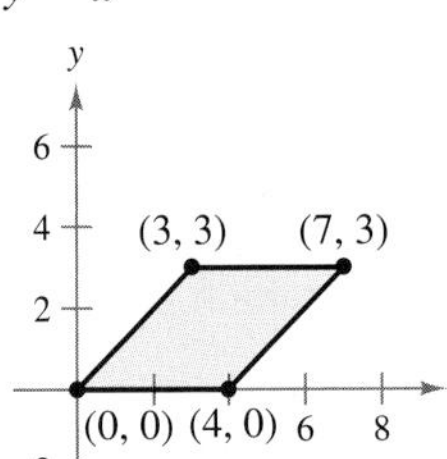

18. $\displaystyle\int_R\int 4(x + y)e^{x-y}\,dA$
    $x = \frac{1}{2}(u + v)$
    $y = \frac{1}{2}(u - v)$

19. $\displaystyle\int_R\int e^{-xy/2}\,dA$
    $x = \sqrt{\dfrac{v}{u}},\ y = \sqrt{uv}$

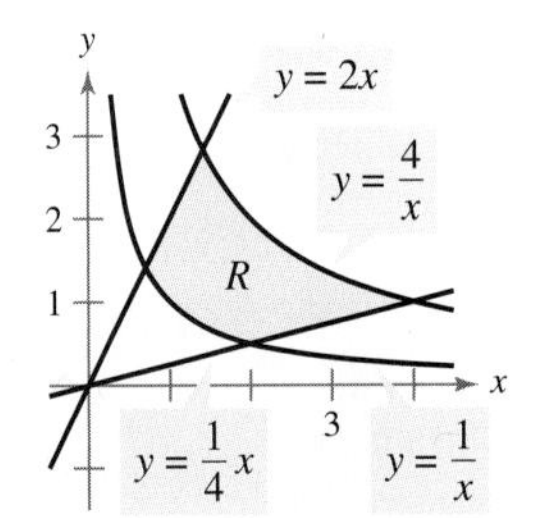

20. $\displaystyle\int_R\int y\sin xy\,dA$
    $x = \dfrac{u}{v},\ y = v$

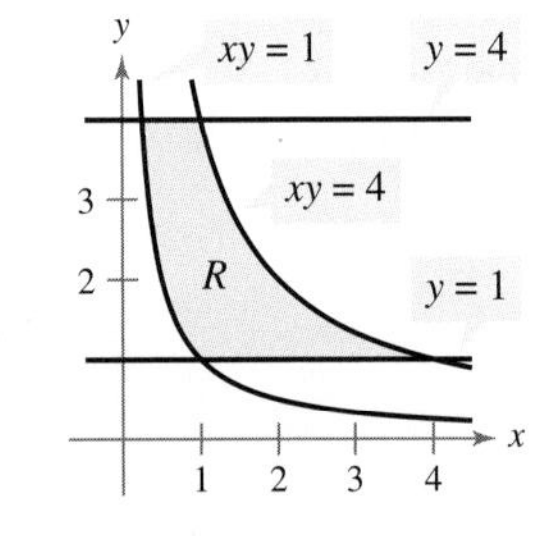

In Exercises 21–28, use a change of variables to find the volume of the solid region lying below the surface $z = f(x, y)$ and above the plane region R.

21. $f(x, y) = 48xy$
    R: region bounded by the square with vertices $(1, 0)$, $(0, 1)$, $(1, 2)$, $(2, 1)$
22. $f(x, y) = (3x + 2y)^2\sqrt{2y - x}$
    R: region bounded by the parallelogram with vertices $(0, 0)$, $(-2, 3)$, $(2, 5)$, $(4, 2)$
23. $f(x, y) = (x + y)e^{x-y}$
    R: region bounded by the square with vertices $(4, 0)$, $(6, 2)$, $(4, 4)$, $(2, 2)$

24. $f(x, y) = (x + y)^2 \sin^2(x - y)$

R: region bounded by the square with vertices $(\pi, 0)$, $(3\pi/2, \pi/2)$, (π, π), $(\pi/2, \pi/2)$

25. $f(x, y) = \sqrt{(x - y)(x + 4y)}$

R: region bounded by the parallelogram with vertices $(0, 0)$, $(1, 1)$, $(5, 0)$, $(4, -1)$

26. $f(x, y) = (3x + 2y)(2y - x)^{3/2}$

R: region bounded by the parallelogram with vertices $(0, 0)$, $(-2, 3)$, $(2, 5)$, $(4, 2)$

27. $f(x, y) = \sqrt{x + y}$

R: region bounded by the triangle with vertices $(0, 0)$, $(a, 0)$, $(0, a)$, where $a > 0$

28. $f(x, y) = \dfrac{xy}{1 + x^2y^2}$

R: region bounded by the graphs of $xy = 1$, $xy = 4$, $x = 1$, $x = 4$ (*Hint:* Let $x = u$, $y = v/u$.)

29. The substitutions $u = 2x - y$ and $v = x + y$ make the region R (see figure) into a simpler region S in the uv-plane. Determine the total number of sides of S that are parallel to either the u-axis or the v-axis.

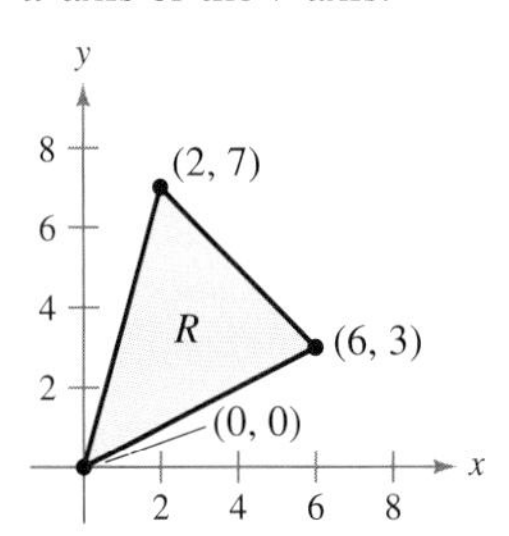

CAPSTONE

30. Find a transformation $T(u, v) = (x, y) = (g(u, v), h(u, v))$ that when applied to the region R will result in the image S (see figure). Explain your reasoning.

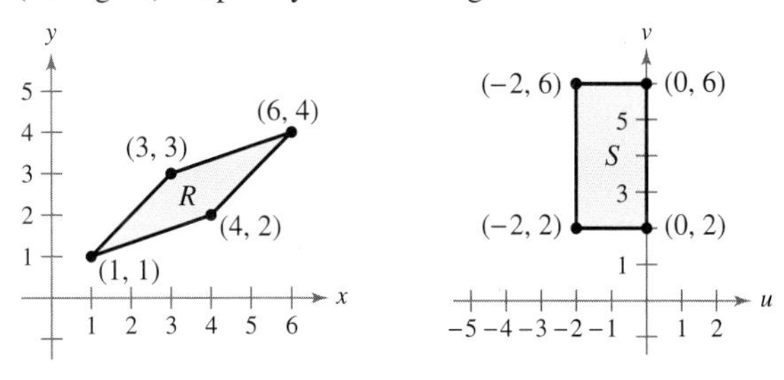

31. Consider the region R in the xy-plane bounded by the ellipse

$$\frac{x^2}{a^2} + \frac{y^2}{b^2} = 1$$

and the transformations $x = au$ and $y = bv$.

(a) Sketch the graph of the region R and its image S under the given transformation.

(b) Find $\dfrac{\partial(x, y)}{\partial(u, v)}$.

(c) Find the area of the ellipse.

32. Use the result of Exercise 31 to find the volume of each dome-shaped solid lying below the surface $z = f(x, y)$ and above the elliptical region R. (*Hint:* After making the change of variables given by the results in Exercise 31, make a second change of variables to polar coordinates.)

(a) $f(x, y) = 16 - x^2 - y^2$

R: $\dfrac{x^2}{16} + \dfrac{y^2}{9} \le 1$

(b) $f(x, y) = A \cos\left(\dfrac{\pi}{2}\sqrt{\dfrac{x^2}{a^2} + \dfrac{y^2}{b^2}}\right)$

R: $\dfrac{x^2}{a^2} + \dfrac{y^2}{b^2} \le 1$

WRITING ABOUT CONCEPTS

33. State the definition of the Jacobian.

34. Describe how to use the Jacobian to change variables in double integrals.

In Exercises 35–40, find the Jacobian $\partial(x, y, z)/\partial(u, v, w)$ for the indicated change of variables. If $x = f(u, v, w)$, $y = g(u, v, w)$, and $z = h(u, v, w)$, then the Jacobian of x, y, and z with respect to u, v, and w is

$$\frac{\partial(x, y, z)}{\partial(u, v, w)} = \begin{vmatrix} \dfrac{\partial x}{\partial u} & \dfrac{\partial x}{\partial v} & \dfrac{\partial x}{\partial w} \\ \dfrac{\partial y}{\partial u} & \dfrac{\partial y}{\partial v} & \dfrac{\partial y}{\partial w} \\ \dfrac{\partial z}{\partial u} & \dfrac{\partial z}{\partial v} & \dfrac{\partial z}{\partial w} \end{vmatrix}.$$

35. $x = u(1 - v)$, $y = uv(1 - w)$, $z = uvw$

36. $x = 4u - v$, $y = 4v - w$, $z = u + w$

37. $x = \frac{1}{2}(u + v)$, $y = \frac{1}{2}(u - v)$, $z = 2uvw$

38. $x = u - v + w$, $y = 2uv$, $z = u + v + w$

39. *Spherical Coordinates*

$x = \rho \sin \phi \cos \theta$, $y = \rho \sin \phi \sin \theta$, $z = \rho \cos \phi$

40. *Cylindrical Coordinates*

$x = r \cos \theta$, $y = r \sin \theta$, $z = z$

PUTNAM EXAM CHALLENGE

41. Let A be the area of the region in the first quadrant bounded by the line $y = \frac{1}{2}x$, the x-axis, and the ellipse $\frac{1}{9}x^2 + y^2 = 1$. Find the positive number m such that A is equal to the area of the region in the first quadrant bounded by the line $y = mx$, the y-axis, and the ellipse $\frac{1}{9}x^2 + y^2 = 1$.

This problem was composed by the Committee on the Putnam Prize Competition.

14 REVIEW EXERCISES

See www.CalcChat.com for worked-out solutions to odd-numbered exercises.

In Exercises 1 and 2, evaluate the integral.

1. $\displaystyle\int_1^{x^2} x \ln y \, dy$

2. $\displaystyle\int_y^{2y} (x^2 + y^2) \, dx$

In Exercises 3–6, sketch the region of integration. Then evaluate the iterated integral. Change the coordinate system when convenient.

3. $\displaystyle\int_0^1 \int_0^{1+x} (3x + 2y) \, dy \, dx$

4. $\displaystyle\int_0^2 \int_{x^2}^{2x} (x^2 + 2y) \, dy \, dx$

5. $\displaystyle\int_0^3 \int_0^{\sqrt{9-x^2}} 4x \, dy \, dx$

6. $\displaystyle\int_0^{\sqrt{3}} \int_{2-\sqrt{4-y^2}}^{2+\sqrt{4-y^2}} dx \, dy$

Area **In Exercises 7–14, write the limits for the double integral**

$$\int_R\int f(x, y) \, dA$$

for both orders of integration. Compute the area of R by letting $f(x, y) = 1$ and integrating.

7. Triangle: vertices $(0, 0)$, $(3, 0)$, $(0, 1)$
8. Triangle: vertices $(0, 0)$, $(3, 0)$, $(2, 2)$
9. The larger area between the graphs of $x^2 + y^2 = 25$ and $x = 3$
10. Region bounded by the graphs of $y = 6x - x^2$ and $y = x^2 - 2x$
11. Region enclosed by the graph of $y^2 = x^2 - x^4$
12. Region bounded by the graphs of $x = y^2 + 1$, $x = 0$, $y = 0$, and $y = 2$
13. Region bounded by the graphs of $x = y + 3$ and $x = y^2 + 1$
14. Region bounded by the graphs of $x = -y$ and $x = 2y - y^2$

Think About It **In Exercises 15 and 16, give a geometric argument for the given equality. Verify the equality analytically.**

15. $\displaystyle\int_0^1 \int_{2y}^{2\sqrt{2-y^2}} (x + y) \, dx \, dy = \int_0^2 \int_0^{x/2} (x + y) \, dy \, dx + \int_2^{2\sqrt{2}} \int_0^{\sqrt{8-x^2}/2} (x + y) \, dy \, dx$

16. $\displaystyle\int_0^2 \int_{3y/2}^{5-y} e^{x+y} \, dx \, dy = \int_0^3 \int_0^{2x/3} e^{x+y} \, dy \, dx + \int_3^5 \int_0^{5-x} e^{x+y} \, dy \, dx$

Volume **In Exercises 17 and 18, use a multiple integral and a convenient coordinate system to find the volume of the solid.**

17. Solid bounded by the graphs of $z = x^2 - y + 4$, $z = 0$, $y = 0$, $x = 0$, and $x = 4$
18. Solid bounded by the graphs of $z = x + y$, $z = 0$, $y = 0$, $x = 3$, and $y = x$

Average Value **In Exercises 19 and 20, find the average of $f(x, y)$ over the region R.**

19. $f(x) = 16 - x^2 - y^2$

    R: rectangle with vertices $(2, 2)$, $(-2, 2)$, $(-2, -2)$, $(2, -2)$

20. $f(x) = 2x^2 + y^2$

    R: square with vertices $(0, 0)$, $(3, 0)$, $(3, 3)$, $(0, 3)$

21. *Average Temperature* The temperature in degrees Celsius on the surface of a metal plate is

    $T(x, y) = 40 - 6x^2 - y^2$

    where x and y are measured in centimeters. Estimate the average temperature if x varies between 0 and 3 centimeters and y varies between 0 and 5 centimeters.

CAS 22. *Average Profit* A firm's profit P from marketing two soft drinks is

    $P = 192x + 576y - x^2 - 5y^2 - 2xy - 5000$

    where x and y represent the numbers of units of the two soft drinks. Use a computer algebra system to evaluate the double integral yielding the average weekly profit if x varies between 40 and 50 units and y varies between 45 and 60 units.

Probability **In Exercises 23 and 24, find k such that the function is a joint density function and find the required probability, where**

$$P(a \le x \le b, c \le y \le d) = \int_c^d \int_a^b f(x, y) \, dx \, dy.$$

23. $f(x, y) = \begin{cases} kxye^{-(x+y)}, & x \ge 0, y \ge 0 \\ 0, & \text{elsewhere} \end{cases}$

    $P(0 \le x \le 1, 0 \le y \le 1)$

24. $f(x, y) = \begin{cases} kxy, & 0 \le x \le 1, 0 \le y \le x \\ 0, & \text{elsewhere} \end{cases}$

    $P(0 \le x \le 0.5, 0 \le y \le 0.25)$

Approximation **In Exercises 25 and 26, determine which value best approximates the volume of the solid between the xy-plane and the function over the region. (Make your selection on the basis of a sketch of the solid and *not* by performing any calculations.)**

25. $f(x, y) = x + y$

    R: triangle with vertices $(0, 0)$, $(3, 0)$, $(3, 3)$

    (a) $\frac{9}{2}$ (b) 5 (c) 13 (d) 100 (e) -100

26. $f(x, y) = 10x^2y^2$

    R: circle bounded by $x^2 + y^2 = 1$

    (a) π (b) -15 (c) $\frac{2}{3}$ (d) 3 (e) 15

True or False? **In Exercises 27–30, determine whether the statement is true or false. If it is false, explain why or give an example that shows it is false.**

27. $\int_a^b \int_c^d f(x)g(y)\,dy\,dx = \left[\int_a^b f(x)\,dx\right]\left[\int_c^d g(y)\,dy\right]$

28. If f is continuous over R_1 and R_2, and

$$\int_{R_1}\int dA = \int_{R_2}\int dA$$

then

$$\int_{R_1}\int f(x, y)\,dA = \int_{R_2}\int f(x, y)\,dA.$$

29. $\int_{-1}^1 \int_{-1}^1 \cos(x^2 + y^2)\,dx\,dy = 4\int_0^1 \int_0^1 \cos(x^2 + y^2)\,dx\,dy$

30. $\int_0^1 \int_0^1 \frac{1}{1 + x^2 + y^2}\,dx\,dy < \frac{\pi}{4}$

In Exercises 31 and 32, evaluate the iterated integral by converting to polar coordinates.

31. $\int_0^h \int_0^x \sqrt{x^2 + y^2}\,dy\,dx$

32. $\int_0^4 \int_0^{\sqrt{16-y^2}} (x^2 + y^2)\,dx\,dy$

Area **In Exercises 33 and 34, use a double integral to find the area of the shaded region.**

33.

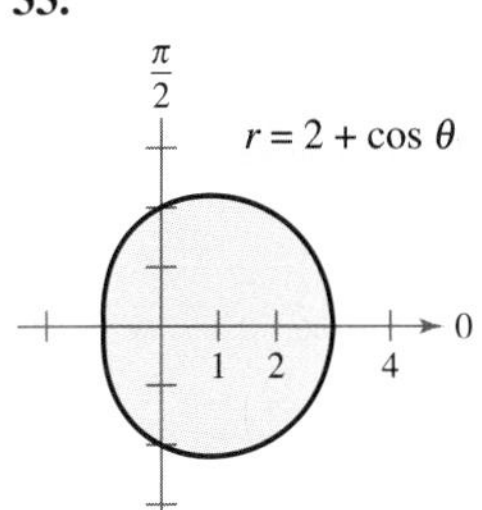

34.

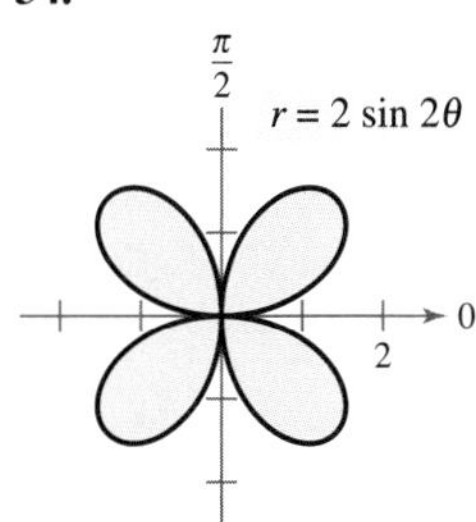

Volume **In Exercises 35 and 36, use a multiple integral and a convenient coordinate system to find the volume of the solid.**

35. Solid bounded by the graphs of $z = 0$ and $z = h$, outside the cylinder $x^2 + y^2 = 1$ and inside the hyperboloid $x^2 + y^2 - z^2 = 1$

36. Solid that remains after drilling a hole of radius b through the center of a sphere of radius R $(b < R)$

37. Consider the region R in the xy-plane bounded by the graph of the equation

$$(x^2 + y^2)^2 = 9(x^2 - y^2).$$

(a) Convert the equation to polar coordinates. Use a graphing utility to graph the equation.

(b) Use a double integral to find the area of the region R.

CAS (c) Use a computer algebra system to determine the volume of the solid over the region R and beneath the hemisphere $z = \sqrt{9 - x^2 - y^2}$.

38. Combine the sum of the two iterated integrals into a single iterated integral by converting to polar coordinates. Evaluate the resulting iterated integral.

$$\int_0^{8/\sqrt{13}} \int_0^{3x/2} xy\,dy\,dx + \int_{8/\sqrt{13}}^4 \int_0^{\sqrt{16-x^2}} xy\,dy\,dx$$

CAS ***Mass and Center of Mass*** **In Exercises 39 and 40, find the mass and center of mass of the lamina bounded by the graphs of the equations for the given density or densities. Use a computer algebra system to evaluate the multiple integrals.**

39. $y = 2x$, $y = 2x^3$, first quadrant

(a) $\rho = kxy$

(b) $\rho = k(x^2 + y^2)$

40. $y = \frac{h}{2}\left(2 - \frac{x}{L} - \frac{x^2}{L^2}\right)$, $\rho = k$, first quadrant

CAS **In Exercises 41 and 42, find I_x, I_y, I_0, $\bar{\bar{x}}$, and $\bar{\bar{y}}$ for the lamina bounded by the graphs of the equations. Use a computer algebra system to evaluate the double integrals.**

41. $y = 0$, $y = b$, $x = 0$, $x = a$, $\rho = kx$

42. $y = 4 - x^2$, $y = 0$, $x > 0$, $\rho = ky$

Surface Area **In Exercises 43–46, find the area of the surface given by $z = f(x, y)$ over the region R.**

43. $f(x, y) = 25 - x^2 - y^2$

$R = \{(x, y): x^2 + y^2 \le 25\}$

CAS **44.** $f(x, y) = 16 - x - y^2$

$R = \{(x, y): 0 \le x \le 2, 0 \le y \le x\}$

Use a computer algebra system to evaluate the integral.

45. $f(x, y) = 9 - y^2$

R: triangle bounded by the graphs of the equations $y = x$, $y = -x$, and $y = 3$

46. $f(x, y) = 4 - x^2$

R: triangle bounded by the graphs of the equations $y = x$, $y = -x$, and $y = 2$

47. ***Building Design*** A new auditorium is built with a foundation in the shape of one-fourth of a circle of radius 50 feet. So, it forms a region R bounded by the graph of

$$x^2 + y^2 = 50^2$$

with $x \ge 0$ and $y \ge 0$. The following equations are models for the floor and ceiling.

Floor: $z = \frac{x + y}{5}$

Ceiling: $z = 20 + \frac{xy}{100}$

(a) Calculate the volume of the room, which is needed to determine the heating and cooling requirements.

(b) Find the surface area of the ceiling.

CAS **48.** ***Surface Area*** The roof over the stage of an open air theater at a theme park is modeled by

$$f(x, y) = 25\left[1 + e^{-(x^2+y^2)/1000}\cos^2\left(\frac{x^2 + y^2}{1000}\right)\right]$$

where the stage is a semicircle bounded by the graphs of $y = \sqrt{50^2 - x^2}$ and $y = 0$.

(a) Use a computer algebra system to graph the surface.

(b) Use a computer algebra system to approximate the number of square feet of roofing required to cover the surface.

In Exercises 49–52, evaluate the iterated integral.

49. $\displaystyle\int_{-3}^{3}\int_{-\sqrt{9-x^2}}^{\sqrt{9-x^2}}\int_{x^2+y^2}^{9}\sqrt{x^2 + y^2}\,dz\,dy\,dx$

50. $\displaystyle\int_{-2}^{2}\int_{-\sqrt{4-x^2}}^{\sqrt{4-x^2}}\int_{0}^{(x^2+y^2)/2}(x^2 + y^2)\,dz\,dy\,dx$

51. $\displaystyle\int_{0}^{a}\int_{0}^{b}\int_{0}^{c}(x^2 + y^2 + z^2)\,dx\,dy\,dz$

52. $\displaystyle\int_{0}^{5}\int_{0}^{\sqrt{25-x^2}}\int_{0}^{\sqrt{25-x^2-y^2}}\frac{1}{1 + x^2 + y^2 + z^2}\,dz\,dy\,dx$

CAS **In Exercises 53 and 54, use a computer algebra system to evaluate the iterated integral.**

53. $\displaystyle\int_{-1}^{1}\int_{-\sqrt{1-x^2}}^{\sqrt{1-x^2}}\int_{-\sqrt{1-x^2-y^2}}^{\sqrt{1-x^2-y^2}}(x^2 + y^2)\,dz\,dy\,dx$

54. $\displaystyle\int_{0}^{2}\int_{0}^{\sqrt{4-x^2}}\int_{0}^{\sqrt{4-x^2-y^2}}xyz\,dz\,dy\,dx$

Volume **In Exercises 55 and 56, use a multiple integral to find the volume of the solid.**

55. Solid inside the graphs of $r = 2\cos\theta$ and $r^2 + z^2 = 4$

56. Solid inside the graphs of $r^2 + z = 16$, $z = 0$, and $r = 2\sin\theta$

Center of Mass **In Exercises 57–60, find the center of mass of the solid of uniform density bounded by the graphs of the equations.**

57. Solid inside the hemisphere $\rho = \cos\phi$, $\pi/4 \le \phi \le \pi/2$, and outside the cone $\phi = \pi/4$

58. Wedge: $x^2 + y^2 = a^2$, $z = cy\ (c > 0)$, $y \ge 0$, $z \ge 0$

59. $x^2 + y^2 + z^2 = a^2$, first octant

60. $x^2 + y^2 + z^2 = 25$, $z = 4$ (the larger solid)

Moment of Inertia **In Exercises 61 and 62, find the moment of inertia I_z of the solid of given density.**

61. The solid of uniform density inside the paraboloid $z = 16 - x^2 - y^2$, and outside the cylinder $x^2 + y^2 = 9$, $z \ge 0$.

62. $x^2 + y^2 + z^2 = a^2$, density proportional to the distance from the center

63. ***Investigation*** Consider a spherical segment of height h from a sphere of radius a, where $h \le a$, and constant density $\rho(x, y, z) = k$ (see figure).

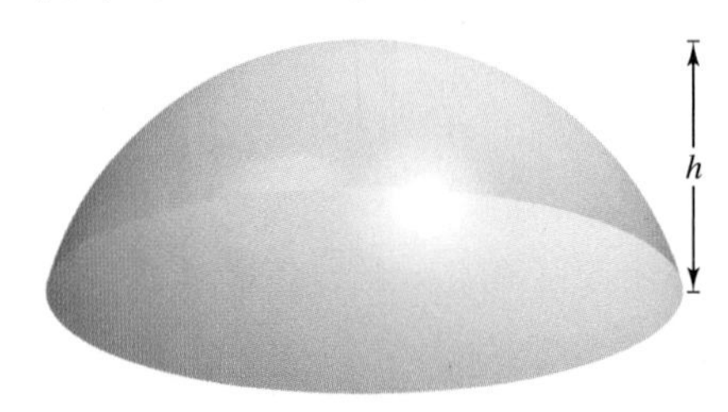

(a) Find the volume of the solid.

(b) Find the centroid of the solid.

(c) Use the result of part (b) to find the centroid of a hemisphere of radius a.

(d) Find $\lim_{h\to 0} \bar{z}$.

(e) Find I_z.

(f) Use the result of part (e) to find I_z for a hemisphere.

64. ***Moment of Inertia*** Find the moment of inertia about the z-axis of the ellipsoid $x^2 + y^2 + \dfrac{z^2}{a^2} = 1$, where $a > 0$.

In Exercises 65 and 66, give a geometric interpretation of the iterated integral.

65. $\displaystyle\int_{0}^{2\pi}\int_{0}^{\pi}\int_{0}^{6\sin\phi}\rho^2\sin\phi\,d\rho\,d\phi\,d\theta$

66. $\displaystyle\int_{0}^{\pi}\int_{0}^{2}\int_{0}^{1+r^2}r\,dz\,dr\,d\theta$

In Exercises 67 and 68, find the Jacobian $\partial(x, y)/\partial(u, v)$ for the indicated change of variables.

67. $x = u + 3v$, $y = 2u - 3v$

68. $x = u^2 + v^2$, $y = u^2 - v^2$

In Exercises 69 and 70, use the indicated change of variables to evaluate the double integral.

69. $\displaystyle\int_R\int \ln(x + y)\,dA$

$x = \frac{1}{2}(u + v)$, $y = \frac{1}{2}(u - v)$

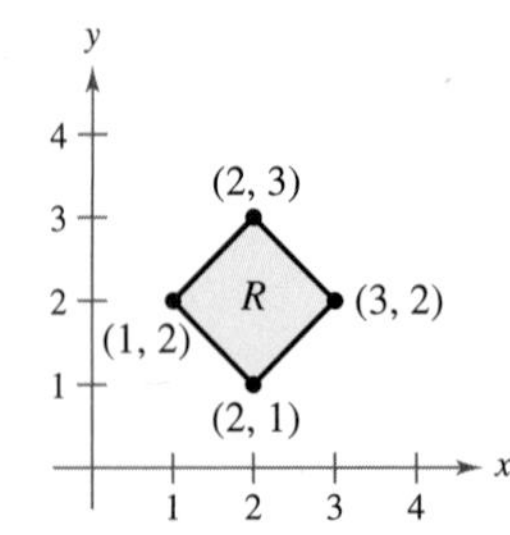

70. $\displaystyle\int_R\int \frac{x}{1 + x^2y^2}\,dA$

$x = u$, $y = \dfrac{v}{u}$

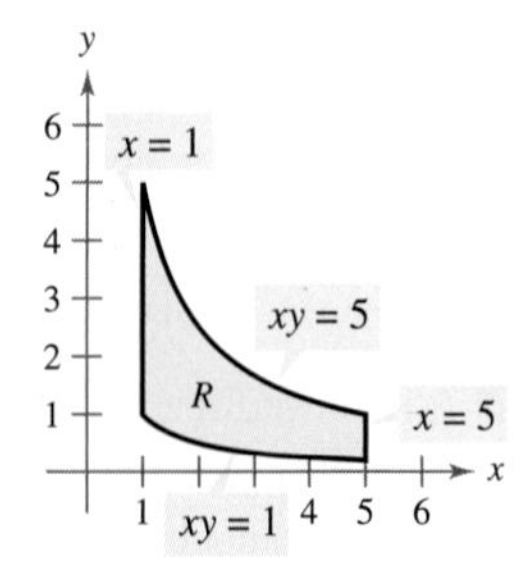

P.S. PROBLEM SOLVING

1. Find the volume of the solid of intersection of the three cylinders $x^2 + z^2 = 1$, $y^2 + z^2 = 1$, and $x^2 + y^2 = 1$ (see figure).

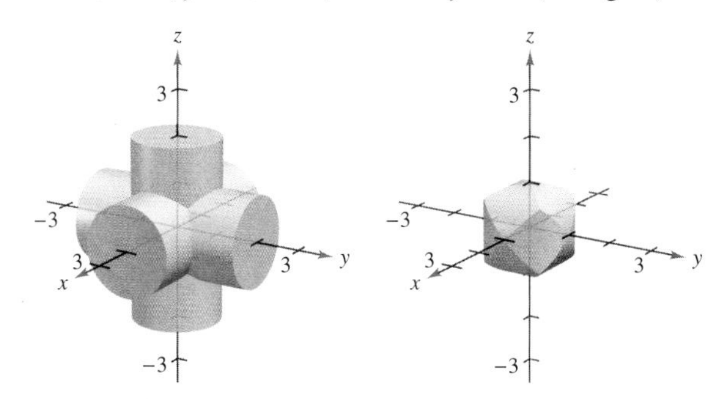

2. Let a, b, c, and d be positive real numbers. The first octant of the plane $ax + by + cz = d$ is shown in the figure. Show that the surface area of this portion of the plane is equal to

$$\frac{A(R)}{c}\sqrt{a^2 + b^2 + c^2}$$

where $A(R)$ is the area of the triangular region R in the xy-plane, as shown in the figure.

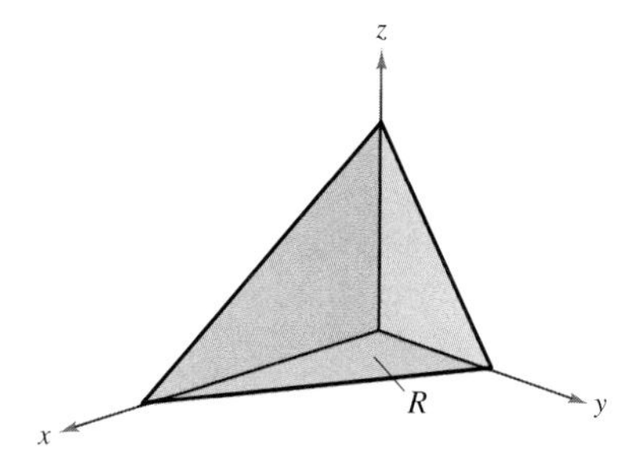

3. Derive Euler's famous result that was mentioned in Section 9.3, $\sum_{n=1}^{\infty} \frac{1}{n^2} = \frac{\pi^2}{6}$, by completing each step.

(a) Prove that $\int \frac{dv}{2 - u^2 + v^2} = \frac{1}{\sqrt{2 - u^2}} \arctan \frac{v}{\sqrt{2 - u^2}} + C$.

(b) Prove that $I_1 = \int_0^{\sqrt{2}/2} \int_{-u}^{u} \frac{2}{2 - u^2 + v^2}\, dv\, du = \frac{\pi^2}{18}$ by using the substitution $u = \sqrt{2} \sin \theta$.

(c) Prove that

$$I_2 = \int_{\sqrt{2}/2}^{\sqrt{2}} \int_{u - \sqrt{2}}^{-u + \sqrt{2}} \frac{2}{2 - u^2 + v^2}\, dv\, du$$

$$= 4\int_{\pi/6}^{\pi/2} \arctan \frac{1 - \sin \theta}{\cos \theta}\, d\theta$$

by using the substitution $u = \sqrt{2} \sin \theta$.

(d) Prove the trigonometric identity $\frac{1 - \sin \theta}{\cos \theta} = \tan\left(\frac{(\pi/2) - \theta}{2}\right)$.

(e) Prove that $I_2 = \int_{\sqrt{2}/2}^{\sqrt{2}} \int_{u - \sqrt{2}}^{-u + \sqrt{2}} \frac{2}{2 - u^2 + v^2}\, dv\, du = \frac{\pi^2}{9}$.

(f) Use the formula for the sum of an infinite geometric series to verify that $\sum_{n=1}^{\infty} \frac{1}{n^2} = \int_0^1 \int_0^1 \frac{1}{1 - xy}\, dx\, dy$.

(g) Use the change of variables $u = \frac{x + y}{\sqrt{2}}$ and $v = \frac{y - x}{\sqrt{2}}$ to prove that $\sum_{n=1}^{\infty} \frac{1}{n^2} = \int_0^1 \int_0^1 \frac{1}{1 - xy}\, dx\, dy = I_1 + I_2 = \frac{\pi^2}{6}$.

4. Consider a circular lawn with a radius of 10 feet, as shown in the figure. Assume that a sprinkler distributes water in a radial fashion according to the formula

$$f(r) = \frac{r}{16} - \frac{r^2}{160}$$

(measured in cubic feet of water per hour per square foot of lawn), where r is the distance in feet from the sprinkler. Find the amount of water that is distributed in 1 hour in the following two annular regions.

$A = \{(r, \theta): 4 \le r \le 5, 0 \le \theta \le 2\pi\}$

$B = \{(r, \theta): 9 \le r \le 10, 0 \le \theta \le 2\pi\}$

Is the distribution of water uniform? Determine the amount of water the entire lawn receives in 1 hour.

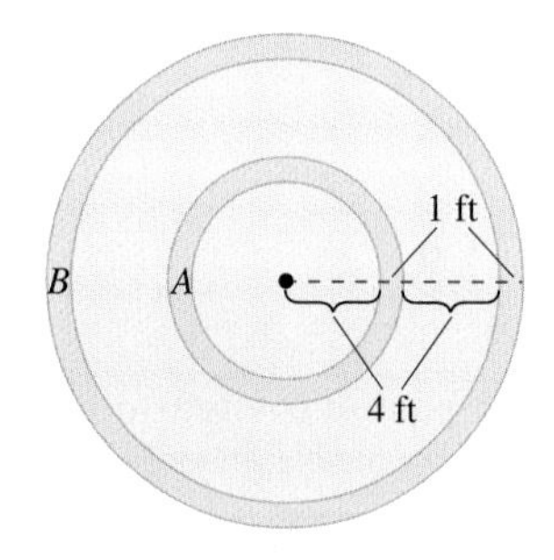

5. The figure shows the region R bounded by the curves $y = \sqrt{x}$, $y = \sqrt{2x}$, $y = \frac{x^2}{3}$, and $y = \frac{x^2}{4}$. Use the change of variables $x = u^{1/3}v^{2/3}$ and $y = u^{2/3}v^{1/3}$ to find the area of the region R.

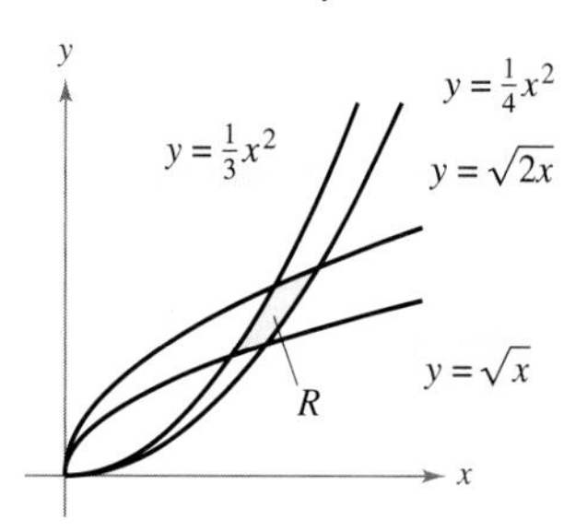

6. The figure shows a solid bounded below by the plane $z = 2$ and above by the sphere $x^2 + y^2 + z^2 = 8$.

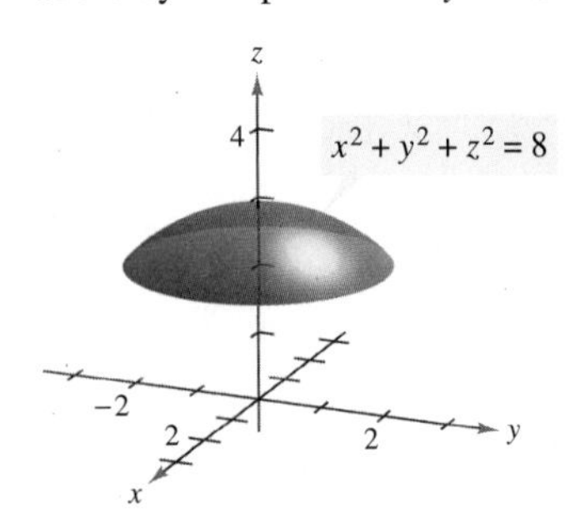

(a) Find the volume of the solid using cylindrical coordinates.

(b) Find the volume of the solid using spherical coordinates.

7. Sketch the solid whose volume is given by the sum of the iterated integrals

$$\int_0^6 \int_{z/2}^3 \int_{z/2}^y dx\, dy\, dz + \int_0^6 \int_3^{(12-z)/2} \int_{z/2}^{6-y} dx\, dy\, dz.$$

Then write the volume as a single iterated integral in the order $dy\, dz\, dx$.

8. Prove that $\lim_{n\to\infty} \int_0^1 \int_0^1 x^n y^n\, dx\, dy = 0.$

In Exercises 9 and 10, evaluate the integral. (*Hint:* See Exercise 69 in Section 14.3.)

9. $\int_0^\infty x^2 e^{-x^2}\, dx$

10. $\int_0^1 \sqrt{\ln \frac{1}{x}}\, dx$

11. Consider the function

$$f(x, y) = \begin{cases} ke^{-(x+y)/a}, & x \ge 0,\ y \ge 0 \\ 0, & \text{elsewhere.} \end{cases}$$

Find the relationship between the positive constants a and k such that f is a joint density function of the continuous random variables x and y.

12. Find the volume of the solid generated by revolving the region in the first quadrant bounded by $y = e^{-x^2}$ about the y-axis. Use this result to find

$$\int_{-\infty}^{\infty} e^{-x^2}\, dx.$$

13. From 1963 to 1986, the volume of the Great Salt Lake approximately tripled while its top surface area approximately doubled. Read the article "Relations between Surface Area and Volume in Lakes" by Daniel Cass and Gerald Wildenberg in *The College Mathematics Journal.* Then give examples of solids that have "water levels" a and b such that $V(b) = 3V(a)$ and $A(b) = 2A(a)$ (see figure), where V is volume and A is area.

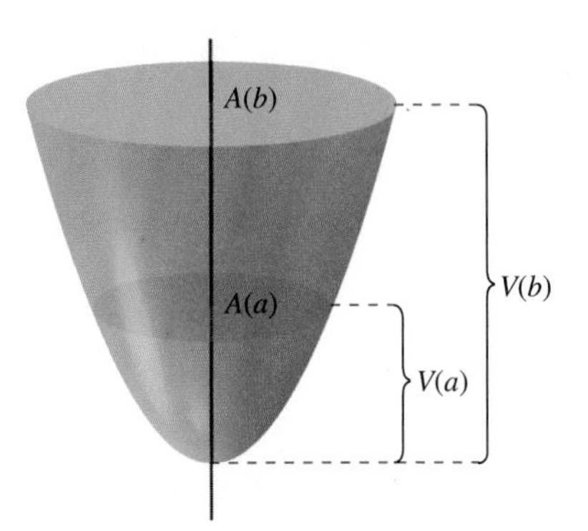

Figure for 13

14. The angle between a plane P and the xy-plane is θ, where $0 \le \theta < \pi/2$. The projection of a rectangular region in P onto the xy-plane is a rectangle whose sides have lengths Δx and Δy, as shown in the figure. Prove that the area of the rectangular region in P is $\sec \theta\, \Delta x\, \Delta y$.

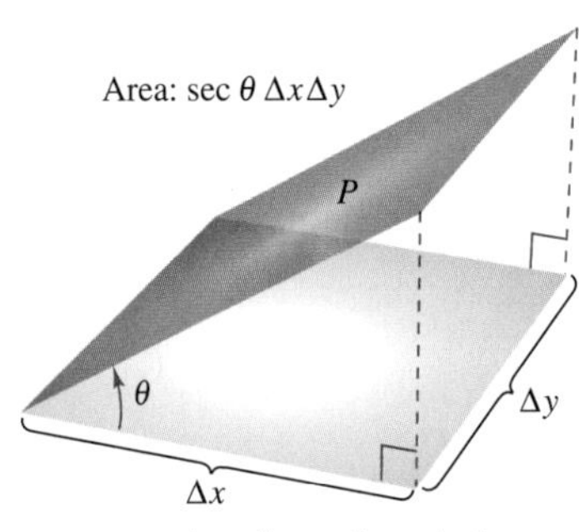

15. Use the result of Exercise 14 to order the planes in ascending order of their surface areas for a fixed region R in the xy-plane. Explain your ordering without doing any calculations.

(a) $z_1 = 2 + x$

(b) $z_2 = 5$

(c) $z_3 = 10 - 5x + 9y$

(d) $z_4 = 3 + x - 2y$

16. Evaluate the integral $\int_0^\infty \int_0^\infty \frac{1}{(1 + x^2 + y^2)^2}\, dx\, dy.$

17. Evaluate the integrals

$$\int_0^1 \int_0^1 \frac{x - y}{(x + y)^3}\, dx\, dy$$

and

$$\int_0^1 \int_0^1 \frac{x - y}{(x + y)^3}\, dy\, dx.$$

Are the results the same? Why or why not?

18. Show that the volume of a spherical block can be approximated by

$\Delta V \approx \rho^2 \sin \phi\, \Delta\rho\, \Delta\phi\, \Delta\theta.$

15 Vector Analysis

In this chapter, you will study vector fields, line integrals, and surface integrals. You will learn to use these to determine real-life quantities such as surface area, mass, flux, work, and energy.

In this chapter, you should learn the following.

- How to sketch a vector field, determine whether a vector field is conservative, find a potential function, find curl, and find divergence. (15.1)
- How to find a piecewise smooth parametrization, write and evaluate a line integral, and use Green's Theorem. (15.2, 15.4)
- How to use the Fundamental Theorem of Line Integrals, independence of path, and conservation of energy. (15.3)
- How to sketch a parametric surface, find a set of parametric equations to represent a surface, find a normal vector, find a tangent plane, and find the area of a parametric surface. (15.5)
- How to evaluate a surface integral, determine the orientation of a surface, evaluate a flux integral, and use the Divergence Theorem. (15.6, 15.7)
- How to use Stokes's Theorem to evaluate a line integral or a surface integral and how to use curl to analyze the motion of a rotating liquid. (15.8)

NASA

While on the ground awaiting liftoff, space shuttle astronauts have access to a basket and slide wire system that is designed to move them as far away from the shuttle as possible in an emergency situation. Does the amount of work done by the gravitational force field vary for different slide wire paths between two fixed points? (See Section 15.3, Exercise 39.)

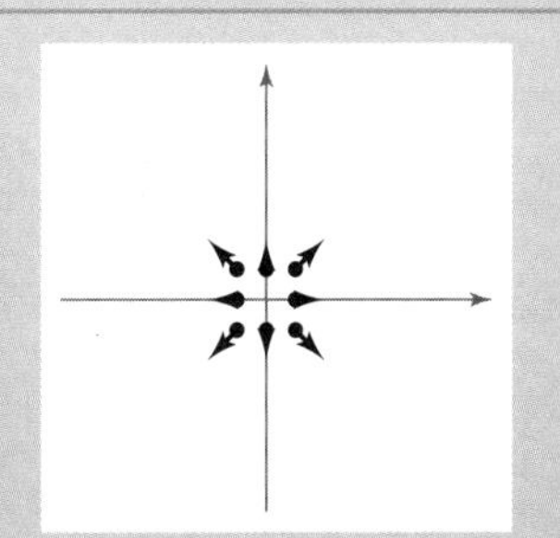

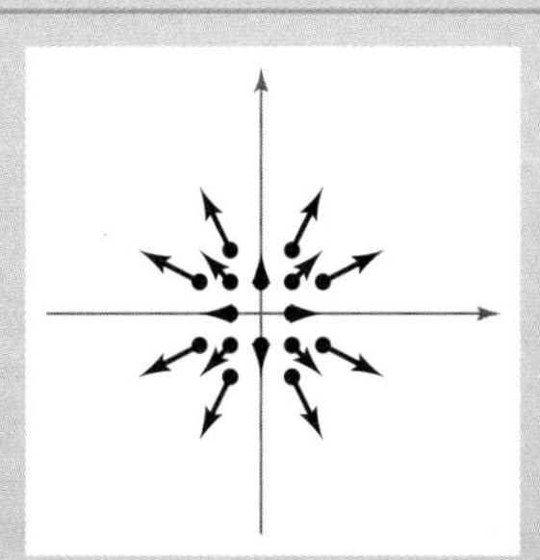

In Chapter 15, you will combine your knowledge of vectors with your knowledge of integral calculus. Section 15.1 introduces *vector fields,* such as those shown above. Examples of vector fields include velocity fields, electromagnetic fields, and gravitational fields.

15.1 Vector Fields

- **Understand the concept of a vector field.**
- **Determine whether a vector field is conservative.**
- **Find the curl of a vector field.**
- **Find the divergence of a vector field.**

Vector Fields

In Chapter 12, you studied vector-valued functions—functions that assign a vector to a *real number*. There you saw that vector-valued functions of real numbers are useful in representing curves and motion along a curve. In this chapter, you will study two other types of vector-valued functions—functions that assign a vector to a *point in the plane* or a *point in space*. Such functions are called **vector fields,** and they are useful in representing various types of **force fields** and **velocity fields.**

DEFINITION OF VECTOR FIELD

A **vector field over a plane region** R is a function $\mathbf{F}$ that assigns a vector $\mathbf{F}(x, y)$ to each point in R.

A **vector field over a solid region** Q **in space** is a function $\mathbf{F}$ that assigns a vector $\mathbf{F}(x, y, z)$ to each point in Q.

NOTE Although a vector field consists of infinitely many vectors, you can get a good idea of what the vector field looks like by sketching several representative vectors $\mathbf{F}(x, y)$ whose initial points are (x, y). ■

The *gradient* is one example of a vector field. For example, if

$$f(x, y) = x^2y + 3xy^3$$

then the gradient of f

$$\begin{aligned}\nabla f(x, y) &= f_x(x, y)\mathbf{i} + f_y(x, y)\mathbf{j}\\ &= (2xy + 3y^3)\mathbf{i} + (x^2 + 9xy^2)\mathbf{j}\end{aligned}$$

Vector field in the plane

is a vector field in the plane. From Chapter 13, the graphical interpretation of this field is a family of vectors, each of which points in the direction of maximum increase along the surface given by $z = f(x, y)$.

Similarly, if

$$f(x, y, z) = x^2 + y^2 + z^2$$

then the gradient of f

$$\begin{aligned}\nabla f(x, y, z) &= f_x(x, y, z)\mathbf{i} + f_y(x, y, z)\mathbf{j} + f_z(x, y, z)\mathbf{k}\\ &= 2x\mathbf{i} + 2y\mathbf{j} + 2z\mathbf{k}\end{aligned}$$

Vector field in space

is a vector field in space. Note that the component functions for this particular vector field are $2x$, $2y$, and $2z$.

A vector field

$$\mathbf{F}(x, y, z) = M(x, y, z)\mathbf{i} + N(x, y, z)\mathbf{j} + P(x, y, z)\mathbf{k}$$

is **continuous** at a point if and only if each of its component functions M, N, and P is continuous at that point.

Some common *physical* examples of vector fields are **velocity fields, gravitational fields,** and **electric force fields.**

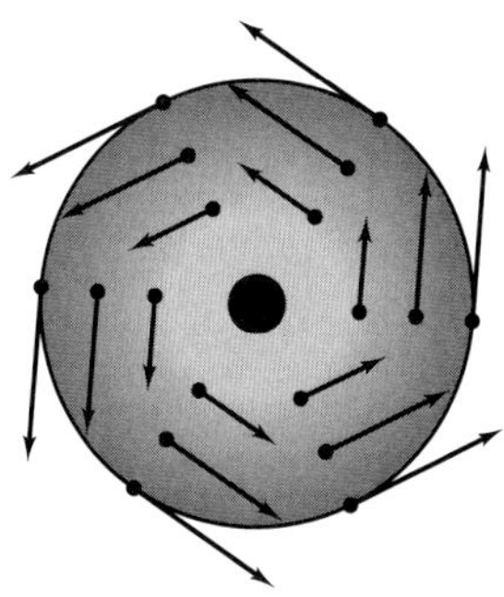

Rotating wheel
Figure 15.1

1. *Velocity fields* describe the motions of systems of particles in the plane or in space. For instance, Figure 15.1 shows the vector field determined by a wheel rotating on an axle. Notice that the velocity vectors are determined by the locations of their initial points—the farther a point is from the axle, the greater its velocity. Velocity fields are also determined by the flow of liquids through a container or by the flow of air currents around a moving object, as shown in Figure 15.2.

2. *Gravitational fields* are defined by **Newton's Law of Gravitation,** which states that the force of attraction exerted on a particle of mass m_1 located at (x, y, z) by a particle of mass m_2 located at $(0, 0, 0)$ is given by

$$\mathbf{F}(x, y, z) = \frac{-Gm_1m_2}{x^2 + y^2 + z^2}\mathbf{u}$$

where G is the gravitational constant and $\mathbf{u}$ is the unit vector in the direction from the origin to (x, y, z). In Figure 15.3, you can see that the gravitational field $\mathbf{F}$ has the properties that $\mathbf{F}(x, y, z)$ always points toward the origin, and that the magnitude of $\mathbf{F}(x, y, z)$ is the same at all points equidistant from the origin. A vector field with these two properties is called a **central force field.** Using the position vector

$$\mathbf{r} = x\mathbf{i} + y\mathbf{j} + z\mathbf{k}$$

for the point (x, y, z), you can write the gravitational field $\mathbf{F}$ as

$$\begin{aligned}\mathbf{F}(x, y, z) &= \frac{-Gm_1m_2}{\|\mathbf{r}\|^2}\left(\frac{\mathbf{r}}{\|\mathbf{r}\|}\right)\\ &= \frac{-Gm_1m_2}{\|\mathbf{r}\|^2}\mathbf{u}.\end{aligned}$$

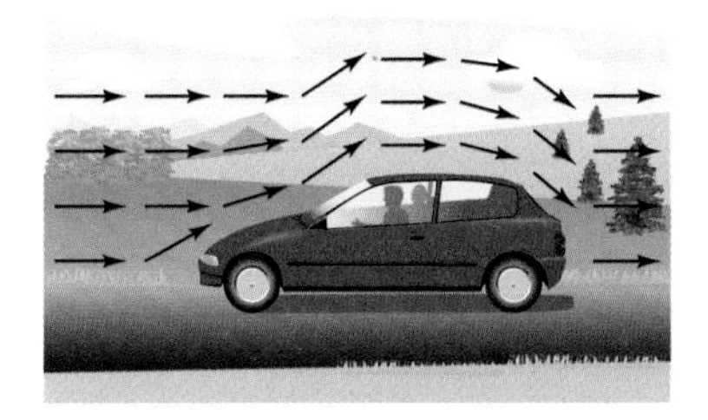

Air flow vector field
Figure 15.2

3. *Electric force fields* are defined by **Coulomb's Law,** which states that the force exerted on a particle with electric charge q_1 located at (x, y, z) by a particle with electric charge q_2 located at $(0, 0, 0)$ is given by

$$\mathbf{F}(x, y, z) = \frac{cq_1q_2}{\|\mathbf{r}\|^2}\mathbf{u}$$

where $\mathbf{r} = x\mathbf{i} + y\mathbf{j} + z\mathbf{k}$, $\mathbf{u} = \mathbf{r}/\|\mathbf{r}\|$, and c is a constant that depends on the choice of units for $\|\mathbf{r}\|$, q_1, and q_2.

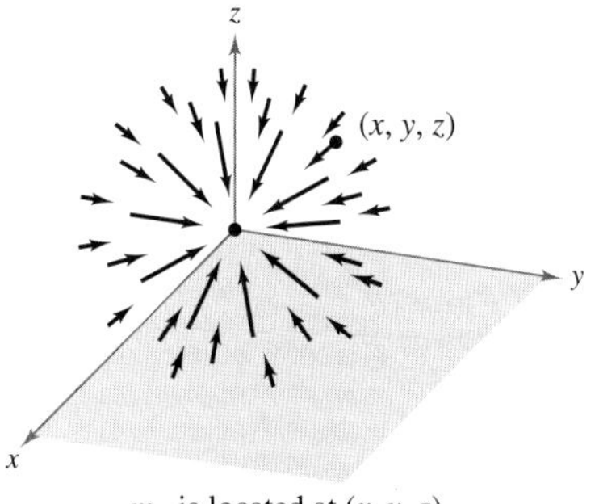

Gravitational force field
Figure 15.3

Note that an electric force field has the same form as a gravitational field. That is,

$$\mathbf{F}(x, y, z) = \frac{k}{\|\mathbf{r}\|^2}\mathbf{u}.$$

Such a force field is called an **inverse square field.**

DEFINITION OF INVERSE SQUARE FIELD

Let $\mathbf{r}(t) = x(t)\mathbf{i} + y(t)\mathbf{j} + z(t)\mathbf{k}$ be a position vector. The vector field $\mathbf{F}$ is an **inverse square field** if

$$\mathbf{F}(x, y, z) = \frac{k}{\|\mathbf{r}\|^2}\mathbf{u}$$

where k is a real number and $\mathbf{u} = \mathbf{r}/\|\mathbf{r}\|$ is a unit vector in the direction of $\mathbf{r}$.

Because vector fields consist of infinitely many vectors, it is not possible to create a sketch of the entire field. Instead, when you sketch a vector field, your goal is to sketch representative vectors that help you visualize the field.

EXAMPLE 1 Sketching a Vector Field

Sketch some vectors in the vector field given by

$$\mathbf{F}(x, y) = -y\mathbf{i} + x\mathbf{j}.$$

Solution You could plot vectors at several random points in the plane. However, it is more enlightening to plot vectors of equal magnitude. This corresponds to finding level curves in scalar fields. In this case, vectors of equal magnitude lie on circles.

$$\|\mathbf{F}\| = c \qquad \text{Vectors of length } c$$

$$\sqrt{x^2 + y^2} = c$$

$$x^2 + y^2 = c^2 \qquad \text{Equation of circle}$$

To begin making the sketch, choose a value for c and plot several vectors on the resulting circle. For instance, the following vectors occur on the unit circle.

Point	*Vector*
$(1, 0)$	$\mathbf{F}(1, 0) = \mathbf{j}$
$(0, 1)$	$\mathbf{F}(0, 1) = -\mathbf{i}$
$(-1, 0)$	$\mathbf{F}(-1, 0) = -\mathbf{j}$
$(0, -1)$	$\mathbf{F}(0, -1) = \mathbf{i}$

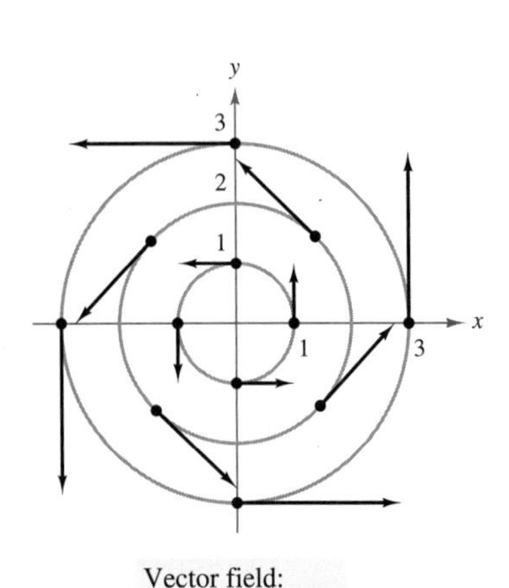

Vector field:
$\mathbf{F}(x, y) = -y\mathbf{i} + x\mathbf{j}$

Figure 15.4

These and several other vectors in the vector field are shown in Figure 15.4. Note in the figure that this vector field is similar to that given by the rotating wheel shown in Figure 15.1.

EXAMPLE 2 Sketching a Vector Field

Sketch some vectors in the vector field given by

$$\mathbf{F}(x, y) = 2x\mathbf{i} + y\mathbf{j}.$$

Solution For this vector field, vectors of equal length lie on ellipses given by

$$\|\mathbf{F}\| = \sqrt{(2x)^2 + (y)^2} = c$$

which implies that

$$4x^2 + y^2 = c^2.$$

For $c = 1$, sketch several vectors $2x\mathbf{i} + y\mathbf{j}$ of magnitude 1 at points on the ellipse given by

$$4x^2 + y^2 = 1.$$

For $c = 2$, sketch several vectors $2x\mathbf{i} + y\mathbf{j}$ of magnitude 2 at points on the ellipse given by

$$4x^2 + y^2 = 4.$$

These vectors are shown in Figure 15.5. ■

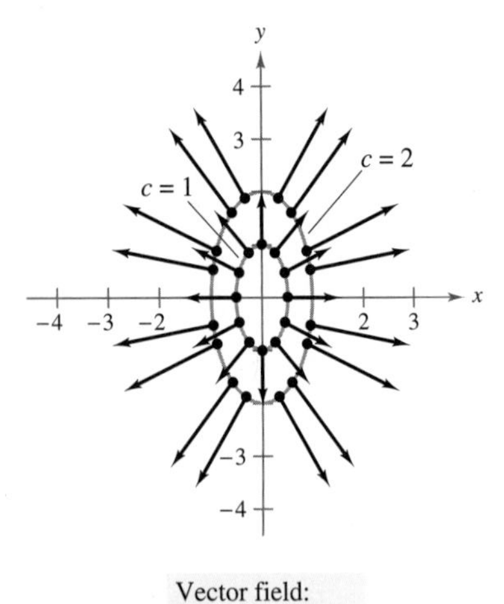

Vector field:
$\mathbf{F}(x, y) = 2x\mathbf{i} + y\mathbf{j}$

Figure 15.5

EXAMPLE 3 Sketching a Velocity Field

Sketch some vectors in the velocity field given by

$$\mathbf{v}(x, y, z) = (16 - x^2 - y^2)\mathbf{k}$$

where $x^2 + y^2 \leq 16$.

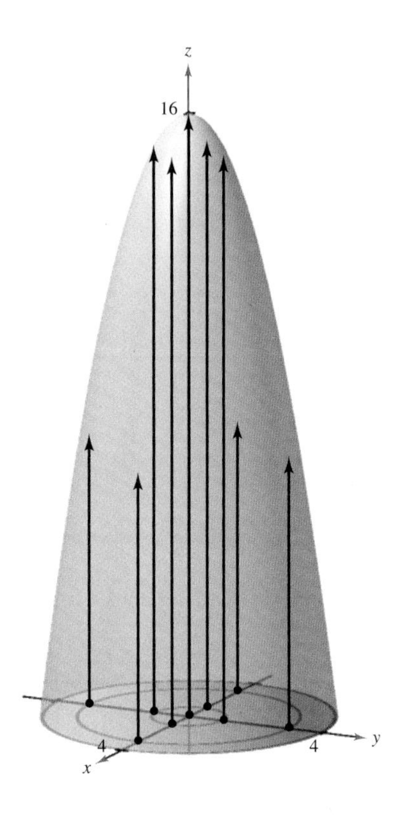

Velocity field:
$\mathbf{v}(x, y, z) = (16 - x^2 - y^2)\mathbf{k}$

Figure 15.6

Solution You can imagine that $\mathbf{v}$ describes the velocity of a liquid flowing through a tube of radius 4. Vectors near the z-axis are longer than those near the edge of the tube. For instance, at the point $(0, 0, 0)$, the velocity vector is $\mathbf{v}(0, 0, 0) = 16\mathbf{k}$, whereas at the point $(0, 3, 0)$, the velocity vector is $\mathbf{v}(0, 3, 0) = 7\mathbf{k}$. Figure 15.6 shows these and several other vectors for the velocity field. From the figure, you can see that the speed of the liquid is greater near the center of the tube than near the edges of the tube. ■

Conservative Vector Fields

Notice in Figure 15.5 that all the vectors appear to be normal to the level curve from which they emanate. Because this is a property of gradients, it is natural to ask whether the vector field given by $\mathbf{F}(x, y) = 2x\mathbf{i} + y\mathbf{j}$ is the *gradient* of some differentiable function f. The answer is that some vector fields can be represented as the gradients of differentiable functions and some cannot—those that can are called **conservative** vector fields.

DEFINITION OF CONSERVATIVE VECTOR FIELD

A vector field $\mathbf{F}$ is called **conservative** if there exists a differentiable function f such that $\mathbf{F} = \nabla f$. The function f is called the **potential function** for $\mathbf{F}$.

EXAMPLE 4 Conservative Vector Fields

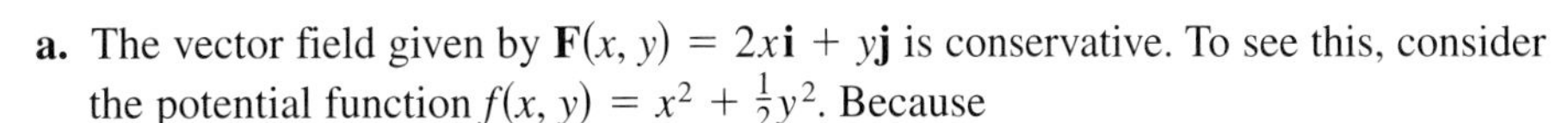

a. The vector field given by $\mathbf{F}(x, y) = 2x\mathbf{i} + y\mathbf{j}$ is conservative. To see this, consider the potential function $f(x, y) = x^2 + \frac{1}{2}y^2$. Because

$$\nabla f = 2x\mathbf{i} + y\mathbf{j} = \mathbf{F}$$

it follows that $\mathbf{F}$ is conservative.

b. Every inverse square field is conservative. To see this, let

$$\mathbf{F}(x, y, z) = \frac{k}{\|\mathbf{r}\|^2}\mathbf{u} \quad \text{and} \quad f(x, y, z) = \frac{-k}{\sqrt{x^2 + y^2 + z^2}}$$

where $\mathbf{u} = \mathbf{r}/\|\mathbf{r}\|$. Because

$$\begin{aligned}\nabla f &= \frac{kx}{(x^2 + y^2 + z^2)^{3/2}}\mathbf{i} + \frac{ky}{(x^2 + y^2 + z^2)^{3/2}}\mathbf{j} + \frac{kz}{(x^2 + y^2 + z^2)^{3/2}}\mathbf{k} \\ &= \frac{k}{x^2 + y^2 + z^2}\left(\frac{x\mathbf{i} + y\mathbf{j} + z\mathbf{k}}{\sqrt{x^2 + y^2 + z^2}}\right) \\ &= \frac{k}{\|\mathbf{r}\|^2}\frac{\mathbf{r}}{\|\mathbf{r}\|} \\ &= \frac{k}{\|\mathbf{r}\|^2}\mathbf{u}\end{aligned}$$

it follows that $\mathbf{F}$ is conservative. ■

As can be seen in Example 4(b), many important vector fields, including gravitational fields and electric force fields, are conservative. Most of the terminology in this chapter comes from physics. For example, the term "conservative" is derived from the classic physical law regarding the conservation of energy. This law states that the sum of the kinetic energy and the potential energy of a particle moving in a conservative force field is constant. (The kinetic energy of a particle is the energy due to its motion, and the potential energy is the energy due to its position in the force field.)

The following important theorem gives a necessary and sufficient condition for a vector field *in the plane* to be conservative.

THEOREM 15.1 TEST FOR CONSERVATIVE VECTOR FIELD IN THE PLANE

Let M and N have continuous first partial derivatives on an open disk R. The vector field given by $\mathbf{F}(x, y) = M\mathbf{i} + N\mathbf{j}$ is conservative if and only if

$$\frac{\partial N}{\partial x} = \frac{\partial M}{\partial y}.$$

PROOF To prove that the given condition is necessary for $\mathbf{F}$ to be conservative, suppose there exists a potential function f such that

$$\mathbf{F}(x, y) = \nabla f(x, y) = M\mathbf{i} + N\mathbf{j}.$$

Then you have

$$f_x(x, y) = M \quad \Longrightarrow \quad f_{xy}(x, y) = \frac{\partial M}{\partial y}$$

$$f_y(x, y) = N \quad \Longrightarrow \quad f_{yx}(x, y) = \frac{\partial N}{\partial x}$$

and, by the equivalence of the mixed partials f_{xy} and f_{yx}, you can conclude that $\partial N/\partial x = \partial M/\partial y$ for all (x, y) in R. The sufficiency of this condition is proved in Section 15.4. ∎

NOTE Theorem 15.1 is valid on *simply connected* domains. A plane region R is simply connected if every simple closed curve in R encloses only points that are in R. See Figure 15.26 in Section 15.4. ∎

EXAMPLE 5 Testing for Conservative Vector Fields in the Plane

Decide whether the vector field given by $\mathbf{F}$ is conservative.

a. $\mathbf{F}(x, y) = x^2y\mathbf{i} + xy\mathbf{j}$ **b.** $\mathbf{F}(x, y) = 2x\mathbf{i} + y\mathbf{j}$

Solution

a. The vector field given by $\mathbf{F}(x, y) = x^2y\mathbf{i} + xy\mathbf{j}$ is not conservative because

$$\frac{\partial M}{\partial y} = \frac{\partial}{\partial y}[x^2y] = x^2 \quad \text{and} \quad \frac{\partial N}{\partial x} = \frac{\partial}{\partial x}[xy] = y.$$

b. The vector field given by $\mathbf{F}(x, y) = 2x\mathbf{i} + y\mathbf{j}$ is conservative because

$$\frac{\partial M}{\partial y} = \frac{\partial}{\partial y}[2x] = 0 \quad \text{and} \quad \frac{\partial N}{\partial x} = \frac{\partial}{\partial x}[y] = 0.$$

∎

Theorem 15.1 tells you whether a vector field is conservative. It does not tell you how to find a potential function of **F**. The problem is comparable to antidifferentiation. Sometimes you will be able to find a potential function by simple inspection. For instance, in Example 4 you observed that

$$f(x, y) = x^2 + \frac{1}{2}y^2$$

has the property that $\nabla f(x, y) = 2x\mathbf{i} + y\mathbf{j}$.

EXAMPLE 6 Finding a Potential Function for F(*x*, *y*)

Find a potential function for

$$\mathbf{F}(x, y) = 2xy\mathbf{i} + (x^2 - y)\mathbf{j}.$$

Solution From Theorem 15.1 it follows that **F** is conservative because

$$\frac{\partial}{\partial y}[2xy] = 2x \quad \text{and} \quad \frac{\partial}{\partial x}[x^2 - y] = 2x.$$

If f is a function whose gradient is equal to $\mathbf{F}(x, y)$, then

$$\nabla f(x, y) = 2xy\mathbf{i} + (x^2 - y)\mathbf{j}$$

which implies that

$$f_x(x, y) = 2xy$$

and

$$f_y(x, y) = x^2 - y.$$

To reconstruct the function f from these two partial derivatives, integrate $f_x(x, y)$ with respect to x and integrate $f_y(x, y)$ with respect to y, as follows.

$$f(x, y) = \int f_x(x, y)\, dx = \int 2xy\, dx = x^2y + g(y)$$

$$f(x, y) = \int f_y(x, y)\, dy = \int (x^2 - y)\, dy = x^2y - \frac{y^2}{2} + h(x)$$

Notice that $g(y)$ is constant with respect to x and $h(x)$ is constant with respect to y. To find a single expression that represents $f(x, y)$, let

$$g(y) = -\frac{y^2}{2} \quad \text{and} \quad h(x) = K.$$

Then, you can write

$$\begin{aligned} f(x, y) &= x^2y + g(y) + K \\ &= x^2y - \frac{y^2}{2} + K. \end{aligned}$$

You can check this result by forming the gradient of f. You will see that it is equal to the original function **F**. ■

NOTE Notice that the solution in Example 6 is comparable to that given by an indefinite integral. That is, the solution represents a family of potential functions, any two of which differ by a constant. To find a unique solution, you would have to be given an initial condition satisfied by the potential function. ■

Curl of a Vector Field

Theorem 15.1 has a counterpart for vector fields in space. Before stating that result, the definition of the **curl of a vector field** in space is given.

DEFINITION OF CURL OF A VECTOR FIELD

The curl of $\mathbf{F}(x, y, z) = M\mathbf{i} + N\mathbf{j} + P\mathbf{k}$ is

$$\begin{aligned}\text{curl } \mathbf{F}(x, y, z) &= \nabla \times \mathbf{F}(x, y, z)\\ &= \left(\frac{\partial P}{\partial y} - \frac{\partial N}{\partial z}\right)\mathbf{i} - \left(\frac{\partial P}{\partial x} - \frac{\partial M}{\partial z}\right)\mathbf{j} + \left(\frac{\partial N}{\partial x} - \frac{\partial M}{\partial y}\right)\mathbf{k}.\end{aligned}$$

NOTE If curl $\mathbf{F} = \mathbf{0}$, then $\mathbf{F}$ is said to be **irrotational.** ■

The cross product notation used for curl comes from viewing the gradient ∇f as the result of the **differential operator** ∇ acting on the function f. In this context, you can use the following determinant form as an aid in remembering the formula for curl.

$$\begin{aligned}\text{curl } \mathbf{F}(x, y, z) &= \nabla \times \mathbf{F}(x, y, z)\\ &= \begin{vmatrix} \mathbf{i} & \mathbf{j} & \mathbf{k} \\ \frac{\partial}{\partial x} & \frac{\partial}{\partial y} & \frac{\partial}{\partial z} \\ M & N & P \end{vmatrix}\\ &= \left(\frac{\partial P}{\partial y} - \frac{\partial N}{\partial z}\right)\mathbf{i} - \left(\frac{\partial P}{\partial x} - \frac{\partial M}{\partial z}\right)\mathbf{j} + \left(\frac{\partial N}{\partial x} - \frac{\partial M}{\partial y}\right)\mathbf{k}\end{aligned}$$

EXAMPLE 7 Finding the Curl of a Vector Field

Find curl $\mathbf{F}$ of the vector field given by

$$\mathbf{F}(x, y, z) = 2xy\mathbf{i} + (x^2 + z^2)\mathbf{j} + 2yz\mathbf{k}.$$

Is $\mathbf{F}$ irrotational?

Solution The curl of $\mathbf{F}$ is given by

$$\begin{aligned}\text{curl } \mathbf{F}(x, y, z) &= \nabla \times \mathbf{F}(x, y, z)\\ &= \begin{vmatrix} \mathbf{i} & \mathbf{j} & \mathbf{k} \\ \frac{\partial}{\partial x} & \frac{\partial}{\partial y} & \frac{\partial}{\partial z} \\ 2xy & x^2 + z^2 & 2yz \end{vmatrix}\\ &= \begin{vmatrix} \frac{\partial}{\partial y} & \frac{\partial}{\partial z} \\ x^2 + z^2 & 2yz \end{vmatrix}\mathbf{i} - \begin{vmatrix} \frac{\partial}{\partial x} & \frac{\partial}{\partial z} \\ 2xy & 2yz \end{vmatrix}\mathbf{j} + \begin{vmatrix} \frac{\partial}{\partial x} & \frac{\partial}{\partial y} \\ 2xy & x^2 + z^2 \end{vmatrix}\mathbf{k}\\ &= (2z - 2z)\mathbf{i} - (0 - 0)\mathbf{j} + (2x - 2x)\mathbf{k}\\ &= \mathbf{0}.\end{aligned}$$

Because curl $\mathbf{F} = \mathbf{0}$, $\mathbf{F}$ is irrotational. ■

The icon indicates that you will find a CAS Investigation on the book's website. The CAS Investigation is a collaborative exploration of this example using the computer algebra systems Maple *and* Mathematica.

Later in this chapter, you will assign a physical interpretation to the curl of a vector field. But for now, the primary use of curl is shown in the following test for conservative vector fields in space. The test states that for a vector field in space, the curl is $\mathbf{0}$ at every point in its domain if and only if $\mathbf{F}$ is conservative. The proof is similar to that given for Theorem 15.1.

THEOREM 15.2 TEST FOR CONSERVATIVE VECTOR FIELD IN SPACE

Suppose that M, N, and P have continuous first partial derivatives in an open sphere Q in space. The vector field given by $\mathbf{F}(x, y, z) = M\mathbf{i} + N\mathbf{j} + P\mathbf{k}$ is conservative if and only if

$$\text{curl } \mathbf{F}(x, y, z) = \mathbf{0}.$$

That is, $\mathbf{F}$ is conservative if and only if

$$\frac{\partial P}{\partial y} = \frac{\partial N}{\partial z}, \quad \frac{\partial P}{\partial x} = \frac{\partial M}{\partial z}, \quad \text{and} \quad \frac{\partial N}{\partial x} = \frac{\partial M}{\partial y}.$$

NOTE Theorem 15.2 is valid for *simply connected* domains in space. A simply connected domain in space is a domain D for which every simple closed curve in D (see Section 15.4) can be shrunk to a point in D without leaving D.

From Theorem 15.2, you can see that the vector field given in Example 7 is conservative because $\text{curl } \mathbf{F}(x, y, z) = \mathbf{0}$. Try showing that the vector field

$$\mathbf{F}(x, y, z) = x^3y^2z\mathbf{i} + x^2z\mathbf{j} + x^2y\mathbf{k}$$

is not conservative—you can do this by showing that its curl is

$$\text{curl } \mathbf{F}(x, y, z) = (x^3y^2 - 2xy)\mathbf{j} + (2xz - 2x^3yz)\mathbf{k} \neq \mathbf{0}.$$

For vector fields in space that pass the test for being conservative, you can find a potential function by following the same pattern used in the plane (as demonstrated in Example 6).

NOTE Examples 6 and 8 are illustrations of a type of problem called *recovering a function from its gradient*. If you go on to take a course in differential equations, you will study other methods for solving this type of problem. One popular method gives an interplay between successive "partial integrations" and partial differentiations.

EXAMPLE 8 Finding a Potential Function for F(x, y, z)

Find a potential function for $\mathbf{F}(x, y, z) = 2xy\mathbf{i} + (x^2 + z^2)\mathbf{j} + 2yz\mathbf{k}$.

Solution From Example 7, you know that the vector field given by $\mathbf{F}$ is conservative. If f is a function such that $\mathbf{F}(x, y, z) = \nabla f(x, y, z)$, then

$$f_x(x, y, z) = 2xy, \quad f_y(x, y, z) = x^2 + z^2, \quad \text{and} \quad f_z(x, y, z) = 2yz$$

and integrating with respect to x, y, and z separately produces

$$f(x, y, z) = \int M\,dx = \int 2xy\,dx = x^2y + g(y, z)$$

$$f(x, y, z) = \int N\,dy = \int (x^2 + z^2)\,dy = x^2y + yz^2 + h(x, z)$$

$$f(x, y, z) = \int P\,dz = \int 2yz\,dz = yz^2 + k(x, y).$$

Comparing these three versions of $f(x, y, z)$, you can conclude that

$$g(y, z) = yz^2 + K, \quad h(x, z) = K, \quad \text{and} \quad k(x, y) = x^2y + K.$$

So, $f(x, y, z)$ is given by

$$f(x, y, z) = x^2y + yz^2 + K.$$

■

Divergence of a Vector Field

NOTE Divergence can be viewed as a type of derivative of **F** in that, for vector fields representing velocities of moving particles, the divergence measures the rate of particle flow per unit volume at a point. In hydrodynamics (the study of fluid motion), a velocity field that is divergence free is called **incompressible.** In the study of electricity and magnetism, a vector field that is divergence free is called **solenoidal.**

You have seen that the curl of a vector field **F** is itself a vector field. Another important function defined on a vector field is **divergence,** which is a scalar function.

DEFINITION OF DIVERGENCE OF A VECTOR FIELD

The **divergence** of $\mathbf{F}(x, y) = M\mathbf{i} + N\mathbf{j}$ is

$$\operatorname{div} \mathbf{F}(x, y) = \nabla \cdot \mathbf{F}(x, y) = \frac{\partial M}{\partial x} + \frac{\partial N}{\partial y}. \qquad \text{Plane}$$

The **divergence** of $\mathbf{F}(x, y, z) = M\mathbf{i} + N\mathbf{j} + P\mathbf{k}$ is

$$\operatorname{div} \mathbf{F}(x, y, z) = \nabla \cdot \mathbf{F}(x, y, z) = \frac{\partial M}{\partial x} + \frac{\partial N}{\partial y} + \frac{\partial P}{\partial z}. \qquad \text{Space}$$

If $\operatorname{div} \mathbf{F} = 0$, then **F** is said to be **divergence free.**

The dot product notation used for divergence comes from considering ∇ as a **differential operator,** as follows.

$$\begin{aligned}\nabla \cdot \mathbf{F}(x, y, z) &= \left[\left(\frac{\partial}{\partial x}\right)\mathbf{i} + \left(\frac{\partial}{\partial y}\right)\mathbf{j} + \left(\frac{\partial}{\partial z}\right)\mathbf{k}\right] \cdot (M\mathbf{i} + N\mathbf{j} + P\mathbf{k}) \\ &= \frac{\partial M}{\partial x} + \frac{\partial N}{\partial y} + \frac{\partial P}{\partial z}\end{aligned}$$

EXAMPLE 9 Finding the Divergence of a Vector Field

Find the divergence at $(2, 1, -1)$ for the vector field

$$\mathbf{F}(x, y, z) = x^3y^2z\mathbf{i} + x^2z\mathbf{j} + x^2y\mathbf{k}.$$

Solution The divergence of **F** is

$$\operatorname{div} \mathbf{F}(x, y, z) = \frac{\partial}{\partial x}[x^3y^2z] + \frac{\partial}{\partial y}[x^2z] + \frac{\partial}{\partial z}[x^2y] = 3x^2y^2z.$$

At the point $(2, 1, -1)$, the divergence is

$$\operatorname{div} \mathbf{F}(2, 1, -1) = 3(2^2)(1^2)(-1) = -12.$$

■

There are many important properties of the divergence and curl of a vector field **F** (see Exercises 83–89). One that is used often is described in Theorem 15.3. You are asked to prove this theorem in Exercise 90.

THEOREM 15.3 DIVERGENCE AND CURL

If $\mathbf{F}(x, y, z) = M\mathbf{i} + N\mathbf{j} + P\mathbf{k}$ is a vector field and M, N, and P have continuous second partial derivatives, then

$$\operatorname{div}(\operatorname{curl} \mathbf{F}) = 0.$$

15.1 Exercises

See www.CalcChat.com for worked-out solutions to odd-numbered exercises.

In Exercises 1–6, match the vector field with its graph. [The graphs are labeled (a), (b), (c), (d), (e), and (f).]

(a)

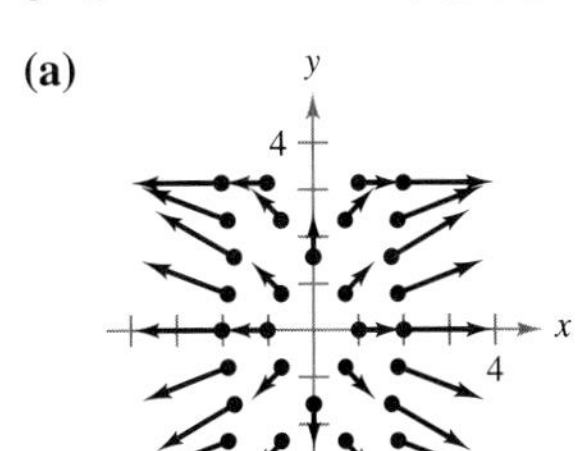

(b)

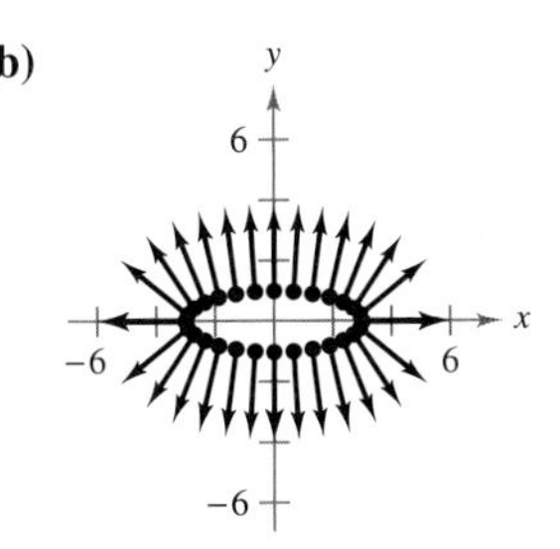

(c)

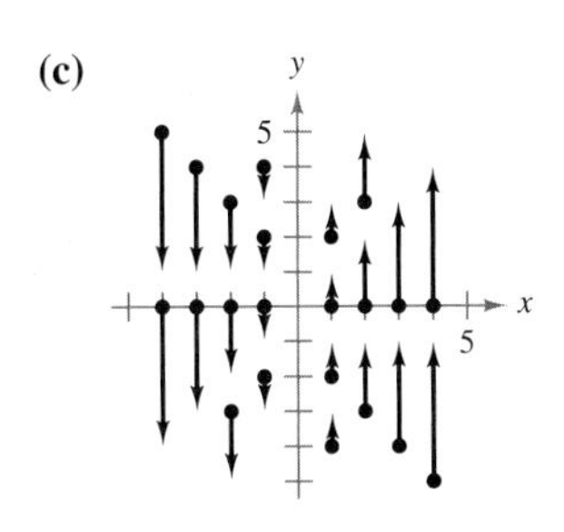

(d)

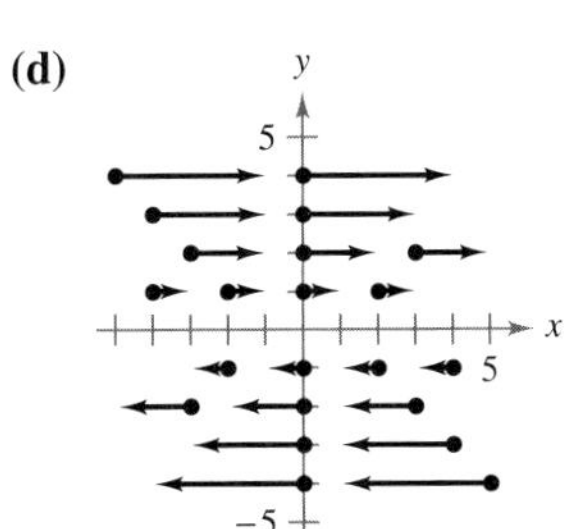

(e)

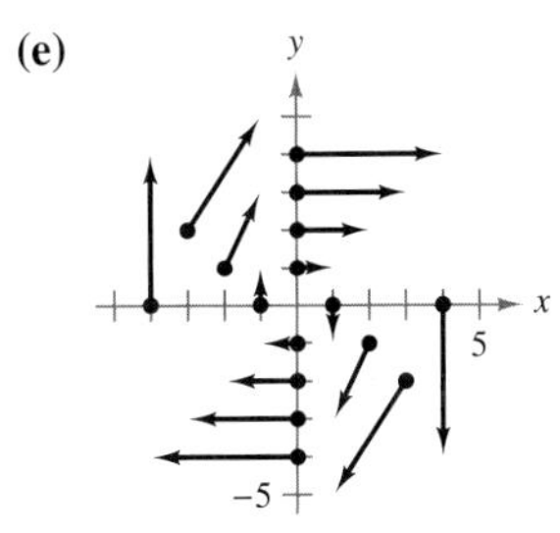

(f)

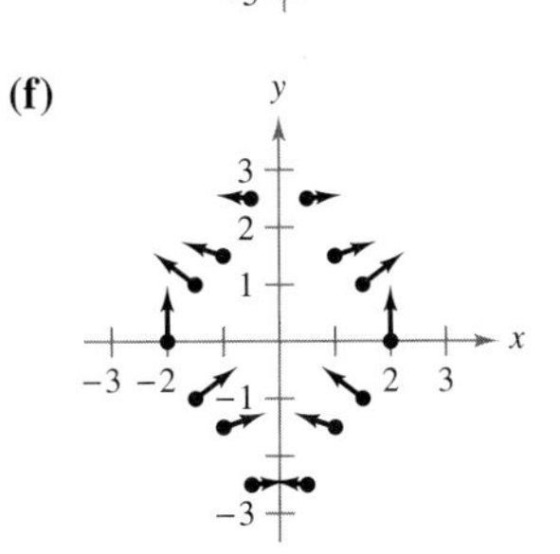

1. $\mathbf{F}(x, y) = y\mathbf{i}$
2. $\mathbf{F}(x, y) = x\mathbf{j}$
3. $\mathbf{F}(x, y) = y\mathbf{i} - x\mathbf{j}$
4. $\mathbf{F}(x, y) = x\mathbf{i} + 3y\mathbf{j}$
5. $\mathbf{F}(x, y) = \langle x, \sin y \rangle$
6. $\mathbf{F}(x, y) = \left\langle \frac{1}{2}xy, \frac{1}{4}x^2 \right\rangle$

In Exercises 7–16, compute $\|\mathbf{F}\|$ and sketch several representative vectors in the vector field.

7. $\mathbf{F}(x, y) = \mathbf{i} + \mathbf{j}$
8. $\mathbf{F}(x, y) = 2\mathbf{i}$
9. $\mathbf{F}(x, y) = y\mathbf{i} + x\mathbf{j}$
10. $\mathbf{F}(x, y) = y\mathbf{i} - 2x\mathbf{j}$
11. $\mathbf{F}(x, y, z) = 3y\mathbf{j}$
12. $\mathbf{F}(x, y) = x\mathbf{i}$
13. $\mathbf{F}(x, y) = 4x\mathbf{i} + y\mathbf{j}$
14. $\mathbf{F}(x, y) = (x^2 + y^2)\mathbf{i} + \mathbf{j}$
15. $\mathbf{F}(x, y, z) = \mathbf{i} + \mathbf{j} + \mathbf{k}$
16. $\mathbf{F}(x, y, z) = x\mathbf{i} + y\mathbf{j} + z\mathbf{k}$

CAS **In Exercises 17–20, use a computer algebra system to graph several representative vectors in the vector field.**

17. $\mathbf{F}(x, y) = \frac{1}{8}(2xy\mathbf{i} + y^2\mathbf{j})$
18. $\mathbf{F}(x, y) = (2y - 3x)\mathbf{i} + (2y + 3x)\mathbf{j}$
19. $\mathbf{F}(x, y, z) = \dfrac{x\mathbf{i} + y\mathbf{j} + z\mathbf{k}}{\sqrt{x^2 + y^2 + z^2}}$
20. $\mathbf{F}(x, y, z) = x\mathbf{i} - y\mathbf{j} + z\mathbf{k}$

In Exercises 21–30, find the conservative vector field for the potential function by finding its gradient.

21. $f(x, y) = x^2 + 2y^2$
22. $f(x, y) = x^2 - \frac{1}{4}y^2$
23. $g(x, y) = 5x^2 + 3xy + y^2$
24. $g(x, y) = \sin 3x \cos 4y$
25. $f(x, y, z) = 6xyz$
26. $f(x, y, z) = \sqrt{x^2 + 4y^2 + z^2}$
27. $g(x, y, z) = z + ye^{x^2}$
28. $g(x, y, z) = \dfrac{y}{z} + \dfrac{z}{x} - \dfrac{xz}{y}$
29. $h(x, y, z) = xy \ln(x + y)$
30. $h(x, y, z) = x \arcsin yz$

In Exercises 31–34, verify that the vector field is conservative.

31. $\mathbf{F}(x, y) = xy^2\mathbf{i} + x^2y\mathbf{j}$
32. $\mathbf{F}(x, y) = \dfrac{1}{x^2}(y\mathbf{i} - x\mathbf{j})$
33. $\mathbf{F}(x, y) = \sin y\mathbf{i} + x \cos y\mathbf{j}$
34. $\mathbf{F}(x, y) = \dfrac{1}{xy}(y\mathbf{i} - x\mathbf{j})$

In Exercises 35–38, determine whether the vector field is conservative. Justify your answer.

35. $\mathbf{F}(x, y) = 5y^2(y\mathbf{i} + 3x\mathbf{j})$
36. $\mathbf{F}(x, y) = \dfrac{2}{y^2}e^{2x/y}(y\mathbf{i} - x\mathbf{j})$
37. $\mathbf{F}(x, y) = \dfrac{1}{\sqrt{x^2 + y^2}}(\mathbf{i} + \mathbf{j})$
38. $\mathbf{F}(x, y) = \dfrac{1}{\sqrt{1 + xy}}(y\mathbf{i} + x\mathbf{j})$

In Exercises 39–48, determine whether the vector field is conservative. If it is, find a potential function for the vector field.

39. $\mathbf{F}(x, y) = y\mathbf{i} + x\mathbf{j}$
40. $\mathbf{F}(x, y) = 3x^2y^2\mathbf{i} + 2x^3y\mathbf{j}$
41. $\mathbf{F}(x, y) = 2xy\mathbf{i} + x^2\mathbf{j}$
42. $\mathbf{F}(x, y) = xe^{x^2y}(2y\mathbf{i} + x\mathbf{j})$
43. $\mathbf{F}(x, y) = 15y^3\mathbf{i} - 5xy^2\mathbf{j}$
44. $\mathbf{F}(x, y) = \dfrac{1}{y^2}(y\mathbf{i} - 2x\mathbf{j})$
45. $\mathbf{F}(x, y) = \dfrac{2y}{x}\mathbf{i} - \dfrac{x^2}{y^2}\mathbf{j}$
46. $\mathbf{F}(x, y) = \dfrac{x\mathbf{i} + y\mathbf{j}}{x^2 + y^2}$
47. $\mathbf{F}(x, y) = e^x(\cos y\mathbf{i} - \sin y\mathbf{j})$
48. $\mathbf{F}(x, y) = \dfrac{2x\mathbf{i} + 2y\mathbf{j}}{(x^2 + y^2)^2}$

In Exercises 49–52, find curl F for the vector field at the given point.

	Vector Field	*Point*
49.	$\mathbf{F}(x, y, z) = xyz\mathbf{i} + xyz\mathbf{j} + xyz\mathbf{k}$	$(2, 1, 3)$
50.	$\mathbf{F}(x, y, z) = x^2z\mathbf{i} - 2xz\mathbf{j} + yz\mathbf{k}$	$(2, -1, 3)$
51.	$\mathbf{F}(x, y, z) = e^x \sin y\mathbf{i} - e^x \cos y\mathbf{j}$	$(0, 0, 1)$
52.	$\mathbf{F}(x, y, z) = e^{-xyz}(\mathbf{i} + \mathbf{j} + \mathbf{k})$	$(3, 2, 0)$

CAS In Exercises 53–56, use a computer algebra system to find the curl F for the vector field.

53. $\mathbf{F}(x, y, z) = \arctan\left(\frac{x}{y}\right)\mathbf{i} + \ln\sqrt{x^2 + y^2}\,\mathbf{j} + \mathbf{k}$

54. $\mathbf{F}(x, y, z) = \frac{yz}{y - z}\mathbf{i} + \frac{xz}{x - z}\mathbf{j} + \frac{xy}{x - y}\mathbf{k}$

55. $\mathbf{F}(x, y, z) = \sin(x - y)\mathbf{i} + \sin(y - z)\mathbf{j} + \sin(z - x)\mathbf{k}$

56. $\mathbf{F}(x, y, z) = \sqrt{x^2 + y^2 + z^2}\,(\mathbf{i} + \mathbf{j} + \mathbf{k})$

In Exercises 57–62, determine whether the vector field F is conservative. If it is, find a potential function for the vector field.

57. $\mathbf{F}(x, y, z) = xy^2z^2\mathbf{i} + x^2yz^2\mathbf{j} + x^2y^2z\mathbf{k}$

58. $\mathbf{F}(x, y, z) = y^2z^3\mathbf{i} + 2xyz^3\mathbf{j} + 3xy^2z^2\mathbf{k}$

59. $\mathbf{F}(x, y, z) = \sin z\mathbf{i} + \sin x\mathbf{j} + \sin y\mathbf{k}$

60. $\mathbf{F}(x, y, z) = ye^z\mathbf{i} + ze^x\mathbf{j} + xe^y\mathbf{k}$

61. $\mathbf{F}(x, y, z) = \frac{z}{y}\mathbf{i} - \frac{xz}{y^2}\mathbf{j} + \frac{x}{y}\mathbf{k}$

62. $\mathbf{F}(x, y, z) = \frac{x}{x^2 + y^2}\mathbf{i} + \frac{y}{x^2 + y^2}\mathbf{j} + \mathbf{k}$

In Exercises 63–66, find the divergence of the vector field F.

63. $\mathbf{F}(x, y) = x^2\mathbf{i} + 2y^2\mathbf{j}$

64. $\mathbf{F}(x, y) = xe^x\mathbf{i} + ye^y\mathbf{j}$

65. $\mathbf{F}(x, y, z) = \sin x\mathbf{i} + \cos y\mathbf{j} + z^2\mathbf{k}$

66. $\mathbf{F}(x, y, z) = \ln(x^2 + y^2)\mathbf{i} + xy\mathbf{j} + \ln(y^2 + z^2)\mathbf{k}$

In Exercises 67–70, find the divergence of the vector field F at the given point.

Vector Field	*Point*
67. $\mathbf{F}(x, y, z) = xyz\mathbf{i} + xy\mathbf{j} + z\mathbf{k}$	$(2, 1, 1)$
68. $\mathbf{F}(x, y, z) = x^2z\mathbf{i} - 2xz\mathbf{j} + yz\mathbf{k}$	$(2, -1, 3)$
69. $\mathbf{F}(x, y, z) = e^x \sin y\mathbf{i} - e^x \cos y\mathbf{j} + z^2\mathbf{k}$	$(3, 0, 0)$
70. $\mathbf{F}(x, y, z) = \ln(xyz)(\mathbf{i} + \mathbf{j} + \mathbf{k})$	$(3, 2, 1)$

WRITING ABOUT CONCEPTS

71. Define a vector field in the plane and in space. Give some physical examples of vector fields.

72. What is a conservative vector field, and how do you test for it in the plane and in space?

73. Define the curl of a vector field.

74. Define the divergence of a vector field in the plane and in space.

In Exercises 75 and 76, find curl(F × G) = ∇ × (F × G).

75. $\mathbf{F}(x, y, z) = \mathbf{i} + 3x\mathbf{j} + 2y\mathbf{k}$
$\mathbf{G}(x, y, z) = x\mathbf{i} - y\mathbf{j} + z\mathbf{k}$

76. $\mathbf{F}(x, y, z) = x\mathbf{i} - z\mathbf{k}$
$\mathbf{G}(x, y, z) = x^2\mathbf{i} + y\mathbf{j} + z^2\mathbf{k}$

In Exercises 77 and 78, find curl(curl F) = ∇ × (∇ × F).

77. $\mathbf{F}(x, y, z) = xyz\mathbf{i} + y\mathbf{j} + z\mathbf{k}$

78. $\mathbf{F}(x, y, z) = x^2z\mathbf{i} - 2xz\mathbf{j} + yz\mathbf{k}$

In Exercises 79 and 80, find div(F × G) = ∇ · (F × G).

79. $\mathbf{F}(x, y, z) = \mathbf{i} + 3x\mathbf{j} + 2y\mathbf{k}$
$\mathbf{G}(x, y, z) = x\mathbf{i} - y\mathbf{j} + z\mathbf{k}$

80. $\mathbf{F}(x, y, z) = x\mathbf{i} - z\mathbf{k}$
$\mathbf{G}(x, y, z) = x^2\mathbf{i} + y\mathbf{j} + z^2\mathbf{k}$

In Exercises 81 and 82, find div(curl F) = ∇ · (∇ × F).

81. $\mathbf{F}(x, y, z) = xyz\mathbf{i} + y\mathbf{j} + z\mathbf{k}$

82. $\mathbf{F}(x, y, z) = x^2z\mathbf{i} - 2xz\mathbf{j} + yz\mathbf{k}$

In Exercises 83–90, prove the property for vector fields F and G and scalar function *f*. (Assume that the required partial derivatives are continuous.)

83. $\text{curl}(\mathbf{F} + \mathbf{G}) = \text{curl}\,\mathbf{F} + \text{curl}\,\mathbf{G}$

84. $\text{curl}(\nabla f) = \nabla \times (\nabla f) = \mathbf{0}$

85. $\text{div}(\mathbf{F} + \mathbf{G}) = \text{div}\,\mathbf{F} + \text{div}\,\mathbf{G}$

86. $\text{div}(\mathbf{F} \times \mathbf{G}) = (\text{curl}\,\mathbf{F}) \cdot \mathbf{G} - \mathbf{F} \cdot (\text{curl}\,\mathbf{G})$

87. $\nabla \times [\nabla f + (\nabla \times \mathbf{F})] = \nabla \times (\nabla \times \mathbf{F})$

88. $\nabla \times (f\mathbf{F}) = f(\nabla \times \mathbf{F}) + (\nabla f) \times \mathbf{F}$

89. $\text{div}(f\mathbf{F}) = f\,\text{div}\,\mathbf{F} + \nabla f \cdot \mathbf{F}$

90. $\text{div}(\text{curl}\,\mathbf{F}) = 0$ (Theorem 15.3)

In Exercises 91–93, let $\mathbf{F}(x, y, z) = x\mathbf{i} + y\mathbf{j} + z\mathbf{k}$, and let $f(x, y, z) = \|\mathbf{F}(x, y, z)\|$.

91. Show that $\nabla(\ln f) = \frac{\mathbf{F}}{f^2}$.

92. Show that $\nabla\left(\frac{1}{f}\right) = -\frac{\mathbf{F}}{f^3}$.

93. Show that $\nabla f^n = nf^{n-2}\mathbf{F}$.

CAPSTONE

94. (a) Sketch several representative vectors in the vector field given by

$$\mathbf{F}(x, y) = \frac{x\mathbf{i} + y\mathbf{j}}{\sqrt{x^2 + y^2}}.$$

(b) Sketch several representative vectors in the vector field given by

$$\mathbf{G}(x, y) = \frac{x\mathbf{i} - y\mathbf{j}}{\sqrt{x^2 + y^2}}.$$

(c) Explain any similarities or differences in the vector fields $\mathbf{F}(x, y)$ and $\mathbf{G}(x, y)$.

True or False? **In Exercises 95–98, determine whether the statement is true or false. If it is false, explain why or give an example that shows it is false.**

95. If $\mathbf{F}(x, y) = 4x\mathbf{i} - y^2\mathbf{j}$, then $\|\mathbf{F}(x, y)\| \to 0$ as $(x, y) \to (0, 0)$.

96. If $\mathbf{F}(x, y) = 4x\mathbf{i} - y^2\mathbf{j}$ and (x, y) is on the positive y-axis, then the vector points in the negative y-direction.

97. If f is a scalar field, then curl f is a meaningful expression.

98. If $\mathbf{F}$ is a vector field and $\text{curl}\,\mathbf{F} = \mathbf{0}$, then $\mathbf{F}$ is irrotational but not conservative.

15.2 Line Integrals

- **Understand and use the concept of a piecewise smooth curve.**
- **Write and evaluate a line integral.**
- **Write and evaluate a line integral of a vector field.**
- **Write and evaluate a line integral in differential form.**

The Granger Collection

JOSIAH WILLARD GIBBS (1839–1903)

Many physicists and mathematicians have contributed to the theory and applications described in this chapter—Newton, Gauss, Laplace, Hamilton, and Maxwell, among others. However, the use of vector analysis to describe these results is attributed primarily to the American mathematical physicist Josiah Willard Gibbs.

Piecewise Smooth Curves

A classic property of gravitational fields is that, subject to certain physical constraints, the work done by gravity on an object moving between two points in the field is independent of the path taken by the object. One of the constraints is that the **path** must be a piecewise smooth curve. Recall that a plane curve C given by

$$\mathbf{r}(t) = x(t)\mathbf{i} + y(t)\mathbf{j}, \quad a \le t \le b$$

is **smooth** if

$$\frac{dx}{dt} \quad \text{and} \quad \frac{dy}{dt}$$

are continuous on $[a, b]$ and not simultaneously 0 on (a, b). Similarly, a space curve C given by

$$\mathbf{r}(t) = x(t)\mathbf{i} + y(t)\mathbf{j} + z(t)\mathbf{k}, \quad a \le t \le b$$

is **smooth** if

$$\frac{dx}{dt}, \quad \frac{dy}{dt}, \quad \text{and} \quad \frac{dz}{dt}$$

are continuous on $[a, b]$ and not simultaneously 0 on (a, b). A curve C is **piecewise smooth** if the interval $[a, b]$ can be partitioned into a finite number of subintervals, on each of which C is smooth.

EXAMPLE 1 Finding a Piecewise Smooth Parametrization

Find a piecewise smooth parametrization of the graph of C shown in Figure 15.7.

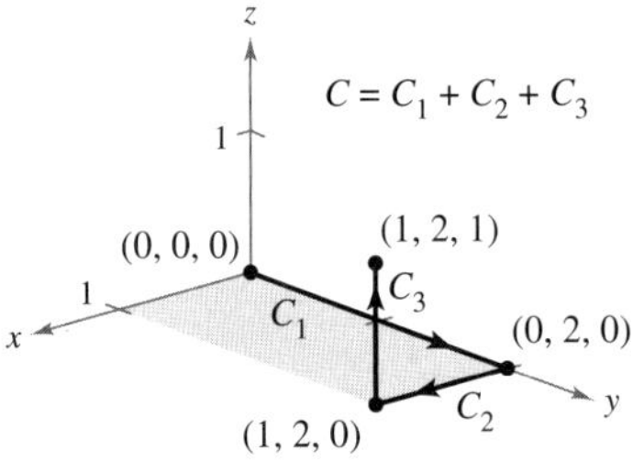

Figure 15.7

Solution Because C consists of three line segments C_1, C_2, and C_3, you can construct a smooth parametrization for each segment and piece them together by making the last t-value in C_i correspond to the first t-value in C_{i+1}, as follows.

$$\begin{array}{llll} C_1\colon x(t) = 0, & y(t) = 2t, & z(t) = 0, & 0 \le t \le 1 \\ C_2\colon x(t) = t - 1, & y(t) = 2, & z(t) = 0, & 1 \le t \le 2 \\ C_3\colon x(t) = 1, & y(t) = 2, & z(t) = t - 2, & 2 \le t \le 3 \end{array}$$

So, C is given by

$$\mathbf{r}(t) = \begin{cases} 2t\mathbf{j}, & 0 \le t \le 1 \\ (t-1)\mathbf{i} + 2\mathbf{j}, & 1 \le t \le 2. \\ \mathbf{i} + 2\mathbf{j} + (t-2)\mathbf{k}, & 2 \le t \le 3 \end{cases}$$

Because C_1, C_2, and C_3 are smooth, it follows that C is piecewise smooth. ■

Recall that parametrization of a curve induces an **orientation** to the curve. For instance, in Example 1, the curve is oriented such that the positive direction is from $(0, 0, 0)$, following the curve to $(1, 2, 1)$. Try finding a parametrization that induces the opposite orientation.

Line Integrals

Up to this point in the text, you have studied various types of integrals. For a single integral

$$\int_a^b f(x)\,dx \qquad \text{Integrate over interval } [a, b].$$

you integrated over the interval $[a, b]$. Similarly, for a double integral

$$\int_R\int f(x, y)\,dA \qquad \text{Integrate over region } R.$$

you integrated over the region R in the plane. In this section, you will study a new type of integral called a **line integral**

$$\int_C f(x, y)\,ds \qquad \text{Integrate over curve } C.$$

for which you integrate over a piecewise smooth curve C. (The terminology is somewhat unfortunate—this type of integral might be better described as a "curve integral.")

To introduce the concept of a line integral, consider the mass of a wire of finite length, given by a curve C in space. The density (mass per unit length) of the wire at the point (x, y, z) is given by $f(x, y, z)$. Partition the curve C by the points

$$P_0, P_1, \ldots, P_n$$

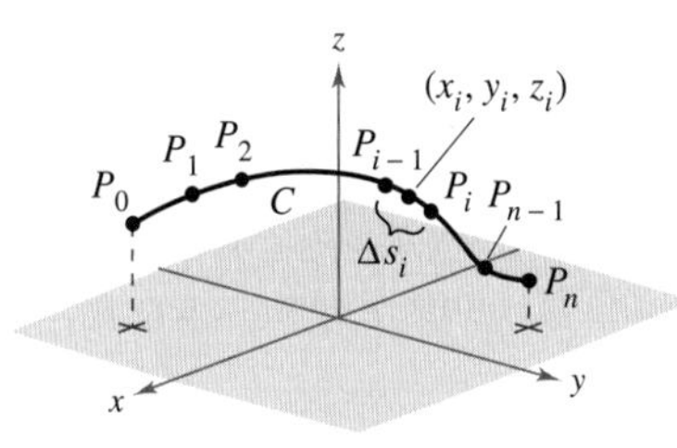

Partitioning of curve C
Figure 15.8

producing n subarcs, as shown in Figure 15.8. The length of the ith subarc is given by Δs_i. Next, choose a point (x_i, y_i, z_i) in each subarc. If the length of each subarc is small, the total mass of the wire can be approximated by the sum

$$\text{Mass of wire} \approx \sum_{i=1}^{n} f(x_i, y_i, z_i)\,\Delta s_i.$$

If you let $\|\Delta\|$ denote the length of the longest subarc and let $\|\Delta\|$ approach 0, it seems reasonable that the limit of this sum approaches the mass of the wire. This leads to the following definition.

DEFINITION OF LINE INTEGRAL

If f is defined in a region containing a smooth curve C of finite length, then the **line integral of f along C** is given by

$$\int_C f(x, y)\,ds = \lim_{\|\Delta\|\to 0} \sum_{i=1}^{n} f(x_i, y_i)\,\Delta s_i \qquad \text{Plane}$$

or

$$\int_C f(x, y, z)\,ds = \lim_{\|\Delta\|\to 0} \sum_{i=1}^{n} f(x_i, y_i, z_i)\,\Delta s_i \qquad \text{Space}$$

provided this limit exists.

As with the integrals discussed in Chapter 14, evaluation of a line integral is best accomplished by converting it to a definite integral. It can be shown that if f is *continuous*, the limit given above exists and is the same for all smooth parametrizations of C.

To evaluate a line integral over a plane curve C given by $\mathbf{r}(t) = x(t)\mathbf{i} + y(t)\mathbf{j}$, use the fact that

$$ds = \|\mathbf{r}'(t)\|\, dt = \sqrt{[x'(t)]^2 + [y'(t)]^2}\, dt.$$

A similar formula holds for a space curve, as indicated in Theorem 15.4.

THEOREM 15.4 EVALUATION OF A LINE INTEGRAL AS A DEFINITE INTEGRAL

Let f be continuous in a region containing a smooth curve C. If C is given by $\mathbf{r}(t) = x(t)\mathbf{i} + y(t)\mathbf{j}$, where $a \le t \le b$, then

$$\int_C f(x, y)\, ds = \int_a^b f(x(t), y(t))\sqrt{[x'(t)]^2 + [y'(t)]^2}\, dt.$$

If C is given by $\mathbf{r}(t) = x(t)\mathbf{i} + y(t)\mathbf{j} + z(t)\mathbf{k}$, where $a \le t \le b$, then

$$\int_C f(x, y, z)\, ds = \int_a^b f(x(t), y(t), z(t))\sqrt{[x'(t)]^2 + [y'(t)]^2 + [z'(t)]^2}\, dt.$$

Note that if $f(x, y, z) = 1$, the line integral gives the arc length of the curve C, as defined in Section 12.5. That is,

$$\int_C 1\, ds = \int_a^b \|\mathbf{r}'(t)\|\, dt = \text{length of curve } C.$$

EXAMPLE 2 Evaluating a Line Integral

Evaluate

$$\int_C (x^2 - y + 3z)\, ds$$

where C is the line segment shown in Figure 15.9.

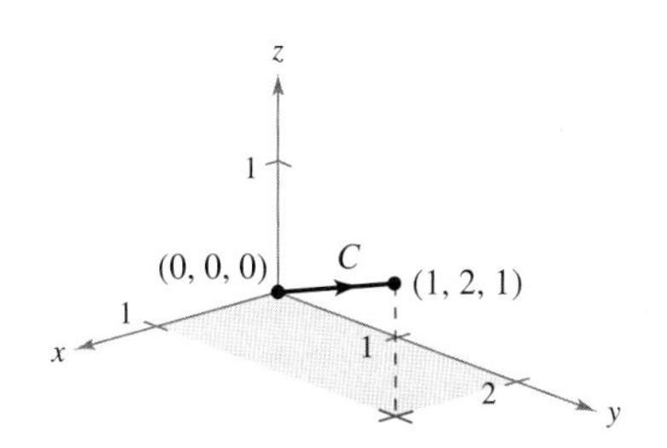

Figure 15.9

Solution Begin by writing a parametric form of the equation of the line segment:

$$x = t, \quad y = 2t, \quad \text{and} \quad z = t, \quad 0 \le t \le 1.$$

Therefore, $x'(t) = 1$, $y'(t) = 2$, and $z'(t) = 1$, which implies that

$$\sqrt{[x'(t)]^2 + [y'(t)]^2 + [z'(t)]^2} = \sqrt{1^2 + 2^2 + 1^2} = \sqrt{6}.$$

So, the line integral takes the following form.

$$\begin{aligned}\int_C (x^2 - y + 3z)\, ds &= \int_0^1 (t^2 - 2t + 3t)\sqrt{6}\, dt \\ &= \sqrt{6}\int_0^1 (t^2 + t)\, dt \\ &= \sqrt{6}\left[\frac{t^3}{3} + \frac{t^2}{2}\right]_0^1 \\ &= \frac{5\sqrt{6}}{6}\end{aligned}$$

NOTE The value of the line integral in Example 2 does not depend on the parametrization of the line segment C (any smooth parametrization will produce the same value). To convince yourself of this, try some other parametrizations, such as $x = 1 + 2t$, $y = 2 + 4t$, $z = 1 + 2t$, $-\frac{1}{2} \le t \le 0$, or $x = -t$, $y = -2t$, $z = -t$, $-1 \le t \le 0$.

Suppose C is a path composed of smooth curves $C_1, C_2, \ldots, C_n$. If f is continuous on C, it can be shown that

$$\int_C f(x, y)\, ds = \int_{C_1} f(x, y)\, ds + \int_{C_2} f(x, y)\, ds + \cdots + \int_{C_n} f(x, y)\, ds.$$

This property is used in Example 3.

EXAMPLE 3 Evaluating a Line Integral Over a Path

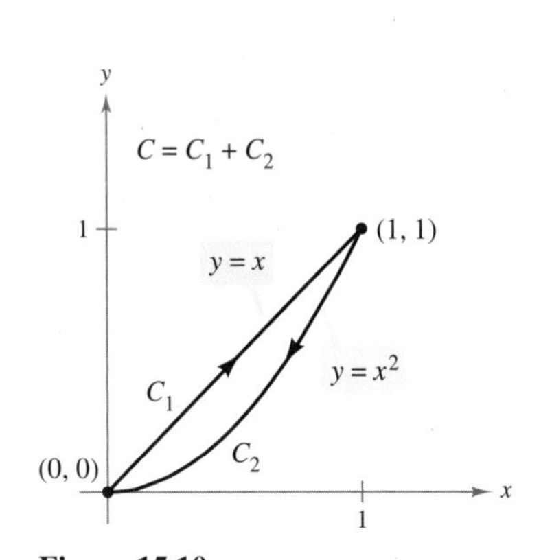

Figure 15.10

Evaluate $\int_C x\, ds$, where C is the piecewise smooth curve shown in Figure 15.10.

Solution Begin by integrating up the line $y = x$, using the following parametrization.

$$C_1\colon\ x = t,\ y = t,\quad 0 \le t \le 1$$

For this curve, $\mathbf{r}(t) = t\mathbf{i} + t\mathbf{j}$, which implies that $x'(t) = 1$ and $y'(t) = 1$. So,

$$\sqrt{[x'(t)]^2 + [y'(t)]^2} = \sqrt{2}$$

and you have

$$\int_{C_1} x\, ds = \int_0^1 t\sqrt{2}\, dt = \frac{\sqrt{2}}{2} t^2 \Big]_0^1 = \frac{\sqrt{2}}{2}.$$

Next, integrate down the parabola $y = x^2$, using the parametrization

$$C_2\colon\ x = 1 - t,\quad y = (1 - t)^2,\quad 0 \le t \le 1.$$

For this curve, $\mathbf{r}(t) = (1 - t)\mathbf{i} + (1 - t)^2\mathbf{j}$, which implies that $x'(t) = -1$ and $y'(t) = -2(1 - t)$. So,

$$\sqrt{[x'(t)]^2 + [y'(t)]^2} = \sqrt{1 + 4(1 - t)^2}$$

and you have

$$\begin{aligned}\int_{C_2} x\, ds &= \int_0^1 (1 - t)\sqrt{1 + 4(1 - t)^2}\, dt \\ &= -\frac{1}{8}\left[\frac{2}{3}[1 + 4(1 - t)^2]^{3/2}\right]_0^1 \\ &= \frac{1}{12}(5^{3/2} - 1).\end{aligned}$$

Consequently,

$$\int_C x\, ds = \int_{C_1} x\, ds + \int_{C_2} x\, ds = \frac{\sqrt{2}}{2} + \frac{1}{12}(5^{3/2} - 1) \approx 1.56.$$ ■

For parametrizations given by $\mathbf{r}(t) = x(t)\mathbf{i} + y(t)\mathbf{j} + z(t)\mathbf{k}$, it is helpful to remember the form of ds as

$$ds = \|\mathbf{r}'(t)\|\, dt = \sqrt{[x'(t)]^2 + [y'(t)]^2 + [z'(t)]^2}\, dt.$$

This is demonstrated in Example 4.

EXAMPLE 4 Evaluating a Line Integral

Evaluate $\int_C (x + 2)\, ds$, where C is the curve represented by

$$\mathbf{r}(t) = t\mathbf{i} + \frac{4}{3}t^{3/2}\mathbf{j} + \frac{1}{2}t^2\mathbf{k}, \quad 0 \le t \le 2.$$

Solution Because $\mathbf{r}'(t) = \mathbf{i} + 2t^{1/2}\mathbf{j} + t\mathbf{k}$ and

$$\|\mathbf{r}'(t)\| = \sqrt{[x'(t)]^2 + [y'(t)]^2 + [z'(t)]^2} = \sqrt{1 + 4t + t^2}$$

it follows that

$$\begin{aligned}\int_C (x + 2)\, ds &= \int_0^2 (t + 2)\sqrt{1 + 4t + t^2}\, dt \\ &= \frac{1}{2}\int_0^2 2(t + 2)(1 + 4t + t^2)^{1/2}\, dt \\ &= \frac{1}{3}\Big[(1 + 4t + t^2)^{3/2}\Big]_0^2 \\ &= \frac{1}{3}\left(13\sqrt{13} - 1\right) \\ &\approx 15.29.\end{aligned}$$

The next example shows how a line integral can be used to find the mass of a spring whose density varies. In Figure 15.11, note that the density of this spring increases as the spring spirals up the z-axis.

EXAMPLE 5 Finding the Mass of a Spring

Find the mass of a spring in the shape of the circular helix

$$\mathbf{r}(t) = \frac{1}{\sqrt{2}}(\cos t\mathbf{i} + \sin t\mathbf{j} + t\mathbf{k}), \quad 0 \le t \le 6\pi$$

where the density of the spring is $\rho(x, y, z) = 1 + z$, as shown in Figure 15.11.

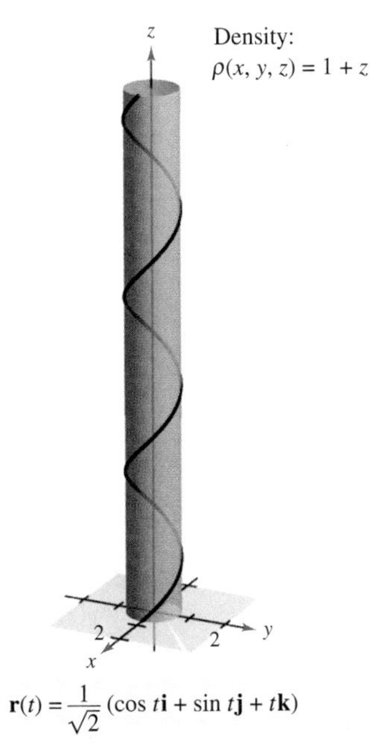

Figure 15.11

Solution Because

$$\|\mathbf{r}'(t)\| = \frac{1}{\sqrt{2}}\sqrt{(-\sin t)^2 + (\cos t)^2 + (1)^2} = 1$$

it follows that the mass of the spring is

$$\begin{aligned}\text{Mass} = \int_C (1 + z)\, ds &= \int_0^{6\pi}\left(1 + \frac{t}{\sqrt{2}}\right) dt \\ &= \left[t + \frac{t^2}{2\sqrt{2}}\right]_0^{6\pi} \\ &= 6\pi\left(1 + \frac{3\pi}{\sqrt{2}}\right) \\ &\approx 144.47.\end{aligned}$$

The mass of the spring is approximately 144.47.

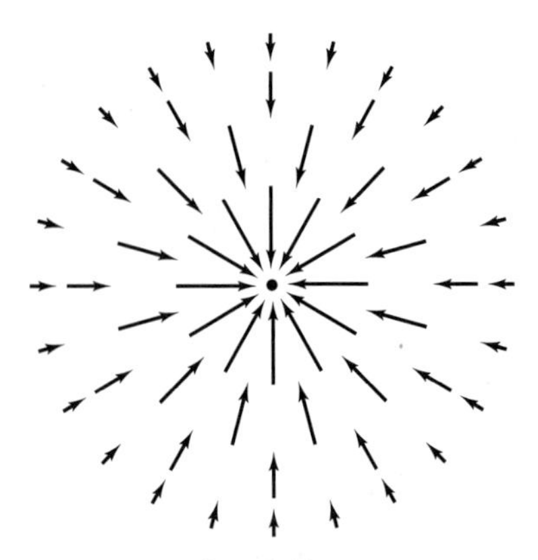

Inverse square force field **F**

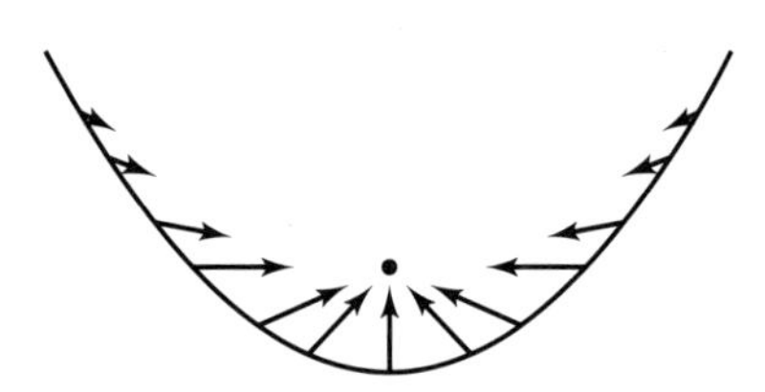

Vectors along a parabolic path in the force field **F**
Figure 15.12

Line Integrals of Vector Fields

One of the most important physical applications of line integrals is that of finding the **work** done on an object moving in a force field. For example, Figure 15.12 shows an inverse square force field similar to the gravitational field of the sun. Note that the magnitude of the force along a circular path about the center is constant, whereas the magnitude of the force along a parabolic path varies from point to point.

To see how a line integral can be used to find work done in a force field $\mathbf{F}$, consider an object moving along a path C in the field, as shown in Figure 15.13. To determine the work done by the force, you need consider only that part of the force that is acting in the same direction as that in which the object is moving (or the opposite direction). This means that at each point on C, you can consider the projection $\mathbf{F} \cdot \mathbf{T}$ of the force vector $\mathbf{F}$ onto the unit tangent vector $\mathbf{T}$. On a small subarc of length Δs_i, the increment of work is

$$\begin{aligned}\Delta W_i &= (\text{force})(\text{distance})\\ &\approx [\mathbf{F}(x_i, y_i, z_i) \cdot \mathbf{T}(x_i, y_i, z_i)]\,\Delta s_i\end{aligned}$$

where (x_i, y_i, z_i) is a point in the ith subarc. Consequently, the total work done is given by the following integral.

$$W = \int_C \mathbf{F}(x, y, z) \cdot \mathbf{T}(x, y, z)\, ds$$

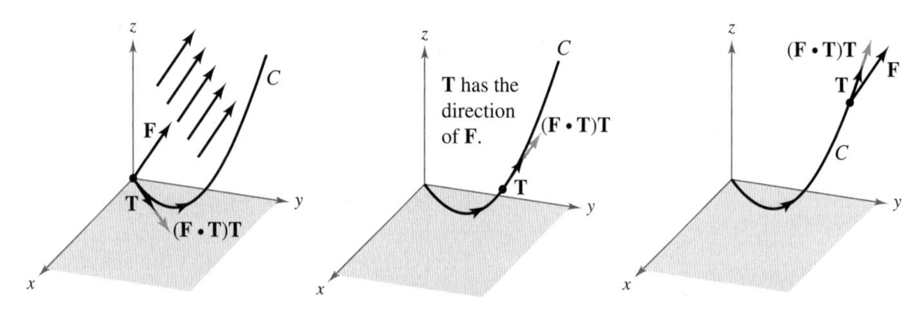

At each point on C, the force in the direction of motion is $(\mathbf{F} \cdot \mathbf{T})\mathbf{T}$.
Figure 15.13

This line integral appears in other contexts and is the basis of the following definition of the **line integral of a vector field.** Note in the definition that

$$\begin{aligned}\mathbf{F} \cdot \mathbf{T}\, ds &= \mathbf{F} \cdot \frac{\mathbf{r}'(t)}{\|\mathbf{r}'(t)\|}\|\mathbf{r}'(t)\|\, dt\\ &= \mathbf{F} \cdot \mathbf{r}'(t)\, dt\\ &= \mathbf{F} \cdot d\mathbf{r}.\end{aligned}$$

DEFINITION OF THE LINE INTEGRAL OF A VECTOR FIELD

Let $\mathbf{F}$ be a continuous vector field defined on a smooth curve C given by $\mathbf{r}(t)$, $a \le t \le b$. The **line integral** of $\mathbf{F}$ on C is given by

$$\int_C \mathbf{F} \cdot d\mathbf{r} = \int_C \mathbf{F} \cdot \mathbf{T}\, ds = \int_a^b \mathbf{F}(x(t),\ y(t), z(t)) \cdot \mathbf{r}'(t)\, dt.$$

EXAMPLE 6 Work Done by a Force

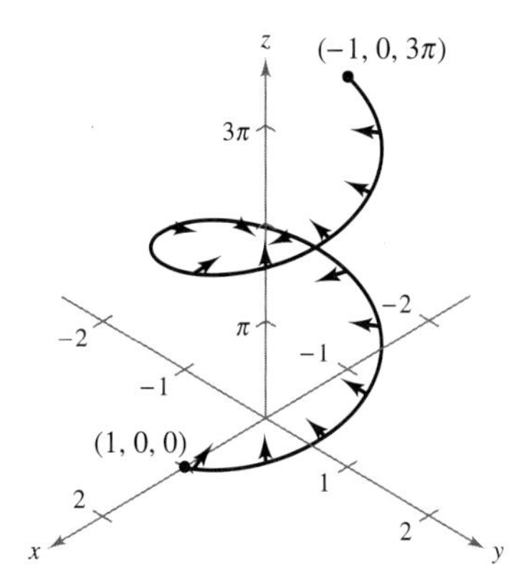

Figure 15.14

Find the work done by the force field

$$\mathbf{F}(x, y, z) = -\frac{1}{2}x\mathbf{i} - \frac{1}{2}y\mathbf{j} + \frac{1}{4}\mathbf{k} \qquad \text{Force field } \mathbf{F}$$

on a particle as it moves along the helix given by

$$\mathbf{r}(t) = \cos t\mathbf{i} + \sin t\mathbf{j} + t\mathbf{k} \qquad \text{Space curve } C$$

from the point $(1, 0, 0)$ to $(-1, 0, 3\pi)$, as shown in Figure 15.14.

Solution Because

$$\begin{aligned}\mathbf{r}(t) &= x(t)\mathbf{i} + y(t)\mathbf{j} + z(t)\mathbf{k}\\ &= \cos t\mathbf{i} + \sin t\mathbf{j} + t\mathbf{k}\end{aligned}$$

it follows that $x(t) = \cos t$, $y(t) = \sin t$, and $z(t) = t$. So, the force field can be written as

$$\mathbf{F}(x(t), y(t), z(t)) = -\frac{1}{2}\cos t\mathbf{i} - \frac{1}{2}\sin t\mathbf{j} + \frac{1}{4}\mathbf{k}.$$

To find the work done by the force field in moving a particle along the curve C, use the fact that

$$\mathbf{r}'(t) = -\sin t\mathbf{i} + \cos t\mathbf{j} + \mathbf{k}$$

and write the following.

$$\begin{aligned}W &= \int_C \mathbf{F} \cdot d\mathbf{r}\\ &= \int_a^b \mathbf{F}(x(t), y(t), z(t)) \cdot \mathbf{r}'(t)\,dt\\ &= \int_0^{3\pi} \left(-\frac{1}{2}\cos t\mathbf{i} - \frac{1}{2}\sin t\mathbf{j} + \frac{1}{4}\mathbf{k}\right) \cdot (-\sin t\mathbf{i} + \cos t\mathbf{j} + \mathbf{k})\,dt\\ &= \int_0^{3\pi} \left(\frac{1}{2}\sin t\cos t - \frac{1}{2}\sin t\cos t + \frac{1}{4}\right) dt\\ &= \int_0^{3\pi} \frac{1}{4}\,dt\\ &= \frac{1}{4}t\Big]_0^{3\pi}\\ &= \frac{3\pi}{4}\end{aligned}$$

■

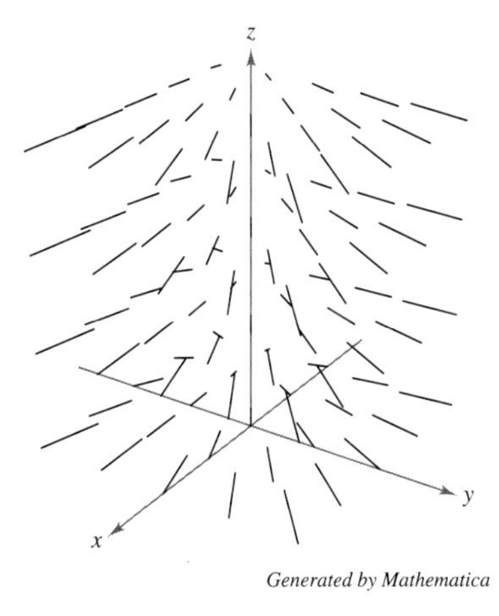

Figure 15.15

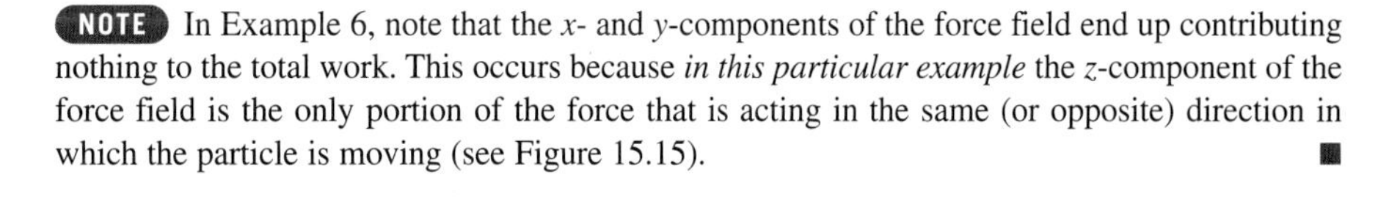

NOTE In Example 6, note that the x- and y-components of the force field end up contributing nothing to the total work. This occurs because *in this particular example* the z-component of the force field is the only portion of the force that is acting in the same (or opposite) direction in which the particle is moving (see Figure 15.15). ■

TECHNOLOGY The computer-generated view of the force field in Example 6 shown in Figure 15.15 indicates that each vector in the force field points toward the z-axis.

For line integrals of vector functions, the orientation of the curve C is important. If the orientation of the curve is reversed, the unit tangent vector $\mathbf{T}(t)$ is changed to $-\mathbf{T}(t)$, and you obtain

$$\int_{-C} \mathbf{F} \cdot d\mathbf{r} = -\int_C \mathbf{F} \cdot d\mathbf{r}.$$

EXAMPLE 7 Orientation and Parametrization of a Curve

Let $\mathbf{F}(x, y) = y\mathbf{i} + x^2\mathbf{j}$ and evaluate the line integral $\int_C \mathbf{F} \cdot d\mathbf{r}$ for each parabolic curve shown in Figure 15.16.

a. C_1: $\mathbf{r}_1(t) = (4 - t)\mathbf{i} + (4t - t^2)\mathbf{j}, \quad 0 \le t \le 3$

b. C_2: $\mathbf{r}_2(t) = t\mathbf{i} + (4t - t^2)\mathbf{j}, \quad 1 \le t \le 4$

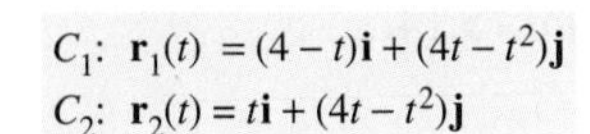

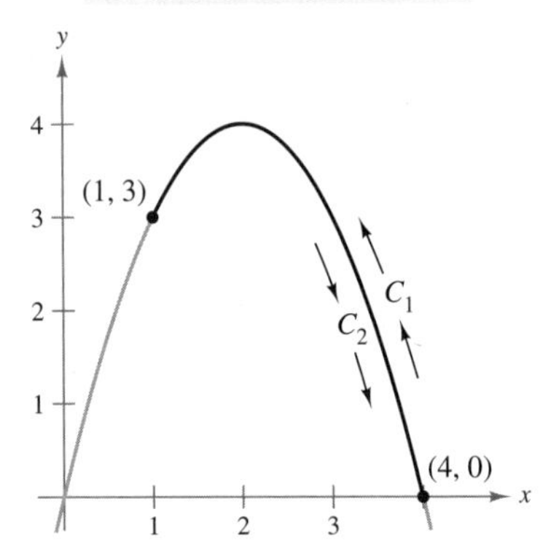

Figure 15.16

Solution

a. Because $\mathbf{r}_1'(t) = -\mathbf{i} + (4 - 2t)\mathbf{j}$ and

$$\mathbf{F}(x(t), y(t)) = (4t - t^2)\mathbf{i} + (4 - t)^2\mathbf{j}$$

the line integral is

$$\begin{aligned}
\int_{C_1} \mathbf{F} \cdot d\mathbf{r} &= \int_0^3 [(4t - t^2)\mathbf{i} + (4 - t)^2\mathbf{j}] \cdot [-\mathbf{i} + (4 - 2t)\mathbf{j}]\, dt \\
&= \int_0^3 (-4t + t^2 + 64 - 64t + 20t^2 - 2t^3)\, dt \\
&= \int_0^3 (-2t^3 + 21t^2 - 68t + 64)\, dt \\
&= \left[-\frac{t^4}{2} + 7t^3 - 34t^2 + 64t\right]_0^3 \\
&= \frac{69}{2}.
\end{aligned}$$

b. Because $\mathbf{r}_2'(t) = \mathbf{i} + (4 - 2t)\mathbf{j}$ and

$$\mathbf{F}(x(t), y(t)) = (4t - t^2)\mathbf{i} + t^2\mathbf{j}$$

the line integral is

$$\begin{aligned}
\int_{C_2} \mathbf{F} \cdot d\mathbf{r} &= \int_1^4 [(4t - t^2)\mathbf{i} + t^2\mathbf{j}] \cdot [\mathbf{i} + (4 - 2t)\mathbf{j}]\, dt \\
&= \int_1^4 (4t - t^2 + 4t^2 - 2t^3)\, dt \\
&= \int_1^4 (-2t^3 + 3t^2 + 4t)\, dt \\
&= \left[-\frac{t^4}{2} + t^3 + 2t^2\right]_1^4 \\
&= -\frac{69}{2}.
\end{aligned}$$

The answer in part (b) is the negative of that in part (a) because C_1 and C_2 represent opposite orientations of the same parabolic segment. ■

NOTE Although the value of the line integral in Example 7 depends on the orientation of C, it does not depend on the parametrization of C. To see this, let C_3 be represented by

$$\mathbf{r}_3 = (t + 2)\mathbf{i} + (4 - t^2)\mathbf{j}$$

where $-1 \le t \le 2$. The graph of this curve is the same parabolic segment shown in Figure 15.16. Does the value of the line integral over C_3 agree with the value over C_1 or C_2? Why or why not?

Line Integrals in Differential Form

A second commonly used form of line integrals is derived from the vector field notation used in the preceding section. If $\mathbf{F}$ is a vector field of the form $\mathbf{F}(x, y) = M\mathbf{i} + N\mathbf{j}$, and C is given by $\mathbf{r}(t) = x(t)\mathbf{i} + y(t)\mathbf{j}$, then $\mathbf{F} \cdot d\mathbf{r}$ is often written as $M\,dx + N\,dy$.

$$\begin{aligned}\int_C \mathbf{F} \cdot d\mathbf{r} &= \int_C \mathbf{F} \cdot \frac{d\mathbf{r}}{dt}\,dt \\ &= \int_a^b (M\mathbf{i} + N\mathbf{j}) \cdot (x'(t)\mathbf{i} + y'(t)\mathbf{j})\,dt \\ &= \int_a^b \left(M\frac{dx}{dt} + N\frac{dy}{dt}\right) dt \\ &= \int_C (M\,dx + N\,dy)\end{aligned}$$

This **differential form** can be extended to three variables. The parentheses are often omitted, as follows.

$$\int_C M\,dx + N\,dy \qquad \text{and} \qquad \int_C M\,dx + N\,dy + P\,dz$$

Notice how this differential notation is used in Example 8.

EXAMPLE 8 Evaluating a Line Integral in Differential Form

Let C be the circle of radius 3 given by

$$\mathbf{r}(t) = 3\cos t\mathbf{i} + 3\sin t\mathbf{j}, \quad 0 \le t \le 2\pi$$

as shown in Figure 15.17. Evaluate the line integral

$$\int_C y^3\,dx + (x^3 + 3xy^2)\,dy.$$

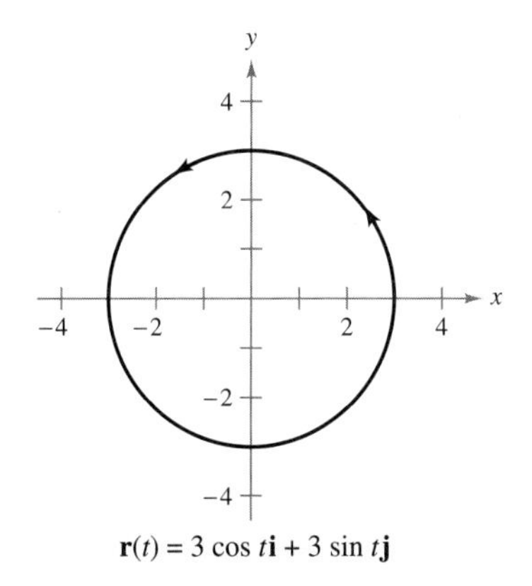

Figure 15.17

Solution Because $x = 3\cos t$ and $y = 3\sin t$, you have $dx = -3\sin t\,dt$ and $dy = 3\cos t\,dt$. So, the line integral is

$$\begin{aligned}&\int_C M\,dx + N\,dy \\ &= \int_C y^3\,dx + (x^3 + 3xy^2)\,dy \\ &= \int_0^{2\pi} \left[(27\sin^3 t)(-3\sin t) + (27\cos^3 t + 81\cos t\sin^2 t)(3\cos t)\right] dt \\ &= 81\int_0^{2\pi} (\cos^4 t - \sin^4 t + 3\cos^2 t\sin^2 t)\,dt \\ &= 81\int_0^{2\pi} \left(\cos^2 t - \sin^2 t + \frac{3}{4}\sin^2 2t\right) dt \\ &= 81\int_0^{2\pi} \left[\cos 2t + \frac{3}{4}\left(\frac{1 - \cos 4t}{2}\right)\right] dt \\ &= 81\left[\frac{\sin 2t}{2} + \frac{3}{8}t - \frac{3\sin 4t}{32}\right]_0^{2\pi} \\ &= \frac{243\pi}{4}.\end{aligned}$$

NOTE The orientation of C affects the value of the differential form of a line integral. Specifically, if $-C$ has the orientation opposite to that of C, then

$$\int_{-C} M\,dx + N\,dy = -\int_C M\,dx + N\,dy.$$

So, of the three line integral forms presented in this section, the orientation of C does not affect the form $\int_C f(x, y)\,ds$, but it does affect the vector form and the differential form.

For curves represented by $y = g(x)$, $a \leq x \leq b$, you can let $x = t$ and obtain the parametric form

$$x = t \quad \text{and} \quad y = g(t), \quad a \leq t \leq b.$$

Because $dx = dt$ for this form, you have the option of evaluating the line integral in the variable x or the variable t. This is demonstrated in Example 9.

EXAMPLE 9 Evaluating a Line Integral in Differential Form

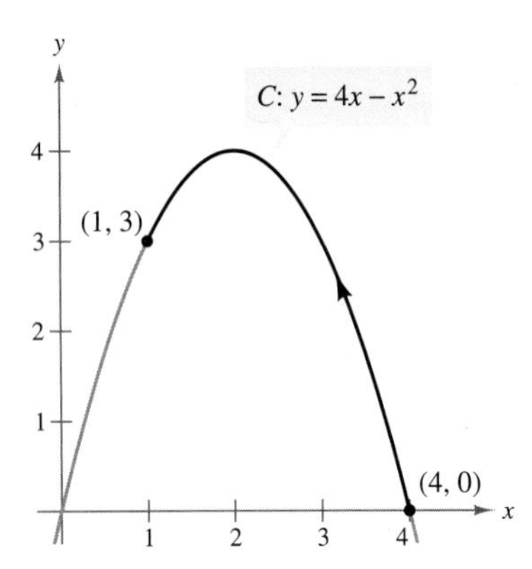

Figure 15.18

Evaluate

$$\int_C y\,dx + x^2\,dy$$

where C is the parabolic arc given by $y = 4x - x^2$ from $(4, 0)$ to $(1, 3)$, as shown in Figure 15.18.

Solution Rather than converting to the parameter t, you can simply retain the variable x and write

$$y = 4x - x^2 \quad \Longrightarrow \quad dy = (4 - 2x)\,dx.$$

Then, in the direction from $(4, 0)$ to $(1, 3)$, the line integral is

$$\begin{aligned}
\int_C y\,dx + x^2\,dy &= \int_4^1 [(4x - x^2)\,dx + x^2(4 - 2x)\,dx] \\
&= \int_4^1 (4x + 3x^2 - 2x^3)\,dx \\
&= \left[2x^2 + x^3 - \frac{x^4}{2}\right]_4^1 = \frac{69}{2}.
\end{aligned}$$

See Example 7. ■

EXPLORATION

Finding Lateral Surface Area The figure below shows a piece of tin that has been cut from a circular cylinder. The base of the circular cylinder is modeled by $x^2 + y^2 = 9$. At any point (x, y) on the base, the height of the object is given by

$$f(x, y) = 1 + \cos\frac{\pi x}{4}.$$

Explain how to use a line integral to find the surface area of the piece of tin.

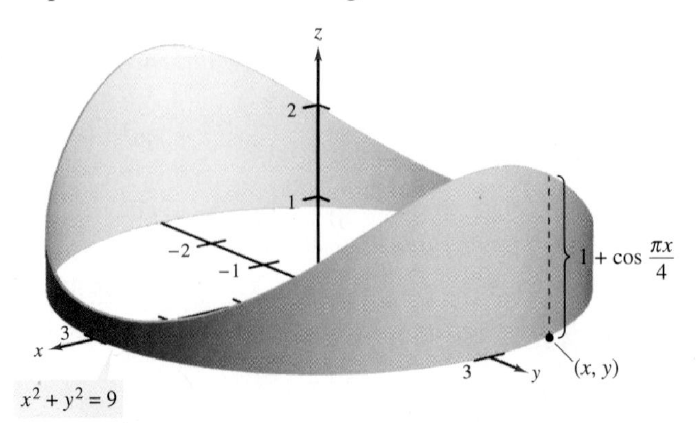

15.2 Exercises

See www.CalcChat.com for worked-out solutions to odd-numbered exercises.

In Exercises 1–6, find a piecewise smooth parametrization of the path C. (Note that there is more than one correct answer.)

1.

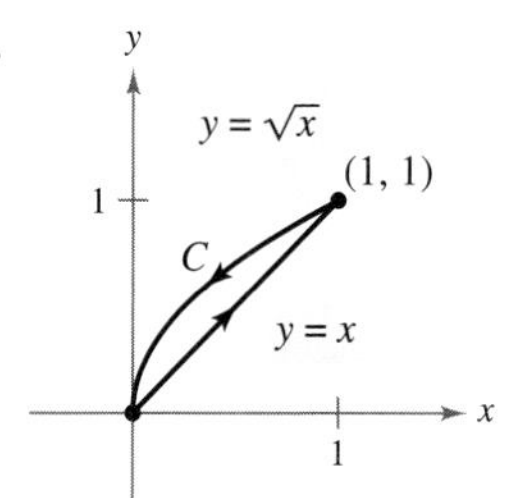

2.

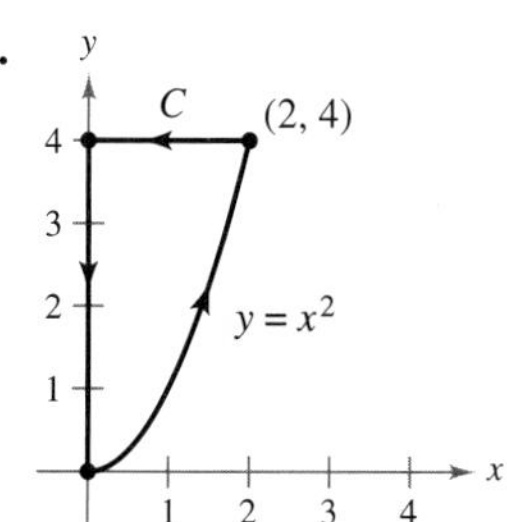

3.

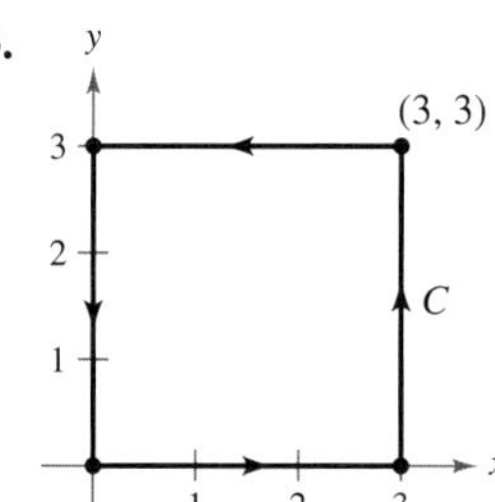

4.

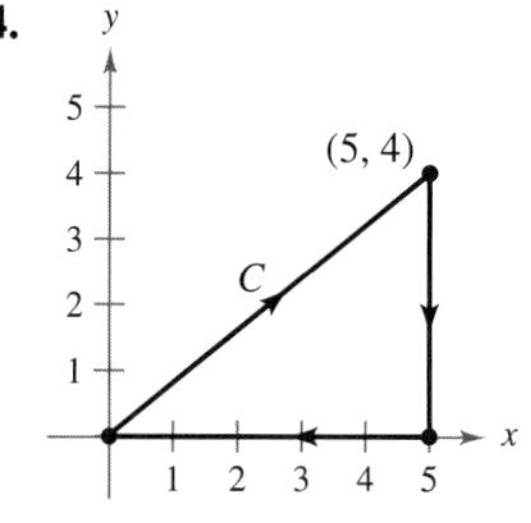

5.

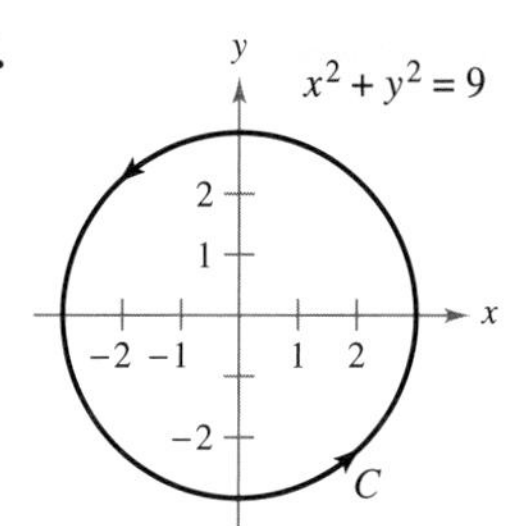

6.

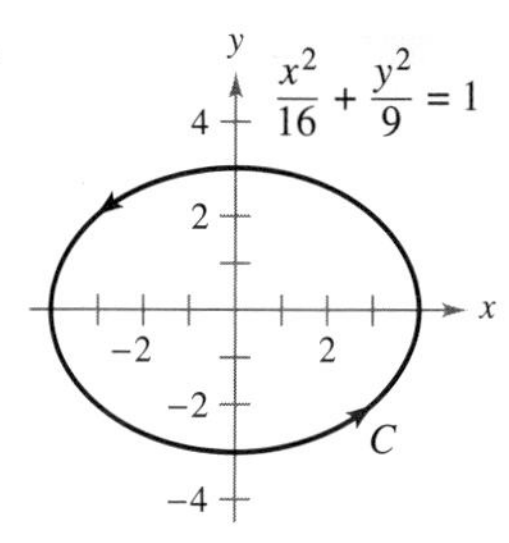

In Exercises 7–10, evaluate the line integral along the given path.

7. $\int_C xy\, ds$

$C: \mathbf{r}(t) = 4t\mathbf{i} + 3t\mathbf{j}$

$0 \le t \le 1$

8. $\int_C 3(x - y)\, ds$

$C: \mathbf{r}(t) = t\mathbf{i} + (2 - t)\mathbf{j}$

$0 \le t \le 2$

9. $\int_C (x^2 + y^2 + z^2)\, ds$

$C: \mathbf{r}(t) = \sin t\mathbf{i} + \cos t\mathbf{j} + 2\mathbf{k}$

$0 \le t \le \pi/2$

10. $\int_C 2xyz\, ds$

$C: \mathbf{r}(t) = 12t\mathbf{i} + 5t\mathbf{j} + 84t\mathbf{k}$

$0 \le t \le 1$

In Exercises 11–14, (a) find a parametrization of the path C, and (b) evaluate

$$\int_C (x^2 + y^2)\, ds$$

along C.

11. C: line segment from $(0, 0)$ to $(1, 1)$

12. C: line segment from $(0, 0)$ to $(2, 4)$

13. C: counterclockwise around the circle $x^2 + y^2 = 1$ from $(1, 0)$ to $(0, 1)$

14. C: counterclockwise around the circle $x^2 + y^2 = 4$ from $(2, 0)$ to $(0, 2)$

In Exercises 15–18, (a) find a parametrization of the path C, and (b) evaluate

$$\int_C (x + 4\sqrt{y})\, ds$$

along C.

15. C: x-axis from $x = 0$ to $x = 1$

16. C: y-axis from $y = 1$ to $y = 9$

17. C: counterclockwise around the triangle with vertices $(0, 0)$, $(1, 0)$, and $(0, 1)$

18. C: counterclockwise around the square with vertices $(0, 0)$, $(2, 0)$, $(2, 2)$, and $(0, 2)$

In Exercises 19 and 20, (a) find a piecewise smooth parametrization of the path C shown in the figure, and (b) evaluate

$$\int_C (2x + y^2 - z)\, ds$$

along C.

19. **20.**

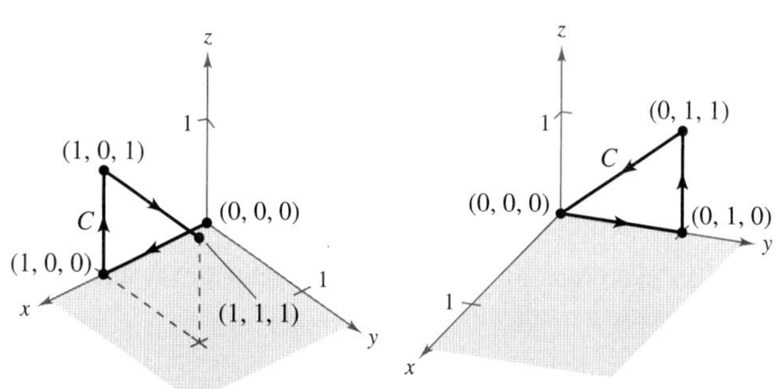

Mass **In Exercises 21 and 22, find the total mass of two turns of a spring with density ρ in the shape of the circular helix**

$\mathbf{r}(t) = 2\cos t\mathbf{i} + 2\sin t\mathbf{j} + t\mathbf{k},\ 0 \le t \le 4\pi.$

21. $\rho(x, y, z) = \frac{1}{2}(x^2 + y^2 + z^2)$

22. $\rho(x, y, z) = z$

Mass **In Exercises 23–26, find the total mass of the wire with density ρ.**

23. $\mathbf{r}(t) = \cos t\mathbf{i} + \sin t\mathbf{j},\quad \rho(x, y) = x + y,\quad 0 \le t \le \pi$

24. $\mathbf{r}(t) = t^2\mathbf{i} + 2t\mathbf{j},\quad \rho(x, y) = \frac{3}{4}y,\quad 0 \le t \le 1$

25. $\mathbf{r}(t) = t^2\mathbf{i} + 2t\mathbf{j} + t\mathbf{k},\quad \rho(x, y, z) = kz\quad (k > 0),\quad 1 \le t \le 3$

26. $\mathbf{r}(t) = 2\cos t\mathbf{i} + 2\sin t\mathbf{j} + 3t\mathbf{k},\quad \rho(x, y, z) = k + z$ $(k > 0),\quad 0 \le t \le 2\pi$

In Exercises 27–32, evaluate

$$\int_C \mathbf{F} \cdot d\mathbf{r}$$

where C is represented by $\mathbf{r}(t)$.

27. $\mathbf{F}(x, y) = x\mathbf{i} + y\mathbf{j}$

C: $\mathbf{r}(t) = t\mathbf{i} + t\mathbf{j}, \quad 0 \le t \le 1$

28. $\mathbf{F}(x, y) = xy\mathbf{i} + y\mathbf{j}$

C: $\mathbf{r}(t) = 4 \cos t\mathbf{i} + 4 \sin t\mathbf{j}, \quad 0 \le t \le \pi/2$

29. $\mathbf{F}(x, y) = 3x\mathbf{i} + 4y\mathbf{j}$

C: $\mathbf{r}(t) = \cos t\mathbf{i} + \sin t\mathbf{j}, \quad 0 \le t \le \pi/2$

30. $\mathbf{F}(x, y) = 3x\mathbf{i} + 4y\mathbf{j}$

C: $\mathbf{r}(t) = t\mathbf{i} + \sqrt{4 - t^2}\mathbf{j}, \quad -2 \le t \le 2$

31. $\mathbf{F}(x, y, z) = xy\mathbf{i} + xz\mathbf{j} + yz\mathbf{k}$

C: $\mathbf{r}(t) = t\mathbf{i} + t^2\mathbf{j} + 2t\mathbf{k}, \quad 0 \le t \le 1$

32. $\mathbf{F}(x, y, z) = x^2\mathbf{i} + y^2\mathbf{j} + z^2\mathbf{k}$

C: $\mathbf{r}(t) = 2 \sin t\mathbf{i} + 2 \cos t\mathbf{j} + \frac{1}{2}t^2\mathbf{k}, \quad 0 \le t \le \pi$

CAS **In Exercises 33 and 34, use a computer algebra system to evaluate the integral**

$$\int_C \mathbf{F} \cdot d\mathbf{r}$$

where C is represented by $\mathbf{r}(t)$.

33. $\mathbf{F}(x, y, z) = x^2z\mathbf{i} + 6y\mathbf{j} + yz^2\mathbf{k}$

C: $\mathbf{r}(t) = t\mathbf{i} + t^2\mathbf{j} + \ln t\mathbf{k}, \quad 1 \le t \le 3$

34. $\mathbf{F}(x, y, z) = \dfrac{x\mathbf{i} + y\mathbf{j} + z\mathbf{k}}{\sqrt{x^2 + y^2 + z^2}}$

C: $\mathbf{r}(t) = t\mathbf{i} + t\mathbf{j} + e^t\mathbf{k}, \quad 0 \le t \le 2$

Work **In Exercises 35–40, find the work done by the force field F on a particle moving along the given path.**

35. $\mathbf{F}(x, y) = x\mathbf{i} + 2y\mathbf{j}$

C: $x = t, y = t^3$ from $(0, 0)$ to $(2, 8)$

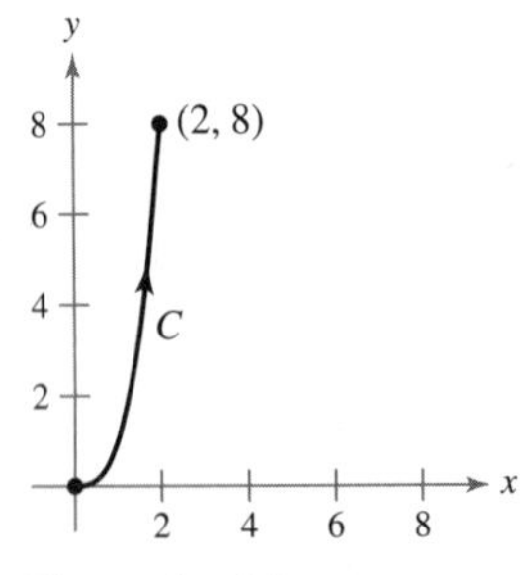

Figure for 35

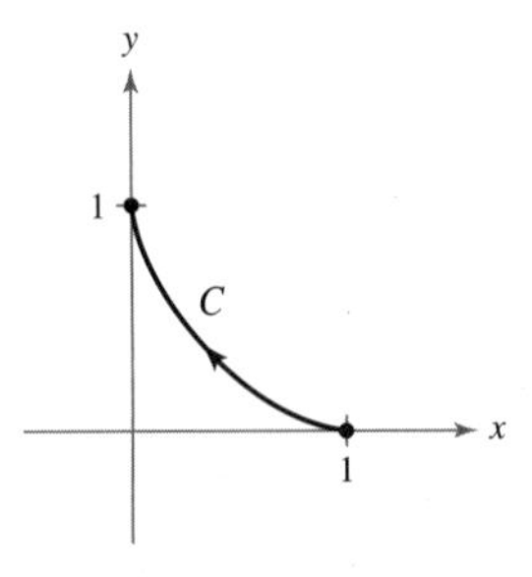

Figure for 36

36. $\mathbf{F}(x, y) = x^2\mathbf{i} - xy\mathbf{j}$

C: $x = \cos^3 t, y = \sin^3 t$ from $(1, 0)$ to $(0, 1)$

37. $\mathbf{F}(x, y) = x\mathbf{i} + y\mathbf{j}$

C: counterclockwise around the triangle with vertices $(0, 0)$, $(1, 0)$, and $(0, 1)$ (*Hint:* See Exercise 17a.)

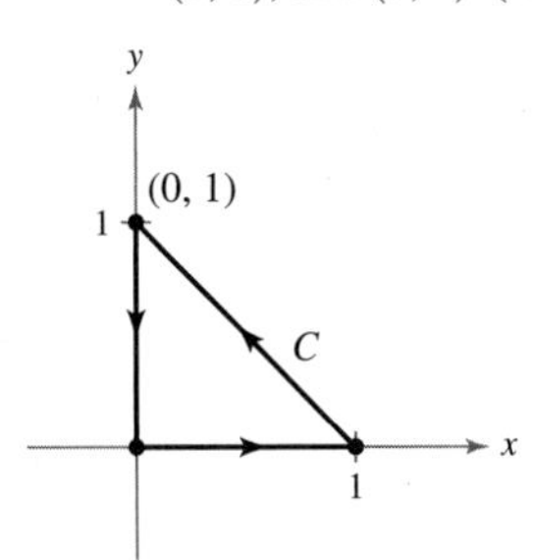

Figure for 37

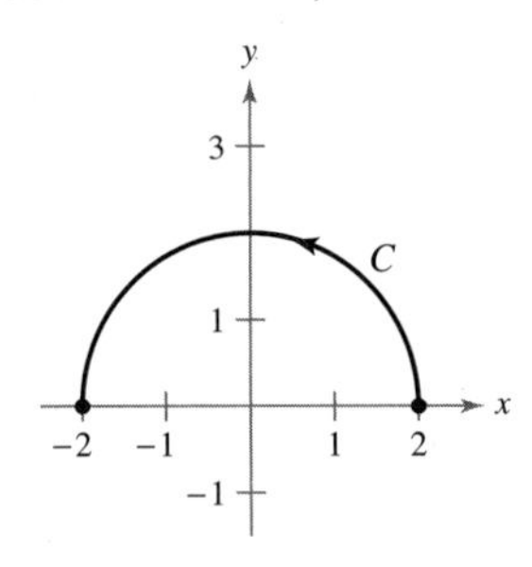

Figure for 38

38. $\mathbf{F}(x, y) = -y\mathbf{i} - x\mathbf{j}$

C: counterclockwise along the semicircle $y = \sqrt{4 - x^2}$ from $(2, 0)$ to $(-2, 0)$

39. $\mathbf{F}(x, y, z) = x\mathbf{i} + y\mathbf{j} - 5z\mathbf{k}$

C: $\mathbf{r}(t) = 2 \cos t\mathbf{i} + 2 \sin t\mathbf{j} + t\mathbf{k}, \quad 0 \le t \le 2\pi$

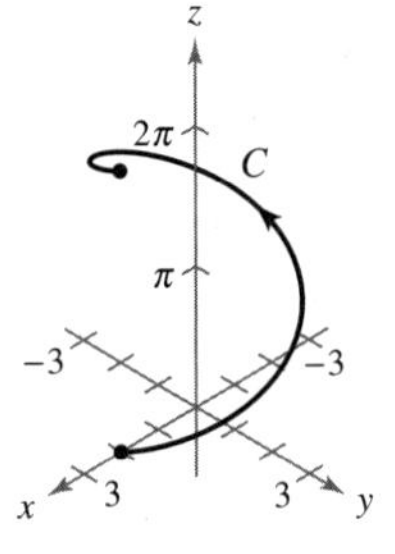

Figure for 39

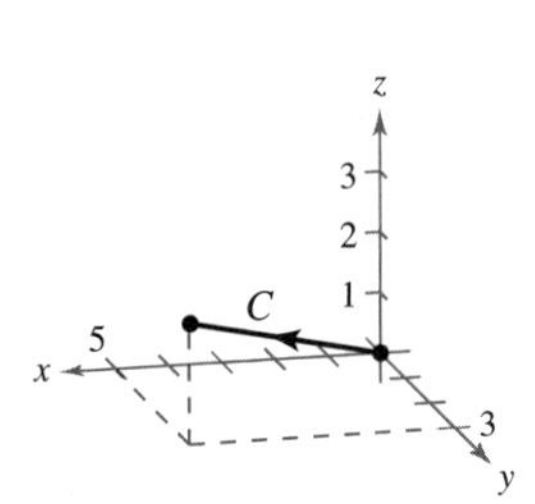

Figure for 40

40. $\mathbf{F}(x, y, z) = yz\mathbf{i} + xz\mathbf{j} + xy\mathbf{k}$

C: line from $(0, 0, 0)$ to $(5, 3, 2)$

In Exercises 41–44, determine whether the work done along the path C is positive, negative, or zero. Explain.

41.

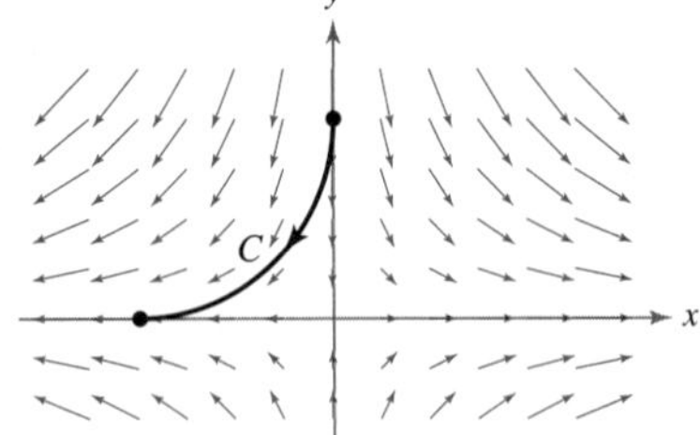

42.

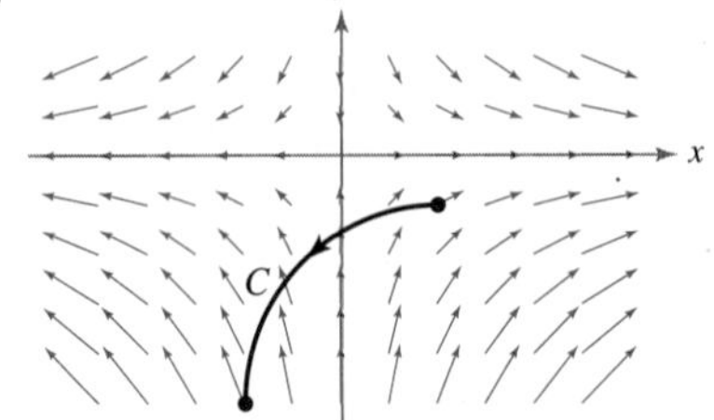

43.

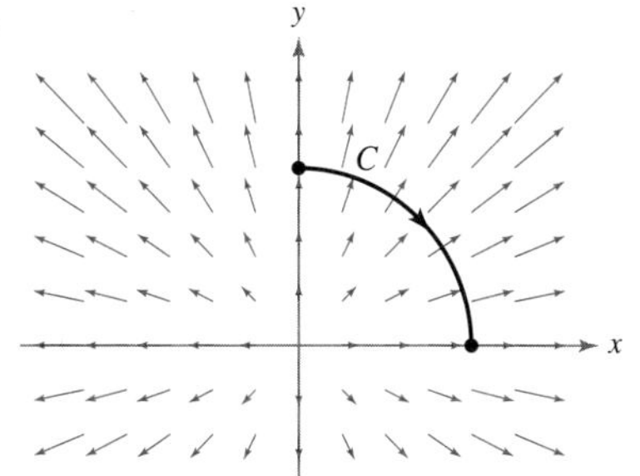

44.

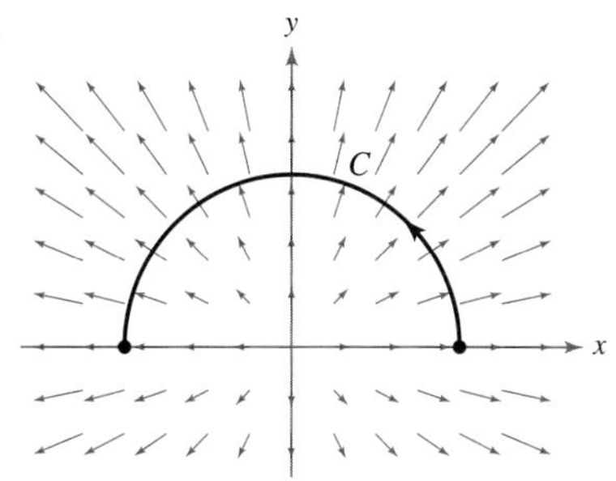

In Exercises 45 and 46, evaluate $\int_C \mathbf{F} \cdot d\mathbf{r}$ for each curve. Discuss the orientation of the curve and its effect on the value of the integral.

45. $\mathbf{F}(x, y) = x^2\mathbf{i} + xy\mathbf{j}$

(a) $\mathbf{r}_1(t) = 2t\mathbf{i} + (t - 1)\mathbf{j}, \quad 1 \le t \le 3$

(b) $\mathbf{r}_2(t) = 2(3 - t)\mathbf{i} + (2 - t)\mathbf{j}, \quad 0 \le t \le 2$

46. $\mathbf{F}(x, y) = x^2y\mathbf{i} + xy^{3/2}\mathbf{j}$

(a) $\mathbf{r}_1(t) = (t + 1)\mathbf{i} + t^2\mathbf{j}, \quad 0 \le t \le 2$

(b) $\mathbf{r}_2(t) = (1 + 2\cos t)\mathbf{i} + (4\cos^2 t)\mathbf{j}, \quad 0 \le t \le \pi/2$

In Exercises 47–50, demonstrate the property that

$$\int_C \mathbf{F} \cdot d\mathbf{r} = 0$$

regardless of the initial and terminal points of C, if the tangent vector $\mathbf{r}'(t)$ is orthogonal to the force field $\mathbf{F}$.

47. $\mathbf{F}(x, y) = y\mathbf{i} - x\mathbf{j}$

C: $\mathbf{r}(t) = t\mathbf{i} - 2t\mathbf{j}$

48. $\mathbf{F}(x, y) = -3y\mathbf{i} + x\mathbf{j}$

C: $\mathbf{r}(t) = t\mathbf{i} - t^3\mathbf{j}$

49. $\mathbf{F}(x, y) = (x^3 - 2x^2)\mathbf{i} + \left(x - \dfrac{y}{2}\right)\mathbf{j}$

C: $\mathbf{r}(t) = t\mathbf{i} + t^2\mathbf{j}$

50. $\mathbf{F}(x, y) = x\mathbf{i} + y\mathbf{j}$

C: $\mathbf{r}(t) = 3\sin t\mathbf{i} + 3\cos t\mathbf{j}$

In Exercises 51–54, evaluate the line integral along the path C given by $x = 2t$, $y = 10t$, where $0 \le t \le 1$.

51. $\displaystyle\int_C (x + 3y^2)\, dy$ **52.** $\displaystyle\int_C (x + 3y^2)\, dx$

53. $\displaystyle\int_C xy\, dx + y\, dy$ **54.** $\displaystyle\int_C (3y - x)\, dx + y^2\, dy$

In Exercises 55–62, evaluate the integral

$$\int_C (2x - y)\, dx + (x + 3y)\, dy$$

along the path C.

55. C: x-axis from $x = 0$ to $x = 5$

56. C: y-axis from $y = 0$ to $y = 2$

57. C: line segments from $(0, 0)$ to $(3, 0)$ and $(3, 0)$ to $(3, 3)$

58. C: line segments from $(0, 0)$ to $(0, -3)$ and $(0, -3)$ to $(2, -3)$

59. C: arc on $y = 1 - x^2$ from $(0, 1)$ to $(1, 0)$

60. C: arc on $y = x^{3/2}$ from $(0, 0)$ to $(4, 8)$

61. C: parabolic path $x = t$, $y = 2t^2$, from $(0, 0)$ to $(2, 8)$

62. C: elliptic path $x = 4\sin t$, $y = 3\cos t$, from $(0, 3)$ to $(4, 0)$

Lateral Surface Area **In Exercises 63–70, find the area of the lateral surface (see figure) over the curve C in the xy-plane and under the surface $z = f(x, y)$, where**

$$\textbf{Lateral surface area} = \int_C f(x, y)\, ds.$$

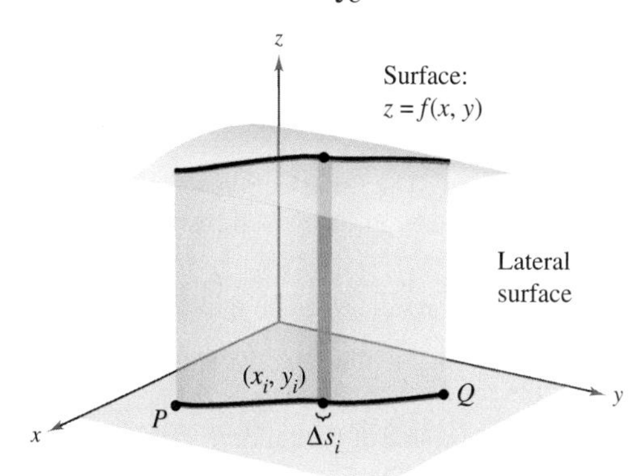

C: Curve in xy-plane

63. $f(x, y) = h$, C: line from $(0, 0)$ to $(3, 4)$

64. $f(x, y) = y$, C: line from $(0, 0)$ to $(4, 4)$

65. $f(x, y) = xy$, C: $x^2 + y^2 = 1$ from $(1, 0)$ to $(0, 1)$

66. $f(x, y) = x + y$, C: $x^2 + y^2 = 1$ from $(1, 0)$ to $(0, 1)$

67. $f(x, y) = h$, C: $y = 1 - x^2$ from $(1, 0)$ to $(0, 1)$

68. $f(x, y) = y + 1$, C: $y = 1 - x^2$ from $(1, 0)$ to $(0, 1)$

69. $f(x, y) = xy$, C: $y = 1 - x^2$ from $(1, 0)$ to $(0, 1)$

70. $f(x, y) = x^2 - y^2 + 4$, C: $x^2 + y^2 = 4$

71. ***Engine Design*** A tractor engine has a steel component with a circular base modeled by the vector-valued function $\mathbf{r}(t) = 2\cos t\mathbf{i} + 2\sin t\mathbf{j}$. Its height is given by $z = 1 + y^2$. (All measurements of the component are in centimeters.)

(a) Find the lateral surface area of the component.

(b) The component is in the form of a shell of thickness 0.2 centimeter. Use the result of part (a) to approximate the amount of steel used in its manufacture.

(c) Draw a sketch of the component.

72. ***Building Design*** The ceiling of a building has a height above the floor given by $z = 20 + \frac{1}{4}x$, and one of the walls follows a path modeled by $y = x^{3/2}$. Find the surface area of the wall if $0 \le x \le 40$. (All measurements are in feet.)

Moments of Inertia **Consider a wire of density $\rho(x, y)$ given by the space curve**

$$C: \mathbf{r}(t) = x(t)\mathbf{i} + y(t)\mathbf{j}, \quad 0 \le t \le b.$$

The moments of inertia about the x- and y-axes are given by

$$I_x = \int_C y^2\rho(x, y)\, ds$$

$$I_y = \int_C x^2\rho(x, y)\, ds.$$

In Exercises 73 and 74, find the moments of inertia for the wire of density ρ.

73. A wire lies along $\mathbf{r}(t) = a \cos t\mathbf{i} + a \sin t\mathbf{j}$, $0 \le t \le 2\pi$ and $a > 0$, with density $\rho(x, y) = 1$.

74. A wire lies along $\mathbf{r}(t) = a \cos t\mathbf{i} + a \sin t\mathbf{j}$, $0 \le t \le 2\pi$ and $a > 0$, with density $\rho(x, y) = y$.

CAS **75.** ***Investigation*** The top outer edge of a solid with vertical sides and resting on the xy-plane is modeled by $\mathbf{r}(t) = 3 \cos t\mathbf{i} + 3 \sin t\mathbf{j} + (1 + \sin^2 2t)\mathbf{k}$, where all measurements are in centimeters. The intersection of the plane $y = b$ $(-3 < b < 3)$ with the top of the solid is a horizontal line.

(a) Use a computer algebra system to graph the solid.

(b) Use a computer algebra system to approximate the lateral surface area of the solid.

(c) Find (if possible) the volume of the solid.

76. ***Work*** A particle moves along the path $y = x^2$ from the point (0, 0) to the point (1, 1). The force field $\mathbf{F}$ is measured at five points along the path, and the results are shown in the table. Use Simpson's Rule or a graphing utility to approximate the work done by the force field.

(x, y)	$(0, 0)$	$\left(\frac{1}{4}, \frac{1}{16}\right)$	$\left(\frac{1}{2}, \frac{1}{4}\right)$	$\left(\frac{3}{4}, \frac{9}{16}\right)$	$(1, 1)$
$\mathbf{F}(x, y)$	$\langle 5, 0\rangle$	$\langle 3.5, 1\rangle$	$\langle 2, 2\rangle$	$\langle 1.5, 3\rangle$	$\langle 1, 5\rangle$

77. ***Work*** Find the work done by a person weighing 175 pounds walking exactly one revolution up a circular helical staircase of radius 3 feet if the person rises 10 feet.

78. ***Investigation*** Determine the value of c such that the work done by the force field

$$\mathbf{F}(x, y) = 15[(4 - x^2y)\mathbf{i} - xy\mathbf{j}]$$

on an object moving along the parabolic path $y = c(1 - x^2)$ between the points $(-1, 0)$ and $(1, 0)$ is a minimum. Compare the result with the work required to move the object along the straight-line path connecting the points.

WRITING ABOUT CONCEPTS

79. Define a line integral of a function f along a smooth curve C in the plane and in space. How do you evaluate the line integral as a definite integral?

80. Define a line integral of a continuous vector field $\mathbf{F}$ on a smooth curve C. How do you evaluate the line integral as a definite integral?

81. Order the surfaces in ascending order of the lateral surface area under the surface and over the curve $y = \sqrt{x}$ from (0, 0) to (4, 2) in the xy-plane. Explain your ordering without doing any calculations.

(a) $z_1 = 2 + x$ (b) $z_2 = 5 + x$

(c) $z_3 = 2$ (d) $z_4 = 10 + x + 2y$

CAPSTONE

82. For each of the following, determine whether the work done in moving an object from the first to the second point through the force field shown in the figure is positive, negative, or zero. Explain your answer.

(a) From $(-3, -3)$ to $(3, 3)$

(b) From $(-3, 0)$ to $(0, 3)$

(c) From $(5, 0)$ to $(0, 3)$

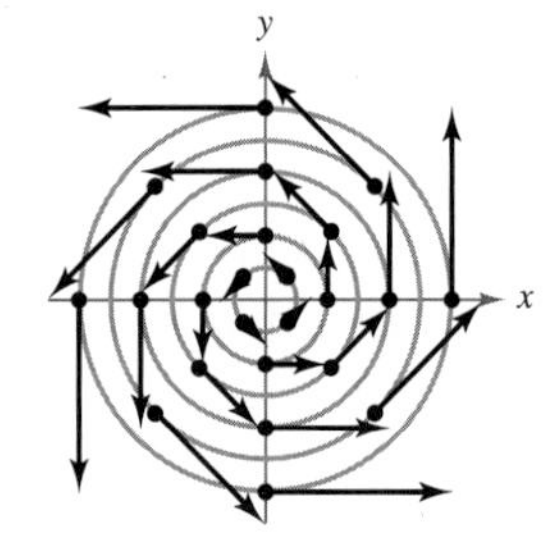

True or False? **In Exercises 83–86, determine whether the statement is true or false. If it is false, explain why or give an example that shows it is false.**

83. If C is given by $x(t) = t$, $y(t) = t$, $0 \le t \le 1$, then

$$\int_C xy\, ds = \int_0^1 t^2\, dt.$$

84. If $C_2 = -C_1$, then $\int_{C_1} f(x, y)\, ds + \int_{C_2} f(x, y)\, ds = 0.$

85. The vector functions $\mathbf{r}_1 = t\mathbf{i} + t^2\mathbf{j}$, $0 \le t \le 1$, and $\mathbf{r}_2 = (1 - t)\mathbf{i} + (1 - t)^2\mathbf{j}$, $0 \le t \le 1$, define the same curve.

86. If $\int_C \mathbf{F} \cdot \mathbf{T}\, ds = 0$, then $\mathbf{F}$ and $\mathbf{T}$ are orthogonal.

87. ***Work*** Consider a particle that moves through the force field $\mathbf{F}(x, y) = (y - x)\mathbf{i} + xy\mathbf{j}$ from the point (0, 0) to the point (0, 1) along the curve $x = kt(1 - t)$, $y = t$. Find the value of k such that the work done by the force field is 1.

15.3 Conservative Vector Fields and Independence of Path

- **Understand and use the Fundamental Theorem of Line Integrals.**
- **Understand the concept of independence of path.**
- **Understand the concept of conservation of energy.**

Fundamental Theorem of Line Integrals

The discussion at the beginning of the preceding section pointed out that in a gravitational field the work done by gravity on an object moving between two points in the field is independent of the path taken by the object. In this section, you will study an important generalization of this result—it is called the **Fundamental Theorem of Line Integrals.**

To begin, an example is presented in which the line integral of a *conservative vector field* is evaluated over three different paths.

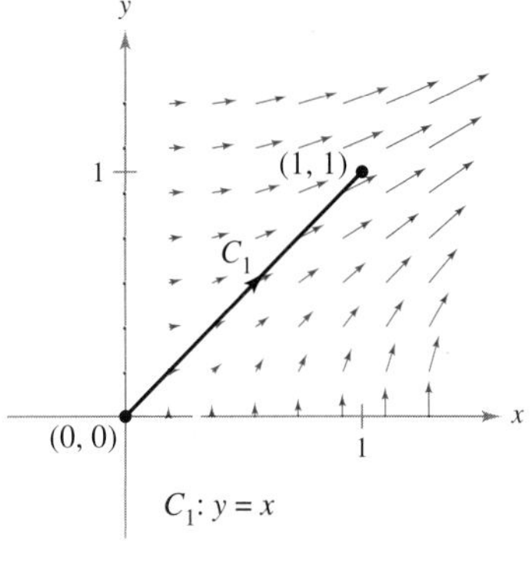

(a)

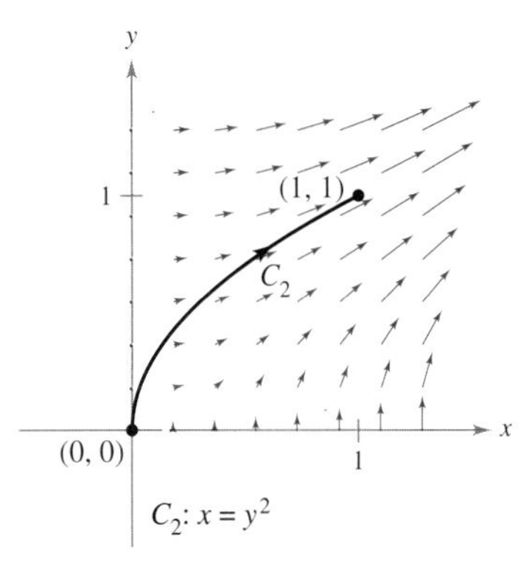

(b)

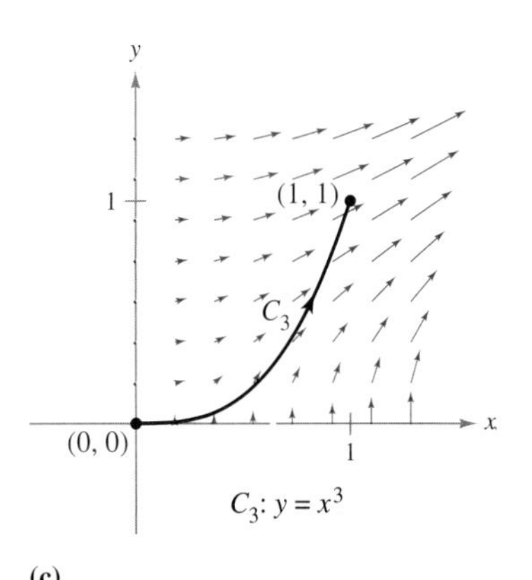

(c)

Figure 15.19

EXAMPLE 1 Line Integral of a Conservative Vector Field

Find the work done by the force field

$$\mathbf{F}(x, y) = \frac{1}{2}xy\mathbf{i} + \frac{1}{4}x^2\mathbf{j}$$

on a particle that moves from $(0, 0)$ to $(1, 1)$ along each path, as shown in Figure 15.19.

a. C_1: $y = x$ **b.** C_2: $x = y^2$ **c.** C_3: $y = x^3$

Solution

a. Let $\mathbf{r}(t) = t\mathbf{i} + t\mathbf{j}$ for $0 \le t \le 1$, so that

$$d\mathbf{r} = (\mathbf{i} + \mathbf{j})\,dt \quad \text{and} \quad \mathbf{F}(x, y) = \frac{1}{2}t^2\mathbf{i} + \frac{1}{4}t^2\mathbf{j}.$$

Then, the work done is

$$W = \int_{C_1} \mathbf{F} \cdot d\mathbf{r} = \int_0^1 \frac{3}{4}t^2\,dt = \frac{1}{4}t^3\Big]_0^1 = \frac{1}{4}.$$

b. Let $\mathbf{r}(t) = t\mathbf{i} + \sqrt{t}\mathbf{j}$ for $0 \le t \le 1$, so that

$$d\mathbf{r} = \left(\mathbf{i} + \frac{1}{2\sqrt{t}}\mathbf{j}\right) dt \quad \text{and} \quad \mathbf{F}(x, y) = \frac{1}{2}t^{3/2}\mathbf{i} + \frac{1}{4}t^2\mathbf{j}.$$

Then, the work done is

$$W = \int_{C_2} \mathbf{F} \cdot d\mathbf{r} = \int_0^1 \frac{5}{8}t^{3/2}\,dt = \frac{1}{4}t^{5/2}\Big]_0^1 = \frac{1}{4}.$$

c. Let $\mathbf{r}(t) = \frac{1}{2}t\mathbf{i} + \frac{1}{8}t^3\mathbf{j}$ for $0 \le t \le 2$, so that

$$d\mathbf{r} = \left(\frac{1}{2}\mathbf{i} + \frac{3}{8}t^2\mathbf{j}\right) dt \quad \text{and} \quad \mathbf{F}(x, y) = \frac{1}{32}t^4\mathbf{i} + \frac{1}{16}t^2\mathbf{j}.$$

Then, the work done is

$$W = \int_{C_3} \mathbf{F} \cdot d\mathbf{r} = \int_0^2 \frac{5}{128}t^4\,dt = \frac{1}{128}t^5\Big]_0^2 = \frac{1}{4}.$$

So, the work done by a conservative vector field is the same for all paths. ■

In Example 1, note that the vector field $\mathbf{F}(x, y) = \frac{1}{2}xy\mathbf{i} + \frac{1}{4}x^2\mathbf{j}$ is conservative because $\mathbf{F}(x, y) = \nabla f(x, y)$, where $f(x, y) = \frac{1}{4}x^2y$. In such cases, the following theorem states that the value of $\int_C \mathbf{F} \cdot d\mathbf{r}$ is given by

$$\begin{aligned}\int_C \mathbf{F} \cdot d\mathbf{r} &= f(x(1), y(1)) - f(x(0), y(0)) \\ &= \frac{1}{4} - 0 \\ &= \frac{1}{4}.\end{aligned}$$

NOTE Notice how the Fundamental Theorem of Line Integrals is similar to the Fundamental Theorem of Calculus (Section 4.4), which states that

$$\int_a^b f(x)\, dx = F(b) - F(a)$$

where $F'(x) = f(x)$.

THEOREM 15.5 FUNDAMENTAL THEOREM OF LINE INTEGRALS

Let C be a piecewise smooth curve lying in an open region R and given by

$$\mathbf{r}(t) = x(t)\mathbf{i} + y(t)\mathbf{j}, \quad a \le t \le b.$$

If $\mathbf{F}(x, y) = M\mathbf{i} + N\mathbf{j}$ is conservative in R, and M and N are continuous in R, then

$$\int_C \mathbf{F} \cdot d\mathbf{r} = \int_C \nabla f \cdot d\mathbf{r} = f(x(b), y(b)) - f(x(a), y(a))$$

where f is a potential function of $\mathbf{F}$. That is, $\mathbf{F}(x, y) = \nabla f(x, y)$.

PROOF A proof is provided only for a smooth curve. For piecewise smooth curves, the procedure is carried out separately on each smooth portion. Because $\mathbf{F}(x, y) = \nabla f(x, y) = f_x(x, y)\mathbf{i} + f_y(x, y)\mathbf{j}$, it follows that

$$\begin{aligned}\int_C \mathbf{F} \cdot d\mathbf{r} &= \int_a^b \mathbf{F} \cdot \frac{d\mathbf{r}}{dt}\, dt \\ &= \int_a^b \left[f_x(x, y)\frac{dx}{dt} + f_y(x, y)\frac{dy}{dt} \right] dt\end{aligned}$$

and, by the Chain Rule (Theorem 13.6), you have

$$\begin{aligned}\int_C \mathbf{F} \cdot d\mathbf{r} &= \int_a^b \frac{d}{dt}[f(x(t), y(t))]\, dt \\ &= f(x(b), y(b)) - f(x(a), y(a)).\end{aligned}$$

The last step is an application of the Fundamental Theorem of Calculus. ■

In space, the Fundamental Theorem of Line Integrals takes the following form. Let C be a piecewise smooth curve lying in an open region Q and given by

$$\mathbf{r}(t) = x(t)\mathbf{i} + y(t)\mathbf{j} + z(t)\mathbf{k}, \quad a \le t \le b.$$

If $\mathbf{F}(x, y, z) = M\mathbf{i} + N\mathbf{j} + P\mathbf{k}$ is conservative and M, N, and P are continuous, then

$$\begin{aligned}\int_C \mathbf{F} \cdot d\mathbf{r} &= \int_C \nabla f \cdot d\mathbf{r} \\ &= f(x(b), y(b), z(b)) - f(x(a), y(a), z(a))\end{aligned}$$

where $\mathbf{F}(x, y, z) = \nabla f(x, y, z)$.

The Fundamental Theorem of Line Integrals states that if the vector field $\mathbf{F}$ is conservative, then the line integral between any two points is simply the difference in the values of the *potential* function f at these points.

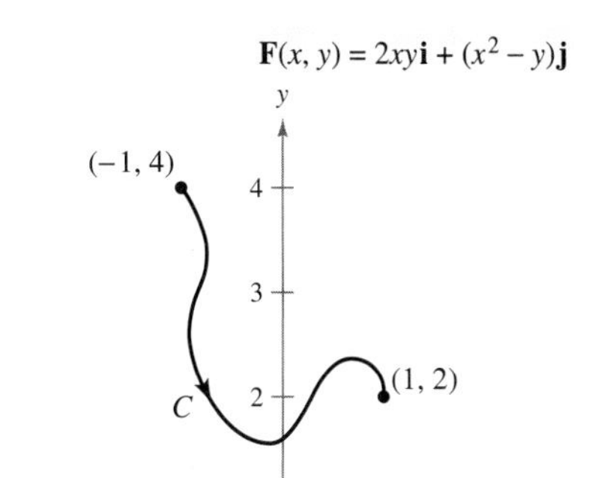

Using the Fundamental Theorem of Line Integrals, $\int_C \mathbf{F} \cdot d\mathbf{r}$.

Figure 15.20

EXAMPLE 2 Using the Fundamental Theorem of Line Integrals

Evaluate $\int_C \mathbf{F} \cdot d\mathbf{r}$, where C is a piecewise smooth curve from $(-1, 4)$ to $(1, 2)$ and

$$\mathbf{F}(x, y) = 2xy\mathbf{i} + (x^2 - y)\mathbf{j}$$

as shown in Figure 15.20.

Solution From Example 6 in Section 15.1, you know that $\mathbf{F}$ is the gradient of f where

$$f(x, y) = x^2y - \frac{y^2}{2} + K.$$

Consequently, $\mathbf{F}$ is conservative, and by the Fundamental Theorem of Line Integrals, it follows that

$$\begin{aligned}\int_C \mathbf{F} \cdot d\mathbf{r} &= f(1, 2) - f(-1, 4)\\ &= \left[1^2(2) - \frac{2^2}{2}\right] - \left[(-1)^2(4) - \frac{4^2}{2}\right]\\ &= 4.\end{aligned}$$

Note that it is unnecessary to include a constant K as part of f, because it is canceled by subtraction.

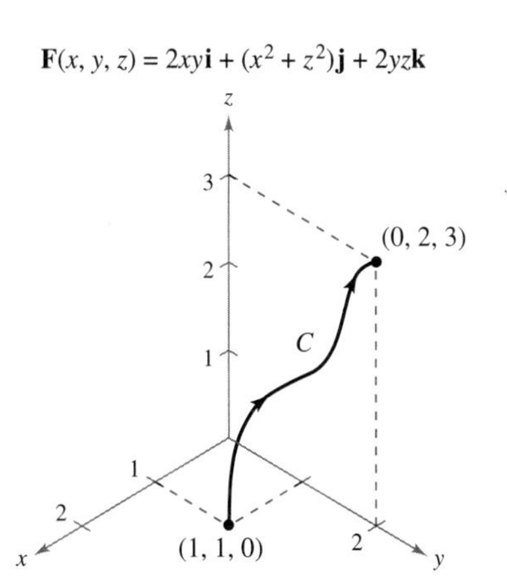

Using the Fundamental Theorem of Line Integrals, $\int_C \mathbf{F} \cdot d\mathbf{r}$.

Figure 15.21

EXAMPLE 3 Using the Fundamental Theorem of Line Integrals

Evaluate $\int_C \mathbf{F} \cdot d\mathbf{r}$, where C is a piecewise smooth curve from $(1, 1, 0)$ to $(0, 2, 3)$ and

$$\mathbf{F}(x, y, z) = 2xy\mathbf{i} + (x^2 + z^2)\mathbf{j} + 2yz\mathbf{k}$$

as shown in Figure 15.21.

Solution From Example 8 in Section 15.1, you know that $\mathbf{F}$ is the gradient of f where $f(x, y, z) = x^2y + yz^2 + K$. Consequently, $\mathbf{F}$ is conservative, and by the Fundamental Theorem of Line Integrals, it follows that

$$\begin{aligned}\int_C \mathbf{F} \cdot d\mathbf{r} &= f(0, 2, 3) - f(1, 1, 0)\\ &= [(0)^2(2) + (2)(3)^2] - [(1)^2(1) + (1)(0)^2]\\ &= 17.\end{aligned}$$

■

In Examples 2 and 3, be sure you see that the value of the line integral is the same for any smooth curve C that has the given initial and terminal points. For instance, in Example 3, try evaluating the line integral for the curve given by

$$\mathbf{r}(t) = (1 - t)\mathbf{i} + (1 + t)\mathbf{j} + 3t\mathbf{k}.$$

You should obtain

$$\begin{aligned}\int_C \mathbf{F} \cdot d\mathbf{r} &= \int_0^1 (30t^2 + 16t - 1)\,dt\\ &= 17.\end{aligned}$$

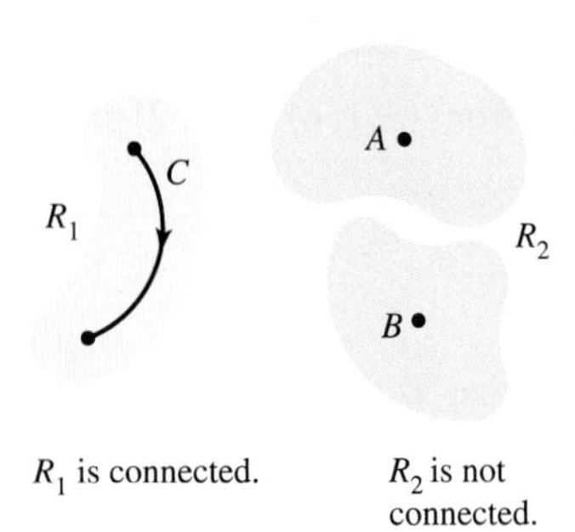

Figure 15.22

Independence of Path

From the Fundamental Theorem of Line Integrals it is clear that if **F** is continuous and conservative in an open region R, the value of $\int_C \mathbf{F} \cdot d\mathbf{r}$ is the same for every piecewise smooth curve C from one fixed point in R to another fixed point in R. This result is described by saying that the line integral $\int_C \mathbf{F} \cdot d\mathbf{r}$ is **independent of path** in the region R.

A region in the plane (or in space) is **connected** if any two points in the region can be joined by a piecewise smooth curve lying entirely within the region, as shown in Figure 15.22. In open regions that are *connected*, the path independence of $\int_C \mathbf{F} \cdot d\mathbf{r}$ is equivalent to the condition that **F** is conservative.

THEOREM 15.6 INDEPENDENCE OF PATH AND CONSERVATIVE VECTOR FIELDS

If **F** is continuous on an open connected region, then the line integral

$$\int_C \mathbf{F} \cdot d\mathbf{r}$$

is independent of path if and only if **F** is conservative.

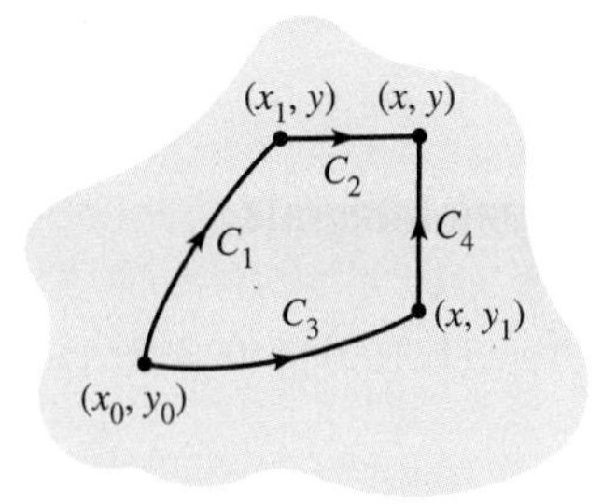

Figure 15.23

PROOF If **F** is conservative, then, by the Fundamental Theorem of Line Integrals, the line integral is independent of path. Now establish the converse for a plane region R. Let $\mathbf{F}(x, y) = M\mathbf{i} + N\mathbf{j}$, and let (x_0, y_0) be a fixed point in R. If (x, y) is any point in R, choose a piecewise smooth curve C running from (x_0, y_0) to (x, y), and define f by

$$f(x, y) = \int_C \mathbf{F} \cdot d\mathbf{r} = \int_C M\,dx + N\,dy.$$

The existence of C in R is guaranteed by the fact that R is connected. You can show that f is a potential function of **F** by considering two different paths between (x_0, y_0) and (x, y). For the *first* path, choose (x_1, y) in R such that $x \neq x_1$. This is possible because R is open. Then choose C_1 and C_2, as shown in Figure 15.23. Using the independence of path, it follows that

$$\begin{aligned} f(x, y) &= \int_C M\,dx + N\,dy \\ &= \int_{C_1} M\,dx + N\,dy + \int_{C_2} M\,dx + N\,dy. \end{aligned}$$

Because the first integral does not depend on x, and because $dy = 0$ in the second integral, you have

$$f(x, y) = g(y) + \int_{C_2} M\,dx$$

and it follows that the partial derivative of f with respect to x is $f_x(x, y) = M$. For the *second* path, choose a point (x, y_1). Using reasoning similar to that used for the first path, you can conclude that $f_y(x, y) = N$. Therefore,

$$\begin{aligned} \nabla f(x, y) &= f_x(x, y)\mathbf{i} + f_y(x, y)\mathbf{j} \\ &= M\mathbf{i} + N\mathbf{j} \\ &= \mathbf{F}(x, y) \end{aligned}$$

and it follows that **F** is conservative. ■

EXAMPLE 4 Finding Work in a Conservative Force Field

For the force field given by

$$\mathbf{F}(x, y, z) = e^x \cos y\mathbf{i} - e^x \sin y\mathbf{j} + 2\mathbf{k}$$

show that $\int_C \mathbf{F} \cdot d\mathbf{r}$ is independent of path, and calculate the work done by $\mathbf{F}$ on an object moving along a curve C from $(0, \pi/2, 1)$ to $(1, \pi, 3)$.

Solution Writing the force field in the form $\mathbf{F}(x, y, z) = M\mathbf{i} + N\mathbf{j} + P\mathbf{k}$, you have $M = e^x \cos y$, $N = -e^x \sin y$, and $P = 2$, and it follows that

$$\frac{\partial P}{\partial y} = 0 = \frac{\partial N}{\partial z}$$

$$\frac{\partial P}{\partial x} = 0 = \frac{\partial M}{\partial z}$$

$$\frac{\partial N}{\partial x} = -e^x \sin y = \frac{\partial M}{\partial y}.$$

So, $\mathbf{F}$ is conservative. If f is a potential function of $\mathbf{F}$, then

$$f_x(x, y, z) = e^x \cos y$$

$$f_y(x, y, z) = -e^x \sin y$$

$$f_z(x, y, z) = 2.$$

By integrating with respect to x, y, and z separately, you obtain

$$f(x, y, z) = \int f_x(x, y, z)\, dx = \int e^x \cos y\, dx = e^x \cos y + g(y, z)$$

$$f(x, y, z) = \int f_y(x, y, z)\, dy = \int -e^x \sin y\, dy = e^x \cos y + h(x, z)$$

$$f(x, y, z) = \int f_z(x, y, z)\, dz = \int 2\, dz = 2z + k(x, y).$$

By comparing these three versions of $f(x, y, z)$, you can conclude that

$$f(x, y, z) = e^x \cos y + 2z + K.$$

Therefore, the work done by $\mathbf{F}$ along *any* curve C from $(0, \pi/2, 1)$ to $(1, \pi, 3)$ is

$$\begin{aligned} W &= \int_C \mathbf{F} \cdot d\mathbf{r} \\ &= \Big[e^x \cos y + 2z \Big]_{(0, \pi/2, 1)}^{(1, \pi, 3)} \\ &= (-e + 6) - (0 + 2) \\ &= 4 - e. \end{aligned}$$

■

How much work would be done if the object in Example 4 moved from the point $(0, \pi/2, 1)$ to $(1, \pi, 3)$ and then back to the starting point $(0, \pi/2, 1)$? The Fundamental Theorem of Line Integrals states that there is zero work done. Remember that, by definition, work can be negative. So, by the time the object gets back to its starting point, the amount of work that registers positively is canceled out by the amount of work that registers negatively.

A curve C given by $\mathbf{r}(t)$ for $a \le t \le b$ is **closed** if $\mathbf{r}(a) = \mathbf{r}(b)$. By the Fundamental Theorem of Line Integrals, you can conclude that if $\mathbf{F}$ is continuous and conservative on an open region R, then the line integral over every closed curve C is 0.

NOTE Theorem 15.7 gives you options for evaluating a line integral involving a conservative vector field. You can use a potential function, or it might be more convenient to choose a particularly simple path, such as a straight line.

THEOREM 15.7 EQUIVALENT CONDITIONS

Let $\mathbf{F}(x, y, z) = M\mathbf{i} + N\mathbf{j} + P\mathbf{k}$ have continuous first partial derivatives in an open connected region R, and let C be a piecewise smooth curve in R. The following conditions are equivalent.

1. $\mathbf{F}$ is conservative. That is, $\mathbf{F} = \nabla f$ for some function f.
2. $\displaystyle\int_C \mathbf{F} \cdot d\mathbf{r}$ is independent of path.
3. $\displaystyle\int_C \mathbf{F} \cdot d\mathbf{r} = 0$ for every *closed* curve C in R.

EXAMPLE 5 Evaluating a Line Integral

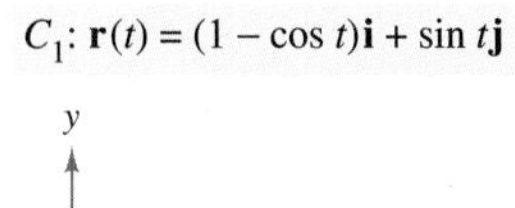

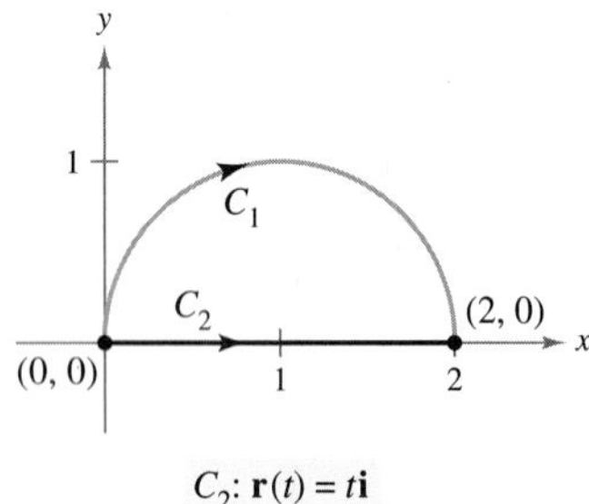

Figure 15.24

Evaluate $\displaystyle\int_{C_1} \mathbf{F} \cdot d\mathbf{r}$, where

$$\mathbf{F}(x, y) = (y^3 + 1)\mathbf{i} + (3xy^2 + 1)\mathbf{j}$$

and C_1 is the semicircular path from $(0, 0)$ to $(2, 0)$, as shown in Figure 15.24.

Solution You have the following three options.

a. You can use the method presented in the preceding section to evaluate the line integral along the *given curve*. To do this, you can use the parametrization $\mathbf{r}(t) = (1 - \cos t)\mathbf{i} + \sin t\mathbf{j}$, where $0 \le t \le \pi$. For this parametrization, it follows that $d\mathbf{r} = \mathbf{r}'(t)\,dt = (\sin t\mathbf{i} + \cos t\mathbf{j})\,dt$, and

$$\int_{C_1} \mathbf{F} \cdot d\mathbf{r} = \int_0^{\pi} (\sin t + \sin^4 t + \cos t + 3\sin^2 t \cos t - 3\sin^2 t \cos^2 t)\,dt.$$

This integral should dampen your enthusiasm for this option.

b. You can try to find a *potential function* and evaluate the line integral by the Fundamental Theorem of Line Integrals. Using the technique demonstrated in Example 4, you can find the potential function to be $f(x, y) = xy^3 + x + y + K$, and, by the Fundamental Theorem,

$$W = \int_{C_1} \mathbf{F} \cdot d\mathbf{r} = f(2, 0) - f(0, 0) = 2.$$

c. Knowing that $\mathbf{F}$ is conservative, you have a third option. Because the value of the line integral is independent of path, you can replace the semicircular path with a *simpler path*. Suppose you choose the straight-line path C_2 from $(0, 0)$ to $(2, 0)$. Then, $\mathbf{r}(t) = t\mathbf{i}$, where $0 \le t \le 2$. So, $d\mathbf{r} = \mathbf{i}\,dt$ and $\mathbf{F}(x, y) = (y^3 + 1)\mathbf{i} + (3xy^2 + 1)\mathbf{j} = \mathbf{i} + \mathbf{j}$, so that

$$\int_{C_1} \mathbf{F} \cdot d\mathbf{r} = \int_{C_2} \mathbf{F} \cdot d\mathbf{r} = \int_0^2 1\,dt = t\Big]_0^2 = 2.$$

Of the three options, obviously the third one is the easiest. ■

Conservation of Energy

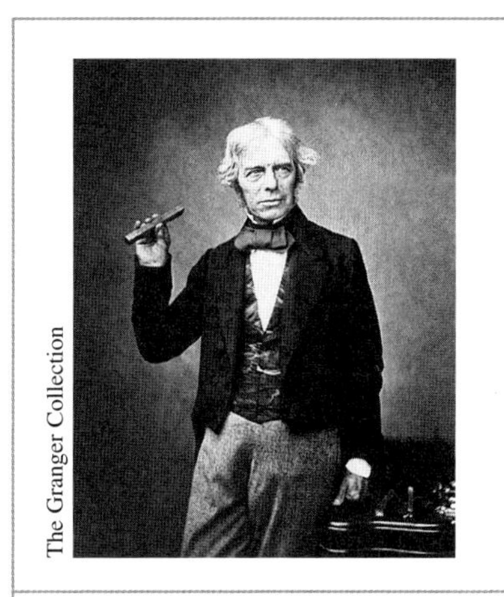

MICHAEL FARADAY (1791–1867)

Several philosophers of science have considered Faraday's Law of Conservation of Energy to be the greatest generalization ever conceived by humankind. Many physicists have contributed to our knowledge of this law. Two early and influential ones were James Prescott Joule (1818–1889) and Hermann Ludwig Helmholtz (1821–1894).

In 1840, the English physicist Michael Faraday wrote, "Nowhere is there a pure creation or production of power without a corresponding exhaustion of something to supply it." This statement represents the first formulation of one of the most important laws of physics—**the Law of Conservation of Energy.** In modern terminology, the law is stated as follows: *In a conservative force field, the sum of the potential and kinetic energies of an object remains constant from point to point.*

You can use the Fundamental Theorem of Line Integrals to derive this law. From physics, the **kinetic energy** of a particle of mass m and speed v is $k = \frac{1}{2}mv^2$. The **potential energy** p of a particle at point (x, y, z) in a conservative vector field $\mathbf{F}$ is defined as $p(x, y, z) = -f(x, y, z)$, where f is the potential function for $\mathbf{F}$. Consequently, the work done by $\mathbf{F}$ along a smooth curve C from A to B is

$$\begin{aligned} W = \int_C \mathbf{F} \cdot d\mathbf{r} &= f(x, y, z)\Big]_A^B \\ &= -p(x, y, z)\Big]_A^B \\ &= p(A) - p(B) \end{aligned}$$

as shown in Figure 15.25. In other words, work W is equal to the difference in the potential energies of A and B. Now, suppose that $\mathbf{r}(t)$ is the position vector for a particle moving along C from $A = \mathbf{r}(a)$ to $B = \mathbf{r}(b)$. At any time t, the particle's velocity, acceleration, and speed are $\mathbf{v}(t) = \mathbf{r}'(t)$, $\mathbf{a}(t) = \mathbf{r}''(t)$, and $v(t) = \|\mathbf{v}(t)\|$, respectively. So, by Newton's Second Law of Motion, $\mathbf{F} = m\mathbf{a}(t) = m(\mathbf{v}'(t))$, and the work done by $\mathbf{F}$ is

$$\begin{aligned} W = \int_C \mathbf{F} \cdot d\mathbf{r} &= \int_a^b \mathbf{F} \cdot \mathbf{r}'(t)\, dt \\ &= \int_a^b \mathbf{F} \cdot \mathbf{v}(t)\, dt = \int_a^b [m\mathbf{v}'(t)] \cdot \mathbf{v}(t)\, dt \\ &= \int_a^b m[\mathbf{v}'(t) \cdot \mathbf{v}(t)]\, dt \\ &= \frac{m}{2}\int_a^b \frac{d}{dt}[\mathbf{v}(t) \cdot \mathbf{v}(t)]\, dt \\ &= \frac{m}{2}\int_a^b \frac{d}{dt}[\|\mathbf{v}(t)\|^2]\, dt \\ &= \frac{m}{2}\Big[\|\mathbf{v}(t)\|^2\Big]_a^b \\ &= \frac{m}{2}\Big[[v(t)]^2\Big]_a^b \\ &= \frac{1}{2}m[v(b)]^2 - \frac{1}{2}m[v(a)]^2 \\ &= k(B) - k(A). \end{aligned}$$

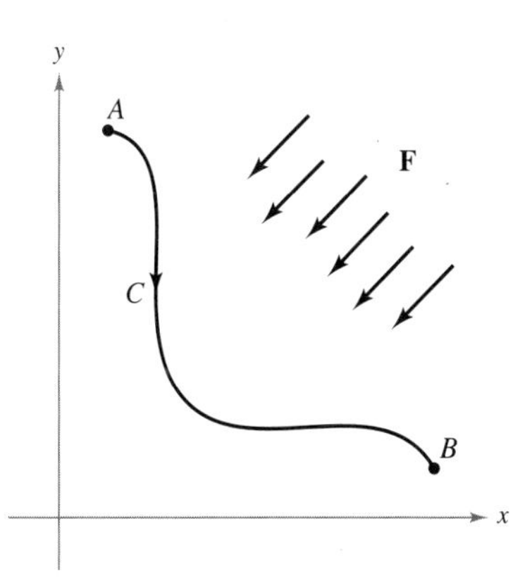

The work done by $\mathbf{F}$ along C is $W = \int_C \mathbf{F} \cdot d\mathbf{r} = p(A) - p(B)$.

Figure 15.25

Equating these two results for W produces

$$p(A) - p(B) = k(B) - k(A)$$
$$p(A) + k(A) = p(B) + k(B)$$

which implies that the sum of the potential and kinetic energies remains constant from point to point.

15.3 Exercises

See www.CalcChat.com for worked-out solutions to odd-numbered exercises.

In Exercises 1–4, show that the value of $\int_C \mathbf{F} \cdot d\mathbf{r}$ is the same for each parametric representation of C.

1. $\mathbf{F}(x, y) = x^2\mathbf{i} + xy\mathbf{j}$
   (a) $\mathbf{r}_1(t) = t\mathbf{i} + t^2\mathbf{j}, \quad 0 \le t \le 1$
   (b) $\mathbf{r}_2(\theta) = \sin\theta\mathbf{i} + \sin^2\theta\mathbf{j}, \quad 0 \le \theta \le \dfrac{\pi}{2}$
2. $\mathbf{F}(x, y) = (x^2 + y^2)\mathbf{i} - x\mathbf{j}$
   (a) $\mathbf{r}_1(t) = t\mathbf{i} + \sqrt{t}\mathbf{j}, \quad 0 \le t \le 4$
   (b) $\mathbf{r}_2(w) = w^2\mathbf{i} + w\mathbf{j}, \quad 0 \le w \le 2$
3. $\mathbf{F}(x, y) = y\mathbf{i} - x\mathbf{j}$
   (a) $\mathbf{r}_1(\theta) = \sec\theta\mathbf{i} + \tan\theta\mathbf{j}, \quad 0 \le \theta \le \dfrac{\pi}{3}$
   (b) $\mathbf{r}_2(t) = \sqrt{t+1}\mathbf{i} + \sqrt{t}\mathbf{j}, \quad 0 \le t \le 3$
4. $\mathbf{F}(x, y) = y\mathbf{i} + x^2\mathbf{j}$
   (a) $\mathbf{r}_1(t) = (2 + t)\mathbf{i} + (3 - t)\mathbf{j}, \quad 0 \le t \le 3$
   (b) $\mathbf{r}_2(w) = (2 + \ln w)\mathbf{i} + (3 - \ln w)\mathbf{j}, \quad 1 \le w \le e^3$

In Exercises 5–10, determine whether the vector field is conservative.

5. $\mathbf{F}(x, y) = e^x(\sin y\mathbf{i} + \cos y\mathbf{j})$
6. $\mathbf{F}(x, y) = 15x^2y^2\mathbf{i} + 10x^3y\mathbf{j}$
7. $\mathbf{F}(x, y) = \dfrac{1}{y^2}(y\mathbf{i} + x\mathbf{j})$
8. $\mathbf{F}(x, y, z) = y\ln z\mathbf{i} - x\ln z\mathbf{j} + \dfrac{xy}{z}\mathbf{k}$
9. $\mathbf{F}(x, y, z) = y^2z\mathbf{i} + 2xyz\mathbf{j} + xy^2\mathbf{k}$
10. $\mathbf{F}(x, y, z) = \sin yz\mathbf{i} + xz\cos yz\mathbf{j} + xy\sin yz\mathbf{k}$

In Exercises 11–24, find the value of the line integral

$$\int_C \mathbf{F} \cdot d\mathbf{r}.$$

(*Hint:* If F is conservative, the integration may be easier on an alternative path.)

11. $\mathbf{F}(x, y) = 2xy\mathbf{i} + x^2\mathbf{j}$
    (a) $\mathbf{r}_1(t) = t\mathbf{i} + t^2\mathbf{j}, \quad 0 \le t \le 1$
    (b) $\mathbf{r}_2(t) = t\mathbf{i} + t^3\mathbf{j}, \quad 0 \le t \le 1$
12. $\mathbf{F}(x, y) = ye^{xy}\mathbf{i} + xe^{xy}\mathbf{j}$
    (a) $\mathbf{r}_1(t) = t\mathbf{i} - (t - 3)\mathbf{j}, \quad 0 \le t \le 3$
    (b) The closed path consisting of line segments from $(0, 3)$ to $(0, 0)$, from $(0, 0)$ to $(3, 0)$, and then from $(3, 0)$ to $(0, 3)$
13. $\mathbf{F}(x, y) = y\mathbf{i} - x\mathbf{j}$
    (a) $\mathbf{r}_1(t) = t\mathbf{i} + t\mathbf{j}, \quad 0 \le t \le 1$
    (b) $\mathbf{r}_2(t) = t\mathbf{i} + t^2\mathbf{j}, \quad 0 \le t \le 1$
    (c) $\mathbf{r}_3(t) = t\mathbf{i} + t^3\mathbf{j}, \quad 0 \le t \le 1$
14. $\mathbf{F}(x, y) = xy^2\mathbf{i} + 2x^2y\mathbf{j}$
    (a) $\mathbf{r}_1(t) = t\mathbf{i} + \dfrac{1}{t}\mathbf{j}, \quad 1 \le t \le 3$
    (b) $\mathbf{r}_2(t) = (t + 1)\mathbf{i} - \frac{1}{3}(t - 3)\mathbf{j}, \quad 0 \le t \le 2$
15. $\displaystyle\int_C y^2\,dx + 2xy\,dy$

(a)

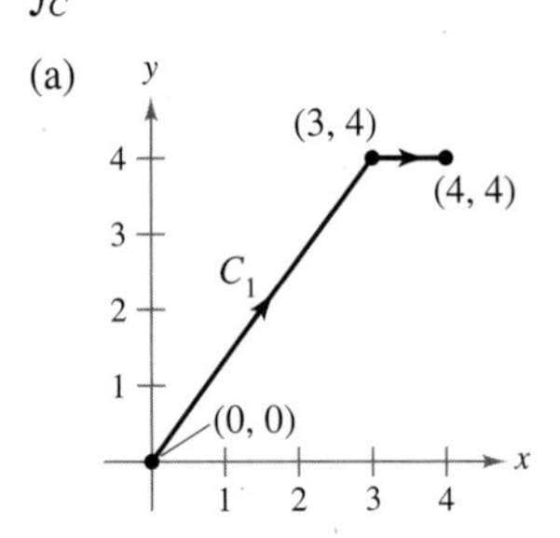

(b)

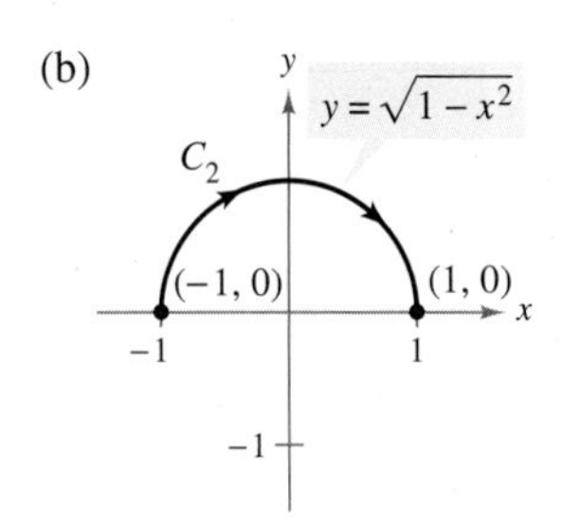

(c)

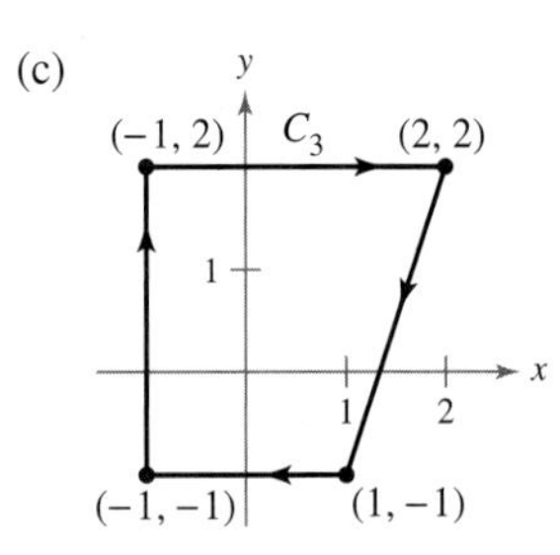

(d)

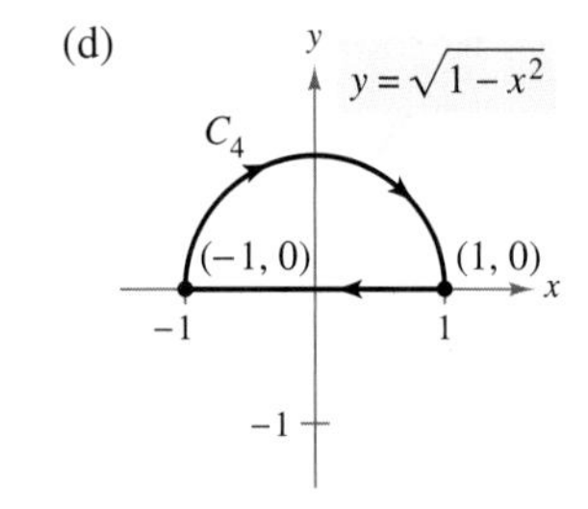

16. $\displaystyle\int_C (2x - 3y + 1)\,dx - (3x + y - 5)\,dy$

(a)

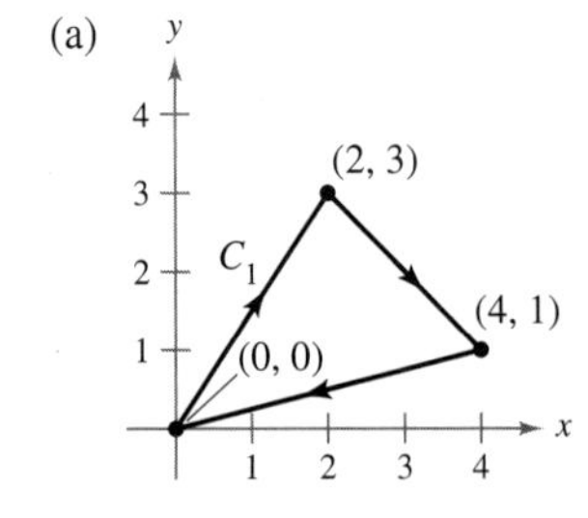

(b)

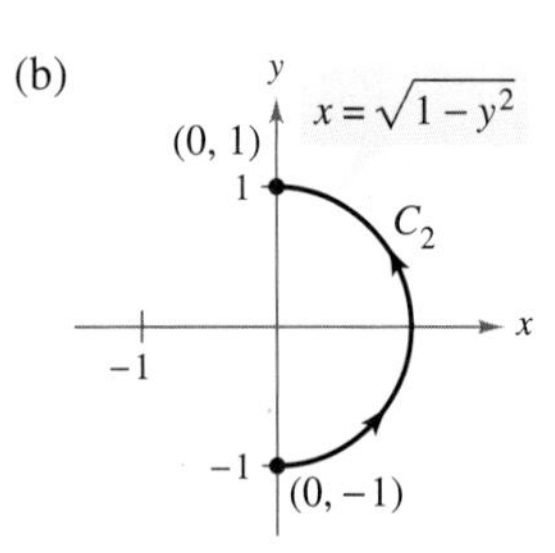

(c)

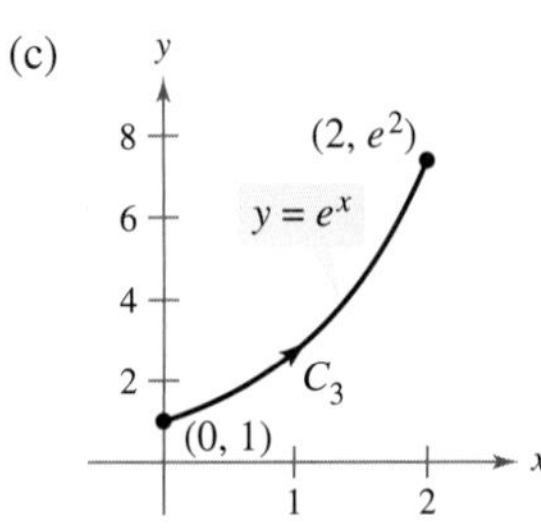

(d)

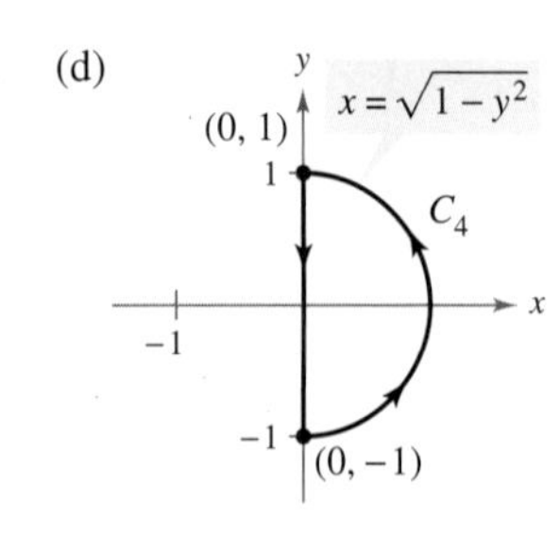

17. $\displaystyle\int_C 2xy\,dx + (x^2 + y^2)\,dy$
    (a) C: ellipse $\dfrac{x^2}{25} + \dfrac{y^2}{16} = 1$ from $(5, 0)$ to $(0, 4)$
    (b) C: parabola $y = 4 - x^2$ from $(2, 0)$ to $(0, 4)$

18. $\int_C (x^2 + y^2)\,dx + 2xy\,dy$

(a) $\mathbf{r}_1(t) = t^3\mathbf{i} + t^2\mathbf{j}, \quad 0 \le t \le 2$

(b) $\mathbf{r}_2(t) = 2\cos t\,\mathbf{i} + 2\sin t\,\mathbf{j}, \quad 0 \le t \le \frac{\pi}{2}$

19. $\mathbf{F}(x, y, z) = yz\mathbf{i} + xz\mathbf{j} + xy\mathbf{k}$

(a) $\mathbf{r}_1(t) = t\mathbf{i} + 2\mathbf{j} + t\mathbf{k}, \quad 0 \le t \le 4$

(b) $\mathbf{r}_2(t) = t^2\mathbf{i} + t\mathbf{j} + t^2\mathbf{k}, \quad 0 \le t \le 2$

20. $\mathbf{F}(x, y, z) = \mathbf{i} + z\mathbf{j} + y\mathbf{k}$

(a) $\mathbf{r}_1(t) = \cos t\,\mathbf{i} + \sin t\,\mathbf{j} + t^2\mathbf{k}, \quad 0 \le t \le \pi$

(b) $\mathbf{r}_2(t) = (1 - 2t)\mathbf{i} + \pi^2 t\mathbf{k}, \quad 0 \le t \le 1$

21. $\mathbf{F}(x, y, z) = (2y + x)\mathbf{i} + (x^2 - z)\mathbf{j} + (2y - 4z)\mathbf{k}$

(a) $\mathbf{r}_1(t) = t\mathbf{i} + t^2\mathbf{j} + \mathbf{k}, \quad 0 \le t \le 1$

(b) $\mathbf{r}_2(t) = t\mathbf{i} + t\mathbf{j} + (2t - 1)^2\mathbf{k}, \quad 0 \le t \le 1$

22. $\mathbf{F}(x, y, z) = -y\mathbf{i} + x\mathbf{j} + 3xz^2\mathbf{k}$

(a) $\mathbf{r}_1(t) = \cos t\,\mathbf{i} + \sin t\,\mathbf{j} + t\mathbf{k}, \quad 0 \le t \le \pi$

(b) $\mathbf{r}_2(t) = (1 - 2t)\mathbf{i} + \pi t\mathbf{k}, \quad 0 \le t \le 1$

23. $\mathbf{F}(x, y, z) = e^z(y\mathbf{i} + x\mathbf{j} + xy\mathbf{k})$

(a) $\mathbf{r}_1(t) = 4\cos t\,\mathbf{i} + 4\sin t\,\mathbf{j} + 3\mathbf{k}, \quad 0 \le t \le \pi$

(b) $\mathbf{r}_2(t) = (4 - 8t)\mathbf{i} + 3\mathbf{k}, \quad 0 \le t \le 1$

24. $\mathbf{F}(x, y, z) = y\sin z\,\mathbf{i} + x\sin z\,\mathbf{j} + xy\cos x\,\mathbf{k}$

(a) $\mathbf{r}_1(t) = t^2\mathbf{i} + t^2\mathbf{j}, \quad 0 \le t \le 2$

(b) $\mathbf{r}_2(t) = 4t\mathbf{i} + 4t\mathbf{j}, \quad 0 \le t \le 1$

In Exercises 25–34, evaluate the line integral using the Fundamental Theorem of Line Integrals. Use a computer algebra system to verify your results.

25. $\int_C (3y\mathbf{i} + 3x\mathbf{j}) \cdot d\mathbf{r}$

C: smooth curve from $(0, 0)$ to $(3, 8)$

26. $\int_C [2(x + y)\mathbf{i} + 2(x + y)\mathbf{j}] \cdot d\mathbf{r}$

C: smooth curve from $(-1, 1)$ to $(3, 2)$

27. $\int_C \cos x \sin y\,dx + \sin x \cos y\,dy$

C: line segment from $(0, -\pi)$ to $\left(\frac{3\pi}{2}, \frac{\pi}{2}\right)$

28. $\int_C \frac{y\,dx - x\,dy}{x^2 + y^2}$

C: line segment from $(1, 1)$ to $(2\sqrt{3}, 2)$

29. $\int_C e^x \sin y\,dx + e^x \cos y\,dy$

C: cycloid $x = \theta - \sin\theta$, $y = 1 - \cos\theta$ from $(0, 0)$ to $(2\pi, 0)$

30. $\int_C \frac{2x}{(x^2 + y^2)^2}\,dx + \frac{2y}{(x^2 + y^2)^2}\,dy$

C: circle $(x - 4)^2 + (y - 5)^2 = 9$ clockwise from $(7, 5)$ to $(1, 5)$

31. $\int_C (z + 2y)\,dx + (2x - z)\,dy + (x - y)\,dz$

(a) C: line segment from $(0, 0, 0)$ to $(1, 1, 1)$

(b) C: line segments from $(0, 0, 0)$ to $(0, 0, 1)$ to $(1, 1, 1)$

(c) C: line segments from $(0, 0, 0)$ to $(1, 0, 0)$ to $(1, 1, 0)$ to $(1, 1, 1)$

32. Repeat Exercise 31 using the integral

$$\int_C zy\,dx + xz\,dy + xy\,dz.$$

33. $\int_C -\sin x\,dx + z\,dy + y\,dz$

C: smooth curve from $(0, 0, 0)$ to $\left(\frac{\pi}{2}, 3, 4\right)$

34. $\int_C 6x\,dx - 4z\,dy - (4y - 20z)\,dz$

C: smooth curve from $(0, 0, 0)$ to $(3, 4, 0)$

Work **In Exercises 35 and 36, find the work done by the force field F in moving an object from *P* to *Q*.**

35. $\mathbf{F}(x, y) = 9x^2y^2\mathbf{i} + (6x^3y - 1)\mathbf{j}$; $P(0, 0)$, $Q(5, 9)$

36. $\mathbf{F}(x, y) = \frac{2x}{y}\mathbf{i} - \frac{x^2}{y^2}\mathbf{j}$; $P(-1, 1)$, $Q(3, 2)$

37. ***Work*** A stone weighing 1 pound is attached to the end of a two-foot string and is whirled horizontally with one end held fixed. It makes 1 revolution per second. Find the work done by the force **F** that keeps the stone moving in a circular path. [*Hint:* Use Force = (mass)(centripetal acceleration).]

38. ***Work*** If $\mathbf{F}(x, y, z) = a_1\mathbf{i} + a_2\mathbf{j} + a_3\mathbf{k}$ is a constant force vector field, show that the work done in moving a particle along any path from P to Q is $W = \mathbf{F} \cdot \overrightarrow{PQ}$.

39. ***Work*** To allow a means of escape for workers in a hazardous job 50 meters above ground level, a slide wire is installed. It runs from their position to a point on the ground 50 meters from the base of the installation where they are located. Show that the work done by the gravitational force field for a 175-pound worker moving the length of the slide wire is the same for each path.

(a) $\mathbf{r}(t) = t\mathbf{i} + (50 - t)\mathbf{j}$

(b) $\mathbf{r}(t) = t\mathbf{i} + \frac{1}{50}(50 - t)^2\mathbf{j}$

40. ***Work*** Can you find a path for the slide wire in Exercise 39 such that the work done by the gravitational force field would differ from the amounts of work done for the two paths given? Explain why or why not.

WRITING ABOUT CONCEPTS

41. State the Fundamental Theorem of Line Integrals.

42. What does it mean that a line integral is independent of path? State the method for determining if a line integral is independent of path.

43. ***Think About It*** Let $\mathbf{F}(x, y) = \dfrac{y}{x^2 + y^2}\mathbf{i} - \dfrac{x}{x^2 + y^2}\mathbf{j}$. Find the value of the line integral

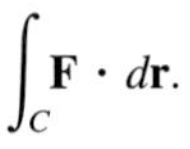

$$\int_C \mathbf{F} \cdot d\mathbf{r}.$$

(a)

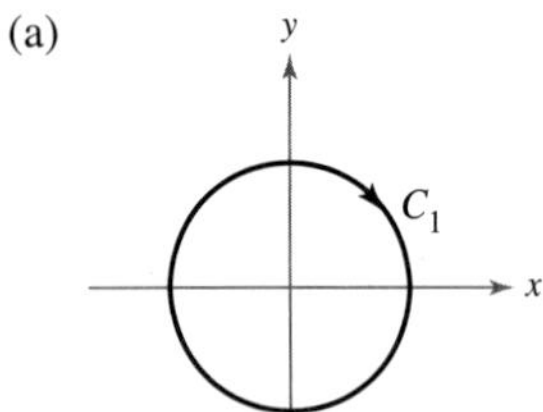

(b)

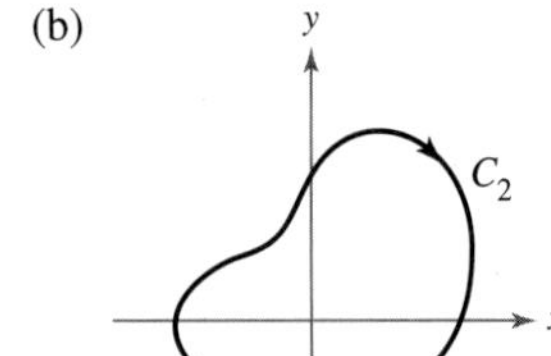

(c)

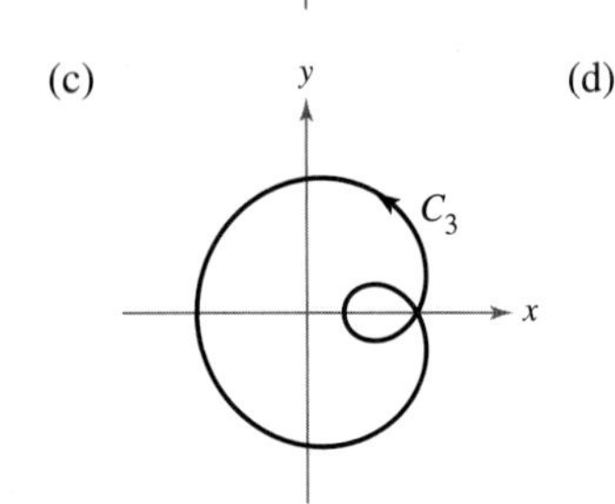

(d)

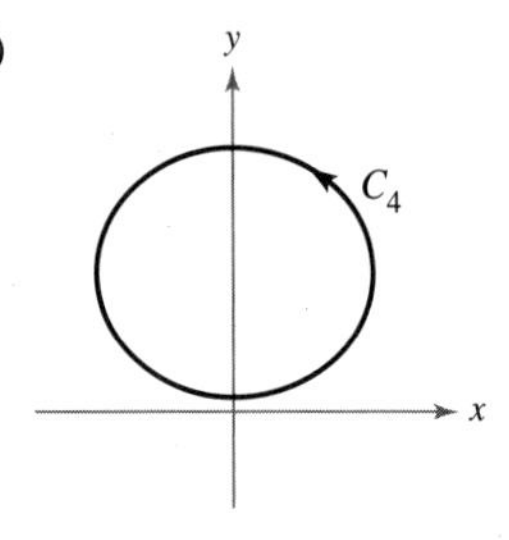

CAPSTONE

44. Consider the force field shown in the figure.

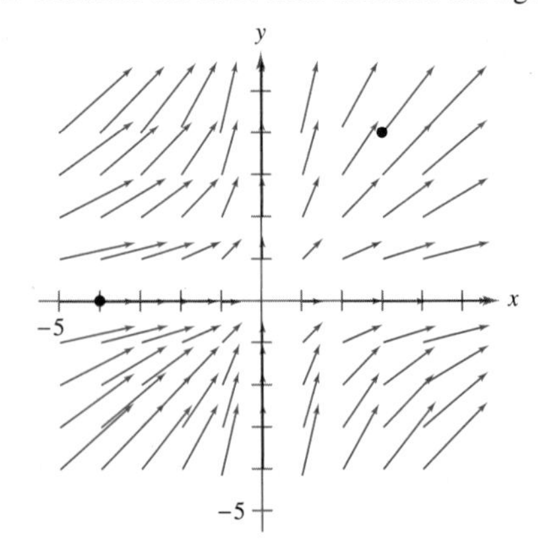

(a) Give a verbal argument that the force field is not conservative because you can identify two paths that require different amounts of work to move an object from $(-4, 0)$ to $(3, 4)$. Identify two paths and state which requires the greater amount of work. To print an enlarged copy of the graph, go to the website *www.mathgraphs.com*.

(b) Give a verbal argument that the force field is not conservative because you can find a closed curve C such that

$$\int_C \mathbf{F} \cdot d\mathbf{r} \neq 0.$$

In Exercises 45 and 46, consider the force field shown in the figure. Is the force field conservative? Explain why or why not.

45.

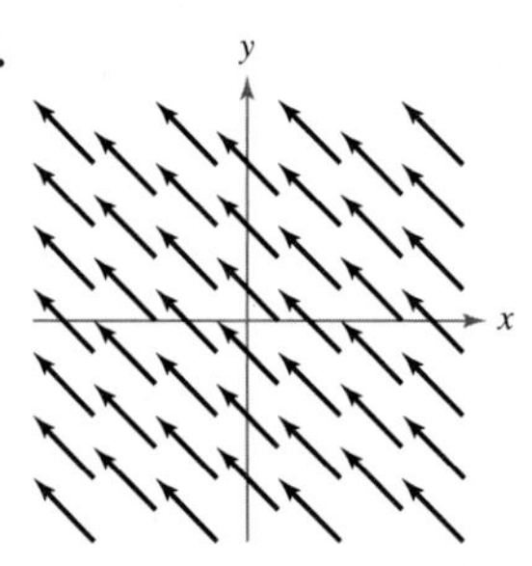

46.

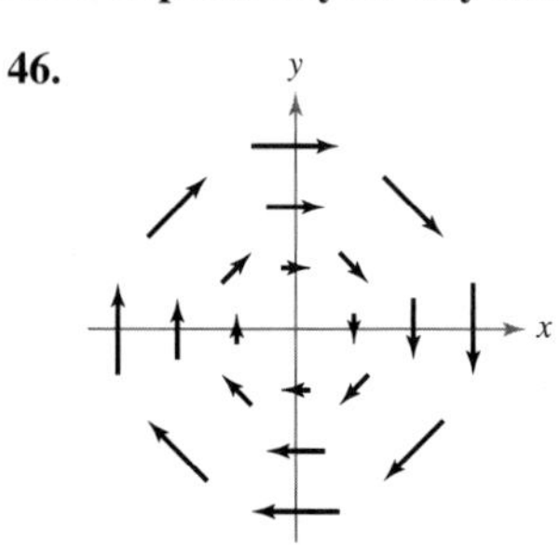

True or False? **In Exercises 47–50, determine whether the statement is true or false. If it is false, explain why or give an example that shows it is false.**

47. If C_1, C_2, and C_3 have the same initial and terminal points and $\int_{C_1} \mathbf{F} \cdot d\mathbf{r}_1 = \int_{C_2} \mathbf{F} \cdot d\mathbf{r}_2$, then $\int_{C_1} \mathbf{F} \cdot d\mathbf{r}_1 = \int_{C_3} \mathbf{F} \cdot d\mathbf{r}_3$.

48. If $\mathbf{F} = y\mathbf{i} + x\mathbf{j}$ and C is given by $\mathbf{r}(t) = (4 \sin t)\mathbf{i} + (3 \cos t)\mathbf{j}$, $0 \le t \le \pi$, then $\int_C \mathbf{F} \cdot d\mathbf{r} = 0$.

49. If $\mathbf{F}$ is conservative in a region R bounded by a simple closed path and C lies within R, then $\int_C \mathbf{F} \cdot d\mathbf{r}$ is independent of path.

50. If $\mathbf{F} = M\mathbf{i} + N\mathbf{j}$ and $\partial M/\partial x = \partial N/\partial y$, then $\mathbf{F}$ is conservative.

51. A function f is called *harmonic* if $\dfrac{\partial^2 f}{\partial x^2} + \dfrac{\partial^2 f}{\partial y^2} = 0$. Prove that if f is harmonic, then

$$\int_C \left(\frac{\partial f}{\partial y}\,dx - \frac{\partial f}{\partial x}\,dy\right) = 0$$

where C is a smooth closed curve in the plane.

52. ***Kinetic and Potential Energy*** The kinetic energy of an object moving through a conservative force field is decreasing at a rate of 15 units per minute. At what rate is the potential energy changing?

53. Let $\mathbf{F}(x, y) = \dfrac{y}{x^2 + y^2}\mathbf{i} - \dfrac{x}{x^2 + y^2}\mathbf{j}$.

(a) Show that

$$\frac{\partial N}{\partial x} = \frac{\partial M}{\partial y}$$

where

$$M = \frac{y}{x^2 + y^2} \text{ and } N = \frac{-x}{x^2 + y^2}.$$

(b) If $\mathbf{r}(t) = \cos t\mathbf{i} + \sin t\mathbf{j}$ for $0 \le t \le \pi$, find $\int_C \mathbf{F} \cdot d\mathbf{r}$.

(c) If $\mathbf{r}(t) = \cos t\mathbf{i} - \sin t\mathbf{j}$ for $0 \le t \le \pi$, find $\int_C \mathbf{F} \cdot d\mathbf{r}$.

(d) If $\mathbf{r}(t) = \cos t\mathbf{i} + \sin t\mathbf{j}$ for $0 \le t \le 2\pi$, find $\int_C \mathbf{F} \cdot d\mathbf{r}$. Why doesn't this contradict Theorem 15.7?

(e) Show that $\nabla\left(\arctan \dfrac{x}{y}\right) = \mathbf{F}$.

15.4 Green's Theorem

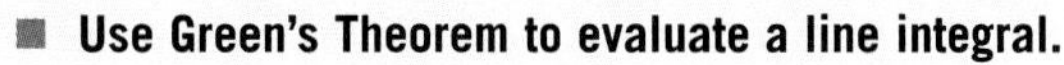

- **Use Green's Theorem to evaluate a line integral.**
- **Use alternative forms of Green's Theorem.**

Green's Theorem

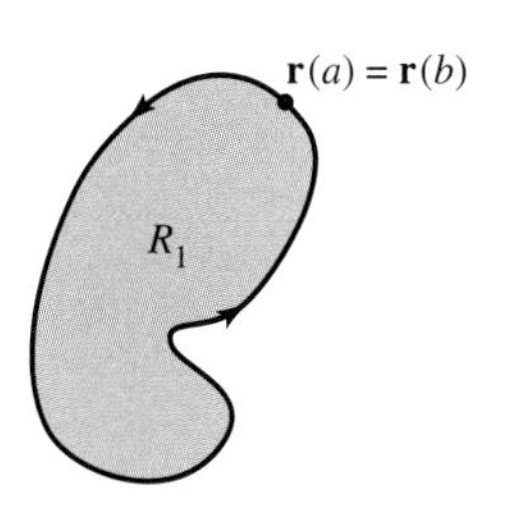

Simply connected

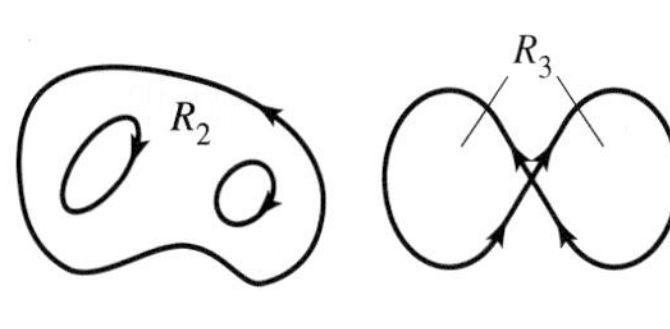

Not simply connected

Figure 15.26

In this section, you will study **Green's Theorem,** named after the English mathematician George Green (1793–1841). This theorem states that the value of a double integral over a *simply connected* plane region R is determined by the value of a line integral around the boundary of R.

A curve C given by $\mathbf{r}(t) = x(t)\mathbf{i} + y(t)\mathbf{j}$, where $a \le t \le b$, is **simple** if it does not cross itself—that is, $\mathbf{r}(c) \ne \mathbf{r}(d)$ for all c and d in the open interval (a, b). A plane region R is **simply connected** if every simple closed curve in R encloses only points that are in R (see Figure 15.26).

> THEOREM 15.8 **GREEN'S THEOREM**
>
> Let R be a simply connected region with a piecewise smooth boundary C, oriented counterclockwise (that is, C is traversed *once* so that the region R always lies to the *left*). If M and N have continuous first partial derivatives in an open region containing R, then
>
> $$\int_C M\,dx + N\,dy = \int_R\!\!\int \left(\frac{\partial N}{\partial x} - \frac{\partial M}{\partial y}\right) dA.$$

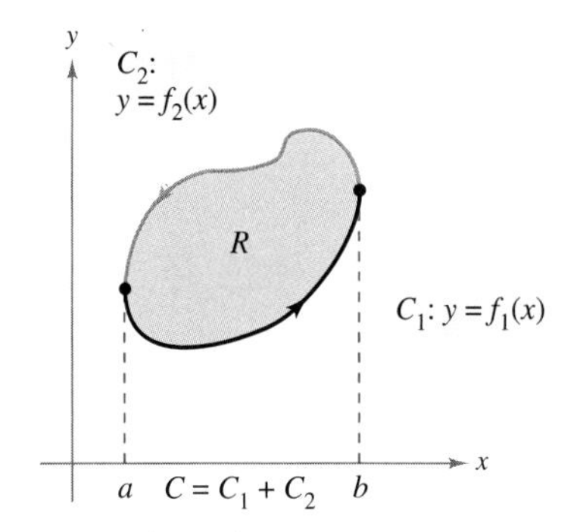

R is vertically simple.

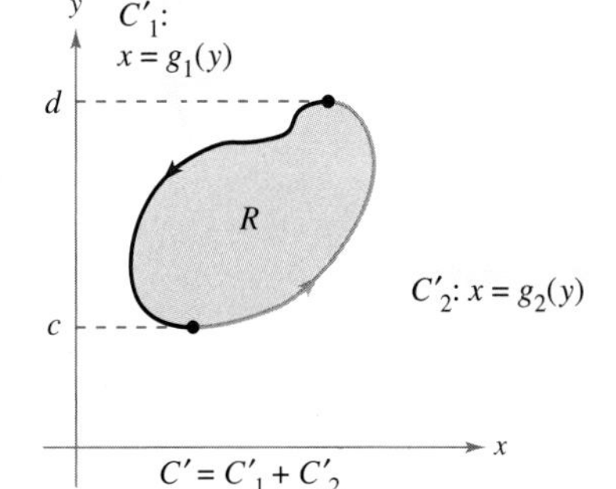

R is horizontally simple.

Figure 15.27

PROOF A proof is given only for a region that is both vertically simple and horizontally simple, as shown in Figure 15.27.

$$\begin{aligned}\int_C M\,dx &= \int_{C_1} M\,dx + \int_{C_2} M\,dx \\ &= \int_a^b M(x, f_1(x))\,dx + \int_b^a M(x, f_2(x))\,dx \\ &= \int_a^b [M(x, f_1(x)) - M(x, f_2(x))]\,dx\end{aligned}$$

On the other hand,

$$\begin{aligned}\int_R\!\!\int \frac{\partial M}{\partial y}\,dA &= \int_a^b \int_{f_1(x)}^{f_2(x)} \frac{\partial M}{\partial y}\,dy\,dx \\ &= \int_a^b M(x, y)\Big]_{f_1(x)}^{f_2(x)}\,dx \\ &= \int_a^b [M(x, f_2(x)) - M(x, f_1(x))]\,dx.\end{aligned}$$

Consequently,

$$\int_C M\,dx = -\int_R\!\!\int \frac{\partial M}{\partial y}\,dA.$$

Similarly, you can use $g_1(y)$ and $g_2(y)$ to show that $\int_C N\,dy = \int_R\!\int \partial N/\partial x\,dA$. By adding the integrals $\int_C M\,dx$ and $\int_C N\,dy$, you obtain the conclusion stated in the theorem. ■

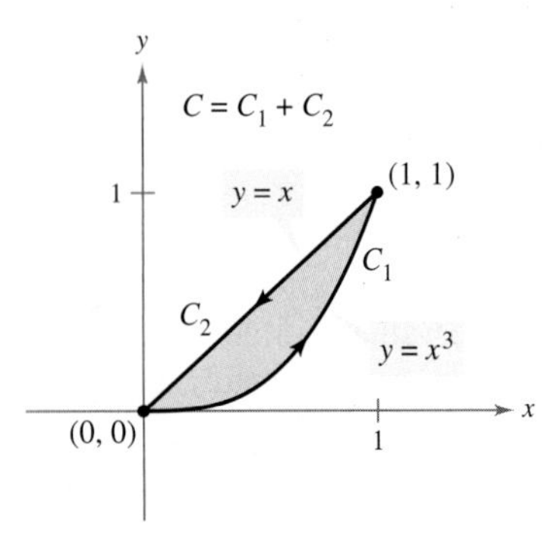

C is simple and closed, and the region R always lies to the left of C.
Figure 15.28

EXAMPLE 1 Using Green's Theorem

Use Green's Theorem to evaluate the line integral

$$\int_C y^3\,dx + (x^3 + 3xy^2)\,dy$$

where C is the path from $(0, 0)$ to $(1, 1)$ along the graph of $y = x^3$ and from $(1, 1)$ to $(0, 0)$ along the graph of $y = x$, as shown in Figure 15.28.

Solution Because $M = y^3$ and $N = x^3 + 3xy^2$, it follows that

$$\frac{\partial N}{\partial x} = 3x^2 + 3y^2 \quad \text{and} \quad \frac{\partial M}{\partial y} = 3y^2.$$

Applying Green's Theorem, you then have

$$\begin{aligned}
\int_C y^3\,dx + (x^3 + 3xy^2)\,dy &= \int_R\int \left(\frac{\partial N}{\partial x} - \frac{\partial M}{\partial y}\right) dA \\
&= \int_0^1 \int_{x^3}^{x} [(3x^2 + 3y^2) - 3y^2]\,dy\,dx \\
&= \int_0^1 \int_{x^3}^{x} 3x^2\,dy\,dx \\
&= \int_0^1 3x^2y\Big]_{x^3}^{x}\,dx \\
&= \int_0^1 (3x^3 - 3x^5)\,dx \\
&= \left[\frac{3x^4}{4} - \frac{x^6}{2}\right]_0^1 \\
&= \frac{1}{4}.
\end{aligned}$$

■

> **GEORGE GREEN (1793–1841)**
>
> Green, a self-educated miller's son, first published the theorem that bears his name in 1828 in an essay on electricity and magnetism. At that time there was almost no mathematical theory to explain electrical phenomena. "Considering how desirable it was that a power of universal agency, like electricity, should, as far as possible, be submitted to calculation, . . . I was induced to try whether it would be possible to discover any general relations existing between this function and the quantities of electricity in the bodies producing it."

Green's Theorem cannot be applied to every line integral. Among other restrictions stated in Theorem 15.8, the curve C must be simple and closed. When Green's Theorem does apply, however, it can save time. To see this, try using the techniques described in Section 15.2 to evaluate the line integral in Example 1. To do this, you would need to write the line integral as

$$\int_C y^3\,dx + (x^3 + 3xy^2)\,dy =$$
$$\int_{C_1} y^3\,dx + (x^3 + 3xy^2)\,dy + \int_{C_2} y^3\,dx + (x^3 + 3xy^2)\,dy$$

where C_1 is the cubic path given by

$$\mathbf{r}(t) = t\mathbf{i} + t^3\mathbf{j}$$

from $t = 0$ to $t = 1$, and C_2 is the line segment given by

$$\mathbf{r}(t) = (1 - t)\mathbf{i} + (1 - t)\mathbf{j}$$

from $t = 0$ to $t = 1$.

EXAMPLE 2 Using Green's Theorem to Calculate Work

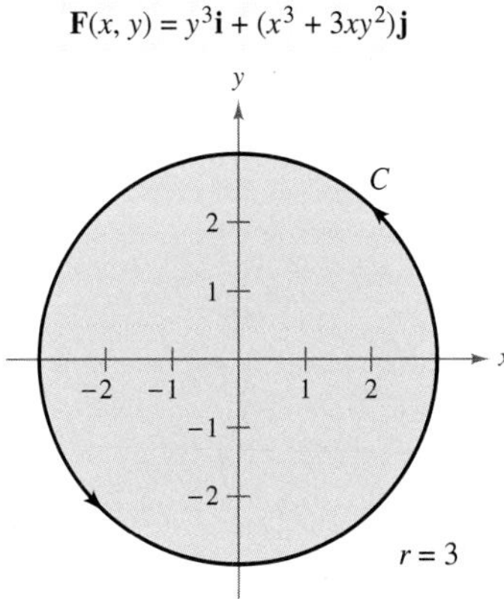

Figure 15.29

While subject to the force

$$\mathbf{F}(x, y) = y^3\mathbf{i} + (x^3 + 3xy^2)\mathbf{j}$$

a particle travels once around the circle of radius 3 shown in Figure 15.29. Use Green's Theorem to find the work done by **F**.

Solution From Example 1, you know by Green's Theorem that

$$\int_C y^3\,dx + (x^3 + 3xy^2)\,dy = \int\!\!\int_R 3x^2\,dA.$$

In polar coordinates, using $x = r\cos\theta$ and $dA = r\,dr\,d\theta$, the work done is

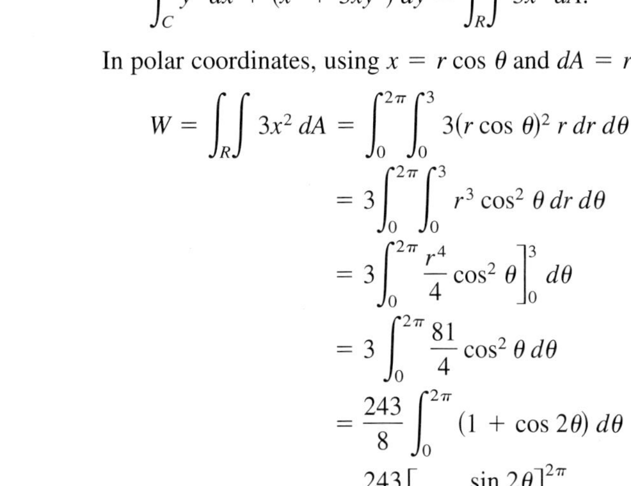

$$\begin{aligned}
W = \int\!\!\int_R 3x^2\,dA &= \int_0^{2\pi}\!\!\int_0^3 3(r\cos\theta)^2\,r\,dr\,d\theta \\
&= 3\int_0^{2\pi}\!\!\int_0^3 r^3\cos^2\theta\,dr\,d\theta \\
&= 3\int_0^{2\pi} \frac{r^4}{4}\cos^2\theta\Big]_0^3\,d\theta \\
&= 3\int_0^{2\pi} \frac{81}{4}\cos^2\theta\,d\theta \\
&= \frac{243}{8}\int_0^{2\pi}(1 + \cos 2\theta)\,d\theta \\
&= \frac{243}{8}\left[\theta + \frac{\sin 2\theta}{2}\right]_0^{2\pi} \\
&= \frac{243\pi}{4}.
\end{aligned}$$

When evaluating line integrals over closed curves, remember that for conservative vector fields (those for which $\partial N/\partial x = \partial M/\partial y$), the value of the line integral is 0. This is easily seen from the statement of Green's Theorem:

$$\int_C M\,dx + N\,dy = \int\!\!\int_R \left(\frac{\partial N}{\partial x} - \frac{\partial M}{\partial y}\right) dA = 0.$$

EXAMPLE 3 Green's Theorem and Conservative Vector Fields

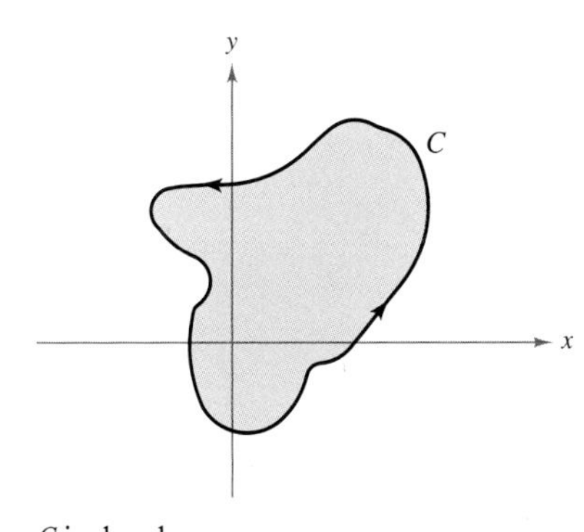

C is closed.
Figure 15.30

Evaluate the line integral

$$\int_C y^3\,dx + 3xy^2\,dy$$

where C is the path shown in Figure 15.30.

Solution From this line integral, $M = y^3$ and $N = 3xy^2$. So, $\partial N/\partial x = 3y^2$ and $\partial M/\partial y = 3y^2$. This implies that the vector field $\mathbf{F} = M\mathbf{i} + N\mathbf{j}$ is conservative, and because C is closed, you can conclude that

$$\int_C y^3\,dx + 3xy^2\,dy = 0.$$

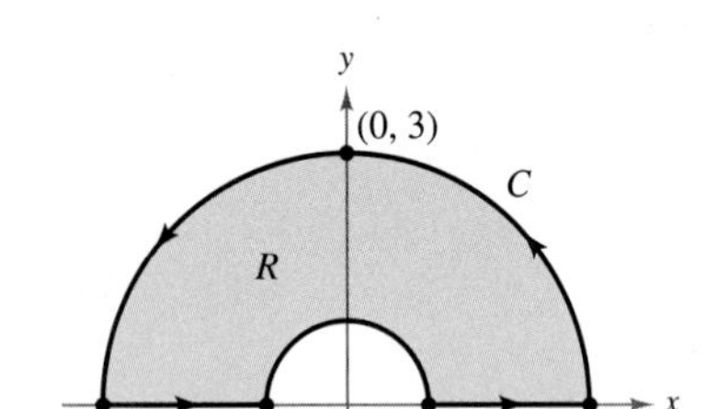

C is piecewise smooth.
Figure 15.31

EXAMPLE 4 Using Green's Theorem for a Piecewise Smooth Curve

Evaluate

$$\int_C (\arctan x + y^2)\, dx + (e^y - x^2)\, dy$$

where C is the path enclosing the annular region shown in Figure 15.31.

Solution In polar coordinates, R is given by $1 \le r \le 3$ for $0 \le \theta \le \pi$. Moreover,

$$\frac{\partial N}{\partial x} - \frac{\partial M}{\partial y} = -2x - 2y = -2(r\cos\theta + r\sin\theta).$$

So, by Green's Theorem,

$$\begin{aligned}\int_C (\arctan x + y^2)\, dx + (e^y - x^2)\, dy &= \int_R\int -2(x + y)\, dA \\ &= \int_0^{\pi}\int_1^3 -2r(\cos\theta + \sin\theta)r\, dr\, d\theta \\ &= \int_0^{\pi} -2(\cos\theta + \sin\theta)\frac{r^3}{3}\Big]_1^3\, d\theta \\ &= \int_0^{\pi}\left(-\frac{52}{3}\right)(\cos\theta + \sin\theta)\, d\theta \\ &= -\frac{52}{3}\Big[\sin\theta - \cos\theta\Big]_0^{\pi} \\ &= -\frac{104}{3}.\end{aligned}$$

In Examples 1, 2, and 4, Green's Theorem was used to evaluate line integrals as double integrals. You can also use the theorem to evaluate double integrals as line integrals. One useful application occurs when $\partial N/\partial x - \partial M/\partial y = 1$.

$$\begin{aligned}\int_C M\, dx + N\, dy &= \int_R\int \left(\frac{\partial N}{\partial x} - \frac{\partial M}{\partial y}\right) dA \\ &= \int_R\int 1\, dA \qquad \frac{\partial N}{\partial x} - \frac{\partial M}{\partial y} = 1 \\ &= \text{area of region } R\end{aligned}$$

Among the many choices for M and N satisfying the stated condition, the choice of $M = -y/2$ and $N = x/2$ produces the following line integral for the area of region R.

THEOREM 15.9 LINE INTEGRAL FOR AREA

If R is a plane region bounded by a piecewise smooth simple closed curve C, oriented counterclockwise, then the area of R is given by

$$A = \frac{1}{2}\int_C x\, dy - y\, dx.$$

EXAMPLE 5 Finding Area by a Line Integral

Use a line integral to find the area of the ellipse

$$\frac{x^2}{a^2} + \frac{y^2}{b^2} = 1.$$

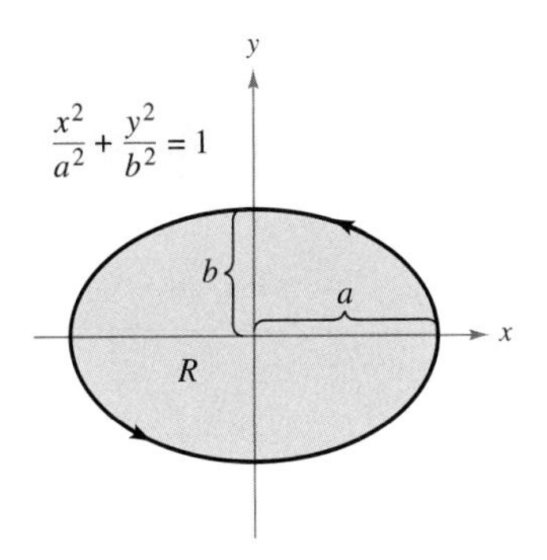

Figure 15.32

Solution Using Figure 15.32, you can induce a counterclockwise orientation to the elliptical path by letting

$$x = a \cos t \quad \text{and} \quad y = b \sin t, \quad 0 \leq t \leq 2\pi.$$

So, the area is

$$\begin{aligned}
A = \frac{1}{2}\int_C x\,dy - y\,dx &= \frac{1}{2}\int_0^{2\pi} [(a\cos t)(b\cos t)\,dt - (b\sin t)(-a\sin t)\,dt] \\
&= \frac{ab}{2}\int_0^{2\pi} (\cos^2 t + \sin^2 t)\,dt \\
&= \frac{ab}{2}\Big[t\Big]_0^{2\pi} \\
&= \pi ab.
\end{aligned}$$

■

Green's Theorem can be extended to cover some regions that are not simply connected. This is demonstrated in the next example.

EXAMPLE 6 Green's Theorem Extended to a Region with a Hole

Let R be the region inside the ellipse $(x^2/9) + (y^2/4) = 1$ and outside the circle $x^2 + y^2 = 1$. Evaluate the line integral

$$\int_C 2xy\,dx + (x^2 + 2x)\,dy$$

where $C = C_1 + C_2$ is the boundary of R, as shown in Figure 15.33.

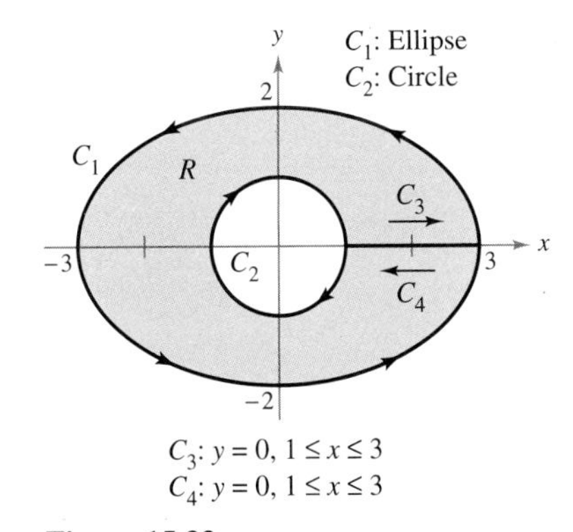

Figure 15.33

Solution To begin, you can introduce the line segments C_3 and C_4, as shown in Figure 15.33. Note that because the curves C_3 and C_4 have opposite orientations, the line integrals over them cancel. Furthermore, you can apply Green's Theorem to the region R using the boundary $C_1 + C_4 + C_2 + C_3$ to obtain

$$\begin{aligned}
\int_C 2xy\,dx + (x^2 + 2x)\,dy &= \int_R\!\!\int \left(\frac{\partial N}{\partial x} - \frac{\partial M}{\partial y}\right) dA \\
&= \int_R\!\!\int (2x + 2 - 2x)\,dA \\
&= 2\int_R\!\!\int dA \\
&= 2(\text{area of } R) \\
&= 2(\pi ab - \pi r^2) \\
&= 2[\pi(3)(2) - \pi(1^2)] \\
&= 10\pi.
\end{aligned}$$

■

In Section 15.1, a necessary and sufficient condition for conservative vector fields was listed. There, only one direction of the proof was shown. You can now outline the other direction, using Green's Theorem. Let $\mathbf{F}(x, y) = M\mathbf{i} + N\mathbf{j}$ be defined on an open disk R. You want to show that if M and N have continuous first partial derivatives and

$$\frac{\partial M}{\partial y} = \frac{\partial N}{\partial x}$$

then $\mathbf{F}$ is conservative. Suppose that C is a closed path forming the boundary of a connected region lying in R. Then, using the fact that $\partial M/\partial y = \partial N/\partial x$, you can apply Green's Theorem to conclude that

$$\begin{aligned}\int_C \mathbf{F} \cdot d\mathbf{r} &= \int_C M\,dx + N\,dy \\ &= \int\!\!\int_R \left(\frac{\partial N}{\partial x} - \frac{\partial M}{\partial y}\right) dA \\ &= 0.\end{aligned}$$

This, in turn, is equivalent to showing that $\mathbf{F}$ is conservative (see Theorem 15.7).

Alternative Forms of Green's Theorem

This section concludes with the derivation of two vector forms of Green's Theorem for regions in the plane. The extension of these vector forms to three dimensions is the basis for the discussion in the remaining sections of this chapter. If $\mathbf{F}$ is a vector field in the plane, you can write

$$\mathbf{F}(x, y, z) = M\mathbf{i} + N\mathbf{j} + 0\mathbf{k}$$

so that the curl of $\mathbf{F}$, as described in Section 15.1, is given by

$$\begin{aligned}\text{curl } \mathbf{F} = \nabla \times \mathbf{F} &= \begin{vmatrix} \mathbf{i} & \mathbf{j} & \mathbf{k} \\ \dfrac{\partial}{\partial x} & \dfrac{\partial}{\partial y} & \dfrac{\partial}{\partial z} \\ M & N & 0 \end{vmatrix} \\ &= -\frac{\partial N}{\partial z}\mathbf{i} + \frac{\partial M}{\partial z}\mathbf{j} + \left(\frac{\partial N}{\partial x} - \frac{\partial M}{\partial y}\right)\mathbf{k}.\end{aligned}$$

Consequently,

$$\begin{aligned}(\text{curl } \mathbf{F}) \cdot \mathbf{k} &= \left[-\frac{\partial N}{\partial z}\mathbf{i} + \frac{\partial M}{\partial z}\mathbf{j} + \left(\frac{\partial N}{\partial x} - \frac{\partial M}{\partial y}\right)\mathbf{k}\right] \cdot \mathbf{k} \\ &= \frac{\partial N}{\partial x} - \frac{\partial M}{\partial y}.\end{aligned}$$

With appropriate conditions on $\mathbf{F}$, C, and R, you can write Green's Theorem in the vector form

$$\begin{aligned}\int_C \mathbf{F} \cdot d\mathbf{r} &= \int\!\!\int_R \left(\frac{\partial N}{\partial x} - \frac{\partial M}{\partial y}\right) dA \\ &= \int\!\!\int_R (\text{curl } \mathbf{F}) \cdot \mathbf{k}\,dA. \qquad \text{First alternative form}\end{aligned}$$

The extension of this vector form of Green's Theorem to surfaces in space produces **Stokes's Theorem,** discussed in Section 15.8.

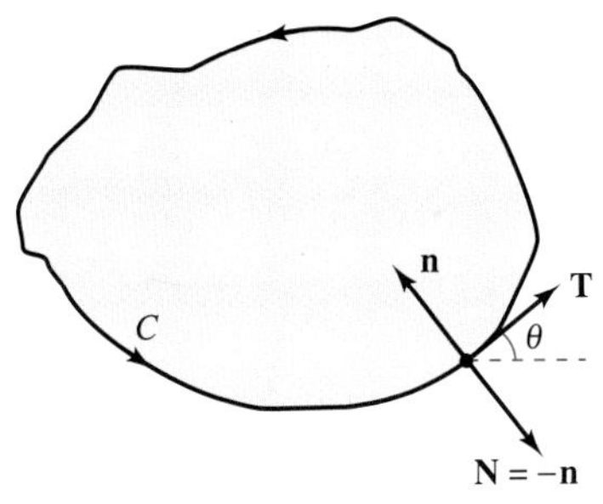

$\mathbf{T} = \cos\theta\mathbf{i} + \sin\theta\mathbf{j}$

$\mathbf{n} = \cos\left(\theta + \frac{\pi}{2}\right)\mathbf{i} + \sin\left(\theta + \frac{\pi}{2}\right)\mathbf{j}$

$\quad = -\sin\theta\mathbf{i} + \cos\theta\mathbf{j}$

$\mathbf{N} = \sin\theta\mathbf{i} - \cos\theta\mathbf{j}$

Figure 15.34

For the second vector form of Green's Theorem, assume the same conditions for $\mathbf{F}$, C, and R. Using the arc length parameter s for C, you have $\mathbf{r}(s) = x(s)\mathbf{i} + y(s)\mathbf{j}$. So, a unit tangent vector $\mathbf{T}$ to curve C is given by $\mathbf{r}'(s) = \mathbf{T} = x'(s)\mathbf{i} + y'(s)\mathbf{j}$. From Figure 15.34 you can see that the *outward* unit normal vector $\mathbf{N}$ can then be written as

$$\mathbf{N} = y'(s)\mathbf{i} - x'(s)\mathbf{j}.$$

Consequently, for $\mathbf{F}(x, y) = M\mathbf{i} + N\mathbf{j}$, you can apply Green's Theorem to obtain

$$\begin{aligned}\int_C \mathbf{F}\cdot\mathbf{N}\,ds &= \int_a^b (M\mathbf{i} + N\mathbf{j})\cdot(y'(s)\mathbf{i} - x'(s)\mathbf{j})\,ds \\ &= \int_a^b \left(M\frac{dy}{ds} - N\frac{dx}{ds}\right)ds \\ &= \int_C M\,dy - N\,dx \\ &= \int_C -N\,dx + M\,dy \\ &= \int_R\int \left(\frac{\partial M}{\partial x} + \frac{\partial N}{\partial y}\right)dA \qquad \text{Green's Theorem} \\ &= \int_R\int \text{div}\,\mathbf{F}\,dA.\end{aligned}$$

Therefore,

$$\int_C \mathbf{F}\cdot\mathbf{N}\,ds = \int_R\int \text{div}\,\mathbf{F}\,dA. \qquad \text{Second alternative form}$$

The extension of this form to three dimensions is called the **Divergence Theorem,** discussed in Section 15.7. The physical interpretations of divergence and curl will be discussed in Sections 15.7 and 15.8.

15.4 Exercises

See www.CalcChat.com for worked-out solutions to odd-numbered exercises.

In Exercises 1–4, verify Green's Theorem by evaluating both integrals

$$\int_C y^2\,dx + x^2\,dy = \int_R\int \left(\frac{\partial N}{\partial x} - \frac{\partial M}{\partial y}\right)dA$$

for the given path.

1. C: boundary of the region lying between the graphs of $y = x$ and $y = x^2$
2. C: boundary of the region lying between the graphs of $y = x$ and $y = \sqrt{x}$
3. C: square with vertices $(0, 0)$, $(1, 0)$, $(1, 1)$, $(0, 1)$
4. C: rectangle with vertices $(0, 0)$, $(3, 0)$, $(3, 4)$, and $(0, 4)$

CAS **In Exercises 5 and 6, verify Green's Theorem by using a computer algebra system to evaluate both integrals**

$$\int_C xe^y\,dx + e^x\,dy = \int_R\int \left(\frac{\partial N}{\partial x} - \frac{\partial M}{\partial y}\right)dA$$

for the given path.

5. C: circle given by $x^2 + y^2 = 4$
6. C: boundary of the region lying between the graphs of $y = x$ and $y = x^3$ in the first quadrant

In Exercises 7–10, use Green's Theorem to evaluate the integral

$$\int_C (y - x)\,dx + (2x - y)\,dy$$

for the given path.

7. C: boundary of the region lying between the graphs of $y = x$ and $y = x^2 - 2x$
8. C: $x = 2\cos\theta$, $y = \sin\theta$
9. C: boundary of the region lying inside the rectangle bounded by $x = -5$, $x = 5$, $y = -3$, and $y = 3$, and outside the square bounded by $x = -1$, $x = 1$, $y = -1$, and $y = 1$
10. C: boundary of the region lying inside the semicircle $y = \sqrt{25 - x^2}$ and outside the semicircle $y = \sqrt{9 - x^2}$

In Exercises 11–20, use Green's Theorem to evaluate the line integral.

11. $\int_C 2xy\,dx + (x + y)\,dy$

C: boundary of the region lying between the graphs of $y = 0$ and $y = 1 - x^2$

12. $\int_C y^2\,dx + xy\,dy$

C: boundary of the region lying between the graphs of $y = 0$, $y = \sqrt{x}$, and $x = 9$

13. $\int_C (x^2 - y^2)\,dx + 2xy\,dy$

C: $x^2 + y^2 = 16$

14. $\int_C (x^2 - y^2)\,dx + 2xy\,dy$

C: $r = 1 + \cos\theta$

15. $\int_C e^x \cos 2y\,dx - 2e^x \sin 2y\,dy$

C: $x^2 + y^2 = a^2$

16. $\int_C 2\arctan\frac{y}{x}\,dx + \ln(x^2 + y^2)\,dy$

C: $x = 4 + 2\cos\theta$, $y = 4 + \sin\theta$

17. $\int_C \cos y\,dx + (xy - x\sin y)\,dy$

C: boundary of the region lying between the graphs of $y = x$ and $y = \sqrt{x}$

18. $\int_C (e^{-x^2/2} - y)\,dx + (e^{-y^2/2} + x)\,dy$

C: boundary of the region lying between the graphs of the circle $x = 6\cos\theta$, $y = 6\sin\theta$ and the ellipse $x = 3\cos\theta$, $y = 2\sin\theta$

19. $\int_C (x - 3y)\,dx + (x + y)\,dy$

C: boundary of the region lying between the graphs of $x^2 + y^2 = 1$ and $x^2 + y^2 = 9$

20. $\int_C 3x^2 e^y\,dx + e^y\,dy$

C: boundary of the region lying between the squares with vertices $(1, 1)$, $(-1, 1)$, $(-1, -1)$, and $(1, -1)$, and $(2, 2)$, $(-2, 2)$, $(-2, -2)$, and $(2, -2)$

Work **In Exercises 21–24, use Green's Theorem to calculate the work done by the force F on a particle that is moving counterclockwise around the closed path *C*.**

21. $\mathbf{F}(x, y) = xy\mathbf{i} + (x + y)\mathbf{j}$

C: $x^2 + y^2 = 1$

22. $\mathbf{F}(x, y) = (e^x - 3y)\mathbf{i} + (e^y + 6x)\mathbf{j}$

C: $r = 2\cos\theta$

23. $\mathbf{F}(x, y) = (x^{3/2} - 3y)\mathbf{i} + (6x + 5\sqrt{y})\mathbf{j}$

C: boundary of the triangle with vertices $(0, 0)$, $(5, 0)$, and $(0, 5)$

24. $\mathbf{F}(x, y) = (3x^2 + y)\mathbf{i} + 4xy^2\,\mathbf{j}$

C: boundary of the region lying between the graphs of $y = \sqrt{x}$, $y = 0$, and $x = 9$

Area **In Exercises 25–28, use a line integral to find the area of the region *R*.**

25. R: region bounded by the graph of $x^2 + y^2 = a^2$

26. R: triangle bounded by the graphs of $x = 0$, $3x - 2y = 0$, and $x + 2y = 8$

27. R: region bounded by the graphs of $y = 5x - 3$ and $y = x^2 + 1$

28. R: region inside the loop of the folium of Descartes bounded by the graph of

$$x = \frac{3t}{t^3 + 1}, \quad y = \frac{3t^2}{t^3 + 1}$$

WRITING ABOUT CONCEPTS

29. State Green's Theorem.

30. Give the line integral for the area of a region R bounded by a piecewise smooth simple curve C.

In Exercises 31 and 32, use Green's Theorem to verify the line integral formulas.

31. The centroid of the region having area A bounded by the simple closed path C is

$$\bar{x} = \frac{1}{2A}\int_C x^2\,dy, \quad \bar{y} = -\frac{1}{2A}\int_C y^2\,dx.$$

32. The area of a plane region bounded by the simple closed path C given in polar coordinates is $A = \frac{1}{2}\int_C r^2\,d\theta$.

CAS ***Centroid*** **In Exercises 33–36, use a computer algebra system and the results of Exercise 31 to find the centroid of the region.**

33. R: region bounded by the graphs of $y = 0$ and $y = 4 - x^2$

34. R: region bounded by the graphs of $y = \sqrt{a^2 - x^2}$ and $y = 0$

35. R: region bounded by the graphs of $y = x^3$ and $y = x$, $0 \le x \le 1$

36. R: triangle with vertices $(-a, 0)$, $(a, 0)$, and (b, c), where $-a \le b \le a$

CAS ***Area*** **In Exercises 37–40, use a computer algebra system and the results of Exercise 32 to find the area of the region bounded by the graph of the polar equation.**

37. $r = a(1 - \cos\theta)$

38. $r = a\cos 3\theta$

39. $r = 1 + 2\cos\theta$ (inner loop)

40. $r = \dfrac{3}{2 - \cos\theta}$

41. (a) Evaluate $\int_{C_1} y^3\,dx + (27x - x^3)\,dy$, where C_1 is the unit circle given by $\mathbf{r}(t) = \cos t\mathbf{i} + \sin t\mathbf{j}$, $0 \le t \le 2\pi$.

(b) Find the maximum value of $\int_C y^3\,dx + (27x - x^3)\,dy$, where C is any closed curve in the xy-plane, oriented counterclockwise.

CAPSTONE

42. For each given path, verify Green's Theorem by showing that

$$\int_C y^2\,dx + x^2\,dy = \int_R\int\left(\frac{\partial N}{\partial x} - \frac{\partial M}{\partial y}\right) dA.$$

For each path, which integral is easier to evaluate? Explain.

(a) C: triangle with vertices $(0, 0), (4, 0), (4, 4)$

(b) C: circle given by $x^2 + y^2 = 1$

43. ***Think About It*** Let

$$I = \int_C \frac{y\,dx - x\,dy}{x^2 + y^2}$$

where C is a circle oriented counterclockwise. Show that $I = 0$ if C does not contain the origin. What is I if C does contain the origin?

44. (a) Let C be the line segment joining (x_1, y_1) and (x_2, y_2). Show that $\int_C -y\,dx + x\,dy = x_1 y_2 - x_2 y_1$.

(b) Let $(x_1, y_1), (x_2, y_2), \ldots, (x_n, y_n)$ be the vertices of a polygon. Prove that the area enclosed is

$$\tfrac{1}{2}[(x_1y_2 - x_2y_1) + (x_2y_3 - x_3y_2) + \cdots + (x_{n-1}y_n - x_ny_{n-1}) + (x_ny_1 - x_1y_n)].$$

Area **In Exercises 45 and 46, use the result of Exercise 44(b) to find the area enclosed by the polygon with the given vertices.**

45. Pentagon: $(0, 0), (2, 0), (3, 2), (1, 4), (-1, 1)$

46. Hexagon: $(0, 0), (2, 0), (3, 2), (2, 4), (0, 3), (-1, 1)$

In Exercises 47 and 48, prove the identity where R is a simply connected region with boundary C. Assume that the required partial derivatives of the scalar functions f and g are continuous. The expressions $D_{\mathbf{N}}f$ and $D_{\mathbf{N}}g$ are the derivatives in the direction of the outward normal vector $\mathbf{N}$ of C, and are defined by $D_{\mathbf{N}}f = \nabla f \cdot \mathbf{N}$, and $D_{\mathbf{N}}g = \nabla g \cdot \mathbf{N}$.

47. Green's first identity:

$$\int_R\int (f\nabla^2 g + \nabla f \cdot \nabla g)\,dA = \int_C f D_{\mathbf{N}}g\,ds$$

[*Hint:* Use the second alternative form of Green's Theorem and the property div $(f\mathbf{G}) = f$ div $\mathbf{G} + \nabla f \cdot \mathbf{G}$.]

48. Green's second identity:

$$\int_R\int (f\nabla^2 g - g\nabla^2 f)\,dA = \int_C (f D_{\mathbf{N}}g - g\,D_{\mathbf{N}}f)\,ds$$

(*Hint:* Use Green's first identity, given in Exercise 47, twice.)

49. Use Green's Theorem to prove that

$$\int_C f(x)\,dx + g(y)\,dy = 0$$

if f and g are differentiable functions and C is a piecewise smooth simple closed path.

50. Let $\mathbf{F} = M\mathbf{i} + N\mathbf{j}$, where M and N have continuous first partial derivatives in a simply connected region R. Prove that if C is simple, smooth, and closed, and $N_x = M_y$, then

$$\int_C \mathbf{F} \cdot d\mathbf{r} = 0.$$

SECTION PROJECT

Hyperbolic and Trigonometric Functions

(a) Sketch the plane curve represented by the vector-valued function $\mathbf{r}(t) = \cosh t\,\mathbf{i} + \sinh t\,\mathbf{j}$ on the interval $0 \le t \le 5$. Show that the rectangular equation corresponding to $\mathbf{r}(t)$ is the hyperbola $x^2 - y^2 = 1$. Verify your sketch by using a graphing utility to graph the hyperbola.

(b) Let $P = (\cosh \phi, \sinh \phi)$ be the point on the hyperbola corresponding to $\mathbf{r}(\phi)$ for $\phi > 0$. Use the formula for area

$$A = \frac{1}{2}\int_C x\,dy - y\,dx$$

to verify that the area of the region shown in the figure is $\frac{1}{2}\phi$.

(c) Show that the area of the indicated region is also given by the integral

$$A = \int_0^{\sinh \phi} \left[\sqrt{1 + y^2} - (\coth \phi)y\right] dy.$$

Confirm your answer in part (b) by numerically approximating this integral for $\phi = 1, 2, 4$, and 10.

(d) Consider the unit circle given by $x^2 + y^2 = 1$. Let θ be the angle formed by the x-axis and the radius to (x, y). The area of the corresponding sector is $\frac{1}{2}\theta$. That is, the trigonometric functions $f(\theta) = \cos\theta$ and $g(\theta) = \sin\theta$ could have been defined as the coordinates of that point $(\cos\theta, \sin\theta)$ on the unit circle that determines a sector of area $\frac{1}{2}\theta$. Write a short paragraph explaining how you could define the hyperbolic functions in a similar manner, using the "unit hyperbola" $x^2 - y^2 = 1$.

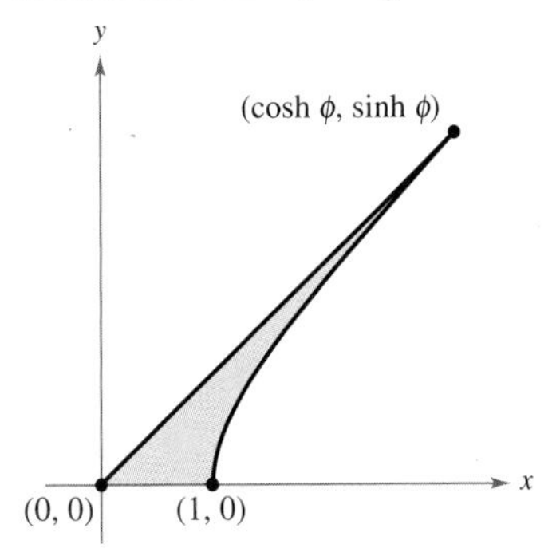

15.5 Parametric Surfaces

- **Understand the definition of a parametric surface, and sketch the surface.**
- **Find a set of parametric equations to represent a surface.**
- **Find a normal vector and a tangent plane to a parametric surface.**
- **Find the area of a parametric surface.**

Parametric Surfaces

You already know how to represent a curve in the plane or in space by a set of parametric equations—or, equivalently, by a vector-valued function.

$$\mathbf{r}(t) = x(t)\mathbf{i} + y(t)\mathbf{j} \qquad \text{Plane curve}$$

$$\mathbf{r}(t) = x(t)\mathbf{i} + y(t)\mathbf{j} + z(t)\mathbf{k} \qquad \text{Space curve}$$

In this section, you will learn how to represent a surface in space by a set of parametric equations—or by a vector-valued function. For curves, note that the vector-valued function $\mathbf{r}$ is a function of a *single* parameter t. For surfaces, the vector-valued function is a function of *two* parameters u and v.

DEFINITION OF PARAMETRIC SURFACE

Let x, y, and z be functions of u and v that are continuous on a domain D in the uv-plane. The set of points (x, y, z) given by

$$\mathbf{r}(u, v) = x(u, v)\mathbf{i} + y(u, v)\mathbf{j} + z(u, v)\mathbf{k} \qquad \text{Parametric surface}$$

is called a **parametric surface.** The equations

$$x = x(u, v), \quad y = y(u, v), \quad \text{and} \quad z = z(u, v) \qquad \text{Parametric equations}$$

are the **parametric equations** for the surface.

If S is a parametric surface given by the vector-valued function $\mathbf{r}$, then S is traced out by the position vector $\mathbf{r}(u, v)$ as the point (u, v) moves throughout the domain D, as shown in Figure 15.35.

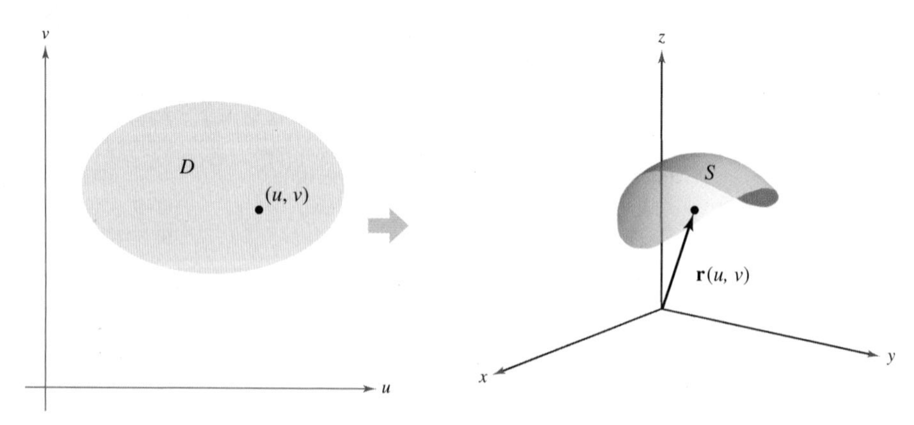

Figure 15.35

TECHNOLOGY Some computer algebra systems are capable of graphing surfaces that are represented parametrically. If you have access to such software, use it to graph some of the surfaces in the examples and exercises in this section.

EXAMPLE 1 Sketching a Parametric Surface

Identify and sketch the parametric surface S given by

$$\mathbf{r}(u, v) = 3 \cos u\mathbf{i} + 3 \sin u\mathbf{j} + v\mathbf{k}$$

where $0 \leq u \leq 2\pi$ and $0 \leq v \leq 4$.

Solution Because $x = 3 \cos u$ and $y = 3 \sin u$, you know that for each point (x, y, z) on the surface, x and y are related by the equation $x^2 + y^2 = 3^2$. In other words, each cross section of S taken parallel to the xy-plane is a circle of radius 3, centered on the z-axis. Because $z = v$, where $0 \leq v \leq 4$, you can see that the surface is a right circular cylinder of height 4. The radius of the cylinder is 3, and the z-axis forms the axis of the cylinder, as shown in Figure 15.36. ■

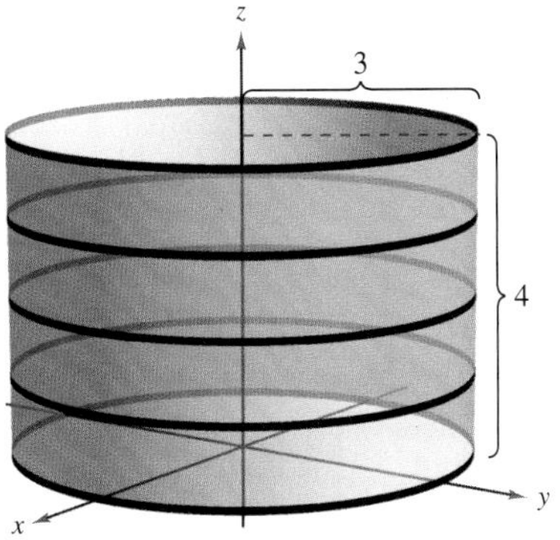

Figure 15.36

As with parametric representations of curves, parametric representations of surfaces are not unique. That is, there are many other sets of parametric equations that could be used to represent the surface shown in Figure 15.36.

EXAMPLE 2 Sketching a Parametric Surface

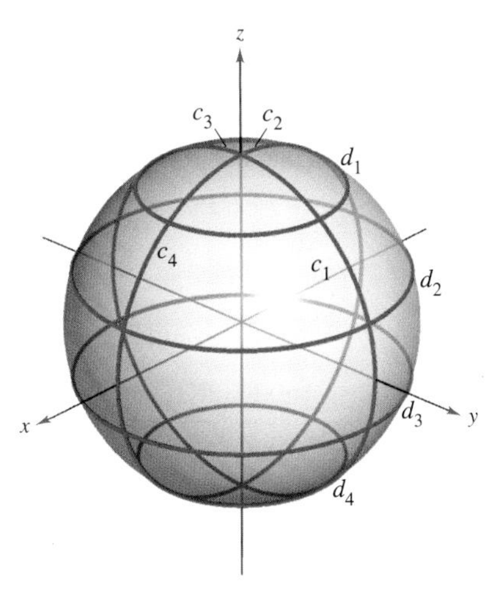

Figure 15.37

Identify and sketch the parametric surface S given by

$$\mathbf{r}(u, v) = \sin u \cos v\mathbf{i} + \sin u \sin v\mathbf{j} + \cos u\mathbf{k}$$

where $0 \leq u \leq \pi$ and $0 \leq v \leq 2\pi$.

Solution To identify the surface, you can try to use trigonometric identities to eliminate the parameters. After some experimentation, you can discover that

$$\begin{aligned} x^2 + y^2 + z^2 &= (\sin u \cos v)^2 + (\sin u \sin v)^2 + (\cos u)^2 \\ &= \sin^2 u \cos^2 v + \sin^2 u \sin^2 v + \cos^2 u \\ &= \sin^2 u(\cos^2 v + \sin^2 v) + \cos^2 u \\ &= \sin^2 u + \cos^2 u \\ &= 1. \end{aligned}$$

So, each point on S lies on the unit sphere, centered at the origin, as shown in Figure 15.37. For fixed $u = d_i$, $\mathbf{r}(u, v)$ traces out latitude circles

$$x^2 + y^2 = \sin^2 d_i, \quad 0 \leq d_i \leq \pi$$

that are parallel to the xy-plane, and for fixed $v = c_i$, $\mathbf{r}(u, v)$ traces out longitude (or meridian) half-circles. ■

NOTE To convince yourself further that the vector-valued function in Example 2 traces out the entire unit sphere, recall that the parametric equations

$$x = \rho \sin \phi \cos \theta, \quad y = \rho \sin \phi \sin \theta, \quad \text{and} \quad z = \rho \cos \phi$$

where $0 \leq \theta \leq 2\pi$ and $0 \leq \phi \leq \pi$, describe the conversion from spherical to rectangular coordinates, as discussed in Section 11.7. ■

Finding Parametric Equations for Surfaces

In Examples 1 and 2, you were asked to identify the surface described by a given set of parametric equations. The reverse problem—that of writing a set of parametric equations for a given surface—is generally more difficult. One type of surface for which this problem is straightforward, however, is a surface that is given by $z = f(x, y)$. You can parametrize such a surface as

$$\mathbf{r}(x, y) = x\mathbf{i} + y\mathbf{j} + f(x, y)\mathbf{k}.$$

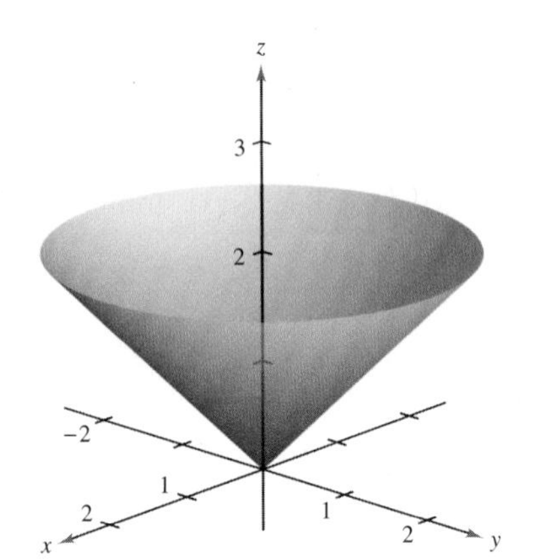

Figure 15.38

EXAMPLE 3 Representing a Surface Parametrically

Write a set of parametric equations for the cone given by

$$z = \sqrt{x^2 + y^2}$$

as shown in Figure 15.38.

Solution Because this surface is given in the form $z = f(x, y)$, you can let x and y be the parameters. Then the cone is represented by the vector-valued function

$$\mathbf{r}(x, y) = x\mathbf{i} + y\mathbf{j} + \sqrt{x^2 + y^2}\,\mathbf{k}$$

where (x, y) varies over the entire xy-plane. ■

A second type of surface that is easily represented parametrically is a surface of revolution. For instance, to represent the surface formed by revolving the graph of $y = f(x)$, $a \le x \le b$, about the x-axis, use

$$x = u, \quad y = f(u)\cos v, \quad \text{and} \quad z = f(u)\sin v$$

where $a \le u \le b$ and $0 \le v \le 2\pi$.

EXAMPLE 4 Representing a Surface of Revolution Parametrically

Write a set of parametric equations for the surface of revolution obtained by revolving

$$f(x) = \frac{1}{x}, \quad 1 \le x \le 10$$

about the x-axis.

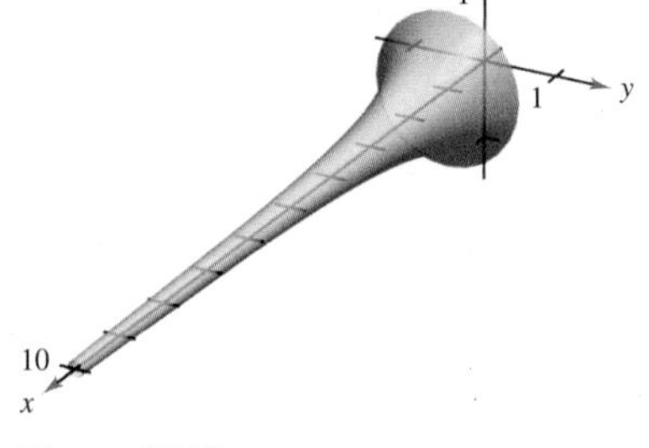

Figure 15.39

Solution Use the parameters u and v as described above to write

$$x = u, \quad y = f(u)\cos v = \frac{1}{u}\cos v, \quad \text{and} \quad z = f(u)\sin v = \frac{1}{u}\sin v$$

where $1 \le u \le 10$ and $0 \le v \le 2\pi$. The resulting surface is a portion of *Gabriel's Horn*, as shown in Figure 15.39. ■

The surface of revolution in Example 4 is formed by revolving the graph of $y = f(x)$ about the x-axis. For other types of surfaces of revolution, a similar parametrization can be used. For instance, to parametrize the surface formed by revolving the graph of $x = f(z)$ about the z-axis, you can use

$$z = u, \quad x = f(u)\cos v, \quad \text{and} \quad y = f(u)\sin v.$$

Normal Vectors and Tangent Planes

Let S be a parametric surface given by

$$\mathbf{r}(u, v) = x(u, v)\mathbf{i} + y(u, v)\mathbf{j} + z(u, v)\mathbf{k}$$

over an open region D such that x, y, and z have continuous partial derivatives on D. The **partial derivatives of r** with respect to u and v are defined as

$$\mathbf{r}_u = \frac{\partial x}{\partial u}(u, v)\mathbf{i} + \frac{\partial y}{\partial u}(u, v)\mathbf{j} + \frac{\partial z}{\partial u}(u, v)\mathbf{k}$$

and

$$\mathbf{r}_v = \frac{\partial x}{\partial v}(u, v)\mathbf{i} + \frac{\partial y}{\partial v}(u, v)\mathbf{j} + \frac{\partial z}{\partial v}(u, v)\mathbf{k}.$$

Each of these partial derivatives is a vector-valued function that can be interpreted geometrically in terms of tangent vectors. For instance, if $v = v_0$ is held constant, then $\mathbf{r}(u, v_0)$ is a vector-valued function of a single parameter and defines a curve C_1 that lies on the surface S. The tangent vector to C_1 at the point $(x(u_0, v_0), y(u_0, v_0), z(u_0, v_0))$ is given by

$$\mathbf{r}_u(u_0, v_0) = \frac{\partial x}{\partial u}(u_0, v_0)\mathbf{i} + \frac{\partial y}{\partial u}(u_0, v_0)\mathbf{j} + \frac{\partial z}{\partial u}(u_0, v_0)\mathbf{k}$$

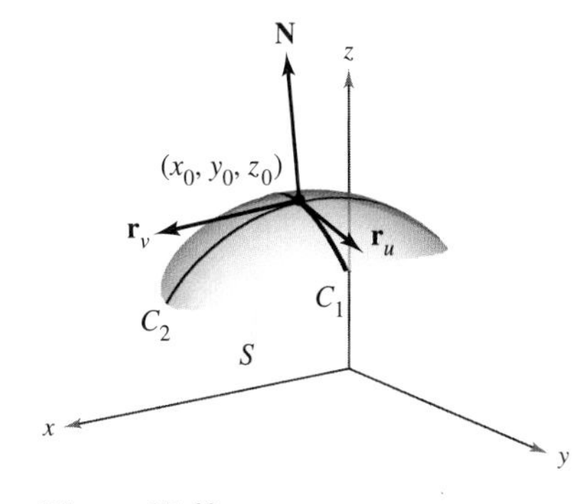

Figure 15.40

as shown in Figure 15.40. In a similar way, if $u = u_0$ is held constant, then $\mathbf{r}(u_0, v)$ is a vector-valued function of a single parameter and defines a curve C_2 that lies on the surface S. The tangent vector to C_2 at the point $(x(u_0, v_0), y(u_0, v_0), z(u_0, v_0))$ is given by

$$\mathbf{r}_v(u_0, v_0) = \frac{\partial x}{\partial v}(u_0, v_0)\mathbf{i} + \frac{\partial y}{\partial v}(u_0, v_0)\mathbf{j} + \frac{\partial z}{\partial v}(u_0, v_0)\mathbf{k}.$$

If the normal vector $\mathbf{r}_u \times \mathbf{r}_v$ is not $\mathbf{0}$ for any (u, v) in D, the surface S is called **smooth** and will have a tangent plane. Informally, a smooth surface is one that has no sharp points or cusps. For instance, spheres, ellipsoids, and paraboloids are smooth, whereas the cone given in Example 3 is not smooth.

NORMAL VECTOR TO A SMOOTH PARAMETRIC SURFACE

Let S be a smooth parametric surface

$$\mathbf{r}(u, v) = x(u, v)\mathbf{i} + y(u, v)\mathbf{j} + z(u, v)\mathbf{k}$$

defined over an open region D in the uv-plane. Let (u_0, v_0) be a point in D. A normal vector at the point

$$(x_0, y_0, z_0) = (x(u_0, v_0), y(u_0, v_0), z(u_0, v_0))$$

is given by

$$\mathbf{N} = \mathbf{r}_u(u_0, v_0) \times \mathbf{r}_v(u_0, v_0) = \begin{vmatrix} \mathbf{i} & \mathbf{j} & \mathbf{k} \\ \dfrac{\partial x}{\partial u} & \dfrac{\partial y}{\partial u} & \dfrac{\partial z}{\partial u} \\ \dfrac{\partial x}{\partial v} & \dfrac{\partial y}{\partial v} & \dfrac{\partial z}{\partial v} \end{vmatrix}.$$

NOTE Figure 15.40 shows the normal vector $\mathbf{r}_u \times \mathbf{r}_v$. The vector $\mathbf{r}_v \times \mathbf{r}_u$ is also normal to S and points in the opposite direction. ■

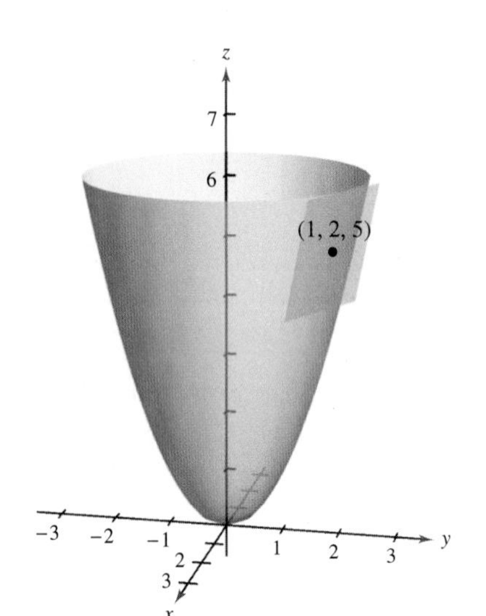

Figure 15.41

EXAMPLE 5 Finding a Tangent Plane to a Parametric Surface

Find an equation of the tangent plane to the paraboloid given by

$$\mathbf{r}(u, v) = u\mathbf{i} + v\mathbf{j} + (u^2 + v^2)\mathbf{k}$$

at the point $(1, 2, 5)$.

Solution The point in the uv-plane that is mapped to the point $(x, y, z) = (1, 2, 5)$ is $(u, v) = (1, 2)$. The partial derivatives of $\mathbf{r}$ are

$$\mathbf{r}_u = \mathbf{i} + 2u\mathbf{k} \quad \text{and} \quad \mathbf{r}_v = \mathbf{j} + 2v\mathbf{k}.$$

The normal vector is given by

$$\mathbf{r}_u \times \mathbf{r}_v = \begin{vmatrix} \mathbf{i} & \mathbf{j} & \mathbf{k} \\ 1 & 0 & 2u \\ 0 & 1 & 2v \end{vmatrix} = -2u\mathbf{i} - 2v\mathbf{j} + \mathbf{k}$$

which implies that the normal vector at $(1, 2, 5)$ is $\mathbf{r}_u \times \mathbf{r}_v = -2\mathbf{i} - 4\mathbf{j} + \mathbf{k}$. So, an equation of the tangent plane at $(1, 2, 5)$ is

$$-2(x - 1) - 4(y - 2) + (z - 5) = 0$$

$$-2x - 4y + z = -5.$$

The tangent plane is shown in Figure 15.41. ■

Area of a Parametric Surface

To define the area of a parametric surface, you can use a development that is similar to that given in Section 14.5. Begin by constructing an inner partition of D consisting of n rectangles, where the area of the ith rectangle D_i is $\Delta A_i = \Delta u_i \, \Delta v_i$, as shown in Figure 15.42. In each D_i let (u_i, v_i) be the point that is closest to the origin. At the point $(x_i, y_i, z_i) = (x(u_i, v_i), y(u_i, v_i), z(u_i, v_i))$ on the surface S, construct a tangent plane T_i. The area of the portion of S that corresponds to D_i, ΔT_i, can be approximated by a parallelogram in the tangent plane. That is, $\Delta T_i \approx \Delta S_i$. So, the surface of S is given by $\Sigma \, \Delta S_i \approx \Sigma \, \Delta T_i$. The area of the parallelogram in the tangent plane is

$$\|\Delta u_i \mathbf{r}_u \times \Delta v_i \mathbf{r}_v\| = \|\mathbf{r}_u \times \mathbf{r}_v\| \, \Delta u_i \, \Delta v_i$$

which leads to the following definition.

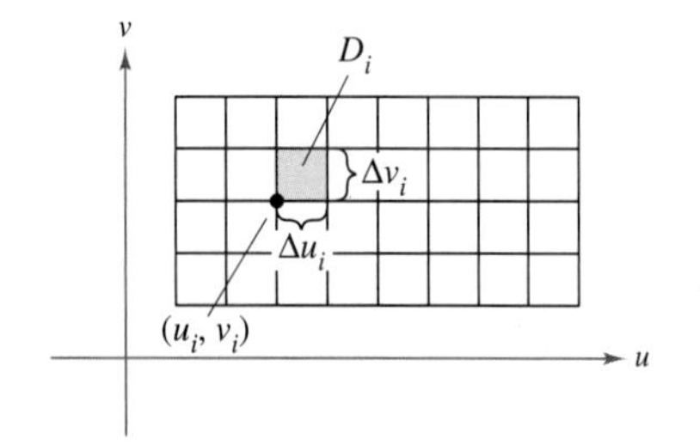

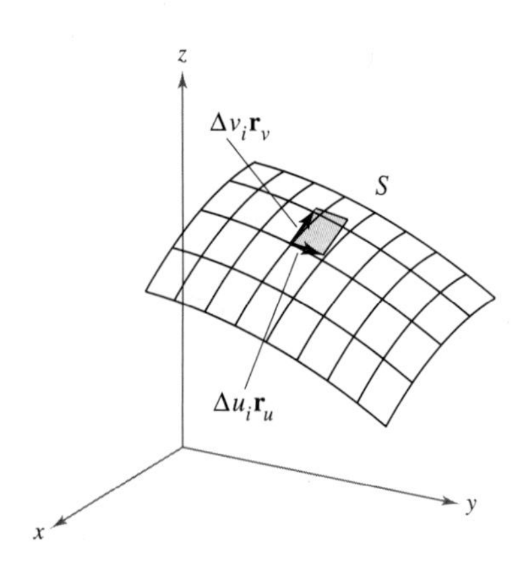

Figure 15.42

AREA OF A PARAMETRIC SURFACE

Let S be a smooth parametric surface

$$\mathbf{r}(u, v) = x(u, v)\mathbf{i} + y(u, v)\mathbf{j} + z(u, v)\mathbf{k}$$

defined over an open region D in the uv-plane. If each point on the surface S corresponds to exactly one point in the domain D, then the **surface area** of S is given by

$$\text{Surface area} = \int\!\!\int_S dS = \int\!\!\int_D \|\mathbf{r}_u \times \mathbf{r}_v\| \, dA$$

where $\mathbf{r}_u = \dfrac{\partial x}{\partial u}\mathbf{i} + \dfrac{\partial y}{\partial u}\mathbf{j} + \dfrac{\partial z}{\partial u}\mathbf{k}$ and $\mathbf{r}_v = \dfrac{\partial x}{\partial v}\mathbf{i} + \dfrac{\partial y}{\partial v}\mathbf{j} + \dfrac{\partial z}{\partial v}\mathbf{k}$.

For a surface S given by $z = f(x, y)$, this formula for surface area corresponds to that given in Section 14.5. To see this, you can parametrize the surface using the vector-valued function

$$\mathbf{r}(x, y) = x\mathbf{i} + y\mathbf{j} + f(x, y)\mathbf{k}$$

defined over the region R in the xy-plane. Using

$$\mathbf{r}_x = \mathbf{i} + f_x(x, y)\mathbf{k} \quad \text{and} \quad \mathbf{r}_y = \mathbf{j} + f_y(x, y)\mathbf{k}$$

you have

$$\mathbf{r}_x \times \mathbf{r}_y = \begin{vmatrix} \mathbf{i} & \mathbf{j} & \mathbf{k} \\ 1 & 0 & f_x(x, y) \\ 0 & 1 & f_y(x, y) \end{vmatrix} = -f_x(x, y)\mathbf{i} - f_y(x, y)\mathbf{j} + \mathbf{k}$$

and $\|\mathbf{r}_x \times \mathbf{r}_y\| = \sqrt{[f_x(x, y)]^2 + [f_y(x, y)]^2 + 1}$. This implies that the surface area of S is

$$\begin{aligned} \text{Surface area} &= \iint_R \|\mathbf{r}_x \times \mathbf{r}_y\| \, dA \\ &= \iint_R \sqrt{1 + [f_x(x, y)]^2 + [f_y(x, y)]^2} \, dA. \end{aligned}$$

EXAMPLE 6 Finding Surface Area

NOTE The surface in Example 6 does not quite fulfill the hypothesis that each point on the surface corresponds to exactly one point in D. For this surface, $\mathbf{r}(u, 0) = \mathbf{r}(u, 2\pi)$ for any fixed value of u. However, because the overlap consists of only a semicircle (which has no area), you can still apply the formula for the area of a parametric surface.

Find the surface area of the unit sphere given by

$$\mathbf{r}(u, v) = \sin u \cos v\mathbf{i} + \sin u \sin v\mathbf{j} + \cos u\mathbf{k}$$

where the domain D is given by $0 \le u \le \pi$ and $0 \le v \le 2\pi$.

Solution Begin by calculating $\mathbf{r}_u$ and $\mathbf{r}_v$.

$$\begin{aligned} \mathbf{r}_u &= \cos u \cos v\mathbf{i} + \cos u \sin v\mathbf{j} - \sin u\mathbf{k} \\ \mathbf{r}_v &= -\sin u \sin v\mathbf{i} + \sin u \cos v\mathbf{j} \end{aligned}$$

The cross product of these two vectors is

$$\begin{aligned} \mathbf{r}_u \times \mathbf{r}_v &= \begin{vmatrix} \mathbf{i} & \mathbf{j} & \mathbf{k} \\ \cos u \cos v & \cos u \sin v & -\sin u \\ -\sin u \sin v & \sin u \cos v & 0 \end{vmatrix} \\ &= \sin^2 u \cos v\mathbf{i} + \sin^2 u \sin v\mathbf{j} + \sin u \cos u\mathbf{k} \end{aligned}$$

which implies that

$$\begin{aligned} \|\mathbf{r}_u \times \mathbf{r}_v\| &= \sqrt{(\sin^2 u \cos v)^2 + (\sin^2 u \sin v)^2 + (\sin u \cos u)^2} \\ &= \sqrt{\sin^4 u + \sin^2 u \cos^2 u} \\ &= \sqrt{\sin^2 u} \\ &= \sin u. \qquad \sin u > 0 \text{ for } 0 \le u \le \pi \end{aligned}$$

Finally, the surface area of the sphere is

$$\begin{aligned} A = \iint_D \|\mathbf{r}_u \times \mathbf{r}_v\| \, dA &= \int_0^{2\pi} \int_0^{\pi} \sin u \, du \, dv \\ &= \int_0^{2\pi} 2 \, dv \\ &= 4\pi. \end{aligned}$$

■

EXPLORATION

For the torus in Example 7, describe the function $\mathbf{r}(u, v)$ for fixed u. Then describe the function $\mathbf{r}(u, v)$ for fixed v.

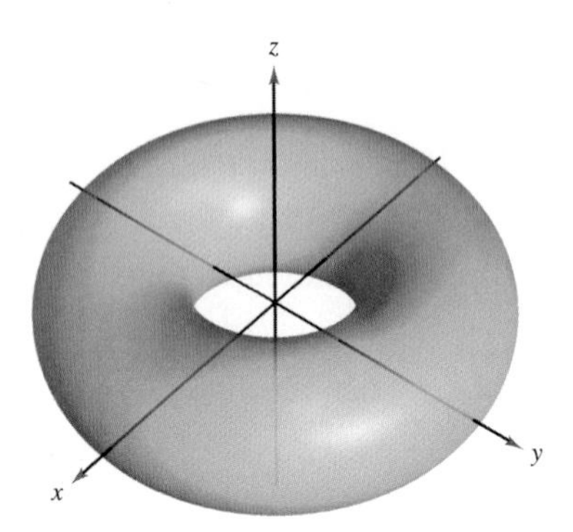

Figure 15.43

EXAMPLE 7 Finding Surface Area

Find the surface area of the torus given by

$$\mathbf{r}(u, v) = (2 + \cos u)\cos v\mathbf{i} + (2 + \cos u)\sin v\mathbf{j} + \sin u\mathbf{k}$$

where the domain D is given by $0 \le u \le 2\pi$ and $0 \le v \le 2\pi$. (See Figure 15.43.)

Solution Begin by calculating $\mathbf{r}_u$ and $\mathbf{r}_v$.

$$\mathbf{r}_u = -\sin u\cos v\mathbf{i} - \sin u\sin v\mathbf{j} + \cos u\mathbf{k}$$
$$\mathbf{r}_v = -(2 + \cos u)\sin v\mathbf{i} + (2 + \cos u)\cos v\mathbf{j}$$

The cross product of these two vectors is

$$\begin{aligned}\mathbf{r}_u \times \mathbf{r}_v &= \begin{vmatrix} \mathbf{i} & \mathbf{j} & \mathbf{k} \\ -\sin u\cos v & -\sin u\sin v & \cos u \\ -(2+\cos u)\sin v & (2+\cos u)\cos v & 0 \end{vmatrix} \\ &= -(2+\cos u)(\cos v\cos u\mathbf{i} + \sin v\cos u\mathbf{j} + \sin u\mathbf{k})\end{aligned}$$

which implies that

$$\begin{aligned}\|\mathbf{r}_u \times \mathbf{r}_v\| &= (2+\cos u)\sqrt{(\cos v\cos u)^2 + (\sin v\cos u)^2 + \sin^2 u} \\ &= (2+\cos u)\sqrt{\cos^2 u(\cos^2 v + \sin^2 v) + \sin^2 u} \\ &= (2+\cos u)\sqrt{\cos^2 u + \sin^2 u} \\ &= 2 + \cos u.\end{aligned}$$

Finally, the surface area of the torus is

$$\begin{aligned}A = \iint_D \|\mathbf{r}_u \times \mathbf{r}_v\|\, dA &= \int_0^{2\pi}\int_0^{2\pi} (2+\cos u)\, du\, dv \\ &= \int_0^{2\pi} 4\pi\, dv \\ &= 8\pi^2.\end{aligned}$$

■

If the surface S is a surface of revolution, you can show that the formula for surface area given in Section 7.4 is equivalent to the formula given in this section. For instance, suppose f is a nonnegative function such that f' is continuous over the interval $[a, b]$. Let S be the surface of revolution formed by revolving the graph of f, where $a \le x \le b$, about the x-axis. From Section 7.4, you know that the surface area is given by

$$\text{Surface area} = 2\pi\int_a^b f(x)\sqrt{1 + [f'(x)]^2}\, dx.$$

To represent S parametrically, let $x = u$, $y = f(u)\cos v$, and $z = f(u)\sin v$, where $a \le u \le b$ and $0 \le v \le 2\pi$. Then,

$$\mathbf{r}(u, v) = u\mathbf{i} + f(u)\cos v\mathbf{j} + f(u)\sin v\mathbf{k}.$$

Try showing that the formula

$$\text{Surface area} = \iint_D \|\mathbf{r}_u \times \mathbf{r}_v\|\, dA$$

is equivalent to the formula given above (see Exercise 58).

15.5 Exercises

See www.CalcChat.com for worked-out solutions to odd-numbered exercises.

In Exercises 1–6, match the vector-valued function with its graph. [The graphs are labeled (a), (b), (c), (d), (e), and (f).]

(a)

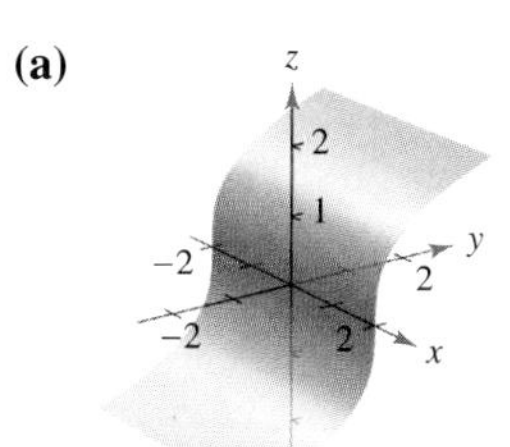

(b)

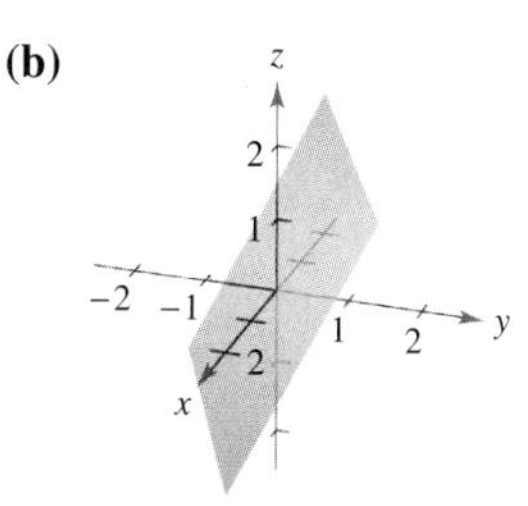

(c)

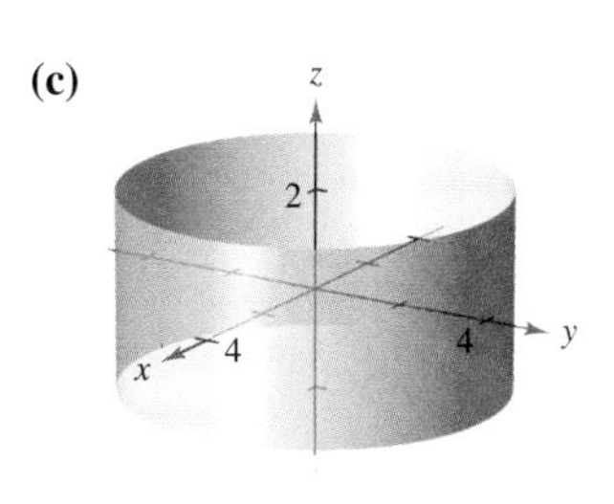

(d)

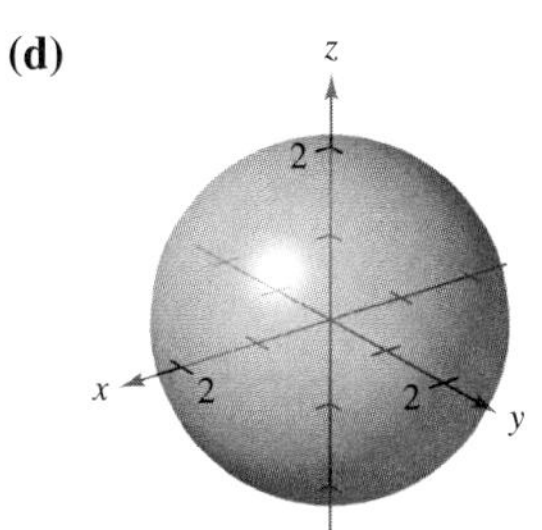

(e)

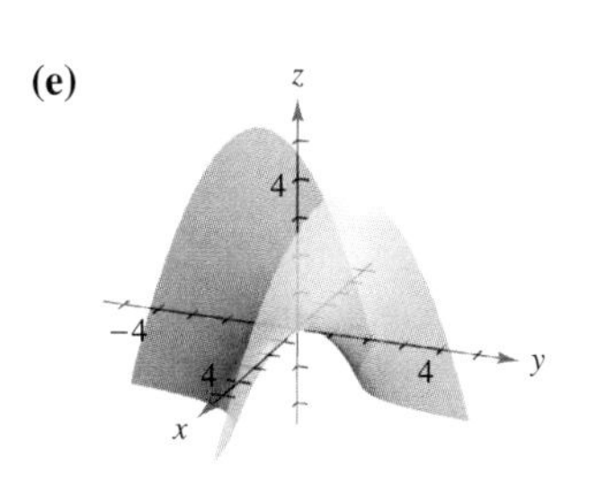

(f)

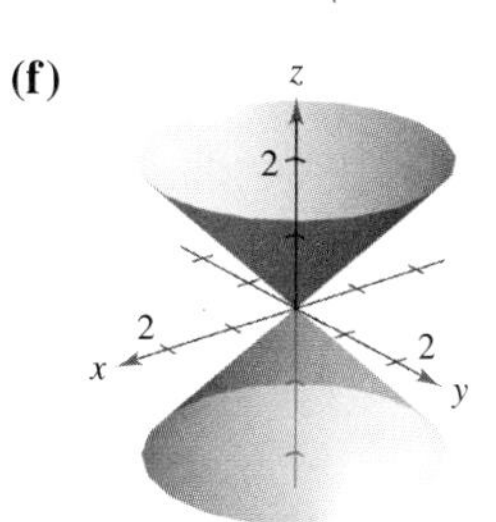

1. $\mathbf{r}(u, v) = u\mathbf{i} + v\mathbf{j} + uv\mathbf{k}$
2. $\mathbf{r}(u, v) = u \cos v\mathbf{i} + u \sin v\mathbf{j} + u\mathbf{k}$
3. $\mathbf{r}(u, v) = u\mathbf{i} + \frac{1}{2}(u + v)\mathbf{j} + v\mathbf{k}$
4. $\mathbf{r}(u, v) = u\mathbf{i} + \frac{1}{4}v^3\mathbf{j} + v\mathbf{k}$
5. $\mathbf{r}(u, v) = 2 \cos v \cos u\mathbf{i} + 2 \cos v \sin u\mathbf{j} + 2 \sin v\mathbf{k}$
6. $\mathbf{r}(u, v) = 4 \cos u\mathbf{i} + 4 \sin u\mathbf{j} + v\mathbf{k}$

In Exercises 7–10, find the rectangular equation for the surface by eliminating the parameters from the vector-valued function. Identify the surface and sketch its graph.

7. $\mathbf{r}(u, v) = u\mathbf{i} + v\mathbf{j} + \dfrac{v}{2}\mathbf{k}$
8. $\mathbf{r}(u, v) = 2u \cos v\mathbf{i} + 2u \sin v\mathbf{j} + \frac{1}{2}u^2\mathbf{k}$
9. $\mathbf{r}(u, v) = 2 \cos u\mathbf{i} + v\mathbf{j} + 2 \sin u\mathbf{k}$
10. $\mathbf{r}(u, v) = 3 \cos v \cos u\mathbf{i} + 3 \cos v \sin u\mathbf{j} + 5 \sin v\mathbf{k}$

CAS **In Exercises 11–16, use a computer algebra system to graph the surface represented by the vector-valued function.**

11. $\mathbf{r}(u, v) = 2u \cos v\mathbf{i} + 2u \sin v\mathbf{j} + u^4\mathbf{k}$
    $0 \le u \le 1, \quad 0 \le v \le 2\pi$
12. $\mathbf{r}(u, v) = 2 \cos v \cos u\mathbf{i} + 4 \cos v \sin u\mathbf{j} + \sin v\mathbf{k}$
    $0 \le u \le 2\pi, \quad 0 \le v \le 2\pi$
13. $\mathbf{r}(u, v) = 2 \sinh u \cos v\mathbf{i} + \sinh u \sin v\mathbf{j} + \cosh u\mathbf{k}$
    $0 \le u \le 2, \quad 0 \le v \le 2\pi$
14. $\mathbf{r}(u, v) = 2u \cos v\mathbf{i} + 2u \sin v\mathbf{j} + v\mathbf{k}$
    $0 \le u \le 1, \quad 0 \le v \le 3\pi$
15. $\mathbf{r}(u, v) = (u - \sin u)\cos v\mathbf{i} + (1 - \cos u)\sin v\mathbf{j} + u\mathbf{k}$
    $0 \le u \le \pi, \quad 0 \le v \le 2\pi$
16. $\mathbf{r}(u, v) = \cos^3 u \cos v\mathbf{i} + \sin^3 u \sin v\mathbf{j} + u\mathbf{k}$
    $0 \le u \le \dfrac{\pi}{2}, \quad 0 \le v \le 2\pi$

Think About It **In Exercises 17–20, determine how the graph of the surface $\mathbf{s}(u, v)$ differs from the graph of $\mathbf{r}(u, v) = u \cos v\mathbf{i} + u \sin v\mathbf{j} + u^2\mathbf{k}$ (see figure), where $0 \le u \le 2$ and $0 \le v \le 2\pi$. (It is not necessary to graph s.)**

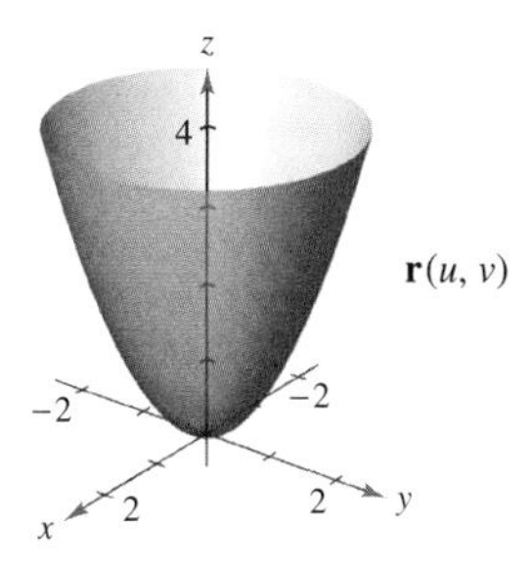

17. $\mathbf{s}(u, v) = u \cos v\mathbf{i} + u \sin v\mathbf{j} - u^2\mathbf{k}$
    $0 \le u \le 2, \quad 0 \le v \le 2\pi$
18. $\mathbf{s}(u, v) = u \cos v\mathbf{i} + u^2\mathbf{j} + u \sin v\mathbf{k}$
    $0 \le u \le 2, \quad 0 \le v \le 2\pi$
19. $\mathbf{s}(u, v) = u \cos v\mathbf{i} + u \sin v\mathbf{j} + u^2\mathbf{k}$
    $0 \le u \le 3, \quad 0 \le v \le 2\pi$
20. $\mathbf{s}(u, v) = 4u \cos v\mathbf{i} + 4u \sin v\mathbf{j} + u^2\mathbf{k}$
    $0 \le u \le 2, \quad 0 \le v \le 2\pi$

In Exercises 21–30, find a vector-valued function whose graph is the indicated surface.

21. The plane $z = y$
22. The plane $x + y + z = 6$
23. The cone $y = \sqrt{4x^2 + 9z^2}$
24. The cone $x = \sqrt{16y^2 + z^2}$
25. The cylinder $x^2 + y^2 = 25$
26. The cylinder $4x^2 + y^2 = 16$
27. The cylinder $z = x^2$
28. The ellipsoid $\dfrac{x^2}{9} + \dfrac{y^2}{4} + \dfrac{z^2}{1} = 1$
29. The part of the plane $z = 4$ that lies inside the cylinder $x^2 + y^2 = 9$
30. The part of the paraboloid $z = x^2 + y^2$ that lies inside the cylinder $x^2 + y^2 = 9$

Surface of Revolution **In Exercises 31–34, write a set of parametric equations for the surface of revolution obtained by revolving the graph of the function about the given axis.**

	Function	*Axis of Revolution*
31.	$y = \frac{x}{2}, \quad 0 \le x \le 6$	x-axis
32.	$y = \sqrt{x}, \quad 0 \le x \le 4$	x-axis
33.	$x = \sin z, \quad 0 \le z \le \pi$	z-axis
34.	$z = y^2 + 1, \quad 0 \le y \le 2$	y-axis

Tangent Plane **In Exercises 35–38, find an equation of the tangent plane to the surface represented by the vector-valued function at the given point.**

35. $\mathbf{r}(u, v) = (u + v)\mathbf{i} + (u - v)\mathbf{j} + v\mathbf{k}, \quad (1, -1, 1)$

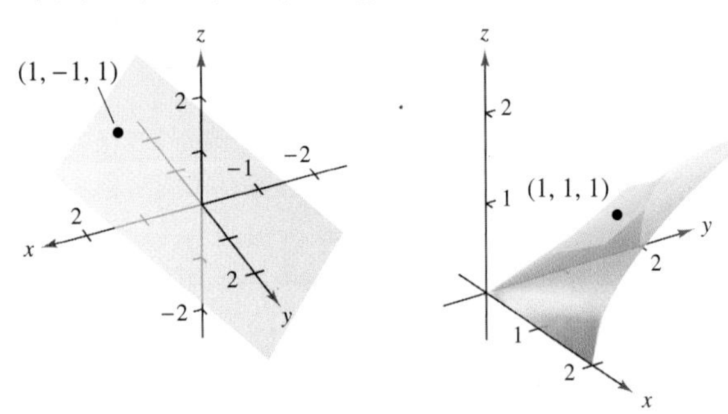

Figure for 35 **Figure for 36**

36. $\mathbf{r}(u, v) = u\mathbf{i} + v\mathbf{j} + \sqrt{uv}\,\mathbf{k}, \quad (1, 1, 1)$

37. $\mathbf{r}(u, v) = 2u \cos v\mathbf{i} + 3u \sin v\mathbf{j} + u^2\mathbf{k}, \quad (0, 6, 4)$

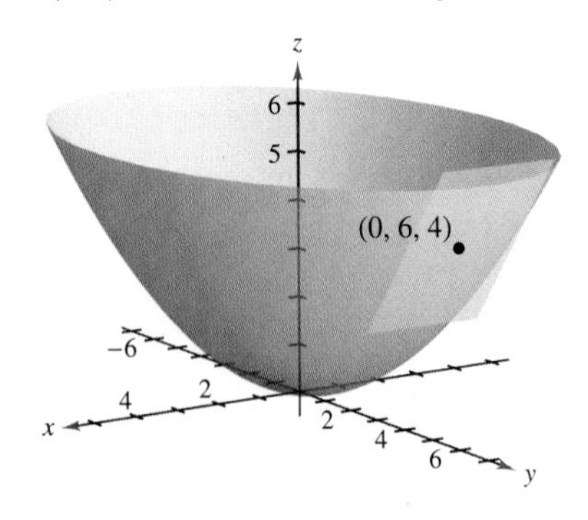

38. $\mathbf{r}(u, v) = 2u \cosh v\mathbf{i} + 2u \sinh v\mathbf{j} + \frac{1}{2}u^2\mathbf{k}, \quad (-4, 0, 2)$

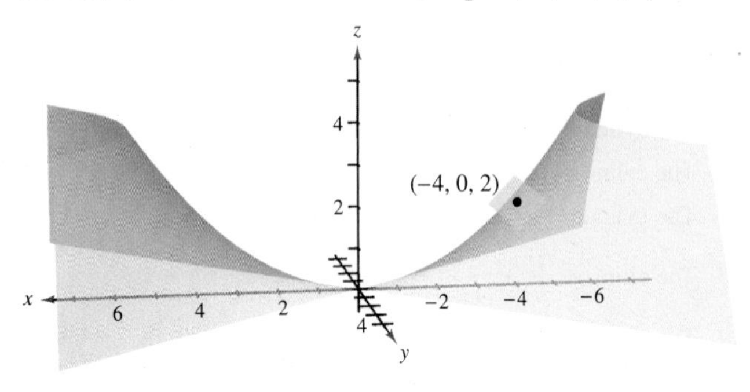

Area **In Exercises 39–46, find the area of the surface over the given region. Use a computer algebra system to verify your results.**

39. The part of the plane $\mathbf{r}(u, v) = 4u\mathbf{i} - v\mathbf{j} + v\mathbf{k}$, where $0 \le u \le 2$ and $0 \le v \le 1$

40. The part of the paraboloid $\mathbf{r}(u, v) = 2u \cos v\mathbf{i} + 2u \sin v\mathbf{j} + u^2\mathbf{k}$, where $0 \le u \le 2$ and $0 \le v \le 2\pi$

41. The part of the cylinder $\mathbf{r}(u, v) = a \cos u\mathbf{i} + a \sin u\mathbf{j} + v\mathbf{k}$, where $0 \le u \le 2\pi$ and $0 \le v \le b$

42. The sphere $\mathbf{r}(u, v) = a \sin u \cos v\mathbf{i} + a \sin u \sin v\mathbf{j} + a \cos u\mathbf{k}$, where $0 \le u \le \pi$ and $0 \le v \le 2\pi$

43. The part of the cone $\mathbf{r}(u, v) = au \cos v\mathbf{i} + au \sin v\mathbf{j} + u\mathbf{k}$, where $0 \le u \le b$ and $0 \le v \le 2\pi$

44. The torus $\mathbf{r}(u, v) = (a + b \cos v)\cos u\mathbf{i} + (a + b \cos v)\sin u\mathbf{j} + b \sin v\mathbf{k}$, where $a > b$, $0 \le u \le 2\pi$, and $0 \le v \le 2\pi$

45. The surface of revolution $\mathbf{r}(u, v) = \sqrt{u} \cos v\mathbf{i} + \sqrt{u} \sin v\mathbf{j} + u\mathbf{k}$, where $0 \le u \le 4$ and $0 \le v \le 2\pi$

46. The surface of revolution $\mathbf{r}(u, v) = \sin u \cos v\mathbf{i} + u\mathbf{j} + \sin u \sin v\mathbf{k}$, where $0 \le u \le \pi$ and $0 \le v \le 2\pi$

WRITING ABOUT CONCEPTS

47. Define a parametric surface.

48. Give the double integral that yields the surface area of a parametric surface over an open region D.

49. Show that the cone in Example 3 can be represented parametrically by $\mathbf{r}(u, v) = u \cos v\mathbf{i} + u \sin v\mathbf{j} + u\mathbf{k}$, where $0 \le u$ and $0 \le v \le 2\pi$.

CAPSTONE

50. The four figures below are graphs of the surface $\mathbf{r}(u, v) = u\mathbf{i} + \sin u \cos v\mathbf{j} + \sin u \sin v\mathbf{k}$,

$0 \le u \le \pi/2, \quad 0 \le v \le 2\pi.$

Match each of the four graphs with the point in space from which the surface is viewed. The four points are $(10, 0, 0)$, $(-10, 10, 0)$, $(0, 10, 0)$, and $(10, 10, 10)$.

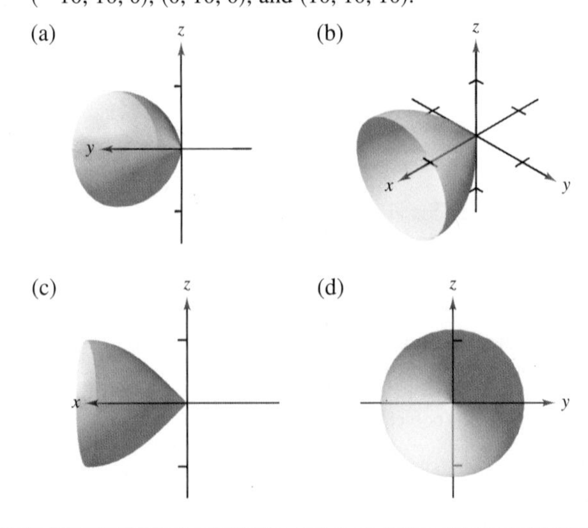

51. ***Astroidal Sphere*** An equation of an **astroidal sphere** in x, y, and z is

$x^{2/3} + y^{2/3} + z^{2/3} = a^{2/3}$.

A graph of an astroidal sphere is shown below. Show that this surface can be represented parametrically by

$\mathbf{r}(u, v) = a \sin^3 u \cos^3 v\mathbf{i} + a \sin^3 u \sin^3 v\mathbf{j} + a \cos^3 u\mathbf{k}$

where $0 \le u \le \pi$ and $0 \le v \le 2\pi$.

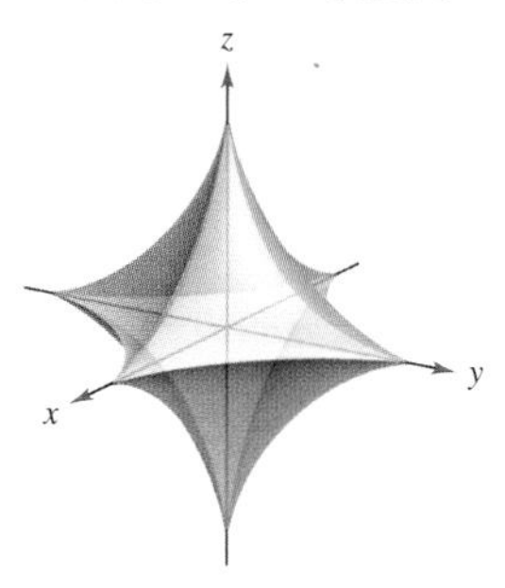

CAS **52.** Use a computer algebra system to graph three views of the graph of the vector-valued function

$$\mathbf{r}(u, v) = u \cos v\mathbf{i} + u \sin v\mathbf{j} + v\mathbf{k}, \quad 0 \le u \le \pi, \quad 0 \le v \le \pi$$

from the points $(10, 0, 0)$, $(0, 0, 10)$, and $(10, 10, 10)$.

CAS **53.** ***Investigation*** Use a computer algebra system to graph the torus

$$\mathbf{r}(u, v) = (a + b \cos v) \cos u\mathbf{i} + (a + b \cos v) \sin u\mathbf{j} + b \sin v\mathbf{k}$$

for each set of values of a and b, where $0 \le u \le 2\pi$ and $0 \le v \le 2\pi$. Use the results to describe the effects of a and b on the shape of the torus.

(a) $a = 4, \quad b = 1$ (b) $a = 4, \quad b = 2$

(c) $a = 8, \quad b = 1$ (d) $a = 8, \quad b = 3$

54. ***Investigation*** Consider the function in Exercise 14.

(a) Sketch a graph of the function where u is held constant at $u = 1$. Identify the graph.

(b) Sketch a graph of the function where v is held constant at $v = 2\pi/3$. Identify the graph.

(c) Assume that a surface is represented by the vector-valued function $\mathbf{r} = \mathbf{r}(u, v)$. What generalization can you make about the graph of the function if one of the parameters is held constant?

55. ***Surface Area*** The surface of the dome on a new museum is given by

$$\mathbf{r}(u, v) = 20 \sin u \cos v\mathbf{i} + 20 \sin u \sin v\mathbf{j} + 20 \cos u\mathbf{k}$$

where $0 \le u \le \pi/3$, $0 \le v \le 2\pi$, and $\mathbf{r}$ is in meters. Find the surface area of the dome.

56. Find a vector-valued function for the hyperboloid

$$x^2 + y^2 - z^2 = 1$$

and determine the tangent plane at $(1, 0, 0)$.

57. Graph and find the area of one turn of the spiral ramp

$$\mathbf{r}(u, v) = u \cos v\mathbf{i} + u \sin v\mathbf{j} + 2v\mathbf{k}$$

where $0 \le u \le 3$ and $0 \le v \le 2\pi$.

58. Let f be a nonnegative function such that f' is continuous over the interval $[a, b]$. Let S be the surface of revolution formed by revolving the graph of f, where $a \le x \le b$, about the x-axis. Let $x = u$, $y = f(u) \cos v$, and $z = f(u) \sin v$, where $a \le u \le b$ and $0 \le v \le 2\pi$. Then, S is represented parametrically by $\mathbf{r}(u, v) = u\mathbf{i} + f(u) \cos v\mathbf{j} + f(u) \sin v\mathbf{k}$. Show that the following formulas are equivalent.

$$\text{Surface area} = 2\pi \int_a^b f(x)\sqrt{1 + [f'(x)]^2}\, dx$$

$$\text{Surface area} = \iint_D \|\mathbf{r}_u \times \mathbf{r}_v\|\, dA$$

CAS **59.** ***Open-Ended Project*** The parametric equations

$$x = 3 + \sin u[7 - \cos(3u - 2v) - 2 \cos(3u + v)]$$

$$y = 3 + \cos u[7 - \cos(3u - 2v) - 2 \cos(3u + v)]$$

$$z = \sin(3u - 2v) + 2 \sin(3u + v)$$

where $-\pi \le u \le \pi$ and $-\pi \le v \le \pi$, represent the surface shown below. Try to create your own parametric surface using a computer algebra system.

CAS **60.** ***Möbius Strip*** The surface shown in the figure is called a **Möbius strip** and can be represented by the parametric equations

$$x = \left(a + u \cos\frac{v}{2}\right) \cos v, \quad y = \left(a + u \cos\frac{v}{2}\right) \sin v, \quad z = u \sin\frac{v}{2}$$

where $-1 \le u \le 1$, $0 \le v \le 2\pi$, and $a = 3$. Try to graph other Möbius strips for different values of a using a computer algebra system.

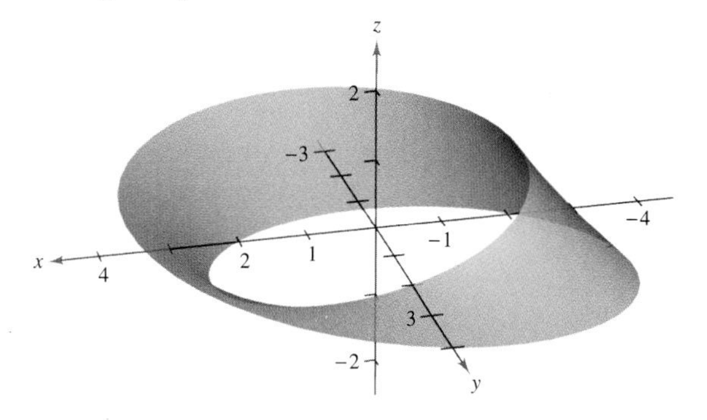

15.6 Surface Integrals

- **Evaluate a surface integral as a double integral.**
- **Evaluate a surface integral for a parametric surface.**
- **Determine the orientation of a surface.**
- **Understand the concept of a flux integral.**

Surface Integrals

The remainder of this chapter deals primarily with **surface integrals.** You will first consider surfaces given by $z = g(x, y)$. Later in this section you will consider more general surfaces given in parametric form.

Let S be a surface given by $z = g(x, y)$ and let R be its projection onto the xy-plane, as shown in Figure 15.44. Suppose that g, g_x, and g_y are continuous at all points in R and that f is defined on S. Employing the procedure used to find surface area in Section 14.5, evaluate f at (x_i, y_i, z_i) and form the sum

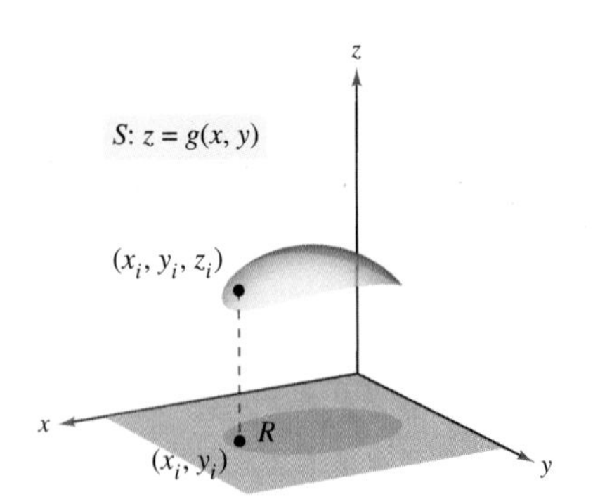

Scalar function f assigns a number to each point of S.
Figure 15.44

$$\sum_{i=1}^{n} f(x_i, y_i, z_i)\,\Delta S_i$$

where $\Delta S_i \approx \sqrt{1 + [g_x(x_i, y_i)]^2 + [g_y(x_i, y_i)]^2}\,\Delta A_i$. Provided the limit of this sum as $\|\Delta\|$ approaches 0 exists, the **surface integral of f over S** is defined as

$$\int_S\int f(x, y, z)\,dS = \lim_{\|\Delta\|\to 0} \sum_{i=1}^{n} f(x_i, y_i, z_i)\,\Delta S_i.$$

This integral can be evaluated by a double integral.

THEOREM 15.10 EVALUATING A SURFACE INTEGRAL

Let S be a surface with equation $z = g(x, y)$ and let R be its projection onto the xy-plane. If g, g_x, and g_y are continuous on R and f is continuous on S, then the surface integral of f over S is

$$\int_S\int f(x, y, z)\,dS = \int_R\int f(x, y, g(x, y))\sqrt{1 + [g_x(x, y)]^2 + [g_y(x, y)]^2}\,dA.$$

For surfaces described by functions of x and z (or y and z), you can make the following adjustments to Theorem 15.10. If S is the graph of $y = g(x, z)$ and R is its projection onto the xz-plane, then

$$\int_S\int f(x, y, z)\,dS = \int_R\int f(x, g(x, z), z)\sqrt{1 + [g_x(x, z)]^2 + [g_z(x, z)]^2}\,dA.$$

If S is the graph of $x = g(y, z)$ and R is its projection onto the yz-plane, then

$$\int_S\int f(x, y, z)\,dS = \int_R\int f(g(y, z), y, z)\sqrt{1 + [g_y(y, z)]^2 + [g_z(y, z)]^2}\,dA.$$

If $f(x, y, z) = 1$, the surface integral over S yields the surface area of S. For instance, suppose the surface S is the plane given by $z = x$, where $0 \le x \le 1$ and $0 \le y \le 1$. The surface area of S is $\sqrt{2}$ square units. Try verifying that $\int_S\int f(x, y, z)\,dS = \sqrt{2}$.

EXAMPLE 1 Evaluating a Surface Integral

Evaluate the surface integral

$$\int_S\int (y^2 + 2yz)\, dS$$

where S is the first-octant portion of the plane $2x + y + 2z = 6$.

Solution Begin by writing S as

$$z = \frac{1}{2}(6 - 2x - y)$$

$$g(x, y) = \frac{1}{2}(6 - 2x - y).$$

Using the partial derivatives $g_x(x, y) = -1$ and $g_y(x, y) = -\frac{1}{2}$, you can write

$$\sqrt{1 + [g_x(x, y)]^2 + [g_y(x, y)]^2} = \sqrt{1 + 1 + \frac{1}{4}} = \frac{3}{2}.$$

Using Figure 15.45 and Theorem 15.10, you obtain

$$\begin{aligned}\int_S\int (y^2 + 2yz)\, dS &= \int_R\int f(x, y, g(x, y))\sqrt{1 + [g_x(x, y)]^2 + [g_y(x, y)]^2}\, dA\\ &= \int_R\int \left[y^2 + 2y\left(\frac{1}{2}\right)(6 - 2x - y)\right]\left(\frac{3}{2}\right) dA\\ &= 3\int_0^3\int_0^{2(3-x)} y(3 - x)\, dy\, dx\\ &= 6\int_0^3 (3 - x)^3\, dx\\ &= -\frac{3}{2}(3 - x)^4\Big]_0^3\\ &= \frac{243}{2}.\end{aligned}$$

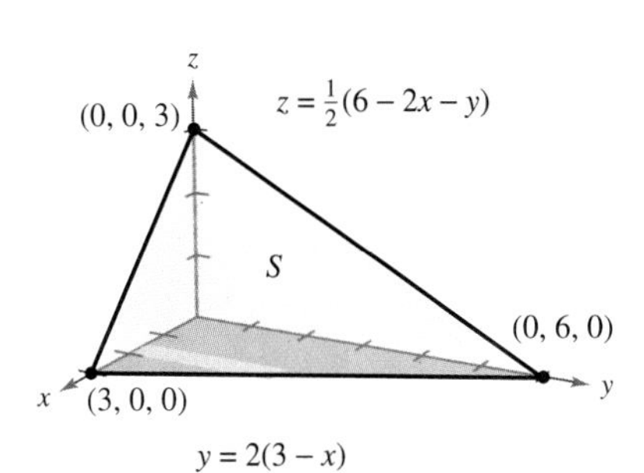

Figure 15.45

An alternative solution to Example 1 would be to project S onto the yz-plane, as shown in Figure 15.46. Then, $x = \frac{1}{2}(6 - y - 2z)$, and

$$\sqrt{1 + [g_y(y, z)]^2 + [g_z(y, z)]^2} = \sqrt{1 + \frac{1}{4} + 1} = \frac{3}{2}.$$

So, the surface integral is

$$\begin{aligned}\int_S\int (y^2 + 2yz)\, dS &= \int_R\int f(g(y, z), y, z)\sqrt{1 + [g_y(y, z)]^2 + [g_z(y, z)]^2}\, dA\\ &= \int_0^6\int_0^{(6-y)/2} (y^2 + 2yz)\left(\frac{3}{2}\right) dz\, dy\\ &= \frac{3}{8}\int_0^6 (36y - y^3)\, dy\\ &= \frac{243}{2}.\end{aligned}$$

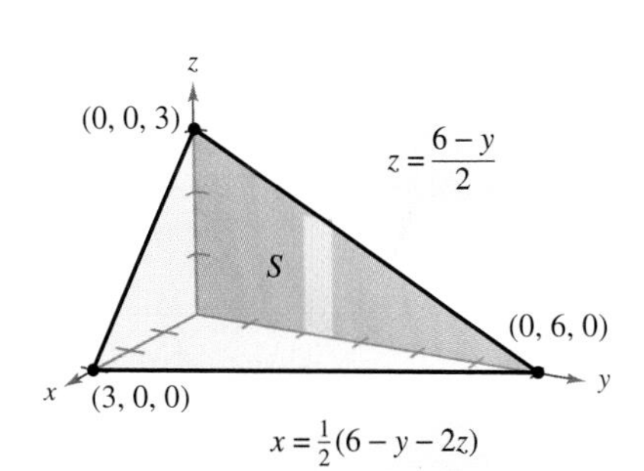

Figure 15.46

Try reworking Example 1 by projecting S onto the xz-plane.

In Example 1, you could have projected the surface S onto any one of the three coordinate planes. In Example 2, S is a portion of a cylinder centered about the x-axis, and you can project it onto either the xz-plane or the xy-plane.

EXAMPLE 2 Evaluating a Surface Integral

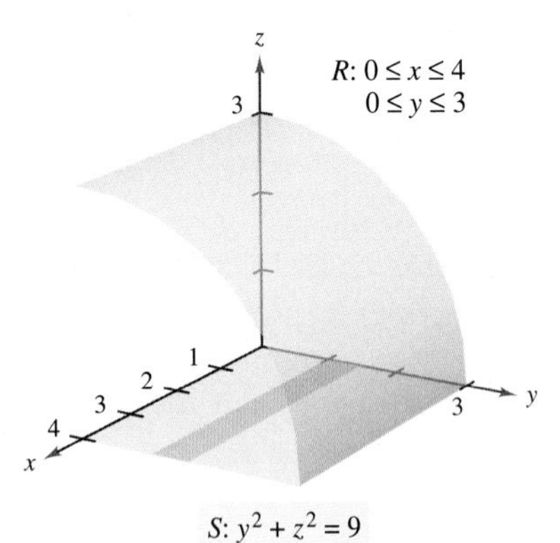

Figure 15.47

Evaluate the surface integral

$$\iint_S (x + z)\, dS$$

where S is the first-octant portion of the cylinder $y^2 + z^2 = 9$ between $x = 0$ and $x = 4$, as shown in Figure 15.47.

Solution Project S onto the xy-plane, so that $z = g(x, y) = \sqrt{9 - y^2}$, and obtain

$$\sqrt{1 + [g_x(x, y)]^2 + [g_y(x, y)]^2} = \sqrt{1 + \left(\frac{-y}{\sqrt{9 - y^2}}\right)^2}$$
$$= \frac{3}{\sqrt{9 - y^2}}.$$

Theorem 15.10 does not apply directly, because g_y is not continuous when $y = 3$. However, you can apply Theorem 15.10 for $0 \le b < 3$ and then take the limit as b approaches 3, as follows.

$$\begin{aligned}
\iint_S (x + z)\, dS &= \lim_{b \to 3^-} \int_0^b \int_0^4 \left(x + \sqrt{9 - y^2}\right)\frac{3}{\sqrt{9 - y^2}}\, dx\, dy \\
&= \lim_{b \to 3^-} 3 \int_0^b \int_0^4 \left(\frac{x}{\sqrt{9 - y^2}} + 1\right) dx\, dy \\
&= \lim_{b \to 3^-} 3 \int_0^b \left[\frac{x^2}{2\sqrt{9 - y^2}} + x\right]_0^4 dy \\
&= \lim_{b \to 3^-} 3 \int_0^b \left(\frac{8}{\sqrt{9 - y^2}} + 4\right) dy \\
&= \lim_{b \to 3^-} 3\left[4y + 8 \arcsin \frac{y}{3}\right]_0^b \\
&= \lim_{b \to 3^-} 3\left(4b + 8 \arcsin \frac{b}{3}\right) \\
&= 36 + 24\left(\frac{\pi}{2}\right) \\
&= 36 + 12\pi
\end{aligned}$$

■

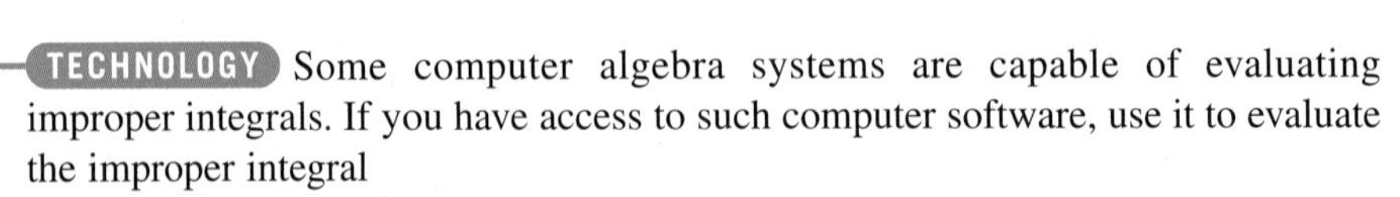

TECHNOLOGY Some computer algebra systems are capable of evaluating improper integrals. If you have access to such computer software, use it to evaluate the improper integral

$$\int_0^3 \int_0^4 \left(x + \sqrt{9 - y^2}\right)\frac{3}{\sqrt{9 - y^2}}\, dx\, dy.$$

Do you obtain the same result as in Example 2?

You have already seen that if the function f defined on the surface S is simply $f(x, y, z) = 1$, the surface integral yields the *surface area* of S.

$$\text{Area of surface} = \iint_S 1\, dS$$

On the other hand, if S is a lamina of variable density and $\rho(x, y, z)$ is the density at the point (x, y, z), then the *mass* of the lamina is given by

$$\text{Mass of lamina} = \iint_S \rho(x, y, z)\, dS.$$

EXAMPLE 3 Finding the Mass of a Surface Lamina

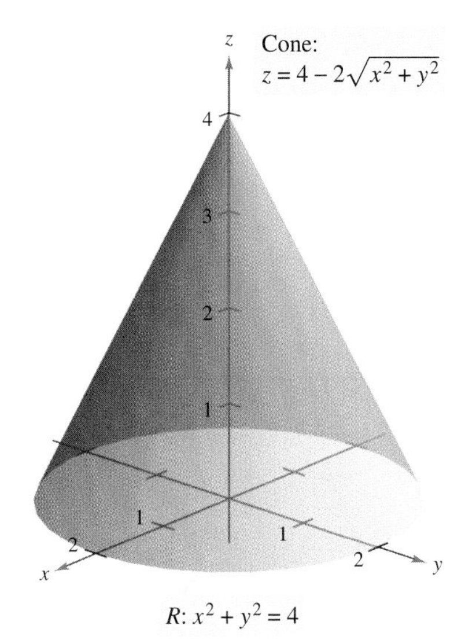

Figure 15.48

A cone-shaped surface lamina S is given by

$$z = 4 - 2\sqrt{x^2 + y^2}, \quad 0 \le z \le 4$$

as shown in Figure 15.48. At each point on S, the density is proportional to the distance between the point and the z-axis. Find the mass m of the lamina.

Solution Projecting S onto the xy-plane produces

$$S: z = 4 - 2\sqrt{x^2 + y^2} = g(x, y), \quad 0 \le z \le 4$$
$$R: x^2 + y^2 \le 4$$

with a density of $\rho(x, y, z) = k\sqrt{x^2 + y^2}$. Using a surface integral, you can find the mass to be

$$\begin{aligned}
m &= \iint_S \rho(x, y, z)\, dS \\
&= \iint_R k\sqrt{x^2 + y^2}\sqrt{1 + [g_x(x, y)]^2 + [g_y(x, y)]^2}\, dA \\
&= k\iint_R \sqrt{x^2 + y^2}\sqrt{1 + \frac{4x^2}{x^2 + y^2} + \frac{4y^2}{x^2 + y^2}}\, dA \\
&= k\iint_R \sqrt{5}\sqrt{x^2 + y^2}\, dA \\
&= k\int_0^{2\pi}\int_0^2 \left(\sqrt{5}r\right)r\, dr\, d\theta \qquad \text{Polar coordinates} \\
&= \frac{\sqrt{5}k}{3}\int_0^{2\pi} r^3\Big]_0^2\, d\theta \\
&= \frac{8\sqrt{5}k}{3}\int_0^{2\pi} d\theta \\
&= \frac{8\sqrt{5}k}{3}\Big[\theta\Big]_0^{2\pi} = \frac{16\sqrt{5}k\pi}{3}.
\end{aligned}$$

■

TECHNOLOGY Use a computer algebra system to confirm the result shown in Example 3. The computer algebra system *Maple* evaluated the integral as follows.

$$k\int_{-2}^{2}\int_{-\sqrt{4-y^2}}^{\sqrt{4-y^2}} \sqrt{5}\sqrt{x^2 + y^2}\, dx\, dy = k\int_0^{2\pi}\int_0^2 \left(\sqrt{5}r\right)r\, dr\, d\theta = \frac{16\sqrt{5}k\pi}{3}$$

Parametric Surfaces and Surface Integrals

For a surface S given by the vector-valued function

$$\mathbf{r}(u, v) = x(u, v)\mathbf{i} + y(u, v)\mathbf{j} + z(u, v)\mathbf{k} \qquad \text{Parametric surface}$$

defined over a region D in the uv-plane, you can show that the surface integral of $f(x, y, z)$ over S is given by

$$\int_S\int f(x, y, z)\, dS = \int_D\int f(x(u, v), y(u, v), z(u, v))\, \|\mathbf{r}_u(u, v) \times \mathbf{r}_v(u, v)\|\, dA.$$

Note the similarity to a line integral over a space curve C.

$$\int_C f(x, y, z)\, ds = \int_a^b f(x(t), y(t), z(t))\, \|\mathbf{r}'(t)\|\, dt \qquad \text{Line integral}$$

NOTE Notice that ds and dS can be written as $ds = \|\mathbf{r}'(t)\|\, dt$ and $dS = \|\mathbf{r}_u(u, v) \times \mathbf{r}_v(u, v)\|\, dA$. ■

EXAMPLE 4 Evaluating a Surface Integral

Example 2 demonstrated an evaluation of the surface integral

$$\int_S\int (x + z)\, dS$$

where S is the first-octant portion of the cylinder $y^2 + z^2 = 9$ between $x = 0$ and $x = 4$ (see Figure 15.49). Reevaluate this integral in parametric form.

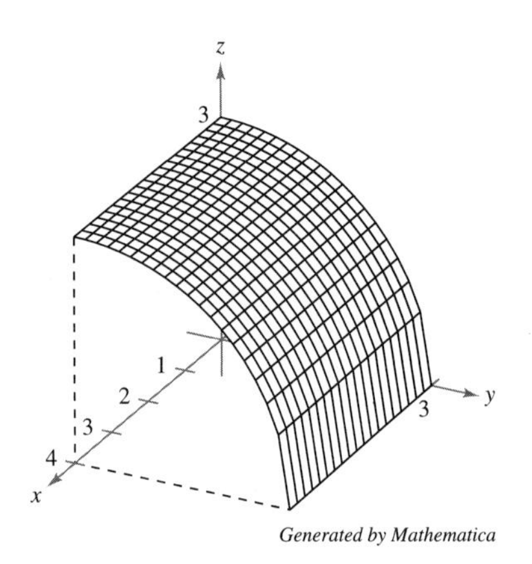

Figure 15.49

Solution In parametric form, the surface is given by

$$\mathbf{r}(x, \theta) = x\mathbf{i} + 3\cos\theta\mathbf{j} + 3\sin\theta\mathbf{k}$$

where $0 \le x \le 4$ and $0 \le \theta \le \pi/2$. To evaluate the surface integral in parametric form, begin by calculating the following.

$$\mathbf{r}_x = \mathbf{i}$$

$$\mathbf{r}_\theta = -3\sin\theta\mathbf{j} + 3\cos\theta\mathbf{k}$$

$$\mathbf{r}_x \times \mathbf{r}_\theta = \begin{vmatrix} \mathbf{i} & \mathbf{j} & \mathbf{k} \\ 1 & 0 & 0 \\ 0 & -3\sin\theta & 3\cos\theta \end{vmatrix} = -3\cos\theta\mathbf{j} - 3\sin\theta\mathbf{k}$$

$$\|\mathbf{r}_x \times \mathbf{r}_\theta\| = \sqrt{9\cos^2\theta + 9\sin^2\theta} = 3$$

So, the surface integral can be evaluated as follows.

$$\begin{aligned}
\int_D\int (x + 3\sin\theta)3\, dA &= \int_0^4\int_0^{\pi/2} (3x + 9\sin\theta)\, d\theta\, dx \\
&= \int_0^4 \Big[3x\theta - 9\cos\theta\Big]_0^{\pi/2} dx \\
&= \int_0^4 \left(\frac{3\pi}{2}x + 9\right) dx \\
&= \left[\frac{3\pi}{4}x^2 + 9x\right]_0^4 \\
&= 12\pi + 36
\end{aligned}$$

■

Orientation of a Surface

Unit normal vectors are used to induce an orientation to a surface S in space. A surface is called **orientable** if a unit normal vector $\mathbf{N}$ can be defined at every nonboundary point of S in such a way that the normal vectors vary continuously over the surface S. If this is possible, S is called an **oriented surface.**

An orientable surface S has two distinct sides. So, when you orient a surface, you are selecting one of the two possible unit normal vectors. If S is a closed surface such as a sphere, it is customary to choose the unit normal vector $\mathbf{N}$ to be the one that points outward from the sphere.

Most common surfaces, such as spheres, paraboloids, ellipses, and planes, are orientable. (See Exercise 43 for an example of a surface that is *not* orientable.) Moreover, for an orientable surface, the gradient vector provides a convenient way to find a unit normal vector. That is, for an orientable surface S given by

$$z = g(x, y) \qquad \text{Orientable surface}$$

let

$$G(x, y, z) = z - g(x, y).$$

Then, S can be oriented by either the unit normal vector

$$\begin{aligned}\mathbf{N} &= \frac{\nabla G(x, y, z)}{\|\nabla G(x, y, z)\|}\\ &= \frac{-g_x(x, y)\mathbf{i} - g_y(x, y)\mathbf{j} + \mathbf{k}}{\sqrt{1 + [g_x(x, y)]^2 + [g_y(x, y)]^2}} \qquad \text{Upward unit normal vector}\end{aligned}$$

or the unit normal vector

$$\begin{aligned}\mathbf{N} &= \frac{-\nabla G(x, y, z)}{\|\nabla G(x, y, z)\|}\\ &= \frac{g_x(x, y)\mathbf{i} + g_y(x, y)\mathbf{j} - \mathbf{k}}{\sqrt{1 + [g_x(x, y)]^2 + [g_y(x, y)]^2}} \qquad \text{Downward unit normal vector}\end{aligned}$$

as shown in Figure 15.50. If the smooth orientable surface S is given in parametric form by

$$\mathbf{r}(u, v) = x(u, v)\mathbf{i} + y(u, v)\mathbf{j} + z(u, v)\mathbf{k} \qquad \text{Parametric surface}$$

the unit normal vectors are given by

$$\mathbf{N} = \frac{\mathbf{r}_u \times \mathbf{r}_v}{\|\mathbf{r}_u \times \mathbf{r}_v\|}$$

and

$$\mathbf{N} = \frac{\mathbf{r}_v \times \mathbf{r}_u}{\|\mathbf{r}_v \times \mathbf{r}_u\|}.$$

NOTE Suppose that the orientable surface is given by $y = g(x, z)$ or $x = g(y, z)$. Then you can use the gradient vector

$$\nabla G(x, y, z) = -g_x(x, z)\mathbf{i} + \mathbf{j} - g_z(x, z)\mathbf{k} \qquad G(x, y, z) = y - g(x, z)$$

or

$$\nabla G(x, y, z) = \mathbf{i} - g_y(y, z)\mathbf{j} - g_z(y, z)\mathbf{k} \qquad G(x, y, z) = x - g(y, z)$$

to orient the surface. ■

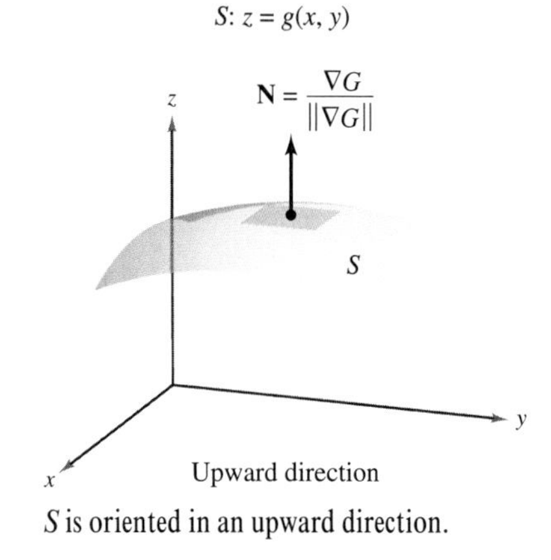

S is oriented in an upward direction.

S: $z = g(x, y)$

$\mathbf{N} = \dfrac{-\nabla G}{\|\nabla G\|}$

z

S

y

x

Downward direction

S is oriented in a downward direction.

Figure 15.50

Flux Integrals

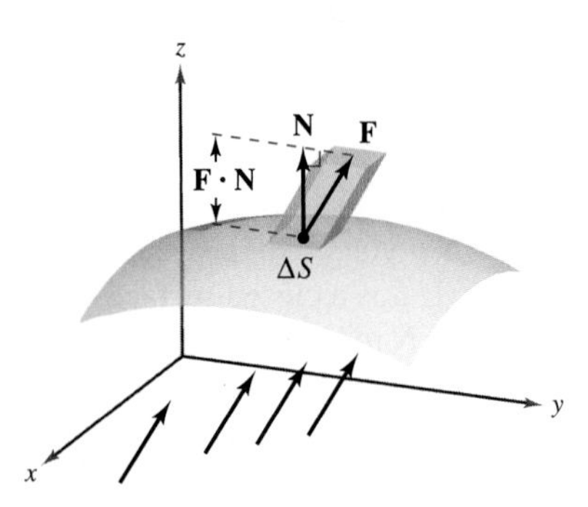

The velocity field **F** indicates the direction of the fluid flow.
Figure 15.51

One of the principal applications involving the vector form of a surface integral relates to the flow of a fluid through a surface S. Suppose an oriented surface S is submerged in a fluid having a continuous velocity field **F**. Let ΔS be the area of a small patch of the surface S over which **F** is nearly constant. Then the amount of fluid crossing this region per unit of time is approximated by the volume of the column of height $\mathbf{F} \cdot \mathbf{N}$, as shown in Figure 15.51. That is,

$$\Delta V = (\text{height})(\text{area of base}) = (\mathbf{F} \cdot \mathbf{N})\Delta S.$$

Consequently, the volume of fluid crossing the surface S per unit of time (called the **flux of F across** S) is given by the surface integral in the following definition.

DEFINITION OF FLUX INTEGRAL

Let $\mathbf{F}(x, y, z) = M\mathbf{i} + N\mathbf{j} + P\mathbf{k}$, where M, N, and P have continuous first partial derivatives on the surface S oriented by a unit normal vector **N**. The **flux integral of F across** S is given by

$$\int_S\int \mathbf{F} \cdot \mathbf{N}\, dS.$$

Geometrically, a flux integral is the surface integral over S of the *normal component* of **F**. If $\rho(x, y, z)$ is the density of the fluid at (x, y, z), the flux integral

$$\int_S\int \rho\mathbf{F} \cdot \mathbf{N}\, dS$$

represents the *mass* of the fluid flowing across S per unit of time.

To evaluate a flux integral for a surface given by $z = g(x, y)$, let

$$G(x, y, z) = z - g(x, y).$$

Then, $\mathbf{N}\, dS$ can be written as follows.

$$\begin{aligned}\mathbf{N}\, dS &= \frac{\nabla G(x, y, z)}{\|\nabla G(x, y, z)\|}\, dS \\ &= \frac{\nabla G(x, y, z)}{\sqrt{(g_x)^2 + (g_y)^2 + 1}}\sqrt{(g_x)^2 + (g_y)^2 + 1}\, dA \\ &= \nabla G(x, y, z)\, dA\end{aligned}$$

THEOREM 15.11 EVALUATING A FLUX INTEGRAL

Let S be an oriented surface given by $z = g(x, y)$ and let R be its projection onto the xy-plane.

$$\int_S\int \mathbf{F} \cdot \mathbf{N}\, dS = \int_R\int \mathbf{F} \cdot [-g_x(x, y)\mathbf{i} - g_y(x, y)\mathbf{j} + \mathbf{k}]\, dA \quad \text{Oriented upward}$$

$$\int_S\int \mathbf{F} \cdot \mathbf{N}\, dS = \int_R\int \mathbf{F} \cdot [g_x(x, y)\mathbf{i} + g_y(x, y)\mathbf{j} - \mathbf{k}]\, dA \quad \text{Oriented downward}$$

For the first integral, the surface is oriented upward, and for the second integral, the surface is oriented downward.

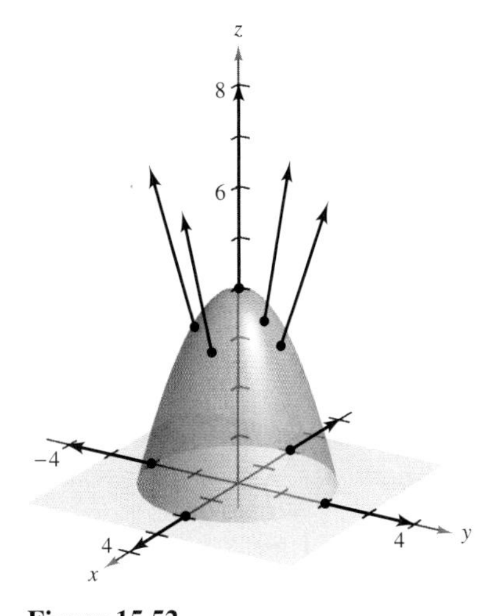

Figure 15.52

EXAMPLE 5 Using a Flux Integral to Find the Rate of Mass Flow

Let S be the portion of the paraboloid

$$z = g(x, y) = 4 - x^2 - y^2$$

lying above the xy-plane, oriented by an upward unit normal vector, as shown in Figure 15.52. A fluid of constant density ρ is flowing through the surface S according to the vector field

$$\mathbf{F}(x, y, z) = x\mathbf{i} + y\mathbf{j} + z\mathbf{k}.$$

Find the rate of mass flow through S.

Solution Begin by computing the partial derivatives of g.

$$g_x(x, y) = -2x$$

and

$$g_y(x, y) = -2y$$

The rate of mass flow through the surface S is

$$\begin{aligned}\int_S\!\!\int \rho\mathbf{F}\cdot\mathbf{N}\,dS &= \rho\int_R\!\!\int \mathbf{F}\cdot\left[-g_x(x, y)\mathbf{i} - g_y(x, y)\mathbf{j} + \mathbf{k}\right]dA\\ &= \rho\int_R\!\!\int \left[x\mathbf{i} + y\mathbf{j} + (4 - x^2 - y^2)\mathbf{k}\right]\cdot(2x\mathbf{i} + 2y\mathbf{j} + \mathbf{k})\,dA\\ &= \rho\int_R\!\!\int \left[2x^2 + 2y^2 + (4 - x^2 - y^2)\right]dA\\ &= \rho\int_R\!\!\int (4 + x^2 + y^2)\,dA\\ &= \rho\int_0^{2\pi}\!\!\int_0^2 (4 + r^2)r\,dr\,d\theta \qquad \text{Polar coordinates}\\ &= \rho\int_0^{2\pi} 12\,d\theta\\ &= 24\pi\rho.\end{aligned}$$

■

For an oriented surface S given by the vector-valued function

$$\mathbf{r}(u, v) = x(u, v)\mathbf{i} + y(u, v)\mathbf{j} + z(u, v)\mathbf{k} \qquad \text{Parametric surface}$$

defined over a region D in the uv-plane, you can define the flux integral of $\mathbf{F}$ across S as

$$\begin{aligned}\int_S\!\!\int \mathbf{F}\cdot\mathbf{N}\,dS &= \int_D\!\!\int \mathbf{F}\cdot\left(\frac{\mathbf{r}_u\times\mathbf{r}_v}{\|\mathbf{r}_u\times\mathbf{r}_v\|}\right)\|\mathbf{r}_u\times\mathbf{r}_v\|\,dA\\ &= \int_D\!\!\int \mathbf{F}\cdot(\mathbf{r}_u\times\mathbf{r}_v)\,dA.\end{aligned}$$

Note the similarity of this integral to the line integral

$$\int_C \mathbf{F}\cdot d\mathbf{r} = \int_C \mathbf{F}\cdot\mathbf{T}\,ds.$$

A summary of formulas for line and surface integrals is presented on page 1121.

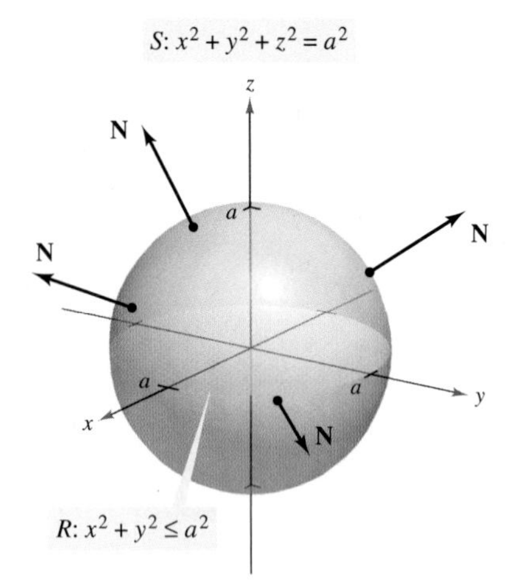

Figure 15.53

EXAMPLE 6 Finding the Flux of an Inverse Square Field

Find the flux over the sphere S given by

$$x^2 + y^2 + z^2 = a^2 \qquad \text{Sphere } S$$

where $\mathbf{F}$ is an inverse square field given by

$$\mathbf{F}(x, y, z) = \frac{kq}{\|\mathbf{r}\|^2}\frac{\mathbf{r}}{\|\mathbf{r}\|} = \frac{kq\mathbf{r}}{\|\mathbf{r}\|^3} \qquad \text{Inverse square field } \mathbf{F}$$

and $\mathbf{r} = x\mathbf{i} + y\mathbf{j} + z\mathbf{k}$. Assume S is oriented outward, as shown in Figure 15.53.

Solution The sphere is given by

$$\begin{aligned}\mathbf{r}(u, v) &= x(u, v)\mathbf{i} + y(u, v)\mathbf{j} + z(u, v)\mathbf{k}\\ &= a \sin u \cos v\mathbf{i} + a \sin u \sin v\mathbf{j} + a \cos u\mathbf{k}\end{aligned}$$

where $0 \le u \le \pi$ and $0 \le v \le 2\pi$. The partial derivatives of $\mathbf{r}$ are

$$\mathbf{r}_u(u, v) = a \cos u \cos v\,\mathbf{i} + a \cos u \sin v\mathbf{j} - a \sin u\mathbf{k}$$

and

$$\mathbf{r}_v(u, v) = -a \sin u \sin v\mathbf{i} + a \sin u \cos v\mathbf{j}$$

which implies that the normal vector $\mathbf{r}_u \times \mathbf{r}_v$ is

$$\begin{aligned}\mathbf{r}_u \times \mathbf{r}_v &= \begin{vmatrix}\mathbf{i} & \mathbf{j} & \mathbf{k}\\ a\cos u\cos v & a\cos u\sin v & -a\sin u\\ -a\sin u\sin v & a\sin u\cos v & 0\end{vmatrix}\\ &= a^2(\sin^2 u \cos v\mathbf{i} + \sin^2 u \sin v\mathbf{j} + \sin u \cos u\mathbf{k}).\end{aligned}$$

Now, using

$$\begin{aligned}\mathbf{F}(x, y, z) &= \frac{kq\mathbf{r}}{\|\mathbf{r}\|^3}\\ &= kq\frac{x\mathbf{i} + y\mathbf{j} + z\mathbf{k}}{\|x\mathbf{i} + y\mathbf{j} + z\mathbf{k}\|^3}\\ &= \frac{kq}{a^3}(a \sin u \cos v\mathbf{i} + a \sin u \sin v\mathbf{j} + a \cos u\mathbf{k})\end{aligned}$$

it follows that

$$\begin{aligned}\mathbf{F} \cdot (\mathbf{r}_u \times \mathbf{r}_v) &= \frac{kq}{a^3}[(a \sin u \cos v\mathbf{i} + a \sin u \sin v\mathbf{j} + a \cos u\mathbf{k}) \cdot\\ &\qquad a^2(\sin^2 u \cos v\mathbf{i} + \sin^2 u \sin v\mathbf{j} + \sin u \cos u\mathbf{k})]\\ &= kq(\sin^3 u \cos^2 v + \sin^3 u \sin^2 v + \sin u \cos^2 u)\\ &= kq \sin u.\end{aligned}$$

Finally, the flux over the sphere S is given by

$$\begin{aligned}\int_S\!\!\int \mathbf{F} \cdot \mathbf{N}\, dS &= \int_D\!\!\int (kq \sin u)\, dA\\ &= \int_0^{2\pi}\!\!\int_0^{\pi} kq \sin u\, du\, dv\\ &= 4\pi kq.\end{aligned}$$

■

The result in Example 6 shows that the flux across a sphere S in an inverse square field is independent of the radius of S. In particular, if $\mathbf{E}$ is an electric field, the result in Example 6, along with Coulomb's Law, yields one of the basic laws of electrostatics, known as **Gauss's Law:**

$$\int_S\int \mathbf{E} \cdot \mathbf{N}\, dS = 4\pi kq \qquad \text{Gauss's Law}$$

where q is a point charge located at the center of the sphere and k is the Coulomb constant. Gauss's Law is valid for more general closed surfaces that enclose the origin, and relates the flux out of the surface to the total charge q inside the surface.

This section concludes with a summary of different forms of line integrals and surface integrals.

SUMMARY OF LINE AND SURFACE INTEGRALS

Line Integrals

$$ds = \|\mathbf{r}'(t)\|\, dt = \sqrt{[x'(t)]^2 + [y'(t)]^2 + [z'(t)]^2}\, dt$$

$$\int_C f(x, y, z)\, ds = \int_a^b f(x(t), y(t), z(t))\, ds \qquad \text{Scalar form}$$

$$\int_C \mathbf{F} \cdot d\mathbf{r} = \int_C \mathbf{F} \cdot \mathbf{T}\, ds = \int_a^b \mathbf{F}(x(t), y(t), z(t)) \cdot \mathbf{r}'(t)\, dt \qquad \text{Vector form}$$

Surface Integrals $[z = g(x, y)]$

$$dS = \sqrt{1 + [g_x(x, y)]^2 + [g_y(x, y)]^2}\, dA$$

$$\int_S\int f(x, y, z)\, dS = \int_R\int f(x, y, g(x, y))\sqrt{1 + [g_x(x, y)]^2 + [g_y(x, y)]^2}\, dA \qquad \text{Scalar form}$$

$$\int_S\int \mathbf{F} \cdot \mathbf{N}\, dS = \int_R\int \mathbf{F} \cdot [-g_x(x, y)\mathbf{i} - g_y(x, y)\mathbf{j} + \mathbf{k}]\, dA \qquad \text{Vector form (upward normal)}$$

Surface Integrals (parametric form)

$$dS = \|\mathbf{r}_u(u, v) \times \mathbf{r}_v(u, v)\|\, dA$$

$$\int_S\int f(x, y, z)\, dS = \int_D\int f(x(u, v), y(u, v), z(u, v))\, dS \qquad \text{Scalar form}$$

$$\int_S\int \mathbf{F} \cdot \mathbf{N}\, dS = \int_D\int \mathbf{F} \cdot (\mathbf{r}_u \times \mathbf{r}_v)\, dA \qquad \text{Vector form}$$

15.6 Exercises

See www.CalcChat.com for worked-out solutions to odd-numbered exercises.

In Exercises 1–4, evaluate $\iint_S (x - 2y + z)\, dS$.

1. S: $z = 4 - x$, $0 \le x \le 4$, $0 \le y \le 3$
2. S: $z = 15 - 2x + 3y$, $0 \le x \le 2$, $0 \le y \le 4$
3. S: $z = 2$, $x^2 + y^2 \le 1$
4. S: $z = \frac{2}{3}x^{3/2}$, $0 \le x \le 1$, $0 \le y \le x$

In Exercises 5 and 6, evaluate $\iint_S xy\, dS$.

5. S: $z = 3 - x - y$, first octant
6. S: $z = h$, $0 \le x \le 2$, $0 \le y \le \sqrt{4 - x^2}$

CAS **In Exercises 7 and 8, use a computer algebra system to evaluate**

$$\iint_S xy\, dS.$$

7. S: $z = 9 - x^2$, $0 \le x \le 2$, $0 \le y \le x$
8. S: $z = \frac{1}{2}xy$, $0 \le x \le 4$, $0 \le y \le 4$

CAS **In Exercises 9 and 10, use a computer algebra system to evaluate**

$$\iint_S (x^2 - 2xy)\, dS.$$

9. S: $z = 10 - x^2 - y^2$, $0 \le x \le 2$, $0 \le y \le 2$
10. S: $z = \cos x$, $0 \le x \le \dfrac{\pi}{2}$, $0 \le y \le \dfrac{1}{2}x$

Mass **In Exercises 11 and 12, find the mass of the surface lamina S of density ρ.**

11. S: $2x + 3y + 6z = 12$, first octant, $\rho(x, y, z) = x^2 + y^2$
12. S: $z = \sqrt{a^2 - x^2 - y^2}$, $\rho(x, y, z) = kz$

In Exercises 13–16, evaluate $\iint_S f(x, y)\, dS$.

13. $f(x, y) = y + 5$
    S: $\mathbf{r}(u, v) = u\mathbf{i} + v\mathbf{j} + 2v\mathbf{k}$, $0 \le u \le 1$, $0 \le v \le 2$
14. $f(x, y) = xy$
    S: $\mathbf{r}(u, v) = 2\cos u\mathbf{i} + 2\sin u\mathbf{j} + v\mathbf{k}$
    $0 \le u \le \dfrac{\pi}{2}$, $0 \le v \le 1$
15. $f(x, y) = x + y$
    S: $\mathbf{r}(u, v) = 2\cos u\mathbf{i} + 2\sin u\mathbf{j} + v\mathbf{k}$
    $0 \le u \le \dfrac{\pi}{2}$, $0 \le v \le 1$
16. $f(x, y) = x + y$
    S: $\mathbf{r}(u, v) = 4u\cos v\mathbf{i} + 4u\sin v\mathbf{j} + 3u\mathbf{k}$
    $0 \le u \le 4$, $0 \le v \le \pi$

In Exercises 17–22, evaluate $\iint_S f(x, y, z)\, dS$.

17. $f(x, y, z) = x^2 + y^2 + z^2$
    S: $z = x + y$, $x^2 + y^2 \le 1$
18. $f(x, y, z) = \dfrac{xy}{z}$
    S: $z = x^2 + y^2$, $4 \le x^2 + y^2 \le 16$
19. $f(x, y, z) = \sqrt{x^2 + y^2 + z^2}$
    S: $z = \sqrt{x^2 + y^2}$, $x^2 + y^2 \le 4$
20. $f(x, y, z) = \sqrt{x^2 + y^2 + z^2}$
    S: $z = \sqrt{x^2 + y^2}$, $(x - 1)^2 + y^2 \le 1$
21. $f(x, y, z) = x^2 + y^2 + z^2$
    S: $x^2 + y^2 = 9$, $0 \le x \le 3$, $0 \le y \le 3$, $0 \le z \le 9$
22. $f(x, y, z) = x^2 + y^2 + z^2$
    S: $x^2 + y^2 = 9$, $0 \le x \le 3$, $0 \le z \le x$

In Exercises 23–28, find the flux of F through S,

$$\iint_S \mathbf{F} \cdot \mathbf{N}\, dS$$

where N is the upward unit normal vector to S.

23. $\mathbf{F}(x, y, z) = 3z\mathbf{i} - 4\mathbf{j} + y\mathbf{k}$
    S: $z = 1 - x - y$, first octant
24. $\mathbf{F}(x, y, z) = x\mathbf{i} + y\mathbf{j}$
    S: $z = 6 - 3x - 2y$, first octant
25. $\mathbf{F}(x, y, z) = x\mathbf{i} + y\mathbf{j} + z\mathbf{k}$
    S: $z = 1 - x^2 - y^2$, $z \ge 0$
26. $\mathbf{F}(x, y, z) = x\mathbf{i} + y\mathbf{j} + z\mathbf{k}$
    S: $x^2 + y^2 + z^2 = 36$, first octant
27. $\mathbf{F}(x, y, z) = 4\mathbf{i} - 3\mathbf{j} + 5\mathbf{k}$
    S: $z = x^2 + y^2$, $x^2 + y^2 \le 4$
28. $\mathbf{F}(x, y, z) = x\mathbf{i} + y\mathbf{j} - 2z\mathbf{k}$
    S: $z = \sqrt{a^2 - x^2 - y^2}$

In Exercises 29 and 30, find the flux of F over the closed surface. (Let N be the outward unit normal vector of the surface.)

29. $\mathbf{F}(x, y, z) = (x + y)\mathbf{i} + y\mathbf{j} + z\mathbf{k}$
    S: $z = 16 - x^2 - y^2$, $z = 0$
30. $\mathbf{F}(x, y, z) = 4xy\mathbf{i} + z^2\mathbf{j} + yz\mathbf{k}$
    S: unit cube bounded by $x = 0$, $x = 1$, $y = 0$, $y = 1$, $z = 0$, $z = 1$

31. ***Electrical Charge*** Let $\mathbf{E} = yz\mathbf{i} + xz\mathbf{j} + xy\mathbf{k}$ be an electrostatic field. Use Gauss's Law to find the total charge enclosed by the closed surface consisting of the hemisphere $z = \sqrt{1 - x^2 - y^2}$ and its circular base in the xy-plane.

32. ***Electrical Charge*** Let $\mathbf{E} = x\mathbf{i} + y\mathbf{j} + 2z\mathbf{k}$ be an electrostatic field. Use Gauss's Law to find the total charge enclosed by the closed surface consisting of the hemisphere $z = \sqrt{1 - x^2 - y^2}$ and its circular base in the xy-plane.

Moment of Inertia **In Exercises 33 and 34, use the following formulas for the moments of inertia about the coordinate axes of a surface lamina of density ρ.**

$$I_x = \iint_S (y^2 + z^2)\rho(x, y, z)\, dS$$

$$I_y = \iint_S (x^2 + z^2)\rho(x, y, z)\, dS$$

$$I_z = \iint_S (x^2 + y^2)\rho(x, y, z)\, dS$$

33. Verify that the moment of inertia of a conical shell of uniform density about its axis is $\frac{1}{2}ma^2$, where m is the mass and a is the radius and height.

34. Verify that the moment of inertia of a spherical shell of uniform density about its diameter is $\frac{2}{3}ma^2$, where m is the mass and a is the radius.

Moment of Inertia **In Exercises 35 and 36, find I_z for the given lamina with uniform density of 1. Use a computer algebra system to verify your results.**

35. $x^2 + y^2 = a^2, \quad 0 \le z \le h$

36. $z = x^2 + y^2, \quad 0 \le z \le h$

CAS ***Flow Rate*** **In Exercises 37 and 38, use a computer algebra system to find the rate of mass flow of a fluid of density ρ through the surface S oriented upward if the velocity field is given by $\mathbf{F}(x, y, z) = 0.5z\mathbf{k}$.**

37. $S\colon z = 16 - x^2 - y^2, \quad z \ge 0$

38. $S\colon z = \sqrt{16 - x^2 - y^2}$

WRITING ABOUT CONCEPTS

39. Define a surface integral of the scalar function f over a surface $z = g(x, y)$. Explain how to evaluate the surface integral.

40. Describe an orientable surface.

41. Define a flux integral and explain how it is evaluated.

42. Is the surface shown in the figure orientable? Explain.

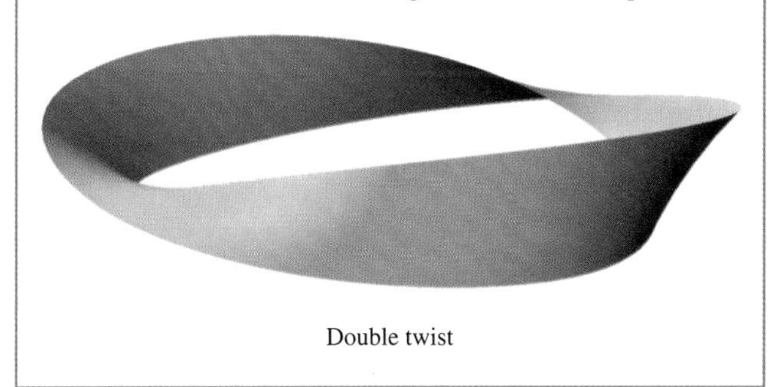

Double twist

CAS **43.** ***Investigation***

(a) Use a computer algebra system to graph the vector-valued function

$$\mathbf{r}(u, v) = (4 - v \sin u)\cos(2u)\mathbf{i} + (4 - v \sin u)\sin(2u)\mathbf{j} + v \cos u\mathbf{k}, \quad 0 \le u \le \pi, \quad -1 \le v \le 1.$$

This surface is called a Möbius strip.

(b) Explain why this surface is not orientable.

(c) Use a computer algebra system to graph the space curve represented by $\mathbf{r}(u, 0)$. Identify the curve.

(d) Construct a Möbius strip by cutting a strip of paper, making a single twist, and pasting the ends together.

(e) Cut the Möbius strip along the space curve graphed in part (c), and describe the result.

CAPSTONE

44. Consider the vector field

$\mathbf{F}(x, y, z) = z\mathbf{i} + x\mathbf{j} + y\mathbf{k}$

and the orientable surface S given in parametric form by

$\mathbf{r}(u, v) = (u + v^2)\mathbf{i} + (u - v)\mathbf{j} + u^2\mathbf{k},$

$0 \le u \le 2, -1 \le v \le 1.$

(a) Find and interpret $\mathbf{r}_u \times \mathbf{r}_v$.

(b) Find $\mathbf{F} \cdot (\mathbf{r}_u \times \mathbf{r}_v)$ as a function of u and v.

(c) Find u and v at the point $P(3, 1, 4)$.

(d) Explain how to find the normal component of $\mathbf{F}$ to the surface at P. Then find this value.

(e) Evaluate the flux integral $\iint_S \mathbf{F} \cdot \mathbf{N}\, dS$.

SECTION PROJECT

Hyperboloid of One Sheet

Consider the parametric surface given by the function

$\mathbf{r}(u, v) = a \cosh u \cos v\mathbf{i} + a \cosh u \sin v\mathbf{j} + b \sinh u\mathbf{k}.$

(a) Use a graphing utility to graph $\mathbf{r}$ for various values of the constants a and b. Describe the effect of the constants on the shape of the surface.

(b) Show that the surface is a hyperboloid of one sheet given by

$$\frac{x^2}{a^2} + \frac{y^2}{a^2} - \frac{z^2}{b^2} = 1.$$

(c) For fixed values $u = u_0$, describe the curves given by

$\mathbf{r}(u_0, v) = a \cosh u_0 \cos v\mathbf{i} + a \cosh u_0 \sin v\mathbf{j} + b \sinh u_0\mathbf{k}.$

(d) For fixed values $v = v_0$, describe the curves given by

$\mathbf{r}(u, v_0) = a \cosh u \cos v_0\mathbf{i} + a \cosh u \sin v_0\mathbf{j} + b \sinh u\mathbf{k}.$

(e) Find a normal vector to the surface at $(u, v) = (0, 0)$.

15.7 Divergence Theorem

- **Understand and use the Divergence Theorem.**
- **Use the Divergence Theorem to calculate flux.**

CARL FRIEDRICH GAUSS (1777–1855)

The **Divergence Theorem** is also called **Gauss's Theorem,** after the famous German mathematician Carl Friedrich Gauss. Gauss is recognized, with Newton and Archimedes, as one of the three greatest mathematicians in history. One of his many contributions to mathematics was made at the age of 22, when, as part of his doctoral dissertation, he proved the *Fundamental Theorem of Algebra.*

Divergence Theorem

Recall from Section 15.4 that an alternative form of Green's Theorem is

$$\int_C \mathbf{F} \cdot \mathbf{N}\, ds = \int_R\!\!\int \left(\frac{\partial M}{\partial x} + \frac{\partial N}{\partial y}\right) dA$$
$$= \int_R\!\!\int \operatorname{div} \mathbf{F}\, dA.$$

In an analogous way, the **Divergence Theorem** gives the relationship between a triple integral over a solid region Q and a surface integral over the surface of Q. In the statement of the theorem, the surface S is **closed** in the sense that it forms the complete boundary of the solid Q. Regions bounded by spheres, ellipsoids, cubes, tetrahedrons, or combinations of these surfaces are typical examples of closed surfaces. Assume that Q is a solid region on which a triple integral can be evaluated, and that the closed surface S is oriented by *outward* unit normal vectors, as shown in Figure 15.54. With these restrictions on S and Q, the Divergence Theorem is as follows.

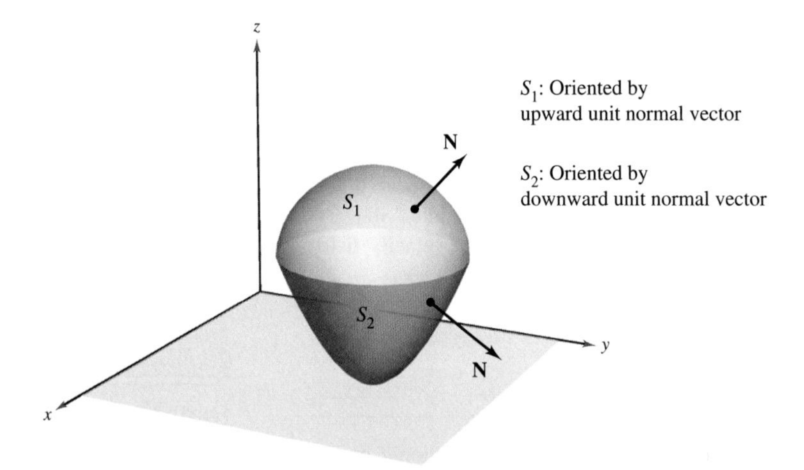

Figure 15.54

> **THEOREM 15.12 THE DIVERGENCE THEOREM**
>
> Let Q be a solid region bounded by a closed surface S oriented by a unit normal vector directed outward from Q. If $\mathbf{F}$ is a vector field whose component functions have continuous first partial derivatives in Q, then
>
> $$\int_S\!\!\int \mathbf{F} \cdot \mathbf{N}\, dS = \iiint_Q \operatorname{div} \mathbf{F}\, dV.$$

NOTE As noted at the left above, the Divergence Theorem is sometimes called Gauss's Theorem. It is also sometimes called Ostrogradsky's Theorem, after the Russian mathematician Michel Ostrogradsky (1801–1861). ■

NOTE This proof is restricted to a *simple* solid region. The general proof is best left to a course in advanced calculus.

PROOF If you let $\mathbf{F}(x, y, z) = M\mathbf{i} + N\mathbf{j} + P\mathbf{k}$, the theorem takes the form

$$\int_S\!\!\int \mathbf{F} \cdot \mathbf{N}\, dS = \int_S\!\!\int (M\mathbf{i} \cdot \mathbf{N} + N\mathbf{j} \cdot \mathbf{N} + P\mathbf{k} \cdot \mathbf{N})\, dS$$

$$= \iiint_Q \left(\frac{\partial M}{\partial x} + \frac{\partial N}{\partial y} + \frac{\partial P}{\partial z}\right) dV.$$

You can prove this by verifying that the following three equations are valid.

$$\int_S\!\!\int M\mathbf{i} \cdot \mathbf{N}\, dS = \iiint_Q \frac{\partial M}{\partial x}\, dV$$

$$\int_S\!\!\int N\mathbf{j} \cdot \mathbf{N}\, dS = \iiint_Q \frac{\partial N}{\partial y}\, dV$$

$$\int_S\!\!\int P\mathbf{k} \cdot \mathbf{N}\, dS = \iiint_Q \frac{\partial P}{\partial z}\, dV$$

Because the verifications of the three equations are similar, only the third is discussed. Restrict the proof to a **simple solid** region with upper surface

$z = g_2(x, y)$ Upper surface

and lower surface

$z = g_1(x, y)$ Lower surface

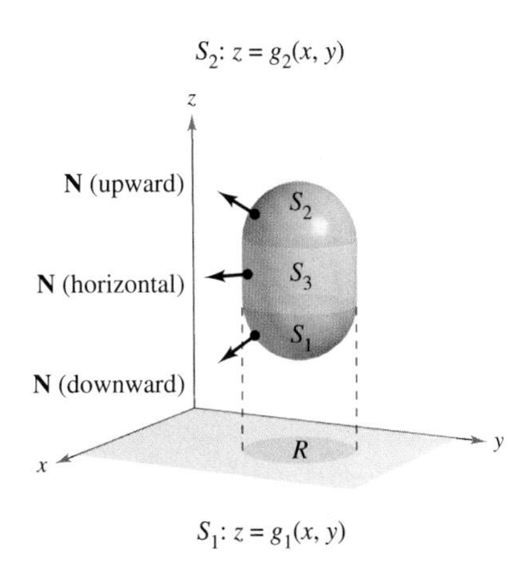

Figure 15.55

whose projections onto the xy-plane coincide and form region R. If Q has a lateral surface like S_3 in Figure 15.55, then a normal vector is horizontal, which implies that $P\mathbf{k} \cdot \mathbf{N} = 0$. Consequently, you have

$$\int_S\!\!\int P\mathbf{k} \cdot \mathbf{N}\, dS = \int_{S_1}\!\!\int P\mathbf{k} \cdot \mathbf{N}\, dS + \int_{S_2}\!\!\int P\mathbf{k} \cdot \mathbf{N}\, dS + 0.$$

On the upper surface S_2, the outward normal vector is upward, whereas on the lower surface S_1, the outward normal vector is downward. So, by Theorem 15.11, you have the following.

$$\int_{S_1}\!\!\int P\mathbf{k} \cdot \mathbf{N}\, dS = \int_R\!\!\int P(x, y, g_1(x, y))\mathbf{k} \cdot \left(\frac{\partial g_1}{\partial x}\mathbf{i} + \frac{\partial g_1}{\partial y}\mathbf{j} - \mathbf{k}\right) dA$$

$$= -\int_R\!\!\int P(x, y, g_1(x, y))\, dA$$

$$\int_{S_2}\!\!\int P\mathbf{k} \cdot \mathbf{N}\, dS = \int_R\!\!\int P(x, y, g_2(x, y))\mathbf{k} \cdot \left(-\frac{\partial g_2}{\partial x}\mathbf{i} - \frac{\partial g_2}{\partial y}\mathbf{j} + \mathbf{k}\right) dA$$

$$= \int_R\!\!\int P(x, y, g_2(x, y))\, dA$$

Adding these results, you obtain

$$\int_S\!\!\int P\mathbf{k} \cdot \mathbf{N}\, dS = \int_R\!\!\int [P(x, y, g_2(x, y)) - P(x, y, g_1(x, y))]\, dA$$

$$= \int_R\!\!\int \left[\int_{g_1(x, y)}^{g_2(x, y)} \frac{\partial P}{\partial z}\, dz\right] dA$$

$$= \iiint_Q \frac{\partial P}{\partial z}\, dV.$$

■

EXAMPLE 1 Using the Divergence Theorem

Let Q be the solid region bounded by the coordinate planes and the plane $2x + 2y + z = 6$, and let $\mathbf{F} = x\mathbf{i} + y^2\mathbf{j} + z\mathbf{k}$. Find

$$\int_S\int \mathbf{F} \cdot \mathbf{N}\, dS$$

where S is the surface of Q.

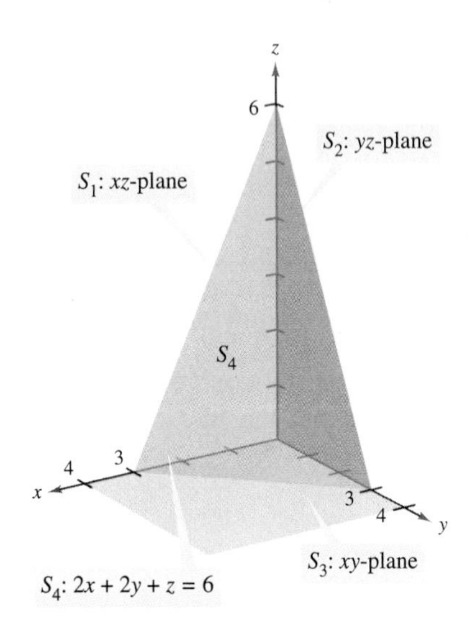

Figure 15.56

Solution From Figure 15.56, you can see that Q is bounded by four subsurfaces. So, you would need four *surface integrals* to evaluate

$$\int_S\int \mathbf{F} \cdot \mathbf{N}\, dS.$$

However, by the Divergence Theorem, you need only one triple integral. Because

$$\begin{aligned}\text{div}\,\mathbf{F} &= \frac{\partial M}{\partial x} + \frac{\partial N}{\partial y} + \frac{\partial P}{\partial z}\\ &= 1 + 2y + 1\\ &= 2 + 2y\end{aligned}$$

you have

$$\begin{aligned}\int_S\int \mathbf{F} \cdot \mathbf{N}\, dS &= \iiint_Q \text{div}\,\mathbf{F}\, dV\\ &= \int_0^3\int_0^{3-y}\int_0^{6-2x-2y} (2 + 2y)\, dz\, dx\, dy\\ &= \int_0^3\int_0^{3-y} (2z + 2yz)\Big]_0^{6-2x-2y} dx\, dy\\ &= \int_0^3\int_0^{3-y} (12 - 4x + 8y - 4xy - 4y^2)\, dx\, dy\\ &= \int_0^3 \Big[12x - 2x^2 + 8xy - 2x^2y - 4xy^2\Big]_0^{3-y} dy\\ &= \int_0^3 (18 + 6y - 10y^2 + 2y^3)\, dy\\ &= \Big[18y + 3y^2 - \frac{10y^3}{3} + \frac{y^4}{2}\Big]_0^3\\ &= \frac{63}{2}.\end{aligned}$$

■

TECHNOLOGY If you have access to a computer algebra system that can evaluate triple-iterated integrals, use it to verify the result in Example 1. When you are using such a utility, note that the first step is to convert the triple integral to an iterated integral—this step must be done by hand. To give yourself some practice with this important step, find the limits of integration for the following iterated integrals. Then use a computer to verify that the value is the same as that obtained in Example 1.

$$\int_?^?\int_?^?\int_?^? (2 + 2y)\, dy\, dz\, dx, \qquad \int_?^?\int_?^?\int_?^? (2 + 2y)\, dx\, dy\, dz$$

EXAMPLE 2 Verifying the Divergence Theorem

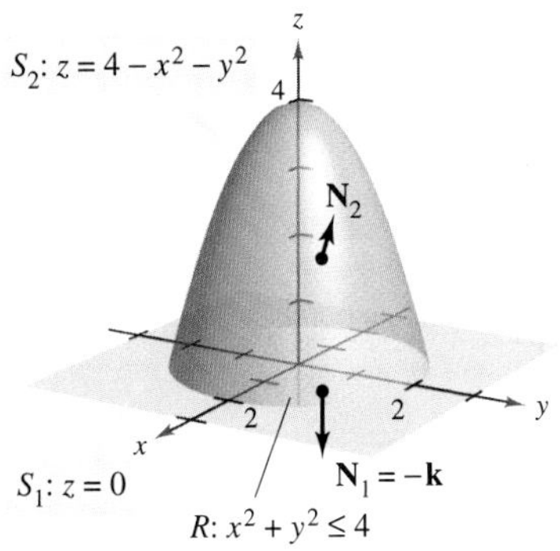

Let Q be the solid region between the paraboloid

$$z = 4 - x^2 - y^2$$

and the xy-plane. Verify the Divergence Theorem for

$$\mathbf{F}(x, y, z) = 2z\mathbf{i} + x\mathbf{j} + y^2\mathbf{k}.$$

S_2: $z = 4 - x^2 - y^2$

S_1: $z = 0$

$\mathbf{N}_2$

$\mathbf{N}_1 = -\mathbf{k}$

R: $x^2 + y^2 \le 4$

Figure 15.57

Solution From Figure 15.57 you can see that the outward normal vector for the surface S_1 is $\mathbf{N}_1 = -\mathbf{k}$, whereas the outward normal vector for the surface S_2 is

$$\mathbf{N}_2 = \frac{2x\mathbf{i} + 2y\mathbf{j} + \mathbf{k}}{\sqrt{4x^2 + 4y^2 + 1}}.$$

So, by Theorem 15.11, you have

$$\begin{aligned}
\int_S\!\!\int \mathbf{F} \cdot \mathbf{N}\, dS
&= \int_{S_1}\!\!\int \mathbf{F} \cdot \mathbf{N}_1\, dS + \int_{S_2}\!\!\int \mathbf{F} \cdot \mathbf{N}_2\, dS \\
&= \int_{S_1}\!\!\int \mathbf{F} \cdot (-\mathbf{k})\, dS + \int_{S_2}\!\!\int \mathbf{F} \cdot \frac{(2x\mathbf{i} + 2y\mathbf{j} + \mathbf{k})}{\sqrt{4x^2 + 4y^2 + 1}}\, dS \\
&= \int_R\!\!\int -y^2\, dA + \int_R\!\!\int (4xz + 2xy + y^2)\, dA \\
&= -\int_{-2}^{2}\int_{-\sqrt{4-y^2}}^{\sqrt{4-y^2}} y^2\, dx\, dy + \int_{-2}^{2}\int_{-\sqrt{4-y^2}}^{\sqrt{4-y^2}} (4xz + 2xy + y^2)\, dx\, dy \\
&= \int_{-2}^{2}\int_{-\sqrt{4-y^2}}^{\sqrt{4-y^2}} (4xz + 2xy)\, dx\, dy \\
&= \int_{-2}^{2}\int_{-\sqrt{4-y^2}}^{\sqrt{4-y^2}} [4x(4 - x^2 - y^2) + 2xy]\, dx\, dy \\
&= \int_{-2}^{2}\int_{-\sqrt{4-y^2}}^{\sqrt{4-y^2}} (16x - 4x^3 - 4xy^2 + 2xy)\, dx\, dy \\
&= \int_{-2}^{2}\left[8x^2 - x^4 - 2x^2y^2 + x^2y\right]_{-\sqrt{4-y^2}}^{\sqrt{4-y^2}} dy \\
&= \int_{-2}^{2} 0\, dy \\
&= 0.
\end{aligned}$$

On the other hand, because

$$\text{div}\,\mathbf{F} = \frac{\partial}{\partial x}[2z] + \frac{\partial}{\partial y}[x] + \frac{\partial}{\partial z}[y^2] = 0 + 0 + 0 = 0$$

you can apply the Divergence Theorem to obtain the equivalent result

$$\begin{aligned}
\int_S\!\!\int \mathbf{F} \cdot \mathbf{N}\, dS &= \iiint_Q \text{div}\,\mathbf{F}\, dV \\
&= \iiint_Q 0\, dV = 0.
\end{aligned}$$

■

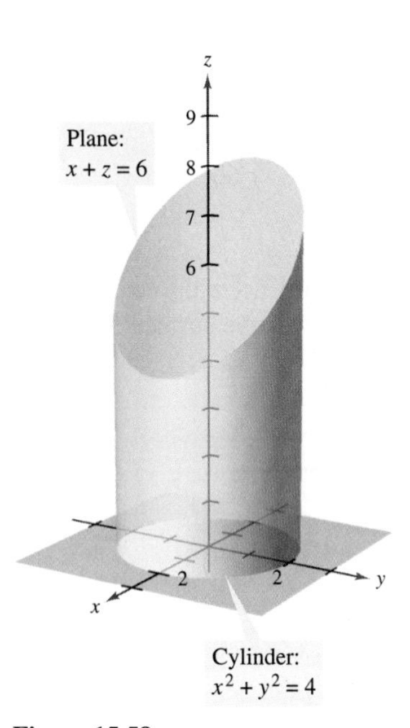

Figure 15.58

EXAMPLE 3 Using the Divergence Theorem

Let Q be the solid bounded by the cylinder $x^2 + y^2 = 4$, the plane $x + z = 6$, and the xy-plane, as shown in Figure 15.58. Find

$$\int_S\int \mathbf{F} \cdot \mathbf{N}\, dS$$

where S is the surface of Q and

$$\mathbf{F}(x, y, z) = (x^2 + \sin z)\mathbf{i} + (xy + \cos z)\mathbf{j} + e^y\mathbf{k}.$$

Solution Direct evaluation of this surface integral would be difficult. However, by the Divergence Theorem, you can evaluate the integral as follows.

$$\begin{aligned}
\int_S\int \mathbf{F} \cdot \mathbf{N}\, dS &= \iiint_Q \operatorname{div} \mathbf{F}\, dV \\
&= \iiint_Q (2x + x + 0)\, dV \\
&= \iiint_Q 3x\, dV \\
&= \int_0^{2\pi}\int_0^2\int_0^{6 - r\cos\theta} (3r\cos\theta)r\, dz\, dr\, d\theta \\
&= \int_0^{2\pi}\int_0^2 (18r^2\cos\theta - 3r^3\cos^2\theta)\, dr\, d\theta \\
&= \int_0^{2\pi} (48\cos\theta - 12\cos^2\theta)d\theta \\
&= \left[48\sin\theta - 6\left(\theta + \frac{1}{2}\sin 2\theta\right)\right]_0^{2\pi} \\
&= -12\pi
\end{aligned}$$

Notice that cylindrical coordinates with $x = r\cos\theta$ and $dV = r\, dz\, dr\, d\theta$ were used to evaluate the triple integral. ■

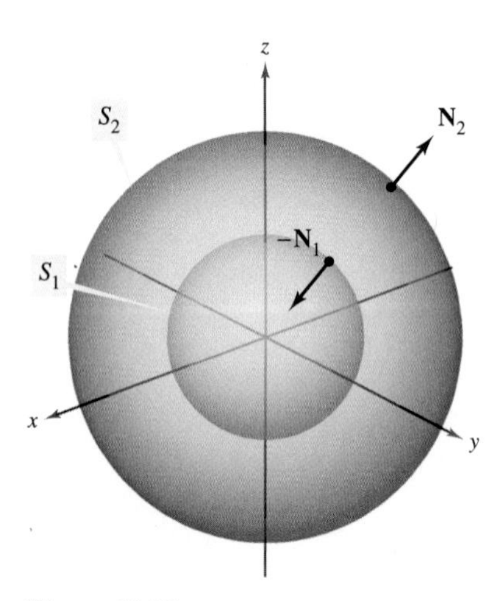

Figure 15.59

Even though the Divergence Theorem was stated for a simple solid region Q bounded by a closed surface, the theorem is also valid for regions that are the finite unions of simple solid regions. For example, let Q be the solid bounded by the closed surfaces S_1 and S_2, as shown in Figure 15.59. To apply the Divergence Theorem to this solid, let $S = S_1 \cup S_2$. The normal vector $\mathbf{N}$ to S is given by $-\mathbf{N}_1$ on S_1 and by $\mathbf{N}_2$ on S_2. So, you can write

$$\begin{aligned}
\iiint_Q \operatorname{div} \mathbf{F}\, dV &= \int_S\int \mathbf{F} \cdot \mathbf{N}\, dS \\
&= \int_{S_1}\int \mathbf{F} \cdot (-\mathbf{N}_1)\, dS + \int_{S_2}\int \mathbf{F} \cdot \mathbf{N}_2\, dS \\
&= -\int_{S_1}\int \mathbf{F} \cdot \mathbf{N}_1\, dS + \int_{S_2}\int \mathbf{F} \cdot \mathbf{N}_2\, dS.
\end{aligned}$$

Flux and the Divergence Theorem

To help understand the Divergence Theorem, consider the two sides of the equation

$$\int\!\!\int_S \mathbf{F} \cdot \mathbf{N}\, dS = \int\!\!\int\!\!\int_Q \text{div}\, \mathbf{F}\, dV.$$

Figure 15.60

You know from Section 15.6 that the flux integral on the left determines the total fluid flow across the surface S per unit of time. This can be approximated by summing the fluid flow across small patches of the surface. The triple integral on the right measures this same fluid flow across S, but from a very different perspective—namely, by calculating the flow of fluid into (or out of) small *cubes* of volume ΔV_i. The flux of the ith cube is approximately

$$\text{Flux of } i\text{th cube} \approx \text{div}\, \mathbf{F}(x_{i,}\, y_i, z_i)\, \Delta V_i$$

for some point (x_i, y_i, z_i) in the ith cube. Note that for a cube in the interior of Q, the gain (or loss) of fluid through any one of its six sides is offset by a corresponding loss (or gain) through one of the sides of an adjacent cube. After summing over all the cubes in Q, the only fluid flow that is not canceled by adjoining cubes is that on the outside edges of the cubes on the boundary. So, the sum

$$\sum_{i=1}^{n} \text{div}\, \mathbf{F}(x_i, y_i, z_i)\, \Delta V_i$$

approximates the total flux into (or out of) Q, and therefore through the surface S.

To see what is meant by the divergence of $\mathbf{F}$ at a point, consider ΔV_α to be the volume of a small sphere S_α of radius α and center (x_0, y_0, z_0), contained in region Q, as shown in Figure 15.60. Applying the Divergence Theorem to S_α produces

$$\begin{aligned} \text{Flux of } \mathbf{F} \text{ across } S_\alpha &= \int\!\!\int\!\!\int_{Q_\alpha} \text{div}\, \mathbf{F}\, dV \\ &\approx \text{div}\, \mathbf{F}(x_0, y_0, z_0,)\, \Delta V_\alpha \end{aligned}$$

(a) Source: div $\mathbf{F} > 0$

where Q_α is the interior of S_α. Consequently, you have

$$\text{div}\, \mathbf{F}(x_0, y_0, z_0) \approx \frac{\text{flux of } \mathbf{F} \text{ across } S_\alpha}{\Delta V_\alpha}$$

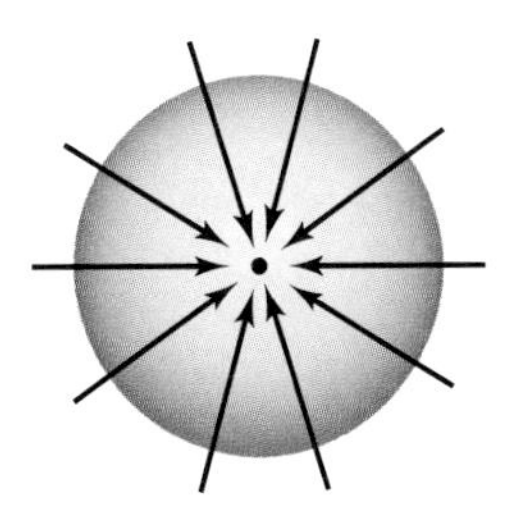

(b) Sink: div $\mathbf{F} < 0$

and, by taking the limit as $\alpha \to 0$, you obtain the divergence of $\mathbf{F}$ at the point (x_0, y_0, z_0).

$$\begin{aligned} \text{div}\, \mathbf{F}(x_0, y_0, z_0) &= \lim_{\alpha \to 0} \frac{\text{flux of } \mathbf{F} \text{ across } S_\alpha}{\Delta V_\alpha} \\ &= \text{flux per unit volume at } (x_0, y_0, z_0) \end{aligned}$$

The point (x_0, y_0, z_0) in a vector field is classified as a source, a sink, or incompressible, as follows.

1. **Source,** if div $\mathbf{F} > 0$ See Figure 15.61(a).
2. **Sink,** if div $\mathbf{F} < 0$ See Figure 15.61(b).
3. **Incompressible,** if div $\mathbf{F} = 0$ See Figure 15.61(c).

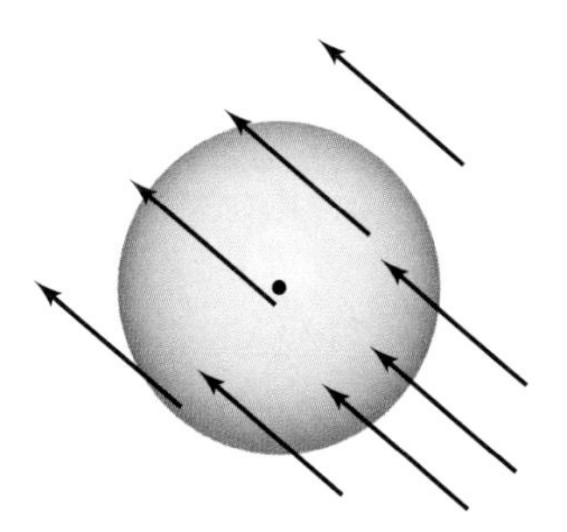

(c) Incompressible: div $\mathbf{F} = 0$

Figure 15.61

NOTE In hydrodynamics, a *source* is a point at which additional fluid is considered as being introduced to the region occupied by the fluid. A *sink* is a point at which fluid is considered as being removed. ■

EXAMPLE 4 Calculating Flux by the Divergence Theorem

Let Q be the region bounded by the sphere $x^2 + y^2 + z^2 = 4$. Find the outward flux of the vector field $\mathbf{F}(x, y, z) = 2x^3\mathbf{i} + 2y^3\mathbf{j} + 2z^3\mathbf{k}$ through the sphere.

Solution By the Divergence Theorem, you have

$$\begin{aligned}
\text{Flux across } S &= \int_S\!\!\int \mathbf{F} \cdot \mathbf{N}\, dS = \iiint_Q \text{div } \mathbf{F}\, dV \\
&= \iiint_Q 6(x^2 + y^2 + z^2)\, dV \\
&= 6\int_0^2\int_0^{\pi}\int_0^{2\pi} \rho^4 \sin\phi\, d\theta\, d\phi\, d\rho \qquad \text{Spherical coordinates} \\
&= 6\int_0^2\int_0^{\pi} 2\pi\rho^4 \sin\phi\, d\phi\, d\rho \\
&= 12\pi\int_0^2 2\rho^4\, d\rho \\
&= 24\pi\left(\frac{32}{5}\right) \\
&= \frac{768\pi}{5}.
\end{aligned}$$

■

15.7 Exercises

See www.CalcChat.com for worked-out solutions to odd-numbered exercises.

In Exercises 1–6, verify the Divergence Theorem by evaluating

$$\int_S\!\!\int \mathbf{F} \cdot \mathbf{N}\, dS$$

as a surface integral and as a triple integral.

1. $\mathbf{F}(x, y, z) = 2x\mathbf{i} - 2y\mathbf{j} + z^2\mathbf{k}$

   S: cube bounded by the planes $x = 0$, $x = a$, $y = 0$, $y = a$, $z = 0$, $z = a$

2. $\mathbf{F}(x, y, z) = 2x\mathbf{i} - 2y\mathbf{j} + z^2\mathbf{k}$

   S: cylinder $x^2 + y^2 = 4$, $0 \le z \le h$

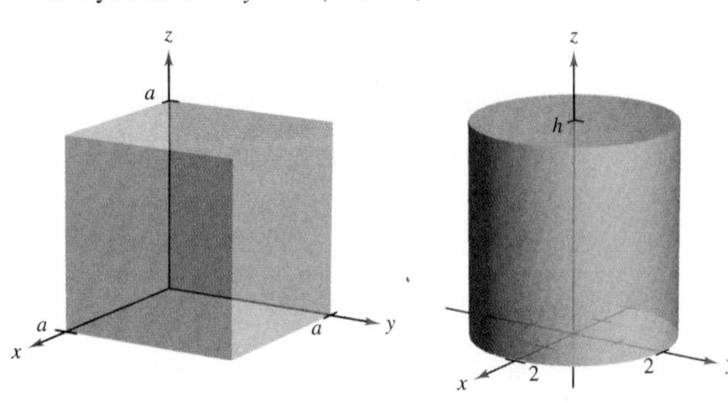

Figure for 1 **Figure for 2**

3. $\mathbf{F}(x, y, z) = (2x - y)\mathbf{i} - (2y - z)\mathbf{j} + z\mathbf{k}$

   S: surface bounded by the plane $2x + 4y + 2z = 12$ and the coordinate planes

4. $\mathbf{F}(x, y, z) = xy\mathbf{i} + z\mathbf{j} + (x + y)\mathbf{k}$

   S: surface bounded by the planes $y = 4$ and $z = 4 - x$ and the coordinate planes

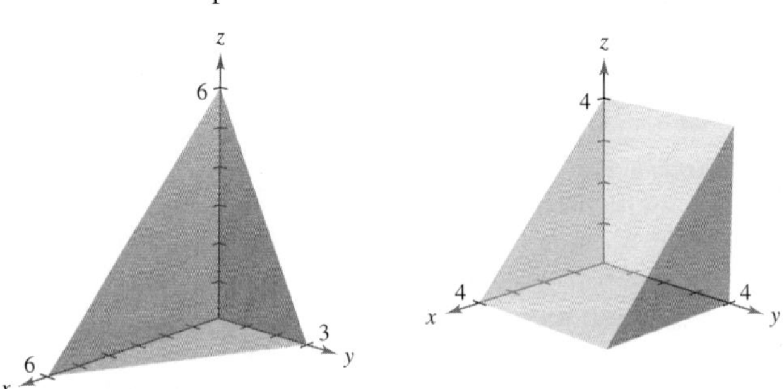

Figure for 3 **Figure for 4**

5. $\mathbf{F}(x, y, z) = xz\mathbf{i} + zy\mathbf{j} + 2z^2\mathbf{k}$

   S: surface bounded by $z = 1 - x^2 - y^2$ and $z = 0$

6. $\mathbf{F}(x, y, z) = xy^2\mathbf{i} + yx^2\mathbf{j} + e\mathbf{k}$

   S: surface bounded by $z = \sqrt{x^2 + y^2}$ and $z = 4$

In Exercises 7–18, use the Divergence Theorem to evaluate

$$\int_S\int \mathbf{F} \cdot \mathbf{N}\, dS$$

and find the outward flux of F through the surface of the solid bounded by the graphs of the equations. Use a computer algebra system to verify your results.

7. $\mathbf{F}(x, y, z) = x^2\mathbf{i} + y^2\mathbf{j} + z^2\mathbf{k}$

S: $x = 0, x = a, y = 0, y = a, z = 0, z = a$

8. $\mathbf{F}(x, y, z) = x^2z^2\mathbf{i} - 2y\mathbf{j} + 3xyz\mathbf{k}$

S: $x = 0, x = a, y = 0, y = a, z = 0, z = a$

9. $\mathbf{F}(x, y, z) = x^2\mathbf{i} - 2xy\mathbf{j} + xyz^2\mathbf{k}$

S: $z = \sqrt{a^2 - x^2 - y^2}, z = 0$

10. $\mathbf{F}(x, y, z) = xy\mathbf{i} + yz\mathbf{j} - yz\mathbf{k}$

S: $z = \sqrt{a^2 - x^2 - y^2}, z = 0$

11. $\mathbf{F}(x, y, z) = x\mathbf{i} + y\mathbf{j} + z\mathbf{k}$

S: $x^2 + y^2 + z^2 = 9$

12. $\mathbf{F}(x, y, z) = xyz\mathbf{j}$

S: $x^2 + y^2 = 4, z = 0, z = 5$

13. $\mathbf{F}(x, y, z) = x\mathbf{i} + y^2\mathbf{j} - z\mathbf{k}$

S: $x^2 + y^2 = 25, z = 0, z = 7$

14. $\mathbf{F}(x, y, z) = (xy^2 + \cos z)\mathbf{i} + (x^2y + \sin z)\mathbf{j} + e^z\mathbf{k}$

S: $z = \frac{1}{2}\sqrt{x^2 + y^2}, z = 8$

15. $\mathbf{F}(x, y, z) = x^3\mathbf{i} + x^2y\mathbf{j} + x^2e^y\mathbf{k}$

S: $z = 4 - y, z = 0, x = 0, x = 6, y = 0$

16. $\mathbf{F}(x, y, z) = xe^z\mathbf{i} + ye^z\mathbf{j} + e^z\mathbf{k}$

S: $z = 4 - y, z = 0, x = 0, x = 6, y = 0$

17. $\mathbf{F}(x, y, z) = xy\mathbf{i} + 4y\mathbf{j} + xz\mathbf{k}$

S: $x^2 + y^2 + z^2 = 16$

18. $\mathbf{F}(x, y, z) = 2(x\mathbf{i} + y\mathbf{j} + z\mathbf{k})$

S: $z = \sqrt{4 - x^2 - y^2}, z = 0$

In Exercises 19 and 20, evaluate

$$\int_S\int \text{curl}\, \mathbf{F} \cdot \mathbf{N}\, dS$$

where S is the closed surface of the solid bounded by the graphs of $x = 4$ and $z = 9 - y^2$, and the coordinate planes.

19. $\mathbf{F}(x, y, z) = (4xy + z^2)\mathbf{i} + (2x^2 + 6yz)\mathbf{j} + 2xz\mathbf{k}$

20. $\mathbf{F}(x, y, z) = xy\cos z\mathbf{i} + yz\sin x\mathbf{j} + xyz\mathbf{k}$

WRITING ABOUT CONCEPTS

21. State the Divergence Theorem.

22. How do you determine if a point (x_0, y_0, z_0) in a vector field is a source, a sink, or incompressible?

23. (a) Use the Divergence Theorem to verify that the volume of the solid bounded by a surface S is

$$\int_S\int x\, dy\, dz = \int_S\int y\, dz\, dx = \int_S\int z\, dx\, dy.$$

(b) Verify the result of part (a) for the cube bounded by $x = 0$, $x = a$, $y = 0$, $y = a$, $z = 0$, and $z = a$.

CAPSTONE

24. Let $\mathbf{F}(x, y, z) = x\mathbf{i} + y\mathbf{j} + z\mathbf{k}$ and let S be the cube bounded by the planes $x = 0$, $x = 1$, $y = 0$, $y = 1$, $z = 0$, and $z = 1$. Verify the Divergence Theorem by evaluating

$$\int_S\int \mathbf{F} \cdot \mathbf{N}\, dS$$

as a surface integral and as a triple integral.

25. Verify that

$$\int_S\int \text{curl}\, \mathbf{F} \cdot \mathbf{N}\, dS = 0$$

for any closed surface S.

26. For the constant vector field $\mathbf{F}(x, y, z) = a_1\mathbf{i} + a_2\mathbf{j} + a_3\mathbf{k}$, verify that

$$\int_S\int \mathbf{F} \cdot \mathbf{N}\, dS = 0$$

where V is the volume of the solid bounded by the closed surface S.

27. Given the vector field $\mathbf{F}(x, y, z) = x\mathbf{i} + y\mathbf{j} + z\mathbf{k}$, verify that

$$\int_S\int \mathbf{F} \cdot \mathbf{N}\, dS = 3V$$

where V is the volume of the solid bounded by the closed surface S.

28. Given the vector field $\mathbf{F}(x, y, z) = x\mathbf{i} + y\mathbf{j} + z\mathbf{k}$, verify that

$$\frac{1}{\|\mathbf{F}\|}\int_S\int \mathbf{F} \cdot \mathbf{N}\, dS = \frac{3}{\|\mathbf{F}\|}\iiint_Q dV.$$

In Exercises 29 and 30, prove the identity, assuming that Q, S, and N meet the conditions of the Divergence Theorem and that the required partial derivatives of the scalar functions f and g are continuous. The expressions $D_{\mathbf{N}}f$ and $D_{\mathbf{N}}g$ are the derivatives in the direction of the vector N and are defined by

$$D_{\mathbf{N}}f = \nabla f \cdot \mathbf{N}, \quad D_{\mathbf{N}}g = \nabla g \cdot \mathbf{N}.$$

29. $$\iiint_Q (f\nabla^2 g + \nabla f \cdot \nabla g)\, dV = \int_S\int f D_{\mathbf{N}}g\, dS$$

[*Hint:* Use div $(f\mathbf{G}) = f$ div $\mathbf{G} + \nabla f \cdot \mathbf{G}$.]

30. $$\iiint_Q (f\nabla^2 g - g\nabla^2 f)\, dV = \int_S\int (f D_{\mathbf{N}}g - g D_{\mathbf{N}}f)\, dS$$

(*Hint:* Use Exercise 29 twice.)

15.8 Stokes's Theorem

- **Understand and use Stokes's Theorem.**
- **Use curl to analyze the motion of a rotating liquid.**

GEORGE GABRIEL STOKES (1819–1903)

Stokes became a Lucasian professor of mathematics at Cambridge in 1849. Five years later, he published the theorem that bears his name as a prize examination question there.

Stokes's Theorem

A second higher-dimension analog of Green's Theorem is called **Stokes's Theorem,** after the English mathematical physicist George Gabriel Stokes. Stokes was part of a group of English mathematical physicists referred to as the Cambridge School, which included William Thomson (Lord Kelvin) and James Clerk Maxwell. In addition to making contributions to physics, Stokes worked with infinite series and differential equations, as well as with the integration results presented in this section.

Stokes's Theorem gives the relationship between a surface integral over an oriented surface S and a line integral along a closed space curve C forming the boundary of S, as shown in Figure 15.62. The positive direction along C is counterclockwise relative to the normal vector $\mathbf{N}$. That is, if you imagine grasping the normal vector $\mathbf{N}$ with your right hand, with your thumb pointing in the direction of $\mathbf{N}$, your fingers will point in the positive direction C, as shown in Figure 15.63.

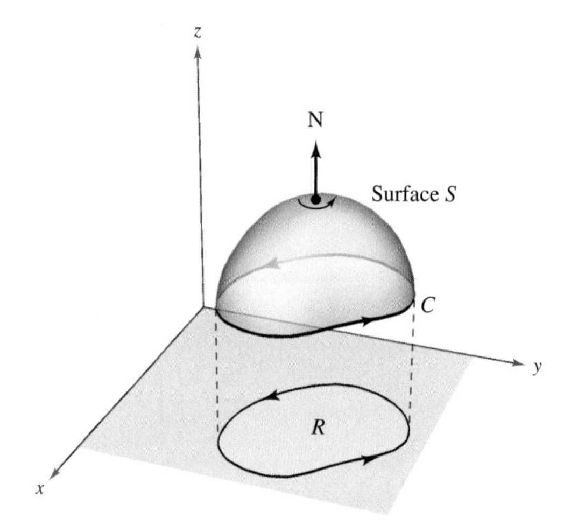

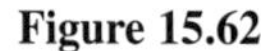

Figure 15.62

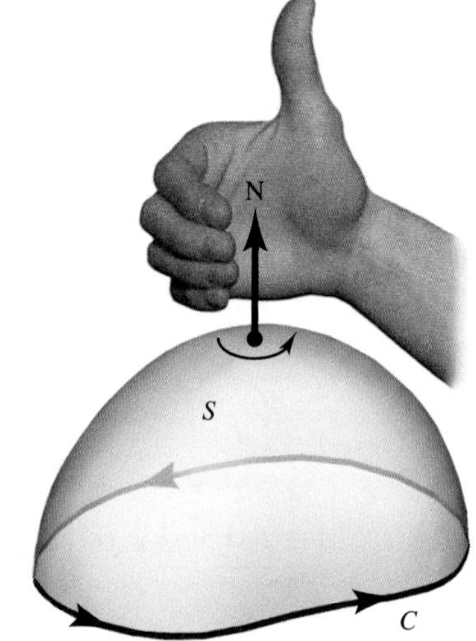

Direction along C is counterclockwise relative to $\mathbf{N}$.
Figure 15.63

> **THEOREM 15.13 STOKES'S THEOREM**
>
> Let S be an oriented surface with unit normal vector $\mathbf{N}$, bounded by a piecewise smooth simple closed curve C with a positive orientation. If $\mathbf{F}$ is a vector field whose component functions have continuous first partial derivatives on an open region containing S and C, then
>
> $$\int_C \mathbf{F} \cdot d\mathbf{r} = \int\int_S (\text{curl } \mathbf{F}) \cdot \mathbf{N}\, dS.$$

NOTE The line integral may be written in the differential form $\int_C M\,dx + N\,dy + P\,dz$ or in the vector form $\int_C \mathbf{F} \cdot \mathbf{T}\,ds$. ■

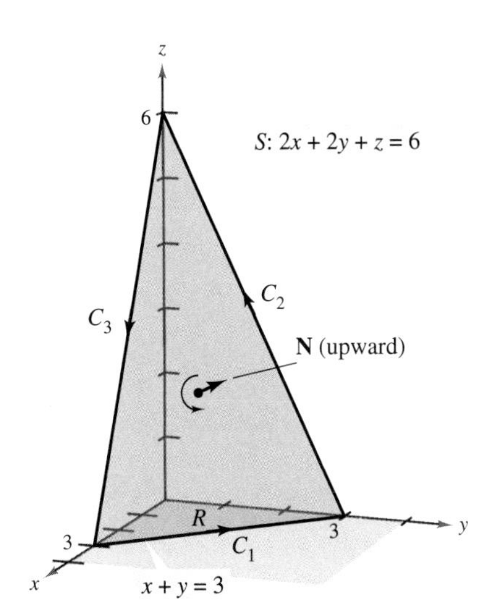

Figure 15.64

EXAMPLE 1 Using Stokes's Theorem

Let C be the oriented triangle lying in the plane $2x + 2y + z = 6$, as shown in Figure 15.64. Evaluate

$$\int_C \mathbf{F} \cdot d\mathbf{r}$$

where $\mathbf{F}(x, y, z) = -y^2\mathbf{i} + z\mathbf{j} + x\mathbf{k}$.

Solution Using Stokes's Theorem, begin by finding the curl of $\mathbf{F}$.

$$\text{curl } \mathbf{F} = \begin{vmatrix} \mathbf{i} & \mathbf{j} & \mathbf{k} \\ \dfrac{\partial}{\partial x} & \dfrac{\partial}{\partial y} & \dfrac{\partial}{\partial z} \\ -y^2 & z & x \end{vmatrix} = -\mathbf{i} - \mathbf{j} + 2y\mathbf{k}$$

Considering $z = 6 - 2x - 2y = g(x, y)$, you can use Theorem 15.11 for an upward normal vector to obtain

$$\begin{aligned}
\int_C \mathbf{F} \cdot d\mathbf{r} &= \int_S\int (\text{curl } \mathbf{F}) \cdot \mathbf{N}\, dS \\
&= \int_R\int (-\mathbf{i} - \mathbf{j} + 2y\mathbf{k}) \cdot [-g_x(x, y)\mathbf{i} - g_y(x, y)\mathbf{j} + \mathbf{k}]\, dA \\
&= \int_R\int (-\mathbf{i} - \mathbf{j} + 2y\mathbf{k}) \cdot (2\mathbf{i} + 2\mathbf{j} + \mathbf{k})\, dA \\
&= \int_0^3 \int_0^{3-y} (2y - 4)\, dx\, dy \\
&= \int_0^3 (-2y^2 + 10y - 12)\, dy \\
&= \left[-\frac{2y^3}{3} + 5y^2 - 12y\right]_0^3 \\
&= -9.
\end{aligned}$$

■

Try evaluating the line integral in Example 1 directly, *without* using Stokes's Theorem. One way to do this would be to consider C as the union of C_1, C_2, and C_3, as follows.

$$C_1: \mathbf{r}_1(t) = (3 - t)\mathbf{i} + t\mathbf{j}, \quad 0 \le t \le 3$$
$$C_2: \mathbf{r}_2(t) = (6 - t)\mathbf{j} + (2t - 6)\mathbf{k}, \quad 3 \le t \le 6$$
$$C_3: \mathbf{r}_3(t) = (t - 6)\mathbf{i} + (18 - 2t)\mathbf{k}, \quad 6 \le t \le 9$$

The value of the line integral is

$$\begin{aligned}
\int_C \mathbf{F} \cdot d\mathbf{r} &= \int_{C_1} \mathbf{F} \cdot \mathbf{r}_1'(t)\, dt + \int_{C_2} \mathbf{F} \cdot \mathbf{r}_2'(t)\, dt + \int_{C_3} \mathbf{F} \cdot \mathbf{r}_3'(t)\, dt \\
&= \int_0^3 t^2\, dt + \int_3^6 (-2t + 6)\, dt + \int_6^9 (-2t + 12)\, dt \\
&= 9 - 9 - 9 \\
&= -9.
\end{aligned}$$

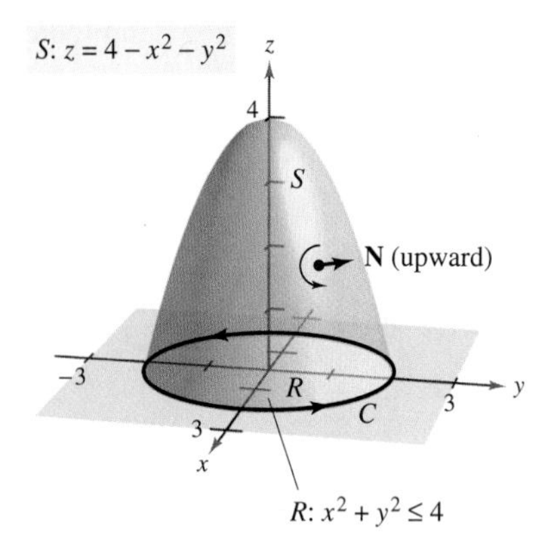

Figure 15.65

EXAMPLE 2 Verifying Stokes's Theorem

Let S be the portion of the paraboloid $z = 4 - x^2 - y^2$ lying above the xy-plane, oriented upward (see Figure 15.65). Let C be its boundary curve in the xy-plane, oriented counterclockwise. Verify Stokes's Theorem for

$$\mathbf{F}(x, y, z) = 2z\mathbf{i} + x\mathbf{j} + y^2\mathbf{k}$$

by evaluating the surface integral and the equivalent line integral.

Solution As a *surface integral*, you have $z = g(x, y) = 4 - x^2 - y^2$, $g_x = -2x$, $g_y = -2y$, and

$$\text{curl } \mathbf{F} = \begin{vmatrix} \mathbf{i} & \mathbf{j} & \mathbf{k} \\ \dfrac{\partial}{\partial x} & \dfrac{\partial}{\partial y} & \dfrac{\partial}{\partial z} \\ 2z & x & y^2 \end{vmatrix} = 2y\mathbf{i} + 2\mathbf{j} + \mathbf{k}.$$

By Theorem 15.11, you obtain

$$\begin{aligned}
\int_S\!\!\int (\text{curl } \mathbf{F}) \cdot \mathbf{N}\, dS &= \int_R\!\!\int (2y\mathbf{i} + 2\mathbf{j} + \mathbf{k}) \cdot (2x\mathbf{i} + 2y\mathbf{j} + \mathbf{k})\, dA \\
&= \int_{-2}^{2} \int_{-\sqrt{4-x^2}}^{\sqrt{4-x^2}} (4xy + 4y + 1)\, dy\, dx \\
&= \int_{-2}^{2} \Big[2xy^2 + 2y^2 + y) \Big]_{-\sqrt{4-x^2}}^{\sqrt{4-x^2}} dx \\
&= \int_{-2}^{2} 2\sqrt{4 - x^2}\, dx \\
&= \text{Area of circle of radius } 2 = 4\pi.
\end{aligned}$$

As a *line integral*, you can parametrize C as

$$\mathbf{r}(t) = 2 \cos t\mathbf{i} + 2 \sin t\mathbf{j} + 0\mathbf{k}, \quad 0 \leq t \leq 2\pi.$$

For $\mathbf{F}(x, y, z) = 2z\mathbf{i} + x\mathbf{j} + y^2\mathbf{k}$, you obtain

$$\begin{aligned}
\int_C \mathbf{F} \cdot d\mathbf{r} &= \int_C M\, dx + N\, dy + P\, dz \\
&= \int_C 2z\, dx + x\, dy + y^2\, dz \\
&= \int_0^{2\pi} [0 + 2 \cos t(2 \cos t) + 0]\, dt \\
&= \int_0^{2\pi} 4 \cos^2 t\, dt \\
&= 2\int_0^{2\pi} (1 + \cos 2t)\, dt \\
&= 2\Big[t + \frac{1}{2} \sin 2t \Big]_0^{2\pi} \\
&= 4\pi.
\end{aligned}$$

■

Physical Interpretation of Curl

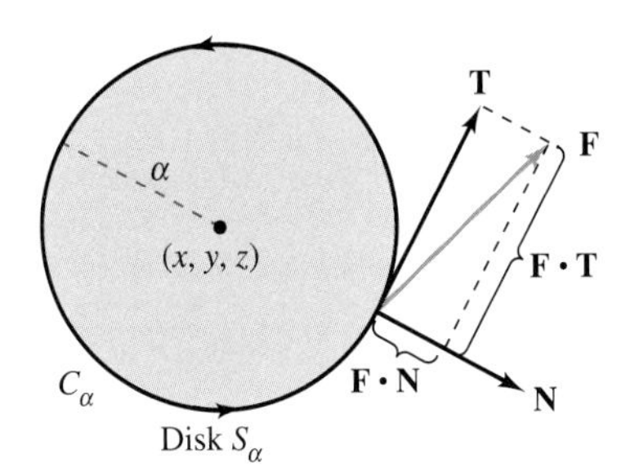

Figure 15.66

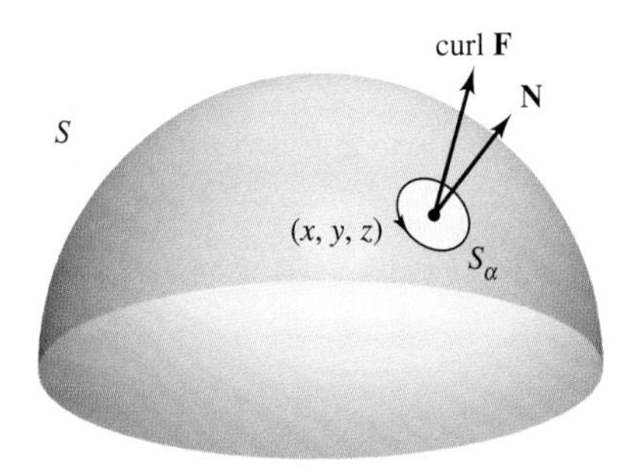

Figure 15.67

Stokes's Theorem provides insight into a physical interpretation of curl. In a vector field **F**, let S_α be a *small* circular disk of radius α, centered at (x, y, z) and with boundary C_α, as shown in Figure 15.66. At each point on the circle C_α, **F** has a normal component $\mathbf{F} \cdot \mathbf{N}$ and a tangential component $\mathbf{F} \cdot \mathbf{T}$. The more closely **F** and **T** are aligned, the greater the value of $\mathbf{F} \cdot \mathbf{T}$. So, a fluid tends to move along the circle rather than across it. Consequently, you say that the line integral around C_α measures the **circulation of F around C_α.** That is,

$$\int_{C_\alpha} \mathbf{F} \cdot \mathbf{T}\, ds = \text{circulation of } \mathbf{F} \text{ around } C_\alpha.$$

Now consider a small disk S_α to be centered at some point (x, y, z) on the surface S, as shown in Figure 15.67. On such a small disk, curl **F** is nearly constant, because it varies little from its value at (x, y, z). Moreover, curl $\mathbf{F} \cdot \mathbf{N}$ is also nearly constant on S_α, because all unit normals to S_α are about the same. Consequently, Stokes's Theorem yields

$$\begin{aligned}\int_{C_\alpha} \mathbf{F} \cdot \mathbf{T}\, ds &= \int_{S_\alpha}\int (\text{curl } \mathbf{F}) \cdot \mathbf{N}\, dS \\ &\approx (\text{curl } \mathbf{F}) \cdot \mathbf{N} \int_{S_\alpha}\int dS \\ &\approx (\text{curl } \mathbf{F}) \cdot \mathbf{N}(\pi\alpha^2).\end{aligned}$$

So,

$$\begin{aligned}(\text{curl } \mathbf{F}) \cdot \mathbf{N} &\approx \frac{\int_{C_\alpha} \mathbf{F} \cdot \mathbf{T}\, ds}{\pi\alpha^2} \\ &= \frac{\text{circulation of } \mathbf{F} \text{ around } C_\alpha}{\text{area of disk } S_\alpha} \\ &= \text{rate of circulation.}\end{aligned}$$

Assuming conditions are such that the approximation improves for smaller and smaller disks $(\alpha \to 0)$, it follows that

$$(\text{curl } \mathbf{F}) \cdot \mathbf{N} = \lim_{\alpha \to 0} \frac{1}{\pi\alpha^2} \int_{C_\alpha} \mathbf{F} \cdot \mathbf{T}\, ds$$

which is referred to as the **rotation of F about N.** That is,

$$\text{curl } \mathbf{F}(x, y, z) \cdot \mathbf{N} = \text{rotation of } \mathbf{F} \text{ about } \mathbf{N} \text{ at } (x, y, z).$$

In this case, the rotation of **F** is maximum when curl **F** and **N** have the same direction. Normally, this tendency to rotate will vary from point to point on the surface S, and Stokes's Theorem

$$\underbrace{\int_S\int (\text{curl } \mathbf{F}) \cdot \mathbf{N}\, dS}_{\text{Surface integral}} = \underbrace{\int_C \mathbf{F} \cdot d\mathbf{r}}_{\text{Line integral}}$$

says that the collective measure of this *rotational* tendency taken over the entire surface S (surface integral) is equal to the tendency of a fluid to *circulate* around the boundary C (line integral).

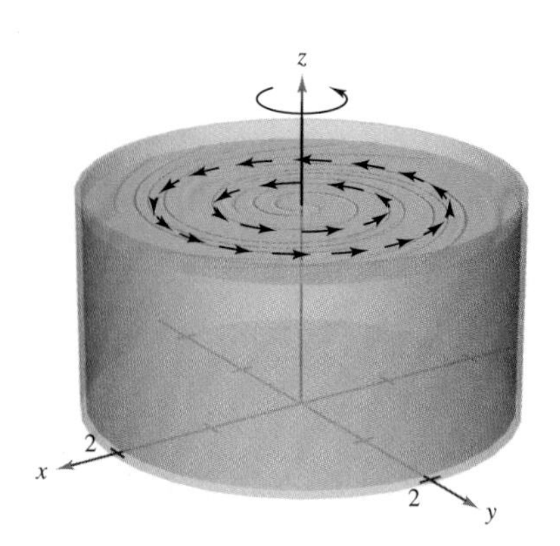

Figure 15.68

EXAMPLE 3 An Application of Curl

A liquid is swirling around in a cylindrical container of radius 2, so that its motion is described by the velocity field

$$\mathbf{F}(x, y, z) = -y\sqrt{x^2 + y^2}\mathbf{i} + x\sqrt{x^2 + y^2}\mathbf{j}$$

as shown in Figure 15.68. Find

$$\int_S\int (\text{curl }\mathbf{F}) \cdot \mathbf{N}\, dS$$

where S is the upper surface of the cylindrical container.

Solution The curl of $\mathbf{F}$ is given by

$$\text{curl }\mathbf{F} = \begin{vmatrix} \mathbf{i} & \mathbf{j} & \mathbf{k} \\ \dfrac{\partial}{\partial x} & \dfrac{\partial}{\partial y} & \dfrac{\partial}{\partial z} \\ -y\sqrt{x^2 + y^2} & x\sqrt{x^2 + y^2} & 0 \end{vmatrix} = 3\sqrt{x^2 + y^2}\mathbf{k}.$$

Letting $\mathbf{N} = \mathbf{k}$, you have

$$\begin{aligned} \int_S\int (\text{curl }\mathbf{F}) \cdot \mathbf{N}\, dS &= \int_R\int 3\sqrt{x^2 + y^2}\, dA \\ &= \int_0^{2\pi}\int_0^2 (3r)r\, dr\, d\theta \\ &= \int_0^{2\pi} r^3\Big]_0^2\, d\theta \\ &= \int_0^{2\pi} 8\, d\theta \\ &= 16\pi. \end{aligned}$$

■

NOTE If curl $\mathbf{F} = \mathbf{0}$ throughout region Q, the rotation of $\mathbf{F}$ about each unit normal $\mathbf{N}$ is 0. That is, $\mathbf{F}$ is irrotational. From earlier work, you know that this is a characteristic of conservative vector fields. ■

SUMMARY OF INTEGRATION FORMULAS

Fundamental Theorem of Calculus:

$$\int_a^b F'(x)\, dx = F(b) - F(a)$$

Fundamental Theorem of Line Integrals:

$$\int_C \mathbf{F} \cdot d\mathbf{r} = \int_C \nabla f \cdot d\mathbf{r} = f(x(b), y(b)) - f(x(a), y(a))$$

Green's Theorem:

$$\int_C M\, dx + N\, dy = \int_R\int \left(\frac{\partial N}{\partial x} - \frac{\partial M}{\partial y}\right) dA = \int_C \mathbf{F} \cdot \mathbf{T}\, ds = \int_C \mathbf{F} \cdot d\mathbf{r} = \int_R\int (\text{curl }\mathbf{F}) \cdot \mathbf{k}\, dA$$

$$\int_C \mathbf{F} \cdot \mathbf{N}\, ds = \int_R\int \text{div }\mathbf{F}\, dA$$

Divergence Theorem:

$$\int_S\int \mathbf{F} \cdot \mathbf{N}\, dS = \int\int_Q\int \text{div }\mathbf{F}\, dV$$

Stokes's Theorem:

$$\int_C \mathbf{F} \cdot d\mathbf{r} = \int_S\int (\text{curl }\mathbf{F}) \cdot \mathbf{N}\, dS$$

15.8 Exercises

See www.CalcChat.com for worked-out solutions to odd-numbered exercises.

In Exercises 1–6, find the curl of the vector field F.

1. $\mathbf{F}(x, y, z) = (2y - z)\mathbf{i} + e^z\mathbf{j} + xyz\mathbf{k}$
2. $\mathbf{F}(x, y, z) = x^2\mathbf{i} + y^2\mathbf{j} + x^2\mathbf{k}$
3. $\mathbf{F}(x, y, z) = 2z\mathbf{i} - 4x^2\mathbf{j} + \arctan x\mathbf{k}$
4. $\mathbf{F}(x, y, z) = x \sin y\mathbf{i} - y \cos x\mathbf{j} + yz^2\mathbf{k}$
5. $\mathbf{F}(x, y, z) = e^{x^2+y^2}\mathbf{i} + e^{y^2+z^2}\mathbf{j} + xyz\mathbf{k}$
6. $\mathbf{F}(x, y, z) = \arcsin y\mathbf{i} + \sqrt{1 - x^2}\,\mathbf{j} + y^2\mathbf{k}$

In Exercises 7–10, verify Stokes's Theorem by evaluating $\int_C \mathbf{F} \cdot \mathbf{T}\, ds = \int_C \mathbf{F} \cdot d\mathbf{r}$ as a line integral and as a double integral.

7. $\mathbf{F}(x, y, z) = (-y + z)\mathbf{i} + (x - z)\mathbf{j} + (x - y)\mathbf{k}$
   S: $z = 9 - x^2 - y^2, \quad z \geq 0$
8. $\mathbf{F}(x, y, z) = (-y + z)\mathbf{i} + (x - z)\mathbf{j} + (x - y)\mathbf{k}$
   S: $z = \sqrt{1 - x^2 - y^2}$
9. $\mathbf{F}(x, y, z) = xyz\mathbf{i} + y\mathbf{j} + z\mathbf{k}$
   S: $6x + 6y + z = 12$, first octant
10. $\mathbf{F}(x, y, z) = z^2\mathbf{i} + x^2\mathbf{j} + y^2\mathbf{k}$
    S: $z = y^2, \quad 0 \leq x \leq a, \quad 0 \leq y \leq a$

In Exercises 11–20, use Stokes's Theorem to evaluate $\int_C \mathbf{F} \cdot d\mathbf{r}$. Use a computer algebra system to verify your results. In each case, C is oriented counterclockwise as viewed from above.

11. $\mathbf{F}(x, y, z) = 2y\mathbf{i} + 3z\mathbf{j} + x\mathbf{k}$
    C: triangle with vertices $(2, 0, 0)$, $(0, 2, 0)$, $(0, 0, 2)$
12. $\mathbf{F}(x, y, z) = \arctan \dfrac{x}{y}\mathbf{i} + \ln\sqrt{x^2 + y^2}\,\mathbf{j} + \mathbf{k}$
    C: triangle with vertices $(0, 0, 0)$, $(1, 1, 1)$, $(0, 0, 2)$
13. $\mathbf{F}(x, y, z) = z^2\mathbf{i} + 2x\mathbf{j} + y^2\mathbf{k}$
    S: $z = 1 - x^2 - y^2, \quad z \geq 0$
14. $\mathbf{F}(x, y, z) = 4xz\mathbf{i} + y\mathbf{j} + 4xy\mathbf{k}$
    S: $z = 9 - x^2 - y^2, \quad z \geq 0$
15. $\mathbf{F}(x, y, z) = z^2\mathbf{i} + y\mathbf{j} + z\mathbf{k}$
    S: $z = \sqrt{4 - x^2 - y^2}$
16. $\mathbf{F}(x, y, z) = x^2\mathbf{i} + z^2\mathbf{j} - xyz\mathbf{k}$
    S: $z = \sqrt{4 - x^2 - y^2}$
17. $\mathbf{F}(x, y, z) = -\ln\sqrt{x^2 + y^2}\,\mathbf{i} + \arctan \dfrac{x}{y}\mathbf{j} + \mathbf{k}$
    S: $z = 9 - 2x - 3y$ over $r = 2 \sin 2\theta$ in the first octant
18. $\mathbf{F}(x, y, z) = yz\mathbf{i} + (2 - 3y)\mathbf{j} + (x^2 + y^2)\mathbf{k}, \quad x^2 + y^2 \leq 16$
    S: the first-octant portion of $x^2 + z^2 = 16$ over $x^2 + y^2 = 16$
19. $\mathbf{F}(x, y, z) = xyz\mathbf{i} + y\mathbf{j} + z\mathbf{k}$
    S: $z = x^2, \quad 0 \leq x \leq a, \quad 0 \leq y \leq a$
    $\mathbf{N}$ is the downward unit normal to the surface.
20. $\mathbf{F}(x, y, z) = xyz\mathbf{i} + y\mathbf{j} + z\mathbf{k}, \quad x^2 + y^2 \leq a^2$
    S: the first-octant portion of $z = x^2$ over $x^2 + y^2 = a^2$

Motion of a Liquid **In Exercises 21 and 22, the motion of a liquid in a cylindrical container of radius 1 is described by the velocity field $\mathbf{F}(x, y, z)$. Find $\int_S\int (\text{curl } \mathbf{F}) \cdot \mathbf{N}\, dS$, where S is the upper surface of the cylindrical container.**

21. $\mathbf{F}(x, y, z) = \mathbf{i} + \mathbf{j} - 2\mathbf{k}$
22. $\mathbf{F}(x, y, z) = -z\mathbf{i} + y\mathbf{k}$

WRITING ABOUT CONCEPTS

23. State Stokes's Theorem.
24. Give a physical interpretation of curl.

25. Let f and g be scalar functions with continuous partial derivatives, and let C and S satisfy the conditions of Stokes's Theorem. Verify each identity.
    (a) $\int_C (f\nabla g) \cdot d\mathbf{r} = \int_S\int (\nabla f \times \nabla g) \cdot \mathbf{N}\, dS$
    (b) $\int_C (f\nabla f) \cdot d\mathbf{r} = 0$
    (c) $\int_C (f\nabla g + g\nabla f) \cdot d\mathbf{r} = 0$
26. Demonstrate the results of Exercise 25 for the functions $f(x, y, z) = xyz$ and $g(x, y, z) = z$. Let S be the hemisphere $z = \sqrt{4 - x^2 - y^2}$.
27. Let $\mathbf{C}$ be a constant vector. Let S be an oriented surface with a unit normal vector $\mathbf{N}$, bounded by a smooth curve C. Prove that

$$\int_S\int \mathbf{C} \cdot \mathbf{N}\, dS = \frac{1}{2}\int_C (\mathbf{C} \times \mathbf{r}) \cdot d\mathbf{r}.$$

CAPSTONE

28. Verify Stokes's Theorem for each given vector field and upward oriented surface. Is the line integral or the double integral easier to set up? to evaluate? Explain.
    (a) $\mathbf{F}(x, y, z) = e^{y+z}\mathbf{i}$
    C: square with vertices $(0, 0, 0)$, $(1, 0, 0)$, $(1, 1, 0)$, $(0, 1, 0)$
    (b) $\mathbf{F}(x, y, z) = z^2\mathbf{i} + x^2\mathbf{j} + y^2\mathbf{k}$
    S: the portion of the paraboloid $z = x^2 + y^2$ that lies below the plane $z = 4$

PUTNAM EXAM CHALLENGE

29. Let $\mathbf{G}(x, y) = \left(\dfrac{-y}{x^2 + 4y^2}, \dfrac{x}{x^2 + 4y^2}, 0\right)$.
    Prove or disprove that there is a vector-valued function $\mathbf{F}(x, y, z) = (M(x, y, z),\ N(x, y, z),\ P(x, y, z))$ with the following properties.
    (i) M, N, P have continuous partial derivatives for all $(x, y, z) \neq (0, 0, 0)$;
    (ii) Curl $\mathbf{F} = \mathbf{0}$ for all $(x, y, z) \neq (0, 0, 0)$;
    (iii) $\mathbf{F}(x, y, 0) = \mathbf{G}(x, y)$.

This problem was composed by the Committee on the Putnam Prize Competition.

15 REVIEW EXERCISES

See www.CalcChat.com for worked-out solutions to odd-numbered exercises.

In Exercises 1 and 2, compute $\|\mathbf{F}\|$ and sketch several representative vectors in the vector field. Use a computer algebra system to verify your results.

1. $\mathbf{F}(x, y, z) = x\mathbf{i} + \mathbf{j} + 2\mathbf{k}$
2. $\mathbf{F}(x, y) = \mathbf{i} - 2y\mathbf{j}$

In Exercises 3 and 4, find the gradient vector field for the scalar function.

3. $f(x, y, z) = 2x^2 + xy + z^2$
4. $f(x, y, z) = x^2 e^{yz}$

In Exercises 5–12, determine whether the vector field is conservative. If it is, find a potential function for the vector field.

5. $\mathbf{F}(x, y) = -\dfrac{y}{x^2}\mathbf{i} + \dfrac{1}{x}\mathbf{j}$
6. $\mathbf{F}(x, y) = \dfrac{1}{y}\mathbf{i} - \dfrac{y}{x^2}\mathbf{j}$
7. $\mathbf{F}(x, y) = (xy^2 - x^2)\mathbf{i} + (x^2y + y^2)\mathbf{j}$
8. $\mathbf{F}(x, y) = (-2y^3 \sin 2x)\mathbf{i} + 3y^2(1 + \cos 2x)\mathbf{j}$
9. $\mathbf{F}(x, y, z) = 4xy^2\mathbf{i} + 2x^2\mathbf{j} + 2z\mathbf{k}$
10. $\mathbf{F}(x, y, z) = (4xy + z^2)\mathbf{i} + (2x^2 + 6yz)\mathbf{j} + 2xz\mathbf{k}$
11. $\mathbf{F}(x, y, z) = \dfrac{yz\mathbf{i} - xz\mathbf{j} - xy\mathbf{k}}{y^2z^2}$
12. $\mathbf{F}(x, y, z) = \sin z(y\mathbf{i} + x\mathbf{j} + \mathbf{k})$

In Exercises 13–20, find (a) the divergence of the vector field F and (b) the curl of the vector field F.

13. $\mathbf{F}(x, y, z) = x^2\mathbf{i} + xy^2\mathbf{j} + x^2z\mathbf{k}$
14. $\mathbf{F}(x, y, z) = y^2\mathbf{j} - z^2\mathbf{k}$
15. $\mathbf{F}(x, y, z) = (\cos y + y\cos x)\mathbf{i} + (\sin x - x\sin y)\mathbf{j} + xyz\mathbf{k}$
16. $\mathbf{F}(x, y, z) = (3x - y)\mathbf{i} + (y - 2z)\mathbf{j} + (z - 3x)\mathbf{k}$
17. $\mathbf{F}(x, y, z) = \arcsin x\mathbf{i} + xy^2\mathbf{j} + yz^2\mathbf{k}$
18. $\mathbf{F}(x, y, z) = (x^2 - y)\mathbf{i} - (x + \sin^2 y)\mathbf{j}$
19. $\mathbf{F}(x, y, z) = \ln(x^2 + y^2)\mathbf{i} + \ln(x^2 + y^2)\mathbf{j} + z\mathbf{k}$
20. $\mathbf{F}(x, y, z) = \dfrac{z}{x}\mathbf{i} + \dfrac{z}{y}\mathbf{j} + z^2\mathbf{k}$

In Exercises 21–26, evaluate the line integral along the given path(s).

21. $\displaystyle\int_C (x^2 + y^2)\,ds$
   (a) C: line segment from $(0, 0)$ to $(3, 4)$
   (b) C: $x^2 + y^2 = 1$, one revolution counterclockwise, starting at $(1, 0)$

22. $\displaystyle\int_C xy\,ds$
   (a) C: line segment from $(0, 0)$ to $(5, 4)$
   (b) C: counterclockwise around the triangle with vertices $(0, 0)$, $(4, 0)$, $(0, 2)$

23. $\displaystyle\int_C (x^2 + y^2)\,ds$
   C: $\mathbf{r}(t) = (1 - \sin t)\mathbf{i} + (1 - \cos t)\mathbf{j}$, $0 \le t \le 2\pi$

24. $\displaystyle\int_C (x^2 + y^2)\,ds$
   C: $\mathbf{r}(t) = (\cos t + t\sin t)\mathbf{i} + (\sin t - t\cos t)\mathbf{j}$, $0 \le t \le 2\pi$

25. $\displaystyle\int_C (2x - y)\,dx + (x + 2y)\,dy$
   (a) C: line segment from $(0, 0)$ to $(3, -3)$
   (b) C: one revolution counterclockwise around the circle $x = 3\cos t$, $y = 3\sin t$

26. $\displaystyle\int_C (2x - y)\,dx + (x + 3y)\,dy$
   C: $\mathbf{r}(t) = (\cos t + t\sin t)\mathbf{i} + (\sin t - t\sin t)\mathbf{j}$, $0 \le t \le \pi/2$

CAS **In Exercises 27 and 28, use a computer algebra system to evaluate the line integral over the given path.**

27. $\displaystyle\int_C (2x + y)\,ds$
   $\mathbf{r}(t) = a\cos^3 t\,\mathbf{i} + a\sin^3 t\,\mathbf{j}$, $0 \le t \le \pi/2$

28. $\displaystyle\int_C (x^2 + y^2 + z^2)\,ds$
   $\mathbf{r}(t) = t\mathbf{i} + t^2\mathbf{j} + t^{3/2}\mathbf{k}$, $0 \le t \le 4$

Lateral Surface Area **In Exercises 29 and 30, find the lateral surface area over the curve C in the xy-plane and under the surface $z = f(x, y)$.**

29. $f(x, y) = 3 + \sin(x + y)$
   C: $y = 2x$ from $(0, 0)$ to $(2, 4)$
30. $f(x, y) = 12 - x - y$
   C: $y = x^2$ from $(0, 0)$ to $(2, 4)$

In Exercises 31–36, evaluate $\displaystyle\int_C \mathbf{F} \cdot d\mathbf{r}$.

31. $\mathbf{F}(x, y) = xy\mathbf{i} + 2xy\mathbf{j}$
   C: $\mathbf{r}(t) = t^2\mathbf{i} + t^2\mathbf{j}$, $0 \le t \le 1$
32. $\mathbf{F}(x, y) = (x - y)\mathbf{i} + (x + y)\mathbf{j}$
   C: $\mathbf{r}(t) = 4\cos t\mathbf{i} + 3\sin t\mathbf{j}$, $0 \le t \le 2\pi$
33. $\mathbf{F}(x, y, z) = x\mathbf{i} + y\mathbf{j} + z\mathbf{k}$
   C: $\mathbf{r}(t) = 2\cos t\mathbf{i} + 2\sin t\mathbf{j} + t\mathbf{k}$, $0 \le t \le 2\pi$
34. $\mathbf{F}(x, y, z) = (2y - z)\mathbf{i} + (z - x)\mathbf{j} + (x - y)\mathbf{k}$
   C: curve of intersection of $x^2 + z^2 = 4$ and $y^2 + z^2 = 4$ from $(2, 2, 0)$ to $(0, 0, 2)$
35. $\mathbf{F}(x, y, z) = (y + z)\mathbf{i} + (x + z)\mathbf{j} + (x + y)\mathbf{k}$
   C: curve of intersection of $z = x^2 + y^2$ and $y = x$ from $(0, 0, 0)$ to $(2, 2, 8)$
36. $\mathbf{F}(x, y, z) = (x^2 - z)\mathbf{i} + (y^2 + z)\mathbf{j} + x\mathbf{k}$
   C: curve of intersection of $z = x^2$ and $x^2 + y^2 = 4$ from $(0, -2, 0)$ to $(0, 2, 0)$

CAS **In Exercises 37 and 38, use a computer algebra system to evaluate the line integral.**

37. $\int_C xy\,dx + (x^2 + y^2)\,dy$

C: $y = x^2$ from $(0, 0)$ to $(2, 4)$ and $y = 2x$ from $(2, 4)$ to $(0, 0)$

38. $\int_C \mathbf{F} \cdot d\mathbf{r}$

$\mathbf{F}(x, y) = (2x - y)\mathbf{i} + (2y - x)\mathbf{j}$

C: $\mathbf{r}(t) = (2\cos t + 2t\sin t)\mathbf{i} + (2\sin t - 2t\cos t)\mathbf{j},$
$0 \le t \le \pi$

39. ***Work*** Find the work done by the force field $\mathbf{F} = x\mathbf{i} - \sqrt{y}\mathbf{j}$ along the path $y = x^{3/2}$ from $(0, 0)$ to $(4, 8)$.

40. ***Work*** A 20-ton aircraft climbs 2000 feet while making a 90° turn in a circular arc of radius 10 miles. Find the work done by the engines.

In Exercises 41 and 42, evaluate the integral using the Fundamental Theorem of Line Integrals.

41. $\int_C 2xyz\,dx + x^2z\,dy + x^2y\,dz$

C: smooth curve from $(0, 0, 0)$ to $(1, 3, 2)$

42. $\int_C y\,dx + x\,dy + \frac{1}{z}\,dz$

C: smooth curve from $(0, 0, 1)$ to $(4, 4, 4)$

43. Evaluate the line integral $\int_C y^2\,dx + 2xy\,dy$.

(a) C: $\mathbf{r}(t) = (1 + 3t)\mathbf{i} + (1 + t)\mathbf{j}, \quad 0 \le t \le 1$

(b) C: $\mathbf{r}(t) = t\mathbf{i} + \sqrt{t}\mathbf{j}, \quad 1 \le t \le 4$

(c) Use the Fundamental Theorem of Line Integrals, where C is a smooth curve from $(1, 1)$ to $(4, 2)$.

44. ***Area and Centroid*** Consider the region bounded by the x-axis and one arch of the cycloid with parametric equations $x = a(\theta - \sin\theta)$ and $y = a(1 - \cos\theta)$. Use line integrals to find (a) the area of the region and (b) the centroid of the region.

In Exercises 45–50, use Green's Theorem to evaluate the line integral.

45. $\int_C y\,dx + 2x\,dy$

C: boundary of the square with vertices $(0, 0)$, $(0, 1)$, $(1, 0)$, $(1, 1)$

46. $\int_C xy\,dx + (x^2 + y^2)\,dy$

C: boundary of the square with vertices $(0, 0)$, $(0, 2)$, $(2, 0)$, $(2, 2)$

47. $\int_C xy^2\,dx + x^2y\,dy$

C: $x = 4\cos t, \quad y = 4\sin t$

48. $\int_C (x^2 - y^2)\,dx + 2xy\,dy$

C: $x^2 + y^2 = a^2$

49. $\int_C xy\,dx + x^2\,dy$

C: boundary of the region between the graphs of $y = x^2$ and $y = 1$

50. $\int_C y^2\,dx + x^{4/3}\,dy$

C: $x^{2/3} + y^{2/3} = 1$

CAS **In Exercises 51 and 52, use a computer algebra system to graph the surface represented by the vector-valued function.**

51. $\mathbf{r}(u, v) = \sec u\cos v\,\mathbf{i} + (1 + 2\tan u)\sin v\,\mathbf{j} + 2u\,\mathbf{k}$

$0 \le u \le \frac{\pi}{3}, \quad 0 \le v \le 2\pi$

52. $\mathbf{r}(u, v) = e^{-u/4}\cos v\,\mathbf{i} + e^{-u/4}\sin v\,\mathbf{j} + \frac{u}{6}\mathbf{k}$

$0 \le u \le 4, \quad 0 \le v \le 2\pi$

CAS **53.** ***Investigation*** Consider the surface represented by the vector-valued function

$\mathbf{r}(u, v) = 3\cos v\cos u\,\mathbf{i} + 3\cos v\sin u\,\mathbf{j} + \sin v\,\mathbf{k}.$

Use a computer algebra system to do the following.

(a) Graph the surface for $0 \le u \le 2\pi$ and $-\frac{\pi}{2} \le v \le \frac{\pi}{2}$.

(b) Graph the surface for $0 \le u \le 2\pi$ and $\frac{\pi}{4} \le v \le \frac{\pi}{2}$.

(c) Graph the surface for $0 \le u \le \frac{\pi}{4}$ and $0 \le v \le \frac{\pi}{2}$.

(d) Graph and identify the space curve for $0 \le u \le 2\pi$ and $v = \frac{\pi}{4}$.

(e) Approximate the area of the surface graphed in part (b).

(f) Approximate the area of the surface graphed in part (c).

54. Evaluate the surface integral $\int_S\int z\,dS$ over the surface S:

$\mathbf{r}(u, v) = (u + v)\mathbf{i} + (u - v)\mathbf{j} + \sin v\,\mathbf{k}$

where $0 \le u \le 2$ and $0 \le v \le \pi$.

CAS **55.** Use a computer algebra system to graph the surface S and approximate the surface integral

$\int_S\int (x + y)\,dS$

where S is the surface

S: $\mathbf{r}(u, v) = u\cos v\,\mathbf{i} + u\sin v\,\mathbf{j} + (u - 1)(2 - u)\mathbf{k}$

over $0 \le u \le 2$ and $0 \le v \le 2\pi$.

56. ***Mass*** A cone-shaped surface lamina S is given by

$$z = a\left(a - \sqrt{x^2 + y^2}\right), \quad 0 \le z \le a^2.$$

At each point on S, the density is proportional to the distance between the point and the z-axis.

(a) Sketch the cone-shaped surface.

(b) Find the mass m of the lamina.

In Exercises 57 and 58, verify the Divergence Theorem by evaluating

$$\int_S\int \mathbf{F} \cdot \mathbf{N}\, dS$$

as a surface integral and as a triple integral.

57. $\mathbf{F}(x, y, z) = x^2\mathbf{i} + xy\mathbf{j} + z\mathbf{k}$

Q: solid region bounded by the coordinate planes and the plane $2x + 3y + 4z = 12$

58. $\mathbf{F}(x, y, z) = x\mathbf{i} + y\mathbf{j} + z\mathbf{k}$

Q: solid region bounded by the coordinate planes and the plane $2x + 3y + 4z = 12$

In Exercises 59 and 60, verify Stokes's Theorem by evaluating

$$\int_C \mathbf{F} \cdot d\mathbf{r}$$

as a line integral and as a double integral.

59. $\mathbf{F}(x, y, z) = (\cos y + y\cos x)\mathbf{i} + (\sin x - x\sin y)\mathbf{j} + xyz\mathbf{k}$

S: portion of $z = y^2$ over the square in the xy-plane with vertices $(0, 0)$, $(a, 0)$, (a, a), $(0, a)$

$\mathbf{N}$ is the upward unit normal vector to the surface.

60. $\mathbf{F}(x, y, z) = (x - z)\mathbf{i} + (y - z)\mathbf{j} + x^2\mathbf{k}$

S: first-octant portion of the plane $3x + y + 2z = 12$

61. Prove that it is not possible for a vector field with twice-differentiable components to have a curl of $x\mathbf{i} + y\mathbf{j} + z\mathbf{k}$.

SECTION PROJECT

The Planimeter

You have learned many calculus techniques for finding the area of a planar region. Engineers use a mechanical device called a *planimeter* for measuring planar areas, which is based on the area formula given in Theorem 15.9 (page 1096). As you can see in the figure, the planimeter is fixed at point O (but free to pivot) and has a hinge at A. The end of the tracer arm AB moves counterclockwise around the region R. A small wheel at B is perpendicular to $\overline{AB}$ and is marked with a scale to measure how much it rolls as B traces out the boundary of region R. In this project you will show that the area of R is given by the length L of the tracer arm $\overline{AB}$ multiplied by the distance D that the wheel rolls.

Assume that point B traces out the boundary of R for $a \le t \le b$. Point A will move back and forth along a circular arc around the origin O. Let $\theta(t)$ denote the angle in the figure and let $(x(t), y(t))$ denote the coordinates of A.

(a) Show that the vector $\overrightarrow{OB}$ is given by the vector-valued function

$$\mathbf{r}(t) = [x(t) + L\cos\theta(t)]\mathbf{i} + [y(t) + L\sin\theta(t)]\mathbf{j}.$$

(b) Show that the following two integrals are equal to zero.

$$I_1 = \int_a^b \frac{1}{2}L^2\frac{d\theta}{dt}\,dt$$

$$I_2 = \int_a^b \frac{1}{2}\left(x\frac{dy}{dt} - y\frac{dx}{dt}\right)dt$$

(c) Use the integral $\int_a^b [x(t)\sin\theta(t) - y(t)\cos\theta(t)]'dt$ to show that the following two integrals are equal.

$$I_3 = \int_a^b \frac{1}{2}L\left(y\sin\theta\frac{d\theta}{dt} + x\cos\theta\frac{d\theta}{dt}\right)dt$$

$$I_4 = \int_a^b \frac{1}{2}L\left(-\sin\theta\frac{dx}{dt} + \cos\theta\frac{dy}{dt}\right)dt$$

(d) Let $\mathbf{N} = -\sin\theta\mathbf{i} + \cos\theta\mathbf{j}$. Explain why the distance D that the wheel rolls is given by

$$D = \int_C \mathbf{N} \cdot \mathbf{T}\,ds.$$

(e) Show that the area of region R is given by $I_1 + I_2 + I_3 + I_4 = DL$.

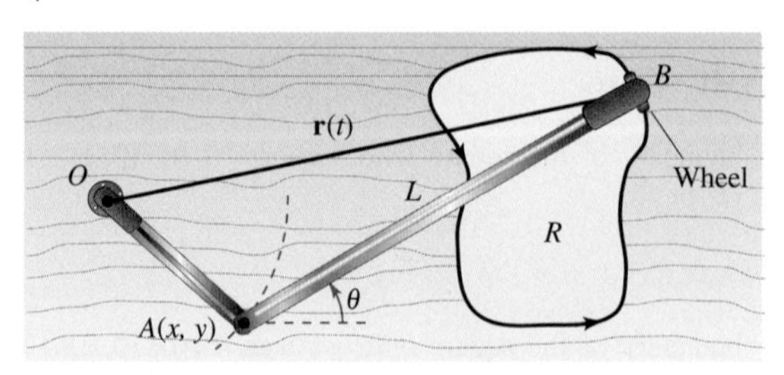

■ **FOR FURTHER INFORMATION** For more information about using calculus to find irregular areas, see "The Amateur Scientist" by C. L. Strong in the August 1958 issue of *Scientific American*.

P.S. PROBLEM SOLVING

1. Heat flows from areas of higher temperature to areas of lower temperature in the direction of greatest change. As a result, measuring heat flux involves the gradient of the temperature. The flux depends on the area of the surface. It is the normal direction to the surface that is important, because heat that flows in directions tangential to the surface will produce no heat loss. So, assume that the heat flux across a portion of the surface of area ΔS is given by $\Delta H \approx -k\nabla T \cdot \mathbf{N}\, dS$, where T is the temperature, $\mathbf{N}$ is the unit normal vector to the surface in the direction of the heat flow, and k is the thermal diffusivity of the material. The heat flux across the surface S is given by

$$H = \iint_S -k\nabla T \cdot \mathbf{N}\, dS.$$

Consider a single heat source located at the origin with temperature

$$T(x, y, z) = \frac{25}{\sqrt{x^2 + y^2 + z^2}}.$$

(a) Calculate the heat flux across the surface

$$S = \left\{(x, y, z): z = \sqrt{1 - x^2},\ -\frac{1}{2} \le x \le \frac{1}{2},\ 0 \le y \le 1\right\}$$

as shown in the figure.

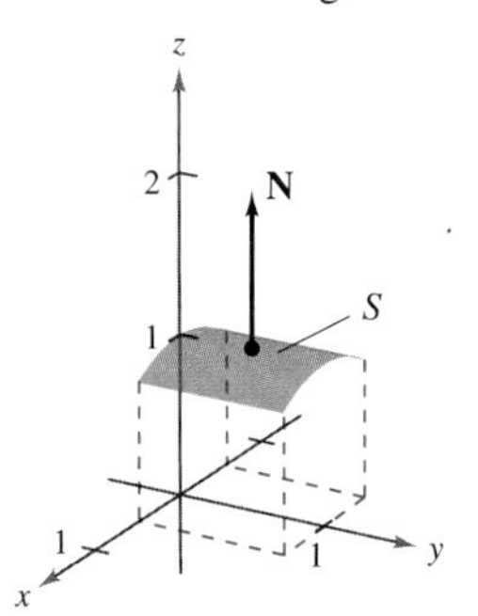

(b) Repeat the calculation in part (a) using the parametrization

$$x = \cos u,\quad y = v,\quad z = \sin u,\quad \frac{\pi}{3} \le u \le \frac{2\pi}{3},\quad 0 \le v \le 1.$$

2. Consider a single heat source located at the origin with temperature

$$T(x, y, z) = \frac{25}{\sqrt{x^2 + y^2 + z^2}}.$$

(a) Calculate the heat flux across the surface

$$S = \left\{(x, y, z): z = \sqrt{1 - x^2 - y^2},\ x^2 + y^2 \le 1\right\}$$

as shown in the figure.

(b) Repeat the calculation in part (a) using the parametrization

$$x = \sin u \cos v,\quad y = \sin u \sin v,\quad z = \cos u,\quad 0 \le u \le \frac{\pi}{2},$$
$$0 \le v \le 2\pi.$$

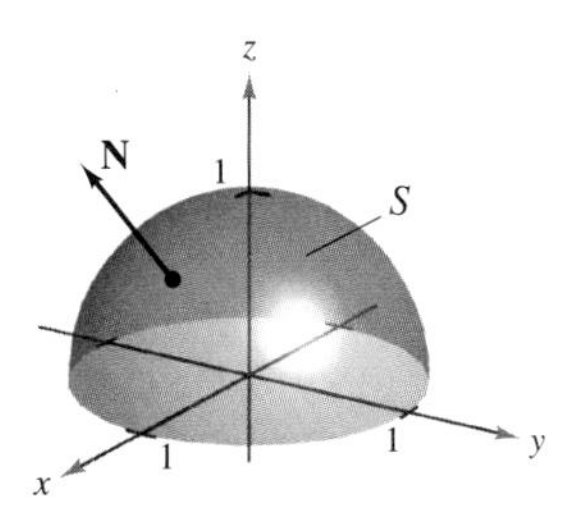

Figure for 2

3. Consider a wire of density $\rho(x, y, z)$ given by the space curve

$$C: \mathbf{r}(t) = x(t)\mathbf{i} + y(t)\mathbf{j} + z(t)\mathbf{k},\quad a \le t \le b.$$

The **moments of inertia** about the x-, y-, and z-axes are given by

$$I_x = \int_C (y^2 + z^2)\rho(x, y, z)\, ds$$
$$I_y = \int_C (x^2 + z^2)\rho(x, y, z)\, ds$$
$$I_z = \int_C (x^2 + y^2)\rho(x, y, z)\, ds.$$

Find the moments of inertia for a wire of uniform density $\rho = 1$ in the shape of the helix

$$\mathbf{r}(t) = 3\cos t\mathbf{i} + 3\sin t\mathbf{j} + 2t\mathbf{k},\quad 0 \le t \le 2\pi \text{ (see figure).}$$

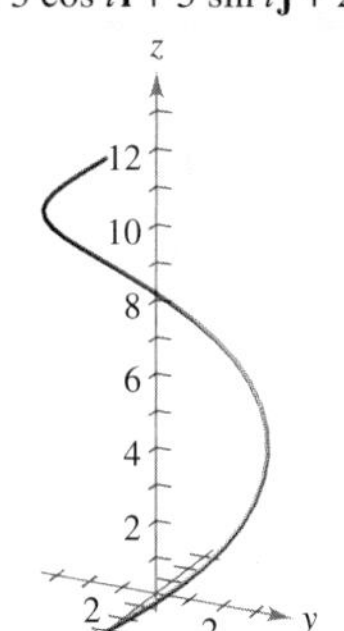

Figure for 3

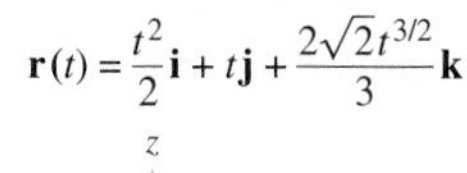
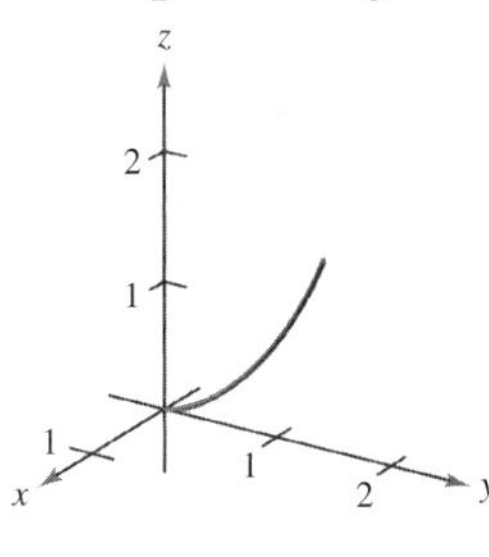

Figure for 4

4. Find the moments of inertia for the wire of density $\rho = \dfrac{1}{1 + t}$ given by the curve

$$C: \mathbf{r}(t) = \frac{t^2}{2}\mathbf{i} + t\mathbf{j} + \frac{2\sqrt{2}\,t^{3/2}}{3}\mathbf{k},\quad 0 \le t \le 1 \text{ (see figure).}$$

5. The **Laplacian** is the differential operator

$$\nabla^2 = \nabla \cdot \nabla = \frac{\partial^2}{\partial x^2} + \frac{\partial^2}{\partial y^2} + \frac{\partial^2}{\partial z^2}$$

and **Laplace's equation** is

$$\nabla^2 w = \frac{\partial^2 w}{\partial x^2} + \frac{\partial^2 w}{\partial y^2} + \frac{\partial^2 w}{\partial z^2} = 0.$$

Any function that satisfies this equation is called **harmonic.** Show that the function $w = 1/f$ is harmonic.

CAS **6.** Consider the line integral

$$\int_C y^n\,dx + x^n\,dy$$

where C is the boundary of the region lying between the graphs of $y = \sqrt{a^2 - x^2}$ $(a > 0)$ and $y = 0$.

(a) Use a computer algebra system to verify Green's Theorem for n, an odd integer from 1 through 7.

(b) Use a computer algebra system to verify Green's Theorem for n, an even integer from 2 through 8.

(c) For n an odd integer, make a conjecture about the value of the integral.

7. Use a line integral to find the area bounded by one arch of the cycloid $x(\theta) = a(\theta - \sin\theta)$, $y(\theta) = a(1 - \cos\theta)$, $0 \le \theta \le 2\pi$, as shown in the figure.

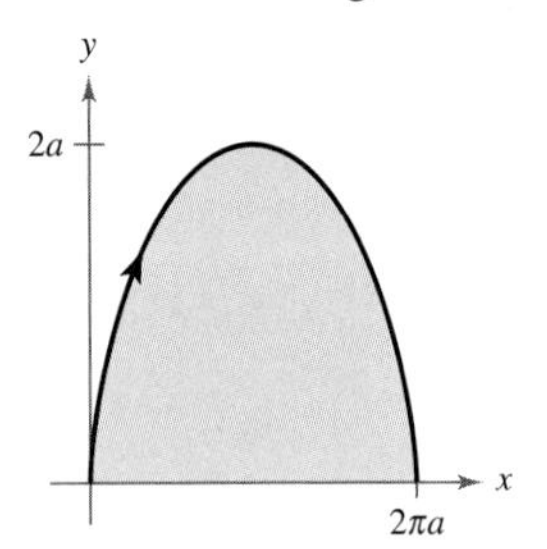

Figure for 7

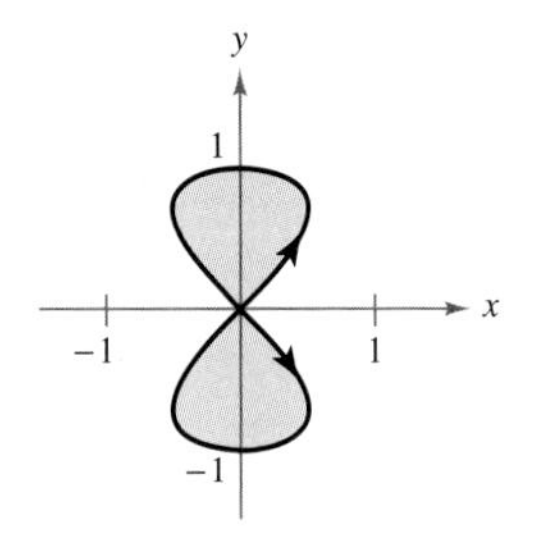

Figure for 8

8. Use a line integral to find the area bounded by the two loops of the eight curve

$$x(t) = \frac{1}{2}\sin 2t,\quad y(t) = \sin t,\quad 0 \le t \le 2\pi$$

as shown in the figure.

9. The force field $\mathbf{F}(x, y) = (x + y)\mathbf{i} + (x^2 + 1)\mathbf{j}$ acts on an object moving from the point $(0, 0)$ to the point $(0, 1)$, as shown in the figure.

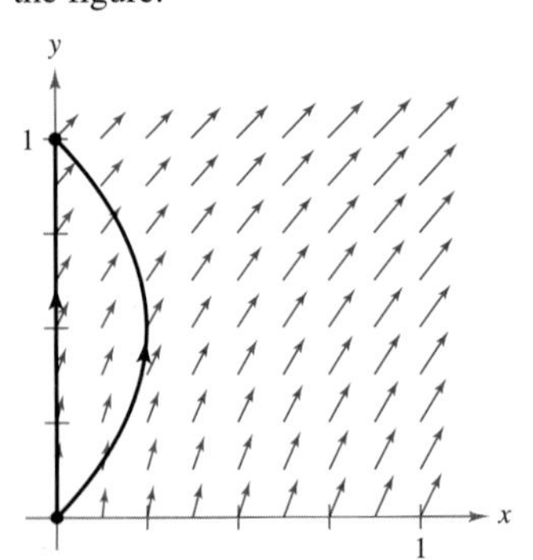

(a) Find the work done if the object moves along the path $x = 0$, $0 \le y \le 1$.

(b) Find the work done if the object moves along the path $x = y - y^2$, $0 \le y \le 1$.

(c) Suppose the object moves along the path $x = c(y - y^2)$, $0 \le y \le 1$, $c > 0$. Find the value of the constant c that minimizes the work.

10. The force field $\mathbf{F}(x, y) = (3x^2y^2)\mathbf{i} + (2x^3y)\mathbf{j}$ is shown in the figure below. Three particles move from the point $(1, 1)$ to the point $(2, 4)$ along different paths. Explain why the work done is the same for each particle, and find the value of the work.

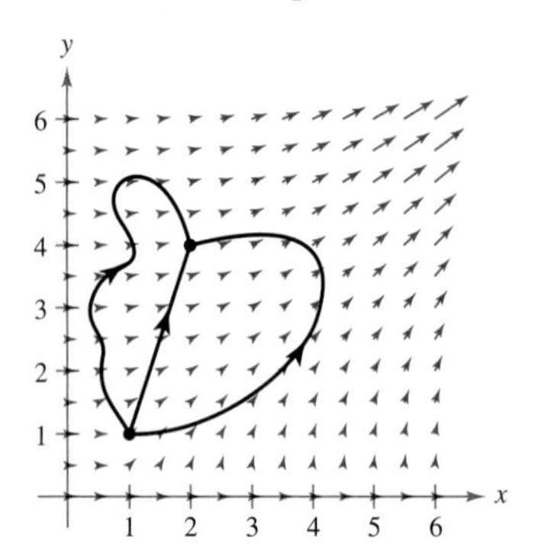

11. Let S be a smooth oriented surface with normal vector $\mathbf{N}$, bounded by a smooth simple closed curve C. Let $\mathbf{v}$ be a constant vector, and prove that

$$\int_S\int (2\mathbf{v} \cdot \mathbf{N})\,dS = \int_C (\mathbf{v} \times \mathbf{r}) \cdot d\mathbf{r}.$$

12. How does the area of the ellipse $\dfrac{x^2}{a^2} + \dfrac{y^2}{b^2} = 1$ compare with the magnitude of the work done by the force field

$$\mathbf{F}(x, y) = -\frac{1}{2}y\mathbf{i} + \frac{1}{2}x\mathbf{j}$$

on a particle that moves once around the ellipse (see figure)?

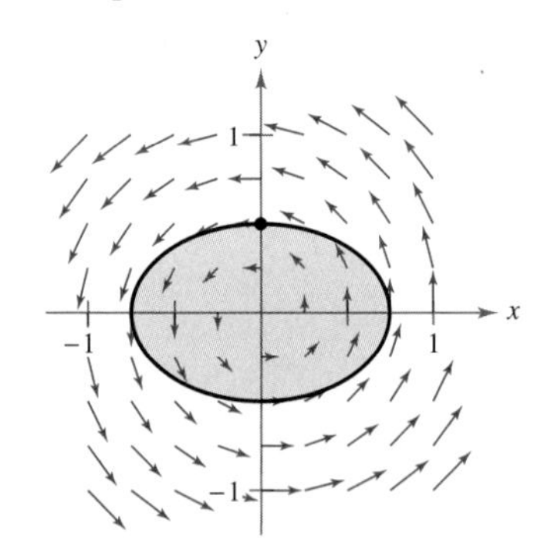

13. A cross section of Earth's magnetic field can be represented as a vector field in which the center of Earth is located at the origin and the positive y-axis points in the direction of the magnetic north pole. The equation for this field is

$$\begin{aligned}\mathbf{F}(x, y) &= M(x, y)\mathbf{i} + N(x, y)\mathbf{j}\\ &= \frac{m}{(x^2 + y^2)^{5/2}}[3xy\mathbf{i} + (2y^2 - x^2)\mathbf{j}]\end{aligned}$$

where m is the magnetic moment of Earth. Show that this vector field is conservative.

16 Additional Topics in Differential Equations

In Chapter 6, you studied differential equations. In this chapter, you will learn additional techniques for solving differential equations.

In this chapter, you should learn the following.

- How to recognize and solve exact differential equations. (16.1)
- How to solve second-order homogeneous linear differential equations and higher-order homogeneous linear differential equations. (16.2)
- How to solve second-order nonhomogeneous linear differential equations. (16.3)
- How to use power series to solve differential equations. (16.4)

Differential equations can be used to model many real-life applications. How can you use a differential equation to describe the fall of a parachutist? (See Section 16.3, Section Project.)

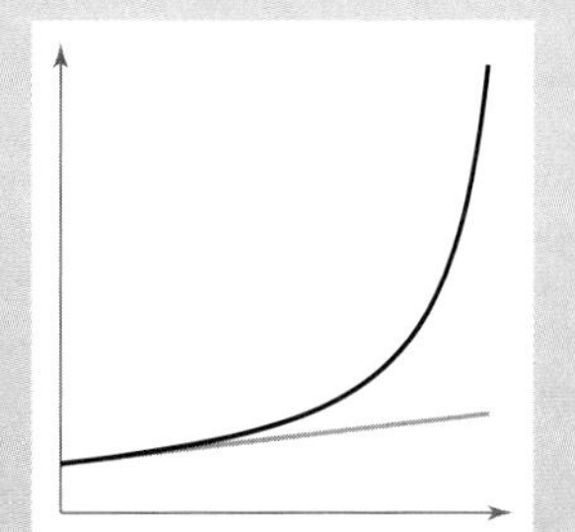

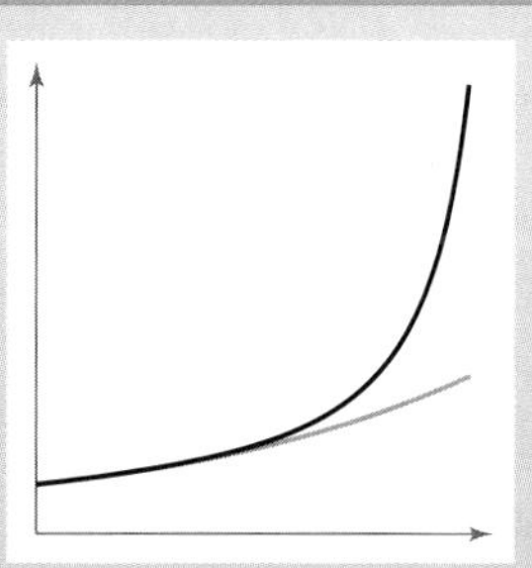

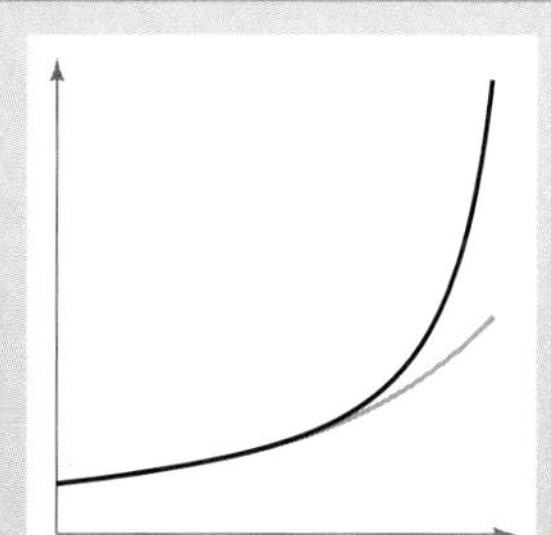

You can use power series to solve certain types of differential equations. A Taylor series was used to find the series solution of $y' = y^2 - x$. You can use n terms of the series to approximate y. As more terms of the series are used, the closer the approximation gets to y. In the graphs, the series solution is shown using 2 terms, 4 terms, and 6 terms, along with y. Can you identify the graphs? (See Section 16.4.)

16.1 Exact First-Order Equations

- **Solve an exact differential equation.**
- **Use an integrating factor to make a differential equation exact.**

Exact Differential Equations

In Chapter 6, you studied applications of differential equations to growth and decay problems. You also learned more about the basic ideas of differential equations and studied the solution technique known as separation of variables. In this chapter, you will learn more about solving differential equations and using them in real-life applications. This section introduces you to a method for solving the first-order differential equation

$$M(x, y)\,dx + N(x, y)\,dy = 0$$

for the special case in which this equation represents the exact differential of a function $z = f(x, y)$.

DEFINITION OF AN EXACT DIFFERENTIAL EQUATION

The equation $M(x, y)\,dx + N(x, y)\,dy = 0$ is an **exact differential equation** if there exists a function f of two variables x and y having continuous partial derivatives such that

$$f_x(x, y) = M(x, y) \qquad \text{and} \qquad f_y(x, y) = N(x, y).$$

The general solution of the equation is $f(x, y) = C$.

From Section 13.3, you know that if f has continuous second partials, then

$$\frac{\partial M}{\partial y} = \frac{\partial^2 f}{\partial y\,\partial x} = \frac{\partial^2 f}{\partial x\,\partial y} = \frac{\partial N}{\partial x}.$$

This suggests the following test for exactness.

THEOREM 16.1 TEST FOR EXACTNESS

Let M and N have continuous partial derivatives on an open disk R. The differential equation $M(x, y)\,dx + N(x, y)\,dy = 0$ is exact if and only if

$$\frac{\partial M}{\partial y} = \frac{\partial N}{\partial x}.$$

NOTE Every differential equation of the form

$$M(x)\,dx + N(y)\,dy = 0$$

is exact. In other words, a separable differential equation is actually a special type of an exact equation. ■

Exactness is a fragile condition in the sense that seemingly minor alterations in an exact equation can destroy its exactness. This is demonstrated in the following example.

EXAMPLE 1 Testing for Exactness

a. The differential equation $(xy^2 + x)\,dx + yx^2\,dy = 0$ is exact because

$$\frac{\partial M}{\partial y} = \frac{\partial}{\partial y}[xy^2 + x] = 2xy \qquad \text{and} \qquad \frac{\partial N}{\partial x} = \frac{\partial}{\partial x}[yx^2] = 2xy.$$

But the equation $(y^2 + 1)\,dx + xy\,dy = 0$ is not exact, even though it is obtained by dividing each side of the first equation by x.

b. The differential equation $\cos y\,dx + (y^2 - x\sin y)\,dy = 0$ is exact because

$$\frac{\partial M}{\partial y} = \frac{\partial}{\partial y}[\cos y] = -\sin y \qquad \text{and} \qquad \frac{\partial N}{\partial x} = \frac{\partial}{\partial x}[y^2 - x\sin y] = -\sin y.$$

But the equation $\cos y\,dx + (y^2 + x\sin y)\,dy = 0$ is not exact, even though it differs from the first equation only by a single sign. ■

Note that the test for exactness of $M(x, y)\,dx + N(x, y)\,dy = 0$ is the same as the test for determining whether $\mathbf{F}(x, y) = M(x, y)\mathbf{i} + N(x, y)\mathbf{j}$ is the gradient of a potential function (Theorem 15.1). This means that a general solution $f(x, y) = C$ to an exact differential equation can be found by the method used to find a potential function for a conservative vector field.

EXAMPLE 2 Solving an Exact Differential Equation

Solve the differential equation $(2xy - 3x^2)\,dx + (x^2 - 2y)\,dy = 0$.

Solution The given differential equation is exact because

$$\frac{\partial M}{\partial y} = \frac{\partial}{\partial y}[2xy - 3x^2] = 2x = \frac{\partial N}{\partial x} = \frac{\partial}{\partial x}[x^2 - 2y].$$

The general solution, $f(x, y) = C$, is given by

$$f(x, y) = \int M(x, y)\,dx = \int (2xy - 3x^2)\,dx = x^2y - x^3 + g(y).$$

In Section 15.1, you determined $g(y)$ by integrating $N(x, y)$ with respect to y and reconciling the two expressions for $f(x, y)$. An alternative method is to partially differentiate this version of $f(x, y)$ with respect to y and compare the result with $N(x, y)$. In other words,

$$f_y(x, y) = \frac{\partial}{\partial y}[x^2y - x^3 + g(y)] = x^2 + g'(y) = \overbrace{x^2 - 2y}^{N(x,\,y)}.$$

$g'(y) = -2y$

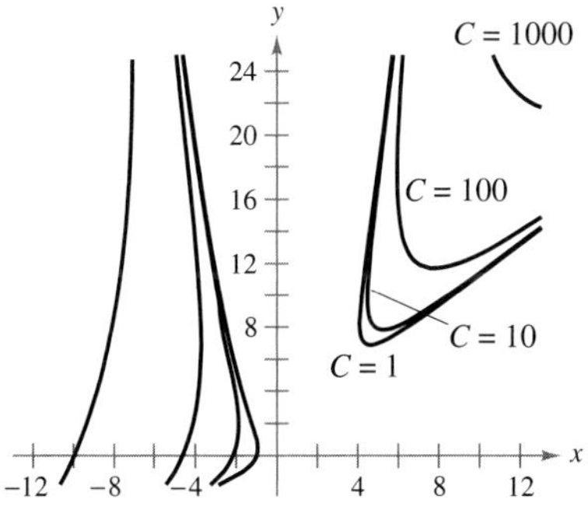

So, $g'(y) = -2y$, and it follows that $g(y) = -y^2 + C_1$. Therefore,

$$f(x, y) = x^2y - x^3 - y^2 + C_1$$

and the general solution is $x^2y - x^3 - y^2 = C$. Figure 16.1 shows the solution curves that correspond to $C = 1, 10, 100,$ and 1000. ■

Figure 16.1

The icon indicates that you will find a CAS Investigation on the book's website. The CAS Investigation is a collaborative exploration of this example using the computer algebra systems Maple *and* Mathematica.

EXAMPLE 3 Solving an Exact Differential Equation

TECHNOLOGY You can use a graphing utility to graph a particular solution that satisfies the initial condition of a differential equation. In Example 3, the differential equation and initial conditions are satisfied when $xy^2 + x\cos x = 0$, which implies that the particular solution can be written as $x = 0$ or $y = \pm\sqrt{-\cos x}$. On a graphing utility screen, the solution would be represented by Figure 16.2 together with the y-axis.

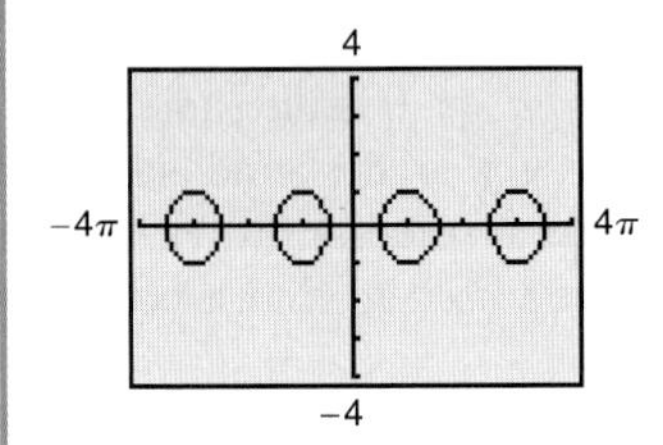

Figure 16.2

Find the particular solution of

$$(\cos x - x\sin x + y^2)\,dx + 2xy\,dy = 0$$

that satisfies the initial condition $y = 1$ when $x = \pi$.

Solution The differential equation is exact because

$$\overbrace{\frac{\partial}{\partial y}[\cos x - x\sin x + y^2]}^{\frac{\partial M}{\partial y}} = 2y = \overbrace{\frac{\partial}{\partial x}[2xy]}^{\frac{\partial N}{\partial x}}.$$

Because $N(x, y)$ is simpler than $M(x, y)$, it is better to begin by integrating $N(x, y)$.

$$f(x, y) = \int N(x, y)\,dy = \int 2xy\,dy = xy^2 + g(x)$$

$$f_x(x, y) = \frac{\partial}{\partial x}[xy^2 + g(x)] = y^2 + g'(x) = \overbrace{\cos x - x\sin x + y^2}^{M(x,y)}$$

$$g'(x) = \cos x - x\sin x$$

So, $g'(x) = \cos x - x\sin x$ and

$$\begin{aligned} g(x) &= \int (\cos x - x\sin x)\,dx \\ &= x\cos x + C_1 \end{aligned}$$

which implies that $f(x, y) = xy^2 + x\cos x + C_1$, and the general solution is

$$xy^2 + x\cos x = C. \qquad \text{General solution}$$

Applying the given initial condition produces

$$\pi(1)^2 + \pi\cos\pi = C$$

which implies that $C = 0$. So, the particular solution is

$$xy^2 + x\cos x = 0. \qquad \text{Particular solution}$$

The graph of the particular solution is shown in Figure 16.3. Notice that the graph consists of two parts: the ovals are given by $y^2 + \cos x = 0$, and the y-axis is given by $x = 0$. ■

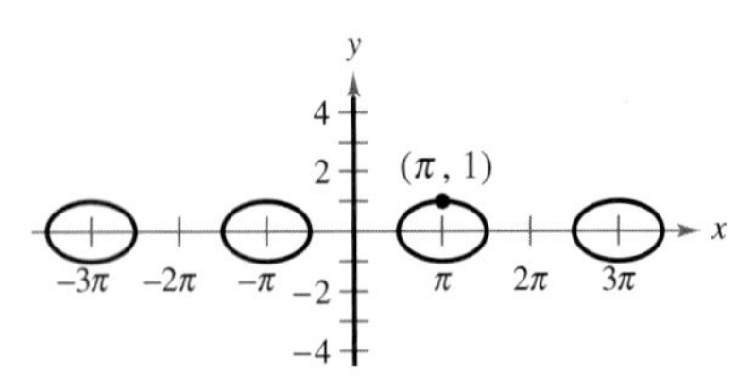

Figure 16.3

In Example 3, note that if $z = f(x, y) = xy^2 + x\cos x$, the total differential of z is given by

$$\begin{aligned} dz &= f_x(x, y)\,dx + f_y(x, y)\,dy \\ &= (\cos x - x\sin x + y^2)\,dx + 2xy\,dy \\ &= M(x, y)\,dx + N(x, y)\,dy. \end{aligned}$$

In other words, $M\,dx + N\,dy = 0$ is called an *exact* differential equation because $M\,dx + N\,dy$ is exactly the differential of $f(x, y)$.

Integrating Factors

If the differential equation $M(x, y)\,dx + N(x, y)\,dy = 0$ is not exact, it may be possible to make it exact by multiplying by an appropriate factor $u(x, y)$, which is called an **integrating factor** for the differential equation.

EXAMPLE 4 Multiplying by an Integrating Factor

a. If the differential equation

$$2y\,dx + x\,dy = 0 \qquad \text{Not an exact equation}$$

is multiplied by the integrating factor $u(x, y) = x$, the resulting equation

$$2xy\,dx + x^2\,dy = 0 \qquad \text{Exact equation}$$

is exact—the left side is the total differential of x^2y.

b. If the equation

$$y\,dx - x\,dy = 0 \qquad \text{Not an exact equation}$$

is multiplied by the integrating factor $u(x, y) = 1/y^2$, the resulting equation

$$\frac{1}{y}\,dx - \frac{x}{y^2}\,dy = 0 \qquad \text{Exact equation}$$

is exact—the left side is the total differential of x/y. ■

EXPLORATION

In Example 4, show that the differential equations

$$2xy\,dx + x^2\,dy = 0$$

and

$$\frac{1}{y}\,dx - \frac{x}{y^2}\,dy = 0$$

are exact.

Finding an integrating factor can be difficult. However, there are two classes of differential equations whose integrating factors can be found routinely—namely, those that possess integrating factors that are functions of either x alone or y alone. The following theorem, which is presented without proof, outlines a procedure for finding these two special categories of integrating factors.

THEOREM 16.2 INTEGRATING FACTORS

Consider the differential equation $M(x, y)\,dx + N(x, y)\,dy = 0$.

1. If

$$\frac{1}{N(x, y)}\left[M_y(x, y) - N_x(x, y)\right] = h(x)$$

is a function of x alone, then $e^{\int h(x)\,dx}$ is an integrating factor.

2. If

$$\frac{1}{M(x, y)}\left[N_x(x, y) - M_y(x, y)\right] = k(y)$$

is a function of y alone, then $e^{\int k(y)\,dy}$ is an integrating factor.

STUDY TIP If either $h(x)$ or $k(y)$ is constant, Theorem 16.2 still applies. As an aid to remembering these formulas, note that the subtracted partial derivative identifies both the denominator and the variable for the integrating factor. ■

EXAMPLE 5 Finding an Integrating Factor

Solve the differential equation $(y^2 - x)\,dx + 2y\,dy = 0$.

Solution The given equation is not exact because $M_y(x, y) = 2y$ and $N_x(x, y) = 0$. However, because

$$\frac{M_y(x, y) - N_x(x, y)}{N(x, y)} = \frac{2y - 0}{2y} = 1 = h(x)$$

it follows that $e^{\int h(x)\,dx} = e^{\int dx} = e^x$ is an integrating factor. Multiplying the given differential equation by e^x produces the exact differential equation

$$(y^2e^x - xe^x)\,dx + 2ye^x\,dy = 0$$

whose solution is obtained as follows.

$$f(x, y) = \int N(x, y)\,dy = \int 2ye^x\,dy = y^2e^x + g(x)$$

$$f_x(x, y) = y^2e^x + g'(x) = \overbrace{y^2e^x - xe^x}^{M(x,y)}$$

$$\boxed{g'(x) = -xe^x}$$

Therefore, $g'(x) = -xe^x$ and $g(x) = -xe^x + e^x + C_1$, which implies that

$$f(x, y) = y^2e^x - xe^x + e^x + C_1.$$

The general solution is $y^2e^x - xe^x + e^x = C$, or $y^2 - x + 1 = Ce^{-x}$. ■

The next example shows how a differential equation can help in sketching a force field given by $\mathbf{F}(x, y) = M(x, y)\mathbf{i} + N(x, y)\mathbf{j}$.

EXAMPLE 6 An Application to Force Fields

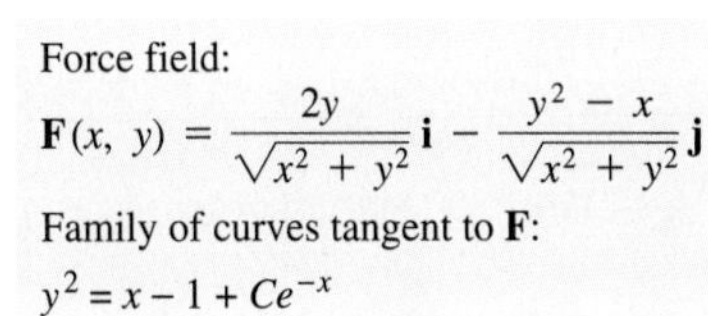

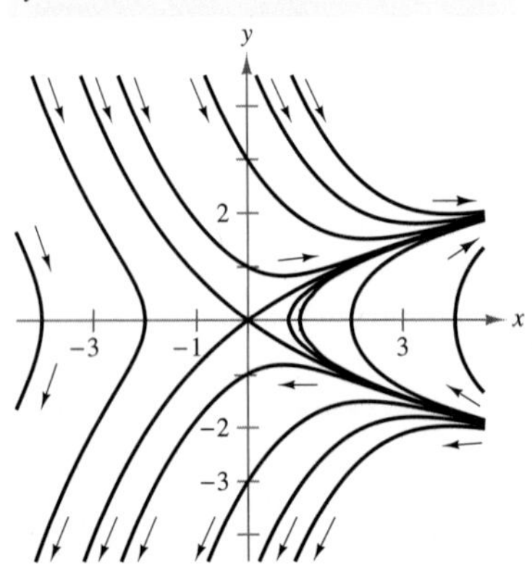

Figure 16.4

Sketch the force field given by

$$\mathbf{F}(x, y) = \frac{2y}{\sqrt{x^2 + y^2}}\mathbf{i} - \frac{y^2 - x}{\sqrt{x^2 + y^2}}\mathbf{j}$$

by finding and sketching the family of curves tangent to **F**.

Solution At the point (x, y) in the plane, the vector $\mathbf{F}(x, y)$ has a slope of

$$\frac{dy}{dx} = \frac{-(y^2 - x)/\sqrt{x^2 + y^2}}{2y/\sqrt{x^2 + y^2}} = \frac{-(y^2 - x)}{2y}$$

which, in differential form, is

$$2y\,dy = -(y^2 - x)\,dx$$
$$(y^2 - x)\,dx + 2y\,dy = 0.$$

From Example 5, you know that the general solution of this differential equation is $y^2 - x + 1 = Ce^{-x}$, or $y^2 = x - 1 + Ce^{-x}$. Figure 16.4 shows several representative curves from this family. Note that the force vector at (x, y) is tangent to the curve passing through (x, y). ■

16.1 Exercises

See www.CalcChat.com for worked-out solutions to odd-numbered exercises.

In Exercises 1–4, determine whether the differential equation is exact. Explain your reasoning.

1. $(2x + xy^2)\,dx + (3 + x^2y)\,dy = 0$
2. $(1 - xy)\,dx + (y - xy)\,dy = 0$
3. $x \sin y\,dx + x \cos y\,dy = 0$
4. $ye^{xy}\,dx + xe^{xy}\,dy = 0$

In Exercises 5–14, determine whether the differential equation is exact. If it is, find the general solution.

5. $(2x - 3y)\,dx + (2y - 3x)\,dy = 0$
6. $ye^x\,dx + e^x\,dy = 0$
7. $(3y^2 + 10xy^2)\,dx + (6xy - 2 + 10x^2y)\,dy = 0$
8. $2\cos(2x - y)\,dx - \cos(2x - y)\,dy = 0$
9. $(4x^3 - 6xy^2)\,dx + (4y^3 - 6xy)\,dy = 0$
10. $2y^2e^{xy^2}\,dx + 2xye^{xy^2}\,dy = 0$
11. $\dfrac{1}{x^2 + y^2}(x\,dy - y\,dx) = 0$
12. $e^{-(x^2+y^2)}(x\,dx + y\,dy) = 0$
13. $\dfrac{1}{(x - y)^2}(y^2\,dx + x^2\,dy) = 0$
14. $e^y \cos xy\,[y\,dx + (x + \tan xy)\,dy] = 0$

In Exercises 15–18, (a) sketch an approximate solution of the differential equation satisfying the initial condition on the slope field, (b) find the particular solution that satisfies the initial condition, and (c) use a graphing utility to graph the particular solution. Compare the graph with the sketch in part (a).

	Differential Equation	*Initial Condition*
15.	$(2x \tan y + 5)\,dx + (x^2 \sec^2 y)\,dy = 0$	$\left(\dfrac{1}{2}, \dfrac{\pi}{4}\right)$
16.	$2xy\,dx + (x^2 + \cos y)\,dy = 0$	$(1, \pi)$
17.	$\dfrac{1}{\sqrt{x^2 + y^2}}(x\,dx + y\,dy) = 0$	$(4, 3)$
18.	$(x^2 - y)\,dx + (y^2 - x)\,dy = 0$	$(4, 5)$

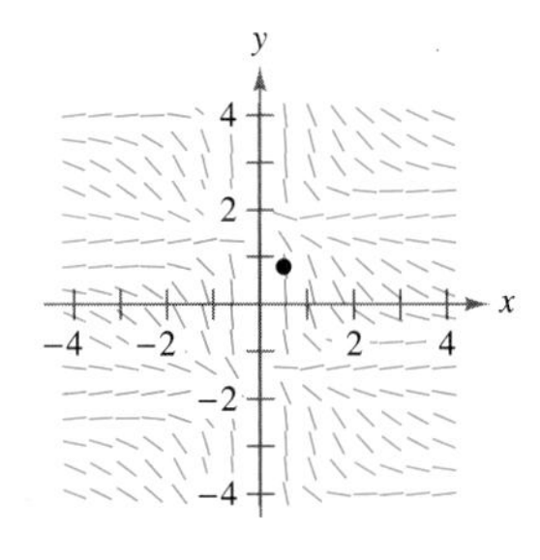

Figure for 15

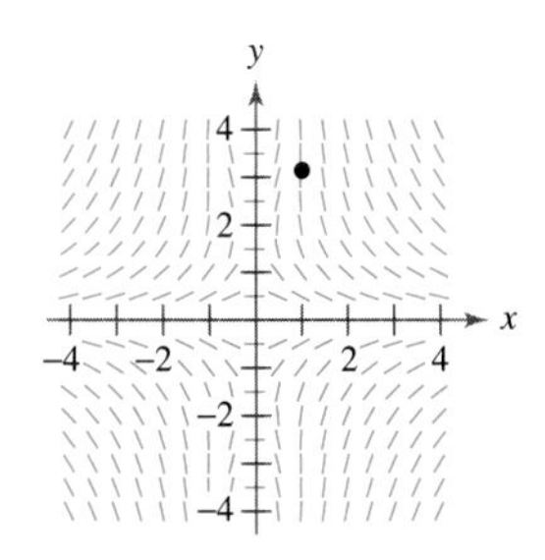

Figure for 16

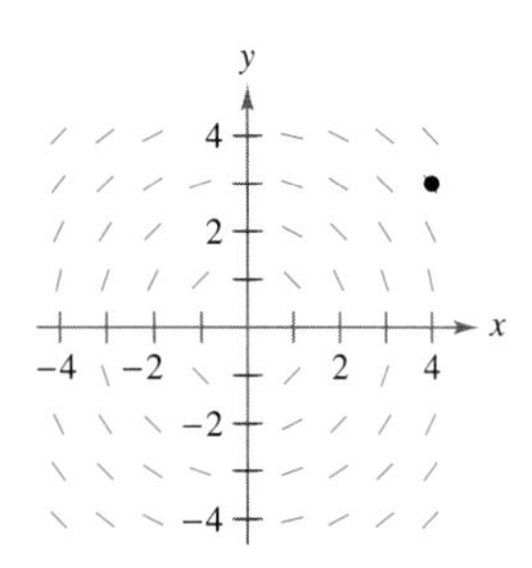

Figure for 17

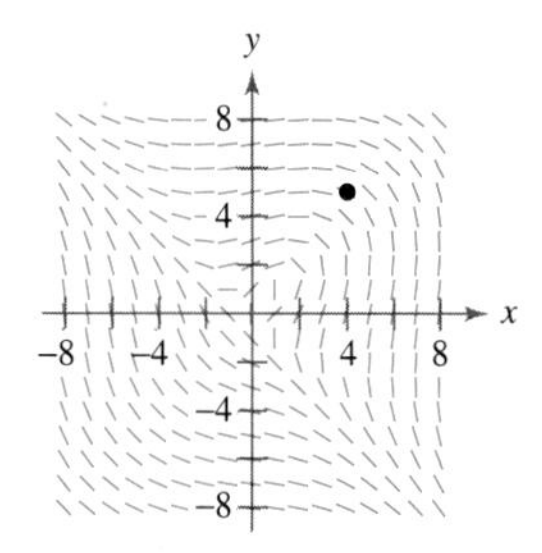

Figure for 18

In Exercises 19–24, find the particular solution that satisfies the initial condition.

	Differential Equation	*Initial Condition*
19.	$\dfrac{y}{x - 1}\,dx + [\ln(x - 1) + 2y]\,dy = 0$	$y(2) = 4$
20.	$\dfrac{1}{x^2 + y^2}(x\,dx + y\,dy) = 0$	$y(0) = 4$
21.	$e^{3x}(\sin 3y\,dx + \cos 3y\,dy) = 0$	$y(0) = \pi$
22.	$(x^2 + y^2)\,dx + 2xy\,dy = 0$	$y(3) = 1$
23.	$(2xy - 9x^2)\,dx + (2y + x^2 + 1)\,dy = 0$	$y(0) = -3$
24.	$(2xy^2 + 4)\,dx + (2x^2y - 6)\,dy = 0$	$y(-1) = 8$

In Exercises 25–34, find the integrating factor that is a function of x or y alone and use it to find the general solution of the differential equation.

25. $y\,dx - (x + 6y^2)\,dy = 0$
26. $(2x^3 + y)\,dx - x\,dy = 0$
27. $(5x^2 - y)\,dx + x\,dy = 0$
28. $(5x^2 - y^2)\,dx + 2y\,dy = 0$
29. $(x + y)\,dx + \tan x\,dy = 0$
30. $(2x^2y - 1)\,dx + x^3\,dy = 0$
31. $y^2\,dx + (xy - 1)\,dy = 0$
32. $(x^2 + 2x + y)\,dx + 2\,dy = 0$
33. $2y\,dx + (x - \sin\sqrt{y})\,dy = 0$
34. $(-2y^3 + 1)\,dx + (3xy^2 + x^3)\,dy = 0$

In Exercises 35–38, use the integrating factor to find the general solution of the differential equation.

	Integrating Factor	*Differential Equation*
35.	$u(x, y) = xy^2$	$(4x^2y + 2y^2)\,dx + (3x^3 + 4xy)\,dy = 0$
36.	$u(x, y) = x^2y$	$(3y^2 + 5x^2y)\,dx + (3xy + 2x^3)\,dy = 0$
37.	$u(x, y) = x^{-2}y^{-3}$	$(-y^5 + x^2y)\,dx + (2xy^4 - 2x^3)\,dy = 0$
38.	$u(x, y) = x^{-2}y^{-2}$	$-y^3\,dx + (xy^2 - x^2)\,dy = 0$

39. Show that each expression is an integrating factor for the differential equation

$$y\,dx - x\,dy = 0.$$

(a) $\frac{1}{x^2}$ (b) $\frac{1}{y^2}$ (c) $\frac{1}{xy}$ (d) $\frac{1}{x^2 + y^2}$

40. Show that the differential equation

$$(axy^2 + by)\,dx + (bx^2y + ax)\,dy = 0$$

is exact only if $a = b$. If $a \neq b$, show that $x^m y^n$ is an integrating factor, where

$$m = -\frac{2b + a}{a + b}, \qquad n = -\frac{2a + b}{a + b}.$$

In Exercises 41–44, use a graphing utility to graph the family of curves tangent to the given force field.

41. $\mathbf{F}(x, y) = \frac{y}{\sqrt{x^2 + y^2}}\mathbf{i} - \frac{x}{\sqrt{x^2 + y^2}}\mathbf{j}$

42. $\mathbf{F}(x, y) = \frac{x}{\sqrt{x^2 + y^2}}\mathbf{i} - \frac{y}{\sqrt{x^2 + y^2}}\mathbf{j}$

43. $\mathbf{F}(x, y) = 4x^2y\mathbf{i} - \left(2xy^2 + \frac{x}{y^2}\right)\mathbf{j}$

44. $\mathbf{F}(x, y) = (1 + x^2)\,\mathbf{i} - 2xy\mathbf{j}$

In Exercises 45 and 46, find an equation of the curve with the specified slope passing through the given point.

	Slope	*Point*
45.	$\frac{dy}{dx} = \frac{y - x}{3y - x}$	$(2, 1)$
46.	$\frac{dy}{dx} = \frac{-2xy}{x^2 + y^2}$	$(0, 2)$

47. ***Cost*** If $y = C(x)$ represents the cost of producing x units in a manufacturing process, the **elasticity of cost** is defined as

$$E(x) = \frac{\text{marginal cost}}{\text{average cost}} = \frac{C'(x)}{C(x)/x} = \frac{x}{y}\frac{dy}{dx}.$$

Find the cost function if the elasticity function is

$$E(x) = \frac{20x - y}{2y - 10x}$$

where $C(100) = 500$ and $x \geq 100$.

CAPSTONE

48. In Chapter 6, you solved the first-order linear differential equation

$$\frac{dy}{dx} + P(x)y = Q(x)$$

by using the integrating factor

$$u(x) = e^{\int P(x)\,dx}.$$

Show that you can obtain this integrating factor by using the methods of this section.

Euler's Method **In Exercises 49 and 50, (a) use Euler's Method and a graphing utility to graph the particular solution of the initial value problem over the indicated interval with the specified value of h and initial condition, (b) find the exact solution of the differential equation analytically, and (c) use a graphing utility to graph the particular solution and compare the result with the graph in part (a).**

	Differential Equation	*Interval*	*h*	*Initial Condition*
49.	$y' = \frac{-xy}{x^2 + y^2}$	$[2, 4]$	0.05	$y(2) = 1$
50.	$y' = \frac{6x + y^2}{y(3y - 2x)}$	$[0, 5]$	0.2	$y(0) = 1$

51. ***Euler's Method*** Repeat Exercise 49 for $h = 1$ and discuss how the accuracy of the result changes.

52. ***Euler's Method*** Repeat Exercise 50 for $h = 0.5$ and discuss how the accuracy of the result changes.

WRITING ABOUT CONCEPTS

53. Explain how to determine whether a differential equation is exact.

54. Outline the procedure for finding an integrating factor for the differential equation $M(x, y)\,dx + N(x, y)\,dy = 0$.

True or False? **In Exercises 55–58, determine whether the statement is true or false. If it is false, explain why or give an example that shows it is false.**

55. The differential equation $2xy\,dx + (y^2 - x^2)\,dy = 0$ is exact.

56. If $M\,dx + N\,dy = 0$ is exact, then $xM\,dx + xN\,dy = 0$ is also exact.

57. If $M\,dx + N\,dy = 0$ is exact, then $[f(x) + M]\,dx + [g(y) + N]\,dy = 0$ is also exact.

58. The differential equation $f(x)\,dx + g(y)\,dy = 0$ is exact.

In Exercises 59 and 60, find all values of k such that the differential equation is exact.

59. $(xy^2 + kx^2y + x^3)\,dx + (x^3 + x^2y + y^2)\,dy = 0$

60. $(ye^{2xy} + 2x)\,dx + (kxe^{2xy} - 2y)\,dy = 0$

61. Find all nonzero functions f and g such that

$$g(y)\sin x\,dx + y^2 f(x)\,dy = 0$$

is exact.

62. Find all nonzero functions g such that

$$g(y)\,e^y\,dx + xy\,dy = 0$$

is exact.

16.2 Second-Order Homogeneous Linear Equations

- **Solve a second-order linear differential equation.**
- **Solve a higher-order linear differential equation.**
- **Use a second-order linear differential equation to solve an applied problem.**

Second-Order Linear Differential Equations

In this section and the following section, you will learn methods for solving higher-order linear differential equations.

DEFINITION OF LINEAR DIFFERENTIAL EQUATION OF ORDER n

Let $g_1, g_2, \ldots, g_n$ and f be functions of x with a common (interval) domain. An equation of the form

$$y^{(n)} + g_1(x)y^{(n-1)} + g_2(x)y^{(n-2)} + \cdots + g_{n-1}(x)y' + g_n(x)y = f(x)$$

is called a **linear differential equation of order n.** If $f(x) = 0$, the equation is **homogeneous**; otherwise, it is **nonhomogeneous.**

NOTE Notice that this use of the term *homogeneous* differs from that in Section 6.3. ■

Homogeneous equations are discussed in this section, and the nonhomogeneous case is discussed in the next section.

The functions $y_1, y_2, \ldots, y_n$ are **linearly independent** if the *only* solution of the equation

$$C_1y_1 + C_2y_2 + \cdots + C_ny_n = 0$$

is the trivial one, $C_1 = C_2 = \cdots = C_n = 0$. Otherwise, this set of functions is **linearly dependent.**

EXAMPLE 1 Linearly Independent and Dependent Functions

a. The functions

$$y_1(x) = \sin x \quad \text{and} \quad y_2(x) = x$$

are linearly independent because the only values of C_1 and C_2 for which

$$C_1 \sin x + C_2x = 0$$

for all x are $C_1 = 0$ and $C_2 = 0$.

b. It can be shown that two functions form a linearly dependent set if and only if one is a constant multiple of the other. For example,

$$y_1(x) = x \quad \text{and} \quad y_2(x) = 3x$$

are linearly dependent because

$$C_1x + C_2(3x) = 0$$

has the nonzero solutions $C_1 = -3$ and $C_2 = 1$. ■

The following theorem points out the importance of linear independence in constructing the general solution of a second-order linear homogeneous differential equation with constant coefficients.

THEOREM 16.3 LINEAR COMBINATIONS OF SOLUTIONS

If y_1 and y_2 are linearly independent solutions of the differential equation $y'' + ay' + by = 0$, then the general solution is

$$y = C_1y_1 + C_2y_2$$

where C_1 and C_2 are constants.

PROOF This theorem is proved in only one direction. If y_1 and y_2 are solutions, you obtain the following system of equations.

$$y_1''(x) + ay_1'(x) + by_1(x) = 0$$
$$y_2''(x) + ay_2'(x) + by_2(x) = 0$$

Multiplying the first equation by C_1, multiplying the second by C_2, and adding the resulting equations together produces

$$[C_1y_1''(x) + C_2y_2''(x)] + a[C_1y_1'(x) + C_2y_2'(x)] + b[C_1y_1(x) + C_2y_2(x)] = 0$$

which means that

$$y = C_1y_1 + C_2y_2$$

is a solution, as desired. The proof that all solutions are of this form is best left to a full course on differential equations. ■

Theorem 16.3 states that if you can find two linearly independent solutions, you can obtain the general solution by forming a **linear combination** of the two solutions.

To find two linearly independent solutions, note that the nature of the equation $y'' + ay' + by = 0$ suggests that it may have solutions of the form $y = e^{mx}$. If so, then $y' = me^{mx}$ and $y'' = m^2e^{mx}$. So, by substitution, $y = e^{mx}$ is a solution if and only if

$$y'' + ay' + by = 0$$
$$m^2e^{mx} + ame^{mx} + be^{mx} = 0$$
$$e^{mx}(m^2 + am + b) = 0.$$

Because e^{mx} is never 0, $y = e^{mx}$ is a solution if and only if

$$m^2 + am + b = 0. \qquad \text{Characteristic equation}$$

This is the **characteristic equation** of the differential equation

$$y'' + ay' + by = 0.$$

Note that the characteristic equation can be determined from its differential equation simply by replacing y'' with m^2, y' with m, and y with 1.

EXPLORATION

For each differential equation below, find the characteristic equation. Solve the characteristic equation for m, and use the values of m to find a general solution to the differential equation. Using your results, develop a general solution to differential equations with characteristic equations that have distinct real roots.

(a) $y'' - 9y = 0$

(b) $y'' - 6y' + 8y = 0$

EXAMPLE 2 Characteristic Equation with Distinct Real Zeros

Solve the differential equation

$$y'' - 4y = 0.$$

Solution In this case, the characteristic equation is

$$m^2 - 4 = 0 \qquad \text{Characteristic equation}$$

so, $m = \pm 2$. Therefore, $y_1 = e^{m_1 x} = e^{2x}$ and $y_2 = e^{m_2 x} = e^{-2x}$ are particular solutions of the given differential equation. Furthermore, because these two solutions are linearly independent, you can apply Theorem 16.3 to conclude that the general solution is

$$y = C_1 e^{2x} + C_2 e^{-2x}. \qquad \text{General solution}$$

■

The characteristic equation in Example 2 has two distinct real zeros. From algebra, you know that this is only one of *three* possibilities for quadratic equations. In general, the quadratic equation $m^2 + am + b = 0$ has zeros

$$m_1 = \frac{-a + \sqrt{a^2 - 4b}}{2} \quad \text{and} \quad m_2 = \frac{-a - \sqrt{a^2 - 4b}}{2}$$

which fall into one of three cases.

1. Two distinct real zeros, $m_1 \neq m_2$
2. Two equal real zeros, $m_1 = m_2$
3. Two complex conjugate zeros, $m_1 = \alpha + \beta i$ and $m_2 = \alpha - \beta i$

In terms of the differential equation $y'' + ay' + by = 0$, these three cases correspond to three different types of general solutions.

THEOREM 16.4 SOLUTIONS OF $y'' + ay' + by = 0$

The solutions of

$$y'' + ay' + by = 0$$

fall into one of the following three cases, depending on the solutions of the characteristic equation, $m^2 + am + b = 0$.

1. *Distinct Real Zeros* If $m_1 \neq m_2$ are distinct real zeros of the characteristic equation, then the general solution is
$$y = C_1 e^{m_1 x} + C_2 e^{m_2 x}.$$
2. *Equal Real Zeros* If $m_1 = m_2$ are equal real zeros of the characteristic equation, then the general solution is
$$y = C_1 e^{m_1 x} + C_2 x e^{m_1 x} = (C_1 + C_2 x) e^{m_1 x}.$$
3. *Complex Zeros* If $m_1 = \alpha + \beta i$ and $m_2 = \alpha - \beta i$ are complex zeros of the characteristic equation, then the general solution is
$$y = C_1 e^{\alpha x} \cos \beta x + C_2 e^{\alpha x} \sin \beta x.$$

■ **FOR FURTHER INFORMATION** For more information on Theorem 16.4, see "A Note on a Differential Equation" by Russell Euler in the 1989 winter issue of the *Missouri Journal of Mathematical Sciences.*

EXAMPLE 3 Characteristic Equation with Complex Zeros

Find the general solution of the differential equation

$$y'' + 6y' + 12y = 0.$$

Solution The characteristic equation

$$m^2 + 6m + 12 = 0$$

has two complex zeros, as follows.

$$\begin{aligned} m &= \frac{-6 \pm \sqrt{36 - 48}}{2} \\ &= \frac{-6 \pm \sqrt{-12}}{2} \\ &= \frac{-6 \pm 2\sqrt{-3}}{2} \\ &= -3 \pm \sqrt{-3} \\ &= -3 \pm \sqrt{3}i \end{aligned}$$

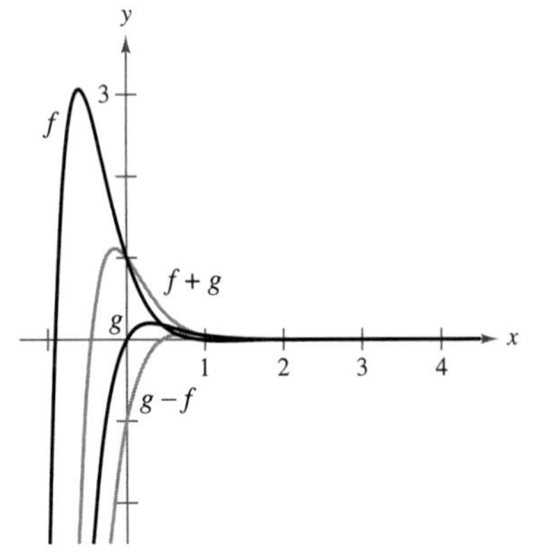

The basic solutions in Example 3, $f(x) = e^{-3x}\cos\sqrt{3}x$ and $g(x) = e^{-3x}\sin\sqrt{3}x$, are shown in the graph along with other members of the family of solutions. Notice that as $x \to \infty$, all of these solutions approach 0.
Figure 16.5

So, $\alpha = -3$ and $\beta = \sqrt{3}$, and the general solution is

$$y = C_1e^{-3x}\cos\sqrt{3}x + C_2e^{-3x}\sin\sqrt{3}x.$$

The graphs of the basic solutions $f(x) = e^{-3x}\cos\sqrt{3}x$ and $g(x) = e^{-3x}\sin\sqrt{3}x$, along with other members of the family of solutions, are shown in Figure 16.5. ■

NOTE In Example 3, note that although the characteristic equation has two *complex* zeros, the solution of the differential equation is *real*. ■

EXAMPLE 4 Characteristic Equation with Repeated Zeros

Solve the differential equation

$$y'' + 4y' + 4y = 0$$

subject to the initial conditions $y(0) = 2$ and $y'(0) = 1$.

Solution The characteristic equation

$$m^2 + 4m + 4 = (m + 2)^2 = 0$$

has two equal zeros given by $m = -2$. So, the general solution is

$$y = C_1e^{-2x} + C_2xe^{-2x}. \qquad \text{General solution}$$

Now, because $y = 2$ when $x = 0$, you have

$$2 = C_1(1) + C_2(0)(1) = C_1.$$

Furthermore, because $y' = 1$ when $x = 0$, you have

$$\begin{aligned} y' &= -2C_1e^{-2x} + C_2(-2xe^{-2x} + e^{-2x}) \\ 1 &= -2(2)(1) + C_2[-2(0)(1) + 1] \\ 5 &= C_2. \end{aligned}$$

Therefore, the solution is

$$y = 2e^{-2x} + 5xe^{-2x}. \qquad \text{Particular solution}$$

Try checking this solution in the original differential equation. ■

Higher-Order Linear Differential Equations

For higher-order homogeneous linear differential equations, you can find the general solution in much the same way as you do for second-order equations. That is, you begin by determining the n zeros of the characteristic equation. Then, based on these n zeros, you form a linearly independent collection of n solutions. The major difference is that with equations of third or higher order, zeros of the characteristic equation may occur more than twice. When this happens, the linearly independent solutions are formed by multiplying by increasing powers of x, as demonstrated in Examples 6 and 7.

EXAMPLE 5 Solving a Third-Order Equation

Find the general solution of $y''' - y' = 0$.

Solution The characteristic equation is

$$m^3 - m = 0$$
$$m(m - 1)(m + 1) = 0$$
$$m = 0, 1, -1.$$

Because the characteristic equation has three distinct zeros, the general solution is

$$y = C_1 + C_2e^{-x} + C_3e^x. \quad \text{General solution}$$

EXAMPLE 6 Solving a Third-Order Equation

Find the general solution of $y''' + 3y'' + 3y' + y = 0$.

Solution The characteristic equation is

$$m^3 + 3m^2 + 3m + 1 = 0$$
$$(m + 1)^3 = 0$$
$$m = -1.$$

Because the zero $m = -1$ occurs three times, the general solution is

$$y = C_1e^{-x} + C_2xe^{-x} + C_3x^2e^{-x}. \quad \text{General solution}$$

EXAMPLE 7 Solving a Fourth-Order Equation

Find the general solution of $y^{(4)} + 2y'' + y = 0$.

Solution The characteristic equation is as follows.

$$m^4 + 2m^2 + 1 = 0$$
$$(m^2 + 1)^2 = 0$$
$$m = \pm i$$

Because each of the zeros $m_1 = \alpha + \beta i = 0 + i$ and $m_2 = \alpha - \beta i = 0 - i$ occurs twice, the general solution is

$$y = C_1 \cos x + C_2 \sin x + C_3x \cos x + C_4x \sin x. \quad \text{General solution}$$

■

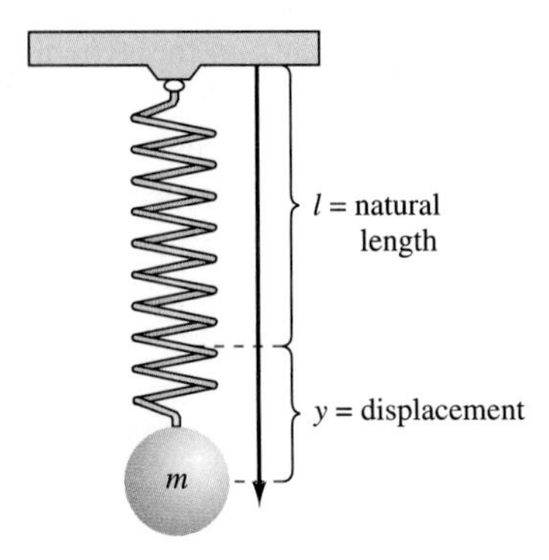

A rigid object of mass m attached to the end of the spring causes a displacement of y.
Figure 16.6

Application

One of the many applications of linear differential equations is describing the motion of an oscillating spring. According to Hooke's Law, a spring that is stretched (or compressed) y units from its natural length l tends to *restore* itself to its natural length by a force F that is proportional to y. That is, $F(y) = -ky$, where k is the **spring constant** and indicates the stiffness of the given spring.

Suppose a rigid object of mass m is attached to the end of a spring and causes a displacement, as shown in Figure 16.6. Assume that the mass of the spring is negligible compared with m. If the object is pulled downward and released, the resulting oscillations are a product of two opposing forces—the spring force $F(y) = -ky$ and the weight mg of the object. Under such conditions, you can use a differential equation to find the position y of the object as a function of time t. According to Newton's Second Law of Motion, the force acting on the weight is $F = ma$, where $a = d^2y/dt^2$ is the acceleration. Assuming that the motion is **undamped**—that is, there are no other external forces acting on the object—it follows that $m(d^2y/dt^2) = -ky$, and you have

$$\frac{d^2y}{dt^2} + \left(\frac{k}{m}\right)y = 0.$$

Undamped motion of a spring

EXAMPLE 8 Undamped Motion of a Spring

A four-pound weight stretches a spring 8 inches from its natural length. The weight is pulled downward an additional 6 inches and released with an initial upward velocity of 8 feet per second. Find a formula for the position of the weight as a function of time t.

Solution By Hooke's Law, $4 = k\left(\frac{2}{3}\right)$, so $k = 6$. Moreover, because the weight w is given by mg, it follows that $m = w/g = \frac{4}{32} = \frac{1}{8}$. So, the resulting differential equation for this undamped motion is

$$\frac{d^2y}{dt^2} + 48y = 0.$$

Because the characteristic equation $m^2 + 48 = 0$ has complex zeros $m = 0 \pm 4\sqrt{3}i$, the general solution is

$$\begin{aligned} y &= C_1e^0 \cos 4\sqrt{3}\,t + C_2e^0 \sin 4\sqrt{3}\,t \\ &= C_1 \cos 4\sqrt{3}\,t + C_2 \sin 4\sqrt{3}\,t. \end{aligned}$$

Using the initial conditions, you have

$$\frac{1}{2} = C_1(1) + C_2(0) \quad \Longrightarrow \quad C_1 = \frac{1}{2} \qquad y(0) = \tfrac{1}{2}$$

$$y'(t) = -4\sqrt{3}\,C_1 \sin 4\sqrt{3}\,t + 4\sqrt{3}\,C_2 \cos 4\sqrt{3}\,t$$

$$8 = -4\sqrt{3}\left(\frac{1}{2}\right)(0) + 4\sqrt{3}\,C_2(1) \quad \Longrightarrow \quad C_2 = \frac{2\sqrt{3}}{3}. \qquad y'(0) = 8$$

Consequently, the position at time t is given by

$$y = \frac{1}{2}\cos 4\sqrt{3}\,t + \frac{2\sqrt{3}}{3}\sin 4\sqrt{3}\,t.$$

■

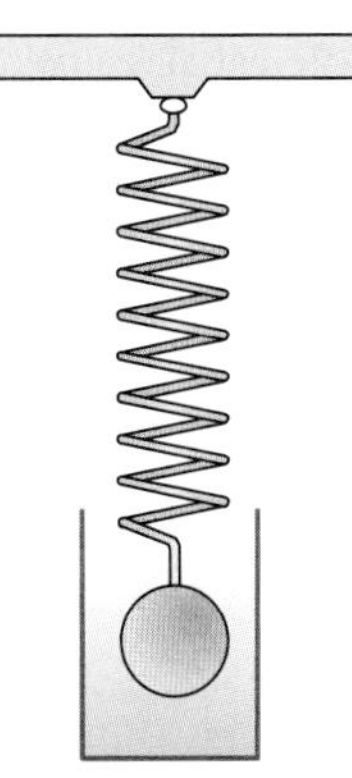

A damped vibration could be caused by friction and movement through a liquid.
Figure 16.7

Suppose the object in Figure 16.7 undergoes an additional damping or frictional force that is proportional to its velocity. A case in point would be the damping force resulting from friction and movement through a fluid. Considering this damping force, $-p(dy/dt)$, the differential equation for the oscillation is

$$m\frac{d^2y}{dt^2} = -ky - p\frac{dy}{dt}$$

or, in standard linear form,

$$\frac{d^2y}{dt^2} + \frac{p}{m}\left(\frac{dy}{dt}\right) + \frac{k}{m}y = 0.$$

Damped motion of a spring

16.2 Exercises

See www.CalcChat.com for worked-out solutions to odd-numbered exercises.

In Exercises 1–4, verify the solution of the differential equation. Then use a graphing utility to graph the particular solutions for several different values of C_1 and C_2. What do you observe?

Solution	*Differential Equation*
1. $y = (C_1 + C_2x)e^{-3x}$	$y'' + 6y' + 9y = 0$
2. $y = C_1e^{2x} + C_2e^{-2x}$	$y'' - 4y = 0$
3. $y = C_1 \cos 2x + C_2 \sin 2x$	$y'' + 4y = 0$
4. $y = C_1e^{-x} \cos 3x + C_2e^{-x} \sin 3x$	$y'' + 2y' + 10y = 0$

In Exercises 5–30, find the general solution of the linear differential equation.

5. $y'' - y' = 0$
6. $y'' + 2y' = 0$
7. $y'' - y' - 6y = 0$
8. $y'' + 6y' + 5y = 0$
9. $2y'' + 3y' - 2y = 0$
10. $16y'' - 16y' + 3y = 0$
11. $y'' + 6y' + 9y = 0$
12. $y'' - 10y' + 25y = 0$
13. $16y'' - 8y' + y = 0$
14. $9y'' - 12y' + 4y = 0$
15. $y'' + y = 0$
16. $y'' + 4y = 0$
17. $y'' - 9y = 0$
18. $y'' - 2y = 0$
19. $y'' - 2y' + 4y = 0$
20. $y'' - 4y' + 21y = 0$
21. $y'' - 3y' + y = 0$
22. $3y'' + 4y' - y = 0$
23. $9y'' - 12y' + 11y = 0$
24. $2y'' - 6y' + 7y = 0$
25. $y^{(4)} - y = 0$
26. $y^{(4)} - y'' = 0$
27. $y''' - 6y'' + 11y' - 6y = 0$
28. $y''' - y'' - y' + y = 0$
29. $y''' - 3y'' + 7y' - 5y = 0$
30. $y''' - 3y'' + 3y' - y = 0$

31. Consider the differential equation $y'' + 100y = 0$ and the solution $y = C_1 \cos 10x + C_2 \sin 10x$. Find the particular solution satisfying each of the following initial conditions.

(a) $y(0) = 2,\ \ y'(0) = 0$
(b) $y(0) = 0,\ \ y'(0) = 2$
(c) $y(0) = -1,\ \ y'(0) = 3$

32. Determine C and ω such that $y = C \sin\sqrt{3}\,t$ is a particular solution of the differential equation $y'' + \omega y = 0$, where $y'(0) = -5$.

In Exercises 33–38, find the particular solution of the linear differential equation that satisfies the initial conditions.

33. $y'' - y' - 30y = 0$
$y(0) = 1,\ \ y'(0) = -4$
34. $y'' - 7y' + 12y = 0$
$y(0) = 3,\ \ y'(0) = 3$
35. $y'' + 16y = 0$
$y(0) = 0,\ \ y'(0) = 2$
36. $9y'' - 6y' + y = 0$
$y(0) = 2,\ \ y'(0) = 1$
37. $y'' + 2y' + 3y = 0$
$y(0) = 2,\ \ y'(0) = 1$
38. $4y'' + 4y' + y = 0$
$y(0) = 3,\ \ y'(0) = -1$

In Exercises 39–44, find the particular solution of the linear differential equation that satisfies the boundary conditions, if possible.

39. $y'' - 4y' + 3y = 0$
$y(0) = 1,\ \ y(1) = 3$
40. $4y'' + y = 0$
$y(0) = 2,\ \ y(\pi) = -5$
41. $y'' + 9y = 0$
$y(0) = 3,\ \ y(\pi) = 5$
42. $4y'' + 20y' + 21y = 0$
$y(0) = 3,\ \ y(2) = 0$
43. $4y'' - 28y' + 49y = 0$
$y(0) = 2,\ \ y(1) = -1$
44. $y'' + 6y' + 45y = 0$
$y(0) = 4,\ \ y(\pi) = 8$

WRITING ABOUT CONCEPTS

45. Is the differential equation $y'' - y' - 5y = \sin x$ homogeneous? Why or why not?

46. The solutions of the differential equation $y'' + ay' + by = 0$ fall into what three cases? What is the relationship of these solutions to the characteristic equation of the differential equation?

47. Two functions are said to be linearly independent provided what?

CAPSTONE

48. Find all values of k for which the differential equation $y'' + 2ky' + ky = 0$ has a general solution of the indicated form.

(a) $y = C_1e^{m_1x} + C_2e^{m_2x}$

(b) $y = C_1e^{m_1x} + C_2xe^{m_1x}$

(c) $y = C_1e^{\alpha x}\cos \beta x + C_2e^{\alpha x}\sin \beta x$

Vibrating Spring **In Exercises 49–54, describe the motion of a 32-pound weight suspended on a spring. Assume that the weight stretches the spring $\frac{2}{3}$ foot from its natural position.**

49. The weight is pulled $\frac{1}{2}$ foot below the equilibrium position and released.

50. The weight is raised $\frac{2}{3}$ foot above the equilibrium position and released.

51. The weight is raised $\frac{2}{3}$ foot above the equilibrium position and started off with a downward velocity of $\frac{1}{2}$ foot per second.

52. The weight is pulled $\frac{1}{2}$ foot below the equilibrium position and started off with an upward velocity of $\frac{1}{2}$ foot per second.

53. The weight is pulled $\frac{1}{2}$ foot below the equilibrium position and released. The motion takes place in a medium that furnishes a damping force of magnitude $\frac{1}{8}$ speed at all times.

54. The weight is pulled $\frac{1}{2}$ foot below the equilibrium position and released. The motion takes place in a medium that furnishes a damping force of magnitude $\frac{1}{4}|v|$ at all times.

Vibrating Spring **In Exercises 55–58, match the differential equation with the graph of a particular solution. [The graphs are labeled (a), (b), (c), and (d).] The correct match can be made by comparing the frequency of the oscillations or the rate at which the oscillations are being damped with the appropriate coefficient in the differential equation.**

(a)

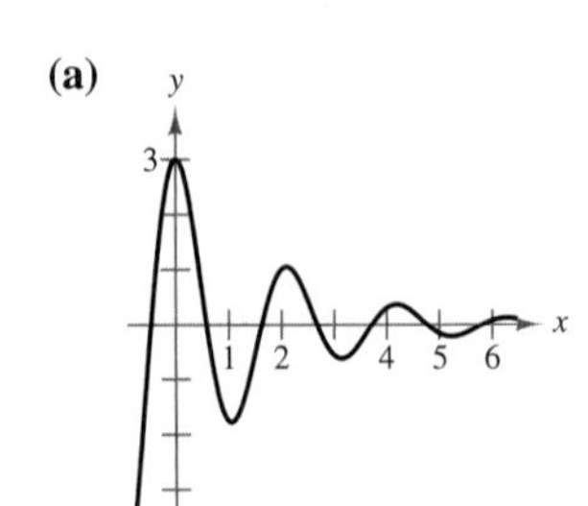

(b)

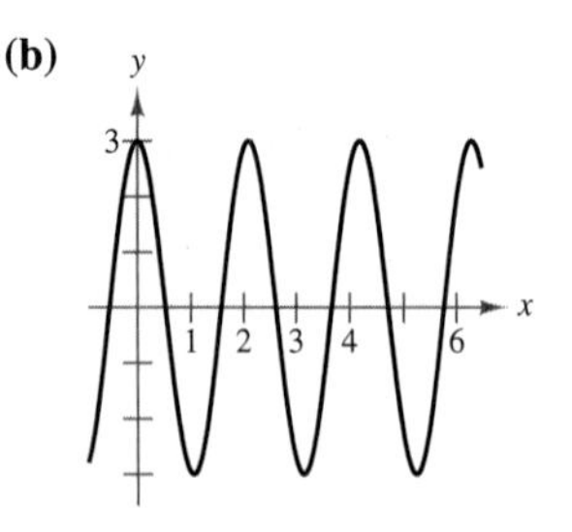

(c)

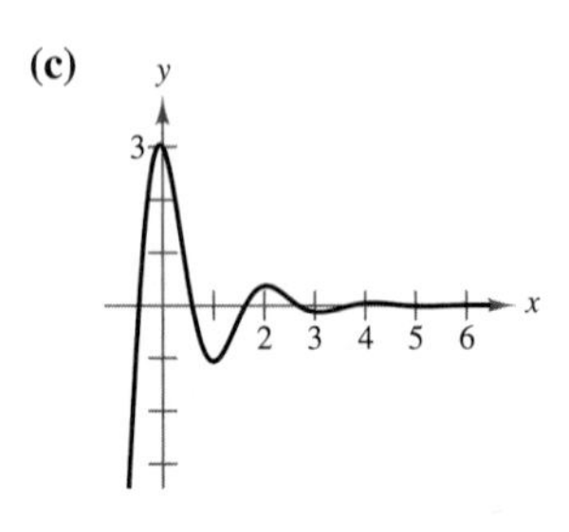

(d)

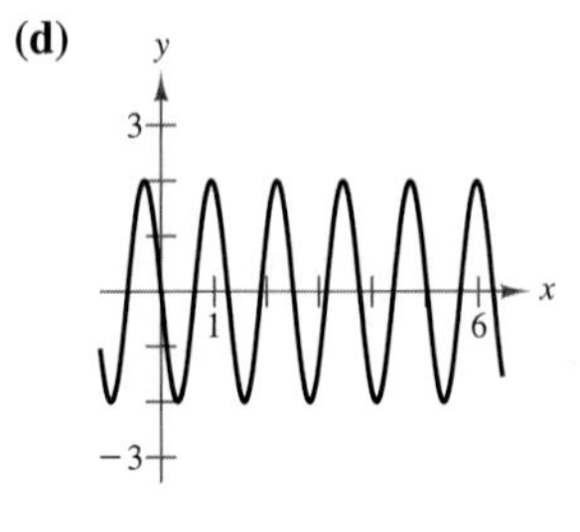

55. $y'' + 9y = 0$

56. $y'' + 25y = 0$

57. $y'' + 2y' + 10y = 0$

58. $y'' + y' + \frac{37}{4}y = 0$

59. If the characteristic equation of the differential equation $y'' + ay' + by = 0$ has two equal real zeros given by $m = r$, show that $y = C_1e^{rx} + C_2xe^{rx}$ is a solution.

60. If the characteristic equation of the differential equation

$$y'' + ay' + by = 0$$

has complex zeros given by $m_1 = \alpha + \beta i$ and $m_2 = \alpha - \beta i$, show that

$$y = C_1e^{\alpha x}\cos \beta x + C_2e^{\alpha x}\sin \beta x$$

is a solution.

True or False? **In Exercises 61–64, determine whether the statement is true or false. If it is false, explain why or give an example that shows it is false.**

61. $y = C_1e^{3x} + C_2e^{-3x}$ is the general solution of $y'' - 6y' + 9 = 0$.

62. $y = (C_1 + C_2x)\sin x + (C_3 + C_4x)\cos x$ is the general solution of $y^{(4)} + 2y'' + y = 0$.

63. $y = x$ is a solution of $a_ny^{(n)} + a_{n-1}y^{(n-1)} + \cdots + a_1y' + a_0y = 0$ if and only if $a_1 = a_0 = 0$.

64. It is possible to choose a and b such that $y = x^2e^x$ is a solution of $y'' + ay' + by = 0$.

The *Wronskian* of two differentiable functions f and g, denoted by $W(f, g)$, is defined as the function given by the determinant

$$W(f, g) = \begin{vmatrix} f & g \\ f' & g' \end{vmatrix}.$$

The functions f and g are linearly independent if there exists at least one value of x for which $W(f, g) \neq 0$. In Exercises 65–68, use the Wronskian to verify the linear independence of the two functions.

65. $y_1 = e^{ax}$
$y_2 = e^{bx}, \quad a \neq b$

66. $y_1 = e^{ax}$
$y_2 = xe^{ax}$

67. $y_1 = e^{ax}\sin bx$
$y_2 = e^{ax}\cos bx, \quad b \neq 0$

68. $y_1 = x$
$y_2 = x^2$

69. Euler's differential equation is of the form

$$x^2y'' + axy' + by = 0, \quad x > 0$$

where a and b are constants.

(a) Show that this equation can be transformed into a second-order linear equation with constant coefficients by using the substitution $x = e^t$.

(b) Solve $x^2y'' + 6xy' + 6y = 0$.

70. Solve

$$y'' + Ay = 0$$

where A is constant, subject to the conditions $y(0) = 0$ and $y(\pi) = 0$.

16.3 Second-Order Nonhomogeneous Linear Equations

- **Recognize the general solution of a second-order nonhomogeneous linear differential equation.**
- **Use the method of undetermined coefficients to solve a second-order nonhomogeneous linear differential equation.**
- **Use the method of variation of parameters to solve a second-order nonhomogeneous linear differential equation.**

Nonhomogeneous Equations

Patrick Zimmermann/AFP/Getty Images

SOPHIE GERMAIN (1776–1831)

Many of the early contributors to calculus were interested in forming mathematical models for vibrating strings and membranes, oscillating springs, and elasticity. One of these was the French mathematician Sophie Germain, who in 1816 was awarded a prize by the French Academy for a paper entitled "Memoir on the Vibrations of Elastic Plates."

In the preceding section, damped oscillations of a spring were represented by the *homogeneous* second-order linear equation

$$\frac{d^2y}{dt^2} + \frac{p}{m}\left(\frac{dy}{dt}\right) + \frac{k}{m}y = 0. \qquad \text{Free motion}$$

This type of oscillation is called **free** because it is determined solely by the spring and gravity and is free of the action of other external forces. If such a system is also subject to an external periodic force such as $a \sin bt$, caused by vibrations at the opposite end of the spring, the motion is called **forced,** and it is characterized by the *nonhomogeneous* equation

$$\frac{d^2y}{dt^2} + \frac{p}{m}\left(\frac{dy}{dt}\right) + \frac{k}{m}y = a \sin bt. \qquad \text{Forced motion}$$

In this section, you will study two methods for finding the general solution of a nonhomogeneous linear differential equation. In both methods, the first step is to find the general solution of the corresponding homogeneous equation.

$$y = y_h \qquad \text{General solution of homogeneous equation}$$

Having done this, you try to find a particular solution of the nonhomogeneous equation.

$$y = y_p \qquad \text{Particular solution of nonhomogeneous equation}$$

By combining these two results, you can conclude that the general solution of the nonhomogeneous equation is $y = y_h + y_p$, as stated in the following theorem.

THEOREM 16.5 SOLUTION OF NONHOMOGENEOUS LINEAR EQUATION

Let

$$y'' + ay' + by = F(x)$$

be a second-order nonhomogeneous linear differential equation. If y_p is a particular solution of this equation and y_h is the general solution of the corresponding homogeneous equation, then

$$y = y_h + y_p$$

is the general solution of the nonhomogeneous equation.

Method of Undetermined Coefficients

You already know how to find the solution y_h of a linear *homogeneous* differential equation. The remainder of this section looks at ways to find the particular solution y_p. If $F(x)$ in

$$y'' + ay' + by = F(x)$$

consists of sums or products of x^n, e^{mx}, $\cos \beta x$, or $\sin \beta x$, you can find a particular solution y_p by the method of **undetermined coefficients.** The object of this method is to guess that the solution y_p is a generalized form of $F(x)$. Here are some examples.

1. If $F(x) = 3x^2$, choose $y_p = Ax^2 + Bx + C$.
2. If $F(x) = 4xe^x$, choose $y_p = Axe^x + Be^x$.
3. If $F(x) = x + \sin 2x$, choose $y_p = (Ax + B) + C \sin 2x + D \cos 2x$.

Then, by substitution, determine the coefficients for the generalized solution.

EXAMPLE 1 Method of Undetermined Coefficients

Find the general solution of the equation

$$y'' - 2y' - 3y = 2 \sin x.$$

Solution To find y_h, solve the characteristic equation.

$$\begin{aligned} m^2 - 2m - 3 &= 0 \\ (m + 1)(m - 3) &= 0 \\ m = -1 \quad &\text{or} \quad m = 3 \end{aligned}$$

So, $y_h = C_1e^{-x} + C_2e^{3x}$. Next, let y_p be a generalized form of $2 \sin x$.

$$\begin{aligned} y_p &= A \cos x + B \sin x \\ y_p' &= -A \sin x + B \cos x \\ y_p'' &= -A \cos x - B \sin x \end{aligned}$$

Substitution into the original differential equation yields

$$\begin{aligned} y'' - 2y' - 3y &= 2 \sin x \\ -A \cos x - B \sin x + 2A \sin x - 2B \cos x - 3A \cos x - 3B \sin x &= 2 \sin x \\ (-4A - 2B)\cos x + (2A - 4B)\sin x &= 2 \sin x. \end{aligned}$$

By equating coefficients of like terms, you obtain

$$-4A - 2B = 0 \quad \text{and} \quad 2A - 4B = 2$$

with solutions $A = \frac{1}{5}$ and $B = -\frac{2}{5}$. Therefore,

$$y_p = \frac{1}{5} \cos x - \frac{2}{5} \sin x$$

and the general solution is

$$\begin{aligned} y &= y_h + y_p \\ &= C_1e^{-x} + C_2e^{3x} + \frac{1}{5} \cos x - \frac{2}{5} \sin x. \end{aligned}$$

■

In Example 1, the form of the homogeneous solution

$$y_h = C_1 e^{-x} + C_2 e^{3x}$$

has no overlap with the function $F(x)$ in the equation

$$y'' + ay' + by = F(x).$$

However, suppose the given differential equation in Example 1 were of the form

$$y'' - 2y' - 3y = e^{-x}.$$

Now it would make no sense to guess that the particular solution was $y = Ae^{-x}$, because you know that this solution would yield 0. In such cases, you should alter your guess by multiplying by the lowest power of x that removes the duplication. For this particular problem, you would guess

$$y_p = Axe^{-x}.$$

EXAMPLE 2 Method of Undetermined Coefficients

Find the general solution of

$$y'' - 2y' = x + 2e^x.$$

Solution The characteristic equation $m^2 - 2m = 0$ has solutions $m = 0$ and $m = 2$. So,

$$y_h = C_1 + C_2 e^{2x}.$$

Because $F(x) = x + 2e^x$, your first choice for y_p would be $(A + Bx) + Ce^x$. However, because y_h *already* contains a constant term C_1, you should multiply the *polynomial part* by x and use

$$y_p = Ax + Bx^2 + Ce^x$$
$$y_p' = A + 2Bx + Ce^x$$
$$y_p'' = 2B + Ce^x.$$

Substitution into the differential equation produces

$$\begin{aligned} y'' - 2y' &= x + 2e^x \\ 2B + Ce^x - 2(A + 2Bx + Ce^x) &= x + 2e^x \\ (2B - 2A) - 4Bx - Ce^x &= x + 2e^x. \end{aligned}$$

Equating coefficients of like terms yields the system

$$2B - 2A = 0, \qquad -4B = 1, \qquad -C = 2$$

with solutions $A = B = -\frac{1}{4}$ and $C = -2$. Therefore,

$$y_p = -\frac{1}{4}x - \frac{1}{4}x^2 - 2e^x$$

and the general solution is

$$\begin{aligned} y &= y_h + y_p \\ &= C_1 + C_2 e^{2x} - \frac{1}{4}x - \frac{1}{4}x^2 - 2e^x. \end{aligned}$$

■

In Example 2, the polynomial part of the initial guess

$$(A + Bx) + Ce^x$$

for y_p overlapped by a constant term with $y_h = C_1 + C_2e^{2x}$, and it was necessary to multiply the polynomial part by a power of x that removed the overlap. The next example further illustrates some choices for y_p that eliminate overlap with y_h. Remember that in all cases the first guess for y_p should match the types of functions occurring in $F(x)$.

EXAMPLE 3 Choosing the Form of the Particular Solution

Determine a suitable choice for y_p for each differential equation, given its general solution of the homogeneous equation.

$y'' + ay' + by = F(x)$	y_h
a. $y'' = x^2$	$C_1 + C_2x$
b. $y'' + 2y' + 10y = 4 \sin 3x$	$C_1e^{-x} \cos 3x + C_2e^{-x} \sin 3x$
c. $y'' - 4y' + 4 = e^{2x}$	$C_1e^{2x} + C_2xe^{2x}$

Solution

a. Because $F(x) = x^2$, the normal choice for y_p would be $A + Bx + Cx^2$. However, because $y_h = C_1 + C_2x$ already contains a linear term, you should multiply by x^2 to obtain

$$y_p = Ax^2 + Bx^3 + Cx^4.$$

b. Because $F(x) = 4 \sin 3x$ and each term in y_h contains a factor of e^{-x}, you can simply let

$$y_p = A \cos 3x + B \sin 3x.$$

c. Because $F(x) = e^{2x}$, the normal choice for y_p would be Ae^{2x}. However, because $y_h = C_1e^{2x} + C_2xe^{2x}$ already contains an xe^{2x} term, you should multiply by x^2 to get

$$y_p = Ax^2e^{2x}.$$

EXAMPLE 4 Solving a Third-Order Equation

Find the general solution of

$$y''' + 3y'' + 3y' + y = x.$$

Solution From Example 6 in the preceding section, you know that the homogeneous solution is

$$y_h = C_1e^{-x} + C_2xe^{-x} + C_3x^2e^{-x}.$$

Because $F(x) = x$, let $y_p = A + Bx$ and obtain $y_p' = B$ and $y_p'' = 0$. So, by substitution, you have

$$0 + 3(0) + 3(B) + A + Bx = (3B + A) + Bx = x.$$

So, $B = 1$ and $A = -3$, which implies that $y_p = -3 + x$. Therefore, the general solution is

$$\begin{aligned} y &= y_h + y_p \\ &= C_1e^{-x} + C_2xe^{-x} + C_3x^2e^{-x} - 3 + x. \end{aligned}$$

■

Variation of Parameters

The method of undetermined coefficients works well if $F(x)$ is made up of polynomials or functions whose successive derivatives have a cyclical pattern. For functions such as $1/x$ and $\tan x$, which do not have such characteristics, it is better to use a more general method called **variation of parameters.** In this method, you assume that y_p has the same *form* as y_h, except that the constants in y_h are replaced by variables.

VARIATION OF PARAMETERS

To find the general solution of the equation $y'' + ay' + by = F(x)$, use the following steps.

1. Find $y_h = C_1y_1 + C_2y_2$.
2. Replace the constants by variables to form $y_p = u_1y_1 + u_2y_2$.
3. Solve the following system for u_1' and u_2'.

$$u_1'y_1 + u_2'y_2 = 0$$
$$u_1'y_1' + u_2'y_2' = F(x)$$

4. Integrate to find u_1 and u_2. The general solution is $y = y_h + y_p$.

EXAMPLE 5 Variation of Parameters

Solve the differential equation

$$y'' - 2y' + y = \frac{e^x}{2x}, \quad x > 0.$$

Solution The characteristic equation $m^2 - 2m + 1 = (m - 1)^2 = 0$ has one repeated solution, $m = 1$. So, the homogeneous solution is

$$y_h = C_1y_1 + C_2y_2 = C_1e^x + C_2xe^x.$$

Replacing C_1 and C_2 by u_1 and u_2 produces

$$y_p = u_1y_1 + u_2y_2 = u_1e^x + u_2xe^x.$$

The resulting system of equations is

$$u_1'e^x + u_2'xe^x = 0$$
$$u_1'e^x + u_2'(xe^x + e^x) = \frac{e^x}{2x}.$$

Subtracting the second equation from the first produces $u_2' = 1/(2x)$. Then, by substitution in the first equation, you have $u_1' = -\frac{1}{2}$. Finally, integration yields

$$u_1 = -\int \frac{1}{2}\,dx = -\frac{x}{2} \quad \text{and} \quad u_2 = \frac{1}{2}\int \frac{1}{x}\,dx = \frac{1}{2}\ln x = \ln\sqrt{x}.$$

From this result it follows that a particular solution is

$$y_p = -\frac{1}{2}xe^x + \left(\ln\sqrt{x}\right)xe^x$$

and the general solution is

$$y = C_1e^x + C_2xe^x - \frac{1}{2}xe^x + xe^x\ln\sqrt{x}.$$

■

EXPLORATION

Notice in Example 5 that the constants of integration were not introduced when finding u_1 and u_2. Show that if

$$u_1 = -\frac{x}{2} + a_1 \text{ and}$$

$$u_2 = \ln\sqrt{x} + a_2$$

then the general solution

$$y = y_h + y_p = C_1e^x + C_2xe^x - \tfrac{1}{2}xe^x + xe^x\ln\sqrt{x}$$

yields the same result as that obtained in the example.

EXAMPLE 6 Variation of Parameters

Solve the differential equation

$$y'' + y = \tan x.$$

Solution Because the characteristic equation $m^2 + 1 = 0$ has solutions $m = \pm i$, the homogeneous solution is

$$y_h = C_1 \cos x + C_2 \sin x.$$

Replacing C_1 and C_2 by u_1 and u_2 produces

$$y_p = u_1 \cos x + u_2 \sin x.$$

The resulting system of equations is

$$\begin{aligned} u_1' \cos x + u_2' \sin x &= 0 \\ -u_1' \sin x + u_2' \cos x &= \tan x. \end{aligned}$$

Multiplying the first equation by $\sin x$ and the second by $\cos x$ produces

$$\begin{aligned} u_1' \sin x \cos x + u_2' \sin^2 x &= 0 \\ -u_1' \sin x \cos x + u_2' \cos^2 x &= \sin x. \end{aligned}$$

Adding these two equations produces $u_2' = \sin x$, which implies that

$$\begin{aligned} u_1' &= -\frac{\sin^2 x}{\cos x} \\ &= \frac{\cos^2 x - 1}{\cos x} \\ &= \cos x - \sec x. \end{aligned}$$

Integration yields

$$\begin{aligned} u_1 &= \int (\cos x - \sec x)\, dx \\ &= \sin x - \ln|\sec x + \tan x| \end{aligned}$$

and

$$\begin{aligned} u_2 &= \int \sin x\, dx \\ &= -\cos x \end{aligned}$$

so that

$$\begin{aligned} y_p &= \sin x \cos x - \cos x \ln|\sec x + \tan x| - \sin x \cos x \\ &= -\cos x \ln|\sec x + \tan x| \end{aligned}$$

and the general solution is

$$\begin{aligned} y &= y_h + y_p \\ &= C_1 \cos x + C_2 \sin x - \cos x \ln|\sec x + \tan x|. \end{aligned}$$

■

16.3 Exercises

See www.CalcChat.com for worked-out solutions to odd-numbered exercises.

In Exercises 1–4, verify the solution of the differential equation.

Solution	*Differential Equation*
1. $y = 2(e^{2x} - \cos x)$	$y'' + y = 10e^{2x}$
2. $y = \left(2 + \frac{1}{2}x\right)\sin x$	$y'' + y = \cos x$
3. $y = 3\sin x - \cos x \ln\lvert\sec x + \tan x\rvert$	$y'' + y = \tan x$
4. $y = \left(5 - \ln\lvert\sin x\rvert\right)\cos x - x\sin x$	$y'' + y = \csc x \cot x$

In Exercises 5–10, find a particular solution of the differential equation.

5. $y'' + 7y' + 12y = 3x + 1$

6. $y'' - y' - 6y = 4$

7. $y'' - 8y' + 16y = e^{3x}$

8. $y'' + y' + 3y = e^{2x}$

9. $y'' - 2y' - 15y = \sin x$

10. $y'' + 4y' + 5y = e^x \cos x$

In Exercises 11–18, solve the differential equation by the method of undetermined coefficients.

11. $y'' - 3y' + 2y = 2x$

12. $y'' - 2y' - 3y = x^2 - 1$

13. $y'' + 2y' = 2e^x$

14. $y'' - 9y = 5e^{3x}$

15. $y'' - 10y' + 25y = 5 + 6e^x$

16. $16y'' - 8y' + y = 4(x + e^x)$

17. $y'' + 9y = \sin 3x$

18. $y''' - 3y' + 2y = 2e^{-2x}$

In Exercises 19–24, solve the differential equation by the method of undetermined coefficients.

19. $y'' + y = x^3$
$y(0) = 1,\ y'(0) = 0$

20. $y'' + 4y = 4$
$y(0) = 1,\ y'(0) = 6$

21. $y'' + y' = 2\sin x$
$y(0) = 0,\ y'(0) = -3$

22. $y'' + y' - 2y = 3\cos 2x$
$y(0) = -1,\ y'(0) = 2$

23. $y' - 4y = xe^x - xe^{4x}$
$y(0) = \dfrac{1}{3}$

24. $y' + 2y = \sin x$
$y\left(\dfrac{\pi}{2}\right) = \dfrac{2}{5}$

In Exercises 25–30, solve the differential equation by the method of variation of parameters.

25. $y'' + y = \sec x$

26. $y'' + y = \sec x \tan x$

27. $y'' + 4y = \csc 2x$

28. $y'' - 4y' + 4y = x^2 e^{2x}$

29. $y'' - 2y' + y = e^x \ln x$

30. $y'' - 4y' + 4y = \dfrac{e^{2x}}{x}$

WRITING ABOUT CONCEPTS

31. Using the method of undetermined coefficients, determine a suitable choice for y_p given $y'' - y' - 12y = x^2$. Explain your reasoning. (You do not need to solve the differential equation.)

32. Using the method of undetermined coefficients, determine a suitable choice for y_p given $y'' - y' - 12y = e^{4x}$. Explain your reasoning. (You do not need to solve the differential equation.)

WRITING ABOUT CONCEPTS (continued)

33. Describe the steps for solving a differential equation by the method of variation of parameters.

CAPSTONE

34. *Think About It*

(a) Explain how, by observation, you know that a particular solution of the differential equation $y'' + 3y = 12$ is $y_p = 4$.

(b) Use your explanation in part (a) to give a particular solution of the differential equation $y'' + 5y = 10$.

(c) Use your explanation in part (a) to give a particular solution of the differential equation $y'' + 2y' + 2y = 8$.

Electrical Circuits **In Exercises 35 and 36, use the electrical circuit differential equation**

$$\frac{d^2q}{dt^2} + \left(\frac{R}{L}\right)\frac{dq}{dt} + \left(\frac{1}{LC}\right)q = \left(\frac{1}{L}\right)E(t)$$

where R is the resistance (in ohms), C is the capacitance (in farads), L is the inductance (in henrys), $E(t)$ is the electromotive force (in volts), and q is the charge on the capacitor (in coulombs). Find the charge q as a function of time for the electrical circuit described. Assume that $q(0) = 0$ and $q'(0) = 0$.

35. $R = 20,\ C = 0.02,\ L = 2,\ E(t) = 12\sin 5t$

36. $R = 20,\ C = 0.02,\ L = 1,\ E(t) = 10\sin 5t$

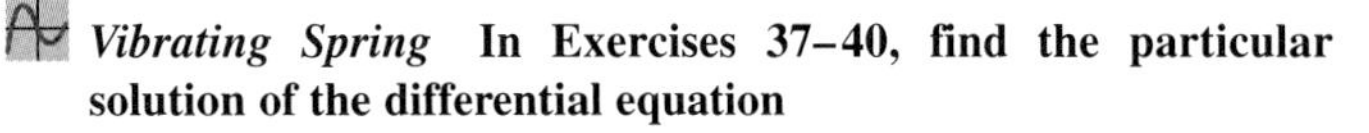

Vibrating Spring **In Exercises 37–40, find the particular solution of the differential equation**

$$\frac{w}{g}y''(t) + by'(t) + ky(t) = \frac{w}{g}F(t)$$

for the oscillating motion of an object on the end of a spring. Use a graphing utility to graph the solution. In the equation, y is the displacement from equilibrium (positive direction is downward) measured in feet, and t is time in seconds (see figure). The constant w is the weight of the object, g is the acceleration due to gravity, b is the magnitude of the resistance to the motion, k is the spring constant from Hooke's Law, and $F(t)$ is the acceleration imposed on the system.

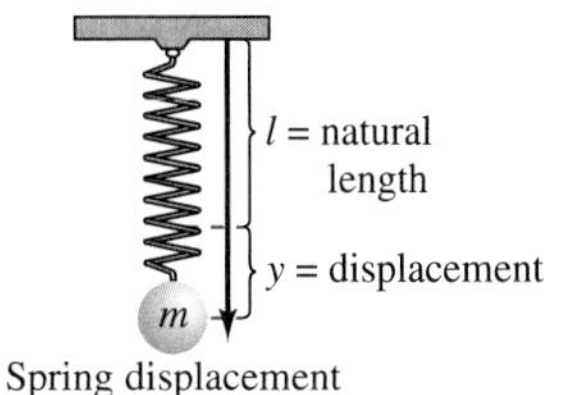

Spring displacement

37. $\frac{24}{32}y'' + 48y = \frac{24}{32}(48 \sin 4t)$
$y(0) = \frac{1}{4}, y'(0) = 0$

38. $\frac{2}{32}y'' + 4y = \frac{2}{32}(4 \sin 8t)$
$y(0) = \frac{1}{4}, y'(0) = 0$

39. $\frac{2}{32}y'' + y' + 4y = \frac{2}{32}(4 \sin 8t)$
$y(0) = \frac{1}{4}, y'(0) = -3$

40. $\frac{4}{32}y'' + \frac{1}{2}y' + \frac{25}{2}y = 0$
$y(0) = \frac{1}{2}, y'(0) = -4$

41. ***Vibrating Spring*** Rewrite y_h in the solution for Exercise 37 by using the identity

$$a \cos \omega t + b \sin \omega t = \sqrt{a^2 + b^2} \sin(\omega t + \phi)$$

where $\phi = \arctan a/b$.

42. ***Vibrating Spring*** The figure shows the particular solution of the differential equation

$$\frac{4}{32}y'' + by' + \frac{25}{2}y = 0$$

$$y(0) = \frac{1}{2}, y'(0) = -4$$

for values of the resistance component b in the interval $[0, 1]$. (Note that when $b = \frac{1}{2}$, the problem is identical to that of Exercise 40.)

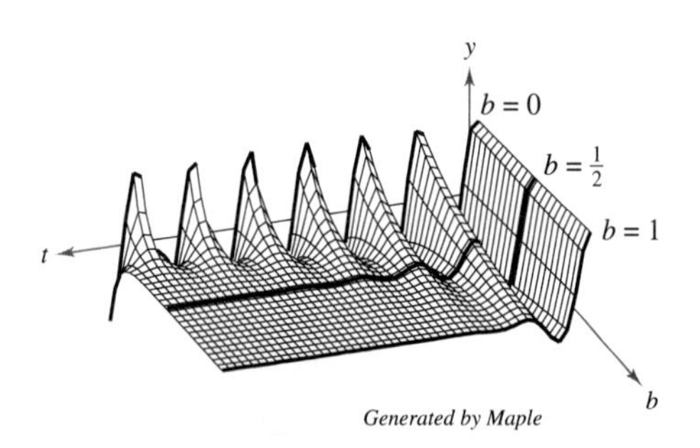

(a) If there is no resistance to the motion ($b = 0$), describe the motion.

(b) If $b > 0$, what is the ultimate effect of the retarding force?

(c) Is there a real number M such that there will be no oscillations of the spring if $b > M$? Explain your answer.

43. Solve the differential equation

$$x^2y'' - xy' + y = 4x \ln x$$

given that $y_1 = x$ and $y_2 = x \ln x$ are solutions of the corresponding homogeneous equation.

44. Solve the differential equation

$$x^2y'' + xy' + 4y = \sin(\ln x)$$

given that $y_1 = \sin(\ln x^2)$ and $y_2 = \cos(\ln x^2)$ are solutions of the corresponding homogeneous equation.

True or False? **In Exercises 45 and 46, determine whether the statement is true or false. If it is false, explain why or give an example that shows it is false.**

45. $y_p = -e^{2x} \cos e^{-x}$ is a particular solution of the differential equation

$$y'' - 3y' + 2y = \cos e^{-x}.$$

46. $y_p = -\frac{1}{8}e^{2x}$ is a particular solution of the differential equation

$$y'' - 6y' = e^{2x}.$$

PUTNAM EXAM CHALLENGE

47. For all real x, the real-valued function $y = f(x)$ satisfies

$$y'' - 2y' + y = 2e^x.$$

(a) If $f(x) > 0$ for all real x, must $f'(x) > 0$ for all real x? Explain.

(b) If $f'(x) > 0$ for all real x, must $f(x) > 0$ for all real x? Explain.

This problem was composed by the Committee on the Putnam Prize Competition.

SECTION PROJECT

Parachute Jump

The fall of a parachutist is described by the second-order linear differential equation

$$\frac{w}{g}\frac{d^2y}{dt^2} - k\frac{dy}{dt} = w$$

where w is the weight of the parachutist, y is the height at time t, g is the acceleration due to gravity, and k is the drag factor of the parachute.

(a) If the parachute is opened at 2000 feet, $y(0) = 2000$, and at that time the velocity is $y'(0) = -100$ feet per second, then for a 160-pound parachutist, using $k = 8$, the differential equation is

$$-5y'' - 8y' = 160.$$

Using the given initial conditions, verify that the solution of the differential equation is

$$y = 1950 + 50e^{-1.6t} - 20t.$$

(b) Consider a 192-pound parachutist who has a parachute with a drag factor of $k = 9$. Using the initial conditions given in part (a), write and solve a differential equation that describes the fall of the parachutist.

16.4 Series Solutions of Differential Equations

- **Use a power series to solve a differential equation.**
- **Use a Taylor series to find the series solution of a differential equation.**

Power Series Solution of a Differential Equation

Power series can be used to solve certain types of differential equations. This section begins with the general **power series solution** method.

Recall from Chapter 9 that a power series represents a function f on an interval of convergence, and you can successively differentiate the power series to obtain a series for f', f'', and so on. These properties are used in the power series solution method demonstrated in the first two examples.

EXAMPLE 1 Power Series Solution

Use a power series to solve the differential equation $y' - 2y = 0$.

Solution Assume that $y = \Sigma a_n x^n$ is a solution. Then, $y' = \Sigma n a_n x^{n-1}$. Substituting for y' and $-2y$, you obtain the following series form of the differential equation. (Note that, from the third step to the fourth, the index of summation is changed to ensure that x^n occurs in both sums.)

$$y' - 2y = 0$$

$$\sum_{n=1}^{\infty} n a_n x^{n-1} - 2\sum_{n=0}^{\infty} a_n x^n = 0$$

$$\sum_{n=1}^{\infty} n a_n x^{n-1} = \sum_{n=0}^{\infty} 2a_n x^n$$

$$\sum_{n=0}^{\infty} (n+1)a_{n+1} x^n = \sum_{n=0}^{\infty} 2a_n x^n$$

Now, by equating coefficients of like terms, you obtain the **recursion formula** $(n+1)a_{n+1} = 2a_n$, which implies that

$$a_{n+1} = \frac{2a_n}{n+1}, \qquad n \geq 0.$$

This formula generates the following results.

a_0	a_1	a_2	a_3	a_4	a_5	$\cdots$
a_0	$2a_0$	$\frac{2^2 a_0}{2}$	$\frac{2^3 a_0}{3!}$	$\frac{2^4 a_0}{4!}$	$\frac{2^5 a_0}{5!}$	$\cdots$

Using these values as the coefficients for the *solution* series, you have

$$\begin{aligned} y &= \sum_{n=0}^{\infty} \frac{2^n a_0}{n!} x^n \\ &= a_0 \sum_{n=0}^{\infty} \frac{(2x)^n}{n!} \\ &= a_0 e^{2x}. \end{aligned}$$

■

EXPLORATION

In Example 1, the differential equation could be solved easily without using a series. Determine which method should be used to solve the differential equation

$y' - 2y = 0$

and show that the result is the same as that obtained in the example.

In Example 1, the differential equation could be solved easily without using a series. The differential equation in Example 2 cannot be solved by any of the methods discussed in previous sections.

EXAMPLE 2 Power Series Solution

Use a power series to solve the differential equation $y'' + xy' + y = 0$.

Solution Assume that $\sum_{n=0}^{\infty} a_n x^n$ is a solution. Then you have

$$y' = \sum_{n=1}^{\infty} na_n x^{n-1}, \qquad xy' = \sum_{n=1}^{\infty} na_n x^n, \qquad y'' = \sum_{n=2}^{\infty} n(n-1)a_n x^{n-2}.$$

Substituting for y'', xy', and y in the given differential equation, you obtain the following series.

$$\sum_{n=2}^{\infty} n(n-1)a_n x^{n-2} + \sum_{n=0}^{\infty} na_n x^n + \sum_{n=0}^{\infty} a_n x^n = 0$$

$$\sum_{n=2}^{\infty} n(n-1)a_n x^{n-2} = -\sum_{n=0}^{\infty} (n+1)a_n x^n$$

To obtain equal powers of x, adjust the summation indices by replacing n by $n + 2$ in the left-hand sum, to obtain

$$\sum_{n=0}^{\infty} (n+2)(n+1)a_{n+2} x^n = -\sum_{n=0}^{\infty} (n+1)a_n x^n.$$

By equating coefficients, you have $(n+2)(n+1)a_{n+2} = -(n+1)a_n$, from which you obtain the recursion formula

$$a_{n+2} = -\frac{(n+1)}{(n+2)(n+1)}a_n = -\frac{a_n}{n+2}, \qquad n \ge 0,$$

and the coefficients of the solution series are as follows.

$$a_2 = -\frac{a_0}{2} \qquad\qquad a_3 = -\frac{a_1}{3}$$

$$a_4 = -\frac{a_2}{4} = \frac{a_0}{2 \cdot 4} \qquad\qquad a_5 = -\frac{a_3}{5} = \frac{a_1}{3 \cdot 5}$$

$$a_6 = -\frac{a_4}{6} = -\frac{a_0}{2 \cdot 4 \cdot 6} \qquad\qquad a_7 = -\frac{a_5}{7} = -\frac{a_1}{3 \cdot 5 \cdot 7}$$

$$\vdots \qquad\qquad \vdots$$

$$a_{2k} = \frac{(-1)^k a_0}{2 \cdot 4 \cdot 6 \cdots (2k)} = \frac{(-1)^k a_0}{2^k(k!)} \qquad\qquad a_{2k+1} = \frac{(-1)^k a_1}{3 \cdot 5 \cdot 7 \cdots (2k+1)}$$

So, you can represent the general solution as the sum of two series—one for the even-powered terms with coefficients in terms of a_0 and one for the odd-powered terms with coefficients in terms of a_1.

$$\begin{aligned} y &= a_0\left(1 - \frac{x^2}{2} + \frac{x^4}{2 \cdot 4} - \cdots\right) + a_1\left(x - \frac{x^3}{3} + \frac{x^5}{3 \cdot 5} - \cdots\right) \\ &= a_0 \sum_{k=0}^{\infty} \frac{(-1)^k x^{2k}}{2^k(k!)} + a_1 \sum_{k=0}^{\infty} \frac{(-1)^k x^{2k+1}}{3 \cdot 5 \cdot 7 \cdots (2k+1)} \end{aligned}$$

The solution has two arbitrary constants, a_0 and a_1, as you would expect in the general solution of a second-order differential equation. ■

Approximation by Taylor Series

A second type of series solution method involves a differential equation *with initial conditions* and makes use of Taylor series, as given in Section 9.10.

EXAMPLE 3 Approximation by Taylor Series

Use Taylor's Theorem to find the first six terms of the series solution of

$$y' = y^2 - x$$

given the initial condition $y = 1$ when $x = 0$. Then, use this polynomial to approximate values of y for $0 \le x \le 1$.

Solution Recall from Section 9.10 that, for $c = 0$,

$$y = y(0) + y'(0)x + \frac{y''(0)}{2!}x^2 + \frac{y'''(0)}{3!}x^3 + \cdot \cdot \cdot .$$

Because $y(0) = 1$ and $y' = y^2 - x$, you obtain the following.

$$\begin{aligned} & & y(0) &= 1 \\ y' &= y^2 - x & y'(0) &= 1 \\ y'' &= 2yy' - 1 & y''(0) &= 2 - 1 = 1 \\ y''' &= 2yy'' + 2(y')^2 & y'''(0) &= 2 + 2 = 4 \\ y^{(4)} &= 2yy''' + 6y'y'' & y^{(4)}(0) &= 8 + 6 = 14 \\ y^{(5)} &= 2yy^{(4)} + 8y'y''' + 6(y'')^2 & y^{(5)}(0) &= 28 + 32 + 6 = 66 \end{aligned}$$

Therefore, the first six terms of the series solution are

$$\begin{aligned} y &= y(0) + y'(0)x + \frac{y''(0)}{2!}x^2 + \frac{y'''(0)}{3!}x^3 + \frac{y^{(4)}(0)}{4!}x^4 + \frac{y^{(5)}(0)}{5!}x^5 \\ &= 1 + x + \frac{1}{2}x^2 + \frac{4}{3!}x^3 + \frac{14}{4!}x^4 + \frac{66}{5!}x^5. \end{aligned}$$

Using this polynomial, you can compute values for y in the interval $0 \le x \le 1$, as shown in the table below.

x	0.0	0.1	0.2	0.3	0.4	0.5	0.6	0.7	0.8	0.9	1.0
y	1.0000	1.1057	1.2264	1.3691	1.5432	1.7620	2.0424	2.4062	2.8805	3.4985	4.3000

■

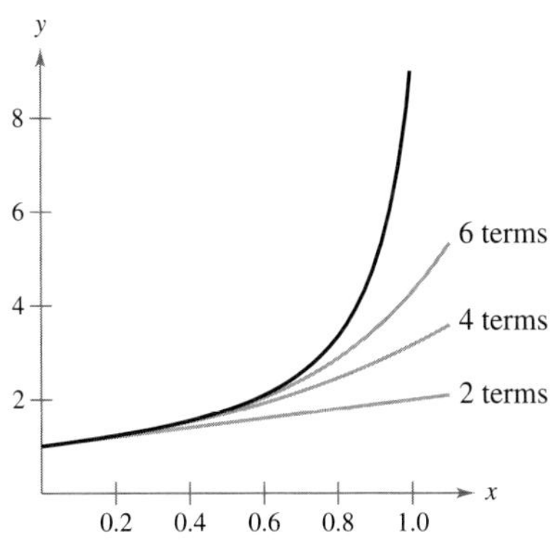

Figure 16.8

In addition to approximating values of a function, you can also use the series solution to sketch a graph. In Figure 16.8, the series solution of $y' = y^2 - x$ using the first two, four, and six terms are shown, along with an approximation found using a computer algebra system. The approximations are nearly the same for values of x close to 0. As x approaches 1, however, there is a noticeable difference between the approximations. For a series solution that is more accurate near $x = 1$, repeat Example 3 using $c = 1$.

16.4 Exercises

See www.CalcChat.com for worked-out solutions to odd-numbered exercises.

In Exercises 1–6, verify that the power series solution of the differential equation is equivalent to the solution found using previously learned solution techniques.

1. $y' - y = 0$
2. $y' - ky = 0$
3. $y'' - 9y = 0$
4. $y'' - k^2y = 0$
5. $y'' + 4y = 0$
6. $y'' + k^2y = 0$

In Exercises 7–10, use power series to solve the differential equation and find the interval of convergence of the series.

7. $y' + 3xy = 0$
8. $y' - 2xy = 0$
9. $y'' - xy' = 0$
10. $y'' - xy' - y = 0$

In Exercises 11 and 12, find the first three terms of each of the power series representing independent solutions of the differential equation.

11. $(x^2 + 4)y'' + y = 0$
12. $y'' + x^2y = 0$

In Exercises 13 and 14, use Taylor's Theorem to find the first n terms of the series solution of the differential equation under the specified initial conditions. Use this polynomial to approximate y for the given value of x and compare the result with the approximation given by Euler's Method for $h = 0.1$.

13. $y' + (2x - 1)y = 0,\ y(0) = 2,\ n = 5,\ x = \frac{1}{2}$
14. $y' - 2xy = 0,\ y(0) = 1,\ n = 4,\ x = 1$

WRITING ABOUT CONCEPTS

15. Describe how to use power series to solve a differential equation.
16. What is a recursion formula? Give an example.

17. ***Investigation*** Consider the differential equation $y'' - xy' = 0$ with the initial conditions $y(0) = 0$ and $y'(0) = 2$. (See Exercise 9.)
    (a) Find the series solution satisfying the initial conditions.
    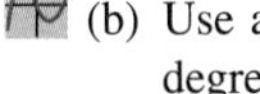
    (b) Use a graphing utility to graph the third-degree and fifth-degree series approximations of the solution. Identify the approximations.
    (c) Identify the symmetry of the solution.

CAPSTONE

18. ***Investigation*** Consider the differential equation

$$y'' + 9y = 0$$

with initial conditions $y(0) = 2$ and $y'(0) = 6$.
    (a) Find the solution of the differential equation using the techniques presented in Section 16.2.

CAPSTONE (continued)

    (b) Find the series solution of the differential equation.
    (c) The figure shows the graph of the solution of the differential equation and the third-degree and fifth-degree polynomial approximations of the solution. Identify each.

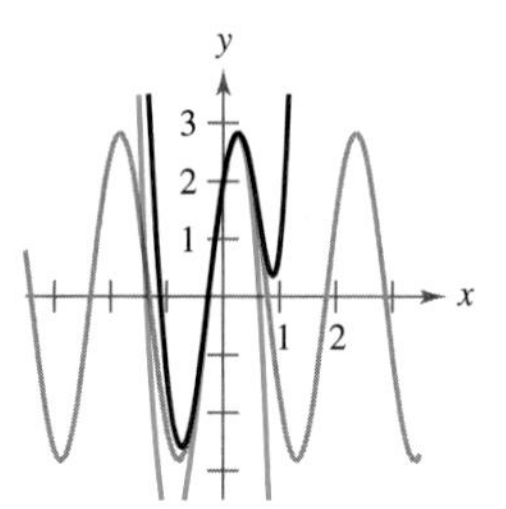

In Exercises 19–22, use Taylor's Theorem to find the first n terms of the series solution of the differential equation under the specified initial conditions. Use this polynomial to approximate y for the given value of x.

19. $y'' - 2xy = 0,\ y(0) = 1,\ y'(0) = -3,\ n = 6,\ x = \frac{1}{4}$
20. $y'' - 2xy' + y = 0,\ y(0) = 1,\ y'(0) = 2,\ n = 8,\ x = \frac{1}{2}$
21. $y'' + x^2y' - (\cos x)y = 0,\ y(0) = 3,\ y'(0) = 2,\ n = 4,\ x = \frac{1}{3}$
22. $y'' + e^xy' - (\sin x)y = 0,\ y(0) = -2,\ y'(0) = 1,\ n = 4,\ x = \frac{1}{5}$

In Exercises 23–26, verify that the series converges to the given function on the indicated interval. (*Hint:* Use the given differential equation.)

23. $\displaystyle\sum_{n=0}^{\infty} \frac{x^n}{n!} = e^x,\ (-\infty, \infty)$
    Differential equation: $y' - y = 0$
24. $\displaystyle\sum_{n=0}^{\infty} \frac{(-1)^n x^{2n}}{(2n)!} = \cos x,\ (-\infty, \infty)$
    Differential equation: $y'' + y = 0$
25. $\displaystyle\sum_{n=0}^{\infty} \frac{(-1)^n x^{2n+1}}{2n + 1} = \arctan x,\ (-1, 1)$
    Differential equation: $(x^2 + 1)y'' + 2xy' = 0$
26. $\displaystyle\sum_{n=0}^{\infty} \frac{(2n)!x^{2n+1}}{(2^n n!)^2(2n + 1)} = \arcsin x,\ (-1, 1)$
    Differential equation: $(1 - x^2)y'' - xy' = 0$

27. ***Airy's Equation*** Find the first six terms in the series solution of Airy's equation, $y'' - xy = 0$.

16 REVIEW EXERCISES

See www.CalcChat.com for worked-out solutions to odd-numbered exercises.

In Exercises 1 and 2, determine whether the differential equation is exact. Explain your reasoning.

1. $(y + x^3 + xy^2)\,dx - x\,dy = 0$

2. $(5x - y)\,dx + (5y - x)\,dy = 0$

In Exercises 3–8, determine whether the differential equation is exact. If it is, find the general solution.

3. $(10x + 8y + 2)\,dx + (8x + 5y + 2)\,dy = 0$

4. $(2x - 2y^3 + y)\,dx + (x - 6xy^2)\,dy = 0$

5. $(x - y - 5)\,dx - (x + 3y - 2)\,dy = 0$

6. $(3x^2 - 5xy^2)\,dx + (2y^3 - 5xy^2)\,dy = 0$

7. $\frac{x}{y}\,dx - \frac{x}{y^2}\,dy = 0$

8. $y\sin(xy)\,dx + [x\sin(xy) + y]\,dy = 0$

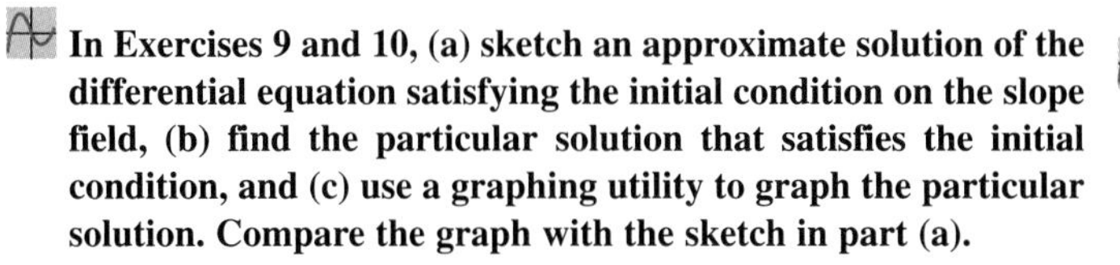

In Exercises 9 and 10, (a) sketch an approximate solution of the differential equation satisfying the initial condition on the slope field, (b) find the particular solution that satisfies the initial condition, and (c) use a graphing utility to graph the particular solution. Compare the graph with the sketch in part (a).

9. $(2x - y)\,dx + (2y - x)\,dy = 0,\; y(2) = 2$

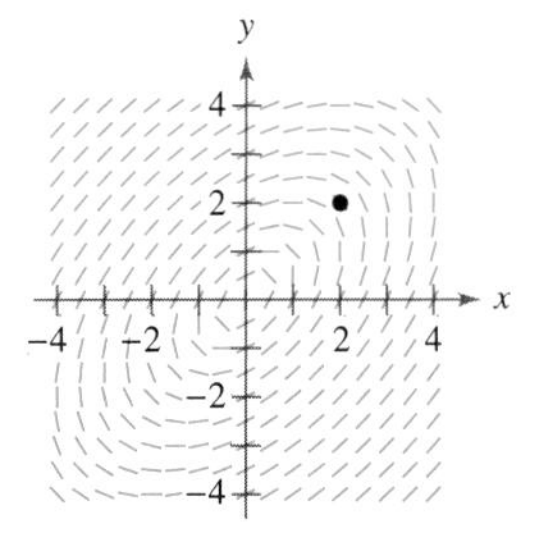

10. $(6xy - y^3)\,dx + (4y + 3x^2 - 3xy^2)\,dy = 0,\; y(0) = 1$

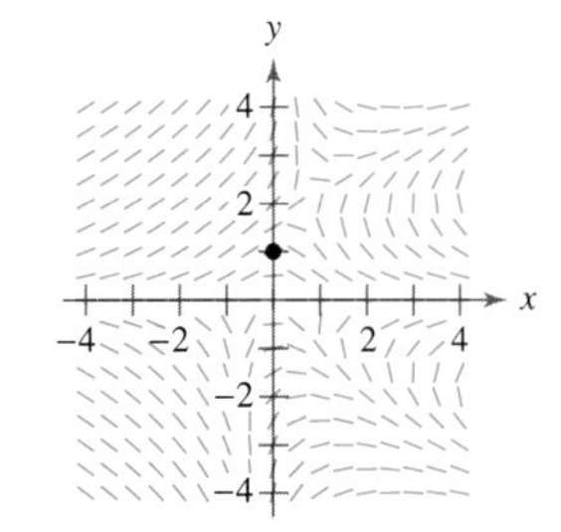

In Exercises 11 and 12, find the particular solution that satisfies the initial condition.

11. $(2x + y - 3)\,dx + (x - 3y + 1)\,dy = 0,\; y(2) = 0$

12. $3x^2y^2\,dx + (2x^3y - 3y^2)\,dy = 0,\; y(1) = 2$

In Exercises 13–16, find the integrating factor that is a function of x or y alone and use it to find the general solution of the differential equation.

13. $(3x^2 - y^2)\,dx + 2xy\,dy = 0$

14. $2xy\,dx + (y^2 - x^2)\,dy = 0$

15. $dx + (3x - e^{-2y})\,dy = 0$

16. $\cos y\,dx - [2(x - y)\sin y + \cos y]\,dy = 0$

In Exercises 17 and 18, verify the solution of the differential equation. Then use a graphing utility to graph the particular solutions for several different values of C_1 and C_2. What do you observe?

	Solution	*Differential Equation*
17.	$y = C_1e^{2x} + C_2e^{-2x}$	$y'' - 4y = 0$
18.	$y = C_1\cos 2x + C_2\sin 2x$	$y'' + 4y = 0$

In Exercises 19–22, find the particular solution of the differential equation that satisfies the initial conditions. Use a graphing utility to graph the solution.

	Differential Equation	*Initial Conditions*
19.	$y'' - y' - 2y = 0$	$y(0) = 0,\; y'(0) = 3$
20.	$y'' + 4y' + 5y = 0$	$y(0) = 2,\; y'(0) = -7$
21.	$y'' + 2y' - 3y = 0$	$y(0) = 2,\; y'(0) = 0$
22.	$y'' + 12y' + 36y = 0$	$y(0) = 2,\; y'(0) = 1$

In Exercises 23 and 24, find the particular solution of the differential equation that satisfies the boundary conditions. Use a graphing utility to graph the solution.

	Differential Equation	*Boundary Conditions*
23.	$y'' + 2y' + 5y = 0$	$y(1) = 4,\; y(2) = 0$
24.	$y'' + y = 0$	$y(0) = 2,\; y(\pi/2) = 1$

Think About It **In Exercises 25 and 26, give a geometric argument to explain why the graph cannot be a solution of the differential equation. It is not necessary to solve the differential equation.**

25. $y'' = y'$

26. $y'' = -\frac{1}{2}y'$

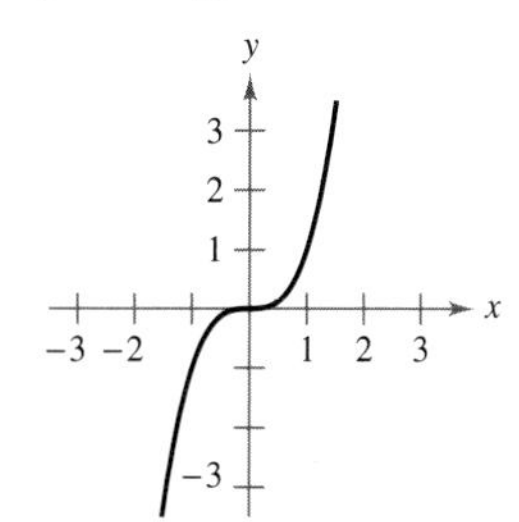

In Exercises 27–32, find the general solution of the second-order differential equation.

27. $y'' + y = x^3 + x$

28. $y'' + 2y = e^{2x} + x$

29. $y'' + y = 2 \cos x$

30. $y'' + 5y' + 4y = x^2 + \sin 2x$

31. $y'' - 2y' + y = 2xe^x$

32. $y'' + 2y' + y = \dfrac{1}{x^2 e^x}$

In Exercises 33–38, find the particular solution of the differential equation that satisfies the initial conditions.

	Differential Equation	*Initial Conditions*
33.	$y'' - y' - 6y = 54$	$y(0) = 2, y'(0) = 0$
34.	$y'' + 25y = e^x$	$y(0) = 0, y'(0) = 0$
35.	$y'' + 4y = \cos x$	$y(0) = 6, y'(0) = -6$
36.	$y'' + 3y' = 6x$	$y(0) = 2, y'(0) = \frac{10}{3}$
37.	$y'' - y' - 2y = 1 + xe^{-x}$	$y(0) = 1, y'(0) = 3$
38.	$y''' - y'' = 4x^2$	$y(0) = 1, y'(0) = 1, y''(0) = 1$

Vibrating Spring **In Exercises 39 and 40, describe the motion of a 64-pound weight suspended on a spring. Assume that the weight stretches the spring $\frac{4}{3}$ feet from its natural position.**

39. The weight is pulled $\frac{1}{2}$ foot below the equilibrium position and released.

40. The weight is pulled $\frac{1}{2}$ foot below the equilibrium position and released. The motion takes place in a medium that furnishes a damping force of magnitude $\frac{1}{8}$ speed at all times.

41. ***Investigation*** The differential equation

$$\frac{8}{32}y'' + by' + ky = \frac{8}{32}F(t), \quad y(0) = \frac{1}{2}, \; y'(0) = 0$$

models the motion of a weight suspended on a spring.

(a) Solve the differential equation and use a graphing utility to graph the solution for each of the assigned quantities for b, k, and $F(t)$.

(i) $b = 0, k = 1, F(t) = 24 \sin \pi t$

(ii) $b = 0, k = 2, F(t) = 24 \sin\left(2\sqrt{2}t\right)$

(iii) $b = 0.1, k = 2, F(t) = 0$

(iv) $b = 1, k = 2, F(t) = 0$

(b) Describe the effect of increasing the resistance to motion b.

(c) Explain how the motion of the object would change if a stiffer spring (increased k) were used.

(d) Matching the input and natural frequencies of a system is known as resonance. In which case of part (a) does this occur, and what is the result?

42. ***True or False?*** The function

$$y_p = \tfrac{1}{4} \cos x$$

is a particular solution of the differential equation

$$y'' + 4y' + 5y = \sin x + \cos x.$$

43. ***Think About It***

(a) Explain how, by observation, you know that a form of a particular solution of the differential equation

$$y'' + 3y = 12 \sin x$$

is

$$y_p = A \sin x.$$

(b) Use your explanation in part (a) to find a particular solution of the differential equation

$$y'' + 5y = 10 \cos x.$$

(c) Compare the algebra required to find particular solutions in parts (a) and (b) with that required if the form of the particular solution were

$$y_p = A \cos x + B \sin x.$$

44. ***Think About It*** Explain how you can find a particular solution of the differential equation

$$y'' + 4y' + 6y = 30$$

by observation.

In Exercises 45 and 46, find the series solution of the differential equation.

45. $(x - 4)y' + y = 0$

46. $y'' + 3xy' - 3y = 0$

In Exercises 47 and 48, use Taylor's Theorem to find the first n terms of the series solution of the differential equation under the specified initial conditions. Use this polynomial to approximate y for the given value of x.

47. $y'' + y' - e^x y = 0, \; y(0) = 2, y'(0) = 0, n = 4, x = \dfrac{1}{4}$

48. $y'' + xy = 0, \; y(0) = 1, \; y'(0) = 1, \; n = 6, \; x = \dfrac{1}{2}$

P.S. PROBLEM SOLVING

1. Find the value of k that makes the differential equation

$(3x^2 + kxy^2)\,dx - (5x^2y + ky^2)\,dy = 0$

exact. Using this value of k, find the general solution.

2. The differential equation

$(kx^2 + y^2)\,dx - kxy\,dy = 0$

is not exact, but the integrating factor $1/x^2$ makes it exact.

(a) Use this information to find the value of k.

(b) Using this value of k, find the general solution.

3. Find the general solution of the differential equation

$y'' - a^2y = 0, \ a > 0.$

Show that the general solution can be written in the form

$y = C_1 \cosh ax + C_2 \sinh ax.$

4. Find the general solution of the differential equation

$y'' + \beta^2 y = 0.$

Show that the general solution can be written in the form

$y = C \sin(\beta x + \phi), \ 0 \le \phi < 2\pi.$

5. Given that the characteristic equation of the differential equation

$y'' + ay' + by = 0$

has two distinct real zeros, $m_1 = r + s$ and $m_2 = r - s$, where r and s are real, show that the general solution of the differential equation can be written in the form

$y = e^{rx}(C_1 \cosh sx + C_2 \sinh sx).$

6. Given that a and b are positive and that $y(x)$ is a solution of the differential equation

$y'' + ay' + by = 0$

show that $\lim_{x\to\infty} y(x) = 0.$

7. Consider the differential equation

$y'' + ay = 0$

with boundary conditions $y(0) = 0$ and $y(L) = 0$ for some nonzero real number L.

(a) If $a = 0$, show that the differential equation has only the trivial solution $y = 0$.

(b) If $a < 0$, show that the differential equation has only the trivial solution $y = 0$.

8. For the differential equation and boundary conditions given in Exercise 7, and with $a > 0$, find the value(s) of a for which the solution is nontrivial. Then find the corresponding solution(s).

9. Consider a pendulum of length L that swings by the force of gravity only.

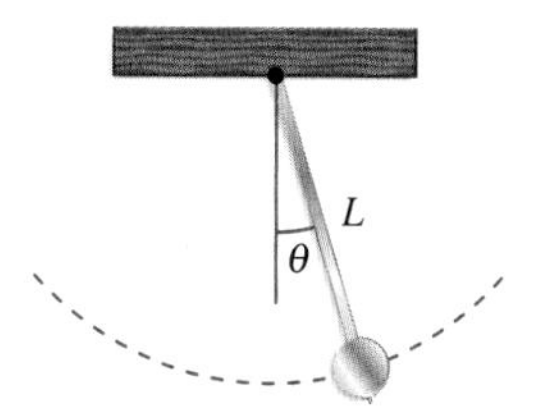

For small values of $\theta = \theta(t)$, the motion of the pendulum can be approximated by the differential equation

$$\frac{d^2\theta}{dt^2} + \frac{g}{L}\theta = 0$$

where g is the acceleration due to gravity.

(a) Find the general solution of the differential equation and show that it can be written in the form

$$\theta(t) = A \cos\left[\sqrt{\frac{g}{L}}(t + \phi)\right].$$

(b) Find the particular solution for a pendulum of length 0.25 meter if the initial conditions are $\theta(0) = 0.1$ radian and $\theta'(0) = 0.5$ radian per second. (Use $g = 9.8$ meters per second per second.)

(c) Determine the period of the pendulum.

(d) Determine the maximum value of θ.

(e) How much time from $t = 0$ does it take for θ to be 0 the first time? the second time?

(f) What is the angular velocity θ' when $\theta = 0$ the first time? the second time?

10. A horizontal beam with a length of 2 meters rests on supports located at the ends of the beam.

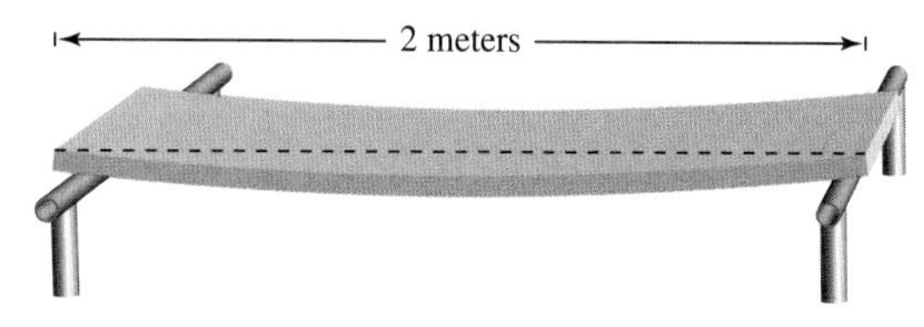

The beam is supporting a load of W kilograms per meter. The resulting deflection y of the beam at a horizontal distance of x meters from the left end can be modeled by

$$A\frac{d^2y}{dx^2} = 2Wx - \frac{1}{2}Wx^2$$

where A is a positive constant.

(a) Solve the differential equation to find the deflection y as a function of the horizontal distance x.

 (b) Use a graphing utility to determine the location and value of the maximum deflection.

In Exercises 11–14, consider a damped mass-spring system whose motion is described by the differential equation

$$\frac{d^2y}{dt^2} + 2\lambda\frac{dy}{dt} + \omega^2 y = 0.$$

The zeros of its characteristic equation are

$$m_1 = -\lambda + \sqrt{\lambda^2 - \omega^2} \text{ and } m_2 = -\lambda - \sqrt{\lambda^2 - \omega^2}.$$

If $\lambda^2 - \omega^2 > 0$, the system is *overdamped*; if $\lambda^2 - \omega^2 = 0$, it is *critically damped*; and if $\lambda^2 - \omega^2 < 0$, it is *underdamped*.

(a) Determine whether the differential equation represents an overdamped, critically damped, or underdamped system.

(b) Find the particular solution corresponding to the given initial conditions.

(c) Use a graphing utility to graph the particular solution found in part (b). Explain how the graph illustrates the type of damping in the system.

11. $\frac{d^2y}{dt^2} + 8\frac{dy}{dt} + 16y = 0$

$y(0) = 1, y'(0) = 1$

12. $\frac{d^2y}{dt^2} + 2\frac{dy}{dt} + 26y = 0$

$y(0) = 1, y'(0) = 4$

13. $\frac{d^2y}{dt^2} + 20\frac{dy}{dt} + 64y = 0$

$y(0) = 2, y'(0) = -20$

14. $\frac{d^2y}{dt^2} + 2\frac{dy}{dt} + y = 0$

$y(0) = 2, y'(0) = -1$

15. Consider Airy's equation given in Section 16.4, Exercise 27. Rewrite the equation as $y'' - (x - 1)y - y = 0$. Then use a power series of the form

$$y = \sum_{n=0}^{\infty} a_n(x-1)^n$$

to find the first *eight* terms of the solution. Compare your result with that of Exercise 27 in Section 16.4.

16. Consider **Chebyshev's equation**

$$(1 - x^2)y'' - xy' + k^2y = 0.$$

Polynomial solutions of this differential equation are called *Chebyshev polynomials* and are denoted by $T_k(x)$. They satisfy the recursion equation

$$T_{n+1}(x) = 2xT_n(x) - T_{n-1}(x).$$

(a) Given that $T_0(x) = 1$ and $T_1(x) = x$, determine the Chebyshev polynomials $T_2(x)$, $T_3(x)$, and $T_4(x)$.

(b) Verify that $T_0(x)$, $T_1(x)$, $T_2(x)$, $T_3(x)$, and $T_4(x)$ are solutions of the given differential equation.

(c) Show that $T_5(x) = 16x^5 - 20x^3 + 5x$,
$T_6(x) = 32x^6 - 48x^4 + 18x^2 - 1$, and
$T_7(x) = 64x^7 - 112x^5 + 56x^3 - 7x$.

17. The differential equation

$$x^2y'' + xy' + x^2y = 0$$

is known as **Bessel's equation of order zero.**

(a) Use a power series of the form

$$y = \sum_{n=0}^{\infty} a_n x^n$$

to find the solution.

(b) Compare your result with that of the function $J_0(x)$ given in Section 9.8, Exercise 71.

18. The differential equation

$$x^2y'' + xy' + (x^2 - 1)y = 0$$

is known as **Bessel's equation of order one.**

(a) Use a power series of the form

$$y = \sum_{n=0}^{\infty} a_n x^n$$

to find the solution.

(b) Compare your result with that of the function $J_1(x)$ given in Section 9.8, Exercise 72.

19. Consider **Hermite's equation**

$$y'' - 2xy' + 2ky = 0.$$

(a) Use a power series of the form

$$y = \sum_{n=0}^{\infty} a_n x^n$$

to find the solution when $k = 4$.

[*Hint:* Choose the arbitrary constants such that the leading term is $(2x)^k$.]

(b) Polynomial solutions of Hermite's equation are called *Hermite polynomials* and are denoted by $H_k(x)$. The general form for $H_k(x)$ can be written as

$$H_k(x) = \sum_{n=0}^{P} \frac{(-1)^n k!(2x)^{k-2n}}{n!(k-2n)!}$$

where P is the greatest integer less than or equal to $k/2$. Use this formula to determine the Hermite polynomials $H_0(x)$, $H_1(x)$, $H_2(x)$, $H_3(x)$, and $H_4(x)$.

20. Consider **Laguerre's equation**

$$xy'' + (1 - x)y' + ky = 0.$$

(a) Polynomial solutions of Laguerre's equation are called *Laguerre polynomials* and are denoted by $L_k(x)$. Use a power series of the form

$$y = \sum_{n=0}^{\infty} a_n x^n$$

to show that

$$L_k(x) = \sum_{n=0}^{k} \frac{(-1)^n k! x^n}{(k-n)!(n!)^2}.$$

Assume that $a_0 = 1$.

(b) Determine the Laguerre polynomials $L_0(x)$, $L_1(x)$, $L_2(x)$, $L_3(x)$, and $L_4(x)$.

Appendices

A Proofs of Selected Theorems

THEOREM 1.2 PROPERTIES OF LIMITS (PROPERTIES 2, 3, 4, AND 5) (PAGE 59)

Let b and c be real numbers, let n be a positive integer, and let f and g be functions with the following limits.

$$\lim_{x \to c} f(x) = L \quad \text{and} \quad \lim_{x \to c} g(x) = K$$

2. Sum or difference: $\lim_{x \to c} [f(x) \pm g(x)] = L \pm K$

3. Product: $\lim_{x \to c} [f(x)g(x)] = LK$

4. Quotient: $\lim_{x \to c} \dfrac{f(x)}{g(x)} = \dfrac{L}{K}$, provided $K \neq 0$

5. Power: $\lim_{x \to c} [f(x)]^n = L^n$

PROOF To prove Property 2, choose $\varepsilon > 0$. Because $\varepsilon/2 > 0$, you know that there exists $\delta_1 > 0$ such that $0 < |x - c| < \delta_1$ implies $|f(x) - L| < \varepsilon/2$. You also know that there exists $\delta_2 > 0$ such that $0 < |x - c| < \delta_2$ implies $|g(x) - K| < \varepsilon/2$. Let δ be the smaller of δ_1 and δ_2; then $0 < |x - c| < \delta$ implies that

$$|f(x) - L| < \frac{\varepsilon}{2} \quad \text{and} \quad |g(x) - K| < \frac{\varepsilon}{2}.$$

So, you can apply the triangle inequality to conclude that

$$|[f(x) + g(x)] - (L + K)| \leq |f(x) - L| + |g(x) - K| < \frac{\varepsilon}{2} + \frac{\varepsilon}{2} = \varepsilon$$

which implies that

$$\lim_{x \to c} [f(x) + g(x)] = L + K = \lim_{x \to c} f(x) + \lim_{x \to c} g(x).$$

The proof that

$$\lim_{x \to c} [f(x) - g(x)] = L - K$$

is similar.

To prove Property 3, given that

$$\lim_{x \to c} f(x) = L \quad \text{and} \quad \lim_{x \to c} g(x) = K$$

you can write

$$f(x)g(x) = [f(x) - L][g(x) - K] + [Lg(x) + Kf(x)] - LK.$$

Because the limit of $f(x)$ is L, and the limit of $g(x)$ is K, you have

$$\lim_{x \to c} [f(x) - L] = 0 \quad \text{and} \quad \lim_{x \to c} [g(x) - K] = 0.$$

Let $0 < \varepsilon < 1$. Then there exists $\delta > 0$ such that if $0 < |x - c| < \delta$, then

$$|f(x) - L - 0| < \varepsilon \quad \text{and} \quad |g(x) - K - 0| < \varepsilon$$

which implies that

$$|[f(x) - L][g(x) - K] - 0| = |f(x) - L|\,|g(x) - K| < \varepsilon\varepsilon < \varepsilon.$$

So,

$$\lim_{x \to c} [f(x) - L][g(x) - K] = 0.$$

Furthermore, by Property 1, you have

$$\lim_{x \to c} Lg(x) = LK \quad \text{and} \quad \lim_{x \to c} Kf(x) = KL.$$

Finally, by Property 2, you obtain

$$\begin{aligned} \lim_{x \to c} f(x)g(x) &= \lim_{x \to c} [f(x) - L][g(x) - K] + \lim_{x \to c} Lg(x) + \lim_{x \to c} Kf(x) - \lim_{x \to c} LK \\ &= 0 + LK + KL - LK \\ &= LK. \end{aligned}$$

To prove Property 4, note that it is sufficient to prove that

$$\lim_{x \to c} \frac{1}{g(x)} = \frac{1}{K}.$$

Then you can use Property 3 to write

$$\lim_{x \to c} \frac{f(x)}{g(x)} = \lim_{x \to c} f(x) \frac{1}{g(x)} = \lim_{x \to c} f(x) \cdot \lim_{x \to c} \frac{1}{g(x)} = \frac{L}{K}.$$

Let $\varepsilon > 0$. Because $\lim\limits_{x \to c} g(x) = K$, there exists $\delta_1 > 0$ such that if

$$0 < |x - c| < \delta_1, \text{ then } |g(x) - K| < \frac{|K|}{2}$$

which implies that

$$|K| = |g(x) + [|K| - g(x)]| \le |g(x)| + ||K| - g(x)| < |g(x)| + \frac{|K|}{2}.$$

That is, for $0 < |x - c| < \delta_1$,

$$\frac{|K|}{2} < |g(x)| \quad \text{or} \quad \frac{1}{|g(x)|} < \frac{2}{|K|}.$$

Similarly, there exists a $\delta_2 > 0$ such that if $0 < |x - c| < \delta_2$, then

$$|g(x) - K| < \frac{|K|^2}{2} \varepsilon.$$

Let δ be the smaller of δ_1 and δ_2. For $0 < |x - c| < \delta$, you have

$$\left|\frac{1}{g(x)} - \frac{1}{K}\right| = \left|\frac{K - g(x)}{g(x)K}\right| = \frac{1}{|K|} \cdot \frac{1}{|g(x)|} |K - g(x)| < \frac{1}{|K|} \cdot \frac{2}{|K|} \frac{|K|^2}{2} \varepsilon = \varepsilon.$$

So, $\lim\limits_{x \to c} \dfrac{1}{g(x)} = \dfrac{1}{K}$.

Finally, the proof of Property 5 can be obtained by a straightforward application of mathematical induction coupled with Property 3. ■

THEOREM 1.4 THE LIMIT OF A FUNCTION INVOLVING A RADICAL (PAGE 60)

Let n be a positive integer. The following limit is valid for all c if n is odd, and is valid for $c > 0$ if n is even.

$$\lim_{x\to c} \sqrt[n]{x} = \sqrt[n]{c}.$$

PROOF Consider the case for which $c > 0$ and n is any positive integer. For a given $\varepsilon > 0$, you need to find $\delta > 0$ such that

$$\left|\sqrt[n]{x} - \sqrt[n]{c}\right| < \varepsilon \quad \text{whenever} \quad 0 < |x - c| < \delta$$

which is the same as saying

$$-\varepsilon < \sqrt[n]{x} - \sqrt[n]{c} < \varepsilon \quad \text{whenever} \quad -\delta < x - c < \delta.$$

Assume $\varepsilon < \sqrt[n]{c}$, which implies that $0 < \sqrt[n]{c} - \varepsilon < \sqrt[n]{c}$. Now, let δ be the smaller of the two numbers.

$$c - \left(\sqrt[n]{c} - \varepsilon\right)^n \quad \text{and} \quad \left(\sqrt[n]{c} + \varepsilon\right)^n - c$$

Then you have

$$\begin{aligned}
-\delta < x - c &< \delta \\
-\left[c - \left(\sqrt[n]{c} - \varepsilon\right)^n\right] < x - c &< \left(\sqrt[n]{c} + \varepsilon\right)^n - c \\
\left(\sqrt[n]{c} - \varepsilon\right)^n - c < x - c &< \left(\sqrt[n]{c} + \varepsilon\right)^n - c \\
\left(\sqrt[n]{c} - \varepsilon\right)^n < x &< \left(\sqrt[n]{c} + \varepsilon\right)^n \\
\sqrt[n]{c} - \varepsilon < \sqrt[n]{x} &< \sqrt[n]{c} + \varepsilon \\
-\varepsilon < \sqrt[n]{x} - \sqrt[n]{c} &< \varepsilon.
\end{aligned}$$

■

THEOREM 1.5 THE LIMIT OF A COMPOSITE FUNCTION (PAGE 61)

If f and g are functions such that $\lim_{x\to c} g(x) = L$ and $\lim_{x\to L} f(x) = f(L)$, then

$$\lim_{x\to c} f(g(x)) = f\left(\lim_{x\to c} g(x)\right) = f(L).$$

PROOF For a given $\varepsilon > 0$, you must find $\delta > 0$ such that

$$|f(g(x)) - f(L)| < \varepsilon \quad \text{whenever} \quad 0 < |x - c| < \delta.$$

Because the limit of $f(x)$ as $x \to L$ is $f(L)$, you know there exists $\delta_1 > 0$ such that

$$|f(u) - f(L)| < \varepsilon \quad \text{whenever} \quad |u - L| < \delta_1.$$

Moreover, because the limit of $g(x)$ as $x \to c$ is L, you know there exists $\delta > 0$ such that

$$|g(x) - L| < \delta_1 \quad \text{whenever} \quad 0 < |x - c| < \delta.$$

Finally, letting $u = g(x)$, you have

$$|f(g(x)) - f(L)| < \varepsilon \quad \text{whenever} \quad 0 < |x - c| < \delta.$$

■

THEOREM 1.7 FUNCTIONS THAT AGREE AT ALL BUT ONE POINT (PAGE 62)

Let c be a real number and let $f(x) = g(x)$ for all $x \neq c$ in an open interval containing c. If the limit of $g(x)$ as x approaches c exists, then the limit of $f(x)$ also exists and

$$\lim_{x \to c} f(x) = \lim_{x \to c} g(x).$$

PROOF Let L be the limit of $g(x)$ as $x \to c$. Then, for each $\varepsilon > 0$ there exists a $\delta > 0$ such that $f(x) = g(x)$ in the open intervals $(c - \delta, c)$ and $(c, c + \delta)$, and

$$|g(x) - L| < \varepsilon \quad \text{whenever} \quad 0 < |x - c| < \delta.$$

Because $f(x) = g(x)$ for all x in the open interval other than $x = c$, it follows that

$$|f(x) - L| < \varepsilon \quad \text{whenever} \quad 0 < |x - c| < \delta.$$

So, the limit of $f(x)$ as $x \to c$ is also L. ■

THEOREM 1.8 THE SQUEEZE THEOREM (PAGE 65)

If $h(x) \leq f(x) \leq g(x)$ for all x in an open interval containing c, except possibly at c itself, and if

$$\lim_{x \to c} h(x) = L = \lim_{x \to c} g(x)$$

then $\lim_{x \to c} f(x)$ exists and is equal to L.

PROOF For $\varepsilon > 0$ there exist $\delta_1 > 0$ and $\delta_2 > 0$ such that

$$|h(x) - L| < \varepsilon \quad \text{whenever} \quad 0 < |x - c| < \delta_1$$

and

$$|g(x) - L| < \varepsilon \quad \text{whenever} \quad 0 < |x - c| < \delta_2.$$

Because $h(x) \leq f(x) \leq g(x)$ for all x in an open interval containing c, except possibly at c itself, there exists $\delta_3 > 0$ such that $h(x) \leq f(x) \leq g(x)$ for $0 < |x - c| < \delta_3$. Let δ be the smallest of δ_1, δ_2, and δ_3. Then, if $0 < |x - c| < \delta$, it follows that $|h(x) - L| < \varepsilon$ and $|g(x) - L| < \varepsilon$, which implies that

$$-\varepsilon < h(x) - L < \varepsilon \quad \text{and} \quad -\varepsilon < g(x) - L < \varepsilon$$
$$L - \varepsilon < h(x) \quad \text{and} \quad g(x) < L + \varepsilon.$$

Now, because $h(x) \leq f(x) \leq g(x)$, it follows that $L - \varepsilon < f(x) < L + \varepsilon$, which implies that $|f(x) - L| < \varepsilon$. Therefore,

$$\lim_{x \to c} f(x) = L.$$ ■

THEOREM 1.11 PROPERTIES OF CONTINUITY (PAGE 75)

If b is a real number and f and g are continuous at $x = c$, then the following functions are also continuous at c.

1. Scalar multiple: bf
2. Sum or difference: $f \pm g$
3. Product: fg
4. Quotient: $\frac{f}{g}$, if $g(c) \neq 0$

PROOF Because f and g are continuous at $x = c$, you can write

$$\lim_{x \to c} f(x) = f(c) \quad \text{and} \quad \lim_{x \to c} g(x) = g(c).$$

For Property 1, when b is a real number, it follows from Theorem 1.2 that

$$\lim_{x \to c} [(bf)(x)] = \lim_{x \to c} [bf(x)] = b \lim_{x \to c} [f(x)] = b\,f(c) = (bf)(c).$$

Thus, bf is continuous at $x = c$.

For Property 2, it follows from Theorem 1.2 that

$$\begin{aligned} \lim_{x \to c} (f \pm g)(x) &= \lim_{x \to c} [f(x) \pm g(x)] \\ &= \lim_{x \to c} [f(x)] \pm \lim_{x \to c} [g(x)] \\ &= f(c) \pm g(c) \\ &= (f \pm g)(c). \end{aligned}$$

Thus, $f \pm g$ is continuous at $x = c$.

For Property 3, it follows from Theorem 1.2 that

$$\begin{aligned} \lim_{x \to c} (fg)(x) &= \lim_{x \to c} [f(x)g(x)] \\ &= \lim_{x \to c} [f(x)] \lim_{x \to c} [g(x)] \\ &= f(c)g(c) \\ &= (fg)(c). \end{aligned}$$

Thus, fg is continuous at $x = c$.

For Property 4, when $g(c) \neq 0$, it follows from Theorem 1.2 that

$$\begin{aligned} \lim_{x \to c} \frac{f}{g}(x) &= \lim_{x \to c} \frac{f(x)}{g(x)} \\ &= \frac{\lim_{x \to c} f(x)}{\lim_{x \to c} g(x)} \\ &= \frac{f(c)}{g(c)} \\ &= \frac{f}{g}(c). \end{aligned}$$

Thus, $\frac{f}{g}$ is continuous at $x = c$. ■

THEOREM 1.14 VERTICAL ASYMPTOTES (PAGE 85)

Let f and g be continuous on an open interval containing c. If $f(c) \neq 0$, $g(c) = 0$, and there exists an open interval containing c such that $g(x) \neq 0$ for all $x \neq c$ in the interval, then the graph of the function given by

$$h(x) = \frac{f(x)}{g(x)}$$

has a vertical asymptote at $x = c$.

PROOF Consider the case for which $f(c) > 0$, and there exists $b > c$ such that $c < x < b$ implies $g(x) > 0$. Then for $M > 0$, choose δ_1 such that

$$0 < x - c < \delta_1 \quad \text{implies that} \quad \frac{f(c)}{2} < f(x) < \frac{3f(c)}{2}$$

and δ_2 such that

$$0 < x - c < \delta_2 \quad \text{implies that} \quad 0 < g(x) < \frac{f(c)}{2M}.$$

Now let δ be the smaller of δ_1 and δ_2. Then it follows that

$$0 < x - c < \delta \quad \text{implies that} \quad \frac{f(x)}{g(x)} > \frac{f(c)}{2}\left[\frac{2M}{f(c)}\right] = M.$$

So, it follows that

$$\lim_{x \to c^+} \frac{f(x)}{g(x)} = \infty$$

and the line $x = c$ is a vertical asymptote of the graph of h. ■

ALTERNATIVE FORM OF THE DERIVATIVE (PAGE 101)

The derivative of f at c is given by

$$f'(c) = \lim_{x \to c} \frac{f(x) - f(c)}{x - c}$$

provided this limit exists.

PROOF The derivative of f at c is given by

$$f'(c) = \lim_{\Delta x \to 0} \frac{f(c + \Delta x) - f(c)}{\Delta x}.$$

Let $x = c + \Delta x$. Then $x \to c$ as $\Delta x \to 0$. So, replacing $c + \Delta x$ by x, you have

$$f'(c) = \lim_{\Delta x \to 0} \frac{f(c + \Delta x) - f(c)}{\Delta x} = \lim_{x \to c} \frac{f(x) - f(c)}{x - c}.$$ ■

THEOREM 2.10 THE CHAIN RULE (PAGE 131)

If $y = f(u)$ is a differentiable function of u, and $u = g(x)$ is a differentiable function of x, then $y = f(g(x))$ is a differentiable function of x and

$$\frac{dy}{dx} = \frac{dy}{du} \cdot \frac{du}{dx}$$

or, equivalently,

$$\frac{d}{dx}[f(g(x))] = f'(g(x))g'(x).$$

PROOF In Section 2.4, you let $h(x) = f(g(x))$ and used the alternative form of the derivative to show that $h'(c) = f'(g(c))g'(c)$, provided $g(x) \neq g(c)$ for values of x other than c. Now consider a more general proof. Begin by considering the derivative of f.

$$f'(x) = \lim_{\Delta x \to 0} \frac{f(x + \Delta x) - f(x)}{\Delta x} = \lim_{\Delta x \to 0} \frac{\Delta y}{\Delta x}$$

For a fixed value of x, define a function η such that

$$\eta(\Delta x) = \begin{cases} 0, & \Delta x = 0 \\ \dfrac{\Delta y}{\Delta x} - f'(x), & \Delta x \neq 0. \end{cases}$$

Because the limit of $\eta(\Delta x)$ as $\Delta x \to 0$ doesn't depend on the value of $\eta(0)$, you have

$$\lim_{\Delta x \to 0} \eta(\Delta x) = \lim_{\Delta x \to 0} \left[\frac{\Delta y}{\Delta x} - f'(x)\right] = 0$$

and you can conclude that η is continuous at 0. Moreover, because $\Delta y = 0$ when $\Delta x = 0$, the equation

$$\Delta y = \Delta x \eta(\Delta x) + \Delta x f'(x)$$

is valid whether Δx is zero or not. Now, by letting $\Delta u = g(x + \Delta x) - g(x)$, you can use the continuity of g to conclude that

$$\lim_{\Delta x \to 0} \Delta u = \lim_{\Delta x \to 0} [g(x + \Delta x) - g(x)] = 0$$

which implies that

$$\lim_{\Delta x \to 0} \eta(\Delta u) = 0.$$

Finally,

$$\Delta y = \Delta u \eta(\Delta u) + \Delta u f'(u) \to \frac{\Delta y}{\Delta x} = \frac{\Delta u}{\Delta x} \eta(\Delta u) + \frac{\Delta u}{\Delta x} f'(u), \quad \Delta x \neq 0$$

and taking the limit as $\Delta x \to 0$, you have

$$\begin{aligned} \frac{dy}{dx} = \frac{du}{dx}\left[\lim_{\Delta x \to 0} \eta(\Delta u)\right] + \frac{du}{dx} f'(u) &= \frac{dy}{dx}(0) + \frac{du}{dx} f'(u) \\ &= \frac{du}{dx} f'(u) \\ &= \frac{du}{dx} \cdot \frac{dy}{du}. \end{aligned}$$

■

CONCAVITY INTERPRETATION (PAGE 190)

1. Let f be differentiable on an open interval I. If the graph of f is concave *upward* on I, then the graph of f lies *above* all of its tangent lines on I.

2. Let f be differentiable on an open interval I. If the graph of f is concave *downward* on I, then the graph of f lies *below* all of its tangent lines on I.

PROOF Assume that f is concave upward on $I = (a, b)$. Then, f' is increasing on (a, b). Let c be a point in the interval $I = (a, b)$. The equation of the tangent line to the graph of f at c is given by

$$g(x) = f(c) + f'(c)(x - c).$$

If x is in the open interval (c, b), then the directed distance from point $(x, f(x))$ (on the graph of f) to the point $(x, g(x))$ (on the tangent line) is given by

$$\begin{aligned} d &= f(x) - [f(c) + f'(c)(x - c)] \\ &= f(x) - f(c) - f'(c)(x - c). \end{aligned}$$

Moreover, by the Mean Value Theorem there exists a number z in (c, x) such that

$$f'(z) = \frac{f(x) - f(c)}{x - c}.$$

So, you have

$$\begin{aligned} d &= f(x) - f(c) - f'(c)(x - c) \\ &= f'(z)(x - c) - f'(c)(x - c) \\ &= [f'(z) - f'(c)](x - c). \end{aligned}$$

The second factor $(x - c)$ is positive because $c < x$. Moreover, because f' is increasing, it follows that the first factor $[f'(z) - f'(c)]$ is also positive. Therefore, $d > 0$ and you can conclude that the graph of f lies above the tangent line at x. If x is in the open interval (a, c), a similar argument can be given. This proves the first statement. The proof of the second statement is similar. ■

THEOREM 3.7 **TEST FOR CONCAVITY (PAGE 191)**

Let f be a function whose second derivative exists on an open interval I.

1. If $f''(x) > 0$ for all x in I, then the graph of f is concave upward in I.

2. If $f''(x) < 0$ for all x in I, then the graph of f is concave downward in I.

PROOF For Property 1, assume $f''(x) > 0$ for all x in (a, b). Then, by Theorem 3.5, f' is increasing on $[a, b]$. Thus, by the definition of concavity, the graph of f is concave upward on (a, b).

For Property 2, assume $f''(x) < 0$ for all x in (a, b). Then, by Theorem 3.5, f' is decreasing on $[a, b]$. Thus, by the definition of concavity, the graph of f is concave downward on (a, b). ■

THEOREM 3.10 LIMITS AT INFINITY (PAGE 199)

If r is a positive rational number and c is any real number, then

$$\lim_{x\to\infty} \frac{c}{x^r} = 0.$$

Furthermore, if x^r is defined when $x < 0$, then $\lim_{x\to-\infty} \frac{c}{x^r} = 0.$

PROOF Begin by proving that

$$\lim_{x\to\infty} \frac{1}{x} = 0.$$

For $\varepsilon > 0$, let $M = 1/\varepsilon$. Then, for $x > M$, you have

$$x > M = \frac{1}{\varepsilon} \implies \frac{1}{x} < \varepsilon \implies \left|\frac{1}{x} - 0\right| < \varepsilon.$$

So, by the definition of a limit at infinity, you can conclude that the limit of $1/x$ as $x \to \infty$ is 0. Now, using this result, and letting $r = m/n$, you can write the following.

$$\begin{aligned}
\lim_{x\to\infty} \frac{c}{x^r} &= \lim_{x\to\infty} \frac{c}{x^{m/n}} \\
&= c\left[\lim_{x\to\infty}\left(\frac{1}{\sqrt[n]{x}}\right)^m\right] \\
&= c\left(\lim_{x\to\infty} \sqrt[n]{\frac{1}{x}}\right)^m \\
&= c\left(\sqrt[n]{\lim_{x\to\infty} \frac{1}{x}}\right)^m \\
&= c\left(\sqrt[n]{0}\right)^m \\
&= 0
\end{aligned}$$

The proof of the second part of the theorem is similar. ■

THEOREM 4.2 SUMMATION FORMULAS (PAGE 260)

1. $\displaystyle\sum_{i=1}^{n} c = cn$
2. $\displaystyle\sum_{i=1}^{n} i = \frac{n(n+1)}{2}$
3. $\displaystyle\sum_{i=1}^{n} i^2 = \frac{n(n+1)(2n+1)}{6}$
4. $\displaystyle\sum_{i=1}^{n} i^3 = \frac{n^2(n+1)^2}{4}$

PROOF The proof of Property 1 is straightforward. By adding c to itself n times, you obtain a sum of cn.

To prove Property 2, write the sum in increasing and decreasing order and add corresponding terms, as follows.

$$\begin{array}{rcccccccccc}
\sum_{i=1}^{n} i = & 1 & + & 2 & + & 3 & + \cdots + & (n-1) & + & n \\
 & \downarrow & & & & \downarrow & & \downarrow & & \downarrow \\
\sum_{i=1}^{n} i = & n & + & (n-1) & + & (n-2) & + \cdots + & 2 & + & 1 \\
 & \downarrow & & \downarrow & & \downarrow & & \downarrow & & \downarrow
\end{array}$$

$$2\sum_{i=1}^{n} i = \underbrace{(n+1) + (n+1) + (n+1) + \cdots + (n+1) + (n+1)}_{n \text{ terms}}$$

So,

$$\sum_{i=1}^{n} i = \frac{n(n+1)}{2}.$$

To prove Property 3, use mathematical induction. First, if $n = 1$, the result is true because

$$\sum_{i=1}^{1} i^2 = 1^2 = 1 = \frac{1(1+1)(2+1)}{6}.$$

Now, assuming the result is true for $n = k$, you can show that it is true for $n = k + 1$, as follows.

$$\begin{aligned}
\sum_{i=1}^{k+1} i^2 &= \sum_{i=1}^{k} i^2 + (k+1)^2 \\
&= \frac{k(k+1)(2k+1)}{6} + (k+1)^2 \\
&= \frac{k+1}{6}(2k^2 + k + 6k + 6) \\
&= \frac{k+1}{6}[(2k+3)(k+2)] \\
&= \frac{(k+1)(k+2)[2(k+1)+1]}{6}
\end{aligned}$$

Property 4 can be proved using a similar argument with mathematical induction. ■

THEOREM 4.8 PRESERVATION OF INEQUALITY (PAGE 278)

1. If f is integrable and nonnegative on the closed interval $[a, b]$, then

$$0 \le \int_a^b f(x)\,dx.$$

2. If f and g are integrable on the closed interval $[a, b]$ and $f(x) \le g(x)$ for every x in $[a, b]$, then

$$\int_a^b f(x)\,dx \le \int_a^b g(x)\,dx.$$

PROOF To prove Property 1, suppose, on the contrary, that

$$\int_a^b f(x)\,dx = I < 0.$$

Then, let $a = x_0 < x_1 < x_2 < \cdot \cdot \cdot < x_n = b$ be a partition of $[a, b]$, and let

$$R = \sum_{i=1}^{n} f(c_i)\,\Delta x_i$$

be a Riemann sum. Because $f(x) \geq 0$, it follows that $R \geq 0$. Now, for $\|\Delta\|$ sufficiently small, you have $|R - I| < -I/2$, which implies that

$$\sum_{i=1}^{n} f(c_i)\,\Delta x_i = R < I - \frac{I}{2} < 0$$

which is not possible. From this contradiction, you can conclude that

$$0 \leq \int_a^b f(x)\,dx.$$

To prove Property 2 of the theorem, note that $f(x) \leq g(x)$ implies that $g(x) - f(x) \geq 0$. So, you can apply the result of Property 1 to conclude that

$$0 \leq \int_a^b [g(x) - f(x)]\,dx$$

$$0 \leq \int_a^b g(x)\,dx - \int_a^b f(x)\,dx$$

$$\int_a^b f(x)\,dx \leq \int_a^b g(x)\,dx.$$

■

PROPERTIES OF THE NATURAL LOGARITHMIC FUNCTION (PAGE 325)

The natural logarithmic function is one-to-one.

$$\lim_{x \to 0^+} \ln x = -\infty \qquad \text{and} \qquad \lim_{x \to \infty} \ln x = \infty$$

PROOF Recall from Section P.3 that a function f is one-to-one if for x_1 and x_2 in its domain

$$x_1 \neq x_2 \quad \Rightarrow \quad f(x_1) \neq f(x_2).$$

Let $f(x) = \ln x$. Then $f'(x) = \dfrac{1}{x} > 0$ for $x > 0$. So f is increasing on its entire domain $(0, \infty)$ and therefore is strictly monotonic (see Section 3.3). Choose x_1 and x_2 in the domain of f such that $x_1 \neq x_2$. Because f is strictly monotonic, it follows that either

$$f(x_1) < f(x_2) \quad \text{or} \quad f(x_1) > f(x_2).$$

In either case, $f(x_1) \neq f(x_2)$. So, $f(x) = \ln x$ is one-to-one. To verify the limits, begin by showing that $\ln 2 \geq \frac{1}{2}$. From the Mean Value Theorem for Integrals, you can write

$$\ln 2 = \int_1^2 \frac{1}{x}\,dx = \frac{1}{c}(2 - 1) = \frac{1}{c}$$

where c is in $[1, 2]$.

This implies that

$$\begin{aligned} 1 \le \quad c \quad &\le 2 \\ 1 \ge \quad \frac{1}{c} \quad &\ge \frac{1}{2} \\ 1 \ge \ln 2 &\ge \frac{1}{2}. \end{aligned}$$

Now, let N be any positive (large) number. Because $\ln x$ is increasing, it follows that if $x > 2^{2N}$, then

$$\ln x > \ln 2^{2N} = 2N \ln 2.$$

However, because $\ln 2 \ge \frac{1}{2}$, it follows that

$$\ln x > 2N \ln 2 \ge 2N\left(\frac{1}{2}\right) = N.$$

This verifies the second limit. To verify the first limit, let $z = 1/x$. Then, $z \to \infty$ as $x \to 0^+$, and you can write

$$\begin{aligned} \lim_{x\to 0^+} \ln x &= \lim_{x\to 0^+} \left(-\ln \frac{1}{x}\right) \\ &= \lim_{z\to\infty} (-\ln z) \\ &= -\lim_{z\to\infty} \ln z \\ &= -\infty. \end{aligned}$$

■

THEOREM 5.8 CONTINUITY AND DIFFERENTIABILITY OF INVERSE FUNCTIONS (PAGE 347)

Let f be a function whose domain is an interval I. If f has an inverse function, then the following statements are true.

1. If f is continuous on its domain, then f^{-1} is continuous on its domain.
2. If f is increasing on its domain, then f^{-1} is increasing on its domain.
3. If f is decreasing on its domain, then f^{-1} is decreasing on its domain.
4. If f is differentiable on an interval containing c and $f'(c) \neq 0$, then f^{-1} is differentiable at $f(c)$.

PROOF To prove Property 1, first show that if f is continuous on I and has an inverse function, then f is strictly monotonic on I. Suppose that f were not strictly monotonic. Then there would exist numbers x_1, x_2, x_3 in I such that $x_1 < x_2 < x_3$, but $f(x_2)$ is not between $f(x_1)$ and $f(x_3)$. Without loss of generality, assume $f(x_1) < f(x_3) < f(x_2)$. By the Intermediate Value Theorem, there exists a number x_0 between x_1 and x_2 such that $f(x_0) = f(x_3)$. So, f is not one-to-one and cannot have an inverse function. So, f must be strictly monotonic.

Because f is continuous, the Intermediate Value Theorem implies that the set of values of f

$$\{f(x) \colon x \in I\}$$

forms an interval J. Assume that a is an interior point of J. From the previous argument, $f^{-1}(a)$ is an interior point of I. Let $\varepsilon > 0$. There exists $0 < \varepsilon_1 < \varepsilon$ such that

$$I_1 = (f^{-1}(a) - \varepsilon_1, f^{-1}(a) + \varepsilon_1) \subseteq I.$$

Because f is strictly monotonic on I_1, the set of values $\{f(x): x \in I_1\}$ forms an interval $J_1 \subseteq J$. Let $\delta > 0$ such that $(a - \delta, a + \delta) \subseteq J_1$. Finally, if

$$|y - a| < \delta, \text{ then } |f^{-1}(y) - f^{-1}(a)| < \varepsilon_1 < \varepsilon.$$

So, f^{-1} is continuous at a. A similar proof can be given if a is an endpoint.

To prove Property 2, let y_1 and y_2 be in the domain of f^{-1}, with $y_1 < y_2$. Then, there exist x_1 and x_2 in the domain of f such that

$$f(x_1) = y_1 < y_2 = f(x_2).$$

Because f is increasing, $f(x_1) < f(x_2)$ holds precisely when $x_1 < x_2$. Therefore,

$$f^{-1}(y_1) = x_1 < x_2 = f^{-1}(y_2)$$

which implies that f^{-1} is increasing. (Property 3 can be proved in a similar way.)

Finally, to prove Property 4, consider the limit

$$(f^{-1})'(a) = \lim_{y \to a} \frac{f^{-1}(y) - f^{-1}(a)}{y - a}$$

where a is in the domain of f^{-1} and $f^{-1}(a) = c$. Because f is differentiable on an interval containing c, f is continuous on that interval, and so is f^{-1} at a. So, $y \to a$ implies that $x \to c$, and you have

$$\begin{aligned}
(f^{-1})'(a) &= \lim_{x \to c} \frac{x - c}{f(x) - f(c)} \\
&= \lim_{x \to c} \frac{1}{\left(\dfrac{f(x) - f(c)}{x - c}\right)} \\
&= \frac{1}{\displaystyle\lim_{x \to c} \frac{f(x) - f(c)}{x - c}} \\
&= \frac{1}{f'(c)}.
\end{aligned}$$

So, $(f^{-1})'(a)$ exists, and f^{-1} is differentiable at $f(c)$. ■

THEOREM 5.9 THE DERIVATIVE OF AN INVERSE FUNCTION (PAGE 347)

Let f be a function that is differentiable on an interval I. If f has an inverse function g, then g is differentiable at any x for which $f'(g(x)) \neq 0$. Moreover,

$$g'(x) = \frac{1}{f'(g(x))}, \quad f'(g(x)) \neq 0.$$

PROOF From the proof of Theorem 5.8, letting $a = x$, you know that g is differentiable. Using the Chain Rule, differentiate both sides of the equation $x = f(g(x))$ to obtain

$$1 = f'(g(x)) \frac{d}{dx}[g(x)].$$

Because $f'(g(x)) \neq 0$, you can divide by this quantity to obtain

$$\frac{d}{dx}[g(x)] = \frac{1}{f'(g(x))}.$$

■

THEOREM 5.10 OPERATIONS WITH EXPONENTIAL FUNCTIONS (PROPERTY 2) (PAGE 353)

2. $\dfrac{e^a}{e^b} = e^{a-b}$ (Let a and b be any real numbers.)

PROOF To prove Property 2, you can write

$$\ln\left(\frac{e^a}{e^b}\right) = \ln e^a - \ln e^b = a - b = \ln(e^{a-b})$$

Because the natural logarithmic function is one-to-one, you can conclude that

$$\frac{e^a}{e^b} = e^{a-b}.$$

■

THEOREM 5.15 A LIMIT INVOLVING *e* (PAGE 366)

$$\lim_{x\to\infty}\left(1 + \frac{1}{x}\right)^x = \lim_{x\to\infty}\left(\frac{x+1}{x}\right)^x = e$$

PROOF Let $y = \displaystyle\lim_{x\to\infty}\left(1 + \frac{1}{x}\right)^x$. Taking the natural logarithm of each side, you have

$$\ln y = \ln\left[\lim_{x\to\infty}\left(1 + \frac{1}{x}\right)^x\right].$$

Because the natural logarithmic function is continuous, you can write

$$\ln y = \lim_{x\to\infty}\left[x \ln\left(1 + \frac{1}{x}\right)\right] = \lim_{x\to\infty}\left\{\frac{\ln[1 + (1/x)]}{1/x}\right\}.$$

Letting $x = \dfrac{1}{t}$, you have

$$\begin{aligned}
\ln y &= \lim_{t\to 0^+}\frac{\ln(1+t)}{t} = \lim_{t\to 0^+}\frac{\ln(1+t) - \ln 1}{t}\\
&= \frac{d}{dx}\ln x \text{ at } x = 1\\
&= \frac{1}{x} \text{ at } x = 1\\
&= 1.
\end{aligned}$$

Finally, because $\ln y = 1$, you know that $y = e$, and you can conclude that

$$\lim_{x\to\infty}\left(1 + \frac{1}{x}\right)^x = e.$$

■

THEOREM 5.16 DERIVATIVES OF INVERSE TRIGONOMETRIC FUNCTIONS (arcsin *u* and arccos *u*) (PAGE 376)

Let u be a differentiable function of x.

$$\frac{d}{dx}[\arcsin u] = \frac{u'}{\sqrt{1-u^2}} \qquad \frac{d}{dx}[\arccos u] = \frac{-u'}{\sqrt{1-u^2}}$$

PROOF
Method 1: Apply Theorem 5.9.

Let $f(x) = \sin x$ and $g(x) = \arcsin x$. Because f is differentiable on $-\pi/2 \le y \le \pi/2$, you can apply Theorem 5.9.

$$g'(x) = \frac{1}{f'(g(x))} = \frac{1}{\cos(\arcsin x)} = \frac{1}{\sqrt{1 - \sin^2(\arcsin x)}} = \frac{1}{\sqrt{1 - x^2}}$$

If u is a differentiable function of x, then you can use the Chain Rule to write

$$\frac{d}{dx}[\arcsin u] = \frac{u'}{\sqrt{1 - u^2}}, \quad \text{where} \quad u' = \frac{du}{dx}.$$

Method 2: Use implicit differentiation.

Let $y = \arccos x$, $0 \le y \le \pi$. So, $\cos y = x$, and you can use implicit differentiation as follows.

$$\cos y = x$$

$$-\sin y \frac{dy}{dx} = 1$$

$$\frac{dy}{dx} = \frac{-1}{\sin y} = \frac{-1}{\sqrt{1 - \cos^2 y}} = \frac{-1}{\sqrt{1 - x^2}}$$

If u is a differentiable function of x, then you can use the Chain Rule to write

$$\frac{d}{dx}[\arccos u] = \frac{-u'}{\sqrt{1 - u^2}}, \quad \text{where} \quad u' = \frac{du}{dx}.$$

■

THEOREM 8.3 THE EXTENDED MEAN VALUE THEOREM (PAGE 570)

If f and g are differentiable on an open interval (a, b) and continuous on $[a, b]$ such that $g'(x) \ne 0$ for any x in (a, b), then there exists a point c in (a, b) such that $\dfrac{f'(c)}{g'(c)} = \dfrac{f(b) - f(a)}{g(b) - g(a)}$.

PROOF You can assume that $g(a) \ne g(b)$, because otherwise, by Rolle's Theorem, it would follow that $g'(x) = 0$ for some x in (a, b). Now, define $h(x)$ as

$$h(x) = f(x) - \left[\frac{f(b) - f(a)}{g(b) - g(a)}\right] g(x).$$

Then

$$h(a) = f(a) - \left[\frac{f(b) - f(a)}{g(b) - g(a)}\right] g(a) = \frac{f(a)g(b) - f(b)g(a)}{g(b) - g(a)}$$

and

$$h(b) = f(b) - \left[\frac{f(b) - f(a)}{g(b) - g(a)}\right] g(b) = \frac{f(a)g(b) - f(b)g(a)}{g(b) - g(a)}$$

and by Rolle's Theorem there exists a point c in (a, b) such that

$$h'(c) = f'(c) - \frac{f(b) - f(a)}{g(b) - g(a)} g'(c) = 0$$

which implies that $\dfrac{f'(c)}{g'(c)} = \dfrac{f(b) - f(a)}{g(b) - g(a)}$.

■

THEOREM 8.4 L'HÔPITAL'S RULE (PAGE 570)

Let f and g be functions that are differentiable on an open interval (a, b) containing c, except possibly at c itself. Assume that $g'(x) \neq 0$ for all x in (a, b), except possibly at c itself. If the limit of $f(x)/g(x)$ as x approaches c produces the indeterminate form $0/0$, then

$$\lim_{x\to c} \frac{f(x)}{g(x)} = \lim_{x\to c} \frac{f'(x)}{g'(x)}$$

provided the limit on the right exists (or is infinite). This result also applies if the limit of $f(x)/g(x)$ as x approaches c produces any one of the indeterminate forms ∞/∞, $(-\infty)/\infty$, $\infty/(-\infty)$, or $(-\infty)/(-\infty)$.

You can use the Extended Mean Value Theorem to prove L'Hôpital's Rule. Of the several different cases of this rule, the proof of only one case is illustrated. The remaining cases where $x \to c^-$ and $x \to c$ are left for you to prove.

PROOF Consider the case for which $\lim_{x\to c^+} f(x) = 0$ and $\lim_{x\to c^+} g(x) = 0$. Define the following new functions:

$$F(x) = \begin{cases} f(x), & x \neq c \\ 0, & x = c \end{cases} \quad \text{and} \quad G(x) = \begin{cases} g(x), & x \neq c \\ 0, & x = c \end{cases}.$$

For any x, $c < x < b$, F and G are differentiable on $(c, x]$ and continuous on $[c, x]$. You can apply the Extended Mean Value Theorem to conclude that there exists a number z in (c, x) such that

$$\frac{F'(z)}{G'(z)} = \frac{F(x) - F(c)}{G(x) - G(c)} = \frac{F(x)}{G(x)} = \frac{f'(z)}{g'(z)} = \frac{f(x)}{g(x)}.$$

Finally, by letting x approach c from the right, $x \to c^+$, you have $z \to c^+$ because $c < z < x$, and

$$\lim_{x\to c^+} \frac{f(x)}{g(x)} = \lim_{x\to c^+} \frac{f'(z)}{g'(z)} = \lim_{z\to c^+} \frac{f'(z)}{g'(z)} = \lim_{x\to c^+} \frac{f'(x)}{g'(x)}.$$

■

THEOREM 9.19 TAYLOR'S THEOREM (PAGE 656)

If a function f is differentiable through order $n + 1$ in an interval I containing c, then, for each x in I, there exists z between x and c such that

$$f(x) = f(c) + f'(c)(x - c) + \frac{f''(c)}{2!}(x - c)^2 + \cdots + \frac{f^{(n)}(c)}{n!}(x - c)^n + R_n(x)$$

where $R_n(x) = \dfrac{f^{(n+1)}(z)}{(n + 1)!}(x - c)^{n+1}$.

PROOF To find $R_n(x)$, fix x in I $(x \neq c)$ and write $R_n(x) = f(x) - P_n(x)$ where $P_n(x)$ is the nth Taylor polynomial for $f(x)$. Then let g be a function of t defined by

$$g(t) = f(x) - f(t) - f'(t)(x - t) - \cdots - \frac{f^{(n)}(t)}{n!}(x - t)^n - R_n(x)\frac{(x - t)^{n+1}}{(x - c)^{n+1}}.$$

The reason for defining g in this way is that differentiation with respect to t has a telescoping effect. For example, you have

$$\frac{d}{dt}[-f(t) - f'(t)(x - t)] = -f'(t) + f'(t) - f''(t)(x - t) = -f''(t)(x - t).$$

The result is that the derivative $g'(t)$ simplifies to

$$g'(t) = -\frac{f^{(n+1)}(t)}{n!}(x - t)^n + (n + 1)R_n(x)\frac{(x - t)^n}{(x - c)^{n+1}}$$

for all t between c and x. Moreover, for a fixed x,

$$g(c) = f(x) - [P_n(x) + R_n(x)] = f(x) - f(x) = 0$$

and

$$g(x) = f(x) - f(x) - 0 - \cdot \cdot \cdot - 0 = f(x) - f(x) = 0.$$

Therefore, g satisfies the conditions of Rolle's Theorem, and it follows that there is a number z between c and x such that $g'(z) = 0$. Substituting z for t in the equation for $g'(t)$ and then solving for $R_n(x)$, you obtain

$$g'(z) = -\frac{f^{(n+1)}(z)}{n!}(x - z)^n + (n + 1)R_n(x)\frac{(x - z)^n}{(x - c)^{n+1}} = 0$$

$$R_n(x) = \frac{f^{(n+1)}(z)}{(n + 1)!}(x - c)^{n+1}.$$

Finally, because $g(c) = 0$, you have

$$0 = f(x) - f(c) - f'(c)(x - c) - \cdot \cdot \cdot - \frac{f^{(n)}(c)}{n!}(x - c)^n - R_n(x)$$

$$f(x) = f(c) + f'(c)(x - c) + \cdot \cdot \cdot + \frac{f^{(n)}(c)}{n!}(x - c)^n + R_n(x).$$

■

THEOREM 9.20 CONVERGENCE OF A POWER SERIES (PAGE 662)

For a power series centered at c, precisely one of the following is true.

1. The series converges only at c.
2. There exists a real number $R > 0$ such that the series converges absolutely for $|x - c| < R$, and diverges for $|x - c| > R$.
3. The series converges absolutely for all x.

The number R is the **radius of convergence** of the power series. If the series converges only at c, the radius of convergence is $R = 0$, and if the series converges for all x, the radius of convergence is $R = \infty$. The set of all values of x for which the power series converges is the **interval of convergence** of the power series.

PROOF In order to simplify the notation, the theorem for the power series $\Sigma\, a_n x^n$ centered at $x = 0$ will be proved. The proof for a power series centered at $x = c$ follows easily. A key step in this proof uses the completeness property of the set of real numbers: If a nonempty set S of real numbers has an upper bound, then it must have a least upper bound (see page 603).

It must be shown that if a power series $\Sigma\, a_n x^n$ converges at $x = d$, $d \neq 0$, then it converges for all b satisfying $|b| < |d|$. Because $\Sigma\, a_n x^n$ converges, $\lim_{x \to \infty} a_n d^n = 0$.

So, there exists $N > 0$ such that $a_n d^n < 1$ for all $n \geq N$. Then for $n \geq N$,

$$|a_n b^n| = \left|a_n b^n \frac{d^n}{d^n}\right| = |a_n d^n|\left|\frac{b^n}{d^n}\right| < \left|\frac{b^n}{d^n}\right|.$$

So, for $|b| < |d|$, $\left|\frac{b}{d}\right| < 1$, which implies that

$$\sum \left|\frac{b^n}{d^n}\right|$$

is a convergent geometric series. By the Comparison Test, the series $\sum a_n b^n$ converges.

Similarly, if the power series $\sum a_n x^n$ diverges at $x = b$, where $b \neq 0$, then it diverges for all d satisfying $|d| > |b|$. If $\sum a_n d^n$ converged, then the argument above would imply that $\sum a_n b^n$ converged as well.

Finally, to prove the theorem, suppose that neither Case 1 nor Case 3 is true. Then there exist points b and d such that $\sum a_n x^n$ converges at b and diverges at d. Let $S = \{x: \sum a_n x^n \text{ converges}\}$. S is nonempty because $b \in S$. If $x \in S$, then $|x| \leq |d|$, which shows that $|d|$ is an upper bound for the nonempty set S. By the completeness property, S has a least upper bound, R.

Now, if $|x| > R$, then $x \notin S$, so $\sum a_n x^n$ diverges. And if $|x| < R$, then $|x|$ is not an upper bound for S, so there exists b in S satisfying $|b| > |x|$. Since $b \in S$, $\sum a_n b^n$ converges, which implies that $\sum a_n x^n$ converges. ■

THEOREM 10.16 CLASSIFICATION OF CONICS BY ECCENTRICITY (PAGE 750)

Let F be a fixed point (*focus*) and let D be a fixed line (*directrix*) in the plane. Let P be another point in the plane and let e (*eccentricity*) be the ratio of the distance between P and F to the distance between P and D. The collection of all points P with a given eccentricity is a conic.

1. The conic is an ellipse if $0 < e < 1$.

2. The conic is a parabola if $e = 1$.

3. The conic is a hyperbola if $e > 1$.

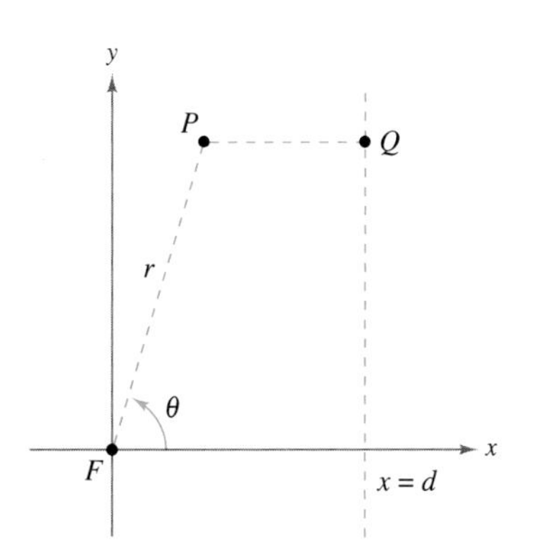

Figure A.1

PROOF If $e = 1$, then, by definition, the conic must be a parabola. If $e \neq 1$, then you can consider the focus F to lie at the origin and the directrix $x = d$ to lie to the right of the origin, as shown in Figure A.1. For the point $P = (r, \theta) = (x, y)$, you have $|PF| = r$ and $|PQ| = d - r\cos\theta$. Given that $e = |PF|/|PQ|$, it follows that

$$|PF| = |PQ|e \quad \Longrightarrow \quad r = e(d - r\cos\theta).$$

By converting to rectangular coordinates and squaring each side, you obtain

$$x^2 + y^2 = e^2(d - x)^2 = e^2(d^2 - 2dx + x^2).$$

Completing the square produces

$$\left(x + \frac{e^2 d}{1 - e^2}\right)^2 + \frac{y^2}{1 - e^2} = \frac{e^2 d^2}{(1 - e^2)^2}.$$

If $e < 1$, this equation represents an ellipse. If $e > 1$, then $1 - e^2 < 0$, and the equation represents a hyperbola. ■

THEOREM 13.4 SUFFICIENT CONDITION FOR DIFFERENTIABILITY (PAGE 919)

If f is a function of x and y, where f_x and f_y are continuous in an open region R, then f is differentiable on R.

PROOF Let S be the surface defined by $z = f(x, y)$, where f, f_x, and f_y are continuous at (x, y). Let A, B, and C be points on surface S, as shown in Figure A.2. From this figure, you can see that the change in f from point A to point C is given by

$$\begin{aligned}\Delta z &= f(x + \Delta x, y + \Delta y) - f(x, y)\\ &= [f(x + \Delta x, y) - f(x, y)] + [f(x + \Delta x, y + \Delta y) - f(x + \Delta x, y)]\\ &= \Delta z_1 + \Delta z_2.\end{aligned}$$

Between A and B, y is fixed and x changes. So, by the Mean Value Theorem, there is a value x_1 between x and $x + \Delta x$ such that

$$\Delta z_1 = f(x + \Delta x, y) - f(x, y) = f_x(x_1, y)\,\Delta x.$$

Similarly, between B and C, x is fixed and y changes, and there is a value y_1 between y and $y + \Delta y$ such that

$$\Delta z_2 = f(x + \Delta x, y + \Delta y) - f(x + \Delta x, y) = f_y(x + \Delta x, y_1)\,\Delta y.$$

By combining these two results, you can write

$$\Delta z = \Delta z_1 + \Delta z_2 = f_x(x_1, y)\Delta x + f_y(x + \Delta x, y_1)\,\Delta y.$$

If you define ε_1 and ε_2 as $\varepsilon_1 = f_x(x_1, y) - f_x(x, y)$ and $\varepsilon_2 = f_y(x + \Delta x, y_1) - f_y(x, y)$, it follows that

$$\begin{aligned}\Delta z = \Delta z_1 + \Delta z_2 &= [\varepsilon_1 + f_x(x, y)]\,\Delta x + [\varepsilon_2 + f_y(x, y)]\,\Delta y\\ &= [f_x(x, y)\,\Delta x + f_y(x, y)\,\Delta y] + \varepsilon_1\Delta x + \varepsilon_2\Delta y.\end{aligned}$$

By the continuity of f_x and f_y and the fact that $x \le x_1 \le x + \Delta x$ and $y \le y_1 \le y + \Delta y$, it follows that $\varepsilon_1 \to 0$ and $\varepsilon_2 \to 0$ as $\Delta x \to 0$ and $\Delta y \to 0$. Therefore, by definition, f is differentiable. ■

THEOREM 13.6 CHAIN RULE: ONE INDEPENDENT VARIABLE (PAGE 925)

Let $w = f(x, y)$, where f is a differentiable function of x and y. If $x = g(t)$ and $y = h(t)$, where g and h are differentiable functions of t, then w is a differentiable function of t, and

$$\frac{dw}{dt} = \frac{\partial w}{\partial x}\frac{dx}{dt} + \frac{\partial w}{\partial y}\frac{dy}{dt}.$$

PROOF Because g and h are differentiable functions of t, you know that both Δx and Δy approach zero as Δt approaches zero. Moreover, because f is a differentiable function of x and y, you know that $\Delta w = (\partial w/\partial x)\,\Delta x + (\partial w/\partial y)\,\Delta y + \varepsilon_1\Delta x + \varepsilon_2\Delta y$, where both ε_1 and $\varepsilon_2 \to 0$ as $(\Delta x, \Delta y) \to (0, 0)$. So, for $\Delta t \ne 0$

$$\frac{\Delta w}{\Delta t} = \frac{\partial w}{\partial x}\frac{\Delta x}{\Delta t} + \frac{\partial w}{\partial y}\frac{\Delta y}{\Delta t} + \varepsilon_1\frac{\Delta x}{\Delta t} + \varepsilon_2\frac{\Delta y}{\Delta t}$$

from which it follows that

$$\frac{dw}{dt} = \lim_{\Delta t\to 0}\frac{\Delta w}{\Delta t} = \frac{\partial w}{\partial x}\frac{dx}{dt} + \frac{\partial w}{\partial y}\frac{dy}{dt} + 0\left(\frac{dx}{dt}\right) + 0\left(\frac{dy}{dt}\right) = \frac{\partial w}{\partial x}\frac{dx}{dt} + \frac{\partial w}{\partial y}\frac{dy}{dt}.$$ ■

B Integration Tables

Forms Involving u^n

1. $\displaystyle\int u^n\,du = \frac{u^{n+1}}{n+1} + C,\ n \neq -1$

2. $\displaystyle\int \frac{1}{u}\,du = \ln|u| + C$

Forms Involving $a + bu$

3. $\displaystyle\int \frac{u}{a+bu}\,du = \frac{1}{b^2}\left(bu - a\ln|a+bu|\right) + C$

4. $\displaystyle\int \frac{u}{(a+bu)^2}\,du = \frac{1}{b^2}\left(\frac{a}{a+bu} + \ln|a+bu|\right) + C$

5. $\displaystyle\int \frac{u}{(a+bu)^n}\,du = \frac{1}{b^2}\left[\frac{-1}{(n-2)(a+bu)^{n-2}} + \frac{a}{(n-1)(a+bu)^{n-1}}\right] + C,\quad n \neq 1, 2$

6. $\displaystyle\int \frac{u^2}{a+bu}\,du = \frac{1}{b^3}\left[-\frac{bu}{2}(2a - bu) + a^2\ln|a+bu|\right] + C$

7. $\displaystyle\int \frac{u^2}{(a+bu)^2}\,du = \frac{1}{b^3}\left(bu - \frac{a^2}{a+bu} - 2a\ln|a+bu|\right) + C$

8. $\displaystyle\int \frac{u^2}{(a+bu)^3}\,du = \frac{1}{b^3}\left[\frac{2a}{a+bu} - \frac{a^2}{2(a+bu)^2} + \ln|a+bu|\right] + C$

9. $\displaystyle\int \frac{u^2}{(a+bu)^n}\,du = \frac{1}{b^3}\left[\frac{-1}{(n-3)(a+bu)^{n-3}} + \frac{2a}{(n-2)(a+bu)^{n-2}} - \frac{a^2}{(n-1)(a+bu)^{n-1}}\right] + C,\quad n \neq 1, 2, 3$

10. $\displaystyle\int \frac{1}{u(a+bu)}\,du = \frac{1}{a}\ln\left|\frac{u}{a+bu}\right| + C$

11. $\displaystyle\int \frac{1}{u(a+bu)^2}\,du = \frac{1}{a}\left(\frac{1}{a+bu} + \frac{1}{a}\ln\left|\frac{u}{a+bu}\right|\right) + C$

12. $\displaystyle\int \frac{1}{u^2(a+bu)}\,du = -\frac{1}{a}\left(\frac{1}{u} + \frac{b}{a}\ln\left|\frac{u}{a+bu}\right|\right) + C$

13. $\displaystyle\int \frac{1}{u^2(a+bu)^2}\,du = -\frac{1}{a^2}\left[\frac{a+2bu}{u(a+bu)} + \frac{2b}{a}\ln\left|\frac{u}{a+bu}\right|\right] + C$

Forms Involving $a + bu + cu^2,\ b^2 \neq 4ac$

14. $$\int \frac{1}{a+bu+cu^2}\,du = \begin{cases} \dfrac{2}{\sqrt{4ac-b^2}}\arctan\dfrac{2cu+b}{\sqrt{4ac-b^2}} + C, & b^2 < 4ac \\[2ex] \dfrac{1}{\sqrt{b^2-4ac}}\ln\left|\dfrac{2cu+b-\sqrt{b^2-4ac}}{2cu+b+\sqrt{b^2-4ac}}\right| + C, & b^2 > 4ac \end{cases}$$

15. $\displaystyle\int \frac{u}{a+bu+cu^2}\,du = \frac{1}{2c}\left(\ln|a+bu+cu^2| - b\int \frac{1}{a+bu+cu^2}\,du\right)$

Forms Involving $\sqrt{a+bu}$

16. $\displaystyle\int u^n\sqrt{a+bu}\,du = \frac{2}{b(2n+3)}\left[u^n(a+bu)^{3/2} - na\int u^{n-1}\sqrt{a+bu}\,du\right]$

17. $\displaystyle\int \frac{1}{u\sqrt{a+bu}}\,du = \begin{cases} \dfrac{1}{\sqrt{a}} \ln\left|\dfrac{\sqrt{a+bu}-\sqrt{a}}{\sqrt{a+bu}+\sqrt{a}}\right| + C, & a > 0 \\ \dfrac{2}{\sqrt{-a}} \arctan \sqrt{\dfrac{a+bu}{-a}} + C, & a < 0 \end{cases}$

18. $\displaystyle\int \frac{1}{u^n\sqrt{a+bu}}\,du = \frac{-1}{a(n-1)}\left[\frac{\sqrt{a+bu}}{u^{n-1}} + \frac{(2n-3)b}{2}\int \frac{1}{u^{n-1}\sqrt{a+bu}}\,du\right],\ n \neq 1$

19. $\displaystyle\int \frac{\sqrt{a+bu}}{u}\,du = 2\sqrt{a+bu} + a\int \frac{1}{u\sqrt{a+bu}}\,du$

20. $\displaystyle\int \frac{\sqrt{a+bu}}{u^n}\,du = \frac{-1}{a(n-1)}\left[\frac{(a+bu)^{3/2}}{u^{n-1}} + \frac{(2n-5)b}{2}\int \frac{\sqrt{a+bu}}{u^{n-1}}\,du\right],\ n \neq 1$

21. $\displaystyle\int \frac{u}{\sqrt{a+bu}}\,du = \frac{-2(2a-bu)}{3b^2}\sqrt{a+bu} + C$

22. $\displaystyle\int \frac{u^n}{\sqrt{a+bu}}\,du = \frac{2}{(2n+1)b}\left(u^n\sqrt{a+bu} - na\int \frac{u^{n-1}}{\sqrt{a+bu}}\,du\right)$

Forms Involving $a^2 \pm u^2,\ a > 0$

23. $\displaystyle\int \frac{1}{a^2+u^2}\,du = \frac{1}{a}\arctan\frac{u}{a} + C$

24. $\displaystyle\int \frac{1}{u^2-a^2}\,du = -\int \frac{1}{a^2-u^2}\,du = \frac{1}{2a}\ln\left|\frac{u-a}{u+a}\right| + C$

25. $\displaystyle\int \frac{1}{(a^2\pm u^2)^n}\,du = \frac{1}{2a^2(n-1)}\left[\frac{u}{(a^2\pm u^2)^{n-1}} + (2n-3)\int \frac{1}{(a^2\pm u^2)^{n-1}}\,du\right],\ n \neq 1$

Forms Involving $\sqrt{u^2 \pm a^2},\ a > 0$

26. $\displaystyle\int \sqrt{u^2\pm a^2}\,du = \frac{1}{2}\left(u\sqrt{u^2\pm a^2} \pm a^2\ln\left|u+\sqrt{u^2\pm a^2}\right|\right) + C$

27. $\displaystyle\int u^2\sqrt{u^2\pm a^2}\,du = \frac{1}{8}\left[u(2u^2\pm a^2)\sqrt{u^2\pm a^2} - a^4\ln\left|u+\sqrt{u^2\pm a^2}\right|\right] + C$

28. $\displaystyle\int \frac{\sqrt{u^2+a^2}}{u}\,du = \sqrt{u^2+a^2} - a\ln\left|\frac{a+\sqrt{u^2+a^2}}{u}\right| + C$

29. $\displaystyle\int \frac{\sqrt{u^2-a^2}}{u}\,du = \sqrt{u^2-a^2} - a\,\mathrm{arcsec}\,\frac{|u|}{a} + C$

30. $\displaystyle\int \frac{\sqrt{u^2\pm a^2}}{u^2}\,du = \frac{-\sqrt{u^2\pm a^2}}{u} + \ln\left|u+\sqrt{u^2\pm a^2}\right| + C$

31. $\displaystyle\int \frac{1}{\sqrt{u^2\pm a^2}}\,du = \ln\left|u+\sqrt{u^2\pm a^2}\right| + C$

32. $\displaystyle\int \frac{1}{u\sqrt{u^2+a^2}}\,du = \frac{-1}{a}\ln\left|\frac{a+\sqrt{u^2+a^2}}{u}\right| + C$

33. $\displaystyle\int \frac{1}{u\sqrt{u^2-a^2}}\,du = \frac{1}{a}\,\mathrm{arcsec}\,\frac{|u|}{a} + C$

34. $\displaystyle\int \frac{u^2}{\sqrt{u^2\pm a^2}}\,du = \frac{1}{2}\left(u\sqrt{u^2\pm a^2} \mp a^2\ln\left|u+\sqrt{u^2\pm a^2}\right|\right) + C$

35. $\displaystyle\int \frac{1}{u^2\sqrt{u^2\pm a^2}}\,du = \mp\frac{\sqrt{u^2\pm a^2}}{a^2u} + C$

36. $\displaystyle\int \frac{1}{(u^2\pm a^2)^{3/2}}\,du = \frac{\pm u}{a^2\sqrt{u^2\pm a^2}} + C$

Forms Involving $\sqrt{a^2 - u^2},\ a > 0$

37. $\displaystyle\int \sqrt{a^2-u^2}\,du = \frac{1}{2}\left(u\sqrt{a^2-u^2} + a^2\arcsin\frac{u}{a}\right) + C$

38. $\displaystyle\int u^2\sqrt{a^2-u^2}\,du = \frac{1}{8}\left[u(2u^2-a^2)\sqrt{a^2-u^2} + a^4\arcsin\frac{u}{a}\right] + C$

39. $\displaystyle\int \frac{\sqrt{a^2 - u^2}}{u}\,du = \sqrt{a^2 - u^2} - a\ln\left|\frac{a + \sqrt{a^2 - u^2}}{u}\right| + C$

40. $\displaystyle\int \frac{\sqrt{a^2 - u^2}}{u^2}\,du = \frac{-\sqrt{a^2 - u^2}}{u} - \arcsin\frac{u}{a} + C$

41. $\displaystyle\int \frac{1}{\sqrt{a^2 - u^2}}\,du = \arcsin\frac{u}{a} + C$

42. $\displaystyle\int \frac{1}{u\sqrt{a^2 - u^2}}\,du = \frac{-1}{a}\ln\left|\frac{a + \sqrt{a^2 - u^2}}{u}\right| + C$

43. $\displaystyle\int \frac{u^2}{\sqrt{a^2 - u^2}}\,du = \frac{1}{2}\left(-u\sqrt{a^2 - u^2} + a^2\arcsin\frac{u}{a}\right) + C$

44. $\displaystyle\int \frac{1}{u^2\sqrt{a^2 - u^2}}\,du = \frac{-\sqrt{a^2 - u^2}}{a^2u} + C$

45. $\displaystyle\int \frac{1}{(a^2 - u^2)^{3/2}}\,du = \frac{u}{a^2\sqrt{a^2 - u^2}} + C$

Forms Involving $\sin u$ or $\cos u$

46. $\displaystyle\int \sin u\,du = -\cos u + C$

47. $\displaystyle\int \cos u\,du = \sin u + C$

48. $\displaystyle\int \sin^2 u\,du = \frac{1}{2}(u - \sin u\cos u) + C$

49. $\displaystyle\int \cos^2 u\,du = \frac{1}{2}(u + \sin u\cos u) + C$

50. $\displaystyle\int \sin^n u\,du = -\frac{\sin^{n-1}u\cos u}{n} + \frac{n-1}{n}\int \sin^{n-2}u\,du$

51. $\displaystyle\int \cos^n u\,du = \frac{\cos^{n-1}u\sin u}{n} + \frac{n-1}{n}\int \cos^{n-2}u\,du$

52. $\displaystyle\int u\sin u\,du = \sin u - u\cos u + C$

53. $\displaystyle\int u\cos u\,du = \cos u + u\sin u + C$

54. $\displaystyle\int u^n\sin u\,du = -u^n\cos u + n\int u^{n-1}\cos u\,du$

55. $\displaystyle\int u^n\cos u\,du = u^n\sin u - n\int u^{n-1}\sin u\,du$

56. $\displaystyle\int \frac{1}{1 \pm \sin u}\,du = \tan u \mp \sec u + C$

57. $\displaystyle\int \frac{1}{1 \pm \cos u}\,du = -\cot u \pm \csc u + C$

58. $\displaystyle\int \frac{1}{\sin u\cos u}\,du = \ln|\tan u| + C$

Forms Involving $\tan u$, $\cot u$, $\sec u$, $\csc u$

59. $\displaystyle\int \tan u\,du = -\ln|\cos u| + C$

60. $\displaystyle\int \cot u\,du = \ln|\sin u| + C$

61. $\displaystyle\int \sec u\,du = \ln|\sec u + \tan u| + C$

62. $\displaystyle\int \csc u\,du = \ln|\csc u - \cot u| + C$ or $\displaystyle\int \csc u\,du = -\ln|\csc u + \cot u| + C$

63. $\displaystyle\int \tan^2 u\,du = -u + \tan u + C$

64. $\displaystyle\int \cot^2 u\,du = -u - \cot u + C$

65. $\displaystyle\int \sec^2 u\,du = \tan u + C$

66. $\displaystyle\int \csc^2 u\,du = -\cot u + C$

67. $\displaystyle\int \tan^n u\,du = \frac{\tan^{n-1}u}{n-1} - \int \tan^{n-2}u\,du,\ n \neq 1$

68. $\displaystyle\int \cot^n u\,du = -\frac{\cot^{n-1}u}{n-1} - \int (\cot^{n-2}u)\,du,\ n \neq 1$

69. $\displaystyle\int \sec^n u\,du = \frac{\sec^{n-2}u\tan u}{n-1} + \frac{n-2}{n-1}\int \sec^{n-2}u\,du,\ n \neq 1$

70. $\displaystyle\int \csc^n u\,du = -\frac{\csc^{n-2}u\cot u}{n-1} + \frac{n-2}{n-1}\int \csc^{n-2}u\,du,\ n \neq 1$

71. $\displaystyle\int \frac{1}{1 \pm \tan u}\,du = \frac{1}{2}\left(u \pm \ln|\cos u \pm \sin u|\right) + C$

72. $\displaystyle\int \frac{1}{1 \pm \cot u}\,du = \frac{1}{2}\left(u \mp \ln|\sin u \pm \cos u|\right) + C$

73. $\displaystyle\int \frac{1}{1 \pm \sec u}\,du = u + \cot u \mp \csc u + C$

74. $\displaystyle\int \frac{1}{1 \pm \csc u}\,du = u - \tan u \pm \sec u + C$

Forms Involving Inverse Trigonometric Functions

75. $\displaystyle\int \arcsin u\,du = u \arcsin u + \sqrt{1 - u^2} + C$

76. $\displaystyle\int \arccos u\,du = u \arccos u - \sqrt{1 - u^2} + C$

77. $\displaystyle\int \arctan u\,du = u \arctan u - \ln\sqrt{1 + u^2} + C$

78. $\displaystyle\int \operatorname{arccot} u\,du = u \operatorname{arccot} u + \ln\sqrt{1 + u^2} + C$

79. $\displaystyle\int \operatorname{arcsec} u\,du = u \operatorname{arcsec} u - \ln\left|u + \sqrt{u^2 - 1}\right| + C$

80. $\displaystyle\int \operatorname{arccsc} u\,du = u \operatorname{arccsc} u + \ln\left|u + \sqrt{u^2 - 1}\right| + C$

Forms Involving e^u

81. $\displaystyle\int e^u\,du = e^u + C$

82. $\displaystyle\int ue^u\,du = (u - 1)e^u + C$

83. $\displaystyle\int u^n e^u\,du = u^n e^u - n\int u^{n-1}e^u\,du$

84. $\displaystyle\int \frac{1}{1 + e^u}\,du = u - \ln(1 + e^u) + C$

85. $\displaystyle\int e^{au}\sin bu\,du = \frac{e^{au}}{a^2 + b^2}(a \sin bu - b \cos bu) + C$

86. $\displaystyle\int e^{au}\cos bu\,du = \frac{e^{au}}{a^2 + b^2}(a \cos bu + b \sin bu) + C$

Forms Involving $\ln u$

87. $\displaystyle\int \ln u\,du = u(-1 + \ln u) + C$

88. $\displaystyle\int u \ln u\,du = \frac{u^2}{4}(-1 + 2\ln u) + C$

89. $\displaystyle\int u^n \ln u\,du = \frac{u^{n+1}}{(n + 1)^2}[-1 + (n + 1)\ln u] + C,\ n \neq -1$

90. $\displaystyle\int (\ln u)^2\,du = u[2 - 2\ln u + (\ln u)^2] + C$

91. $\displaystyle\int (\ln u)^n\,du = u(\ln u)^n - n\int (\ln u)^{n-1}\,du$

Forms Involving Hyperbolic Functions

92. $\displaystyle\int \cosh u\,du = \sinh u + C$

93. $\displaystyle\int \sinh u\,du = \cosh u + C$

94. $\displaystyle\int \operatorname{sech}^2 u\,du = \tanh u + C$

95. $\displaystyle\int \operatorname{csch}^2 u\,du = -\coth u + C$

96. $\displaystyle\int \operatorname{sech} u \tanh u\,du = -\operatorname{sech} u + C$

97. $\displaystyle\int \operatorname{csch} u \coth u\,du = -\operatorname{csch} u + C$

Forms Involving Inverse Hyperbolic Functions (in logarithmic form)

98. $\displaystyle\int \frac{du}{\sqrt{u^2 \pm a^2}} = \ln\left(u + \sqrt{u^2 \pm a^2}\right) + C$

99. $\displaystyle\int \frac{du}{a^2 - u^2} = \frac{1}{2a}\ln\left|\frac{a + u}{a - u}\right| + C$

100. $\displaystyle\int \frac{du}{u\sqrt{a^2 \pm u^2}} = -\frac{1}{a}\ln\frac{a + \sqrt{a^2 \pm u^2}}{|u|} + C$

Answers to Odd-Numbered Exercises

Chapter 11

Section 11.1 (page 771)

1. 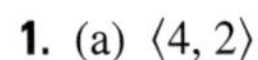 (a) $\langle 4, 2 \rangle$

(b)

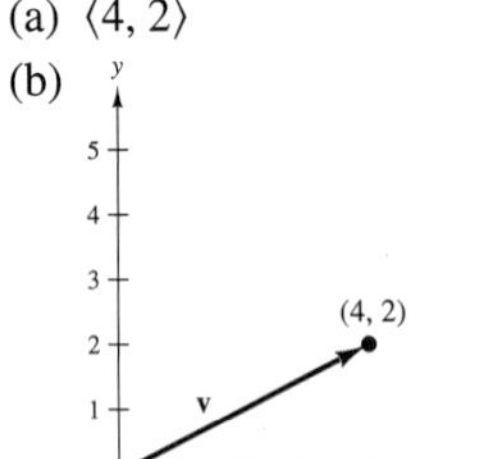

3. (a) $\langle -6, 0 \rangle$

 (b)

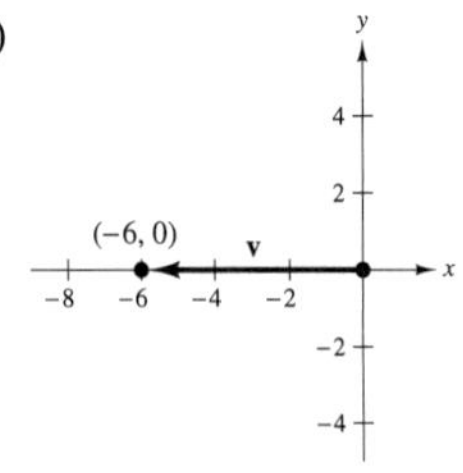

5. $\mathbf{u} = \mathbf{v} = \langle 2, 4 \rangle$ **7.** $\mathbf{u} = \mathbf{v} = \langle 6, -5 \rangle$

9. (a) and (d)

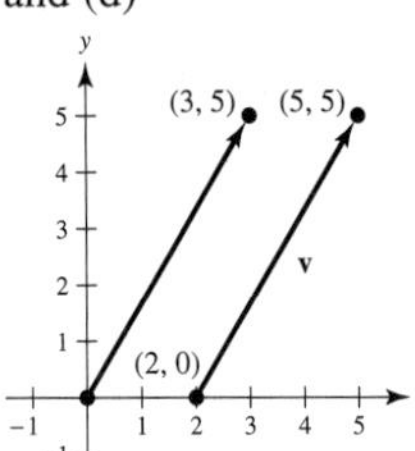

(b) $\langle 3, 5 \rangle$

(c) $\mathbf{v} = 3\mathbf{i} + 5\mathbf{j}$

11. (a) and (d)

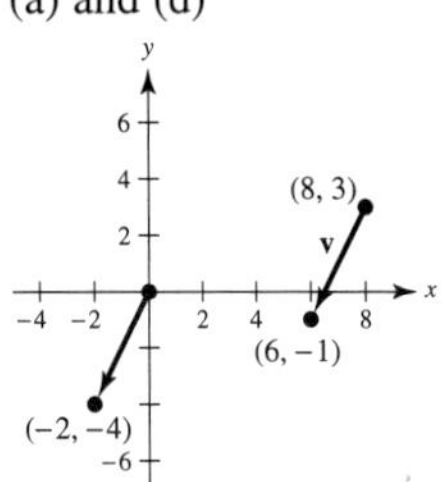

(b) $\langle -2, -4 \rangle$

(c) $\mathbf{v} = -2\mathbf{i} - 4\mathbf{j}$

13. (a) and (d)

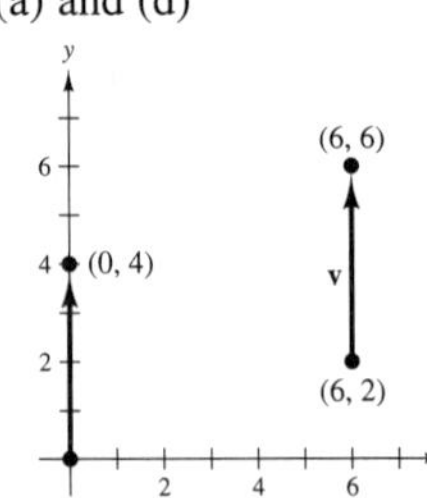

(b) $\langle 0, 4 \rangle$ (c) $\mathbf{v} = 4\mathbf{j}$

15. (a) and (d)

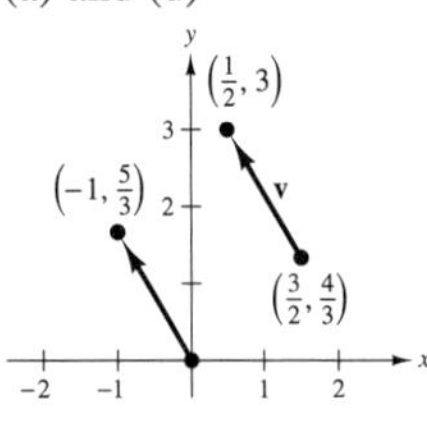

(b) $\left\langle -1, \frac{5}{3} \right\rangle$

(c) $\mathbf{v} = -\mathbf{i} + \frac{5}{3}\mathbf{j}$

17. (a) $\langle 6, 10 \rangle$

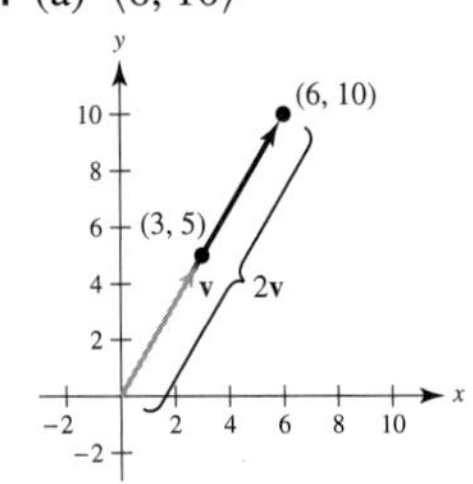

(b) $\langle -9, -15 \rangle$

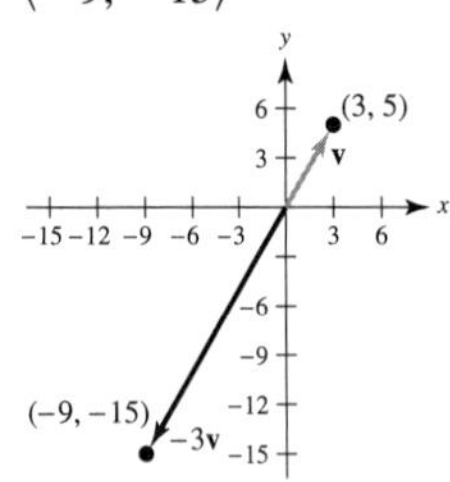

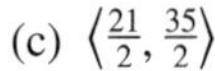 (c) $\left\langle \frac{21}{2}, \frac{35}{2} \right\rangle$

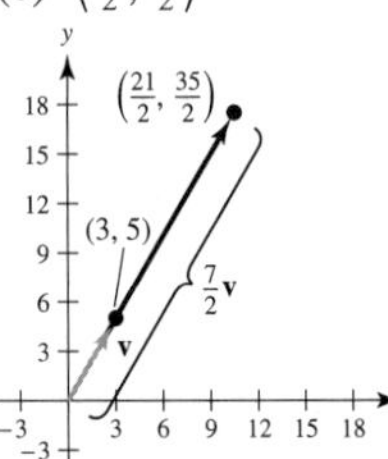

(d) $\left\langle 2, \frac{10}{3} \right\rangle$

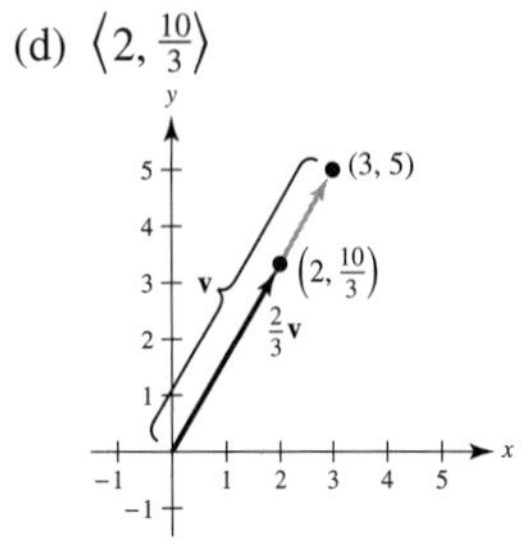

19.

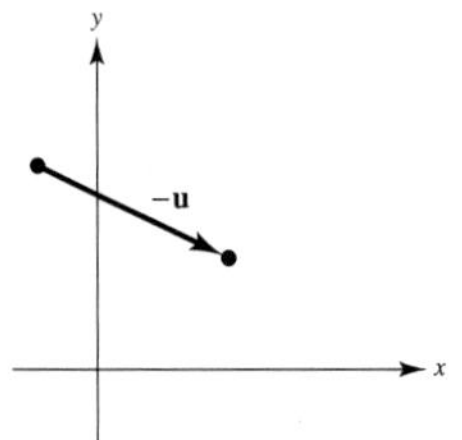

21.

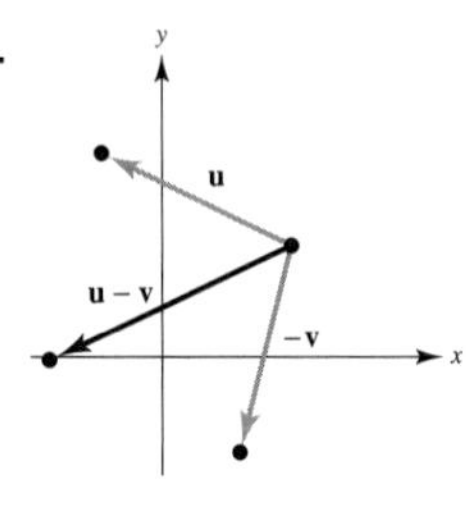

23. (a) $\left\langle \frac{8}{3}, 6 \right\rangle$ (b) $\langle -2, -14 \rangle$ (c) $\langle 18, -7 \rangle$

25. $\left\langle 3, -\frac{3}{2} \right\rangle$

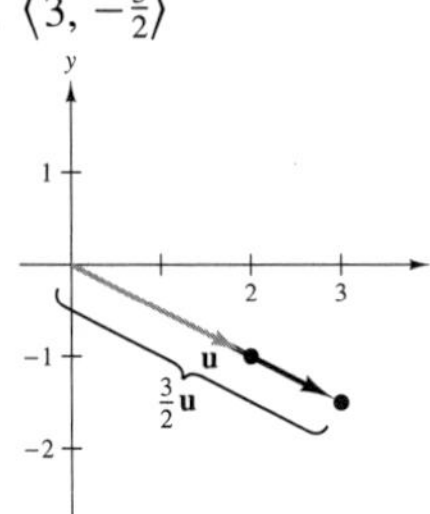

27. $\langle 4, 3 \rangle$

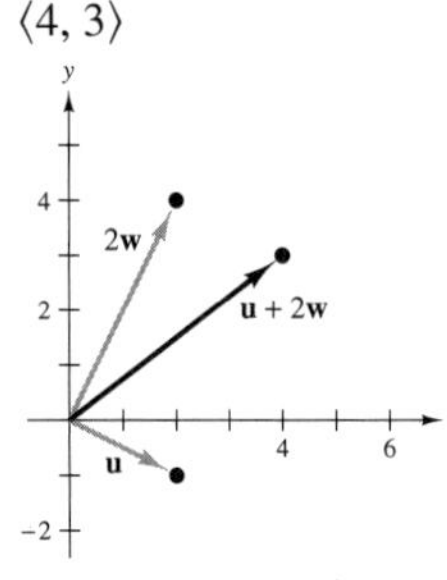

29. $(3, 5)$ **31.** 7 **33.** 5 **35.** $\sqrt{61}$

37. $\left\langle \sqrt{17}/17, 4\sqrt{17}/17 \right\rangle$ **39.** $\left\langle 3\sqrt{34}/34, 5\sqrt{34}/34 \right\rangle$

41. (a) $\sqrt{2}$ (b) $\sqrt{5}$ (c) 1 (d) 1 (e) 1 (f) 1

43. (a) $\sqrt{5}/2$ (b) $\sqrt{13}$ (c) $\sqrt{85}/2$ (d) 1 (e) 1 (f) 1

45.

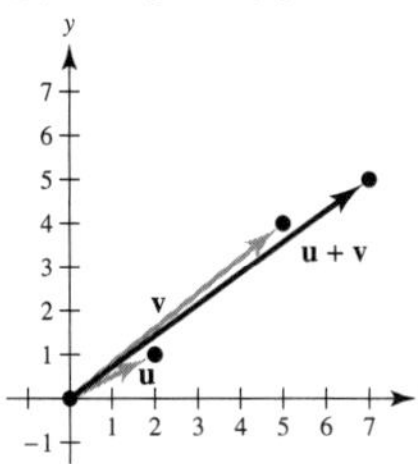

$\|\mathbf{u}\| + \|\mathbf{v}\| = \sqrt{5} + \sqrt{41}$ and $\|\mathbf{u} + \mathbf{v}\| = \sqrt{74}$

$\sqrt{74} \le \sqrt{5} + \sqrt{41}$

47. $\langle 0, 6 \rangle$ **49.** $\left\langle -\sqrt{5}, 2\sqrt{5} \right\rangle$ **51.** $\langle 3, 0 \rangle$ **53.** $\left\langle -\sqrt{3}, 1 \right\rangle$

55. $\left\langle \dfrac{2 + 3\sqrt{2}}{2}, \dfrac{3\sqrt{2}}{2} \right\rangle$ **57.** $\langle 2\cos 4 + \cos 2, 2\sin 4 + \sin 2 \rangle$

59. Answers will vary. Example: A scalar is a single real number such as 2. A vector is a line segment having both direction and magnitude. The vector $\left\langle \sqrt{3}, 1 \right\rangle$, given in component form, has a direction of $\pi/6$ and a magnitude of 2.

61. (a) Vector; has magnitude and direction
(b) Scalar; has only magnitude

63. $a = 1, b = 1$ **65.** $a = 1, b = 2$ **67.** $a = \frac{2}{3}, b = \frac{1}{3}$

69. (a) $\pm\left(1/\sqrt{37}\right)\langle 1, 6 \rangle$
(b) $\pm\left(1/\sqrt{37}\right)\langle 6, -1 \rangle$

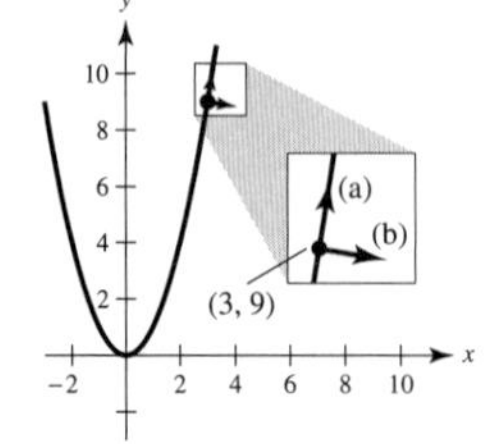

71. (a) $\pm\left(1/\sqrt{10}\right)\langle 1, 3 \rangle$
(b) $\pm\left(1/\sqrt{10}\right)\langle 3, -1 \rangle$

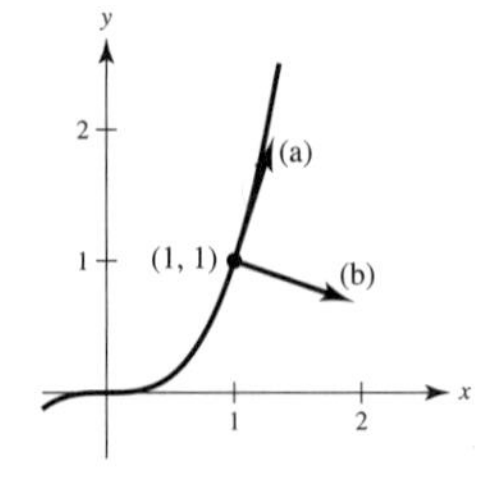

73. (a) $\pm\frac{1}{5}\langle -4, 3\rangle$ (b) $\pm\frac{1}{5}\langle 3, 4\rangle$

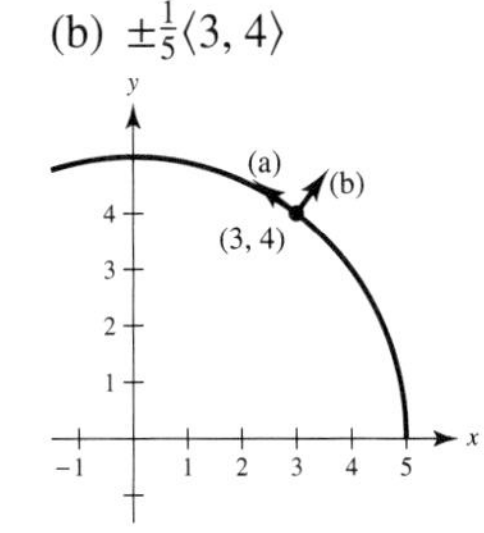

75. $\langle -\sqrt{2}/2, \sqrt{2}/2\rangle$

77. (a)–(c) Answers will vary.
(d) Magnitude ≈ 63.5, direction $\approx -8.26°$

79. 1.33, 132.5° **81.** 10.7°, 584.6 lb **83.** 71.3°, 228.5 lb

85. (a) $\theta = 0°$ (b) $\theta = 180°$
(c) No, the resultant can only be less than or equal to the sum.

87. $(-4, -1), (6, 5), (10, 3)$

89. Tension in cable $AC \approx 2638.2$ lb
Tension in cable $BC \approx 1958.1$ lb

91. Horizontal: 1193.43 ft/sec
Vertical: 125.43 ft/sec

93. 38.3° north of west
882.9 km/h

95. True **97.** True **99.** False. $\|a\mathbf{i} + b\mathbf{j}\| = \sqrt{2}|a|$

101–103. Proofs **105.** $x^2 + y^2 = 25$

Section 11.2 (page 780)

1. $A(2, 3, 4)$, $B(-1, -2, 2)$

3.

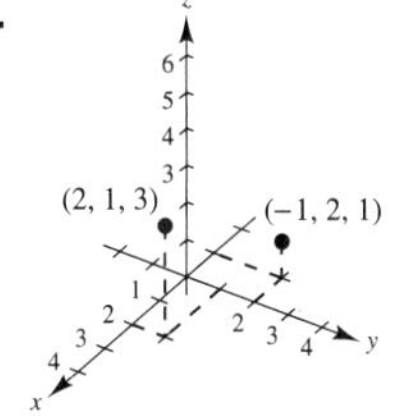

5.

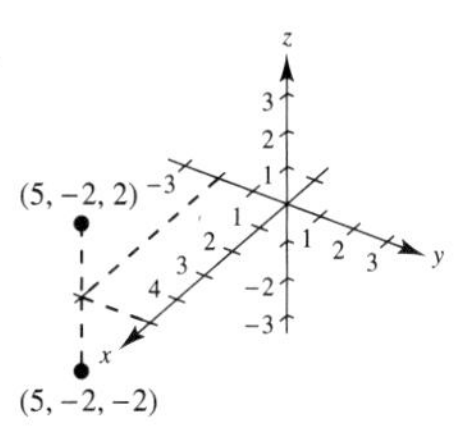

7. $(-3, 4, 5)$ **9.** $(12, 0, 0)$ **11.** 0

13. Six units above the xy-plane

15. Three units behind the yz-plane

17. To the left of the xz-plane

19. Within three units of the xz-plane

21. Three units below the xy-plane, and below either quadrant I or III

23. Above the xy-plane and above quadrants II or IV, *or* below the xy-plane and below quadrants I or III

25. $\sqrt{69}$ **27.** $\sqrt{61}$ **29.** $7, 7\sqrt{5}, 14$; Right triangle

31. $\sqrt{41}, \sqrt{41}, \sqrt{14}$; Isosceles triangle

33. $(0, 0, 9), (2, 6, 12), (6, 4, -3)$

35. $\left(\frac{3}{2}, -3, 5\right)$ **37.** $(x - 0)^2 + (y - 2)^2 + (z - 5)^2 = 4$

39. $(x - 1)^2 + (y - 3)^2 + (z - 0)^2 = 10$

41. $(x - 1)^2 + (y + 3)^2 + (z + 4)^2 = 25$
Center: $(1, -3, -4)$
Radius: 5

43. $\left(x - \frac{1}{3}\right)^2 + (y + 1)^2 + z^2 = 1$
Center: $\left(\frac{1}{3}, -1, 0\right)$
Radius: 1

45. A solid sphere with center $(0, 0, 0)$ and radius 6

47. Interior of sphere of radius 4 centered at $(2, -3, 4)$

49. (a) $\langle -2, 2, 2\rangle$
(b) $\mathbf{v} = -2\mathbf{i} + 2\mathbf{j} + 2\mathbf{k}$
(c)

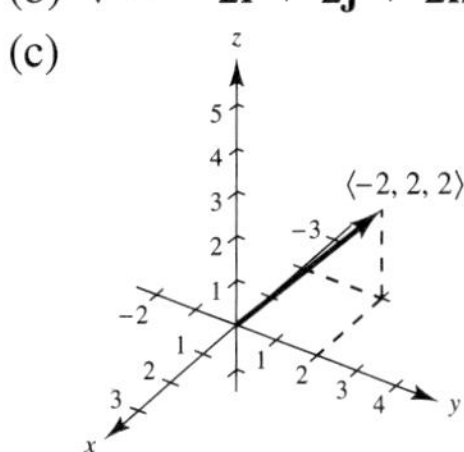

51. (a) $\langle -3, 0, 3\rangle$
(b) $\mathbf{v} = -3\mathbf{i} + 3\mathbf{k}$
(c)

53. $\mathbf{v} = \langle 1, -1, 6\rangle$
$\|\mathbf{v}\| = \sqrt{38}$
$\mathbf{u} = \dfrac{1}{\sqrt{38}}\langle 1, -1, 6\rangle$

55. $\mathbf{v} = \langle -1, 0, -1\rangle$
$\|\mathbf{v}\| = \sqrt{2}$
$\mathbf{u} = \dfrac{1}{\sqrt{2}}\langle -1, 0, -1\rangle$

57. (a) and (d)

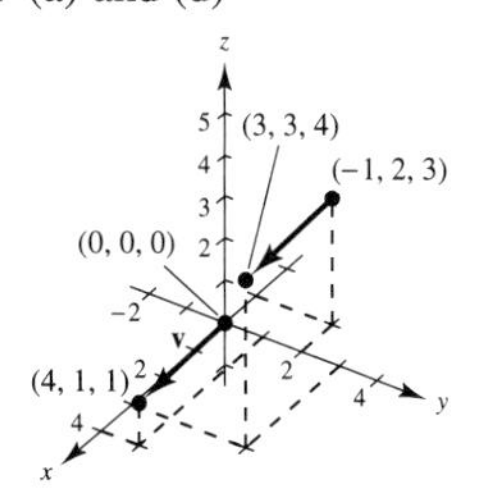

(b) $\langle 4, 1, 1\rangle$ (c) $\mathbf{v} = 4\mathbf{i} + \mathbf{j} + \mathbf{k}$

59. $(3, 1, 8)$

61. (a) (b)

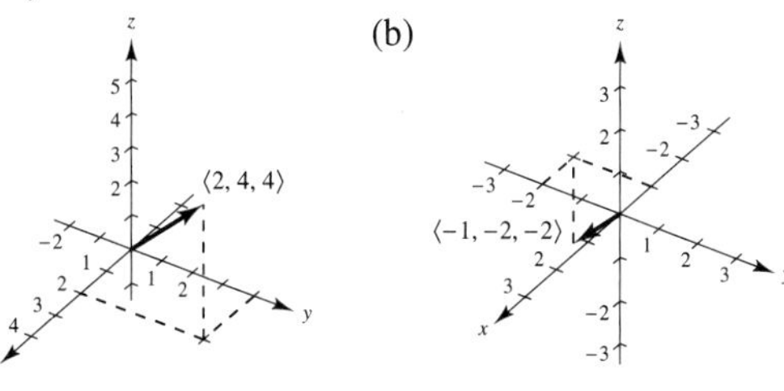

(c)

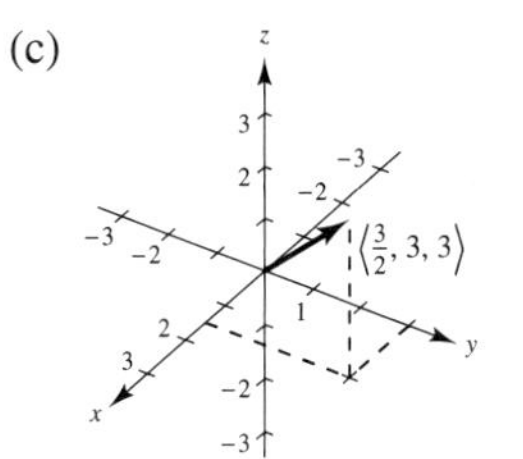

(d)

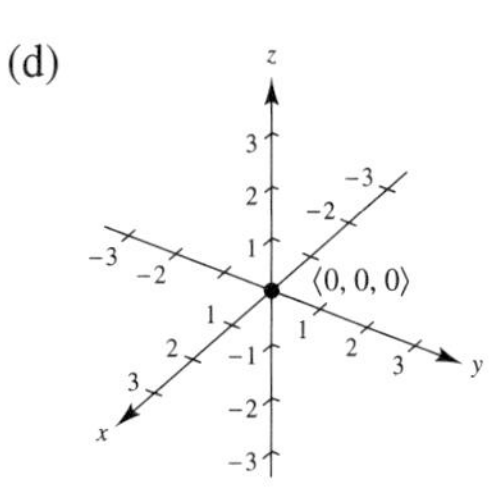

63. $\langle -1, 0, 4\rangle$ **65.** $\langle 6, 12, 6\rangle$ **67.** $\left\langle \frac{7}{2}, 3, \frac{5}{2}\right\rangle$

69. a and b **71.** a **73.** Collinear **75.** Not collinear

77. $\overrightarrow{AB} = \langle 1, 2, 3\rangle$
$\overrightarrow{CD} = \langle 1, 2, 3\rangle$
$\overrightarrow{BD} = \langle -2, 1, 1\rangle$
$\overrightarrow{AC} = \langle -2, 1, 1\rangle$
Because $\overrightarrow{AB} = \overrightarrow{CD}$ and $\overrightarrow{BD} = \overrightarrow{AC}$, the given points form the vertices of a parallelogram.

79. 0 **81.** $\sqrt{34}$ **83.** $\sqrt{14}$

85. (a) $\frac{1}{3}\langle 2, -1, 2\rangle$ (b) $-\frac{1}{3}\langle 2, -1, 2\rangle$

87. (a) $(1/\sqrt{38})\langle 3, 2, -5\rangle$ (b) $-(1/\sqrt{38})\langle 3, 2, -5\rangle$

89. (a)–(d) Answers will vary.
(e) $\mathbf{u} + \mathbf{v} = \langle 4, 7.5, -2\rangle$
$\|\mathbf{u} + \mathbf{v}\| \approx 8.732$
$\|\mathbf{u}\| \approx 5.099$
$\|\mathbf{v}\| \approx 9.014$

91. $\pm\frac{7}{3}$ **93.** $\langle 0, 10/\sqrt{2}, 10/\sqrt{2}\rangle$ **95.** $\langle 1, -1, \frac{1}{2}\rangle$

97.

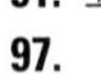

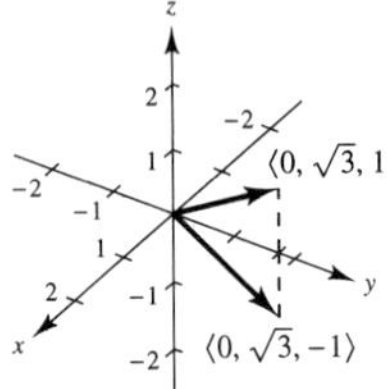

$\langle 0, \sqrt{3}, \pm 1\rangle$

99. $(2, -1, 2)$

101. (a)

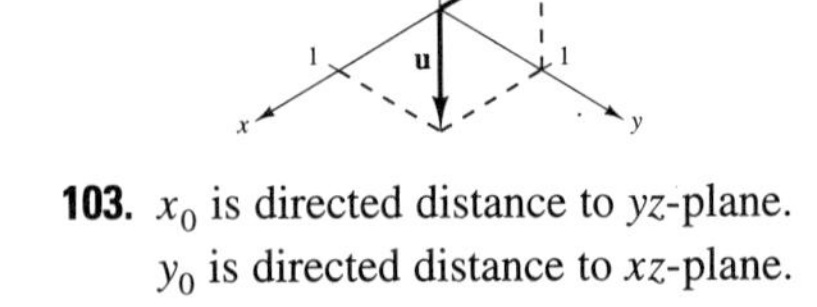

(b) $a = 0, a + b = 0, b = 0$
(c) $a = 1, a + b = 2, b = 1$
(d) Not possible

103. x_0 is directed distance to yz-plane.
y_0 is directed distance to xz-plane.
z_0 is directed distance to xy-plane.

105. $(x - x_0)^2 + (y - y_0)^2 + (z - z_0)^2 = r^2$ **107.** 0

109. (a) $T = 8L/\sqrt{L^2 - 18^2}, L > 18$

(b)

L	20	25	30	35	40	45	50
T	18.4	11.5	10	9.3	9.0	8.7	8.6

(c)

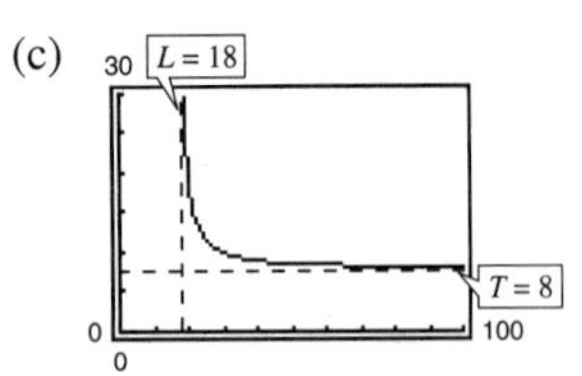

(d) Proof (e) 30 in.

111. $(\sqrt{3}/3)\langle 1, 1, 1\rangle$

113. Tension in cable AB: 202.919 N
Tension in cable AC: 157.909 N
Tension in cable AD: 226.521 N

115. $\left(x - \frac{4}{3}\right)^2 + (y - 3)^2 + \left(z + \frac{1}{3}\right)^2 = \frac{44}{9}$

Section 11.3 (page 789)

1. (a) 17 (b) 25 (c) 25 (d) $\langle -17, 85\rangle$ (e) 34

3. (a) -26 (b) 52 (c) 52 (d) $\langle 78, -52\rangle$ (e) -52

5. (a) 2 (b) 29 (c) 29 (d) $\langle 0, 12, 10\rangle$ (e) 4

7. (a) 1 (b) 6 (c) 6 (d) $\mathbf{i} - \mathbf{k}$ (e) 2

9. 20 **11.** $\pi/2$ **13.** $\arccos\left[-1/(5\sqrt{2})\right] \approx 98.1°$

15. $\arccos(\sqrt{2}/3) \approx 61.9°$ **17.** $\arccos(-8\sqrt{13}/65) \approx 116.3°$

19. Neither **21.** Orthogonal **23.** Neither

25. Orthogonal **27.** Right triangle; answers will vary.

29. Acute triangle; answers will vary.

31. $\cos\alpha = \frac{1}{3}$
$\cos\beta = \frac{2}{3}$
$\cos\gamma = \frac{2}{3}$

33. $\cos\alpha = 0$
$\cos\beta = 3/\sqrt{13}$
$\cos\gamma = -2/\sqrt{13}$

35. $\alpha \approx 43.3°, \beta \approx 61.0°, \gamma \approx 119.0°$

37. $\alpha \approx 100.5°, \beta \approx 24.1°, \gamma \approx 68.6°$

39. Magnitude: 124.310 lb
$\alpha \approx 29.48°, \beta \approx 61.39°, \gamma \approx 96.53°$

41. $\alpha = 90°, \beta = 45°, \gamma = 45°$ **43.** (a) $\langle 2, 8\rangle$ (b) $\langle 4, -1\rangle$

45. (a) $\left\langle \frac{5}{2}, \frac{1}{2}\right\rangle$ (b) $\left\langle -\frac{1}{2}, \frac{5}{2}\right\rangle$ **47.** (a) $\langle -2, 2, 2\rangle$ (b) $\langle 2, 1, 1\rangle$

49. (a) $\left\langle 0, \frac{33}{25}, \frac{44}{25}\right\rangle$ (b) $\left\langle 2, -\frac{8}{25}, \frac{6}{25}\right\rangle$

51. See "Definition of Dot Product," page 783.

53. (a) and (b) are defined. (c) and (d) are not defined because it is not possible to find the dot product of a scalar and a vector or to add a scalar to a vector.

55. See Figure 11.29 on page 787.

57. Yes.

$$\left\|\frac{\mathbf{u}\cdot\mathbf{v}}{\|\mathbf{v}\|^2}\mathbf{v}\right\| = \left\|\frac{\mathbf{v}\cdot\mathbf{u}}{\|\mathbf{u}\|^2}\mathbf{u}\right\|$$
$$|\mathbf{u}\cdot\mathbf{v}|\frac{\|\mathbf{v}\|}{\|\mathbf{v}\|^2} = |\mathbf{v}\cdot\mathbf{u}|\frac{\|\mathbf{u}\|}{\|\mathbf{u}\|^2}$$
$$\frac{1}{\|\mathbf{v}\|} = \frac{1}{\|\mathbf{u}\|}$$
$$\|\mathbf{u}\| = \|\mathbf{v}\|$$

59. \$12,351.25; Total revenue **61.** (a)–(c) Answers will vary.

63. Answers will vary. **65.** $\mathbf{u}$

67. Answers will vary. Example: $\langle 12, 2\rangle$ and $\langle -12, -2\rangle$

69. Answers will vary. Example: $\langle 2, 0, 3\rangle$ and $\langle -2, 0, -3\rangle$

71. (a) 8335.1 lb (b) 47,270.8 lb

73. 425 ft-lb **75.** 2900.2 km-N

77. False. For example, $\langle 1, 1\rangle \cdot \langle 2, 3\rangle = 5$ and $\langle 1, 1\rangle \cdot \langle 1, 4\rangle = 5$, but $\langle 2, 3\rangle \neq \langle 1, 4\rangle$.

79. $\arccos(1/\sqrt{3}) \approx 54.7°$

81. (a) $(0, 0), (1, 1)$
(b) To $y = x^2$ at $(1, 1)$: $\langle \pm\sqrt{5}/5, \pm 2\sqrt{5}/5\rangle$
To $y = x^{1/3}$ at $(1, 1)$: $\langle \pm 3\sqrt{10}/10, \pm\sqrt{10}/10\rangle$
To $y = x^2$ at $(0, 0)$: $\langle \pm 1, 0\rangle$
To $y = x^{1/3}$ at $(0, 0)$: $\langle 0, \pm 1\rangle$
(c) At $(1, 1)$: $\theta = 45°$
At $(0, 0)$: $\theta = 90°$

83. (a) $(-1, 0), (1, 0)$
(b) To $y = 1 - x^2$ at $(1, 0)$: $\langle \pm\sqrt{5}/5, \mp 2\sqrt{5}/5 \rangle$
To $y = x^2 - 1$ at $(1, 0)$: $\langle \pm\sqrt{5}/5, \pm 2\sqrt{5}/5 \rangle$
To $y = 1 - x^2$ at $(-1, 0)$: $\langle \pm\sqrt{5}/5, \pm 2\sqrt{5}/5 \rangle$
To $y = x^2 - 1$ at $(-1, 0)$: $\langle \pm\sqrt{5}/5, \mp 2\sqrt{5}/5 \rangle$
(c) At $(1, 0)$: $\theta = 53.13°$
At $(-1, 0)$: $\theta = 53.13°$

85. Proof

87. (a)

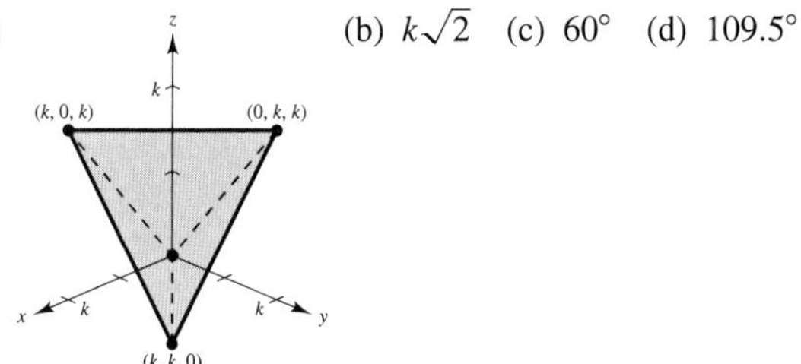

(b) $k\sqrt{2}$ (c) $60°$ (d) $109.5°$

89–91. Proofs

Section 11.4 (page 798)

1. $-\mathbf{k}$

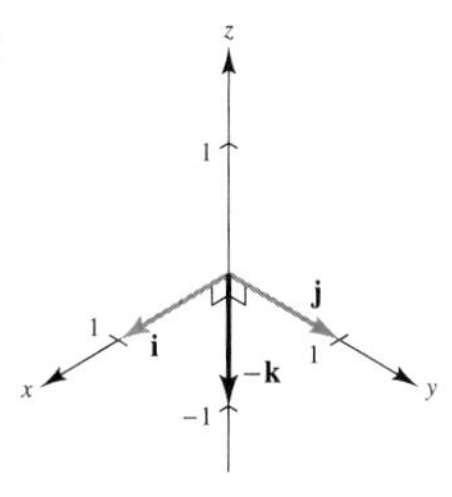

3. $\mathbf{i}$

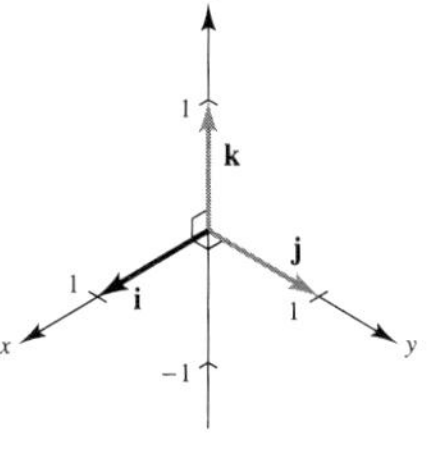

5. $-\mathbf{j}$

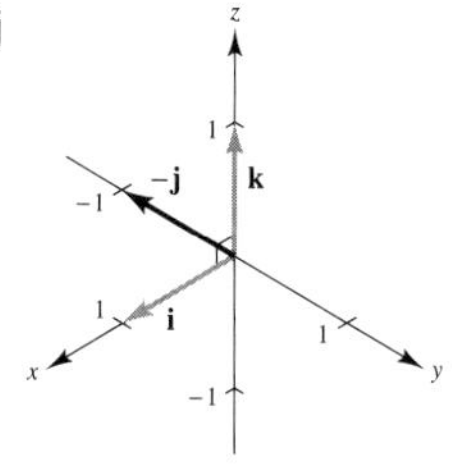

7. (a) $20\mathbf{i} + 10\mathbf{j} - 16\mathbf{k}$
(b) $-20\mathbf{i} - 10\mathbf{j} + 16\mathbf{k}$
(c) $\mathbf{0}$

9. (a) $17\mathbf{i} - 33\mathbf{j} - 10\mathbf{k}$
(b) $-17\mathbf{i} + 33\mathbf{j} + 10\mathbf{k}$
(c) $\mathbf{0}$

11. $\langle 0, 0, 54 \rangle$ **13.** $\langle -1, -1, -1 \rangle$ **15.** $\langle -2, 3, -1 \rangle$

17.

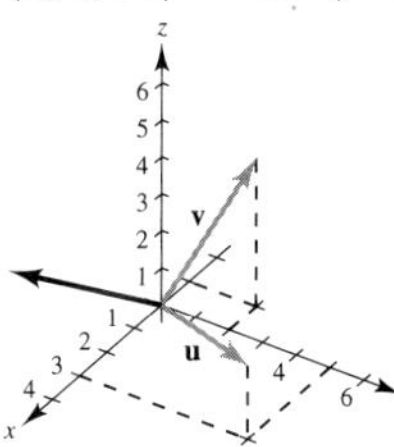

19.

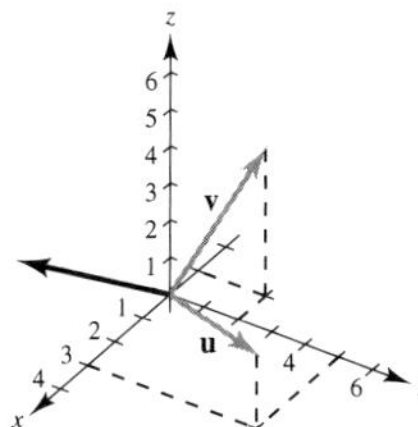

21. $\langle -73.5, 5.5, 44.75 \rangle, \left\langle -\frac{2.94}{\sqrt{11.8961}}, \frac{0.22}{\sqrt{11.8961}}, \frac{1.79}{\sqrt{11.8961}} \right\rangle$

23. $\langle -3.6, -1.4, 1.6 \rangle, \left\langle -\frac{1.8}{\sqrt{4.37}}, -\frac{0.7}{\sqrt{4.37}}, \frac{0.8}{\sqrt{4.37}} \right\rangle$

25. Answers will vary. **27.** 1 **29.** $6\sqrt{5}$ **31.** $9\sqrt{5}$

33. $\frac{11}{2}$ **35.** $\frac{\sqrt{16{,}742}}{2}$ **37.** $10 \cos 40° \approx 7.66$ ft-lb

39. (a) $84 \sin \theta$

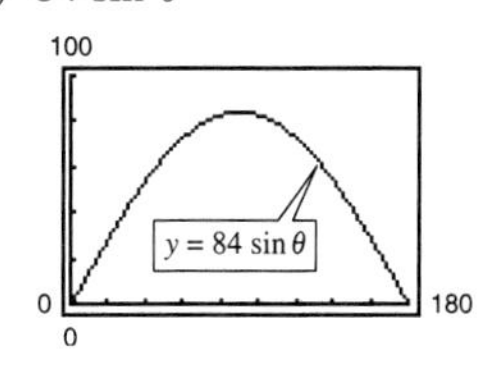

(b) $42\sqrt{2} \approx 59.40$
(c) $\theta = 90°$; This is what should be expected. When $\theta = 90°$, the pipe wrench is horizontal.

41. 1 **43.** 6 **45.** 2 **47.** 75

49. At least one of the vectors is the zero vector.

51. See "Definition of Cross Product of Two Vectors in Space," page 792.

53. The magnitude of the cross product will increase by a factor of 4.

55. False. The cross product of two vectors is not defined in a two-dimensional coordinate system.

57. False. Let $\mathbf{u} = \langle 1, 0, 0 \rangle$, $\mathbf{v} = \langle 1, 0, 0 \rangle$, and $\mathbf{w} = \langle -1, 0, 0 \rangle$. Then $\mathbf{u} \times \mathbf{v} = \mathbf{u} \times \mathbf{w} = \mathbf{0}$, but $\mathbf{v} \neq \mathbf{w}$.

59–67. Proofs

Section 11.5 (page 807)

1. (a)

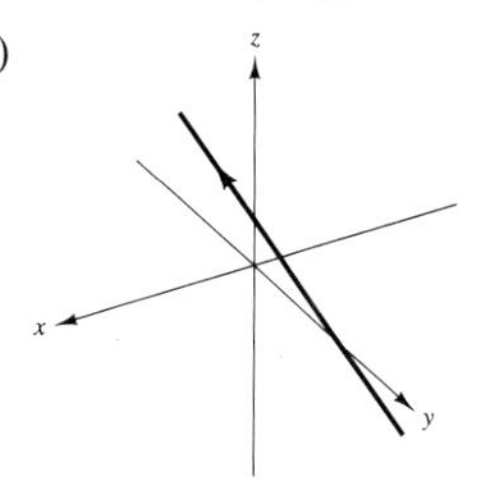

(b) $P = (1, 2, 2), Q = (10, -1, 17), \overrightarrow{PQ} = \langle 9, -3, 15 \rangle$
(There are many correct answers.) The components of the vector and the coefficients of t are proportional because the line is parallel to $\overrightarrow{PQ}$.
(c) $\left(-\frac{1}{5}, \frac{12}{5}, 0\right), (7, 0, 12), \left(0, \frac{7}{3}, \frac{1}{3}\right)$

3. (a) Yes (b) No

	Parametric Equations (a)	*Symmetric Equations (b)*	*Direction Numbers*
5.	$x = 3t$, $y = t$, $z = 5t$	$\frac{x}{3} = y = \frac{z}{5}$	3, 1, 5
7.	$x = -2 + 2t$, $y = 4t$, $z = 3 - 2t$	$\frac{x+2}{2} = \frac{y}{4} = \frac{z-3}{-2}$	2, 4, −2
9.	$x = 1 + 3t$, $y = -2t$, $z = 1 + t$	$\frac{x-1}{3} = \frac{y}{-2} = \frac{z-1}{1}$	3, −2, 1

	Parametric Equations (a)	*Symmetric Equations (b)*	*Direction Numbers*
11.	$x = 5 + 17t$ $y = -3 - 11t$ $z = -2 - 9t$	$\frac{x-5}{17} = \frac{y+3}{-11} = \frac{z+2}{-9}$	$17, -11, -9$
13.	$x = 7 - 10t$ $y = -2 + 2t$ $z = 6$	Not possible	$-10, 2, 0$

15. $x = 2$
$y = 3$
$z = 4 + t$

17. $x = 2 + 3t$
$y = 3 + 2t$
$z = 4 - t$

19. $x = 5 + 2t$
$y = -3 - t$
$z = -4 + 3t$

21. $x = 2 - t$
$y = 1 + t$
$z = 2 + t$

23. $P(3, -1, -2)$; $\mathbf{v} = \langle -1, 2, 0 \rangle$

25. $P(7, -6, -2)$; $\mathbf{v} = \langle 4, 2, 1 \rangle$

27. $L_1 = L_2$ and is parallel to L_3. **29.** L_1 and L_3 are identical.

31. $(2, 3, 1)$; $\cos \theta = 7\sqrt{17}/51$ **33.** Not intersecting

35.

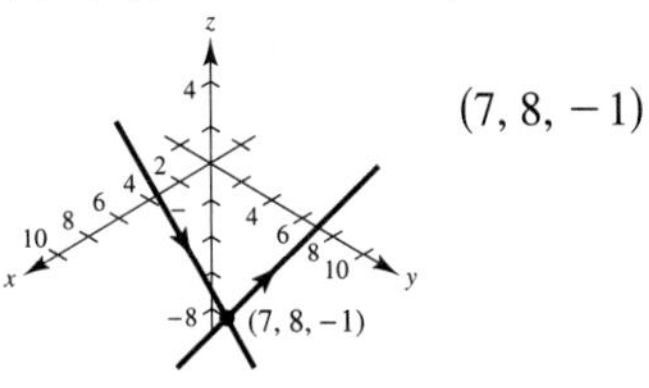

$(7, 8, -1)$

37. (a) $P = (0, 0, -1), Q = (0, -2, 0), R = (3, 4, -1)$
$\overrightarrow{PQ} = \langle 0, -2, 1 \rangle, \overrightarrow{PR} = \langle 3, 4, 0 \rangle$
(There are many correct answers.)
(b) $\overrightarrow{PQ} \times \overrightarrow{PR} = \langle -4, 3, 6 \rangle$
The components of the cross product are proportional to the coefficients of the variables in the equation. The cross product is parallel to the normal vector.

39. (a) Yes (b) Yes **41.** $y - 3 = 0$

43. $2x + 3y - z = 10$ **45.** $2x - y - 2z + 6 = 0$

47. $3x - 19y - 2z = 0$ **49.** $4x - 3y + 4z = 10$ **51.** $z = 3$

53. $x + y + z = 5$ **55.** $7x + y - 11z = 5$ **57.** $y - z = -1$

59.

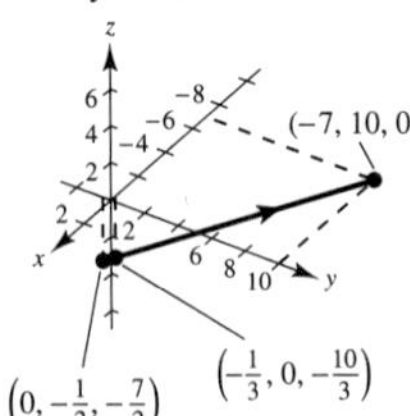

61. $x - z = 0$ **63.** $9x - 3y + 2z - 21 = 0$

65. Orthogonal **67.** Neither; 83.5° **69.** Parallel

71.

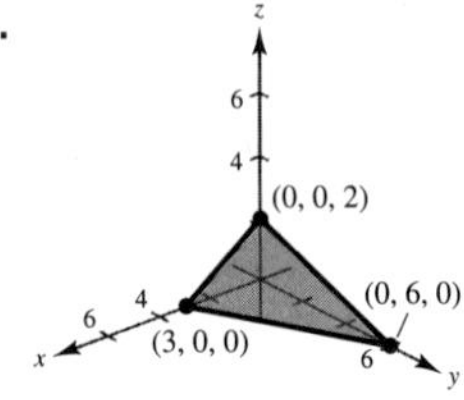

73.

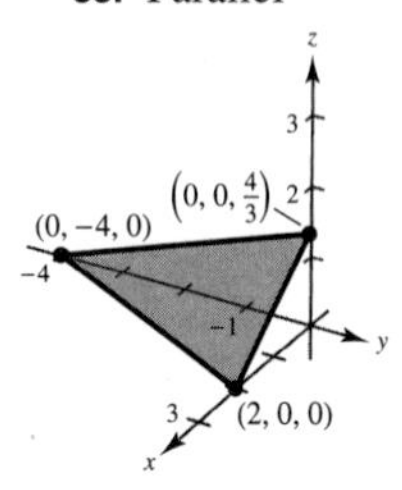

75.

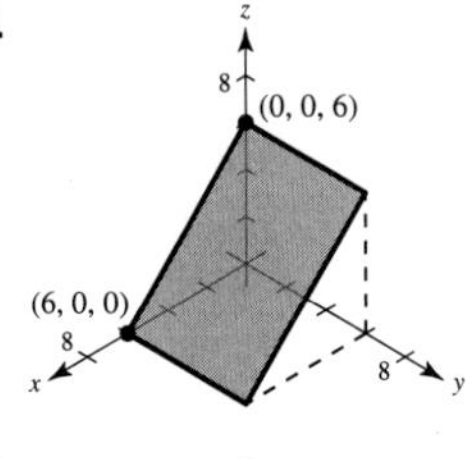

77.

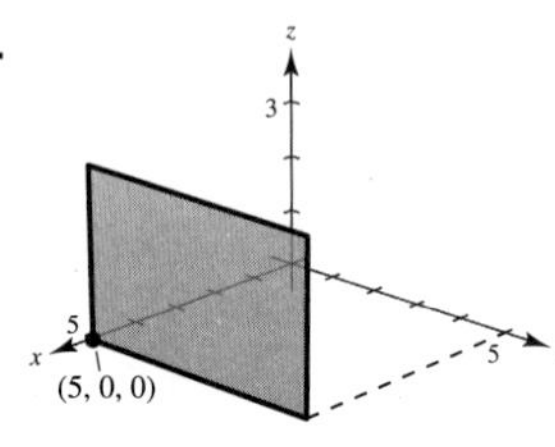

79.

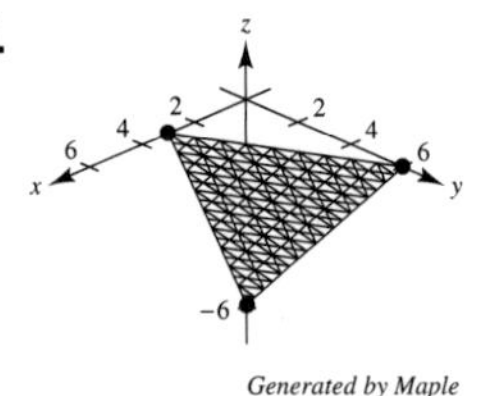

81.

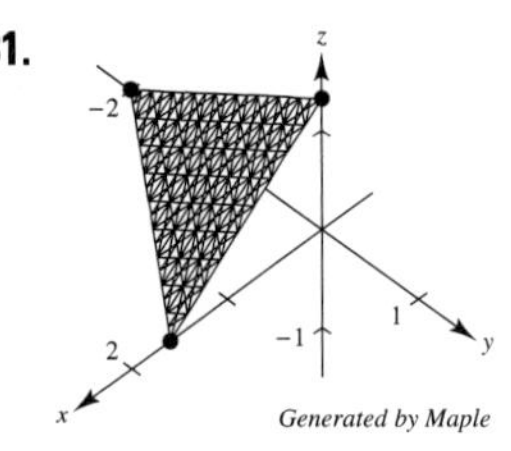

83. P_1 and P_2 are parallel. **85.** $P_1 = P_4$ and is parallel to P_2.

87. The planes have intercepts at $(c, 0, 0)$, $(0, c, 0)$, and $(0, 0, c)$ for each value of c.

89. If $c = 0$, $z = 0$ is the xy-plane; If $c \neq 0$, the plane is parallel to the x-axis and passes through $(0, 0, 0)$ and $(0, 1, -c)$.

91. (a) $\theta \approx 65.91°$
(b) $x = 2$
$y = 1 + t$
$z = 1 + 2t$

93. $(2, -3, 2)$; The line does not lie in the plane.

95. Not intersecting **97.** $6\sqrt{14}/7$ **99.** $11\sqrt{6}/6$

101. $2\sqrt{26}/13$ **103.** $27\sqrt{94}/188$ **105.** $\sqrt{2533}/17$

107. $7\sqrt{3}/3$ **109.** $\sqrt{66}/3$

111. Parametric equations: $x = x_1 + at$, $y = y_1 + bt$, and $z = z_1 + ct$
Symmetric equations: $\frac{x - x_1}{a} = \frac{y - y_1}{b} = \frac{z - z_1}{c}$
You need a vector $\mathbf{v} = \langle a, b, c \rangle$ parallel to the line and a point $P(x_1, y_1, z_1)$ on the line.

113. Simultaneously solve the two linear equations representing the planes and substitute the values back into one of the original equations. Then choose a value for t and form the corresponding parametric equations for the line of intersection.

115. (a) Parallel if vector $\langle a_1, b_1, c_1 \rangle$ is a scalar multiple of $\langle a_2, b_2, c_2 \rangle$; $\theta = 0$.
(b) Perpendicular if $a_1a_2 + b_1b_2 + c_1c_2 = 0$; $\theta = \pi/2$.

117. $cbx + acy + abz = abc$

119. Sphere: $(x - 3)^2 + (y + 2)^2 + (z - 5)^2 = 16$

121. (a)

Year	1999	2000	2001	2002
z (approx.)	6.25	6.05	5.94	5.76

Year	2003	2004	2005
z (approx.)	5.66	5.56	5.56

The approximations are close to the actual values.
(b) Answers will vary.

123. (a) $\sqrt{70}$ in.
(b) [graph]
(c) The distance is never zero.
(d) 5 in.

125. $\left(\frac{77}{13}, \frac{48}{13}, -\frac{23}{13}\right)$ **127.** $\left(-\frac{1}{2}, -\frac{9}{4}, \frac{1}{4}\right)$ **129.** True **131.** True

133. False. Plane $7x + y - 11z = 5$ and plane $5x + 2y - 4z = 1$ are perpendicular to plane $2x - 3y + z = 3$ but are not parallel.

Section 11.6 (page 820)

1. c **2.** e **3.** f **4.** b **5.** d **6.** a

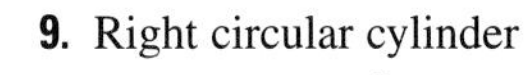

7. Plane

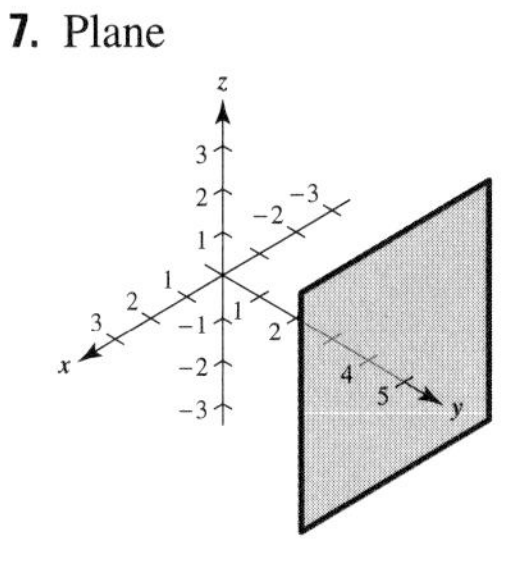

9. Right circular cylinder

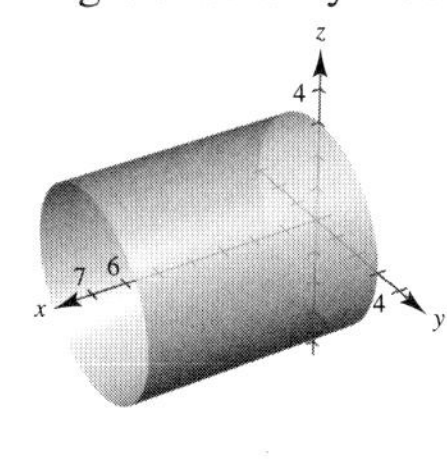

11. Parabolic cylinder

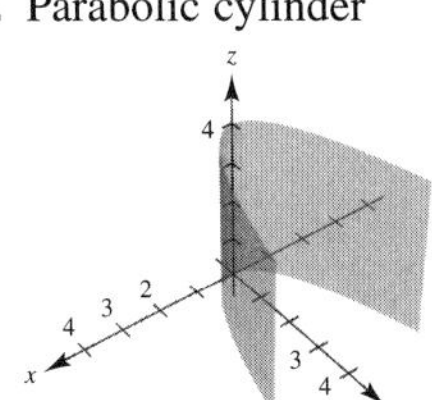

13. Elliptic cylinder

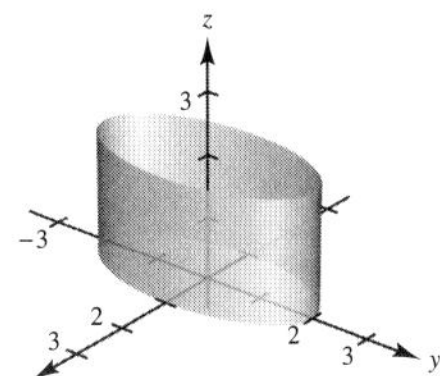

15. Cylinder

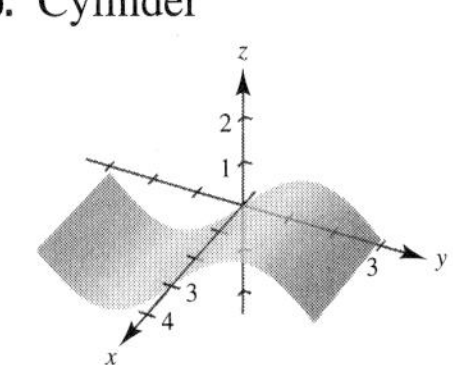

17. (a) $(20, 0, 0)$
(b) $(10, 10, 20)$
(c) $(0, 0, 20)$
(d) $(0, 20, 0)$

19. Ellipsoid

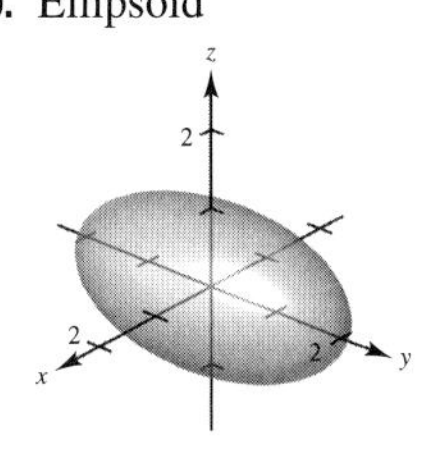

21. Hyperboloid of one sheet

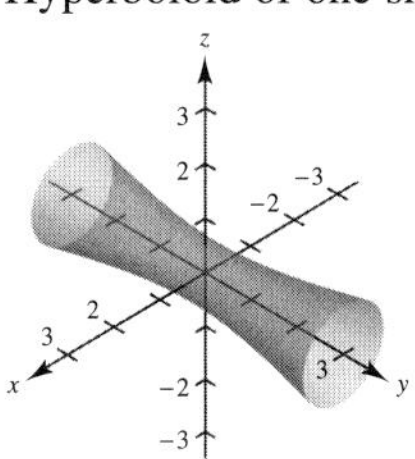

23. Hyperboloid of two sheets

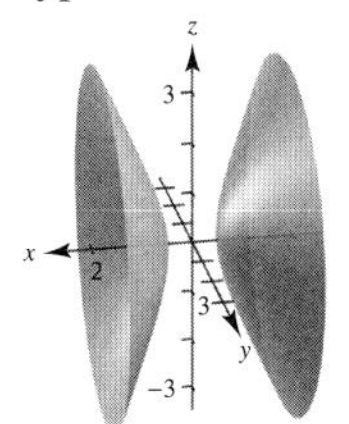

25. Elliptic paraboloid

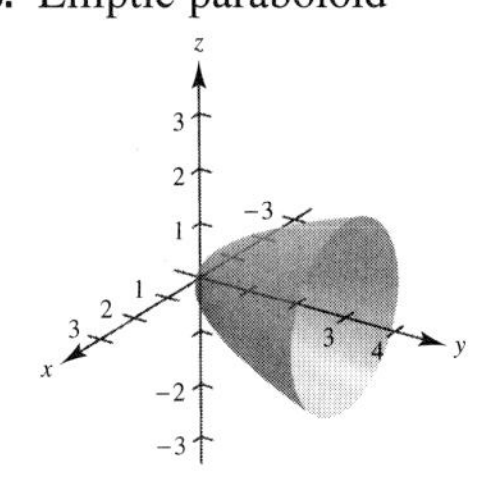

27. Hyperbolic paraboloid

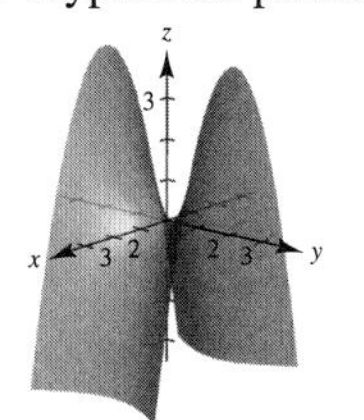

29. Elliptic cone

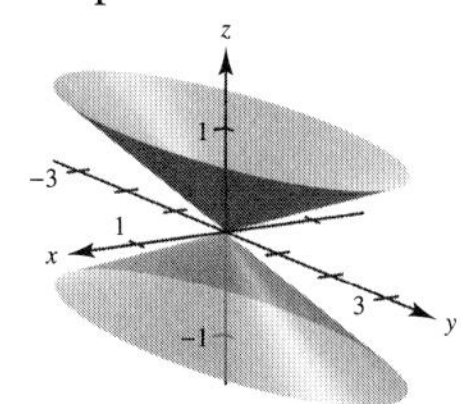

31. Ellipsoid

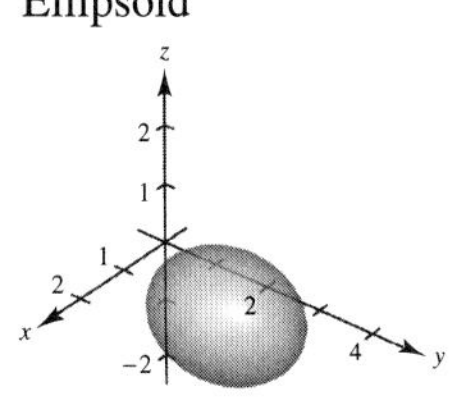

33.

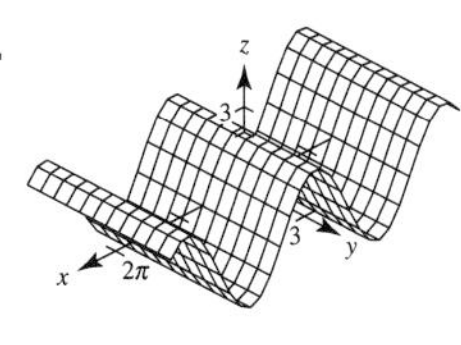

35.

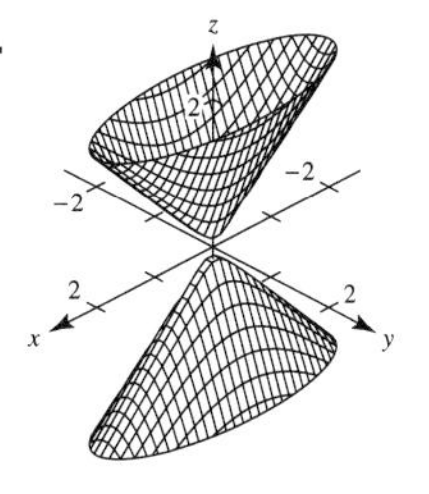

37.

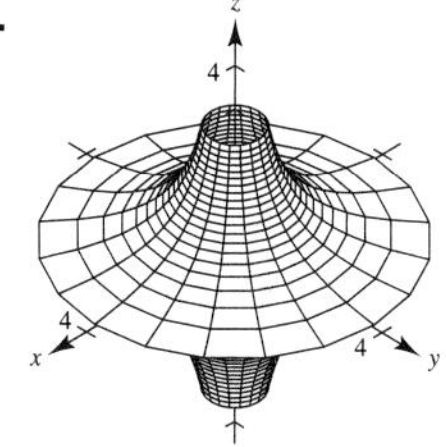

39.

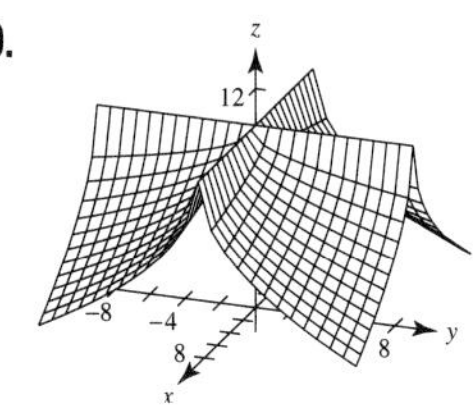

41.

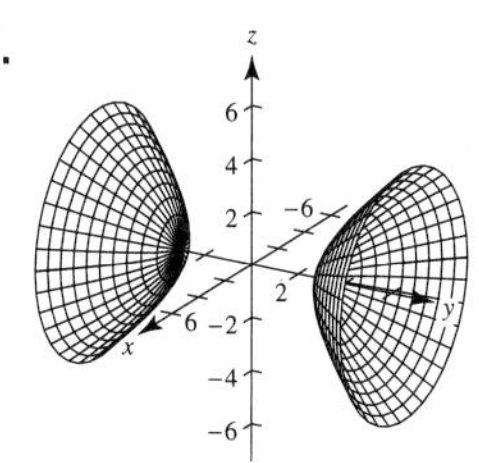

43.

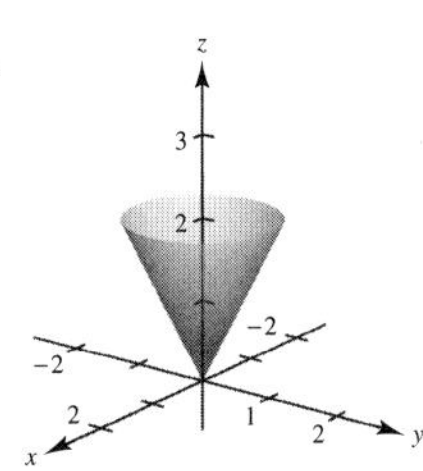

45.

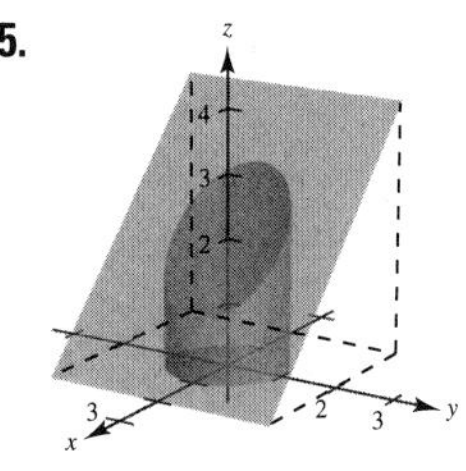

47. $x^2 + z^2 = 4y$ **49.** $4x^2 + 4y^2 = z^2$ **51.** $y^2 + z^2 = 4/x^2$

53. $y = \sqrt{2z}$ $\left(\text{or } x = \sqrt{2z}\right)$

55. Let C be a curve in a plane and let L be a line not in a parallel plane. The set of all lines parallel to L and intersecting C is called a cylinder. C is called the generating curve of the cylinder, and the parallel lines are called rulings.

57. See pages 814 and 815. **59.** $128\pi/3$

61. (a) Major axis: $4\sqrt{2}$
Minor axis: 4
Foci: $(0, \pm 2, 2)$

(b) Major axis: $8\sqrt{2}$
Minor axis: 8
Foci: $(0, \pm 4, 8)$

63. $x^2 + z^2 = 8y$; Elliptic paraboloid

65. $x^2/3963^2 + y^2/3963^2 + z^2/3950^2 = 1$

67. $x = at, y = -bt, z = 0$;
$x = at, y = bt + ab^2, z = 2abt + a^2b^2$

69. True

71. False. A trace of an ellipsoid can be a single point.

73. The Klein bottle does not have both an "inside" and an "outside." It is formed by inserting the small open end through the side of the bottle and making it contiguous with the top of the bottle.

Section 11.7 (page 827)

1. $(-7, 0, 5)$ **3.** $\left(3\sqrt{2}/2, 3\sqrt{2}/2, 1\right)$ **5.** $\left(-2\sqrt{3}, -2, 3\right)$

7. $(5, \pi/2, 1)$ **9.** $\left(2\sqrt{2}, -\pi/4, -4\right)$ **11.** $(2, \pi/3, 4)$

13. $z = 4$ **15.** $r^2 + z^2 = 17$ **17.** $r = \sec\theta\tan\theta$

19. $r^2\sin^2\theta = 10 - z^2$

21. $x^2 + y^2 = 9$

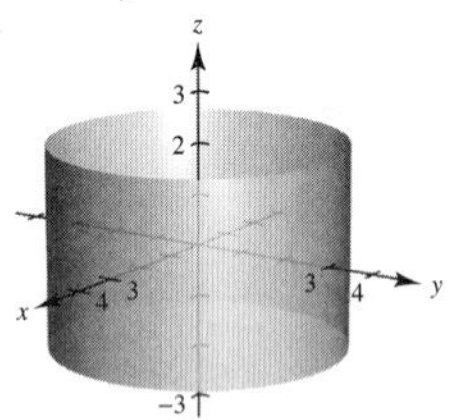

23. $x - \sqrt{3}y = 0$

25. $x^2 + y^2 + z^2 = 5$

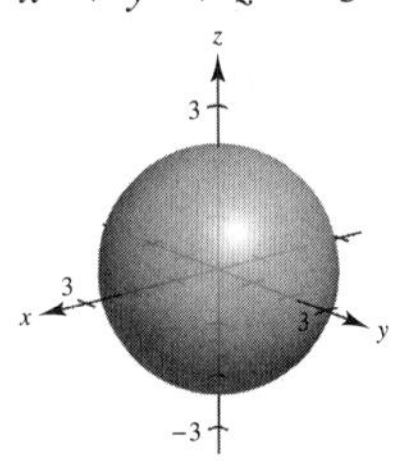

27. $x^2 + y^2 - 2y = 0$

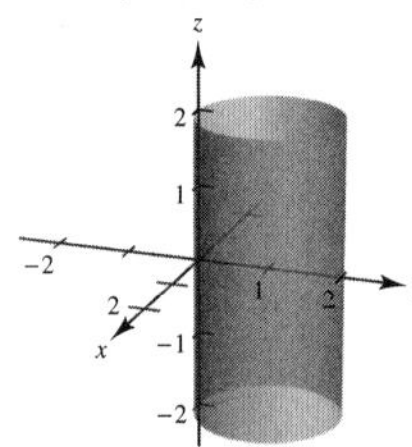

29. $(4, 0, \pi/2)$ **31.** $\left(4\sqrt{2}, 2\pi/3, \pi/4\right)$ **33.** $(4, \pi/6, \pi/6)$

35. $\left(\sqrt{6}, \sqrt{2}, 2\sqrt{2}\right)$ **37.** $(0, 0, 12)$ **39.** $\left(\frac{5}{2}, \frac{5}{2}, -5\sqrt{2}/2\right)$

41. $\rho = 2\csc\phi\csc\theta$ **43.** $\rho = 7$

45. $\rho = 4\csc\phi$ **47.** $\tan^2\phi = 2$

49. $x^2 + y^2 + z^2 = 25$

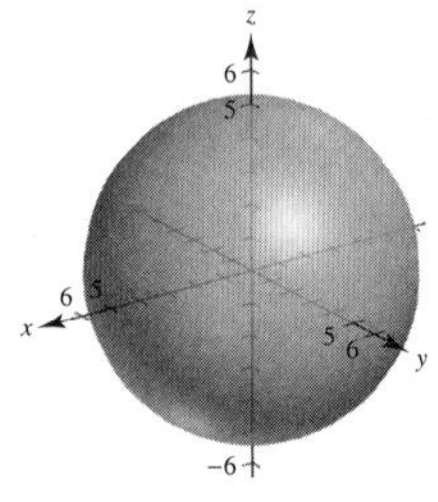

51. $3x^2 + 3y^2 - z^2 = 0$

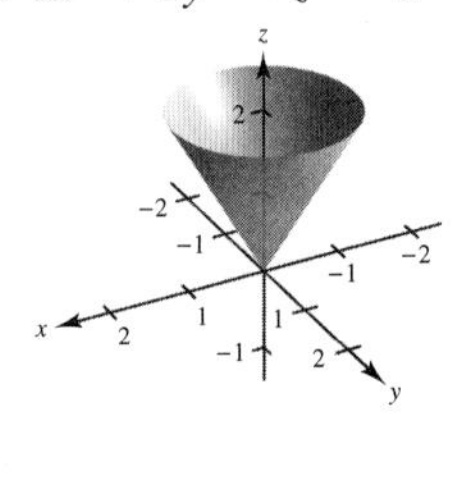

53. $x^2 + y^2 + (z - 2)^2 = 4$

55. $x^2 + y^2 = 1$

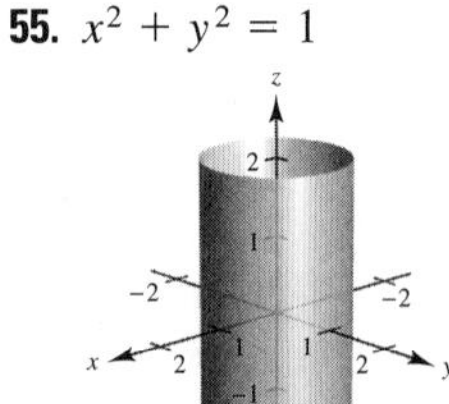

57. $(4, \pi/4, \pi/2)$ **59.** $\left(4\sqrt{2}, \pi/2, \pi/4\right)$

61. $\left(2\sqrt{13}, -\pi/6, \arccos[3/\sqrt{13}]\right)$ **63.** $(13, \pi, \arccos[5/13])$

65. $(10, \pi/6, 0)$ **67.** $(36, \pi, 0)$

69. $\left(3\sqrt{3}, -\pi/6, 3\right)$ **71.** $\left(4, 7\pi/6, 4\sqrt{3}\right)$

	Rectangular	*Cylindrical*	*Spherical*
73.	$(4, 6, 3)$	$(7.211, 0.983, 3)$	$(7.810, 0.983, 1.177)$
75.	$(4.698, 1.710, 8)$	$(5, \pi/9, 8)$	$(9.434, 0.349, 0.559)$
77.	$(-7.071, 12.247, 14.142)$	$(14.142, 2.094, 14.142)$	$(20, 2\pi/3, \pi/4)$
79.	$(3, -2, 2)$	$(3.606, -0.588, 2)$	$(4.123, -0.588, 1.064)$
81.	$\left(\frac{5}{2}, \frac{4}{3}, -\frac{3}{2}\right)$	$(2.833, 0.490, -1.5)$	$(3.206, 0.490, 2.058)$
83.	$(-3.536, 3.536, -5)$	$(5, 3\pi/4, -5)$	$(7.071, 2.356, 2.356)$
85.	$(2.804, -2.095, 6)$	$(-3.5, 2.5, 6)$	$(6.946, 5.642, 0.528)$
87.	$(-1.837, 1.837, 1.5)$	$(2.598, 2.356, 1.5)$	$(3, 3\pi/4, \pi/3)$

89. d **90.** e **91.** c **92.** a **93.** f **94.** b

95. Rectangular to cylindrical:
$r^2 = x^2 + y^2$, $\tan\theta = y/x$, $z = z$
Cylindrical to rectangular:
$x = r\cos\theta$, $y = r\sin\theta$, $z = z$

97. Rectangular to spherical:
$\rho^2 = x^2 + y^2 + z^2$, $\tan\theta = y/x$, $\phi = \arccos\left(z/\sqrt{x^2 + y^2 + z^2}\right)$
Spherical to rectangular:
$x = \rho\sin\phi\cos\theta$, $y = \rho\sin\phi\sin\theta$, $z = \rho\cos\phi$

99. (a) $r^2 + z^2 = 25$ (b) $\rho = 5$

101. (a) $r^2 + (z - 1)^2 = 1$ (b) $\rho = 2\cos\phi$

103. (a) $r = 4\sin\theta$ (b) $\rho = 4\sin\theta/\sin\phi = 4\sin\theta\csc\phi$

105. (a) $r^2 = 9/(\cos^2\theta - \sin^2\theta)$
(b) $\rho^2 = 9\csc^2\phi/(\cos^2\theta - \sin^2\theta)$

107.

109.

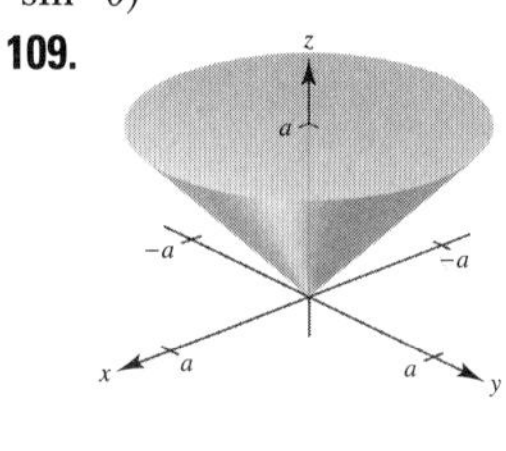

111.

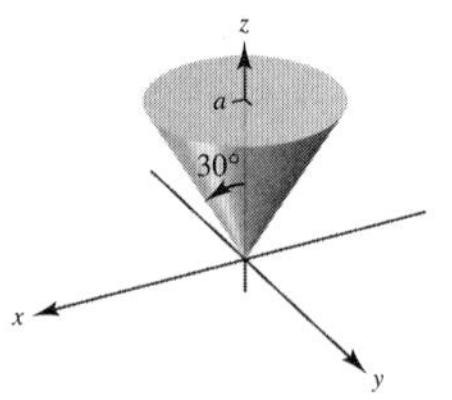

113.

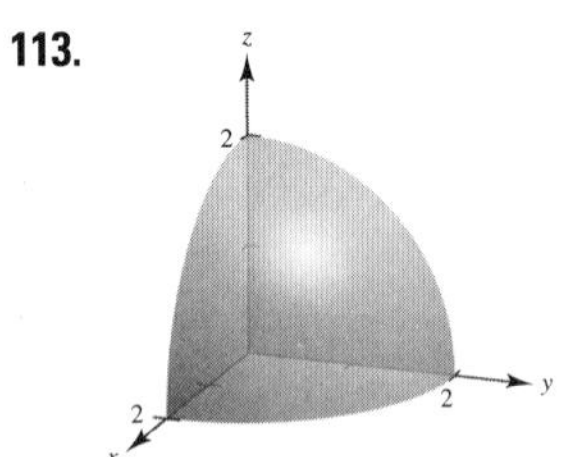

115. Rectangular: $0 \le x \le 10$
$0 \le y \le 10$
$0 \le z \le 10$

117. Spherical: $4 \le \rho \le 6$

119. Cylindrical: $r^2 + z^2 \le 9$, $r \le 3\cos\theta$, $0 \le \theta \le \pi$

121. False. $r = z$ represents a cone.

123. False. See page 823. **125.** Ellipse

Review Exercises for Chapter 11 (page 829)

1. (a) $\mathbf{u} = \langle 3, -1\rangle$ (b) $\mathbf{u} = 3\mathbf{i} - \mathbf{j}$ (c) $2\sqrt{5}$ (d) $10\mathbf{i}$
$\mathbf{v} = \langle 4, 2\rangle$

3. $\mathbf{v} = \left\langle 4, 4\sqrt{3}\right\rangle$ **5.** $(-5, 4, 0)$

7. Above the xy-plane and to the right of the xz-plane *or* below the xy-plane and to the left of the xz-plane

9. $(x - 3)^2 + (y + 2)^2 + (z - 6)^2 = \frac{225}{4}$

11. $(x - 2)^2 + (y - 3)^2 + z^2 = 9$
Center: $(2, 3, 0)$
Radius: 3

13. (a) and (d) (b) $\mathbf{u} = \langle 2, 5, -10 \rangle$
(c) $\mathbf{u} = 2\mathbf{i} + 5\mathbf{j} - 10\mathbf{k}$

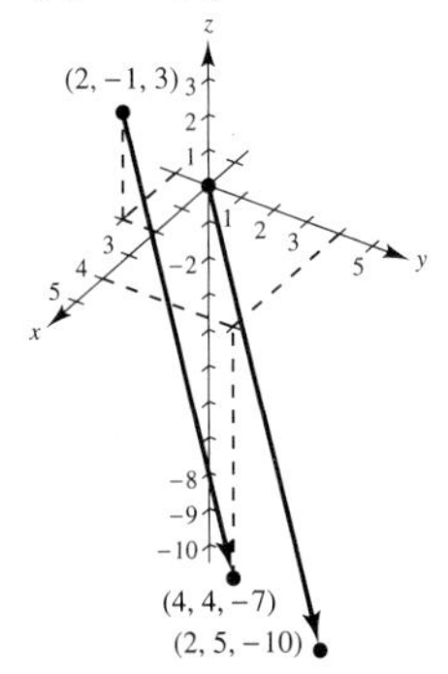

15. Collinear **17.** $(1/\sqrt{38})\langle 2, 3, 5 \rangle$

19. (a) $\mathbf{u} = \langle -1, 4, 0 \rangle, \mathbf{v} = \langle -3, 0, 6 \rangle$ (b) 3 (c) 45

21. Orthogonal **23.** $\theta = \arccos\left(\frac{\sqrt{2} + \sqrt{6}}{4}\right) = 15°$ **25.** π

27. Answers will vary. Example: $\langle -6, 5, 0 \rangle, \langle 6, -5, 0 \rangle$

29. $\mathbf{u} \cdot \mathbf{u} = 14 = \|\mathbf{u}\|^2$ **31.** $\left\langle -\frac{15}{14}, \frac{5}{7}, -\frac{5}{14} \right\rangle$

33. $(1/\sqrt{5})(-2\mathbf{i} - \mathbf{j})$ or $(1/\sqrt{5})(2\mathbf{i} + \mathbf{j})$

35. 4 **37.** $\sqrt{285}$ **39.** $100 \sec 20° \approx 106.4$ lb

41. (a) $x = 3 + 6t, y = 11t, z = 2 + 4t$
(b) $(x - 3)/6 = y/11 = (z - 2)/4$

43. (a) $x = 1, y = 2 + t, z = 3$ (b) None
(c)

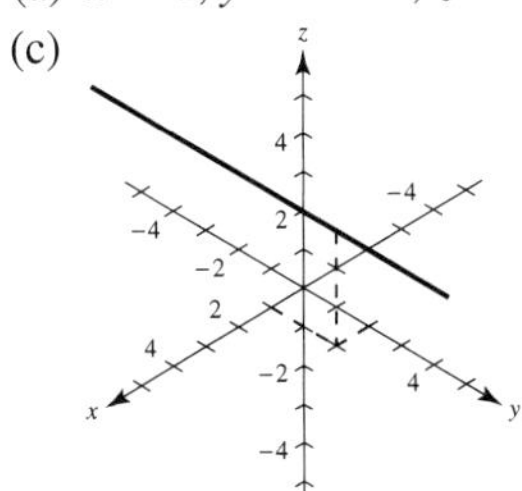

45. (a) $x = t, y = -1 + t, z = 1$ (b) $x = y + 1, z = 1$
(c)

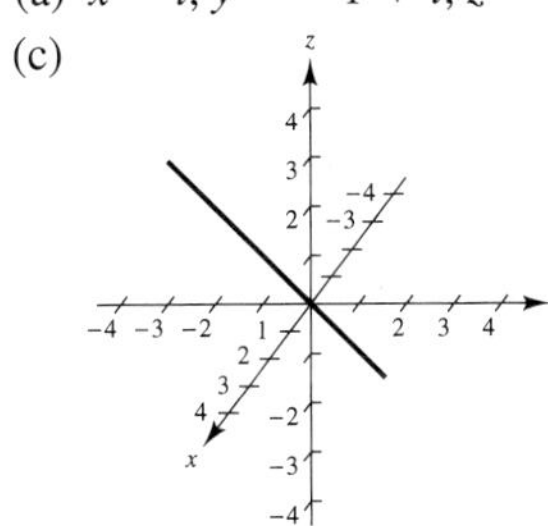

47. $27x + 4y + 32z + 33 = 0$ **49.** $x + 2y = 1$

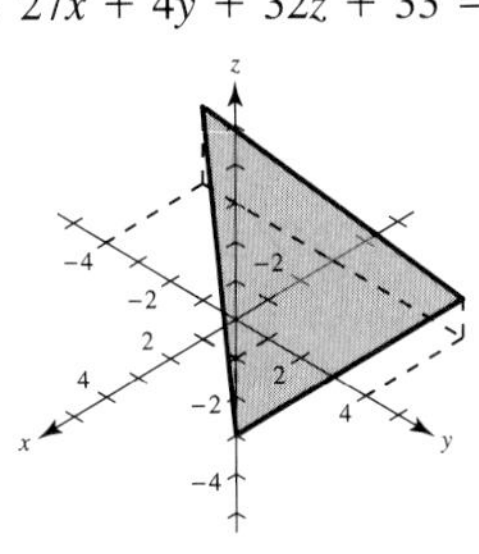

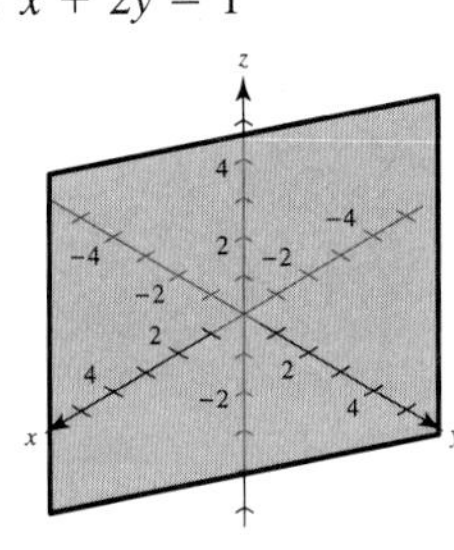

51. $\frac{8}{7}$ **53.** $\sqrt{35}/7$

55. Plane

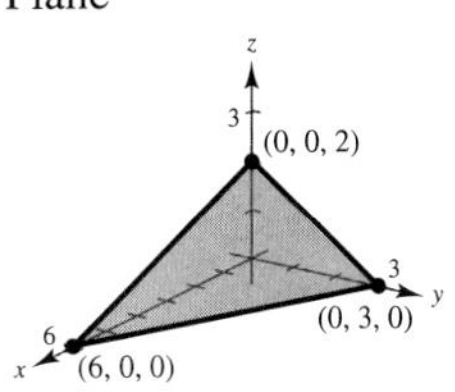

57. Plane

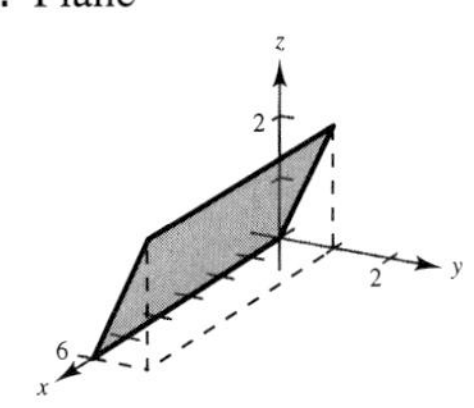

59. Ellipsoid

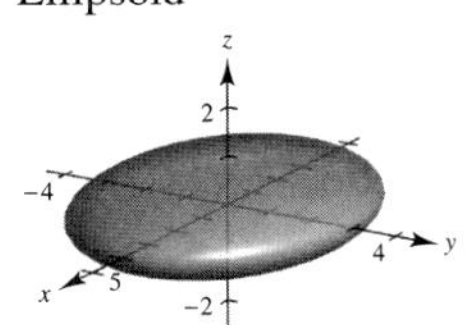

61. Hyperboloid of two sheets

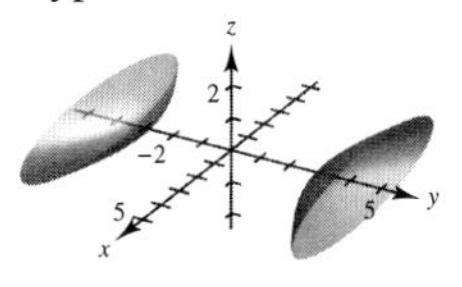

63. Cylinder

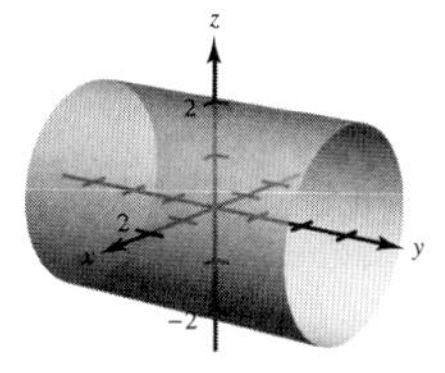

65. Let $y = 2\sqrt{x}$ and revolve around the x-axis.

67. $x^2 + z^2 = 2y$

69. (a) $(4, 3\pi/4, 2)$ (b) $\left(2\sqrt{5}, 3\pi/4, \arccos\left[\sqrt{5}/5\right]\right)$

71. $\left(50\sqrt{5}, -\pi/6, \arccos\left[1/\sqrt{5}\right]\right)$

73. $\left(25\sqrt{2}/2, -\pi/4, -25\sqrt{2}/2\right)$

75. (a) $r^2 \cos 2\theta = 2z$ (b) $\rho = 2 \sec 2\theta \cos \phi \csc^2 \phi$

77. $\left(x - \frac{5}{2}\right)^2 + y^2 = \frac{25}{4}$ **79.** $x = y$

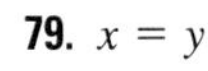

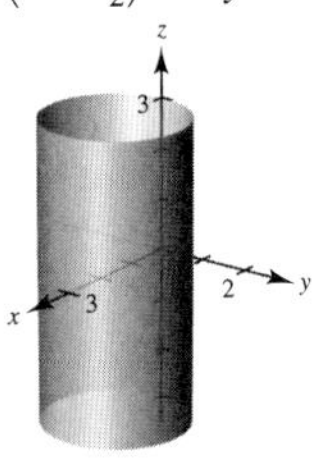

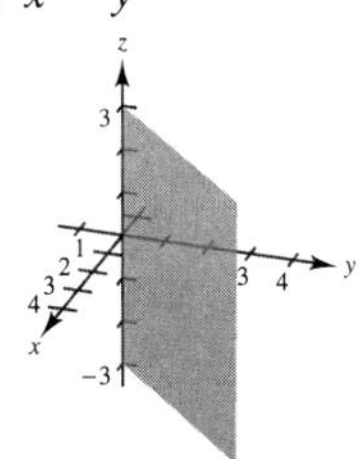

P.S. Problem Solving (page 831)

1–3. Proofs **5.** (a) $3\sqrt{2}/2 \approx 2.12$ (b) $\sqrt{5} \approx 2.24$

7. (a) $\pi/2$ (b) $\frac{1}{2}(\pi abk)k$
(c) $V = \frac{1}{2}(\pi ab)k^2$
$V = \frac{1}{2}$(area of base)height

9. (a) (b)

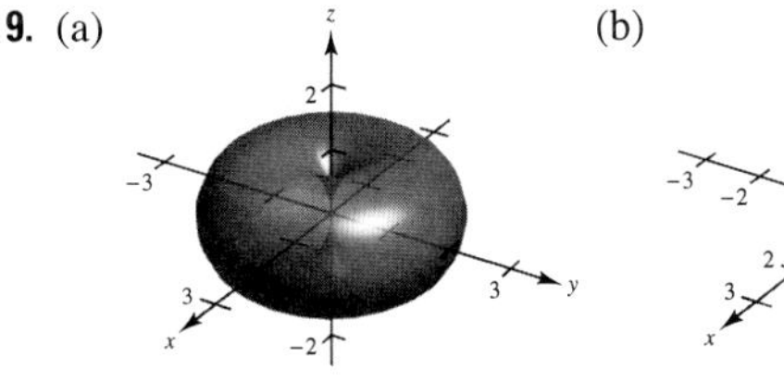

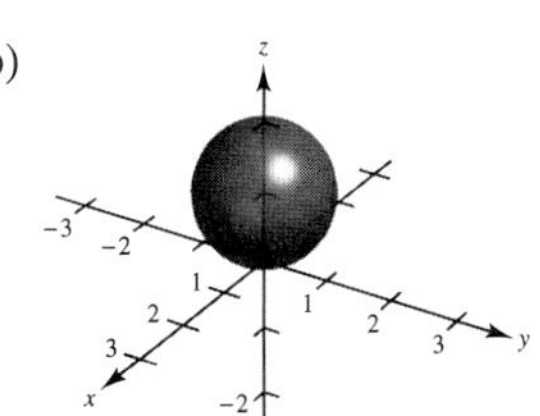

11. Proof

13. (a) Tension: $2\sqrt{3}/3 \approx 1.1547$ lb
Magnitude of $\mathbf{u}$: $\sqrt{3}/3 \approx 0.5774$ lb
(b) $T = \sec \theta$; $\|\mathbf{u}\| = \tan \theta$; Domain: $0° \le \theta \le 90°$
(c)

θ	0°	10°	20°	30°
T	1	1.0154	1.0642	1.1547
$\|\mathbf{u}\|$	0	0.1763	0.3640	0.5774

θ	40°	50°	60°
T	1.3054	1.5557	2
$\|\mathbf{u}\|$	0.8391	1.1918	1.7321

(d) 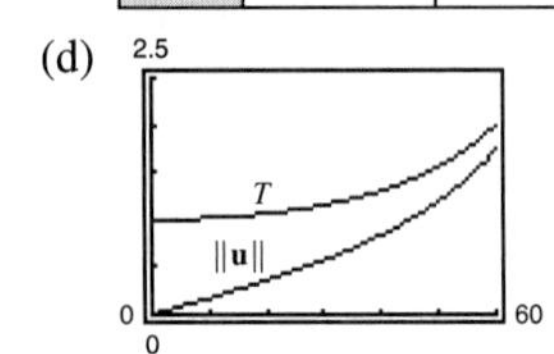

(e) Both are increasing functions.
(f) $\lim_{\theta \to \pi/2^-} T = \infty$ and $\lim_{\theta \to \pi/2^-} \|\mathbf{u}\| = \infty$
Yes. As θ increases, both T and $\|\mathbf{u}\|$ increase.

15. $\langle 0, 0, \cos \alpha \sin \beta - \cos \beta \sin \alpha \rangle$; Proof

17. $D = \dfrac{|\overrightarrow{PQ} \cdot \mathbf{n}|}{\|\mathbf{n}\|}$
$= \dfrac{|\mathbf{w} \cdot (\mathbf{u} \times \mathbf{v})|}{\|\mathbf{u} \times \mathbf{v}\|} = \dfrac{|(\mathbf{u} \times \mathbf{v}) \cdot \mathbf{w}|}{\|\mathbf{u} \times \mathbf{v}\|} = \dfrac{|\mathbf{u} \cdot (\mathbf{v} \times \mathbf{w})|}{\|\mathbf{u} \times \mathbf{v}\|}$

19. Proof

Chapter 12

Section 12.1 (page 839)

1. $(-\infty, -1) \cup (-1, \infty)$ **3.** $(0, \infty)$
5. $[0, \infty)$ **7.** $(-\infty, \infty)$
9. (a) $\frac{1}{2}\mathbf{i}$ (b) $\mathbf{j}$ (c) $\frac{1}{2}(s+1)^2\mathbf{i} - s\mathbf{j}$ (d) $\frac{1}{2}\Delta t(\Delta t + 4)\mathbf{i} - \Delta t\mathbf{j}$
11. (a) $\ln 2\mathbf{i} + \dfrac{1}{2}\mathbf{j} + 6\mathbf{k}$ (b) Not possible
(c) $\ln(t-4)\mathbf{i} + \dfrac{1}{t-4}\mathbf{j} + 3(t-4)\mathbf{k}$
(d) $\ln(1+\Delta t)\mathbf{i} - \dfrac{\Delta t}{1+\Delta t}\mathbf{j} + 3\Delta t\mathbf{k}$
13. $\sqrt{t(1+25t)}$
15. $\mathbf{r}(t) = 3t\mathbf{i} + t\mathbf{j} + 2t\mathbf{k}$
$x = 3t,\ y = t,\ z = 2t$
17. $\mathbf{r}(t) = (-2+t)\mathbf{i} + (5-t)\mathbf{j} + (-3+12t)\mathbf{k}$
$x = -2 + t,\ y = 5 - t,\ z = -3 + 12t$
19. $t^2(5t-1)$; No, the dot product is a scalar.
21. b **22.** c **23.** d **24.** a
25. (a) $(-20, 0, 0)$ (b) $(10, 20, 10)$
(c) $(0, 0, 20)$ (d) $(20, 0, 0)$

27. 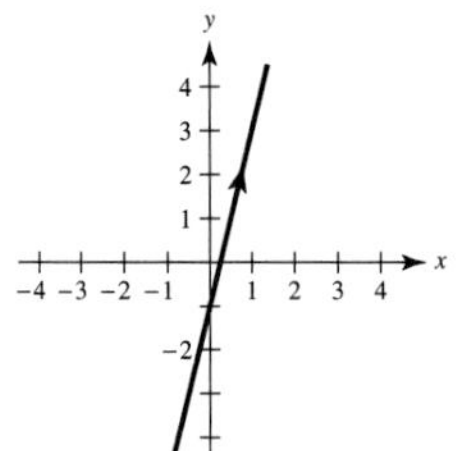

29. 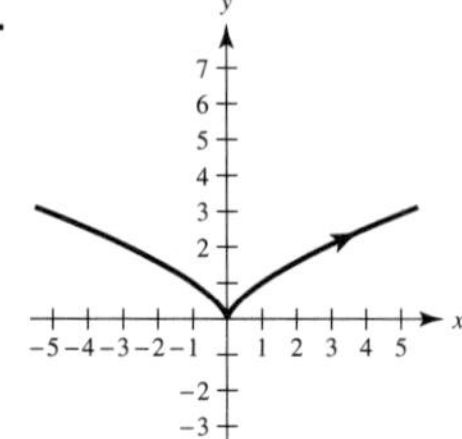

31. 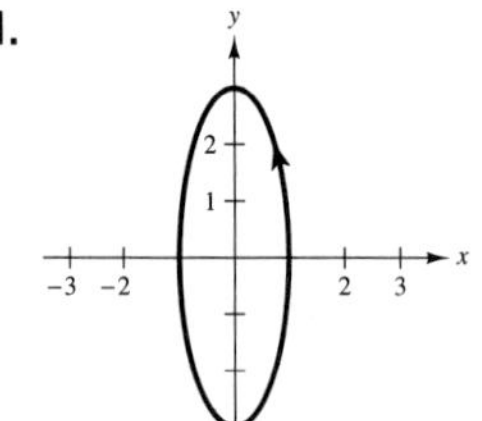

33. 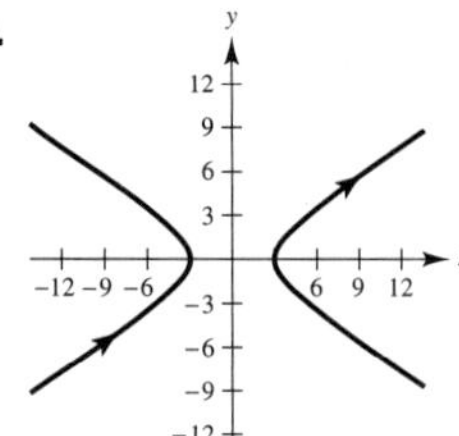

35. 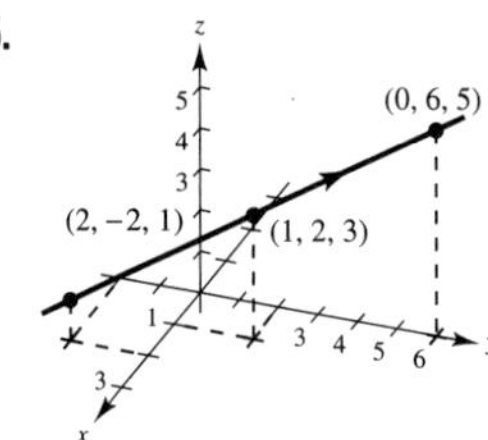

37. 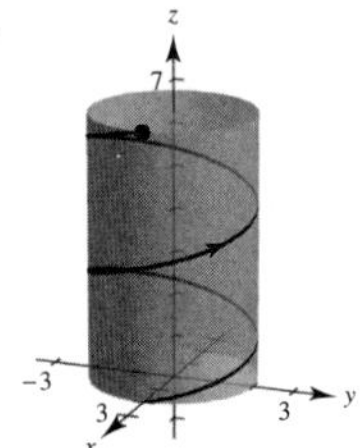

39. 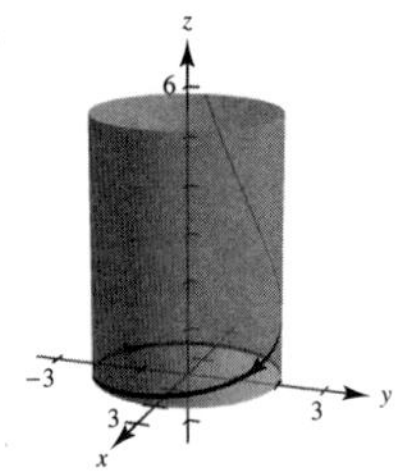

41. 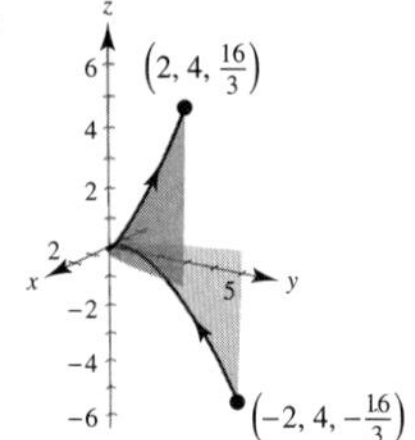

43. 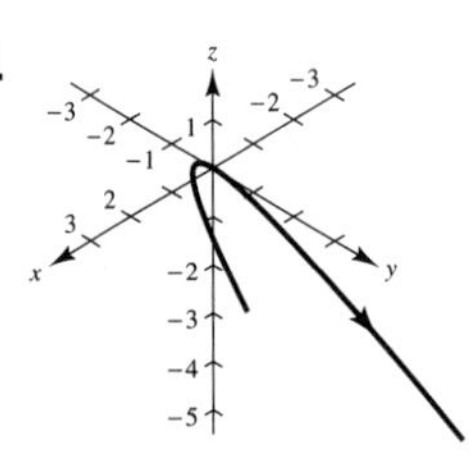

45. 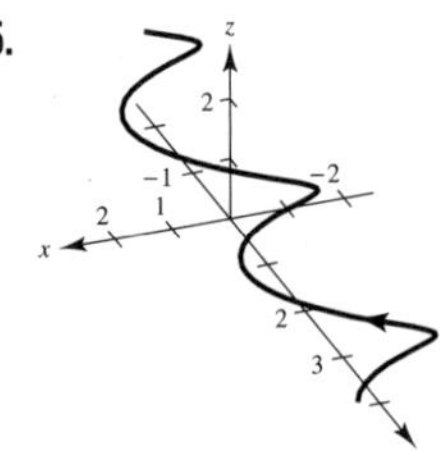

47. Parabola
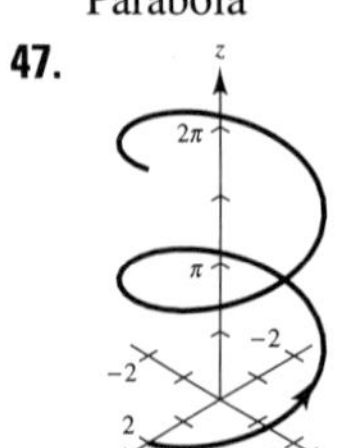

Helix
(a) The helix is translated two units back on the x-axis.
(b) The height of the helix increases at a greater rate.
(c) The orientation of the graph is reversed.
(d) The axis of the helix is the x-axis.
(e) The radius of the helix is increased from 2 to 6.

49–55. Answers will vary.
57. Answers will vary. Sample answer:
$\mathbf{r}_1(t) = t\mathbf{i} + t^2\mathbf{j},\quad 0 \le t \le 2$
$\mathbf{r}_2(t) = (2-t)\mathbf{i} + 4\mathbf{j},\quad 0 \le t \le 2$
$\mathbf{r}_3(t) = (4-t)\mathbf{j},\quad 0 \le t \le 4$

59.

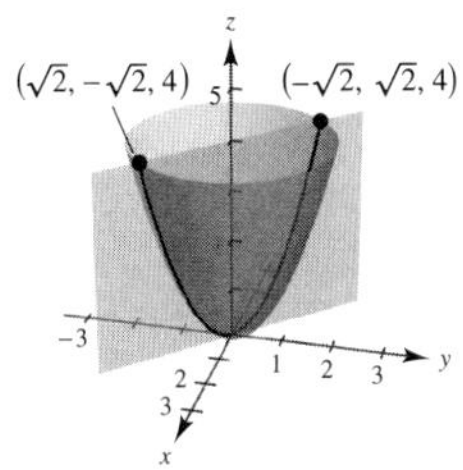

$\mathbf{r}(t) = t\mathbf{i} - t\mathbf{j} + 2t^2\mathbf{k}$

61.

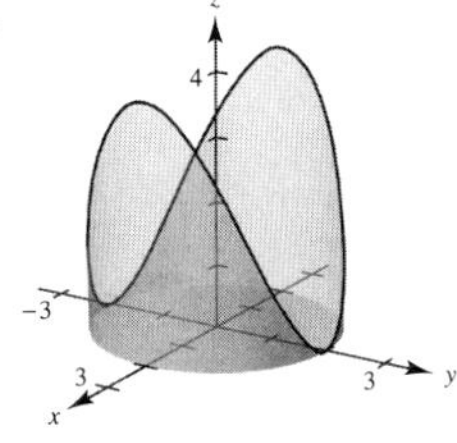

$\mathbf{r}(t) = 2 \sin t\mathbf{i} + 2 \cos t\mathbf{j} + 4 \sin^2 t\mathbf{k}$

63.

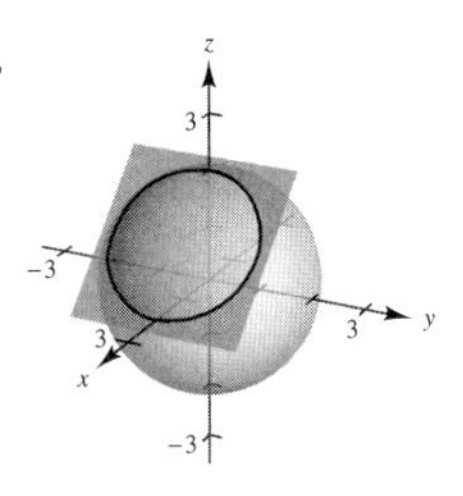

$\mathbf{r}(t) = (1 + \sin t)\mathbf{i} + \sqrt{2} \cos t\mathbf{j} + (1 - \sin t)\mathbf{k}$ and
$\mathbf{r}(t) = (1 + \sin t)\mathbf{i} - \sqrt{2} \cos t\mathbf{j} + (1 - \sin t)\mathbf{k}$

65.

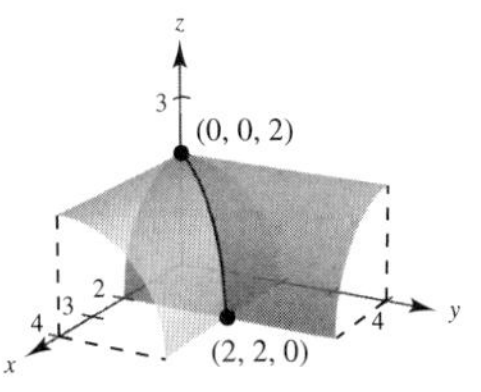

$\mathbf{r}(t) = t\mathbf{i} + t\mathbf{j} + \sqrt{4 - t^2}\,\mathbf{k}$

67. Let $x = t$, $y = 2t \cos t$, and $z = 2t \sin t$. Then
$y^2 + z^2 = (2t \cos t)^2 + (2t \sin t)^2 = 4t^2 \cos^2 t + 4t^2 \sin^2 t$
$= 4t^2(\cos^2 t + \sin^2 t) = 4t^2.$
Because $x = t$, $y^2 + z^2 = 4x^2$.

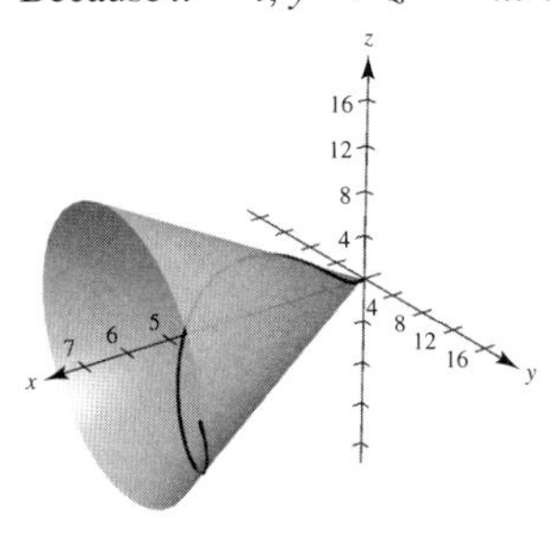

69. $\pi\mathbf{i} - \mathbf{j}$ **71.** $\mathbf{0}$ **73.** $\mathbf{i} + \mathbf{j} + \mathbf{k}$
75. $(-\infty, 0), (0, \infty)$ **77.** $[-1, 1]$
79. $(-\pi/2 + n\pi, \pi/2 + n\pi)$, n is an integer.
81. (a) $\mathbf{s}(t) = t^2\mathbf{i} + (t - 3)\mathbf{j} + (t + 3)\mathbf{k}$
(b) $\mathbf{s}(t) = (t^2 - 2)\mathbf{i} + (t - 3)\mathbf{j} + t\mathbf{k}$
(c) $\mathbf{s}(t) = t^2\mathbf{i} + (t + 2)\mathbf{j} + t\mathbf{k}$
83. Answers will vary. Sample answer:

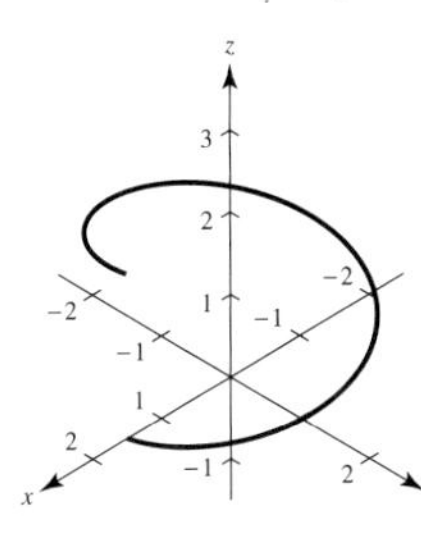

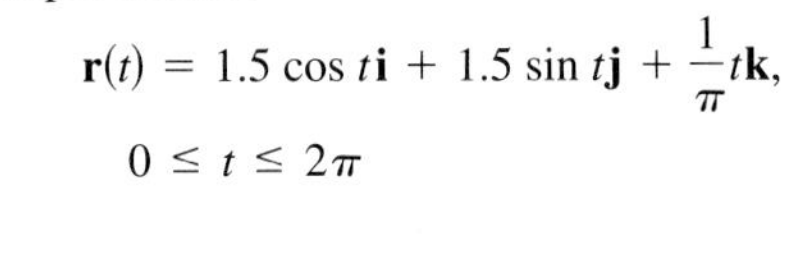

$\mathbf{r}(t) = 1.5 \cos t\mathbf{i} + 1.5 \sin t\mathbf{j} + \dfrac{1}{\pi}t\mathbf{k}$, $0 \le t \le 2\pi$

85–87. Proofs **89.** Yes; Yes **91.** Not necessarily
93. True **95.** True

Section 12.2 (page 848)

1. $\mathbf{r}(2) = 4\mathbf{i} + 2\mathbf{j}$
$\mathbf{r}'(2) = 4\mathbf{i} + \mathbf{j}$

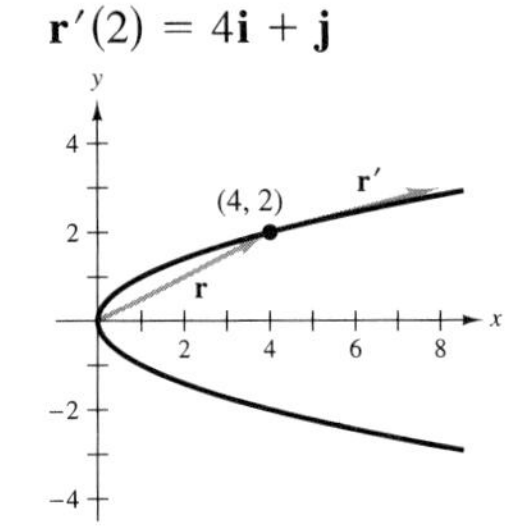

$\mathbf{r}'(t_0)$ is tangent to the curve at t_0.

3.

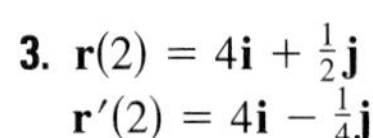

$\mathbf{r}(2) = 4\mathbf{i} + \frac{1}{2}\mathbf{j}$
$\mathbf{r}'(2) = 4\mathbf{i} - \frac{1}{4}\mathbf{j}$

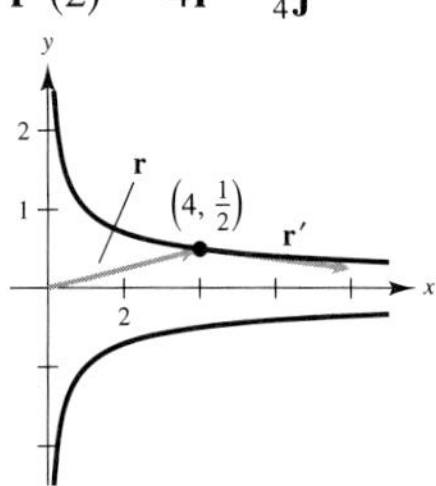

$\mathbf{r}'(t_0)$ is tangent to the curve at t_0.

5. $\mathbf{r}(\pi/2) = \mathbf{j}$
$\mathbf{r}'(\pi/2) = -\mathbf{i}$

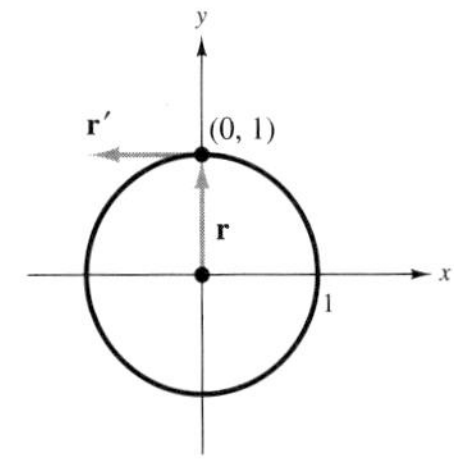

$\mathbf{r}'(t_0)$ is tangent to the curve at t_0.

7. $\mathbf{r}(0) = \mathbf{i} + \mathbf{j}$
$\mathbf{r}'(0) = \mathbf{i} + 2\mathbf{j}$

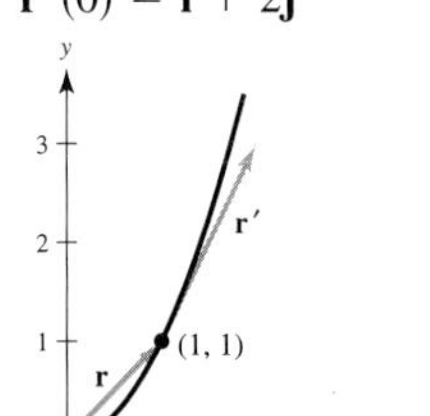

$\mathbf{r}'(t_0)$ is tangent to the curve at t_0.

9. $\mathbf{r}\left(\dfrac{3\pi}{2}\right) = -2\mathbf{j} + \left(\dfrac{3\pi}{2}\right)\mathbf{k}$
$\mathbf{r}'\left(\dfrac{3\pi}{2}\right) = 2\mathbf{i} + \mathbf{k}$

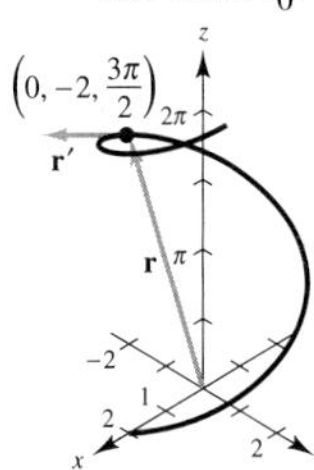

11. $3t^2\mathbf{i} - 3\mathbf{j}$ **13.** $-2 \sin t\mathbf{i} + 5 \cos t\mathbf{j}$
15. $6\mathbf{i} - 14t\mathbf{j} + 3t^2\mathbf{k}$ **17.** $-3a \sin t \cos^2 t\mathbf{i} + 3a \sin^2 t \cos t\mathbf{j}$
19. $-e^{-t}\mathbf{i} + (5te^t + 5e^t)\mathbf{k}$
21. $\langle \sin t + t \cos t, \cos t - t \sin t, 1\rangle$
23. (a) $3t^2\mathbf{i} + t\mathbf{j}$ (b) $6t\mathbf{i} + \mathbf{j}$ (c) $18t^3 + t$
25. (a) $-4 \sin t\mathbf{i} + 4 \cos t\mathbf{j}$ (b) $-4 \cos t\mathbf{i} - 4 \sin t\mathbf{j}$ (c) 0
27. (a) $t\mathbf{i} - \mathbf{j} + \frac{1}{2}t^2\mathbf{k}$ (b) $\mathbf{i} + t\mathbf{k}$ (c) $t^3/2 + t$
29. (a) $\langle t \cos t, t \sin t, 1\rangle$
(b) $\langle \cos t - t \sin t, \sin t + t \cos t, 0\rangle$ (c) t

31. $\dfrac{\mathbf{r}'(-1/4)}{\|\mathbf{r}'(-1/4)\|} = \dfrac{1}{\sqrt{4\pi^2 + 1}}\left(\sqrt{2}\pi\mathbf{i} + \sqrt{2}\pi\mathbf{j} - \mathbf{k}\right)$
$\dfrac{\mathbf{r}''(-1/4)}{\|\mathbf{r}''(-1/4)\|} = \dfrac{1}{2\sqrt{\pi^4 + 4}}\left(-\sqrt{2}\pi^2\mathbf{i} + \sqrt{2}\pi^2\mathbf{j} + 4\mathbf{k}\right)$

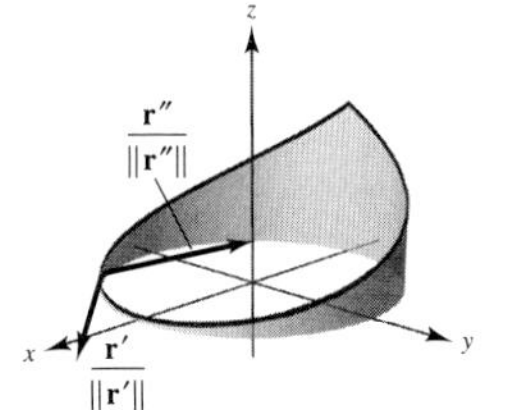

33. $(-\infty, 0), (0, \infty)$ **35.** $(n\pi/2, (n + 1)\pi/2)$

37. $(-\infty, \infty)$ **39.** $(-\infty, 0), (0, \infty)$

41. $(-\pi/2 + n\pi, \pi/2 + n\pi)$, n is an integer.

43. (a) $\mathbf{i} + 3\mathbf{j} + 2t\mathbf{k}$ (b) $2\mathbf{k}$ (c) $8t + 9t^2 + 5t^4$
(d) $-\mathbf{i} + (9 - 2t)\mathbf{j} + (6t - 3t^2)\mathbf{k}$
(e) $8t^3\mathbf{i} + (12t^2 - 4t^3)\mathbf{j} + (3t^2 - 24t)\mathbf{k}$
(f) $(10 + 2t^2)/\sqrt{10 + t^2}$

45. (a) $7t^6$ (b) $12t^5\mathbf{i} - 5t^4\mathbf{j}$

47. $\theta(t) = \arccos\left(\dfrac{-7 \sin t \cos t}{\sqrt{9 \sin^2 t + 16 \cos^2 t}\sqrt{9 \cos^2 t + 16 \sin^2 t}}\right)$

Maximum: $\theta\left(\dfrac{\pi}{4}\right) = \theta\left(\dfrac{5\pi}{4}\right) \approx 1.855$

Minimum: $\theta\left(\dfrac{3\pi}{4}\right) = \theta\left(\dfrac{7\pi}{4}\right) \approx 1.287$

Orthogonal: $\dfrac{n\pi}{2}$, n is an integer.

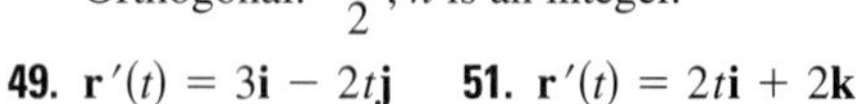

49. $\mathbf{r}'(t) = 3\mathbf{i} - 2t\mathbf{j}$ **51.** $\mathbf{r}'(t) = 2t\mathbf{i} + 2\mathbf{k}$

53. $t^2\mathbf{i} + t\mathbf{j} + t\mathbf{k} + \mathbf{C}$ **55.** $\ln t\mathbf{i} + t\mathbf{j} - \frac{2}{5}t^{5/2}\mathbf{k} + \mathbf{C}$

57. $(t^2 - t)\mathbf{i} + t^4\mathbf{j} + 2t^{3/2}\mathbf{k} + \mathbf{C}$

59. $\tan t\mathbf{i} + \arctan t\mathbf{j} + \mathbf{C}$ **61.** $4\mathbf{i} + \frac{1}{2}\mathbf{j} - \mathbf{k}$

63. $a\mathbf{i} + a\mathbf{j} + (\pi/2)\mathbf{k}$

65. $2\mathbf{i} + (e^2 - 1)\mathbf{j} - (e^2 + 1)\mathbf{k}$

67. $2e^{2t}\mathbf{i} + 3(e^t - 1)\mathbf{j}$ **69.** $600\sqrt{3}t\mathbf{i} + (-16t^2 + 600t)\mathbf{j}$

71. $((2 - e^{-t^2})/2)\mathbf{i} + (e^{-t} - 2)\mathbf{j} + (t + 1)\mathbf{k}$

73. See "Definition of the Derivative of a Vector-Valued Function" and Figure 12.8 on page 842.

75. The three components of $\mathbf{u}$ are increasing functions of t at $t = t_0$.

77–83. Proofs

85. (a) The curve is a cycloid.

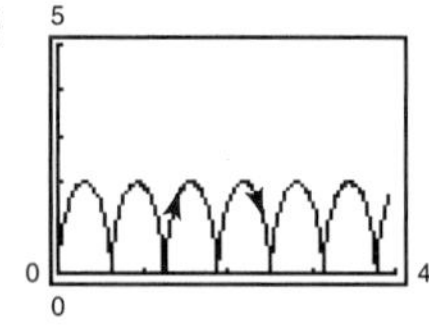

(b) The maximum of $\|\mathbf{r}'\|$ is 2; the minimum of $\|\mathbf{r}'\|$ is 0. The maximum and the minimum of $\|\mathbf{r}''\|$ is 1.

87. Proof **89.** True

91. False: Let $\mathbf{r}(t) = \cos t\mathbf{i} + \sin t\mathbf{j} + \mathbf{k}$, then $d/dt[\|\mathbf{r}(t)\|] = 0$, but $\|\mathbf{r}'(t)\| = 1$.

Section 12.3 (page 856)

1. $\mathbf{v}(1) = 3\mathbf{i} + \mathbf{j}$
$\mathbf{a}(1) = \mathbf{0}$

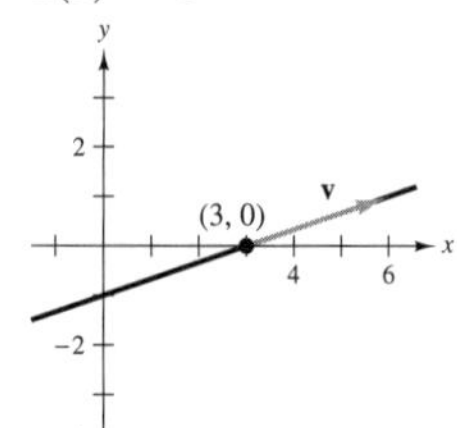

3. $\mathbf{v}(2) = 4\mathbf{i} + \mathbf{j}$
$\mathbf{a}(2) = 2\mathbf{i}$

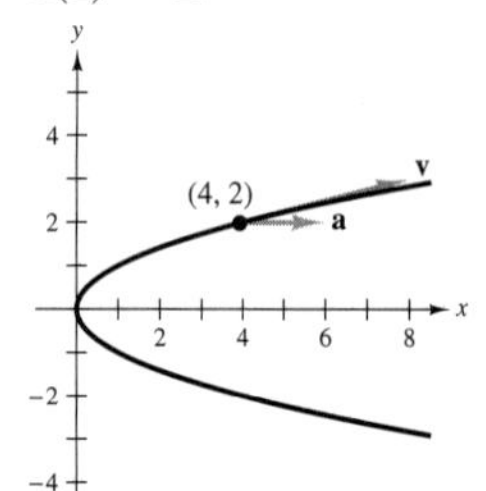

5. $\mathbf{v}(1) = 2\mathbf{i} + 3\mathbf{j}$
$\mathbf{a}(1) = 2\mathbf{i} + 6\mathbf{j}$

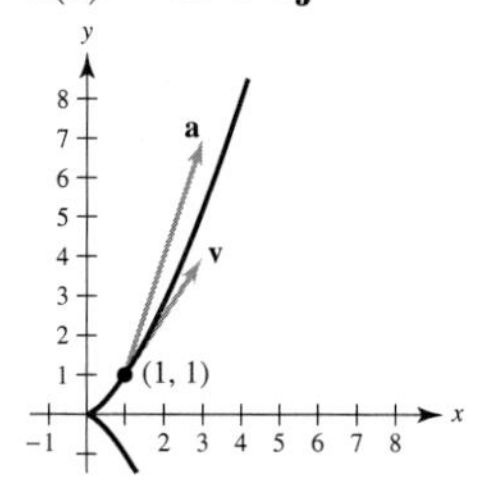

7. $\mathbf{v}(\pi/4) = -\sqrt{2}\mathbf{i} + \sqrt{2}\mathbf{j}$
$\mathbf{a}(\pi/4) = -\sqrt{2}\mathbf{i} - \sqrt{2}\mathbf{j}$

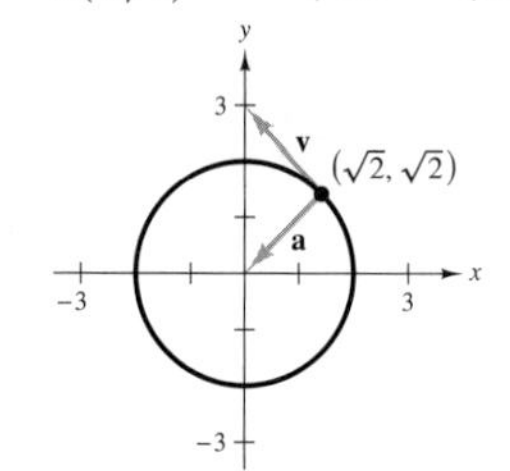

9. $\mathbf{v}(\pi) = 2\mathbf{i}$
$\mathbf{a}(\pi) = -\mathbf{j}$

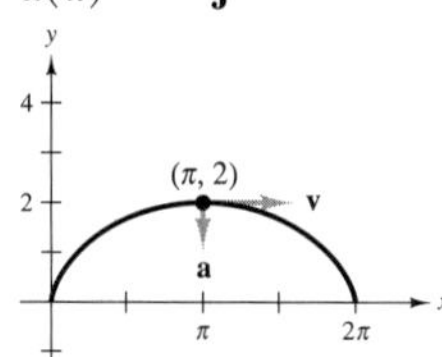

11. $\mathbf{v}(t) = \mathbf{i} + 5\mathbf{j} + 3\mathbf{k}$
$\|\mathbf{v}(t)\| = \sqrt{35}$
$\mathbf{a}(t) = \mathbf{0}$

13. $\mathbf{v}(t) = \mathbf{i} + 2t\mathbf{j} + t\mathbf{k}$
$\|\mathbf{v}(t)\| = \sqrt{1 + 5t^2}$
$\mathbf{a}(t) = 2\mathbf{j} + \mathbf{k}$

15. $\mathbf{v}(t) = \mathbf{i} + \mathbf{j} - \left(t/\sqrt{9 - t^2}\right)\mathbf{k}$
$\|\mathbf{v}(t)\| = \sqrt{(18 - t^2)/(9 - t^2)}$
$\mathbf{a}(t) = (-9/(9 - t^2)^{3/2})\mathbf{k}$

17. $\mathbf{v}(t) = 4\mathbf{i} - 3 \sin t\mathbf{j} + 3 \cos t\mathbf{k}$
$\|\mathbf{v}(t)\| = 5$
$\mathbf{a}(t) = -3 \cos t\mathbf{j} - 3 \sin t\mathbf{k}$

19. $\mathbf{v}(t) = (e^t \cos t - e^t \sin t)\mathbf{i} + (e^t \sin t + e^t \cos t)\mathbf{j} + e^t\mathbf{k}$
$\|\mathbf{v}(t)\| = e^t\sqrt{3}$
$\mathbf{a}(t) = -2e^t \sin t\mathbf{i} + 2e^t \cos t\mathbf{j} + e^t\mathbf{k}$

21. (a) $x = 1 + t$
$y = -1 - 2t$
$z = \frac{1}{4} + \frac{3}{4}t$
(b) $(1.100, -1.200, 0.325)$

23. $\mathbf{v}(t) = t(\mathbf{i} + \mathbf{j} + \mathbf{k})$
$\mathbf{r}(t) = (t^2/2)(\mathbf{i} + \mathbf{j} + \mathbf{k})$
$\mathbf{r}(2) = 2(\mathbf{i} + \mathbf{j} + \mathbf{k})$

25. $\mathbf{v}(t) = \left(t^2/2 + \frac{9}{2}\right)\mathbf{j} + \left(t^2/2 - \frac{1}{2}\right)\mathbf{k}$
$\mathbf{r}(t) = \left(t^3/6 + \frac{9}{2}t - \frac{14}{3}\right)\mathbf{j} + \left(t^3/6 - \frac{1}{2}t + \frac{1}{3}\right)\mathbf{k}$
$\mathbf{r}(2) = \frac{17}{3}\mathbf{j} + \frac{2}{3}\mathbf{k}$

27. $\mathbf{v}(t) = -\sin t\mathbf{i} + \cos t\mathbf{j} + \mathbf{k}$
$\mathbf{r}(t) = \cos t\mathbf{i} + \sin t\mathbf{j} + t\mathbf{k}$
$\mathbf{r}(2) = (\cos 2)\mathbf{i} + (\sin 2)\mathbf{j} + 2\mathbf{k}$

29. $\mathbf{r}(t) = 44\sqrt{3}t\mathbf{i} + (10 + 44t - 16t^2)\mathbf{j}$

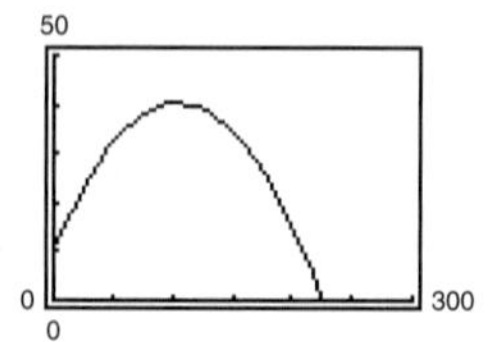

31. $v_0 = 40\sqrt{6}$ ft/sec; 78 ft **33.** Proof

35. (a) $y = -0.004x^2 + 0.37x + 6$
$\mathbf{r}(t) = t\mathbf{i} + (-0.004t^2 + 0.37t + 6)\mathbf{j}$

(b) 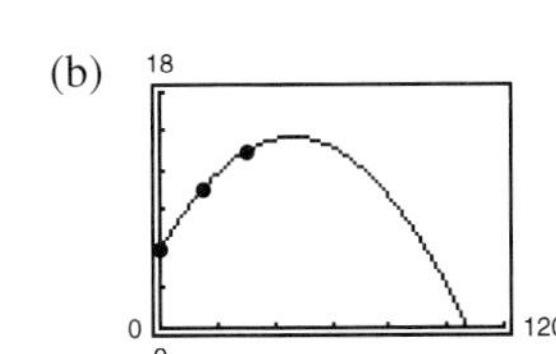

(c) 14.56 ft
(d) Initial velocity: 67.4 ft/sec; $\theta \approx 20.14°$

37. (a) $\mathbf{r}(t) = \left(\frac{440}{3}\cos\theta_0\right)t\mathbf{i} + \left[3 + \left(\frac{440}{3}\sin\theta_0\right)t - 16t^2\right]\mathbf{j}$
(b)

The minimum angle appears to be $\theta_0 = 20°$.
(c) $\theta_0 \approx 19.38°$

39. (a) $v_0 = 28.78$ ft/sec; $\theta = 58.28°$ (b) $v_0 \approx 32$ ft/sec

41. 1.91°

43. (a) 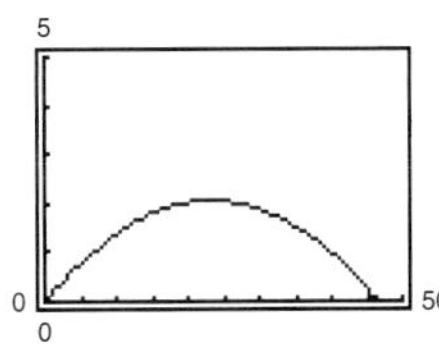

Maximum height: 2.1 ft
Range: 46.6 ft

(b) 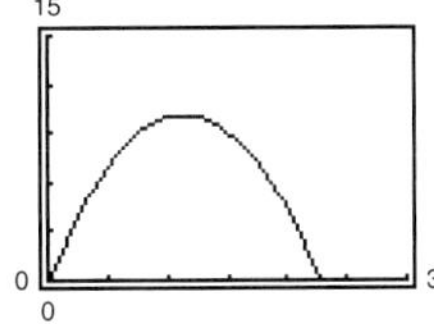

Maximum height: 10.0 ft
Range: 227.8 ft

(c) 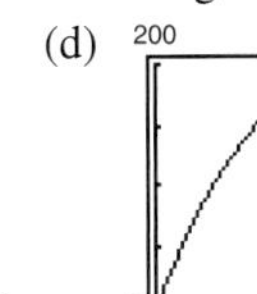

Maximum height: 34.0 ft
Range: 136.1 ft

(d) 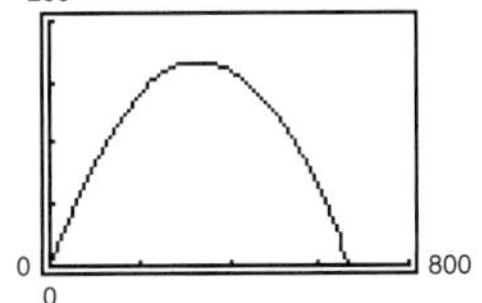

Maximum height: 166.5 ft
Range: 666.1 ft

(e)

Maximum height: 51.0 ft
Range: 117.9 ft

(f) 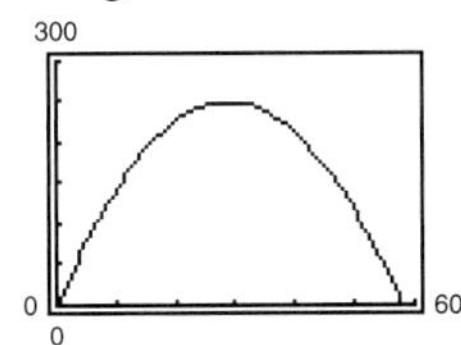

Maximum height: 249.8 ft
Range: 576.9 ft

45. Maximum height: 129.1 m
Range: 886.3 m

47. $\mathbf{v}(t) = b\omega[(1 - \cos\omega t)\mathbf{i} + \sin\omega t\mathbf{j}]$
$\mathbf{a}(t) = b\omega^2(\sin\omega t\mathbf{i} + \cos\omega t\mathbf{j})$
(a) $\|\mathbf{v}(t)\| = 0$ when $\omega t = 0, 2\pi, 4\pi, \ldots$
(b) $\|\mathbf{v}(t)\|$ is maximum when $\omega t = \pi, 3\pi, \ldots$

49. $\mathbf{v}(t) = -b\omega\sin\omega t\mathbf{i} + b\omega\cos\omega t\mathbf{j}$
$\mathbf{v}(t) \cdot \mathbf{r}(t) = 0$

51. $\mathbf{a}(t) = -b\omega^2(\cos\omega t\mathbf{i} + \sin\omega t\mathbf{j}) = -\omega^2\mathbf{r}(t)$; $\mathbf{a}(t)$ is a negative multiple of a unit vector from $(0, 0)$ to $(\cos\omega t, \sin\omega t)$, so $\mathbf{a}(t)$ is directed toward the origin.

53. $8\sqrt{10}$ ft/sec **55–57.** Proofs

59. (a) $\mathbf{v}(t) = -6\sin t\mathbf{i} + 3\cos t\mathbf{j}$
$\|\mathbf{v}(t)\| = 3\sqrt{3\sin^2 t + 1}$
$\mathbf{a}(t) = -6\cos t\mathbf{i} - 3\sin t\mathbf{j}$

(b)

t	0	$\pi/4$	$\pi/2$	$2\pi/3$	π
Speed	3	$3\sqrt{10}/2$	6	$3\sqrt{13}/2$	3

(c) 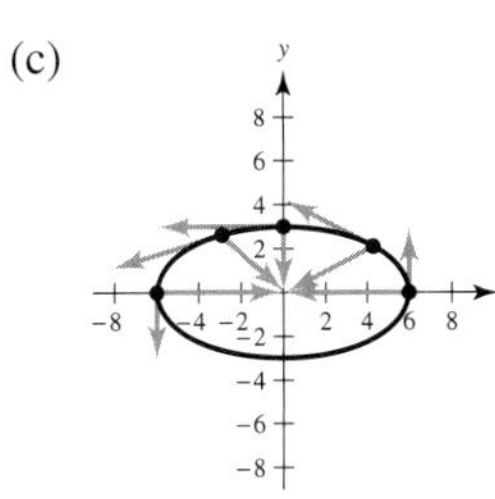

(d) The speed is increasing when the angle between $\mathbf{v}$ and $\mathbf{a}$ is in the interval $[0, \pi/2)$, and decreasing when the angle is in the interval $(\pi/2, \pi]$.

61. The velocity of an object involves both magnitude and direction of motion, whereas speed involves only magnitude.

63. (a) Velocity: $\mathbf{r}_2'(t) = 2\mathbf{r}_1'(2t)$
Acceleration: $\mathbf{r}_2''(t) = 4\mathbf{r}_1''(2t)$
(b) In general, if $\mathbf{r}_3(t) = \mathbf{r}_1(\omega t)$, then:
Velocity: $\mathbf{r}_3'(t) = \omega\mathbf{r}_1'(\omega t)$
Acceleration: $\mathbf{r}_3''(t) = \omega^2\mathbf{r}_1''(\omega t)$

65. False; acceleration is the derivative of the velocity.

67. True

Section 12.4 (page 865)

1. 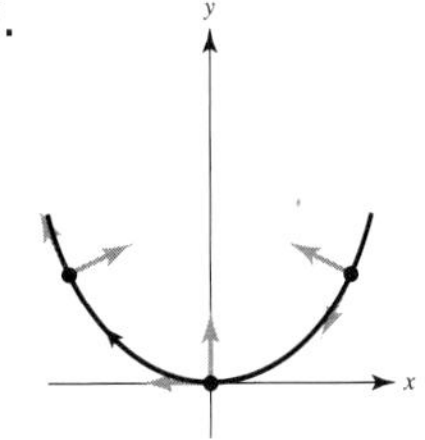

3. 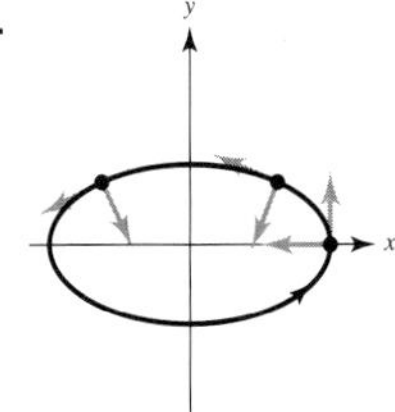

5. $\mathbf{T}(1) = (\sqrt{2}/2)(\mathbf{i} + \mathbf{j})$ **7.** $\mathbf{T}(\pi/4) = (\sqrt{2}/2)(-\mathbf{i} + \mathbf{j})$

9. $\mathbf{T}(e) = (3e\mathbf{i} - \mathbf{j})/\sqrt{9e^2 + 1} \approx 0.9926\mathbf{i} - 0.1217\mathbf{j}$

11. $\mathbf{T}(0) = (\sqrt{2}/2)(\mathbf{i} + \mathbf{k})$
$x = t$
$y = 0$
$z = t$

13. $\mathbf{T}(0) = (\sqrt{10}/10)(3\mathbf{j} + \mathbf{k})$
$x = 3$
$y = 3t$
$z = t$

15. $\mathbf{T}(\pi/4) = \frac{1}{2}\langle -\sqrt{2}, \sqrt{2}, 0\rangle$
$x = \sqrt{2} - \sqrt{2}t$
$y = \sqrt{2} + \sqrt{2}t$
$z = 4$

17. $\mathbf{T}(3) = \frac{1}{19}\langle 1, 6, 18\rangle$
$x = 3 + t$
$y = 9 + 6t$
$z = 18 + 18t$

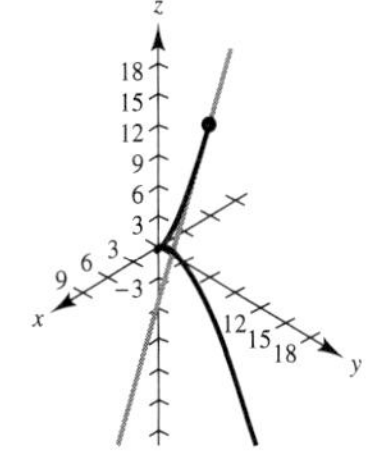

19. Tangent line: $x = 1 + t$, $y = t$, $z = 1 + \frac{1}{2}t$
$\mathbf{r}(1.1) \approx \langle 1.1, 0.1, 1.05\rangle$

21. 1.2° **23.** $\mathbf{N}(2) = (\sqrt{5}/5)(-2\mathbf{i} + \mathbf{j})$

25. $\mathbf{N}(2) = (-\sqrt{5}/5)(2\mathbf{i} - \mathbf{j})$

27. $\mathbf{N}(1) = (-\sqrt{14}/14)(\mathbf{i} - 2\mathbf{j} + 3\mathbf{k})$

29. $\mathbf{N}(3\pi/4) = (\sqrt{2}/2)(\mathbf{i} - \mathbf{j})$

31. $\mathbf{v}(t) = 4\mathbf{i}$
$\mathbf{a}(t) = \mathbf{0}$
$\mathbf{T}(t) = \mathbf{i}$
$\mathbf{N}(t)$ is undefined. The path is a line and the speed is constant.

33. $\mathbf{v}(t) = 8t\mathbf{i}$
$\mathbf{a}(t) = 8\mathbf{i}$
$\mathbf{T}(t) = \mathbf{i}$
$\mathbf{N}(t)$ is undefined. The path is a line and the speed is variable.

35. $\mathbf{T}(1) = (\sqrt{2}/2)(\mathbf{i} - \mathbf{j})$
$\mathbf{N}(1) = (\sqrt{2}/2)(\mathbf{i} + \mathbf{j})$
$a_{\mathbf{T}} = -\sqrt{2}$
$a_{\mathbf{N}} = \sqrt{2}$

37. $\mathbf{T}(1) = (-\sqrt{5}/5)(\mathbf{i} - 2\mathbf{j})$
$\mathbf{N}(1) = (-\sqrt{5}/5)(2\mathbf{i} + \mathbf{j})$
$a_{\mathbf{T}} = 14\sqrt{5}/5$
$a_{\mathbf{N}} = 8\sqrt{5}/5$

39. $\mathbf{T}(0) = (\sqrt{5}/5)(\mathbf{i} - 2\mathbf{j})$
$\mathbf{N}(0) = (\sqrt{5}/5)(2\mathbf{i} + \mathbf{j})$
$a_{\mathbf{T}} = -7\sqrt{5}/5$
$a_{\mathbf{N}} = 6\sqrt{5}/5$

41. $\mathbf{T}(\pi/2) = (\sqrt{2}/2)(-\mathbf{i} + \mathbf{j})$
$\mathbf{N}(\pi/2) = (-\sqrt{2}/2)(\mathbf{i} + \mathbf{j})$
$a_{\mathbf{T}} = \sqrt{2}e^{\pi/2}$
$a_{\mathbf{N}} = \sqrt{2}e^{\pi/2}$

43. $\mathbf{T}(t_0) = (\cos \omega t_0)\mathbf{i} + (\sin \omega t_0)\mathbf{j}$
$\mathbf{N}(t_0) = (-\sin \omega t_0)\mathbf{i} + (\cos \omega t_0)\mathbf{j}$
$a_{\mathbf{T}} = \omega^2$
$a_{\mathbf{N}} = \omega^3 t_0$

45. $\mathbf{T}(t) = -\sin(\omega t)\mathbf{i} + \cos(\omega t)\mathbf{j}$
$\mathbf{N}(t) = -\cos(\omega t)\mathbf{i} - \sin(\omega t)\mathbf{j}$
$a_{\mathbf{T}} = 0$
$a_{\mathbf{N}} = a\omega^2$

47. $\|\mathbf{v}(t)\| = a\omega$; The speed is constant because $a_{\mathbf{T}} = 0$.

49. $\mathbf{r}(2) = 2\mathbf{i} + \frac{1}{2}\mathbf{j}$
$\mathbf{T}(2) = (\sqrt{17}/17)(4\mathbf{i} - \mathbf{j})$
$\mathbf{N}(2) = (\sqrt{17}/17)(\mathbf{i} + 4\mathbf{j})$

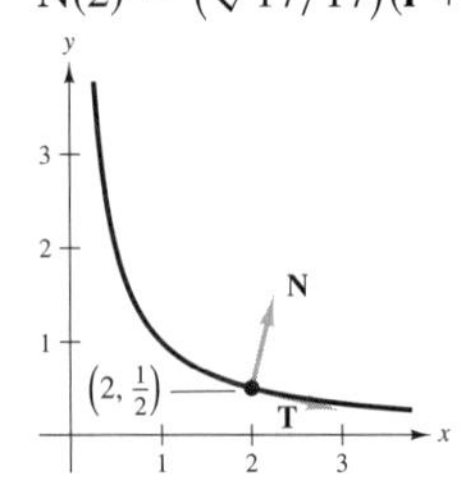

51. $\mathbf{r}(1/4) = \mathbf{i} + (1/4)\mathbf{j}$
$\mathbf{T}(1/4) = (\sqrt{5}/5)(2\mathbf{i} + \mathbf{j})$
$\mathbf{N}(1/4) = (2\sqrt{5}/5)[-(1/2)\mathbf{i} + \mathbf{j}]$

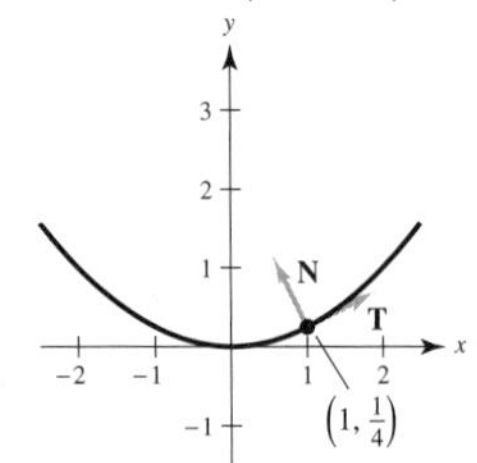

53. $\mathbf{r}(\pi/4) = \sqrt{2}\mathbf{i} + \sqrt{2}\mathbf{j}$
$\mathbf{T}(\pi/4) = (\sqrt{2}/2)(-\mathbf{i} + \mathbf{j})$
$\mathbf{N}(\pi/4) = (\sqrt{2}/2)(-\mathbf{i} - \mathbf{j})$

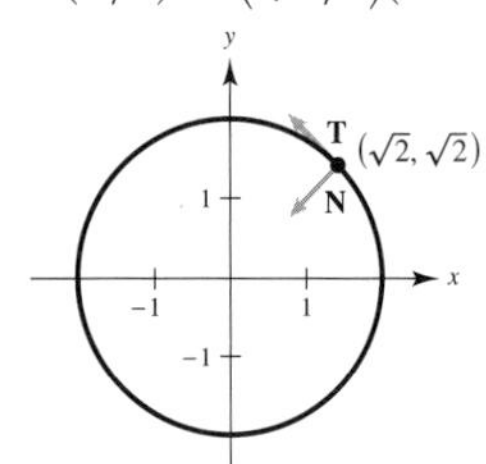

55. $\mathbf{T}(1) = (\sqrt{14}/14)(\mathbf{i} + 2\mathbf{j} - 3\mathbf{k})$
$\mathbf{N}(1)$ is undefined.
$a_{\mathbf{T}}$ is undefined.
$a_{\mathbf{N}}$ is undefined.

57. $\mathbf{T}(\pi/3) = (\sqrt{5}/5)[-(\sqrt{3}/2)\mathbf{i} + (1/2)\mathbf{j} + 2\mathbf{k}]$
$\mathbf{N}(\pi/3) = -(1/2)\mathbf{i} - (\sqrt{3}/2)\mathbf{j}$
$a_{\mathbf{T}} = 0$
$a_{\mathbf{N}} = 1$

59. $\mathbf{T}(1) = (\sqrt{6}/6)(\mathbf{i} + 2\mathbf{j} + \mathbf{k})$
$\mathbf{N}(1) = (\sqrt{30}/30)(-5\mathbf{i} + 2\mathbf{j} + \mathbf{k})$
$a_{\mathbf{T}} = 5\sqrt{6}/6$
$a_{\mathbf{N}} = \sqrt{30}/6$

61. $\mathbf{T}(0) = (\sqrt{3}/3)(\mathbf{i} + \mathbf{j} + \mathbf{k})$
$\mathbf{N}(0) = (\sqrt{2}/2)(\mathbf{i} - \mathbf{j})$
$a_{\mathbf{T}} = \sqrt{3}$
$a_{\mathbf{N}} = \sqrt{2}$

63. $\mathbf{T}(\pi/2) = \frac{1}{5}(4\mathbf{i} - 3\mathbf{j})$
$\mathbf{N}(\pi/2) = -\mathbf{k}$
$a_{\mathbf{T}} = 0$
$a_{\mathbf{N}} = 3$

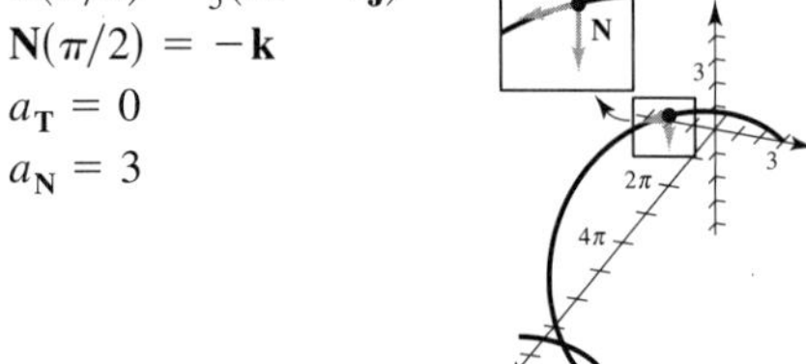

65. $\mathbf{T}(2) = (\sqrt{149}/149)(\mathbf{i} + 12\mathbf{j} + 2\mathbf{k})$
$\mathbf{N}(2) = (\sqrt{5513}/5513)(-74\mathbf{i} + 6\mathbf{j} + \mathbf{k})$
$a_{\mathbf{T}} = 74\sqrt{149}/149$
$a_{\mathbf{N}} = \sqrt{5513}/149$

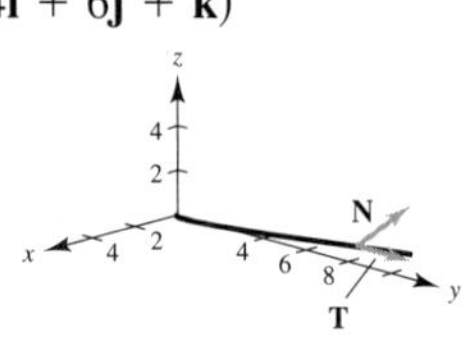

67. Let C be a smooth curve represented by $\mathbf{r}$ on an open interval I. The unit tangent vector $\mathbf{T}(t)$ at t is defined as

$$\mathbf{T}(t) = \frac{\mathbf{r}'(t)}{\|\mathbf{r}'(t)\|}, \ \mathbf{r}'(t) \neq 0.$$

The principal unit normal vector $\mathbf{N}(t)$ at t is defined as

$$\mathbf{N}(t) = \frac{\mathbf{T}'(t)}{\|\mathbf{T}'(t)\|}, \ \mathbf{T}'(t) \neq 0.$$

The tangential and normal components of acceleration are defined as $\mathbf{a}(t) = a_{\mathbf{T}}\mathbf{T}(t) + a_{\mathbf{N}}\mathbf{N}(t)$.

69. (a) The particle's motion is in a straight line.
(b) The particle's speed is constant.

71. (a) $t = \frac{1}{2}$: $a_{\mathbf{T}} = \sqrt{2}\pi^2/2$, $a_{\mathbf{N}} = \sqrt{2}\pi^2/2$
$t = 1$: $a_{\mathbf{T}} = 0$, $a_{\mathbf{N}} = \pi^2$
$t = \frac{3}{2}$: $a_{\mathbf{T}} = -\sqrt{2}\pi^2/2$, $a_{\mathbf{N}} = \sqrt{2}\pi^2/2$
(b) $t = \frac{1}{2}$: Increasing because $a_{\mathbf{T}} > 0$.
$t = 1$: Maximum because $a_{\mathbf{T}} = 0$.
$t = \frac{3}{2}$: Decreasing because $a_{\mathbf{T}} < 0$.

73. $\mathbf{T}(\pi/2) = (\sqrt{17}/17)(-4\mathbf{i} + \mathbf{k})$
$\mathbf{N}(\pi/2) = -\mathbf{j}$
$\mathbf{B}(\pi/2) = (\sqrt{17}/17)(\mathbf{i} + 4\mathbf{k})$

75. $\mathbf{T}(\pi/4) = (\sqrt{2}/2)(\mathbf{j} - \mathbf{k})$
$\mathbf{N}(\pi/4) = -(\sqrt{2}/2)(\mathbf{j} + \mathbf{k})$
$\mathbf{B}(\pi/4) = -\mathbf{i}$

77. $\mathbf{T}(\pi/3) = (\sqrt{5}/5)(\mathbf{i} - \sqrt{3}\mathbf{j} + \mathbf{k})$
$\mathbf{N}(\pi/3) = -\frac{1}{2}(\sqrt{3}\mathbf{i} + \mathbf{j})$
$\mathbf{B}(\pi/3) = (\sqrt{5}/10)(\mathbf{i} - \sqrt{3}\mathbf{j} - 4\mathbf{k})$

79. $a_{\mathbf{T}} = \dfrac{-32(v_0 \sin\theta - 32t)}{\sqrt{v_0^2\cos^2\theta + (v_0\sin\theta - 32t)^2}}$
$a_{\mathbf{N}} = \dfrac{32v_0\cos\theta}{\sqrt{v_0^2\cos^2\theta + (v_0\sin\theta - 32t)^2}}$
At maximum height, $a_{\mathbf{T}} = 0$ and $a_{\mathbf{N}} = 32$.

81. (a) $\mathbf{r}(t) = 60\sqrt{3}t\mathbf{i} + (5 + 60t - 16t^2)\mathbf{j}$
(b)
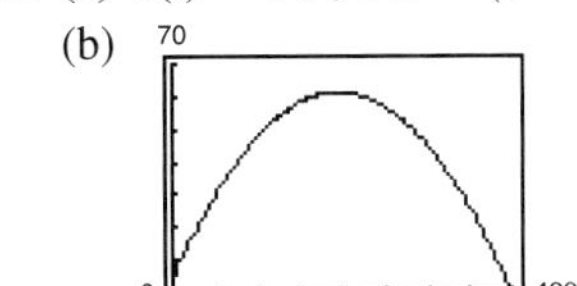

Maximum height ≈ 61.245 ft
Range ≈ 398.186 ft
(c) $\mathbf{v}(t) = 60\sqrt{3}\mathbf{i} + (60 - 32t)\mathbf{j}$
$\|\mathbf{v}(t)\| = 8\sqrt{16t^2 - 60t + 225}$
$\mathbf{a}(t) = -32\mathbf{j}$
(d)

t	0.5	1.0	1.5
Speed	112.85	107.63	104.61

t	2.0	2.5	3.0
Speed	104	105.83	109.98

(e)
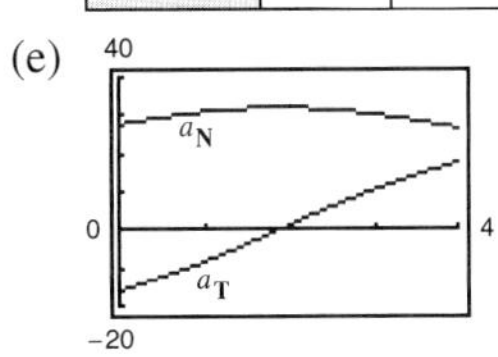

The speed is decreasing when $a_{\mathbf{T}}$ and $a_{\mathbf{N}}$ have opposite signs.

83. (a) $4\sqrt{625\pi^2 + 1} \approx 314$ mi/h
(b) $a_{\mathbf{T}} = 0$, $a_{\mathbf{N}} = 1000\pi^2$
$a_{\mathbf{T}} = 0$ because the speed is constant.

85. (a) The centripetal component is quadrupled.
(b) The centripetal component is halved.

87. 4.82 mi/sec **89.** 4.67 mi/sec

91. False; centripetal acceleration may occur with constant speed.

93. (a) Proof (b) Proof **95–97.** Proofs

Section 12.5 (page 877)

1.
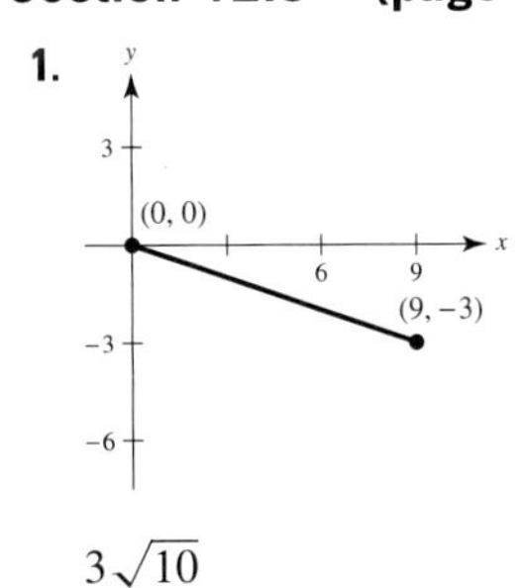

$3\sqrt{10}$

3.
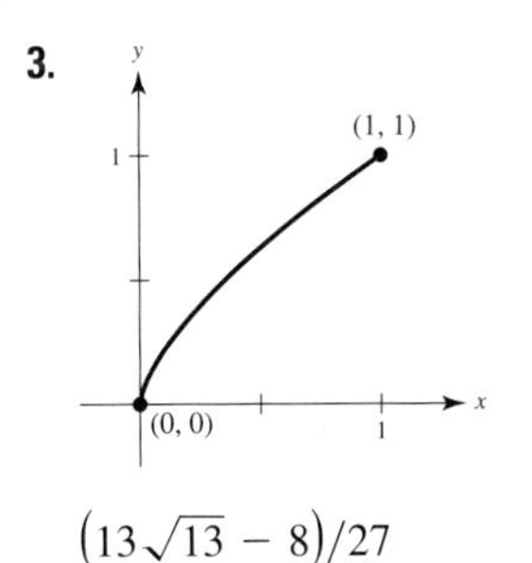

$(13\sqrt{13} - 8)/27$

5.
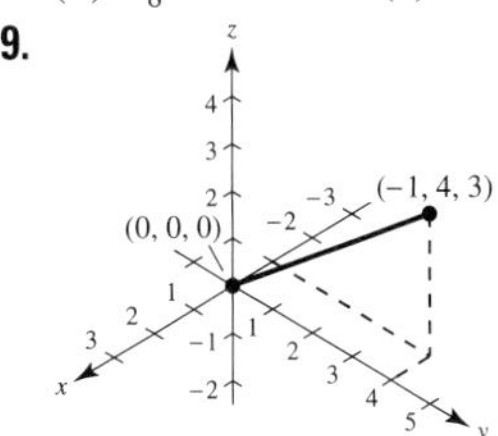

$6a$

7. (a) $\mathbf{r}(t) = (50t\sqrt{2})\mathbf{i} + (3 + 50t\sqrt{2} - 16t^2)\mathbf{j}$
(b) $\frac{649}{8} \approx 81$ ft (c) 315.5 ft (d) 362.9 ft

9.
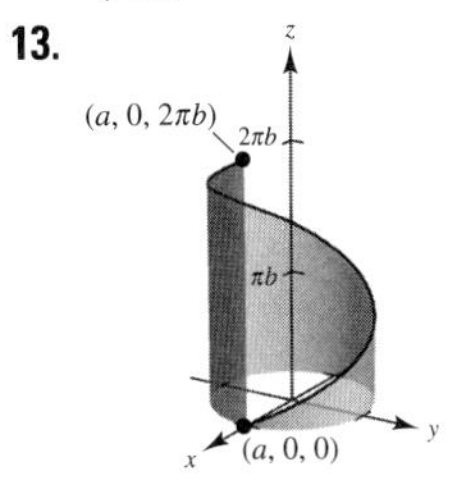

$\sqrt{26}$

11.
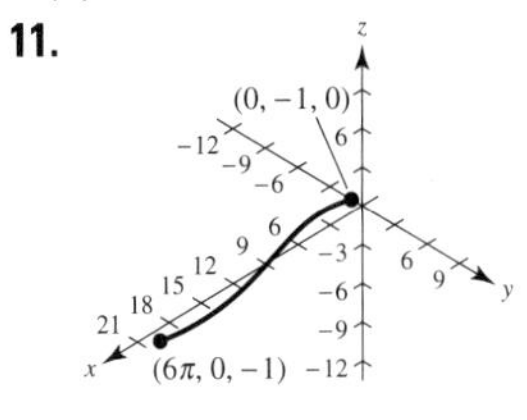

$3\sqrt{17}\pi/2$

13.

$2\pi\sqrt{a^2 + b^2}$

15. 8.37

17. (a) $2\sqrt{21} \approx 9.165$ (b) 9.529
(c) Increase the number of line segments. (d) 9.571

19. (a) $s = \sqrt{5}t$ (b) $\mathbf{r}(s) = 2\cos\dfrac{s}{\sqrt{5}}\mathbf{i} + 2\sin\dfrac{s}{\sqrt{5}}\mathbf{j} + \dfrac{s}{\sqrt{5}}\mathbf{k}$
(c) $s = \sqrt{5}$: (1.081, 1.683, 1.000)
$s = 4$: (−0.433, 1.953, 1.789)
(d) Proof

21. 0 **23.** $\frac{2}{5}$ **25.** 0 **27.** $\sqrt{2}/2$ **29.** 1

31. $\frac{1}{4}$ **33.** $1/a$ **35.** $\sqrt{2}/(4a\sqrt{1 - \cos\omega t})$

37. $\sqrt{5}/(1 + 5t^2)^{3/2}$ **39.** $\frac{3}{25}$ **41.** $\frac{12}{125}$ **43.** $7\sqrt{26}/676$

45. $K = 0$, $1/K$ is undefined.

47. $K = 4/17^{3/2}$, $1/K = 17^{3/2}/4$ **49.** $K = 4$, $1/K = 1/4$

51. $K = 1/a$, $1/K = a$

53. $K = 12/145^{3/2}$, $1/K = 145^{3/2}/12$

55. (a) $(x - \pi/2)^2 + y^2 = 1$
(b) Because the curvature is not as great, the radius of the curvature is greater.

57. $(x - 1)^2 + \left(y - \frac{5}{2}\right)^2 = \left(\frac{1}{2}\right)^2$

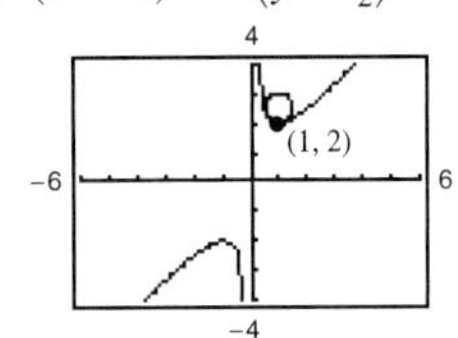

59. $(x + 2)^2 + (y - 3)^2 = 8$

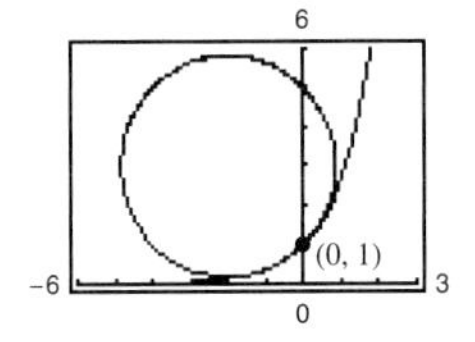

61.

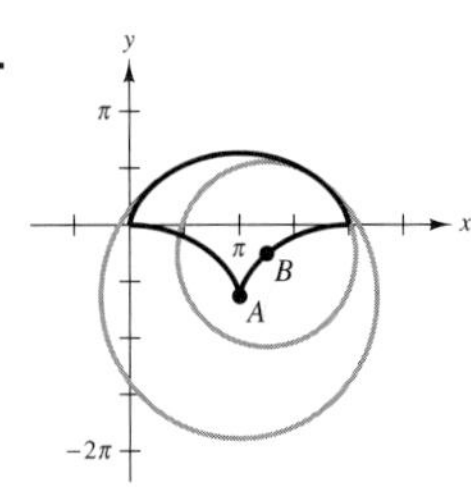

63. (a) $(1, 3)$ (b) 0

65. (a) $K \to \infty$ as $x \to 0$ (No maximum) (b) 0

67. (a) $(1/\sqrt{2}, -\ln 2/2)$ (b) 0

69. (a) $(\pm \text{arcsinh}(1), 1)$ (b) 0

71. $(0, 1)$ **73.** $(\pi/2 + K\pi, 0)$

75. (a) $s = \int_a^b \sqrt{[x'(t)]^2 + [y'(t)]^2 + [z'(t)]^2}\, dt = \int_a^b \|\mathbf{r}'(t)\|\, dt$

(b) Plane: $K = \left\|\dfrac{d\mathbf{T}}{ds}\right\| = \|\mathbf{T}'(s)\|$

Space: $K = \dfrac{\|\mathbf{T}'(t)\|}{\|\mathbf{r}'(t)\|} = \dfrac{\|\mathbf{r}'(t) \times \mathbf{r}''(t)\|}{\|\mathbf{r}'(t)\|^3}$

77. $K = |y''|$; Yes, for example, $y = x^4$ has a curvature of 0 at its relative minimum $(0, 0)$. The curvature is positive at any other point on the curve.

79. Proof

81. (a) $K = \dfrac{2|6x^2 - 1|}{(16x^6 - 16x^4 + 4x^2 + 1)^{3/2}}$

(b) $x = 0$: $x^2 + \left(y + \dfrac{1}{2}\right)^2 = \dfrac{1}{4}$

$x = 1$: $x^2 + \left(y - \dfrac{1}{2}\right)^2 = \dfrac{5}{4}$

(c)

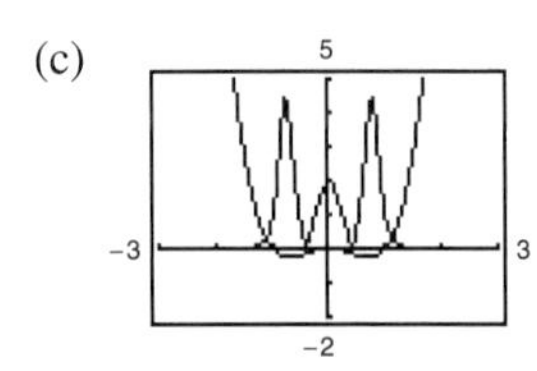

The curvature tends to be greatest near the extrema of the function and decreases as $x \to \pm\infty$. However, f and K do not have the same critical numbers.

Critical numbers of f: $x = 0, \pm\sqrt{2}/2 \approx \pm 0.7071$

Critical numbers of K: $x = 0, \pm 0.7647, \pm 0.4082$

83. (a) 12.25 units (b) $\frac{1}{2}$ **85–87.** Proofs

89. (a) 0 (b) 0 **91.** $\frac{1}{4}$ **93.** Proof

95. $K = [1/(4a)]|\csc(\theta/2)|$ **97.** 3327.5 lb

Minimum: $K = 1/(4a)$

There is no maximum.

99. Proof

101. False. See Exploration on page 869. **103.** True

105–111. Proofs

Review Exercises for Chapter 12 (page 881)

1. (a) All reals except $(\pi/2) + n\pi$, n is an integer

(b) Continuous except at $t = (\pi/2) + n\pi$, n is an integer

3. (a) $(0, \infty)$ (b) Continuous for all $t > 0$

5. (a) $\mathbf{i} - \sqrt{2}\mathbf{k}$ (b) $-3\mathbf{i} + 4\mathbf{j}$

(c) $(2c - 1)\mathbf{i} + (c - 1)^2\mathbf{j} - \sqrt{c + 1}\,\mathbf{k}$

(d) $2\Delta t\mathbf{i} + \Delta t(\Delta t + 2)\mathbf{j} - \left(\sqrt{\Delta t + 3} - \sqrt{3}\right)\mathbf{k}$

7.

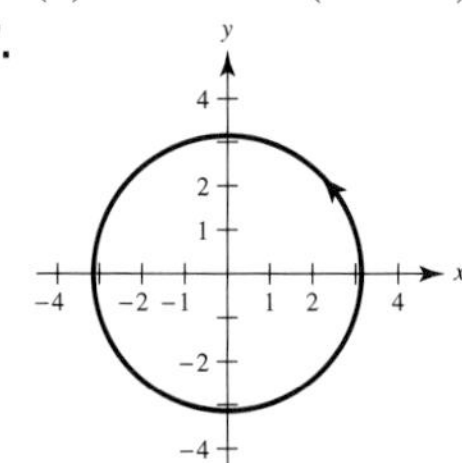

9.

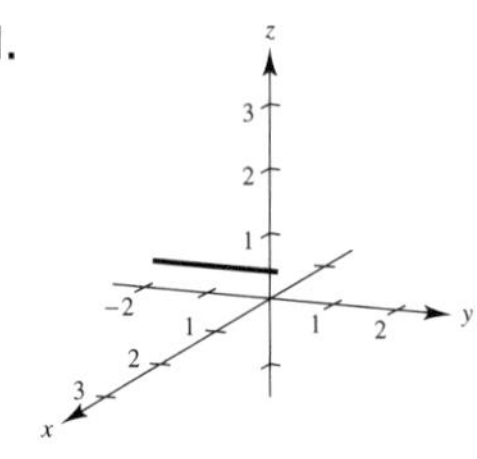

11.

13.

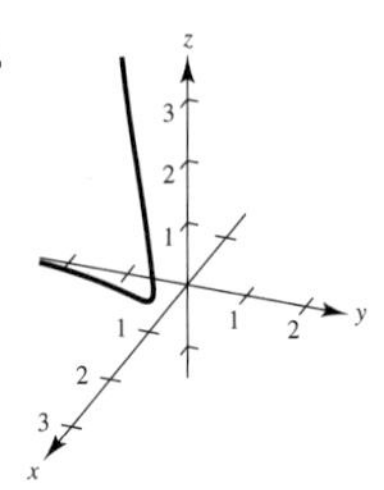

15. $\mathbf{r}_1(t) = 3t\mathbf{i} + 4t\mathbf{j}$, $0 \le t \le 1$

$\mathbf{r}_2(t) = 3\mathbf{i} + (4 - t)\mathbf{j}$, $0 \le t \le 4$

$\mathbf{r}_3(t) = (3 - t)\mathbf{i}$, $0 \le t \le 3$

17. $\mathbf{r}(t) = \langle -2 + 7t, -3 + 4t, 8 - 10t \rangle$

(Answer is not unique.)

19.

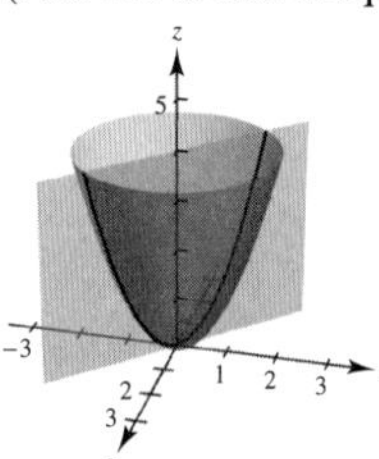

$x = t, y = -t, z = 2t^2$

21. $4\mathbf{i} + \mathbf{k}$

23. (a) $3\mathbf{i} + \mathbf{j}$ (b) $\mathbf{0}$ (c) $4t + 3t^2$

(d) $-5\mathbf{i} + (2t - 2)\mathbf{j} + 2t^2\mathbf{k}$

(e) $(10t - 1)/\sqrt{10t^2 - 2t + 1}$

(f) $\left(\frac{8}{3}t^3 - 2t^2\right)\mathbf{i} - 8t^3\mathbf{j} + (9t^2 - 2t + 1)\mathbf{k}$

25. $x(t)$ and $y(t)$ are increasing functions at $t = t_0$, and $z(t)$ is a decreasing function at $t = t_0$.

27. $\sin t\mathbf{i} + (t \sin t + \cos t)\mathbf{j} + \mathbf{C}$

29. $\frac{1}{2}\left(t\sqrt{1 + t^2} + \ln\left|t + \sqrt{1 + t^2}\right|\right) + \mathbf{C}$

31. $\frac{32}{3}\mathbf{j}$ **33.** $2(e - 1)\mathbf{i} - 8\mathbf{j} - 2\mathbf{k}$

35. $\mathbf{r}(t) = (t^2 + 1)\mathbf{i} + (e^t + 2)\mathbf{j} - (e^{-t} + 4)\mathbf{k}$

37. $\mathbf{v}(t) = 4\mathbf{i} + 3t^2\mathbf{j} - \mathbf{k}$

$\|\mathbf{v}(t)\| = \sqrt{17 + 9t^4}$

$\mathbf{a}(t) = 6t\mathbf{j}$

39. $\mathbf{v}(t) = \langle -3\cos^2 t \sin t, 3\sin^2 t \cos t, 3 \rangle$

$\|\mathbf{v}(t)\| = 3\sqrt{\sin^2 t \cos^2 t + 1}$

$\mathbf{a}(t) = \langle 3\cos t(2\sin^2 t - \cos^2 t), 3\sin t(2\cos^2 t - \sin^2 t), 0 \rangle$

41. $x(t) = t, y(t) = 16 + 8t, z(t) = 2 + \frac{1}{2}t$

$\mathbf{r}(4.1) \approx \langle 0.1, 16.8, 2.05 \rangle$

43. About 191.0 ft **45.** About 38.1 m/sec

47. $\mathbf{v} = -\mathbf{i} + 3\mathbf{j}$
$\|\mathbf{v}\| = \sqrt{10}$
$\mathbf{a} = \mathbf{0}$
$\mathbf{a} \cdot \mathbf{T} = 0$
$\mathbf{a} \cdot \mathbf{N}$ does not exist.

49. $\mathbf{v} = \mathbf{i} + \left[1/\left(2\sqrt{t}\right)\right]\mathbf{j}$
$\|\mathbf{v}\| = \sqrt{4t+1}/\left(2\sqrt{t}\right)$
$\mathbf{a} = \left[-1/\left(4t\sqrt{t}\right)\right]\mathbf{j}$
$\mathbf{a} \cdot \mathbf{T} = -1/\left(4t\sqrt{t}\sqrt{4t+1}\right)$
$\mathbf{a} \cdot \mathbf{N} = 1/\left(2t\sqrt{4t+1}\right)$

51. $\mathbf{v} = e^t\mathbf{i} - e^{-t}\mathbf{j}$
$\|\mathbf{v}\| = \sqrt{e^{2t} + e^{-2t}}$
$\mathbf{a} = e^t\mathbf{i} + e^{-t}\mathbf{j}$
$\mathbf{a} \cdot \mathbf{T} = \dfrac{e^{2t} - e^{-2t}}{\sqrt{e^{2t} + e^{-2t}}}$
$\mathbf{a} \cdot \mathbf{N} = \dfrac{2}{\sqrt{e^{2t} + e^{-2t}}}$

53. $\mathbf{v} = \mathbf{i} + 2t\mathbf{j} + t\mathbf{k}$
$\|\mathbf{v}\| = \sqrt{1 + 5t^2}$
$\mathbf{a} = 2\mathbf{j} + \mathbf{k}$
$\mathbf{a} \cdot \mathbf{T} = \dfrac{5t}{\sqrt{1 + 5t^2}}$
$\mathbf{a} \cdot \mathbf{N} = \dfrac{\sqrt{5}}{\sqrt{1 + 5t^2}}$

55. $x = -\sqrt{3}t + 1$
$y = t + \sqrt{3}$
$z = t + (\pi/3)$

57. 4.58 mi/sec

59.

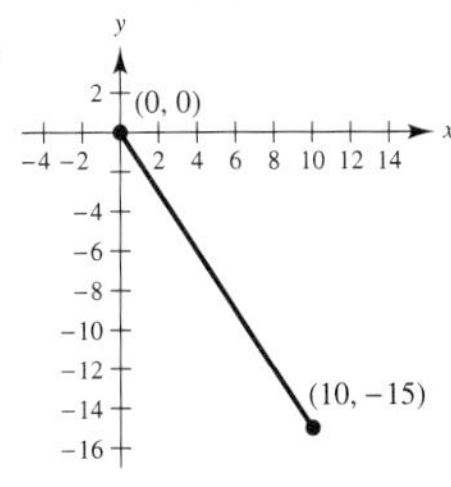

$5\sqrt{13}$

61.

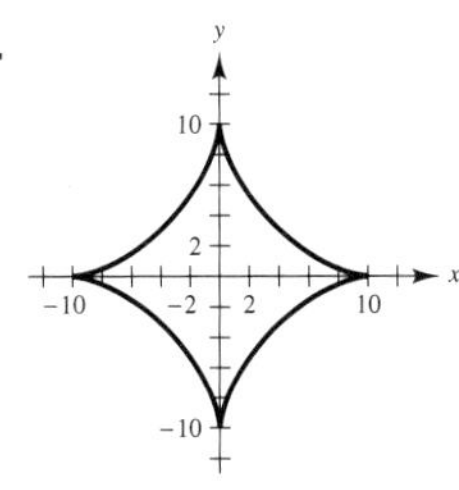

60

63.

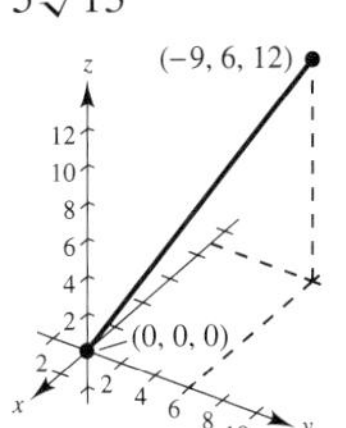

$3\sqrt{29}$

65.

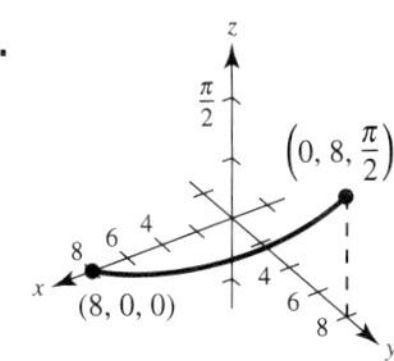

$\sqrt{65}\pi/2$

67. 0 **69.** $\left(2\sqrt{5}\right)/(4 + 5t^2)^{3/2}$ **71.** $\sqrt{2}/3$

73. $K = \sqrt{17}/289$; $r = 17\sqrt{17}$ **75.** $K = \sqrt{2}/4$; $r = 2\sqrt{2}$

77. The curvature changes abruptly from zero to a nonzero constant at the points B and C.

P.S. Problem Solving (page 883)

1. (a) a (b) πa (c) $K = \pi a$

3. Initial speed: 447.21 ft/sec; $\theta \approx 63.43°$

5–7. Proofs

9. Unit tangent: $\left\langle -\frac{4}{5}, 0, \frac{3}{5} \right\rangle$
Unit normal: $\langle 0, -1, 0 \rangle$
Binormal: $\left\langle \frac{3}{5}, 0, \frac{4}{5} \right\rangle$

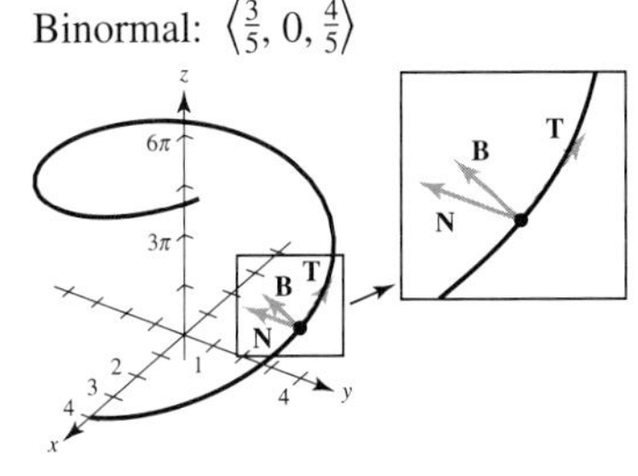

11. (a) Proof (b) Proof

13. (a)

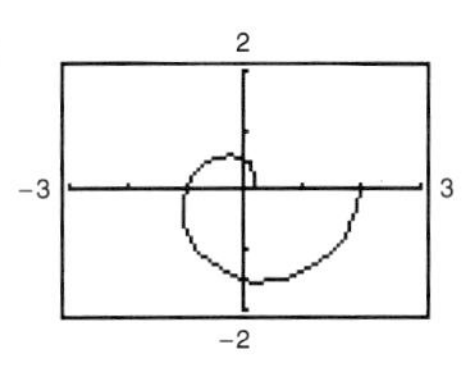

(b) 6.766

(c) $K = [\pi(\pi^2t^2 + 2)]/(\pi^2t^2 + 1)^{3/2}$
$K(0) = 2\pi$
$K(1) = [\pi(\pi^2 + 2)]/(\pi^2 + 1)^{3/2} \approx 1.04$
$K(2) \approx 0.51$

(d)

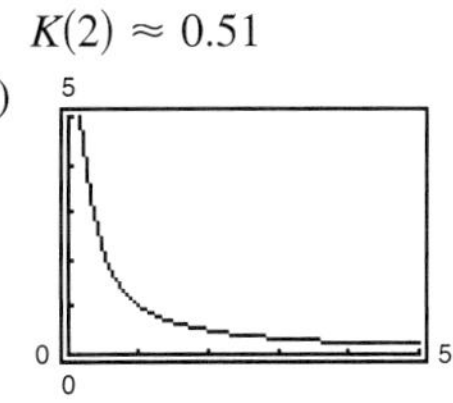

(e) $\lim_{t\to\infty} K = 0$

(f) As $t \to \infty$, the graph spirals outward and the curvature decreases.

Chapter 13

Section 13.1 (page 894)

1. Not a function because for some values of x and y (for example $x = y = 0$), there are two z-values.

3. z is a function of x and y. **5.** z is not a function of x and y.

7. (a) 6 (b) -4 (c) 150 (d) $5y$ (e) $2x$ (f) $5t$

9. (a) 5 (b) $3e^2$ (c) $2/e$ (d) $5e^y$ (e) xe^2 (f) te^t

11. (a) $\frac{2}{3}$ (b) 0 (c) $-\frac{3}{2}$ (d) $-\frac{10}{3}$

13. (a) $\sqrt{2}$ (b) $3 \sin 1$ (c) $-3\sqrt{3}/2$ (d) 4

15. (a) -4 (b) -6 (c) $-\frac{25}{4}$ (d) $\frac{9}{4}$

17. (a) $2, \Delta x \neq 0$ (b) $2y + \Delta y, \Delta y \neq 0$

19. Domain: $\{(x, y)$: x is any real number, y is any real number$\}$
Range: $z \geq 0$

21. Domain: $\{(x, y): y \geq 0\}$
Range: all real numbers

23. Domain: $\{(x, y): x \neq 0, y \neq 0\}$
Range: all real numbers

25. Domain: $\{(x, y): x^2 + y^2 \leq 4\}$
Range: $0 \leq z \leq 2$

27. Domain: $\{(x, y): -1 \leq x + y \leq 1\}$
Range: $0 \leq z \leq \pi$

29. Domain: $\{(x, y): y < -x + 4\}$
Range: all real numbers

31. (a) (20, 0, 0) (b) (−15, 10, 20)
(c) (20, 15, 25) (d) (20, 20, 0)

33.

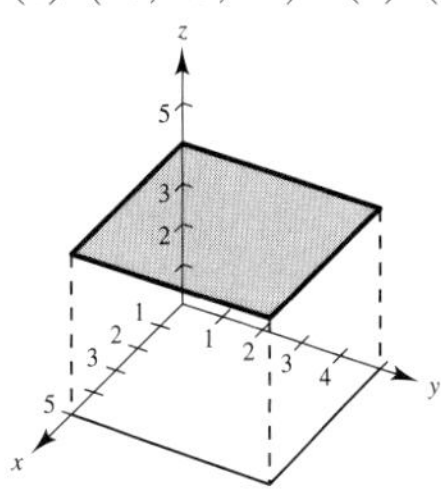

35.

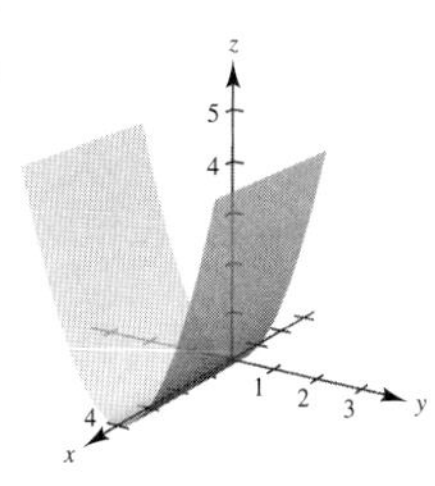

37.

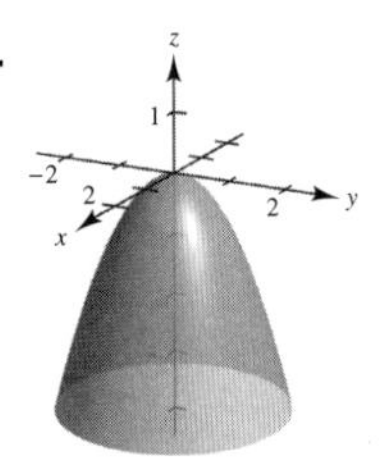

39.

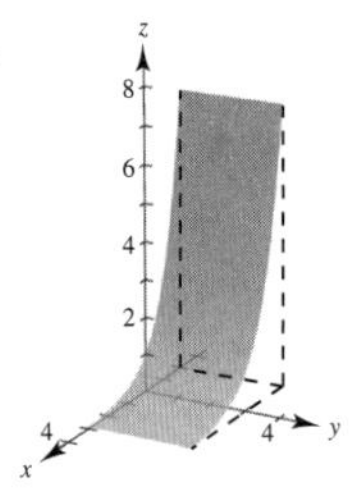

41.

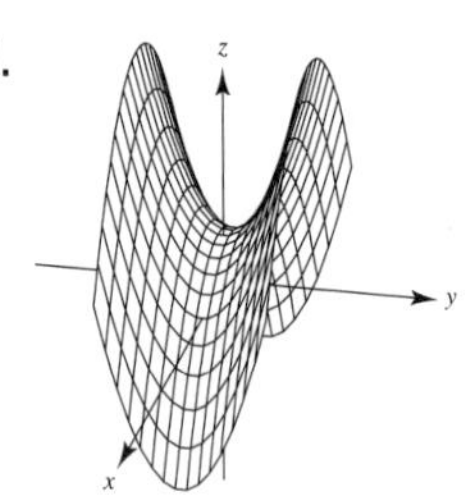

43.

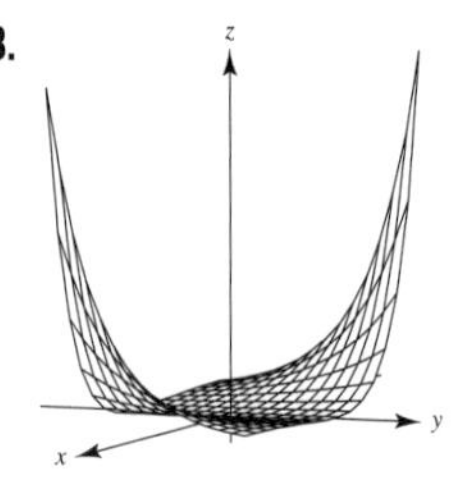

45. c **46.** d **47.** b **48.** a

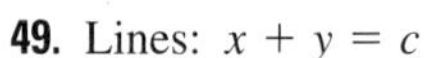

49. Lines: $x + y = c$

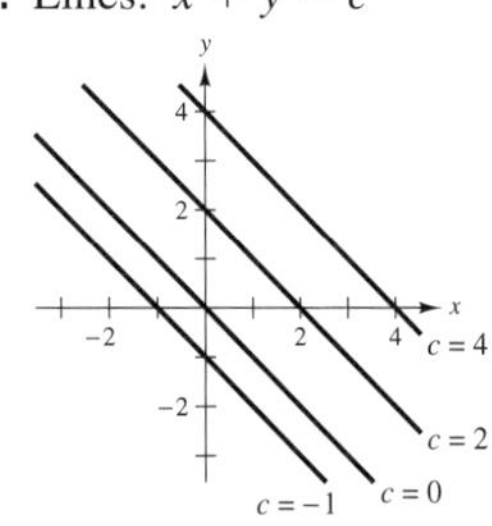

51. Ellipses: $x^2 + 4y^2 = c$
(except $x^2 + 4y^2 = 0$ is the point $(0, 0)$.)

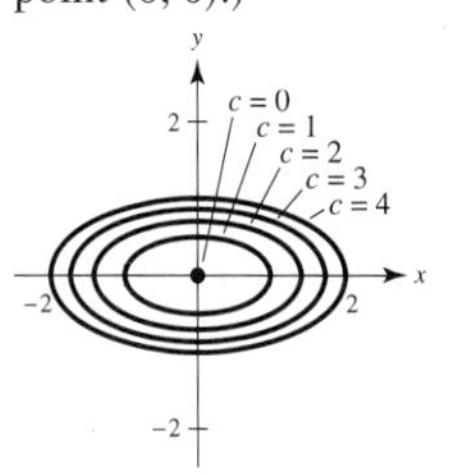

53. Hyperbolas: $xy = c$

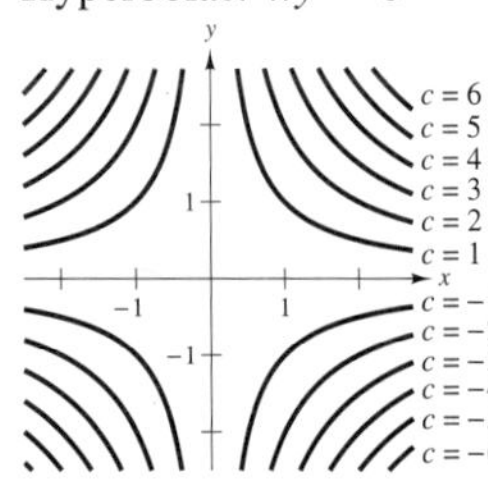

55. Circles passing through $(0, 0)$
Centered at $(1/(2c), 0)$

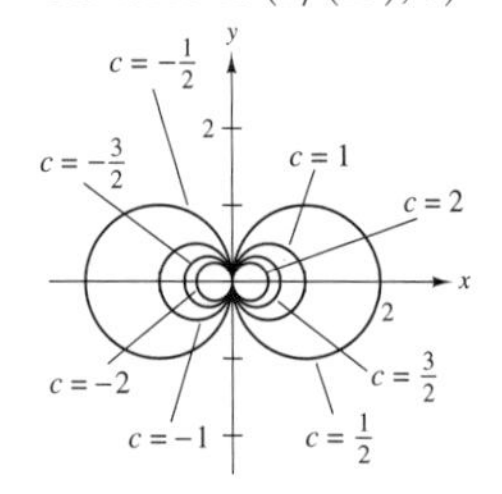

57.

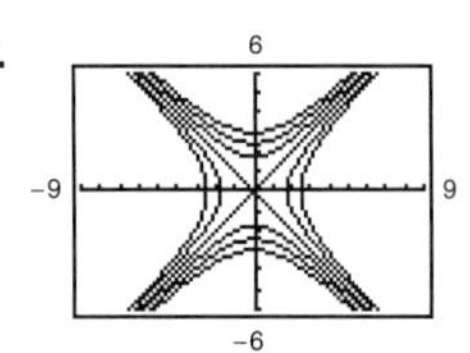

59.

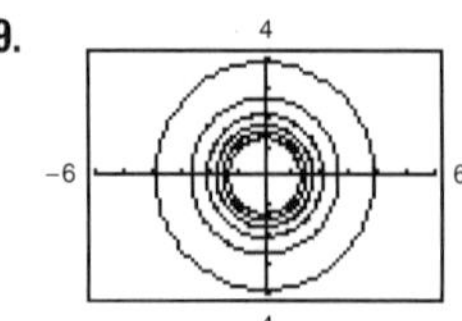

61. The graph of a function of two variables is the set of all points (x, y, z) for which $z = f(x, y)$ and (x, y) is in the domain of f. The graph can be interpreted as a surface in space. Level curves are the scalar fields $f(x, y) = c$, where c is a constant.

63. $f(x, y) = x/y$; the level curves are the lines $y = (1/c)x$.

65. The surface may be shaped like a saddle. For example, let $f(x, y) = xy$. The graph is not unique; any vertical translation will produce the same level curves.

67.

	Inflation Rate		
Tax Rate	**0**	**0.03**	**0.05**
0	\$1790.85	\$1332.56	\$1099.43
0.28	\$1526.43	\$1135.80	\$937.09
0.35	\$1466.07	\$1090.90	\$900.04

69.

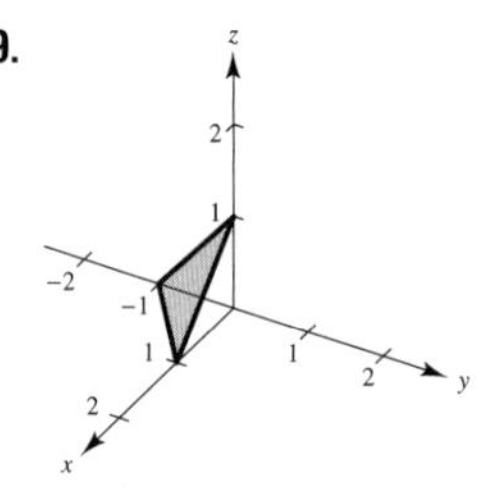

71.

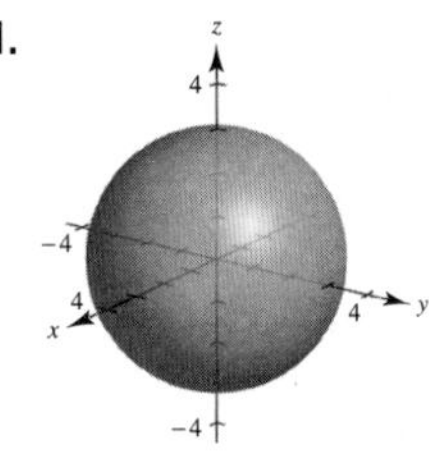

73.

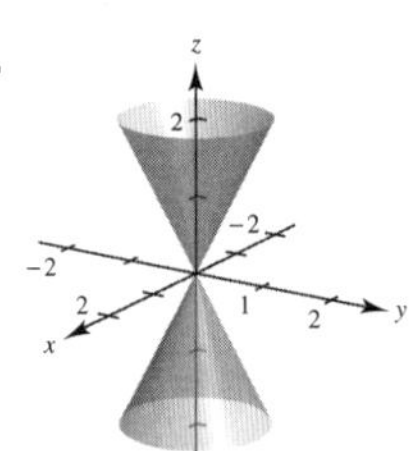

75. (a) 243 board-ft
(b) 507 board-ft

77.

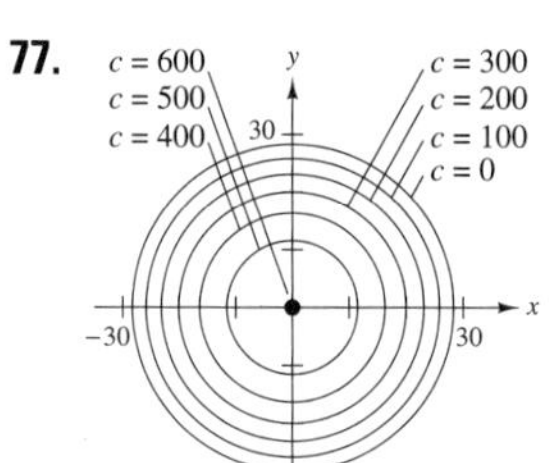

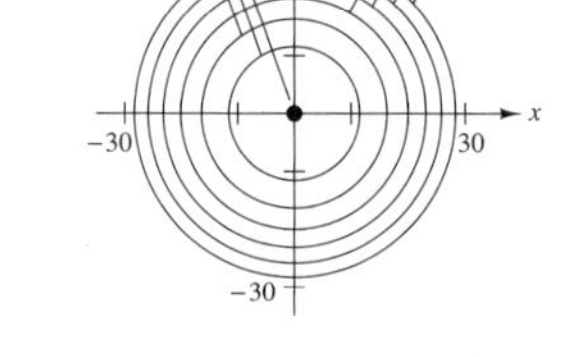

79. Proof

81. $C = 1.20xy + 1.50(xz + yz)$

83. (a) $k = \frac{520}{3}$
(b) $P = 520T/(3V)$
The level curves are lines.

85. (a) C (b) A (c) B

87. (a) No; the level curves are uneven and sporadically spaced.
(b) Use more colors.

89. False: let $f(x, y) = 4$. **91.** True

Section 13.2 (page 904)

1–3. Proofs **5.** 1 **7.** 12 **9.** 9, continuous
11. e^2, continuous **13.** 0, continuous for $y \neq 0$
15. $\frac{1}{2}$, continuous except at $(0, 0)$ **17.** 0, continuous
19. 0, continuous for $xy \neq 1$, $|xy| \leq 1$
21. $2\sqrt{2}$, continuous for $x + y + z \geq 0$ **23.** 0
25. Limit does not exist. **27.** 4 **29.** Limit does not exist.
31. Limit does not exist. **33.** 0
35. Limit does not exist. **37.** Continuous, 1

39.

(x, y)	$(1, 0)$	$(0.5, 0)$	$(0.1, 0)$	$(0.01, 0)$	$(0.001, 0)$
$f(x, y)$	0	0	0	0	0

$y = 0$: 0

(x, y)	$(1, 1)$	$(0.5, 0.5)$	$(0.1, 0.1)$
$f(x, y)$	$\frac{1}{2}$	$\frac{1}{2}$	$\frac{1}{2}$

(x, y)	$(0.01, 0.01)$	$(0.001, 0.001)$
$f(x, y)$	$\frac{1}{2}$	$\frac{1}{2}$

$y = x$: $\frac{1}{2}$

Limit does not exist.

Continuous except at $(0, 0)$

41.

(x, y)	$(1, 1)$	$(0.25, 0.5)$	$(0.01, 0.1)$
$f(x, y)$	$-\frac{1}{2}$	$-\frac{1}{2}$	$-\frac{1}{2}$

(x, y)	$(0.0001, 0.01)$	$(0.000001, 0.001)$
$f(x, y)$	$-\frac{1}{2}$	$-\frac{1}{2}$

$x = y^2$: $-\frac{1}{2}$

(x, y)	$(-1, 1)$	$(-0.25, 0.5)$	$(-0.01, 0.1)$
$f(x, y)$	$\frac{1}{2}$	$\frac{1}{2}$	$\frac{1}{2}$

(x, y)	$(-0.0001, 0.01)$	$(-0.000001, 0.001)$
$f(x, y)$	$\frac{1}{2}$	$\frac{1}{2}$

$x = -y^2$: $\frac{1}{2}$

Limit does not exist.

Continuous except at $(0, 0)$

43. f is continuous. g is continuous except at $(0, 0)$. g has a removable discontinuity at $(0, 0)$.

45. f is continuous. g is continuous except at $(0, 0)$. g has a removable discontinuity at $(0, 0)$.

47. 0

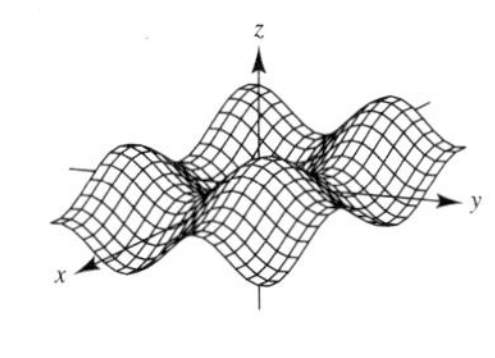

49. Limit does not exist.

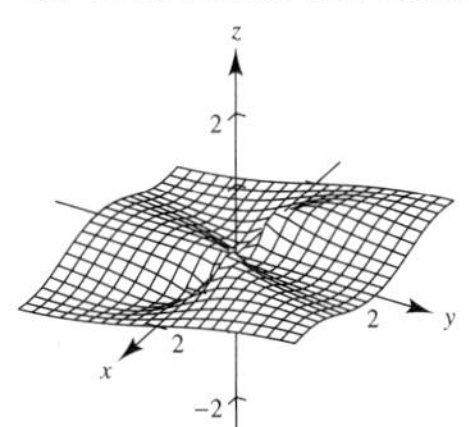

51. Limit does not exist.

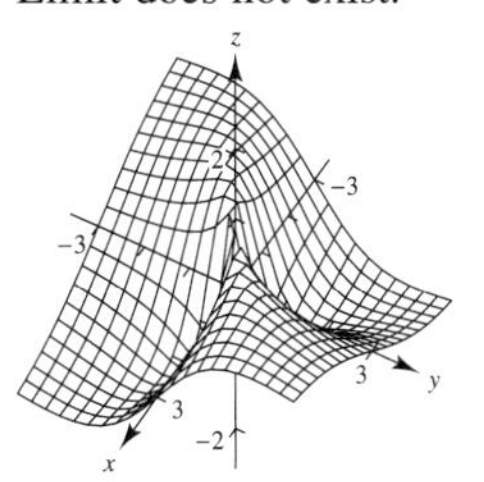

53. 0 **55.** 0 **57.** 1 **59.** 1 **61.** 0

63. Continuous except at $(0, 0, 0)$ **65.** Continuous

67. Continuous **69.** Continuous

71. Continuous for $y \neq 2x/3$ **73.** (a) $2x$ (b) -4

75. (a) $1/y$ (b) $-x/y^2$ **77.** (a) $3 + y$ (b) $x - 2$

79. True **81.** False: let $f(x, y) = \begin{cases} \ln(x^2 + y^2), & x \neq 0, y \neq 0 \\ 0, & x = 0, y = 0 \end{cases}$.

83. (a) $(1 + a^2)/a, \quad a \neq 0$ (b) Limit does not exist.

(c) No, the limit does not exist. Different paths result in different limits.

85. 0 **87.** $\pi/2$ **89.** Proof

91. See "Definition of the Limit of a Function of Two Variables," on page 899; show that the value of $\lim_{(x, y)\to(x_0, y_0)} f(x, y)$ is not the same for two different paths to (x_0, y_0).

93. (a) True. To find the first limit, you substitute $(2, 3)$ for (x, y). To find the second limit, you first substitute 3 for y to find a function of x. Then you substitute 2 for x.

(b) False. The convergence of one path does not imply the convergence of all paths.

(c) False. Let $f(x, y) = 4\left(\dfrac{(x - 2)^2 - (y - 3)^2}{(x - 2)^2 + (y - 3)^2}\right)^2$.

(d) True. When you multiply 0 by any real number, you always get 0.

Section 13.3 (page 914)

1. $f_x(4, 1) < 0$ **3.** $f_y(4, 1) > 0$

5. No. Because you are finding the partial derivative with respect to y, you consider x to be constant. So, the denominator is considered a constant and does not contain any variables.

7. Yes. Because you are finding the partial derivative with respect to x, you consider y to be constant. So, both the numerator and denominator contain variables.

9. $f_x(x, y) = 2$
$f_y(x, y) = -5$

11. $f_x(x, y) = 2xy^3$
$f_y(x, y) = 3x^2y^2$

13. $\partial z/\partial x = \sqrt{y}$
$\partial z/\partial y = x/(2\sqrt{y})$

15. $\partial z/\partial x = 2x - 4y$
$\partial z/\partial y = -4x + 6y$

17. $\partial z/\partial x = ye^{xy}$
$\partial z/\partial y = xe^{xy}$

19. $\partial z/\partial x = 2xe^{2y}$
$\partial z/\partial y = 2x^2e^{2y}$

21. $\partial z/\partial x = 1/x$
$\partial z/\partial y = -1/y$

23. $\partial z/\partial x = 2x/(x^2 + y^2)$
$\partial z/\partial y = 2y/(x^2 + y^2)$

25. $\partial z/\partial x = (x^3 - 3y^3)/(x^2y)$
$\partial z/\partial y = (-x^3 + 12y^3)/(2xy^2)$

27. $h_x(x, y) = -2xe^{-(x^2+y^2)}$
$h_y(x, y) = -2ye^{-(x^2+y^2)}$

29. $f_x(x, y) = x/\sqrt{x^2 + y^2}$
$f_y(x, y) = y/\sqrt{x^2 + y^2}$

31. $\partial z/\partial x = -y \sin xy$
$\partial z/\partial y = -x \sin xy$

33. $\partial z/\partial x = 2 \sec^2(2x - y)$
$\partial z/\partial y = -\sec^2(2x - y)$

35. $\partial z/\partial x = ye^y \cos xy$
$\partial z/\partial y = e^y(x \cos xy + \sin xy)$

37. $\partial z/\partial x = 2\cosh(2x + 3y)$
$\partial z/\partial y = 3\cosh(2x + 3y)$

39. $f_x(x, y) = 1 - x^2$
$f_y(x, y) = y^2 - 1$

41. $f_x(x, y) = 3$
$f_y(x, y) = 2$

43. $f_x(x, y) = 1/(2\sqrt{x + y})$
$f_y(x, y) = 1/(2\sqrt{x + y})$

45. $\partial z/\partial x = -1$
$\partial z/\partial y = 0$

47. $\partial z/\partial x = -1$
$\partial z/\partial y = \frac{1}{2}$

49. $\partial z/\partial x = \frac{1}{4}$
$\partial z/\partial y = \frac{1}{4}$

51. $\partial z/\partial x = -\frac{1}{4}$
$\partial z/\partial y = \frac{1}{4}$

53. $g_x(1, 1) = -2$
$g_y(1, 1) = -2$

55.

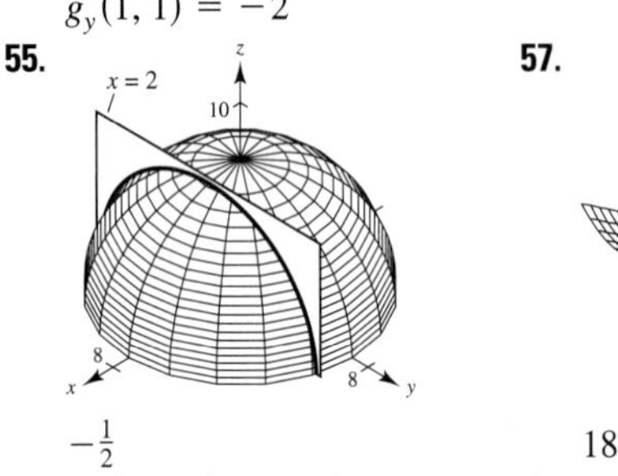

$-\frac{1}{2}$

57. 18

59. $H_x(x, y, z) = \cos(x + 2y + 3z)$
$H_y(x, y, z) = 2\cos(x + 2y + 3z)$
$H_z(x, y, z) = 3\cos(x + 2y + 3z)$

61. $\dfrac{\partial w}{\partial x} = \dfrac{x}{\sqrt{x^2 + y^2 + z^2}}$
$\dfrac{\partial w}{\partial y} = \dfrac{y}{\sqrt{x^2 + y^2 + z^2}}$
$\dfrac{\partial w}{\partial z} = \dfrac{z}{\sqrt{x^2 + y^2 + z^2}}$

63. $F_x(x, y, z) = \dfrac{x}{x^2 + y^2 + z^2}$
$F_y(x, y, z) = \dfrac{y}{x^2 + y^2 + z^2}$
$F_z(x, y, z) = \dfrac{z}{x^2 + y^2 + z^2}$

65. $f_x = 3; f_y = 1; f_z = 2$

67. $f_x = 1; f_y = 1; f_z = 1$

69. $f_x = 0; f_y = 0; f_z = 1$

71. $\dfrac{\partial^2 z}{\partial x^2} = 0$
$\dfrac{\partial^2 z}{\partial y^2} = 6x$
$\dfrac{\partial^2 z}{\partial y \partial x} = \dfrac{\partial^2 z}{\partial x \partial y} = 6y$

73. $\dfrac{\partial^2 z}{\partial x^2} = 2$
$\dfrac{\partial^2 z}{\partial y^2} = 6$
$\dfrac{\partial^2 z}{\partial y \partial x} = \dfrac{\partial^2 z}{\partial x \partial y} = -2$

75. $\dfrac{\partial^2 z}{\partial x^2} = \dfrac{y^2}{(x^2 + y^2)^{3/2}}$
$\dfrac{\partial^2 z}{\partial y^2} = \dfrac{x^2}{(x^2 + y^2)^{3/2}}$
$\dfrac{\partial^2 z}{\partial y \partial x} = \dfrac{\partial^2 z}{\partial x \partial y} = \dfrac{-xy}{(x^2 + y^2)^{3/2}}$

77. $\dfrac{\partial^2 z}{\partial x^2} = e^x \tan y$
$\dfrac{\partial^2 z}{\partial y^2} = 2e^x \sec^2 y \tan y$
$\dfrac{\partial^2 z}{\partial y \partial x} = \dfrac{\partial^2 z}{\partial x \partial y} = e^x \sec^2 y$

79. $\dfrac{\partial^2 z}{\partial x^2} = -y^2 \cos xy$
$\dfrac{\partial^2 z}{\partial y^2} = -x^2 \cos xy$
$\dfrac{\partial^2 z}{\partial y \partial x} = \dfrac{\partial^2 z}{\partial x \partial y} = -xy \cos xy - \sin xy$

81. $x = 2, y = -2$ **83.** $x = -6, y = 4$

85. $x = 1, y = 1$ **87.** $x = 0, y = 0$

89. $\partial z/\partial x = \sec y$
$\partial z/\partial y = x \sec y \tan y$
$\partial^2 z/\partial x^2 = 0$
$\partial^2 z/\partial y^2 = x \sec y(\sec^2 y + \tan^2 y)$
$\partial^2 z/\partial y \partial x = \partial^2 z/\partial x \partial y = \sec y \tan y$
No values of x and y exist such that $f_x(x, y) = f_y(x, y) = 0$.

91. $\partial z/\partial x = (y^2 - x^2)/[x(x^2 + y^2)]$
$\partial z/\partial y = -2y/(x^2 + y^2)$
$\partial^2 z/\partial x^2 = (x^4 - 4x^2y^2 - y^4)/[x^2(x^2 + y^2)^2]$
$\partial^2 z/\partial y^2 = 2(y^2 - x^2)/(x^2 + y^2)^2$
$\partial^2 z/\partial y \partial x = \partial^2 z/\partial x \partial y = 4xy/(x^2 + y^2)^2$
No values of x and y exist such that $f_x(x, y) = f_y(x, y) = 0$.

93. $f_{xyy}(x, y, z) = f_{yxy}(x, y, z) = f_{yyx}(x, y, z) = 0$

95. $f_{xyy}(x, y, z) = f_{yxy}(x, y, z) = f_{yyx}(x, y, z) = z^2e^{-x} \sin yz$

97. $\partial^2 z/\partial x^2 + \partial^2 z/\partial y^2 = 0 + 0 = 0$

99. $\partial^2 z/\partial x^2 + \partial^2 z/\partial y^2 = e^x \sin y - e^x \sin y = 0$

101. $\partial^2 z/\partial t^2 = -c^2 \sin(x - ct) = c^2(\partial^2 z/\partial x^2)$

103. $\partial^2 z/\partial t^2 = -c^2/(x + ct)^2 = c^2(\partial^2 z/\partial x^2)$

105. $\partial z/\partial t = -e^{-t} \cos x/c = c^2(\partial^2 z/\partial x^2)$

107. Yes, $f(x, y) = \cos(3x - 2y)$. **109.** 0

111. If $z = f(x, y)$, then to find f_x you consider y constant and differentiate with respect to x. Similarly, to find f_y, you consider x constant and differentiate with respect to y.

113.

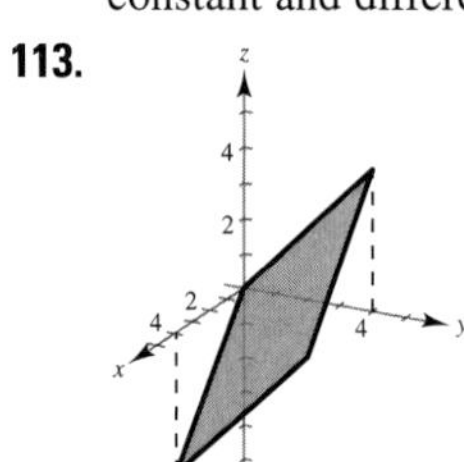

115. The mixed partial derivatives are equal. See Theorem 13.3.

117. (a) 72 (b) 72

119. $IQ_M = \dfrac{100}{C}, IQ_M(12, 10) = 10$

IQ increases at a rate of 10 points per year of mental age when the mental age is 12 and the chronological age is 10.

$IQ_C = -\dfrac{100M}{C^2}, IQ_C(12, 10) = -12$

IQ decreases at a rate of 12 points per year of chronological age when the mental age is 12 and the chronological age is 10.

121. An increase in either the charge for food and housing or the tuition will cause a decrease in the number of applicants.

123. $\partial T/\partial x = -2.4°/\text{m}, \partial T/\partial y = -9°/\text{m}$

125. $T = PV/(nR) \implies \partial T/\partial P = V/(nR)$
$P = nRT/V \implies \partial P/\partial V = -nRT/V^2$
$V = nRT/P \implies \partial V/\partial T = nR/P$
$\partial T/\partial P \cdot \partial P/\partial V \cdot \partial V/\partial T = -nRT/(VP) = -nRT/(nRT) = -1$

127. (a) $\partial z/\partial x = -0.92; \partial z/\partial y = 1.03$
(b) As the consumption of flavored milk (x) increases, the consumption of plain light and skim milks (z) decreases. As the consumption of plain reduced-fat milk (y) decreases, the consumption of plain light and skim milks also decreases.

129. False; Let $z = x + y + 1$. **131.** True

133. (a) $f_x(x, y) = \dfrac{y(x^4 + 4x^2y^2 - y^4)}{(x^2 + y^2)^2}$

$f_y(x, y) = \dfrac{x(x^4 - 4x^2y^2 - y^4)}{(x^2 + y^2)^2}$

(b) $f_x(0, 0) = 0,\ f_y(0, 0) = 0$

(c) $f_{xy}(0, 0) = -1,\ f_{yx}(0, 0) = 1$

(d) f_{xy} or f_{yx} or both are not continuous at $(0, 0)$.

135. (a) $f_x(0, 0) = 1, f_y(0, 0) = 1$

(b) $f_x(x, y)$ and $f_y(x, y)$ do not exist when $y = -x$.

Section 13.4 (page 923)

1. $dz = 4xy^3\,dx + 6x^2y^2\,dy$

3. $dz = 2(x\,dx + y\,dy)/(x^2 + y^2)^2$

5. $dz = (\cos y + y \sin x)\,dx - (x \sin y + \cos x)\,dy$

7. $dz = (e^x \sin y)\,dx + (e^x \cos y)\,dy$

9. $dw = 2z^3y \cos x\,dx + 2z^3 \sin x\,dy + 6z^2 y \sin x\,dz$

11. (a) $f(2, 1) = 1, f(2.1, 1.05) = 1.05, \Delta z = 0.05$

(b) $dz = 0.05$

13. (a) $f(2, 1) = 11, f(2.1, 1.05) = 10.4875, \Delta z = -0.5125$

(b) $dz = -0.5$

15. (a) $f(2, 1) = e^2 \approx 7.3891, f(2.1, 1.05) = 1.05e^{2.1} \approx 8.5745,$ $\Delta z \approx 1.1854$

(b) $dz \approx 1.1084$

17. 0.44 **19.** -0.012

21. If $z = f(x, y)$ and Δx and Δy are increments of x and y, and x and y are independent variables, then the total differential of the dependent variable z is

$dz = (\partial z/\partial x)\,dx + (\partial z/\partial y)\,dy = f_x(x, y)\,\Delta x + f_y(x, y)\,\Delta y.$

23. The approximation of Δz by dz is called a linear approximation, where dz represents the change in height of a plane that is tangent to the surface at the point $P(x_0, y_0)$.

25. $dA = h\,dl + l\,dh$

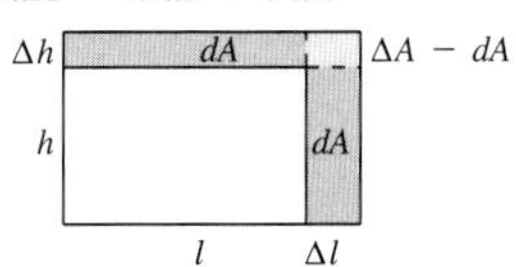

$\Delta A - dA = dl\,dh$

27.

Δr	Δh	dV	ΔV	$\Delta V - dV$
0.1	0.1	8.3776	8.5462	0.1686
0.1	-0.1	5.0265	5.0255	-0.0010
0.001	0.002	0.1005	0.1006	0.0001
-0.0001	0.0002	-0.0034	-0.0034	0.0000

29. (a) $dz = -0.92\,dx + 1.03\,dy$

(b) $dz = \pm 0.4875;\ dz/z \approx 8.1\%$

31. 10% **33.** $dC = \pm 2.4418;\ dC/C = 19\%$

35. (a) $V = 18 \sin \theta$ ft³; $\theta = \pi/2$

(b) 1.047 ft³

37. 10% **39.** $L \approx 8.096 \times 10^{-4} \pm 6.6 \times 10^{-6}$ microhenrys

41. Answers will vary.
Example:
$\varepsilon_1 = \Delta x$
$\varepsilon_2 = 0$

43. Answers will vary.
Example:
$\varepsilon_1 = y\,\Delta x$
$\varepsilon_2 = 2x\,\Delta x + (\Delta x)^2$

45–47. Proofs

Section 13.5 (page 931)

1. $26t$ **3.** $e^t(\sin t + \cos t)$ **5.** (a) and (b) $-e^{-t}$

7. (a) and (b) $2e^{2t}$ **9.** (a) and (b) $3(2t^2 - 1)$

11. $-11\sqrt{29}/29 \approx -2.04$ **13.** $\dfrac{4}{(e^t + e^{-t})^2}$; 1

15. $\partial w/\partial s = 4s, 4$
$\partial w/\partial t = 4t, 0$

17. $\partial w/\partial s = 5 \cos(5s - t), 0$
$\partial w/\partial t = -\cos(5s - t), 0$

19. $\partial w/\partial r = 2r/\theta^2$
$\partial w/\partial \theta = -2r^2/\theta^3$

21. $\partial w/\partial r = 0$
$\partial w/\partial \theta = 1$

23. $\dfrac{\partial w}{\partial s} = t^2(3s^2 - t^2)$
$\dfrac{\partial w}{\partial t} = 2st(s^2 - 2t^2)$

25. $\dfrac{\partial w}{\partial s} = te^{s^2 - t^2}(2s^2 + 1)$
$\dfrac{\partial w}{\partial t} = se^{s^2 - t^2}(1 - 2t^2)$

27. $\dfrac{y - 2x + 1}{2y - x + 1}$ **29.** $-\dfrac{x^2 + y^2 + x}{x^2 + y^2 + y}$

31. $\dfrac{\partial z}{\partial x} = \dfrac{-x}{z}$
$\dfrac{\partial z}{\partial y} = \dfrac{-y}{z}$

33. $\dfrac{\partial z}{\partial x} = -\dfrac{x}{y + z}$
$\dfrac{\partial z}{\partial y} = -\dfrac{z}{y + z}$

35. $\dfrac{\partial z}{\partial x} = \dfrac{-\sec^2(x + y)}{\sec^2(y + z)}$
$\dfrac{\partial z}{\partial y} = -1 - \dfrac{\sec^2(x + y)}{\sec^2(y + z)}$

37. $\dfrac{\partial z}{\partial x} = -\dfrac{(ze^{xz} + y)}{xe^{xz}}$
$\dfrac{\partial z}{\partial y} = -e^{-xz}$

39. $\dfrac{\partial w}{\partial x} = -\dfrac{y + w}{x - z}$
$\dfrac{\partial w}{\partial y} = -\dfrac{x + z}{x - z}$
$\dfrac{\partial w}{\partial z} = \dfrac{w - y}{x - z}$

41. $\dfrac{\partial w}{\partial x} = \dfrac{y \sin xy}{z}$
$\dfrac{\partial w}{\partial y} = \dfrac{x \sin xy - z \cos yz}{z}$
$\dfrac{\partial w}{\partial z} = -\dfrac{y \cos yz + w}{z}$

43. (a) $f(tx, ty) = \dfrac{(tx)(ty)}{\sqrt{(tx)^2 + (ty)^2}} = t\left(\dfrac{xy}{\sqrt{x^2 + y^2}}\right) = tf(x, y);\ n = 1$

(b) $xf_x(x, y) + yf_y(x, y) = \dfrac{xy}{\sqrt{x^2 + y^2}} = 1f(x, y)$

45. (a) $f(tx, ty) = e^{tx/ty} = e^{x/y} = f(x, y);\ n = 0$

(b) $xf_x(x, y) + yf_y(x, y) = \dfrac{xe^{x/y}}{y} - \dfrac{xe^{x/y}}{y} = 0$

47. 47 **49.** $dw/dt = (\partial w/\partial x \cdot dx/dt) + (\partial w/\partial y \cdot dy/dt)$

51. $\dfrac{dy}{dx} = -\dfrac{f_x(x, y)}{f_y(x, y)}$

$\dfrac{\partial z}{\partial x} = -\dfrac{f_x(x, y, z)}{f_z(x, y, z)}$

$\dfrac{\partial z}{\partial y} = -\dfrac{f_y(x, y, z)}{f_z(x, y, z)}$

53. 4608π in.³/min; 624π in.²/min

55. $\dfrac{dT}{dt} = \dfrac{1}{mR}\left[V\dfrac{dp}{dt} + p\dfrac{dV}{dt}\right]$ **57.** $28m$ cm²/sec

59. Proof **61.** (a) Proof (b) Proof **63–65.** Proofs

Section 13.6 (page 942)

1. 1 **3.** -1 **5.** $-e$ **7.** $-\frac{7}{25}$ **9.** $2\sqrt{3}/3$ **11.** $\frac{8}{3}$

13. $\sqrt{2}(x+y)$ **15.** $\left[(2+\sqrt{3})/2\right]\cos(2x+y)$ **17.** 6

19. $-8/\sqrt{5}$ **21.** $3\mathbf{i}+10\mathbf{j}$ **23.** $4\mathbf{i}-\mathbf{j}$ **25.** $6\mathbf{i}-10\mathbf{j}-8\mathbf{k}$

27. $3\sqrt{2}$ **29.** $2\sqrt{5}/5$ **31.** $2[(x+y)\mathbf{i}+x\mathbf{j}];\ 2\sqrt{2}$

33. $\tan y\mathbf{i}+x\sec^2 y\mathbf{j},\ \sqrt{17}$ **35.** $e^{-x}(-y\mathbf{i}+\mathbf{j});\ \sqrt{26}$

37. $\dfrac{x\mathbf{i}+y\mathbf{j}+z\mathbf{k}}{\sqrt{x^2+y^2+z^2}},\ 1$ **39.** $yz(yz\mathbf{i}+2xz\mathbf{j}+2xy\mathbf{k});\ \sqrt{33}$

41.

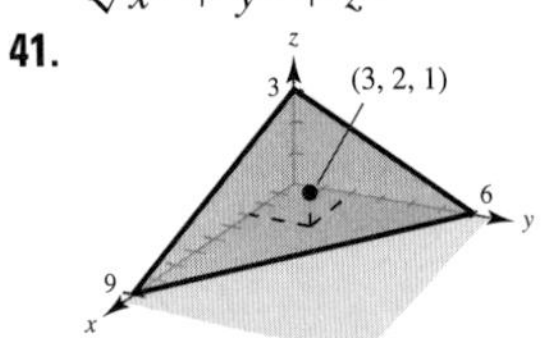

43. (a) $-5\sqrt{2}/12$ (b) $\frac{3}{5}$ (c) $-\frac{1}{5}$ (d) $-11\sqrt{10}/60$

45. $\sqrt{13}/6$

47. (a) Answers will vary. Example: $-4\mathbf{i}+\mathbf{j}$

(b) $-\frac{2}{5}\mathbf{i}+\frac{1}{10}\mathbf{j}$ (c) $\frac{2}{5}\mathbf{i}-\frac{1}{10}\mathbf{j}$

The direction opposite that of the gradient

49. (a)

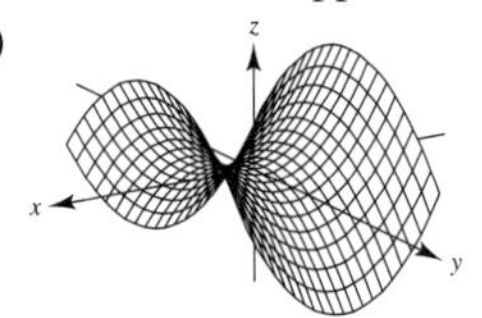

(b) $D_{\mathbf{u}}f(4,-3)=8\cos\theta+6\sin\theta$

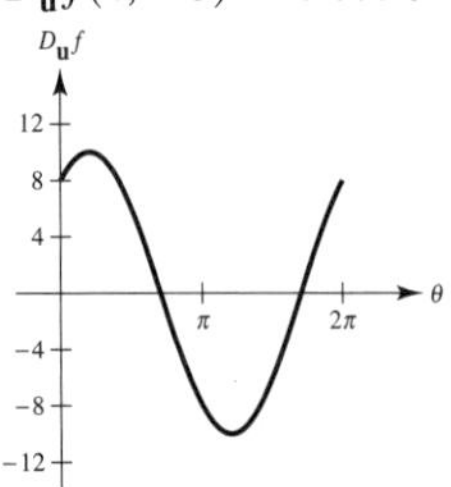

Generated by Mathematica

(c) $\theta\approx 2.21,\ \theta\approx 5.36$
Directions in which there is no change in f

(d) $\theta\approx 0.64,\ \theta\approx 3.79$
Directions of greatest rate of change in f

(e) 10; Magnitude of the greatest rate of change

(f)

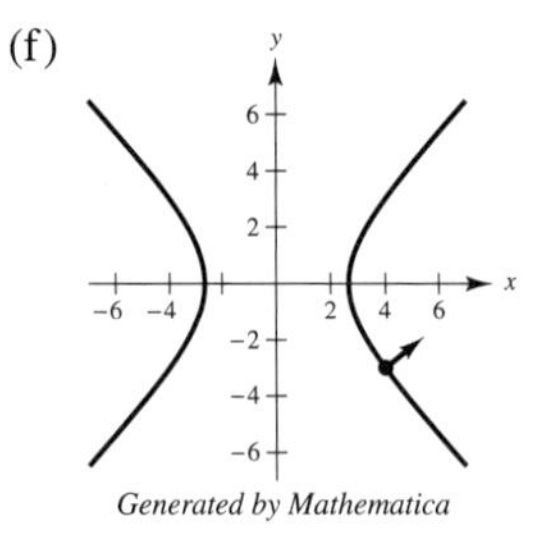

Generated by Mathematica

Orthogonal to the level curve

51. $-2\mathbf{i}-3\mathbf{j}$ **53.** $3\mathbf{i}-\mathbf{j}$

55. (a) $16\mathbf{i}-\mathbf{j}$ (b) $(\sqrt{257}/257)(16\mathbf{i}-\mathbf{j})$ (c) $y=16x-22$

(d)

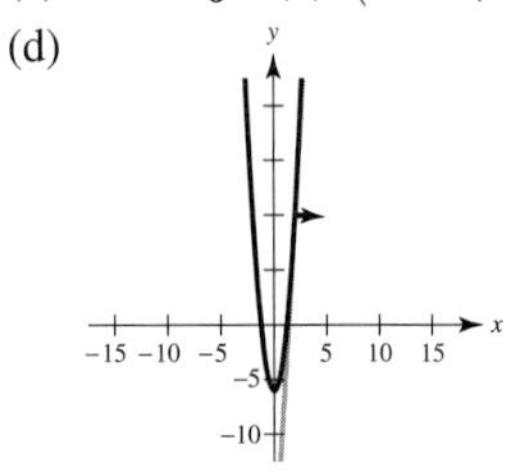

57. (a) $6\mathbf{i}-4\mathbf{j}$ (b) $(\sqrt{13}/13)(3\mathbf{i}-2\mathbf{j})$ (c) $y=\frac{3}{2}x-\frac{1}{2}$

(d)

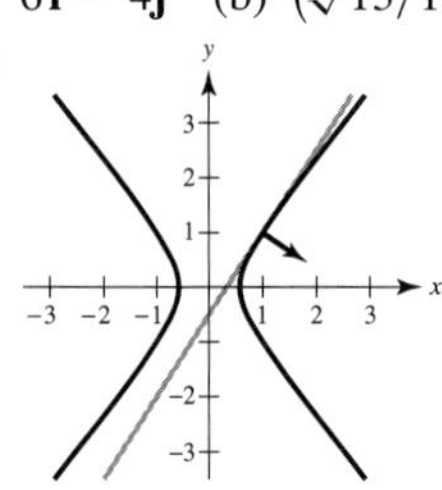

59. The directional derivative of $z=f(x,y)$ in the direction of $\mathbf{u}=\cos\theta\mathbf{i}+\sin\theta\mathbf{j}$ is

$$D_{\mathbf{u}}f(x,y)=\lim_{t\to 0}\frac{f(x+t\cos\theta,\ y+t\sin\theta)-f(x,y)}{t}$$

if the limit exists.

61. See the definition on page 936. See the properties on page 937.

63. The gradient vector is normal to the level curves.

65. $\frac{1}{625}(7\mathbf{i}-24\mathbf{j})$

67. [map] **69.** $y^2=10x$

71. (a)

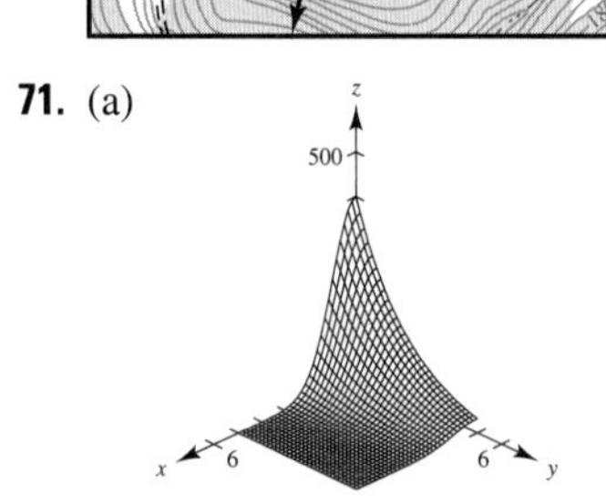

(b) There is no change in heat in directions perpendicular to the gradient: $\pm(\mathbf{i}-6\mathbf{j})$.

(c) The greatest increase is in the direction of the gradient: $-3\mathbf{i}-\frac{1}{2}\mathbf{j}$.

73. True **75.** True **77.** $f(x,y,z)=e^x\cos y+\frac{1}{2}z^2+C$

79. (a) Proof (b) Proof

(c)

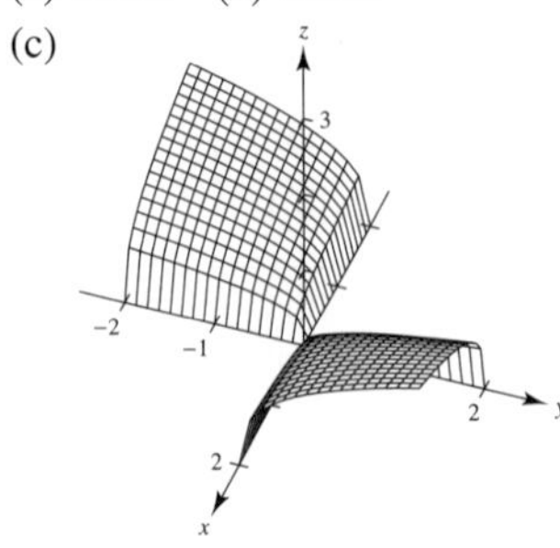

Section 13.7 (page 951)

1. The level surface can be written as $3x-5y+3z=15$, which is an equation of a plane in space.

3. The level surface can be written as $4x^2+9y^2-4z^2=0$, which is an elliptic cone that lies on the z-axis.

5. $\frac{1}{13}(3\mathbf{i} + 4\mathbf{j} + 12\mathbf{k})$ **7.** $(\sqrt{6}/6)(\mathbf{i} + \mathbf{j} + 2\mathbf{k})$
9. $(\sqrt{145}/145)(12\mathbf{i} - \mathbf{k})$ **11.** $\frac{1}{13}(4\mathbf{i} + 3\mathbf{j} + 12\mathbf{k})$
13. $(\sqrt{3}/3)(\mathbf{i} - \mathbf{j} + \mathbf{k})$ **15.** $(\sqrt{113}/113)(-\mathbf{i} - 6\sqrt{3}\mathbf{j} + 2\mathbf{k})$
17. $4x + 2y - z = 2$ **19.** $3x + 4y - 5z = 0$
21. $2x - 2y - z = 2$ **23.** $2x + 3y + 3z = 6$
25. $3x + 4y - 25z = 25(1 - \ln 5)$ **27.** $x - 4y + 2z = 18$
29. $6x - 3y - 2z = 11$ **31.** $x + y + z = 9$
$x - 3 = y - 3 = z - 3$
33. $2x + 4y + z = 14$
$\frac{x-1}{2} = \frac{y-2}{4} = \frac{z-4}{1}$
35. $6x - 4y - z = 5$
$\frac{x-3}{6} = \frac{y-2}{-4} = \frac{z-5}{-1}$
37. $10x + 5y + 2z = 30$
$\frac{x-1}{10} = \frac{y-2}{5} = \frac{z-5}{2}$
39. $x - y + 2z = \pi/2$
$\frac{(x-1)}{1} = \frac{(y-1)}{-1} = \frac{z-(\pi/4)}{2}$
41. (a) $\frac{x-1}{1} = \frac{y-1}{-1} = \frac{z-1}{1}$ (b) $\frac{1}{2}$, not orthogonal
43. (a) $\frac{x-3}{4} = \frac{y-3}{4} = \frac{z-4}{-3}$ (b) $\frac{16}{25}$, not orthogonal
45. (a) $\frac{x-3}{1} = \frac{y-1}{5} = \frac{z-2}{-4}$ (b) 0, orthogonal
47. 86.0° **49.** 77.4° **51.** (0, 3, 12) **53.** (2, 2, −4)
55. (0, 0, 0) **57.** Proof **59.** (a) Proof (b) Proof
61. (−2, 1, −1) or (2, −1, 1)
63. $F_x(x_0, y_0, z_0)(x - x_0) + F_y(x_0, y_0, z_0)(y - y_0) + F_z(x_0, y_0, z_0)(z - z_0) = 0$
65. Answers will vary.
67. (a) Line: $x = 1, y = 1, z = 1 - t$
Plane: $z = 1$
(b) Line: $x = -1, y = 2 + \frac{6}{25}t, z = -\frac{4}{5} - t$
Plane: $6y - 25z - 32 = 0$
(c)

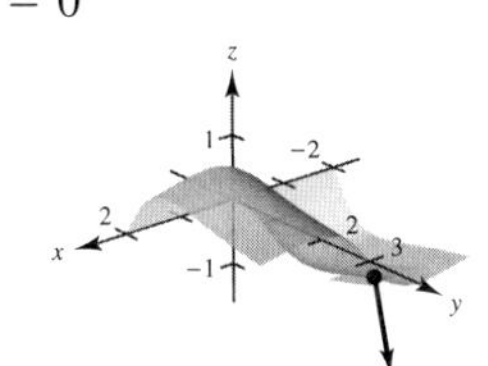

69. (a) $x = 1 + t$
$y = 2 - 2t$
$z = 4$
$\theta \approx 48.2°$
(b)

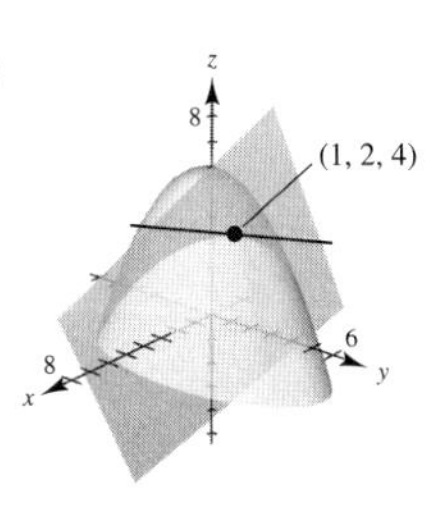

71. $F(x, y, z) = \frac{x^2}{a^2} + \frac{y^2}{b^2} + \frac{z^2}{c^2} - 1$
$F_x(x, y, z) = 2x/a^2$
$F_y(x, y, z) = 2y/b^2$
$F_z(x, y, z) = 2z/c^2$
Plane: $\frac{2x_0}{a^2}(x - x_0) + \frac{2y_0}{b^2}(y - y_0) + \frac{2z_0}{c^2}(z - z_0) = 0$
$\frac{x_0 x}{a^2} + \frac{y_0 y}{b^2} + \frac{z_0 z}{c^2} = 1$
73. $F(x, y, z) = a^2x^2 + b^2y^2 - z^2$
$F_x(x, y, z) = 2a^2x$
$F_y(x, y, z) = 2b^2y$
$F_z(x, y, z) = -2z$
Plane: $2a^2x_0(x - x_0) + 2b^2y_0(y - y_0) - 2z_0(z - z_0) = 0$
$a^2x_0x + b^2y_0y - z_0z = 0$
Therefore, the plane passes through the origin.
75. (a) $P_1(x, y) = 1 + x - y$
(b) $P_2(x, y) = 1 + x - y + \frac{1}{2}x^2 - xy + \frac{1}{2}y^2$
(c) If $x = 0$, $P_2(0, y) = 1 - y + \frac{1}{2}y^2$.
This is the second-degree Taylor polynomial for e^{-y}.
If $y = 0$, $P_2(x, 0) = 1 + x + \frac{1}{2}x^2$.
This is the second-degree Taylor polynomial for e^x.
(d)

x	y	$f(x, y)$	$P_1(x, y)$	$P_2(x, y)$
0	0	1	1	1
0	0.1	0.9048	0.9000	0.9050
0.2	0.1	1.1052	1.1000	1.1050
0.2	0.5	0.7408	0.7000	0.7450
1	0.5	1.6487	1.5000	1.6250

(e)

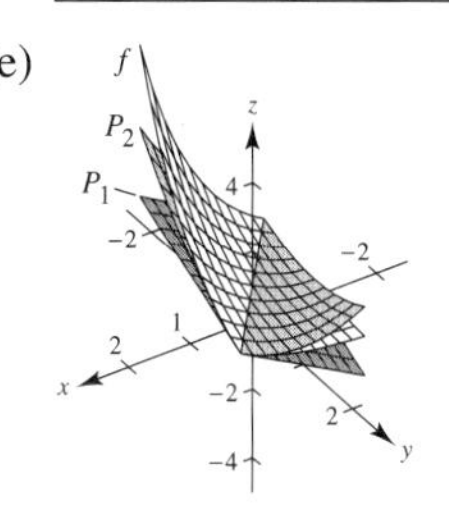

77. Proof

Section 13.8 (page 960)

1. Relative minimum: (1, 3, 0) **3.** Relative minimum: (0, 0, 1)
5. Relative minimum: (−1, 3, −4)
7. Relative minimum: (1, 1, 11)
9. Relative maximum: (5, −1, 2)
11. Relative minimum: (3, −4, −5)
13. Relative minimum: (0, 0, 0)
15. Relative maximum: (0, 0, 4)

17. Relative maximum: $(-1, 0, 2)$
Relative minimum: $(1, 0, -2)$

19. Relative minimum: $(0, 0, 0)$
Relative maxima: $(0, \pm 1, 4)$
Saddle points: $(\pm 1, 0, 1)$

21. Relative maximum: $(40, 40, 3200)$

23. Saddle point: $(0, 0, 0)$ **25.** Saddle point: $(1, -1, -1)$

27. There are no critical numbers.

29. z is never negative. Minimum: $z = 0$ when $x = y \neq 0$.

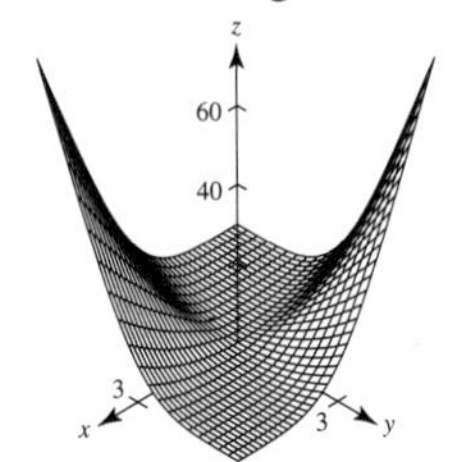

31. Insufficient information **33.** Saddle point

35. $-4 < f_{xy}(3, 7) < 4$

37. (a) $(0, 0)$ (b) Saddle point: $(0, 0, 0)$ (c) $(0, 0)$
(d)

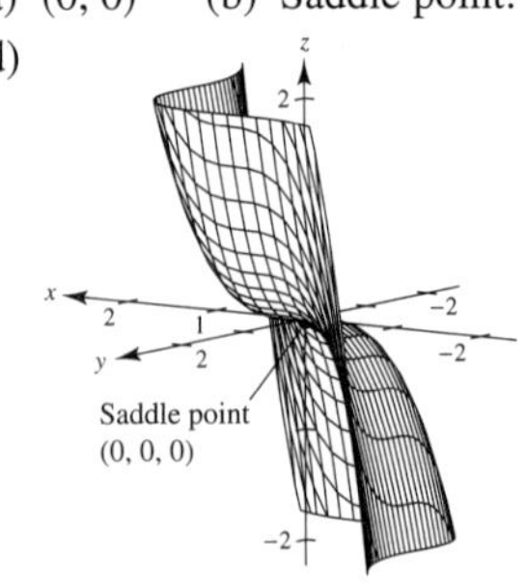

39. (a) $(1, a), (b, -4)$ (b) Absolute minima: $(1, a, 0), (b, -4, 0)$
(c) $(1, a), (b, -4)$ (d)

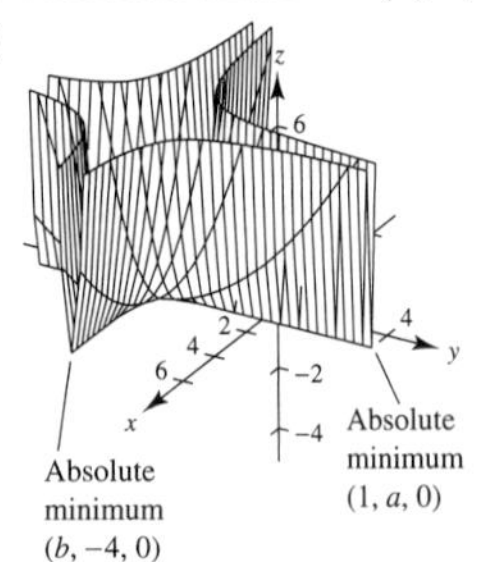

41. (a) $(0, 0)$ (b) Absolute minimum: $(0, 0, 0)$ (c) $(0, 0)$
(d)

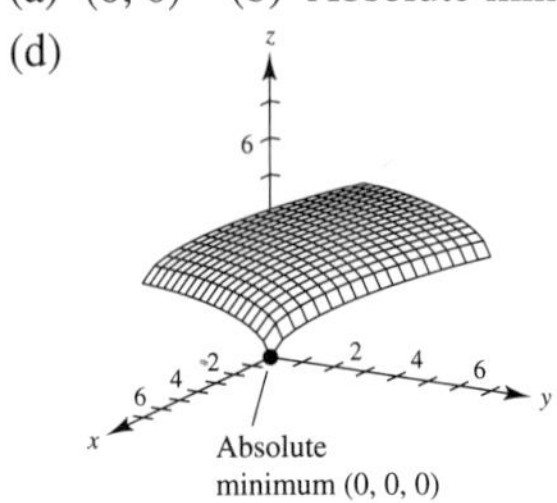

43. Relative minimum: $(0, 3, -1)$

45. Absolute maximum: $(4, 0, 21)$
Absolute minimum: $(4, 2, -11)$

47. Absolute maximum: $(0, 1, 10)$
Absolute minimum: $(1, 2, 5)$

49. Absolute maxima: $(\pm 2, 4, 28)$
Absolute minimum: $(0, 1, -2)$

51. Absolute maxima: $(-2, -1, 9), (2, 1, 9)$
Absolute minima: $(x, -x, 0),\ |x| \leq 1$

53. Absolute maximum: $(1, 1, 1)$
Absolute minimum: $(0, 0, 0)$

55. Point A is a saddle point.

57. Answers will vary.
Sample answer:

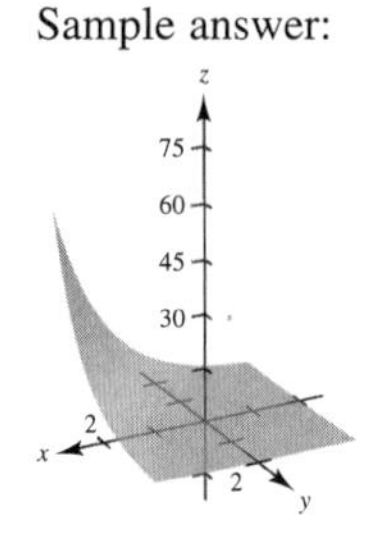

No extrema

59. Answers will vary.
Sample answer:

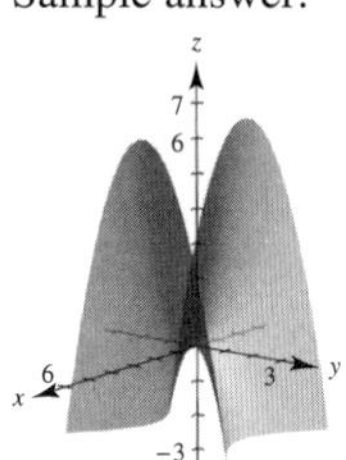

Saddle point

61. False. Let $f(x, y) = 1 - |x| - |y|$ at the point $(0, 0, 1)$.

63. False. Let $f(x, y) = x^2y^2$ (see Example 4 on page 958).

Section 13.9 (page 966)

1. $\sqrt{3}$ **3.** $\sqrt{7}$ **5.** $x = y = z = 3$ **7.** 10, 10, 10

9. 9 ft × 9 ft × 8.25 ft; \$26.73

11. Let $a + b + c = k$.
$V = 4\pi abc/3 = \frac{4}{3}\pi ab(k - a - b) = \frac{4}{3}\pi(kab - a^2b - ab^2)$

$$\left.\begin{aligned} V_a &= \tfrac{4}{3}\pi(kb - 2ab - b^2) = 0 \\ V_b &= \tfrac{4}{3}\pi(ka - a^2 - 2ab) = 0 \end{aligned}\right\} \begin{aligned} kb - 2ab - b^2 &= 0 \\ kb - a^2 - 2ab &= 0 \end{aligned}$$

So, $a = b$ and $b = k/3$. Thus, $a = b = c = k/3$.

13. Let x, y, and z be the length, width, and height, respectively, and let V_0 be the given volume. Then $V_0 = xyz$ and $z = V_0/xy$. The surface area is
$S = 2xy + 2yz + 2xz = 2(xy + V_0/x + V_0/y)$.

$$\left.\begin{aligned} S_x &= 2(y - V_0/x^2) = 0 \\ S_y &= 2(x - V_0/y^2) = 0 \end{aligned}\right\} \begin{aligned} x^2y - V_0 &= 0 \\ xy^2 - V_0 &= 0 \end{aligned}$$

So, $x = \sqrt[3]{V_0}$, $y = \sqrt[3]{V_0}$, and $z = \sqrt[3]{V_0}$.

15. $x_1 = 3$; $x_2 = 6$ **17.** Proof

19. $x = \sqrt{2}/2 \approx 0.707$ km
$y = (3\sqrt{2} + 2\sqrt{3})/6 \approx 1.284$ km

21. (a) $S = \sqrt{x^2 + y^2} + \sqrt{(x + 2)^2 + (y - 2)^2} + \sqrt{(x - 4)^2 + (y - 2)^2}$

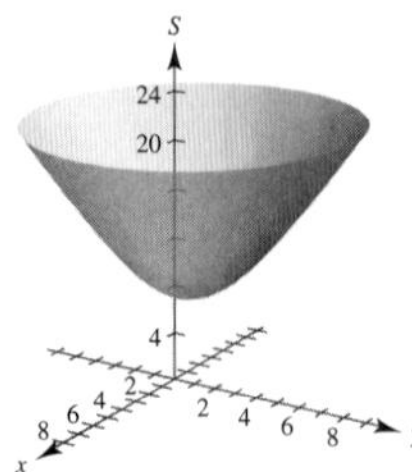

The surface has a minimum.

(b) $S_x = \dfrac{x}{\sqrt{x^2+y^2}} + \dfrac{x+2}{\sqrt{(x+2)^2+(y-2)^2}} + \dfrac{x-4}{\sqrt{(x-4)^2+(y-2)^2}}$

$S_y = \dfrac{y}{\sqrt{x^2+y^2}} + \dfrac{y-2}{\sqrt{(x+2)^2+(y-2)^2}} + \dfrac{y-2}{\sqrt{(x-4)^2+(y-2)^2}}$

(c) $-\dfrac{1}{\sqrt{2}}\mathbf{i} - \left(\dfrac{1}{\sqrt{2}} - \dfrac{2}{\sqrt{10}}\right)\mathbf{j}$

$\theta \approx 186.0°$

(d) $t = 1.344$; $(x_2, y_2) \approx (0.05, 0.90)$

(e) $(x_4, y_4) \approx (0.06, 0.45)$; $S = 7.266$

(f) $-\nabla S(x, y)$ gives the direction of greatest rate of decrease of S. Use $\nabla S(x, y)$ when finding a maximum.

23. Write the equation to be maximized or minimized as a function of two variables. Take the partial derivatives and set them equal to zero or undefined to obtain the critical points. Use the Second Partials Test to test for relative extrema using the critical points. Check the boundary points.

25. (a) $y = \frac{3}{4}x + \frac{4}{3}$ (b) $\frac{1}{6}$ **27.** (a) $y = -2x + 4$ (b) 2

29. $y = \frac{37}{43}x + \frac{7}{43}$

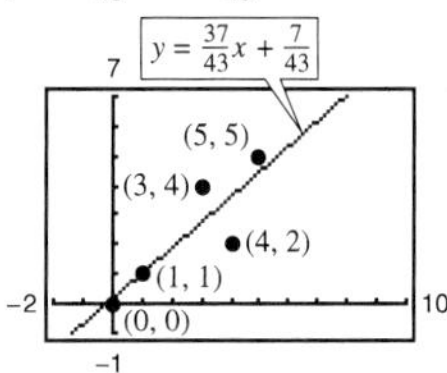

31. $y = -\frac{175}{148}x + \frac{945}{148}$

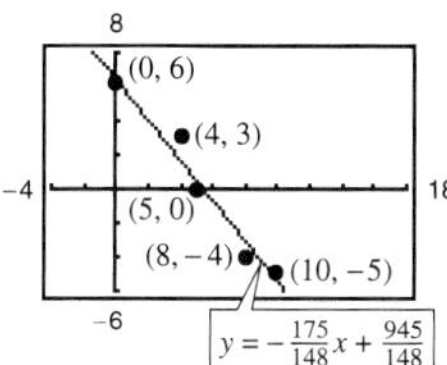

33. (a) $y = 1.6x + 84$

(b)

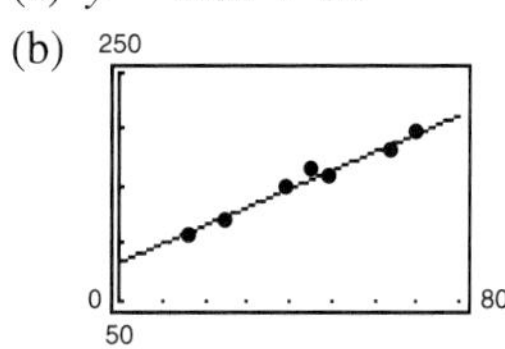

(c) 1.6

35. $y = 14x + 19$

41.4 bushels per acre

37. $a\sum_{i=1}^{n} x_i^4 + b\sum_{i=1}^{n} x_i^3 + c\sum_{i=1}^{n} x_i^2 = \sum_{i=1}^{n} x_i^2 y_i$

$a\sum_{i=1}^{n} x_i^3 + b\sum_{i=1}^{n} x_i^2 + c\sum_{i=1}^{n} x_i = \sum_{i=1}^{n} x_i y_i$

$a\sum_{i=1}^{n} x_i^2 + b\sum_{i=1}^{n} x_i + cn = \sum_{i=1}^{n} y_i$

39. $y = \frac{3}{7}x^2 + \frac{6}{5}x + \frac{26}{35}$

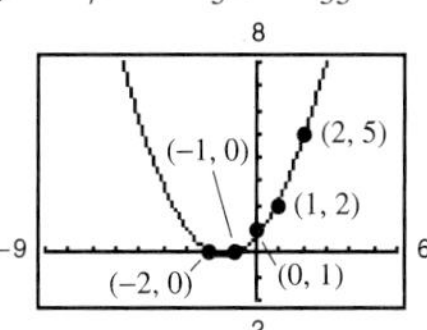

41. $y = x^2 - x$

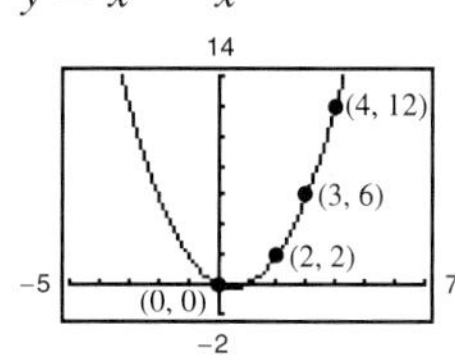

43. (a) $y = -0.22x^2 + 9.66x - 1.79$

(b)

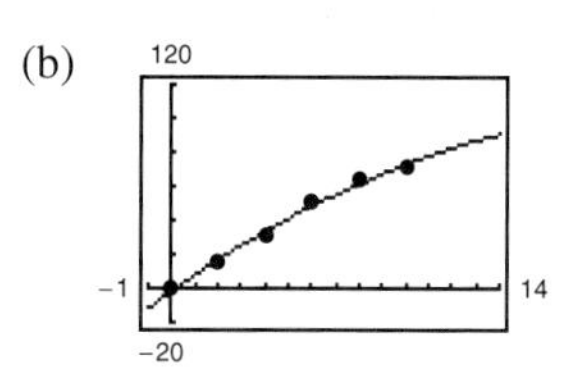

45. (a) $\ln P = -0.1499h + 9.3018$ (b) $P = 10{,}957.7e^{-0.1499h}$

(c)

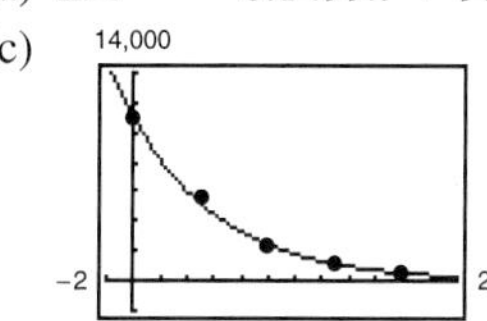

(d) Proof

47. Proof

Section 13.10 (page 976)

1.

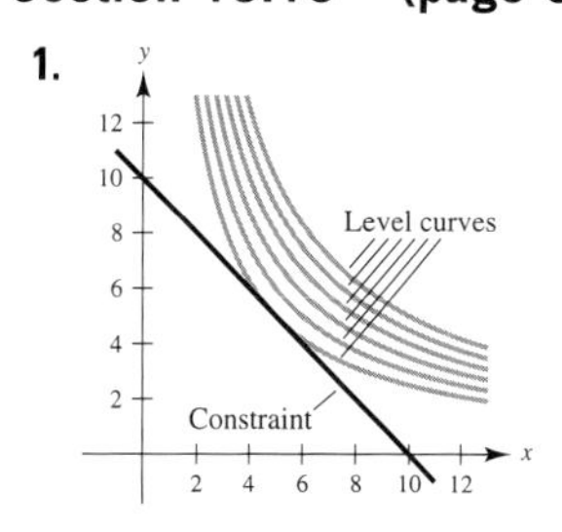

$f(5, 5) = 25$

3.

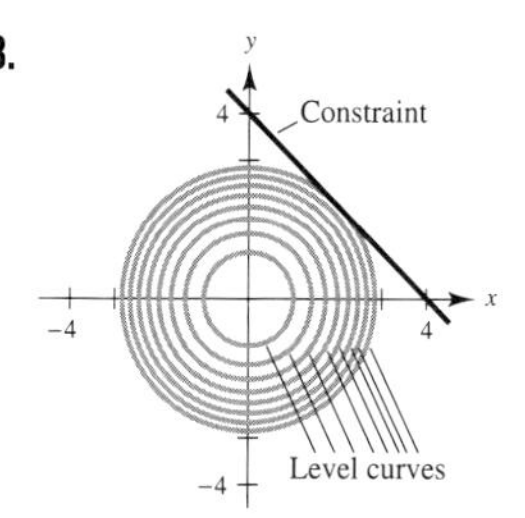

$f(2, 2) = 8$

5. $f(1, 2) = 5$ **7.** $f(25, 50) = 2600$

9. $f(1, 1) = 2$ **11.** $f(3, 3, 3) = 27$ **13.** $f\left(\frac{1}{3}, \frac{1}{3}, \frac{1}{3}\right) = \frac{1}{3}$

15. Maxima: $f\left(\sqrt{2}/2, \sqrt{2}/2\right) = \frac{5}{2}$

$f\left(-\sqrt{2}/2, -\sqrt{2}/2\right) = \frac{5}{2}$

Minima: $f\left(-\sqrt{2}/2, \sqrt{2}/2\right) = -\frac{1}{2}$

$f\left(\sqrt{2}/2, -\sqrt{2}/2\right) = -\frac{1}{2}$

17. $f(8, 16, 8) = 1024$ **19.** $\sqrt{2}/2$ **21.** $3\sqrt{2}$ **23.** $\sqrt{11}/2$

25. 0.188 **27.** $\sqrt{3}$ **29.** $(-4, 0, 4)$

31. Optimization problems that have restrictions or constraints on the values that can be used to produce the optimal solutions are called constrained optimization problems.

33. $\sqrt{3}$ **35.** $x = y = z = 3$

37. 9 ft × 9 ft × 8.25 ft; \$26.73 **39.** $a = b = c = k/3$

41. Proof **43.** $2\sqrt{3}a/3 \times 2\sqrt{3}b/3 \times 2\sqrt{3}c/3$

45. $\sqrt[3]{360} \times \sqrt[3]{360} \times \frac{4}{3}\sqrt[3]{360}$ ft

47. $r = \sqrt[3]{\dfrac{v_0}{2\pi}}$ and $h = 2\sqrt[3]{\dfrac{v_0}{2\pi}}$ **49.** Proof

51. $P(15{,}625/18, 3125) \approx 226{,}869$

53. $x \approx 191.3$

$y \approx 688.7$

Cost ≈ \$55,095.60

55. (a) $g(\pi/3, \pi/3, \pi/3) = \frac{1}{8}$

(b)

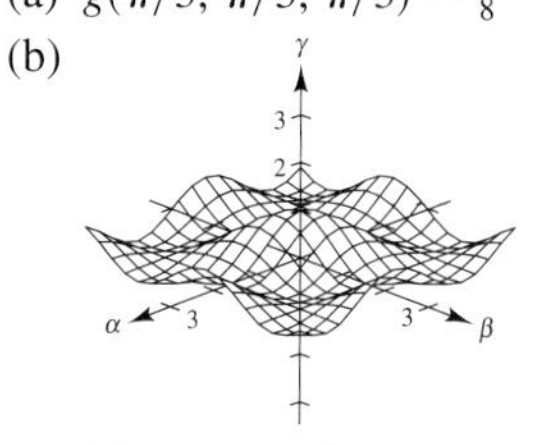

Maximum values occur when $\alpha = \beta$.

Review Exercises for Chapter 13 (page 978)

1.

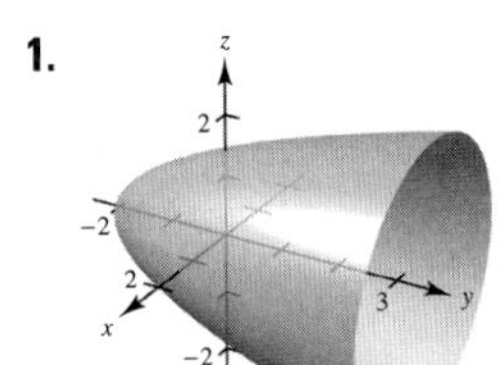

3. (a)

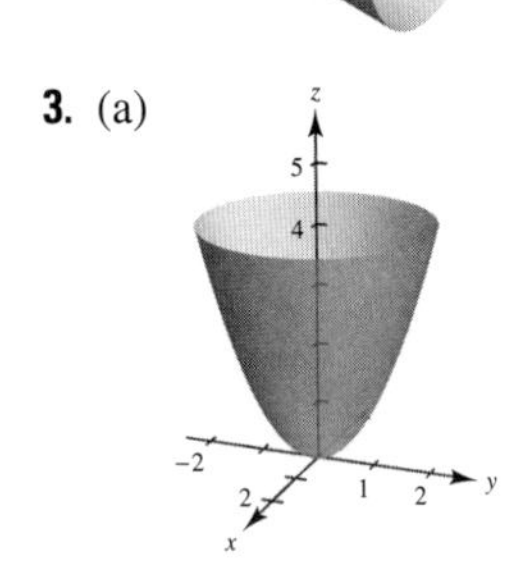

(b) g is a vertical translation of f two units upward.

(c) g is a horizontal translation of f two units to the right.

(d)

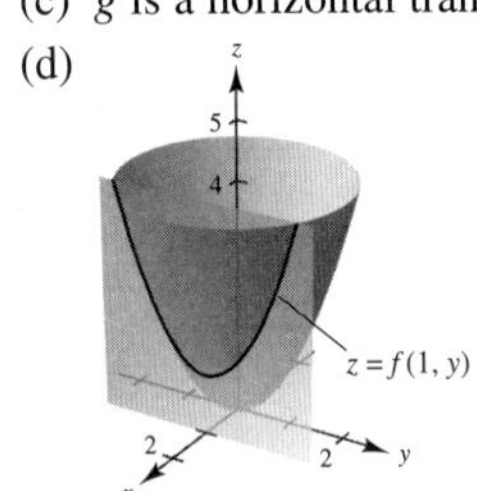

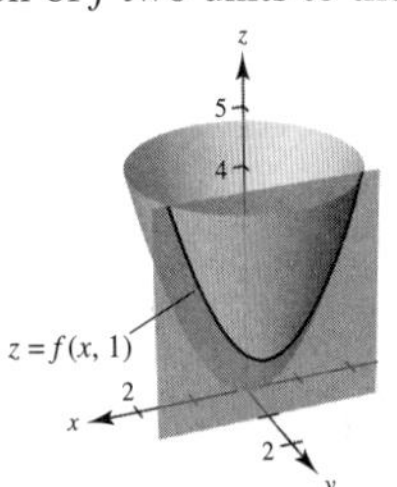

5.

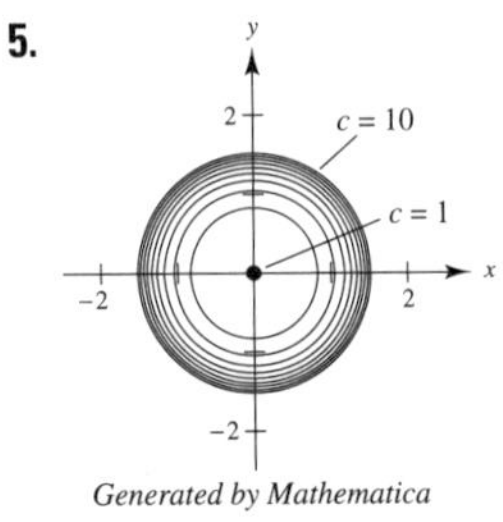

Generated by Mathematica

7.

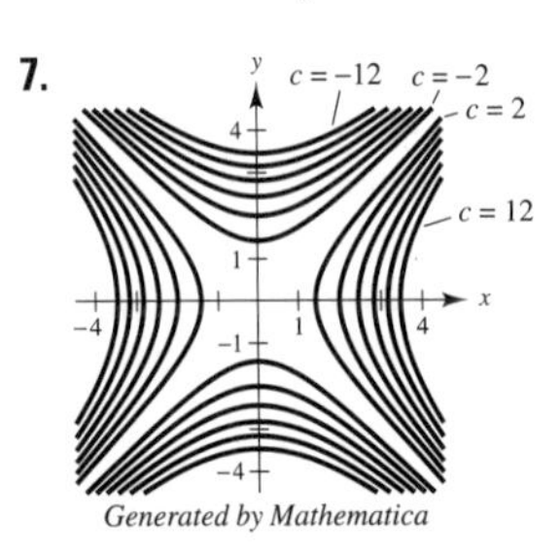

Generated by Mathematica

9.

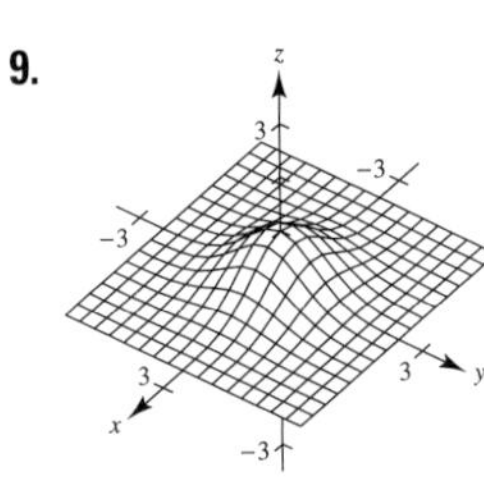

11. Limit: $\frac{1}{2}$
Continuous except at $(0, 0)$

13. Limit: 0
Continuous

15. $f_x(x, y) = e^x \cos y$
$f_y(x, y) = -e^x \sin y$

17. $\partial z/\partial x = -e^{-x}$
$\partial z/\partial y = -e^{-y}$

19. $g_x(x, y) = [y(y^2 - x^2)]/(x^2 + y^2)^2$
$g_y(x, y) = [x(x^2 - y^2)]/(x^2 + y^2)^2$

21. $f_x(x, y, z) = -yz/(x^2 + y^2)$
$f_y(x, y, z) = xz/(x^2 + y^2)$
$f_z(x, y, z) = \arctan y/x$

23. $u_x(x, t) = cne^{-n^2t} \cos nx$
$u_t(x, t) = -cn^2e^{-n^2t} \sin nx$

25. Answers will vary. Example:

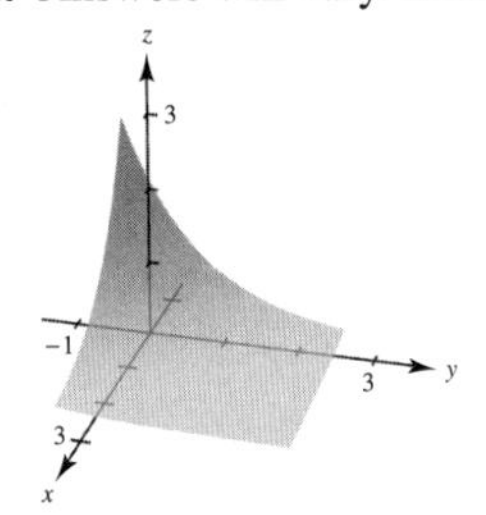

27. $f_{xx}(x, y) = 6$
$f_{yy}(x, y) = 12y$
$f_{xy}(x, y) = f_{yx}(x, y) = -1$

29. $h_{xx}(x, y) = -y \cos x$
$h_{yy}(x, y) = -x \sin y$
$h_{xy}(x, y) = h_{yx}(x, y) = \cos y - \sin x$

31. $\partial^2z/\partial x^2 + \partial^2z/\partial y^2 = 2 + (-2) = 0$

33. $\dfrac{\partial^2 z}{\partial x^2} + \dfrac{\partial^2 z}{\partial y^2} = \dfrac{6x^2y - 2y^3}{(x^2 + y^2)^3} + \dfrac{-6x^2y + 2y^3}{(x^2 + y^2)^3} = 0$

35. $(xy \cos xy + \sin xy)\, dx + (x^2 \cos xy)\, dy$

37. 0.6538 cm, 5.03% **39.** $\pm\pi$ in.^{3}

41. $dw/dt = (8t - 1)/(4t^2 - t + 4)$

43. $\partial w/\partial r = (4r^2t - 4rt^2 - t^3)/(2r - t)^2$
$\partial w/\partial t = (4r^2t - rt^2 + 4r^3)/(2r - t)^2$

45. $\partial z/\partial x = (-2x - y)/(y + 2z)$
$\partial z/\partial y = (-x - 2y - z)/(y + 2z)$

47. -50 **49.** $\frac{2}{3}$ **51.** $\langle 4, 4\rangle, 4\sqrt{2}$ **53.** $\langle -\frac{1}{2}, 0\rangle, \frac{1}{2}$

55. (a) $54\mathbf{i} - 16\mathbf{j}$ (b) $\dfrac{27}{\sqrt{793}}\mathbf{i} - \dfrac{8}{\sqrt{793}}\mathbf{j}$ (c) $y = \dfrac{27}{8}x - \dfrac{65}{8}$

(d)

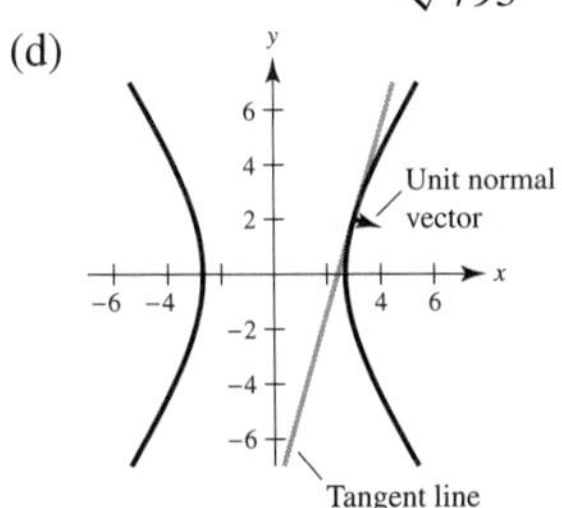

57. Tangent plane: $4x + 4y - z = 8$
Normal line: $x = 2 + 4t,\ y = 1 + 4t,\ z = 4 - t$

59. Tangent plane: $z = 4$
Normal line: $x = 2,\ y = -3,\ z = 4 + t$

61. $(x - 2)/1 = (y - 2)/1 = (z - 5)/(-4)$ **63.** $\theta \approx 36.7°$

65. Relative minimum: $\left(-4, \frac{4}{3}, -2\right)$

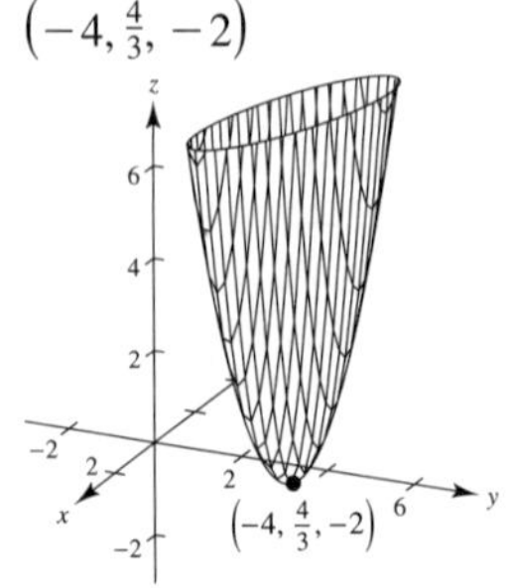

67. Relative minimum: $(1, 1, 3)$

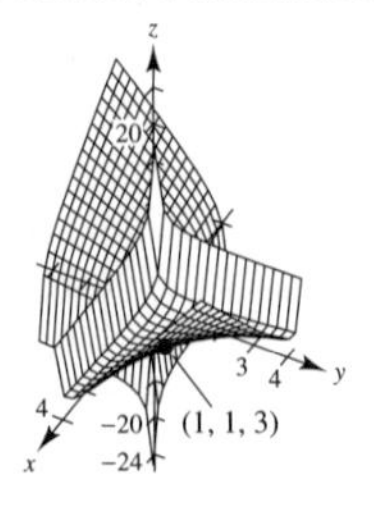

69. The level curves are hyperbolas. The critical point $(0, 0)$ may be a saddle point or an extremum.

71. $x_1 = 94, x_2 = 157$ **73.** $f(49.4, 253) = 13{,}201.8$

75. (a) $y = 0.004x^2 + 0.07x + 19.4$ (b) 50.6 kg

77. Maximum: $f\left(\frac{1}{3}, \frac{1}{3}, \frac{1}{3}\right) = \frac{1}{3}$

79. $x = \sqrt{2}/2 \approx 0.707$ km; $y = \sqrt{3}/3 \approx 0.577$ km; $z = \left(60 - 3\sqrt{2} - 2\sqrt{3}\right)/6 \approx 8.716$ km

P.S. Problem Solving (page 981)

1. (a) 12 square units (b) Proof (c) Proof

3. (a) $y_0z_0(x - x_0) + x_0z_0(y - y_0) + x_0y_0(z - z_0) = 0$

(b) $x_0y_0z_0 = 1 \Rightarrow z_0 = 1/x_0y_0$

Then the tangent plane is

$$y_0\left(\frac{1}{x_0y_0}\right)(x - x_0) + x_0\left(\frac{1}{x_0y_0}\right)(y - y_0) + x_0y_0\left(z - \frac{1}{x_0y_0}\right) = 0.$$

Intercepts: $(3x_0, 0, 0), (0, 3y_0, 0), \left(0, 0, \dfrac{3}{x_0y_0}\right)$

$V = \frac{1}{3}bh = \frac{9}{2}$

5. (a)

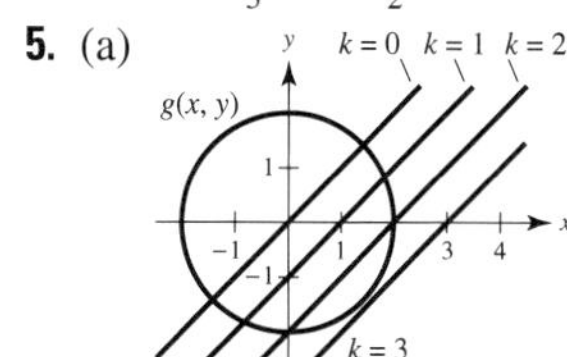

Maximum value: $2\sqrt{2}$

(b)

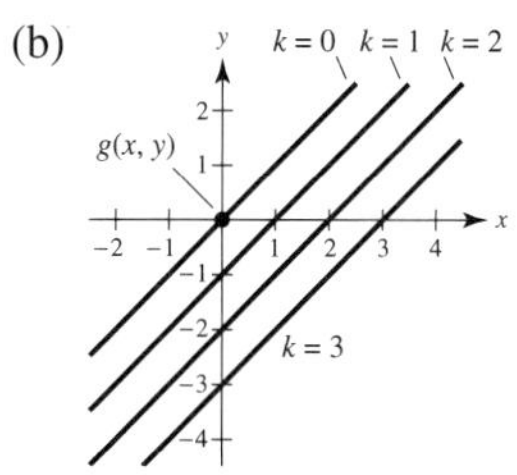

Maximum and minimum value: 0

The method of Lagrange multipliers does not work because $\nabla g(x_0, y_0) = \mathbf{0}$.

7. $2\sqrt[3]{150} \times 2\sqrt[3]{150} \times 5\sqrt[3]{150}/3$

9. (a)
$$\begin{aligned} x\frac{\partial f}{\partial x} + y\frac{\partial f}{\partial y} &= xCy^{1-a}ax^{a-1} + yCx^a(1 - a)y^{1-a-1} \\ &= ax^aCy^{1-a} + (1 - a)x^aC(y^{1-a}) \\ &= Cx^ay^{1-a}[a + (1 - a)] \\ &= Cx^ay^{1-a} \\ &= f(x, y) \end{aligned}$$

(b)
$$\begin{aligned} f(tx, ty) &= C(tx)^a(ty)^{1-a} \\ &= Ctx^ay^{1-a} \\ &= tCx^ay^{1-a} \\ &= tf(x, y) \end{aligned}$$

11. (a) $x = 32\sqrt{2}t$

$y = 32\sqrt{2}t - 16t^2$

(b) $\alpha = \arctan\left(\dfrac{y}{x + 50}\right) = \arctan\left(\dfrac{32\sqrt{2}t - 16t^2}{32\sqrt{2}t + 50}\right)$

(c) $\dfrac{d\alpha}{dt} = \dfrac{-16\left(8\sqrt{2}t^2 + 25t - 25\sqrt{2}\right)}{64t^4 - 256\sqrt{2}t^3 + 1024t^2 + 800\sqrt{2}t + 625}$

(d)

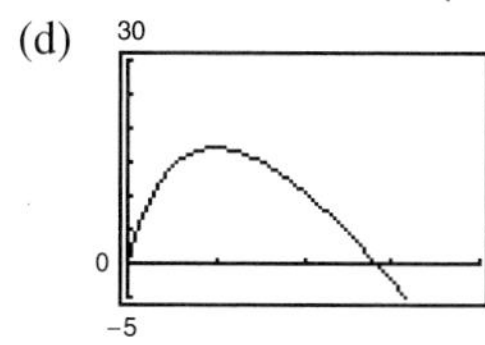

No; The rate of change of α is greatest when the projectile is closest to the camera.

(e) α is maximum when $t = 0.98$ second.

No; the projectile is at its maximum height when $t = \sqrt{2} \approx 1.41$ seconds.

13. (a)

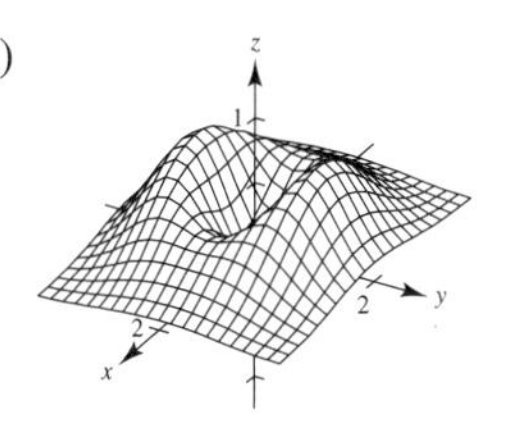

Minimum: $(0, 0, 0)$

Maxima: $(0, \pm1, 2e^{-1})$

Saddle points: $(\pm1, 0, e^{-1})$

(b)

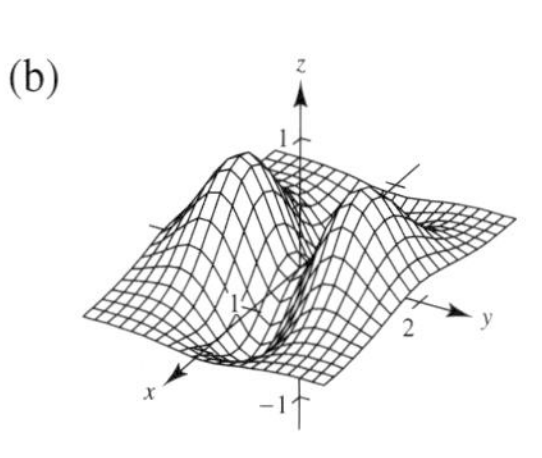

Minima: $(\pm1, 0, -e^{-1})$

Maxima: $(0, \pm1, 2e^{-1})$

Saddle point: $(0, 0, 0)$

(c) $\alpha > 0$

Minimum: $(0, 0, 0)$

Maxima: $(0, \pm1, \beta e^{-1})$

Saddle points: $(\pm1, 0, \alpha e^{-1})$

$\alpha < 0$

Minima: $(\pm1, 0, \alpha e^{-1})$

Maxima: $(0, \pm1, \beta e^{-1})$

Saddle point: $(0, 0, 0)$

15. (a)

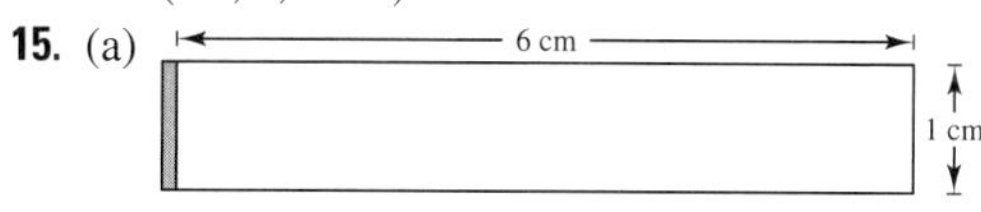

(b)

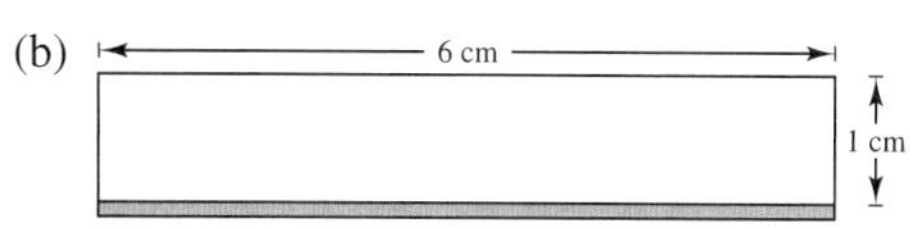

(c) Height

(d) $dl = 0.01, dh = 0$: $dA = 0.01$

$dl = 0, dh = 0.01$: $dA = 0.06$

17–19. Proofs

Chapter 14

Section 14.1 (page 990)

1. $2x^2$ **3.** $y\ln(2y)$ **5.** $(4x^2 - x^4)/2$

7. $(y/2)[(\ln y)^2 - y^2]$ **9.** $x^2(1 - e^{-x^2} - x^2e^{-x^2})$ **11.** 3

13. $\frac{8}{3}$ **15.** $\frac{1}{2}$ **17.** 2 **19.** $\frac{1}{3}$ **21.** 1629 **23.** $\frac{2}{3}$ **25.** 4

27. $\pi/2$ **29.** $\pi^2/32 + \frac{1}{8}$ **31.** $\frac{1}{2}$ **33.** Diverges **35.** 24

37. $\frac{16}{3}$ **39.** $\frac{8}{3}$ **41.** 5 **43.** πab **45.** $\frac{9}{2}$

47.

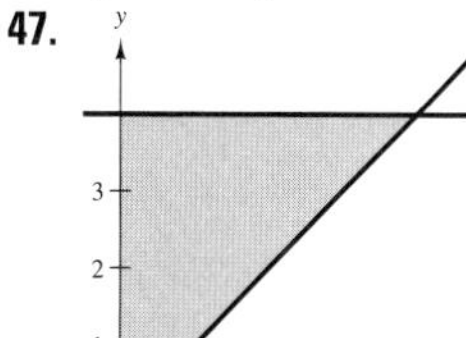

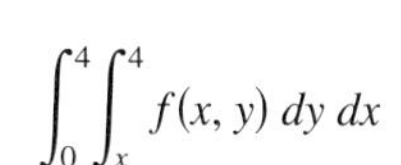

$$\int_0^4 \int_x^4 f(x, y)\, dy\, dx$$

49.

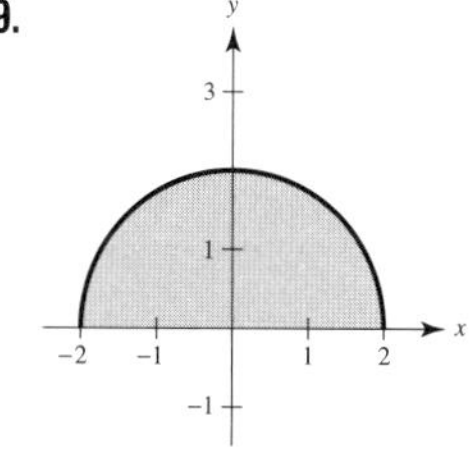

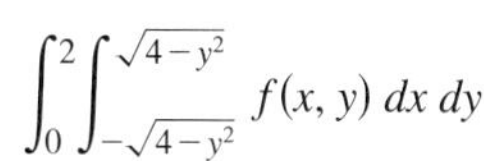

$$\int_0^2 \int_{-\sqrt{4-y^2}}^{\sqrt{4-y^2}} f(x, y)\, dx\, dy$$

51.

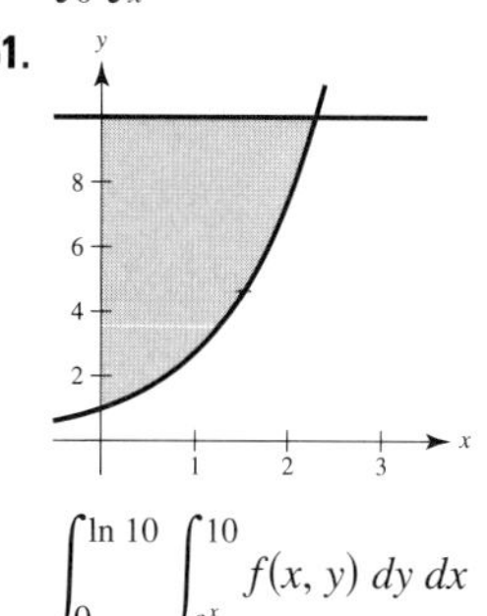

$$\int_0^{\ln 10} \int_{e^x}^{10} f(x, y)\, dy\, dx$$

53.

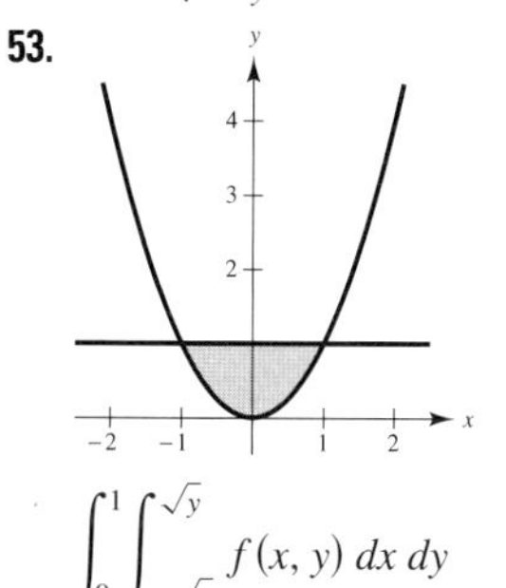

$$\int_0^1 \int_{-\sqrt{y}}^{\sqrt{y}} f(x, y)\, dx\, dy$$

55.

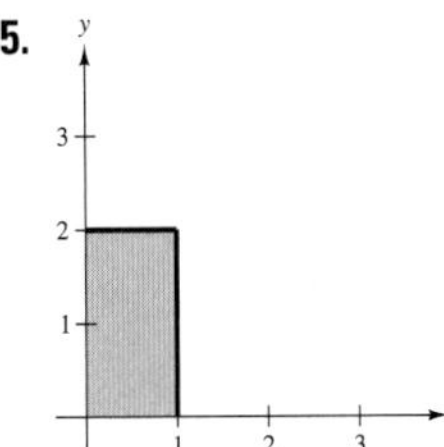

$$\int_0^1\int_0^2 dy\,dx = \int_0^2\int_0^1 dx\,dy = 2$$

57.

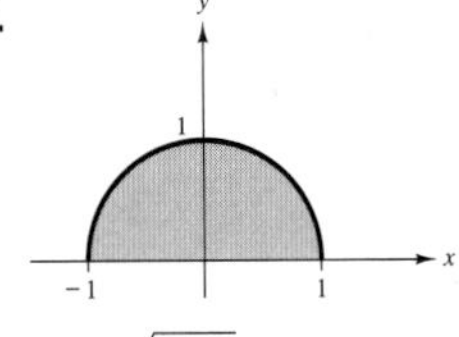

$$\int_0^1\int_{-\sqrt{1-y^2}}^{\sqrt{1-y^2}} dx\,dy = \int_{-1}^1\int_0^{\sqrt{1-x^2}} dy\,dx = \frac{\pi}{2}$$

59.

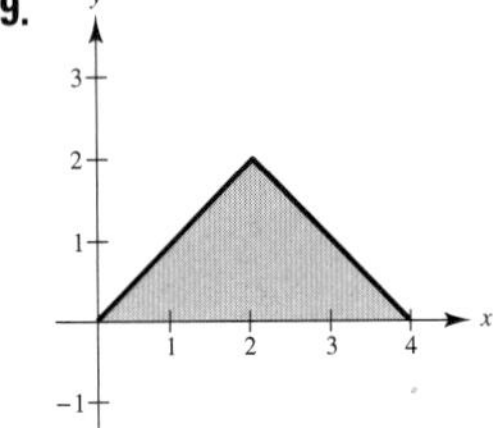

$$\int_0^2\int_0^x dy\,dx + \int_2^4\int_0^{4-x} dy\,dx = \int_0^2\int_y^{4-y} dx\,dy = 4$$

61.

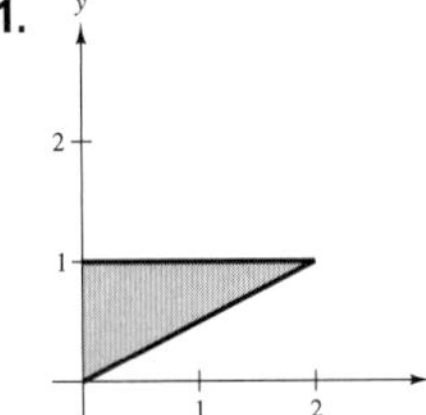

$$\int_0^2\int_{x/2}^1 dy\,dx = \int_0^1\int_0^{2y} dx\,dy = 1$$

63.

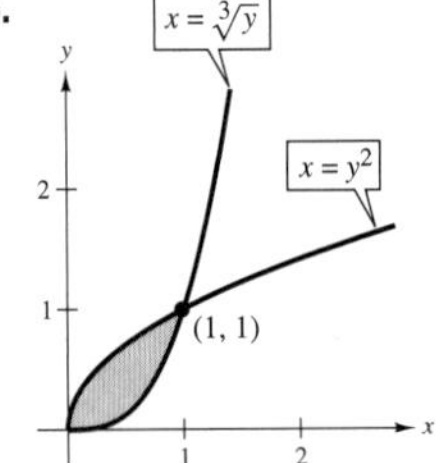

$$\int_0^1\int_{y^2}^{\sqrt[3]{y}} dx\,dy = \int_0^1\int_{x^3}^{\sqrt{x}} dy\,dx = \frac{5}{12}$$

65. The first integral arises using vertical representative rectangles. The second two integrals arise using horizontal representative rectangles.
Value of the integrals: $15{,}625\pi/24$

67.

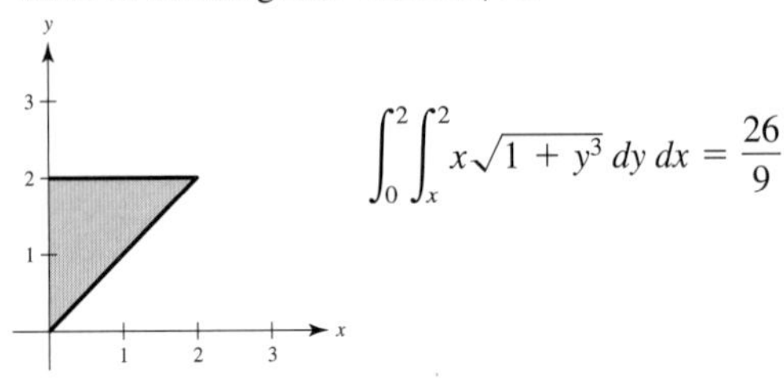

$$\int_0^2\int_x^2 x\sqrt{1+y^3}\,dy\,dx = \frac{26}{9}$$

69.

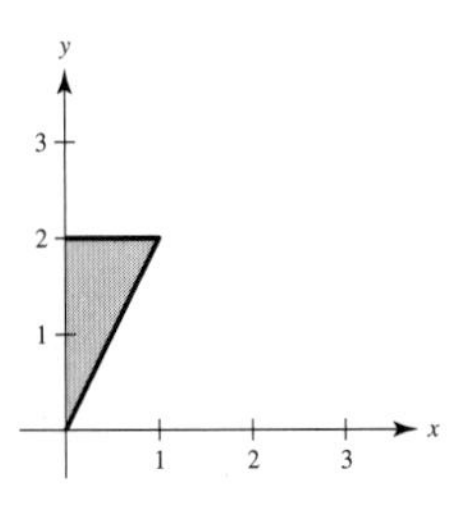

$$\int_0^1\int_{2x}^2 4e^{y^2}\,dy\,dx = e^4 - 1 \approx 53.598$$

71.

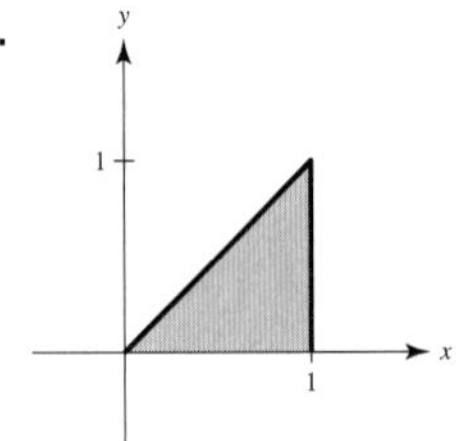

$$\int_0^1\int_y^1 \sin x^2\,dx\,dy = \frac{1}{2}(1 - \cos 1) \approx 0.230$$

73. $\frac{1664}{105}$ **75.** $(\ln 5)^2$

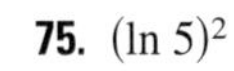

77. (a)

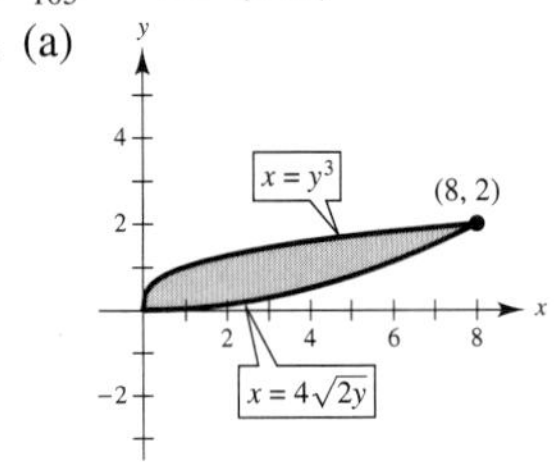

(b) $\displaystyle\int_0^8\int_{x^2/32}^{\sqrt[3]{x}} (x^2y - xy^2)\,dy\,dx$ (c) $67{,}520/693$

79. 20.5648 **81.** $15\pi/2$

83. An iterated integral is an integral of a function of several variables. Integrate with respect to one variable while holding the other variables constant.

85. If all four limits of integration are constant, the region of integration is rectangular.

87. True

Section 14.2 (page 1000)

1. 24 (approximation is exact)

3. Approximation: 52; Exact: $\frac{160}{3}$ **5.** 400; 272

7.

8

9.

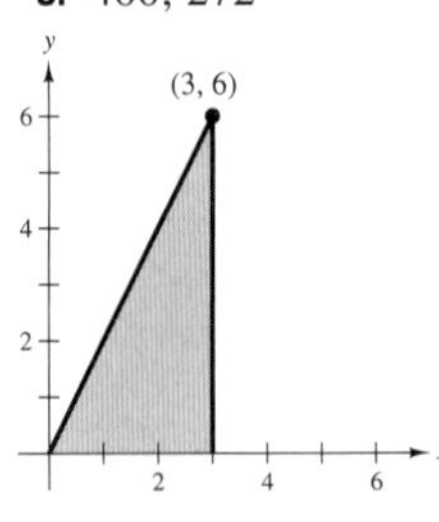

36

11.

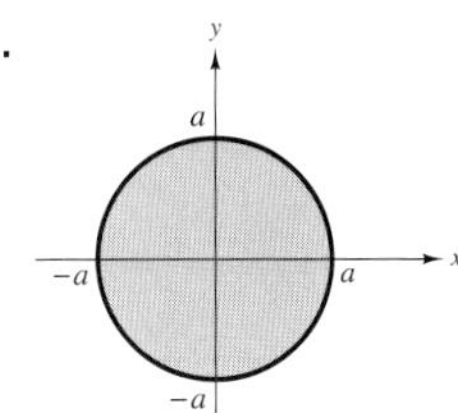

0

13. $\int_0^3\int_0^5 xy\,dy\,dx = \frac{225}{4}$

$\int_0^5\int_0^3 xy\,dx\,dy = \frac{225}{4}$

15. $\int_1^2\int_x^{2x} \frac{y}{x^2+y^2}\,dy\,dx = \frac{1}{2}\ln\frac{5}{2}$

$\int_1^2\int_1^y \frac{y}{x^2+y^2}\,dx\,dy + \int_2^4\int_{y/2}^2 \frac{y}{x^2+y^2}\,dx\,dy = \frac{1}{2}\ln\frac{5}{2}$

17. $\int_0^1\int_{4-x}^{4-x^2} -2y\,dy\,dx = -\frac{6}{5}$

$\int_3^4\int_{4-y}^{\sqrt{4-y}} -2y\,dx\,dy = -\frac{6}{5}$

19. $\int_0^3\int_{4y/3}^{\sqrt{25-y^2}} x\,dx\,dy = 25$

$\int_0^4\int_0^{3x/4} x\,dy\,dx + \int_4^5\int_0^{\sqrt{25-x^2}} x\,dy\,dx = 25$

21. 4 **23.** 4 **25.** 12 **27.** $\frac{3}{8}$ **29.** 1 **31.** $32\sqrt{2}\pi/3$

33. $\int_0^1\int_0^x xy\,dy\,dx = \frac{1}{8}$ **35.** $\int_0^2\int_0^4 x^2\,dy\,dx = \frac{32}{3}$

37. $2\int_0^1\int_0^x \sqrt{1-x^2}\,dy\,dx = \frac{2}{3}$

39. $\int_0^2\int_0^{\sqrt{4-x^2}} (x+y)\,dy\,dx = \frac{16}{3}$

41. $2\int_0^2\int_0^{\sqrt{1-(x-1)^2}} (2x - x^2 - y^2)\,dy\,dx$

43. $4\int_0^2\int_0^{\sqrt{4-x^2}} (x^2+y^2)\,dy\,dx$

45. $\int_0^2\int_{-\sqrt{2-2(y-1)^2}}^{\sqrt{2-2(y-1)^2}} (4y - x^2 - 2y^2)\,dx\,dy$

47. $81\pi/2$ **49.** 1.2315 **51.** Proof

53.

$\int_0^1\int_{y/2}^{1/2} e^{-x^2}\,dx\,dy = 1 - e^{-1/4} \approx 0.221$

55.

$\int_{-2}^2\int_{-\sqrt{4-x^2}}^{\sqrt{4-x^2}} \sqrt{4-y^2}\,dy\,dx = \frac{64}{3}$

57.

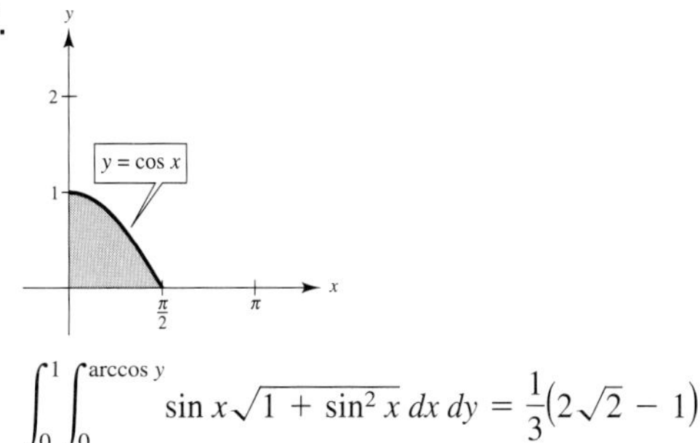

$\int_0^1\int_0^{\arccos y} \sin x\sqrt{1+\sin^2 x}\,dx\,dy = \frac{1}{3}(2\sqrt{2} - 1)$

59. 2 **61.** $\frac{8}{3}$ **63.** $(e-1)^2$ **65.** 25,645.24

67. See "Definition of Double Integral" on page 994. The double integral of a function $f(x, y) \geq 0$ over the region of integration yields the volume of that region.

69. (a) The total snowfall in county R

(b) The average snowfall in county R

71. No; 6π is the greatest possible value. **73.** Proof; $\frac{1}{5}$

75. Proof; $\frac{7}{27}$ **77.** 2500 m^3 **79.** (a) 1.784 (b) 1.788

81. (a) 11.057 (b) 11.041 **83.** d

85. False. $V = 8\int_0^1\int_0^{\sqrt{1-y^2}} \sqrt{1-x^2-y^2}\,dx\,dy.$

87. $\frac{1}{2}(1-e)$ **89.** R: $x^2 + y^2 \leq 9$ **91.** About 0.82736

93. Putnam Problem A2, 1989

Section 14.3 (page 1009)

1. Rectangular **3.** Polar

5. The region R is a half-circle of radius 8. It can be described in polar coordinates as

$R = \{(r, \theta): 0 \leq r \leq 8, 0 \leq \theta \leq \pi\}.$

7. The region R is a cardioid with $a = b = 3$. It can be described in polar coordinates as

$R = \{(r, \theta): 0 \leq r \leq 3 + 3\sin\theta, 0 \leq \theta \leq 2\pi\}.$

9. $\pi/4$

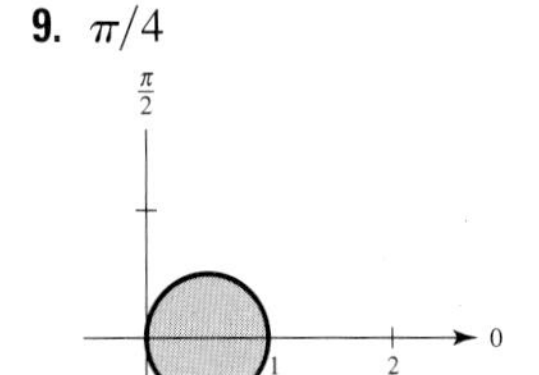

11. 0

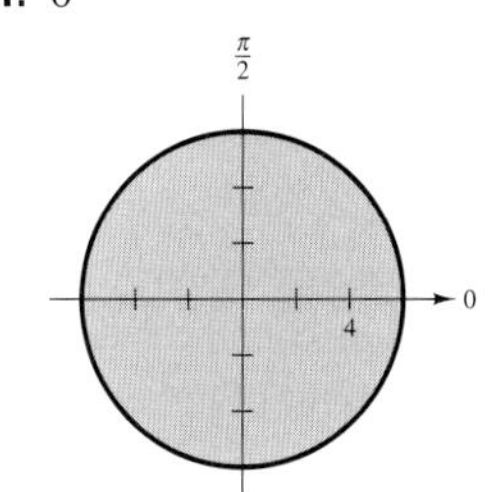

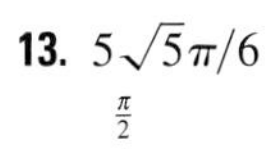

13. $5\sqrt{5}\pi/6$

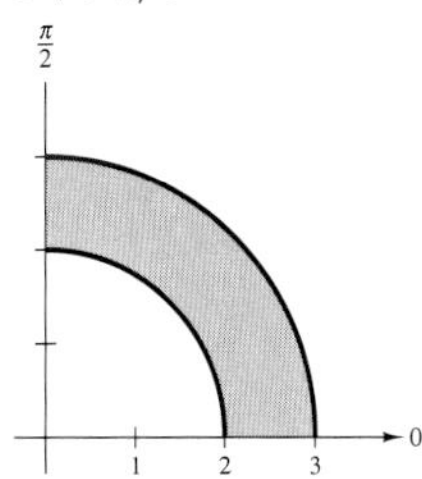

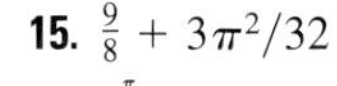

15. $\frac{9}{8} + 3\pi^2/32$

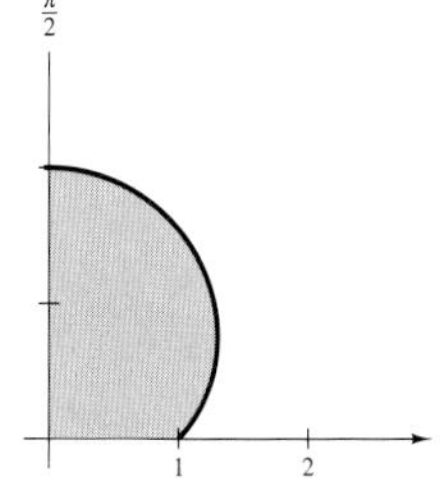

17. $a^3/3$ **19.** 4π **21.** $243\pi/10$ **23.** $\frac{2}{3}$ **25.** $(\pi/2)\sin 1$

27. $\int_0^{\pi/4}\int_0^{2\sqrt{2}} r^2\,dr\,d\theta = \frac{4\sqrt{2}\pi}{3}$

29. $\int_0^{\pi/2}\int_0^2 r^2(\cos\theta + \sin\theta)\, dr\, d\theta = \frac{16}{3}$

31. $\int_0^{\pi/4}\int_1^2 r\theta\, dr\, d\theta = \frac{3\pi^2}{64}$ **33.** $\frac{1}{8}$ **35.** $\frac{250\pi}{3}$

37. $\frac{64}{9}(3\pi - 4)$ **39.** $2\sqrt{4 - 2\sqrt[3]{2}}$ **41.** 1.2858

43. 9π **45.** $3\pi/2$ **47.** π

49.

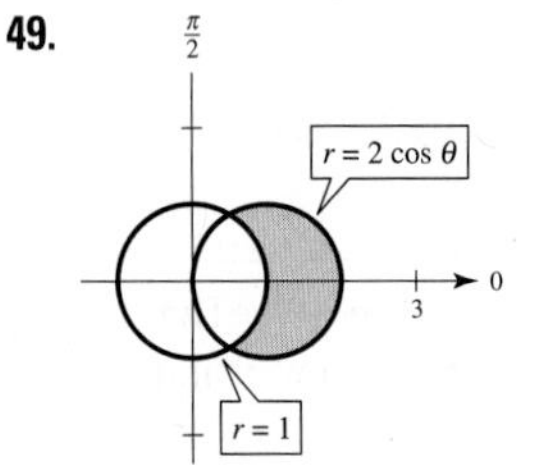

$\frac{\pi}{3} + \frac{\sqrt{3}}{2}$

51.

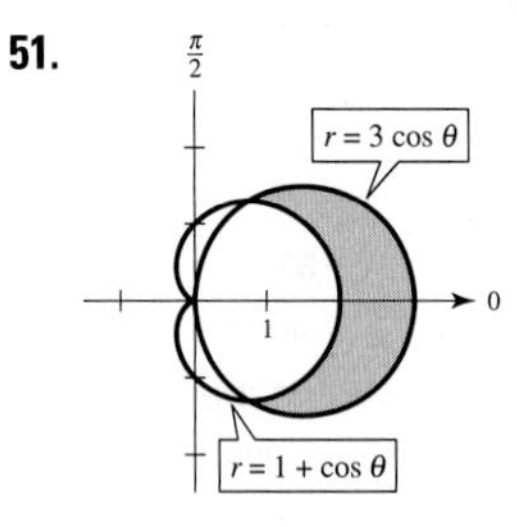

π

53.

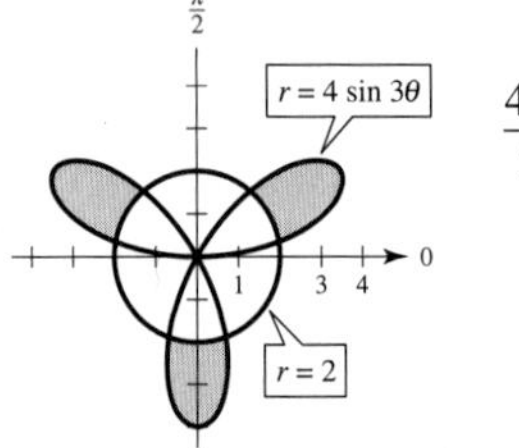

$\frac{4\pi}{3} + 2\sqrt{3}$

55. Let R be a region bounded by the graphs of $r = g_1(\theta)$ and $r = g_2(\theta)$ and the lines $\theta = a$ and $\theta = b$. When using polar coordinates to evaluate a double integral over R, R can be partitioned into small polar sectors.

57. r-simple regions have fixed bounds for θ and variable bounds for r.

θ-simple regions have variable bounds for θ and fixed bounds for r.

59. (a) $\int_{-3}^{3}\int_{-\sqrt{9-x^2}}^{\sqrt{9-x^2}} f(x, y)\, dy\, dx$

(b) $\int_0^{2\pi}\int_0^3 f(r\cos\theta, r\sin\theta)\, r\, dr\, d\theta$

(c) Choose the integral in part (b) because the limits of integration are less complicated.

61. Insert a factor of r; Sector of a circle **63.** 56.051 **65.** c

67. False. Let $f(r, \theta) = r - 1$ and let R be a sector where $0 \le r \le 6$ and $0 \le \theta \le \pi$.

69. (a) 2π (b) $\sqrt{2\pi}$ **71.** 486,788

73. (a) $\int_2^4\int_{y/\sqrt{3}}^{y} f\, dx\, dy$

(b) $\int_{2/\sqrt{3}}^{2}\int_2^{\sqrt{3}x} f\, dy\, dx + \int_2^{4/\sqrt{3}}\int_x^{\sqrt{3}x} f\, dy\, dx + \int_{4/\sqrt{3}}^{4}\int_x^4 f\, dy\, dx$

(c) $\int_{\pi/4}^{\pi/3}\int_{2\csc\theta}^{4\csc\theta} f\, r\, dr\, d\theta$

75. $A = \frac{\Delta\theta r_2^2}{2} - \frac{\Delta\theta r_1^2}{2} = \Delta\theta\left(\frac{r_1 + r_2}{2}\right)(r_2 - r_1) = r\,\Delta r\,\Delta\theta$

Section 14.4 (page 1018)

1. $m = 4$ **3.** $m = \frac{1}{8}$

5. (a) $m = ka^2, (a/2, a/2)$ (b) $m = ka^3/2, (a/2, 2a/3)$
(c) $m = ka^3/2, (2a/3, a/2)$

7. (a) $m = ka^2/2, (a/3, 2a/3)$ (b) $m = ka^3/3, (3a/8, 3a/4)$
(c) $m = ka^3/6, (a/2, 3a/4)$

9. (a) $\left(\frac{a}{2} + 5, \frac{a}{2}\right)$ (b) $\left(\frac{a}{2} + 5, \frac{2a}{3}\right)$

(c) $\left(\frac{2(a^2 + 15a + 75)}{3(a + 10)}, \frac{a}{2}\right)$

11. $m = k/4, (2/3, 8/15)$ **13.** $m = 30k, (14/5, 4/5)$

15. (a) $m = k(e - 1), \left(\frac{1}{e - 1}, \frac{e + 1}{4}\right)$

(b) $m = \frac{k}{4}(e^2 - 1), \left(\frac{e^2 + 1}{2(e^2 - 1)}, \frac{4(e^3 - 1)}{9(e^2 - 1)}\right)$

17. $m = 256k/15, (0, 16/7)$ **19.** $m = \frac{2kL}{\pi}, \left(\frac{L}{2}, \frac{\pi}{8}\right)$

21. $m = \frac{k\pi a^2}{8}, \left(\frac{4\sqrt{2}a}{3\pi}, \frac{4a(2 - \sqrt{2})}{3\pi}\right)$

23. $m = \frac{k}{8}(1 - 5e^{-4}), \left(\frac{e^4 - 13}{e^4 - 5}, \frac{8}{27}\left[\frac{e^6 - 7}{e^6 - 5e^2}\right]\right)$

25. $m = k\pi/3, \left(81\sqrt{3}/(40\pi), 0\right)$

27. $\bar{\bar{x}} = \sqrt{3}b/3$, $\bar{\bar{y}} = \sqrt{3}h/3$ **29.** $\bar{\bar{x}} = a/2$, $\bar{\bar{y}} = a/2$ **31.** $\bar{\bar{x}} = a/2$, $\bar{\bar{y}} = a/2$

33. $I_x = kab^4/4$
$I_y = kb^2a^3/6$
$I_0 = (3kab^4 + 2ka^3b^2)/12$
$\bar{\bar{x}} = \sqrt{3}a/3$
$\bar{\bar{y}} = \sqrt{2}b/2$

35. $I_x = 32k/3$
$I_y = 16k/3$
$I_0 = 16k$
$\bar{\bar{x}} = 2\sqrt{3}/3$
$\bar{\bar{y}} = 2\sqrt{6}/3$

37. $I_x = 16k$
$I_y = 512k/5$
$I_0 = 592k/5$
$\bar{\bar{x}} = 4\sqrt{15}/5$
$\bar{\bar{y}} = \sqrt{6}/2$

39. $I_x = 3k/56$
$I_y = k/18$
$I_0 = 55k/504$
$\bar{\bar{x}} = \sqrt{30}/9$
$\bar{\bar{y}} = \sqrt{70}/14$

41. $2k\int_{-b}^{b}\int_0^{\sqrt{b^2-x^2}} (x - a)^2\, dy\, dx = \frac{k\pi b^2}{4}(b^2 + 4a^2)$

43. $\int_0^4\int_0^{\sqrt{x}} kx(x - 6)^2\, dy\, dx = \frac{42{,}752k}{315}$

45. $\int_0^a\int_0^{\sqrt{a^2-x^2}} k(a - y)(y - a)^2\, dy\, dx = ka^5\left(\frac{7\pi}{16} - \frac{17}{15}\right)$

47. See definitions on page 1014. **49.** Answers will vary.

51. $L/3$ **53.** $L/2$ **55.** Proof

Section 14.5 (page 1025)

1. 24 **3.** 12π **5.** $\frac{1}{2}\left[4\sqrt{17} + \ln\left(4 + \sqrt{17}\right)\right]$

7. $\frac{4}{27}\left(31\sqrt{31} - 8\right)$ **9.** $\sqrt{2} - 1$ **11.** $\sqrt{2}\pi$

13. $2\pi a\left(a - \sqrt{a^2 - b^2}\right)$ **15.** $48\sqrt{14}$ **17.** 20π

19. $\int_0^1\int_0^x \sqrt{5 + 4x^2}\, dy\, dx = \frac{27 - 5\sqrt{5}}{12} \approx 1.3183$

21. $\int_{-3}^{3}\int_{-\sqrt{9-x^2}}^{\sqrt{9-x^2}} \sqrt{1 + 4x^2 + 4y^2}\, dy\, dx$

$= \frac{\pi}{6}\left(37\sqrt{37} - 1\right) \approx 117.3187$

23. $\int_0^1\int_0^1 \sqrt{1 + 4x^2 + 4y^2}\, dy\, dx \approx 1.8616$ **25.** e

27. 2.0035 **29.** $\int_{-1}^{1}\int_{-1}^{1} \sqrt{1 + 9(x^2 - y)^2 + 9(y^2 - x)^2}\, dy\, dx$

31. $\int_{-2}^{2}\int_{-\sqrt{4-x^2}}^{\sqrt{4-x^2}} \sqrt{1 + e^{-2x}}\, dy\, dx$

33. $\int_{0}^{4}\int_{0}^{10} \sqrt{1 + e^{2xy}(x^2 + y^2)}\, dy\, dx$

35. If f and its first partial derivatives are continuous on the closed region R in the xy-plane, then the area of the surface S given by $z = f(x, y)$ over R is

$$\int_R\int \sqrt{1 + [f_x(x, y)]^2 + [f_y(x, y)]^2}\, dA.$$

37. No. The size and shape of the graph stay the same, just the position is changed. So, the surface area does not increase.

39. 16 **41.** (a) $812\pi\sqrt{609}$ cm^3 (b) $100\pi\sqrt{609}$ cm^2

Section 14.6 (page 1035)

1. 18 **3.** $\frac{1}{10}$ **5.** $\frac{15}{2}(1 - 1/e)$ **7.** $-\frac{40}{3}$ **9.** $\frac{324}{5}$

11. 2.44167 **13.** $V = \int_0^5\int_0^{5-x}\int_0^{5-x-y} dz\, dy\, dx$

15. $V = \int_{-\sqrt{6}}^{\sqrt{6}}\int_{-\sqrt{6-y^2}}^{\sqrt{6-y^2}}\int_0^{6-x^2-y^2} dz\, dx\, dy$

17. $V = \int_{-4}^{4}\int_{-\sqrt{16-x^2}}^{\sqrt{16-x^2}}\int_{(x^2+y^2)/2}^{\sqrt{80-x^2-y^2}} dz\, dy\, dx$

19. $\frac{256}{15}$ **21.** $4\pi a^3/3$ **23.** $\frac{256}{15}$ **25.** 10

27.

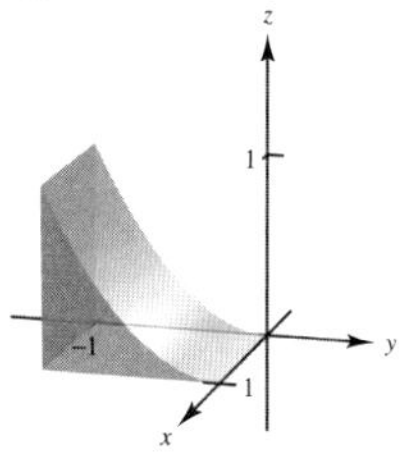

$$\int_0^1\int_0^1\int_{-1}^{-\sqrt{z}} dy\, dz\, dx$$

29.

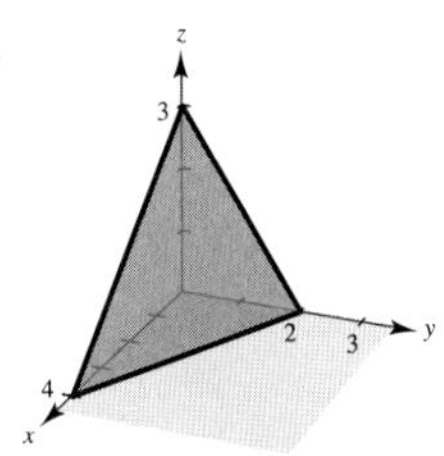

$$\int_0^3\int_0^{(12-4z)/3}\int_0^{(12-4z-3x)/6} dy\, dx\, dz$$

31.

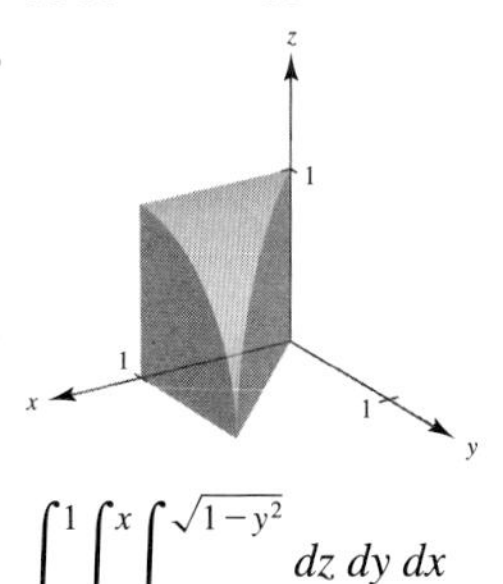

$$\int_0^1\int_0^x\int_0^{\sqrt{1-y^2}} dz\, dy\, dx$$

33. $\int_0^1\int_0^x\int_0^3 xyz\, dz\, dy\, dx,\ \int_0^1\int_y^1\int_0^3 xyz\, dz\, dx\, dy,$

$\int_0^1\int_0^3\int_0^x xyz\, dy\, dz\, dx,\ \int_0^3\int_0^1\int_0^x xyz\, dy\, dx\, dz,$

$\int_0^3\int_0^1\int_y^1 xyz\, dx\, dy\, dz,\ \int_0^1\int_0^3\int_y^1 xyz\, dx\, dz\, dy$

35. $\int_{-3}^{3}\int_{-\sqrt{9-x^2}}^{\sqrt{9-x^2}}\int_0^4 xyz\, dz\, dy\, dx,\ \int_{-3}^{3}\int_{-\sqrt{9-y^2}}^{\sqrt{9-y^2}}\int_0^4 xyz\, dz\, dx\, dy,$

$\int_{-3}^{3}\int_0^4\int_{-\sqrt{9-x^2}}^{\sqrt{9-x^2}} xyz\, dy\, dz\, dx,\ \int_0^4\int_{-3}^{3}\int_{-\sqrt{9-x^2}}^{\sqrt{9-x^2}} xyz\, dy\, dx\, dz,$

$\int_0^4\int_{-3}^{3}\int_{-\sqrt{9-y^2}}^{\sqrt{9-y^2}} xyz\, dx\, dy\, dz,\ \int_{-3}^{3}\int_0^4\int_{-\sqrt{9-y^2}}^{\sqrt{9-y^2}} xyz\, dx\, dz\, dy$

37. $\int_0^1\int_0^{1-z}\int_0^{1-y^2} dx\, dy\, dz,\ \int_0^1\int_0^{1-y}\int_0^{1-y^2} dx\, dz\, dy,$

$\int_0^1\int_0^{2z-z^2}\int_0^{1-z} 1\, dy\, dx\, dz + \int_0^1\int_{2z-z^2}^{1}\int_0^{\sqrt{1-x}} 1\, dy\, dx\, dz,$

$\int_0^1\int_{1-\sqrt{1-x}}^{1}\int_0^{1-z} 1\, dy\, dz\, dx + \int_0^1\int_0^{1-\sqrt{1-x}}\int_0^{\sqrt{1-x}} 1\, dy\, dz\, dx,$

$\int_0^1\int_0^{\sqrt{1-x}}\int_0^{1-y} dz\, dy\, dx$

39. $m = 8k$ **41.** $m = 128k/3$
$\bar{x} = \frac{3}{2}$ $\qquad \bar{z} = 1$

43. $m = k\int_0^b\int_0^b\int_0^b xy\, dz\, dy\, dx$

$M_{yz} = k\int_0^b\int_0^b\int_0^b x^2y\, dz\, dy\, dx$

$M_{xz} = k\int_0^b\int_0^b\int_0^b xy^2\, dz\, dy\, dx$

$M_{xy} = k\int_0^b\int_0^b\int_0^b xyz\, dz\, dy\, dx$

45. $\bar{x}$ will be greater than 2, and $\bar{y}$ and $\bar{z}$ will be unchanged.

47. $\bar{x}$ and $\bar{z}$ will be unchanged, and $\bar{y}$ will be greater than 0.

49. $(0, 0, 3h/4)$ **51.** $\left(0, 0, \frac{3}{2}\right)$ **53.** $\left(5, 6, \frac{5}{4}\right)$

55. (a) $I_x = 2ka^5/3$ (b) $I_x = ka^8/8$
$I_y = 2ka^5/3$ $\qquad I_y = ka^8/8$
$I_z = 2ka^5/3$ $\qquad I_z = ka^8/8$

57. (a) $I_x = 256k$ (b) $I_x = 2048k/3$
$I_y = 512k/3$ $\qquad I_y = 1024k/3$
$I_z = 256k$ $\qquad I_z = 2048k/3$

59. Proof **61.** $\int_{-1}^{1}\int_{-1}^{1}\int_0^{1-x} (x^2 + y^2)\sqrt{x^2 + y^2 + z^2}\, dz\, dy\, dx$

63. (a) $m = \int_{-2}^{2}\int_{-\sqrt{4-x^2}}^{\sqrt{4-x^2}}\int_0^{4-x^2-y^2} kz\, dz\, dy\, dx$

(b) $\bar{x} = \bar{y} = 0$, by symmetry.

$\bar{z} = \frac{1}{m}\int_{-2}^{2}\int_{-\sqrt{4-x^2}}^{\sqrt{4-x^2}}\int_0^{4-x^2-y^2} kz^2\, dz\, dy\, dx$

(c) $I_z = \int_{-2}^{2}\int_{-\sqrt{4-x^2}}^{\sqrt{4-x^2}}\int_0^{4-x^2-y^2} kz(x^2 + y^2)\, dz\, dy\, dx$

65. See "Definition of Triple Integral" on page 1027 and Theorem 14.4, "Evaluation by Iterated Integrals," on page 1028.

67. (a) Solid B

(b) Solid B has the greater moment of inertia because it is more dense.

(c) Solid A will reach the bottom first. Because Solid B has a greater moment of inertia, it has a greater resistance to rotational motion.

69. $\frac{13}{3}$ **71.** $\frac{3}{2}$

73. Q: $3z^2 + y^2 + 2x^2 \le 1; 4\sqrt{6}\pi/45 \approx 0.684$

75. $a = 2, \frac{16}{3}$ **77.** Putnam Problem B1, 1965

Section 14.7 (page 1043)

1. 27 **3.** $\frac{52}{45}$ **5.** $\pi/8$ **7.** $\pi(e^4 + 3)$

9.

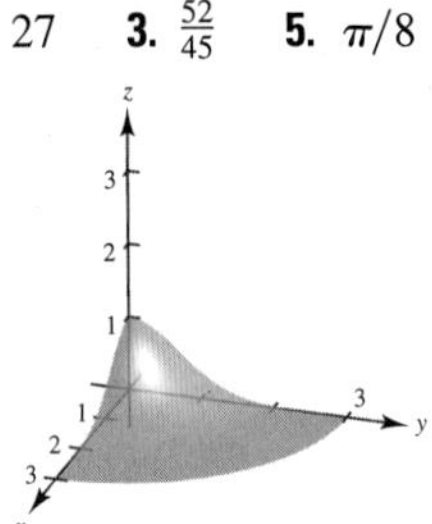

$(1 - e^{-9})\pi/4$

11.

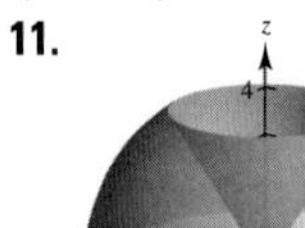
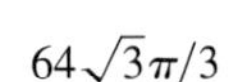

$64\sqrt{3}\pi/3$

13. Cylindrical: $\displaystyle\int_0^{2\pi}\int_0^2\int_{r^2}^4 r^2 \cos\theta \, dz \, dr \, d\theta = 0$

Spherical: $\displaystyle\int_0^{2\pi}\int_0^{\arctan(1/2)}\int_0^{4\sec\phi} \rho^3 \sin^2\phi \cos\theta \, d\rho \, d\phi \, d\theta$

$\displaystyle + \int_0^{2\pi}\int_{\arctan(1/2)}^{\pi/2}\int_0^{\cot\phi\csc\phi} \rho^3 \sin^2\phi \cos\phi \, d\rho \, d\phi \, d\theta = 0$

15. Cylindrical: $\displaystyle\int_0^{2\pi}\int_0^a\int_a^{a+\sqrt{a^2-r^2}} r^2 \cos\theta \, dz \, dr \, d\theta = 0$

Spherical: $\displaystyle\int_0^{\pi/4}\int_0^{2\pi}\int_{a\sec\phi}^{2a\cos\phi} \rho^3 \sin^2\phi \cos\theta \, d\rho \, d\theta \, d\phi = 0$

17. $(2a^3/9)(3\pi - 4)$ **19.** $\pi/16$ **21.** $(2a^3/9)(3\pi - 4)$

23. $48k\pi$ **25.** $\pi r_0^2 h/3$ **27.** $(0, 0, h/5)$

29. $\displaystyle I_z = 4k\int_0^{\pi/2}\int_0^{r_0}\int_0^{h(r_0-r)/r_0} r^3 \, dz \, dr \, d\theta = 3mr_0^2/10$

31. Proof **33.** $9\pi\sqrt{2}$ **35.** $16\pi^2$

37. $k\pi a^4$ **39.** $(0, 0, 3r/8)$ **41.** $k\pi/192$

43. Rectangular to cylindrical: $r^2 = x^2 + y^2$
$\tan\theta = y/x$
$z = z$

Cylindrical to rectangular: $x = r\cos\theta$
$y = r\sin\theta$
$z = z$

45. $\displaystyle\int_{\theta_1}^{\theta_2}\int_{g_1(\theta)}^{g_2(\theta)}\int_{h_1(r\cos\theta, r\sin\theta)}^{h_2(r\cos\theta, r\sin\theta)} f(r\cos\theta, r\sin\theta, z) r \, dz \, dr \, d\theta$

47. (a) r constant: right circular cylinder about z-axis
θ constant: plane parallel to z-axis
z constant: plane parallel to xy-plane

(b) ρ constant: sphere
θ constant: plane parallel to z-axis
ϕ constant: cone

49. $\frac{1}{2}\pi^2 a^4$ **51.** Putnam Problem A1, 2006

Section 14.8 (page 1050)

1. $-\frac{1}{2}$ **3.** $1 + 2v$ **5.** 1 **7.** $-e^{2u}$

9.

11.

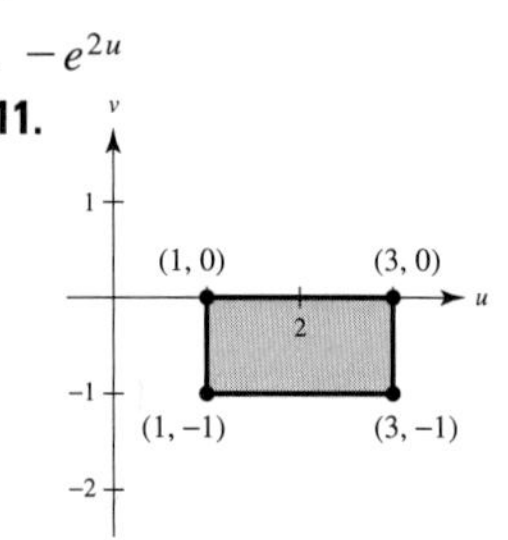

13. $\displaystyle\iint_R 3xy \, dA = \int_{-2/3}^{2/3}\int_{1-x}^{(1/2)x+2} 3xy \, dy \, dx$

$\displaystyle + \int_{2/3}^{4/3}\int_{(1/2)x}^{(1/2)x+2} 3xy \, dy \, dx + \int_{4/3}^{8/3}\int_{(1/2)x}^{4-x} 3xy \, dy \, dx = \frac{164}{9}$

15. $\frac{8}{3}$ **17.** 36 **19.** $(e^{-1/2} - e^{-2}) \ln 8 \approx 0.9798$ **21.** 96

23. $12(e^4 - 1)$ **25.** $\frac{100}{9}$ **27.** $\frac{2}{5}a^{5/2}$ **29.** One

31. (a)

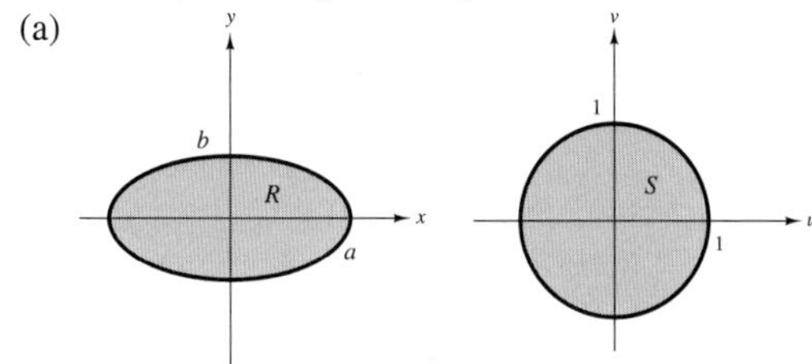

(b) ab (c) πab

33. See "Definition of the Jacobian" on page 1045. **35.** $u^2 v$

37. $-uv$ **39.** $-\rho^2 \sin\phi$ **41.** Putnam Problem A2, 1994

Review Exercises for Chapter 14 (page 1052)

1. $x - x^3 + x^3 \ln x^2$

3.

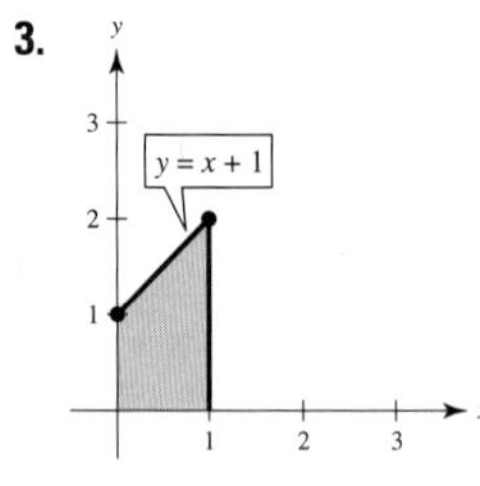

$\frac{29}{6}$

5.

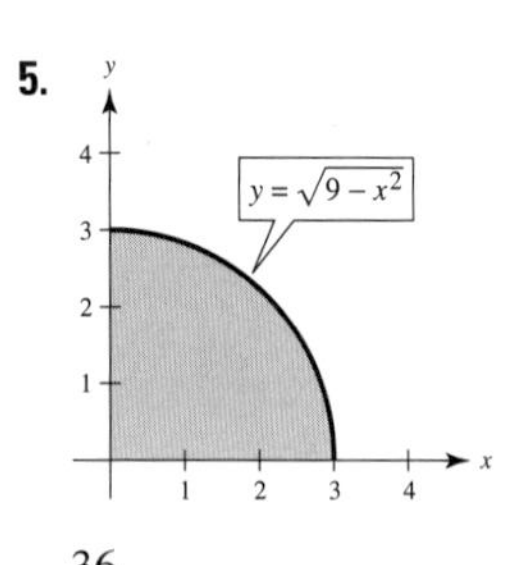

36

7. $\displaystyle\int_0^3\int_0^{(3-x)/3} dy \, dx = \int_0^1\int_0^{3-3y} dx \, dy = \frac{3}{2}$

9. $\displaystyle\int_{-5}^{3}\int_{-\sqrt{25-x^2}}^{\sqrt{25-x^2}} dy \, dx$

$\displaystyle = \int_{-5}^{-4}\int_{-\sqrt{25-y^2}}^{\sqrt{25-y^2}} dx \, dy + \int_{-4}^{4}\int_{-\sqrt{25-y^2}}^{3} dx \, dy$

$\displaystyle + \int_4^5\int_{-\sqrt{25-y^2}}^{\sqrt{25-y^2}} dx \, dy$

$= 25\pi/2 + 12 + 25 \arcsin\frac{3}{5} \approx 67.36$

11. $\displaystyle 4\int_0^1\int_0^{x\sqrt{1-x^2}} dy \, dx = 4\int_0^{1/2}\int_{\sqrt{(1-\sqrt{1-4y^2})/2}}^{\sqrt{(1+\sqrt{1-4y^2})/2}} dx \, dy = \frac{4}{3}$

13. $\int_2^5 \int_{x-3}^{\sqrt{x-1}} dy\,dx + 2\int_1^2 \int_0^{\sqrt{x-1}} dy\,dx = \int_{-1}^2 \int_{y^2+1}^{y+3} dx\,dy = \frac{9}{2}$

15. Both integrations are over the common region R, as shown in the figure. Both integrals yield $\frac{4}{3} + \frac{4}{3}\sqrt{2}$.

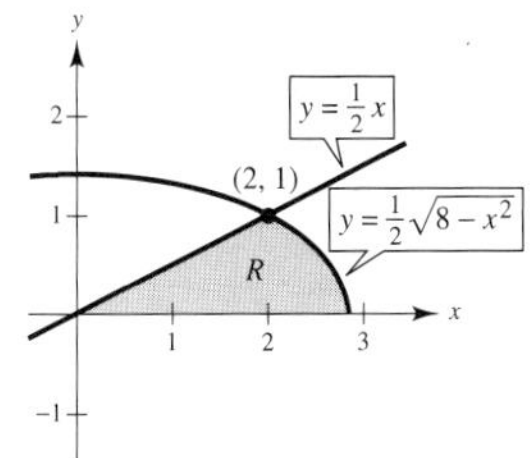

17. $\frac{3296}{15}$ **19.** $\frac{40}{3}$ **21.** 13.67°C **23.** $k = 1, 0.070$ **25.** c

27. True **29.** True **31.** $(h^3/6)\left[\ln(\sqrt{2} + 1) + \sqrt{2}\right]$

33. $9\pi/2$ **35.** $\pi h^3/3$

37. (a) $r = 3\sqrt{\cos 2\theta}$ (b) 9 (c) $3(3\pi - 16\sqrt{2} + 20) \approx 20.392$

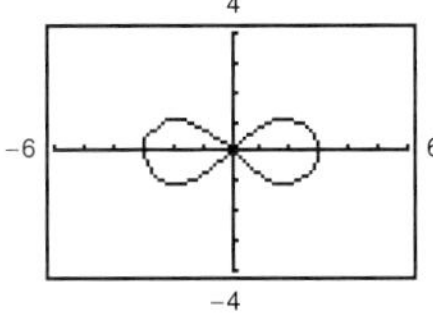

(b) 9 (c) $3(3\pi - 16\sqrt{2} + 20) \approx 20.392$

39. (a) $m = k/4, \left(\frac{32}{45}, \frac{64}{55}\right)$ (b) $m = 17k/30, \left(\frac{936}{1309}, \frac{784}{663}\right)$

41. $I_x = ka^2b^3/6$
$I_y = ka^4b/4$
$I_0 = (2ka^2b^3 + 3ka^4b)/12$
$\bar{\bar{x}} = a/\sqrt{2}$
$\bar{\bar{y}} = b/\sqrt{3}$

43. $\frac{(101\sqrt{101} - 1)\pi}{6}$ **45.** $\frac{1}{6}(37\sqrt{37} - 1)$

47. (a) 30,415.74 ft³ (b) 2081.53 ft² **49.** $324\pi/5$

51. $(abc/3)(a^2 + b^2 + c^2)$ **53.** $8\pi/15$ **55.** $\frac{32}{3}\left(\pi/2 - \frac{2}{3}\right)$

57. $\left(0, 0, \frac{1}{4}\right)$ **59.** $(3a/8, 3a/8, 3a/8)$ **61.** $833k\pi/3$

63. (a) $\frac{1}{3}\pi h^2(3a - h)$ (b) $\left(0, 0, \frac{3(2a - h)^2}{4(3a - h)}\right)$ (c) $\left(0, 0, \frac{3}{8}a\right)$

(d) a (e) $(\pi/30)h^3(20a^2 - 15ah + 3h^2)$ (f) $4\pi a^5/15$

65. Volume of a torus formed by a circle of radius 3, centered at (0, 3, 0) and revolved about the z-axis

67. −9 **69.** $5 \ln 5 - 3 \ln 3 - 2 \approx 2.751$

P.S. Problem Solving (page 1055)

1. $8(2 - \sqrt{2})$ **3.** (a)–(g) Proofs **5.** $\frac{1}{3}$

7.

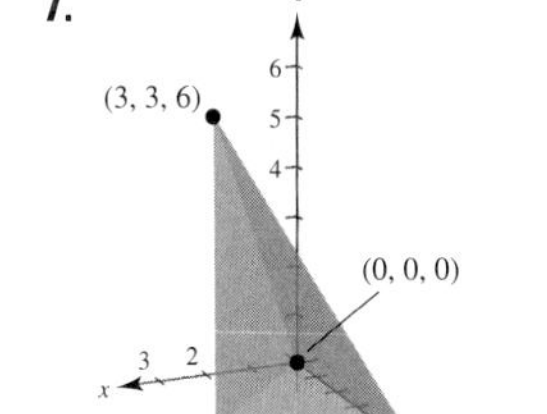

$\int_0^3 \int_0^{2x} \int_x^{6-x} dy\,dz\,dx = 18$

9. $\sqrt{\pi}/4$

11. If $a, k > 0$, then $1 = ka^2$ or $a = 1/\sqrt{k}$.

13. Answers will vary.

15. The greater the angle between the given plane and the xy-plane, the greater the surface area. So, $z_2 < z_1 < z_4 < z_3$.

17. The results are not the same. Fubini's Theorem is not valid because f is not continuous on the region $0 \le x \le 1$, $0 \le y \le 1$.

Chapter 15

Section 15.1 (page 1067)

1. d **2.** c **3.** e **4.** b **5.** a **6.** f

7. $\sqrt{2}$

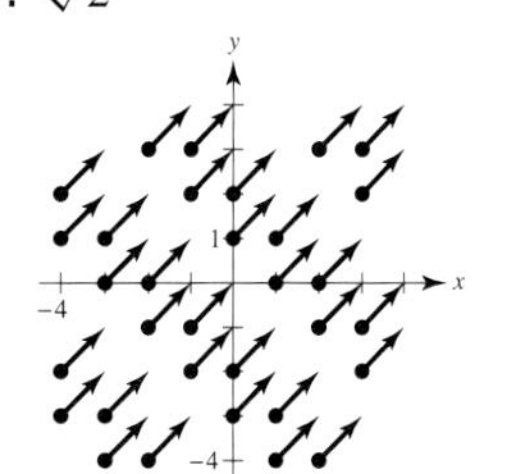

9. $\sqrt{x^2 + y^2}$

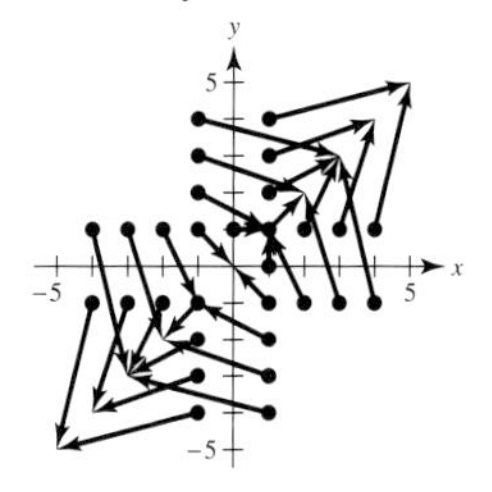

11. $3|y|$

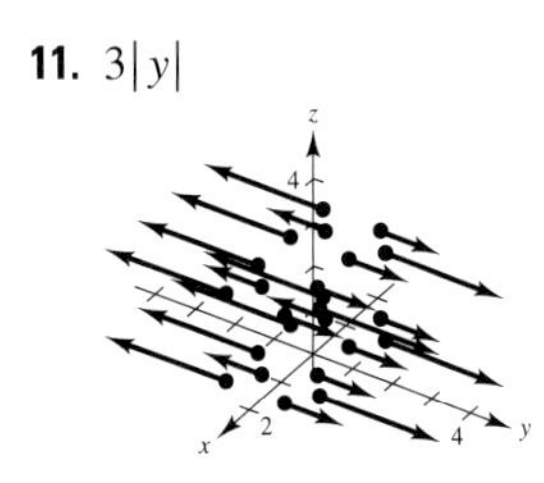

13. $\sqrt{16x^2 + y^2}$

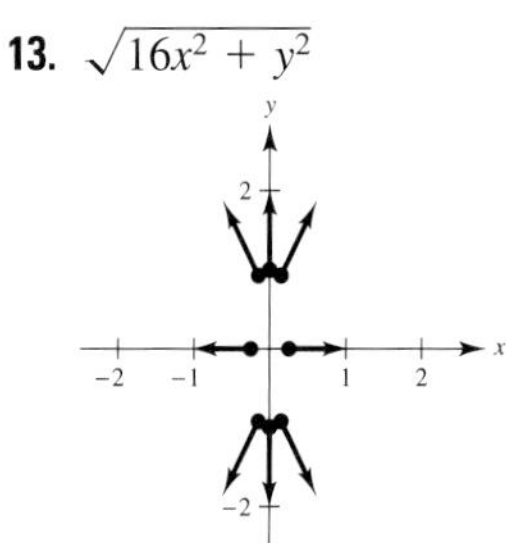

15. $\sqrt{3}$

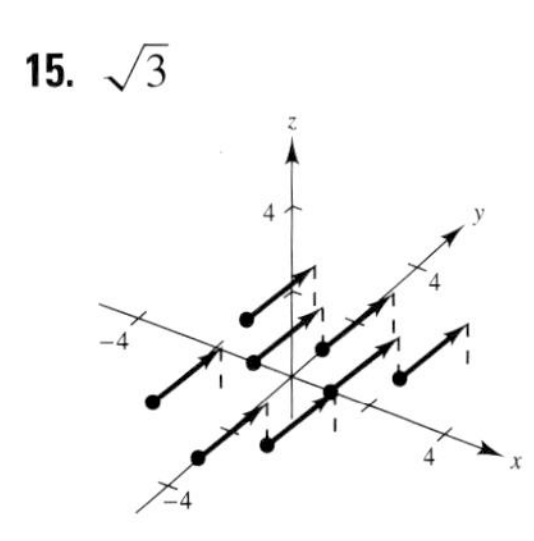

17.

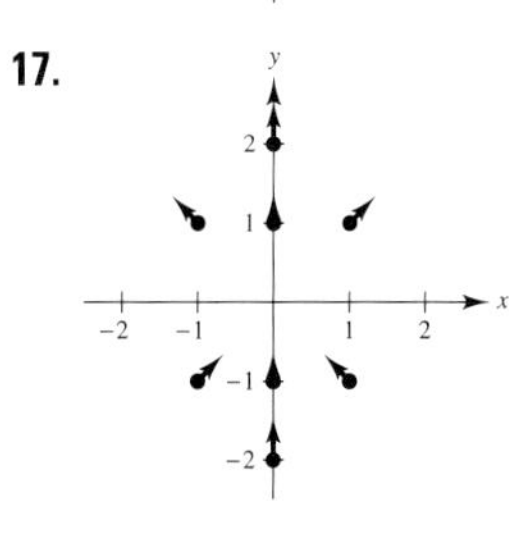

19.

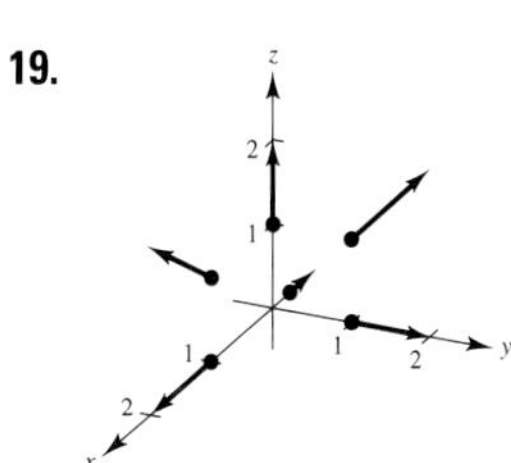

21. $2x\mathbf{i} + 4y\mathbf{j}$

23. $(10x + 3y)\mathbf{i} + (3x + 2y)\mathbf{j}$ **25.** $6yz\mathbf{i} + 6xz\mathbf{j} + 6xy\mathbf{k}$

27. $2xye^{x^2}\mathbf{i} + e^{x^2}\mathbf{j} + \mathbf{k}$

29. $[xy/(x + y) + y\ln(x + y)]\mathbf{i} + [xy/(x + y) + x\ln(x + y)]\mathbf{j}$

31–33. Proofs **35.** Conservative because $\partial N/\partial x = \partial M/\partial y$.

37. Not conservative because $\partial N/\partial x \ne \partial M/\partial y$.

39. Conservative: $f(x, y) = xy + K$

41. Conservative: $f(x, y) = x^2y + K$

43. Not conservative **45.** Not conservative
47. Conservative: $f(x, y) = e^x \cos y + K$ **49.** $4\mathbf{i} - \mathbf{j} - 3\mathbf{k}$
51. $-2\mathbf{k}$ **53.** $2x/(x^2 + y^2)\mathbf{k}$
55. $\cos(y - z)\mathbf{i} + \cos(z - x)\mathbf{j} + \cos(x - y)\mathbf{k}$
57. Conservative: $f(x, y, z) = \frac{1}{2}(x^2y^2z^2) + K$
59. Not conservative **61.** Conservative: $f(x, y, z) = xz/y + K$
63. $2x + 4y$ **65.** $\cos x - \sin y + 2z$ **67.** 4 **69.** 0
71. See "Definition of Vector Field" on page 1058. Some physical examples of vector fields include velocity fields, gravitational fields, and electric force fields.
73. See "Definition of Curl of a Vector Field" on page 1064.
75. $9x\mathbf{j} - 2y\mathbf{k}$ **77.** $z\mathbf{j} + y\mathbf{k}$ **79.** $3z + 2x$ **81.** 0
83–89. Proofs
91. $f(x, y, z) = \|\mathbf{F}(x, y, z)\| = \sqrt{x^2 + y^2 + z^2}$

$$\ln f = \frac{1}{2}\ln(x^2 + y^2 + z^2)$$
$$\nabla \ln f = \frac{x}{x^2 + y^2 + z^2}\mathbf{i} + \frac{y}{x^2 + y^2 + z^2}\mathbf{j} + \frac{z}{x^2 + y^2 + z^2}\mathbf{k} = \frac{\mathbf{F}}{f^2}$$

93. $f^n = \|\mathbf{F}(x, y, z)\|^n = \left(\sqrt{x^2 + y^2 + z^2}\right)^n$

$$\nabla f^n = n\left(\sqrt{x^2 + y^2 + z^2}\right)^{n-1}\left(\frac{x\mathbf{i} + y\mathbf{j} + z\mathbf{k}}{\sqrt{x^2 + y^2 + z^2}}\right) = nf^{n-2}\mathbf{F}$$

95. True
97. False. Curl f is meaningful only for vector fields, when direction is involved.

Section 15.2 (page 1079)

1. $\mathbf{r}(t) = \begin{cases} t\mathbf{i} + t\mathbf{j}, & 0 \le t \le 1 \\ (2 - t)\mathbf{i} + \sqrt{2 - t}\,\mathbf{j}, & 1 \le t \le 2 \end{cases}$

3. $\mathbf{r}(t) = \begin{cases} t\mathbf{i}, & 0 \le t \le 3 \\ 3\mathbf{i} + (t - 3)\mathbf{j}, & 3 \le t \le 6 \\ (9 - t)\mathbf{i} + 3\mathbf{j}, & 6 \le t \le 9 \\ (12 - t)\mathbf{j}, & 9 \le t \le 12 \end{cases}$

5. $\mathbf{r}(t) = 3\cos t\mathbf{i} + 3\sin t\mathbf{j},\ 0 \le t \le 2\pi$ **7.** 20 **9.** $5\pi/2$
11. (a) C: $\mathbf{r}(t) = t\mathbf{i} + t\mathbf{j},\ 0 \le t \le 1$ (b) $2\sqrt{2}/3$
13. (a) C: $\mathbf{r}(t) = \cos t\mathbf{i} + \sin t\mathbf{j},\ 0 \le t \le \pi/2$ (b) $\pi/2$
15. (a) C: $\mathbf{r}(t) = t\mathbf{i},\ 0 \le t \le 1$ (b) $1/2$
17. (a) C: $\mathbf{r}(t) = \begin{cases} t\mathbf{i}, & 0 \le t \le 1 \\ (2 - t)\mathbf{i} + (t - 1)\mathbf{j}, & 1 \le t \le 2 \\ (3 - t)\mathbf{j}, & 2 \le t \le 3 \end{cases}$
(b) $\frac{19}{6}(1 + \sqrt{2})$
19. (a) C: $\mathbf{r}(t) = \begin{cases} t\mathbf{i}, & 0 \le t \le 1 \\ \mathbf{i} + t\mathbf{k}, & 0 \le t \le 1 \\ \mathbf{i} + t\mathbf{j} + \mathbf{k}, & 0 \le t \le 1 \end{cases}$ (b) $\frac{23}{6}$
21. $8\sqrt{5}\pi(1 + 4\pi^2/3) \approx 795.7$ **23.** 2
25. $(k/12)(41\sqrt{41} - 27)$ **27.** 1 **29.** $\frac{1}{2}$ **31.** $\frac{9}{4}$
33. About 249.49 **35.** 66 **37.** 0 **39.** $-10\pi^2$
41. Positive **43.** Zero
45. (a) $\frac{236}{3}$; Orientation is from left to right, so the value is positive.
(b) $-\frac{236}{3}$; Orientation is from right to left, so the value is negative.

47. $\mathbf{F}(t) = -2t\mathbf{i} - t\mathbf{j}$
$\mathbf{r}'(t) = \mathbf{i} - 2\mathbf{j}$
$\mathbf{F}(t) \cdot \mathbf{r}'(t) = -2t + 2t = 0$
$\int_C \mathbf{F} \cdot d\mathbf{r} = 0$
49. $\mathbf{F}(t) = (t^3 - 2t^2)\mathbf{i} + (t - t^2/2)\mathbf{j}$
$\mathbf{r}'(t) = \mathbf{i} + 2t\mathbf{j}$
$\mathbf{F}(t) \cdot \mathbf{r}'(t) = t^3 - 2t^2 + 2t^2 - t^3 = 0$
$\int_C \mathbf{F} \cdot d\mathbf{r} = 0$
51. 1010 **53.** $\frac{190}{3}$ **55.** 25 **57.** $\frac{63}{2}$ **59.** $-\frac{11}{6}$ **61.** $\frac{316}{3}$
63. $5h$ **65.** $\frac{1}{2}$ **67.** $(h/4)\left[2\sqrt{5} + \ln(2 + \sqrt{5})\right]$
69. $\frac{1}{120}(25\sqrt{5} - 11)$
71. (a) $12\pi \approx 37.70\ \text{cm}^2$ (b) $12\pi/5 \approx 7.54\ \text{cm}^3$
(c)

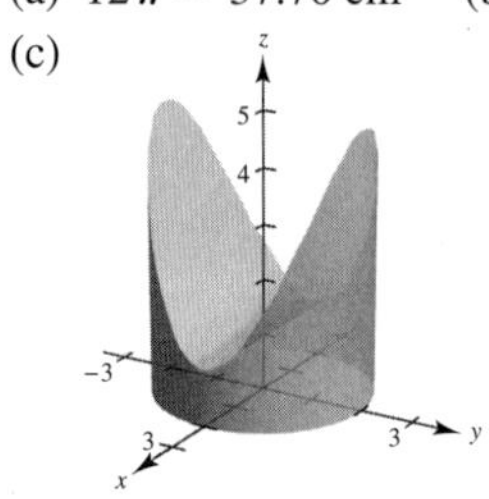

73. $I_x = I_y = a^3\pi$
75. (a)

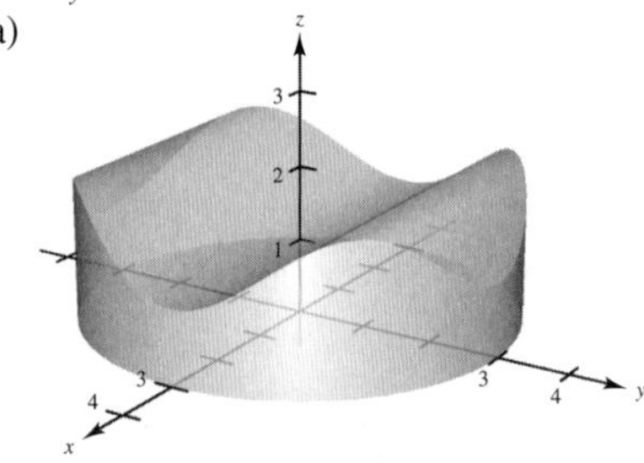

(b) $9\pi\ \text{cm}^2 \approx 28.274\ \text{cm}^2$
(c) $\text{Volume} = 2\int_0^3 2\sqrt{9 - y^2}\left[1 + 4\frac{y^2}{9}\left(1 - \frac{y^2}{9}\right)\right] dy = 27\pi/2 \approx 42.412\ \text{cm}^3$
77. 1750 ft-lb
79. See "Definition of Line Integral" on page 1070 and Theorem 15.4, "Evaluation of a Line Integral as a Definite Integral" on page 1071.
81. z_3, z_1, z_2, z_4; The greater the height of the surface over the curve $y = \sqrt{x}$, the greater the lateral surface area.
83. False: $\int_C xy\, ds = \sqrt{2}\int_0^1 t^2\, dt$.
85. False: the orientations are different. **87.** -12

Section 15.3 (page 1090)

1. (a) $\int_0^1 (t^2 + 2t^4)\, dt = \frac{11}{15}$
(b) $\int_0^{\pi/2} (\sin^2\theta\cos\theta + 2\sin^4\theta\cos\theta)\, d\theta = \frac{11}{15}$
3. (a) $\int_0^{\pi/3} (\sec\theta\tan^2\theta - \sec^3\theta)\, d\theta \approx -1.317$
(b) $\int_0^3 \left[\frac{\sqrt{t}}{2\sqrt{t + 1}} - \frac{\sqrt{t + 1}}{2\sqrt{t}}\right] dt \approx -1.317$

5. Conservative **7.** Not conservative **9.** Conservative
11. (a) 1 (b) 1 **13.** (a) 0 (b) $-\frac{1}{3}$ (c) $-\frac{1}{2}$
15. (a) 64 (b) 0 (c) 0 (d) 0 **17.** (a) $\frac{64}{3}$ (b) $\frac{64}{3}$
19. (a) 32 (b) 32 **21.** (a) $\frac{2}{3}$ (b) $\frac{17}{6}$ **23.** (a) 0 (b) 0
25. 72 **27.** -1 **29.** 0 **31.** (a) 2 (b) 2 (c) 2
33. 11 **35.** 30,366 **37.** 0
39. (a) $d\mathbf{r} = (\mathbf{i} - \mathbf{j})\,dt \Longrightarrow \int_0^{50} 175\,dt = 8750$ ft-lb
(b) $d\mathbf{r} = \left(\mathbf{i} - \frac{1}{25}(50 - t)\mathbf{j}\right)dt \Longrightarrow 7\int_0^{50} (50 - t)\,dt$
$= 8750$ ft-lb
41. See Theorem 15.5, "Fundamental Theorem of Line Integrals," on page 1084.
43. (a) 2π (b) 2π (c) -2π (d) 0
45. Yes, because the work required to get from point to point is independent of the path taken.
47. False. It would be true if $\mathbf{F}$ were conservative.
49. True **51.** Proof
53. (a) Proof (b) $-\pi$ (c) π
(d) -2π; does not contradict Theorem 15.7 because $\mathbf{F}$ is not continuous at $(0, 0)$ in R enclosed by C.
(e) $\nabla\left(\arctan\dfrac{x}{y}\right) = \dfrac{1/y}{1 + (x/y)^2}\mathbf{i} + \dfrac{-x/y^2}{1 + (x/y)^2}\mathbf{j}$

Section 15.4 (page 1099)

1. $\frac{1}{30}$ **3.** 0 **5.** About 19.99 **7.** $\frac{9}{2}$ **9.** 56 **11.** $\frac{4}{3}$ **13.** 0
15. 0 **17.** $\frac{1}{12}$ **19.** 32π **21.** π **23.** $\frac{225}{2}$ **25.** πa^2 **27.** $\frac{9}{2}$
29. See Theorem 15.8 on page 1093. **31.** Proof **33.** $\left(0, \frac{8}{5}\right)$
35. $\left(\frac{8}{15}, \frac{8}{21}\right)$ **37.** $3\pi a^2/2$ **39.** $\pi - 3\sqrt{3}/2$
41. (a) $51\pi/2$ (b) $243\pi/2$
43. $\displaystyle\int_C \mathbf{F}\cdot d\mathbf{r} = \int_C M\,dx + N\,dy = \iint_R \left(\frac{\partial N}{\partial x} - \frac{\partial M}{\partial y}\right) dA = 0;$
$I = -2\pi$ when C is a circle that contains the origin.
45. $\frac{19}{2}$ **47–49.** Proofs

Section 15.5 (page 1109)

1. e **2.** f **3.** b **4.** a **5.** d **6.** c
7. $y - 2z = 0$
Plane

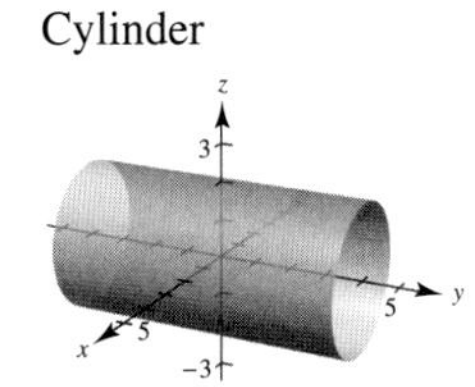

9. $x^2 + z^2 = 4$
Cylinder

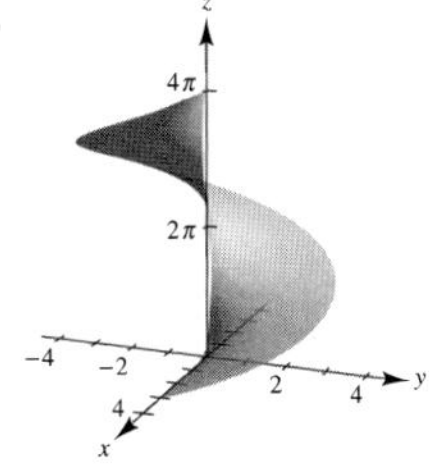

11.

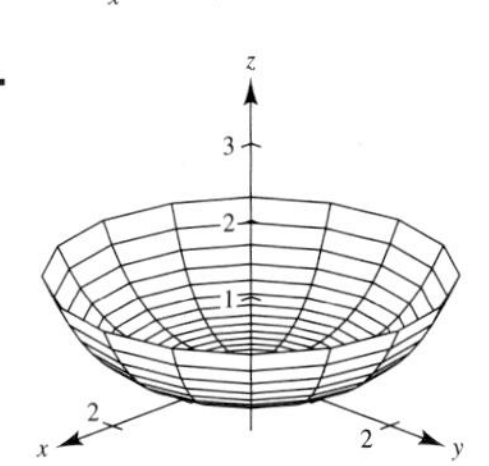

13.

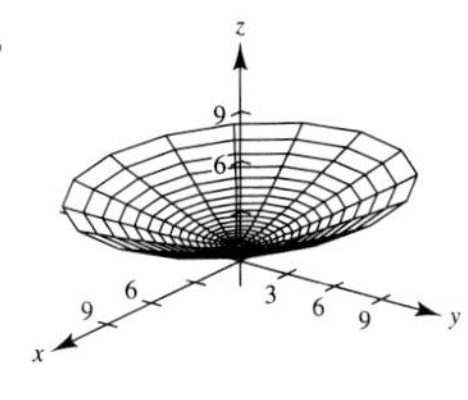

15.

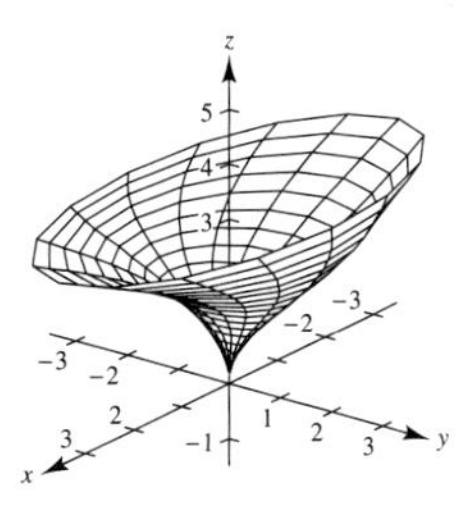

17. The paraboloid is reflected (inverted) through the xy-plane.
19. The height of the paraboloid is increased from 4 to 9.
21. $\mathbf{r}(u, v) = u\mathbf{i} + v\mathbf{j} + v\mathbf{k}$
23. $\mathbf{r}(u, v) = \frac{1}{2}u\cos v\mathbf{i} + u\mathbf{j} + \frac{1}{3}u\sin v\mathbf{k},\ u \geq 0,\ 0 \leq v \leq 2\pi$ *or*
$\mathbf{r}(x, y) = x\mathbf{i} + \sqrt{4x^2 + 9y^2}\mathbf{j} + z\mathbf{k}$
25. $\mathbf{r}(u, v) = 5\cos u\mathbf{i} + 5\sin u\mathbf{j} + v\mathbf{k}$
27. $\mathbf{r}(u, v) = u\mathbf{i} + v\mathbf{j} + u^2\mathbf{k}$
29. $\mathbf{r}(u, v) = v\cos u\mathbf{i} + v\sin u\mathbf{j} + 4\mathbf{k},\quad 0 \leq v \leq 3$
31. $x = u,\ y = \dfrac{u}{2}\cos v,\ z = \dfrac{u}{2}\sin v,\quad 0 \leq u \leq 6,\ 0 \leq v \leq 2\pi$
33. $x = \sin u\cos v,\ y = \sin u\sin v,\ z = u$
$0 \leq u \leq \pi,\ 0 \leq v \leq 2\pi$
35. $x - y - 2z = 0$ **37.** $4y - 3z = 12$ **39.** $8\sqrt{2}$
41. $2\pi ab$ **43.** $\pi ab^2\sqrt{a^2 + 1}$
45. $(\pi/6)\left(17\sqrt{17} - 1\right) \approx 36.177$
47. See "Definition of Parametric Surface" on page 1102.
49–51. Proofs
53. (a) (b)

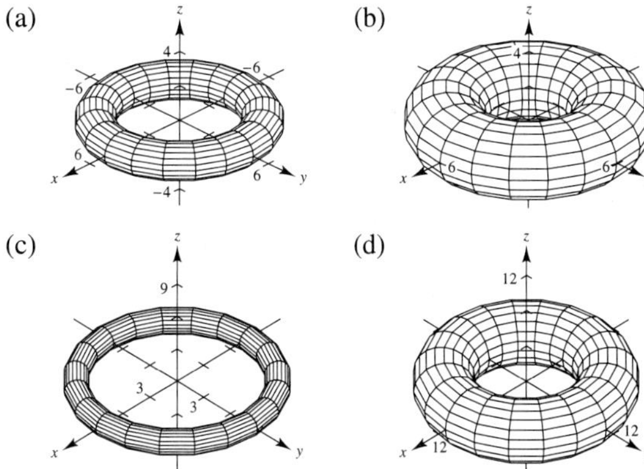

(c) (d)

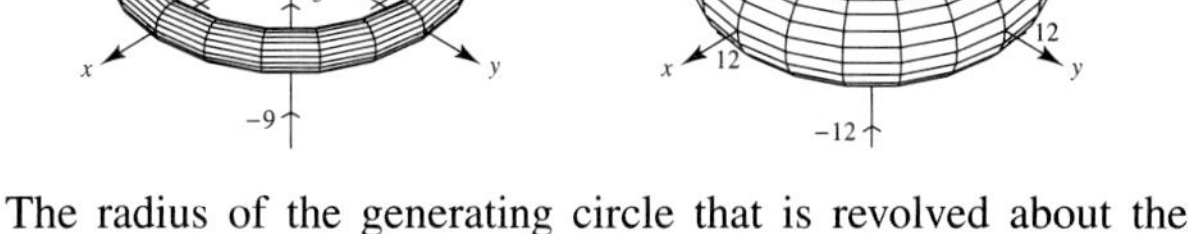

The radius of the generating circle that is revolved about the z-axis is b, and its center is a units from the axis of revolution.
55. 400π m^2
57.

$2\pi\left[\frac{3}{2}\sqrt{13} + 2\ln\left(3 + \sqrt{13}\right) - 2\ln 2\right]$

59. Answers will vary. Sample answer: Let
$x = (2 - u)(5 + \cos v) \cos 3\pi u$
$y = (2 - u)(5 + \cos v) \sin 3\pi u$
$z = 5u + (2 - u) \sin v$
where $-\pi \le u \le \pi$ and $-\pi \le v \le \pi$.

Section 15.6 (page 1122)

1. $12\sqrt{2}$ **3.** 2π **5.** $27\sqrt{3}/8$ **7.** $(391\sqrt{17} + 1)/240$
9. About -11.47 **11.** $\frac{364}{3}$ **13.** $12\sqrt{5}$ **15.** 8 **17.** $\sqrt{3}\pi$
19. $32\pi/3$ **21.** 486π **23.** $-\frac{4}{3}$ **25.** $3\pi/2$ **27.** 20π
29. 384π **31.** 0 **33.** Proof **35.** $2\pi a^3 h$ **37.** $64\pi\rho$
39. See Theorem 15.10, "Evaluating a Surface Integral," on page 1112.
41. See "Definition of Flux Integral," on page 1118; see Theorem 15.11, "Evaluating a Flux Integral," on page 1118.
43. (a)

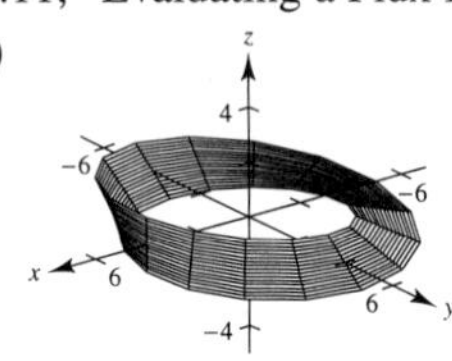

(b) If a normal vector at a point P on the surface is moved around the Möbius strip once, it will point in the opposite direction.

(c)

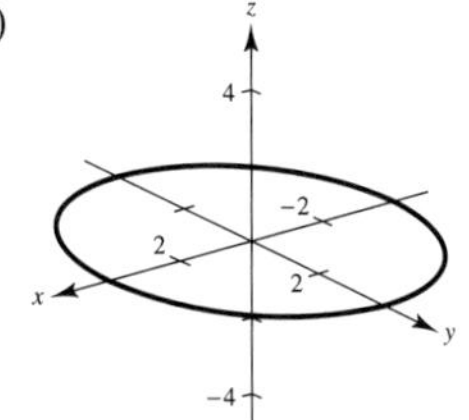

Circle

(d) Construction
(e) A strip with a double twist that is twice as long as the Möbius strip.

Section 15.7 (page 1130)

1. a^4 **3.** 18 **5.** π **7.** $3a^4$ **9.** 0 **11.** 108π
13. 0 **15.** 2304 **17.** $1024\pi/3$ **19.** 0
21. See Theorem 15.12, "The Divergence Theorem," on page 1124.
23–29. Proofs

Section 15.8 (page 1137)

1. $(xz - e^z)\mathbf{i} - (yz + 1)\mathbf{j} - 2\mathbf{k}$ **3.** $[2 - 1/(1 + x^2)]\mathbf{j} - 8x\mathbf{k}$
5. $z(x - 2e^{y^2+z^2})\mathbf{i} - yz\mathbf{j} - 2ye^{x^2+y^2}\mathbf{k}$ **7.** 18π **9.** 0
11. -12 **13.** 2π **15.** 0 **17.** $\frac{8}{3}$ **19.** $a^5/4$ **21.** 0
23. See Theorem 15.13, "Stokes's Theorem," on page 1132.
25–27. Proofs **29.** Putnam Problem A5, 1987

Review Exercises for Chapter 15 (page 1138)

1. $\sqrt{x^2 + 5}$

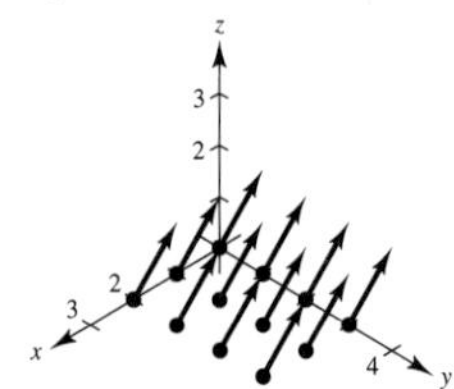

3. $(4x + y)\mathbf{i} + x\mathbf{j} + 2z\mathbf{k}$
5. Conservative: $f(x, y) = y/x + K$
7. Conservative: $f(x, y) = \frac{1}{2}x^2y^2 - \frac{1}{3}x^3 + \frac{1}{3}y^3 + K$
9. Not conservative **11.** Conservative: $f(x, y, z) = x/(yz) + K$
13. (a) $\text{div}\,\mathbf{F} = 2x + 2xy + x^2$ (b) $\text{curl}\,\mathbf{F} = -2xz\mathbf{j} + y^2\mathbf{k}$
15. (a) $\text{div}\,\mathbf{F} = -y \sin x - x \cos y + xy$
(b) $\text{curl}\,\mathbf{F} = xz\mathbf{i} - yz\mathbf{j}$
17. (a) $\text{div}\,\mathbf{F} = 1/\sqrt{1 - x^2} + 2xy + 2yz$
(b) $\text{curl}\,\mathbf{F} = z^2\mathbf{i} + y^2\mathbf{k}$
19. (a) $\text{div}\,\mathbf{F} = \dfrac{2x + 2y}{x^2 + y^2} + 1$
(b) $\text{curl}\,\mathbf{F} = \dfrac{2x - 2y}{x^2 + y^2}\mathbf{k}$
21. (a) $\frac{125}{3}$ (b) 2π **23.** 6π **25.** (a) 18 (b) 18π
27. $9a^2/5$ **29.** $(\sqrt{5}/3)(19 - \cos 6) \approx 13.446$ **31.** 1
33. $2\pi^2$ **35.** 36 **37.** $\frac{4}{3}$ **39.** $\frac{8}{3}(3 - 4\sqrt{2}) \approx -7.085$
41. 6 **43.** (a) 15 (b) 15 (c) 15
45. 1 **47.** 0 **49.** 0
51.

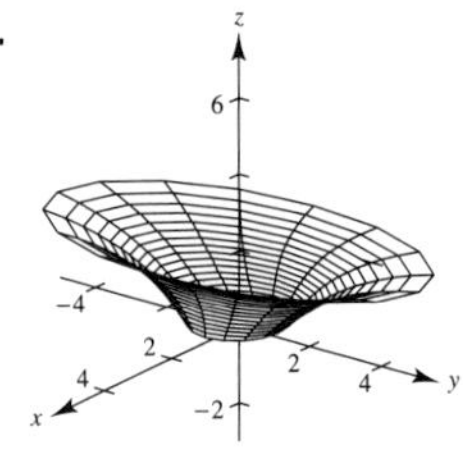

53. (a)

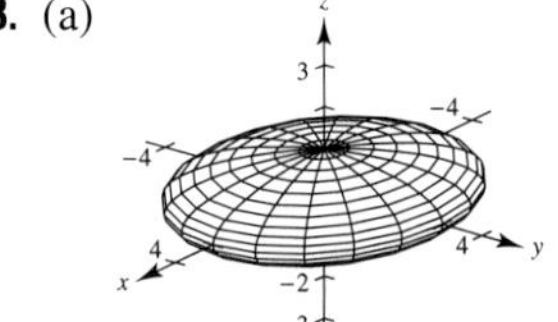

(b)

(c)

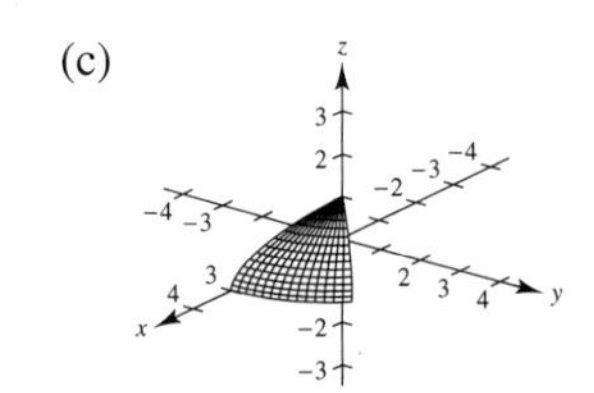

(d)

Circle

(e) About 14.436 (f) About 4.269

55.

0

57. 66 **59.** $2a^6/5$ **61.** Proof

P.S. Problem Solving (page 1141)

1. (a) $(25\sqrt{2}/6)k\pi$ (b) $(25\sqrt{2}/6)k\pi$
3. $I_x = (\sqrt{13}\pi/3)(27 + 32\pi^2)$; $I_y = (\sqrt{13}\pi/3)(27 + 32\pi^2)$; $I_z = 18\sqrt{13}\pi$
5. Proof **7.** $3a^2\pi$ **9.** (a) 1 (b) $\frac{13}{15}$ (c) $\frac{5}{2}$ **11.** Proof
13. $M = 3mxy(x^2 + y^2)^{-5/2}$
$\partial M/\partial y = 3mx(x^2 - 4y^2)/(x^2 + y^2)^{7/2}$
$N = m(2y^2 - x^2)(x^2 + y^2)^{-5/2}$
$\partial N/\partial x = 3mx(x^2 - 4y^2)/(x^2 + y^2)^{7/2}$
Therefore, $\partial N/\partial x = \partial M/\partial y$ and $\mathbf{F}$ is conservative.

Chapter 16

Section 16.1 (page 1149)

1. Exact; $M_y = 2xy = N_x$

3. Not exact; $M_y = x \cos y$, $N_x = \cos y$

5. $x^2 - 3xy + y^2 = C$ **7.** $3xy^2 + 5x^2y^2 - 2y = C$

9. Not exact **11.** $\arctan(x/y) = C$ **13.** Not exact

15. (a) Answers will vary.

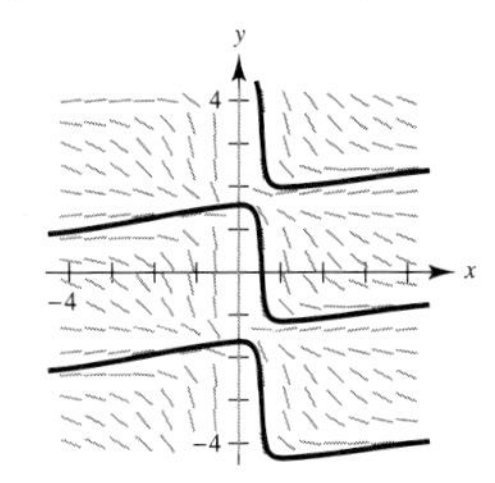

(b) $x^2 \tan y + 5x = \dfrac{11}{4}$

(c)

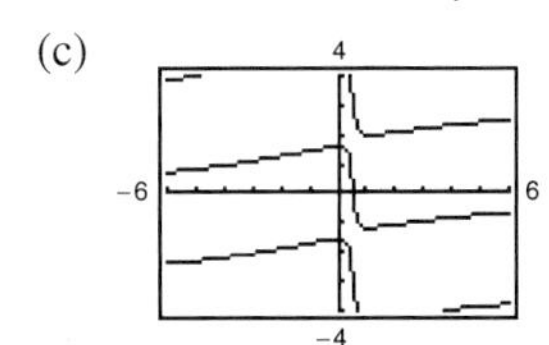

17. (a) Answers will vary.

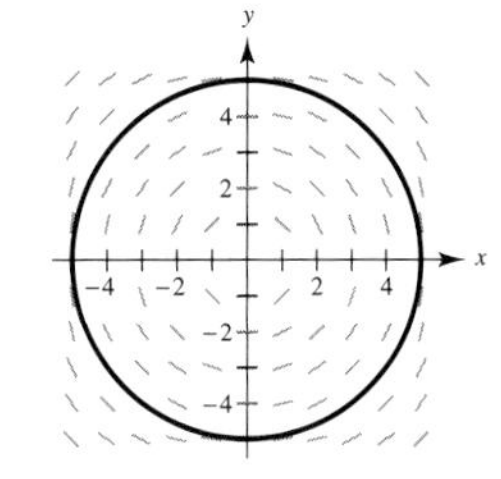

(b) $x^2 + y^2 = 25$

(c)

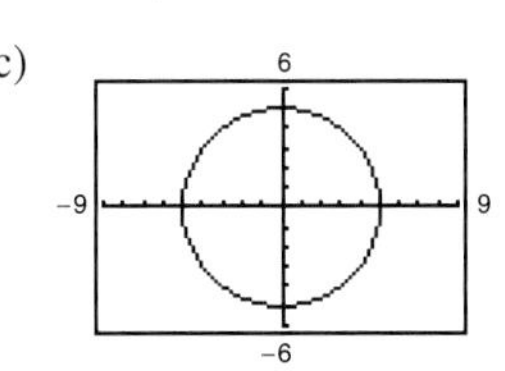

19. $y \ln(x - 1) + y^2 = 16$ **21.** $e^{3x} \sin 3y = 0$

23. $x^2y - 3x^3 + y^2 + y = 6$

25. Integrating factor: $1/y^2$

$\dfrac{x}{y} - 6y = C$

27. Integrating factor: $1/x^2$

$\dfrac{y}{x} + 5x = C$

29. Integrating factor: $\cos x$

$y \sin x + x \sin x + \cos x = C$

31. Integrating factor: $1/y$

$xy - \ln|y| = C$

33. Integrating factor: $1/\sqrt{y}$

$x\sqrt{y} + \cos\sqrt{y} = C$

35. $x^4y^3 + x^2y^4 = C$ **37.** $\dfrac{y^2}{x} + \dfrac{x}{y^2} = C$ **39.** Proof

41. $x^2 + y^2 = C$

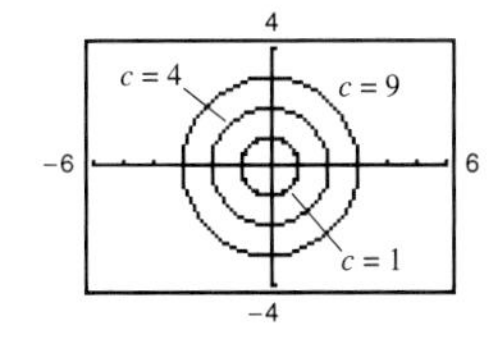

43. $2x^2y^4 + x^2 = C$

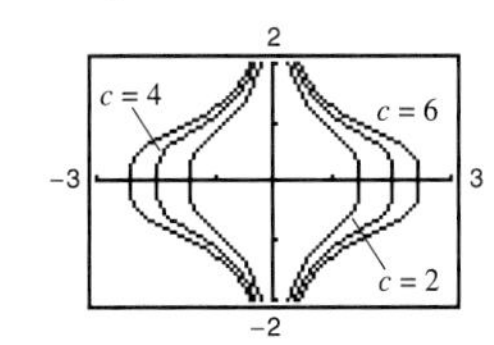

45. $x^2 - 2xy + 3y^2 = 3$ **47.** $C = \dfrac{5\left(x^2 + \sqrt{x^4 - 1{,}000{,}000x}\right)}{x}$

49. (a)

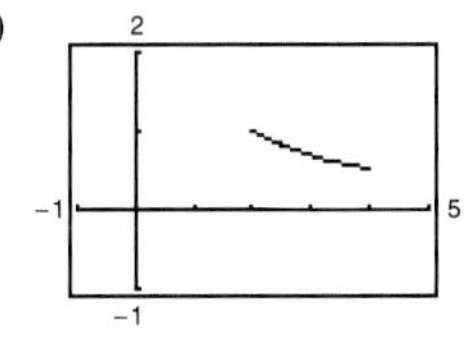

(c)

(b) $y^2(2x^2 + y^2) = 9$

51. (a)

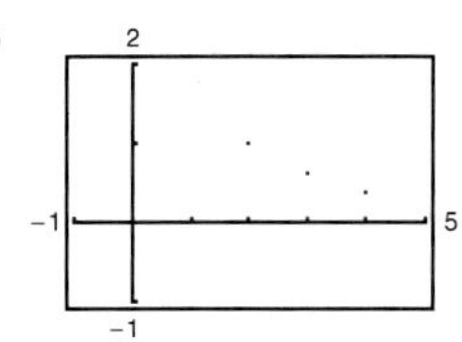

(c)

(b) $y^2(2x^2 + y^2) = 9$ Less accurate

53. See Theorem 16.1 on page 1144.

55. False; $\dfrac{\partial M}{\partial y} = 2x$, $\dfrac{\partial N}{\partial x} = -2x$. **57.** True **59.** $k = 3$

61. $f(x) = -\cos x + C_1$, $g(y) = \frac{1}{3}y^3 + C_2$

Section 16.2 (page 1157)

1.

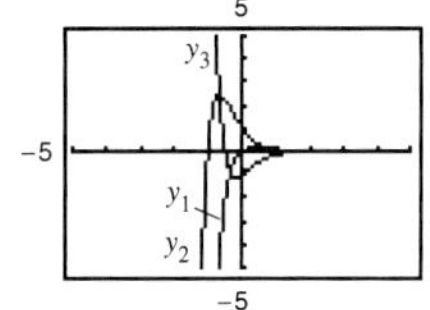

y_1: $C_1 = 0, C_2 = 1$
y_2: $C_1 = 1, C_2 = 1$
y_3: $C_1 = -1, C_2 = -2$

y approaches zero as $x \to \infty$.

3. y_1: $C_1 = 1, C_2 = -1$
y_2: $C_1 = -1, C_2 = 1$
y_3: $C_1 = 2, C_2 = 3$

The graphs are basically the same shape, with left and right shifts and varying ranges.

5. $y = C_1 + C_2e^x$ **7.** $y = C_1e^{3x} + C_2e^{-2x}$

9. $y = C_1e^{x/2} + C_2e^{-2x}$ **11.** $y = C_1e^{-3x} + C_2xe^{-3x}$

13. $y = C_1e^{x/4} + C_2xe^{x/4}$ **15.** $y = C_1 \sin x + C_2 \cos x$

17. $y = C_1e^{3x} + C_2e^{-3x}$ **19.** $y = e^x\left(C_1 \sin\sqrt{3}x + C_2 \cos\sqrt{3}x\right)$

21. $y = C_1e^{(3+\sqrt{5})x/2} + C_2e^{(3-\sqrt{5})x/2}$

23. $y = e^{2x/3}\left(C_1 \sin\dfrac{\sqrt{7}x}{3} + C_2 \cos\dfrac{\sqrt{7}x}{3}\right)$

25. $y = C_1e^x + C_2e^{-x} + C_3 \sin x + C_4 \cos x$

27. $y = C_1e^x + C_2e^{2x} + C_3e^{3x}$

29. $y = C_1e^x + e^x(C_2 \sin 2x + C_3 \cos 2x)$

31. (a) $y = 2 \cos 10x$ (b) $y = \frac{1}{5} \sin 10x$
(c) $y = -\cos 10x + \frac{3}{10} \sin 10x$

33. $y = \frac{1}{11}(e^{6x} + 10e^{-5x})$ **35.** $y = \frac{1}{2} \sin 4x$

37. $y = 2e^{x/3} + \frac{1}{3}xe^{x/3}$

39. $y = \left(\dfrac{e - 3}{e - e^3}\right)e^{3x} + \left(\dfrac{3 - e^3}{e - e^3}\right)e^x$ **41.** No Solution

43. $y = 2e^{7x/2} + \left(-\dfrac{1}{e^{7/2}} - 2\right)xe^{7x/2}$ **45.** No; $f(x) \neq 0$.

47. The functions y_1 and y_2 are linearly independent if the only solution of $C_1 y_1 + C_2 y_2 = 0$ is $C_1 = C_2 = 0$.

49. $y = \frac{1}{2} \cos 4\sqrt{3}t$

51. $y = -\dfrac{2}{3} \cos 4\sqrt{3}t + \dfrac{\sqrt{3}}{24} \sin 4\sqrt{3}t$

53. $y = \dfrac{e^{-t/16}}{2}\left(\cos\dfrac{\sqrt{12{,}287}\,t}{16} + \dfrac{\sqrt{12{,}287}}{12{,}287} \sin\dfrac{\sqrt{12{,}287}\,t}{16}\right)$

55. b **56.** d **57.** c **58.** a **59.** Proof

61. False; the general solution is $y = C_1e^{3x} + C_2xe^{3x}$.

63. True **65.** Proof **67.** Proof

69. (a) Proof (b) $y = \dfrac{C_1}{x^3} + \dfrac{C_2}{x^2}$

Section 16.3 (page 1165)

1–3. Proofs **5.** $y_p = \frac{1}{4}x - \frac{1}{16}$ **7.** $y_p = e^{3x}$

9. $y_p = -\frac{4}{65}\sin x + \frac{1}{130}\cos x$

11. $y = C_1e^x + C_2e^{2x} + x + \frac{3}{2}$ **13.** $y = C_1 + C_2e^{-2x} + \frac{2}{3}e^x$

15. $y = (C_1 + C_2x)e^{5x} + \frac{3}{8}e^x + \frac{1}{5}$

17. $y = \left(C_1 - \dfrac{x}{6}\right)\cos 3x + C_2 \sin 3x$

19. $y = 6\sin x + \cos x + x^3 - 6x$

21. $y = -1 + 2e^{-x} - \sin x - \cos x$

23. $y = \left(\dfrac{4}{9} - \dfrac{1}{2}x^2\right)e^{4x} - \dfrac{1}{9}(1 + 3x)e^x$

25. $y = (C_1 + \ln|\cos x|)\cos x + (C_2 + x)\sin x$

27. $y = \left(C_1 - \dfrac{x}{2}\right)\cos 2x + \left(C_2 + \dfrac{1}{4}\ln|\sin 2x|\right)\sin 2x$

29. $y = (C_1 + C_2x)e^x + \dfrac{x^2e^x}{4}(\ln x^2 - 3)$

31. $y_p = Ax^2 + Bx + C$; This is a generalized form of $F(x) = x^2$.

33. See "Variation of Parameters" box on page 1163.

35. $q = \frac{3}{25}(e^{-5t} + 5te^{-5t} - \cos 5t)$

37. $y = \frac{1}{4}\cos 8t - \frac{1}{2}\sin 8t + \sin 4t$

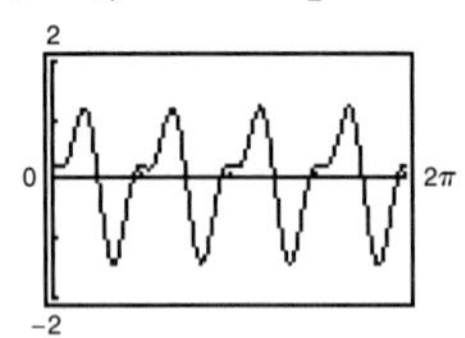

39. $y = \left(\frac{9}{32} - \frac{3}{4}t\right)e^{-8t} - \frac{1}{32}\cos 8t$

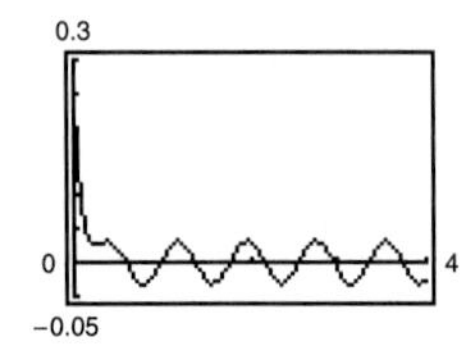

41. $y = \dfrac{\sqrt{5}}{4}\sin\left(8t + \pi - \arctan\dfrac{1}{2}\right) \approx \dfrac{\sqrt{5}}{4}\sin(8t + 2.6779)$

43. $y = C_1x + C_2x\ln x + \frac{2}{3}x(\ln x)^3$ **45.** True

47. (a) No (b) Yes

Putnam Problem A3, 1987

Section 16.4 (page 1170)

1–5. Proofs

7. $y = a_0\displaystyle\sum_{k=0}^{\infty}\frac{(-3)^k}{2^kk!}x^{2k}$

Interval of convergence: $(-\infty, \infty)$

9. $y = a_0 + a_1\displaystyle\sum_{k=0}^{\infty}\frac{x^{2k+1}}{2^k(k!)(2k+1)}$

Interval of convergence: $(-\infty, \infty)$

11. $y = a_0\left(1 - \dfrac{x^2}{8} + \dfrac{x^4}{128} - \cdots\right) + a_1\left(x - \dfrac{x^3}{24} + \dfrac{7x^5}{1920} - \cdots\right)$

13. Taylor's Theorem: $y = 2 + \dfrac{2x}{1!} - \dfrac{2x^2}{2!} - \dfrac{10x^3}{3!} + \dfrac{2x^4}{4!} + \cdots$

$y\left(\dfrac{1}{2}\right) \approx 2.547$

Euler's Method: $y\left(\dfrac{1}{2}\right) \approx 2.672$

15. Given a differential equation, assume that the solution is of the form $y = \sum a_nx^n$. Then substitute y and its derivatives into the differential equation. You should then be able to determine the coefficients $(a_0, a_1, \ldots)$ for the solution series.

17. (a) $y = 2x + \dfrac{x^3}{3} + \dfrac{x^5}{20} + \cdots$

(b)

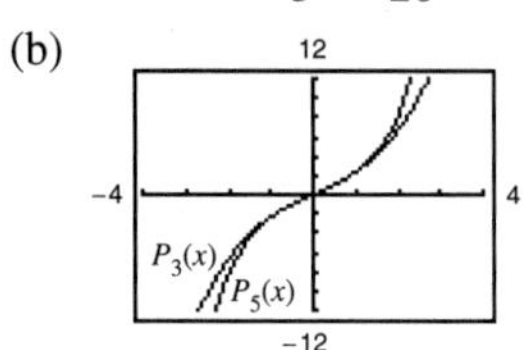

(c) The solution is symmetric about the origin.

19. $y = 1 - \dfrac{3x}{1!} + \dfrac{2x^3}{3!} - \dfrac{12x^4}{4!} + \dfrac{16x^6}{6!} - \dfrac{120x^7}{7!}$

$y\left(\dfrac{1}{4}\right) \approx 0.253$

21. $y = 3 + \dfrac{2x}{1!} + \dfrac{3x^2}{2!} + \dfrac{2x^3}{3!}$; $y\left(\dfrac{1}{3}\right) \approx 3.846$

23–25. Proofs

27. $y = a_0 + a_1x + \dfrac{a_0}{6}x^3 + \dfrac{a_1}{12}x^4 + \dfrac{a_0}{180}x^6 + \dfrac{a_1}{504}x^7$

Review Exercises for Chapter 16 (page 1171)

1. Not exact; $\dfrac{\partial M}{\partial y} \neq \dfrac{\partial N}{\partial x}$

3. Exact; $16xy + 10x^2 + 4x + 5y^2 + 4y = C$

5. Exact; $-2xy - 3y^2 + 4y + x^2 - 10x = C$ **7.** Not exact

9. (a) Answers will vary. (b) $x^2 + y^2 - xy = 4$

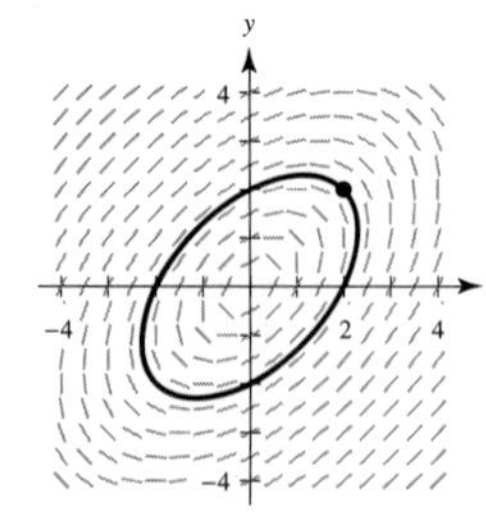

(c)

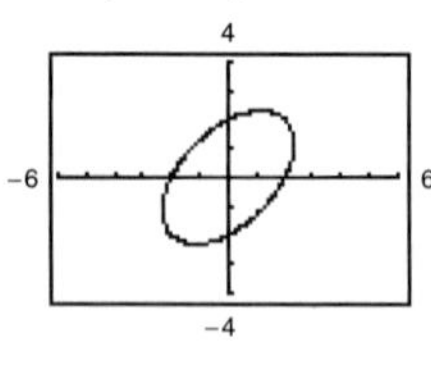

11. $2xy + 2x^2 - 6x - 3y^2 + 2y = -4$

13. $3x + \dfrac{y^2}{x} = C$ **15.** $xe^{3y} - e^y = C$

17.

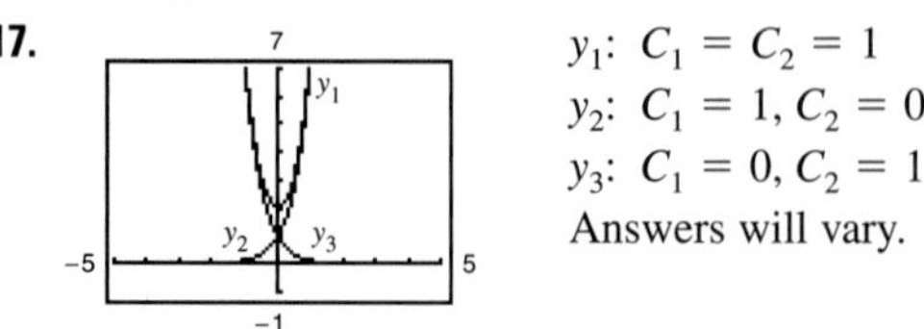

y_1: $C_1 = C_2 = 1$
y_2: $C_1 = 1, C_2 = 0$
y_3: $C_1 = 0, C_2 = 1$
Answers will vary.

19. $y = e^{2x} - e^{-x}$

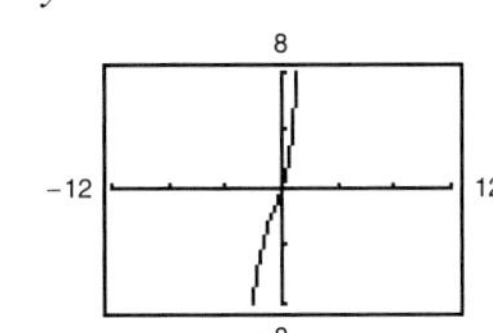

21. $y = \frac{1}{2}e^{-3x} + \frac{3}{2}e^{x}$

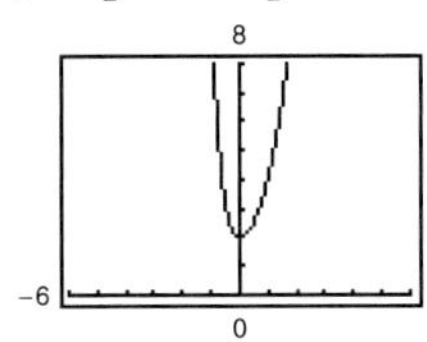

23. $y = \left(\dfrac{4e}{\sin 2 - \tan 4 \cos 2}\right)(e^{-x} \sin 2x - \tan 4e^{-x} \cos 2x)$

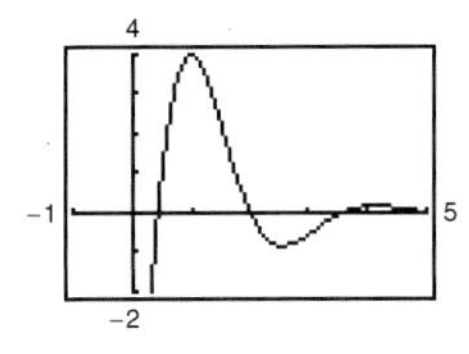

25. y'' is always positive according to the graph but y' is negative when $x < 0$, so $y'' \neq y'$.

27. $y = C_1 \sin x + C_2 \cos x - 5x + x^3$

29. $y = (C_1 + x)\sin x + C_2 \cos x$

31. $y = \left(C_1 + C_2x + \frac{1}{3}x^3\right)e^x$ **33.** $y = \frac{11}{5}(2e^{3x} + 3e^{-2x}) - 9$

35. $y = \frac{17}{3}\cos 2x - 3 \sin 2x + \frac{1}{3}\cos x$

37. $y = -\frac{1}{2} - \frac{1}{27}e^{-x} - \frac{1}{9}xe^{-x} - \frac{1}{6}x^2e^{-x} + \frac{83}{54}e^{2x}$

39. $y = \frac{1}{2}\cos\left(2\sqrt{6}t\right)$

41. (a) (i) $y = \dfrac{1}{2}\cos 2t + \dfrac{12\pi}{\pi^2 - 4}\sin 2t + \dfrac{24}{4 - \pi^2}\sin \pi t$

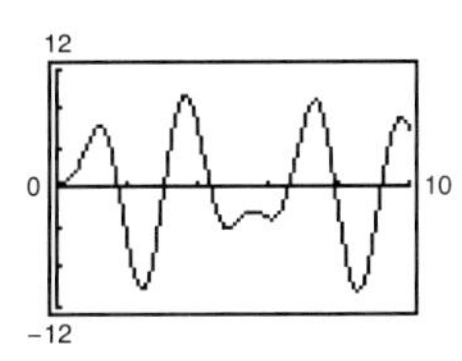

(ii) $y = \frac{1}{2}\left[\left(1 - 6\sqrt{2}t\right)\cos\left(2\sqrt{2}t\right) + 3\sin\left(2\sqrt{2}t\right)\right]$

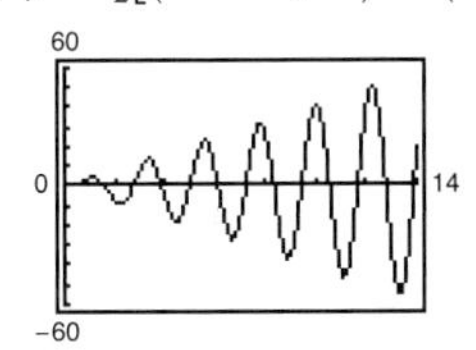

(iii) $y = \dfrac{e^{-t/5}}{398}\left[199 \cos \dfrac{\sqrt{199}\,t}{5} + \sqrt{199}\sin\dfrac{\sqrt{199}\,t}{5}\right]$

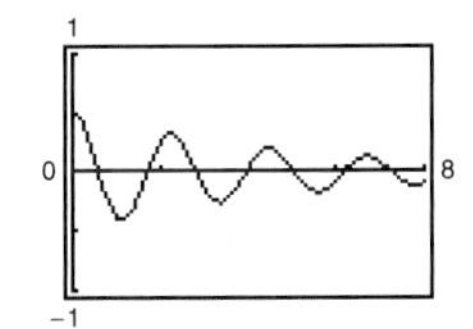

(iv) $y = \frac{1}{2}e^{-2t}(\cos 2t + \sin 2t)$

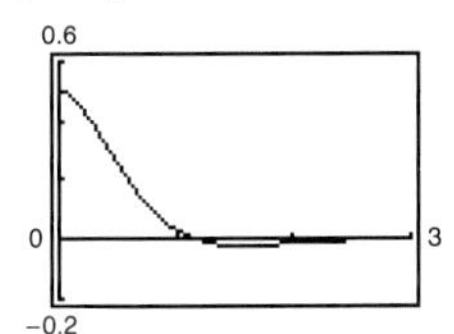

(b) The object would come to rest more quickly. It might not oscillate at all, as in part (iv).

(c) The object would oscillate more rapidly.

(d) Part (ii). The amplitude becomes increasingly large.

43. (a) Only a second derivative is used, so a cosine is unnecessary.

(b) $y_p = \frac{5}{2}\cos x$

(c) If $y_p = A \cos x + B \sin x$, then $y_p'' = -A \cos x - B \sin x$. So it would be more difficult to solve for A and B.

45. $y = a_0 \displaystyle\sum_{n=0}^{\infty} \frac{x^n}{4^n}$

47. $y = 2 + \dfrac{2x^2}{2!} + \dfrac{4x^4}{4!} + \dfrac{4x^5}{5!}$; $y\left(\dfrac{1}{4}\right) \approx 2.063$

P.S. Problem Solving (page 1173)

1. $k = -5$; $6x^3 + 10y^3 - 15x^2y^2 = C$

3. $y = B_1e^{ax} + B_2e^{-ax}$; Proof **5.** Proof

7. (a) and (b) Proofs

9. (a) $\theta(t) = C_1 \cos\left(\sqrt{\dfrac{g}{L}}t\right) + C_2 \sin\left(\sqrt{\dfrac{g}{L}}t\right)$; Proof

(b) $\theta(t) = 0.128 \cos\left[\sqrt{39.2}(t - 0.108)\right]$ (c) Period $\approx$ 1 sec

(d) 0.128 (e) 0.358 sec; 0.860 sec

(f) $\theta'(0.358) = -0.8012$; $\theta'(0.860) = 0.8012$

11. (a) Critically damped (b) $y = e^{-4t} + 5te^{-4t}$

(c)

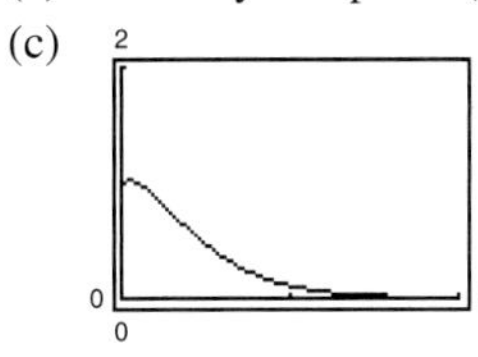

Answers will vary.

13. (a) Overdamped (b) $y = e^{-16t} + e^{-4t}$

(c)

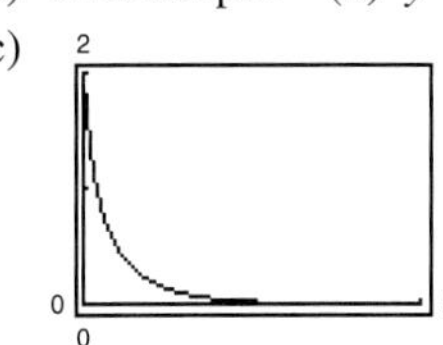

Answers will vary.

15. $y = a_0 + a_1(x - 1) + \dfrac{a_0}{2}(x - 1)^2 + \dfrac{(a_0 + a_1)}{6}(x - 1)^3$
$+ \dfrac{(a_0 + 2a_1)}{24}(x - 1)^4 + \dfrac{(4a_0 + a_1)}{120}(x - 1)^5$
$+ \dfrac{(5a_0 + 6a_1)}{720}(x - 1)^6 + \dfrac{(9a_0 + 11a_1)}{5040}(x - 1)^7$

Answers will vary.

17. (a) $y = a_0 \displaystyle\sum_{k=0}^{\infty} \frac{(-1)^k x^{2k}}{2^{2k}(k!)^2}$ (b) $y = a_0 J_0(x)$

19. (a) $y = 16x^4 - 48x^2 + 12$

(b) $H_0(x) = 1$

$H_1(x) = 2x$

$H_2(x) = 4x^2 - 2$

$H_3(x) = 8x^3 - 12x$

$H_4(x) = 16x^4 - 48x^2 + 12$

Index

INDEX

C

D

G

H

I

J

K

INDEX

L

M

N

O

P

V

W

X

Y

Z

ALGEBRA

Factors and Zeros of Polynomials

Let $p(x) = a_n x^n + a_{n-1}x^{n-1} + \cdots + a_1 x + a_0$ be a polynomial. If $p(a) = 0$, then a is a *zero* of the polynomial and a solution of the equation $p(x) = 0$. Furthermore, $(x - a)$ is a *factor* of the polynomial.

Fundamental Theorem of Algebra

An nth degree polynomial has n (not necessarily distinct) zeros. Although all of these zeros may be imaginary, a real polynomial of odd degree must have at least one real zero.

Quadratic Formula

If $p(x) = ax^2 + bx + c$, and $0 \le b^2 - 4ac$, then the real zeros of p are $x = \left(-b \pm \sqrt{b^2 - 4ac}\right)/2a$.

Special Factors

$x^2 - a^2 = (x - a)(x + a)$

$x^3 - a^3 = (x - a)(x^2 + ax + a^2)$

$x^3 + a^3 = (x + a)(x^2 - ax + a^2)$

$x^4 - a^4 = (x^2 - a^2)(x^2 + a^2)$

Binomial Theorem

$(x + y)^2 = x^2 + 2xy + y^2$

$(x - y)^2 = x^2 - 2xy + y^2$

$(x + y)^3 = x^3 + 3x^2y + 3xy^2 + y^3$

$(x - y)^3 = x^3 - 3x^2y + 3xy^2 - y^3$

$(x + y)^4 = x^4 + 4x^3y + 6x^2y^2 + 4xy^3 + y^4$

$(x - y)^4 = x^4 - 4x^3y + 6x^2y^2 - 4xy^3 + y^4$

$$(x + y)^n = x^n + nx^{n-1}y + \frac{n(n-1)}{2!}x^{n-2}y^2 + \cdots + nxy^{n-1} + y^n$$

$$(x - y)^n = x^n - nx^{n-1}y + \frac{n(n-1)}{2!}x^{n-2}y^2 - \cdots \pm nxy^{n-1} \mp y^n$$

Rational Zero Theorem

If $p(x) = a_n x^n + a_{n-1}x^{n-1} + \cdots + a_1 x + a_0$ has integer coefficients, then every *rational zero* of p is of the form $x = r/s$, where r is a factor of a_0 and s is a factor of a_n.

Factoring by Grouping

$acx^3 + adx^2 + bcx + bd = ax^2(cx + d) + b(cx + d) = (ax^2 + b)(cx + d)$

Arithmetic Operations

$ab + ac = a(b + c)$

$\frac{a}{b} + \frac{c}{d} = \frac{ad + bc}{bd}$

$\frac{a + b}{c} = \frac{a}{c} + \frac{b}{c}$

$\frac{\left(\frac{a}{b}\right)}{\left(\frac{c}{d}\right)} = \left(\frac{a}{b}\right)\left(\frac{d}{c}\right) = \frac{ad}{bc}$

$\frac{\left(\frac{a}{b}\right)}{c} = \frac{a}{bc}$

$\frac{a}{\left(\frac{b}{c}\right)} = \frac{ac}{b}$

$a\left(\frac{b}{c}\right) = \frac{ab}{c}$

$\frac{a - b}{c - d} = \frac{b - a}{d - c}$

$\frac{ab + ac}{a} = b + c$

Exponents and Radicals

$a^0 = 1, \quad a \ne 0$

$(ab)^x = a^x b^x$

$a^x a^y = a^{x+y}$

$\sqrt{a} = a^{1/2}$

$\frac{a^x}{a^y} = a^{x-y}$

$\sqrt[n]{a} = a^{1/n}$

$\left(\frac{a}{b}\right)^x = \frac{a^x}{b^x}$

$\sqrt[n]{a^m} = a^{m/n}$

$a^{-x} = \frac{1}{a^x}$

$\sqrt[n]{ab} = \sqrt[n]{a}\,\sqrt[n]{b}$

$(a^x)^y = a^{xy}$

$\sqrt[n]{\frac{a}{b}} = \frac{\sqrt[n]{a}}{\sqrt[n]{b}}$

FORMULAS FROM GEOMETRY

Triangle

$h = a \sin \theta$

Area $= \frac{1}{2}bh$

(Law of Cosines)

$c^2 = a^2 + b^2 - 2ab \cos \theta$

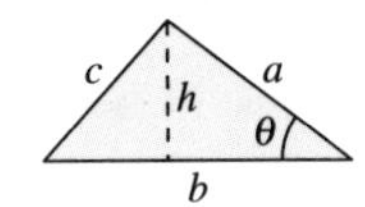

Sector of Circular Ring

(p = average radius,

w = width of ring,

θ in radians)

Area $= \theta p w$

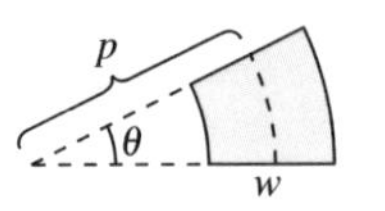

Right Triangle

(Pythagorean Theorem)

$c^2 = a^2 + b^2$

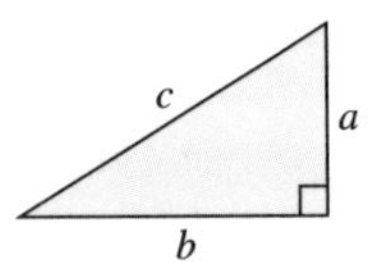

Ellipse

Area $= \pi ab$

Circumference $\approx 2\pi\sqrt{\frac{a^2 + b^2}{2}}$

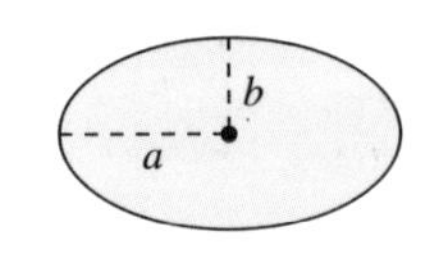

Equilateral Triangle

$h = \frac{\sqrt{3}s}{2}$

Area $= \frac{\sqrt{3}s^2}{4}$

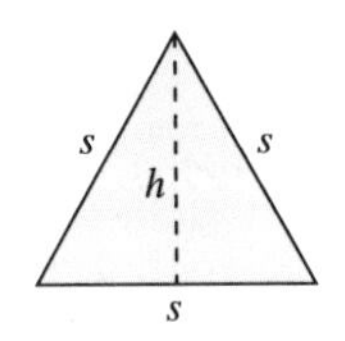

Cone

(A = area of base)

Volume $= \frac{Ah}{3}$

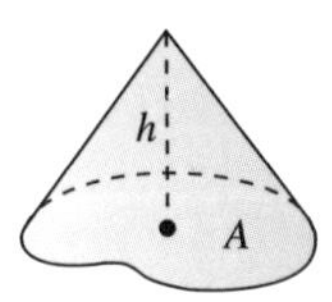

Parallelogram

Area $= bh$

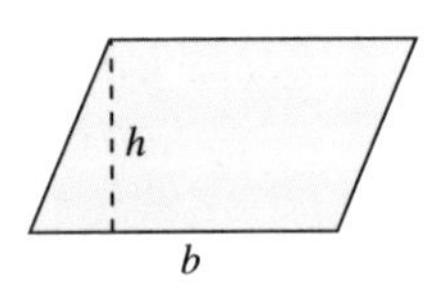

Right Circular Cone

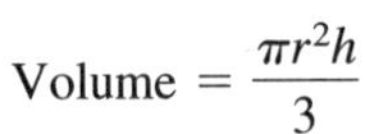

Volume $= \frac{\pi r^2 h}{3}$

Lateral Surface Area $= \pi r\sqrt{r^2 + h^2}$

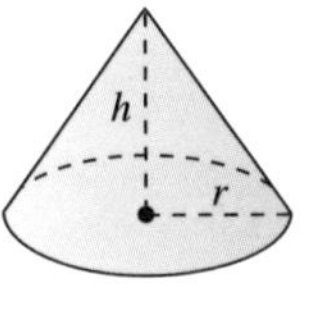

Trapezoid

Area $= \frac{h}{2}(a + b)$

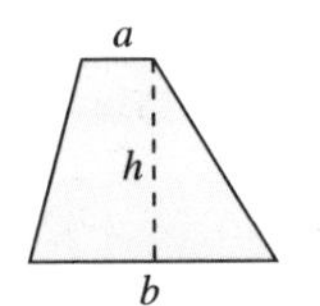

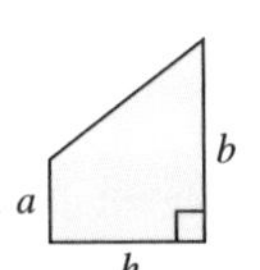

Frustum of Right Circular Cone

Volume $= \frac{\pi(r^2 + rR + R^2)h}{3}$

Lateral Surface Area $= \pi s(R + r)$

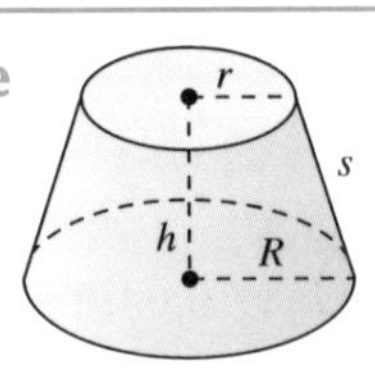

Circle

Area $= \pi r^2$

Circumference $= 2\pi r$

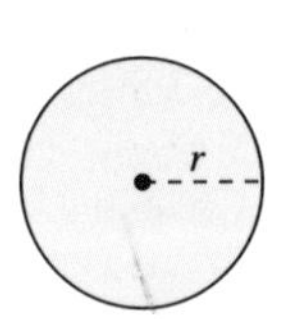

Right Circular Cylinder

Volume $= \pi r^2 h$

Lateral Surface Area $= 2\pi rh$

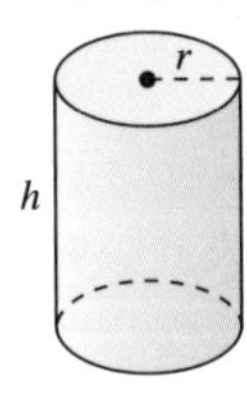

Sector of Circle

(θ in radians)

Area $= \frac{\theta r^2}{2}$

$s = r\theta$

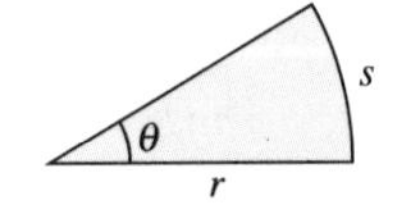

Sphere

Volume $= \frac{4}{3}\pi r^3$

Surface Area $= 4\pi r^2$

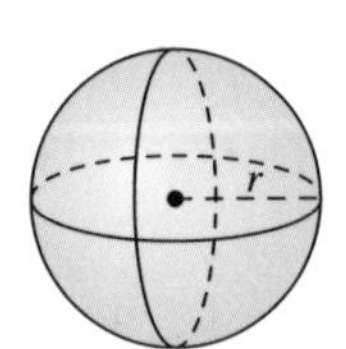

Circular Ring

(p = average radius,

w = width of ring)

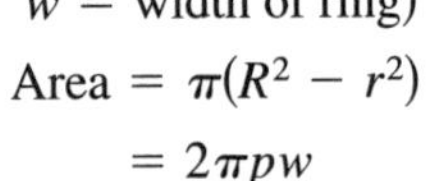

Area $= \pi(R^2 - r^2)$

$= 2\pi pw$

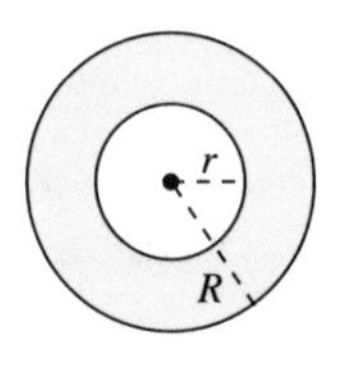

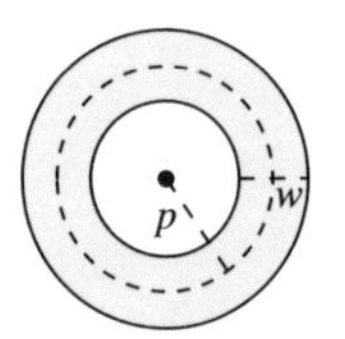

Wedge

(A = area of upper face,

B = area of base)

$A = B \sec \theta$

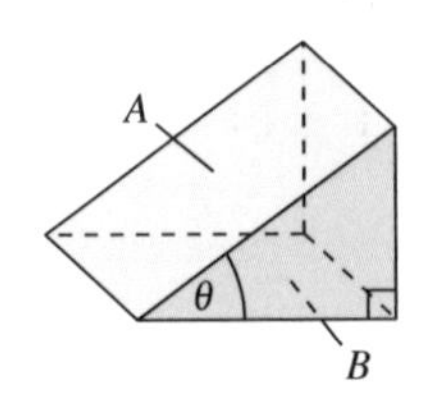